ENCYCLOPEDIA OF STATISTICS IN QUALITY AND RELIABILITY

ENCYCLOPEDIA OF STATISTICS IN QUALITY AND RELIABILITY

Volume 3

M–Q

Editors-in-Chief

FABRIZIO RUGGERI

CNR-IMATI, Milano, Italy

RON S. KENETT

KPA Ltd., Raanana, Israel

FREDERICK W. FALTIN

The Faltin Group, Cody, WY, USA

The Atrium,
Southern Gate,
Chichester,
West Sussex,
PO19 8SQ, England

Telephone: (+44) 1243 779777
Email (for orders and customer service enquiries): cs-books@wiley.co.uk
Visit our Home Page on www.wiley.com

Other Wiley Editorial Offices

John Wiley & Sons Inc., 111 River Street,
Hoboken, NJ 07030, USA

Jossey-Bass, 989 Market Street,
San Francisco, CA 94103-1741, USA

Wiley-VCH Verlag GmbH, Boschstr. 12,
D-69469 Weinheim, Germany

John Wiley & Sons Australia Ltd, 42 McDougall Street,
Milton, Queensland 4064, Australia

John Wiley & Sons (Asia) Pte Ltd, 2 Clementi Loop #02-01,
Jin Xing Distripark, Singapore 129809

John Wiley & Sons Canada Ltd, 6045 Freemont Blvd,
Mississauga, ONT, L5R 4J3, Canada

Wiley also publishes its books in a variety of electronic formats. Some content that appears in print may not be available in electronic books.

Anniversary Logo Design: Richard J. Pacifico

Library of Congress Cataloging-in-Publication Data

Encyclopedia statistics in quality and reliability / editors-in-chief, Fabrizio Ruggeri, Ron S. Kenett, Frederick W. Faltin.
p. cm.
Includes bibliographical references and index.
ISBN 978-0-470-01861-3 (cloth)
1. Production management – Quality control – Encyclopedias. 2. Reliability (Engineering) – Encyclopedias. I. Ruggeri, Fabrizio. II. Kenett, Ron. III. Faltin, Frederick W.
TS156.E54 2007
658.4′013 – dc22

2007037127

British Library Cataloguing in Publication Data

A catalogue record for this book is available from the British Library

978-0-470-01861-3

Typeset in 9½/11½ pt Times by Laserwords Private Limited, Chennai, India.
Printed and bound in Spain by Grafos S.A.
This book is printed on acid-free paper responsibly manufactured from sustainable forestry in which at least two trees are planted for each one used for paper production.

To

Anna, Giacomo and Lorenzo

FR

Jonathan, Alma and Tomer

RSK

Otto, Mary and Donna

FWF

Editorial Board

International Advisory Board

Contents

VOLUME 1

VOLUME 2

VOLUME 3

VOLUME 4

Contributors

Abbas, Ismail *Universitat Politècnica de Catalunya, Barcelona, Spain*

Adams, Benjamin M. *University of Alabama, Tuscaloosa, AL, USA*

Adhikari, Sondipon *University of Wales Swansea, Bristol, UK*

Agustin, Ma. Zenia N. *Southern Illinois University, Edwardsville, IL, USA*

Agustin, Marcus A. *Southern Illinois University, Edwardsville, IL, USA*

Ahmed, Abdulwaheed S. *FRS Marine Laboratory and Robert Gordon University, Aberdeen, UK*

Allen, Elaine *Babson College, Babson Park, MA, USA*

Allen, Yvonne *Burnham Laboratory, Burnham-on-Crouch, UK*

Altham, P.M.E. *University of Cambridge, Cambridge, UK*

Anderson, Chris K. *Cornell University, Ithaca, NY, USA*

Anderson-Cook, Christine *Los Alamos National Laboratory, Los Alamos, NM, USA*

Andrews, John D. *Loughborough University, Loughborough, UK*

Ankenman, Bruce E. *Northwestern University, Evanston, IL, USA*

Arvidsson, Martin *Chalmers University of Technology, Göteborg, Sweden*

Ascher, Harold E. *Harold E. Ascher & Associates, Potomac, MD, USA*

Atkinson, Anthony C. *London School of Economics and Political Science, London, UK*

Ausín, M. Concepción *Universidade da Coruña, Coruña, Spain*

Aven, Terje *University of Stavanger, Stavanger, Norway*

Bae, Suk J. *Hanyang University, Seoul, Korea*

Bai, Do Sun *Korea Advanced Institute of Science and Technology (KAIST), Daejon, Korea*

Baker, Rose D. *University of Salford, Salford, UK*

Balakrishnan, Narayanaswamy *McMaster University, Hamilton, Ontario, Canada*

Balamurali, Saminathan *Pohang University of Science and Technology, Pohang, Korea*

Barlović, Robert *AMD Saxony LLC & Co. KG, Dresden, Germany*

Barry, Don *University of Limerick, Limerick, Ireland*

Barry, Jon *Burnham Laboratory, Burnham-on-Crouch, UK*

Bartlett, Lisa M. *Loughborough University, Loughborough, UK*

Basu, Asit P. *University of Missouri-Columbia, Columbia, MO, USA*

Basu, Sanjib *Northern Illinois University, Dekalb, IL, USA*

Beder, Jay H. *University of Wisconsin, Milwaukee, WI, USA*

Bedford, Tim *University of Strathclyde, Glasgow, UK*

Belyaev, Yuri K. *Umeå University, Umeå, Sweden*

Belzunce, Félix *Universidad de Murcia, Murcia, Spain*

Ben-Gal, Irad *Tel-Aviv University, Tel-Aviv, Israel*

Benmerzouga, Ali *Sultan Qaboos University, Muscat, Sultanate of Oman*

Bergman, Bo *Chalmers University of Technology, Göteborg, Sweden*

Betrò, Bruno *CNR-IMATI, Milano, Italy*
Bevan, Richard G. *London School of Economics and Political Science, London, UK*
Bhattacharjee, Manish C. *New Jersey Institute of Technology, Newark, NJ, USA*
Biemer, Paul P. *RTI International and University of North Carolina, Chapel Hill, NC, USA*
Bingham, Derek *Simon Fraser University, Burnaby, British Columbia, Canada*
Blischke, Wallace R. *University of Southern California, Sherman Oaks, CA, USA*
Borkowski, John J. *Montana State University, Bozeman, MT, USA*
Borror, Connie M. *Arizona State University, Phoenix, AZ, USA*
Bothe, Davis Ross *International Quality Institute, Cedarburg, WI, USA*
Bowles, John B. *University of South Carolina, Columbia, SC, USA*
Boxma, Onno J. *Eindhoven University of Technology, Eindhoven, The Netherlands*
Brenneman, William A. *Procter & Gamble Company, Mason, OH, USA*
Brockwell, P.J. *Colorado State University, Fort Collins, CO, USA*
Broemeling, Lyle D. *M.D. Anderson Cancer Center, Houston, TX, USA*
Bunea, Cornel *George Washington University, Washington, DC, USA*
Burdick, Richard K. *Amgen, Inc., Thousand Oaks, CA, USA*
Bzik, Thomas J. *Air Products and Chemicals Inc., Allentown, PA, USA*

Casanovas, Josep *Universitat Politècnica de Catalunya, Barcelona, Spain*
Çekyay, Bora *Koç University, Sarıyer-Istanbul, Turkey*
Chakraborti, Subhabrata *University of Alabama, Tuscaloosa, AL, USA*
Champ, Charles W. *Georgia Southern University, Statesboro, GA, USA*
Chang, Young Soon *Myongji University, Seoul, Korea*
Chaubey, Yogendra P. *Concordia University, Montréal, Quebec, Canada*
Chen, Gilad *Texas A&M University, College Station, TX, USA*
Cheng, Smiley *University of Manitoba, Winnipeg, Manitoba, Canada*
Chiang, Andy Kok Leong *National University of Singapore, Singapore*
Chien, Yu-hung *National Taiwan University of Science and Technology, Taipei and National Taichung Institute of Technology, Taichung, Taiwan*
Chou, Youn-Min *University of Texas, San Antonio, TX, USA*
Chun, Young H. *Louisiana State University, Baton Rouge, LA, USA*
Çinlar, Erhan *Princeton University, Princeton, NJ, USA*
Clapp, Timothy G. *North Carolina State University, Raleigh, NC, USA*
Clark-Carter, David *Staffordshire University, Stoke on Trent, UK*
Cobo, Erik *Universitat Politècnica de Catalunya, Barcelona, Spain*
Cofino, Wim P. *Wageningen University and Research Centre, Wageningen, The Netherlands*
Coit, David W. *Rutgers, The State University of New Jersey, Piscataway, NJ, USA*
Colbourn, Charles J. *Arizona State University, Tempe, Arizona, USA*
Coleman, Shirley Y. *Newcastle University, Newcastle, UK*
Collani, Elart von *University of Würzburg, Würzburg, Germany*
Colosimo, Bianca M. *Politecnico di Milano, Milano, Italy*
Conerly, Michael D. *University of Alabama, Tuscaloosa, AL, USA*
Connor, Stephen B. *University of Warwick, Coventry, UK*
Conti, Tito A. *Organizational Assessment Management, Ivrea, Italy*
Cooke, Roger M. *Resources for the Future, Washington, DC, USA and Delft University of Technology, Delft, The Netherlands*
Coolen, Frank P.A. *Durham University, Durham, UK*

Coolen-Schrijner, Pauline *Durham University, Durham, UK*
Cooper, Tony *Six Sigma Associates, Knoxville, TN, USA*
Cox, David R. *University of Oxford, Oxford, UK*
Cox, Sue *Lancaster University Management School, Lancaster, UK*
Crowder, Stephen V. *Sandia National Laboratories, Albuquerque, NM, USA*

Daniels, Lorraine *Tempe, AZ, USA*
Davidov, Ori *University of Haifa, Mount Carmel, Haifa, Israel*
Davies, Ian M. *FRS Marine Laboratory, Aberdeen, UK*
Dekker, Rommert *Erasmus University, Rotterdam, The Netherlands*
Demidenko, Eugene *Dartmouth Medical School, Hanover, NH, USA*
Deng, Lih-Yuan *University of Memphis, Memphis, TN, USA*
De Blasi, Pierpaolo *University of Turin, Turin, Italy*
De Iorio, maria *Imperial College London, London, UK*
de Mast, Jeroen *University of Amsterdam, Amsterdam, The Netherlands*
Di Bucchianico, Alessandro *Eindhoven University of Technology, Eindhoven, The Netherlands*
Djerf, Kari *Statistics Finland, Helsinki, Finland*
Dracup, Chris *Northumbria University, Newcastle, UK*
Draper, Norman R. *University of Wisconsin, Madison, WI, USA*

Ebrahimi, Nader *Northern Illinois University, DeKalb, IL, USA*
Elam, Matthew E. *University of Alabama, Tuscaloosa, AL, USA*
Elsayed, Elsayed A. *Rutgers, The State University of New Jersey, Piscataway, NJ, USA*
El-Haik, Basem *Six Sigma Professionals, Inc., Canton, MI, USA*
Erkanli, Alaattin *Duke University Medical Center, Durham, NC, USA*
Evandt, Øystein *Det Norske Veritas, Høvik, Norway and Newcastle University, Newcastle, UK*
Everitt, Brian S. *King's College London, London, UK*

Faltin, Donna M. *The Faltin Group, Cody, WY, USA*
Faltin, Frederick W. *The Faltin Group, Cody, WY, USA*
Fang, Kai-Tai *BNU-HKBU United International College, Zhuhai, China*
Farrington, C. Paddy *Open University, Milton Keynes, UK*
Fatt Gan, Fah *National University of Singapore, Singapore*
Federer, Walter T. *Cornell University, Ithaca, NY, USA*
Fisher, Dennis G. *California State University, Long Beach, CA, USA*
Fuchs, Camil *Tel Aviv University, Tel Aviv, Israel*

Gabler, Siegfried *Centre for Survey Research and Methodology (GESIS-ZUMA), Mannheim, Germany*
Garcidueñas, Jorge E. *George Washington University, Washington, DC, USA*
Gaver, Donald P. *Naval Postgraduate School, Monterey, CA, USA*
Ghosh, Kaushik *University of Nevada at Las Vegas, Las Vegas, NV and National Cancer Institute, Bethesda, MD, USA*
Gilmour, Steven G. *Queen Mary, University of London, London, UK*
Giorgio, Massimiliano *Seconda Università di Napoli, Aversa, Italy*
Glaz, Joseph *University of Connecticut, Storrs, CT, USA*
Göb, Rainer *University of Würzburg, Würzburg, Germany*
Godfrey, A. Blanton *North Carolina State University, Raleigh, NC, USA*
Goos, Peter *Universiteit Antwerpen, Antwerpen, Belgium*
Gormley, Isobel C. *University College Dublin, Dublin, Ireland*

GRAHAM, MARIEN A. *University of Pretoria, Pretoria, South Africa*

GRAYESKI, FRANK *Cohn Consulting Group, Roseland, NJ, USA*

GRAYSON, KEVIN *North Carolina State University, Raleigh, NC, USA*

GREENBERG, BETSY S. *University of Texas at Austin, Austin, TX, USA*

GRIGG, OLIVIA A.J. *Institute of Public Health, Cambridge, UK*

GUNTER, BERT *Genentech Inc., South San Francisco, CA, USA*

GÜR ALI, ÖZDEN F. *Koç University, Istanbul, Turkey*

GUTHRIE, WILLIAM F. *National Institute of Standards and Technology, Gaithersburg, MD, USA*

HANSON, TIMOTHY E. *University of Minnesota, Minneapolis, MN, USA*

HARDMAN, GAVIN *University of Strathclyde, Glasgow, UK*

HAWKINS, DOUGLAS M. *University of Minnesota, Minneapolis, MN, USA*

HAYES, KEVIN *University of Limerick, Limerick, Ireland*

HE, BIN *University of Albama, Tuscaloosa, AL, USA*

HEERWEGH, DIRK *Katholieke Universiteit Leuven, Leuven, Belgium*

HERMAN, AMIR *University of Haifa, Mount Carmel, Haifa, Israel*

HERSHBERGER, SCOTT L. *California State University, Long Beach, CA, USA*

HEYDE, CHRISTOPHER. C. *The Australian National University, Canberra, Australia*

HICKERNELL, FRED J. *Illinois Institute of Technology, Chicago, IL, USA*

HIGDON, DAVID M. *Los Alamos National Laboratory, Los Alamos, NM, USA*

HILD, CHERYL *University of Tennessee, Knoxville, TN, USA*

HOCINE, MOUNIA N. *Open University, Milton Keynes, UK and INSERM, Villejuif, France*

HOERL, ROGER W. *GE Global Research, Niskayuna, NY, USA*

HOLFELD, ANDRE *AMD Saxony LLC & Co. KG, Dresden, Germany*

HOLLANDER, MYLES *Florida State University, Tallahassee, FL, USA*

HOWELL, DAVID C. *University of Vermont, Burlington, VT, USA*

HRYNIEWICZ, OLGIERD *Systems Research Institute, Warsaw, Poland*

HU, PING *National Cancer Institute, Bethesda, MD, USA*

HULTQUIST, ROBERT *Pennsylvania State University, University Park, PN, USA*

HUMAN, SCHALK W. *University of Pretoria, Pretoria, South Africa*

HUZURBAZAR, APARNA V. *University of New Mexico, Albuquerque, NM, USA*

HWARNG, H. BRIAN *National University of Singapore, Singapore*

JACROUX, MIKE *Washington State University, Pullman, WA, USA*

JENSEN, FINN VERNER *Aalborg University, Aalborg, Denmark*

JENSEN, UWE *Universität Hohenheim, Stuttgart, Germany*

JOHNSON, RICHARD A. *University of Wisconsin, Madison, WI, USA*

JONES, JEFF *University of Warwick, Coventry, UK*

JONES-FARMER, L. ALLISON *Auburn University College of Business, Auburn, AL, USA*

JORDAN, MICHAEL E. *Jones Lang LaSalle Strategic Consulting Organization, Chicago, IL, USA*

JOSEPH, V. ROSHAN *Georgia Institute of Technology, Atlanta, GA, USA*

JUN, CHI-HYUCK *Pohang University of Science and Technology, Pohang, Republic of Korea*

KALLEN, MAARTEN-JAN *HKV Consultants, Lelystad and Delft University of Technology, Delft, The Netherlands*

KALLENBERG, LODEWIJK C.M. *Leiden University, Leiden, The Netherlands*

KARIM, M. REZAUL *University of Rajshahi, Rajshahi, Bangladesh*

KAY, JIM W. *University of Glasgow, Glasgow, UK*

KELLA, OFFER *Hebrew University of Jerusalem, Jerusalem, Israel*

Kenett, Ron S. *KPA Ltd., Raanana, Israel*
Klebanov, Lev B. *Charles University, Prague, Czech Republic*
Klefsjö, B. *Luleå University of Technology, Luleå, Sweden*
Koehler Wiedlea, Andrew *Los Alamos National Laboratory, Los Alamos, NM, USA*
Koning, Alex J. *Erasmus University Rotterdam, Rotterdam, The Netherlands*
Koroliuk, Vladimir S. *Ukrainian National Academy of Science, Kiev, Ukraine*
Kottas, Athanasios *University of California, Santa Cruz, CA, USA*
Kumar, Naveen *Intel Corporation, Hillsboro, OR, USA*
Kunert, Joachim *Universität Dortmund, Dortmund, Germany*
Kuo, Lynn *University of Connecticut, Storrs, CT, USA*
Kuo, Way *University of Tennessee, Knoxville, TN, USA*
Kvam, Paul H. *Georgia Institute of Technology, Atlanta, Georgia, USA*

Lakhani, Azim D.H. *Universities of London and Oxford, London and Oxford, UK*
Langseth, Helge *Norwegian University of Science and Technology, Trondheim, Norway*
Larsen, Greg A. *Agilent Technologies, Loveland, CO, USA*
Laud, Purushottam W. *Medical College of Wisconsin, Milwaukee, WI, USA*
Lawson, Andrew B. *University of South Carolina, Columbia, SC, USA*
Ledolter, Johannes *University of Iowa, Iowa City, IA, USA*
Leemis, Lawrence M. *College of William & Mary, Williamsburg, VA, USA*
Lehmann, Axel *Otto-von Guericke-University, Magdeburg, Germany*
Lehmann, Erich L. *University of California, Berkley, CA, USA*
Leitnaker, Mary *University of Tennessee, Knoxville, TN, USA*
Lenth, Russell V. *University of Iowa, Iowa City, IA, USA*
Lenz, Hans-J. *Freie Universität Berlin, Berlin, Germany*
Lewis, Donald K. *Lewis Consulting, Lake Oswego, OR, USA*
Li, William *University of Minnesota, Minneapolis, MN, USA*
Limnios, Nikolaos *University of Technology of Compiègne, Compiègne, France*
Lin, Dennis K.J. *Pennsylvania State University, University Park, PA, USA*
Lin, Pi-Chuan *National Chin-Yi Institute of Technology, Taiwan, Republic of China*
Lindqvist, Bo Henry *Norwegian University of Science and Technology, Trondheim, Norway*
Lipsitz, Stuart R. *Dana-Farber Cancer Institute, Boston, MA, USA*
Loh, Wei-Yin *University of Wisconsin, Madison, WI, USA*
Lovelace, Cynthia R. *University of Alabama, Huntsville, AL, USA*
Lovie, Pat *Keele University, Keele, UK*
Lovie, Sandy *University of Liverpool, Liverpool, UK*
Lu, Jye-Chyi *Georgia Institute of Technology, Atlanta, Georgia, USA*
Luceño, Alberto *University of Cantabria, Santander, Spain*
Lütkebohmert, Constanze *Universität Hohenheim, Stuttgart, Germany*

Maass, Eric C. *Motorola Inc., Scottsdale, AZ, USA*
Malaiya, Yashwant K. *Colorado State University, Fort Collins, CO, USA*
Marco, Lluis *Technical University of Catalonia (UPC), Barcelona, Spain*
Marseguerra, Marzio *Polytechnic of Milan, Milan, Italy*
Martin, Michael A. *Australian National University, Canberra, Australia*
Martz, Harry F. *Los Alamos National Laboratory, Los Alamos, NM, USA*

Mason, Robert L. *Southwest Research Institute, San Antonio, TX, USA*
Mastrangelo, Christina *University of Washington, Seattle, WA, USA*
Maxwell, David L. *Centre for Environment, Fisheries and Aquaculture Science (CEFAS), Lowestoft, UK*
Mazzuchi, Thomas A. *George Washington University, Washington, DC, USA*
McCarty, Thomas *Jones Lang LaSalle Strategic Consulting Organization, Chicago, IL, USA*
McCollin, Christopher *Nottingham Trent University, Nottingham, UK*
McGrath, Richard N. *Bowling Green State University, Bowling Green, OH, USA*
McShane-Vaughn, Mary *Southern Polytechnic State University, Marietta, GA, USA*
Meeker, William Q. *Iowa State University, Ames, IA, USA*
Melgaard, Henrik *Novo Nordisk A/S, Allé, Bagsværd, Denmark*
Mendel, Max *Boston Fund Services, Salem, MA, USA*
Merrick, Jason R. *Virginia Commonwealth University, Richmond, VA, USA*
Meyer, R. Daniel *Pfizer Inc., New London, CT, USA*
Mi, Jie *Florida International University, Miami, FL, USA*
Michelson, Diane K. *International SEMATECH Manufacturing Initiative, Austin, TX, USA*
Miller, Arden *University of Auckland, Auckland, New Zealand*
Mira, Antonietta *University of Insubria, Varese, Italy*
Mitra, Amitava *Auburn University, Auburn, AL, USA*
Moffat, Colin F. *FRS Marine Laboratory and Robert Gordon University, Aberdeen, UK*
Momenkhani, Kourosh *George Washington University, Washington, DC, USA*
Montgomery, Douglas C. *Arizona State University, Tempe, AZ, USA*
Moore, Joseph D. *North Carolina State University, Raleigh, NC, USA*
Mordoch, Avraham *TOC Solutions, Emek Hefer, Israel*
Morgan, J.P. *Virginia Tech, Blacksburg, VA, USA*
Morris, Max D. *Iowa State University, Ames, IA, USA*
Morrison, Donald F. *University of Pennsylvania, Philadelphia, PA, USA*
Müller, Hans-Georg *University of California, Davis, CA, USA*
Murphy, Terrence E. *Yale University School of Medicine, New Haven, CT, USA*
Murphy, Thomas B. *Trinity College Dublin, Dublin, Ireland*
Murthy, D.N. Prabhakar *University of Queensland, Queensland, Australia*

Nakajo, Takeshi *Chuo University, Tokyo, Japan*
Nakamura, Tsutomu *Japan Institute of Plant Maintenance, Tokyo, Japan*
Natvig, Bent *University of Oslo, Oslo, Norway*
Nat, Jack *University of Abertay Dundee, Dundee, UK*
Neath, Andrew A. *Southern Illinois University, Edwardsville, IL, USA*
Nelson, Wayne B. *Consultant, Schenectady, NY, USA*
Nicolai, Robin P. *Erasmus University, Rotterdam, The Netherlands*
Nikulin, Mikhail *Université Victor Segalen Bordeaux 2, Bordeaux, France*
Nishina, Ken *Nagoya Institute of Technology, Nagoya, Japan*
Noguera, John *SigmaXL, Toronto, Ontario, Canada*
Notz, William I. *Ohio State University, Columbus, OH, USA*

O'Brien, Margaret G. *North Carolina State University, Raleigh, NC, USA*
Ograjenšek, Irena *University of Ljubljana, Ljubljana, Slovenia*
Onar, Arzu *St. Jude Children's Research Hospital, Memphis, TN, USA*

Özekici, Süleyman *Koç University, Sarıyer-İstanbul, Turkey*

Padgett, William J. *Clemson University, Clemson and University of South Carolina, Columbia, SC, USA*

Pajewski, Nicholas M. *Medical College of Wisconsin, Milwaukee, WI, USA*

Palcat, Frank *Industry Canada, Ottawa, Ontario, Canada*

Pall, Gabriel A. *College of William & Mary, Williamsburg, VA, USA*

Park, Chanseok *Clemson University, Clemson, SC, USA*

Parr, William C. *China Europe International Business School, Shanghai, China*

Pearn, Wen Lea *National Chiao Tung University, Hsinchu, Taiwan, Feng Chia University, Taichung, Taiwan*

Pecht, Michael *University of Maryland, College Park, MD, USA*

Peña, Daniel *Universidad Carlos III de Madrid, Getafe, Spain*

Peña, Edsel A. *Southern Illinois University, Edwardsville, IL and University of South Carolina, Columbia, SC, USA*

Pensky, Marianna *University of Central Florida, Orlando, FL, USA*

Pfeifer, Charles G. *E.I. du Pont de Nemours and Company, Inc., Wilmington, DE, USA*

Pham, Hoang *Rutgers, The State University of New Jersey, Piscataway, NJ, USA*

Piazza, Roberto *European Railway Agency (ERA), Valenciennes, France*

Piepel, Greg F. *Pacific Northwest National Laboratory, Richland, WA, USA*

Pievatolo, Antonio *CNR-IMATI, Milano, Italy*

Pilling, Graham M. *Centre for Environment, Fisheries and Aquaculture Science (CEFAS), Lowestoft, UK*

Pira, Angelo *European Railway Agency (ERA), Valenciennes, France*

Plsek, Paul E. *DirectedCreativity.com, Roswell, GA, USA*

Polansky, Alan M. *Northern Illinois University, De Kalb, IL, USA*

Pollard, Pat *Robert Gordon University, Aberdeen, UK*

Popova, Elmira *University of Texas at Austin, Austin, TX, USA*

Popova, Ivilina *University of Seattle, Seattle, WA, USA*

Post, Richard I. *Retired, formerly San Jose State University and formerly Intel Corporation, Santa Clara, CA, USA*

Prescott, Philip *University of Southampton, Southampton, UK*

Pulcini, Gianpaolo *CNR, Istituto Motori, Naples, Italy*

Qiu, Peihua *University of Minnesota, Minneapolis, MN, USA*

Quigley, John *University of Strathclyde, Glasgow, UK*

Räbiger, Jan *AMD Saxony LLC & Co. KG, Dresden, Germany*

Rajan, Jane *European Railway Agency (ERA), Valenciennes, France*

Redman, Thomas C. *Navesink Consulting Group, Little Silver, NJ, USA*

Reese, C. Shane *Brigham Young University, Provo, UT, USA*

Reiter, Jerome P. *Duke University, Durham, NC, USA*

Reynolds Jr, Marion R. *Virginia Tech, Blacksburg, VA, USA*

Rigdon, Christopher J. *Arizona State University, Tempe, AZ, USA*

Rigdon, Steven E. *Southern Illinois University, Edwardsville, IL, USA*

Roberts, Steven *Australian National University, Canberra, Australia*

Robinson, Jeffrey A. *General Motors R&D Center, Warren, MI, USA*

Robinson, Timothy J. *University of Wyoming, Laramie, WY, USA*

Rodebaugh Jr, William F. *GlaxoSmithKline, Philadelphia, PA, USA*

Romano, Daniele *University of Cagliari, Cagliari, Italy*
Romeu, Joan *Hospital Universitari Germans Trias i Pujol, Badalona, Spain*
Rosiński, Jan *University of Tennessee, Knoxville, TN, USA*
Rovira, Joan *Universitat de Barcelona, Barcelona, Spain*
Runeson, Per *Lund University, Lund, Sweden*
Runger, George C. *Arizona State University, Tempe, AZ, USA*
Russell, Marie *FRS Marine Laboratory, Aberdeen, UK*
Ryan, Thomas P. *Acworth, GA, USA*

Saithanu, Kidakan *Burapha University, Chonburi, Thailand*
Samaniego, Francisco J. *University of California, Davis, CA, USA*
Sanders, Doug *University of Tennessee, Knoxville, TN, USA*
Saniga, Erwin M. *University of Delaware, Newark, DE, USA*
Sarkani, Shahram *George Washington University, Washington, DC, USA*
Schueremans, Luc *Katholieke Universiteit Leuven, Heverlee, Belgium*
Seaman, Chris *City University of New York, New York, NY, USA*
Segal, Eran *Tel Aviv University, Tel Aviv, Israel*
Seidel, Wilfried *Helmut Schmidt University, Hamburg, Germany*
Sen, Ananda *University of Michigan, Ann Arbor, MI, USA*
Sen, Arusharka *Concordia University, Montréal, Quebec, Canada*
Sen, Pranab K. *University of North Carolina, Chapel Hill, NC, USA*
Sethuraman, Jayaram *Florida State University, Tallahassee, FL, USA*
Shainin, Richard D. *Shainin LLC, Livonia, MI, USA*
Shaked, Moshe *University of Arizona, Tucson, AZ, USA*
Shenk, Debra *Agilent Technologies Inc., Wilmington, DE, USA*
Shepherd, Deborah K. *Louisiana State University, Shreveport, LA, USA*
Sheu, Shey-Huei *National Taiwan University of Science and Technology, Taipei and National Taichung Institute of Technology, Taichung, Taiwan*
Shier, Douglas R. *Clemson University, Clemson, South Carolina, USA*
Shore, Haim *Ben-Gurion University of the Negev, Beer-Sheva, Israel*
Shu, Lianjie *University of Macau, Taipa, Macau*
Sims, Benjamin H. *Los Alamos National Laboratory, Los Alamos, NM, USA*
Singpurwalla, Nozer D. *George Washington University, Washington, DC, USA*
Snee, Ronald D. *Tunnell Consulting, King of Prussia, PA, USA*
Soofi, Ehsan S. *University of Wisconsin-Milwaukee, Milwaukee, WI, USA*
Soyer, Refik *George Washington University, Washington, DC, USA*
Spiring, Fred *University of Manitoba Winnipeg, Manitoba, Canada*
Spizzichino, Fabio *University "La Sapienza", Rome, Italy*
Srinivasan, Mandyam M. *University of Tennessee, Knoxville, TN, USA*
Stadje, Wolfgang *Universität Osnabrück, Osnabrück, Germany*
Stavish, Len *Cohn Consulting Group, Roseland, NJ, USA*
Steinberg, David M. *Tel Aviv University, Tel Aviv, Israel*
Stephens, Ken *University of South Florida, Tampa, FL, USA*
Stoumbos, Zachary G. *Rutgers, The State University of New Jersey, Piscataway, NJ, USA*
Sullivan, Joe H. *Mississippi State University, Starkville, MS, USA*
Suresh, Kodakanallur Krishnaswamy *Bharathiar University, Coimbatore, Tamilnadu, India*
Suzuki, Kazuyuki *University of Electro-Communications, Tokyo, Japan*

Taam, Winson *The Boeing Company, Seattle, WA, USA*
Taboada, Heidi A. *University of Texas, El Paso, TX, USA*
Tang, Boxin *Simon Fraser University, Burnaby, British Columbia, Canada*
Testik, Murat C. *Hacettepe University, Ankara, Turkey*
Teugels, Jozef L. *Katholieke Universiteit Leuven, Leuven, Belgium and Technische Universiteit Eindhoven, Eindhoven, The Netherlands*
Thomas, Marlin U. *Air Force Institute of Technology, Dayton, OH, USA*
Thyregod, Poul *Technical University of Denmark, Herlev, Denmark*
Tiwari, Ram C. *University of Nevada at Las Vegas, Las Vegas, NV, and National Cancer Institute, Bethesda, MD, USA*
Tong, Y.L. *Georgia Institue of Technology, Atlanta, GA, USA*
Tony Ng, Hon Keung *Southern Methodist University, Dallas, TX, USA*
Tort-Martorell, Xavier *Technical University of Catalonia (UPC), Barcelona, Spain*
Tortorella, Michael *Rutgers, The State University of New Jersey, Piscataway, NJ, USA*
Tosteson, Tor D. *Dartmouth Medical School, Hanover, NH, USA*
Trabert, Thomas F. *Cinteger LLC, Chapel Hill, NC, USA*
Trip, Albert *University of Amsterdam, Amsterdam, The Netherlands*
Tsiamyrtzis, Panagiotis *Athens University of Economics and Business, Athens, Greece*
Tsui, Kwok-Leung *Georgia Institute of Technology, Atlanta, GA, USA*
Tsung, Fugee *Hong Kong University of Science and Technology, Kowloon, Hong Kong*
Tyssedal, John *Norwegian University of Science and Technology, Trondheim, Norway*
Vander Wiel, Scott A. *Los Alamos National Laboratories, Los Alamos, NM, USA*
Vännman, Kerstin *Luleå University of Technology, Luleå, Switzerland*
van Dorp, Johan René *George Washington University, Washington, DC, USA*
Van Gemert, Dionys *Katholieke Universiteit Leuven, Heverlee, Belgium*
van Noortwijk, Jan M. *HKV Consultants, Lelystad, The Netherlands, and Delft University of Technology, Delft, The Netherlands*
van de Geer, Sara A. *University of Leiden, Leiden, The Netherlands*
van der Weijden, Trudy *Maastricht University, Maastricht, The Netherlands*
Vardeman, Stephen B. *Iowa State University, Ames, IA, USA*
Voelkel, Joseph G. *Rochester Institute of Technology, Rochester, NY, USA*

Walls, Lesley *University of Strathclyde, Glasgow, UK*
Walsh, Cathal D. *Trinity College Dublin, Dublin, Ireland*
Walwyn, Rebecca *King's College London, London, UK*
Wan, Rui *University of Tennessee, Knoxville, TN, USA*
Wang, Fu-Kwun *National Taiwan University of Science and Technology, Taipei, Republic of China*
Wang, Jane-Ling *University of California, Davis, CA, USA*
Wang, Xiao *University of Maryland, Baltimore, MD, USA*
Wardell, Don G. *University of Utah, Salt Lake City, UT, USA*
Watson, Gregory H. *Business Excellence Solutions, Helsinki, Finland and Oklahoma State University, Stillwater, OK, USA*
Watson, G. S. *Deceased*
Webster, Lynda *FRS Marine Laboratory, Aberdeen, UK*
Weiß, Christian H. *University of Würzburg, Würzburg, Germany*

Wells, David E. *FRS Marine Laboratory, Aberdeen, UK*

Wensing, Michel *Radboud University Nijmegen Medical Centre, Nijmegen, The Netherlands*

Whitaker, Heather J. *Open University, Milton Keynes, UK*

Wichern, Dean W. *Texas A & M University, College Station, TX, USA*

Williams, Brian J. *Los Alamos National Laboratory, Los Alamos, NM, USA*

Wilrich, Peter-Th. *Freie Universität Berlin, Promenadenstrasse, Berlin, Germany*

Wilson, Alyson G. *Los Alamos National Laboratory, Los Alamos, NM, USA*

Wilson, Gregory D. *Los Alamos National Laboratory, Los Alamos, NM, USA*

Wilson, John G. *University of Western Ontario, London, Ontario, Canada*

Wilson, Simon P. *Trinity College Dublin, Dublin, Ireland*

Winkel, Per *Copenhagen University Hospital, Blegdamsvej, Denmark*

Wiper, Michael P. *Universidad Carlos III de Madrid, Madrid, Spain*

Woodall, William H. *Virginia Tech, Blacksburg, VA, USA*

Wu, Chien-Wei *National Chiao Tung University, Hsinchu, and Feng Chia University, Taichung, Taiwan*

Xie, Min *National University of Singapore, Singapore*

Yang, Bo *University of Electronic Science and Technology, Chengdu, China*

Yashchin, Emmanuel *IBM Research Division, Yorktown Heights, NY, USA*

Yechiali, Uri *Tel Aviv University, Tel-Aviv, Israel*

Yeh, Arthur B. *Bowling Green State University, Bowling Green, OH, USA*

Young, John C. *McNeese State University, Lake Charles, LA, USA*

Zacks, Shelemyahu *Binghamton University, Binghamton, NY, USA*

Zamba, Kokou D. *University of Iowa, Iowa City, IA, USA*

Zhang, Nien Fan *National Institute of Standards and Technology, Gaithersburg, MD, USA*

Zhang, Yao *Pfizer Global Research and Development, Groton, CT, USA*

Zhu, Yu *Purdue University, West Lafayette, IN, USA*

Zio, Enrico *Polytechnic of Milan, Milan, Italy*

Zlotin, Boris *Ideation International Inc., Farmington Hills, MI, USA*

Zuo, Ming J. *University of Alberta, Edmonton, Alberta, Canada*

Zusman, Alla *Ideation International Inc., Farmington Hills, MI, USA*

Perspectives on Quality and Reliability

Statistics: the Key to Improvement

Quality and reliability are hot topics today – as well they should be. We still hear too often about the disastrous consequences of inadequate quality or reliability. Non-robust design of space vehicle components, fire hazards of computer batteries, and inadequate assessments of the side effects of new drugs are just a few examples. But these highly publicized cases are only the tip of the proverbial iceberg. Companies throughout the world lose vast sums of money each year due to rejects, rework, and warranty claims; to say nothing about loss of business from dissatisfied customers and expensive lawsuits.

As consumers, we have become sensitive to shoddy products and services as we travel on airplanes, wait in doctor's offices, or try to open containers that will not open. And we have learned to appreciate the many cases when products do perform well and when we are provided good service.

Interest in quality hails back to the days of Shewhart, close to a century ago. General recognition of its criticality did not surface in the Western world, however, until Deming returned from Japan and participated in the now historic 1980 television white paper provocatively entitled "If Japan Can, Why Can't We?" Various quality gurus, such as Crosby, Feigenbaum, Golomski, Juran, and Shainin-as well as Deming-gained rapid prominence. The prestigious Malcolm Baldrige Award was established by the US Congress in 1987 to reward high quality.

As we became more customer-focused, we also became more sensitive to the importance of high reliability-that is quality over time. And most importantly, enlightened management recognized the need to replace a "ship and fix" mentality with the concept of "doing it right the first time" by designing products and services for high quality and reliability.

It rapidly became clear that a disciplined approach to quality and reliability improvement was needed. Inherent in the earlier Shewhart-Deming PDSA (plan, do, study, act) cycle, this became the foundation of various new initiatives. Prominent among these is Six Sigma-which adopted the slogan "In God we trust, all others bring data." Leveraging our data-rich environment and computer-intensive technology is, indeed, a key element of quality and reliability improvement. And modern statistics is at the heart of it all-to which I can bear witness after a half-century career in industry.

So how do we proceed? Most of the quality gurus, unfortunately, are no longer here to guide us. So we are on our own. To address the challenge, statisticians and engineers, mostly from academia, have developed an impressive arsenal of tools. Together with good scientific and applications-area knowledge and an inquisitive mindset, these help us steer successfully through the rough seas of quality and reliability improvement.

But here is the rub. Information about these tools is scattered over scores of technical journals and conference transactions, making it hard for practitioners to locate. The work, moreover, is often presented in a highly technical manner, often intelligible only to fellow researchers.

That is where this Encyclopedia comes in. It aims to present quality and reliability concepts and methods ranging from fundamental to advanced, and do so in a complete, comprehensible, and understandable manner. It provides numerous cross-references to permit seamless movement from one topic to the next. It addresses a key need of today's quality and reliability community who come to the subject with widely varying backgrounds-from students to Ph.D. statisticians and from design engineers to Six Sigma Black Belts and quality managers.

Gerald J. Hahn
Retired
GE Corporate Research and Development
November 2007

Defining Perfection

Perfect quality, perfect delivery, perfect reliability, perfect service-these are achievable. The content of this encyclopedia is a substantial endorsement of these objectives, and the principles and methodologies for the achievement of these results. The quality system that will be embraced by any particular organization that takes the subject very seriously will aim for these goals and be measurable by the appropriate and dedicated use of the statistical systems that are now readily available.

The company that I was with for a lifetime was keenly dedicated to the measurements of our quality efforts. The term Six Sigma describes and emphasizes the fact that the system was a mathematical system, along with many other of its qualities.

At times we may feel as if we are generating an overload of information, but we cannot have too much relevant data to know where we are and what the expectations can be for an even more perfect system. Not knowing how our products and services stack up against competition is a void that must not be accepted. Only with constant monitoring and continual follow-up can an institution have a true picture of where one stands and a clear view of where one needs to go. Sometimes statistical data measures our failures; acknowledge them and improve. They also measure our successes, and the more we operate to the expectation of perfection, the measurements will please us with our successes.

A substantial value of a major reference work like this is to challenge us as to whether we think we know what we know. When we first started with our quality system and our statistical system, many of us assumed that we had already learned all the fundamentals. When we got right down to it, we had a lot to teach ourselves. That is the value of taking a good look in the mirror, taking a good look at these pages, and determining that there is still something more we can do to perfect our systems in order to achieve the perfect results.

Robert W. Galvin
Retired
Motorola Inc.
November 2007

Preface

Quality and Reliability are terms which permeate our daily experience. Statistics is a basic discipline in science and management. The domain of Statistics in Quality and Reliability covers a wide range of applications, from consumer products to health care, manufacturing to software engineering, and electronics to the environment. This Encyclopedia has been designed to provide a comprehensive and integrated knowledge source for anyone having a theoretical or practical interest in applications of statistics to any of these or related fields.

Our lives are filled with daily quests for "good", "reliable" products: we'd like to start our day with a good cup of coffee from a reliable espresso machine, use highly serviceable and functional products throughout the day, and fall asleep in a cozy, quiet bed at night. In fact, the search for high quality, reliable products gets more intense with each passing day. In industry, quality and reliability are properties that can make the difference between success and failure. In a highly competitive global market, the buyer's choice is almost always influenced by the search for a balance between cost on the one hand, and quality/reliability on the other. The general public may have no knowledge of the technical terms our profession applies to these concepts, but they have a clear, intuitive idea of what quality and reliability are. In our parlance, quality denotes a product's performance to given specifications and/or expectations (efficiency), and how these needs and expectations are met (effectiveness), whereas reliability denotes the degree of permanence of such performances over time.

Quality and reliability are not principally deterministic phenomena. They are the result of a series of factors (design, production, operation, environment, etc.) which define and influence them. The variable nature of these factors makes it natural to search for stochastic models that describe, plan for, measure, predict, and track quality and reliability. As a consequence, in the last century, these became important fields of study for applied probabilists and statisticians.

There are any number of communities interested in quality and reliability: business leaders, product and process designers, industrial and quality engineers, and academic researchers, to name a few. When Wiley invited us to serve as Editors-in-Chief of the Encyclopedia of Statistics in Quality and Reliability (EQR), we agreed from the outset on the need to provide a reference source that would be of interest to practitioners and researchers alike. Therefore, we sought a mix of contributions which would present both basic principles and advanced mathematical methods, shown wherever possible in action on real problems. By this route, we hope to illustrate the applicability of quality and reliability concepts and practices throughout business, science, society, and everyday life.

This project started with Wiley contacting leading researchers from all over the world to determine whether there was need for a major reference work in these two related fields. The answers were, in general, very positive, and the project started in earnest with the appointment of three Editors-in-Chief who had a wide range of experience across both academia and business. The next step was the identification of critical topical areas that would span the subject, and the search for outstanding experts in these fields who would commission papers in cooperation with the Editors-in-Chief. Editors were appointed for twelve sections:

- Mike Adams (University of Alabama, USA), *Process Control*
- Connie Borror (Arizona State University, USA), *Process Capability & Measurement Systems Analysis*
- Jeroen de Mast (University of Amsterdam, The Netherlands), *Basics Statistics*
- Rainer Göb (University of Würzburg, Germany), *Sampling*
- Blanton Godfrey (North Carolina State University, USA) and Ramón León (University of Tennessee, USA), *Management of Quality & Business Statistics*
- Tony Greenfield (Greenfield Research, UK), *Health, Safety & Environmental Applications*
- Tom Mazzuchi (George Washington University, USA), *Reliability: Life Distribution Modeling & Accelerated Testing*
- Steven Rigdon (Southern Illinois University, USA), *System Reliability*
- Refik Soyer (George Washington University, USA), *Reliability: Life Cycle & Warranty Cost Prediction*
- David Steinberg (Tel Aviv University, Israel), *Design of Experiments & Robust Design*
- Simon Wilson (Trinity College Dublin, Ireland), *Computationally Intensive Methods & Simulation*
- Shelley Zacks (Binghamton University, USA), *Statistical & Stochastic Modeling*

The Section Editors, in cooperation with the Editors-in-Chief, selected key topics and commissioned leading authors to write authoritative articles on each. It has been a long process since then, but it is worth mentioning that most of the potential contributors contacted worldwide agreed to contribute. We are very proud to present four outstanding volumes comprising over 400 entries, written by some of the world's most prominent authors in their respective fields. We would like to thank all of them for their very substantive and insightful contributions. Among other things, their works provide extensive illustration of the cumulative knowledge attained to date in each designated field, and a richness of examples which will prove invaluable to readers seeking to divine a comprehensive set of concepts and tools in quality and reliability.

We would be remiss were we not to acknowledge the remarkable work of the Section Editors, who not only identified and invited contributing authors, but also acted as referees for the papers submitted. The realization of this project could have not been possible without their contribution, and we wish to extend them our gratitude and admiration. In addition, we applaud the staff of John Wiley & Sons, whose consummate professionalism has been a further key factor in the successful completion of this work. In particular, we would like to recognise the experience and leadership of David Hughes, Liz Smith, Kerry Powell, Tony Carwardine, Jill Hawthorne, and Sian Jones, who defined and guided the relationship of the editorial and production teams throughout the project, providing us direction and assistance all the way. We acknowledge, as well, the contribution of Sangeetha and the team at Laserwords Private Ltd., who were responsible for the copy-editing and typesetting of all the papers.

Fabrizio Ruggeri
Ron S. Kenett
Frederick W. Faltin
November 2007

Abbreviations and Acronyms

ABC	Activity-Based Costing
ACASI	Audio Computer-Assisted Self Interview
ACCP	Accelerated Central Cutting Plane
ACL	Acceptance Control Limit
ACO	Ant Colony Optimization
ACSI	American Customer Satisfaction Index
ADT	Accelerated Degradation Testing
AECM	Alternating Expectation Conditional Maximization
AFD	Anticipatory Failure Determination
AFI	Average Fraction of Items
AFM	Atomic Force Microscope
AFT	Accelerated Failure Time
AIAG	Automotive Industry Action Group
AICPA	American Institute of Certified Public Accountants
AIC	Akaike Information Criterion
ALMs	Accelerated Life Models
ALTs	Accelerated Life Tests
AMOC	At Most One Change
AMR	Average Moving Range
ANOVA	Analysis of Variance
ANSI	American National Standards Institute
AO	Additive Outliers
AOQ	Average Outgoing Quality
AOQL	Average Outgoing Quality Level
AOV	Analysis of Variance
APARs	Authorized Program Analysis Reports
APC	Advanced Process Control
APICS	Association for Operations Management
APIMS	Atmospheric Pressure Ion Mass Spectrometers
APL	Acceptable Process Level
APP	Allocation of Priorities Principle
APPL	A Probability Programming Language
AQL	Acceptable Quality Level
AQL	Average Quality Level
ARCH	Autoregressive Conditional Heteroscedastic
ARDS	Acute Respiratory Distress Syndrome
ARFIMA	Auto-Regressive Fractionally Integrated Moving Average
ARIMA	Autoregressive Integrated Moving Average
ARIZ	Algorithm of Inventive Problem Solving
ARL	Average Run Length

ARMA Auto-Regressive Moving Average
ARS Adaptive Rejection Sampling
ART Adaptive Resonance Theory
ASA American Statistical Association
ASL Actual Significance Level
ASN Average Sample Number
ASPC Algorithmic Statistical Process Control
ASQ American Society for Quality
ATI Average Total Inspection
ATS Average Time to Signal

BBD Box–Behnken Design
BBS Block, Borges, and Savits
BC_a Accelerated Bias-Corrected
BCS Binary Coherent System
BDD Binary Decision Diagram
BEQUALM Biological Effects Quality Assurance in Monitoring Programmes
BHAG Big Hairy Audacious Goal
BIBD Balanced Incomplete Block Design
BIC Bayesian Information Criterion
BINCDF Binomial Cumulative Distribution Function
BLM Bivariate Lack of Memory
BLUEs Best Linear Unbiased Estimators
BLUP Best Linear Unbiased Predictor
BN Bayesian Network
BP Bacteriological Peptone
BP Brown and Proschan
BPR Business Process Reengineering
BRD Bayesian Reliability Demonstration
BS British Standard
BTMSCS Binary Type Multistate Strongly Coherent System
BUGS Bayesian Inference Using Gibbs Sampling
BYM Besag, York, and Mollié

CA Component Amounts
CABG Coronary Artery Bypass Graft
CAI Computer-Assisted Interviewing
CAPI Computer-Administered Personal Interviewing
CaRT Classification and Regression Trees
CAR Combined Age-Based Replacement
CASI Computer-Assisted Self Interview
CASIC Computer-Assisted Survey Information Collection
CATI Computer-Assisted Telephone Interviewing
CAWI Computer-Assisted Web Interviewing
CB Chlorinated Biphenyl
CBM Condition-Based Maintenance
CC Categorized Component
CCADs Central Composite Analog Designs
CCC Core Conflict Cloud
CCD Central Composite Design

CCPM	Critical-Chain Project Management
CCPS/AIChE	Center for Chemical Process Safety/American Institute of Chemical Engineers
CCP	Central Cutting Plane
CCs	Complete Cases
CD	Centered L_2 Discrepancy
CD	Critical Dimension
CDA	Confirmatory Data Analysis
CDC	Centers for Disease Control
CDF	Cumulative Distribution Function
CDR	Critical Design Review
C&E	Cause and Effect
CFR	Combined Failure-Based Replacement
CFTP	Coupling From the Past
CHAID	Chi-Squared Automatic Interaction Detector
CHF	Cumulative Hazard Function
ChSP-1	Chain Sampling Plan
CHSS	Changing Shape and Scale
CI	Confidence Interval
CI	Continuous Improvement
CIAFTP	Coupling Into and From the Past
CIPM	International Committee for Weights and Measures
CIRP	Constant-Interval Replacement Policy
CL	Center Line
CLRT	Central Limit Resampling Theorem
CM	Corrective Maintenance
CMOS	Complementary Metal Oxide Semiconductor
CMP	Chemical Mechanical Polishing
CMR	Comparative Mean Ratio
COD	Close-Out Designs
CP	Cutting Plane
C2P	Constrained Two-Step Procedure
CPD	Conditional Probability Distribution
CPF	Conditional Probability Function
CPO	Conditional Predictive Ordinate
CPP	Compound Poisson Process
CPs	Component Proportions
CPS	Current Population Survey
CPT	Conditional Probability Table
CPU	Central Processing Unit
CPUE	Catch Per Unit Effort
CQL	Consumer's Quality Level
CR	Consumer's Risk
CRAG	Clinical Resource and Audit Group
CRD	Conflict Resolution Diagram
CRPs	Compound Renewal Processes
CRQ	Consumer's Risk Quality
CRT	Current Reality Tree
CSAQ	Computerized Self-Administered Questionnaires
CSEMP	Clean Seas Environmental Monitoring Programme

CSF	Critical Success Factor
CSIs	Common Safety Indicators
CSLM	Component-Slope Linear Mixture
CSM	Chopped Strand Mat
CSP	Continuous Sampling Plan
CSTs	Common Safety Targets
CTQ	Critical-to-Quality
CTS	Critical-to-Satisfaction
CUSCORE	CUmulative SCORE
CUSUM	CUmulative SUM
CVB	Chakraborti, Van Der Laan, and Bakir
CWP	Compound Weibull Process
DACE	Design and Analysis of Computer Experiments
DAG	Directed Acyclic Graph
DBM	Disk by Mail
DBR	Drum–Buffer–Rope
DCS	Digital Control System
DE	Desired Effect
DE	Directed Evolution
DES	Discrete-Event Simulation
DF	Distribution Function
DF	Degrees of Freedom
DFR	Decreasing Failure Rate
DFSS	Design for Six Sigma
DGI	Dirección General de Investigación
DHS	Dorado, Hollander, and Sethuraman
DIC	Deviance Information Criterion
DLM	Dynamic Linear Model
DLS95	Damien, Laud, and Smith Algorithm
DMADV	Define, Measure, Analyze, Design, Verify
DMAIC	Define, Measure, Analyze, Improve, Control
DMAWC	Define, Measure, Analyze, Work-out, Control
DMEA	Damage Modes and Effects Analysis
DMRL	Decreasing Mean Residual Life
DoCDat	Directory of Clinical Databases
DoD	Department of Defense
DOE	Design of Experiments
domCFTP	Dominated Coupling from the Past
DP	Dirichlet Process
DP	Dynamic Programming
DPM	Dirichlet Process Mixture
DPMO	Defect Per Million Opportunities
DPs	Design Parameters
DPU	Defects Per Unit
DR	Discrimination Ratio
D&R	Delete and Revise
DRAM	Delayed Rejection in Adaptive Metropolis
DRD	Dynamic Robust Design
DRI	Directly Riemann Integrable

DS	Directional Sampling
DTP	Diphtheria-Pertussis-Tetanus
DTS	Degradation-Threshold-Shock
DVT	Deep Venous Thrombosis
EB	Empirical Bayes
ECG	Electrocardiogram
ECM	Expectation Conditional Maximization
EDA	Exploratory Data Analysis
EDF	Empirical Distribution Function
EDP	Early Detection Program
EEAC	Expected Equivalent Average Cost
EFQM	European Foundation for Quality Management
EIV	Errors-in-Variables
EM	Expectation Maximization
EMS	Expected Mean Squares
EOS	Equation of State
EPC	Engineering Process Control
ER	Entity-Relationship
ERA	European Railway Agency
ERP	Enterprise Resource Planning
ESP	Extrasensory Perception
E-Step	Expectation Step
ESS	Equivalent Sample Size
ETR	Expected Total Reserve
EU	Experimental Unit
EVI	Extreme Value Index
EWMA	Exponentially Weighted Moving-Average
FA	Failure Analysis
FAR	False Alarm Rate
FCCD	Face-Center Cube Design
FDA	Food and Drug Administration
FDM	Functional Deployment Map
FDS	Fraction of Design Space
FEPA	Food and Environment Protection Act
FF	False Failure
FF	Fractional Factorial
FFS	Fee-for-Service
FIR	Fast Initial Response
FK74	Ferguson and Klass Algorithm
FMEA	Failure Mode and Effect Analysis
FMECA	Failure Mode, Effect and Criticality Analysis
FMR	Forever Minimal Repairs
FOM	Force of Mortality
FORM	First-Order Reliability Method
FPR	Forever Perfect Repairs
FRD	Factor Relationship Diagram
FRP	Fiber Reinforced Plastic
FRT	Future Reality Tree

FRW	Free Replacement Warranty
FS	Fisher Scoring
FSI	Fixed Sampling Interval
FSP	False Signal Probability
FSR	Fail–Safe Redundancy
FSR	Fixed Sampling Rate
FSS	Fail–Safe System
FTA	Fault Tree Analysis
GA	Genetic Algorithm
GARCH	Generalized Auto-Regressive Conditional Heteroscedasticity
GBS	Guillain-Barré Syndrome
GC	Gas Chromatographs
GCIs	Generalized Confidence Intervals
GDA	Great Deluge Algorithm
GEE	Generalized Estimating Equations
GEM	Generalized Expectation Maximization
GET	General Equivalence Theorem
GI	Gini Index
GIDEP	Government-Industry Data Exchange Programme
GLM	Generalized Linear Model
GLPH	Generalized Linear Proportional Hazards
GLR	Generalized Likelihood Ratio
GM	Graphical Model
GMA	Generalized Minimum Aberration
GMP	Good Manufacturing Practice
GOF	Goodness-of-Fit
GP	Gaussian Process
GPH	Generalized Proportional Hazards
GPQ	Generalized Pivotal Quantity
GPRS	General Packet Radio Service
GPs	General Practitioners
GPS	Global Positions System
GRI	Generalized Rule Induction
GRR	Gauge Repeatability and Reproducibility
GSM	Generalized Sedyakin Model
GSR	Grouped Signed-Rank
GSR-CUSUM	Grouped Signed-Rank Cumulative Sum
GSR-EWMA	Grouped Signed-Rank Exponentially Weighted Moving Average
GUM	Guide to the Expression of Uncertainty in Measurement
HALT	Highly Accelerated Life Tests
HAZAN	Hazard Analysis
HBP	High Blood Pressure
HDD	Hard Disk Drives
HDS	Historical Data Set
HEDIS	Health Plan Employer Data and Information Set
HEP	Human Error Probability
HF	Hydrofluoric Acid
HMM	Hidden Markov Models

HNBUE	Harmonic New Better Than Used in Expectation
HOQ	House of Quality
HP	Hewlett–Packard
HPP	Homogeneous Poisson Process
HPS	Hollander, Presnell and Sethuraman
HSD	Honestly Significant Difference
HSE	Health Survey for England
IA	Immune Algorithm
IAR	Integrated Alternator Regulator
IBD	Incomplete Block Design
IBF	Intrinsic Bayes Factor
IC	In-Control
ICC	Intraclass Correlation Coefficient
ICD	International Classification of Diseases
ICI	Imperial Chemical Industries
ICP-MS	Ion-Coupled Plasma Mass Spectrometers
ICU	Intensive Care Unit
IEC	International Engineering Consortium
IFR	Increasing Failure Rate
IFRA	Increasing Failure Rate Average
IG	Inverse Gaussian
IHI	Institute of Healthcare Improvement
IID	Independent and Identically Distributed
IIDRVs	Independent Identically Distributed Random Variables
IMA	Integrated Moving Average
IMEP	Indicated Mean Effective Pressure
IMF	International Monetary Fund
IMRL	Increasing Mean Residual Life
IMSE	Integrated Mean Squared Error
IO	Innovative Outliers
IO	Intermediate Objective
IPF	Iterative Proportional Fitting
IQR	Interquartile Range
IRLS	Iteratively Reweighted Least Squares
IRR	Internal Rate of Return
IRs	Imperfect Repairs
ISMC	Importance Sampling Monte Carlo
ISO	International Organization for Standardization
ISQ	Innovation Situation Questionnaire
IT	Information Technology
ITP	Idiopathic Thrombocytopenic Purpura
ITT	Intent-to-Treat
IVs	Independent Variables
IVHM	Integrated Vehicle Health Management
IVR	Interactive Voice Response
JAN-STD	Joint Army–Navy Standard
JIPM	Japan Institute of Plant Maintenance
JIT	Just-in-Time

JM	Jelinski and Moranda
JPD	Joint Probability Distribution
JQT	Journal of Quality Technology
JSPI	Journal of Statistical Planning and Inference
JUSE	Union of Japanese Scientists and Engineers
KDD	Knowledge Discovery in Databases
KF	Kalman Filter
KME	Kaplan–Meier Estimator
KPI	Key Performance Indicator
KS	Kolmogorov–Smirnov
LA	Local Authority
LC	Liquid Chromatograph
LCC	Life Cycle Cost
LCD	Liquid Crystal Display
LCL	Lower Control Limit
LCVs	Left-Censored Values
LDL	Lower Detectable Limit
LDs	Layered Designs
LHD	Load–Haul–Dump
LL	Lower Limit
LLS	Log-Location Scale
LMP	Locally Most Powerful
LOCF	Last Observation Carried Forward
LOD	Limit of Detection
LOGD	Limit of Guaranteed Detection
LOQ	Limit of Quantization
LOR	Line of Restriction
LORA	Level of Repair Analysis
LP	Linear Programming
LQL	Limiting Quality Level
LR	Likelihood-Ratio
LRT	Likelihood-Ratio Test
LS	Last and Szekli
LSF	Limit State Function
LSFE	Limit State Function Evaluation
LSI	Least Significant Interval
LSIP	Linear Semi-Infinite Programming
LSL	Lower Specification Limit
LST	Laplace–Stieltjes Transform
LT	Laplace Transform
LTPD	Lot Tolerance Percent Defective
LVQ	Learning Vector Quantization
5M1E	Machine, Man, Material, Method, Measurement, and Environment
6-MP	6-mercaptopurine
MA	Minimum Aberration
MA	Mixture Amount
MA	Moving Average
MAD	Median Absolute Deviation

MANOVA	Multivariate Analysis Of Variance
MAPE	Mean Absolute Percentage Error
MAR	Missing at Random
MAR	Multivariate Age Replacement
MAZ	Mixture-Amount-Zero
MB	Malcolm Baldrige
MBOs	Management by Objectives
MBP	Management by Policy
MBRP	Modified Block Replacement
MC	Managed Care
MCAR	Missing Completely at Random
MCF	Mean Cumulative Function
MCHR	Multivariate Conditional Hazard Rate
MCMC	Markov Chain Monte Carlo
MCSs	Multistate Coherent Systems
MCUSUM	Multivariate Cumulative Sum
MDCs	Markov Decision Chains
MDE	Maximum Dynamic Entropy
MDL	Method Detection Limit
MDL	Minimum Description Length
MDP	Mixture of Dirichlet Process
ME	Margin of Error
MEDLARS	MEDical Literature and Retrieval System
MESD	Minimum Engineering Significant Difference
MEs	Main Effects
MEWMA	Multivariate Exponentially Weighted Moving Average
MEWMS	Multivariate Exponentially Weighted Moving Squared-Deviation
MF	Missed Fault
MFM	Multifactor Mixture
MGFs	Moment Generating Functions
MH	Modified Harvey
MI	Multiple Imputation
MI	Myocardial Infarction
MIFR	Multivariate Increasing Failure Rate
MIFRA	Multivariate Increasing Failure Rate Average
MIL-STD	Military Standard
MILP	Mixed Integer Linear Program
MINQUE	Minimum Norm Quadratic Unbiased Estimation
MIS	Management Information Systems
MISL	Mixed MIL-Std 105D and Skip-Lot Sampling System
MIVs	Mathematically Independent Variables
ML	Maximum Likelihood
MLE	Maximum Likelihood Estimate
MLP	Multi-Layer Perceptron
MLPs	Multilevel Plans
MLR	Major Labor Force Recode
MLS	Modified Large Sample
MM	Markov Model
MMRM	Mixed Model for Repeated Measures
MMR	Measles, Mumps, and Rubella

MMS	Multistate Monotone System
MMSE	Minimum Mean Square Error
MNAR	Missing Not at Random
MNBU	Multivariate New Better than Used
MoM	Mixture of Mixtures
MOPs	Multiple-Objective Problems
MOS	Metal-Oxide Semiconductor
MOSDs	Marginally Oversaturated Designs
MP	Maintenance Prevention
MP	Markov Process
MPLP	Modulated Power Law Process
MPT	Mixture of Polya Trees
MPV	Mixture-Process Variable
MR	Mean Ratio
MR	Minimal Repair
MR	Moving Range
MRFF	Model Robust Fractional Factorial
MRP	Minimal Repair Process
MRR	Multivariate Renewal Replacement
MRs	Modification Requests
MRSA	Methicillin-Resistant *Staphylococcus Aureus*
MSA	Measurement Systems Analysis
MSCS	Multistate Strongly Coherent System
MSE	Mean Square Error
MSG	Maintenance Steering Group
MSPC	Multivariate Statistical Process Control
M-Step	Maximization Step
MSs	Mean Squares
MSs	Member States
MSS	Multistate System
MSTr	Treatment Mean Square
MTBF	Mean Time Between Failure
MTP_2	Multivariate Totally Positive of Order 2
MTTF	Mean Time to Failure
MTTR	Mean Time to Repair
MW	Mann–Whitney
MWCS	Multistate Weakly Coherent System
MWW	Mann–Whitney–Wilcoxon
NABU	New Better That Used in Average
NASC	Necessary And Sufficient Condition
NAVORD	Navy Ordnance Standard
NBR	Negative Branch Reservations
NBUC	New Better Than Used in Convex Order
NBUE	New Better Than Used in Expectation
NCHOD	National Center for Health Outcomes Development
NCPPM	Nonconforming Parts Per Million
NDA	Normal Distribution Approximation
NFT	Near Feasible Threshold
NGKF	Nonlinear Kalman Filter

NHPP	Nonhomogeneous Poisson Process
NHS	National Health Service
NIBs	National Investigation Bodies
NIH	National Institutes of Health
NIST	National Institute of Standards and Technology
NMAR	Not Missing at Random
NMR	N-Modular Redundancy
NN	Neural Network
NOAME	Nearly Orthogonal Array Main Effect
NObs	Number of Observations
NPCUSUM	Nonparametric Cumulative Sum
NPM	Nonconforming Items Per Million
NPMLE	Nonparametric Maximum-Likelihood Estimator
NPV	Net Present Value
NR	Newton–Raphson
NSAs	National Safety Authorities
NSGA	Nondominated Sorting Genetic Algorithm
NTB	Nominal-The-Best
NTR	Neutral-To-The-Right
NVP	N-Version Programming
NWUE	New Worse Than Used in Expectation
OA	Orthogonal Array
OAME	Orthogonal Array Main Effect
OCAP	Out-of-Control Action Plan
OC	Operating Characteristic
OEE	Overall Equipment Efficiency
OEs	Operational Expenses
OFAT	One-Factor-at-a-Time
OH	Open Heating
OK	Ordinary Kriging
OLAP	On-Line Analytical Processing
OLS	Ordinary Least Squares
OLTP	On-Line Transaction Processing
OME	Orthogonal Main Effect
OMED	Orthogonal Main Effect Design
OMFs	Observation Measurement Functions
ONS	Office for National Statistics
OOS	Out-of-Specification
OPE	Overall Plant Efficiency
OREDA	Offshore Reliability Data
OSPAR	Oslo and Paris Commissions
OSTD	Ordance Standard
PALT	Partially Accelerated Life Test
PASTA	Poisson Arrivals See Time Averages
PAT	Process Analytical Technology
PAVA	Pool Adjacent Violators Algorithm
PB	Partition-Based
PB	Plackett and Burman

PB	Pressure Bomb
PB	Purely Bayes
PBD	Partition-Based Dirichlet
PBIBD	Partially Balanced Incomplete Block Designs
PCA	Principal Component Analysis
PCB	Printed Circuit Board
PCE	Process Cycle Efficiency
PCI	Process Capability Index
PCO	Primary Care Organization
PDCA	Plan–Do–Check–Act
PDF	Probability Density Function
PDM	Positive Dependence and Multivariate
PDMOSA	Pareto Domination Based Multi-Objective Simulated Annealing
PDR	Preliminary Design Review
PerMIA	Performance Measure Independent of Adjustment
PEXE	Piecewise Exponential Estimator
PGF	Probability Generating Function
PGW	Power Generalized Weibull
PH	Peña and Hollander
PH	Proportional Hazard
PHS	Presnell, Hollander and Sethuraman
PI	Policy Improvement
PI	Profitability Index
PI	Protease Inhibitor
PID	Proportional Integral Derivative
PIM	Proportional Intensity Modeling
PL	Product-Limit
PLMLE	Partial-Likelihood, Maximum-Likelihood Estimator
PLP	Power Law Process
PLS	Partial Least Squares
PLS	Projection to Latent Structures
PM	Preventive Maintenance
PMFs	Population Measurement Functions
PMMs	Pattern-Mixture Models
PMRI	Preventive Maintenance
POOGI	Process of Ongoing Improvement
POP	Persistent Organic Pollutant
POT	Peaks-Over-Threshold
P–P	Probability–Probability
PPM	Product Portfolio Management
PQ	Producer's Quality
PQD	Positive Quadrant Dependence
PQM	Partial Quadratic Mixture
PR	Perfect Repair
PR	Producer's Risk
PRESS	Prediction Error Sum of Squares
PRL	Positive Reinforcement Loop
Proc IML	Interactive Matrix Language Procedure
PRT	Prerequisite Tree
PRW	Pro Rata Warranty

PSA	Pareto Simulated Annealing
PSA	Propensity Score Adjustment
PSA	Prostate-Specific Antigen
PSE	Practical Statistical Efficiency
PSE	Pseudo Standard Error
PSUs	Primary Sampling Units
PT	Proficiency Test
P/T	Precision-to-Tolerance
PTR	Precision-to-Tolerance Ratio
PTV	Percent of Total Variation
PTW	Preston–Tonks–Wallace
PUOD	Positive Upper Orthant Dependence
PVR	Precision Voltage Reference
PVs	Process Variables
PWM	Position Weight Matrix
QA	Quality Assurance
QAR	Quality Audit and Review
QC	Quality Control
QE	Quality Engineering
QFD	Quality Function Deployment
QI	Quality Improvement
QMP	Quality Measurement Plan
QMS	Quality Management System
QN	Quasi-Newton
Q–Q	Quantile–Quantile
RAP	Redundancy Allocation Problem
RB	Recovery Block
RBD	Reliability Block Diagram
RBF	Radial Basis Function
RCBD	Randomized Complete Block Design
RCTs	Randomized Clinical Trials
RD	Robust Design
RDA	Recurrence Data Analysis
RDB	Reliability Databases
REML	Residual Maximum Likelihood
REML	Restricted Maximum Likelihood
RG	Reliability Growth
RI	Relative Incidence
RM	Raw Material
RMS	Root Mean Square
RMSE	Root Mean Square Error
ROC	Receiver Operating Characteristic
ROCOF	Rate of Occurrence of Failures
ROI	Return on Investment
RP	Renewal Process
RPL	Rejectable Process Level
RPM	Revolutions Per Minute
RPN	Risk Priority Number

RQL	Rejectable Quality Level
R&R	Repeatability and Reproducibility
RSM	Response Surface Modeling
RTY	Rolled Throughput Yield
RV	Random Variable
SAE	Society of Automotive Engineering
SAS	Statistical Analysis System
SCAD	Smoothly Clipped Absolute Deviation
SCM	Sugar Cane Molasses
SCRIMP	Seeman Composite Resin Infusion Molding Process
SD	Standard Deviation
SDD	System Downtime Distribution
SDDS	Special Data Dissemination Standard
SDMX	Statistical Data and Metadata Exchange
SE	Standard Error
SEM	Scanning Electron Microscope
SEM	Stochastic Expectation Maximization
SF	Survival Function
S–F	Substance–Field
SH	Sethuraman and Hollander
SHA	Strategic Health Authority
SIC	Schwarz Information Criterion
SID	Stationary Initial Distribution
SIDS	Sudden Infant Death Syndrome
SIL	Smiths Industries Ltd.
SIPOC	Supplier, Inputs, Process, Outputs, and Customer
SIPP	Survey on Income and Program Participation
SIT	Structured Inventive Thinking
SkSP	Skip-Lot Sampling Plan
SM	Sedyakin Model
SMs	Selection Models
SMA	Straight Moving Average
SMART	Self-Monitoring Analysis and Reporting Technology
SMART	Specific, Measurable, Achievable, Realistic, and with a Known Time Scale
SME	Simultaneous Margin of Error
SMOSA	Suppapitnarm Multi-Objective Simulated Annealing
SMR	Standardized Mortality Ratios
SMS	Short Message Service
SN	Signal-to-Noise
SNK	Student–Newman–Keuls
SNR	Signal-to-Noise Ratio
SOC	Standard Occupation Classification
SOMs	Self-Organizing Maps
SOP	Standard Operating Procedure
SORM	Second-Order Reliability Method
SOX	Sarbanes–Oxley
SPC	Statistical Process Control
SPE	Squared Prediction Error
SPI	Statistical Product Inspection

SPICE	Simulation Program with Integrated Circuit Emphasis
SPRT	Sequential Probability Ratio Test
SPs	Splines
SQC	Statistical Quality Control
SR	Signed-Rank
SRGM	Software Reliability Growth Model
SRI	Stanford Research Institute
SRP	Superimposed Renewal Process
SRR	Software Requirements Review
SRS	Simple Random Sampling
SS	Sums of Squares
SSALT	Step-Stress Accelerated Life Tests
SSC	Strong Slater Condition
SSE	Error Sum of Squares
SSM	Structural Safety Measure
SSN	Social Security Number
SSUs	Secondary Sampling Units
SSVS	Stochastic Search Variable Selection
STAM	Software Trouble Assessment Matrix
S&T	Strategy and Tactics
SV	Slack Variable
T-ACASI	Telephone Version of Audio Computer-Assisted Self Interview
TAAF	Test-Analyze-and-Fix
TER	Tolerance of Electric Resistance
TET	Total Elapsed Time
T&EM	Trial-and-Error Method
TFR	Tampered Failure Rate
TMR	Triple-Modular Redundancy
TOC	Theory of Constraints
TOC	Total Organic Carbon
TOCICO	Theory of Constraints International Certification Organization
TOP	Top Event
TPM	Total Productive Maintenance
TQM	Total Quality Management
TRIZ	Theory of Inventive Problem Solving
TRs	Trouble Reports
TS	Tabu Search
TT	Transition Tree
TTF	Time-to-Failure
TTT	Total Time on Test
UBIBD	Unreduced Balanced Incomplete Block Design
UCL	Upper Control Limit
UDE	Undesirable Effects
UFD	Up-Front Designs
UGF	Universal Generating Function
UL	Upper Limit
UMGF	Universal Moment Generating Function
UMOSA	Ulungu Multi-Objective Simulated Annealing

UMVUEs	Uniformly Minimum Variance Unbiased Estimators
USL	Upper Specification Limit
USS	Unweighted Sum of Squares
VA	Veterans Administration
VAERS	Vaccine Adverse Event Reporting System
VAP	Ventilator-Associated Pneumonia
VAT	Value-Added Time
VCASI	Video Computer-Assisted Self Interview
VDG	Variance Dispersion Graph
VFD	Vacuum Fluorescent Display
VIF	Variance Inflation Factor
VLAD	Variable Life-Adjusted Display
VMI	Vendor-Managed Inventory
VOC	Voice of the Customer
VOM	Variable-Order Markov
VPA	Virtual Population Analysis
VSD	Vaccine Safety Datalink
VSI	Variable Sampling Interval
VSM	Value Stream Map
VSR	Variable Sampling Rate
VSS	Variable Sample Size
WAP	Wireless Application Protocol
WBF	Weakened By Failure
WD98	Walker and Damien Algorithm
WI98	Wolpert and Ickstadt Algorithm
WIP	Work-In-Process
WLP	Word Length Pattern
WLS	Weighted Least-Squares
WMOSA	Weight Based Multi-Objective Simulated Annealing
WR	Woven Roving
WS	Whitaker and Samaniego
WSD	Weighted Standard Deviation
WV	Weighted Variance
YEP	Yeast Extract Prodex
YHP	Yokagawa Hewlett–Packard
YSM	Yield Surface Modeling
Z–N	Ziegler–Nichols

Main Effect Designs

Introduction

In the planning stages of any experimental study, an investigator must determine which experimental factors to include in the study. An experimental factor is a variable that is studied in the experiment and believed to potentially have an effect on a response variable of interest. The values of a factor at which experimental runs are conducted are called the *levels of the factor*. During an experimental study, the levels of each factor are usually experimentally changed during different runs of the experiment so that the direct effects of each factor (called *main effects* (MEs)) on the response can be assessed as well as how factors affect the response in combination (called *interactions*, *see* **Interactions**). However, when the list of potential experimental factors is long, it is important in the early stages of an experimental process to identify those experimental factors which do in fact have an influential effect on a response variable and to screen out those that do not. This is the primary purpose of screening experiments (*see* **Screening Designs**). One type of screening experiment which is useful in such situations is called a *main effects design*. ME designs focus on the identification of influential MEs rather than interactions and are particularly useful for experimental studies in which the initial list of experimental factors is large compared to the number of runs that can be made in the initial screening experiment. In this article, we consider the use and construction of ME designs.

In an ME design, a factor may be either quantitative or qualitative. Quantitative factors such as time or age can assume all values on a continuous scale. Qualitative factors such as different machine types or different crop varieties have levels which can only assume discrete values and typically cannot be rank ordered. The selection of levels for factors to include in an experiment is somewhat subjective. For example, if a quantitative factor is believed to have primarily a linear effect on a response, then only two levels of the factor may be selected whereas if it is believed that a quantitative factor may effect a response nonlinearly, then three or more levels may be selected to allow for detection of a factor curvature effect. There is usually much less choice in the selection of levels of a qualitative factor as the levels of such a factor are determined by the qualitative nature of the variable.

The most popular screening experiments in use today are factorial experiments (*see* **Factorial Experiments**). A factorial experiment is one in which the treatments consist of different combinations of experimental factor levels. ME designs are one type of factorial experiment.

Further Notation and Model

In this article we consider experimental situations in which m factors are to be studied using a factorial experiment with N runs. We shall use d to denote a design which can be used in such a situation and interchangeably represent d by an $N \times m$ matrix $X_d = (\mathbf{x}_{d1}, \ldots, \mathbf{x}_{dm}) = (x_{dij})$ whose ith row corresponds to run i and jth column corresponds to factor j. If factor j has s_j levels, then the entries in column j of X_d will consist of symbols $1, 2, \ldots, s_j$ with $x_{dij} = t\epsilon\{1, 2, \ldots, s_j\}$ if factor j occurs at level t in run i.

The model used to analyze the data from a given design d is

$$\mathbf{Y} = \beta_0 \mathbf{1}_N + \sum_{i=1}^{m} X_{di}\boldsymbol{\beta}_i + \boldsymbol{\varepsilon} \tag{1}$$

where $\mathbf{Y}$ is a $N \times 1$ vector of observations, β_0 is an overall mean effect, $\mathbf{1}_N$ is an $N \times 1$ vector of 1's, X_{di} is an $N \times (s_i - 1)$ matrix corresponding to factor i having s_i levels, β_i is an associated set of ME parameters, and $\boldsymbol{\varepsilon}$ is a vector of random error terms whose entries are assumed to be uncorrelated with mean *zero* and constant variance σ^2. The actual entries of X_{di} usually depend upon whether factor i is a quantitative or qualitative factor. If factor i is a quantitative factor with s_i levels, the entries of X_{di} are usually coded (using orthogonal polynomials) so that all polynomial effects up through degree $s_i - 1$ can be estimated. For example, if factor i has $s_i = 3$ equally spaced factors all occurring in $N/3$ runs, then X_{di} has two columns with entries $-1, 0, 1$ and $1, -2, 1$, respectively, depending on whether factor i occurs at its low, middle or high level. If factor i is a qualitative variable with s_i levels then the entries of X_{di} can be coded in several different ways. We shall assume for convenience that the average effect of factor i is zero which implies one parameterization where column p of X_{di} corresponding to level p, $p = 1, \ldots, s_i - 1$, has entries 1 if factor i occurs at level p, -1 if factor i occurs at level s_i and **0** otherwise. Under this parameterization, the parameter corresponding to column p of X_{di} represents the difference in the average response at level p and the overall average. Because we are only interested in estimating MEs for factors, we do not include interaction terms in model (1). We shall say that a design is saturated if $\sum_{i=1}^{m}(s_i - 1) + 1 = N$ or nearly saturated if $\sum_{i=1}^{m}(s_i - 1) + 1$ is "close" to N. When ME designs are used as **screening designs**, they are often saturated or nearly saturated. In these situations, the identification of influential effects in model (1) usually depends upon the form of the matrices X_{di}. If the columns of all the X_{di} are orthogonal to one another and the corresponding parameter estimates can be assumed to be normal random variables with equal variances, then graphical methods such as half-normal plots (*see* **Half-Normal Plot**), can be used to identify significant effects. However, because the use of graphical methods is somewhat subjective in nature, other, more objective, statistical methods have been developed to identify significant effects in these situations, for example, see Hamada and Balakrishnan [1] for a good discussion of available methods such as Lenth's method (*see* **Lenth's Method for the Analysis of Unreplicated Experiments**). When the columns of the X_{di} in model (1) are nonorthogonal or are orthogonal but the corresponding parameter estimates do not satisfy the assumptions required for the use of half-normal plots or other available methods, regression analysis and variable selection techniques can be used to identify influential effects.

When using model (1) to analyze the data from a given design d, a useful property for d to have is that of proportional frequencies. In particular, let factors i and j have s_i and s_j levels, respectively, and let n_{pq} be the number of runs in d at level p of factor i and level q of factor j. We say factors i and j have proportional frequencies if for all $p = 1, \ldots, s_i$ and $q = 1, \ldots, s_j$,

$$n_{pq} = \frac{n_{p.} \cdot n_{.q}}{N} \tag{2}$$

where $n_{p.} = \sum_{l=1}^{s_j} n_{pl}$ and $n_{.q} = \sum_{l=1}^{s_i} n_{lq}$. The property of proportional frequencies was derived by Addelman [2] and insures that any difference between the averages of observations at two levels of factor i is orthogonal (when considered as a linear combination of the observations) to any difference between the averages of any two levels of factor j. Any design in which all pairs of factors have proportional frequencies is called an *orthogonal main effects (OME) design*. The property of proportional frequencies in an OME design thus makes the **estimation** of OME parameters in model (1) relatively easy. We shall denote an OME design d in N runs having m_i factors with s_i levels, $i = 1, \ldots, k$, by OME$(N, s_1^{m_1} \ldots s_k^{m_k})$.

We note that there are no interaction terms included in model (1). This implies the assumption that interaction effects are small compared to MEs. When this assumption is valid, then the model given in equation (1) is useful for identifying significant MEs of factors on a response. When this assumption is invalid, the analysis using model (1) can lead to the identification of some factors as having an influential effect on a response variable when they do not, or, what is worse, lead the experimenter to miss entirely some influential factors. In any case, it is always wise after running an ME screening design to run a follow-up experiment to explore interactions between factors and develop a more detailed model relating the response variable to experimental factors.

Orthogonal Array Main Effect *(OAME)* Designs

Some of the most popular ME designs used today are the two- and three-level regular **fractional factorial** design, and Plackett–Burman (PB) (*see* **Plackett–Burman Designs**) designs. However, because these designs are all special cases of orthogonal array (*OA*) (*see* **Orthogonal Arrays**) designs of strength two defined below and because these designs are considered elsewhere in this volume we do not consider them here. We also note that OA designs serve as the building blocks for other useful and more general ME designs. We shall henceforth denote an OA design d of strength two in N runs having m_i factors with s_i levels, $i = 1, \dots, k$ by $\text{OAME}(N, s_1^{m_1} \dots s_k^{m_k})$.

Definition 1 Let d be an ME design with $N \times m$ matrix X_d.

1. We say factors p and q are orthogonal if in columns p and q of X_d each level of factor p occurs equally often with each level of factor q.
2. We say d is an $\text{OAME}(N, s_1^{m_1} \dots s_k^{m_k})$ design of strength two if every pair of factors in d is orthogonal.

Comment *The definition of an orthogonal array* **main effect** *(OAME) design given above can be extended to that of an OA design of strength t (see* **Orthogonal Arrays***). However, because such designs typically require relatively* **large numbers** *of runs, we do not consider them here. Also, because we will only be considering OAME designs of strength two, we will henceforth omit any reference to the strength of the designs considered. An excellent reference on the properties and construction of general OA designs is the book Orthogonal Arrays: Theory and Applications by Hedayat et al.* [3].

We note that in an ME design d, if factors p and q are orthogonal to one another, then factors p and q have the property of proportional frequencies since each level of factor p occurs equally often with each level of factor q. Thus an *OAME* design d is clearly an *OME* design since all pairs of factors in d are orthogonal. Another useful property regarding OAME designs is that it is possible to determine a lower bound for the number of runs required to accommodate all factors (though the bound is not necessarily achievable in all cases). The following theorem can be derived using the property of orthogonality of factors in an *OAME* design.

Theorem 1 *For an* $\text{OAME}(N, s_1^{m_1} \dots s_k^{m_k})$ *design d, N must be divisible by* $s_i s_j, i, j = 1, \dots, k$.

Example 1 Suppose an experimental situation requires the use of a design having m_1 factors with three levels and m_2 factors with two levels where $m_1 \geq 2, m_2 \geq 2$. Then, using Theorem 1, any number of runs required for an $\text{OAME}(N, 3^{m_1}2^{m_2})$ design must have N divisible by $3^2 = 9, 2^2 = 4$, and $3 \times 2 = 6$. Hence, the minimum possible value for N is $N = 36$.

A large number of *OAME* designs have been constructed using various methods. The largest listing and catalogue of such designs known to the author is available on the web at `www.research.att.com/~njas/oadir`. This website is maintained by N.J.A. Sloane. Some useful OAME designs having 36 or fewer runs are listed in Table 1. For the construction and actual form of the arrays given in that Table, the reader is referred to the website given earlier.

With regard to the usage of OAME designs, there are certain advantages and disadvantages that should be kept in mind.

Advantages

1. Economy of run size. In a number of situations, an OAME design will allow an experimenter to run a screening experiment using the minimum number of runs required to estimate all or nearly all parameter effects of interest, that is, the OAME design will be saturated or nearly saturated. For example, if a screening experiment

Table 1 Orthogonal array main effects designs

Number of observations	Design
4	2^3
6	$3^1 2^1$
8	$2^7, 4^1 2^4$
9	3^4
12	$2^{11}, 3^1 2^4, 6^1 2^2, 4^1 3^1$
15	$5^1 3^1$
16	$2^{15}, 4^1 2^{12}, 4^2 2^9, 8^1 2^8, 4^3 2^6, 4^4 2^3, 4^5$
18	$3^7 2^1, 9^1 2^1, 6^1 3^6$
20	$2^{19}, 5^1 2^8, 10^1 2^2, 5^1 4^1$
24	$2^{23}, 4^1 20^{20}, 3^1 2^{11}, 6^1 2^{14}, 4^1 3^1 2^{13}, 12^1 2^{12}, 6^1 4^1 2^1, 8^1 3^1$
25	5^6
27	$3^{13}, 9^3, 3^9$
28	$2^{27}, 7^1 2^{12}$
30	$5^1 3^1 2^1$
32	$2^{31}, 4^1 2^{28}, 4^2 2^{25}, 8^1 2^{24}, 4^3 2^{22}, 8^1 4^1 2^{21}, 4^4 2^{19}, 8^1 4^1 2^{18}$, $4^5 2^{16}, 16^1 2^{16}, 8^1 4^3 2^{15}, 4^6 2^{13}, 8^1 4^4 2^{12}, 4^7 2^{10}, 8^1 4^5 2^9, 4^8 2^7$, $8^1 4^6 2^6, 4^9 2^4, 8^1 4^7 2^3, 8^1 4^8$
36	$2^{35}, 3^1 2^{27}, 3^2 2^{30}, 6^1 3^1 2^{18}, 3^4 2^{13}, 6^2 2^{13}, 9^1 2^{12}, 3^{12} 2^{11}$, $6^1 3^2 2^{11}, 6^1 3^8 2^{10}, 6^2 3^1 2^{10}, 6^2 3^4 2^9, 6^3 2^8, 3^{13} 2^4, 6^3 3^1 2^4, 6^1 3^9 2^3$, $6^3 3^2 2^3, 6^1 3^{12} 2^2, 6^2 3^5 2^7, 6^2 3^8 2^1, 6^3 3^3 2^1, 4^1 3^{13}, 12^1 3^{12}, 6^3 3^7$

requires 18 runs and has eight factors with seven factors having three levels and one factor having two levels, then an OAME(18, $3^7 2^1$) can be used, or if there are four factors having three levels each to be screened in nine runs, then an OAME(9, 3^4) can be used.

2. Estimates for all ME parameters between different factors can be estimated orthogonally to one another and such estimates will be unbiased as long as higher-order interactions are negligible.
3. Flexibility. OAME designs provide the opportunity to combine factors having differing numbers of levels in the same experiment. This is particularly useful when a study involves only qualitative or both qualitative and quantitative factors requiring differing numbers of levels.

Disadvantages

1. For a given design d, we say two effects are fully aliased (*see* **Aliasing in Fractional Designs**) or confounded if the **correlation** between their parameter estimates under model (1) (when their corresponding terms are included in the model) has absolute value 1 and we say two effects are partially aliased or confounded if the correlation between their estimates has absolute value between 0 and 1. The primary disadvantage with OAME designs, particularly those that are nearly saturated,is that many of the estimates for ME parameters will be fully or partially confounded with low order interactions. For example, in an OAME(18, $3^7 2^1$) design d, because d is nearly saturated, there are at most two lower order interactions between MEs that can be estimated. Thus, most MEs in d will be at least partially aliased with some low order interactions. Whenever a design d has a number of MEs that are at least partially aliased with a number of low order interactions, we say that the design has a complex **aliasing** structure. A complex aliasing structure can sometimes lead to difficulty in the interpretation of a significant ME parameter estimate. For example, in the OAME(18, $3^7 2^1$) design d, if a parameter ME is found to be significant, then it may be difficult to determine (unless all two-factor and higher-order interactions are negligible) whether the significance was due to the associated factor effect or some combination of the partially aliased low order interactions. Hamada and Wu [4] have suggested a

stepwise, regression-based procedure for analyzing data from designs having complex aliasing structures and their method of analysis can be useful when only a few effects are influential and the correlation between partially aliased effects is small to moderate. However, as also pointed out in Hamada and Wu [4], their technique is not foolproof. When the assumption of two-factor and higher-order interactions being negligible is valid, the OAME designs can be of great use in identifying influential factors in a screening process. However, when the assumption is not valid, it is possible to identify some factors which are not truly significant as being influential and completely fail to identify other factors which are truly influential.

Other OME Plans

As mentioned in the section titled "Orthogonal Array Main Effect (OAME) Designs", OAME designs are special cases of OME plans. In fact, OAME designs provide one of the basic tools for constructing OME designs to handle experimental situations for which an OAME design may not exist. The simplest method for constructing OME designs from OAME plans is called the *method of collapsing factors*. We illustrate the technique through an example.

Example 2 Consider the OAME$(9, 3^4)$ design d which exists and has

$$X_d = \begin{pmatrix} 1 & 1 & 1 & 1 \\ 1 & 2 & 2 & 2 \\ 1 & 3 & 3 & 3 \\ 2 & 1 & 2 & 3 \\ 2 & 2 & 3 & 1 \\ 2 & 3 & 1 & 2 \\ 3 & 1 & 3 & 2 \\ 3 & 2 & 1 & 3 \\ 3 & 3 & 2 & 1 \end{pmatrix} \tag{3}$$

Using d, we can now construct an OME$(9, 3^2 2^2)$ design $\tilde{d}$ using the method of "collapsing factors". In particular, we assign all observations occurring at levels 1, 2, and 3 of factors 1 and 2 in d to the same levels for factors 1 and 2 in $\tilde{d}$. However, we also assign those observations occurring at level 1 of factors 3 and 4 in d to the same level of factors 3 and 4 in $\tilde{d}$ but assign those observations occurring at levels 2 and 3 of factors 3 and 4 in d to level 2 of factors 3 and 4 in $\tilde{d}$. Thus $\tilde{d}$ has

$$X_{\tilde{d}} = \begin{pmatrix} 1 & 1 & 1 & 1 \\ 1 & 2 & 2 & 2 \\ 1 & 3 & 2 & 2 \\ 2 & 1 & 2 & 2 \\ 2 & 2 & 2 & 1 \\ 2 & 3 & 1 & 2 \\ 3 & 1 & 2 & 2 \\ 3 & 2 & 1 & 2 \\ 3 & 3 & 2 & 1 \end{pmatrix} \tag{4}$$

and we have "collapsed" factors 3 and 4 in d having three levels each to factors 3 and 4 in $\tilde{d}$ having two levels each. In doing so, we note that level 1 in factors 3 and 4 in $\tilde{d}$ occurs in three runs whereas level 2 of factors 3 and 4 in $\tilde{d}$ occurs in six runs. However, because the levels of each pair of factors in d occur equally often together, the property of proportional frequencies in d carries over to $\tilde{d}$ during the "collapsing" process. We also note that using the method of "collapsing factors", additional OME designs can be constructed from d such as OME$(9, 3^3 2^1)$, OME$(9, 3^1 2^3)$, and OME$(9, 2^4)$.

Using the method of "collapsing factors", a large number of OME designs can be constructed from OAME designs such as those listed in Table 1 or the web site mentioned in the section titled "Orthogonal Array Main Effect (OAME) Designs". A collection of useful OME designs can be found in Dey [5]. We also note that, as in the example above, when applying the method of "collapsing factors" to a given OAME design d, the property of proportional frequencies carries over from d to the OME design $\tilde{d}$ constructed.

Similar to OAME plans, the property of proportional frequencies makes it possible to find a lower bound for the number of observations required for an OME$(N, s_1^{m_1} \ldots s_k^{m_k})$ design. In particular, we have the following theorem which has been proved by Jacroux [6].

Theorem 2 *Suppose that an OME design d requires $m \geq 3$ factors with factor i having s_i levels, $i = 1, \ldots, m$ and assume $s_1 \geq s_2 \geq \cdots \geq s_m$. If $N = \tilde{s}_1 \tilde{s}_2$ is the minimum possible value for integers $\tilde{s}_1$ and $\tilde{s}_2$ satisfying*

1. $\tilde{s}_1 \geq s_2$ and $\tilde{s}_2 \geq s_2$;
2. $\tilde{s}_1\tilde{s}_2 \leq 2s_1s_2$;
3. $s_3 \times \{lcm(\tilde{s}_1, \tilde{s}_2)\} \leq \tilde{s}_1\tilde{s}_2$,

where $lcm(x, y)$ denotes the least common multiple between positive integers x and y, then N is a lower bound for the number of observations required for d.

A design having N satisfying the conditions of Theorem 2 is called a *minimal OME plan*.

Comment *If d requires m factors having level $s_1 \geq s_2 \geq \cdots \geq s_m$, then another obvious lower bound for N is $\sum_{i=1}^{m} s_i - m + 1$ since this is the minimal number of runs required to estimate all MEs in d.*

Example 3 Consider an experimental setting in which a screening experiment is to be run and requires the use of a design having two factors with three levels and two factors with two levels. Using Theorem 1, it is seen that a lower bound for the number of runs required for an OAME design is 36. On the other hand, it is seen using Theorem 2 that the lower bound for the number of observations required for an OME design is 9. We saw in Example 2 that an OME$(9, 3^2 2^2)$ exists which can be used to handle this experimental situation. Thus in this case, we see that the use of an OME design can provide substantial savings in terms of runs required over the best possible OAME design. Many other such situations can be found in which the use of OME designs provides similar savings in terms of runs required over corresponding OAME designs.

For a specific experimental situation requiring m factors with $\tilde{s}_1 \geq \cdots \geq \tilde{s}_m$ levels, the following steps can be used to construct a specific OME design $\tilde{d}$ for use.

1. Use Theorem 2 to find a lower bound for the number of observations required for an OME plan to accommodate all m factors.
2. Consult a table of OAME designs such as given in Table 1 or at the website mentioned previously and find an OAME design d whose number of runs is equal to or as close as possible to the lower bound found in step 1 and having factors $i, i = 1, \ldots, m$ with s_i levels, $s_i \geq \tilde{s}_i, i = 1, \ldots, m$.
3. Collapse the levels of the factors $i = 1, \ldots, m$ in d to the appropriate number of levels for factors $i = 1, \ldots, m$ in $\tilde{d}$.

We illustrate the use of this procedure in the following example.

Example 4 Consider an experimental situation requiring three three-level factors and six two-level factors. vv What is the smallest OME$(N, 3^3 2^6)$ that can be run? Following the stepwise procedure outlined above, we proceed as follows:

1. Using Theorem 2, a lower bound for N is seen to be $N = 9$.
2. Consulting Table 1, we see that the smallest OA design d that has nine factors with three factors having at least three levels is OAME$(16, 4^3 2^6)$.
3. Using the method of collapsing factors, we can construct the desired OME$(16, 3^3 2^6)$ design $\tilde{d}$ by taking the three four-level factors in d and collapsing their levels so that we obtain the required three factors having three levels in $\tilde{d}$.

Once an OME has been selected for use, the actual analysis of data from the design would proceed, using regression and variable selection techniques. The systematic procedure of analysis suggested by Hamada and Wu [4] may also be of use under the same conditions mentioned in the previous section, but the limitations of the method should also be kept in mind.

As in the case of OAME designs, there are certain advantages and disadvantages an experimenter should keep in mind when using OME plans as screening designs.

Advantages over OAME designs

1. Economy of run size. Because OME designs are more general than OAME plans, it is often the case that the number of runs required to construct an OME design to handle a given number of factors having given numbers of factor levels is smaller than the number required to construct an OAME design. This is clearly illustrated in Example 3.
2. Because differences between factor level averages between different factors are orthogonal, this makes the estimation of ME parameters relatively easy in OME designs.

Disadvantages

1. Because most OME designs will be nearly saturated when used as screening experiments,

many of the ME parameter estimates will be at least partially aliased with low order interactions which lead to a complex aliasing structure. All the problems mentioned earlier for complex aliasing structures in OAME designs hold true for OME plans as well.

2. When obtaining an OME design $\tilde{d}$ from an OAME design d through the method of "collapsing factors", it will be the case that some of the factors in $\tilde{d}$ will have levels which occur in differing numbers of runs. Thus some ME parameters associated with these factors will be estimated with greater precision than others. This can result in a loss of efficiency. However, the loss of efficiency can be minimized by collapsing levels of a factor in d so that the levels of the resulting factor in $\tilde{d}$ occur as equally often in runs as possible.

Nonorthogonal ME Designs

In the previous sections, we introduced designs which allow for the orthogonal estimation of ME parameters associated with different factors. In this section, we consider designs which sometimes prove useful as screening designs, but which are nonorthogonal in the sense that ME parameters cannot necessarily be orthogonally estimated. In particular, we consider what we refer to as *nearly orthogonal array main effect (NOAME) designs*. NOAME designs were introduced by Wang and Wu [7]. In OAME designs, for any two factors i and j, each level of factor i occurs equally often with each level of factor j. This latter property implies proportional frequencies between factors which ensure that ME parameters between different factors can be orthogonally estimated. The property of proportional frequencies between factors in both OAME designs and OME designs places some restrictions on the number of factors that can be accommodated in a given set of runs. For example, in a design requiring six runs and a $3^1 2^2$ fractional factorial experiment, it can be shown that neither an OAME design nor an OME design exists. However, by relaxing the condition of orthogonality between ME estimates for different parameters, Wang and Wu [7] give a method for constructing an NOAME design which can be used for such a situation.

Wang and Wu [7] give three different methods for constructing NOAME designs. In some cases, they use their methods to find designs which minimize the number of nonorthogonal pairs of columns of certain types and in other cases, they find designs which minimize the canonical correlations between two sets of effects. A list of parameters for the actual designs constructed by Wang and Wu [7] is given in Table 2, but for the actual construction of the designs given, the reader is referred to Wang and Wu [7].

Table 2 Nearly orthogonal array main effects designs

Number of observations	Design
6	$3^1 2^3$
10	$5^1 2^5$
12	$3^4 2^3, 6^1 2^5, 3^1 2^9, 3^5 2^1, 3^4 2^2$
15	$5^1 3^5$
18	$9^1 2^8, 3^4 2^8$
20	$5^1 2^{15}$
24	$8^1 3^8, 3^8 2^7, 4^1 3^8 2^4, 3^1 2^{21}, 6^1 2^{18}, 3^{11} 2^1$
	$4^7 3^1, 4^1 3^9 2^1, 3^9 2^5, 4^6 3^1 2^3$
36	$4^{11} 3^1, 4^{10} 3^2$

Once again, the statistical analysis of NOAME designs usually relies upon regression analysis and variable selection techniques.

Advantages of NOAME designs over OME designs

1. Economy of run size. NOAME designs can be used to handle experimental situations where OAME and OME designs do not exist for a given number of runs, factors, and factor levels. Thus such designs are used in situations where other options do not exist.

Disadvantages

1. The analysis of NOAME designs is not as straightforward as that of orthogonal designs since parameter ME estimates between different factors are not necessarily orthogonal. The lack of orthogonality also leads to some loss of efficiency in terms of the estimating factor MEs.
2. The alias structure associated with NOAME designs tends to be even more complex than that of OAME and OME designs. Hence, all of the issues mentioned previously, associated with complex alias structures, hold for NOAME designs.

References

[1] Hamada, M and Balakrishnan, N. (1998). Analyzing unreplicated factorial experiments: a review with some new proposals (with discussion), *Statistica Sinica* **8**, 1–41.

[2] Addelman, S. (1962). Orthogonal main effects plans for asymmetrical factorial experiments, *Technometrics* **4**, 21–45.

[3] Hedayat, A., Sloane, N. & Stufken, J. (1999). *Orthogonal Arrays: Theory and Applications*, Springer, New York.

[4] Hamada, M. & Wu, C.F.J. (1992). Analysis of designed experiments with complex aliasing, *Journal of Quality Technology* **24**, 130–137.

[5] Dey, A. (1985). *Orthogonal Fractional Factorial Designs*, Holsted Press, New York.

[6] Jacroux, M. (1992). A note on the determination and construction of minimal orthogonal main effects plans, *Technometrics* **34**, 92–96.

[7] Wang, J.C. & Wu, C.F.J. (1992). Nearly orthogonal arrays with mixed levels, *Technometrics* **34**, 409–422.

Further Reading

John, P.W.M. (1971). *Statistical Design and Analysis of Experiments*, Macmillan, New York.

Plackett, R.L. & Burman, J.P. (1946). The design of optimum multifactorial experiments, *Biometrika* **33**, 305–325.

Related Articles

Aliasing in Fractional Designs; **Factorial Experiments**; **Fractional Factorial Designs**; **Half-Normal Plot**; **Interactions**; **Main Effects**; **Orthogonal Arrays**; **Plackett–Burman Designs**; **Screening Designs**.

MIKE JACROUX

Main Effects

The Cell-Means Definition

In a *factorial experiment* (*see* **Factorial Experiments**), the ***main effect*** of a given factor, say F, is a measure of the average change in expected response that occurs when the level of factor F is changed. Here the "average" is taken over all other factors in the experiment. We will first develop this idea through several examples, and then follow with a formal definition.

The 2 x 2 Experiment. Here there are two factors, say A and B, each measured at two levels, which we may denote by 0 and 1. Let us suppose that the true (theoretical) mean responses to the four treatment combinations are given in the following table:

		Levels of B	
		0	1
Levels	0	4	6
of A	1	8	9

The treatment combination (0, 0) – A and B both at level 0 – is assumed to yield a mean response of 4, and similarly for the other treatment combinations (or *cells*). The change in the average expected response due to change in the level of A is

$$\frac{4+6}{2} - \frac{8+9}{2} = -3.5 \tag{1}$$

(or 3.5 if we change in the opposite direction). Thus a measure of the main effect of A is -3.5 units. Similarly, the main effect of B is -1.5.

Note that if we were to replace the 9 by a 6 in this example, we would calculate the main effect of B to be 0. This would not mean that B has no effect on responses, but merely that it affects responses only through its *interaction* (*see* **Interactions**) with A. We may say in this case that B has no main effect, or that the main effect of B is absent.

A simulated experiment based on the theoretical values given above might yield the following data, with two observations per cell. Cell averages, which are least-squares estimates of the true mean responses above, are given in parentheses:

	0	1
0	4.75, 3.79 (4.270)	6.24, 5.79 (6.015)
1	7.83, 7.82 (7.825)	10.37, 8.83 (9.600)

The main effect of A calculated in equation (1) could thus be estimated by

$$\frac{4.270+6.015}{2} - \frac{7.825+9.600}{2} = -3.57 \tag{2}$$

Table 1 Analysis of variance for the 2 × 2 example

Source	*df*	*SS*	*MS*	*F*	*p*
A	1	25.4898	25.4898	58.33	0.002
B	1	6.1952	6.1952	14.18	0.020
Interaction	1	0.0004	0.0004	0.00	0.976
Error	4	1.7479	0.4370	–	–
Total	7	33.4334			

The corresponding main effect of B would be estimated as -1.76. The *analysis of variance* in Table 1 (*see* **Analysis of Variance**) confirms that the main effects of A and B are significant – that is, that the estimated values are significantly different from 0.

To formulate our ideas more generally, we denote the true mean response to treatment combination $t = (i, j)$ by μ_{ij}. We have described the main effect of factor A as the difference between the average response to level "0", or $(\mu_{00} + \mu_{01})/2$, and that to level "1" of A, namely $(\mu_{10} + \mu_{11})/2$, that is,

$$\frac{\mu_{00} + \mu_{01}}{2} - \frac{\mu_{10} + \mu_{11}}{2} \tag{3}$$

(*see*, e.g. [1, Chapter 10]). Of course, we could subtract in the opposite order as well. In fact, one often describes this effect by ignoring the factors of 1/2 and writing simply:

$$\mu_{00} + \mu_{01} - \mu_{10} - \mu_{11} \tag{4}$$

Expressions such as (3) and (4) are called *contrasts*, as they contrast the responses to (in this case) the different levels of factor A. We say that *A has no main effect*, or that *the main effect of A is absent*, if any of these expressions equals 0, in which case they all do. We also note that the **sum of squares** (SS) for a contrast is not affected if we multiply the contrast by a nonzero constant (such as -1 or 1/2). Thus the main effect of factor A may simply be defined by the contrast (equation (4)). We note that the coefficients in this expression are $+1$ where $i = 0$, and -1 where $i = 1$, independent of the value of j (the level of B).

In similar fashion, the main effect of factor B is defined as the contrast

$$\mu_{00} - \mu_{01} + \mu_{10} - \mu_{11} \tag{5}$$

Here the values of the coefficients are independent of the value of i (the level of A). We note that the interaction of A and B is also given by a contrast, namely

$$\mu_{00} - \mu_{01} - \mu_{10} + \mu_{11} \tag{6}$$

If this equals 0, then $\mu_{00} - \mu_{01} = \mu_{10} - \mu_{11}$, so that changing the level of B alters μ_{ij} by the same amount at each level of A. In this case we say that *interaction is absent*. We see that the coefficients of equation (6) depend on the levels of *both* factors.

The 2 x 2 x 2 Experiment. Here we have three factors, say A, B, and C, each at two levels. The treatment combinations are ordered triples $t = (i, j, k)$, where k denotes the level of factor C. The contrasts for the main effects are often given *via* a table such as the following:

Cell	000	001	010	011	100	101	110	111
A	1	1	1	1	−1	−1	−1	−1
B	1	1	−1	−1	1	1	−1	−1
C	1	−1	1	−1	1	−1	1	−1

The main effect of C, for example, is defined by the contrast

$$\mu_{000} - \mu_{001} + \mu_{010} - \mu_{011} + \mu_{100} - \mu_{101} + \mu_{110} - \mu_{111} \tag{7}$$

Some authors prefer to view this as a difference of averages, analogous to equation (3), and so multiply it by 1/4.

The 2^n Experiment. The extension to four or more two-level factors is handled analogously. Wu and Hamada [2, p. 97] give an example of a $2 \times 2 \times 2 \times 2$ or 2^4 experiment to study the growth of an epitaxial layer on polished silicon wafers. The factors studied are susceptor-rotation method (continuous or oscillating), nozzle position (2 or 6), deposition temperature (1210 or 1220 °C), and deposition time (low or high). From the observations in each treatment combination they calculate an observed average thickness (μm), and use this to estimate the main effect of factor susceptor-rotation method [2 p. 105] by calculating (1/8) (sum of average thicknesses, oscillating − sum of average thicknesses, continuous) = −0.077.

This effect turns out not to be statistically significant ($p = 0.439$).

The 2 x 3 Experiment. This is the simplest example of a factorial experiment with a factor at more than two levels. Here we have two factors, say A and B, such that A has two levels and B has three. The main effect of B must now be described by *two* independent contrasts (we say that it has *2 degrees of freedom* (df)). Denoting the levels of A as before and those of B by 0, 1, and 2, one choice of main effects contrasts is given in the following table, which lists the contrast coefficients:

Cell	00	01	02	10	11	12
A	1	1	1	−1	−1	−1
B	1	−1	0	1	−1	0
	1	0	−1	1	0	−1

Other choices of contrasts are possible. As before, the coefficients of the A contrast do not depend on the level of B, and those of the two B contrasts do not depend on the level of A. As we have noted, the choice of contrasts does not affect the sums of squares for main effects, which thus provide an overall summary of the intensity of each main effect. (This also applies in so-called nonorthogonal designs, although there the sums of squares must be adjusted.)

The general case. If factor F is measured at s levels, then it has $s-1$ *df*. This means that its main effect is described by a set of $s-1$ contrasts in cell means. The main effect of F is said to be absent if all these contrasts are zero. We now give a more precise definition of the contrasts belonging to a main effect.

A factorial experiment consists of n factors, the ith factor measured at s_i levels, so that there are $s_1 \times \ldots \times s_n$ *treatment combinations* or *cells*. We let T be the set of treatment combinations. Each $t \in T$ is a multi-index or n-tuple (e.g., $t = (i, j, k)$ if $n = 3$), and we denote the corresponding *cell mean* by μ_t.

A *contrast* is an expression $\sum_{t \in T} c_t \mu_t$ such that the coefficients c_t add up to zero. The least-squares estimate of a contrast is calculated by replacing cell means by observed sample averages.

We say that a contrast $\sum_{t \in T} c_t \mu_t$ *belongs to the main effect of the ith factor* if the coefficients c_t do not depend on any factors other than the ith [3]. Contrasts belonging to interactions may be defined in similar fashion.

The Factor Effects Parametrization

A different way of defining main effects arises by using a different set of parameters. In the two-factor case, the means μ_{ij} are expressed by the set of equations

$$\mu_{ij} = \mu + \alpha_i + \beta_j + \gamma_{ij} \tag{8}$$

This is often accompanied by a set of side conditions of the form

$$\sum_i v_i \alpha_i = 0, \quad \sum_j w_j \beta_j = 0,$$
$$\sum_i v_i \gamma_{ij} = 0 \quad \text{for each } j,$$
$$\sum_j w_j \gamma_{ij} = 0 \quad \text{for each } i \tag{9}$$

where the weights v_i and w_j are nonnegative and sum respectively to unity. In the presence of these side conditions, μ represents an overall level of response, α_i is the *main effect of the ith level of A*, β_j is the *main effect of the jth level of B*, and γ_{ij} is the *interaction of the ith level of A with the jth level of B*. According to Scheffé, "if the set of constants $\{\mu, \alpha_i, \beta_j, \gamma_{ij}\}$ satisfies (condition (8)) this is not sufficient for them to be the general mean, main effects, and interactions, unless the side conditions (9) are satisfied" [7, p. 92].

The model $E(Y_{ij}) = \mu + \alpha_i + \beta_j + \gamma_{ij}$ is sometimes called a *factor effects model* [5], while $E(Y_{ij}) = \mu_{ij}$ is a *cell means model* [6].

With the constraints (9) in place, one may solve equation (8) for the factor effects parameters. It is not hard to see that the resulting equations express these parameters as contrasts in the cell means μ_{ij}. But do the contrasts expressing α_i belong to the main effect for A, and those for β_j to B, in our earlier sense? The answer is yes if, and only if, one uses equal weights ($v_1 = \cdots = v_a = 1/a$ and $w_1 = \cdots = w_b = 1/b$). In this case equation (9) reduces to the more familiar "zero-sum" constraints

$$\sum_i \alpha_i = 0, \quad \sum_j \beta_j = 0,$$
$$\sum_i \gamma_{ij} = 0 \quad \text{for each } j,$$
$$\sum_j \gamma_{ij} = 0 \quad \text{for each } i \tag{10}$$

For example, solving equation (8) in the 2×3 case using the zero-sum constraints, we find that

$$\begin{aligned}
\mu &= \left(\tfrac{1}{6}\right)\mu_{..} \\
\alpha_1 &= \left(\tfrac{1}{6}\right)(\mu_{1.} - \mu_{2.}) \\
\beta_1 &= \left(\tfrac{1}{6}\right)(2\mu_{.1} - \mu_{.2} - \mu_{.3}) \\
\beta_2 &= \left(\tfrac{1}{6}\right)(-\mu_{.1} + 2\mu_{.2} - \mu_{.3}) \\
\gamma_{11} &= \left(\tfrac{1}{6}\right)(2\mu_{11} - \mu_{12} - \mu_{13} - 2\mu_{21} + \mu_{22} + \mu_{23}) \\
\gamma_{12} &= \left(\tfrac{1}{6}\right)(-\mu_{11} + 2\mu_{12} - \mu_{13} \\
&\quad + \mu_{21} - 2\mu_{22} + \mu_{23})
\end{aligned} \tag{11}$$

where, as usual, $\mu_{i.} = \sum_j \mu_{ij}$, $\mu_{.j} = \sum_i \mu_{ij}$, and $\mu_{..} = \sum_i \sum_j \mu_{ij}$. The other parameters are obtained from equation (11) by applying the constraints; for example, $\alpha_2 = -\alpha_1$. It is evident that the contrasts for α_1 and α_2 belong to the main effect of A, and those of the β_j to B.

Main Effects in Fractional Factorial Experiments

A subset $S \subset T$ of the set of treatment combinations in a factorial experiment is a *fractional factorial* (*see* **Fractional Factorial Designs**) experiment, or *fraction*. Viewed as an *orthogonal array* (*see* **Orthogonal Arrays**), a fraction has a given *strength* t. If $t \geq 1$, then the restriction of each main-effect contrast to the subset S will still be a contrast, and each main effect will still be represented by the same df. For example, in the 2^3 experiment above, suppose $S = \{000, 011, 101, 110\}$, a fraction of strength 2. Then in S the contrast (equation (7)) for the main effect of C becomes

$$\mu_{000} - \mu_{011} - \mu_{101} + \mu_{110} \tag{12}$$

If in addition $t \geq 2$, then contrasts belonging to different main effects will also be mutually orthogonal. We may take these restricted contrasts as defining the main effects in the fraction [7].

One difficulty that arises in a fraction is that restricted contrasts belonging to a main effect may be the same as those belonging to an interaction. We say that the main effect is *aliased* (*see* **Aliasing in Fractional Designs**) with the interaction in the fraction. If one assumes in addition that interactions are absent, then we say that the main effects are *measurable* in the fraction [8].

References

[1] Box, G.E.P., Hunter, J.S. & Hunter, W.G. (2005). *Statistics for Experimenters: Design, Innovation and Discovery*, 2nd Edition, John Wiley & Sons, New York.

[2] Wu, C.F.J. & Hamada, M. (2000). *Experiments: Planning, Analysis, and Parameter Design Optimization*, John Wiley & Sons, New York.

[3] Bose, R.C. (1947). Mathematical theory of the symmetrical factorial design, *Sankhya* **8**, 107–166.

[4] Scheffé, H. (1959). *The Analysis of Variance*, John Wiley & Sons, New York, Reissued in the Wiley Classics series, 1999.

[5] Neter, J., Kutner, M.H., Nachtsheim, C.J. & Wasserman, W. (1996). *Applied Linear Statistical Models*, 4th Edition, Irwin Publishing, Chicago.

[6] Hocking, R.R. (1985). *The Analysis of Linear Models*, Brooks/Cole Publishing, Monterey.

[7] Beder, J.H. (2004). On the definition of effects in fractional factorial designs, *Utilitas Mathematica* **66**, 47–60.

[8] Rao, C.R. (1947). Factorial experiments derivable from combinatorial arrangements of arrays, *Journal of the Royal Statistical Society* Suppl 9, 128–139.

Related Articles

Main Effect Designs; **Factorial Designs, Resolution of**.

Jay H. Beder

Maintenance and Markov Decision Models

Introduction

Maintenance modeling and Markov decision chains (MDCs) have a large joint history. For both areas the main research started in the 1950s. They formed a fruitful partnership, with theory allowing applications and applications guiding the need for theoretical developments. In this chapter we sketch the main ideas behind MDCs and indicate how they are applied in the maintenance area. In particular, we treat the area of maintenance of civil structures. Apart from providing ample references, we state the pros and cons of the Markov decision modeling.

Historical Perspective

Intense scientific interest in problems related to maintenance management originated only a few decades ago. The basis for the scientific support of maintenance is found in reliability engineering. The book *Mathematical Theory of Reliability* (Barlow and Proschan [1]) indicates the start of scientific interest in maintenance problems. In the early stages, maintenance was almost exclusively interpreted as preventive replacement of components (*see* **Replacement Strategies**). The development of some simple but insightful models describing aging phenomena of technical components, as well as the choice of appropriate actions to cope with these phenomena, showed that a scientific approach could be useful. One such model is the MDC. In the 1970s and early 1980s, the basic models were extended in several directions. In the 1980s, scientists made a step forward in bridging the gap between the analytical models and practice by a systematic analysis of systems. Moreover, similar to many other areas of applied (stochastic) optimization, the major advances in information technology brought the use of quantitative analytic models within computational reach.

During the last 15 years, substantial progress has been made in developing quantitative decision support systems for maintenance management of complex systems. Sophisticated decision support systems are in operation nowadays in the oil industry (both for maintenance of refineries as well as offshore installations), the civil infrastructure sector (road, bridges, and railway), and electric power generation. The number of organizations that make use of some kind of quantitative tools to support inspection and replacement of equipment is increasing rapidly.

Maintenance

Maintenance can be defined as the combination of all technical and associated administrative actions intended to retain an item or system in, or restore it to, a state in which it can perform its required function. The maintenance objectives can be summarized under four headings – ensuring system function (availability, efficiency, and product quality); ensuring system life (asset management), ensuring safety, and ensuring human well-being.

For production equipment, ensuring the system function is often the prime objective. Here, maintenance has to provide the right (but not the maximum) reliability, availability, efficiency, and capability (i.e., producing at the right quality) of production systems in accordance with the need for these characteristics. In principle it is possible to give an economic value to the maintenance results, and a cost balance can be done. Ensuring system life and asset management refers to keeping systems as such in proper condition, while there are only indirect links to a possible production of goods or services. This objective is appropriate for civil structures like buildings, dams, offshore platforms, and roads, as their function is complex and not easy to measure. Often norms have to be set to define failure, and the benefits of maintenance are therefore more difficult to quantify. In this case, one has to minimize maintenance costs to meet the norms or conditions on states. Safety plays a role, as failures can have dramatic consequences, e.g., in the case of airplanes, and nuclear and chemical plants. In this case, testing and inspection activities constitute an important part of the maintenance work. Here, costs of maintenance have to be minimized while keeping the risks within strict limits and meeting statutory requirements. Finally, we refer to human well-being or shine as an objective; though there is no direct economic or technical necessity, it has primarily a psychological one (which indirectly may be economical). An example is painting, which is not done for protective reasons.

Many models for maintenance consider replacing of parts or systems. Typical models are the age- and block-replacement models (*see* **Multivariate Age and Multivariate Renewal Replacement**). These models consider only one state of the system, *viz* working or failed. Next to that there are condition-based models that consider multiple states and these can be modeled by MDCs. For mechanical equipment, deterioration can be quick, and the economic importance of failures can be clear. In the case of civil infrastructures, deterioration can be very slow, and failure does not always have direct economic consequences. As a result, it is much more important to decide when to do maintenance for such systems and to allocate the budget to those assets with the highest needs. Markov decision models can be helpful in this respect.

Markov Decision Models

A Markov model is a special type of dynamic model with which the probabilistic evolution of a system can be modeled in time. The main assumption underlying such a model is that all information about the future behavior is captured in the state description. In other words, the present state provides all relevant information about the future behavior, and knowledge about previous states is not necessary. More formally, a Markov chain is a discrete-time process governed by a discrete state space E (observed at discrete time points) and transition matrix P, for which the Markov property holds, i.e.,

$$\begin{aligned} P_{ij} &= P(X_{t+1} = j | X_0 = i_0, X_1 = i_1, \ldots, X_t = i) \\ &= P(X_{t+1} = j | X_t = i) \end{aligned} \quad (1)$$

If the transition probabilities do not depend on t, then the Markov chain is said to be stationary. An MDC is an extension of a Markov chain, where the Markov chain can be steered by actions and with which optimal actions can be determined. An alternative name is *Markov decision process* as it basically describes a stochastic process. The term *chain* is used to indicate that it gives rise to a chain of states in time. It is defined by a four-tuple E, A, P, and r – where E denotes the state space; $A(i)$ the action set in state $i \in E$; $P_{ij}(a), i, j \in E$ the transition probabilities; and $r_i(a), i \in E, a \in A(i)$, the immediate rewards in state i when action a is chosen. The control is defined through policies and decision rules. A policy π is a sequence of decision rules $(\pi^1, \pi^2, \ldots)$ where π^t is a function that assigns to each action a the probability of that action being taken at time t. For a memoryless or Markov policy, the decision rule at time t is independent of the states and actions before time t. A policy is said to be stationary and deterministic if all decision rules are identical and nonrandomized. For any decision rule π, we denote by $P(\pi)$, $r(\pi)$ the matrix and the vector with $P_{ij}(\pi(i))$ and $r_i(\pi(i))$ as ith element, respectively. Let $P^k(R) = P(\pi^k) \cdots P(\pi^1)$ and $P^0(R) = I$, the identity matrix. The evolution of the MDC under a given stationary and deterministic policy f is a Markov chain with transition probability matrix $P(f)$. Let C be the class of all policies and C_M be the class of Markov policies. In policy optimization the issue is to find the best action a in each state such that some criterion is optimized. The expected total α-discounted rewards are defined by

$$v^\alpha(R) = \sum_{k=0}^{\infty} \alpha^k P^k(R) r\left(\pi^{k+1}\right) \quad (2)$$

Here α is the discount factor, which is taken from [0, 1). For a stationary policy f, one can also obtain the discounted rewards $v^\alpha(f)$ by solving a set of equations, i.e. (in vector notation)

$$v^\alpha(f) = r(f) + \alpha P(f) v^\alpha(f) \quad (3)$$

Again if $\alpha < 1$, this set has a unique solution. Note that the size of this set of equations is equal to the size of the state space. Normally solving such a set takes a number of operations, which increases with the cube of the size of the state space. A policy R_α is called *α-discounted optimal* if for all $i \in E, R \in C$ we have

$$v_i^\alpha(R_\alpha) \geq v_i^\alpha(R) \quad (4)$$

A second criterion is the long-run expected average rewards, defined by

$$g_i(R) = \liminf_{N\to\infty} \frac{1}{N+1} \sum_{k=0}^{N} \sum_{j\in E} P_{ij}^k(R) r_j\left(\pi^{k+1}\right) \quad (5)$$

Let $g_i^* = \sup_{R\in C} g_i(R)$. In the finite-state and finite-action case, there exists an average optimal policy and the supremum can be replaced by a maximum. In the case of a denumerable state space, an average optimal policy may exist only under certain conditions (like unichain and a certain recurrence). A policy R is long-run average optimal if $g_i(R) = g_i^*$, for all $i \in E$. For both criteria it can be shown that one can restrict oneself to the class of Markov policies. The first step in optimization is the derivation of the so-called optimality equations. For the α-discounted rewards they are defined as follows:

$$v_i^\alpha = \max_{a\in A(i)} \left\{ r_i(a) + \sum_{j\in E} \alpha P_{ij}(a) v_j^\alpha \right\}, \quad i \in E \quad (6)$$

It appears that there exists a unique solution to them and that any policy that takes maximizing actions in each state is α-discounted optimal. The optimality equations for the long-run average costs consist of two equations, *viz*

$$g_i = \max_{a \in A(i)} \left\{ \sum_{j \in E} P_{ij}(a) g_j \right\}, i \in E \quad (7)$$

$$g_i + v_i = \max_{a \in A^0(i)} \left\{ r_i(a) + \sum_{j \in E} P_{ij}(a) v_j \right\},$$

$$i \in E \quad (8)$$

$$\text{where } A^0(i) = \left\{ a \in A(i) | g_i = \sum_{j \in E} P_{ij}(a) g_j \right\}$$

Again a solution g, v to this set of equations is unique up to a constant vector for **v** and the solution g equals g^*. Any policy that takes maximizing actions in both equations in all states is average optimal. These equations can be simplified in case the MDC is *unichain* (which means that under all policies the Markov chain has one minimal closed set) or even weaker, if it is *communicating*, i.e., for every two pairs of states i and j, there is a policy f such that $P_{ij}^k(f) > 0$ for some $k > 0$. In this case the optimal average reward vector is a constant vector **g** and the first equation in the optimality equation is automatically satisfied. The next problem is to solve the equations.

Solution Methods

In principle there are three solution methods, i.e., policy improvement (PI), value iteration, and linear programming (LP). PI works with a starting policy f_0 and determines in each step for each state an improving action. It appears that the stationary policy belonging to these actions also improves the objective function. As there are only a finite number of policies, the algorithm is finite. An algorithm for the α-discounted rewards is as follows:

Step 0: let $k = 0$ and choose a starting policy f_0.
Step 1: policy evaluation: determine $v^\alpha(f_k)$ from equation (3).
Step 2: PI: determine for each state $i : \max_{a \in A(i)} r_i(a) + \alpha \sum_{j \in E} P_{ij}(a) v_j^\alpha(f_k)$.
Let $f_{k+1}(i)$ be the policy consisting of the maximizing action for each state i. Choose $f_{k+1}(i) = f_k(i)$ when possible.
Step 3: if $f_{k+1} = f_k$, then stop, else increase k by 1 and return to step 1.

The successive approximation algorithm works with value iteration. It tries to find a solution to the optimality equations by repeatedly evaluating them. An algorithm is as follows:

Step 1: choose a starting vector $v_i^0, i \in E$, e.g., by taking $v_i^0 = \max_{a \in A(i)} r_i(a)$ and set $k = 0$.
Step 2: let $v_i^{k+1} = \max_{a \in A(i)} r_i(a) + \alpha \sum_{j \in E} P_{ij}(a) v_j^k$, $i \in E$.
Step 3: check if $|v_i^{k+1} - v_i^k| < \varepsilon$ for all $i \in E$, if not go back to step 2.
Step 4: determine the policy that takes maximizing actions in the last step.

The resulting policy has α-discounted rewards, which differ by less than $2\alpha(1-\alpha)^{-1}\varepsilon$ from the maximal α-discounted rewards. The algorithm may be speeded up by eliminating nonoptimal actions. It is faster than PI if the transition matrix is sparse and only few transitions are possible. Finally the LP approach consists of formulating an LP for the optimality criterion. One can prove the following formulation

$$\min \sum_{j \in E} \beta_j v_j,$$

$$\text{s.t.} \quad v_i \geq r_i(a) + \alpha \sum_{j \in E} P_{ij}(a) v_j, a \in A(i), i \in E$$

where $\beta_j, j \in E$, is just a vector of positive components. The solution equals the optimal discounted rewards, and the policy consisting of actions that are binding in the LP restrictions is α-discount optimal. The advantage is that standard LP solvers can be used. From the LP formulation it follows that an MDC can be solved in polynomial time. More complex results are given in Papadimitriou and Tsitsiklis [2].

Issues in Applying MDCs

In general, there are two problems with applying the Markov decision model. The first is identifying states and establishing the Markov property and the second, which is related to the first, is the issue that state spaces can be very large, with the consequence that computation times can be prohibitive.

New techniques concentrate on finding a functional form for the value function. Remember that if one has an approximation for the value function, one can apply the PI step from the PI algorithm to obtain a better- or near-optimal policy. Several approaches have been suggested to this end. We would like to mention neurodynamic programming from Bertsekas and Tsitsiklis [3], reinforcement learning from Das *et al.* [4] and simulation-based approaches by Marbach and Tsitsiklis [5].

Extensions of the Markov Decision Chain Model

In a *semi-MDC* the time between observations of the system is not only fixed, but it is also a random variable, with a distribution that depends on the state and action chosen. Most parts of the analysis can be extended to semi-Markov chains. It is also possible to define a corresponding discrete-time Markov chain, by discretizing the transition time distribution, but this does increase the state space considerably.

In a *continuous-time Markov process*, the time between consecutive transitions is exponentially distributed. These **Markov processes** can be converted into discrete-time Markov chains. We refer to **Markov Processes** for an extensive treatment of Markov processes.

In a *partially observable MDC*, the problem is that not all states are observable and hence no specific actions can be taken once the system enters these states. A separate branch of theory is devoted to these types of Markov chains. A problem is that it is no longer optimal to restrict oneself to Markov polices, because information on the past history also plays a role. This makes both analysis and solution methods inherently more difficult. Overviews on this type of models have been given by Monahan [6] and Lovejoy [7].

MDC Applications to Maintenance Problems

Maintenance problems have been among the first applications of MDCs, as the model allows both the modeling of the deterioration as well as the determination of the structure of optimal policies. Early examples were given by Sasieni [8] and Howard [9] in his car replacement problem. Let us describe the latter.

Consider a car, the age of which is expressed in periods of 3 months; so an age of j indicates an age of $3j$ months. In every period, the car could either age with another period or have a catastrophic failure/accident after which it is replaced with another car. Instead of letting the car age, we can also decide to replace it preventively with another car. In both cases, one needs to choose what kind of age the replacement car should have. The system can be modeled by taking the present age of the car as a state. Transitions to other states occur because of either the aging of the car or a catastrophic failure, in which case we replace the car. From a lifetime distribution, the transition probabilities can be derived. Next we need the cost figures for operating (e.g., due to some small maintenance), for failures, and for acquiring a car of a certain age j, for all values of j. We would like to remark that in the Markov chain modeling with expected rewards criteria, we do not need to bother whether costs were actually incurred before or after a transition. One just takes an expectation over all possible transitions in a state to define the expected immediate rewards. One of the results of MDC theory is that the optimal policy is of the control-limit type, i.e., replace the car if its age is beyond a critical age threshold. This can be proved by showing that the α-discounted costs increase in the states. Other approaches would not give such results.

The Markov decision approach has particularly been followed in condition-based maintenance, where different conditions of equipment are considered and maintenance actions are based on these. Next, there are many studies in maintenance that make use of Markov models (*see* **Markov Processes**), but they use other methods to optimize rather than Markov decision theory.

Real applications have been very much in the sector of civil infrastructure, *viz* roads, bridges, and buildings. Nice overviews of applications have been given by Hontelez *et al.* [10] and Frangopol *et al.* [11]. Applications in other maintenance areas are less common, and as an example we would like to mention Amari *et al.* [12] for inspection planning of condition-based maintenance and Stengos and Thomas [13], who consider the replacement of blast furnaces in a steelworks. It appears that preventively replacing both the furnaces at the same time is not beneficial and that a specific cycle should be followed to reduce the probability that both furnaces fail together. In general, there are two different ways to model states. In the first, one takes the age of a structure as central and takes that as the basis to derive transition probabilities. An alternative is that one takes the remaining life span as a state. In the other modeling, one classifies the condition of the object according to some scale. Table 1 below shows a classification given by Scherer and Glagola [14] for bridge components.

Van Winden and Dekker [15] also use such a condition scale, albeit with five levels. These conditions are typically identified on inspection by classified personnel. One problem however with such scales is that the time to transition is not constant and that a semi-Markov modeling should be used. The latter can be converted by taking the residence time as an extra state variable, but it would make the state space two dimensional.

Another practical issue is that often the state of multiple components or different deterioration mechanisms has to be taken into account. Van Winden and Dekker [15] consider various elements in a building, like window frames, masonry, pointing, and painting. Dekker *et al.* [16] consider for a road segment longitudinal and transversal unevenness, cracking, and raveling. All these aspects can be modeled but they make the state space multidimensional. A decomposition is not always possible because there can be actions which address multiple deterioration aspects or multiple components and which are cheaper than addressing these components separately.

The final issue is that often multiple objects need to be considered, because of either economies of scale in maintenance or budget restrictions that allow only some maintenance to be carried out. We now would like to discuss approaches for roads, bridges, and buildings.

A seminal paper on road maintenance was published by Golabi [17] and it discusses a pavement maintenance system for Arizona's highways. In the system the whole state highway system was modeled. LP was used as the technique; it has the advantage that restrictions on the fraction of road segments in a certain state can be taken into account. A problem is that no costs are involved with bad states; so one has to find a driver to stay out of these states. A special modeling advantage of roads is that they are more or less similar objects and that large sums of money are involved in their maintenance. The disadvantage is that one needs to consider road segments of 100 m and build up a quite large state space. The MDC approach is typically split 1 into several phases. First for a single segment, a Markov decision model is formulated and an optimal policy is derived. From this policy it follows what kind of budget is needed to keep the road in a good condition. Next one considers a whole network of road segments and determines priorities to work on the segments in case the budget is restricted. The advantage of this approach is that the state of all the highways can be taken into account, which gives a much fairer budget allocation. Many papers extended this approach, e.g., Chen *et al.* [18], Abaza and Ashur [19], Bako *et al.* [20], and Dekker *et al.* [16]. Later on, pavement management systems using Markov decision models came into use in many US states and other countries.

For bridges, the first paper was written by Scherer and Glagola [14], who have investigated the applicability and discussed the problems to keep the state space limited. The first real application, Pontis, was again made by a team led by Golabi [21]. Later on more applications followed (see Hawk and Small [22]). A problem with bridges, compared to roads, is that the bridges are more or less unique, which makes the modeling quite time consuming.

Table 1 Condition state descriptions[(a)]

Condition state	Description
9	New condition
8	Good condition: no repairs needed
7	Generally good condition: potential exists for minor maintenance
6	Fair condition: potential exists for major maintenance
5	Fair condition: potential exists for minor rehabilitation
4	Marginal condition: potential exists for major rehabilitation
3	Poor condition: repair or rehabilitation required immediately

(a) © American Society of Civil Engineers, 1994

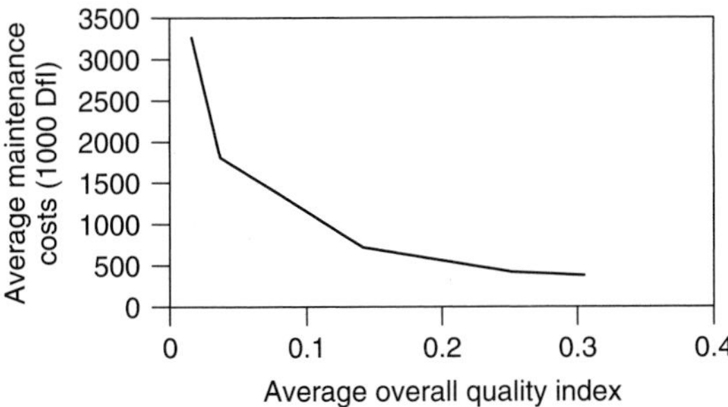

Figure 1 Trade-off between quality and costs

Building maintenance can be approached in a way similar to bridge maintenance. One trick with buildings is that one can work with generic elements and determine policies for these. Next an upscaling is made. For example, pointing is measured in square meters and the total costs can be determined by considering the total number of square meters per building multiplied by the cost per square meter. Sometimes a subdivision is necessary, but otherwise most of the same parts have a similar aging. This limits somewhat the number of components in a building, but considering a whole inventory of the building does give a large state space. One of the interesting results of MDC in this respect is that a trade-off between quality and costs can be obtained, which facilitates decision making, as shown in Figure 1, derived by Van Winden and Dekker [15]. Again, this is difficult to achieve with other techniques.

Finally we would like to mention quite a few papers on partially observable MDCs and maintenance. Again this is a natural combination as quite often deterioration is not visible, but it can be modeled. Frangopol *et al.* [11] give a number of examples for bridge maintenance.

A natural extension of these is the application of MDCs in health maintenance. Here again the modeling is on the evolution of a disease within a human with several stages. The MDCs are used to determine the effectiveness of early screening for diseases and an example is Singer [23].

Conclusions

Markov decision chains have been a very successful tool in modeling maintenance. As such they are the main models in civil infrastructure maintenance. The main problem lies however in the size of the state space. But new techniques allow for much larger state spaces. So far no real alternative models have emerged which allow the determinination of optimal policies. Hence, the main future issue is to develop new techniques to tackle the large multidimensional state spaces.

References

[1] Barlow, R.E. & Proschan, F. (1965). *Mathematical Theory of Reliability*, John Wiley & Sons, New York.

[2] Papadimitriou, C. & Tsitsiklis, J.N. (1987). The complexity of Markov decision processes, *Mathematics of Operations Research* **12**, 441–450.

[3] Bertsekas, W. & Tsitsiklis, J.N. (1996). *Neuro-Dynamic Programming*, Athena Scientific, Nashua.

[4] Das, T.K., Gosavi, A., Mahadevan, S. & Marchalleck, N. (1999). Solving semi-Markov decision problems using average reward reinforcement learning, *Management Science* **45**(4), 560–574.

[5] Marbach, P. & Tsitsiklis, J.N. (2001). Simulation-based optimization of Markov reward processes, *IEEE Transactions on Automatic Control* **46**(2), 191–209.

[6] Monahan, G.E. (1982). A survey of partially observable Markov decision processes: theory, models and algorithms, *Management Science* **28**(1), 1–16.

[7] Lovejoy, W.S. (1991). A survey of algorithmic methods for partially observed Markov decision processes, *Annals of Operations Research* **28**, 47–66.

[8] Sasieni, M. (1956). A Markov chain process in industrial replacement, *Operational Research Quarterly* **7**, 148–155.

[9] Howard, R.A. (1960). *Dynamic Programming and Markov Processes*, Technology Press of MIT.

[10] Hontelez, J.A.M., Burger, H.H. & Wijnmalen, D.J.D. (1996). Optimum condition-based maintenance policies for deteriorating systems with partial information, *Reliability Engineering and System Safety* **98**(1), 52–63.

[11] Frangopol, D.M., Kallen, M.-J. & Van Noortwijk, J.M. (2004). Probabilistic models for life-cycle performance of deteriorating structures: review and future directions, *Progress in Structural Engineering and Materials* **6**, 197–212.

[12] Amari, S.V., McLaughlin, L. & Pham, H. (2006). Cost-effective condition-based maintenance using Markov decision processes, *Proceedings Annual Reliability and Maintainability Symposium, RAMS 2006*, IEEE, 464–469.

[13] Stengos, D. & Thomas, L. (1980). The blast furnaces problem, *European Journal of Operational Research* **4**, 330–336.

[14] Scherer, W.T. & Glagola, D.M. (1994). Markovian models for bridge maintenance management, *Journal of Transportation Engineering* **120**(1), 37–51.

[15] Van Winden, C. & Dekker, R. (1998). Markov decision models for building maintenance, *Journal of the Operational Research Society* **49**(9), 928–935.
[16] Dekker, R., Plasmeijer, R. & Swart, J. (1998). Evaluation of a new maintenance concept for the preservation of highways, *IMA Journal of Mathematics Applied in Business and Industry* **9**, 109–156.
[17] Golabi, K. (1982). A statewide pavement management system, *Interfaces* **6**, 5–21.
[18] Chen, X., Hudson, S., Pajoh, M. & Dickinson, M. (1996). Development of new network optimization model for Oklahoma department of transportation, *Annual Meeting TRB*, Washington DC, No 1524, 103–108.
[19] Abaza, K.A. & Ashur, S.A. (1999). *Optimum Decision Policy for Management of Pavement Maintenance and Rehabilitation, Transportation Research Record 1655*, Journal of Transportation Research Board, 8–15.
[20] Bako, A., Klafszky, E., Szantai, T. & Gaspar, L. (1995). Optimization techniques for planning highway pavement improvements, *Annals of Operations Research* **58**(1), 55–66.
[21] Golabi, K. & Shepard, R. (1997). Pontis: a system for maintenance optimization and improvement of US bridge networks, *Interfaces* **27**, 71–88.
[22] Hawk, H. & Small, E.P. (1998). The BRIDGIT bridge management system, *Structural Engineering International* **8**(4), 309–314.
[23] Singer, M. (2001). Cost effectiveness of screening for hepatitis C virus in asymptotic, average risk adults, *The American Journal of Medicine* **111**(8), 614–621.

Related Articles

Markov Processes; **Replacement Strategies**; **Multivariate Age and Multivariate Renewal Replacement**.

ROMMERT DEKKER, ROBIN P. NICOLAI, AND LODEWIJK C.M. KALLENBERG

Maintenance Optimization

Introduction

As the cost of maintenance is continuously increasing [1], a scientific approach to maintenance optimization is of considerable interest. Maintenance optimization is the problem of determining cost-optimal maintenance decisions for an object (system or structure or one of its components) to ensure a safe and economic operation. An important concept in maintenance optimization is that of life-cycle costing (*see* **Life Cycle Costs and Reliability Engineering**), where the total cost of design, building, maintenance, and demolition is considered over the entire life span of the object in question. In optimizing maintenance, the uncertainties in the time to failure and/or the deteriorating condition should be taken into account (*see* **Stochastic Deterioration**). Many maintenance models, under uncertain deterioration, have been developed in various fields of engineering and a large number of articles, mainly focusing on the mathematical aspects, have been published [1–12]. Herein, a number of probabilistic models for optimizing the life-cycle cost of deteriorating objects are reviewed.

Maintenance optimization techniques are applicable in the design phase and the application phase of the life span of an object. In the design phase, the initial cost of investment has to be balanced against the future cost of maintenance by optimizing the design: the larger the initial resistance, the higher the cost of investment, but the lower the cost of maintenance. In the application phase, the costs of inspection and preventive maintenance (PM) have to be balanced against the costs of corrective maintenance (CM) and failure, by optimizing intervals of inspection and PM: the more PM is carried out, the higher the costs of inspection and PM, but the smaller the costs of CM and failure.

Maintenance

Maintenance Strategies

Usually, maintenance is defined as a combination of actions carried out to restore a component to, or to "renew" it to, a specified condition in which the component can perform its required functions. Inspections, replacements, perfect repairs, partial repairs (lifetime-extending maintenance), and minimal repairs (restoring the component to its prefailure state) are possible maintenance actions. Roughly, there are two types of maintenance: corrective maintenance (after failure) and preventive maintenance (mainly before failure). The decision diagram for the choice between CM and PM is given in Figure 1.

PM can be further subdivided into: time-based maintenance carried out at regular intervals of time,

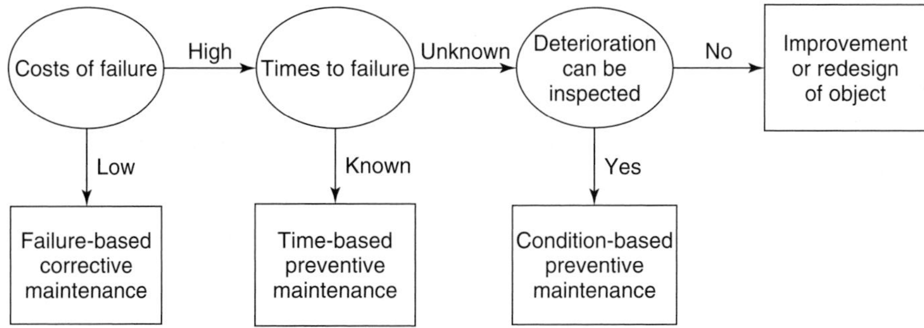

Figure 1 Decision diagram for corrective and preventive maintenance

use-based maintenance carried out after a fixed cumulative use, operation or load, and condition-based maintenance (CBM) carried out at times determined by (non)periodic inspection or continuous monitoring of a component's condition. According to Moubray [13], industrial maintenance changed from corrective to preventive in 1950s; initially time-based maintenance in 1960s (such as equipment overhauls at fixed intervals) and from 1970s on more CBM (such as condition monitoring). The above maintenance strategies can be applied to a single component or groups of components.

To determine optimal maintenance decisions under stochastic deterioration (*see* **Stochastic Deterioration**), we distinguish two approaches [14, Chapter 1]: the actuarial approach based on the notion of lifetime (time to failure) and the physical approach based on the notion of failure due the stress exceeding the resistance (*see* **Stress–Strength Model**). The former is used to model time-based maintenance and the latter to model CBM.

Assumptions, Notations, and Abbreviations

In the material reviewed below, the following assumptions are made: inspection is perfect in the sense that deterioration will be observed with certainty; inspection, maintenance, and replacement take negligible time and do not degrade the component at hand.

In addition, the following notation will be used in the description of the models: c_P is the cost of preventive replacement or maintenance, c_F is the cost of corrective replacement or maintenance and failure, c_I is the cost of a single inspection, and c_U is the cost of unavailability per unit time in the event of an unrevealed failure.

Finally, the following standard abbreviations will be used for brevity of presentation: PM (preventive maintenance), CM (corrective maintenance), CBM (condition-based maintenance), and EEAC (expected equivalent average cost).

Renewal Reward Processes

Maintenance can often be modeled as a renewal process, whereby the renewals are the maintenance actions that bring a component back into its original condition or "good as new" state (*see* **Renewal Theory**). A renewal process $\{N(t), t \geq 0\}$ is a nonnegative integer-valued stochastic process that registers the successive renewals in the time interval $(0, t]$. Let $F(t)$ be the cumulative distribution function of the renewal time $T \geq 0$ and let $c(t)$ be the cost associated with a renewal at time t.

There are three cost-based criteria that can be used to compare maintenance decisions [15, Chapter 11]: (a) the expected average cost per unit time, (b) the expected discounted cost over an unbounded time horizon, and (c) the EEAC per unit time. Because the planned lifetime of most systems and structures is very long, maintenance decisions may be compared over an unbounded time horizon. For an overview on maintenance optimization over a bounded horizon, see [16].

Using renewal reward theory [17, Chapter 3], the expected average cost per unit time is

$$\lim_{t\to\infty} \frac{E(K(t))}{t} = \frac{\displaystyle\int_0^\infty c(t)\,\mathrm{d}F(t)}{\displaystyle\int_0^\infty t\,\mathrm{d}F(t)} = \frac{E(\text{cycle cost})}{E(\text{cycle length})} \tag{1}$$

where $E(K(t))$ represents the expected nondiscounted cost in the bounded time interval $(0, t]$. Let a renewal cycle be the time period between two renewals, and recognize the numerator of equation (1) as the expected cycle cost and the denominator as the expected cycle length (mean lifetime).

To account for the time value of money, Samuelson [18] proposed to use the exponential discount function e^{-rt} with constant discount rate $r > 0$ for time $t \geq 0$. Using renewal reward theory with discounting [19, 20], the expected discounted cost over an unbounded horizon can be written as

$$c_0 + \lim_{t\to\infty} E(K(t,r)) = c_0 + \frac{\displaystyle\int_0^\infty \mathrm{e}^{-rt}c(t)\,\mathrm{d}F(t)}{1 - \displaystyle\int_0^\infty \mathrm{e}^{-rt}\,\mathrm{d}F(t)} = L(r) \quad (2)$$

where c_0 is the investment cost and $E(K(t, r))$ is the expected discounted cost in the bounded time interval $(0, t]$, $t > 0$. Similar results can be obtained for discrete-time **renewal processes** [21]. For renewal theory with other types of discounting (such as generalized hyperbolic or gamma discounting), see van der Weide *et al.* [20].

For the purpose of reserving budget for performing future maintenance actions, it is important to determine the amount of money these actions cost per unit time while taking the discounting into account. This cost is known as the *equivalent average cost per unit time* [15, Chapter 11]. The EEAC per unit time computed over an unbounded time horizon is defined as

$$EEAC = \lim_{t\to\infty} \frac{c_0 + E(K(t,r))}{\displaystyle\int_0^t \mathrm{e}^{-rx}\,\mathrm{d}x} = rL(r) \quad (3)$$

The EEAC per unit time can also be interpreted as a stream of fixed identical cost per unit time sufficient to recover all the necessary discounted cost. As r tends to zero from above, the EEAC approaches the expected average cost per unit time [21]; that is,

$$\lim_{r\downarrow 0} \frac{E(K(t,r))}{\displaystyle\int_0^t \mathrm{e}^{-rx}\,\mathrm{d}x} = \frac{E(K(t,0))}{t} \quad (4)$$

The cost-based criteria of discounted cost and EEAC are most suitable for balancing the initial building cost optimally against the future maintenance cost, because the contribution of the initial investment cost is not ignored (as opposed to the expected average cost per unit time). The criterion of the expected average cost per unit time can be used in situations in which no large investments are made (like inspections) and in which the time value of money is of no consequence. Often, it is preferable to spread the cost of maintenance over time and to use discounting. The area of optimizing maintenance through mathematical models based on lifetime distributions was founded in the early 1960s [2, 3]. Well-known decision models of this period are the age-replacement model and the block-replacement model.

Age Replacement

Under an age-replacement policy [2, Chapters 3–4], a replacement is carried out at age $k > 0$ (preventive replacement) *or* at failure (corrective replacement), whichever occurs first. Let the time to failure T have a cumulative distribution function $F(t)$. According to equation (1), the expected average cost of age replacement per unit time is

$$\lim_{t\to\infty} \frac{E(K(t))}{t} = \frac{c_F F(k) + c_P \bar{F}(k)}{\displaystyle\int_0^k t\,\mathrm{d}F(t) + k\bar{F}(k)} \quad (5)$$

where k is the age-replacement interval and $0 < c_P \leq c_F$. The optimal age-replacement interval k^* is an interval for which the expected average cost per unit time is minimal. Note that PM only makes sense for deteriorating components (i.e., having increasing failure rates). According to equation (2) and Fox [22], the expected discounted cost of age replacement over an unbounded horizon is

$$\lim_{t\to\infty} E(K(t,r)) = \frac{c_F \displaystyle\int_0^k \mathrm{e}^{-rt}\,\mathrm{d}F(t) + c_P \mathrm{e}^{-rk}\bar{F}(k)}{1 - \left[\displaystyle\int_0^k \mathrm{e}^{-rt}\,\mathrm{d}F(t) + \mathrm{e}^{-rk}\bar{F}(k)\right]} = L(r) \quad (6)$$

Note that the replacement model can also be applied for determining an optimal component in the design phase, which balances the initial cost of investment c_P optimally against the future cost of maintenance, by adding c_P to equation (6). The age-replacement model is one of the maintenance optimization models that has been applied most [1, 9]. For example, it has been extended with the possibility of lifetime-extending maintenance in [23].

Block Replacement

Under a block-replacement policy (*see* **Block Replacement**), a replacement is carried out at failure (corrective replacement) *and* periodically at the times $k, 2k, 3k, \ldots$ (preventive replacement). Let the failure times $T_1, T_2, T_3, \ldots$ be nonnegative, independent, identically distributed, random quantities having the cumulative distribution function $F(t)$. The expected average cost of block replacement per unit time when the decision maker chooses block-replacement interval k is

$$\lim_{t\to\infty} \frac{E(K(t))}{t} = \frac{c_F E(N(k)) + c_P}{k} \tag{7}$$

where $E(N(t))$ is the expected number of failures in $(0, t]$:

$$N(t) = \max\{j \mid S_j \le t\} = \sum_{i=1}^{\infty} 1_{\{S_i \le t\}}, \quad t \ge 0 \tag{8}$$

where $S_j = T_1 + \cdots + T_j$, $j = 1, 2, \ldots$, and 1_A denotes the indicator function of the set A. The expected discounted cost of block replacement over an unbounded horizon when the decision maker chooses block-replacement interval k is

$$\lim_{t\to\infty} E(K(t, r)) = \frac{c_F E(N(k, r)) + c_P e^{-rk}}{1 - e^{-rk}} \tag{9}$$

where $E(N(t, r))$ is the expected number of "discounted failures" in $(0, t]$; that is,

$$E(N(t, r)) = E\left(\sum_{i=1}^{\infty} e^{-rS_i} 1_{\{S_i \le t\}}\right), \quad t \ge 0 \tag{10}$$

The block-replacement policy may be modified by performing a **minimal repair** when a component fails in the block interval [24]. Assuming the failure rate h before and after the minimal repair to be identical, then $E(N(k)) = \int_0^k h(t)\,dt$ and $E(N(k, r)) = \int_0^k e^{-rt} h(t)\,dt$.

Inspection to Detect Failures

The age- and block-replacement models assume failures to be detected immediately (revealed failures). When failures are detected only by inspection (unrevealed failures), then a different maintenance model should be used. For this purpose, we assume that the inspection times are $0 = t_0 < t_1 < t_2 < \ldots$ A renewal is assumed when inspection reveals failure. According to Barlow and Proschan [2, Chapter 4], the expected costs of inspection and unavailability can be written as

$$E(\text{cycle cost}) = \sum_{j=1}^{\infty} \int_{t_{j-1}}^{t_j} \left[c_I j + c_U(t_j - t)\right] dF(t) + c_F \tag{11}$$

and the expected renewal time as

$$E(\text{cycle length}) = \sum_{j=1}^{\infty} \int_{t_{j-1}}^{t_j} t_j\, dF(t) \tag{12}$$

Another inspection model is the delay-time model [25, 26], which assumes the failure process to have two stages: the first stage at which the component is functioning well and the second stage at which the component is still functioning but defective, in the sense that failure can be expected soon. Periodic inspection is performed to reveal whether a component is defective or not with inspection interval τ. The time interval between the first moment at which a defect can be noticed until the time of failure is called the *delay time* and is assumed to have cumulative distribution function G. The occurrence process of defects is assumed to be a Poisson with a rate λ. Using equation (1), the expected average cost per unit time can be written as

$$\frac{E(\text{cycle cost})}{E(\text{cycle length})} = \frac{c_I + c_F \int_0^{\tau} \lambda G(\tau - x)\,dx}{\tau} \tag{13}$$

Multicomponent Maintenance

When production downtime occurs or when other system component failures occur requiring downtime, maintenance opportunities may be presented for all components. When such is the case, opportunity-based maintenance models are used. These approaches use either age [27–31] or block [32, 33]

replacement strategies with opportunity times described by a renewal, Markov or **Poisson process**.

Dependence among components can arise due to economies of scale for maintenance cost, structural relationships, or stochastic dependence of failure times [34]. Accounting for these types of dependencies has also been an important research area in maintenance optimization. Reviews of models for multicomponent maintenance can be found in [8, 35] (*see* **Multicomponent Maintenance**).

Condition-based Maintenance

Due to the usual lack of failure data, a reliability approach solely based on lifetime distributions is unsatisfactory. If possible, it is recommended to model deterioration in terms of a time-dependent stochastic process $\{X(t),\ t \geq 0\}$ where $X(t)$ is a random quantity for all $t \geq 0$. The deterioration as a function of time t, $X(t)$, is usually assumed to be a Markov process [2, Chapter 5] (*see* **Markov Processes**). Classes of Markov processes which are useful for modeling stochastic deterioration are discrete-time Markov processes having a finite or countable state space called *Markov chains* and continuous-time nondecreasing Markov processes with independent increments such as the compound Poisson process and the gamma process. Compound Poisson processes (gamma processes) are jump processes with a finite (infinite) number of jumps in a bounded time interval (for a sample path of the gamma process, see Figure 2 taken from [36]). Compound Poisson processes are suitable for modeling damage due to sporadic shocks and gamma processes for describing gradual damage by continuous use [36].

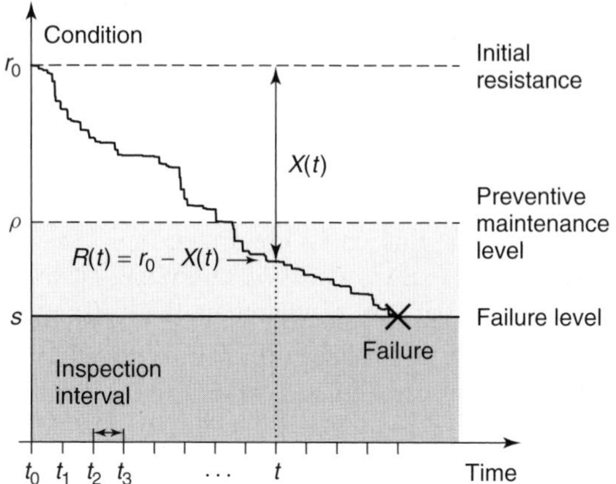

Figure 2 Condition-based maintenance model under gamma process deterioration

Monotone Jump Process. Modeling the deterioration as a monotone stochastic jump process is especially suited when inspections are involved [36]. In optimizing periodic inspection, the two decision variables are the inspection interval τ and the PM level ρ (Figure 2). Such a policy is called a *control-limit policy* with the PM level called the *control limit* [11]. The deterioration is inspected periodically (i.e., $t_j = j\tau,\ j = 1, 2, \ldots$) and is regarded as a gamma process. At an inspection, the object can be in a functional (good), marginal, or failed state. If the object is found to be in a functional state, no maintenance is required and only the cost of the inspection is incurred. If it is found to be in a marginal state, the cost of PM is added to the cost of the inspection. For an object in failed state, there are two scenarios: either the failure is immediately noticed at the time of occurrence [37–41], without the necessity of an inspection, or failure is only detected at the next planned inspection [42–44]. In the first case, only the cost of CM is incurred and in the second case, the cost of the inspection and unavailability has to be included as well. A renewal can be PM or CM and it brings the object back to its "good as new" condition. The optimal maintenance decision is determined by minimizing the long-term expected average cost per unit time in equation (1).

When a failure is detected immediately, the expected renewal-cycle cost is the sum of the costs of all inspections during the cycle and either PM or CM:

$$E(\text{cycle cost}) = \sum_{j=1}^{\infty} \Big[(jc_{\mathrm{I}} + c_{\mathrm{P}}) \Pr\{\text{PM in } (t_{j-1}, t_j]\} + ((j-1)c_{\mathrm{I}} + c_{\mathrm{F}}) \times \Pr\{\text{CM in } (t_{j-1}, t_j]\} \Big] \quad (14)$$

In terms of the deterioration process $X(t)$, the event $\{\text{PM in } (t_{j-1}, t_j]\}$ means PM at the jth inspection and is equivalent to $\{r_0 - X(t_{j-1}) \geq \rho, s \leq r_0 - X(t_j) < \rho\}$. The event $\{\text{CM in } (t_{j-1}, t_j]\}$ means CM during the jth inspection interval, which is equivalent to $\{r_0 - X(t_{j-1}) \geq \rho, r_0 - X(t_j) < s\}$. In a similar way, the expected renewal-cycle length is

$$E(\text{cycle length}) = \sum_{j=1}^{\infty} \left[t_j \Pr\left\{\text{PM in } (t_{j-1}, t_j]\right\} + E\left(T_F; \text{CM in } (t_{j-1}, t_j]\right) \right] \quad (15)$$

where T_F is the time to failure. Extensions and variations of the CBM model with failures detected immediately include discounting [39, 40], nonperiodic inspection [40, 45], and partial or **imperfect repair** [46].

When a failure is detected only by inspection, the object is renewed when an inspection reveals either that the PM level ρ is crossed while no failure has occurred (PM) or that the failure level s is crossed (CM). The expected cycle cost in equation (14) should now be extended with the cost of the inspection at which failure is detected and a penalization for the unavailability of the object due to the unrevealed failure being

$$\sum_{j=1}^{\infty} \left[c_I \Pr\left\{\text{CM in } (t_{j-1}, t_j]\right\} + c_U E\left(t_j - T_F; \text{CM in } (t_{j-1}, t_j]\right) \right] \quad (16)$$

where c_U is the cost of unavailability per unit time. The expected cycle length is

$$E(\text{cycle length}) = \sum_{j=1}^{\infty} t_j \Pr\left\{\text{PM or CM in } (t_{j-1}, t_j]\right\} \quad (17)$$

Extensions and variations of the CBM model with failures detected only by inspection include nonperiodic inspection [47, 48], **optimal design** [49], and damage initiation [50]. A CBM model with deterioration described as a compound Poisson process can be found in Zuckerman [44].

Markov Decision Process. When a finite or countable state space is assumed, implying that the condition of a component can be in any one of $n \geq 0$ discrete states, a Markov-chain model can be used (*see* **Maintenance and Markov Decision Models**). The condition-based inspection model described in the previous section can also be applied when deterioration is modeled by a finite-state Markov process. See [51] for an application to bridge inspections. However, a more common optimization framework for these processes are the so-called Markov decision processes.

Decisions about an optimal policy for maintenance actions are made on a finite set of actions A and costs $C(i, a)$, which are incurred when the process is in state i and action $a \in A$ is taken. The costs are assumed to be bounded and a policy is defined to be any rule for choosing actions. When the process is in state i at time $t = 0, 1, 2, \ldots$ and an action a is taken, the process moves into state j after one unit of time with probability

$$P_{ij}(a) = \Pr\{X_{t+1} = j | X_t = i, a_t = a\} \quad (18)$$

Like for regular Markov chains, this transition probability does not depend on the state history. If a stationary policy is selected, then this process is a Markov decision process. A stationary policy arises when the decision for an action only depends on the current state of the process and not on the time at which the action is performed.

Given that the state of the process at times $0, 1, \ldots$ is modeled by a Markov decision process $X_0, X_1, \ldots$ governed by the transition probabilities $P_{ij}(a)$, the optimization of inspection and/or maintenance policies using this process can be performed. For example, when the object is in state i the expected discounted costs over an unbounded horizon are given by the recurrent relation

$$V_\alpha(i) = C(i, a) + \alpha \sum_{j=1}^{n} P_{ij}(a) V_\alpha(j) \quad (19)$$

where $\alpha = 1/(1 + r)$ is the discount factor for one unit of time, r the discount rate, and V_α the value function using α. Starting from state i, $V_\alpha(i)$ gives us the cost of performing an action a given by $C(i, a)$ and adds the expected discounted costs of moving into another state one unit of time later with probability $P_{ij}(a)$. The discounted costs over an unbounded horizon associated with a start in state j are given by $V_\alpha(j)$, therefore, equation (19) is a recursive equation. The choice for the action a is determined by the maintenance policy and also includes no repair.

A cost-optimal decision can now be found by minimizing equation (19) with respect to the action a. There are a number of ways to find this optimal solution. One of these is the so-called policy improvement algorithm, where equation (19) is calculated for increasingly better policies until no more improvement can be made. Also, it is possible to formulate the minimization problem as a linear programming problem. This is used in the Arizona pavement management system [52] and the PONTIS bridge management system [53]. As Golabi [54] illustrates, we can choose to maximize the condition of the road system under a budget constraint or we can minimize the maintenance cost under a minimum safety constraint. This can be achieved by using the original linear programming formulation or with its dual formulation.

The difference between the use of the condition-based inspection model from the previous section and Markov decision processes lies primarily in the fact that the latter approach is used to make a decision about what to do at fixed times $t = 0, 1, 2, \ldots$. The condition-based inspection model has a fixed policy, namely, do nothing if the condition is functional, do a preventative repair if the condition is marginal, and do a corrective repair if the object is in a failed condition. Given this policy, the inspection interval and the thresholds for marginal and failed conditions may be optimized. The decision is therefore not what to do, but when to do it.

Adaptive Maintenance Policies

In most cases, the maintenance optimization approaches are developed under the assumption that the parameters of the process are known (or estimated) within a degree of certainty. Adaptive methods for maintenance optimization using a Bayes approach where **prior distributions** are formulated for distribution parameters and updated as data becomes available have been proposed for only a few basic models. Wilson and Benmerzouga [55] have used this approach for analysis of group replacement strategies for parallel components with exponential failure times. Mazzuchi and Soyer [56] propose a Bayes approach for the standard age- and block-replacement model where the underlying failure time distribution is Weibull. There are several extensions in the literature. In Markov decision processes, Bayes theorem can be applied to update prior distributions on transition probabilities elicited from engineers with inspections [57]. Finally, Kallen and van Noortwijk [41] provide a Bayes analysis for a periodic (imperfect) inspection and maintenance plan where the deterioration process is modeled by a gamma process.

References

[1] Dekker, R. & Scarf, P.A. (1998). On the impact of optimisation models in maintenance decision making: the state of the art, *Reliability Engineering and System Safety* **60**(2), 111–119.

[2] Barlow, R.E. & Proschan, F. (1965). *Mathematical Theory of Reliability*, John Wiley & Sons, New York.

[3] McCall, J.J. (1965). Maintenance policies for stochastically failing equipment: a survey, *Management Science* **11**, 493–524.

[4] Pierskalla, W.P. & Voelker, J.A. (1976). A survey of maintenance models: the control and surveillance of deteriorating systems, *Naval Research Logistics Quarterly* **23**, 353–388.

[5] Sherif, Y.S. & Smith, M.L. (1981). Optimal maintenance models for systems subject to failure - a review, *Naval Research Logistics Quarterly* **28**, 47–74.

[6] Sherif, Y.S. (1982). Reliability analysis: optimal inspection and maintenance schedules of failing systems, *Microelectronics Reliability* **22**(1), 59–115.

[7] Valdez-Flores, C. & Feldman, R.M. (1989). A survey of preventive maintenance models for stochastically deteriorating single-unit systems, *Naval Research Logistics* **36**, 419–446.

[8] Cho, D.I. & Parlar, M. (1991). A survey of maintenance models for multi-unit systems, *European Journal of Operational Research* **51**(1), 1–23.

[9] Dekker, R. (1996). Applications of maintenance optimization models: a review and analysis, *Reliability Engineering and System Safety* **51**(3), 229–240.

[10] Pintelon, L.M. & Gelders, L.F. (1992). Maintenance management decision making, *European Journal of Operational Research* **58**(3), 301–317.

[11] Wang, H. (2002). A survey of maintenance policies of deteriorating systems, *European Journal of Operational Research* **139**(3), 469–489.

[12] Frangopol, D.M., Kallen, M.J. & van Noortwijk, J.M. (2004). Probabilistic models for life-cycle performance of deteriorating structures: review and future directions, *Progress in Structural Engineering and Materials* **6**(4), 197–212.

[13] Moubray, J. (1997). *Reliability-Centered Maintenance*, 2nd Edition, Industrial Press, New York.

[14] Rausand, R. & Høyland, A. (2004). *System Reliability Theory; Models, Statistical Methods, and Applications*, 2nd edition, John Wiley & Sons, Hoboken.

[15] Wagner, H.M. (1975). *Principles of Operations Research*, 2nd Edition, Prentice-Hall, Englewood Cliffs.

[16] Nakagawa, T. & Mizutani, S. (2007). A summary of maintenance policies for a finite interval, *Reliability Engineering and System Safety* In Press; doi: 10.1016/j.ress.2007.04.004.

[17] Ross, S.M. (1970). *Applied Probability Models with Optimization Applications*, Dover Publications, New York.

[18] Samuelson, P.A. (1937). A note on measurement of utility, *The Review of Economic Studies* **4**(2), 155–161.

[19] Rackwitz, R. (2001). Optimizing systematically renewed structures, *Reliability Engineering and System Safety* **73**(3), 269–279.

[20] van der Weide, J.A.M., Suyono, & van Noortwijk, J.M. (2008). Renewal theory with exponential and hyperbolic discounting, *Probability in the Engineering and Informational Sciences* **22**(1), (In Press).

[21] van Noortwijk, J.M. (2003). Explicit formulas for the variance of discounted life-cycle cost, *Reliability Engineering and System Safety* **80**(2), 185–195.

[22] Fox, B. (1966). Age replacement with discounting, *Operations Research* **14**(3), 533–537.

[23] van Noortwijk, J.M. & Frangopol, D.M. (2004). Two probabilistic life-cycle maintenance models for deteriorating civil infrastructures, *Probabilistic Engineering Mechanics* **19**(4), 345–359.

[24] Barlow, R.E. & Hunter, L.C. (1960). Optimum preventive maintenance policies, *Operations Research* **8**(1), 90–100.

[25] Christer, A.H. (1982). Modeling inspection policies for building maintenance, *Journal of the Operational Research Society* **33**(8), 723–732.

[26] Baker, R.D. & Christer, A.H. (1994). Review of delay-time OR modelling of engineering aspects of maintenance, *European Journal of Operational Research* **73**(3), 407–422.

[27] Dekker, R. & Dijkstra, M.C. (1992). Opportunity-based age replacement: exponentially distributed times between opportunities, *Naval Research Logistics* **39**(2), 175–190.

[28] Coolen-Schrijner, P., Coolen, F.P.A. & Shaw, S.C. (2006). Nonparametric adaptive opportunity-based age replacement strategies, *Journal of the Operational Research Society* **57**(1), 63–81.

[29] Satow, T. & Osaki, S. (2003). Opportunity-based age replacement with different intensity rates, *Mathematical and Computer Modelling* **38**(11–13), 1419–1426.

[30] Jhang, J.P. & Sheu, S.H. (1999). Opportunity-based age replacement policy with minimal repair, *Reliability Engineering and System Safety* **64**(3), 339–344.

[31] Iskandar, B.P. & Sandoh, H. (2000). An extended opportunity-based age replacement policy, *RAIRO Recherche Opérationnelle – Operations Research* **34**(2), 145–154.

[32] Dekker, R. & Smeitink, E. (1991). Opportunity-based block replacement, *European Journal of Operational Research* **53**(1), 46–63.

[33] Dekker, R. & Smeitink, E. (1994). Preventive maintenance at opportunities of restricted duration, *Naval Research Logistics* **41**(3), 335–353.

[34] Thomas, L.C. (1986). A survey of maintenance and replacement models for maintainability and reliability of multi-item systems, *Reliability Engineering* **16**(4), 297–309.

[35] Nicolai, R.P. & Dekker, R. (2007). Optimal maintenance of multi-component systems: a review, in *Complex System Maintenance Handbook*, K.A.H. Kobbacy & D.N.P. Murthy, eds, Springer Verlag.

[36] van Noortwijk, J.M. (2007). A survey of the application of gamma processes in maintenance, *Reliability Engineering and System Safety* In Press; doi:10.1016/j.ress.2007.03.019.

[37] Kong, M.B. & Park, K.S. (1997). Optimal replacement of an item subject to cumulative damage under periodic inspections, *Microelectronics Reliability* **37**(3), 467–472.

[38] Park, K.S. (1988). Optimal continuous-wear limit replacement under periodic inspections, *IEEE Transactions on Reliability* **37**(1), 97–102.

[39] van Noortwijk, J.M., Kok, M. & Cooke, R.M. (1997). Optimal maintenance decisions for the sea-bed protection of the Eastern-Scheldt barrier, in *Engineering Probabilistic Design and Maintenance for Flood Protection*, R. Cooke, M., Mendel & H., Vrijling, eds, Kluwer Academic Publishers, Dordrecht, pp 25–56.

[40] Newby, M. & Dagg, R. (2004). Optimal inspection and perfect repair, *IMA Journal of Management Mathematics* **15**(2), 175–192.

[41] Kallen, M.J. & van Noortwijk, J.M. (2005). Optimal maintenance decisions under imperfect inspection, *Reliability Engineering and System Safety* **90**(2–3), 177–185.

[42] Abdel-Hameed, M. (1987). Inspection and maintenance policies of devices subject to deterioration, *Advances in Applied Probability* **19**, 917–931.

[43] Abdel-Hameed, M. (1995). Correction to: "Inspection and maintenance policies of devices subject to deterioration", *Advances in Applied Probability* **27**, 584.

[44] Zuckerman, D. (1980). Inspection and replacement policies, *Journal of Applied Probability* **17**, 168–177.

[45] Grall, A., Bérenguer, C. & Dieulle, L. (2002). A condition-based maintenance policy for stochastically deteriorating systems, *Reliability Engineering and System Safety* **76**(2), 167–180.

[46] Castanier, B., Bérenguer, C. & Grall, A. (2003). A sequential condition-based repair/replacement policy with non-periodic inspections for a system subject to continuous wear, *Applied Stochastic Models in Business and Industry* **19**(4), 327–347.

[47] Dieulle, L., Bérenguer, C., Grall, A. & Roussignol, M. (2003). Sequential condition-based maintenance scheduling for a deteriorating system, *European Journal of Operational Research* **150**(2), 451–461.

[48] Grall, A., Dieulle, L., Bérenguer, C. & Roussignol, M. (2002). Continuous-time predictive-maintenance scheduling for a deteriorating system, *IEEE Transactions on Reliability* **51**(2), 141–150.

[49] Speijker, L.J.P., van Noortwijk, J.M., Kok, M. & Cooke, R.M. (2000). Optimal maintenance decisions for dikes, *Probability in the Engineering and Informational Sciences* **14**(1), 101–121.

[50] van Noortwijk, J.M. & Klatter, H.E. (1999). Optimal inspection decisions for the block mats of the Eastern-Scheldt barrier, *Reliability Engineering and System Safety* **65**(3), 203–211.

[51] Kallen, M.J. & van Noortwijk, J.M. (2006). Optimal periodic inspection of a deterioration process with sequential condition states, *International Journal of Pressure Vessels and Piping* **83**(4), 249–255.

[52] Golabi, K., Kulkarni, R.B. & Way, G.B. (1982). A statewise pavement management system, *Interfaces* **12**(6), 5–21.

[53] Golabi, K. & Shepard, R. (1997). Pontis: a system for maintenance optimization and improvement of US bridge networks, *Interfaces* **27**(1), 71–88.

[54] Golabi, K. (1983). A Markov decision modeling approach to a multi-objective maintenance problem, in *Essays and Surveys on Multiple Criteria Decision Making*, P. Hansen, ed Springer, Berlin, Vol. 209, pp. 115–125.

[55] Wilson, J.G. & Benmerzouga, A. (1995). Bayesian group replacement policies, *Operations Research* **43**(3), 471–476.

[56] Mazzuchi, T.A. & Soyer, R. (1996). A Bayesian perspective on some replacement strategies, *Reliability Engineering and System Safety* **51**(3), 295–303.

[57] Lee, T.C., Judge, G.G. & Zellner, A. (1970). *Estimating the Parameters of the Markov Probability Model from Aggregate Time Series Data*, North-Holland, Amsterdam.

Related Articles

Age-Dependent Minimal Repair and Maintenance; **Block Replacement**; **Analysis of Recurrent Events from Repairable Systems**; **General Minimal Repair Models**; **Group Maintenance Policies**; **Imperfect Repair**; **Inspection Policies for Reliability**; **Life Cycle Costs and Reliability Engineering**; **Maintenance and Markov Decision Models**; **Maintenance Optimization in Random Environments**; **Markov Processes**; **Multicomponent Maintenance**; **Multivariate Age and Multivariate Renewal Replacement**; **Multivariate Imperfect Repair Models**; **Nonparametric Methods for Analysis of Repair Data**; **Renewal Theory**; **Repairable Systems Reliability**; **Replacement Strategies**; **Stochastic Deterioration**; **Stress–Strength Model**.

THOMAS A. MAZZUCHI, JAN M. VAN NOORTWIJK AND MAARTEN-JAN KALLEN

Maintenance Optimization in Random Environments

Introduction

In this article, we focus on a number of optimal management problems involving reliability and risk models operating in a random environment. The environment is described by a stochastic process, which represents important factors that affect the stochastic and deterministic parameters of the reliability or risk model. The term *environment* is used in the generic sense so that it represents any set of conditions that affect the stochastic structure of the model investigated. The concept of an "environmental" process, in one form or another, has been used in earlier literature for various purposes. Neveu [1] provides an early reference to paired stochastic processes where the first component is a Markov process while the second one has conditionally independent increments given the first. Ezhov and Skorohod [2] refer to this as a Markov process with a homogeneous second component. In a more modern setting, Çınlar [3, 4] introduced Markov additive processes and provided a detailed description on the structure of the additive component. The environment is modeled as a Markov process in all these cases and the additive process represents the stochastic evolution of a quantity of interest. Since the model parameters depend on the state of the environment at any time, the environmental process is also a process that modulates the model. As the state of the environment changes randomly, so do the model parameters and the environmental process indirectly generates the variation in the parameters.

The use of a modulating stochastic process as a source of variation in the model parameters and of dependence among the model components has proved to be quite useful in operations research and management science applications. In modeling hardware reliability for complex devices consisting of many interdependent components, the concept was introduced by Çınlar and Özekici [5] where the failure rate and hazard functions of the components all depend on the prevailing environmental conditions. For example, a complex device like a jet engine consists of a large number of components where the failure structure of each component depends very much on the set of environmental conditions that it is subjected to during flight. The levels of vibration, atmospheric pressure, temperature, and so on, obviously change during take-off, cruising, and landing. Component lifetimes and reliabilities depend on these random environmental variations. Moreover, the components have dependent lifetimes since they operate in the same environment. The concept of random hazard functions is also used in Gaver [6] and Arjas [7]. The intrinsic aging model of Çınlar and Özekici [5] is studied further in Çınlar *et al.* [8] to determine the conditions that lead to associated component lifetimes, as well as multivariate increasing failure rate (IFR) and new better than used (NBU) life distribution characterizations. It was also extended in Shaked and Shanthikumar [9] by discussions on several different models with multicomponent replacement policies. Lindley and Singpurwalla [10] discuss the effects of the random environment on the reliability of a system consisting of components which share the same environment. Although the initial state of the environment is random, they assume that it remains constant in time and components have exponential life distributions in each possible environment. This model is also studied by Lefèvre and Malice [11] to determine partial orderings on the number of functioning components and the reliability of k-out-of-n systems, for different partial orderings of the probability distribution on the environmental state. The association of the lifetimes of components subjected to a randomly varying environment is discussed in Lefèvre and Milhaud [12]. Singpurwalla and Youngren [13] also discuss multivariate distributions that arise in models where a dynamic environment affects the failure rates of the components.

Although we discuss the use of environmental processes in managing hardware as well as **software reliability** and risk models, there is now considerable amount of literature on modulation in a variety of applications. An example in queuing is provided by Prabhu and Zhu [14] where customer arrival and service rates are modulated by a Markov process. Song and Zipkin [15] consider an inventory model with a demand process that fluctuates with respect to stochastically changing economic conditions. A general discussion on the idea can be found in Özekici [16]. The interested reader is referred to Asmussen [17] and Rolski *et al.* [18] for further applications in queuing, insurance, and finance.

The stochastic structure of a rather general environmental process involving a **Markov renewal** process is described in the section titled "The Environmental Process". Issues regarding the maintenance of complex systems under random environments are discussed in the section titled "Maintenance of Complex Systems", and the last section presents some results on the structure of optimal **maintenance policies**.

The Environmental Process

We suppose that the environmental process is quite general as given in Çekyay and Özekici [19] (*see* **Markov Renewal Processes in Reliability Modeling**). Let T_n denote the time of the nth environment change and X_n denote the nth environmental state with $T_0 \equiv 0$. The main assumption is that the process $(X, T) = \{(X_n, T_n); n \geq 0\}$ is a Markov renewal process on the state space E with some semi-Markov kernel Q. Moreover, the environmental process $Y = \{Y_t; t \geq 0\}$ is the minimal semi-Markov process associated with (X, T) so that Y_t is the state of the environment at time t. More precisely, $Y_t = X_n$ whenever $T_n \leq t < T_{n+1}$. For any $i, j \in E$ and $t \geq 0$,

$$Q(i, j, t) = P\{X_{n+1} = j, T_{n+1} - T_n \leq t | X_n = i\} \tag{1}$$

and it is well-known that X is a Markov chain on E with transition matrix $P(i, j) = P\{X_{n+1} = j | X_n = i\} = Q(i, j, +\infty)$. We further assume that the Markov renewal process has infinite lifetime so that $\sup_n T_n = +\infty$. The probabilistic structure of T is given by the conditional distributions

$$G(i, j, t) = P\{T_{n+1} - T_n \leq t | X_n = i, X_{n+1} = j\} = \frac{Q(i, j, t)}{P(i, j)} \tag{2}$$

for $P(i, j) > 0$. Moreover, it is easy to see that

$$F_i(t) = P_i\{T_1 \leq t\} = \sum_{j \in E} Q(i, j, t) \tag{3}$$

denotes the distribution of the duration of state i.

Maintenance of Complex Systems

A complex system consists of many components that have stochastically dependent lifetimes. This dependence is induced by the common environment that they all operate in. Moreover, there is usually some economic dependence among the components due to the economies of scale obtained by maintaining many components at the same time. The most common type of maintenance involves the replacement of some of the functioning and nonfunctional components. The problem is to determine when replacement should be done and which components to replace. This should of course be done by observing the state of the reliability system. The intrinsic aging model introduced by Çınlar and Özekici [5] is a tractable one since it provides a realistic and simple stochastic structure of complex systems with many components. In Çekyay and Özekici [19], the authors consider various issues regarding reliability of such systems under the assumption that there is either no repair or maximal repair when all components are replaced at the beginning of each environmental state. In particular, the system reliability for the two cases are characterized through Markov renewal equations (15) and (37) in Çekyay and Özekici [19]. We will now extend their analysis by considering what happens when one incorporates a decision making process in their setting. We will follow their notation and terminology and the reader should first read that chapter. In particular, we set $R_+ = [0, \infty)$ and $F = R_+^m$ where m is the number of components. Moreover, we assume throughout that the components are serially connected so that the system fails as soon as any one of the components fails.

The intrinsic aging model is described by equations (25)–(33) in Çekyay and Özekici [19]. Recall that for any component k, $H_k(i, \cdot)$ is the cumulative hazard function and $r_k(i, \cdot)$ is the intrinsic aging rate function in environment i. The intrinsic age process is defined through the differential equation $\mathrm{d}A_t(k)/\mathrm{d}t = r_k(Y_t, A_t(k))$. The age of the system is $A_t = (A_t(1), A_2(t), \ldots, A_t(m))$ at time t. It is clear that the process A takes values in F. Whenever the state of the environment changes at time T_n there is a decision made on maintenance based on the observation of the new environment X_n and intrinsic age $B_n = A_{T_n}$ of the system. The primary process is now given by the new Markov renewal process $((X, B), T)$.

We propose to use policies given by sets $M \subseteq E \times F$ so that **preventive maintenance** is done at the first-passage time

$$T_M = \inf\{T_n : (X_n, B_n) \in M\} \tag{4}$$

when the system is maintained if the observed environmental state i and the intrinsic ages $a = (a_1, a_2, \ldots, a_m)$ of the components satisfy $(i, a) \in M$. The stopping time equation (4) identifies the time of maintenance. The type of maintenance to be done is another issue. The most common type involves replacements where one can replace either all components, or only a given a set of components, or only those that have failed. One can also use a repair type of policy where instead of replacing a component by a brand-new one, one repairs it to a lower intrinsic age level. A typical policy in this regard is the **minimal repair** one when a failed component is given a jump start and brought back to the working condition or age just before failure.

For any such policy, Çınlar and Özekici [5] make a Markov renewal argument to obtain the characterization

$$P_{ia}\{L > T_M\} = \sum_{j \in E} \int_F \overline{R}(i, a; j, \mathrm{d}b) \times \sum_{k \in E} \int_{R_+} Q(j, k, \mathrm{d}s) \mathrm{e}^{-\sigma(h(k,b,s)-b)} \times 1_M(k, h(k, b, s)) \tag{5}$$

for $(i, a) \notin M$ where $1_M(y)$ is the indicator function that is equal to 1 if and only if $y \in M$, and $\overline{R} = \sum_n \overline{P}^n$ is the potential kernel corresponding to the defective transition kernel

$$\overline{P}(i, a; j, \mathrm{d}b) = \int_{R_+} Q(i, j, \mathrm{d}s) H(i, a, s, \mathrm{d}b) \times \mathrm{e}^{-\sigma(b-a)} 1_{M^c}(j, b) \tag{6}$$

where M^c denotes the complement of M. Note that equation (5) gives the probability that there is no system failure before the time of maintenance. The probability $P_{ia}\{L > t\}$ that the system does not fail until time t is already determined by equations (15) and (37) in Çekyay and Özekici [19] for the two extreme cases of no repair ($M = \phi$ and no preventive replacement is done) and maximal repair ($M = E \times R_+^m$ and all components are replaced), respectively.

If $L(k)$ is the lifetime of the kth component, and L is the lifetime of the system, then $L = \min\{L(k); k = 1, 2, \ldots, m\}$ and

$$S = \min\{T_M, L\} \tag{7}$$

is the first time either maintenance is done or the system fails, whichever occurs sooner. Then, using Markov renewal theory one obtains the joint distribution

$$\begin{aligned} P_{ia}\{X_S = j, A_S \in \mathrm{d}b, S \leq t, S = L(k)\} \\ = \sum_{l \in E} \int_{F \times [0,t]} \widetilde{R}(i, a; l, \mathrm{d}c; \mathrm{d}s) \int_{[0,t-s]} \mathrm{d}u \overline{F}_l(u) \\ \times I(l, j) H(l, c, u; \mathrm{d}b) \mathrm{e}^{-\sigma(b-c)} r_k(l, b) \end{aligned} \tag{8}$$

for $(i, a) \notin M$ where $\widetilde{R} = \sum_n \widetilde{Q}^n$ is the Markov renewal kernel corresponding to the semi-Markov kernel

$$\begin{aligned} \widetilde{Q}(i, a; j, \mathrm{d}b; \mathrm{d}s) = Q(i, j, \mathrm{d}s) H(i, a, s; \mathrm{d}b) \\ \times \mathrm{e}^{-\sigma(b-a)} 1_{M^c}(j, b) \end{aligned} \tag{9}$$

and r_k is the intrinsic aging rate function given by equation (25) in Çekyay and Özekici [19]. Using a similar analysis, one also obtains

$$\begin{aligned} P_{ia}\{X_S = j, A_S \in \mathrm{d}b, S \leq t, S = T_M\} \\ = \sum_{l \in E} \int_{F \times [0,t]} \widetilde{R}(i, a; l, \mathrm{d}c; \mathrm{d}s) \int_{[0,t-s]} Q(l, j, \mathrm{d}u) \\ \times H(l, c, u; \mathrm{d}b) \mathrm{e}^{-\sigma(b-c)} 1_M(j, b) \end{aligned} \tag{10}$$

for $(i, a) \notin M$. In these expressions, equation (8) gives the probability that the system fails before time t due to the failure of component k in environmental state j at age db, while equation (10) gives the probability that the system is maintained before it fails until time t in environmental state j at age db. Both solutions are of the form $\widetilde{R} * g$ for an appropriate function g. The sum of equation (8) for all k and equation (10) provides another Markov renewal characterization for the probability that preventive or failure maintenance is done before time t in environmental state j at age db. The probabilities (5), (8), and (10) provide important performance measures to the decision maker to choose maintenance policies.

We will not attempt to formulate an optimization problem in our setting to determine optimal maintenance policies. It is indeed an interesting problem which has not been analyzed before for models with intrinsic aging in random environments depicted by a Markov renewal process. This will undoubtedly require various cost figures on system and component failures as well as costs of maintenance. We should point out, however, that the structure of optimal policies is by no means trivial and intuitive. The formulation of the optimization problem, the characterization of optimal policies, and the solution procedure is quite complicated. It is well known that the structure of optimal policies may be quite complex in multicomponent systems even when there are no environmental fluctuations. Preventive replacement is perhaps the most widely used maintenance policy to prevent the device from failure during operation, thus incurring excessive failure costs. The fixed environment case with several cost structures and objectives is discussed extensively in the literature by many authors and, in most cases, an age-replacement policy is optimal if the life distribution is IFR. Özekici [20] provides an example along this direction and a discussion on the optimality conditions for control-limit policies can be found in So [21].

Structure of Optimal Policies

In the optimal replacement problem, for example, perhaps the most widely used policy is the age-replacement policy where a component is replaced when its age is observed to be over a critical level. The problem then is to determine these ages optimally in order to minimize the average cost or total cost. A simple age replacement policy is one in which, in any environment i, there is critical age vector b_i such that component k is replaced if and only if the intrinsic age of the component exceeds $b_i(k)$. This policy is specified by m critical thresholds for each environment. However, such a policy is not necessarily optimal. It is clear that the component lifetimes are stochastically dependent due to their common environment. This implies that a component may be more prone to failure if the age of another component is higher and the critical threshold may depend on the age of other components. Another complication is due to economic dependence among the components. This is the case when there is a fixed cost of replacement such that one is tempted to replace a functioning component when a failed component must be replaced since the fixed cost is already paid. In such cases, so-called opportunistic replacement polices with rather complex structures may be optimal. Özekici [22, 23] demonstrates this through several examples. For simplicity, suppose that there are two components only so that the optimal policy can be identified by the ages at which only component 1 is replaced (1), or only component 2 is replaced (2), or both components are replaced (1,2), or no component is replaced (0). Figure 1 depicts a typical optimal policy.

It is clear that the structure of the optimal policy, in general, is nontrivial and it cannot be identified by a few critical numbers. This is actually due to two reasons: economic and stochastic dependence among the components. Some rather counterintuitive observations can be made on Figure 1. For example, at point B, no component should be replaced but at point A, where there is substantial aging on component 2, both 1 and 2 should be replaced even though component 1 is at the same age as point B. This is called *opportunistic replacement* since 1 is replaced by making use of the opportunity in replacing component 2. Since an "old" component 2 is being replaced optimally, it may be best to replace component 1 at the same time. Opportunistic replacement is due to the fact that $b_1 \leq c_1$ and $b_2 \leq c_2$. This could be further clarified by considering the case where the components are stochastically independent while 1 has an exponentially distributed lifetime and

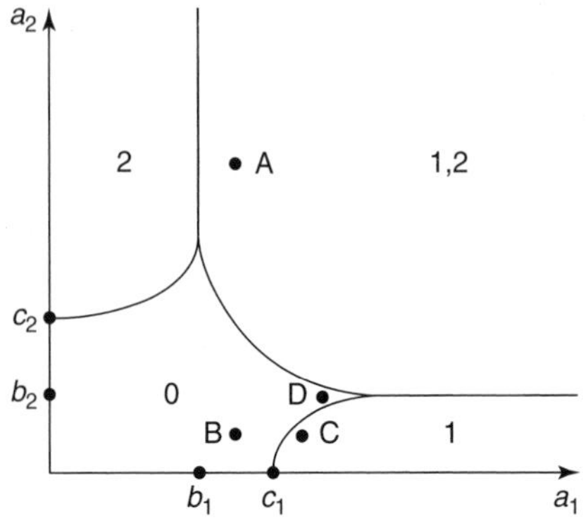

Figure 1 Optimal policy for 2 components

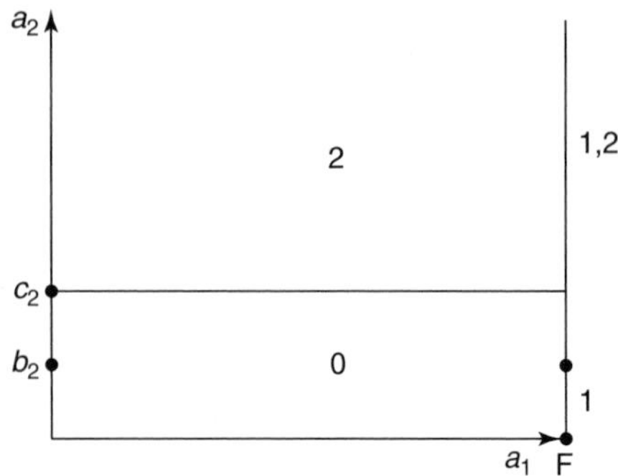

Figure 2 Optimal policy when one component has exponential lifetime

2 has an IFR life distribution. If the components are economically dependent, Radner and Jorgenson [24] showed that the optimal policy has the simple form given in Figure 2 for a fixed environment model. Note that component 1 is replaced only at failure (F) since it has an exponential lifetime. Component 2 is optimally replaced at the critical age c_2, but if component 1 has failed and must be replaced, then 2 can be replaced opportunistically as early as age $b_2 \leq c_2$.

Another interesting observation follows from the comparison of points C and D on Figure 1. Note that component 1 is replaced at C but not at the "older" system age D. The system is not interfered with at the "worse" state D while a replacement decision is given at C. This is due to "opportunistic nonreplacement" since it may be best not to do any replacement at D and wait a little longer to replace both 1 and 2 at the same time.

The complexity in the structure of the optimal replacement policy creates computational as well as practical difficulties since it is not identified by a few critical numbers. Even if the optimal policy is determined by any one of the procedures, its implementation requires extreme precision. For these reasons, one may have to approximate the optimal policy by one that has a much simpler structure. An example is given by Van der Duyn Schouten, and Vanneste [25] analyzed the (n, N) policy given in Figure 3.

Note that the (n, N) policy provides substantial simplification when compared with the optimal policy in Figure 1. The fact that $n_1 \leq N_1$ and $n_2 \leq N_2$ show that, in essence, these policies reflect the existence of opportunistic replacement. However, this does not allow for "opportunistic nonreplacement" since the

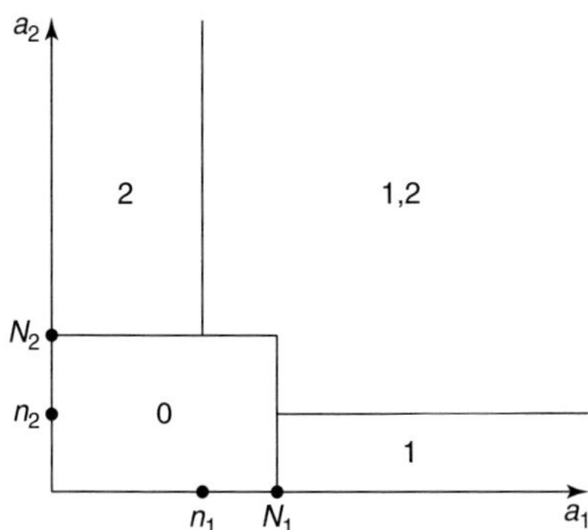

Figure 3 (n, N) policy

number of components replaced in an "older" system is at least as many as that of a "younger" system.

Acknowledgment

This research is supported by the Scientific and Technological Research Council of Turkey through grant 106M044.

References

[1] Neveu, J. (1961). Une généralisation des processus á accroisements positifs indépendants, *Abhandlungen aus den Mathematischen Seminar der Universitat Hamburg* **25**, 36–61.

[2] Ezhov, I.I. & Skorohod, A.V. (1969). Markov processes with homogeneous second component: I, *Theory of Probability and its Application* **14**, 3–14.

[3] Çınlar, E. (1972). Markov additive processes: I, *Zeitschrift für Wahrscheinlichkeitstheorie und verwandte Gebiete* **24**, 85–93.

[4] Çınlar, E. (1972). Markov additive processes: II, *Zeitschrift für Wahrscheinlichkeitstheorie und verwandte Gebiete* **24**, 95–121.

[5] Çınlar, E. & Özekici, S. (1987). Reliability of complex devices in random environments, *Probability in the Engineering and Informational Sciences* **1**, 97–115.

[6] Gaver, D.P. (1963). Random hazard in reliability problems, *Technometrics* **5**, 211–226.

[7] Arjas, E. (1981). The failure and hazard process in multivariate reliability systems, *Mathematics of Operations Research* **6**, 551–562.

[8] Çınlar, E., Shaked, M. & Shanthikumar, J.G. (1989). On lifetimes influenced by a common environment, *Stochastic Processes and their Applications* **33**, 347–359.

[9] Shaked, M. & Shanthikumar, J.G. (1989). Some replacement policies in a random environment, *Probability in the Engineering and Informational Sciences* **3**, 117–134.

[10] Lindley, D.V. & Singpurwalla, N.D. (1986). Multivariate distributions for the lifelenghts of components of a system sharing a common environment, *Journal of Applied Probability* **23**, 418–431.

[11] Lefèvre, C. & Malice, M.-P. (1989). On a system of components with joint lifetimes distributed as a mixture of exponential laws, *Journal of Applied Probability* **26**, 202–208.

[12] Lefèvre, C. & Milhaud, X. (1990). On the association of the lifelenghts of components subjected to a stochastic environment, *Advances in Applied Probability* **22**, 961–964.

[13] Singpurwalla, N.D. & Youngren, M.A. (1993). Multivariate distributions induced by dynamic environments, *Scandinavian Journal of Statistics* **20**, 251–261.

[14] Prabhu, N.U. & Zhu, Y. (1989). Markov-modulated queueing systems, *Queueing Systems* **5**, 215–246.

[15] Song, J.S. & Zipkin, P. (1993). Inventory control in fluctuating demand environment, *Operations Research* **41**, 351–370.

[16] Özekici, S. (1996). Complex systems in random environments, in *Reliability and Maintenance of Complex Systems, NATO ASI Series*, S. Ozekici, ed, Springer-Verlag, Berlin, 137–157 Vol. F154.

[17] Asmussen, S. (2000). *Ruin Probabilities*, World Scientific, Singapore.

[18] Rolski, T., Schmidli, H., Schmidt, V. & Teugels, J. (1999). *Stochastic Processes for Insurance and Finance*, John Wiley & Sons, Chichester.

[19] Çekyay, B. & Özekici, S. (2007). Markov renewal processes in reliability modelling, in *Encyclopedia of Statistics in Quality and Reliability*, F. Ruggeri, F. Faltin & R. Kenett, eds, John Wiley & Sons, Chichester.

[20] Özekici, S. (1985). Optimal replacement of one-unit systems under periodic inspection, *SIAM Journal on Control and Optimization* **23**, 122–128.

[21] So, K.C. (1992). Optimality of control limit policies in replacement models, *Naval Research Logistics* **39**, 685–697.

[22] Özekici, S. (1988). Optimal periodic replacement of multicomponent reliability systems, *Operations Research* **36**, 542–552.

[23] Özekici, S. (1995). Optimal maintenance policies in random environments, *European Journal of Operational Research* **82**, 283–294.

[24] Radner, R. & Jorgenson, D.W. (1963). Opportunistic replacement of a single part in the presence of several monitored parts, *Management Science* **10**, 70–84.

[25] Van der Duyn Schouten, F.A. & Vanneste, S.G. (1990). Analysis and computation of (n, n) strategies for maintenance of a two-component system, *European Journal of Operational Research* **48**, 260–274.

Related Articles

Block Replacement; Group Maintenance Policies; Maintenance and Markov Decision Models; Maintenance Optimization; Markov Processes; Markov

Renewal Processes in Reliability Modeling; **Multicomponent Maintenance**; **Multivariate Age and Multivariate Renewal Replacement**; **Multivariate Imperfect Repair Models**; **Repairable Systems Reliability**; **Replacement Strategies**; **Total Productive Maintenance**.

SÜLEYMAN ÖZEKICI

Maintenance, Age Dependent *see* Age-Dependent Minimal Repair and Maintenance

Management of Quality *see* Quality Management, Overview

Markov Chain Monte Carlo, Hierarchical *see* Hierarchical Markov Chain Monte Carlo (MCMC) for Bayesian System Reliability

Markov Chain Monte Carlo, Introduction

Introduction to Monte Carlo Simulation

High-dimensional integrals and simulation problems are seen in areas such as statistics, reliability, statistical physics, and so on. For example, in Bayesian statistics, given a prior distribution $P(\theta)$ and a likelihood function $l(\theta|\text{data})$, then the posterior mean of θ is calculated *via* Bayes theorem as

$$E[\theta|\text{data}] = \frac{\int \theta P(\theta) l(\theta|\text{data})\,\mathrm{d}\theta}{\int P(\theta) l(\theta|\text{data})\,\mathrm{d}\theta} \tag{1}$$

In many cases, the integrals in the numerator and denominator cannot be evaluated analytically and various alternatives may be considered. One possibility is to use numerical integration techniques (*see* **Quadrature and Numerical Integration**), and an alternative is to use analytic approximations of the integrands such as the Laplace approximation, (*see* e.g. [1] for a general review and **Laplace Approximations in Bayesian Lifetime Analysis** in the context of lifetime data analysis). However, the first method suffers strongly from the curse of dimensionality and approximation methods are only appropriate under restricted conditions. A more general approach is to use simulation.

Suppose then that for a variable X with distribution π that we wish to estimate the expected value of some functional $f(X)$. Then the **Monte Carlo** method, see [2], is to generate a sample of values $x_1, \ldots, x_N$ from the density π and estimate the expectation using the sample mean

$$E_\pi[f(X)] \approx \frac{\sum_{n=1}^{N} f(x_i)\pi(x_i)}{\sum_{n=1}^{N} \pi(x_i)} \tag{2}$$

It is usually difficult or impossible to directly sample from the density π so that certain alternatives such as importance sampling or rejection sampling may be used. A full review of the Monte Carlo approach is given in **Monte Carlo Methods**. See also **System Reliability: Monte Carlo Estimation**.

In a large number of cases, however, exact Monte Carlo samples may not be easily constructed. Therefore, an alternative strategy is to try to generate an approximate Monte Carlo sample. One approach is to construct a discrete-time stochastic process (**Markov chain**) such that the distribution of the states of the process converges over time to the distribution of interest π when a sample can be taken and used in the same way as a standard Monte Carlo sample. This is the basis of the Markov chain Monte Carlo (MCMC) algorithms.

Markov Chains

Before introducing the different MCMC algorithms, it is first useful to review the properties of Markov chains. A full review is given in, for example [3].

A Markov chain, $\{X_t\}$, is defined to be a sequence of variables $X_1, X_2, \ldots$ such that the distribution of X_t, given the previous values $X_0, \ldots, X_{t-1}$ only depends on X_{t-1}, so that

$$P(X_t \in A|X_1, \ldots, X_{t-1}) = P(X_t \in A|X_{t-1}) \quad (3)$$

for any set A, where, $P(\cdot|\cdot)$ represents a conditional probability. A Markov chain is further said to be time homogeneous if

$$P(X_{t+k} \in A|X_t) = P(X_k \in A|X_0) \quad (4)$$

for any k. A time homogeneous Markov chain is thereby determined by the distribution or value of the initial state X_0 and by the transition kernel

$$P(x, y) = P(X_{t+1} = y|X_t = x) \quad (5)$$

For most problems of interest to us in the context of MCMC, the Markov chain will take values in some subset of $\mathbb{R}^m$ for some possibly large m. However, in order to illustrate the most important concepts, it is convenient to assume that the state space is countable so that the possible states of the chain can be represented as the nonnegative integers $\{0, 1, 2, \ldots\}$. We can then denote the t-step transition probabilities as

$$p_{ij}(t) = P(X_t = j|X_0 = i) \quad (6)$$

Then our interest concerns the conditions under which the t-step transition probabilities converge to a stationary distribution π so that

$$p_{ij}(t) \longrightarrow \pi(j) \quad \text{for all} \quad i, j, \quad \text{as} \quad t \longrightarrow \infty \quad (7)$$

so that the distribution of the states of the chain becomes approximately independent of the initial state or distribution as t increases.

A first necessary condition is that the Markov chain is irreducible. A Markov chain is said to be irreducible if every state can be reached from any initial state, that is if $p_{ij}(t) > 0$ for some t and all i, j. Secondly, all states of the Markov chain must be positive recurrent. A state, i, is said to be recurrent, if the Markov chain is certain to return to i, so if we define $\tau_i = \min\{t > 0 : X_t = i|X_0 = i\}$ to be the time taken to return to state i, then $P(\tau_i < \infty) = 1$. A recurrent state i is said to be positive recurrent if its expected recurrence time is finite, that is if $E[\tau_i] < \infty$. For an irreducible Markov chain, it can be demonstrated that if any state is positive recurrent, then all states are positive recurrent. Finally, all states of the chain must be aperiodic. The period of state i is defined to be $d(i)$ where,

$$d(i) = \text{greatest common divider } \{t : p_{ii}(t) > 0\} \quad (8)$$

For an irreducible Markov chain, it can be shown that all states have the same period. The chain is said to be aperiodic if this period is one.

It can then be shown that for an irreducible, positive recurrent, aperiodic Markov chain, a unique stationary distribution π exists so that for all i, j,

$$\pi(j) = \lim_{t \to \infty} p_{ij}(t) \quad (9)$$

Moreover, suppose that X is a random variable with distribution π and that $f(X)$ is a function such that $E_\pi[|f(X)|] < \infty$. Then it follows that if $\{X_t\}$ is a Markov chain with stationary distribution π, then

$$\frac{1}{T}\sum_{t=1}^{T} f(X_t) \longrightarrow E_\pi[f(X)] \quad (10)$$

as $t \to \infty$.

A sufficient condition for the existence of a unique stationary distribution is that of reversibility. A Markov chain with transition probabilities $p_{ij} = P(X_{t+1} = j|X_t = i)$ is said to be (time) reversible, if there exists a probability density π that satisfies the detailed balance equation so that for any i, j, then

$$p_{ij}\pi(i) = p_{ji}\pi(j) \quad (11)$$

For a time-reversible Markov chain, then it can be immediately demonstrated that π is then a stationary distribution of the Markov chain because, in this case,

$$\sum_i p_{ij}\pi(i) = \sum_i p_{ji}\pi(j) = \pi(j) \quad (12)$$

The previous ideas can be generalized to the case of a continuous state space, although the conditions are slightly more technical, for details, see [4, 20]. In this case, given a transition kernel $P(x, y)$ such that

$$\int_y P(x, y)\,\mathrm{d}y = 1 \tag{13}$$

then the stationary distribution π satisfies

$$\pi(y) = \int_x \pi(x)P(x, y)\,\mathrm{d}x \tag{14}$$

The objective of the MCMC approach is therefore to construct a Markov chain with a given stationary distribution π. A general algorithm for doing this is discussed in the following section.

The Metropolis–Hastings Algorithm

The Metropolis–Hastings algorithm of [5, 6] is a general method of constructing a Markov chain with a given equilibrium distribution $X \sim \pi$. A full description is given in [7]. The algorithm generates a Markov chain in which each state only depends on the previous state as follows.

1. Given the current value, $X_t = x$, generate a candidate value, y, from a proposal density $q(y|x)$.
2. Calculate the acceptance probability

$$\alpha(x, y) = \min\left\{1, \frac{\pi(y)q(x|y)}{\pi(x)q(y|x)}\right\} \tag{15}$$

3. With probability $\alpha(x, y)$ define $X_{t+1} = y$ and otherwise reject the proposed value and set $X_{t+1} = x$.
4. Repeat until convergence is judged and a sample of the desired size is obtained.

Firstly, it is clear that the acceptance probability $\alpha(x, y)$ used in the Metropolis–Hastings algorithm only depends on the ratio $\pi(y)/\pi(x)$, which is not dependent upon the integration constant of the density π. Secondly, it is straightforward to demonstrate that π is a stationary distribution of the Markov chain as follows. The transition kernel for the density of a move from x to y is given by

$$P(x, y) = \alpha(x, y)q(y|x) + \left(1 - \int \alpha(x, y)q(y|x)\,\mathrm{d}y\right)\delta_x \tag{16}$$

where δ_x is the Dirac mass at x. Now it is easy to show that

$$\pi(x)q(y|x)\alpha(x, y) = \pi(y)q(x|y)\alpha(y, x) \tag{17}$$

and that

$$\left(1 - \int \alpha(x, y)q(y|x)\,\mathrm{d}y\right)\delta_x = \left(1 - \int \alpha(y, x)q(x|y)\,\mathrm{d}x\right)\delta_y \tag{18}$$

and therefore, we have the detailed balance equation

$$\pi(x)P(x, y) = \pi(y)P(y, x) \tag{19}$$

as in equation (11) which implies that π is the stationary distribution of the chain. Note also that if the proposal density $q(y|x) = \pi(y)$, then the Metropolis–Hastings algorithm simply reduces to standard Monte Carlo sampling.

One might expect that the Metropolis–Hastings algorithm would, in general, be more efficient if the acceptance probability $\alpha(x, y)$ was high. However, this is not the case in general. In [8] it is recommended that for high-dimensional models, the acceptance rate for random walk type Metropolis–Hastings algorithms (see below) should be around 25%, whereas in models of dimension one or two, the acceptance rate should be around 50%. However, general results are not available and the efficiency of a Metropolis–Hastings algorithm will generally be heavily dependent on the proposal density $q(y|x)$. Various possibilities are considered in the following section.

Different Types of MCMC Samplers

Theoretically, the Metropolis–Hastings algorithm can be implemented with almost any proposal density $q(\cdot|\cdot)$, However, the performance of the algorithm can be influenced by the form of this density. Usually, the proposal density is chosen so that it is easy to sample from.

One possibility is to use an independence sampler, see e.g. [9] that is $q(y|x) = q(y)$ independent of x. This will often work very well if the density q is similar to π, although with somewhat heavier tails, in a similar way, to the rejection sampler in standard Monte Carlo sampling.

Another alternative is the Metropolis [5] sampler which has the property that $q(x|y) = q(y|x)$ and has the advantage of simplifying the acceptance probability in equation (15) to $\alpha(x, y) = \pi(y)/\pi(x)$. A special case is the random walk Metropolis algorithm, which assumes that $q(y|x) = q(|y - x|)$.

In many cases, X will be a high-dimensional random variable and it may be much more convenient to implement the Metropolis–Hastings algorithm in blocks $X = (X_1, \ldots, X_k)$. The simplest algorithm of this type is the so-called Gibbs sampler (see [10, 11]). In order to illustrate this approach, suppose initially that $X = (X_1, X_2)$ has joint density $\pi(x_1, x_2)$ and that the marginal densities are $\pi_{X_1}(x_1)$ and $\pi_{X_2}(x_2)$, then the Gibbs sampler proceeds as follows:

1. Set an initial value $X_1 = x_{10}$.
2. For $t = 1, 2, \ldots$ generate:
   (a) $X_{2t} \sim \pi_{X_2|X_1}(\cdot|X_1 = x_{1t-1})$
   (b) $X_{1t} \sim \pi_{X_1|X_2}(\cdot|X_2 = x_{2t})$
   where $\pi_{X_1|X_2}(\cdot|x_2)$ and $\pi_{X_2|X_1}(\cdot|x_1)$ are the associated conditional distributions so that $\pi_{X_1|X_2}(x_1|x_2) = \dfrac{\pi(x_1, x_2)}{\pi_{X_2}(x_2)}$ and $\pi_{X_2|X_1}(x_2|x_1) = \dfrac{\pi(x_1, x_2)}{\pi_{X_1}(x_1)}$.

This algorithm can be easily extended to the case of a higher-dimensional X by successively sampling from the full conditional distributions $\pi_{X_i|X_{-i}}(x_i|x_{-i})$ where $X_{-i} = (X_1, \ldots, X_{i-1}, X_{i+1}, \ldots, X_k)$ of each component at each step.

It is clear that the Gibbs algorithm generates a Markov chain and it can be shown that it corresponds to a composition of Metropolis–Hastings steps with acceptance probabilities equal to 1 and stationary distribution π. Full descriptions of the Gibbs sampler are given in [12, 20].

A number of methods have been developed for the implementation of Gibbs sampling algorithms. Firstly, the adaptive rejection sampler for log concave densities was introduced in [13] and extended to arbitrary densities in [14]. Generic software, e.g. WinBUGS, for implementation of Gibbs sampling based on these methods is available (see [15]), and software for assessing convergence of Gibbs samplers, e.g. CODA, has also been developed (see [16]).

When full Gibbs sampling is impossible, as some of the conditional distributions cannot be sampled directly, it is possible to use Metropolis–Hastings updates in these steps which produces a hybrid, Metropolis–within–Gibbs sampler, see e.g. [9].

Note finally that as well as the various options discussed here, many alternative MCMC samplers have been developed. Of particular importance, we may cite the slice sampler [17] which is a simple approach for sampling relatively low-dimensional distributions and the reversible jump sampler [18], which has been developed to sample over state spaces of various dimensions and the perfect sampler [19] which, for a restricted range of problems, is designed to avoid the problems of assessing MCMC convergence by generating an exact sample from the target distribution π. See e.g. [20] for more details.

Apart from the perfect sampler, although theoretically, the Metropolis–Hastings algorithm guarantees convergence to the stationary distribution. In practice, this may be very slow and the performance of the algorithm will depend on both the initial values chosen to start the chain and the form of the proposal distribution $q(\cdot|\cdot)$. Different convergence monitoring techniques are discussed in the following section.

Monitoring the Markov Chain

The objective of this section is to present the basic techniques for monitoring the output of a Markov chain for convergence. Full reviews of this area are given in, e.g. [21, 23]. Firstly, it is important to assess when the sampled values X_t have approximately converged to the stationary distribution π. This will depend largely on how well the MCMC algorithm is able to explore the state space and also on the levels of correlation between the X_t's. Secondly, it is important to assess the convergence of MCMC averages $\frac{1}{T}\sum_{t=1}^{T} f(X_t)$, as in equation (10), and finally we need to be able to assess how close a given sample is to being independent and identically distributed.

When assessing the performance of an MCMC algorithm, one possibility is to consider running the algorithm various times independently with a variety of widely dispersed starting values. Then, for example, one might assess convergence to equilibrium by examining the time at which the sample averages of functionals $f(X)$ of interest have converged for all the chains. Formal diagnostics for doing this are introduced in, e.g. [22].

An alternative is to use a single run of a Markov chain. In this case, it is always useful to produce simple graphs of the generated values X_t *versus* time, which can be used to assess how quickly or

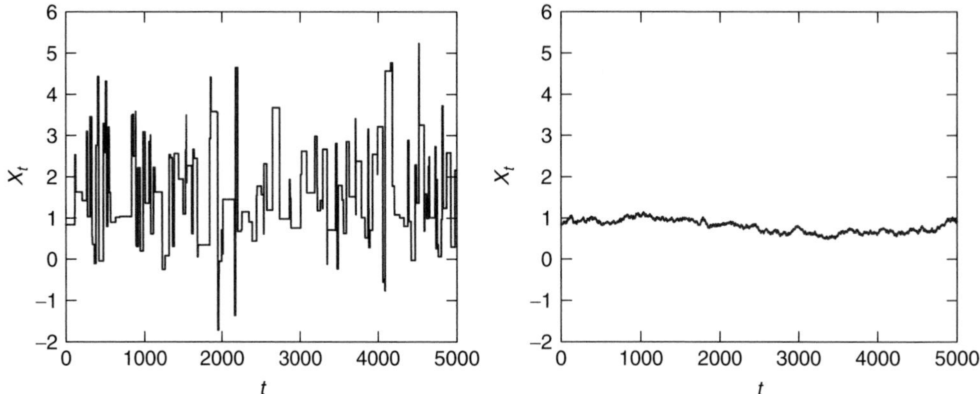

Figure 1 Metropolis samplers showing poor mixing properties

slowly the Markov chain moves or any deviations from stationarity. Furthermore, convergence may be assessed using graphs of sample averages as in equation 10). More formal approaches are discussed in, e.g. [23].

Figure 1 shows the states generated using the Metropolis sampler described in the section titled "Example" when this is run for 5000 iterations starting from the same initial value with standard deviations of 50 and 0.01, respectively.

In the first case, the chain mixes quite badly; only 3% of the proposed values are accepted and the chain spends relatively long periods in the same state before jumping to another value. In the second case, almost all of the proposed values (around 99%) are accepted, but it is clear that the chain has not explored the full sample space. We can improve the mixing of the chain by changing the standard deviation of the proposal distribution. It is often useful, in general, to use the first part of a Metropolis–Hastings run to tune the sampler (for example, by varying the variance of the proposal distribution) to achieve reasonable mixing and movement through the state space so that equilibrium can be reached.

Even when convergence has been assessed, it is important to take into account that for any Markov chain–based sampler, the output is correlated. Although this does not affect the estimation of the mean, a standard Monte Carlo estimate of the variance of X,

$$\frac{1}{T}\sum_{i=1}^{T}(x_i - \bar{x})^2 \tag{20}$$

will underestimate the true variance because of the positive correlation between the sampled data. One method of taking this into account is to calculate the effective sample size T/κ, where $\kappa = 1 + 2\sum_{i=1}^{\infty} \rho_i$, where ρ_i is the autocorrelation (*see* **Autocorrelation Function**) of lag i (see [24]). For example, in the case discussed earlier of running the Metropolis sampler with $\sigma = 50$, for 5000 iterations, the effective sample size is approximately 125.

A full review of the different strategies for improving the convergence and mixing of MCMC algorithms is given in **Convergence and Mixing in Markov Chain Monte Carlo**.

Example

We wish to sample a distribution of the form

$$\pi(x) \propto \phi\left(\frac{x + 0.94}{5/\sqrt{14}}\right) \bar{\Phi}\left(\frac{5 - x}{5}\right)^6 \tag{21}$$

where $\phi(x) = \exp\left(-\frac{1}{2}x^2\right)$ is the standard normal density function and $\bar{\Phi}(x) = \int_x^{\infty} \phi(u)\,du$ is the normal survivor function. This distribution is the posterior distribution derived from observing a sample of 20 normally distributed lifetimes with mean x and standard deviation 5, $Z|x \sim N(x, 25)$, where 14 values less than 5, say $z_1, \ldots, z_{14}$ are observed completely and have mean -0.94 and the remaining 6 values are truncated so that it is only known that they take values greater than 5 and a uniform prior distribution for X is assumed.

We shall consider three different MCMC samplers: an independence sampler $q(y|x) = N(0.85, 5)$, a Metropolis sampler $q(y|x) = N(x, 5)$ with initial values $x_0 = 0.85$ in both cases, and thirdly a Gibbs. The Gibbs sampler is implemented by defining and introducing truncated normal latent variables $Z_i|x \sim TN(x, 5)$ such that $Z_i > 5$, for $= 15, \ldots, 20$ to represent the true values of the truncated normal data. Note that it is straightforward to sample from a truncated normal distribution *via* rejection sampling (*see* e.g. [25]; **Monte Carlo Methods**). Given the full sample $z_1, \ldots, z_{14}, Z_{15} = z_{15}, Z_{20} = z_{20}$, the conditional density of X is normal

$$X|z_1, \ldots, z_{14}, z_{15}, \ldots, z_{20} \sim N\left(\frac{0.94 \times 14 + \sum_{i=15}^{20} z_i}{20}, \frac{5}{\sqrt{20}}\right) \quad (22)$$

This leads to the Gibbs sampling algorithm:

1. Fix $x_0 = 0.85$.
2. For $t = 1, 2, \ldots$
   (a) Generate $z_{ti} \sim TN(x_{t-1}, 5)$ for $i = 15, \ldots, 20$.
   (b) Calculate $\mu_t = (0.94 \times 14 + \sum_{i=15}^{20} z_{ti})/20$.
   (c) Generate $x_t \sim N(\mu_t, 5/\sqrt{20})$.

Figure 2 gives the convergence of the means $\frac{1}{T}\sum_{t=1}^{T} x_i$ for each of the three algorithms. In this case, the mean for the Gibbs algorithm has converged after about 5000 iterations, whereas the Metropolis and independence samplers perform somewhat worse.

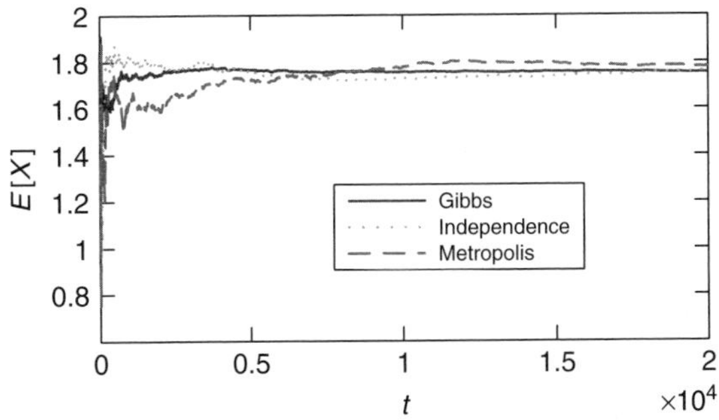

Figure 2 Convergence of the sample estimates of $E[X]$

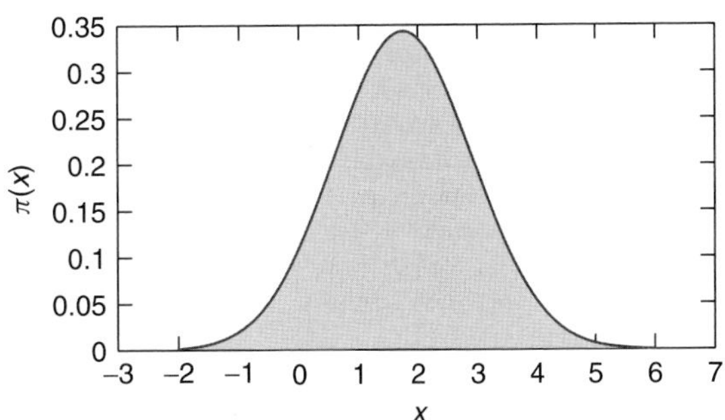

Figure 3 Histogram of the Gibbs–sampled data and fitted density estimate

In Figure 3 a histogram of the data generated from the Gibbs algorithm along with an estimate of the density function of X is shown. The modal value is very close to 2 which was the true value of x used to generate these data.

References

[1] Tierney, L. & Kadane, J. (1986). Accurate approximations for posterior moments and marginal densities, *Journal of the American Statistical Association* **81**, 82–86.

[2] Metropolis, N. & Ulam, S. (1949). The Monte Carlo method, *Journal of the American Statistical Association* **44**, 335–341.

[3] Norris, J. (1998). *Markov Chains*, University Press, Cambridge.

[4] Tierney, L. (1996). Introduction to general state-space Markov Chain theory, in *Markov Chain Monte Carlo in Practice*, W. Gilks, S. Richardson & D.J. Spiegelhalter, eds, Chapman & Hall, London, pp. 59–74.

[5] Metropolis, N., Rosenbluth, A., Rosenbluth, M., Teller, A. & Teller, E. (1953). Equations of state calculations by fast computing machines, *The Journal of Chemical Physics* **21**, 1087–1092.

[6] Hastings, W. (1970). Monte Carlo sampling methods using Markov chains and their application, *Biometrika* **57**, 97–109.

[7] Chib, S. & Greenberg, G. (1995). Understanding the Metropolis Hastings algorithm, *The American Statistician* **49**, 327–335.

[8] Gelman, A. Gilks, W. R. & Roberts, G.O. (1997). Weak convergence and optimal scaling of random walk Metropolis algorithms, *Annals of Applied Probability* **7**, 110–120.

[9] Tierney, L. (1994). Markov chains for exploring posterior distributions (with discussion), *Annals of Statistics* **22**, 1701–1762.

[10] Geman, S. & Geman, D. (1984). Stochastic relaxation, Gibbs distributions and the Bayesian restoration of

images, *IEEE Transactions on Pattern Analysis and Machine Intelligence* **6**, 721–741.
[11] Gelfand, A. & Smith, A. (1990). Sampling based approaches to calculating marginal densities, *Journal of the American Statistical Association* **85**, 398–409.
[12] Casella, G. & George, E. (1992). Explaining the Gibbs sampler, *The American Statistician* **46**, 167–174.
[13] Gilks, W. & Wild, P. (1992). Adaptive rejection sampling for Gibbs sampling, *Applied Statistics* **41**, 337–348.
[14] Gilks, W., Best, N. & Tan, K. (1995). Adaptive rejection metropolis sampling within Gibbs sampling, *Applied Statistics* **44**, 455–472.
[15] Spiegelhalter, D., Thomas, A., Best, N. & Gilks, W. (1995). BUGS: Bayesian inference using Gibbs sampling, Technical Report, MRC Biostatistics Unit, University of Cambridge.
[16] Best, N., Cowles, M. & Vines, K. (1995). CODA: convergence diagnosis and output analysis software for Gibbs sampling output, version 0.30, Technical Report, MRC Biostatistics Unit, University of Cambridge.
[17] Neal, R. (2003). Slice sampling (with discussion), *Annals of Statistics* **31**, 705–767.
[18] Green, P. (1995). Reversible jump MCMC computation and Bayesian model determination, *Biometrika* **82**, 711–732.
[19] Propp, J. & Wilson, D. (1996). Exact sampling with coupled Markov chains and applications to statistical mechanics, *Random Structures and Algorithms* **9**, 223–252.
[20] Robert, C. & Casella, G. (2004). *Monte Carlo Statistical Methods*, 2nd Edition, Springer, New York.
[21] Brooks, S. & Roberts, G. (1998). Assessing convergence of Markov chain Monte Carlo algorithms, *Statistics and Computing* **8**, 319–335.
[22] Gelman, A. & Rubin, D. (1992). Inference from iterative simulation using multiple sequences (with discussion), *Statistical Science* **7**, 457–511.
[23] Mengersen, K., Robert, C. & Guihenneuc Jouyaux, C. (1999). In *MCMC Convergence Diagnostics: A "Reviewww"*, *Bayesian Statistics, Vol. 6*, J.M. Bernardo, J.O. Berger, A.P. Davis & A.F.M. Smith, eds, Oxford University Press, pp. 415–440.
[24] Liu, J. & Chen, R. (1995). Blind deconvolution via sequential imputations, *Journal of the American Statistical Association* **90**, 567–576.
[25] Geweke, J. (1991). Efficient simulation from the multivariate normal and student-t distributions subject to linear constraints, in *Computing Science and Statistics: Proceedings of the Twenty-Third Symposium on the Interface*, E.M. Keramidas, ed., Interface Foundation of North America, Fairfax, pp. 571–578.

Related Articles

Convergence and Mixing in Markov Chain Monte Carlo; Hierarchical Markov Chain Monte Carlo (MCMC) for Bayesian System Reliability; Monte Carlo Methods; System Reliability: Monte Carlo Estimation.

MICHAEL P. WIPER

Markov decision models *see* Maintenance and Markov Decision Models

Markov Processes

Basic Definitions

A *Markov process* (MP) is a stochastic process $(X_t)_{t\in T}$ (where usually $T=\mathbb{Z}_+$ or $T=\mathbb{R}_+$) with values in some state space S that has the *Markov property*: for any $n\in\mathbb{N}$, time indices $t_1<\cdots<t_n<t$ and states $x_1,\ldots,x_n\in S$ the conditional distribution of X_t given that $X_{t_1}=x_1,\ldots,X_{t_n}=x_n$ coincides with that under the condition $X_{t_n}=x_n$. In other words, the probability law of the evolution of X_t after any fixed time instant s depends on its past only through the value of X_s, so that the "present state" contains all the relevant information about the future. The Markov property is equivalent to the following conditional independence property. Let $\mathcal{F}_s$ and $\mathcal{F}^s$ be the σ-algebras generated by $(X_t)_{t\le s}$ and $(X_t)_{t\ge s}$, describing the past up to time s and the future starting at s, respectively. Then for all s and events $A\in\mathcal{F}_s$ and $B\in\mathcal{F}^s$,

$$\mathbb{P}(A\cap B\mid X_s)=\mathbb{P}(A\mid X_s)\,\mathbb{P}(B\mid X_s) \quad (1)$$

Depending on whether the time index set T and the state space S are discrete or continuous, one can distinguish four types of MPs. We will for simplicity only consider Borel subsets $S\subset\mathbb{R}$ and the time ranges $T=\mathbb{R}_+$ or $T=\mathbb{Z}_+$. If either the state space or the time range or both are countable, one also uses the term *Markov chain*.

The feature that the future development depends on the past only through the current state lies behind the ubiquitous occurrence of MPs in applied probability. It is analogous to the well-known property of deterministic dynamical systems that the solution of their motion equations are uniquely determined by the data (i.e., the positions and velocities of the particles involved) at one time instant. Arguably most stochastic models for real-life systems are based on underlying MPs. Important classes of MPs are random walks, birth-and-death processes, many of the processes occurring in classical queuing, dam, storage and reliability theory, Lévy processes (e.g., compound Poisson processes and **Brownian motion**), and diffusions. In this article we can give only a condensed survey of the basic mathematical machinery for the analysis of MPs. In the section titled "Discrete-Time Markov Processes" we deal with a case of discrete time, describing in particular the classical theory for countable state space and the main results for Harris chains. The section titled "Continuous-Time Markov Processes" is devoted to the basics of the continuous-time case. In the section titled "Example: The Performance of a Replacement Policy" we present the computation of performance measures of a replacement policy in a concrete model of a failure system; this is meant as an example to show the general techniques at work. Finally, in the section titled "Continuous-Time MPs with Countable State Space" we discuss the basic constructions and analysis of continuous-time MPs with countable state space.

Among the many excellent monographs and textbooks on different levels we mention [1–4] for the general theory of MPs and [5–8] for an in-depth analysis of Markov chains with a view toward applications. See also [9, 10] for the case of discrete state space and [11] for Harris chains. We do not discuss Lévy processes [12, 13], Brownian motion [4, 14] or diffusions [1, 2, 15]. The example in the section titled "Example: The Performance of a Replacement Policy" is taken from [16]. We start with the case of discrete time.

Discrete-Time Markov Processes

If $T = \mathbb{Z}_+$, the Markov property can be formulated as

$$\mathbb{P}(X_{n+1} \in B \mid X_0, \ldots, X_n) = \mathbb{P}(X_{n+1} \in B \mid X_n) \quad \text{almost surely} \tag{2}$$

for all $n \in \mathbb{Z}_+$ and all $B \in \mathcal{B}(S)$ (the Borel σ-algebra in S). For every n one can choose a stochastic kernel $p_n(x, B)$ which is a version of $\mathbb{P}(X_{n+1} \in B \mid X_n = x)$. These kernels give the *transition probabilities* of the MP. Together with the *initial distribution*,

$$\pi(B) = \mathbb{P}(X_0 \in B) \tag{3}$$

they determine the finite-dimensional distributions and thus the law of the MP: one can show that for all n and $B_0, \ldots, B_n \in \mathcal{B}(S)$,

$$\begin{aligned}\mathbb{P}(X_0 \in B_0, \ldots, X_n \in B_n) = \int_{B_{n-1}} \cdots \int_{B_0} \\ \times\, p_{n-1}(x_{n-1}, B_{n-1}) p_{n-2}(x_{n-2}, \mathrm{d}x_{n-1}) \\ \ldots p_0(x_0, \mathrm{d}x_1)\pi(\mathrm{d}x_0)\end{aligned} \tag{4}$$

Conversely, for any sequence $p_n(x, B)$, $n \in \mathbb{Z}_+$, of stochastic kernels and any distribution π on S one can construct an MP on the infinite product space $(S^{\mathbb{N}}, \mathcal{B}(S)^{\mathbb{N}})$ of countably many factors $(S, \mathcal{B}(S))$ with the X_ns being the coordinate mappings, having the $p_n(x, B)$ as transition probabilities and π as distribution of X_0. Let $\mathbb{P}_\pi$ and $\mathbb{E}_\pi$ be the corresponding probability measure on $(S^{\mathbb{N}}, \mathcal{B}(S)^{\mathbb{N}})$ and expectation, respectively. This standard measure theoretic construction allows to consider simultaneously a whole collection of MPs having the same transition probabilities but variable initial distributions. If π is the point mass at some x, we write $\mathbb{P}_\pi = \mathbb{P}_x$ and $\mathbb{E}_\pi = \mathbb{E}_x$ and speak of the MP starting from the point x.

In the following we will restrict ourselves to the case when $p(x, B) = p_n(x, B)$ can be chosen independently of n so that the time index n can be omitted. The MP is then called *homogeneous*; $p(x, B)$ is the probability of going from x to B in one step. For homogeneous MPs equation (2) takes the form

$$\mathbb{P}_\pi(X_{n+1} \in B \mid X_0, \ldots, X_n) = \mathbb{P}_{X_n}(X_1 \in B) \quad \mathbb{P}_\pi\text{-almost surely} \tag{5}$$

For homogeneous discrete-time MPs a generalization of their defining property (5) from deterministic times n to certain nonanticipatory random times, the so-called *strong Markov property*, holds. A *stopping time* is a random variable with values in $\mathbb{Z}_+ \cup \{\infty\}$ for which $\{\tau = n\} \in \sigma(X_0, \ldots, X_n)$ (the σ-algebra generated by $X_0, \ldots, X_n$) for all $n \in \mathbb{Z}_+$. Then $\mathcal{F}_\tau$ is defined to be the set of all events A for which $A \cap \{\tau = n\} \in \sigma(X_0, \ldots, X_n)$ for all $n \in \mathbb{Z}_+$; $\mathcal{F}_\tau$ is the σ-algebra consisting of all events whose occurrence or nonoccurrence is known at time τ. If $\tau < \infty$, the MP is at state X_τ at time τ. The strong Markov property states that the evolution of the process after time τ is conditionally independent of $\mathcal{F}_\tau$, given X_τ, and follows the same law as the process started at the state X_τ. Formally,

$$\mathbb{P}_\pi((X_\tau, X_{\tau+1}, \ldots) \in C | \mathcal{F}_\tau) = \mathbb{P}_{X_\tau}((X_0, X_1, \ldots) \in C)$$

$$\mathbb{P}_\pi\text{-almost surely on } \{\tau < \infty\} \text{ for every } C \in \mathcal{B}(S)^{\mathbb{N}} \tag{6}$$

For example, the strong Markov property is basic for studying the recurrence behavior of MPs, because it implies that following any visit to some state the process behaves as if starting anew from this state.

An MP is called *asymptotically stationary* if there is a stationary process $(Y_n)_{n \in \mathbb{Z}_+}$ such that the joint distribution of the tail sequence $(X_{k+n})_{n \in \mathbb{Z}_+}$ converges to that of $(Y_n)_{n \in \mathbb{Z}_+}$ in the sense that

$$\lim_{k \to \infty} \mathbb{P}_\pi((X_{k+n})_{n \in \mathbb{Z}_+} \in C) = \mathbb{P}((Y_n)_{n \in \mathbb{Z}_+} \in C)$$

$$\text{for all } C \in \mathcal{B}(S)^\infty \tag{7}$$

under any initial distribution π, as $k \to \infty$. To prove asymptotic stationarity, it turns out to be sufficient to show a one-dimensional convergence result. A probability measure ρ on the state space is called *stationary initial distribution* (s.i.d.) if

$$\rho(B) = \int_S p(x, B)\, \mathrm{d}\rho(x) \quad \text{for all } B \in \mathcal{B}(S) \tag{8}$$

If an MP has an s.i.d. ρ, then it is a stationary process under $\mathbb{P}_\rho$, and if it can be shown that for some probability measure ρ, $\mathbb{P}_x(X_n \in B)$ converges to $\rho(B)$ for all x and B, then the MP is asymptotically stationary (the limit being $(X_n)_{n \in \mathbb{Z}_+}$ under $\mathbb{P}_\rho$). There is at most one s.i.d. if the MP is not decomposable, i.e., if the state space cannot be partitioned into two disjoint sets A_1, A_2 such that $p(x, A_i) = 1$ for all $x \in A_i$, $i = 1, 2$. If an s.i.d. ρ exists, one can show that equation (7) holds with limit $\mathbb{P}_\rho((X_n)_{n \in \mathbb{Z}_+} \in \cdot)$ if

1. for every $n = 1, 2, \ldots$, indecomposability holds under all transition probabilities $p^{(n)}(x, B) = P_x(X_n \in B)$, and
2. the distribution $p(x, \cdot)$ is absolutely continuous with respect to ρ for every x.

In general the problems of the existence of an s.i.d. and of asymptotic stationarity are difficult. A complete solution can be given for countable state spaces, whereas for uncountable state spaces there is the theory of Harris chains. We will now discuss these two special cases.

Countable State Space

In this subsection we take S to be a subset of the integers. The transition probabilities can then be represented as a matrix $P = (p_{ij})_{i,j \in S}$, where $p_{ij} = p(i, \{j\})$, whereas the initial distribution becomes a vector $\pi = (\mathbb{P}(X_0 = i))_{i \in S}$. The *n-step transition probabilities* $p_{ij}^{(n)} = \mathbb{P}(X_n = j \mid X_0 = i)$ satisfy the Chapman–Kolmogorov equations

$$p_{ij}^{(n+1)} = \sum_{k \in S} p_{ik}^{(n)} p_{kj} \tag{9}$$

This means that the matrix $p_{ij}^{(n)}$ is just the nth power P^n of the *transition matrix* P. Thus, the stationarity and asymptotics of MPs can be tackled by studying the limiting behavior of the powers of stochastic matrices. If the state space is finite, this is a classical topic in matrix theory. In the general countable case one has to study the sequences of visits of the MP to its states because by the strong Markov property at each time a state i is visited, the MP restarts anew from i independently of the past.

A state i is called *recurrent* if $\mathbb{P}_i(X_n = i$ for some $n \geq 1) = 1$ and *transient* otherwise. Let N_i denote the number of visits in i. It can be shown that i is recurrent if and only if $\mathbb{P}_i(N_i < \infty) = 0$, which in turn is equivalent to

$$\mathbb{E}_i(N_i) = \sum_{n=1}^{\infty} p_{ii}^{(n)} = \infty \tag{10}$$

If i is recurrent, the ($\mathbb{P}_i$-almost surely nonterminating) sequence of times of visits in i forms a renewal process, i.e., the times between visits are i.i.d. random variables. Let $\tau_i = \inf\{n \geq 1 \mid X_n = i\}$ be the time of the first visit of i (with $\inf \emptyset = \infty$). i is called *positive recurrent* if $\mathbb{E}_i(\tau_i) < \infty$, and *null recurrent* if $\mathbb{E}_i(\tau_i) = \infty$. The largest integer d for which $\mathbb{P}_i(\tau_i \in \{d, 2d, 3d, \ldots\}) = 1$ is called the *period* of i, and if the period is 1, i is called *aperiodic*. Two states i and j are said to *communicate* if there are integers $n, m \geq 1$ such that $p_{ij}^{(n)} > 0$ and $p_{ji}^{(m)} > 0$. If two states communicate, they have the same period and are simultaneously either positive recurrent, or null recurrent, or transient. On the set R of recurrent states communication is an equivalence relation, so that R consists of disjoint communication classes. The states in one class have the same period d, and if $d > 1$ the class can be split into disjoint "periodic" subsets $C_1, \ldots, C_d$ such that for any state $i \in C_l$ the next transition goes to some state in C_{l+1} (where $C_{d+1} = C_1$)). Finally, an MP is called *irreducible* if all states communicate with each other.

On the basis of these classifications of the states a complete description of the asymptotic behavior of the transition probabilities can be given. First, let us call a measure $\rho = (\rho_j)_{j \in S}$ *stationary* if $0 \leq \rho_j < \infty$ and $\rho_j = \sum_{i \in S} \rho_i p_{ij}$ for all $j \in S$. For any fixed recurrent state i one can define a stationary measure by setting ρ_j equal to the expected number of visits of j between two consecutive visits of i. If X_n is irreducible and recurrent, this ρ is the only stationary measure up to a multiplicative constant and $\rho_j > 0$ for all j. In particular, if we additionally assume positive recurrence, there is exactly one stationary distribution π, and it is given by $\pi_j = 1/\mathbb{E}_j(\tau_j)$. Now let us look at the limiting behavior of $p_{ij}^{(n)}$.

1. Let j be null recurrent or transient. Then $\lim_{n\to\infty} p_{ij}^{(n)} = 0$ for every state i.
2. Let j be positive recurrent with period d, and let i communicate with j. Then if $p_{ij}^{(nd+l)} > 0$ for some $n \geq 0$ and $l \in \{1, \ldots, d\}$, we have $p_{ij}^{(N)} = 0$ for $N \not\equiv l \bmod d$ and $\lim_{n\to\infty} p_{ij}^{(nd+l)} = d/\mathbb{E}_j(\tau_j)$.
3. If j is positive recurrent and $p_{ij}^{(n)} > 0$ for some n, then either i and j communicate (so that we are in case (2)), or i is transient. Let i be transient. Then if j is aperiodic, $\lim_{n\to\infty} p_{ij}^{(n)} = r(i)/\mathbb{E}_j(\tau_j)$, where $r(i) = \mathbb{P}_i(X_n = j \text{ for some } n)$. If j has period $d > 1$ the limiting behavior depends on which (if any) of the periodic subsets $C_1, \ldots, C_d$ of the communication class C of j is entered first after the start from i, and after how many steps this happens. Let $\sigma_j = \inf\{n \geq 1 \mid X_n \in C\}$. Then for all $N \geq 1$ and $l \in \{1, \ldots, d\}$,

$$\lim_{n\to\infty} \mathbb{P}_i(X_{N+l+nd} = j \mid \sigma_j = N, X_{\sigma_j} \in C_{d-l+1}) = d/\mathbb{E}_j(\tau_j) \quad (11)$$

$$\mathbb{P}_i(X_{N+l+nd} = j \mid \sigma_j = N, X_{\sigma_j} \in C_{d-l+1}) = 0 \quad \text{if } n \not\equiv N + l \bmod d \quad (12)$$

The limiting behavior of $p_{ij}^{(n)}$ is then obtained by deconditioning with respect to the pair (σ_j, X_{σ_j}).

The main result in this description is the ergodic theorem for discrete-time MPs. If X_n is irreducible, aperiodic and positive recurrent, then

$$\lim_{n\to\infty} p_{ij}^{(n)} = 1/\mathbb{E}_j(\tau_j) = \pi_j \quad \text{for all } i, j \in S \quad (13)$$

Note that the limit is the stationary distribution of X_n. Actually, a stronger result is valid: if all states communicate, are aperiodic and recurrent, we have

$$\lim_{n\to\infty} \sum_{j \in I} |\mathbb{P}_i(X_n = j) - \mathbb{P}_k(X_n = j)| = 0 \quad \text{for all } i, k \in S \quad (14)$$

Equation (13) follows from equation (14) because for any fixed states $i, j_0 \in S$

$$\begin{aligned} 0 &= \lim_{n\to\infty} \sum_{k \in I} \pi_k \sum_{j \in I} |\mathbb{P}_i(X_n = j) - \mathbb{P}_k(X_n = j)| \\ &\geq \lim_{n\to\infty} \sum_{k \in I} \pi_k |\mathbb{P}_i(X_n = j_0) - \mathbb{P}_k(X_n = j_0)| \\ &\geq \lim_{n\to\infty} \left| \sum_{k \in I} \left(\pi_k \mathbb{P}_i(X_n = j_0) - \pi_k \mathbb{P}_k(X_n = j_0) \right) \right| \\ &= \lim_{n\to\infty} |\mathbb{P}_i(X_n = j_0) - \mathbb{P}_\pi(X_n = j_0)| \\ &= \lim_{n\to\infty} |p_{ij_0}^{(n)} - \pi_{j_0}| \end{aligned} \quad (15)$$

Continuous State Space

For a discrete-time MP X_n with state space $\mathbb{R}$ there is in most cases no simple regeneration structure, since single states are usually not revisited with positive probability. Thus one has to consider visits to subsets $R \subset \mathbb{R}$: we call R *recurrent* if $\tau_R = \inf\{n \in \mathbb{N} \mid X_n \in R\}$ satisfies $\mathbb{P}_x(\tau_R < \infty) = 1$ for all $x \in R$. A recurrent set R is said to be a *regeneration set* if for some $n \in \mathbb{N}$ the n-step transition probability measures $p^{(n)}(x,\cdot) = \mathbb{P}_x(X_n \in \cdot)$, $x \in R$, share some component, i.e., if there is an $n_0 \in \mathbb{N}$, a probability measure μ on the state space and an $\varepsilon > 0$ such that

$$p^{(n_0)}(x,\cdot) \geq \varepsilon\mu(\cdot) \quad \text{for all } x \in R \tag{16}$$

If X_n possesses a regeneration set, it is called a *Harris chain*. For given transition probabilities with a regeneration set R we can construct a corresponding Harris chain with an embedded regenerative structure as follows. First take the standard version of the MP from an arbitrary initial state x until time τ_R when R is visited for the first time. Then define $X_{\tau_R+n_0}$ by **randomization**: take a random variable which has either distribution μ with probability ε or distribution $(1-\varepsilon)^{-1}[p^{(n_0)}(X_{\tau_R},\cdot) - \varepsilon\mu(\cdot)]$ with probability $1-\varepsilon$. Next, construct the missing part $(X_{\tau_R+n})_{0<n<n_0}$, according to the conditional distribution of $(X_n)_{0<n<n_0}$, given that the pair of "boundary values" (X_0, X_{n_0}) is equal to the pair of values obtained before for X_{τ_R} and $X_{\tau_R+n_0}$. Finally, continue the construction from the new "initial" value $X_{\tau_R+n_0}$, and so on. The process constructed this way is an MP with transition probability measures $p(x,\cdot)$ which, due to the randomization, has distribution μ with probability ε n_0 steps after any visit to R. Let us start this MP with μ as initial distribution for X_0. The times between two consecutive recurrences of the distribution μ obtained by the randomizations are called *cycles*.

Using this cycle decomposition, the following results can be proved for Harris chains:

1. There is exactly one stationary measure up to a positive factor; this measure ρ can be defined by setting $\rho(A)$ equal to the expected number of visits to A during a cycle started with initial distribution μ.
2. X_n is called *aperiodic* if the distribution of the length of a cycle started with μ is aperiodic. It is called *positive recurrent* (*null recurrent*) if the stationary measure has finite (infinite) total mass. For an aperiodic, positive recurrent Harris chain $p^{(n)}(x,\cdot)$ converges to the stationary probability measure in total variation.

For an aperiodic, null recurrent Harris chain, $\lim_{n\to\infty} p^{(n)}(x,B) = 0$ for all x and all Borel sets B with finite stationary measure.

Continuous-Time Markov Processes

In the case $T = \mathbb{R}_+$, $S \subset \mathbb{R}$ the Markov property reads as

$$\mathbb{P}(X_{t+s} \in B \mid (X_u)_{u\in[0,t]}) = \mathbb{P}(X_{t+s} \in B \mid X_t)$$
$$\mathbb{P}\text{-almost surely for all } t,s \geq 0,\ B \in \mathcal{B}(S) \tag{17}$$

For all $t > s \geq 0$ one can choose a stochastic kernel $p_{s,t}(x,B)$ which is a version of $\mathbb{P}(X_t \in B \mid X_s = x)$. These kernels give the *transition probabilities* of the MP. Together with the *initial distribution*,

$$\pi(B) = \mathbb{P}(X_0 \in B) \tag{18}$$

they determine the finite-dimensional distributions and thus the law of the MP: one can show that for all n, $0 \leq t_0 < t_1 < \cdots < t_n$ and $B_0, \ldots, B_n \in \mathcal{B}(S)$,

$$\mathbb{P}(X_{t_0} \in B_0, \ldots, X_{t_n} \in B_n) = \int_{B_{n-1}} \cdots \int_{B_0}$$
$$\times\, p_{t_{n-1},t_n}(x_{n-1}, \mathrm{d}x_n)\, p_{t_{n-2},t_{n-1}}(x_{n-2}, \mathrm{d}x_{n-1})$$
$$\cdots p_{t_0,t_1}(x_0, \mathrm{d}x_1)\pi(\mathrm{d}x_0) \tag{19}$$

The Chapman–Kolmogorov equations hold in the form

$$p_{t_0,t_1}(x,B) = \int_S p_{s,t_1}(y,B)\, p_{t_0,s}(x,\mathrm{d}y) \tag{20}$$

almost surely with respect to the distribution of X_{t_0} for all $t_1 > s > t_0 \geq 0$.

Usually, the transition probabilities and the initial distribution of an MP are specified, and one wants to study the properties of an MP in these terms. Suppose a probability measure π on S and a collection of stochastic kernels $p_{s,t}(x,B)$, $t > s \geq 0$, are given. The kernels are assumed to satisfy equation (20) for all $x \in S$, $B \in \mathcal{B}(S)$ and all $t_1 > s > t_0 \geq 0$. Then one can show that the finite-dimensional distributions defined by equation (19) are consistent, so that they uniquely determine a probability measure on the product space $(S^{[0,\infty)}, \mathcal{B}(S)^{[0,\infty)})$ such that its coordinate mappings X_t form an MP having the functions $p_{s,t}(x,B)$ as transition probabilities and π as distribution of X_0. Thus the only requirement to create an MP from transition kernels is that they satisfy the Chapman–Kolmogorov equations identically.

From now on we will again only consider homogeneous MPs: these are MPs whose transition probabilities satisfy

$$p_{s,t}(x,B) = p_{0,t-s}(x,B) \quad \text{for all } t > s \geq 0 \quad (21)$$

We write $p_t(x,B) = p_{0,t}(x,B)$. The Chapman–Kolmogorov equations now take the form

$$p_{t+s}(x,B) = \int_S p_s(y,B)p_t(x,\mathrm{d}y) \quad (22)$$

for all $t, s > 0$ and all x and B. In the following all conditional probabilities and expectations are assumed to be obtained from the $p_t(x,B)$.

Let $\mathbb{P}_\pi$ and $\mathbb{E}_\pi$ be the corresponding probability measure and expectation on $(S^{[0,\infty)}, \mathcal{B}(S)^{[0,\infty)})$ constructed from the $p_t(x,B)$ and the initial distribution π. If π is the point mass at some x, we write $\mathbb{P}_\pi = \mathbb{P}_x$ and $\mathbb{E}_\pi = \mathbb{E}_x$ and speak of the MP starting from the point x. We now consider only this coordinate representation process (so that the measurability of sets and functions will not depend on the selection of a starting point) simultaneously for variable initial distributions.

By equation (22), the transition probabilities $p_t(x,B)$ are completely determined for all $t > 0$ by their values for t in an arbitrarily small time interval $(0,\varepsilon)$, $\varepsilon > 0$. Thus it seems that the distribution of an MP can be specified by their infinitesimal behavior at zero. From now on we only consider MPs for which $p_t(x,(x-\varepsilon, x+\varepsilon)) \to 1$ for all $\varepsilon > 0$ as $t \to 0$, i.e., $X_t \to X_0$ in probability. The *generator* of an MP is defined to be the operator $\mathcal{G}$, given by

$$(\mathcal{G}f)(x) = \lim_{t\to 0} t^{-1}\left[\mathbb{E}_x(f(X_t)) - f(x)\right] \quad (23)$$

whose domain $D_\mathcal{G}$ is the set of all bounded measurable functions f on S for which the limit in equation (23) exists for all $x \in S$. For example, if X_t is a compound **Poisson process** with rate λ and jump size distribution function F, then $D_\mathcal{G}$ is the set of all bounded measurable functions on S and

$$(\mathcal{G}f)(x) = \lambda \int_{\mathbb{R}} [f(x+u) - f(x)]\,\mathrm{d}F(u) \quad (24)$$

whereas if X_t is standard Brownian motion,

$$(\mathcal{G}f)(x) = \frac{1}{2}f''(x) \quad (25)$$

for all bounded $f : \mathbb{R} \to \mathbb{R}$ having a bounded continuous second derivative (in this case it is not obvious what $D_\mathcal{G}$ is).

In general, the generator $(\mathcal{G}, D_\mathcal{G})$ does not determine the transition probabilities uniquely. However, it leads to Kolmogorov's so-called *backward equations* and *forward equations*, which are valid under smoothness conditions on $p_t(x,B)$. Setting $f_{t,B}(x) = p_t(x,B)$, the backward equations are

$$(\mathcal{G}f_{t,B})(x) = \frac{\partial f_{t,B}}{\partial t}(x) \quad (26)$$

The forward equations are

$$\int (\mathcal{G}f)(u)p_t(x,\mathrm{d}u) = \int f(u)\frac{\partial}{\partial t}p_t(x,\mathrm{d}u),\ f \in D_\mathcal{G} \quad (27)$$

Uniqueness can be achieved by restricting the domain of $\mathcal{G}$ to the set $\overline{D}_\mathcal{G}$ of all bounded measurable functions on S for which (a) $\lim_{t\to 0}\mathbb{E}_x(f(X_t)) = f(x)$ for all $x \in S$ and (b) the ratio on the right side of equation (23) converges to $(\mathcal{G}f)(x)$ and is uniformly bounded in x for sufficiently small t. Then if two MPs have the same generator $\mathcal{G}$ with the same restricted domain $\overline{D}_\mathcal{G}$, they have the same law.

The strong Markov property does not hold in general for MPs in continuous time. It is valid if the stopping time takes only countably many values. Arbitrary stopping times can be approximated by those with countably many values under regularity conditions on the sample paths, and the strong Markov property can be carried over to more general cases. For example, it holds if X_t is a Feller process and has right-continuous paths. An MP with state space $\mathbb{R}$ is called a *Feller process* if for every continuous function f on the state space satisfying $\lim_{|x|\to\infty} f(x) = 0$ the function $x \mapsto \mathbb{E}_x(f(X_t))$ has the same properties and $\lim_{t\to 0}\mathbb{E}_x(f(X_t)) = f(x)$ for all x. For example, Lévy processes, diffusions, and Markov jump processes are Feller processes. Every Feller process has a version that has right-continuous paths with left-hand limits. A very useful property of Feller processes is that for any $f \in D_\mathcal{G}$ the process

$$f(X_t) - \int_0^t (\mathcal{G}f)(X_s)\,\mathrm{d}s$$

is a martingale. *Dynkin's formula* states that

$$\mathbb{E}_x(f(X_\tau)) = f(x) + \mathbb{E}_x\left(\int_0^\tau (\mathcal{G}f)(X_s)\,\mathrm{d}s\right) \quad (28)$$

for every stopping time τ for which $\mathbb{E}_x(\tau) < \infty$. In the next section we present a typical example of how these methods can be employed in stochastic modeling.

Example: The Performance of a Replacement Policy

Consider a technical system subject to fatal failure at some random time σ. The state of the system X_t, $t \geq 0$, is described by a real number indicating its extent of attrition. The process $(X_t)_{t\geq 0}$ starts at $X_0 = 0$ and increases linearly at rate $c(J_t)$, where $(J_t)_{t\geq 0}$ is a "modulating" irreducible Markov chain having state space $\{1, \ldots, n\}$ and the rates $c(j)$ satisfy $c(1) > c(2) > \cdots > c(n) \geq 0$. The failure rate $r(x)$ is assumed to be a function of the state variable x. The system has to run indefinitely; upon failure it has to be replaced by a new identical one. However, it can also be replaced preventively when its attrition reaches a certain threshold $a > 0$ which has to be specified by the controller. Thus, setting $T_a = \inf\{t > 0 \mid X_t = a\}$, the first replacement takes place at time $\min[T_a, \sigma]$. Suppose that after replacement the modulating chain is also restarted at some fixed state i_0. We want to determine the performance of the replacement strategy T_a as a function of a.

The underlying MP of this system is two dimensional: $Z_t = (X_t, J_t)$. The generator of Z_t is of course defined analogously to the section titled "Continuous-Time Markov Processes" by $(\mathcal{G}f)(x,i) = \lim_{t\to 0} t^{-1}\left[\mathbb{E}_{x,i}(f(Z_t)) - f(x,i)\right]$ for functions $f : \mathbb{R}_+ \times \{1, \ldots, n\} \to \mathbb{R}$. It can be shown that

$$\begin{aligned}(\mathcal{G}f)(x,i) &= c(i)f'(x,i) + \sum_{j\neq i} q_{ij} f(x,j) \\ &\quad - (q_i + r(x))f(x,i) \\ &\quad + r(x)\int_{[0,x)} f(y,i)\mu_x(\mathrm{d}y), \\ i &= 1, \ldots, n\end{aligned} \quad (29)$$

or in matrix form

$$\begin{aligned}(\mathcal{G}f)(x) &= Cf'(x) + (Q - r(x)E)f(x) \\ &\quad + r(x)D_f(x)\end{aligned} \quad (30)$$

where $f'(x,i)$ denotes the derivative of f with respect to x and

1. $f(x) = (f(x,1), \ldots, f(x,n))^t$;
2. $Q = (q_{ij})_{i,j\in\{1,\ldots,n\}}$ is the generator of the Markov chain J_t and $q_i = -q_{ii}$;
3. C and $D_f(x)$ are diagonal matrices with diagonal entries $c(i)$ and $\int_{[0,x)} f(y,i)\,\mu_x(\mathrm{d}y)$, respectively;
4. E is the $n \times n$ identity matrix.

Now assume a replacement of a still functioning system costs C_1 and a replacement upon disaster costs C_2 (where $C_1 < C_2$). Then the long-run average cost of running the system when using the replacement policy T_a is given by

$$C(a) = \frac{C_1\mathbb{P}_{(0,i_0)}(\min[T_a,\sigma] = T_a) + C_2(1 - \mathbb{P}_{(0,i_0)}(\min[T_a,\sigma] = T_a))}{\mathbb{E}_{(0,i_0)}(\min[T_a,\sigma])} \quad (31)$$

Hence, we have to compute

$$\mathbb{P}_{(x,i)}(\min[T_a,\sigma] = T_a) \quad \text{and} \quad \mathbb{E}_{(x,i)}(\min[T_a,\sigma]).$$

Once these quantities are known as *functions of a* one can try to minimize $C(a)$.

This problem can be tackled as follows. Suppose the process is killed at time σ by entering into a "coffin state" ∂. By Dynkin's formula, we have

$$f(x,i) + \mathbb{E}_{(x,i)}\left(\int_0^T (\mathcal{G}f)(Z_t)\,\mathrm{d}t\right) = \mathbb{E}_{(x,i)}(f(Z_T)) \quad (32)$$

for f bounded and in the domain of $\mathcal{G}$ and T any integrable stopping time. Now apply equation (32) to $T = \min[T_a, \sigma]$ in two cases:

1. $f = f_1$ such that $(\mathcal{G}f_1)(x,i) = 0$, $x \in (0,a)$, and $f_1(\partial) = 0$, $f_1(a,i) = 1$.
2. $f = f_2$ such that $(\mathcal{G}f_2)(x,i) = -1$, $x \in (0,a)$, and $f_2(\partial) = 0$, $f_2(a,i) = 0$.

A moment's reflection shows that if f_1, f_2 have these properties, then

$$f_1(x,i) = \mathbb{P}_{(x,i)}(\min[T_a,\sigma] = T_a) \quad (33)$$

$$f_2(x,i) = \mathbb{E}_{(x,i)}(\min[T_a, \sigma]) \tag{34}$$

Hence, we have to solve

$$Cf'(x) + (Q - r(x)E)f(x) = (0, \ldots, 0)^t \tag{35}$$

subject to

$$f(a, \cdot) = (1, \ldots, 1)^t \tag{36}$$

and

$$Cf'(x) + (Q - r(x)E)f(x) = (-1, \ldots, -1)^t \tag{37}$$

subject to

$$f(a, \cdot) = (0, \ldots, 0)^t \tag{38}$$

These are linear systems of differential equations that can be dealt with using standard methods. For special failure rate functions $r(x)$ one can even obtain closed-form solutions and thus determine the long-run average cost $C(a)$ explicitly.

Continuous-Time MPs with Countable State Space

Let the state space of X_t be a set of integers. Then the transition probabilities are given by the set of functions $p_{ij}(t) = \mathbb{P}_i(X_t = j)$, $t > 0$, where $i, j \in S$, which have the following properties:

1. $p_{ij}(t) \geq 0$, $\sum_{j \in S} p_{ij}(t) = 1$;
2. $p_{ij}(t+s) = \sum_{j \in S} p_{ik}(t) p_{kj}(s)$;
3. $\lim_{t \to 0} p_{ij}(t) = \delta_{ij}$.

It follows from 1 to 3 that the limits $\lim_{t \to 0} t^{-1} p_{ij}(t) = \lambda_{ij}$, $i \neq j$, and $\lim_{t \to 0} t^{-1}[1 - p_{ii}(t)] = \lambda_i$ exist, but the latter ones may be infinite. If λ_i is finite for some i, the functions $p_{ij}(t)$, $j \in S$, are continuously differentiable on $(0, \infty)$ and we have the linear approximations

$$p_{ij}(t) = \lambda_{ij} t + o(t), \quad j \neq i \tag{39}$$

$$p_{ij}(t) = 1 - \lambda_i t + o(t) \tag{40}$$

as $t \to 0$. Thus, λ_{ij} can be interpreted as the transition rate from i to j and λ_i as the total transition rate away from i. From now on we assume that $\sum_{j \neq i} \lambda_{ij} = \lambda_i < \infty$ for all states i. Then the backward equations are valid in the form

$$p'_{ij}(t) = -\lambda_i p_{ij}(t) + \sum_{k \neq i} \lambda_{ik} p_{kj}(t) \tag{41}$$

The forward equations are

$$p'_{ij}(t) = -\lambda_j p_{ij}(t) + \sum_{k \neq j} \lambda_{kj} p_{ik}(t) \tag{42}$$

Equation (42) does not hold in general; its derivation additionally requires that

$$\lim_{t \to 0} \sup_{i \neq j} |t^{-1} p_{ij}(t) - \lambda_{ij}| = 0 \tag{43}$$

a uniformity property that is satisfied in most concrete models. Let Λ be the matrix with entries λ_{ij}, $i \neq j$, and $-\lambda_i$ on the diagonal, and let $P(t)$ and $P'(t)$ be the matrices with entries $p_{ij}(t)$ and $p'_{ij}(t)$, respectively. Then equation (41) and equation (42) can be written as the linear matrix differential equations

$$P'(t) = P(t)\Lambda \tag{44}$$

$$P'(t) = \Lambda P(t) \tag{45}$$

with the initial condition $P(0) = (\delta_{ij})$. A formal solution to equation (44) and equation (45) is

$$P(t) = e^{\Lambda t} \tag{46}$$

If the state space is finite, equation (46) is the unique solution of equation (41) (and of equation (42)), and when Λ can be diagonalized, $e^{\Lambda t}$ can be expressed in terms of eigenvectors and eigenvalues. The case of an infinite state space is more complicated.

In many stochastic systems transition probabilities for the successively visited states and total transition rates λ_i are given by the intuitive formulation of the model. The standard construction of MPs with countable state space starts from an arbitrary discrete-time MP $(Y_n)_{n \in \mathbb{Z}_+}$, giving the sequence of visited states, and arbitrary non-negative sojourn time parameters λ_i for the holding times (which turn out to be exponential). Let $Q = (q_{ij})_{i,j \in S}$ be a transition matrix satisfying $q_{ii} = 0$ for all i. We introduce an extra point $\partial \notin S$ to account for possible "explosions" of the MP in finite time. Let $S_\partial = S \cup \{\partial\}$ and extend Q to a transition matrix Q_∂ on S_∂ by making ∂ an absorbing state, i.e., setting $q_{\partial,\partial} = 1$. Define the underlying sample space by

$$\Omega = (0, \infty]^{\mathbb{N}} \times (S_\partial)^{\mathbb{N}} \tag{47}$$

and let $T_0, T_1, \ldots : \Omega \to (0, \infty]$ and $Y_0, Y_1, \ldots : \Omega \to S_\partial$ be the coordinate projections. Then one can, for any probability vector $\pi = (\pi_i)_{i \in S}$, construct probability measures $\mathbb{P}_\pi$ on Ω, endowed with its product σ-algebra, such that, under $\mathbb{P}_\pi$:

1. $(Y_n)_{n\in\mathbb{Z}_+}$ is a discrete-time MP on S_∂ governed by Q_∂ and satisfying $\mathbb{P}_\pi(Y_0 = i) = \pi_i$.
2. Conditional on $(Y_n)_{n\in\mathbb{Z}_+}$, the random variables $T_0, T_1, \ldots,$ are independent and T_n is exponential with rate λ_{Y_n}.

Then

$$X_t = \begin{cases} Y_n, & \text{if } T_0 + \cdots + T_{n-1} \le t < T_0 + \cdots + T_n, \\ & n \in \mathbb{Z}_+ \\ \partial, & \text{if } t \ge \sup_{n\in\mathbb{Z}_+}(T_0 + \cdots + T_n) \end{cases} \quad (48)$$

defines an MP which runs through the states $Y_0, Y_1, \ldots$ and has for every visit to state i an exponential sojourn time (with mean λ_i^{-1}) which is independent of its history before and after the current visit. The rates λ_{ij} in equation (39) are given by $\lambda_{ij} = \lambda_i q_{ij}$, $i \ne j$.

This process can be *explosive*, i.e. $\mathbb{P}_i(M < \infty) > 0$ for some i, where $M = \sup_{n\in\mathbb{Z}_+}(T_0 + \cdots + T_n)$. According to definition (48), it will stay in ∂ forever after time M, but there is an infinity of other possibilities to define the process after an explosion. The MP constructed by equation (48) is called *minimal* as it minimizes $\mathbb{P}_\pi(X_t = i)$ for all $i \in S$, $t > 0$ and π. Instead of introducing an absorbing extra state, one could define that at every explosion time the process moves to state i with some probability $\alpha_i \ge 0$, where $\sum_{i\in S}\alpha_i = 1$ (in this case one does not need ∂). Formally, starting with M and X_t from the above construction,

$$p_{ij}^{(0)}(t) = \mathbb{P}_i(X_t = j, M > t) \quad (49)$$

is the probability to go from i to j before the first explosion. Now define recursively the probability to go from i to j in time t with exactly n explosions in between by setting

$$p_{ij}^{(n)}(t) = \int_0^t \sum_{k\in S} p_{ik}^{(n-1)}(t-s)\alpha_k \mathbb{P}_k(M \in \mathrm{d}s), \\ \times n \ge 1 \quad (50)$$

and let

$$p_{ij}(t) = \sum_{n=0}^{\infty} p_{ij}^{(n)}(t) \quad (51)$$

It can be shown that these $p_{ij}(t)$ define a family of transition probability functions which satisfy the backward but not the forward equations.

The MP constructed by equation (48) is nonexplosive (that is, $\mathbb{P}_i(M < \infty) = 0$ for every $i \in S$) if and only if the vector equation $\Lambda a = a$, $a = (a_i)_{i\in S}$ admits only the trivial solution $a = 0$. Sufficient conditions to rule out explosions are: (a) $\sup_{i\in S}\lambda_i < \infty$ or (b) Y_n is recurrent.

The asymptotic behavior of X_t is slightly simpler than in the discrete-time case because periodicity phenomena are smoothed out by the exponential sojourn times. In the following we consider a minimal MP X_t and summarize the main results. X_t is called *irreducible* if $p_{ij}(t) > 0$ for some $t > 0$ for all $i, j \in S$ (this is equivalent to the irreducibility of Y_n). A state i is called *recurrent* (*transient*) if it is recurrent (transient) for Y_n, and the MP itself is said to be recurrent (transient) if all states have this property.

1. If X_t is irreducible and recurrent, it possesses a stationary measure $\rho = (\rho_i)_{i\in S}$, i.e., $\rho \ne 0$, $0 \le \rho_i < \infty$ and $\rho_i = \sum_{j\in S}\rho_j p_{ji}(t)$ for all $i \in S$ and $t > 0$. Such a ρ is unique up to positive factors and is given by $\tilde{\rho}_i = \mu_i/\lambda_i$, where $(\tilde{\rho}_i)_{i\in S}$ is stationary for Y_n.
2. Alternatively, ρ_i can be characterized (up to a constant factor) as the expected time spent in i between two consecutive visits of j, where j is an arbitrary fixed state, or as solution of the equation $\rho\Lambda = 0$.
3. If the stationary measure has finite total mass, the MP is called *ergodic*. If the MP is irreducible and nonexplosive, it is ergodic if and only if there is a stationary probability distribution; for an irreducible MP to be ergodic it is sufficient that there exists a probability vector $\pi = (\pi_i)_{i\in S}$ solving $\pi\Lambda = 0$ and satisfying $\sum_i \pi_i\lambda_i < \infty$. In this case $\lim_{t\to\infty} p_{ij}(t) = \pi_j$ for all states i and j. If X_t is irreducible and recurrent but not ergodic, the limits are all equal to zero.

Spelled out, the vector equation $\pi\Lambda = 0$ is tantamount to $\pi_j\lambda_j = \sum_{i\ne j}\rho_i\lambda_{ij}$ for all j; this means that in steady state the outflow rate of any state is equal to its inflow rate.

References

[1] Dynkin, E.B. (1965). *Markov Processes*, Springer, Vols 1–2, English translation.

[2] Breiman, L. (1968). *Probability*, Addison-Wesley.
[3] Ethier, S.N. & Kurtz, T.G. (1986). *Markov Processes: Characterization and Convergence*, John Wiley & Sons.
[4] Kallenberg, O. (2001). *Foundations of Modern Probability*, 2nd Edition, Springer.
[5] Feller, W. (1968, 1971). *An Introduction to Probability Theory and its Applications*, Vol. 1 (3rd Edition), Vol. 2 (2nd Edition), John Wiley & Sons.
[6] Karlin, S. & Taylor, H.M. (1975). *A First Course in Stochastic Processes*, 2nd Edition, Academic Press.
[7] Karlin, S. & Taylor, H.M. (1981). *A Second Course in Stochastic Processes*, Academic Press.
[8] Asmussen, S. (2003). *Applied Probability and Queues*, 2nd Edition, Springer.
[9] Chung, K.L. (1960). *Markov Chains with Stationary Transition Probabilities*, Springer.
[10] Freedman, D. (1971). *Markov Chains*, Holden-Day.
[11] Meyn, S.P. & Tweedie, R.L. (1993). *Markov Chains and Stochastic Stability*, Springer.
[12] Bertoin, J. (1996). *Lévy Processes*, Cambridge University Press.
[13] Sato, K. (1999). *Lévy Processes and Infinitely Divisible Distributions*, Cambridge University Press.
[14] Revuz, P. & Yor, M. (1999). *Continuous Martingales and Brownian Motion*, 3rd Edition, Springer.
[15] Rogers, L.C.G. & Williams, D. (2000). *Diffusions, Markov Processes, and Martingales*, Cambridge University Press, Vols 1–2.
[16] Boxma, O., Perry, D., Stadje, W. & Zacks, S. (2006). A Markovian growth-collapse model, *Advances in Applied Probability* **76**, 221–243.

Further Reading

Karatzas, I. & Shreve, S.E. (1991). *Brownian Motion and Stochastic Calculus*, 2nd Edition, Springer.

WOLFGANG STADJE

Markov Renewal Processes in Reliability Modeling

Introduction and Mission Process

In this article, we consider a reliability system that is designed to perform a random sequence of phases or stages with random durations. These are referred to as *phased-mission* or *mission-based systems* where the stochastic structure of the mission process plays a critical role. It is assumed that the mission follows a Markov **renewal process**, and we will analyze various issues related to the reliability of such systems. This line of stochastic modeling and research is often classified under stochastic models in random environments. The idea is used not only in a reliability setting, but in other applications as well. For example, Özekici [1] discusses other applications in inventory and queuing. Asmussen [2] and Rolski *et al.* [3] also give further applications in queuing, insurance, and finance.

In reliability modeling, a device generally consists of a large number of components with stochastically dependent lifetimes. **Random environments** are used to provide a tractable model of dependence since this is taken as an external process that affects the deterioration, aging, and failure of all of the components. Since all components are subjected to the same environmental conditions, their lifetimes are dependent *via* their common environmental process. Thus, the environmental process is actually a factor of variation in the failure structure of the components. An interesting model was introduced by Çınlar and Özekici [4] where stochastic dependence is introduced by a randomly changing common environment that all components of the system are subjected to. This model is based on the simple observation that the aging or deterioration process of any component depends very much on the environment that the component is operating in. They propose to construct an intrinsic clock that ticks differently in different environments to measure the intrinsic age of the device. The environment is modeled by a semi-Markov jump process and the intrinsic age is represented by the cumulative hazard accumulated in time during the operation of the device in the randomly varying environment. This is a rather stylish choice, which envisions that the intrinsic lifetime of any device has an exponential distribution with parameter 1. The concept of random hazard functions is also used in Gaver [5] and Arjas [6]. The intrinsic aging model of Çınlar and Özekici [4] is studied further in Çınlar *et al.* [7] to determine the conditions that lead to associated component lifetimes, as well as multivariate increasing failure rate (IFR) and new better than used (NBU) life distribution characterizations. It was also extended in Shaked and Shanthikumar [8] by discussions on several different models with multicomponent replacement policies. Lindley and Singpurwalla [9] discuss the effects of the random environment on the reliability of a system consisting of components that share the same environment. Although the initial state of the environment is random, they assume that it remains constant in time and components have exponential life distributions in each possible environment. This model is also studied by Lefèvre and Malice [10] to determine partial orderings on the number of functioning components and the reliability of k-out-of-n systems, for different partial orderings of the probability distribution on the environmental state. The association of the lifetimes of components subjected to a randomly varying environment is discussed in Lefèvre and Milhaud [11]. Singpurwalla and Youngren [12] also discuss multivariate distributions that arise in models where a dynamic environment affects the failure rates of the components.

Reliability models in random environments also provide a perfect opportunity to analyze mission-based or so-called phased-mission reliability systems. Our aim in this article is to present an example along this direction. These systems involve devices or machines that are designed to perform or assigned to missions consisting of a number of phases. The sequence as well as the durations of the phases may be random and, as in all random environment models, all stochastic and deterministic failure properties depend on the phases of the mission that is performed at a given time. Therefore, the random environmental process is the mission process in such models. A phased-mission consists of a number of stages or phases that must be successfully completed. These systems were introduced by Esary and Ziehms [13] and a vast literature has accumulated since then. The phases may be deterministic or stochastic as in Kim and Park [14]. The components are either repairable or nonrepairable. Alam and Al-Saggaf [15] discuss a phased-mission system with **repairable components** where the repair activity begins as soon as a component fails. A case study involving spacecraft is provided by Mura and Bondavalli [16] where the mission consists of four basic phases: launch, hibernation, planet, and scientific observation. The reader is referred to these papers and the many references cited in them for various models, techniques, and applications on phased-mission systems.

Let X_n denote the nth phase of the mission and T_n denote the time at which the nth phase starts with $T_0 \equiv 0$. The main assumption is that the process $(X, T) = \{(X_n, T_n); n \geq 0\}$ is a Markov renewal process on the countable state space E with some semi-Markov kernel Q. The state space E is actually that of the process X and it is implicitly understood that the process T always takes values in $R_+ = [0, +\infty)$ since they denote times at which certain events occur. We refer the reader to Çınlar [17] for a more rigorous and detailed treatment of Markov renewal processes and theory. The Markov renewal property states that

$$\begin{aligned} P\{X_{n+1} = j, T_{n+1} - T_n \leq t | X_0, \ldots, X_n; T_0, \ldots, T_n\} \\ = P\{X_{n+1} = j, T_{n+1} - T_n \leq t | X_n\} \end{aligned} \quad (1)$$

where we suppose that the process is time homogeneous with the semi-Markov kernel

$$Q(i, j, t) = P\{X_{n+1} = j, T_{n+1} - T_n \leq t | X_n = i\} \quad (2)$$

for any $i, j \in E$ and $t \geq 0$. It is well known that X is a Markov chain on E with transition matrix $P(i, j) = P\{X_{n+1} = j | X_n = i\} = Q(i, j, +\infty)$. We further assume that the Markov renewal process has infinite lifetime so that $\sup_n T_n = +\infty$. The probabilistic structure of T is given by the conditional distributions

$$\begin{aligned} G(i, j, t) &= P\{T_{n+1} - T_n \leq t | X_n = i, X_{n+1} = j\} \\ &= \frac{Q(i, j, t)}{P(i, j)} \end{aligned} \quad (3)$$

for $P(i, j) > 0$. Moreover, it is easy to see that

$$F_i(t) = P\{T_1 \leq t | X_0 = i\} = \sum_{j \in E} Q(i, j, t) \quad (4)$$

denotes the distribution of the duration of phase i. Finally, the mission process $Y = \{Y_t; t \geq 0\}$ is the minimal semi-Markov process associated with (X, T) so that Y_t is the stage or phase of the mission at time t. More precisely, $Y_t = X_n$ whenever $T_n \leq t < T_{n+1}$.

To simplify the notation, for any event A we will let $P_i\{A\} = P\{A | X_0 = i\}$ denote the conditional probability of A given $X_0 = i$. The reliability of a mission-based system will be analyzed with two extreme repair policies: maximal repair and no repair.

The system is replaced by a brand-new one at the beginning of each phase in the first case whereas no repair or replacement is done in the second case until the system fails. We discuss system and mission reliability with maximal repair in the second section. Mission reliability can be defined in different ways. We consider two alternatives where we focus on the probability of completing the first n phases successfully in the first one, and the probability of completing a given critical phase in the second alternative. These issues are discussed for the no repair case in the third section using intrinsic aging concepts.

Models with Maximal Repair

Suppose that the system performs the mission such that at the beginning of each phase it is replaced by a brand-new one. This simplifying assumption allows us to obtain various reliability measures quite easily by using the renewal property. It is clear that the probability of survival during a given phase depends on it, and we let

$$\overline{P}_i(t) = 1 - P_i(t) = P_i\{L > t | Y_s = i; s \in [0, t]\} \tag{5}$$

denote the survival probability of the system in phase i for t units of time where L is the lifetime of the system.

We can obtain another Markov renewal processes using (X, T), which will be used during the analysis. Define a new Markov renewal process $(\widetilde{X}, \widetilde{T})$ through its minimal semi-Markov process $\widetilde{Y}$ so that

$$\widetilde{Y}_t = \begin{cases} Y_t & \text{if } t < L \\ \Delta & \text{if } t \geq L \end{cases} \tag{6}$$

where Δ is an absorbing state. This also implies that

$$\widetilde{T}_n = \begin{cases} 0 & n = 0 \\ \inf\{t > T_{n-1} : \widetilde{Y}_t \neq \widetilde{Y}_{T_{n-1}}\} & n \geq 1 \end{cases} \tag{7}$$

and $\widetilde{X}_n = \widetilde{Y}_{\widetilde{T}_n}$ for $n \geq 0$. Clearly, the state space of $(\widetilde{X}, \widetilde{T})$ is $\widetilde{E} = E \cup \{\Delta\}$ and its semi-Markov kernel is

$$\widetilde{Q}(i, k, \mathrm{d}s) = \begin{cases} \overline{P_i}(s)\, Q(i, k, \mathrm{d}s) & \text{if } i, k \in E \\ \overline{F_i}(s)\, P_i(\mathrm{d}s) & \text{if } i \in E,\ k = \Delta \\ 0 & \text{if } i, k = \Delta \end{cases} \tag{8}$$

Note that $\widetilde{Q}(\Delta, \Delta, t) = 0$ for every $t \geq 0$ and $\widetilde{P}(\Delta, \Delta) = 1$. In plain words, the new process $\widetilde{Y}$ is obtained by "stopping" or "killing" the process Y whenever the system fails at time L. At the time of failure, the process is dumped to the absorbing state Δ that denotes system failure. The state space is therefore extended by including this new state Δ. We can find the transition matrix of the Markov chain $\widetilde{X}$ as

$$\widetilde{P}(i, j) = \widetilde{Q}(i, j, \infty) = \int_0^\infty \widetilde{Q}(i, j, \mathrm{d}s) = \int_0^\infty \overline{P_i}(s)\, Q(i, j, \mathrm{d}s) \tag{9}$$

for $i, j \in E$, and

$$\widetilde{P}(i, \Delta) = \int_0^\infty \overline{F_i}(s)\, P_i(\mathrm{d}s) \tag{10}$$

for $i \in E$.

Suppose that one is interested in the successful completion of a given critical phase j of the mission. Letting

$$U_j = \inf\{t \geq 0; Y_t \neq Y_{t-} = j\} \tag{11}$$

denote the first time that the mission process leaves state j, we can now define another new Markov renewal process $(\overline{X}, \overline{T})$ as appropriate through its minimal semi-Markov process $\overline{Y}$. The new process is defined by

$$\overline{Y}_t = \begin{cases} Y_t & \text{if } t < \min\{L, U_j\} \\ \Delta & \text{if } L \leq \min\{t, U_j\} \\ S & \text{if } U_j \leq \min\{t, L\} \end{cases} \tag{12}$$

so that we now extend E such that $\overline{E} = E \cup \{\Delta, S\}$. We let the semi-Markov mission process Y go to the absorbing state Δ as soon as the system fails before completing phase j or to the absorbing state S as soon as phase j is completed without failure. Thus, if $\overline{X}_n = \Delta$ or $\overline{X}_n = S$, then $\overline{T}_{n+1} = \infty$. If the minimal semi-Markov process associated with this Markov renewal process is $\overline{Y}$, then the probability of completing the critical phase j until time t is $P_i\{\overline{Y}_t = S\}$ if the initial state is i. The Markov renewal kernel of this process can be obtained as

$$\overline{Q}(i,k,\mathrm{d}s)=\begin{cases}\widetilde{Q}(i,k,\mathrm{d}s) & i\in E-\{j\},k\in E\\ Q(i,\Delta,\mathrm{d}s) & i\in E,k=\Delta\\ F_j(\mathrm{d}s)\overline{P}_j(s) & i=j,k=S\\ 0 & \text{otherwise}\end{cases} \tag{13}$$

Note that $\overline{Q}(S,S,t)=\overline{Q}(\Delta,\Delta,t)=0$ for every $t\geq 0$, and $\overline{P}(S,S)=\overline{P}(\Delta,\Delta)=1$. If $i\in E-\{j\},k\in E$, then phase i is completed successfully before system failure and hence $\overline{Q}(i,k,\mathrm{d}s)=\widetilde{Q}(i,k,\mathrm{d}s)$. If $i\in E$ and $k=\Delta$, then $\overline{Q}(i,k,\mathrm{d}s)$ represents the probability of failure during phase i in the vicinity of time s, which means that the duration of phase i is longer than s units of time and a failure will occur in the vicinity of time s. Therefore, this probability is equal to $\overline{Q}(i,\Delta,\mathrm{d}s)=\overline{F}_i(s)P_i(\mathrm{d}s)$. Moreover, $\overline{Q}(j,S,\mathrm{d}s)$ represents the probability that the system will complete phase j in the vicinity of time s given that the system is in working condition at the beginning of phase j. This situation occurs if the duration of mission j is in the vicinity of time s and the system survives more than s units of time in these conditions; hence, $\overline{Q}(j,S,ds)=F_j(ds)\overline{P}_j(s)$. It is clear from the definition of the process $(\overline{X},\overline{T})$ that the states S and Δ are absorbing states. Therefore, if the process gets into states S or Δ, it remains there forever where S now represents the "success" state and Δ represents the "failure" state.

System Reliability

System reliability is given by the probability that the system will function until time t. Let $f(i,t)=P_i\{L>t\}$ denote the desired probability. Then, conditioning on T_1

$$\begin{aligned}f(i,t)&=P_i\{L>t,T_1>t\}+P_i\{L>t,T_1\leq t\}\\&=P_i\{L>t|T_1>t\}P_i\{T_1>t\}\\&\quad+P_i\{L>t,T_1\leq t\}=\overline{P_i}(t)\overline{F_i}(t)\\&\quad+\sum_{j\in E}\int_0^t Q(i,j,\mathrm{d}s)\overline{P_i}(s)f(j,t-s)\\&=g(i,t)+\widetilde{Q}*f(i,t)\end{aligned} \tag{14}$$

where $g(i,t)=\overline{P_i}(t)\overline{F_i}(t)$ and $\widetilde{Q}(i,j,\mathrm{d}s)=Q(i,j,\mathrm{d}s)\overline{P_i}(s)$ for $i\in E$. Clearly, $f(\Delta,t)=g(\Delta,t)=0$. Then, equation (14) is a Markov renewal equation and it has the unique solution $f=\widetilde{R}*g$ so that

$$\begin{aligned}P_i\{L>t\}&=\sum_{j\in\widetilde{E}}\int_0^t\widetilde{R}(i,j,\mathrm{d}s)g(j,t-s)\\&=\sum_{j\in\widetilde{E}}\int_0^t\widetilde{R}(i,j,\mathrm{d}s)\overline{P_j}(t-s)\overline{F_j}(t-s)\end{aligned} \tag{15}$$

where $\widetilde{R}=\sum_{n=0}^{\infty}\widetilde{Q}^n$ is the Markov renewal kernel corresponding to $\widetilde{Q}$.

Mission Reliability

In a given application, it may be important to determine the probability that the system will complete the first n phases successfully. We will now focus on this issue and show that this probability can be calculated using a recursive formula and then obtain an explicit solution. We first find the probability that the first phase will be completed without failure. This can be done by conditioning on the next phase after the first one so that

$$\begin{aligned}P_i\{L>T_1\}&=\sum_{j\in E}P_i\{L>T_1,X_1=j\}\\&=\sum_{j\in E}\int_0^{\infty}\overline{P_i}(s)Q(i,j,\mathrm{d}s)\\&=\int_0^{\infty}\overline{P_i}(s)F_i(\mathrm{d}s)\end{aligned} \tag{16}$$

Now, we can write,

$$\begin{aligned}&P_i\{L>T_{n+1}\}\\&\quad=\sum_{j\in E}P_i\{L>T_{n+1},X_1=j\}\\&\quad=\sum_{j\in E}P_j\{L>T_n\}P_i\{X_1=j,L>T_1\}\\&\quad=\sum_{j\in E}P_j\{L>T_n\}\int_0^{\infty}\overline{P_i}(s)Q(i,j,\mathrm{d}s)\end{aligned} \tag{17}$$

which is a recursive relationship with the boundary condition $P_i\{L>T_0\}=1$ for $n\geq 0$. Using equations

(9) and (17), we have

$$P_i\{L > T_{n+1}\} = \sum_{j\in E} \widetilde{P}(i,j)\, P_j\{L > T_n\} \quad (18)$$

Letting $f_n(i) = P_i\{L > T_n\}$, we can rewrite equation (18) as

$$f_{n+1} = \widetilde{P} f_n \quad (19)$$

for $n \geq 0$ with the boundary condition $f_0 = 1$. Then, it is clear that $f_1 = \widetilde{P} f_0$, $f_2 = \widetilde{P} f_1 = \widetilde{P}^2 f_0$, and, more generally, $f_n = \widetilde{P} f_{n-1} = \widetilde{P}^n f_0$ so that the solution is

$$f_n(i) = \widetilde{P}^n f_0(i) = \sum_{j\in E} \widetilde{P}^n(i,j) \quad (20)$$

It is also possible to obtain the same solution by noting that

$$P_i\{L > T_n\} = P_i\left\{\widetilde{X}_n \in E\right\} = \sum_{j\in E} \widetilde{P}^n(i,j) \quad (21)$$

since the nth phase of the mission is successfully completed if $\widetilde{X}_n \neq \Delta$.

Critical Phase Reliability

For a complex system, a given phase can be more important than the others because of the overall objective of the mission. Therefore, the probability that this phase will be completed in a fixed period is an important measure to represent the reliability of the system. To analyze this case, define

$$W_j(t) = \begin{cases} 1 & \text{if phase } j \text{ is completed before time } t \\ 0 & \text{otherwise} \end{cases} \quad (22)$$

where j is the critical phase. Then, using the definition of $\left(\overline{X}, \overline{T}\right)$, the desired probability is

$$P_i\left\{W_j(t) = 1\right\} = P_i\left\{\overline{Y}_t = S\right\} \quad (23)$$

and, using Proposition (10.5.4) in [18], we have

$$P_i\left\{W_j(t) = 1\right\} = P_i\left\{\overline{Y}_t = S\right\} = \overline{R}(i,S,t) \quad (24)$$

where $\overline{R} = \sum_n \overline{Q}^n$ is the Markov renewal kernel corresponding to $\overline{Q}$.

Models with No Repair

In the previous section, we considered systems with maximal repair so that we had a brand-new system at the beginning of each phase. In this section, we remove this assumption and, hence, the system will deteriorate or get older in time. Therefore, the concept of "aging" comes into consideration. In this study, we will use the "intrinsic aging" model introduced by Çınlar and Özekici [4]. The analysis will be presented for a series system with m components, which requires that all components must function for the system to be functional.

Before the analysis, some new notation should be introduced. Let $H_k(i,t)$ be the cumulative hazard of component k at time t in phase i. Then, the intrinsic age of component k at time t is defined as $H_k(i,t)$ provided that the system performs phase i throughout $[0,t]$. The intrinsic aging rate of component k during phase i is defined as

$$r_k(i,a) = \frac{\mathrm{d}}{\mathrm{d}t} H_k(i,t)\,|_{t=\tau(x,a)} \quad (25)$$

at any age $a \geq 0$, where $\tau_k(i,a)$ is the time at which the intrinsic age of component k becomes a if the system performs phase i; or

$$\tau_k(i,a) = \inf\{t \geq 0 : H_k(i,t) > a\} \quad (26)$$

Let $A_t(k)$ denote the intrinsic age process of component k. We assume that the intrinsic age process satisfies

$$\frac{\mathrm{d}A_t(k)}{\mathrm{d}t} = r_k(Y_t, A_t(k)) \quad (27)$$

Therefore, if the intrinsic age of component k at time s when phase i starts is $A_s(k) = a$, then after t units of time its age becomes

$$A_{s+t}(k) = h_k(i,a,t) = H_k(i, \tau_k(i,a) + t) \quad (28)$$

Let $B_0(k) = A_0(k)$ and define an embedded process $\{B_n(k)\}$ recursively through

$$B_{n+1}(k) = h(X_n, B_n(k), T_{n+1} - T_n) \quad (29)$$

for $n \geq 0$. Then, the intrinsic age of component k at time t is given by

$$A_t(k) = h(X_n, B_n(k), t - T_n) \quad (30)$$

whenever $T_n \leq t \leq T_{n+1}$. Let $L(k)$ and $\widehat{L}(k)$ denote the lifetime and the intrinsic lifetime of component k

respectively. Then,

$$\widehat{L}(k) = A_{L(k)}(k) \tag{31}$$

for all k. Following Çınlar and Özekici [4], $\widehat{L}(k)$ is exponentially distributed with rate 1 for all k and

$$\begin{aligned} P\{L(k) > t | A_0(k) = a(k)\} &= P\left\{\widehat{L}(k) > A_t(k) | A_0(k) = a(k)\right\} \\ &= E\left[e^{-(A_t(k)-a(k))}\right] \end{aligned} \tag{32}$$

Let A_t, a, $h(i,a,t)$, L, and $\widehat{L}$ denote the vectors with elements $A_t(k)$, $a(k)$, $h_k(i,a,t)$, $L(k)$, and $\widehat{L}(k)$ for all k, respectively. We let $F = R_+^m$ denote the collection of all nonnegative vectors where m is the number of components in the system. Each a in F represents a vector of intrinsic ages of all components. Assume that if b is a vector, then $b > x$ for a scalar x means that $b(k) > x$ for all k; and if a, b are both vectors in F, then $b > a$ means that $b(k) > a(k)$ for all k. We let σ be a row vector with m entries which are all equal to 1, and

$$H(i,a,t;\,db) = \begin{cases} 1 & \text{if } h(i,a,t) = b \\ 0 & \text{otherwise} \end{cases} \tag{33}$$

To simplify our notation, for any event A we will let $P_{ia}\{A\} = P\left\{A | X_0 = i, A_0 = a, \widehat{L} > a\right\}$ denote the conditional probability of A given $X_0 = i$, $A_0 = a$, $\widehat{L} > a$. Our analysis in the previous section used the fact that (X, T) is a Markov renewal process to obtain Markov renewal characterizations on various reliability measures. This analysis was possible because of the maximal repair assumption so that the system was brand new at the beginning of each phase. However, when this is not the case, we need to base our analysis on the new Markov renewal process $((X, B), T)$ that also includes information on the intrinsic ages of the components. The state space is now extended to $E \times F$.

System Reliability of a Series System

We first consider the system reliability of a series system that is performing a mission. In other words, we want to determine the probability that the system will work without failure until time t. Using a Markov renewal characterization, we define $f(i,a,t) = P_{ia}\{L > t\}$ and obtain

$$\begin{aligned} P_{ia}\{L > t\} &= P_{ia}\{L > t, T_1 > t\} \\ &+ P_{ia}\{L > t, T_1 \le t\} = \overline{F_i}(t)\,e^{-\sigma(h(i,a,t)-a)} \\ &+ \sum_{j\in E}\int_{F\times[0,t]} Q(i,j,ds)\,H(i,a,s;\,db) \\ &\times e^{-\sigma(b-a)} f(j,b,t-s) \end{aligned} \tag{34}$$

which is a Markov renewal equation of the form $f = g + \widehat{Q} * f$, where

$$g(i,a,t) = \overline{F_i}(t)\,e^{-\sigma(h(i,a,t)-a)} \tag{35}$$

and

$$\widehat{Q}(i,a,j,db,ds) = Q(i,j,ds)\,H(i,a,s;\,db) \times e^{-\sigma(b-a)} \tag{36}$$

It is clear that $\widehat{Q}$ is a semi-Markov kernel on the extended state space $\widetilde{E} \times F$ and the Markov renewal equation has the unique solution $f = \widehat{R} * g$ such that

$$P_{ia}\{L > t\} = \sum_{j\in\widetilde{E}}\int_{F\times[0,t]} \widehat{R}(i,a;j,db;ds) \times \overline{F_j}(t-s)\,e^{-\sigma(h(j,b,t-s)-b)} \tag{37}$$

where $\widehat{R} = \sum_n \widehat{Q}^n$ is the Markov renewal kernel corresponding to $\widehat{Q}$.

Mission Reliability of a Series System

We now focus on computing mission reliability involving the first n phases of the mission. We first find the probability that the first phase will be completed by conditioning on the next phase so that

$$\begin{aligned} P_{ia}\{L > T_1\} &= \sum_{j\in E} P_{ia}\{L > T_1, X_1 = j\} \\ &= \sum_{j\in E} P_{ia}\{L > T_1 | X_1 = j\}\,P(i,j) \end{aligned} \tag{38}$$

Note that

$$\begin{aligned} &P_{ia}\{L > T_1 | X_1 = j\} \\ &= \int_{F\times[0,\infty)} P_{ia}\{L > T_1, T_1 \in ds, B_1 \in db | X_1 = j\} \\ &= \int_{F\times[0,\infty)} e^{-\sigma(b-a)} G(i,j,ds)\,H(i,a,s;\,db) \end{aligned} \tag{39}$$

Combining equations (38) and (39), we obtain

$$P_{ia}\{L > T_1\} = \sum_{j\in E}\int_{F\times[0,\infty)} Q(i, j, ds) \times e^{-\sigma(b-a)} H(i, a, s; db) \tag{40}$$

Similarly, we can now obtain a recursive relationship

$$\begin{aligned} P_{ia}\{L > T_{n+1}\} &= \sum_{j\in E} P_{ia}\{L > T_{n+1} | X_1 = j\} P(i, j) \\ &= \sum_{j\in E}\int_{F\times[0,\infty)} P_{ia}\{L > T_{n+1}, T_1 \in ds, \\ &\quad B_1 \in db | X_1 = j\} P(i, j) \\ &= \sum_{j\in E}\int_{F\times[0,\infty)} e^{-\sigma(b-a)} Q(i, j, ds) \\ &\quad \times H(i, a, s; db) P_{jb}\{L > T_n\} \end{aligned} \tag{41}$$

for $n \geq 0$. Let $\widehat{Q}(i, a; j, db; ds) = Q(i, j, ds) H(i, a, s; db) e^{-\sigma(b-a)}$, and define

$$\begin{aligned} \widehat{P}(i, a; j, db) &= \widehat{Q}(i, a; j, db; \infty) \\ &= P_{ia}\{X_1 = j, B_1 \in db\} \end{aligned} \tag{42}$$

and

$$\begin{aligned} \widehat{P}^n(i, a; j, db) &= \sum_{j\in E}\int_F \widehat{P}^{n-1}(i, a; k, dc) \widehat{P}(k, c; j, db) \\ &= P_{ia}\{X_n = j, B_n \in db\} \end{aligned} \tag{43}$$

for $n \geq 1$.

Then, it follows from equation (40) that

$$\begin{aligned} P_{ia}\{L > T_1\} &= \sum_{j\in E}\int_{F\times[0,\infty)} \widehat{Q}(i, a; j, db; ds) \\ &= \sum_{j\in E}\int_F \widehat{Q}(i, a; j, db; \infty) \\ &= \sum_{j\in E}\int_F \widehat{P}(i, a; j, db) \end{aligned} \tag{44}$$

Now, using induction, we will show that

$$P_{ia}\{L > T_n\} = \sum_{j\in E}\int_F \widehat{P}^n(i, a; j, db) \tag{45}$$

for every $n \geq 1$. We have already shown that equation (45) holds for $n = 1$. Suppose that

$$P_{ia}\{L > T_k\} = \sum_{j\in E}\int_F \widehat{P}^k(i, a; j, db) \tag{46}$$

for all $k \leq n - 1$. Then, using equations (41) and (46),

$$\begin{aligned} P_{ia}\{L > T_n\} &= \sum_{j\in E}\int_{F\times[0,\infty)} \widehat{Q}(i, a; j, db; ds) P_{jb}\{L > T_{n-1}\} \\ &= \sum_{j\in E}\int_F \widehat{P}(i, a; j, db) \sum_{k\in E}\int_F \widehat{P}^{n-1}(j, b; k, dc) \\ &= \sum_{j\in E}\int_F \widehat{P}^n(i, a; j, db) \end{aligned} \tag{47}$$

Critical Phase Reliability of a Series System

Suppose that there is a critical phase j and we are interested in computing the probability $P_{ia}\{W_j(t) = 1\}$ that phase j will be completed successfully by time t. Using a similar approach as in the critical reliability part of the previous section, we note that the process $\overline{Y}$ defined by equation (12) is a semiregenerative process. In this case, the semi-Markov kernel corresponding to the Markov renewal process $((\overline{X}, B), \overline{T})$ is given by

$$\overline{Q}(i, a; k, db; ds) = \begin{cases} Q(i, k, ds) H(i, a, s; db) \times e^{-\sigma(b-a)} & \text{if } i \in E - \{j\},\ k \in E \\ \overline{F}_i(s) H(i, a, s; db) U_{ia}(ds) & \text{if } i \in E, k = \Delta \\ F_j(ds) H(i, a, s; db) e^{-\sigma(b-a)} & \text{if } i = j, k = S \\ 0 & \text{otherwise} \end{cases} \tag{48}$$

where

$$U_{ia}(s) = 1 - e^{-\sigma(h(i,a,s)-a)}. \tag{49}$$

Note that $\overline{Q}(S, a; S, db; t) = \overline{Q}(\Delta, a; \Delta, db; t) = 0$ trivially for every $t \geq 0$, $a, b \in F$ and $\overline{P}(S, a; S, da) = \overline{P}(\Delta, a; \Delta, da) = 1$ for every $a \in F$. We now find the probability $f(i, a, t) = P_{ia}\{\overline{Y}_t = j, A_t \in db\}$ for some $j \in \overline{E}$ and $b \in F$. Using Markov renewal theory,

$$
\begin{aligned}
f(i,a,t) &= P_{ia}\left\{\overline{Y}_t = j, A_t \in \mathrm{d}b, \overline{T}_1 > t\right\} \\
&\quad + P_{ia}\left\{\overline{Y}_t = j, A_t \in \mathrm{d}b, \overline{T}_1 \le t\right\} \\
&= I(i,j)\, I(a,b)\, q(i,a,t) \\
&\quad + \sum_{l \in \overline{E}} \int_{F\times[0,t]} \overline{Q}(i,a;l,\mathrm{d}c;\mathrm{d}s) \\
&\quad \times P_{lc}\left\{\overline{Y}_{t-s} = j, A_{t-s} \in \mathrm{d}b\right\} \\
&= I(i,j)\, I(a,b)\, q(i,a,t) \\
&\quad + \sum_{l \in \overline{E}} \int_{F\times[0,t]} \overline{Q}(i,a;l,\mathrm{d}c;\mathrm{d}s) \\
&\quad \times f(l,c,t-s)
\end{aligned} \tag{50}
$$

where

$$
\begin{aligned}
q(i,a,t) &= P_{ia}\left\{\overline{T}_1 > t\right\} \\
&= 1 - \sum_{l \in \overline{E}} \int_F \overline{Q}(i,a;l,\mathrm{d}b;t)
\end{aligned} \tag{51}
$$

We have a Markov renewal equation $f = g + \overline{Q} * f$ with

$$
g(i,a,t) = I(i,j)\, I(a,b)\, q(i,a,t) \tag{52}
$$

Following Proposition A.2 in [4], equation (50) has a unique solution

$$
\begin{aligned}
&f(i,a,t) \\
&\quad = P_{ia}\left\{\overline{Y}_t = j, A_t \in \mathrm{d}b\right\} \\
&\quad = \sum_{l \in \overline{E}} \int_{F\times[0,t]} \overline{R}(i,a;l,\mathrm{d}c;\mathrm{d}s)\, g(l,c,t-s) \\
&\quad = \int_{[0,t]} \overline{R}(i,a;j,\mathrm{d}b;\mathrm{d}s)\, q(j,b,t-s)
\end{aligned} \tag{53}
$$

where $\overline{R} = \sum_n \overline{Q}^n$ is the Markov renewal kernel corresponding to $\overline{Q}$.

Then, using equation (53)

$$
\begin{aligned}
P_{ia}\left\{W_j(t) = 1\right\} &= \int_F P_{ia}\left\{\overline{Y}_t = j, A_t \in \mathrm{d}b\right\} \\
&= \int_{F\times[0,t]} \overline{R}(i,a;j,\mathrm{d}b;\mathrm{d}s) \\
&\quad \times q(j,b,t-s)
\end{aligned} \tag{54}
$$

Acknowledgment

This research is supported by the Scientific and Technological Research Council of Turkey through grant 106M044.

References

[1] Özekici, S. (1996). Complex systems in random environments, In *Reliability and Maintenance of Complex Systems, NATO ASI Series*, S. Ozekici, ed, Springer-Verlag, Berlin, Vol. F154, pp. 137–157.

[2] Asmussen, S. (2000). *Ruin Probabilities*, World Scientific, Singapore.

[3] Rolski, T., Schmidli, H., Schmidt, V. & Teugels, J. (1999). *Stochastic Processes for Insurance and Finance*, John Wiley & Sons, Chichester.

[4] Çınlar, E. & Özekici, S. (1987). Reliability of complex devices in random environments, *Probability in the Engineering and Informational Sciences* **1**, 97–115.

[5] Gaver, D.P. (1963). Random hazard in reliability problems, *Technometrics* **5**, 211–226.

[6] Arjas, E. (1981). The failure and hazard process in multivariate reliability systems, *Mathematics of Operations Research* **6**, 551–562.

[7] Çınlar, E., Shaked, M. & Shanthikumar, J.G. (1989). On lifetimes influenced by a common environment, *Stochastic Processes and their Applications* **33**, 347–359.

[8] Shaked, M. & Shanthikumar, J.G. (1989). Some replacement policies in a random environment, *Probability in the Engineering and Informational Sciences* **3**, 117–134.

[9] Lindley, D.V. & Singpurwalla, N.D. (1986). Multivariate distributions for the lifelenghts of components of a system sharing a common environment, *Journal of Applied Probability* **23**, 418–431.

[10] Lefèvre, C. & Malice, M.-P. (1989). On a system of components with joint lifetimes distributed as a mixture of exponential laws, *Journal of Applied Probability* **26**, 202–208.

[11] Lefèvre, C. & Milhaud, X. (1990). On the association of the lifelenghts of components subjected to a stochastic environment, *Advances in Applied Probability* **22**, 961–964.

[12] Singpurwalla, N.D. & Youngren, M.A. (1993). Multivariate distributions induced by dynamic environments, *Scandinavian Journal of Statistics* **20**, 251–261.

[13] Esary, J.D. & Ziehms, H. (1975). Reliability analysis of phased missions, *Proceedings of the Conference on Reliability and Fault Tree Analysis*, SIAM, pp. 213–236.

[14] Kim, K. & Park, K.S. (1994). Phased-mission system reliability under Markov environment, *IEEE Transactions on Reliability* **43**, 301–309.

[15] Alam, M. & Al-Saggaf, U.M. (1986). Quantitative reliability evaluation of repairable phased-mission systems using Markov approach, *IEEE Transactions on Reliability* **35**, 498–503.

[16] Mura, I. & Bondavalli, A. (1999). Hierarchical modelling and evaluation of phased-mission systems, *IEEE Transactions on Reliability* **48**, 360–368.
[17] Çınlar, E. (1969). Markov renewal theory, *Advances in Applied Probability* **1**, 123–187.
[18] Çınlar, E. (1975). *Introduction to Stochastic Processes*, Prentice Hall, Englewood Cliffs.

Related Articles

Aging and Positive Dependence; **General Minimal Repair Models Imperfect Repair**; **Maintenance Optimization in Random Environments**; **Markov Processes**; **Parallel, Series, and Series–Parallel Systems**; **Renewal Theory**; **Repairable Systems Reliability**.

BORA ÇEKYAY AND SÜLEYMAN ÖZEKICI

Masked Failure Data

Introduction

In many life-testing situations one is only in a position to establish a subset of causes responsible for a given failure, but not the actual cause. This phenomenon is termed *masking*. For example, a typical failure of a personal computer results in error codes that enable one to identify the card (printed circuit board) responsible for the failure. However, in the vast majority of failures the actual offending component on the card remains unidentified. Situations of this type occur because of a number of reasons, the leading one being the cost of failure analysis (FA). Typically, the top priority in the wake of a failure is bringing the system back to operational state as quickly as possible. This is usually achieved through a modular system design and modular diagnostic tests. This approach to system design is the leading cause for the prevalence of masked in manufacturing, field, and warranty areas.

The key statistical problems related to masked data are those of modeling, **estimation**, and diagnostics. Most of the literature has been focused on **competing risk models**, which assume that there is only one cause of failure. However, models in which more than one component could be a culprit have also been considered [1]. As the concept of masked failure appears as a natural extension of the concept of completely identified failure, the whole spectrum of statistical models for lifetime data can be extended to accommodate these types of failures. The estimation and inference problems are typically handled *via* a conventional likelihood approach. One of the key estimation problems in this context relates to survival curves of the individual causes of failure. This problem is of special importance in the area of quality control associated with masked data. For example, one can find that there exists enough evidence to alert the manufacturer of a specific component that the reliability of that component is problematic, even though in most cases the component was not directly implicated in causing the system failure.

Another class of problems is related to estimation of parameters that govern the masking phenomenon itself. For example, in many cases one would want to estimate the probability that a failure of a particular component will lead to a given type of masking. In situations where one is interested in setting up an efficient repair operation, the inverse problem is also of importance. For example, if we know that replacing a particular set of five components will take care of the failure, could we assert that replacing only components 1 and 3 will eliminate the failure with probability 0.99 or higher? Furthermore, what is the probability that each one of the components $1, 2, \ldots, 5$ was the cause of the failure? The latter probabilities are not only important in the problem of diagnostics, but they also play a key role in quality reporting where it is frequently necessary to "assign the blame" or "allocate the failure" to several causes. Such reporting, in turn, enables one to use, in conjunction with masked data, a common set of reporting and monitoring tools, such as **Pareto** analysis and **control charts** for attribute data.

Model Specification and Parameters

Consider a system that contains k components in series. For simplicity, let us associate these components with causes of failure: we will say that a particular system failed because of component i (or because of cause i). Then one can typically prespecify the *masking groups*: for example, in the case $k = 5$ we could have just three masking groups, $g_1 = (1, 2)$,

$g_2 = (1, 3)$ and $g_3 = (3, 4, 5)$. In other words, given that the masking group g_2 is feasible, we could have a situation where, in the wake of failure, we are definitely able to establish that the failure is due to either component 1 or 3, but we are not able to establish which of these components is culpable. In what follows, we will denote by G the set of the possible masking groups and use the symbol g to index these groups. If the cause i is part of the masking group g, we will express it as $i \subset g$.

We must note that the problem of analysis of masked data has been studied extensively not only in the engineering and reliability literature, but also in literature related to biostatistics and medical research. Therefore, in what follows we will mention a number of publications from journals related to these areas.

A typical dataset associated with masked failures contains the following information [2]:

N is the number of systems tested.
n_0 is the number of systems that survived.
$n_i, i = 1, \ldots, k$ is the number of failed systems for which the failure is known to be caused by component i.
n_g is the number of failed systems for which it is known that the cause belongs to a masking group g (given for every $g \in G$).
$t_j^{(i)}, j = 1, \ldots, n_i$ are the lifetimes of the n_i failures identified as caused by component i.
$t_j^{(g)}, j = 1, \ldots, n_g$ are the lifetimes of the n_g failures identified as caused by one of the components in the masking group g.
$t_j^{(c)}, j = 1, \ldots, n_0$ are the censored lifetimes.

The above data must satisfy the conditions $n_0 + \sum_{i=1}^{k} n_i + \sum_{g \in G} n_g = N$.

A typical model used in conjunction with the above data is defined in terms of the following functions:

$P_i(t), i = 1, \ldots, k$, is called the ith *identification probability*. This is the probability that a system failure at time t that is due to cause i is identified as such (i.e., no masking occurs).
$P_{g|i}(t)$ is called the *masking probability*. This is the probability that a system failure at time t that is actually due to cause i will result in a masking group g.

Under the above setting, the above functions must satisfy the condition

$$P_i(t) + \sum_{g \supset i} P_{g|i}(t) = 1, \quad i = 1, \ldots, k \tag{1}$$

The lifetime distributions of the individual components (causes of failure) are also of key importance. We will denote the density function and survival function corresponding to the ith component by $f_i(t)$ and $S_i(t)$, respectively ($i = 1, \ldots, k$). A number of publications (for example, [3–5]) consider a *nonparametric* approach, in which no further assumptions are made about the nature of the distributions of individual causes of failure. However, in many problems originating in the area of engineering and reliability, there is a basis for assuming a more refined structure. In this case we could represent the density and survival functions in a parameterized form, $f_i(t; \boldsymbol{\beta}_i)$ and $S_i(t; \boldsymbol{\beta}_i)$, where $\boldsymbol{\beta}_i$ is the vector of parameters corresponding to cause i, and then use a *parametric* approach [6–12].

The primary quantities of interest are the survival curves corresponding to the individual causes, $S_i(t)$, the identification probabilities and the masking probabilities. Some derived parameters are also of interest in many applications. For example, one could be interested in estimating a survival curve of a system of different architecture constructed (wholly or partially) using the k components our system is based on. Problems of this type occur frequently in the process of designing subsequent versions of the system. Another derived quantity is the *diagnostic probability* $\pi_{i|g}(t)$ defined as a probability that a failure at time t that is associated with the masking group g is actually caused by the ith component:

$$\pi_{i|g}(t) = \frac{f_i(t) P_{g|i}(t)}{\sum_{j \subset g} f_j(t) P_{g|j}(t)} \tag{2}$$

Modeling and Analysis of Masked Data

The most popular and productive methods of analysis have been based on the likelihood. The complexity of the likelihood function depends greatly on further assumptions about the nature of failures, and we will only show the form that corresponds to *parametric analysis* under the assumptions that the competing risks are independent. Under this assumption the survival probability $S(t)$ of the system is given by

$$S(t;\boldsymbol{\beta}_1,\ldots,\boldsymbol{\beta}_k)=\prod_{i=1}^{k}S_i(t;\boldsymbol{\beta}_i) \tag{3}$$

and the likelihood function can be written in the form:

$$\begin{aligned}L=&\prod_{j=1}^{n_0}S\big(t_j^{(c)};\boldsymbol{\beta}_1,\ldots,\boldsymbol{\beta}_k\big)\\&\times\prod_{i=1}^{k}\prod_{j=1}^{n_i}\Big\{f_i\big(t_j^{(i)};\boldsymbol{\beta}_i\big)P_i\big(t_j^{(i)}\big)\prod_{m\neq i}S_m\big(t_j^{(i)};\boldsymbol{\beta}_l\big)\Big\}\\&\times\prod_{g}\prod_{j=1}^{n_g}\Big[\sum_{r\subset g}f_r\big(t_j^{(g)};\boldsymbol{\beta}_r\big)P_{g|r}\big(t_j^{(g)}\big)\prod_{m\neq r}S_m\big(t_j^{(g)};\boldsymbol{\beta}_l\big)\Big]\end{aligned} \tag{4}$$

In situations where the identification and masking probabilities are completely specified, the likelihood (equation 4) can be easily analyzed using standard techniques. In other cases, however, one needs to deal with the parameters that determine these probabilities – and the likelihood analysis becomes rather difficult, because of both numeric issues and issues related to the identifiability and strength of the resulting inference. Therefore, a number of simplifying assumptions have been made in the literature. A number of papers discuss the situation where the identification and masking probabilities are not dependent on time, i.e.,

$$P_i(t)\equiv P_i;\quad P_{g|i}(t)\equiv P_{g|i} \tag{5}$$

(for example, see [12, 13]; however, some more general situations have also been discussed [5, 14]).

Estimability issues

Though the assumptions (equation 5) greatly simplify the likelihood analysis, it could still remain quite involved, even in the parametric case with relatively simple component distributions. One of the reasons for this is that in many applications one needs to take into account *covariates* corresponding to items on test that could reflect, for example, vintage characteristics or environmental conditions – and these usually inject additional parameters into the model [15, 16]. Furthermore, in the case of *proportional hazards*, i.e. when

$$S_i(t)=[S(t)]^{\phi_i},\ i=1,\ldots,k \tag{6}$$

with $\phi_i>0$ and $\sum_{i=1}^{k}\phi_i=1$, the problem is not estimable, even in the case where the survival function of the system $S(t)$ is completely identified (see Flehinger *et al.* [2]). For example, in the case $k=2$ the likelihood function is proportional to $(\phi_1P_1)^{n_1}\times(\phi_2P_2)^{n_2}\times[1-(\phi_1P_1+\phi_2P_2)]^{n_{(1,2)}}$; therefore, only the products ϕ_1P_1 and $\phi_2P_2=(1-\phi_1)P_2$ are estimable. Given that in most practical situations the assumption of proportional hazards (possibly involving only a subgroup of competing risks, which would be sufficient to undermine estimability) cannot be *a priori* excluded, the issue of estimability plays a prominent role in the analysis of masked data. In light of this issue, a number of further simplifications have been considered. One of the most popular ones is based on the *symmetry assumption* [17, 18]:

$$P_{g|i}=P_{g|j}\ \text{for every } i,j\subset g \tag{7}$$

Plausibility of the symmetry assumption typically depends on the reasons for masking. It can be justified, for example, in situations where any failed component in the group causes so much destruction that identification of the culprit is not possible without considerable expense (such situations sometimes occur in crashes of hard drives). This assumption could also be relevant in cases when masking occurs at random; in this case the masking phenomenon is quite similar to random censoring. However, in cases where masking patterns are determined by the FA strategy [19] one could find situations where the symmetry assumption could be *a priori* declared irrelevant. For example, when

1. the diagnostic procedure in the wake of a failure proceeds to examine all potential culprits in a given order (represented by a list),
2. the search for the cause of failure is time-censored, and
3. the inspection time of a given component is considerably shorter in the case when this component is faulty than in the case when it is healthy,

one could expect that for any given masking group (whose members are ordered in accordance with the list), the masking probability for the first component is smaller than that for the other components.

When the symmetry assumption is not justified, the estimability issues frequently necessitate use of more elaborate approaches to the problem of inference. In some cases, one could obtain fairly good estimates of the identification and masking probabilities based on external information. For example, in the situation described above where masking is determined by the postfailure diagnostics policy, one could sometimes obtain reasonable estimates based on the characteristics of this policy and information about the duration (or costs) of inspection under various scenarios. The evaluation can be typically carried out *via* a simulation study of various types of failure.

Two-Stage Models

In many cases one could take advantage of the fact that some of the masked cases can actually be resolved. This leads to *two-stage* analysis [2]. The data for the two-stage model is similar to one given in the section titled "Model Specification and Parameters", with addition of the information on how the selected cases were resolved. Use of two-stage models eliminates the estimability issue: one can carry out a likelihood analysis of the expanded dataset in a straightforward way, provided the policy that determines which masked cases are subject to resolution does not depend on the model parameters that we seek to estimate. Research results show that one can obtain a strong boost in statistical power by resolving only a small fraction of cases. Furthermore, one can control the strength of inference *via* a policy that determines which cases are to be resolved.

The two-stage data can be found in a large number of cases involving system manufacturing. The fact is that engineering departments normally do not allow situations generating exclusively masked data to persist for too long: inability to establish in a timely matter that a given component is causing reliability problems could mean exposure to significant financial losses related to warranty and maintenance costs, as large volumes of defective components continue to propagate through the supply chain and the field. However, the databases related to field failure/warranty records and to FA labs are frequently controlled by different organizations, which typically results in problems with compatibility of formats and policies. However, we found that in many cases creation of unified databases suitable for two-stage statistical analysis of masked data could lead to substantial reduction in quality and maintenance costs.

It is also important to note that in some cases, the data for two-stage analysis is potentially available at almost no additional investment and can be obtained *via* better defect documentation procedures. For example, some of the failures that are recorded as nonmasked in the data described in the section titled "Introduction" sometimes represent cases that were originally masked; however, the information about this masking was simply not recorded.

Dependent Causes of Failure

In some cases the basic assumption that the competing hazards are independent cannot be justified. In general, this assumption cannot be overturned on the basis of the observed failure data alone, since the marginal survival probabilities $S_i(t)$ are no longer identifiable when the causes of failure are dependent, even in the case where there is no masking. To see why this is the case, imagine that we have two types of a two-component system, A and B. For a system of type A the components are independent, with survival curves $S_1(t) = S_2(t) = \sqrt{S(t)}$. For a system of type B the survival curves of both components are equal to $S(t)$, but the lifetimes are dependent in such a way that they are both almost identical, and the probability that component 1 has a shorter life (always by a minuscule amount) than component 2 is 0.5. Under such conditions, distribution of data from lifetime tests related to system A is basically indistinguishable from the similar data for system B.

The assumption of independence can, however, be overturned on the basis of the *a priori* engineering knowledge. For example, it may be known that the lifetime of a system is affected by a "frailty" parameter that affects several components, inducing dependence of the corresponding lifetimes. In the case of depending failures, analysis of masked data can be performed on the basis of *cause-specific hazards* defined by:

$$\lambda_i(t) = \lim_{\Delta t \to \infty} \left[\frac{\text{Prob}(t \le T \le t + \Delta t, C = i | T \ge t)}{\Delta t} \right]$$
$$i = 1, \ldots, k \tag{8}$$

where (T, C) denotes the system lifetime and cause of failure. In this case, instead of representing $S(t)$ in the form of a product (equation 3), we represent $1 - S(t)$ as a sum of *cumulative incidence probabilities* $I_i(t)$ defined *via*

$$I_i(t) = \text{Prob}(T \le t, C = i) = \int_0^t \lambda_i(u) S(u)\,\text{d}u \quad i = 1, \ldots, k \qquad (9)$$

Now the likelihood function in the case of dependent causes of failure can be written in the form that is very similar to equation (4), but is based on the cause-specific density functions $g_i(t)$ defined by

$$g_i(t) = \frac{\text{d}}{\text{d}t} I_i(t) = \lambda_i(t) S(t) \qquad (10)$$

see [2]. This approach enables one to estimate the parameters related to the cause-specific hazards, identification, masking probabilities, and other quantities of interest. Craiu and Reiser [20] developed an approach based on this technique for estimation related to a model with depending competing risks and piecewise-constant, cause-specific hazards.

Bayesian Approach and Computational Issues

Another method for incorporating external information about the model parameters (or even internal information, presented in the form of so-called noninformative priors) is to use Bayesian techniques. Literature on this approach to the problem of masking is quite extensive [11, 21–24]. In this context, the authors typically use the **Markov chain** (MCMC) techniques for estimation and inference.

In the frequentist setting, the inference can be carried out by maximizing the likelihood (for example, of type shown in equation (4), or its two-stage versions) *via* conventional nonlinear optimization techniques. Confidence statements and test statistics can then be obtained by using the *profile-likelihood* method. However, since the problem is naturally fitting the framework of estimation with missing data, the expectation-maximization (EM) algorithm is also most useful, e.g. see [5, 25, 26].

References

[1] Reiser, B., Flehinger, B.J. & Conn, A.R. (1996). Estimating component defect probability from masked system success/failure data, *IEEE Transactions in Reliability* **45**, 238–243.

[2] Flehinger, B.J., Reiser, B. & Yashchin, E. (2001). Statistical analysis of masked data, in *Handbook of Statistics, Advances in Reliability*, N. Balakrishnan & C.R. Rao, eds, Elsevier, Amsterdam, Vol. 20, pp. 499–522.

[3] Dinse, G.E. (1986). Nonparametric prevalence and mortality estimators for animal experiments with incomplete cause-of-death data, *Journal of American Statistical Association* **81**, 328–336.

[4] Flehinger, B.J., Reiser, B. & Yashchin, E. (1998). Survival with competing risks and masked causes of failures, *Biometrika* **85**, 151–164.

[5] Craiu, R.V. & Duchesne, T. (2004). Inference based on EM algorithm for the competing risks model with masked causes of failure, *Biometrika* **91**, 543–558.

[6] Miyakawa, M. (1984). Analyses of incomplete data in competing risk model, *IEEE Transactions in Reliability* **33**, 293–296.

[7] Usher, J.S. & Hodgson, T.J. (1988). Maximum likelihood estimation of component reliability using masked system life test data, *IEEE Transactions in Reliability* **37**, 550–555.

[8] Usher, J.S. & Guess, F.M. (1989). An iterative approach for estimating component reliability from masked system life data, *Quality and Reliability Engineering International* **5**, 257–261.

[9] Guess, F.M., Usher, J.S. & Hodgson, T.J. (1991). Estimating system and component reliabilities under partial information on the cause of failure, *Journal of Statistical Planning and Inference* **29**, 75–85.

[10] Usher, J.S. (1996). Weibull component reliability – prediction in the presence of masked data, *IEEE Transactions in Reliability* **45**, 229–232.

[11] Basu, S., Basu, A. & Mukhopadhyay, C. (1999). Bayesian analysis for masked system failure data using non-identical Weibull models, *Journal of Statistical Planning and Inference* **78**, 255–275.

[12] Flehinger, B.J., Reiser, B. & Yashchin, E. (2002). Parametric modeling for survival with competing risks and masked failure causes, *Lifetime Data Analysis* **8**, 177–203.

[13] Lo, S.-H. (1991). Estimating a survival function with incomplete cause-of-death data, *Journal of Multivariate Analysis* **39**, 217–235.

[14] Lin, D.K.J. & Guess, F.M. (1994). System life data analysis with dependent partial knowledge on the exact cause of system failure, *Microelectronics Reliability* **34**, 535–544.

[15] Andersen, J.W., Goetghebeur, E. & Ryan, L.M. (1996). Missing cause of death information in the analysis of survival data, *Statistics in Medicine* **15**, 2191–2201.

[16] Nagai, Y. (2004). Maximum likelihood analysis of masked data in competing risks models with an environmental stress, *IEICE Transactions of Fundamentals of Electronic Communications* **E87A**, 3389–3396.

[17] Guttman, I., Lin, D.K., Reiser, B. & Usher, J.S. (1995). Dependent masking and system life data analysis: Bayesian inference for two-component systems, *Lifetime Data Analysis* **1**, 87–100.

[18] Sen, A., Basu, S. & Banerjee, M. (2001). Analysis of masked failure data under competing risks, in *Handbook of Statistics, Advances in Reliability*, N. Balakrishnan & C.R. Rao, eds, Elsevier, Amsterdam, Vol. 20, pp. 523–522.

[19] Gupta, S.S. & Gastaldi, T. (1996). Life testing for multi-component systems with incomplete information on the cause of failure: a study on some inspection strategies, *Computational Statistics and Data Analysis* **22**, 373–393.

[20] Craiu, R.V. & Reiser, B. (2006). Inference for the dependent competing risks model with masked causes of failure, *Lifetime Data Analysis* **12**, 21–33.

[21] Reiser, B., Guttman, I., Lin, D.K.J., Usher, J.S. & Guess, F.M. (1995). Bayesian inference for masked system life time data, *Applied Statistics* **44**, 79–90.

[22] Lin, D.K.J., Usher, J.S. & Guess, F.M. (1996). Bayes estimation of component reliability from masked system life data, *IEEE Transactions in Reliability* **45**, 233–237.

[23] Mukhopadhyay, C. & Basu, A.P. (1997). Bayesian analysis of incomplete time and cause of failure data, *Journal of Statistical Planning and Inference* **59**, 79–100.

[24] Basu, S., Sen, A. & Banerjee, M. (2003). Bayesian analysis of competing risks with partially masked cause-of-failure, *Journal of the Royal Statistical Society. Series C: Applied Statistics* **52**, 77–93.

[25] Kodell, R.K. & Chen, J.J. (1987). Handling cause of death in equivocal cases using the EM algorithm, *Communications in Statistics A: Theory and Methods* **16**, 2565–2603.

[26] Zhao, M. & Xie, M. (1994). EM algorithms for estimating software reliability based on masked data, *Microelectronics Reliability* **34**, 1027–1038.

Related Articles

Masked Failure Data: Competing Risks; **Masked Failure Data: Bayesian Modeling**; .

EMMANUEL YASHCHIN

Masked Failure Data: Bayesian Modeling

Introduction

Suppose we have a system with J components and a defect in any component causes the system to fail. We test many of these identical systems to assess each component's reliability. In the ideal situation, the exact cause of the system's failure can be identified. Then a detailed **Failure Mode Effect Analysis** can be carried out. However, one often encounters the situation where the exact component causing the system to fail is unknown, nevertheless, it can be narrowed to a subset of components. So we have masked data that is, the data consist of the lifetimes of n systems randomly chosen for life testing and their associated labels identifying the probable components that may cause each system to fail. Masking is usually due to limited resources for diagnosing the causes of failures. More importantly, it is due to the modular nature of present systems. For example, in computer repair shops, the cause of failure is often isolated to a module of several components. The system is brought back into operation by replacing the whole module. The objective in analyzing the masked system **failure data** is to make inferences for the reliability of the components. The **system reliability** is usually derived from the **component reliability** and the system's configuration. Another objective is to make inference for the diagnostic probability that is, the probability of a specific component causing the system to fail given the set of probable causes and the system's failure time.

Recently, several authors studied this problem for the series system using exponential distributions for the component lifetimes. Most authors make an equiprobable assumption for the masking probabilities, where the conditional masking probabilities given each cause of failure in the masked set take the same value. This assumption (called the symmetry assumption by some authors) allows one to consider only the reduced likelihood. The work includes Miyakawa [1], Usher and Hodgson [2], Guess *et al.* [3], Lin *et al.* [4, 5], Doganaksoy [6], Gastaldi [7], Reiser *et al.* [8], Usher [9], Mukhopadhyay and Basu [10], Basu *et al.* [11] and a review article by Sen *et al.* [12].

To relax the equiprobable assumption, Lin and Guess [13], and Guttman *et al.* [14] consider a proportional probability model for a dependent masking probability for a two-component series system. They assume that the masking probability conditional on the first component's failure and the system's failure time is proportional to that of the second component. Although both masking probabilities are functions of the system's failure time, they assume the proportionality is independent of the system's failure time. Craiu and Duchesne [15] propose EM algorithms for models that allow the masking probabilities to depend on time. Craiu and Reiser [16] propose EM based approach for inference for dependent **competing risks** models.

In a similar spirit to relax the equiprobable assumption, Kuo and Yang [17] explore other probability assumptions for the masking probabilities. They explore two different stochastic models for the masking probabilities. One model allows the masking probabilities to have different values depending on the true cause of failure, but they do not depend on the system's failure time. The other model assumes the masking probabilities are further dependent on some monotone functions of the system's failure time. In addition to exponential distributions for the components, they also consider Weibull distributions. To ensure identifiability, they assume the components have independent failure distributions. They show that better predictive power can be achieved by modeling the masking probabilities without the equiprobable assumption. They demonstrate this with a simulated data set and with a real data set based on Dinse [18].

Note that the system's failure occurs at the earliest onset of any component's failure. So we are using the competing risk methodology in our analysis. Furthermore, we are assuming the components are acting independently for identifiability issue. **Markov chain Monte Carlo** (MCMC) methods are typically used for the Bayesian inference for the problems posed here. In particular, an MCMC algorithm augmented with latent variables that simulate the true causes of failures will be used for Bayesian computation of the component reliability. In addition, an **EM algorithm** [19] for this problem will also be illustrated.

Model selection by a predictive approach as in Geisser and Eddy [20] can be developed naturally from the Bayesian framework. Kuo and Yang [17] provide more details on the model determination and show that improved model can be obtained by modeling the masking probabilities.

The sections titled "Full Likelihood" and "Reduced Likelihood" describe the basic models for the equiprobable assumption. The section titled "Relaxing the Equiprobable Assumption" discusses two variations of the basic models that model the masking probabilities. The MCMC algorithms with auxiliary latent variable augmentation are developed for these models. Some numerical results for a simulated data set are given in the section titled "Simulation Study" to illustrate the techniques. A brief discussion on the model determination is given in the section titled "Conclusion".

Full Likelihood

In the full likelihood formulation, we not only consider the system lifetimes but also the labels that identify the possible failed components. Let t_{ij} denote the random lifetime of the jth component in the ith system. Assume we observe a sample of n **series systems** each having J components; therefore we observe $t_i = \min(t_{i1}, \ldots, t_{iJ})$, for $i = 1, \ldots, n$. In addition, for each system, we observe a set of labels of the components that include the failed component. Let S_i denote the set of components (labels) that possibly cause the system i to fail, and let s_i denote the realized set for S_i. The observed data denoted by $\mathbf{d}$ consist of $(t_1, s_1), \ldots, (t_n, s_n)$. If each set s_i is a singleton, then this reduces to the usual competing risk study where the exact cause of failure is identified. The set of labels is often masked because the true cause of failure is hiding with other components. Let K_i denote the index of the component actually causing the ith system to fail. It is a singleton because of the series system and continuous lifetimes. Let f_j denote the density for the jth component's lifetime and $\overline{F}_j$ denote its survival function. We assume the component lifetime distributions are independent. For the ith system, the likelihood contribution coming from the datum (t_i, s_i) consists of $p(t_i, s_i)$. This can be written as $\sum_{j=1}^{J} p(t_i, K_i = j)p(s_i|t_i, K_i = j)$. Note the masking probability $p(s_i|t_i, K_i = j) = 0$, if $j \notin s_i$. Moreover, we have: $p(t_i, K_i = j) = f_j(t_i)\prod_{l=1,l\neq j}^{J} \overline{F}_l(t_i)$. Therefore the full likelihood of this data set is

$$L_{\mathrm{F}} = \prod_{i=1}^{n}\left[\sum_{j\in s_i}\left\{\left(f_j(t_i)\prod_{l=1,l\neq j}^{J}\overline{F}_l(t_i)\right)\right.\right.$$
$$\left.\left.\times\, P(S_i = s_i|T_i = t_i, K_i = j)\right\}\right] \quad (1)$$

It is often of interest to compute the diagnostic probability that is, the probability that the jth component actually causing the system to fail given the possible malfunctioning subset and the system's failure time. For the ith system and $j \in s_i$, the diagnostic probability for the jth cause is

$$p(k_i = j|s_i, t_i) = \frac{p(k_i = j, s_i, t_i)}{p(s_i, t_i)} = \frac{P(S_i = s_i|T_i = t_i, K_i = j) f_j(t_i) \times \prod_{l=1, l\neq j}^{J} \overline{F}_l(t_i)}{\sum_{j\in s_i} P(S_i = s_i|T_i = t_i, K_i = j) \times f_j(t_i) \prod_{l=1, l\neq j}^{J} \overline{F}_l(t_i)} \quad (2)$$

Reduced Likelihood

Several authors assume an equiprobable assumption where the masking probabilities take on the same value irrelevant of the particular cause of failure. That is, they make the following assumption for a fixed $j \in s_i$:

$$P(S_i = s_i|T_i = t_i, K_i = j') = P(S_i = s_i|T_i = t_i, K_i = j) \text{ for all } j' \in s_i \quad (3)$$

Then they proceed with their analysis with a reduced likelihood that does not depend on the masking probabilities:

$$L_R = \prod_{i=1}^{n}\left[\sum_{j\in s_i}\left\{f_j(t_i) \prod_{l=1, l\neq j}^{J} \overline{F}_l(t_i)\right\}\right] \quad (4)$$

Reiser *et al.* [8] provide a Bayesian analysis with this reduced likelihood assuming exponential distributions $f_j(t_i) = \lambda_j e^{-\lambda_j t_i} I\{t_i > 0\}$, and the noninfromative Jeffrey's prior:

$$\pi(\lambda_1, \ldots, \lambda_J) \propto \prod_{j=1}^{J} \lambda_j^{-1} \quad (5)$$

For the system with $J = 2$, let n_1, n_2, and n_{12} denote the numbers of systems in the sample with true causes identified to be the first component, the second component, or masked (either component). Then we can summarize the observed data with $T = \sum_{i=1}^{n} t_i$ (**total time on test**), and n_1, n_2, and n_{12}. Then the posterior density of λ_1, λ_2 given the data is

$$\pi(\lambda_1, \lambda_2|\mathbf{d}) \propto \exp\{-T(\lambda_1 + \lambda_2)\}\lambda_1^{n_1-1}\lambda_2^{n_2-1} \times (\lambda_1 + \lambda_2)^{n_{12}} \quad (6)$$

So they show the Bayes estimators (the posterior means) are given by

$$\tilde{\lambda}_1 = \frac{n_1}{T}\frac{n_1 + n_2 + n_{12}}{n_1 + n_2} \quad (7)$$

$$\tilde{\lambda}_2 = \frac{n_2}{T}\frac{n_1 + n_2 + n_{12}}{n_1 + n_2} \quad (8)$$

They are exactly the maximum likelihood estimators given by Usher and Hodgson [2]. An MCMC algorithm for computing the Bayes estimates can be developed easily that will be given in the subsection titled "Gibbs Sampling for the Reduced Model" as a special case of the algorithm. The MCMC algorithm allows us to compute **quantiles**, other functionals of the posterior distribution, and the highest posterior density credible sets.

Lin *et al.* [5] use a step-function prior to λ_j:

$$\pi(\lambda_j) = \sum_{k=1}^{\infty} \alpha_{j,k} I\{\lambda_j \in [a_{k-1}, a_k)\} \quad (9)$$

where I denotes the indicator function and $\sum_{k=1}^{\infty} \alpha_{j,k}(a_k - a_{k-1}) = 1$ satisfying the probability constraint. They provide a numerical example for the $J = 2$ system. An MCMC algorithm with data augmentation similar to Yang and Kuo [21] can be developed easily for this problem.

Relaxing the Equiprobable Assumption

Most of the Bayesian analysis is based on the equiprobable assumption for the masking probability. So reduced likelihood can be used. There are some treatments on relaxing the equiprobable assumption. For example, Kuo and Yang [17] model the masking probabilities from the simple classification point of view. One of the simple assumptions is to make the probabilities independent of the time and having different values conditional on their components. Let us look at the two-components system. Assume $P(S_i = \{1\}|t_i, K_i = 1) = \theta_1$, and $P(S_i = \{2\}|t_i, K_i = 2) = \theta_2$. Consequently, $P(S_i =$

$\{1,2\}|t_i, K_i = 1) = 1 - \theta_1$, and $P(S_i = \{1,2\}|t_i, K_i = 2) = 1 - \theta_2$. Then the full likelihood in this case is

$$L_F = \prod_{i|s_i=\{1\}} \theta_1 f_1(t_i)\overline{F}_2(t_i) \prod_{i|s_i=\{2\}} \theta_2 f_2(t_i)\overline{F}_1(t_i) \times \prod_{i|s_i=\{1,2\}} \{(1-\theta_1) f_1(t_i)\overline{F}_2(t_i) + (1-\theta_2) f_2(t_i)\overline{F}_1(t_i)\} \quad (10)$$

Let us note the cardinalities of the indices in each product in equation (10) are exactly n_1, n_2, and n_{12}, defined earlier in the section titled "Reduced Likelihood".

The same idea can be generalized to an arbitrary number of components. For example for $J = 3$, we define the following masking probabilities: $P(S_i = \{j\}|t_i, K_i = j) = \theta_j$ for all j, $P(S_i = \{1,2\}|t_i, K_i = 1) = \theta_{12|1}$, and $P(S_i = \{1,3\}|t_i, K_i = 1) = \theta_{13|1}$. Consequently, $P(S_i = \{1,2,3\}|t_i, K_i = 1) = 1 - \theta_1 - \theta_{12|1} - \theta_{13|1}$. Similarly for causes two and three. These statements can be simplified if we know more about the configuration of the system. For example, components one and two are mounted on the same card separately from component three. Then we can assume $\theta_{13|1} = \theta_{23|2} = \theta_{13|3} = \theta_{23|3} = 0$. There will be fewer parameters to be estimated. Aboul-Séoud and Usher [22] describe an example of the modular system structure.

If the replaced modules are further sent to the factories for identifying the exact component causing the failure, we can produce fairly accurate estimates of the θ's from the follow-up study. Therefore, we can estimate f_j by assuming the θ's are known. Alternatively, we can combine the original data $(t_i, s_i), i = 1, \ldots, n$, with the follow-up "autopsy" data to obtain better inferences as done in Flehinger *et al.* [23, 24], and Reiser *et al.* [25].

In addition to exponential distributions for the component lifetimes, we will also consider Weibull distributions for F_j. Gibbs sampling will be developed for Bayesian inference for the θ's and the parameters in f_j. Although we concentrate on the two-components system here for illustration, the Gibbs sampler can be extended straightforwardly to any number of components. The method is akin to the Gibbs sampling used in many aggregated models (cf. [21]).

Gibbs Sampling for the Full Model

Let us define the hazard function $h_j(t) = f_j(t)/\overline{F_j}(t)$ for all j. Then equation (10) can be rewritten as

$$L_F = \left(\prod_{i|s_i=\{1\}} \theta_1 h_1(t_i) \prod_{i|s_i=\{2\}} \theta_2 h_2(t_i) \times \prod_{i|s_i=\{1,2\}} \{(1-\theta_1)h_1(t_i) + (1-\theta_2)h_2(t_i)\} \right) \times \left(\prod_{i=1}^{n} \overline{F}_1(t_i)\overline{F}_2(t_i) \right) \quad (11)$$

The summation in the third product of this likelihood prevents us from updating the parameters in F easily. Therefore, we introduce latent variables z_i for the systems where the cause of failure is masked. Let z_i be a Bernoulli variable with the success ($z_i = 1$) probability to be $p_i = (1-\theta_1)h_1(t_i)/\{(1-\theta_1)h_1(t_i) + (1-\theta_2)h_2(t_i)\}$ for i with $s_i = \{1,2\}$. It is interesting to note that this probability of success p_i is exactly the diagnostic probability $p(k_i = 1|s_i, t_i)$ in equation (2) for the true cause due to the first component conditional on $s_i = \{1,2\}$ and t_i being observed. Let $z_+ = \sum_{i|s_i=\{1,2\}} z_i$. Then the conditional density of $\boldsymbol{\theta} = (\theta_1, \theta_2)$ and the unknown parameters $\boldsymbol{\lambda}$ in F's given $\mathbf{d}$ and the latent variables $\mathbf{z}$ is

$$f(\boldsymbol{\theta}, \boldsymbol{\lambda}|\mathbf{z}, \mathbf{d}) \propto \theta_1^{n_1}(1-\theta_1)^{z_+}\theta_2^{n_2}(1-\theta_2)^{n_{12}-z_+} \times \left(\prod_{i|s_i=\{1\}} h_1(t_i) \prod_{i|s_i=\{2\}} h_2(t_i) \times \prod_{i|s_i=\{1,2\}} h_1(t_i)^{z_i} h_2(t_i)^{1-z_i} \right) \times \left(\prod_{i=1}^{n} \overline{F}_1(t_i)\overline{F}_2(t_i) \right) \pi(\boldsymbol{\theta}, \boldsymbol{\lambda}) \quad (12)$$

where π denotes the prior density of $\boldsymbol{\theta}$ and $\boldsymbol{\lambda}$. Therefore, updating $\boldsymbol{\theta}$ and $\boldsymbol{\lambda}$ can be done easily and independently for independent prior on $\boldsymbol{\theta}$ and $\boldsymbol{\lambda}$. We are assuming that readers are familiar with the basics of MCMC algorithms as described in Gelfand and Smith [26] and Tanner and Wong [27]. In the following, we only describe the transitional kernel in the MCMC algorithm. Now we write the Gibbs steps for two special cases, one for each component to have an exponential distribution and the other to have a Weibull distribution.

(1) The exponential component $\overline{F}_j(t) = \exp(-\lambda_j t)$:

We consider a prior with the following independent components for $j = 1$ and 2: $\theta_j \sim \beta e(\alpha_j, \beta_j)$ and $\lambda_j \sim G(a_j, b_j)$, where $G(a_j, b_j)$ denotes a gamma density with mean a_j/b_j. The conjugate priors for θ_j and λ_j in the augmented likelihood are chosen so that we can easily specify the conditional densities in the Gibbs algorithm. Gamma and beta prior densities are also versatile enough to incorporate various shapes for the distributions of the unknown parameters.

Step a (data augmentation step): given $\boldsymbol{\theta}$ and $\boldsymbol{\lambda}$, we generate $z_i \sim B(1, (1-\theta_1)\lambda_1/\{(1-\theta_1)\lambda_1 + (1-\theta_2)\lambda_2\})$ a Bernoulli distribution, independently for each i with $s_i = \{1, 2\}$. Then we compute $z_+ = \sum_{i|s_i=\{1,2\}} z_i$.

Step b (parameter updating step): given $\mathbf{z}$ and $\mathbf{d}$, we update the parameters by:

$$\theta_1|\mathbf{z}, \mathbf{d}, \theta_2, \boldsymbol{\lambda} \sim Beta(\alpha_1 + n_1, \ \beta_1 + z_+) \tag{13}$$

$$\theta_2|\mathbf{z}, \mathbf{d}, \theta_1, \boldsymbol{\lambda} \sim Beta(\alpha_2 + n_2, \ \beta_2 + n_{12} - z_+) \tag{14}$$

$$\lambda_1|\mathbf{z}, \mathbf{d}, \lambda_2, \boldsymbol{\theta} \sim G\left(a_1 + n_1 + z_+, \ b_1 + \sum_{i=1}^{n} t_i\right) \tag{15}$$

$$\lambda_2|\mathbf{z}, \mathbf{d}, \lambda_1, \ \boldsymbol{\theta} \sim G \times \left(a_2 + n_2 + n_{12} - z_+, \ b_2 + \sum_{i=1}^{n} t_i\right) \tag{16}$$

Note that the parameters $\theta_1, \theta_2, \lambda_1$ and λ_2 are conditionally independent given $\mathbf{z}$, and $\mathbf{d}$ from the above expressions. Therefore, updating the four parameters can be done in parallel.

(2) The Weibull component $\overline{F}_j(t) = \exp(-\lambda_j t^{\rho_j})$:

We consider a prior with the following independent components: $\theta_j \sim \beta e(\alpha_j, \beta_j)$, $\lambda_j \sim G(a_j, b_j)$, and ρ_j has density $\pi(\rho_j)$ that can be arbitrary. The same ideas can be generalized to a system with more than two components. The assumption of independent components in the prior is chosen for convenience. We have chosen the gamma distribution for λ_j for conjugacy. We can choose the prior on ρ_j to be uniformly distributed over $(0, m_j)$ as suggested by Sinha [28, p.160]. We can also explore the choice $\pi(\rho_j, \lambda_j) \propto 1/(\rho_j \lambda_j)$ as given by Sinha and Zellner [29]. Let $\boldsymbol{\rho} = (\rho_1, \rho_2)$.

Step a (data augmentation step): given $\boldsymbol{\theta}$, $\boldsymbol{\lambda}$ and $\boldsymbol{\rho}$, we generate $z_i \sim B(1, (1-\theta_1)\lambda_1\rho_1 t_i^{\rho_1 - 1}/\{(1-\theta_1)\lambda_1\rho_1 t_i^{\rho_1-1} + (1-\theta_2)\lambda_2\rho_2 t_i^{\rho_2-1}\})$, independently for i with $s_i = \{1, 2\}$. Then we compute $z_+ = \sum_{i|s_i=\{1,2\}} z_i$.

Step b (parameter updating step): given $\mathbf{z}$ and $\mathbf{d}$, we update the parameters by:

$$\theta_1|\mathbf{z}, \mathbf{d}, \boldsymbol{\lambda}, \boldsymbol{\rho}, \theta_2 \sim Beta(\alpha_1 + n_1, \ \beta_1 + z_+) \tag{17}$$

$$\theta_2|\mathbf{z}, \mathbf{d}, \boldsymbol{\lambda}, \boldsymbol{\rho}, \theta_1 \sim Beta(\alpha_2 + n_2, \beta_2 + n_{12} - z_+) \tag{18}$$

$$\lambda_1|\mathbf{z}, \mathbf{d}, \boldsymbol{\rho}, \boldsymbol{\theta}, \lambda_2 \sim G\left(a_1 + n_1 + z_+, \ b_1 + \sum_{i=1}^{n} t_i^{\rho_1}\right) \tag{19}$$

$$\lambda_2|\mathbf{z}, \mathbf{d}, \boldsymbol{\rho}, \boldsymbol{\theta}, \lambda_1 \sim G\left(a_2 + n_2 + n_{12} - z_+, \ b_2 + \sum_{i=1}^{n} t_i^{\rho_2}\right) \tag{20}$$

and update the parameters ρ_1 and ρ_2 independently given $\mathbf{z}, \mathbf{d}, \boldsymbol{\lambda}$ and $\boldsymbol{\theta}$, each by the Metropolis law (cf. [30], and [31, 32]) with the following density as the target density:

$$\pi(\rho_1|\mathbf{z}, \mathbf{d}, \boldsymbol{\lambda}, \boldsymbol{\theta}) \propto \rho_1^{n_1+z_+} \left(\prod_{i|s_i=\{1\}} t_i^{\rho_1-1}\right) \times \left(\prod_{i|s_i=\{1,2\}} t_i^{(\rho_1-1)z_i}\right) \exp \times \left(-\lambda_1 \sum_{i=1}^{n} t_i^{\rho_1}\right) \pi(\rho_1) \tag{21}$$

$$\pi(\rho_2|\mathbf{z}, \mathbf{d}, \boldsymbol{\lambda}, \boldsymbol{\theta}) \propto \rho_2^{n_2+n_{12}-z_+} \left(\prod_{i|s_i=\{2\}} t_i^{\rho_2-1} \right) \times \left(\prod_{i|s_i=\{1,2\}} t_i^{(\rho_2-1)(1-z_i)} \right) \times \exp(-\lambda_2 \sum_{i=1}^{n} t_i^{\rho_2}) \pi(\rho_2) \quad (22)$$

Note in the above updating, the generation of $\boldsymbol{\theta}$ and $(\boldsymbol{\lambda}, \boldsymbol{\rho})$ can be done independently.

To generalize the above discussion to three components, we can easily extend the same prior for $\boldsymbol{\lambda}$, $\boldsymbol{\rho}$ in the component lifetimes to more components. For the masking probabilities $\boldsymbol{\theta}_1 = (\theta_1, \theta_{12|1}, \theta_{13|1}, \theta_{123|1})$ with $\theta_1 + \theta_{12|1} + \theta_{13|1} + \theta_{123|1} = 1$, a Dirichlet prior with parameters $(\alpha_1, \alpha_2, \alpha_3, \alpha_4)$ can be considered. Similarly, we can define independent Dirichlet priors for the other masking probabilities $\boldsymbol{\theta}_2$ and $\boldsymbol{\theta}_3$, where $j = 2$ or $j = 3$ is the true cause of failure. Then we can employ the same data augmentation idea as before and updating the parameters $\boldsymbol{\theta}_1, \boldsymbol{\theta}_2$, and $\boldsymbol{\theta}_3$ independently, each with an updated Dirichlet distribution.

EM Algorithm for the Full Model

If we are only interested in the posterior mode or the mode of the likelihood with their error estimates, we can apply the EM algorithm [19]. The algorithm may be simpler than the Gibbs sampler in implementation, it also may converge faster. The EM algorithm essentially converts a complicated optimization problem to many simpler iterative steps. The iteration steps are similar to that in the Gibbs algorithm, except the data augmentation step is replaced by an E-step (expectation) where the missing data (latent variables) are replaced by their expectations, and the parameter updating step is replaced by a maximization step. In the following, we make the same prior assumption as in the previous subsection to obtain the posterior mode. The algorithm can be easily modified to obtain the maximum likelihood estimate using the flat prior.

(1) For the exponential component, we make the following changes in the $l + 1st$ iteration:

Step a:

$$z_+^l = \frac{n_{12}(1-\theta_1^l)\lambda_1^l}{(1-\theta_1^l)\lambda_1^l + (1-\theta_2^l)\lambda_2^l} \quad (23)$$

Step b:

$$\theta_1^{l+1} = \frac{\alpha_1 + n_1 - 1}{\alpha_1 + \beta_1 + n_1 + z_+^l - 2} \quad (24)$$

$$\theta_2^{l+1} = \frac{\alpha_2 + n_2 - 1}{\alpha_2 + \beta_2 + n_2 + n_{12} - z_+^l - 2} \quad (25)$$

$$\lambda_1^{l+1} = \frac{a_1 + n_1 + z_+^l - 1}{b_1 + \sum_{i=1}^n t_i} \quad (26)$$

$$\lambda_2^{l+1} = \frac{a_2 + n_2 + n_{12} - z_+^l - 1}{b_2 + \sum_{i=1}^n t_i} \quad (27)$$

(2) For the Weibull component:

Step a: the generation of z_i is replaced by setting $z_+^l = n_{12}(1-\theta_1^l)\lambda_1^l\rho_1^l t_i^{\rho_1^l-1}/\{(1-\theta_1^l)\lambda_1^l \rho_1^l t_i^{\rho_1^l-1} + (1-\theta_2^l)\lambda_2^l\rho_2^l t_i^{\rho_2^l-1}\}$.

Step b: the generation of θ_1, θ_2, λ_1, and λ_2 is replaced by

$$\theta_1^{l+1} = \frac{\alpha_1 + n_1 - 1}{\alpha_1 + \beta_1 + n_1 + z_+^l - 2} \quad (28)$$

$$\theta_2^{l+1} = \frac{\alpha_2 + n_2 - 1}{\alpha_2 + \beta_2 + n_2 + n_{12} - z_+^l - 2} \quad (29)$$

$$\lambda_1^{l+1} = \frac{a_1 + n_1 + z_+^l - 1}{b_1 + \sum_{i=1}^n t_i^{\rho_1}} \quad (30)$$

$$\lambda_2^{l+1} = \frac{a_2 + n_2 + n_{12} - z_+^l - 1}{b_2 + \sum_{i=1}^n t_i^{\rho_2}} \quad (31)$$

and the Metropolis algorithm for generating ρ_1 and ρ_2 is replaced by finding ρ_1 that maximizes equation (21) and finding ρ_2 that maximizes equation (22).

Gibbs Sampling for the Reduced Model

Now we write the Gibbs sampler for the reduced model with the equiprobable assumption. Observe $1 - \theta_1 = 1 - \theta_2 = 1 - \theta$ in this case. We assume the same prior as the one in the full model except θ is

$\beta e(\alpha_1, \beta_1)$ independent of other parameters. So the likelihood in this situation reduces to

$$L_R = \theta^{n_1+n_2}(1-\theta)^{n_{12}} \left(\prod_{i=1}^{n} \overline{F}_1(t_i)\overline{F}_2(t_i)\right) \prod_{i|s_i=\{1\}} h_1(t_i) \times \prod_{i|s_i=\{2\}} h_2(t_i) \prod_{i|s_i=\{1,2\}} \{h_1(t_i) + h_2(t_i)\} \quad (32)$$

So the Gibbs sampling for the exponential model is similar to the full model except z_i is generated by $B(1, \lambda_1/\{\lambda_1 + \lambda_2\})$, and instead of θ_1 and θ_2, we generate a single $\theta \sim \beta e(\alpha_1 + n_1 + n_2, \beta_1 + n_{12})$. The Gibbs sampling for the Weibull model is the same as the full model except $\theta_1 = \theta_2 = \theta$ and we generate $\theta \sim \beta e(\alpha_1 + n_1 + n_2, \beta_1 + n_{12})$ to replace the generation of θ_1 and θ_2.

Note the Gibbs sampler for the posterior distribution in equation (6) can be developed similarly. That is, we generate $z \sim Bin(n_{12}, \lambda_1/\{\lambda_1 + \lambda_2\})$ a binomial distribution, then generate $\lambda_1 \sim G(n_1 + z - 1, T)$, $\lambda_2 \sim G(n_2 + n_{12} - z - 1, T)$ independently.

EM Algorithm for the Reduced Model

The maximization for θ and other parameters can be done separately, where $\hat{\theta} = (\alpha_1 + n_1 + n_2 - 1)/(\alpha_1 + \beta_1 + n_1 + n_2 + n_{12} - 2)$ with no iteration. For other parameters we follow the same procedure as the EM algorithm for the full model in the subsection titled "EM Algorithm for the Full Model" with $\theta_1 = \theta_2$.

Simulation Study

We consider a series system with two exponential components, where $f_j(t) = \lambda_j \exp(-\lambda_j t)$, $j = 1, 2$, and $\lambda_1 = 0.01$ and $\lambda_2 = 0.005$. For each system, we simulate a pair of failure times from f_1 and f_2 respectively. Then, we take the minimum of the two failure times as the system's life time and note its label, the cause of the system's failure. If the cause is from the first (second) component, we randomly mask 10% (20%) of the observations. That is, the unmasked probabilities are $\theta_1 = 0.9$ and $\theta_2 = 0.8$. The data in Table 1 represent the failure time and cause of failure for a random sample of $n = 50$ systems.

Table 1 Simulated data

Set of labels	System's failure time
$s_i = \{1\}$	3.4, 10.9, 71.5, 56.9, 41.6, 41.6, 38.6, 4.6, 8.5, 27.8, 2.9, 13.3, 37.4, 57.5, 168.1, 123.9, 136.3, 6.5, 21.4, 135.4, 93.1, 14.3, 10.8, 26.1, 95.3, 68.7, 22.3, 13.3, 73.4, 35.2, 28.0, 21.1
$s_i = \{2\}$	2.5, 212.2, 147.3, 63.6, 45.6, 4.2, 123.1, 10.5, 2.6, 27.4
$s_i = \{1, 2\}$	68.6, 2.3, 5.4, 8.8, 63.7, 325.2, 216.4, 132.0

We apply our analysis in the section titled "Relaxing the Equiprobable Assumption" to this simulated data set. We fit the data with the full model in equation (10) and the reduced model in equation (32) with exponentially distributed component lifetimes. Table 2 summarizes the priors, the posterior means, the EM estimates, and the posterior estimates of the diagnostic probability of the cause one for each of the full and reduced models. We choose uniform priors for the unmasked probabilities. We monitor the convergence of the Gibbs sampler using the Gelman and Rubin [33] method that uses the analysis of variance technique to determine whether or not further iterations are needed. We found 3000

Table 2 Priors, posterior means, and Bayesian diagnostic probability estimates

Model	Full exponential	Reduced exponential
Prior	$\theta_i \sim$ *Beta* $(1, 1)$ for $i = 1, 2$ $\lambda_i \sim G(1, 0.001)$ for $i = 1, 2$	$\theta \sim$ *Beta* $(1, 1)$ $\lambda_i \sim G(1, 0.001)$ for $i = 1, 2$
Posterior means Using Gibbs	$\hat{\theta}_1 = 0.866$, $\hat{\theta}_2 = 0.718$ $\hat{\lambda}_1 = 0.0125$, $\hat{\lambda}_2 = 0.0050$	$\hat{\theta} = 0.827$ $\hat{\lambda}_1 = 0.0131$, $\hat{\lambda}_2 = 0.0044$
Estimates Using EM	$\hat{\theta}_1 = 0.889$, $\hat{\theta}_2 = 0.714$ $\hat{\lambda}_1 = 0.0121$, $\hat{\lambda}_2 = 0.00471$	$\hat{\theta} = 0.840$ $\hat{\lambda}_1 = 0.0128$, $\hat{\lambda}_2 = 0.00401$
$\hat{p}(k_i = 1\|t_i, s_i = \{1, 2\})$ for each i with $s_i = \{1, 2\}$	0.53	0.75

iterations to be large enough for the priors being considered. All the following numerical results are obtained with 3 000 iterations and 50 replications in the Gibbs sampler. The table shows that the posterior means using the MCMC algorithm and the EM estimates are comparable. Starting at $\theta_1^0 = 0.5, \theta_2^0 = 0.5, \lambda_1^0 = 0.005, \lambda_2^0 = 0.005$, the EM algorithm converges after the 6th iteration when all the iteration error bounds are met, where each error bound for each parameter is set at the E^{-8} level. The diagnostic probability of the first cause conditional on $s_i = \{1, 2\}$ and t_i for such i is: $p(k_i = 1|t_i, s_i = \{1, 2\}) = (1 - \theta_1)\lambda_1/\{(1 - \theta_1)\lambda_1 + (1 - \theta_2)\lambda_2\}$ for the full exponential model and $p(k_i = 1|t_i, s_i = \{1, 2\}) = \lambda_1/(\lambda_1 + \lambda_2)$ for the reduced exponential model. Note these diagnostic probabilities are the same independent of t_i as long as i is such that $s_i = \{1, 2\}$. Table 2 also lists the posterior means for these two probabilities evaluated from the Gibbs sampler.

Conclusion

So far, we have presented several models for the masked system life times: full model and reduced model, each with exponential and Weibull component lifetimes. Which model is most desirable? We will use a predictive approach based on cross-validated data to address the model selection issue here. We can define the conditional predictive ordinate (CPO) for the ith data point for a model to be the predictive density evaluated at the ith data point given the remaining data. Then we define the pseudo-marginal predictive likelihood [20] to be the product of the CPOs over i. The pseudo-Bayes factor criterion is to select the model with the largest pseudo-marginal predictive likelihood. The computation of the CPOs follows straightforwardly from the Gibbs samplers. Details of this model determination are given in Kuo and Yang [17].

References

[1] Miyakawa, M. (1984). Analysis of incomplete data in a competing risks model, *IEEE Transaction on Reliability* **33**, 293–296.

[2] Usher, J. & Hodgson, T. (1988). Maximum likelihood analysis of component reliability using masked system life-test data, *IEEE Transaction on Reliability* **37**, 550–555.

[3] Guess, F.M., Usher, J.S. & Hodgson, T.J. (1991). Estimating system and component reliabilities under partial information on cause of failure, *Journal of Statistical Planning and Inference* **29**, 75–85.

[4] Lin, D.K.J., Usher, J.S. & Guess, F.M. (1993). Exact maximum likelihood estimation using masked system data, *IEEE Transaction on Reliability* **42**, 631–635.

[5] Lin, D.K.J., Usher, J.S. & Guess, F.M. (1996). Bayes estimation of component-reliability from masked system-life data, *IEEE Transaction on Reliability* **45**, 233–237.

[6] Doganaksoy, N. (1991). Interval estimation from censored & masked system-failure data, *IEEE Transsaction on Reliability* **40**, 280–286.

[7] Gastaldi, T. (1994). Improved maximum likelihood estimation for component reliabilities with Miyakawa-Usher-Hodgson-Guess' estimators under censored search for the cause of failure, *Statistics & Probability Letters* **19**, 5–18.

[8] Reiser, B., Guttman, I., Lin, D.K.J., Guess, F.M. & Usher, J.S. (1995). Bayesian inference for masked system lifetime data, *Applied Statistics* **44**, 79–90.

[9] Usher, J. (1996). Weibull component reliability-prediction in the presence of masked data, *IEEE Transaction on Reliability* **45**, 229–232.

[10] Mukhopadhyay, C. & Basu, A.P. (1997). Bayesian analysis of incomplete time and cause of failure data, *Journal of Statistical Planning and Inference* **59**, 79–100.

[11] Basu, S., Sen, A. & Banerjee, M. (2003). Bayesian analysis of competing risks with partially masked cause of failure, *Applied Statistics* **52**, 77–93.

[12] Sen, A., Basu, S. & Banerjee, M. (2001). Analysis of masked failure data under competing risks, in *Handbook of Statistics*, N. Balakrishnan & C.R. Rao, eds, Elsevier Science, Amsterdam, Vol. 20, pp. 523–540.

[13] Lin, D.K.J. & Guess, F.M. (1994). System life data analysis with dependent partial knowledge on the exact cause of system failure, *Microelectronics Reliability* **34**, 535–544.

[14] Guttman, I., Lin, D.K.J., Reiser, B. & Usher, J.S. (1995). Dependent masking and system life data analysis: Bayesian inference for two-component systems, *Lifetime Data Analysis* **1**, 87–100.

[15] Craiu, R.V. & Duchesne, T. (2004). Inference based on the EM algorithm for the competing risks model with masked causes of failure, *Biometrika* **91**, 543–558.

[16] Craiu, R.V. & Reiser, B. (2006). Inference for the dependent competing risks model with masked causes of failure, *Lifetime Data Analysis* **12**, 21–33.

[17] Kuo, L. & Yang, T. (2000). Bayesian reliability modeling for masked system lifetime data, *Statistics and Probability Letters* **47**, 229–241.

[18] Dinse, G. (1986). Nonparametric prevalence and mortality estimators for animal experiments with incomplete cause-of-death data, *Journal of the American Statistical Association* **81**, 328–336.

[19] Dempster, A., Laird, N. & Rubin, D. (1977). Maximum likelihood from incomplete data via the EM algorithm, *Journal of Royal Statistical Society. Series B* **39**, 1–38.

[20] Geisser, S. & Eddy, W. (1979). A predictive approach to model selection, *Journal of the American Statistical Association* **74**, 153–160.

[21] Yang, T. & Kuo, L. (1999). Bayesian computation for the superposition of nonhomogeneous Poisson processes, *Canadian Journal of Statistics* **27**, 547–556.

[22] Aboul-Séoud, M.H. & Usher, J.S. (1996). The effect of modular system structures on component reliability estimation, in *5th Industrial Engineering Research Conference Proceedings*, Minneapolis, MN, pp. 464–467.

[23] Flehinger, B.J., Reiser, B. & Yashchin, E. (1996). Inference about defects in the presence of masking, *Technometrics* **38**, 247–255.

[24] Flehinger, B.J., Reiser, B. & Yashchin, E. (1998). Survival with competing risks and masked causes of failures, *Biometrika* **85**, 151–164.

[25] Reiser, B., Flehinger, B.J. & Conn, A.R. (1996). Estimating component-defect probability form masked system success/failure data, *IEEE Transaction on Reliability* **45**, 238–243.

[26] Gelfand, A. & Smith, A.F.M. (1990). Sampling based approaches to calculating marginal densities, *Journal of the American Statistical Association* **85**, 398–409.

[27] Tanner, M.A. & Wong, W.H. (1987). The calculation of posterior distributions by data augmentation, *Journal of the American Statistical Association* **82**, 528–550.

[28] Sinha, S. (1986). *Reliability and Life Testing*, John Wiley & Sons, New York.

[29] Sinha, B. & Zellner, A. (1990). A note on the prior distributions of Weibull parameters, *SCIMA* **19**, 5–13.

[30] Metropolis, N., Rosenbluth, A.W., Rosenbluth, M.N., Teller, A.H. & Teller, E. (1953). Equations of state calculations by fast computing machines, *Journal of Chemical Physics* **21**, 1087–1091.

[31] Kuo, L. & Yang, T. (1995). Bayesian computation for software reliability, *Journal of Computational and Graphical Statistics* **4**, 1–18.

[32] Kuo, L. & Yang, T. (1996). Bayesian computation for nonhomogeneous Poisson processes in software reliability, *Journal of the American Statistical Association* **91**, 763–773.

[33] Gelamn, A. & Rubin, D.B. (1992). Inference from iterative simulation using multiple sequences (with discussion), *Statistical Science* **7**, 457–511.

Related Articles

Masked Failure Data: Competing Risks; **Masked Failure Data**.

LYNN KUO

Masked Failure Data: Competing Risks

Introduction

The term **competing-risks** *problem* has come to encompass the study of any failure process in which there is more than one distinct cause or type of failure. Frequent reference is made to possible influences of competing failure types in studies reported in the industrial, engineering as well as clinical, epidemiologic, demographic, and basic science literature. In engineering applications, the causes or risks may signify either multiple modes of failure for a complex unit or multiple components or subsystems which comprise an entire system. Occurrence of a system failure is caused by the earliest onset of any of these component failures. In this respect, the framework is that of a system with components connected in series. An additional complication arises when the causes of failure for a subset of systems are only known to belong to certain subsets of all possible causes; this is generally referred to as the cause of failure being *masked*. Masking is often the manifestation of an attempt to expedite the process of repair by replacing the entire subset of components responsible for failure instead of further investigation toward identifying the specific cause for failure. In practice, one possibility is to carry out a second-stage analysis to uniquely determine the cause; however, statistical inference may be possible even when the cause of failure is masked for a subset of systems.

Examples of failure data obtained under competing risks are abundant in reliability applications in engineering as well as in biomedical applications. In a medical context, the causes of failure (a failure may indicate death or relapse or other time-to-event outcome) typically refer to various potential risk factors for a patient (or animal) observed in a biomedical study. For example, Basu *et al.* [1] consider the risks from breast cancer and heart disease as well as other causes for breast cancer patients.

The literature on competing risks spans the work of Berman [2] to very recent work including the 2006 special issue of (JSPI) Journal of Statistical Planning and Inference [3]; the latter contains a wealth of references. An excellent resource for statistical theory and analysis of competing risks is the book by

Crowder [4]. The more recent book by Pintilie [5] emphasizes practical application of competing-risks theory, mostly in biomedical applications. In engineering applications, Langseth and Lindqvist [6] considered an interesting application of competing-risks models in the repairable systems, whereas Bunea and Mazzuchi [7] studied **competing failure modes in accelerated life testing**. Sun and Tiwari [8] analyzed the failure times of small electric appliances that may fail due to two competing risk whereas Taylor [9] modeled the tensile strength of certain materials known to contain two or more flaw types.

Competing Risks (without masking)

The early approach to analysis of competing-risks data introduced conceptual or latent failure times [2, 10] T_j, $j = 1, \ldots, K$, corresponding to time to failure from risks $j = 1, \ldots, K$, with associated probability density function $f_j(t)$, marginal survival function $S_j(t) = 1 - F_j(t)$, and net hazard function $h_j(t) = f_j(t)/S_j(t)$. Thus T_j is the potential failure time from cause $C = j$ that would be observed if the possibility of failure from causes other than j were removed. The observed lifetime T is then taken to be $T = \min(T_1, \ldots, T_K)$. Let $S(t) = P(T > t)$ be the survival function for the observed lifetime T; under the assumption that the potential failure times $T_1, \ldots, T_K$ from the K competing risks are independent we have $S(t) = \prod_{j=1}^{K} S_j(t)$. The observation triplet is (T, C, δ) where C is the cause of failure and δ is the censoring indicator. The likelihood for an observation at time t can be expressed as

$$L(t,\ \delta = 0) = S(t) \text{ for a censored case} \quad (1)$$

$$\begin{aligned} L(t, C = j, \delta = 1) &= f_j(t) \prod_{l \neq j} S_l(t) \\ &= S(t)\, h_j(t) \text{if cause is identified as } C = j \end{aligned} \quad (2)$$

$$\begin{aligned} L(t, C \in \mathcal{C}, \delta = 1) &= \sum_{j \in \mathcal{C}} f_j(t) \prod_{l \neq j} S_l(t) \\ &= S(t) \sum_{j \in \mathcal{C}} h_j(t) \end{aligned} \quad (3)$$

where the last expression applies when the cause of failure is masked and is narrowed down to a subset $\mathcal{C}$ of $\{1, \ldots, K\}$. This construction, of course, assumes independence of the K competing risks.

The random variable T_j used in the latent formulation is often referred to as the *potential failure time* since it represents the time to failure of the system when all other risks except the jth risk are removed from the system. The usefulness and interpretability of this model have been the subjects of debate [11, 12] since in most cases, it is either not feasible or physically impossible to remove all other risks from the system. The alternative cause-specific hazard formulation [11–13] is based on the joint distribution of the observed survival time T and cause of failure C. In particular, let $S(j, t) = P(C = j, T > t)$ be a "subsurvival" function with $S(t) = \sum_{j=1}^{K} S(j, t)$ denoting the marginal survival function of T. The corresponding subdensity function is denoted as $f(j, t)$. The cause-specific or subhazard function

$$\begin{aligned} h(j, t) &= \lim_{\varepsilon \to 0} \frac{P(C = j, T \leq t + \varepsilon | T > t)}{\varepsilon} \\ &= \frac{f(j, t)}{S(t)} \end{aligned} \quad (4)$$

is the instantaneous failure rate from cause j at time t after surviving from all risks $1, \ldots, K$ up to time t. The cause-specific subhazard functions $h(j, t)$ are thus defined in terms of the joint distribution of the observed time and cause of failure (T, C) and constitute the basis of the cause-specific approach. The overall hazard from all risks combined is $h(t) = \sum_{j=1}^{K} h(j, t)$. The marginal survival can be written in term of the cause-specific hazards as $S(t) = \exp\left[-\sum \int_0^t h(j, s) ds\right]$ and each of the likelihood terms in equations (1–3) can be reexpressed by replacing the net hazard $h_j(t)$ with the cause-specific subhazard $h(j, t)$. The cumulative incidence function is defined as

$$\begin{aligned} P(T < t, J = j) &= \int_0^t f(j, u)\, \mathrm{d}u \\ &= \int_0^t h(j, u) S(u)\, \mathrm{d}u \end{aligned} \quad (5)$$

and may serve as a useful descriptive device [13, 14].

The book by Crowder [4] contains detailed comparison of the latent and the cause-specific approaches. Under the assumption of independence of the potential failure times $T_1, \ldots, T_K$, the net hazard $h_j(t)$ and the cause-specific hazard $h(j,t)$ are identical (see Crowder [4] and Kalbfleisch and Prentice [12]), and the latent and cause-specific approaches lead to identical statistical inference. Independence assumes that the time of failure from cause j under one set of study conditions in which all K causes are operative is precisely the same as under an altered set of conditions in which all causes except the jth have been removed. However, the elimination of certain risks may well alter the hazards from other causes making the independence assumption of questionable validity (see Lagakos [15]). In addition, the all-too-known identifiability problem [16–20] entails that given a competing-risks model with a specific joint survivor function $\overline{F}(t_1, \ldots, t_K)$ and arbitrary dependence between $T_1, \ldots, T_K$, there exists a different joint survivor function in which $T_1, \ldots, T_K$ are independent such that both models reproduce the same set of subsurvival functions $S(j,t)$. In particular, one cannot distinguish between the dependent model and the proxy independent model on the basis of observations on (C, T) alone. In fact, each such a model has a whole class of proxy models (not necessarily independent). Crowder [19] has extended this result to show that (under mild conditions) there exist infinitely many proxy joint survivor functions which will reproduce both the set of subsurvival functions $S(j,t)$ as well as the set of marginal survival functions $S_j(t)$, $j = 1, \ldots, K$. Identifiability, however, can be regained when covariates are present. Heckman and Honore [21] showed that when there are explanatory variables in the model, identification of the joint survivor function $\overline{F}(t_1, \ldots, t_K)$ is possible from the subsurvivor functions within a certain framework. Slud [22] proved a result of similar spirit under a different setup.

Moeschberger and Klein [23] provided a thorough review of analytic approaches for handling competing-risks data when the independence assumption is suspect. One approach is to assume a parametric joint distribution for the latent lifetimes T_j, $j = 1, \ldots, K$. Parametric models are common in reliability applications and the identifiability problem is much less pervasive here (as opposed to non- or semiparametric models, which are more popular in biomedical applications). Moeschberger [24] suggested using multivariate Weibull and normal distributions for the joint distribution of the latent lifetimes $(T_1, \ldots, T_K)$. Basu and Klein [25] established that there is no identifiability problem in this setting. Basu and Ghosh [26, 27] demonstrated that many parametric distributions, including the Marshall–Olkin and the Gumbel distribution, are identifiable (up to a permutation) from (T, C), the time and the cause.

As we noted before, the assumption of independence of the potential failure times $T_1, \ldots, T_K$ implies that the cause-specific hazard $h(j,t)$ and the net hazard $h_j(t)$ are the same for all j and all $t > 0$. The latter equality $h(j,t) = h_j(t)$, $\forall j, t$ is known as the *Makeham assumption* [4]. While independence implies the Makeham condition, Gail [28] showed that the Makeham condition, in turn, implies that the subsurvival functions $S(j,t)$ determine the marginal survival functions $S_j(t)$. This implies $\overline{F}(t, t, \ldots, t) = \prod_{j=1}^{K} S_j(t)$ or "independence on the diagonal", but rather interestingly, the Makeham assumption may not imply independence (see William and Lagakos [29], and Crowder [4, 19]).

Competing Risks for Masked Data

Masking refers to the case when the causes of failure for a subset of systems are not exactly identified, but instead, have been narrowed down to subsets of $\{1, \ldots, K\}$. Dinse [30] and Kodell and Chen [31] considered bioassays for animal carcinogenicity where masking arises owing to the disagreement on the reliability of cause of death information. They focused on nonparametric estimation of survival functions associated with the cause of interest as well as the other causes. Racine-Poon and Hoel [32] formally accounted for the uncertainty in diagnosing the exact cause of death by assigning a score on the diagnostic probability being correct. Dinse [33] obtained nonparametric maximum-likelihood estimates (MLEs) of cause-specific hazards in the setting of two competing risks and indicated a generalization to the multirisk case. Dinse [34] developed nonparametric estimation of the incidence rate. Lagakos and Louis [35] investigated various nonparametric tests for comparing difference in survival functions for two groups under masking. Goetghebeur and Ryan [36] derived a modified log-rank test for comparing the survival of two populations that was later modified by Dewanji [37]. Goetghebeur and Ryan [38] developed proportional hazards structure in the context of two competing risks under the assumption that the baseline hazards for the two competing risks are proportional. Lu and Tsiatis [39] utilized multiple imputations to generalize this to the case when the baselines are not proportional. Flehinger *et al.* [40] proposed maximum-likelihood estimation under a similar model. Dewanji and SenGupta [41] developed an **EM algorithm** in the setting of masked but grouped survival data. Lu and Tsiatis [42] compared the two partial-likelihood approaches in masked data.

The parametric models are typically considered in engineering applications and are mostly based on the latent lifetime approach. One of the earliest parametric models was proposed by Miyakawa [43], who derived closed-form estimates for the case of two competing risks, each following exponential failure time distribution. In the same exponential framework, Usher and Hodgson [44] and Lin *et al.* [45] presented iterative algorithms for obtaining maximum-likelihood estimators when three competing risks are present. Guess *et al.* [46] presented a general framework that included an arbitrary number of competing risks. They introduced the term *minimum random subset* (MRS) to denote the subset $\mathcal{C}$ of the causes $\{1, \ldots, K\}$ to which the cause of failure can be narrowed down. They also noticed that the likelihood term in equations (1–3) for the masked case actually contains additional terms involving the masking probabilities $Q(\mathcal{C}|C = j, t)$ and is given by

$$L(t, C \in \mathcal{C}, \delta = 1) = S(t) \sum_{j \in \mathcal{C}} \{h_j(t) Q(\mathcal{C}|C = j, t)\} \tag{6}$$

where the masking probabilities are defined as

$$Q(\mathcal{C}|C = j, t) = P(\text{cause is maksed in subset } \mathcal{C}|C = j, t) \tag{7}$$

As noted by Flehinger *et al.* [40, 47], Craiu and Reiser [48], Craiu and Lee [49], and Craiu and Duchesne [50], the diagnostic probabilities

$$\pi(C = j|\mathcal{C}, t) = P(\text{actual failure due to cause } j|\text{cause masked in } \mathcal{C},\ t) \tag{8}$$

which can be obtained from $Q(\mathcal{C}|C = j, t)$ and the likelihood model *via* Bayes rule are of interest. Guess *et al.* [46] described two "noninformative masking" and "symmetry" conditions under which one can ignore the $Q(\mathcal{C}|C = j, t)$ terms and proceed with likelihood inference. Guess *et al.* [46] and Lin *et al.* [45] cited industrial examples where it is reasonable to assume that these conditions hold. Similar symmetry conditions are assumed in Schäbe's [51] and Goetghebeur and Ryan's [38] semiparametric formulations. Lin and Guess [52] and Guttman *et al.* [53] use a proportionality assumption to meet the symmetry condition. Flehinger *et al.* [40] consider an interesting case when further data are available from second-stage autopsy on a subset of masked units. Craiu and Duchesne [50] and Craiu and Reiser [48] used EM-based methods to model and analyze these masking probabilities. Craiu and Lee [49] developed model selection for these models, whereas Craiu and Duchesne [54] compared the performances of EM and Bayesian data augmentation methods in this scenario. Sen *et al.* [55] provided a recent review of competing-risks analysis for masked data. Mukhopadhay [56] obtained (MLEs) and bootstrap estimates of these masking probabilities in a general setting and established consistency and as well asymptotic normality of the MLEs under suitable regularity conditions. Kuo and Yang [57] developed different models for the masking probabilities and described Bayesian analysis of these models using Gibbs sampling methods. Mukhopadhyay and Basu [58] described a general Bayesian model for the masking probabilities; the details of this model is discussed in the next section.

Bayesian Approaches

Bayesian analysis of masked competing risks has been mostly limited to parametric models and the latent failure time approach, but see Craiu and Duchesne [54]. Mukhopadhyay and Basu [59], in one of the early Bayesian works, assumed that the component net survival distributions are exponential and obtain closed-form Bayesian estimates under a Dirichlet prior. Reiser *et al.* [60] considered two independent latent risks following exponential distributions, derived closed-form Bayes estimators for a noninformative Jeffreys' prior, and showed that they are identical to the corresponding MLEs. Other prior choices under exponential lifetimes are discussed in Lin *et al.* [61] and Mukhopadhyay and Basu [62]. Guttman *et al.* [53] developed Bayesian analysis relaxing the symmetry condition for two competing risks with associated exponential net survival distributions. Berger and Sun [63], Mukhopadhyay and Basu [62], and Basu *et al.* [64] considered the case of arbitrary number of competing risks where the latent failure time of each risk follows a Weibull distribution. While Mukhopadhyay and Basu [62] provided an EM algorithm, Basu *et al.* [64] used data augmentation and Gibbs sampling in the Bayesian analysis. Basu *et al.* [65] and Sen *et al.* [55] provided a general framework for parametric Bayesian analysis which encompasses location-scale and log-location-scale distributions. They used Markov chain sampling for the posterior analysis where they introduced latent variables C_i to augment the causes of failure for the masked cases. Basu *et al.* [65] also considered the case of interval censoring (when, instead of actually observing the time to event, events are only known to occur within a specified interval), applied these Bayesian parametric models in two engineering applications and provided Bayesian estimates of the diagnostic probability for the masked cases. They further compared different parametric models by Bayes factors and provide model diagnostics using the cross-validated predictive approach.

Mukhopadhyay and Basu [58] described a general approach for Bayesian analysis with the masking probabilities $Q(\mathcal{C}|C = j, t)$. Suppose that $Q(\mathcal{C}|C = j, t) = Q(\mathcal{C}|C = j)$ are free of t and let $\mathcal{F}_j$ denote the collection of all subsets of $\{1, \ldots, K\}$ which includes j. We then have $\sum \{Q(\mathcal{C}|C = j) : \mathcal{C} \in \mathcal{F}_j\} = 1$ and Mukhopadhyay and Basu [58] assigned Dirichlet distribution prior on this simplex. Such a prior specification, in fact, produces a conditionally conjugate structure, which lead to a straightforward updating scheme for the masking probabilities within the Markov chain sampling. Mukhopadhyay and Basu [58] obtained Bayesian estimates of both the masking and the diagnostic probabilities in the data example. As we noted before, Kuo and Yang [57] also considered Bayesian analysis of the masking probabilities, whereas Craiu and Duchesne [54] compared Bayesian data augmentation with the EM algorithm.

Applications

Crowder [4] and Nelson [66] listed a collection of data sets from engineering applications involving competing risks. They include data on breaking strength of wire connections (2 possible causes of failure), data on failure time of electric appliances (many possible causes), data from accelerated life tests of industrial heaters and data from low-cycle fatigue test of a superalloy. Langseth and Lindqvist [6] analyzed an interesting dataset from the Offshore Reliability Database which involved competing risks in repairable systems. Flehinger *et al.* [47] listed data of 172 failure times observed over a period of 4 years during which 10 000 computer hard drives were monitored. There are three possible caused of failure; but they are masked for some of the systems in either $\mathcal{C} = \{1, 3\}$ or $\mathcal{C} = \{1, 2, 3\}$. These data were also analyzed by Craiu and Reiser [48]. Basu *et al.* [65] analyzed data from a 10 000-hr field trial of 4993 circuit-boards reported in Chan and Meeker [67] and considered "infant mortality" and wearout as two competing risks. There is an interval when both modes are simultaneously operative and the cause for a failure within this interval is considered as masked. Reiser *et al.* [60], Basu *et al.* [64, 65], Mukhopadhyay [56], and Mukhopadhyay and Basu [58] analyzed time-to-failure data of computer component systems where failure was attributed to malfunctioning of one of the three components. Out of a total of eight system failures, the causes of three were masked. In addition, 676 systems were progressively censored at various times and this heavy censoring poses an additional challenge to statistical analysis. Basu *et al.* [65] analyzed these data using different parametric models and noted that due to the limited information in the data, the Bayesian inference is expected to depend on the assumed model and prior. Mukhopadhyay [56] and Mukhopadhyay and Basu [58] analyzed these data using a general model for the masking probabilities $Q(\mathcal{C}|C = j)$. The maximum-likelihood analysis in Mukhopadhyay [56], for example, estimate that if the true cause of failure is $C = 3$, then the probability that the cause will be identified $\hat{Q}(3|C = 3) \approx 0.5$, the probability that the cause will be masked in $\{2, 3\}$ is $\hat{Q}(\{2, 3\}|C = 3) \approx 0.5$, whereas $\hat{Q}(\{1, 3\}|C = 3)$ and $\hat{Q}(\{1, 2.3\}|C = 3)$ are ≈ 0. The Bayesian estimates in Mukhopadhyay and Basu [58], however, assign approximately 0.20 probability to each of $Q(\{1, 3\}|C = 3)$ and $\hat{Q}(\{1, 2.3\}|C = 3)$. The diagnostic probabilities $\pi(C = j|\mathcal{C})$ reported in Basu *et al.* [69] also differ from the estimates reported in Mukhopadhyay and Basu [58]. Such diversity in the results, however, are expected due to the limited information available in these data and the diversity in the models used for analysis. Mukhopadhayay and Basu, for example, used noninformative priors whereas Basu *et al.* [65] argued for a rather informative prior.

In their review on analysis of masked data in competing risks, Sen *et al.* [55] cited dependent causes, symmetry assumption, and covariate inclusion as research areas that need further exploration. The recent literature has seen substantial progress in each of these areas. Identifiability concerns from competing risks are more pervasive in masked data (when some causes are not completely known); while individual results are scattered in the literature, we are not aware of a general result. The shortage in the availability of rich data from reliability applications is also often a deterrent to methodological development in this area.

Acknowledgment

This material is based upon work partially supported by the National Science Foundation under Grant No. 0306416. Any opinions, findings, and conclusions or recommendations expressed in this material are those of the author(s) and do not necessarily reflect the views of the National Science Foundation.

References

[1] Basu, S., Tiwari, R.C. & Feuer, E.J. (2007). *Bayesian Analysis of Cancer Survival Data Using Competing Risks and Mixture Cure Models*, Technical Report, Northern Illinois University.

[2] Berman, S.M. (1963). Notes on extreme values, competing risks, and semi-Markov processes, *Annals of Mathematical Statistics* **34**, 1104–1106.

[3] Deshpande, J.V. & Cooke, R.M. (2006). Editors, competing risks: theory and applications, *A Special Volume of the Journal of Statistical Planning and Inference* **136**(5), 1569–1746.

[4] Crowder, M. (2001). *Classical Competing Risks*, Chapman & Hall/CRC, London.

[5] Pintilie, M. (2006). *Competing Risks: A Practical Perspective*, John Wiley & Sons, Hoboken.

[6] Langseth, H. & Lindqvist, B.H. (2006). Competing risks for repairable systems: a data study, *Journal of Statistical Planning and Inference* **136**(5), 1687–1700.

[7] Bunea, C. & Mazzuchi, T.A. (2006). Competing failure modes in accelerated life testing, *Journal of Statistical Planning and Inference* **136**(5), 1608–1620.

[8] Sun, Y. & Tiwari, R.C. (1997). Comparing cumulative incidence functions of a competing-risks model, *IEEE Transactions on Reliability* **46**, 247–253.

[9] Taylor, H.M. (1994). The Poisson-Weibull flaw model for brittle fiber strength, in *Extreme Value Theory*, J. Galambos, J. Lechner & Emil Simiu, eds, Kluwer Academic Publishers, Amsterdam, Vol 1, pp. 43–59.

[10] David, H.A. & Moeschberger, M.L. (1978). *The Theory of Competing Risks, Griffin's Statistical Monographs & Courses, Vol. 39*, Charles W. Griffin, London.

[11] Prentice, R.L., Kalbfleisch, J.D., Peterson, A.V., Flournoy, N., Farewell, V.T. & Breslow, N.E. (1978). The analysis of failure time data in the presence of competing risks, *Biometrics* **34**, 541–554.

[12] Kalbfleisch, J.D. & Prentice, R.L. (2002). *The Statistical Analysis of Failure Time Data*, 2nd Edition, John Wiley & Sons, New York.

[13] Pepe, M.S. & Mori, M. (1993). Kaplan-Meier, marginal or conditional probability curves in summarizing competing risks failure time data? *Statistics in Medicine* **12**, 737–751.

[14] Gray, R.J. (1988). A class of k-sample tests for comparing the cumulative incidences of a competing risk, *Annals of Statistics* **16**, 1141–1154.

[15] Lagakos, S.W. (1979). General right censoring and its impact on the analysis of survival data, *Biometrics* **35**, 139–156.

[16] Cox, D.R. (1959). The analysis of exponentially distributed lifetimes with two types of failures, *Journal of the Royal Statistical Society. Series B* **21**, 411–421.

[17] Tsiatis, A. (1975). A nonidentifiability aspect of the problem of competing risks, *Proceedings of the National Academy of Sciences of the United States of America* **72**, 20–22.

[18] Peterson, A.V. (1976). Bounds for a joint distribution function with fixed sub-distribution functons: applications to competing risks, *Proceedings of the National Academy of Sciences* **73**, 11–13.

[19] Crowder, M. (1991). On the identifiability crisis in competing risks theory, *Scandinavian Journal of Statistics* **18**, 223–233.

[20] Crowder, M. (1994). Identifiability crises in competing risks, *International Statistical Review* **62**, 379–391.

[21] Heckman, J.J. & Honore, B.E. (1989). The identifiability of the competing risks model, *Biometrika* **77**, 893–896.

[22] Slud, E. (1992). Nonparametric identifiability of marginal survival distributions in the presence of dependent competing risks and a prognostic covariate, in *Survival Analysis: State of the Art*, J.P. Klein & P.K. Goel, eds, Kluwer Academic Publishers, Boston, pp. 355–368.

[23] Moeschberger, M.L. & Klein, J.P. (1996). Statistical methods for dependent competing risks, in *Lifetime Data Models in Reliability and Survival Analysis*, N.P. Jewell, A.C. Kimber, M.-L.T. Lee & G.A. Whitmore, eds, Kluwer Academic Publishers, Norwell, pp. 233–242.

[24] Moeschberger, M.L. (1974). Life tests under dependent competing causes of failure, *Technometrics* **16**, 39–47.

[25] Basu, A.P. & Klein, J.P. (1982). Some recent results in competing risks theory, in *Survival Analysis*, J. Crowley & R.A. Johnson, eds, Hayward, California, pp. 216–229.

[26] Basu, A.P. & Ghosh, J.K. (1978). Identifiability of the multinormal distribution under competing risks models, *Journal of Multivariate Analysis* **8**, 413–429.

[27] Basu, A.P. & Ghosh, J.K. (1980). Identifiability of distributions under competing risks and complementary risks models, *Communications in Statistics-Theory and Methods* **9**, 1515–1525.

[28] Gail, M. (1975). A review and critique of some models used in competing risk analyses, *Biometrics* **31**, 209–222.

[29] Williams, J.S. & Lagakos, S.W. (1977). Models for censored survival analysis: constant-sum and variable-sum models, *Biometrika* **64**, 215–224.

[30] Dinse, G.E. (1982). Nonparametric estimates for partially-complete time and type of failure data, *Biometrics* **38**, 417–431.

[31] Kodell, R.J. & Chen, J.J. (1987.) Handling cause of death in equivocal cases using the EM algorithm, *Communications in Statistics – Theory and Methods* **16**, 2565–2585.

[32] Racine-Poon, A.H. & Hoel, D.G. (1984). Nonparametric estimation of survival function when cause of death is uncertain, *Biometrics* **40**, 1151–1158.

[33] Dinse, G.E. (1986). Nonparametric prevalence and mortality estimators for animal experiments with incomplete cause-of-death data, *Journal of the American Statistical Association* **81**, 328–336.

[34] Dinse, G.E. (1988). Estimating tumor incidence rates in animal carcinogenicity experiments, *Biometrics* **44**, 405–415.

[35] Lagakos, S.W. & Louis, T.A. (1988). Use of tumor lethality to interpret tumorigenicity experiments lacking cause-of-death data, *Applied Statistics* **37**, 169–179.

[36] Goetghebeur, E. & Ryan, L. (1990). A modified logrank test for competing risks with missing failure type, *Biometrika* **77**, 207–211.

[37] Dewanji, A. (1992). A note on the test for competing risks with missing failure type, *Biometrika* **79**, 855–857.

[38] Goetghebeur, E. & Ryan, L. (1995). Analysis of competing risks survival data when some failure types are missing, *Biometrika* **82**, 821–833.

[39] Lu, K. & Tsiatis, A. (2001). Multiple imputation methods for estimating regression coefficients in the competing risks model with missing casue of failure, *Biometrics* **57**, 1191–1197.

[40] Flehinger, B.J., Reiser, B. & Yashchin, E. (1998). Survival with competing risks and masked causes of failures, *Biometrika* **85**, 151–164.

[41] Dewanji, A. & Sengupta, D. (2003). Estimation of competing risks with general missing pattern in failure types, *Biometrics* **59**, 1063–1070.

[42] Lu, K. & Tsiatis, A. (2005). Comparison between two partial likelihood approaches for the competing risks model with missing cause of failure, *Lifetime Data Analysis* **11**, 29–40.

[43] Miyakawa, M. (1984). Analysis of incomplete data in competing risks model, *IEEE Transactions on Reliability* **33**, 293–296.

[44] Usher, J.S. & Hodgson, T.J. (1988). Maximum likelihood analysis of component reliability using masked system life-test data, *IEEE Transactions on Reliability* **37**, 550–555.

[45] Lin, D., Usher, J. & Guess, F. (1993). Exact maximum likelihood estimation using masked system data, *IEEE Transactions on Reliability* **42**, 631–635.

[46] Guess, F.M., Usher, J.S. & Hodgson, T.J. (1991). Estimating system and component reliabilities under partial information on cause of failure, *Journal of Statistical Planning and Inference* **29**, 75–85.

[47] Flehinger, B.J., Reiser, B. & Yashchin, E. (2001). Statistical analysis for masked data advances in reliability, in *Handbook of Statistics*, N. Balakrishnan & C.Radhakrishna Rao, eds, Elsevier/North-Holland, Vol. 18, pp. 499–522.

[48] Craiu, R.V. & Reiser, B. (2006). Inference for the dependent competing risks model with masked causes of failure, *Lifetime Data Analysis* **12**(1), 21–33.

[49] Craiu, R.V. & Lee, T.C.M. (2005). Model selection for the competing risks model with and without masking, *Technometrics* **47**(4), 457–467.

[50] Craiu, R.V. & Duchesne, T. (2004a). Inference based on the EM algorithm for the competing risks model with masked causes of failure, *Biometrika* **91**(3), 543–558.

[51] Schäbe, H. (1994). Nonparametric estimation of component lifetime based on masked system life test data, *Journal of the Royal Statistical. Society B* **56**, 251–259.

[52] Lin, D. & Guess, F.M. (1994). System life data analysis with dependent partial knowledge on the exact cause of system failure, *Microelectronics and Reliability* **34**, 535–544.

[53] Guttman, I., Lin, D.K., Reiser, B. & Usher, J.S. (1995). Dependent masking and system life data analysis: Bayesian inference for two-component systems, *Lifetime Data Analysis* **1**, 87–100.

[54] Craiu, R.V. & Duchesne, T. (2004b). Using EM and data augmentation for the competing risks model, in *Applied Bayesian Modeling and Causal Inference from an Incomplete-Data Perspective*, A. Gelman & X.L. Meng, eds, John Wiley & Sons.

[55] Sen, A., Basu, S. & Banerjee, M. (2001). Statistical analysis of life-data with masked cause-of-failure, in *Handbook of Statistics, Advances in Reliability*, A.P. Basu & N. Balakrishnan, eds, Elsevier Sciences, Amsterdam, Vol. 20, pp. 523–540.

[56] Mukhopadhyay, C. (2006). Maximum likelihood analysis of masked series system lifetime data, *Journal of Statistical Planning and Inference* **136**, 803–838.

[57] Kuo, L. & Yang, T.Y. (2000). Bayesian reliability modeling for masked system lifetime data, *Statistics and Probability Letters* **47**, 229–241.

[58] Mukhopadhyay, C. & Basu, S. (2007). Bayesian analysis of masked series system lifetime data, *Communication in Statistics: Theory and Applications*, **36**, 329–348.

[59] Mukhopadhyay, C. & Basu, A.P. (1993). *Bayesian Analysis of Competing Risks: K Independent Exponentials*,

Technical Report #516, Department of Statistics, The Ohio State University.

[60] Reiser, B., Guttman, I., Lin, D.K.J., Guess, F.M. & Usher, J.S. (1995). Bayesian inference for masked system lifetime data, *Applied Statistics* **44**, 79–90.

[61] Lin, D., Usher, J. & Guess, F. (1996). Bayes estimation of component-reliability from masked system-life data, *IEEE Transactions on Reliability* **45**, 233–237.

[62] Mukhopadhyay, C. & Basu, A.P. (1997). Bayesian analysis of incomplete time and cause of failure data, *Journal of Statistical Planning and Inference* **59**, 79–100.

[63] Berger, J.O. & Sun, D. (1993). Bayesian analysis of the ply-Weibull distribution, *Journal of the American Statistical Association* **82**, 1412–1418.

[64] Basu, S., Basu, A.P. & Mukhopadhyay, C. (1999). Bayesian analysis of masked system failure data using non-identical Weibull models, *Journal of Statistical Planning and Inference* **78**, 255–275.

[65] Basu, S., Sen, A. & Banerjee, M. (2003). Bayesian analysis of competing risks with partially masked cause-of-failure, *Journal of Royal Statistical Society. Series C: Applied Statistics* **52**, 77–93.

[66] Nelson, W. (1982). *Applied Life Data Analysis*, John Wiley & Sons, New York.

[67] Chan, V. & Meeker, W.Q. (1999). A failure-time model for infant mortality and wearout failure modes, *IEEE Transactions on Reliability* **TR-48**, 678–682.

Further Reading

Flehinger, B.J., Reiser, B. & Yashchin, E. (1996). Inference about defects in the presence of masking, *Technometrics* **38**, 247–255.

Related Articles

Accelerated Life Tests: Analysis with Competing Failure Modes; **Competing Risks**; **Expectation Maximization Algorithm**; **Masked Failure Data: Bayesian Modeling**.

SANJIB BASU

Maximum Likelihood

Introduction

The term *maximum likelihood* refers to a general method of **Estimation** with important historical and practical significance. Consider a sample $y_1, y_2, \ldots, y_n$ drawn independently from a distribution with density or probability function $f(y;\theta)$ with an unknown vector parameter θ. The likelihood function is defined as

$$L(\theta) = \prod_{i=1}^{n} f(y_i;\theta) \tag{1}$$

The maximum-likelihood estimate (MLE) is the value $\theta = \hat{\theta}$ which maximizes $L(\theta)$ over the set of all possible values for θ. In practice, it is usually more convenient to maximize the logarithm of the likelihood function, $l(\theta) = \ln L(\theta)$. From calculus, it follows that $\hat{\theta}$ satisfies the score equation $\partial l/\partial\theta = 0$.

Early references to the method of maximum likelihood are attributed to Gauss, Laplace, and Edgeworth [1]. However, the prominent English statistician R. A. Fisher is unquestionably responsible for popularizing the technique and for identifying many of its statistical properties [2].

The method has a strong heuristic appeal: choose as your parameter estimate the one which makes the observed data seem most likely. As it turns out, it is often the best estimate possible, particularly in large samples. Inference is made easy by the fact that MLEs are consistent and asymptotically normal under broad regularity assumptions, with a variance that can be estimated from the observed or expected information matrix. The MLE is also invariant under one-to-one transformations of the parameters, so to obtain the MLE of a transformation of the original parameters, one need only apply the transformation to the original MLE.

Optimal Properties

The following is a list of the most important properties of the MLE in large samples (i.e., $n \to \infty$):

1. **Consistency**
   The MLE converges in probability to the true value of the parameter (i.e., for any $\epsilon > 0$, $\lim_{n\to\infty} P(|\hat{\theta}_n - \theta| > \epsilon) = 0$).
2. **Asymptotic normality**
   The distribution of $n^{1/2}(\hat{\boldsymbol{\theta}} - \boldsymbol{\theta})$ converges to a **normal distribution** with zero mean and a **covariance** matrix which is the inverse of the information matrix $\mathbf{I}$. From a practical point of view it is more convenient to say that $\hat{\boldsymbol{\theta}}$ is approximately distributed as $\mathrm{N}(\boldsymbol{\theta}, \mathbf{I}^{-1}/n)$.

3. **Asymptotic efficiency**
   The MLE is the best asymptotically normal estimate in terms of its variance in large samples. Put more precisely, if $\tilde{\boldsymbol{\theta}}$ is another estimate such that $n^{1/2}(\tilde{\boldsymbol{\theta}} - \boldsymbol{\theta}) \xrightarrow{\mathcal{L}} \mathrm{N}(\boldsymbol{\theta}, \mathbf{C})$, where $\mathbf{C}$ is a fixed matrix, then $\mathbf{C} \geq \mathbf{I}^{-1}/n$ in the sense that $\mathbf{C} - \mathbf{I}^{-1}/n$ is a positive semidefinite matrix.

Regularity Conditions

It is important to be aware of the general regularity conditions for the optimal properties of the MLE. Most advanced statistics texts have a discussion of these conditions [3, 4], with the classic reference being Cramér [5]. The following three conditions are commonly given:

1. The observed data points $y_1, y_2, \ldots, y_n$ are independently and identically distributed (i.i.d.) according to a density or probability function $f(y;\boldsymbol{\theta})$, where $\boldsymbol{\theta}$ has finite dimension m. This condition is less restrictive than it may seem, given that y_i may be a vector. In fact, the method of maximum likelihood is popular and appropriate for many regression problems where the distributions of the data points are not identical, and many texts give less restrictive assumptions.
2. The underlying density or probability function is identifiable, that is, $f(y;\boldsymbol{\theta}_1) = f(y;\boldsymbol{\theta}_2)$ for all y implies that $\boldsymbol{\theta}_1 = \boldsymbol{\theta}_2$.
3. The density is "smooth" in the sense that f has derivatives up to the third order with finite expectation, and the information matrix

$$\mathbf{I} = E_\theta\left[\frac{\partial^2 \ln f(y;\theta)}{\partial\theta_j \partial\theta_k}\right], \quad j, k = 1, \ldots, m \quad (2)$$

   exists and is nonsingular. The latter conditions ensure that the asymptotic covariance matrix exists.

These conditions are satisfied for many models of interest, such as the binomial, normal, and Poisson distributions (*see* **Probability Density Functions**).

Maximum-Likelihood Calculations

Maximum-likelihood **estimation** involves finding a global maximum of a function of one or several parameters. For certain models, the solution can be expressed as a simple function of the data. However, more often the solution must be posed as a nonlinear optimization problem. Typically, it involves solving a system of nonlinear equations.

The main methods in use today are the Newton–Raphson (NR) or quasi-Newton (QN) methods, and the Fisher scoring (FS) algorithm [6]. The NR algorithm is based on an approximation of the log likelihood by a quadratic function through a Taylor series expansion of the score functions. To implement the NR algorithm, one has to provide second-order derivatives for the log-likelihood function – the so-called Hessian matrix. Initial values for the parameters are updated and the process is repeated until convergence is obtained. If the initial value for parameters is close enough to the maximum, the NR algorithm usually converges quickly. However, if the initial value is poorly chosen it may fail. In particular, the Hessian matrix can become a nonpositive-definite. Upon convergence, at the final iteration, the inverse of the Hessian matrix provides an approximation to the asymptotic covariance of the MLE. In the QN algorithm only the first derivatives are used, and the second derivatives are estimated on the basis of the results from previous iterations. The QN algorithm involves line search methods, that is, maximization of the log likelihood along a given ray in the parameter space.

The difference between the NR and FS algorithms is that the latter uses the expectation of the Hessian matrix rather than the Hessian matrix itself, as in the NR algorithm. There are two versions of the FS algorithm. In the first version, the expectation of the Hessian matrix is approximated as the sum of cross-products of first derivatives. In the second version, the exact calculated information matrix is used. Thus, to use this version of the FS algorithm one has to have a formula for the information matrix as a function of the parameters calculated prior to the maximization procedure.

Other methods are sometimes used for maximum-likelihood estimation. One that deserves special mention is the EM (expectation–maximization) algorithm. Certain likelihoods may be thought as involving missing data, with the most notable example being random effects models. The EM method works in this setting by maximizing the expectation of the log-likelihood iteratively for the complete data. This may aid in difficult maximization problems by taking advantage of simple closed-form solutions available for the "M" stage. Other advantages of the **EM algorithm** include its natural statistical interpretation and its property of producing an increasing sequence of log likelihood values in the specified parameter space. The principal drawback is that it may be relatively slow.

Example: Estimation of a Proportion

We observe the occurrence of a certain event for n individuals, where y_i is 1 if the event occurs and 0 otherwise. It can be assumed that events are independent among individuals and have the same probability of occurrence θ. The likelihood can be written as

$$L(\theta) = \prod_{i=1}^{n} \theta^{y_i}(1-\theta)^{1-y_i} = \theta^m(1-\theta)^{n-m} \qquad (3)$$

where m is the number of events observed among the n individuals. The log likelihood is

$$l(\theta) = m\ln(\theta) + (n-m)\ln(1-\theta) \qquad (4)$$

To find the maximum, we consider the following score equation:

$$\frac{\mathrm{d}l}{\mathrm{d}\theta} = \frac{m}{\theta} - \frac{n-m}{1-\theta} = 0 \qquad (5)$$

which has the unique solution $\hat{\theta} = m/n$.

Discussion and Extensions

In theory, the method of maximum likelihood is very simple: determine an appropriate sampling distribution for the data, write down the likelihood as a function of the unknown parameters, and solve for the estimate. Of course, in practice it is not always so easy. Solutions to the likelihood score equations usually must be arrived at by numerical methods, and may be computationally intensive or inaccurate. For some models and data sets it may be difficult to demonstrate that the likelihood has a unique global maximum. Many small-sample MLEs can be shown to be biased (*see* **Bias of an Estimator**). The presence of nuisance parameters can exacerbate the computational difficulties. Often the MLE is quite nonrobust to **Outliers** and, unlike competing general methods such as **Least-Squares Estimation** and the method of **moments** (*see* **Estimation**), computation of the MLE requires that the distribution of the data be completely parameterized. Uniformly minimum variance unbiased (UMVU) theory and Bayesian methods compete with maximum-likelihood estimation with regard to optimal properties, but also require complete specification of the distribution, and may be even harder to implement.

To overcome some of the difficulties of maximum-likelihood estimation, modified methods have been proposed such as restricted maximum likelihood (REML), conditional likelihood, pseudolikelihood, quasi-likelihood, partial likelihood, and M estimation. These methods were all inspired by the powerful heuristic appeal and conceptual simplicity of the original formulation of the method of maximum likelihood.

References

[1] Stigler, S.M. (1986). *The History of Statistics: The Measurement of Uncertainty Before 1900*, Harvard University Press, Cambridge.

[2] Rao, C.R. (1992). R.A. Fisher: The founder of modern statistics, *Statistical Science* **7**, 34–48.

[3] Cox, D.R. & Hinckley, D.V. (1974). *Theoretical Statistics*, Chapman & Hall, London.

[4] Stuart, H. & Ord, J.K. (1991). *Kendall's Advanced Theory of Statistics*, Oxford University Press, New York, Vol. 2.

[5] Cramér, H. (1945). *Mathematical Methods of Statistics*, Princeton University Press, Princeton.

[6] Thisted, R.A. (1988). *Elements of Statistical Computing*, Chapman & Hall, New York.

TOR D. TOSTESON AND EUGENE DEMIDENKO

Article originally published in Encyclopedia of Biostatistics, 2nd Edition *(2005, John Wiley & Sons, Inc.). Minor revisions for this publication by Jeroen de Mast.*

Mean Residual Life

Consider a random variable T denoting the failure time of an item. Suppose that the item has been functioning up to time t, then the mean residual life at time t is defined as

$$m(t) = E(T - t | T \geq t) \qquad (1)$$

Given the reliability function (*see* **Reliability Function**) $R(t)$, it follows that $m(t) = \int_t^\infty R(u)\,du/R(t)$. There is a one-to-one correspondence between mean residual life and reliability function since it holds that

$$R(t) = \frac{m(0)}{m(t)} e^{-\int_0^t (1/m(u))\,du} \qquad (2)$$

Suppose that T has an exponential distribution, with density $\lambda e^{-\lambda t}$ and reliability function $e^{-\lambda t}$, $\lambda > 0$, $t \geq 0$. Using the previous result about the link with the reliability function, it follows that $m(t) = 1/\lambda$, as another consequence of the "memoryless"property.

Mean Square Error

In *analysis of variance* (*see* **Analysis of Variance**) the total **sum of squares** is divided into a sum of squares attributed to the factors, and a residual sum of squares (often called the *error sum of squares*). The residual sum of squares divided by the residual **degrees of freedom** is the mean square error (MSE). The square root of the MSE is often used as an estimator of noise in the measurements. For example, in the simple linear regression analysis $y_i = a + bx_i + \epsilon_i$, $i = 1, \ldots, n$, the MSE is given by

$$MSE = \frac{1}{n-2} \sum_{i=1}^{n} (y_i - \hat{y}_i)^2 \qquad (1)$$

with $\hat{y}_i$ the least-squares fit for y_i.

Mean Time Between Failures

Consider a random variable T denoting a time between two failures. The mean time between failures (MTBF) is defined as the expected value of T, provided it exists.

If T has a density function $f(t)$, then the MTBF is given by $\int_0^\infty t f(t)\,dt$ or $\sum_i t_i f(t_i)$, depending on whether T is absolutely continuous or discrete, respectively.

Suppose that T has an exponential distribution, with density $\lambda e^{-\lambda t}$, $\lambda > 0$, $t \geq 0$. Then the MTBF is given by

$$\int_0^\infty t\lambda e^{-\lambda t}\,dt = \frac{1}{\lambda} \qquad (1)$$

Mean Time to Failure

Consider a random variable T denoting a time to failure. The mean time to failure (MTTF) is defined as the expected value of T, provided it exists.

If T has a density function $f(t)$, then the MTTF is given by $\int_0^\infty t f(t)\,dt$ or $\sum_i t_i f(t_i)$, depending on whether T is absolutely continuous or discrete, respectively.

Given the reliability function (*see* **Reliability Function**) $R(t)$, it follows that the MTTF equals $\int_0^\infty R(t)\,dt$ in the absolutely continuous case.

Suppose that T has an exponential distribution, with density $\lambda e^{-\lambda t}$ and reliability function $e^{-\lambda t}$, $\lambda > 0$, $t \geq 0$. Then the MTTF is given by

$$\int_0^\infty t\lambda e^{-\lambda t}\,dt = \int_0^\infty e^{-\lambda t}\,dt = \frac{1}{\lambda} \qquad (1)$$

Mean Time to Repair

Consider a random variable R denoting the repair (or down) time after the failure of an item. The mean time to repair (MTTR) is defined as the expected value of R, provided it exists.

If R has a density function $f(r)$, then the MTTR is given by $\int_0^\infty r f(r)\,dr$ or $\sum_i r_i f(r_i)$, depending on whether R is absolutely continuous or discrete, respectively.

Given the distribution function $F(r)$, it follows that the MTTR equals $\int_0^\infty [1 - F(r)]\,dr$ in the absolutely continuous case.

Suppose that R has an exponential distribution, with density $\lambda e^{-\lambda r}$ and distribution function $1 - e^{-\lambda r}$, $\lambda > 0$, $r \geq 0$. Then the MTTR is given by

$$\int_0^\infty r\lambda e^{-\lambda r}\,dr = \int_0^\infty e^{-\lambda r}\,dr = \frac{1}{\lambda} \qquad (1)$$

Measurement Capability *see* Measurement Systems Analysis, Capability Measures for

Measurement Error *see* Gauge Repeatability and Reproducibility (R&R) Studies

Measurement Error and Uncertainty

Introduction

Measurement systems analysis (MSA) involves understanding the measurement process in place, identifying and estimating the error present in the process or system, and determining the adequacy of the **measurement system** for use in product and **process control** (*see* **Engineering Process Control**). A measurement system itself consists of devices to obtain measurements (gauges), standards, methods, personnel, and so on. That is, it is the entire process for obtaining measurements on some quantity of interest (*measurand*). *Error* and *uncertainty* will be a part of any measurement system or in the results from any measurement system put into place. Unfortunately, the concepts of error and uncertainty are often confused with one another and sometimes used (incorrectly) interchangeably. There are two (if not more) schools of thought with respect to the operational definitions of error and uncertainty. In this chapter, measurement error, uncertainty, and their components are presented as they are often used and defined in the literature [1–6]. In addition, definitions and relationships for precision and accuracy with respect to measurement systems are provided.

Error

Error is by definition the difference between the measured value of the quantity of interest (*measurand*) and the true value of the measurand. In the literature, error has also been referred to and written as *measurement error, error of measurement, experimental error*, and so forth; although depending on the experimental situation, these terms may have specific and therefore, quite different meanings. In addition, issue has sometimes been taken with the use of the term *true* value of the measurand (see [3, 4, 7] for example), since a true value is unknowable. "Reference value" is often used interchangeably with true value although this use is not recommended [8]. The reference value should be considered more as the best *approximation* available for the true value.

There are numerous types of error. These may include random error, systematic error, observer error, repeatability error, discrimination error, total error, and rounding error, just to name a few (see [6]). In this section, two important and common types of error are presented as well as their use in MSA: systematic error and random error.

Systematic Error

Systematic errors can be caused by human activities, inadequate manufacturing methods, imperfections of the measuring device, changes in temperature, pressure, and so on. This error remains the same over repeated measures taken under identical conditions. That is, the error that is induced for one measurement will be identical to the error for the next measurement. The error is *systematic* and results in values that are consistently above or consistently below the true or reference value of the measurand. Systematic error can also be defined as the difference between error and random error (discussed in the next subsection).

Random Error

Suppose that repeated measures are taken on the same item, under identical conditions (same operator, same gauge). Under ideal conditions, even when systematic errors have been identified and accounted for, error will occur. *Random errors* vary randomly over all measurements taken under identical conditions. Even in a well-designed experiment or well-designed MSA, normal random fluctuations will occur and often occur according to some distribution (usually a Gaussian distribution with zero mean). If only random errors exist in the system, then increasing the number of measurements taken will provide a better estimate or better approximation of the measurand's true value. For example, if the average of the measurements taken is the best estimate for the true value of the measurand, then increasing the number of measurements will result in a better estimate.

Measurement Results Modeled by Systematic and Random Errors

Suppose that the following could represent a measurement result

$$Y = X + \varepsilon \quad (1)$$

where Y is the measurement result, X is the true value of the measurand, and ε represents the combination of both systematic and random error. If the error is combined to represent both the systematic and random error, and we let υ represent the mean of ε and σ^2 the variance, then it can be said that $\varepsilon \sim (\upsilon, \sigma^2)$. In this form, if $\upsilon = 0$ then the measurements are considered more *accurate* and there is no "systematic error". Furthermore, if σ is sufficiently small, the measurements are considered more *precise*. (For more details, the reader is encouraged to see [9]). Precision and accuracy play a very important role in MSA and are discussed in the next subsection.

Measurement System Error

Measurement system error as it pertains to MSA (*see* **Measurement Systems Analysis, Overview**) is a combination of variability attributable to gauge ***bias*** *stability, linearity, repeatability*, and *reproducibility* (*see* **Gauge Repeatability and Reproducibility (R&R) Studies**; **Gauge Repeatability and Reproducibility (R&R), Variance Components in**; **Gauge Repeatability and Reproducibility (R&R) Studies, Misclassification Rates**; **Gauge Repeatability and Reproducibility (R&R) Studies, Confidence Intervals for**). Bias, linearity, and stability are the components that make up the *accuracy* of a measurement system. Repeatability and reproducibility are the components that make up the *precision* (measurement variation).

Accuracy. *Accuracy* is defined as the difference between the measurement and the value of the measurand. Accuracy is a qualitative concept and, as a result, is not quantifiable. It does not make sense to describe a measurement system in these terms by saying, for instance, "the accuracy is 0.05 cm". *Uncertainty* (discussed in the section titled "Uncertainty"), however, is quantifiable and can be used to describe the system: "The standard uncertainty is 0.05 cm." In measurement system error, the three components of accuracy and their definitions are

- *Bias* – the difference between the observed average measurement and a reference or master value. It is a measure of the systematic error with respect to the measurement system.
- *Linearity* – measures how changes in the size of the measurand (part being measured) will affect measurement system bias over the expected range of the process.
- *Stability* – measures how well the measurement system performs over time. It reflects the change in bias (if any) over time when the same parts or master value are measured.

In general, accuracy focuses on measures of location or the relationship between the measurement results and value of the measurand.

Precision. *Precision* is defined as the variation identified when the same part or item is measured repeatedly using the same measurement device or gauge. The components of precision are listed below (*see* **Measurement Systems Analysis, Overview**; **Gauge Repeatability and Reproducibility (R&R), Variance Components in**; **Gauge Repeatability and Reproducibility (R&R) Studies, Misclassification Rates**; **Gauge Repeatability and Reproducibility (R&R) Studies, Confidence Intervals for**).

- *Repeatability* – variability due to the gauge or test instrument when used by the same operator to measure the same unit. In other words, this is variability arising from the same operator (or setup) measuring the same part repeatedly using the same measuring device (gauge).
- *Reproducibility* – variability due to different operators or setups measuring the same parts using the same measuring device (gauge). This is considered the variability due to the measurement system.

Summary of Error

Chunovkina [3] among others argues that error is a model concept and as a result cannot be used to directly calculate error. Arguments have been made that error differs conceptually from uncertainty and as a result these terms should be used cautiously. Full and excellent discussions on measurement error can be found in [3, 4, 6, 9]. Measurement system error is discussed in [7].

In the next section, measurement uncertainty is presented. The discussion is based on several articles that discuss the *Guide to the Expression of Uncertainty in Measurement* (1993) also known as *GUM*, as well as the guide itself. In addition, the reader is encouraged to see *Guidelines for Evaluating and Expressing the Uncertainty of NIST Measurement Results* [2].

Uncertainty

The International Committee for Weights and Measures (CIPM) in conjunction with several international organizations, including the International Engineering Consortium (IEC) and International Organization for Standards (ISO) began work on a project that would hopefully result in some common approach for describing the quality of measurement results among international laboratories. The result of this project is the GUM published in 1993. GUM attempts to provide a conceptual framework for assessing and reporting uncertainty in measurements. In this section, uncertainty and its extensions are presented. The reader is encouraged to see GUM as well as [2–5, 10] for full details and discussion of uncertainty as well as some critiques of GUM itself [3].

Standard Uncertainty

As stated previously, MSA as a whole is used to understand the measurement system and attempt to improve it through variation reduction techniques. *Standard uncertainty* on the other hand attempts to *quantify* the range of measurement values by providing intervals over which the measurement results would vary with a certain probability (or in some instances confidence levels). The expectation is the value of the measurand would be captured by this interval. Technically, uncertainties are characterized by standard deviations (*see* **Variance**).

The purpose of the GUM (1993) was to provide comprehensive and consistent information on how to report measurement uncertainty. GUM also provided a common framework for comparison of measurement results internationally. GUM states that the uncertainty of a measurement is a "... parameter, associated with the result of a measurement that characterizes the dispersion of the values that could reasonably be attributed to the measurand". Again, standard uncertainty, u_i, is described by standard deviation. Obviously, u_i is also the positive square root of u_i^2. Components that make up uncertainty (to be discussed in this section) are categorized into one of two categories (this is the categorization presented by [1]). Grouping of components depends on the method used to estimate numerical values. There are two types of uncertainties, type A and type B:

- *Type A* – based on statistical analysis of a series of measurements.
- *Type B* – based on other sources of information. These "other" sources can include, for example, gauge manufacturer's specification or published reference values.

The definitions given here are very simplistic and the reader is encouraged to see [1, 2]. In the remainder of this section, more basic definitions and descriptions for uncertainty are provided.

Combined Standard Uncertainty

Combined standard uncertainty, u_c, is the standard deviation of the combined errors. These errors include both random error and systematic error. The combined uncertainty, u_c can often be obtained using the *law of propagation of uncertainty* when the measurand, Y is a function of several independent variables. For more details, see [9] or [2, Appendix A].

In MSA for example, the combined uncertainty would be a combination of all sources of measurement variation for that particular experimental situation. This would include measurement variation in the measurement process, **calibration** issues, environment errors, and so on. (See [7] for more discussion of measurement uncertainty in MSA.) To illustrate, suppose that the sources of variation for a particular process include repeatability, reproducibility, and stability denoted by $\sigma^2_{\text{repeatability}}$, $\sigma^2_{\text{reproducibility}}$, and $\sigma^2_{\text{stability}}$, respectively. The combined uncertainty would then be

$$u_c = \sqrt{\sigma^2_{\text{repeatability}} + \sigma^2_{\text{reproducibility}} + \sigma^2_{\text{stability}}} \quad (2)$$

There can be many situations where the combined uncertainty can be quite complicated [2, 10] and the law of propagation of uncertainty can be very useful.

Expanded Uncertainty

Often in measurement studies it is not enough to simply report a single combined estimate of uncertainty. An interval about the measurement results that defines the measurement uncertainty is often required. *Expanded uncertainty*, U, is the combined standard uncertainty, u_c multiplied by some "coverage factor", k. In other words, $U = ku_c$. The coverage factor, k, is chosen such that a certain level of confidence (*see* **Confidence Intervals**) (for Gaussian or normally distributed measurements in a measurement system) is attained. In all instances, the factor k results in a certain coverage interval. When the normal distribution applies (*see* **Normal Distribution**; **Normality Tests**), $k = 2$ will result in an approximate 95% **confidence interval**. Suppose a measurand, Y, is estimated by some value y. The combined uncertainty would be given by $u_c(y)$, then $U = ku_c(y)$ and the expanded uncertainty would be written as $Y = y \pm U$. The expanded uncertainty provides more information about the process than a single-number statistic.

Other Considerations and Standards

To attain some level of standardization and agreement among laboratories internationally, GUM (1993) has provided guidance on how to evaluate and express measurement uncertainty. The guide is a high-level reference document with some very broad and sweeping specifications and definitions. Some very basic definitions are given in this section and the reader is encouraged to review [1–5] for more complete discussions and illustrations of the concepts outlined here.

Summary and Conclusions

In this article, some very basic definitions and concepts concerning measurement error and uncertainty are provided. There remains a great deal of room in the literature for further discussion and clarification; not only for the benefit of the measurement community, but for any practitioner or entity that must rely on measurement results, MSA, and the inferences that are to be made from the results obtained.

It is also important to note that many of the terms and concepts presented here and in the literature may take on different meanings and interpretation based on the experimental situation at hand. For instance, a source of random error in one experimental setup may easily be a systematic error in a different setup. The roles of error and uncertainty may be quite different depending on the type of measurements actually recorded, and their use in practice. Again, the reader is encouraged to review the references provided here and the references provided within those documents.

References

[1] ISO. (1993). *Guide to the Expression of Uncertainty in Measurement*, 1st Edition, International Organization of Standards, Switzerland.

[2] Taylor, B.N. & Kuyatt, C.E. (1994). Guidelines for Evaluating and Expressing the Uncertainty of NIST Measurement Results, NIST Technical Note 1297.

[3] Chunovkina, A.G. (2000). Measurement error, measurement uncertainty, and measurand uncertainty, *Measurement Techniques* **43**, 581–586.

[4] Mari, L.P. (2005). *Explanation of Key Error and Uncertainty Concepts and Terms. Handbook of Measuring System Design*, P.H. Sydenham & R. Thorn, eds, John Wiley & Sons.

[5] Cox, M. & Harris, P. (2005). An outline of supplement 1 to the guide to the expression of uncertainty in measurement on numerical methods for the propagation of distributions, *Measurement Techniques* **48**, 336–345.
[6] McGhee, J. (2005). *Calculation and Treatment of Errors. Handbook of Measuring System Design*, P.H. Sydenham & R. Thorn, eds, John Wiley & Sons.
[7] Automotive Industry Action Group. *Measurement Systems Analysis Reference Manual*, 3rd Edition, Automotive Industry Action Group, Southfield.
[8] ASTM. E177-90a (2002). standard practice for use of the terms precision and bias in ASTM test methods, *Book of Standards 14.02.*, ASTM International West Conshohocken.
[9] Gertsbakh, I. (2003). *Measurement Theory for Engineers*, Springer-Verlag, Berlin Heidelberg.
[10] van der Veen, A.M.H. & Cox, M.G. (2003). Error analysis in the evaluation of measurement uncertainty, *Metrologia* **40**, 42–50.

CONNIE M. BORROR

Measurement Systems Analysis, Attribute

Introduction

Methods for assessing the capability of a quantitative or continuous **measurement system** are well documented and include the tabular method and the analysis of variance approach. When the measurement system involves attribute data, the standard quantitative methods are not appropriate. An attribute gauge measurement system is one in which the parts or objects under study are placed into one of two or more possible categories. The classification of the part is the measurement of interest in an attribute measurement system. Several commonly used statistics for "appraiser agreement" and attribute gauge repeatability and reproducibility (GRR) are presented.

Kappa Statistics

Two Appraisers

The κ statistic suggested by [1] measures the agreement between two appraisers (raters, judges) when evaluating or rating the same item. Cohen's kappa statistic is given by

$$\kappa = \frac{p_o - p_e}{1 - p_e} \tag{1}$$

where p_o represents the observed proportion of agreement between appraisers and p_e represents the expected proportion of agreement due to chance. Table 1 illustrates a typical 2×2 table with two appraisers (X, Y) and two categories (pass, fail). The values a, b, c, d represent the number of times the appraisers agreed (a, d) and disagreed (b, c).

Table 1 Standard summary for two appraisers and two categories

		Ratings by appraiser X		
		Pass	Fail	Totals
Ratings by appraiser Y	Pass	a	b	n_1
	Fail	c	d	n_2
	Totals	m_1	m_2	N

Using the notation from Table 1, the observed proportion of agreement for this binary system is $p_o = (a + d)/N$ and the expected proportion of agreement due to chance is $p_e = (m_1n_1 + m_2n_2)/N^2$. If there is perfect agreement between appraisers, then $\kappa = 1$. If all observed agreement is due to chance, then $\kappa = 0$. κ will be negative if the agreement is less than what would be expected purely by chance. Cohen's κ can be calculated for any two appraisers and is a single value summarizing the quality of measurements.

More than Two Appraisers

The κ statistic suggested by [2] is a variant to Cohen's kappa statistic which takes into account more than two appraisers measuring the same number of items. Fleiss [2] suggested that the proportion of agreeing pairs be used to determine the degree of agreement among appraisers. Specifically, p_o and p_e are formulated to account for more than two appraisers by averaging the proportion of agreement for each pair of appraisers. Following the notation of [2], suppose there are N parts, n ratings per part, and k possible ratings (categories). Furthermore, suppose there are n_{ij} appraisers who will assign the ith part to the jth category. The proportion of agreeing pairs out of all possible pairs of category assignments is

$$p_i = \frac{1}{n(n-1)} \sum_{j=1}^{k} n_{ij}(n_{ij} - 1) \quad (2)$$

The average of all p_i's ($i = 1, \ldots, N$) can be used as the measure for overall agreement,

$$p_o = \frac{1}{N} \sum_{i=1}^{N} p_i$$

$$= \frac{1}{Nn(n-1)} \left(\sum_{i=1}^{N} \sum_{j=1}^{k} n_{ij}^2 - Nn \right) \quad (3)$$

If the extent of agreement is purely by chance, then we can expect the average proportion of agreement to be

$$p_e = \sum_{j=1}^{k} p_j^2 \quad (4)$$

The quantities given in equations (3) and (4) are used to calculate κ given in equation (1). The resulting κ statistic is often referred to as *Fleiss's kappa statistic*. It is of interest to note that Fleiss's kappa is useful for determining which category level(s) the appraisers (raters) are in relative agreement on, as well as the one(s) in which they do not have a common understanding (disagreement). In practice, this makes it easier to target remedial training or rewrite operational definitions of the categories in order to improve the measurement system. For more information on κ with multiple appraisers see [2–3] and [4].

Limitations of κ Statistics

κ cannot measure the degree of agreement or the size of disagreement between the two appraisers. Also, the value of κ can be misleading. For example, it would seem intuitive that a large value of p_o (indicating high agreement between appraisers), would result in a large value of κ. But, the resulting κ could be misleadingly small if p_e happens to be a large value. To illustrate, consider the data displayed in Table 2.

Table 2 Sample data for two appraisers and two categories

		Ratings by appraiser X		
		Pass	Fail	Totals
Ratings by appraiser Y	Pass	80	5	80
	Fail	15	5	20
	Totals	95	5	100

In this hypothetical situation, we have $p_o = (80 + 5)/100 = 0.85$, a value indicating a high level of agreement between the appraisers. Next, $p_e = [(95)(80) + (5)(20)]/100^2 = 0.77$. As a result, $\kappa = 0.34$. Therefore, a large value of p_e can result in a small value of κ, even if the proportion of observed agreement between appraisers is quite large. This phenomenon is one of two paradoxes associated with κ (see [5] for details and illustrations of both paradoxes).

Summary of κ Statistics

Overall, the κ statistics are easy to calculate and are often reported in standard statistical software. The κ statistics presented in this section can be used for binary data or ratings with more than two categories. The statistics provide some measure of reproducibility for attribute measurement systems. A measure of repeatability is possible if each appraiser rates each part more than once. Often, repeatability is of less importance than reproducibility in these situations since appraisers tend to give similar (if not the same) ratings to the same part; this results in a high level of repeatability.

There are several drawbacks associated with the use of κ statistics, one of which has been illustrated here. Other problems with the use of Cohen's κ are that it assumes that the two appraisers use the same rating scale. Although this is a desirable condition, there may be situations where one appraiser would use one rating scale (say 1–8) while the other uses a different rating scale (say 1–10). This situation may be appropriate when there is interest in quantifying the consistency of ratings for an appraiser who uses different rating scales. There have been several extensions of Cohen's κ including a weighted κ statistic, statistics for paired data and for situations where **covariate** information is available. See [2–7] for more details concerning drawbacks and alternative statistics for interrater agreement.

Intraclass Correlation

The intraclass correlation coefficient (ICC) measures the correlation among multiple measurements of the same item or part. ICC was developed as an alternative to Cohen's κ [8] and has been used to evaluate reproducibility due to different appraisers. (Note: ICC is sometimes referred to as the *intraclass* κ and could have been discussed in the section titled "Kappa Statistics" (see [9]).)

The intraclass correlation is the correlation between different measurements Y_{ij} and Y_{ik} taken on the ith part or object and can be calculated as

$$ICC = \frac{\mathrm{Cov}(Y_{ij}, Y_{ik})}{\sqrt{\mathrm{Var}(Y_{ij}) \cdot \mathrm{Var}(Y_{ik})}} = \frac{\sigma_p^2}{\sigma_p^2 + \sigma_e^2} \quad (5)$$

σ_p^2 represents the variability between parts σ_e^2 represents the variability within a part or object. An ICC of 0 would indicate lack of consistency among appraisers. An ICC = 1 would indicate perfect consistency or perfect reliability. It should be noted that for continuous measurement systems, ICC can also be written in terms of GRR for a one-way analysis of variance (*see* **Variance Components**; **Gauge Repeatability and Reproducibility (R&R), Variance Components in**; **Gauge Repeatability and Reproducibility (R&R) Studies, Misclassification Rates**). One summary of GRR for a one-factor random model is

$$GRR = \frac{\sigma_e}{\sqrt{\sigma_p^2 + \sigma_e^2}} \quad (6)$$

We can then write $ICC = 1 - (GRR)^2$. Although the ICC can be used to provide a measure of reproducibility, complete distributional assumptions are necessary since it is estimated using random-effects models.

Binary Measurement Systems

Suppose the quality characteristic under study can assume only two possible values (pass/fail, go/no go, and 0/1). Initially, also assume there are only two appraisers. For this binary measurement system, the ICC could be written as (this is identical to the ϕ coefficient defined in [4])

$$ICC_b = \frac{\mathrm{Cov}(Y_{ij}, Y_{ik})}{\sqrt{\mathrm{Var}(Y_{ij}) \cdot \mathrm{Var}(Y_{ik})}}$$

$$= \frac{P(Y_{i1} = 1, Y_{i2} = 1) - p_1 p_2}{\sqrt{p_1(1 - p_1) \cdot p_2(1 - p_2)}} \quad (7)$$

where $Y_{ij} = 1$ represents a successful rating on the ith part by the jth appraiser. The estimates for p_1, p_2, and $P(Y_{i1} = 1, Y_{i2} = 1)$ are

$$\hat{p}_1 = \frac{\sum_{i=1}^{n} Y_{i1}}{n}, \quad \hat{p}_2 = \frac{\sum_{i=1}^{n} Y_{i2}}{n},$$

$$\hat{P}(Y_{i1} = 1, Y_{i2} = 1) = \frac{\sum_{i=1}^{n} Y_{i1} Y_{i2}}{n} \quad (8)$$

and n is the number of parts or items being measured.

If the binary measurement system has m appraisers ($m > 2$), then the recommendation by [10] and [11] (and summarized in [4]) involves averaging the $\mathrm{ICC_b}$ values for all pairs of appraisers. They also assume that $p_j = p$ for all appraisers (see [4] for more discussion).

Analytic Method for Estimating Bias and Repeatability

Gauge studies are carried out to determine the suitability of the measurement system for use in a particular application. The *analytic method* detailed in [12] provides estimates of bias and repeatability for an attribute gauge and involves approximating the naturally discrete distribution with a continuous distribution, namely the normal (or Gaussian) distribution. This method assumes there are multiple parts or objects to be measured and that each part is measured multiple times by a single appraiser or measuring device. Once measured, the part is classified as acceptable (a) or unacceptable (e.g., binary data). Furthermore, it is assumed that a "standard" or reference value is known or can be obtained for the quality characteristic of interest.

The analytic method has the following requirements [12]:

- There must be a known reference value for each part selected.
- At least eight parts are selected at equidistant intervals.
- The minimum and the maximum values must represent the range of the process.

- Each part is measured $m = 20$ times under identical conditions.
- The number of "accepts" (a) is recorded.
  - Smallest part must have no accepts $(a = 0)$. (If this is not satisfied, smaller parts are selected until $a = 0$.)
  - The largest part must have all accepts $(a = 20)$. (If this is not satisfied, then larger parts are selected until $a = 20$.)
  - The remaining six parts must have between 1 and 19 accepts, inclusive $(1 \leq a \leq 19)$. (If this is not satisfied, more parts are selected over the range – the additional parts are selected at the midpoint between the parts that have already been measured.)

In the analytic approach, gauge bias and gauge repeatability are then estimated by calculating probabilities of acceptance using normal probability plotting (*see* **Probability Plots**) – a procedure first put forth by [13]. Reproducibility is not estimable since there is a single gauge measuring the parts. The accuracy of the bias and repeatability estimates has been questioned since it depends on the normal approximation of discrete data. The authors in [14] propose an improved approach for estimating bias and repeatability. In their work, they provide an alternative statistical estimation procedure that takes advantage of modern computational capabilities.

Latent-Class Models

A promising approach to assessing an attribute measurement system – repeatability and reproducibility in particular – is the use of latent-class models. The latent variable in this case represents the true value of the part or object, and as the name implies, is unknown. For interrater agreement, the latent-class model approach assumes that a part being rated or judged belongs to one of two or more possible latent classes. For example, in a binary system there would be two latent classes, namely, pass/fail. The probability that a part receives a particular rating is assumed to be dependent upon the latent class to which the part truly belongs. The joint distribution of the ratings is represented by a latent-class model and consists of a mixture of distributions for the classes of the latent variable. For example, in a binary measurement system, the latent-class model would consist of a mixture of binomial distributions.

Binary measurement systems are commonly encountered in industrial practice. Boyles [15] describes an approach for this situation using latent-class models. Using the notation and method presented in [15], suppose there are p parts and n_i ratings (inspections) of the ith part. Let s_i represent the number of "pass" evaluations of the ith part, i.e. $s_i = \sum_{j=1}^{n} y_{ij}$, where y_{ij} represents the jth rating of the ith part. For a binary measurement system, let $y_{ij} = 1$ if the ith part receives a pass evaluation (rating) on the jth rating, and $y_{ij} = 0$ if the ith part does not receive a pass evaluation (fails) on the jth rating. The distribution of s_i consists of a mixture of binomial distributions, namely

$$\begin{aligned} P(s_i = s) &= E[P(s_i = s|\xi_i)] \\ &= \binom{n_i}{s}[\phi\theta_1^s(1-\theta_1)^{n_i-s} \\ &\quad + (1-\phi)\theta_0^s(1-\theta_0)^{n_i-s}] \end{aligned} \tag{9}$$

where ξ_i represents the latent classes with $\xi_i = 1$ if the ith part is good and $\xi_i = 0$ if the ith part is bad. It is assumed that parts in each category of ξ are homogeneous. As a result, for a given level of ξ_i, the joint ratings of the appraisers are assumed to be statistically independent. If the parts are randomly selected from a population of parts, $\phi = P(\xi_i = 1)$ represents material variation. Boyles [15] further denotes the variability due to the measurement system by

$$\theta_0 = P(y_{ij} = 1|\xi_i = 0) \tag{10}$$

$$\theta_1 = P(y_{ij} = 1|\xi_i = 1) \tag{11}$$

Therefore, θ_0 is the probability that a part passes when the part is bad and θ_1 is the probability that a part passes when the part is good. The misclassification rates are θ_0 and $1 - \theta_1$. θ_0 and θ_1 are considered the gauge parameters. Furthermore, it is assumed that $\theta_0 < \theta_1$, i.e., the probability that a good part is classified as good is greater than the probability that a good part is classified as bad. Under these conditions (which [15] refers to as the basic model), it can be shown that

$$\begin{aligned} \mathrm{Var}(y_{ij}) &= [\phi\theta_1(1-\theta_1) + (1-\phi)\theta_0(1-\theta_0)] \\ &\quad + (\theta_1-\theta_0)^2\phi(1-\phi) \end{aligned} \tag{12}$$

It is desired that estimates be obtained for the gauge parameters (θ_0, θ_1) and the material variation (ϕ). Maximum-likelihood estimates (MLEs) (*see* **Maximum Likelihood**) can be found using the expectation-maximization (EM) algorithm where the likelihood function from equation (9) is

$$L \propto \prod_{i=1}^{p} \{\phi\theta_1^s(1-\theta_1)^{n_i-s} + (1-\phi)\theta_0^s(1-\theta_0)^{n_i-s}\} \tag{13}$$

The resulting MLEs of (θ_0, θ_1, ϕ) provide information about the misclassification levels in the system. In addition, **confidence intervals** can be constructed on the misclassification rates (*see* [15], p. 225; **Gauge Repeatability and Reproducibility (R&R) Studies, Misclassification Rates**). The author in [15] goes on to demonstrate how the reproducibility can be examined using this modeling approach. Detailed information on latent-class models, their development, and applications can be found in [4, 9, 16–20].

Summary and Conclusions

Several commonly used measures of assessment for attribute data have been presented. Attribute GRR studies can be carried out using appraiser agreement statistics (κ statistics, intraclass correlation), the analytic method, and latent-class models. Each approach can provide some measure of reproducibility or repeatability as discussed. The analytic method also provides some measure of system bias.

Commonly used criteria such as the κ statistics provide only a single statistic to describe the capability of the entire system. The latent-class model approach provides a model that can be used to describe the gauge capability for a measurement system.

If summary statistics are to be used, the ICC is a better measure of interrater agreement than Cohen's κ statistic, if the underlying marginal distributions are identical across all appraisers.

The latent-class model approach is extremely useful for the evaluation of attribute measurement systems; in particular, binary (pass/fail) measurement systems. Reference [15] employs latent-class models to estimate misclassification rates for binary measurement systems under a model where a standard is given and a model when no standard is given. The author in [15] describes a method for estimating repeatability and reproducibility effects utilizing misclassification rates. The authors in [4] compare latent-class models with κ statistics and log-linear models. As they have noted, the latent-class model approach provides a model for the outcome of measurement system, whereas the κ statistics and intraclass correlation are statistics – one-number summaries that are supposed to describe an entire measurement system. The authors in [4] recommend the latent-class model approach for dealing with binary measurement systems due to its flexibility and ability to handle various conditions and restrictions.

References

[1] Cohen, J. (1960). A coefficient of agreement for nominal scales, *Educational and Psychological Measurement* **20**, 37–46.

[2] Fleiss, J.L. (1971). Measuring nominal scale agreement among many raters, *Psychological Bulletin* **76**, 378–382.

[3] Conger, A.J. (1980). Integration and generalization of kappas for multiple raters, *Psychological Bulletin* **88**, 322–328.

[4] van Wieringen, W.N. & van Heuvel, E.R. (2005). A comparison of methods for the evaluation of binary measurement systems, *Quality Engineering* **17**, 495–507.

[5] Feinstein, A.R. & Cicchetti, D.V. (1990). High agreement but low kappa: I. The problem of two paradoxes, *Journal of Clinical Epidemiology* **43**, 543–549.

[6] Cicchetti, D.V. & Feinstein, A.R. (1990). High agreement but low kappa: II. Resolving the paradoxes, *Journal of Clinical Epidemiology* **43**, 551–558.

[7] Uebersax, J.S. (1988). Validity inferences from interobserver agreement, *Psychological Bulletin* **104**, 405–416.

[8] Bloch, D.A. & Kraemer, H.C. (1989). 2 × 2 kappa coefficients: measures of agreement or association, *Biometrics* **45**, 269–287.

[9] Banerjee, M., Capozzoli, M., McSweeney, L. & Sinha, D. (1999). Beyond kappa: a review of interrater agreement measures, *The Canadian Journal of Statistics* **27**, 3–23.

[10] Fleiss, J.L. (1965). Estimating the accuracy of dichotomous judgments, *Psychometrika* **30**, 469–479.

[11] Bartko, J.J. & Carpenter, W.T. (1976). On the methods and theory of reliability, *Journal of Nervous and Mental Disease* **163**, 307–317.

[12] Automotive Industry Action Group. (2002). *Measurement Systems Analysis Reference Manual*, 3rd Edition, Automotive Industry Action Group, Southfield.

[13] McCaslin, J.A. & Gruska, G.F. (1976). Analysis of attribute gage systems, *ASQC Technical Conference Transactions* **30**, 392–399.

[14] Sweet, A.L., Tjokrodjojo, S. & Wijaya, P. (2005). An investigation of the measurements systems analysis

"Analytic method" for attribute gages, *Quality Engineering* **17**, 219–226.
[15] Boyles, R.A. (2001). Gauge capability for pass-fail inspection, *Technometrics* **43**, 223–229.
[16] Agresti, A. (1992). Modelling patterns of agreement and disagreement, *Statistical Methods in Medical Research* **1**, 201–218.
[17] Agresti, A. & Lang, J.B. (1993). Quasi-symmetric latent class models, with application to rater agreement, *Biometrics* **49**, 131–139.
[18] Agresti, A. (1988). A model for agreement between ratings on an ordinal scale, *Biometrics* **44**, 539–548.
[19] Aickin, M. (1990). Maximum likelihood estimation of agreement in the constant predictive model, and its relation to Cohen's kappa, *Biometrics* **46**, 293–302.
[20] Uebersax, J.S. & Grove, W.M. (1990). Latent class analysis of diagnostic agreement, *Statistical Methods* **9**, 559–572.

Related Articles

Gauge Repeatability and Reproducibility (R&R) Studies, Destructive Testing; **Gauge Repeatability and Reproducibility (R&R) Studies, Misclassification Rates**; **Gauge Repeatability and Reproducibility (R&R), Variance Components in**; **Maximum Likelihood**; **Measurement Systems Analysis, Capability Measures for**; **Probability Plots**; **Variance Components**.

CONNIE M. BORROR

Measurement Systems Analysis, Capability Measures for

Introduction

There are two "families" of measurement capability metrics in common use. There are those that compare measurement variation to total or part variation, and there are those that compare to the tolerance or specification width. These capability metrics are all functions of the **variance components** in the statistical model used to describe the response of interest. In addition to these capability metrics, it is also often useful to estimate the misclassification rates that arise from using the measurement system to discriminate between "good" and "bad" items being tested. An additional capability metric is obtained from the ratio of the estimated misclassification rates to the rates that would occur purely due to random chance. This section will define several capability measures, how they are related to one another, and give some common criteria for capability. Additional topics discuss the effect of the unit of measure and present a situation where none of the aforementioned metrics are particularly useful in quantifying the capability of a measurement process.

Standard MSA

The standard experiment for conducting a measurement systems analysis (**MSA**) uses a two-factor design with "parts" and "operators". The "parts" can be almost anything for which a measurement is desired – people on whom a medical test is performed, metal rods for which the length is desired, or cell phones to determine if the functionality is working properly. "Operators" can be the people doing the measurement or can be different gauges or measurement systems. The statistical model used to describe the response is the random two-factor model

$$Y_{ijk} = \mu + P_i + O_j + PO_{ij} + E_{ijk} \qquad (1)$$
$$i = 1, \ldots, p; \quad j = 1, \ldots, o; \quad k = 1, \ldots, r$$

where μ is a constant, and P_i, O_j, PO_{ij}, and E_{ijk} are jointly independent normal random variables with means of zero and variances $\sigma_P^2, \sigma_O^2, \sigma_{PO}^2$, and σ_E^2, respectively. These variance components are used to compute the measurement capability measures. A number of authors discuss how to estimate variance components in the random two-factor model (see, e.g. [1]).

Commonly Used Capability Measures

The primary purpose of MSA is to determine whether the measurement process is adequate for its intended use. This is typically done by comparing the estimated measurement variation to either the part or total variation or to the width of any applicable tolerances or specifications.

Table 1 provides parameters of interest for assessing the capability of a measurement process. Table 2 presents some commonly used capability metrics. Perhaps the most common capability metric is the precision-to-tolerance ratio (PTR). The PTR is defined as:

$$PTR = \frac{k\sqrt{\gamma_M}}{USL - LSL} \tag{2}$$

where USL and LSL are upper and lower specification or tolerance limits and k is either 5.15 or 6. The value of $k = 6$ corresponds to the middle 99.73% of a normal probability distribution and $k = 5.15$ corresponds to the middle 99%. PTR compares the total measurement variation to the width of the specifications or tolerances the items being measured are to adhere to. The Automotive Industry Action Group (AIAG) MSA manual [2] recommends the following criteria be used for PTR:

$PTR \leq 0.1$: the measurement process is capable;
$0.1 < PTR \leq 0.3$: the process is marginal;
$PTR > 0.3$: the process is not capable.

It is further recommended that a confidence level be placed on the PTR to reflect the uncertainty associated with estimating variance components. For example, for the measurement process to be capable, the PTR should be less than 0.1 with at least a 90% level of confidence. Burdick *et al.* [3] give methods for computing **confidence intervals** on a variety of MSA metrics.

It has been noted in several places (see, e.g. [4]) that the PTR does not necessarily provide a good indication of how well the measurement system performs for a particular process. This is due to the relationship between the capability of the production process and the capability of the measurement system used to monitor the production process. When the production process is highly capable of meeting the production limits, it can tolerate a higher PTR than a production process which is not capable. It is therefore useful to compute other measurement capability metrics that compare the measurement variation to total or part variation.

Table 1 Parameters of interest in MSA

Parameter	Variance component(s)	Definition
γ_P	σ_P^2	Variance among parts
γ_M	$\sigma_O^2 + \sigma_{PO}^2 + \sigma_E^2$	Total measurement variation
$\gamma_T = \gamma_P + \gamma_M$	$\sigma_P^2 + \sigma_O^2 + \sigma_{PO}^2 + \sigma_E^2$	Total variation
$\gamma_R = \gamma_P/\gamma_M$	$\sigma_P^2/(\sigma_O^2 + \sigma_{PO}^2 + \sigma_E^2)$	Ratio of part to measurement variation
$\rho_P = \gamma_P/\gamma_T$	$\sigma_P^2/(\sigma_P^2 + \sigma_O^2 + \sigma_{PO}^2 + \sigma_E^2)$	Ratio of part to total variation
$\rho_M = \gamma_M/\gamma_T$	$(\sigma_O^2 + \sigma_{PO}^2 + \sigma_E^2)/(\sigma_P^2 + \sigma_O^2 + \sigma_{PO}^2 + \sigma_E^2)$	Ratio of measurement to total variation

Table 2 Common capability metrics

Metric	Definition	Criteria
PTR	$\dfrac{k\sqrt{\gamma_M}}{USL - LSL}$	$k = 5.15$ or 6 $PTR \leq 0.1$: capable $0.1 < PTR \leq 0.3$: marginal $PTR > 0.3$: not capable
SNR	$\sqrt{2\gamma_R}$	$SNR > 10$: excellent $5 < SNR \leq 10$: acceptable $SNR < 5$: unacceptable
$PTV = \%R\&R$	$\sqrt{\rho_M} \times 100\%$	$PTV \leq 10\%$: excellent $10\% < PTV \leq 30\%$: acceptable $PTV > 30\%$: unacceptable
DR	$\sqrt{\dfrac{1+\rho_P}{1-\rho_P}}$	$DR > 4$: acceptable $2 < DR \leq 4$: marginal $DR < 2$: unacceptable

The signal-to-noise ratio (SNR) is defined as:

$$SNR = \sqrt{2\gamma_R} \tag{3}$$

The AIAG MSA manual defines SNR as the number of distinct data categories that can be reliably distinguished by the measurement system. A value of 5 or greater is recommended. Based on experience over a number of years, the author recommends the following criteria be met with at least a 90% level of confidence:

$SNR > 10$: the measurement process is excellent;
$5 < SNR \leq 10$: the process is acceptable;
$SNR < 5$: the process is unacceptable.

Another commonly used metric in the same class is the percent of total variation (PTV) or equivalently the percent of reproducibility and repeatability (%R&R). The PTV is defined as:

$$PTV = \sqrt{\rho_M} \times 100\% \tag{4}$$

AIAG recommends criteria of:

$PTV \leq 10\%$: the measurement process is excellent;
$10\% < PTV \leq 30\%$: the process is acceptable;
$PTV > 30\%$: the process is unacceptable.

Again, the author further recommends that the PTV be bounded by confidence intervals and the criteria be applied using at least a 90% confidence level.

Another metric in the same class is the discrimination ratio (DR) (see [5] or [6]). The DR is defined as:

$$DR = \sqrt{\frac{1+\rho_P}{1-\rho_P}} \tag{5}$$

ρ_P is the intraclass **correlation** defined as the correlation between two measurements taken on the same part. Mader *et al.* [5] suggest the following criteria:

$DR > 4$: the measurement process is acceptable;
$2 < DR \leq 4$: the process is marginal;
$DR < 2$: the process is unacceptable.

An additional recommendation is that DR be bounded by confidence intervals and at least a 90% confidence level be used to judge the capability of the measurement process.

Interrelationships among Capability Metrics

$$PTV = \frac{100}{\sqrt{\dfrac{SNR^2}{2}+1}} \qquad SNR = \sqrt{2\left(\frac{100^2}{PTV^2}-1\right)}$$

$$\rho_P = \frac{SNR^2}{SNR^2+2} \qquad SNR = \sqrt{\frac{2\rho_P}{1-\rho_P}}$$

$$DR = \sqrt{SNR^2+1} \quad SNR = \sqrt{DR^2-1}$$

Using Misclassification Rates to Determine Capability

In a "test" situation where the measurement system is used to discriminate among items such as a medical test for disease or a functional test for cell phones, Burdick *et al.* [3] have suggested a misclassification index for assessing capability. The misclassification rates or test errors are sometimes called *false failures* and *missed faults* or *false positives* and *false negatives*. The misclassification rates are directly analogous to the type I (α) and type II (β) errors in statistical **hypothesis testing**. Methods to estimate the misclassification rates have been described by several authors (see, e.g. [7]) and include numerical integration and computer simulation.

Suppose we wish to test a product to determine whether or not it meets acceptable limits. In the case of two-sided test limits and specifications, the situation appears as in Figure 1. Following the notation in [7], define the following terms:

Y	=	true value $\sim N(\mu_P, \gamma_P)$
X	=	measured value $= Y + E$
E	=	measurement error $\sim N(0, \gamma_M)$
$f(x, y)$	=	joint pdf
$f(x)$	=	marginal pdf for X
$f(y)$	=	marginal pdf for Y

SL and SU are the lower and upper product specifications
TL and TU are the lower and upper test limits.

The test limits may or may not be equal to the product specifications.

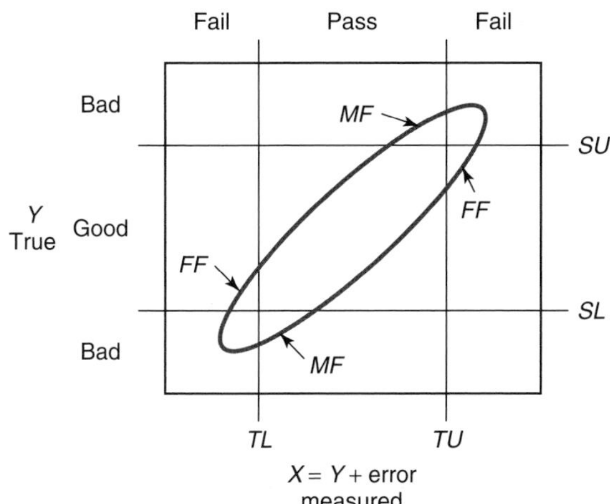

Figure 1 Misclassification rates with two-sided test limits and specifications

The following probabilities are obtained by numerical integration or computer simulation.

$$P(\text{Bad}) = \int_{-\infty}^{SL} f(y)\,\mathrm{d}y + \int_{SU}^{+\infty} f(y)\,\mathrm{d}y = 1 - \pi \tag{6}$$

$$P(\text{Fail}) = \int_{-\infty}^{TL} f(x)\,\mathrm{d}x + \int_{TU}^{+\infty} f(x)\,\mathrm{d}x \tag{7}$$

$$\begin{aligned} P(\text{Pass, Bad}) &= \text{Joint MF} \\ &= \int_{-\infty}^{SL}\int_{TL}^{TU} f(x,y)\,\mathrm{d}x\,\mathrm{d}y \\ &\quad + \int_{SU}^{+\infty}\int_{TL}^{TU} f(x,y)\,\mathrm{d}x\,\mathrm{d}y \\ &= \beta \end{aligned} \tag{8}$$

Both conditional and joint misclassification rates are obtained using Bayes formula and some algebra.

$$\begin{aligned} P(\text{Fail}|\text{Good}) &= \text{FF} \\ &= \frac{P(\text{Fail}) - P(\text{Bad}) + P(\text{Pass, Bad})}{1 - P(\text{Bad})} \\ &= \alpha_c \end{aligned} \tag{9}$$

$$P(\text{Pass}|\text{Bad}) = \text{MF} = \frac{P(\text{Pass, Bad})}{P(\text{Bad})} = \beta_c \tag{10}$$

$$\begin{aligned} P(\text{Fail, Good}) &= \text{Joint FF} \\ &= \text{FF}(1 - P(\text{Bad})) = \alpha \end{aligned} \tag{11}$$

The joint misclassification rates, α and β are typically reported because they are relevant to the whole population of products.

When considering whether the misclassification rates are sufficiently small, Burdick *et al.* [3] have suggested comparing them to what could be obtained by chance. The proposal goes as follows. Suppose the value of π, $P(\text{Good})$, in equation 6 is known. A chance process for classifying parts is to take no measurements and randomly classify π of the parts as good and $1 - \pi$ as bad. For this chance process, the probability of a false failure is

$$\begin{aligned} \alpha^* &= P(\text{Fail, Good}) \\ &= (1 - \pi)\pi \end{aligned} \tag{12}$$

Likewise, the probability of a missed fault for this chance process is

$$\begin{aligned} \beta^* &= P(\text{Pass, Bad}) \\ &= \pi(1 - \pi) \end{aligned} \tag{13}$$

Thus, for a measurement system to be useful, the misclassification rates α and β should be less than what could be obtained by chance, α^* and β^*. A similar argument can be made for the conditional probabilities α_c and β_c. In this case, the conditional probabilities that could be obtained by chance are $\alpha_c^* = 1 - \pi$ and $\beta_c^* = \pi$.

Burdick *et al.* [3] proposed the following indices be used to assess the capability of the measurement system to discriminate good product from bad.

$$\alpha_{\text{index}} = \frac{\alpha}{\alpha^*} \tag{14}$$

$$\beta_{\text{index}} = \frac{\beta}{\beta^*} \tag{15}$$

And since $\alpha_c/\alpha_c^* = \alpha_{\text{index}}$ and $\beta_c/\beta_c^* = \beta_{\text{index}}$, these indices can be used for both the conditional and joint definitions of misclassification rates. The criteria for establishing the capability of the measurement process is for both α_{index} and β_{index} to be less than unity with at least a 90% confidence level. Burdick *et al.* [3] suggested using generalized confidence intervals and Larsen [7] proposed a different computer simulation method which provides similar results.

Effect of the Unit of Measure

Experience has shown that SNR and PTR are largely scale invariant so long as the data do not contain **outliers**. For example, decibel (db) is a common unit for electrical measurement. Decibel can be defined as the log of the ratio of two voltages. Whether the MSA is conducted directly in decibel units or with a linearizing transformation does not matter in so far as the capability measures are concerned so long as the data do not contain outliers. SNR and PTR values will be similar either way. Doing the analysis both with raw and transformed data and finding a large difference in either SNR or PTR often indicates the presence of outliers. And these outliers should be examined for validity and possible exclusion from the MSA.

A Situation Where the Usual Metrics Can Be Noninformative

It is increasingly the case in testing consumer electronic devices (cell phones, PDAs, etc.) that the measurement system is used to "align" the device rather than determine whether it performs within acceptable limits. With alignment, the value the device initially has is unimportant because the measurement system changes the value of the device to lie within a narrow range. When device parameters are aligned, the resulting "part" variation is similar to that found in the measurement system. This implies that the SNR will tend to be well below the usual criteria of 5. Further, parameters which are aligned do not have test limits in the usual sense because the objective is not to determine pass or fail but rather to set the parameter to a predefined value or narrow range of values called *alignment limits*. The alignment limits are a trade-off between providing a sufficient margin with industry standard specifications and aligning to a nominal value that often takes excessive test time. The object of the alignment is not to center the parameter within the limits but just to get within them. Often there are no product specifications warranted to the end customer on aligned parameters of consumer electronic devices. This situation makes the PTR difficult to apply as well. Using the alignment limits in the denominator of the PTR typically provides values too high to meet the usual criteria. Finally, because the measurement system is not used to determine pass or fail, the misclassification indices are not applicable.

The question of how to assess the capability of a measurement system used to align parameters is difficult to answer. Larsen [8] makes several recommendations which include comparing the total measurement system variability found from an MSA to a tolerance stack of component tolerances in the measurement system. If the measurement variability from the MSA is less than the value from the tolerance stack, then the measurement process is at least meeting capability expectations. However, as noted, this is no guarantee that the measurement system is adequate in a particular application. It is also suggested that the measurement equipment be calibrated and traceable to National Institute of Standards and Technology (NIST) so that system-to-system variation is lessened. Finally, the PTR might be applicable if the alignment limits are used with relaxed criteria or if product specifications are developed which reflect acceptable functional performance to the end customer.

References

[1] Vardeman, S.B. & VanValkenburg, E.S. (1999). Two-way random-effects analyses and gauge R&R studies, *Technometrics* **41**(3), 202–211.

[2] Automotive Industry Action Group. (1995). *Measurement Systems Analysis*, 2nd Edition, AIAG, Detroit.

[3] Burdick, R.K., Borror, C.M. & Montgomery, D.C. (2005). *Design and Analysis of Gauge R&R Studies: Making Decisions with Confidence Intervals in Random and Mixed ANOVA Models, SIAM-ASA Series on Statistics and Applied Probability*, SIAM, Philadelphia, ASA, Alexandria.

[4] Montgomery, D.C. & Runger, G.C. (1993). Gauge capability and designed experiments, part I: basic methods, *Quality Engineering* **6**(1), 115–135.

[5] Mader, D.P., Prins, J. & Lampe, R.E. (1999). The economic impact of measurement error, *Quality Engineering* **11**(4), 563–574.

[6] Wheeler, D.J. (1992). Problems with gauge R&R studies, in 46th ASQC Quality Congress Transactions, Nashville, pp. 179–185.

[7] Larsen, G.A. (2003). Measurement system analysis in a production environment with multiple test parameters, *Quality Engineering* **16**(2), 297–306.

[8] Larsen, G.A. (2002). Measurement system analysis: the usual metrics can be noninformative, *Quality Engineering* **15**(2), 293–298.

GREG A. LARSEN

Measurement Systems Analysis, Overview

Introduction

Measurement systems analysis (MSA) is a collection of statistical techniques designed to quantify the variability in a manufacturing process that is attributable to the metrology of that process. No process variability can be measured directly; that is, no process variability can be quantified without including the variability induced by measuring that process. MSA is therefore a study, separate from the processing of the product, that estimates the variability of the measurement process under conditions similar to those of the manufacturing process.

MSA is also called, variously, *measurement capability analysis* (in the context of determining whether or not a measurement system's variability is small enough compared with total process variability to adequately measure the process), *gauge studies*, or *gauge repeatability and reproducibility* (*GR&R*) *studies*.

Historical Perspective

Some of the earliest references to MSA in the statistical literature are found in the 1930s and 1940s. However, measurement precision was of concern even in older times. Shewhart [1] refers readers to a book by Goodwin [2] that was based on lecture notes from a course at MIT in the late 1800s. In 1948, Grubbs [3] laid out the first equations for estimating variability due to measurement repeatability.

The rise of MSA as a fundamental component of a process characterization study seems to have occurred in the 1980s, along with the rise of **statistical process control** (SPC) and other statistical quality methods in American manufacturing. Both the automotive and semiconductor industries mandate the use of gauge studies in quality improvement projects. Gauge studies are an integral part of both the Deming "Plan, Do, Check, Act" cycle, and the Six Sigma "Define, Measure, Analyze, Improve, Control" procedure. Wheeler and Lyday [4] include details on running gauge studies. In the early 1990s, SEMATECH included metrology system characterization as a step in their equipment qualification plan [5]. A good review of methods for MSA can be found in Burdick *et al.* [6].

The design of the gauge study experiment has changed over the years. Early on, use of the standard 10-part, three-operator experimental design was extremely popular. In this design, multiple repeated measurements are performed by each operator, and the study yields estimates of variability due to repeats as well as variability due to operators. Another type of design used was the so-called $30-30$ study, in which two studies are performed, the first with one setup and 30 repeated measurements, the other with 30 setups of one measurement each. The variance of the 30 repeated measurements is an estimate of repeatability, and the variance of the 30 setups is an estimate of reproducibility.

Terminology and methods of MSA have been somewhat standardized over the years; however, they have not been truly standardized. Care must be taken when performing or discussing MSA studies. Terms such as *repeatability* and *reproducibility* have been replaced by *variability due to repeats* or *variability due to operators*, for example. Industrial standard bodies have produced guideline documents that attempt standardization. See [7] for the automotive industry's guide to MSA or SEMI E89 [8] for the semiconductor industry's standard.

Experimental Design

A measurement process consists of the measurement tools themselves, including all hardware and software, as well as the procedures for using the tools: operators, setup and handling procedures, any off-line calculations or data entry, and **calibration** or other **preventive maintenance** frequency and technique.

A good measurement study will address the following questions: How big is the measurement error? What are the sources of the measurement error? Is the tool stable over time? Is the tool capable of making measurements for this process? What needs to be done to improve the measurement process?

Designing a MSA study is similar to designing any experiment (*see* **Gauge Repeatability and Reproducibility (R&R) Studies**). Up-front planning should include listing specific goals, obtaining measurement standards to use in the study, reviewing previous studies, identifying potential sources of variability, estimating time and costs, assigning responsibilities, and securing management buy-in. Some of the potential sources of variability will not be studied, for example, yearly preventive maintenance activities. Others will be included in the study, for example, operator training and techniques or issues related to loading parts into the measurement tool. The factors that are not directly under study should still be documented and possibly added to a corrective action plan in the future.

Most gauge study information assumes the measurement tool gives continuous data and the true value of the standard remains constant over time (*see* **Gauge Repeatability and Reproducibility (R&R) Studies, Destructive Testing**).

The first step is to determine if the measurement tool is calibrated in the ranges of interest (*see* **Calibration**).

The fundamental equation of process variability is $\sigma^2_{\text{process}} = \sigma^2_{\text{product}} + \sigma^2_{\text{measurement}}$. That is, the variability seen in a process is made up of both variability due to the manufacturing process as well as variability due to the measurement process. Most gauge studies are designed to yield an estimate of $\sigma^2_{\text{measurement}}$. Given a list of possible contributors to $\sigma^2_{\text{measurement}}$, designing an experiment to estimate the size of the contributors to the total variability is usually straightforward. For example, consider an ellipsometer, used to measure the thickness of films on silicon wafers. Possible significant sources of variability include day-to-day effects, wafer-handling effects, and the actual measurement itself. The latter factor is often called *repeatability*. A good experiment will, therefore, include multiple days, multiple loads per day, and multiple repeated measurements per load. Note the hierarchical nature of the experimental factors. If the tool is not automated, multiple operators, or at least multiple levels of operator training, might be considered. If significant effects due to changes in ambient temperature and humidity over the day are expected, then measurements might be taken over multiple labor shifts.

Analytical Methods

Most of these effects will be treated as random factors in the ensuing model. Operators are often treated as fixed, especially if the operators in the study are the only operators used in the process or if different operators represent different (fixed) levels of competency or training. Otherwise, Operator is treated as a random effect, and a variance component due to operators is calculated (*see* **Gauge Repeatability and Reproducibility (R&R), Variance Components in**; **Gauge Repeatability and Reproducibility (R&R) Studies, Confidence Intervals for**).

Once the estimate of measurement variability has been calculated, it can be used to answer the question of whether or not the measurement tool is capable of measuring the process under study. One popular metric for determining gauge capability is the precision-to-tolerance ratio, P/T. P/T is defined as the ratio of the measurement distribution to the process specification window, that is $6 \times \sigma_{\text{measurement}}/(USL - LSL)$. In automotive applications, the P/T ratio is often calculated as $5.15 \times \sigma_{\text{measurement}}/(USL - LSL)$. A rule of thumb is that a gauge with $P/T < 10\%$ is considered capable, one with $10\% < P/T < 30\%$ might be suitable, and one with $P/T > 30\%$ is not suitable for manufacturing. A large P/T ratio increases the misclassification probabilities (*see* **Gauge Repeatability and Reproducibility (R&R) Studies, Misclassification Rates**).

When specification limits are poorly defined, a **signal-to-noise** ratio (SNR) can be calculated. The SNR most often used is $SNR = \sigma_{\text{product}}/\sigma_{\text{measurement}}$. An SNR of at least 10 means that the measurement system is adequate for the process. A control system with a small SNR will have reduced sensitivity, meaning a longer time until an out-of-control condition is detected (*see* **Average Run Lengths and Operating Characteristic Curves**).

In addition to the metrology portion of the gauge study, care must be taken when choosing parts to use in the study. As with any statistical study, the results are applicable only to the population from which the sample was drawn. Therefore, it is crucial that the parts used in the study are representative of the process the measurement tool will be used for. Burdick *et al.* recommend using more parts and fewer measurements, as a rule.

Graphical Methods

In addition to the graphs generated for any other modeling effort, such as a time plot of the response, or a scatterplot of the response *versus* the factors, a key graph used in the analysis of gauge study data

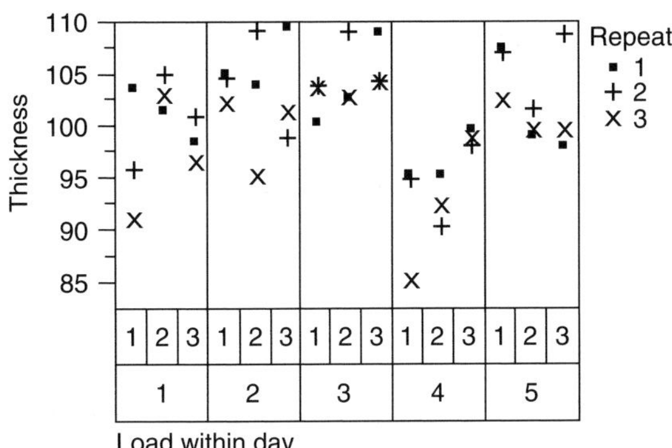

Figure 1 Multi-vari chart of film thickness gauge study data

is the multi-vari plot, also called a *variability graph*. This graph plots the response on the vertical axis, with all hierarchical factors but the lowest on the horizontal axis. Figure 1 illustrates the concept. The data are simulated from a gauge study on a metrology tool measuring wafer film thickness. The study was conducted over 5 days, with three loads (setups) per day and three repeated measurements per load. One part was used in this example.

From the graph, it is easily seen that day-to-day and repeat-to-repeat variability are large, while load-to-load variability is small. The **variance components** and percent contribution of each to the total are found in Table 1.

Stability

It is important for a measurement system to remain stable, as a shift in the measurement distribution can lead to false out-of-control signals or false in-control behavior on the process **control chart**.

Stability of the measurement system is determined in a similar way to stability of a process. Data are collected over time, oversampling from the potential significant sources of variability, and control charts are calculated (*see* **Control Charts, Overview**).

Typically, a pool of standards is used for control charting a measurement system. Control limits are calculated using the entire pool, and then one standard is measured at specified intervals for the control system. When that standard breaks or is otherwise retired, the control limits need only to be shifted to match the mean value of the new standard from the pool. The width of the limits remains the same.

As with any SPC system, an out-of-control action plan should be developed. Often the original brainstorming session that determined the factors to be considered in the gauge study will provide information for the out-of-control action plan. For example, if yearly preventive maintenance activities have the potential to cause measurement system variability, but are not included in the gauge study, one line in the action plan flow chart might read: "Has the tool been returned properly from preventive maintenance activities?"

Table 1 Variance components for the film thickness gauge study

Source	Variance component	Percent of total (%)
Day	13.95	43
Load	1.98	6
Repeat	16.26	51
Total	32.19	100

Summary

MSA is an important set of techniques used to determine if the measurement process is suitable for a manufacturing process. MSA gives an estimate of the measurement process variability and goodness metrics can then be calculated. Verification of the suitability of a measurement process should be the first step in any process improvement activity.

References

[1] Shewhart, W.A. (1986). *Statistical Method from the Viewpoint of Quality Control*, Dover.

[2] Goodwin, H.M. (1908). *Elements of the Precision of Measurement and Graphical Methods*, G.H. Ellis Co.

[3] Grubbs, F.E. (1948). On estimating precision of measuring instruments and product variability, *Journal of the American Statistical Association* **43**, 243–264.

[4] Wheeler, D.J. & Lyday, R.W. (1989). *Evaluating the Measurement Process*, SPC Press.

[5] SEMATECH Qualification Plan Guidelines for Engineers (1992). Technology Transfer Document 92061182A-GEN, and SEMATECH Qualification Plan Overview, Technology Transfer Document 91050538A-GEN.

[6] Burdick, R.K., Borror, C.M. & Montgomery, D.C. (2003). A review of methods for measurement systems capability analysis, *Journal of Quality Technology* **35**, 342–354.

[7] Automotive Industry Action Group (2002). *Measurement Systems Analysis*.
[8] SEMI E89-1106E (2006). *Guide for Measurement System Analysis (MSA)*, http://www.semi.org (accessed 2006).

DIANE K. MICHELSON

Measures of Association

Introduction

Measures of association quantify the statistical dependence of two or more categorical variables. Many measures of association have been proposed; in this article, we restrict our attention to the ones that are most commonly used. We broadly divide the measures into those suitable for (a) unordered (nominal) categorical variables, and (b) ordered categorical variables.

Measures of Association for Unordered Categorical Variables

As an example, consider the contingency table shown in Table 1, with sex (s) as the row variable and employment status (e) as the column variable.

Measures Based on the Odds Ratio

The odds that an event will occur are computed by dividing the probability that the event will occur by the probability that the event will not occur. For males, the odds of being employed equal the probability of being employed divided by the probability of not being employed:

$$odds_{y|m} = \frac{p(y|m)}{1 - p(y|m)} = \frac{p(y|m)}{p(n|m)} = \frac{42/75}{33/75} = \frac{0.56}{0.44} = 1.27 \quad (1)$$

Similarly, for females:

$$odds_{y|f} = \frac{p(y|f)}{1 - p(y|f)} = \frac{p(y|f)}{p(n|f)} = \frac{40/165}{125/165} = \frac{0.242}{0.758} = 0.319 \quad (2)$$

These odds are large when employment is likely, and small when it is unlikely. The odds can be interpreted as the number of times that employment occurs for each time it does not. For example, since the odds of employment for males is 1.27, approximately 1.27 males are employed for every unemployed male, or, in round numbers, 5 males are employed for every four males that are unemployed. For females, there is approximately one employed female for every three unemployed females.

The association between sex and employment status can be expressed by the degree to which the two odds differ. This difference can be summarized by the odds ratio (α):

$$\alpha = \frac{odds_{y|f}}{odds_{y|m}} = \frac{p(y|f)/p(n|f)}{p(y|m)/p(n|m)} = \frac{1.27}{0.32} = 3.98 \quad (3)$$

This odds ratio indicates that for every one employed female, there are four employed males. Although not necessary, it is customary to place the larger of the two odds in the numerator.

Odds ratios are not symmetric around one: An odds ratio larger than one by a given amount indicates a smaller effect than an odds ratio smaller than one by the same amount [1]. While the magnitude of an odds ratio less than one is restricted to the range between zero and one, odds ratios greater than one and are not restricted, allowing the ratio to potentially take on any value. If the natural logarithm (ln) of the odds ratio is taken, the odds ratio is symmetric above and below one, with $\ln(1) = 0$. For example, to take the 2×2 example above, the odds ratio of a male being employed, compared to a female, was

Table 1 Employment status of males and females

		Employment status		
		Employed (y)	Not employed (n)	Total
Sex	Male (m)	42	33	75
	Female (f)	40	125	165
	Total	82	58	40

3.98. If we reverse that and take the odds ratio of a female being employed compared to a male, it is $0.32/1.27 = 0.25$. These ratios are clearly not symmetric about 1.00. However, $\ln(3.98) = 1.38$ and $\ln(0.25) = -1.38$, and these ratios are symmetric.

Yule's Q Coefficient of Association. Both α and $\ln(\alpha)$ range from $-\infty$ to $+\infty$. to restrict the odds ratio within the interval -1 to $+1$, Yule introduced the coefficient of association, Q, for 2×2 tables. Its definition is [2]:

$$Q = \frac{n_{11}n_{22} - n_{12}n_{21}}{n_{11}n_{22} + n_{12}n_{21}} = \frac{\alpha - 1}{\alpha + 1} \quad (4)$$

Therefore, if odds ratio (α) expressing the relationship between sex and employment is 3.98,

$$Q = \frac{(42)(125) - (33)(40)}{(42)(125) + (33)(40)} = \frac{3.98 - 1}{3.98 + 1} = 0.59 \quad (5)$$

Yule's Q is algebraically equal to the 2×2 Goodman–Kruskal γ coefficient (described below), and thus measures the degree of concordance or discordance between two variables.

Yule's Coefficient of Colligation. As an alternative to Q, Yule proposed the *coefficient of colligation* [2]:

$$\widehat{Y} = \frac{\sqrt{n_{11}n_{22}} - \sqrt{n_{12}n_{21}}}{\sqrt{n_{11}n_{22}} + \sqrt{n_{12}n_{21}}} = \frac{\sqrt{a} - 1}{\sqrt{a} + 1} \quad (6)$$

The coefficient of colligation between sex and employment is

$$\widehat{Y} = \frac{\sqrt{(42)(125)} - \sqrt{(33)(40)}}{\sqrt{(42)(125)} + \sqrt{(33)(40)}} = \frac{\sqrt{3.98} - 1}{\sqrt{3.98} + 1} = 0.33 \quad (7)$$

The coefficient of colligation has a different interpretation than the coefficient of association, and is interpreted as a Pearson product–moment **correlation** coefficient r (see below), but is not algebraically equivalent to one.

Both Q and $\widehat{Y}$ are symmetric measures of association, and are invariant under changes in the ordering of rows and columns.

Measures Based on Pearson's χ^2

$$\chi^2 = \sum_i \sum_j \frac{\left(n_{ij} - \widehat{n}_{ij}\right)^2}{n_{ij}} \quad (8)$$

where $\widehat{n}_{ij}$ is the expected frequency in cell ij, and can be usefully transformed into several measures of association [3].

The Φ Coefficient. The ϕ *coefficient* is defined as [4]:

$$\Phi = \sqrt{\frac{\chi^2}{N}} \quad (9)$$

For sex and employment,

$$\Phi = \sqrt{\frac{23.12}{240}} = 0.31 \quad (10)$$

The Φ coefficient can vary between 0 and 1, and is algebraically equivalent to $|r|$. However, the lower and upper limits of Φ in a 2×2 table are dependent on two conditions. In order for Φ to equal -1 or $+1$, (a) $(n_{11} + n_{12}) = (n_{21} + n_{22})$, and (b) $(n_{11} + n_{21}) = (n_{12} + n_{22})$.

The Pearson Product-Moment Correlation Coefficient. The general formula for the *Pearson product–moment correlation coefficient* is

$$r = \frac{\sum_i \left(X_i - \overline{X}\right)\left(Y_i - \overline{Y}\right)}{N s_x s_y} \quad (11)$$

where X and Y are two continuous, interval-level variables. The categories of a dichotomous variable can be coded 0 and 1, and used in the formula for r. In a 2×2 contingency table, the calculations reduce to [5]:

$$r = \frac{n_{11}n_{22} - n_{12}n_{21}}{\sqrt{n_{1+}n_{2+}n_{+1}n_{+2}}} \quad (12)$$

For sex and employment, $r = \{(42)(125) - (33)(40) / \sqrt{[(75)(165)(82)(158)]}\} = 0.31$.

A symmetric measure of association is also invariant to row and column order, r varies between -1 to $+1$. From the formula, it is apparent that $r = 1$ if $n_{12} = n_{21} = 0$, and $r = -1$ if $n_{11} = n_{22} = 0$. In a standardized 2×2 table, where each marginal probability $= 0.5$, $r = \widehat{Y}$; otherwise $|r| < |\widehat{Y}|$ except when the variables are independent or completely related.

Measures of Predictive Association

Another class of measures of association for nominal variables is measures of prediction analogous in concept to the multiple correlation coefficient or R^2 in regression analysis [2](*see* **Coefficient of Determination (R^2)**). When there is an association between two nominal variables X and Y, then knowledge about X allows one to obtain knowledge about Y, more knowledge than would have been available without X. Let Δ_Y be the dispersion of Y and $\Delta_{Y \cdot X}$ be the conditional dispersion of Y given X. A measure of prediction

$$\phi_{Y \cdot X} = 1 - \frac{\Delta_{Y \cdot X}}{\Delta_Y} \tag{13}$$

compares the conditional dispersion of Y given X to the unconditional dispersion of Y, similar to how the multiple correlation coefficient compares the conditional variance of the dependent variable to its unconditional variance [3]. When $\phi_{Y.X} = 0$, X and Y are independently distributed; when $\phi_{Y.X} = 1$, X is a perfect predictor of Y.

We describe four measures that operationalize the idea underlying $\phi_{Y \cdot X} = 0$.

The first measure is the *asymmetric lambda coefficient* of Goodman and Kruskal [6]:

$$\lambda(Y = c|X = r) = \frac{\sum_i \max_j p_{ij} - \max_j p_{+j}}{1 - \max_j p_{+j}}$$
$$\lambda(X = r|Y = c) = \frac{\sum_j \max_i p_{ij} - \max_i p_{+i}}{1 - \max_i p_{+i}} \tag{14}$$

The second is the *symmetric lambda coefficient* (λ) of Goodman and Kruskal [6]:

$$\lambda = \frac{\sum_i \max_j p_{ij} + \sum_j \max_i p_{ij} - \max_j p_{+j} - \max_i p_{i+}}{2 - \max_j p_{+j} - \max_i p_{i+}} \tag{15}$$

The third is the *asymmetric uncertainty coefficient* of Theil [7]:

$$U(Y = c|X = r) = \frac{\left(-\sum_i (p_{i+}) \ln(p_{i+})\right) + \left(-\sum_j (p_{+j}) \ln(p_{+j})\right) - \left(-\sum_i \sum_j (p_{ij}) \ln(p_{ij})\right)}{-\sum_j (p_{+j}) \ln(p_{+j})}$$
$$U(X = r|Y = c) = \frac{\left(-\sum_i (p_{i+}) \ln(p_{i+})\right) + \left(-\sum_j (p_{+j}) \ln(p_{+j})\right) - \left(-\sum_i \sum_j (p_{ij}) \ln(p_{ij})\right)}{-\sum_i (p_{i+}) \ln(p_{i+})} \tag{16}$$

The fourth measure of predictive association is U, Theil's *symmetric uncertainty coefficient* [7]:

$$U = \frac{2\left[\left(-\sum_i (p_{i+}) \ln(p_{i+})\right) + \left(-\sum_j (p_{+j}) \ln(p_{+j})\right) - \left(-\sum_i \sum_j (p_{ij}) \ln(p_{ij})\right)\right]}{\left(-\sum_i (p_{i+}) \ln(p_{i+})\right) + \left(-\sum_j (p_{+j}) \ln(p_{+j})\right)} \tag{17}$$

We use the contingency table shown in Table 2 to illustrate, in turn, the computation of the four measures of predictive association.

The first measure, asymmetric $\lambda(Y|X)$, is interpreted as the relative decrease in the probability of incorrectly predicting the column variable Y between not knowing X and knowing X. As such, lambda can also be interpreted as a proportional reduction in

Table 2 The computation of the four measures of predictive association

	Y_1	Y_2	Y_3	Y_4	Total
X_1	28	32	49	36	145
X_2	44	21	17	12	94
X_3	46	31	53	63	193
Total	118	84	119	111	

variation measure – how much of the variance in one variable is accounted for by the other? In the equation for $\lambda(Y|X)$, $1 - \max p_{ij}$ is the minimum probability of error from the prediction that Y is a function of X, and $1 - \max_j p_{+j}$ is the minimum probability of error from a prediction that Y is a constant over X. For the data we obtain:

$$\lambda(Y|X) = \frac{((49+44+63)/432) - (119/432)}{1-(119/432)} = \frac{0.36-0.28}{1-0.28} = 0.11 \quad (18)$$

Thus, there is an 11% improvement in predicting the column variable Y given the knowledge of the row variable X. Asymmetric λ has the range $0 \le \lambda(Y|X) \le 1$.

In general, $\lambda(Y|X) \neq \lambda(X|Y); \lambda(X|Y)$ is the relative decrease in the probability of incorrectly predicting the row variable X between not knowing Y and knowing Y – hence, the term "asymmetric". For these data,

$$\lambda(X|Y) = \frac{((46+32+53+63)/432) - (193/432)}{1-(193/432)} = \frac{0.45-0.44}{1-0.44} = 0.02 \quad (19)$$

Symmetric λ is the average of the two asymmetric lambdas and has the range $0 \le \lambda \le 1$. For our example,

$$\lambda = \frac{0.36+0.45-0.28-0.44}{2-0.28-0.44} = \frac{0.09}{1.28} = 0.07 \quad (20)$$

Theil's asymmetric uncertainty coefficient, $U(Y|X)$, is the proportion of uncertainty in the column variable Y that is explained by the row variable X, or, alternatively, $U(X|Y)$ is the proportion of uncertainty in the row variable X that is explained by the column variable Y. The asymmetric uncertainty coefficient has the range $0 \le U(Y|X) \le 1$. For $U(Y|X)$, we obtain for these data:

$$U(Y|X) = \frac{(1.06)+(1.38)-(2.40)}{1.38} = 0.03 \quad (21)$$

and for $U(X|Y)$,

$$U(X|Y) = \frac{(1.06)+(1.38)-(2.40)}{1.06} = 0.04 \quad (22)$$

The symmetric U is computed as

$$U = \frac{2\,[(1.06)+(1.38)-(2.40)]}{(1.06)+(1.38)} = 0.03 \quad (23)$$

Both the asymmetric and symmetric uncertainty coefficients have the range $0 \le U \le 1$.

The quantities in the numerators of the equations for the asymmetric lambda and uncertainty coefficients are interpreted as measures of variation for nominal responses: In the case of lambda coefficients, the variation measure is called the *Gini concentration*, and the variation measure used in the uncertainty coefficients is called the *entropy* [2]. More specifically, in $\lambda(Y|X)$, the numerator represents the variance of the Y or column variable:

$$V(Y) = \sum_i \max_j p_{ij} - \max_j p_{+j} \quad (24)$$

and in $\lambda(X|Y)$, the numerator represents the variance of the X or row variable:

$$V(X) = \sum_j \max_i p_{ij} - \max_i p_{i+} \quad (25)$$

Analogously, the numerator of $U(Y|X)$ is the variance of Y,

$$\begin{aligned} V(Y) &= \left(-\Sigma_i(p_{i+})\ln(p_{i+})\right) \\ &\quad + \left(-\Sigma_j(p_{+j})\ln(p_{+j})\right) \\ &\quad - \left(-\Sigma_i\Sigma_j(p_{ij})\ln(p_{ij})\right) \end{aligned} \quad (26)$$

and the numerator of $U(X|Y)$ is the variance of X,

$$\begin{aligned} V(X) &= \left(-\Sigma_i(p_{i+})\ln(p_{i+})\right) \\ &\quad + \left(-\Sigma_j(p_{+j})\ln(p_{+j})\right) \\ &\quad - \left(-\Sigma_i\Sigma_j(p_{ij})\ln(p_{ij})\right) \end{aligned} \quad (27)$$

However, in these measures, there is a positive relationship between the number of categories and the magnitude of the variation, thus introducing ambiguity in evaluating both the variance of the variables and their relationship.

Measures of Association for Ordered Categorical Variables

Measures of Concordance/Discordance

A pair of observations is *concordant* if the observation that ranks higher on X also ranks higher on Y. A pair of observations is *discordant* if the observation that ranks higher on X ranks lower on Y. The number of concordant pairs is

$$C = \sum_{i<i'} \sum_{j<j'} n_{ij} n_{i'j'} \tag{28}$$

where the first summation is over all rows $i < i'$, and the second summation is over all pairs of columns $j < j'$. The number of discordant pairs is

$$D = \sum_{i<i'} \sum_{j>j'} n_{ij} n_{i'j'} \tag{29}$$

where the first summation is over all rows $i < i'$, and the second summation is over all columns $j > j'$.

To illustrate the calculation of C and D, consider the 3×3 contingency table shown in Table 3 cross-classifying income with education.

In this table, the number of concordant pairs is

$$\begin{aligned} C &= 302(331 + 250 + 155 + 185) \\ &\quad + 105(250 + 185) + 409(155 + 185) \\ &\quad + 331(185) = 524\,112 \end{aligned} \tag{30}$$

Note that the process of identifying concordant pairs amounts to taking each cell frequency (e.g., 302) in turn, and multiplying that frequency by each of the cell frequencies "southeast" of it (e.g., 331, 250, 155, 185).

The number of discordant pairs is

$$\begin{aligned} D &= 23(409 + 331 + 15 + 155) \\ &\quad + 105(409 + 15) + 250(15 + 155) \\ &\quad + 331(15) = 112\,915 \end{aligned} \tag{31}$$

Discordant pairs can readily be identified by taking each cell frequency (e.g., 23) in turn, and multiplying that cell frequencies by each of the cell frequencies "southwest" of it (e.g., 409, 331, 15, 155).

Let $n_{i+} = \sum_j n_{ij}$ and $n_{+j} = \sum_i n_{ij}$. We can express the total number of pairs of observations as

$$\frac{n(n-1)}{2} = C + D + T_X + T_Y - T_{XY} \tag{32}$$

where $T_X = \Sigma_i n_i(n_{i+} - 1)/2$ is the number of pairs tied on the row variable X, $T_Y = \Sigma_j n_j(n_{+j} - 1)/2$ is the number of pairs tied on the column variable Y, and $T_{XY} = \Sigma_i \Sigma_j n_{ij}(n_{ij} - 1)/2$ is the number of pairs from a common cell (tied on X and Y). In this formula for $n(n-1)/2$, T_{XY} is subtracted because pairs tied on both X and Y have been counted twice, once in T_X and once in T_Y. Therefore,

$$\begin{aligned} T_X &= \frac{(430 \times 429) + (990 \times 989) + (355 \times 354)}{2} \\ &= 644\,625 \end{aligned} \tag{33}$$

Table 3 Income for three levels of education

		Income				
		Low	Medium	High	Total	Proportion
	LT than high school	302	105	23	430	0.243
	High school	409	331	250	990	0.557
Education	GT than high school	15	155	185	355	0.200
	Total	726	591	458	1775	1.00
	Proportion	0.409	0.333	0.258	1.00	

$$T_Y = \frac{(726 \times 725) + (591 \times 590) + (458 \times 457)}{2} = 542\,173 \tag{34}$$

$$T_{XY} = \frac{(302 \times 301) + (105 \times 104) + \cdots + (185 \times 184)}{2} = 249\,400 \tag{35}$$

$$\frac{n(n-1)}{2} = (524\,112) + (112\,915) + (644\,625) + (542\,173) - (249\,400) = 1\,574\,425 \tag{36}$$

Kendall's τ Statistics. Several measures of concordance/discordance are based on the difference $C - D$. Kendall's three *tau* (τ) statistics are among the most well known of these, and can be applied to $r \times c$ contingency tables with $r \geq 2$ and $c \geq 2$. The row and column variables do not have to have the same number of categories. The (τ) statistics have the range $-1 \leq 0 \leq +1$.

The three (τ) statistics are [8]:

$$\tau_a = \frac{2(C-D)}{N(N-1)} \tag{37}$$

$$\tau_b = \frac{(C-D)}{\sqrt{[(C+D+T_Y-T_{XY})(C+D+T_X-T_{XY})]}} \tag{38}$$

$$\tau_c = \frac{2m(C-D)}{N^2(m-1)} \tag{39}$$

where $m =$ the number of rows or column, whichever is smaller. For our data, the tau statistics are equal to

$$\tau_a = \frac{2(524\,112 - 112\,915)}{1775(1775-1)} = 0.26 \tag{40}$$

$$\tau_b = \frac{(524\,112 - 112\,915)}{\sqrt{\begin{array}{l}[(524\,112 + 112\,915 + 542\,173 - 249\,900)\\ (524\,112 + 112\,915 + 644\,625 - 249\,900)]\end{array}}} = 0.42 \tag{41}$$

$$\tau_c = \frac{3 \times 2(524\,112 - 112\,915)}{1775^2(3-1)} = 0.39 \tag{42}$$

One note of qualification is that $\tau - a$ assumes there are no tied observations; thus, strictly speaking, it is not applicable to contingency tables. It is generally more useful as a measure of rank correlation. For 2×2 tables, τ_b simplifies to the Pearson product–moment correlation obtained by assigning any scores to the rows and columns consistent with their orderings.

Goodman and Kruskal's γ. Goodman and Kruskal suggested yet another measure, *gamma* (γ) based on $C - D$ [6]:

$$\gamma = \frac{C-D}{C+D} \tag{43}$$

For our example data, $\gamma = (524\,112 - 112\,915)/(524\,112 + 112\,915) = 0.65$ γ has the range $-1 \leq 0 \leq +1$, equal to $+1$ when the data are concentrated in the upper-left to lower-right diagonal, and equal to -1 for the converse pattern. Although γ does equal zero when the two variables are independent, two variables can be completely dependent and still have a value of γ less than unity. For 2×2 tables, γ is equal to Yule's Q.

Measures Based on Derived Scores

Some methods for measuring the association between ordinal variables require assigning scores to the levels of the ordinal variables. When a contingency table is involved, scale values are assigned to row and column categories, the data are treated as a grouped frequency distribution, and the Pearson product–moment correlation is computed.

Specifically, let x_i and y_j be the values assigned to the rows and columns. The Pearson product–moment correlation for grouped data is then

$$r = \frac{CP_{XY}}{\sqrt{SS_X SS_Y}} \tag{44}$$

where CP_{XY} is the sum of cross products,

$$CP_{XY} = \sum_{i,j} x_i y_j n_{++} - \frac{\left(\sum_i x_i n_{i+}\right)\left(\sum_j y_j n_{+j}\right)}{n_{++}} \tag{45}$$

and SS_X and SS_Y are the sums of squares,

$$SS_X = \Sigma_i x_i^2 n_{i+} - \frac{\left(\sum_i x_i n_{i+}\right)^2}{n_{++}} \quad (46)$$

$$SS_Y = \Sigma_j y_j^2 n_{+j} - \frac{\left(\sum_j y_j n_{+j}\right)^2}{n_{++}} \quad (47)$$

Therefore, if for our 3×3 contingency table for income and education, we assign the values, "1", "2", and "3" to the three levels of each variable, we have

$$CP_{XY} = 6863 - \frac{(3475)(3282)}{1775} = 437.68 \quad (48)$$

$$SS_X = 7585 - \frac{(3475)^2}{1775} = 781.83 \quad (49)$$

$$SS_Y = 7212 - \frac{(3282)^2}{1775} = 1143.54 \quad (50)$$

$$r = \frac{437.68}{\sqrt{(781.83)(1143.54)}} = 0.46 \quad (51)$$

The value of $r = 0.46$ is also known as the *Spearman rank correlation coefficient* r_s, sometimes referred to as Spearman's rho [9]. The Spearman rank correlation coefficient can also be computed by using:

$$r_S = \frac{6\sum d^2}{n(n^2 - 1)} \quad (52)$$

where d is the difference in the rank scores between X and Y. Computed by using this formula,

$$r_S = \frac{6(20706)^2}{1775(1775^2 - 1)} = 0.46 \quad (53)$$

Instead of using the rank scores of "1", "2", and "3" directly, we can use *ridit scores* [10]. Ridit scores are cumulative probabilities; each ridit score represents the proportion of observations below category j plus half the proportion within j. We illustrate the calculation of ridit scores using the income and education contingency table. Beneath each marginal total is the proportion of observations in the row or column. The ridit score for the low income category is $0.5(0.409) = 0.205$; for medium income, $0.409 + 0.5(0.333) = 0.576$; and for high income, the ridit score is $0.409 + 0.333 + 0.5(0.258) = 0.871$. The ridit scores for less than high school education is $0.5(0.243) = 0.122$; for high school education, $0.243 + 0.5(0.557) = 0.522$; and for greater than high school education, the ridit score is $0.243 + 0.557 + 0.5(0.200) = 0.900$. Applying the Pearson product–moment correlation formula to the ridit scores, we obtain $r_s = 0.47$.

References

[1] Rudas, T. (1998). *Odds Ratios in the Analysis of Contingency Tables*, Sage Publications, Thousand Oaks.

[2] Wickens, T.D. (1986). *Multiway Contingency Tables Analysis for the Social Sciences*, Lawrence Erlbaum, Hillsdale.

[3] Agresti, A. (1984). *Analysis of Ordinal Categorical Data*, John Wiley & Sons, New York.

[4] Fleiss, J.L. (1981). *Statistical Methods for Rates and Proportions*, 2nd Edition, John Wiley & Sons, New York.

[5] Reynolds, H.T. (1984). *Analysis of Nominal Data*, Sage Publications, Newbury Park.

[6] Goodman, L.A. & Kruskal, W.H. (1972). Measures of association for cross-classification, IV, *Journal of the American Statistical Association* **67**, 415–421.

[7] Everitt, B.S. (1992). *The Analysis of Contingency Tables*, 2nd Edition, Chapman & Hall, Boca Raton.

[8] Gibbons, J.D. (1993). *Nonparametric Measures of Association*, Sage Publications, Newbury Park.

[9] Hildebrand, D.K., Laing, J.D. & Rosenthal, H. (1977). *Analysis of Ordinal Data*, Sage Publications, Newbury Park.

[10] Bross, I.D.J. (1958). How to use ridit analysis, *Biometrics* **14**, 18–38.

Further Reading

Liebetrau, A.M. (1983). *Measures of Association*, Sage Publications, Newbury Park.

Scott L. Hershberger and Dennis G. Fisher

Article originally published in Encyclopedia of Statistics in Behavioral Science *(2005, John Wiley & Sons, Ltd). Minor revisions for this publication by Jeroen de Mast.*

Measures of Location

Measures of location are statistics that yield information on the point of gravity or the center of the data. A more formal description is a function of data that is equivariant under affine transformations, i.e., commutes with the group of transformations $x \to ax + b$. Common examples include the sample mean $\sum_{i=1}^{n} x_i/n$, the sample median (the middle value of the data after sorting from small to large in the case of an odd number of observations or the average of the two middle values in the case of an even number of observations), and the sample mode (the value that appears most frequently; this need not be a unique value). When data are normally distributed (*see* **Normal Distribution**), then the sample mean is the optimal estimator for the population mean in the sense that it is the unique unbiased estimator with minimal variance. A disadvantage of the mean is that it is not robust against ***outliers***. Well-known alternatives that have a certain resistance against outliers include the median, the shorth (the sample mean of the shortest intervals containing at least half of the observations), *trimmed means* (sample means when extreme order statistics have been removed), winsorized means (sample means when extreme order statistics have been set to an extreme order statistic), and the Hodges–Lehmann estimator (median of pairwise averages). Many of these sample statistics can be derived from general construction principles or techniques for estimators, like ***Maximum Likelihood***, method of **moments**, M-estimators (extension of Maximum Likelihood), L-estimators (linear combinations of order statistics), or R-estimators (rank-based estimators). We refer to [1] for extensive details.

Reference

[1] Barnett, V. & Lewis, T. (1994). *Outliers in Statistical Data*, John Wiley & Sons, New York.

Related Articles

Maximum Likelihood; **Normal Distribution**; **Outliers**; **Trimmed Mean**.

ALESSANDRO DI BUCCHIANICO

Measures of Scale

Measures of scale are statistics that yield information on the spread of data. A more formal definition is that it is a function of data that changes with a multiplicative factor $|a|$ after applying an affine transformation $x \to ax + b$. The most well-known example is the standard deviation, which is the square root of the sample variance $\sum_{i=1}^{n} (x_i - \bar{x})^2/(n-1)$ (the sample *variance* is more tractable from a computational point of view, but has the drawback that it does not have the right unit when applied to physical data). A disadvantage of the standard deviation is that it is not robust against ***outliers***. This is even more the case for the range (the difference of the largest and the smallest observation), which was a popular alternative to the standard deviation in the precomputer era. The range and the related average moving range (AMR) $\sum_{i=1}^{n-1} |x_{i+1} - x_i|/(n-1)$ are often used in **quality control**, especially for **control charts** (*see* **Moving Range and *R* Charts**). Well-known alternatives that have a certain resistance against *outliers* include the interquartile range (distance between third and first *quartile*), the median absolute deviation (MAD), which is the median of the absolute deviations from the median), and the length of the shorth (the shortest interval containing at least half of the data, *see* **Measures of Location**). Many of these sample statistics can be derived from general construction principles or techniques for estimators, like *Maximum Likelihood*, method of **moments**, M-estimators (extension of maximum likelihood), L-estimators (linear combinations of order statistics), or R-estimators (rank-based estimators). We refer to [1] for extensive details.

Reference

[1] Barnett, V. & Lewis, T. (1994). *Outliers in Statistical Data*, John Wiley & Sons, New York.

Related Articles

Maximum Likelihood; **Measures of Location**; **Outliers**; **Variance**.

ALESSANDRO DI BUCCHIANICO

Measuring Process Yield *see* Process Yield

Metamodels in Reliability Analysis *see* Splines and other Metamodels in Reliability Analysis

Minimal Maintenance and Repair *see* Age-Dependent Minimal Repair and Maintenance

Minimum Aberration

Introduction

Two-level **factorial** and **fractional factorial** designs are most widely used in experimental investigations. When m factors, each with two levels such as high (+1) and low (–1), are to be studied in an experiment, one may choose a full factorial design in which all possible 2^m level combinations of the factors are investigated. Such a level combination of the factors is often referred to as a *run* of the design. However, for a moderately large number of factors, a two-level full factorial design requires a large number of runs, which may be very time consuming, too expensive, and even redundant (see [1, p. 243]). Therefore, investigators often turn to a fractional factorial design, in which only a fraction of runs of a full factorial design are studied. In general, a 2^{m-p} fractional factorial design studies m factors with a 2^{-p}th fraction of 2^m runs of a full factorial design. A good fractional factorial design allows one to investigate many factors with relatively small run size, while still enabling one to estimate important effects. In order to select good fractional factorial designs for a given dimension (i.e., m and p), Box and Hunter [2] introduced the idea of design *resolution*. Fries and Hunter [3] further proposed a refined *minimum aberration* criterion to assess fractional factorial designs with the same resolution.

Minimum Aberration in Regular Designs

In this section we define regular two-level fractional factorial designs, describe how they alias certain effects and explain how the minimum aberration criterion can be helpful in selecting a good design. We illustrate the ideas by considering an experiment in which it is desired to study 7 factors in 32 ($= 2^{7-2}$) runs. Denote the seven factors by numbers $1, 2, 3, 4, 5, 6, 7$, which correspond to seven columns with entries ± 1 in the design matrix. A 2^{7-2} fractional factorial design can be constructed by first writing down a basic design consisting of 32 runs associated with a full factorial design in five factors, say, $1, 2, 3, 4, 5$. Then we associate the two additional factors 6 and 7 with appropriately chosen interactions involving the factors $1, 2, 3, 4, 5$. For example, we may obtain a design D_1 by associating factors $6 = 123$ and $7 = 234$, where the multiplication of letters is the element-wise multiplication of their corresponding columns. Hence, main effects 6 and 7 of design D_1 are aliased by its three-factor interactions 123 and 234, respectively. The assignment of factors 6 and 7 can be rewritten as $I = 1236$ and $I = 2347$, where I is a column of "+1". One can find another relation among the factors as $I = 1236 \times 2347 = 1467$ using the fact that two same letters result in the identity column I. Altogether we have $I = 1236 = 2347 = 1467$, which forms the *defining relation* of the design. The elements 1236, 2347, and 1467 in the defining relation are referred to as *words*. The number of letters in a word is called the *length* of the word or *word length*. The words 1236 and 2347, used to define additional factors 6 and 7, are referred to as the *design generators*. In general, the design generators plus all the possible products of them constitute the complete defining relation of a design. Two-level fractional factorial designs 2^{m-p} constructed as above have the complete defining relation and are known as *regular fractional factorial designs* or *simply regular designs*.

In regular fractional factorial designs, some of the effects are completely entangled, or aliased, with one another. From the defining relation, it is straightforward to find the **aliasing** structures. For example, multiplying the above defining relation on both sides by 1 yields $1 = 236 = 467 = 12347$. That is, the **main effect** 1 is aliased with two three-factor interactions 236, 467 and a five-factor interaction 12347. Similarly, $12 = 36 = 2467 = 1347$. The **estimation** of the two-factor interaction 12 is actually the sum of effects $12 + 36 + 246 + 1347$. In the literature, when an effect A is aliased with another effect B, it is also said that the effect A is confounded with the effect B, which means these two effects are indistinguishable from each other. Design D_1 can estimate all main effects, but cannot estimate some two-factor interactions such as 12 and 36, even if we assume that the three-factor and higher-order interactions are negligible. However, all two-factor interactions involving 5 in D_1 are estimable. Therefore, the aliasing structure is an important information in planning industrial experiments.

Let $A_i(D)$ be the number of words of length i in the defining relation of a design D. The *word length pattern* is then the vector $W(D) = [A_1(D), A_2(D), A_3(D), \ldots]$. Since any two columns of a two-level fractional factorial design are orthogonal, the minimum length of a word will be 3; the maximum length possible is the total number of factors in the design. Thus, the word length pattern of design D_1 above can be shortened to $W(D_1) = [A_3(D_1), A_4(D_1), A_5(D_1), A_6(D_1), A_7(D_1)] = [0, 3, 0, 0, 0]$.

Box and Hunter [2] introduced *resolution* as a criterion for selecting regular fractional factorial designs. The resolution R of a regular fractional factorial design is defined as the length of the shortest word in the defining relation of the design. Therefore, if a regular fractional factorial design is of resolution R, then no k-factor interaction is aliased with any other interaction containing less than $R - k$ factors. The resolution of a regular design is denoted by a Roman numeral subscript. For instance, the design D_1 mentioned above is of resolution IV and is denoted as 2^{7-2}_{IV}. The **design resolution** is a criterion to measure the occurrences of worst aliases of a regular fractional factorial design. In general, a design with a higher resolution is better. However, there may be many regular fractional factorial designs with the same resolution, especially for larger designs with many factors. Resolution alone is sometimes not sufficient to distinguish among regular fractional factorial designs. Suppose a design D_2 is obtained by redefining factors 6 and 7 in design D_1 above as $6 = 1234$ and $7 = 1245$. So design D_2 has the defining relation $I = 12346 = 12457 = 3567$. Both designs D_1 and D_2 are of resolution IV, but they have rather different alias structures. If we assume that interactions involving three or more factors are negligible, which is often the case at the initial stage of the investigation, design D_2 has only three pairs of aliases (i.e., $35 = 67$, $36 = 57$, and $37 = 56$) with respect to the two-factor interactions, whereas there are nine pairs of such aliases in design D_1. Clearly, design D_2 is a better choice for a 2^{7-2}_{IV}, if there is little knowledge about the factors.

In order to discriminate among regular fractional factorial designs with the same resolution, Fries and Hunter [3] proposed the *minimum aberration* criterion based on the word length pattern. Suppose for any two regular fractional factorial designs D and D', r is the smallest positive integer such that $A_r(D) \neq A_r(D')$. Then design D is said to have less aberration than design D' if $A_r(D) < A_r(D')$. If there exists no other regular design having less aberration than design D, then D is said to be a minimum aberration design. Minimum aberration is a criterion to compare the frequency of aliases of regular designs at different levels sequentially, and a regular design having the smallest frequency of worst aliases is the best. For example, the word length pattern for design D_2 is $W(D_2) = [A_3(D_2), A_4(D_2), A_5(D_2), A_6(D_2), A_7(D_2)] = [0, 1, 2, 0, 0]$. Since $A_3(D_1) = A_3(D_2) = 0$ and $A_4(D_2) = 1 < 3 = A_4(D_1)$, design D_2 has less aberration than design D_1. It is easy to check that design D_1 is a minimum aberration design. Chen *et al.* [4] and Wu and Hamada [5, Chapter 4, Appendix] listed minimum aberration (and other selected) two-level regular fractional factorial designs from 8 to 128 runs. Some important contributions related to minimum aberration can be found in Franklin [6], Chen and Wu [7], Chen [8], Tang and Wu [9], Chen and Hedayat [10], Suen *et al.* [11], and Cheng *et al.* [12]. There are many additional works on minimum aberration designs; see of Wu and Hamada [5, Chapter 4, References] for details.

Nonregular Designs and Generalized Minimum Aberration

Nonregular fractional factorial designs, or simply nonregular designs, are commonly obtained from Plackett–Burman designs given in Plackett and Burman [13] or more generally from Hadamard matrices by selecting a subset of the columns. Recall that a Hadamard matrix of order n is an $n \times n$ matrix with the elements ± 1 whose columns (and rows) are orthogonal to each other. The order n is a multiple of 4. Other widely used nonregular designs are selected from orthogonal arrays. See Hedayat *et al.* [14] and Wu and Hamada [5] for further details. While regular designs have a clear defining relation and therefore simple aliasing structures, the gap of runs for regular factorial designs is becoming larger and larger (4, 8, 16, 32, 64, 128, ...). In contrast, nonregular designs have no clear defining relation and thus their aliasing structures are very complicated, but they have a very flexible run size of a multiple of 4. Hamada and Wu [15] argued that the nature of the complex aliasing structure of nonregular designs could be exploited to estimate some interactions without increasing the number of runs. They proposed an analysis strategy to turn the liability of the complex aliasing structure into the virtue of model estimability. Therefore, good nonregular designs not only provide a much wider range of possible experiment sizes but may also have advantages in the estimation of interactions; see the illustrative example in Wang and Wu [16].

The popular minimum aberration criterion was first proposed for selecting regular designs, and it cannot be directly applied to nonregular designs. In order to assess nonregular designs, one needs to find a suitable criterion. There are several significant articles in this direction as reported by Lin and Draper [17], Wang and Wu [16], and Box and Tyssedal [18]. The criteria proposed were reasonable and heuristic. Most of them however, were highly model based. They had no apparent connection with the minimum aberration criterion or no rigorous theoretical justifications. Deng and Tang [19, 20] proposed *generalized resolution* and *generalized minimum aberration* (GMA) for ranking nonregular two-level designs in a systematic approach. These criteria are based on a generalization of the word length pattern and is useful for comparing regular/nonregular designs. We give a more detailed description next.

Let D be a two-level (regular or nonregular) factorial design with n runs and m factors. The design D can also be regarded as a set of m column vectors of length n. For any subset with k columns selected from D, say, $s = \{\mathbf{x}_1, \ldots, \mathbf{x}_k\}$, define $J_k(s) = |\sum_{i=1}^{n} x_{i1}, \ldots, x_{ik}|$, where x_{ij} is the ith component of column vector $\mathbf{x}_j$. $J_k(s)$ is the absolute value of the sum of the interaction column $\mathbf{x}_1, \mathbf{x}_2, \ldots, \mathbf{x}_k$.

For regular designs, $J_k(s)$ equals either 0 or n. When $J_k(s) = n$, it corresponds to a full confounding among these columns whose indices are in s. For nonregular designs, $J_k(s)$ can take values between 0 and n which means a partial confounding. For a design D, let r be the smallest integer such that $\max_{|s|=r} J_r(s) > 0$, where the maximization is over all the subsets of r distinct columns of D. Then the *generalized resolution* of D, proposed in Deng and Tang [19], is defined as $R(D) = r + [1 - \max_{|s|=r} J_r(s)/n]$. Obviously, $r \leq R(D) < r + 1$. For a regular design, $R(D) = r$ since $J_k(s)$ takes either 0 or n. Therefore, $\max_{|s|=r} J_r(s) > 0$ is equivalent to $\max_{|s|=r} J_r(s) = n$ for a regular design.

For nonregular designs, it is important to know the possible values of $J_k(s)$. It is straightforward to see that $J_k(s)$ is a multiple of 4. Deng and Tang [20] derived a general formula for possible values of $J_k(s)$ and proved that the gap between two consecutive possible J values is a multiple of 8. We then list all possible values of $J_k(s)$ in decreasing order and denote the frequency distribution of $J_k(s)$ for all possible s out of k columns in D by $[f_{k1}, f_{k2}, \ldots, f_{k\,g+1}]$, where f_{kj} is the frequency of $J_k(s)$ taking the jth largest possible value and $g = [\frac{n}{8}]$, the integer part of $\frac{n}{8}$. The total number of k-column subdesigns (*projections*) from an m-column design is a fixed number $\binom{m}{k}$ which is equal to $\sum_{j=1}^{g+1} f_{kj}$. Therefore, it is sufficient to consider the first g entries and let

$$F_k(D) = [f_{k1}, f_{k2}, \ldots, f_{kg}] \tag{1}$$

The confounding frequency vector (or generalized word length pattern) of D, proposed by Deng and Tang [19, 20], is given by

$$W(D) = [F_3(D), F_4(D), \ldots, F_m(D)] \tag{2}$$

Since only designs with orthogonal columns are considered, $F_1(D) = F_2(D) = 0$.

Note that for a regular design, f_{k1} is the number of words with length k and $f_{ki} = 0\,(i > 1)$ because

$J_k(s)$ takes either 0 or n. Therefore, the corresponding confounding frequency vector reduces to the word length pattern for regular designs.

Any two fractional factorial designs D_1 and D_2 with the same run size and number of factors can be ranked by comparing their confounding frequency vectors element by element. Specifically, suppose r is the smallest integer such that $F_r(D_1) \neq F_r(D_2)$. Design D_1 is said to have less generalized aberration than design D_2 if $F_r(D_1) < F_r(D_2)$, which are compared element by element from the first entry to the last. In other words, we would sequentially compare the frequency of the components and prefer the one that minimizes the frequency of the largest value. As proposed first by Deng and Tang [19], a design is said to have *generalized minimum aberration* (also referred to as *minimum G aberration*) if there exists no other design having less generalized aberration. Clearly this criterion reduces to the traditional minimum aberration for regular designs. The GMA criterion can compare and assess any two factorial designs, whether or not they are regular.

Using the confounding frequency vector, one can rank various nonregular designs of m columns and n runs. Deng *et al.* [21] and Deng and Tang [20] tabulated and ranked a collection of nonregular designs of run sizes 16, 20, and 24.

Other Related Criteria

Instead of counting the frequency of $J_k(s)$ in the confounding frequency vector $W(D)$, Tang and Deng [22] considered a relaxed version as $B_k(D) = n^{-2}\sum_{|s|=k}[J_k(s)]^2$ and proposed *minimum G_2 aberration* which is based on the vector $W_B(D) = [B_3(D), B_4(D), \ldots, B_m(D)]$. They further developed a complementary design theory for the minimum G_2 aberration criterion. Cheng *et al.* [23] demonstrated that the criterion based on $W_B(D)$ is consistent with some model-dependent efficiency criteria under certain two-factor interaction models. Cheng and Tang [24] further developed a general theory of minimum aberration based on some statistical principles and their theory provided a unified framework for minimum aberration.

While the work on GMA shows a promising direction in the study of nonregular designs, this approach is mostly suitable for two-level designs. By studying treatment contrasts and **ANOVA** models, Xu and Wu [25] proposed a generalization of the minimum aberration criterion that can be used for multilevel designs. To develop an additional theory for the proposed criterion, they successfully explored the connection between the theory of factorial designs and the theory of coding. In other words, instead of studying the relationship between the factors (columns), one can investigate the relationship between the runs (rows). Following this approach, Xu [26] proposed the *minimum moment aberration* criterion, which is based on the power moments, for measuring the similarity among runs (rows). Minimizing the power moments makes runs as dissimilar as possible. Therefore, good designs should have small power moments.

Using power moments of the projection designs and counting their frequency distributions, Xu and Deng [27] proposed the *moment aberration projection* criterion to rank and classify nonregular designs (including multilevel designs). Two fractional factorial designs are *isomorphic* if one can be obtained from the other by permuting rows and columns and/or changing the signs of columns. In general, it is time consuming to perform an isomorphism check between two designs as demonstrated in Chen *et al.* [4]. The confounding frequency vector and the moment aberration projection, although originally proposed to rank and assess designs, are also useful in testing for isomorphism. Clearly, two designs with different confounding frequency vectors or moment aberration projections are not isomorphic. On the other hand, two nonisomorphic designs may have the same confounding frequency vector or moment aberration projection. Xu and Deng [27] demonstrated that moment aberration projection has a better classification power than the confounding frequency vector. The ranking of various designs given by both criteria are fairly consistent, but not identical, with each other.

Fang and Mukerjee [28] found a connection analytically between the minimum aberration criterion and a uniformity criterion in the area of **uniform design**. See Fang *et al.* [29, Chapter 3] for more references.

References

[1] Box, G.E.P., Hunter, W.G. & Hunter, J.S. (1978). *Statistics for Experimenters: An Introduction to Design, Data Analysis, and Model Building*, John Wiley & Sons, New York.

[2] Box, G.E.P. & Hunter, J.S. (1961). The 2^{k-p} fractional factorial designs. *Technometrics* **3**, 311–351 and 449–458.

[3] Fries, A. & Hunter, W.G. (1980). Minimum aberration 2^{k-p} designs, *Technometrics* **22**, 601–608.

[4] Chen, J., Sun, D.X. & Wu, C.F.J. (1993). A catalogue of two-level and three-level fractional factorial designs with small runs, *International Statistical Review* **61**, 131–146.

[5] Wu, C.F.J. & Hamada, M. (2000). *Experiments: Planning, Analysis and Parameter Design Optimization*, John Wiley & Sons, New York.

[6] Franklin, M.F. (1984). Constructing tables of minimum aberration p^{n-m} designs, *Technometrics* **26**, 225–232.

[7] Chen, J. & Wu, C.F.J. (1991). Some results on s^{n-k} fractional factorial designs with minimum aberration or optimal moments, *Annals of Statistics* **19**, 1028–1041.

[8] Chen, J. (1992). Some results on 2^{n-k} fractional factorial designs and search for minimum aberration designs, *Annals of Statistics* **20**, 2124–2141.

[9] Tang, B. & Wu, C.F.J. (1996). Characterization of minimum aberration 2^{n-k} designs in terms of their complementary designs, *Annals of Statistics* **24**, 2549–2559.

[10] Chen, H. & Hedayat, A.S. (1996). 2^{n-m} fractional factorial designs with weak minimum aberration, *Annals of Statistics* **24**, 2536–2548.

[11] Suen, C.-Y., Chen, H. & Wu, C.F.J. (1997). Some identities on q^{n-m} designs with application to minimum aberrations, *Annals of Statistics* **25**, 1176–1188.

[12] Cheng, C.S., Steinberg, D.M. & Sun, D.X. (1999). Minimum aberration and model robustness for two-level fractional factorial designs, *Journal of the Royal Statistical Society, Series B* **61**, 85–91.

[13] Plackett, R.L. & Burman, J.P. (1946). The design of optimum multifactorial experiments, *Biometrika* **33**, 305–325.

[14] Hedayat, A.S., Sloane, N.J.A. & Stufken, J. (1999). *Orthogonal Arrays: Theory and Applications*, Springer-Verlag, New York.

[15] Hamada, M. & Wu, C.F.J. (1992). Analysis of designed experiments with complex aliasing, *Journal of Quality Technology* **24**, 130–137.

[16] Wang, J.C. & Wu, C.F.J. (1995). A hidden projection property of Plackett-Burman and related designs, *Statistica Sinica* **5**, 235–250.

[17] Lin, D.K.J. & Draper, N.R. (1992). Projection properties of Plackett and Burman designs, *Technometrics* **34**, 423–428.

[18] Box, G.E.P. & Tyssedal, J. (1996). Projective properties of certain orthogonal arrays, *Biometrika* **84**, 950–955.

[19] Deng, L.-Y. & Tang, B. (1999). Generalized resolution and minimum aberration criteria for Plackett-Burman and other nonregular factorial designs, *Statistica Sinica* **9**, 1071–1082.

[20] Deng, L.-Y. & Tang, B. (2002). Design selection and classification for Hadamard matrices using generalized minimum aberration criteria, *Technometrics* **44**, 173–184.

[21] Deng, L.-Y., Li, Y. & Tang, B. (2000). Catalogue of small runs nonregular designs from Hadamard matrices with generalized minimum aberrations, *Communications in Statistics: Theory and Methods* **29**, 1379–1395.

[22] Tang, B. & Deng, L.Y. (1999). Minimum G_2-aberration for nonregular fractional factorial designs, *Annals of Statistics* **27**, 1914–1926.

[23] Cheng, C.S., Deng, L.Y. & Tang, B. (2002). Generalized minimum aberration and design efficiency for nonregular fractional factorial designs, *Statistica Sinica* **12**, 991–1000.

[24] Cheng, C.S. & Tang, B. (2005). A general theory of minimum aberration and its applications, *Annals of Statistics* **33**, 944–958.

[25] Xu, H. & Wu, C.F.J. (2001). Generalized minimum aberration for asymmetrical fractional factorial designs, *Annals of Statistics* **29**, 1066–1077.

[26] Xu, H. (2003). Minimum moment aberration for nonregular designs and supersaturated designs, *Statistica Sinica* **13**, 691–708.

[27] Xu, H. & Deng, L.Y. (2005). Moment aberration projection for nonregular fractional factorial designs, *Technometrics* **47**, 121–131.

[28] Fang, K.T. & Mukerjee, R. (2000). A connection between uniformity and aberration in regular fractions of two-level factorials, *Biometrika* **87**, 193–198.

[29] Fang, K.T., Li, R. & Sudjianto, S. (2005). *Design and Modeling for Computer Experiments*, Chapman & Hall/CRC.

LIH-YUAN DENG

Missing Data *see* Expectation Maximization Algorithm

Missing Data and Imputation

Introduction

Missing data is a comprehensive term that comprises various aspects of incomplete information on the studied variables. An example of incomplete information is a longitudinal study of consumer satisfaction, with several surveys performed over a period. The data on consumer satisfaction is obtained by interviewing a sample from the customers' population.

1. Some of the customers from the intended sample may refuse to participate in the study. At the analysis stage, the information from those individuals will be missing on all the items and in all the subsequent surveys. This type of missing data is a *unit nonresponse* for all the surveys in the study.
2. Some customers may participate in initial studies but drop out from the study afterwards. In this attrition case with dropouts, the missing data is a still considered a *unit nonresponse* but only for the late surveys in which they do not participate.
3. In each particular survey, respondents may decide not to respond to specific question, e.g. when questions are sensitive, or the respondent ends the interview before its completion. The resulting data thus contains *item nonresponses*. Item nonresponse may also occur when respondents have no opinion or knowledge on issues, and the questions do not contain categories to cover such eventualities. Furthermore, if the respondents provide information that does not pass the logical tests (e.g. outside allowable range), the data will be again coded as *item nonresponse*.
4. Incomplete information may also result from *partial recording*. Examples of such partial recording are values known to be outside prespecified limits, or data with *interval censoring*, which are known to be between specific values.
5. In the **quality control** settings, *partial recording* may occur when the data are *batched* and the available information pertains to the entire batch rather than to the individual observations. Such partial recordings are missing data by design. A different type of partial recording by design can occur when a subset of the data is subjected to more expensive measurements, while the values for the rest of the data are obtained by a simpler and more error-prone method of measurements.

In sample surveys, unit and item nonresponses can hardly be avoided and are an integral part of any study (*see* **Design and Testing of Questionnaires**; **Data Management in Survey Sampling**). In the quality control settings, efforts to avoid missing data can be more successful and typically rely on improved technology (*see* **Control Charts and Process Capability**). In multivariate process control, the problem is amplified since measurements can be derived from several measurement systems that have different levels of uncertainty and operate at difference cycle times (*see* **Multivariate Control Charts Overview**; **Statistical Process Control, Multivariate**). In clinical trials, participants who drop out of the study for any reason cause missing data (*see* **Case–Control Studies**; **Disease and Clinical Trial Modeling**). In designed experiments, some replicates or experimental runs might have not been executed causing another type of missing data (*see* **Factorial Experiments**; **Screening Designs**; **Response Surface Methodology**). Methods of dealing with missing data typically cannot overcome the impact of missing data and yield less accurate results than what can be achieved with datasets collected on all the intended units and items. Indeed, as mentioned by one of the leaders in the research on missing data "the most important step in dealing with missing data is to try to avoid it during the data-collection stage" [1].

In many cases, the transformation of the sets of data containing missing observations into a complete dataset precedes the analyses aimed at assessing the parameters of the data. The two basic methods for transforming the dataset into a rectangular array to be analyzed are as follows:

1. discard the incomplete cases and then analyze the units with complete information (complete-case analysis), or
2. impute estimated values instead of the missing ones and then analyze the filled in data.

In the longitudinal study of consumer satisfaction example, for each survey separately, missing data have to be considered in the analysis, usually by properly weighting the observed items. Alternatively, incomplete data can be analyzed by more sophisticated methods which do not require a complete dataset. In some cases, the common practice is to perform a basic comparison between the characteristics of the units (say, respondents) with complete observations and the characteristics of the units with incomplete data.

The pattern of the missing data defines the observed and the missing values in the dataset.

Especially in longitudinal studies, when the missing data are due to attrition and dropout, the pattern of missing data contains important relevant information for analysis. Another important pattern is the univariate item nonresponse, i.e. missing data on a single variable with complete data on all the others.

The Process that Causes Missing Data and the Data Analysis

The process that causes missing data is related to the possible relationship between the fact that an observation is missing and the values of observations. When analyzing the data, while we consider the observed pattern of missing data, we almost always ignore the process that caused the data. As Rubin [2] mentioned, "When making sampling distribution inferences about the parameter of the data, θ, it is appropriate to ignore the process that causes missing data if the missing data are 'missing at random' and the observed data are 'observed at random', but these inferences are generally conditional on the observed pattern of missing data".

The terms *missing at random* (MAR) and *missing completely at random* (MCAR) are fundamental in defining the possible relationship between the fact that an observation is missing and the values of observations. In this context, values of observations refer both to the values of the observed data (call them Y_{obs}) as well as to the potential values of the missing data (call them Y_{miss}). Let Y denote all the observations, observed or missing, with Y_{ij} being the variable j on the ith observation. The terms *MAR* and *MCAR* concern the relationships between those values and the distribution of the random matrix M of missing data indicators, (i.e., $m_{ij} = 1$ is variable j on observation i is missing, and 0 otherwise).

Missing data are MCAR, if the conditional distribution of M may depend on the unknown parameters of the data Θ but not on Y_{obs} or Y_{miss}. Then, if the data are MCAR, the conditional distribution of M satisfies

$$f(M|Y, \Theta) = f(M|\Theta) \text{ for all } Y \text{ and } \Theta \qquad (1)$$

In the consumer survey example, data on specific items will be MCAR if, say, individuals may decide by a random drawing whether to reply to some questions, with possible different probabilities of replying in the various subpopulations. Also, unit nonresponse may be completely at random, if again, individuals may decide by a random drawing whether to participate in the study, with possible different probabilities of participation in the various subpopulations.

Missing data are MAR, if the conditional distribution of M may depend on Θ and Y_{obs} but not on Y_{miss}. Thus, if the data are MAR, the conditional distribution of M satisfies

$$f(M|Y, \Theta) = f(M|Y_{\text{obs}}, \Theta) \text{ for all } Y_{\text{miss}} \text{ and } \Theta \quad (2)$$

In the longitudinal consumer study example, MAR will occur if say, the customers' decision regarding participation in the survey at time t, depends on the grade that they assigned in the previous surveys, but not on the grades that they would have assigned at this survey. Again, the probabilities of participations may also vary in the various subpopulations. Obviously, since in the unit nonresponse there are no Y_{obs}, there is no MAR for an entire unit.

All other cases of missing data are not missing at random (NMAR). As an example, let two units have equal observations on all but one variable, with one unit having data and the other not having data on the remaining variable. If missing data are NMAR, the two units differ systematically on that remaining variable. In terms of the conditional distribution of M, the fact that the data is missing is affected by the unobserved value.

Handling of Missing Data – Listwise Deletion and Weighting

In most of the cases, the process that causes missing data is unknown. However, by using various methods of handling missing data, which ignore either only the process that causes missing data or both the process as well as the pattern of the observed data, the investigator implicitly assumes a certain mechanism that caused the missing data.

The simplest method of handling missing data is to discard the incomplete cases and then analyze the units with complete information "as is". This complete-case analysis is thus preceded by what is known in the computer software as a {*listwise deletion*}. This method of handling missing data is indeed simple and quite common. As an example, in a study [3] the investigators collected data on more than 500 companies from the United Kingdom and aimed at creating an algorithm that can be used in the formulation of a knowledge-based decision system. They mentioned that "prior to conducting the statistical analysis, cases with missing data were eliminated from the dataset".

This type of complete-case analysis ignores both the mechanism that created the missing data as well as the observed pattern of missing data. When the observed pattern of missing data is ignored, MAR is not a sufficient assumption. The required assumption in this case is MCAR. Under the MCAR mechanism, for each variable, the complete data can be considered as a random subsample from the intended sample. Moreover, even if we assume that the MCAR assumption holds, deletion of collected data in the complete-case analysis is a waste of information, which is rarely justifiable. Furthermore, if the data originates from a designed balanced experiment, the listwise deletion will create an undesirable unbalanced design.

A different type of complete-case analysis considers the pattern of the observed pattern of missing data and weighs differentially the respondents while still discarding the incomplete cases. The underlying assumption is that the data are MAR. The weights are usually based on poststratification, either direct or based on the response propensity stratification. In the direct poststratification the allocation of the observations to strata is based on the respondents' recorded values (e.g., geographical or socioeconomic data). For each stratum, given the prior sampling probabilities, respondents' weights are the inverse of the product of those probabilities and the estimated probability of participation in that stratum. In the consumer satisfaction example, with respondents grouped by the duration since becoming customers, the weight w_k for an observation in the kth group is given by $w_k = 1/(p_k r_k)$ where p_k is the prior sampling probability and r_k is the response rate. In many cases, in the sample design the sampling rates for strata are proportional to the population rates and equal to the prior sampling probabilities. The corresponding weights will have only to account for the estimated probability of participation in each stratum.

When the weighting is based on the response propensity stratification, the definition of the strata is based on categories of variables ("adjustment cells") with homogeneous estimated rates of response. The rates of response are estimated by regressing the indicator variable m_i (for response–nonresponse) on the background variables. Little [4] showed theoretically and in a simulation study that when weighting adjustments are employed (in contrast to imputation) adjustment cells based on the response propensity are effective in controlling **bias**.

Handling of Missing Data – Available-Case Analysis

The listwise deletion method discards information collected in the study and is thus inefficient. The available-case analysis is a relatively simple method to estimate the means and the correlations matrix from all the information available for estimating each parameter separately. Thus, the means of variable Y_k is estimated from all the cases with observed data on the kth variable, while the **correlation** between Y_k and Y_j is estimated from all the cases with observed data on the kth and the jth variable.

The advantage of making use of all the available information is counterbalanced by the fact that the sample base changes from the **estimation** of one correlation to the other, and especially in regression problems, when the data is highly correlated, simulations performed by Haitovsky [5] found that the available data analysis can yield considerably inferior results than the listwise deletion. Later simulations [6] found the available-case analysis superior to the listwise deletion in regression models with week correlation. Little [1] mentions that the available-case method cannot be generally recommended.

Imputation – Unconditional and Conditional Mean Imputation

Imputation is the replacement of unknown, unmeasured, or missing data with a particular value. As in the available-case analysis, all the incomplete cases are retained, and the dataset is transformed into a complete (rectangular) dataset by imputing values instead of the missing ones.

The simplest form of imputation is to replace all missing values with the average of that variable. It has been shown that, as expected, the unconditional mean imputation yields inconsistent estimates of parameters both for the variances as well as for the parameters based on the **covariance** matrix (e.g.

Affifi and Elashoff [7]). By imputing a single average value for all the missing data, the resulting estimated variance clearly underestimates the actual parameter. Also, the **measures of association** are distorted. For this case as well, Little [1] mentions that the unconditional case imputation cannot be generally recommended.

The conditional mean imputation uses the correlation structure among observed variables. The value imputed for a missing value on a specific variable is the value resulting from the regression prediction equation obtained from the set of complete cases (CCs), with the variable to be predicted as the response variable and all the others as predictor variables. The estimates resulting from the conditional mean imputation improves the estimation of the parameters, which are linear in the data (like regression equations), but can still yield distorted estimates for variances, correlations, **quantiles**, and other parameters, which are not linear in the data.

Imputation – Hot Deck and Predictive Mean Matching

The hot-deck imputation strategy replaces the missing data on one observation with the available data from another matched observation. The method is used by many surveys in the official statistics. Among the versions of hot-deck imputation strategies, the univariate hot-deck strategy is probably still the most commonly used. Under this strategy, given a unit with missing data on one or more items, the method assigns for each variable separately, a value from a record with similar characteristics on the recorded variables. For example, the current population survey (CPS) performed by the Census Bureau [8, 9], use different hot decks for the households, demographic, labor force, industry and occupation, earning, and school enrollment edits. Hot decks are always defined by age, race, and gender. The hot decks for the major labor force recode (MLR), which classifies adults as either employed, unemployed, or not in the labor force, are defined by age, race, and gender and, on occasion, by a related labor force characteristic. The number of cells in labor force item hot decks is relatively small, perhaps less than 100. On the other hand, the weekly earnings hot deck is defined by age, race, gender, usual hours, occupation, and educational attainment. This hot deck has several thousand cells. David *et al.* [10] compared the performance of the hot-deck imputation in the CPS data to the performance of a structured linear regression model.

The univariate hot-deck strategy handles many imputation problems well but handles more complex problems especially in the multivariate context poorly. One of the main problems is the shrunk correlation when the imputation on each variable is processed individually. As an example, Thibaudeau *et al.* [11] mention the distortion of the multivariate properties of the data obtained by using the univariate hot-deck strategy in the Wealth topical module for the Survey on Income and Program Participation (SIPP). In the sixth wave of the 1996 panel, the correlation coefficient between property value (assets) and mortgage(liability) was 0.39 in the reported (not imputed) data, but only 0.16 in the data imputed by the present univariate hot-deck strategy (i.e., less than half).

The joint hot-deck strategy preserves the correlation between jointly missing data, by imputing two or more items from the same donor. For a unit Y with missing data on several items whose correlation is to be analyzed (say, both assets and liabilities) the joint hot-deck imputation strategy finds a corresponding unit Z which matches Y as well as possible on the items on which both Y and Z are recorded. The missing items in Y are then filled in with the corresponding items in Z. The searching method is sequential. The search starts with all the items, and if no matching unit is found, the search continues for a subset of the most important variables or by reducing the number of categories in variables through collapsing.

The predictive mean imputation and the predictive mean matching can be conceptually considered as extensions to the hot-deck strategies by progressively expanding the number of cells while keeping at least one donor in each cell. In the analyses of paired variables (as assets and liabilities), for each missing observation on one out of the two variables, the value imputed by the predictive mean imputation strategy is from a donor, such that both the donor and the recipient have matched values on the other variable. On the basis of results of tests performed, Thibaudeau *et al.* [11] suggest replacing gradually the currently used univariate hot-deck strategy in SIPP, first by a bivariate hot-deck strategy for specific pairs of variables, and then by a predictive mean methodology.

The predictive mean matching is a further extension, which selects the donor whose value is to be imputed instead of the missing data on the basis of a distance function rather than directly on the values of the covariates. For each variable Y and based on all other variables, the predictions on Y are computed from the complete data. For a recipient with missing value on Y, the set of potential donors are the units with recorded values on Y. The imputed value by predictive mean matching method is the recorded value of the nearest-neighbor potential donor, as computed from the distance between the donor and the recipient on the predictions on Y. In ordinary predictive mean matching the predictions are computed through a linear regression model.

The methods presented above implicitly assume that the missing data are MAR. Hot-deck procedures and models have been proposed also for cases when the missing data are NMAR. In the hot-deck procedures, instead of imputing the actual donor's value, for the various variables the values may be multiplied by appropriate inflation or factors. By definition, the observed data in a set with NMAR missing do not provide information to allow us to cross-validate the validity of the models' assumption. A cross-validation method that has been suggested and tested is to fit the models under various assumptions and to test the sensitivity of the resulting conclusions.

Multiple Imputation and Bootstrapping

While the biases in estimation caused by missing data are widely recognized, the imputation is not a universally accepted solution. In a glossary of the National Institute of Standards and Technology (`http://ciks.cbt.nist.gov`), it is mentioned that "... imputation is most common in surveys of human populations. It is also used in certain computer experiment applications". Even in "surveys of human populations" the imputation and estimation is not taken lightly. The international standards code for market, opinion, and social research [12] requires that "... no data shall be imputed/assumed without the knowledge of the project manager. Comparison to the original source data shall be the first step in the process. Any imputation process shall be documented and available to client on request."

It is clear that the imputation raises concern with the practitioners. This concern can be attributed in part to the feeling that imputation is in some sense "inventing data" [1], and that even the sophisticated imputation methods cannot account for all the uncertainty in the actual value of the missing data.

Multiple imputation (MI) developed by Rubin [13] is a practical solution, which can alleviate the problem. For the estimation of an unknown parameter vector $\mathbf{\Theta}$, if we have a model for predicting the missing (or erroneous) values conditional on all observed data, we can use this model to make m independent simulated draws for the missing data, each containing different estimated values of the missing values from their predictive distribution. Rubin [13] showed that for a fixed number of draws, $m \geq 2$, the estimate of $\mathbf{\Theta}$ obtained by averaging the individual estimates is a consistent estimator. Furthermore, a consistent estimate of the covariance matrix can be constructed from the averages of the individual covariance matrices estimated for each draw, and an estimate of the covariance of the estimates between the draws. In particular, the variance of the estimate is the average of the variances from the m draws plus $1 + 1/m$ times the sample variance of the estimates between the m datasets.

The MI can enhance the accuracy of estimated critical values and **confidence intervals** and bands as compared to other methods of dealing with missing data. An additional technique, which was compared in a series of articles with the MIs (e.g. [14]), and which also relies on repeated sampling from empirical datasets and associated estimates is bootstrapping. Both are computationally intensive methods and their development has been supported by the increase in computing **power**. Especially in econometrics, they contribute to the improvement of the estimates of precision of the statistical estimates when standard analytical estimates are biased or even unavailable. Furthermore, the bootstrapping and MIs can be combined, with the **bootstrap** replica being used in the MIs (e.g., Serrat *et al.* [15]).

In a recent industrial design of experiment [16] the bootstrap method was used to impute missing data at **center point** in an industrial **design of experiments**. Indeed, although "imputation is most common in surveys of human populations" the industrial experimental designs are also faced with missing or extra experimental runs, resulting in an unplanned not-rectangular dataset. Imputation can be highly effective in those circumstances.

Despite the concern related to "inventing data" in imputations, both theoretical results and simulation clearly demonstrate that, when performed properly, imputation can considerably improve the accuracy and the efficiency of the analyses. It is our opinion that they probably should be used more often than they are in quality control settings.

Parameter Estimation from Incomplete Data – The EM algorithm

The disadvantage of the imputation techniques mentioned above regarding "invented data" is alleviated in the direct analysis of the nonrectangular, incomplete dataset by the maximum-likelihood analysis. Let us assume that the underlying statistical model whose parameters Θ we have to estimate and the distribution of the data under the model are specified. Furthermore, let us assume that the missing pattern is ignorable, i.e. that in the maximum-likelihood approach, the parameter estimates do not depend on the pattern of the missing values and that pattern can thus be ignored. Little and Rubin [17] called those *ignorable models*.

Under those conditions the expectation-maximization (EM) algorithm [18], is a general and vastly used iterative algorithm, which computes the maximum-likelihoods estimates for Θ and in the intermediate steps performs pseudo-imputations as explained below. Specifically, at the completion of the tth iteration or step of the **EM algorithm**, the current estimated value of Θ is $\Theta^{(t)}$. The expectation step (E-step) $(t+1)$th step computes the expected value of the log-likelihood, over the conditional distribution of the missing data Y_{miss}, given the observed data Y_{obs} and the current estimate $\Theta^{(t)}$ of Θ. Now, if the distribution of the data is from the exponential family, the above expectation is equivalent to computing expected values for the sufficient statistics, given Y_{obs} and $\Theta^{(t)}$ [1, 19]. This is the pseudo-imputation alluded to above. In the M-step of the algorithm, a new estimate $\Theta^{(t+1)}$ of Θ is calculated by maximizing with respect to Θ the expected value of the log-likelihood found in the E-step. The algorithm was proved to converge to the maximum-likelihood estimates of Θ. Since the algorithm was developed, many enhancements, which increase the rate of convergence for various problems with application in industry, have been proposed (e.g. Fessler and Hero [20, 21]).

In many situations, calculating the conditional expectation required in the E-step, may be infeasible. A possible alternative is the stochastic EM algorithm [22] in which the expectations are performed *via* simulation. At the $(t+1)$th iteration the missing data is imputed with a draw from the conditional distribution of the missing data given Y_{obs} and $\Theta^{(t)}$. Given the pseudocomplete data, one can maximize the log-likelihood to obtain $\Theta^{(t+1)}$.

The EM algorithm is widely used in a very large spectrum of applications, either in its direct version or in special versions and enhancements developed for particular situations (as the stochastic EM). Among the applications we can mention biomedical imaging, such as positron emission tomography [20, 21, 23, 24] as a robust inference in the multivariate **normal distribution** with missing data [25, 26]; log-linear models for contingency tables with missing data [27]; models for logistic regression with missing covariates [28]; and **survival analysis** with missing covariates [29].

Repeated Measurements in Longitudinal Studies

In the conclusion of this review we return to the example of a longitudinal consumer satisfaction study and present some issues specific to missing data in longitudinal studies. In many such studies (as in consumer satisfaction with a single response or in some clinical studies), the response variable is univariate, with the predictor variables assumed constant over time, and recorded at the baseline. More complex models assume variable predictors, which have to be recorded at each survey time separately. The missing data in longitudinal studies can come in various forms: from the common case of dropouts after one or more initial surveys, to more complex cases of incidental dropouts at no particular order, or even more complex patterns of missing with unit nonresponse on some surveys and item nonresponse on others. For the sake of simplicity, we focus on the important case in longitudinal studies where the missing data is due to dropout.

The main approaches of dealing with missing data in longitudinal studies are equivalent to those used in single-time studies, with proper and important adjustments. The listwise deletion (or CC) approach performs the analysis on the subset with complete data and ignores the dropouts. The analysis yields valid estimates in the MCAR case, and yields biased estimates in all other cases.

Both single and multiple imputations are used in longitudinal studies. In the clinical trials the entire groups of subjects who started the study are known as the *intent-to-treat* (ITT) group. In those studies, one of the widely used single-imputation techniques (at least until recently) was the last observation carried forward (LOCF) technique. Under this procedure, the observation recorded before the dropout is "carried forward" and imputed for the missing in all subsequent times. LOCF is a model with simple assumptions behind it and it is indeed easy to understand and communicate.

However, the procedure does not give valid analyses if the dropout mechanism is not MCAR. Even then, it can yield biased estimates unless the pattern of response over time is constant, and if groups are compared, the pattern of dropout had to be the same for each group. The LOCF does not give a statistical estimator of any actual population parameter and yields underestimated variances. The procedure was considered conservative (and it is so, but only if a treatment is monotonically beneficial). A frequent argument in its favour is that it can be interpreted as "real world results". However, this argument is not tenable since the results are likely to be misinterpreted as if they were from marginal inference, and the actual interpretation is very difficult to explain (e.g. Lane [30]). The performance of the procedure was compared in many studies under a variety of simulated types of missingness, and it was found to be under par even under most favorable situations (e.g. [30–34]). Indeed as Carpenter *et al.* [35] and Shao and Zhong [36] put it, "It is time to stop carrying it forward."

Other single imputations, mentioned in the context of a single-wave study, as the hot deck and the predicted mean imputation are also used in longitudinal studies. Unlike the single-wave studies, the data used for matching in the hot-deck procedure and for prediction in the predictive mean procedure, include the responses on the previous waves on the same variable.

The MI approach, which was also presented in the context of single-wave studies is effective in longitudinal studies as well. The MI approach carries out analysis under each set of imputation and combines analyses to reflect within-imputation and between-imputation variability. Several techniques involved in MI and used in longitudinal studies are mentioned by Rubin [37], Little and Rubin [38], Schafer and Olsen [39], and Schafer [40].

Assuming ignorable missing data as well as multivariate normal distribution of responses in longitudinal studies, the general mixed model can be used for analyzing the data with or without imputation. The model is also known under the name "mixed model for repeated measures" (MMRM). For the ith unit, $i = 1, \ldots, n$, the model for the $(\tau \times p)$ response vector, $\mathbf{Y}_i$ is $\mathbf{Y}_i = \mathbf{X}_i\boldsymbol{\beta} + \mathbf{Z}_i\mathbf{b}_i + \boldsymbol{\varepsilon}_i$, where $\mathbf{b}_i \sim \mathbf{N}_q(\mathbf{0},\mathbf{D})$, $\boldsymbol{\varepsilon}_i \sim \mathbf{N}_\tau\ (\mathbf{0},\boldsymbol{\Sigma})$, and $\mathbf{b}_i$ and $\boldsymbol{\varepsilon}_i$ are statistically independent. $\mathbf{X}_i$ is a $(\tau \times p)$ design matrix for the fixed effects and $\mathbf{Z}_i$ is a $(\tau \times q)$ design matrix for the random effects. $\boldsymbol{\beta}$ is a $(p \times 1)$ vector of unknown fixed effects and $\mathbf{b}_i$ is an unknown $(q \times 1)$ random-coefficient vector (e.g. [41, 42]). Maximum-likelihood estimates for this model (e.g. [43]) can now be calculated either by standard software programs like PROC MIXED in SAS or by more specialized programs like HLM.

If the assumption of nonignoribility is considered to be untenable, one can model the dropout as well as the response through selection models (SMs) or pattern-mixture models (PMMs). In the SMs, given the matrix of covariates X, the joint conditional distribution for the responses Y and matrix of missing M is

$$f(Y, M|X, \Theta, \Psi) = f(Y|X, \Theta)f(M|Y, X, \Psi) \quad (3)$$

where the model under no missing data is $f(Y|X, \Theta)$ and $f(M|Y, X, \Psi)$ is the model for the pattern of missing data.

In the PMMs the joint conditional distribution is

$$f(Y, M|X, \Theta, \Psi) = f(Y|X, M, \Theta)f(M|X, \Psi) \quad (4)$$

with the two models coinciding when M is independent of Y. The shortcoming of the models for dropout is that the model assumptions cannot be tested from the data.

An issue of considerable practical importance is to assess theoretically or in simulations, the performance and robustness of various procedures both in the case when the underlying assumptions are met as well as when they are not met. Thus, one can generate various patterns of missing not at random (MNAR), and assess the robustness of the methods which implicitly assume a MAR mechanism. When the MMRM model is fitted, the fitting may or may not be preceded by a preliminary imputation step.

Several such studies were performed lately. The comparison between LOCF and MMRM [30] led to the conclusion that the use of the LOCF method is likely to seriously misrepresent the results of the study and that LOCF is "not a good choice for primary analysis". In contrast, except for MNAR cases with substantially unequal dropouts in various groups, the MMRM is unlikely to result in serious misinterpretation. Sensitivity analysis with more complex models is required in dealing with such MNAR patterns of missing data. Simulation results obtained with MMRM, with and without imputations found the MIs to be a valuable technique, and superior to complete-case analysis [44]. The data were simulated under the MAR and MCAR mechanisms. Furthermore, the direct analysis of the incomplete dataset was effective as compared to the imputed data. The effectiveness of adding MIs to the MMRM and to the generalized estimating equations (GEE) was tested recently [45] in a simulated clinical trial setting aimed at determining the treatment effect at specific time epochs. They concluded that in this setting, the additional use of MIs is, in general, counterproductive both in terms of bias as well as in terms of variances.

The imputation procedures and MMRM modeling lately gained acceptance in industry (e.g. [30]). This trend is a positive result of the studies that showed that both imputations (especially MIs) and model fittings are highly beneficial in improving the accuracy of the performed analyses. The benefits to the performed studies far outweigh the shortcomings of the methods, i.e. the fact that imputation seem to be inventing data and the results of MMRM models are more difficult to communicate than analyses based either on CCs or on simple analysis of available data. The trend of increased acceptance of imputation and MMRM model fitting deserves to be encouraged.

References

[1] Little, R.J.A. (1997). Biostatistical analysis with missing data, in *Encyclopedia of Biostatistics*, P. Armitage & T. Colton, eds, John Wiley & Sons, London.

[2] Rubin, D.B. (1976). Inference and missing data, *Biometrika* **63**, 581–592.

[3] Nookabadi, A.S. & Middle, J.E. (2001). A knowledge-based system as a design aid for quality assurance information systems, *International Journal of Quality and Reliability Management* **18**, 657–671.

[4] Little, R.J.A. (1986). Survey nonresponse adjustments for estimates of means, *International Statistical Review* **54**, 137–139.

[5] Haitovsky, Y. (1968). Missing data in regression analysis, *Journal of the Royal Statistical Society, Series B* **30**, 67–81.

[6] Kim, J.O. & Curry, J. (1977). The treatment of missing data in multivariate analysis, *Sociological Methods and Research* **6**, 215–240.

[7] Afifi, A. & Elashoff, R. (1967). Missing observations in: II. Point estimation in simple linear regression, *Journal of the American Statistical Association* **62**(317),, 10–29.

[8] Kostanich, D.L. & Dippo, C.S. (supervisors) (2002). *The Current Population Survey: Design and Methodology*, Technical Paper 63RV, U.S. Bureau of the Census, Washington, DC.

[9] Hanson, R.H. (1978). *The Current Population Survey: Design and Methodology*, Technical Paper 40, U.S. Bureau of the Census, Washington, DC.

[10] David, M., Little, R.J.A., Samuhel, M.E. & Triest, R.K. (1986). Alternative methods for CPS income imputation, *Journal of the American Statistical Association* **81**, 29–41.

[11] Thibaudeau, Y., Gottschalck, A. & Palumbo, T. (2006). *The Predictive-Mean Method of Imputation for Preserving Coupling Between Assets and Liabilities*, U.S. Census Bureau, Statistical Research Division Research Report Series.

[12] ISO 20252 (2006). *Market, Opinion and Social Research – Vocabulary and Service Requirements*, International Organization for Standardization.

[13] Rubin, D.B. (1987). *Multiple Imputation for Nonresponse in Surveys*, Wiley, New York.

[14] Brownstone, D. & Valletta, R. (2001). The bootstrap and multiple imputations: harnessing increased computing power for improved statistical tests, *The Journal of Economic Perspectives* **15**, 129–141.

[15] Serrat, C., Gomez, G., Garcia de Olallab, P. & Caylab, J. (1998). CD4+ lymphocytes and tuberculin skin test as survival predictors in pulmonary tuberculosis HIV-infected patients, *International Journal of Epidemiology* **27**, 703–712.

[16] Kenett, R.S., Rahav, E. & Steinberg, D.M. (2006). Bootstrap analysis of designed, *Experiments Quality and Reliability Engineering International* **22**, 659–667.

[17] Little, R.J.A. & Rubin, D.B. (1989). Missing data in social science data sets, *Sociological Methods and Research* **18**, 292–326.

[18] Dempster, A.P., Laird, N.M. & Rubin, D.B. (1977). Maximum likelihood from incomplete data via the EM algorithm, *Journal of the Royal Statistical Society* **B39**, 1–38.

[19] Sundberg, R. (1974). Maximum likelihood theory for incomplete data from an exponential family, *Scandinavian Journal of Statistics* **1**, 49–58.

[20] Fessler, J.A. & Hero, A.O. (1994). Space-alternating generalized expectation-maximization algorithm, *IEEE Transactions on Signal Processing* **42**, 2664–2677.

[21] Fessler, J.A. & Hero, A.O. (1995). Penalized maximum-likelihood image reconstruction using space alternating

generalized expectation-maximization algorithm, *IEEE Transactions on Image Processing* **4**, 1417–1438.

[22] Celuex, G. & Diebolt, J. (1985). The SEM algorithm: a probabilistic teacher algorithm derived from the EM algorithm for the mixture problems, *Computational Statistics Quarterly* **2**, 73–82.

[23] Lange, F. & Carson, R. (1984). EM reconstruction algorithms for emission and transmission tomography, *Journal of Computer-Assisted Tomography* **8**, 306–316.

[24] Shepp, L.A. & Vardi, Y. (1982). Maximum likelihood reconstruction for emission tomography, *IEEE Transactions on Image Processing* **2**, 113–122.

[25] Lange, K., Little, R.J.A. & Taylor, J. (1989). Robust statistical inference using the *T* distribution, *Journal of the American Statistical Association* **84**, 881–896.

[26] Little, R.J.A. (1988). Robust estimation of the mean and covariance matrix from data with missing values, *Applied Statistics* **37**, 23–38.

[27] Fuchs, C. (1982). Maximum likelihood estimation and model selection in contingency tables with missing data, *Journal of the American Statistical Association* **77**, 270–278.

[28] Vach, W. (1994). *Logistic Regression with Missing Values in the Covariates*, Springer-Verlag, New York.

[29] Schluchter, M.D. & Jackson, K.L. (1989). Loglinear analysis of censored survival data with partially observed covariates, *Journal of the American Statistical Association* **84**, 42–52.

[30] Lane, P. (2007). Handling drop-out in longitudinal clinical trials: a comparison of the LOCF and MMRM approaches, *Pharmaceutical Statistics*, Wiley InterScience.

[31] Heyting, A., Tolboom, J.T.B.M. & Essers, J.G.A. (1992). Statistical handling of drop-outs in longitudinal clinical trials, *Statistics in Medicine* **11**, 2043–2061.

[32] Mallinckrodt, C.H., Clark, W.S. & David, S.R. (2001). Type I error rates from mixed effects model repeated measures compared with fixed effects ANOVA with missing values imputed via LOCF, *Drug Information Journal* **35**, 1214–1225.

[33] Mallinckrodt, C.H., Kaiser, C.J., Watkin, J.G., Molenberghs, G. & Carroll, R.J. (2004). The effect of correlation structure on treatment contrasts estimated from incomplete clinical trial data with likelihood-based repeated measures compared with last observation carried forward ANOVA, *Clinical Trials* **1**, 477–489.

[34] Cook, R.J., Zeng, L. & Yi, G.Y. (2004). Marginal analysis of incomplete longitudinal binary data; a cautionary note on LOCF imputation, *Biometrics* **60**, 820–828.

[35] Carpenter, J., Kenward, M., Evans, S. & White, I. (2004). Last observation carry forward and last observation analysis (letter to the editor), *Statistics in Medicine* **23**, 3241–3244.

[36] Shao, J. & Zhong, B. (2003). Last observation carry forward and last observation analysis, *Statistics in Medicine* **22**, 2429–2441.

[37] Rubin, D.B. (1986). Statistical matching and file concatenation with adjusted weights and multiple imputations, *Journal of Business and Economic Statistics* **4**, 87–94.

[38] Little, R.J.A. & Rubin, D.B. (1987). *Statistical Analysis with Missing Data*, John Wiley & Sons, New York.

[39] Schafer, J.L. & Olsen, M.K. (1998). Multiple imputation for multivariate missing-data problems: a data analyst's perspective, *Multivariate Behavioral Research* **33**, 545–571.

[40] Schafer, J.L. (1997). *Analysis of Incomplete Multivariate Data*, Chapman & Hall, London.

[41] Molenberghs, G., Kenward, M.G. & Lesaffre, E. (1977). The analysis of longitudinal ordinal data with nonrandom drop-out, *Biometrika* **84**(1), 33–44.

[42] Verbeke, G. & Molenberghs, G. (2000). *Linear Mixed Models for Longitudinal Data*, Springer, New York.

[43] Laird, N.M. & Ware, J.H. (1982). Random-effects models for longitudinal data, *Biometrics* **38**, 963–974.

[44] Shieh, Y.-Y. (2003). *Imputation Methods on General Linear Mixed Models of Longitudinal Studies*, Federal Committee on Statistical Methodology (FCSM), Washington, DC.

[45] Fong, D.Y.T., Rai, S.N. & Lam, K.S.L. (2006). *Use of Multiple Imputation on Linear Mixed Model and Generalized Estimating Equations for Longitudinal Data Analysis: A Simulation Study*, SCRA 2006-FIM XIII, Lisbon-Tomar, Portugal.

Further Reading

Cook, R. (1986). Assessment of local influence, *Journal of the Royal Statistical Society. Series B (Statistical Methodology)* **48**, 133–169.

Lange, K. (1995a). A gradient algorithm locally equivalent to the EM algorithm, *Journal of the Royal Statistical Society. Series B* **57**, 425–437.

Lange, K. (1995b). A quasi-Newtonian acceleration of the EM algorithm, *Statistica Sinica* **5**, 1–18.

Molenberghs, G. & Verbeke, G. (2005). *Models for Discrete Longitudinal Data*, Springer, New York.

CAMIL FUCHS AND RON S. KENETT

Mixture Experiments

Introduction

A *mixture experiment* involves combining two or more components in various proportions or amounts and then measuring one or more responses for the resulting end products. Other factors that affect the response(s), such as process variables (PVs) and/or the total amount of the mixture, may also be studied in the experiment. A *mixture experiment design* specifies the combinations of mixture components and other experimental factors (if any) to be studied and the response(s) to be measured. *Mixture experiment data analyses* are then used to achieve the desired goals, which may include (a) understanding the effects of components and other factors on the response(s), (b) identifying components and other factors with significant and nonsignificant effects on the response(s), (c) developing models for predicting the response(s) as functions of the mixture components and any other factors, and (d) developing end products with optimal or desired values and uncertainties of the response(s).

Given a mixture experiment problem, a practitioner must consider the possible approaches for designing the experiment and analyzing the data, and then select the approach best suited to the problem. Eight possible approaches are (a) component proportions (CPs), (b) mathematically independent variables (MIVs), (c) slack variable (SV), (d) mixture amount (MA), (e) component amounts (CA), (f) mixture-process variable (MPV), (g) mixture of mixtures, (MoM) and (h) multifactor mixture (MFM). The article provides an overview of the mixture experiment designs, models, and data analyses appropriate for these approaches. For more discussion and examples of these approaches, see Piepel and Cornell [1]. Also, the books by Cornell [2] and Smith [3] are devoted to the design and data analysis of mixture experiments and are recommended for additional information. Piepel [4] summarizes 50 years of mixture experiment research from 1955 to 2004 along with an extensive bibliography.

Before designing a mixture experiment, a practitioner must choose the model form(s) likely to be adequate to achieve the objectives of the experiment, and then design the experiment to adequately support fitting the model form(s). Hence in the subsequent section for each mixture experiment approach, appropriate models are discussed before experimental designs.

Component Proportions Approach

With the CP approach for mixture experiments, q mixture components and no other factors are studied. The experiment is designed and responses are modeled using the proportions of the mixture components x_i, which are subject to the constraints

$$0 \le x_i \le 1, i = 1, 2, \ldots, q \text{ and } \sum_{i=1}^{q} x_i = 1 \qquad (1)$$

The experimental region defined by these constraints is a $(q-1)$-dimensional simplex. Additional single-component constraints of the form

$$L_i \le x_i \le U_i, \quad i = 1, 2, \ldots, q \qquad (2)$$

and/or multiple-component constraints of the form

$$C_k \le \sum_{i=1}^{q} A_{ki} x_i \le D_k, \quad k = 1, 2, \ldots, K \qquad (3)$$

generally yield a $(q-1)$-dimensional polyhedral experimental region. The constraint in equation (1), that the mixture CPs must sum to one, means that the mixture CPs cannot be varied independently. This fact affects the choice of experimental designs and models for the CP approach (and some other approaches) to mixture experiments.

Scheffé Polynomial Models in Component Proportions

Polynomial models are often successful in modeling response surfaces based on Taylor series theory for approximating functions with polynomial expansions. However, the standard polynomial models must be modified to account for the mixture constraint $\sum_{i=1}^{q} x_i = 1$ in equation (1). Several commonly used Scheffé [5] polynomial models for CP mixture experiments are

Linear:

$$\eta(\boldsymbol{x}) = \sum_{i=1}^{q} \beta_i x_i \qquad (4)$$

Quadratic:

$$\eta(\boldsymbol{x}) = \sum_{i=1}^{q} \beta_i x_i + \sum_{i=1}^{q-1} \sum_{j=i+1}^{q} \beta_{ij} x_i x_j \qquad (5)$$

Special cubic:

$$\eta(\boldsymbol{x}) = \sum_{i=1}^{q} \beta_i x_i + \sum_{i=1}^{q-1} \sum_{j=i+1}^{q} \beta_{ij} x_i x_j + \sum_{i=1}^{q-2} \sum_{j=i+1}^{q-1} \sum_{k=j+1}^{q} \beta_{ijk} x_i x_j x_k \qquad (6)$$

Full cubic:

$$\eta(\mathbf{x}) = \sum_{i=1}^{q} \beta_i x_i + \sum_{i=1}^{q-1} \sum_{j=i+1}^{q} \beta_{ij} x_i x_j + \sum_{i=1}^{q-1} \sum_{j=i+1}^{q} \delta_{ij} x_i x_j (x_i - x_j) + \sum_{i=1}^{q-2} \sum_{j=i+1}^{q-1} \sum_{k=j+1}^{q} \beta_{ijk} x_i x_j x_k \quad (7)$$

Full-quartic and special-quartic models are discussed by Smith [3, Section 3.4]. In all of these model forms, $\eta(\mathbf{x})$ is the expected value of the response at mixture $\mathbf{x} = (x_1, x_2, \ldots, x_q)$. The $\sum_{i=1}^{q} \beta_i x_i$ model terms in equations (4–7) describe *linear blending* of the mixture components, while the remaining terms in equations (5–7) describe *nonlinear blending* of the mixture components in various degrees (i.e., quadratic, special cubic, and full cubic). The nonlinear blending terms are not referred to as *interaction terms*, because curvature and interaction effects of the mixture components are partially confounded as a result of the mixture constraint $\sum_{i=1}^{q} x_i = 1$. Methods for investigating curvature and interaction effects of mixture components are discussed by Cox [6] and Piepel *et al.* [7].

The coefficient β_i in equations (4–7) represents the expected value of the response at the pure mixture with $x_i = 1$ and $x_j = 0$ for $j \neq i$. Cornell [2, Sections 2.2 and 2.3] discusses the interpretations of other coefficients in mixture experiment models. However, all such interpretations are extrapolations when the experimental region is restricted by constraints of the form in equations (2) and/or (3).

Other Polynomial Models in Component Proportions

Cox [6] proposed using ordinary first- and second-degree polynomial models with the parameters constrained to estimate the main, curvature, and interactive effects of components in a way appropriate to mixture experiments. Cornell [2, Sections 6.7 and 6.8] and Smith and Beverly [8] discuss and illustrate the *Cox mixture polynomial models*. Piepel [9, 10] proposed *component-slope linear mixture (CSLM) models* with parameters representing the slopes of the **response surface** along the component effect directions. The advantage of the Cox and CSLM models over the Scheffé models is that their coefficients provide estimates of component effects. The disadvantage is that the models are overparameterized and involve constraints on the coefficients.

Scheffé quadratic models in equation (5) are adequate for many mixture experiment applications, but in practice only a few quadratic terms may be needed. *Partial quadratic mixture* (PQM) models of the form

$$\eta(\mathbf{x}) = \sum_{i=1}^{q} \beta_i x_i + \text{subset of } \left\{ \sum_{i=1}^{q} \beta_{ii} x_i^2 + \sum_{i=1}^{q-1} \sum_{j=i+1}^{q} \beta_{ij} x_i x_j \right\} \quad (8)$$

contain a subset of squared terms and/or cross-product terms. PQM models contain as a subset the set of reduced Scheffé quadratic models, and thus can more efficiently capture curvature as well as interactive effects of mixture components. See Piepel *et al.* [11] and Smith [3, Section 10.3] for discussion and illustrations of PQM models.

Nonpolynomial Models in Component Proportions

The mixture polynomial model forms previously introduced provide for adequately approximating many true response surfaces for mixture experiments, but there are situations when nonpolynomial model forms are more appropriate.

For CP mixture experiments where one or more components have additive effects, three *models homogeneous of degree one* proposed by Becker [12] are appropriate

$$\text{H1: } \eta(\mathbf{x}) = \sum_{i=1}^{q} \beta_i x_i + \sum_{i=1}^{q-1} \sum_{j=i+1}^{q} \beta_{ij} \min(x_i, x_j) + \sum_{i=1}^{q-2} \sum_{j=i+1}^{q-1} \sum_{k=j+1}^{q} \beta_{ijk} \min(x_i, x_j, x_k) \quad (9)$$

$$\text{H2: } \eta(\mathbf{x}) = \sum_{i=1}^{q} \beta_i x_i + \sum_{i=1}^{q-1} \sum_{j=i+1}^{q} \beta_{ij} \frac{x_i x_j}{x_i + x_j} + \sum_{i=1}^{q-2} \sum_{j=i+1}^{q-1} \sum_{k=j+1}^{q} \beta_{ijk} \frac{x_i x_j x_k}{x_i + x_j + x_k} \quad (10)$$

$$\text{H3: } \eta(\boldsymbol{x}) = \sum_{i=1}^{q} \beta_i x_i + \sum_{i=1}^{q-1} \sum_{j=i+1}^{q} \beta_{ij} (x_i x_j)^{1/2} + \sum_{i=1}^{q-2} \sum_{j=i+1}^{q-1} \sum_{k=j+1}^{q} \beta_{ijk} (x_i x_j x_k)^{1/3} \quad (11)$$

The cubic forms of these models are shown, but the quadratic forms (without the last group of terms in each model) are more frequently used in practice. For additional discussion of these models, see Cornell [2, Sections 6.3–6.5] and Snee [13].

For CP mixture experiments where the response variable increases rapidly as CPs approach zero, Scheffé linear and quadratic models with inverse terms

$$\eta(\boldsymbol{x}) = \sum_{i=1}^{q} \beta_i x_i + \sum_{i=1}^{q} \delta_i x_i^{-1} \quad (12)$$

$$\eta(\boldsymbol{x}) = \sum_{i=1}^{q} \beta_i x_i + \sum_{i=1}^{q-1} \sum_{j=i+1}^{q} \beta_{ij} x_i x_j + \sum_{i=1}^{q} \delta_i x_i^{-1} \quad (13)$$

are appropriate. Inverse terms can be added to other mixture models such as in equations (6) and (7) if needed. It is only necessary to add inverse terms for those components needing them. For CPs that can equal zero, the corresponding x_i^{-1} in equations (12) and (13) are replaced with $(x_i + c_i)^{-1}$, where the c_i values are small positive constants. The forms of Scheffé models with inverse terms can also be modified for use when the response increases rapidly as constrained CPs approach their lower limits. See Cornell [2, Sections 6.1 and 6.2] for further discussion of these topics.

Log-contrast models were introduced by Aitchison and Bacon-Shone [14] for situations in which one or more mixture components are inert or have additive effects on the response. The linear and quadratic forms of the log-contrast models are given by

$$\eta(\boldsymbol{x}) = \gamma_0 + \sum_{i=1}^{q} \gamma_i \log x_i \quad (14)$$

$$\eta(\boldsymbol{x}) = \gamma_0 + \sum_{i=1}^{q} \gamma_i \log x_i + \sum_{i=1}^{q-1} \sum_{j=i+1}^{q} \gamma_{ij} (\log x_i - \log x_j)^2 \quad (15)$$

which are overparameterized and thus subject to constraints on the parameters. For CPs that can equal zero, the corresponding $\log x_i$ in equations (14) and (15) are replaced with $\log(x_i + c_i)$. See Aitchison and Bacon-Shone [14] and Cornell [2, Sections 6.9 and 6.10] for additional discussion of the constraints on the parameters, interpretations of the parameters, and examples of these models.

Simplex Designs in Component Proportions

Four types of experimental design are commonly used to explore how responses depend on the proportions of mixture components over simplex experimental regions (either the whole simplex region as defined in equation (1) or a simplex-shaped subregion).

A $\{q, m\}$ *simplex-lattice design* consists of $\binom{q+m-1}{m} = (q+m-1)!/m!(q-1)!$ points with each component having $m+1$ equally spaced values of $0, 1/m, 2/m, \ldots, 1$. For example, the $\{3, 2\}$ simplex-lattice design contains the six points $(x_1, x_2, x_3) = (1, 0, 0)$, (0, 1, 0), (0, 0, 1), (1/2, 1/2, 0), (1/2, 0, 1/2), and (0, 1/2, 1/2). This design is pictured in Figure 1(a). As another example, the $\{4, 3\}$ simplex-lattice design contains 20 points including 4 permutations of (1, 0, 0, 0), 12 permutations of (1/3, 2/3, 0, 0), and 4 permutations of (1/3, 1/3, 1/3, 0). A $\{q, m)$ simplex-lattice design supports fitting a Scheffé polynomial model of degree m (see Cornell [2, Section 2.2]). For example, the $\{3, 3\}$ simplex-lattice design supports fitting the full-cubic model in equation (7).

A *simplex-centroid design* for q components contains $2^q - 1$ points, consisting of all permutations of $(1, 0, 0, \ldots, 0)$, $(1/2, 1/2, 0, \ldots, 0)$, $(1/3, 1/3, 1/3, 0, \ldots, 0)$, ... as well as the centroid $(1/q, 1/q, 1/q, \ldots, 1/q)$. Figure 1(b) illustrates this design for $q = 3$. The simplex-centroid design supports fitting the full Scheffé polynomial model in equation (4), as well as the Scheffé linear, quadratic, and special-cubic polynomial models in equations (5–7).

An *augmented simplex-centroid design* adds to the simplex-centroid design the q interior axial points of the form $((q+1)/2q, 1/2q, 1/2q, \ldots, 1/2q)$. This design for $q = 3$ is shown in Figure 1(c).

Snee and Marquardt [15] proposed a *simplex screening design* for q components that contains

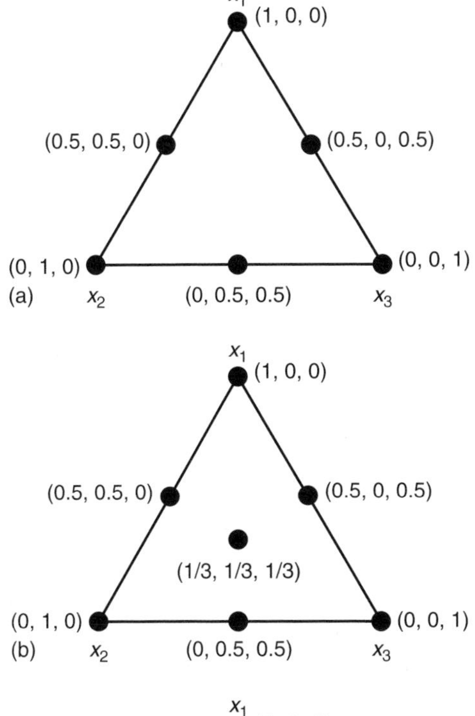

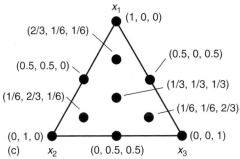

Figure 1 Example designs for three mixture components with the CP approach: (a) {3, 2} simplex-lattice design, (b) simplex-centroid design, and (c) augmented simplex-centroid design

points along the component axes, including the q vertices of the form $(1, 0, 0, \ldots, 0)$, the q axis endpoints of the form $(0, 1/q, \ldots, 1/q)$, the q interior axial points of the form $((q+1)/2q, 1/2q, 1/2q, \ldots, 1/2q)$, and the overall centroid $(1/q, 1/q, 1/q, \ldots, 1/q)$. For $q = 3$, the simplex screening design is the same as the augmented simplex-centroid design shown in Figure 1(c).

When one or more mixture CPs are constrained by lower bounds $x_i \geq L_i$, the constrained experimental region is a simplex-shaped subregion of the whole simplex defined by equation (1). The simplex subregion can be transformed to the whole simplex by L-pseudocomponent transformations

$$x_i' = \frac{x_i - L_i}{1 - \sum_{j=1}^{q} L_j}, i = 1, 2, \ldots, q \qquad (16)$$

In such cases, any of the simplex designs described previously can be used. Designs expressed in proportions of the L-pseudocomponents can be transformed into designs expressed in proportions of the original components via

$$x_i = L_i + \left(1 - \sum_{j=1}^{q} L_j\right) x_i' \qquad (17)$$

Constraints of the form $x_i \leq U_i$ can sometimes lead to a simplex-shaped subregion. Cornell [2, Section 4.6] discusses a U-pseudocomponent transformation that can be applied to adapt the simplex designs discussed previously to such situations.

Constrained Designs in Component Proportions

When there are single- and/or multiple-component constraints on the CPs, as in equations (2) and/or (3), the mixture experimental region is generally a convex polyhedron that can be irregular in shape. Designs for constrained CP mixture experiments are generally constructed by *optimal experimental design* methods. These methods involve selecting design points that satisfy the applicable constraints (possibly from a pregenerated candidate set) so as to optimize a selected design optimality criterion.

By analogy with **factorial** and **fractional factorial** designs for nonmixture variables on cuboidal experimental regions, candidate points often consist of all or some of the vertices and dimensional centroids (i.e., 1-D edge centroids, 2-D face centroids, ..., $(q-2)$-D face centroids, and the $(q-1)$-D overall centroid) of the constrained CP experimental region. Cornell [2, Section 4.9] discusses several algorithms for calculating all or a fraction of the vertices of a polyhedral experimental region defined by single-component constraints as in equation (2). Snee [16] and Piepel [17] discuss algorithms for calculating vertices and dimensional centroids for a constrained region defined by single- and/or multiple-component constraints of the forms (2) and/or (3). Mixtures in the interior of the constrained region, possibly obtained on a grid, are needed as candidate points for some design optimality criteria.

The most commonly used design optimality criteria include

D-optimality: Minimize $|(\mathbf{U}'\mathbf{U})^{-1}|$, the generalized variance of the vector of estimated model coefficients.

G-optimality: Minimize $\max_{\mathbf{x}\in R}\{\mathbf{u}'(\mathbf{U}'\mathbf{U})^{-1}\mathbf{u}\}$, the maximum standardized prediction variance over the constrained experimental region R.

I-optimality: Minimize $\text{avg}_{\mathbf{x}\in R}\{\mathbf{u}'(\mathbf{U}'\mathbf{U})^{-1}\mathbf{u}\}$, the average standardized prediction variance over the constrained experimental region R.

In the preceding notation, $\boldsymbol{x}$ is a vector of predictor variables (e.g., mixture CPs), $\boldsymbol{X}$ is the experimental design matrix expressed in the predictor variables, and $\boldsymbol{u}$ and $\boldsymbol{U}$ are extensions of $\boldsymbol{x}$ and $\boldsymbol{X}$ in the form of the model for which an **optimal design** is desired. Note that I-optimality is sometimes referred to as *V-optimality* or *IV-optimality*. These optimality criteria (and others not listed here) are often called *alphabetic design optimality criteria* because letters of the alphabet are used to identify them.

Mathematically Independent Variables Approach

With the MIVs approach for mixture experiments, q mixture components and no other factors are studied. The MIVs approach can be extended to study nonmixture factors (e.g., total amount or PVs) in addition to the mixture components, but such extensions are not discussed here. An MIVs experiment is designed and responses are modeled using $q-1$ MIVs to which the q mixture CPs ($x_i, i = 1, 2, \ldots, q$) have been transformed. The MIVs are denoted $z_i, i = 1, 2, \ldots, q-1$. Standard designs, models, and other response surface methodology for nonmixture experiments (*see* **Response Surface Methodology**) can be used on the basis of the $q-1$ MIVs. For this reason, the MIV approach was widely used many years ago prior to the development of experimental design, modeling, and other methods specifically for mixture experiments. However, it is now recommended that the MIVs approach only be used if the transformation of q mixture CPs to $q-1$ MIVs is natural or of long-standing practice. For example, in aluminum production the long-term practice has been to use the MIVs approach with bath ratio $= NaF/AlF_3$ in addition to other separate components such as Al_2O_3, MgF_2, and CaF_2. Various forms of ratio transformations and other transformations to MIVs that are natural for a particular problem are possible. Cornell [2, Section 6.6] discusses the MIV approach using ratios of components. Cornell [2, Chapter 3] also discusses general methods for obtaining MIVs, although those methods typically do not yield transformations natural to any particular problem.

Slack Variable Approach

With the SV approach for mixture experiments, only mixture components and no other factors are studied. The experiment is designed and response(s) modeled using proportions of $q-1$ of the q mixture components. For a given design in the $q-1$ components, the proportion of the qth component (called the *slack variable*) is obtained by $x_q = 1 - x_1 - x_2 - \ldots - x_{q-1}$ for each point in the design. In this sense the qth component "takes up the slack". Standard designs, models, and other response surface methods (*see* **Response Surface Methodology**) are used with the SV approach. For example, the linear and quadratic models with the qth component as the SV are

Linear:

$$\eta(\boldsymbol{x}) = \alpha_0 + \sum_{i=1}^{q-1} \alpha_i x_i \tag{18}$$

Quadratic:

$$\eta(\boldsymbol{x}) = \alpha_0 + \sum_{i=1}^{q-1} \alpha_i x_i + \sum_{i=1}^{q-2} \sum_{j=i+1}^{q-1} \alpha_{ij} x_i x_j + \sum_{i=1}^{q-1} \alpha_{ii} x_i^2 \tag{19}$$

These models provide exactly the same fits as would the linear and quadratic Scheffé mixture models in equations (4) and (5). However, reductions of SV models can have different fits than reductions of

Scheffé mixture models and can yield misleading conclusions.

The SV approach is often used when the component chosen as the SV (a) makes up most (e.g., $x_q \geq 0.80$) of the mixture, (b) has a negligible effect on the response(s), or (c) serves as a filler or diluent. In all of these cases applications or adaptations of the CP approach are appropriate, whereas the SV approach can result in misleading conclusions. In situation (a), a component that makes up the majority of a mixture often has large effects on the response variables. However, the effects of this component are confounded with the effects of other components in models such (18) or (19). Hence, the SV approach could lead to incorrect conclusions about the effects of other components as well as the SV component. It is better to use the CP approach in this situation. In situation (b), when a component has no effect on a response over the composition region of interest, the CP approach is appropriate, with the response depending on the proportions of the remaining components (i.e., $X_j = x_j/(1 - \sum_{k=1}^{q-1} x_k)$, with $j = 1, 2, \ldots, q-1$). In situation (c), a filler or diluent component typically would not blend nonlinearly with the remaining components, but still may affect the response. In such cases, the CP approach using a model homogeneous of degree one (Becker H1, H2, and H3) or a log-contrast model should be used.

Mixture-Amount Approach

When a response variable is expected to depend on the total amount of the mixture as well as the proportions of the components, a MA approach should be used. An MA experiment is designed and the response(s) modeled in terms of the proportions (x_i) and total amount (A) of the mixture components. The CPs are subject to the basic mixture constraints in equation (1), and possibly to single-component constraints in equation (2) and/or multicomponent constraints in equation (3). The total amount is typically restricted to fall within a specified range, often with two or three levels specified for study. Typically two levels of A are coded as $A' = -1$ and $+1$, while three levels are coded as $A' = -1, 0$, and $+1$.

Mixture-Amount Models

If the amount of a mixture affects a response but not the blending effects of the mixture components, then MA models have the general form

$$\eta(\boldsymbol{x}) = \{\text{Mixture model}\} + \{\text{Amount model}\} \quad (20)$$

The mixture model can take any permissible form in the CPs (e.g., the forms in equations (4–7)), while the amount model is typically a linear or quadratic polynomial in the amount A (or coded A'). Two example models of this type are

Linear–linear:

$$\eta(\boldsymbol{x}) = \sum_{i=1}^{q} \beta_i^0 x_i + \beta_0^1 A' \quad (21)$$

Quadratic–quadratic:

$$\eta(\boldsymbol{x}) = \sum_{i=1}^{q} \beta_i^0 x_i + \sum_{i=1}^{q-1} \sum_{j=i+1}^{q} \beta_{ij}^0 x_i x_j + \beta_0^1 A' + \beta_0^2 A'^2 \quad (22)$$

In equation (21), the mixture components blend linearly and the amount variable has a linear effect on the response. In equation (22), the mixture components blend quadratically and the amount variable has a quadratic effect on the response. In both equations, the component blending is not affected by the amount of the mixture (i.e., the components and amount do not interact).

If the amount of a mixture affects the component blending properties, then MA models have the general form

$$\eta(\boldsymbol{x}) = (\text{Mixture model})(\text{Amount model}) + \text{Mixture terms} + \text{Amount terms} \quad (23)$$

An example model of this form is

$$\begin{aligned}\eta(\boldsymbol{x}) &= \left(\sum_{i=1}^{q} \beta_i x_i\right)(\alpha_0 + \alpha_1 A') + \sum_{i=1}^{q-1} \sum_{j=i+1}^{q} \beta_{ij}^0 x_i x_j + \beta_0^2 A'^2 \\ &= \sum_{i=1}^{q} \beta_i^0 x_i + \sum_{i=1}^{q} \beta_i^1 x_i A' + \sum_{i=1}^{q-1} \sum_{j=i+1}^{q} \beta_{ij}^0 x_i x_j + \beta_0^2 A'^2 \quad (24)\end{aligned}$$

where (a) the linear blending effects of the components are affected by the amount of the mixture, (b) the components have quadratic blending effects that are not affected by the amount of the mixture, and (c) the amount has the same quadratic effect on the response for all mixtures.

Mixture-Amount Designs

Complete MA designs can be formed by running a CP mixture experiment design at each of two or three (or more, if desired) levels of the amount variable. As an example, Figure 2 illustrates a constrained mixture design in three components performed at each of two levels of the amount variable. This design supports fitting a MA model of the form

$$\begin{aligned}\eta(\boldsymbol{x}) = {} & \beta_1^0 x_1 + \beta_2^0 x_2 + \beta_3^0 x_3 + \beta_{12}^0 x_1 x_2 + \beta_{13}^0 x_1 x_3 \\ & + \beta_{23}^0 x_2 x_3 + \beta_{123}^0 x_1 x_2 x_3 + \beta_1^1 x_1 A' + \beta_2^1 x_2 A' \\ & + \beta_3^1 x_3 A' + \beta_{12}^1 x_1 x_2 A' + \beta_{13}^1 x_1 x_3 A' \\ & + \beta_{23}^1 x_2 x_3 A' + \beta_{123}^1 x_1 x_2 x_3 A' \qquad (25)\end{aligned}$$

With A coded to -1 and $+1$, the terms with 0 superscripts describe the component blending at the average of the two amounts (i.e., at zero-coded amount), while the terms with a 1 superscript describe the effect of the amount variable on the component blending properties relative to the average component blending properties.

Fractional MA designs can be formed with a fraction of a CP mixture design at one or more levels of the total amount variable. For example, the MA design with three mixture components and three levels of total amount shown in Figure 3 has a complete simplex-centroid design at $A' = -1$ and $+1$, but a vertex-only design at $A' = 0$. This design supports fitting the model

$$\begin{aligned}\eta(\boldsymbol{x}) = {} & \beta_1^0 x_1 + \beta_2^0 x_2 + \beta_3^0 x_3 + \beta_{12}^0 x_1 x_2 + \beta_{13}^0 x_1 x_3 \\ & + \beta_{23}^0 x_2 x_3 + \beta_{123}^0 x_1 x_2 x_3 + \beta_1^1 x_1 A' + \beta_2^1 x_2 A' \\ & + \beta_3^1 x_3 A' + \beta_{12}^1 x_1 x_2 A' + \beta_{13}^1 x_1 x_3 A' \\ & + \beta_{23}^1 x_2 x_3 A' + \beta_{123}^1 x_1 x_2 x_3 A' + \beta_1^2 x_1 A'^2 \\ & + \beta_2^2 x_2 A'^2 + \beta_3^2 x_3 A'^2 \qquad (26)\end{aligned}$$

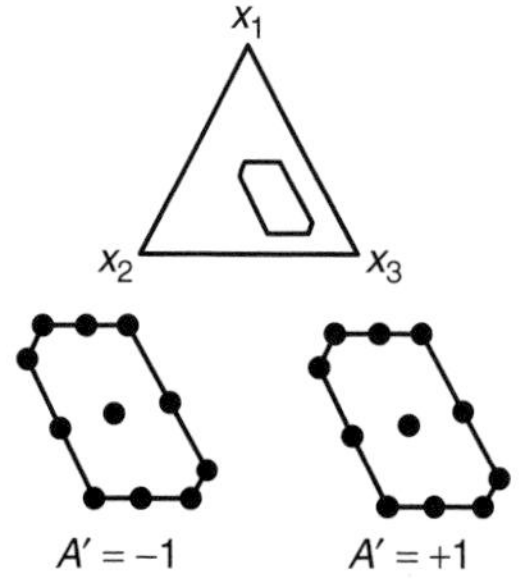

Figure 2 Example of a complete mixture-amount design for three constrained mixture components and two levels of total amount

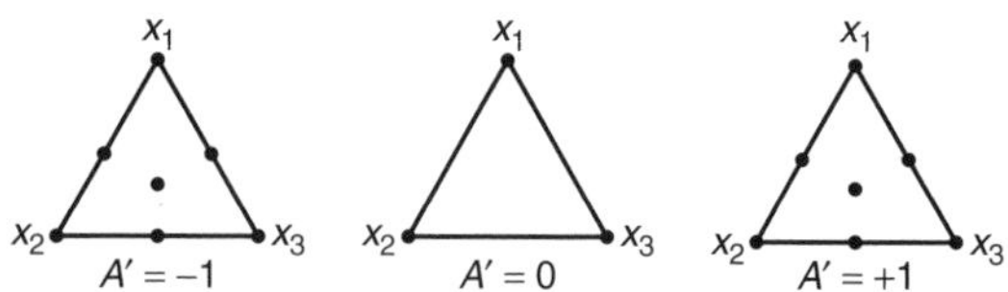

Figure 3 Example of a fractional mixture-amount design with three mixture components and three levels of total amount

which assumes that the total amount has (a) linear effects on the linear and nonlinear blending properties of the mixture components and (b) a quadratic effect only on the linear blending properties of the mixture components.

Piepel *et al.* [18] discusses the special case of an MA experiment and models when the total amount takes a zero value in addition to nonzero values. These experiments are sometimes referred to as *mixture-amount-zero* (MAZ) experiments.

Component Amounts Approach

With the *CA* approach for mixture experiments, q mixture components and no other factors are studied. The CA approach can be extended to study nonmixture factors (e.g., total amount or PVs) in addition to the mixture components, but such extensions are not discussed here. A CA experiment is designed and responses are modeled using amounts (e.g., masses or volumes) of the mixture components, denoted $a_i, i = 1, 2, \ldots, q$. The amounts of mixture components are not subject to the constraints in equation (1) for mixture CPs. Hence, the experimental region of interest can be scaled to be cuboidal or spherical, enabling standard statistical designs such as *factorial, fractional factorial, Plackett–Burman,* **central composite**, *Box–Behnken*, and others to be used. Similarly, standard response surface models, such as

Linear:

$$\eta(\boldsymbol{x}) = \alpha_0 + \sum_{i=1}^{q} \alpha_i a_i \tag{27}$$

Linear plus interaction:

$$\eta(\boldsymbol{x}) = \alpha_0 + \sum_{i=1}^{q} \alpha_i a_i + \sum_{i=1}^{q-1} \sum_{j=i+1}^{q} \alpha_{ij} a_i a_j \tag{28}$$

Quadratic:

$$\eta(\boldsymbol{x}) = \alpha_0 + \sum_{i=1}^{q} \alpha_i a_i + \sum_{i=1}^{q-1} \sum_{j=i+1}^{q} \alpha_{ij} a_i a_j + \sum_{i=1}^{q} \alpha_{ii} a_i^2 \tag{29}$$

and other standard data analysis and graphics methods (*see* **Response Surface Methodology**) can be used with the CA approach.

When using the CA approach, information about effects of CPs (x_i) and effects of the total amount of the mixture ($\sum_{i=1}^{q} a_i = A$) are confounded because $a_i = x_i A$, $i = 1, 2, \ldots, q$. If it is natural or desirable to separately understand the effects of the component (x_i) and total amount (A) variables on the response(s), then the MA approach (discussed previously) should be used. On the other hand, if it is natural or desirable, or to understand the effects of the CAs on the response(s), the CA approach is appropriate. Finally, if the total amount of the mixture turns out to have little or no effect on the response, the CA approach could complicate discovering it, so that the MA approach is preferred for such cases. See Piepel [19] for more discussion of the CA approach and how it compares to the MA approach.

Mixture-Process Variable Approach

With the MPV approach, the effects on a response variable of both mixture components and process variables (a general term for "other factors") are studied. A MPV experiment is usually designed and responses modeled in terms of the proportions of the mixture components and the levels of the PVs. The CPs are subject to the basic mixture constraints in equation (1), and possibly to single-component constraints in equation (2) and/or multicomponent constraints in equation (3). One or more PVs are typically restricted to fall within specified ranges. Often, two or three values of each PV are specified for study. Typically two levels of a PV would be coded as $z = -1$ and $+1$, while three levels would be coded as $z = -1$, 0, and $+1$.

Mixture-Process Variable Models

If the levels of PVs affect a response but not the blending effects of the mixture components, then MPV models have the general form

$$\eta(\boldsymbol{x}) = \{\text{Mixture model}\} + \{\text{PVs model}\} \tag{30}$$

The mixture model can take any permissible form in the CPs (e.g., the forms in equations (4–7)), while the PV model is typically a linear, linear plus interactions, or quadratic polynomial in the PVs. An example model of this type with two PVs (z_1 and z_2) is given by

$$\eta(\boldsymbol{x}) = \sum_{i=1}^{q} \beta_i^0 x_i + \sum_{i=1}^{q-1} \sum_{j=i+1}^{q} \beta_{ij}^0 x_i x_j + \beta_0^1 z_1 + \beta_0^2 z_2 + \beta_0^{12} z_1 z_2 \tag{31}$$

In this model, the mixture components blend quadratically and the PVs have linear and interactive effects. However, the component blending is not affected by the PVs (i.e., the components and PVs do not interact).

If the levels of PVs affect the component blending properties (which happens frequently in practice), then MPV models have the general form

$$\eta(\boldsymbol{x}) = (\text{Mixture model})(\text{PVs model}) + \text{Mixture terms} + \text{PVs terms} \tag{32}$$

An example model of this form is

$$\begin{aligned}\eta(\boldsymbol{x}) &= \left(\sum_{i=1}^{q} \beta_i x_i\right)(\alpha_0 + \alpha_1 z_1 + \alpha_2 z_2) + \sum_{i=1}^{q-1} \sum_{j=i+1}^{q} \beta_{ij}^0 x_i x_j + \beta_0^1 z_1^2 \\ &= \sum_{i=1}^{q} \beta_i^0 x_i + \sum_{i=1}^{q-1} \sum_{j=i+1}^{q} \beta_{ij}^0 x_i x_j + \sum_{i=1}^{q} \beta_i^1 x_i z_1 + \sum_{i=1}^{q} \beta_i^2 x_i z_2 + \beta_0^{11} z_1^2\end{aligned} \tag{33}$$

where (a) the linear blending properties of the components are affected by linear effects of the PVs, (b) the components have quadratic blending properties that are not affected by the PVs, and (c) one PV (z_1) has a quadratic effect on the response that is the same for all mixtures.

Mixture-Process Variable Designs

A *complete MPV design* is formed by running the same CP mixture experiment design at each point in a PV design (or vice versa). A *fractional MPV design* results from running different mixture designs at each PV design point (or vice versa). As an example, Figure 4 illustrates a simplex-centroid mixture design in three components performed at each of the four combinations in a 2^2 design for two PVs. This 28-point complete MPV design supports exactly fitting a 28-term model of the form

$$\begin{aligned}\eta(\boldsymbol{x}) = {} & \beta_1^0 x_1 + \beta_2^0 x_2 + \beta_3^0 x_3 + \beta_{12}^0 x_1 x_2 + \beta_{13}^0 x_1 x_3 \\ & + \beta_{23}^0 x_2 x_3 + \beta_{123}^0 x_1 x_2 x_3 + (\beta_1^1 x_1 + \beta_2^1 x_2 \\ & + \beta_3^1 x_3 + \beta_{12}^1 x_1 x_2 + \beta_{13}^1 x_1 x_3 + \beta_{23}^1 x_2 x_3 \\ & + \beta_{123}^1 x_1 x_2 x_3) z_1 + (\beta^2 x_1 + \beta_2^2 x_2 + \beta_3^2 x_3 \\ & + \beta_{12}^2 x_1 x_2 + \beta_{13}^2 x_1 x_3 + \beta_{23}^2 x_2 x_3 + \beta_{123}^2 x_1 x_2 x_3) z_2 \\ & + (\beta_1^{12} x_1 + \beta_2^{12} x_2 + \beta_3^{12} x_3 + \beta_{12}^{12} x_1 x_2 \\ & + \beta_{13}^{12} x_1 x_3 + \beta_{23}^{12} x_2 x_3 + \beta_{123}^{12} x_1 x_2 x_3) z_1 z_2 \quad (34)\end{aligned}$$

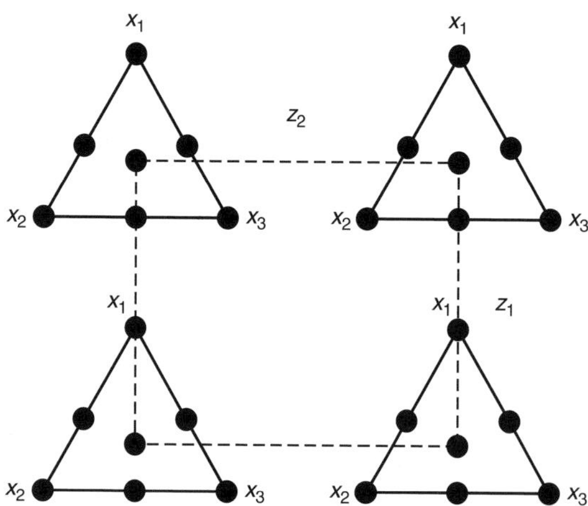

Figure 4 Example of a complete mixture-process variable design with a simplex-centroid mixture design in three components at each of the four combinations of a 2^2 design for two process variables

With z_1 and z_2 coded to -1 and $+1$ the terms with 0 superscripts describe the component blending at the average levels of the PVs (i.e., zero-coded values). Terms with a 1 or a 2 superscript describe the effects of the first or second PV on the component blending properties relative to the average component blending properties. The terms with a 12 superscript describe the interactive effects of the two PVs on the component blending properties relative to the average component blending properties.

Sometimes MPV designs are run as **split-plot** experiments when the mixture blends or the PVs are hard to change. When the mixture components are hard to change, all points in a PV design are run together for each point in a mixture design. When the PVs are hard to change, all points in a mixture design are run together for each point in a PV design. MPV designs run as split-plot experiments require special data analysis methods to account for the restriction on **randomization** of the design points. See Cornell [2, Sections 7.3 and 7.4] and Kowalski *et al.* [20] for additional discussion of the design and analysis of MPV split-plot experiments.

Mixture-of-Mixtures Approach

With the MoM approach, the end product is a mixture of main components (sometimes called *major components*), which are themselves mixtures of subcomponents (sometimes called *minor components*). For example, an ink-jet printer image is a mixture of three to six colors of inks, with each ink a mixture of chemicals. In some cases the main components represent categories of components, and the subcomponents are kinds of ingredients in each category. Such cases are referred to as *categorized component* (CC) mixture experiments.

There are two types of MoM/CC experiments. In one type (type A), the proportions of the main components have fixed values and only the proportions of the subcomponents are varied. In this type of MoM/CC experiment, only the blending among subcomponents is studied. In the other type (type B) of MoM/CC experiments, the proportions of the main components and the subcomponents are varied. In this type of experiment, blending among main components and among subcomponents is studied. Models and designs for both types of MoM/CC experiments are discussed by Cornell [2, Section 4.13]. In an

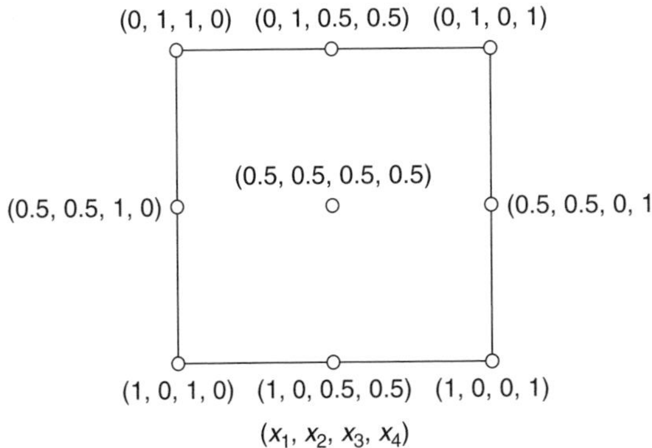

Figure 5 The {2, 2; 2, 2} double-lattice design for a mixture-of-mixtures experiment

example discussed by Cornell [2, Section 4.13.2] the main components are two types of resins, with two subcomponents (x_1 and x_2 for the first type, x_3 and x_4 for the second type). The {2, 2; 2, 2} *double-lattice design* used in this example is shown in Figure 5. The corresponding *double-Scheffé (quadratic–quadratic) model* is

$$\begin{aligned}\eta(\boldsymbol{x}) &= \gamma_{13}x_1x_3 + \gamma_{14}x_1x_4 + \gamma_{23}x_2x_3 + \gamma_{24}x_2x_4 \\ &\quad + \gamma_{123}x_1x_2x_3 + \gamma_{124}x_1x_2x_4 + \gamma_{134}x_1x_3x_4 \\ &\quad + \gamma_{234}x_2x_3x_4 + \gamma_{1234}x_1x_2x_3x_4 \qquad (35)\end{aligned}$$

See Cornell [2, Section 4.13.2] for further discussion of this example. Piepel [21] discusses methods for developing and reducing models for type A MoM experiments with a pharmaceutical tablet formulation example.

Multifactor Mixture Approach

In a *MFM* experiment, mixture experiments are performed for each of two or more independent mixture factors in an experiment. If there are n mixture factors each having $q_k (k = 1, 2, \ldots, n)$ components, then the experimental region is $(\sum_{k=1}^{n} q_k - n)$ dimensional.

Nigam [22] discusses an MFM example in which the goal is to maximize recombination frequency of a mixed bacterial culture (male and female) when irradiated with short exposures of mixtures of mutagens (x-rays and γ-rays). In this example there are two mixture factors, bacterial strains and mutagens, with each having two components. Note that the two mixture factors are independent and are not themselves major components of a mixture, which differentiates MFM experiments from MoM/CC experiments. Nigam [22] discusses designs and models for MFM experiments.

Optimal Experimental Design

Optimal experimental design methods were discussed previously for the CP approach, but can be used with most of the approaches when needed. Exceptions are the MoM and MFM approaches, where the methods are applicable but not currently implemented in available software. Two common situations when optimal experimental design is needed are (a) the experimental region is irregular in shape because of linear constraints on the variables, and (b) the standard designs for a given approach contain too few or too many design points compared to the desired number.

For optimal experimental design methods that require a set of candidate points, the algorithms of Piepel [16] provide for generating the vertices and various dimensional centroids of any polyhedral region defined by linear inequality constraints on mixture variables, nonmixture variables, or both. Those algorithms only allow for at most one equality (mixture) constraint and hence are applicable to all but the MoM and MFM approaches. Even for those approaches candidate points can be generated separately and combined. However, the optimal design algorithms and software for candidate and candidate-free methods would have to be modified to account for more than one equality (mixture) constraint.

Optimal design methods are very useful for generating nonstandard designs, but there are three cautions to keep in mind. First, optimal designs (especially if constructed using vertices and dimensional centroids as candidate points) often have all (or nearly all) design points on or near the boundary of the constrained experimental region. It may be desirable to have a substantial fraction of the design points in the interior of the experimental region. *Space-filling designs* (sometimes referred to as ***uniform designs***) are one alternative to obtain a more even coverage of the interior as well as the exterior of a constrained experimental region. Piepel [4, Table 2] lists several publications that discuss uniform designs for

CP mixture experiments. Other alternatives are the *central composite analog designs* (CCADs) and *layered designs* (LDs) proposed by Piepel *et al.* [23] to assure having design points on different layers of the constrained experimental region. **Ordinary central composite designs** can be used with the MIV, SV, CA approaches, and the PV portion of an MPV experiment. CCADs can be applied with the CP approach or the mixture portion of the MA or MPV approaches. LDs can be applied with any of the mixture approaches.

The second caution is that optimal designs are constructed for a specific model form, and may not be optimal or even good for other model forms. This tends to be the case when a design is optimized for a simpler (smaller) model but a larger model is more appropriate. However, designs optimized for a larger model often tend to be good for reduced forms of that model. Hence, construction of optimal designs for the largest model form that may be needed is generally recommended.

The third caution is that the number of design points must be specified in the optimal design approach. Practitioners should investigate optimal designs with fewer and more points than the desired number and assess the sensitivity of design properties to the number of design points.

Assessing and Comparing Designs

Piepel [4, Section 5] summarizes several analytical and graphical methods for assessing and comparing mixture designs. These include design efficiency measures, collinearity and leverage diagnostics, variance and **bias** dispersion graphs, fraction of design space plots, and other kinds of prediction variance plots. Many of these methods are applicable or can be adapted regardless of the mixture experiment approach used. For discussion of these methods, *see* **Assessment of Experimental Designs** and the associated references in [4].

Data Analyses for Mixture Experiments

The data analysis process for a mixture experiment usually begins by fitting and assessing models (corresponding to the selected approach) until a model is obtained that does not substantially underfit or overfit the data. Traditional regression methods for assessing model **lack of fit** are applicable to all of the mixture experiment approaches. However, appropriate formulas must be used to calculate entries of the analysis of variance table and related statistics (e.g., R^2) associated with models using the CP, MA, MPV, MoM, and MFM approaches (see Cornell [2, Section 5.2]).

After an adequately fitting model is obtained, it can be used to achieve the desired goals of the experiment. If the goal is to identify important (and unimportant) mixture components in the CP approach, numeric estimates of component effects can be calculated or displayed graphically using response trace plots (see Cornell [2, Section 5.9] and Smith [3, Section 11]). These methods can be adapted for the MA and MPV approaches. If the goal is to select settings of the mixture and/or nonmixture variables to optimize one response variable, standard optimization methods can be used for any of the approaches. If there are two or more response variables, multiresponse optimization methods can be applied for any of the approaches. Cornell [2, Sections 8.13 and 8.14] and Smith [3, Section 12.2] discuss these methods for the CP approach.

Piepel [4, Section 7] lists and provides references for many other data analysis methods applicable for the CP, MA, and MPV approaches. Several of these are also discussed throughout the book by Cornell [2]. Standard data analysis methods from *response surface methodology* are applicable to the MIV, CA, and SV approaches, but can result in misleading conclusions for the SV approach as noted previously.

Choosing a Mixture Experiment Approach

Choosing a mixture experiment approach depends primarily on the types of variables whose effects on the response variables are to be studied in the experiment. If only mixture variables (components) affect the response(s) and are to be studied, then the CP, MIV, and SV approaches are possible. However, the CP approach is recommended in general because the experiment is designed, the data are analyzed, and conclusions are made directly in terms of the proportions of the mixture variables. The MIV approach can be appropriate if the MIVs are the natural or long-standing variables of interest for a particular problem. The SV approach is not recommended because of the potential for making wrong interpretations and conclusions.

If the total amount of the mixture as well as the CPs affect the response(s), then the MA and CA approaches are possible. The MA approach is recommended when the effects of the components and the effect of the total amount of the mixture are to be studied, understood, and optimized separately. The CA approach is appropriate when the effects of amounts of the individual components are to be studied, rather than effects of the component proportions and total amount.

The MPV approach should be used when the effects on the response(s) of mixture components and PVs are to be studied simultaneously. While it is possible to use the MIV and SV approaches in a MPV situation, doing so is not recommended because these approaches do not directly study all of the mixture components and PVs.

Finally, the MoM and MFM approaches are used in special cases that involve more than one mixture. The MoM approach is appropriate when the components of a primary mixture are in turn mixtures of subcomponents, hence the terminology of *mixtures-of-mixtures*. The MFM approach is appropriate when there are two or more independent factors to be studied in an experiment, and the factors are mixtures of different sets of components.

References

[1] Piepel, G.F. & Cornell, J.A. (1994). Mixture experiment approaches: examples, discussion, and recommendations, *Journal of Quality Technology* **26**, 177–196.

[2] Cornell, J.A. (2002). *Experiments with Mixtures: Designs, Models, and the Analysis of Mixture Data*, John Wiley & Sons, New York.

[3] Smith, W.F. (2005). *Experimental Design for Formulation, ASA-SIAM Series on Statistics and Applied Probability*, ASA, SIAM, Alexandria, Philadelphia.

[4] Piepel, G.F. (2004). 50 years of mixture experiment research: 1955–2004, in *Response Surface Methodology and Related Topics*, A.I. Khuri, ed, World Scientific Press, Singapore, Chapter 12, pp. 283–327.

[5] Scheffé, H. (1958). Experiments with mixtures, *Journal of the Royal Statistical Society. Series B* **20**, 344–360.

[6] Cox, D.R. (1971). A note on polynomial response functions for mixtures, *Biometrika* **58**, 155–159.

[7] Piepel, G.F., Hicks, R.D., Szychowski, J.M. & Loeppky, J.L. (2002). Methods for assessing curvature and interaction in mixture experiments, *Technometrics* **44**, 161–172.

[8] Smith, W.F. & Beverly, T.A. (1997). Generating linear and quadratic Cox mixture models, *Journal of Quality Technology* **29**, 211–224.

[9] Piepel, G.F. (2006). A note comparing component-slope, Scheffé, and Cox parameterizations of the linear mixture experiment model, *Journal of Applied Statistics* **33**, 397–403.

[10] Piepel, G.F. (2007). A component slope linear model for mixture experiments, *Quality Technology and Quantitative Management* **4**(3).

[11] Piepel, G.F., Szychowski, J.M. & Loeppky, J.L. (2002). Augmenting Scheffé linear mixture models with squared and/or crossproduct terms, *Journal of Quality Technology* **34**, 297–314.

[12] Becker, N.G. (1968). Models for the response of a mixture, *Journal of the Royal Statistical Society. Series B* **30**, 349–358.

[13] Snee, R.D. (1971). Design and analysis of mixture experiments, *Journal of Quality Technology* **3**, 159–169.

[14] Aitchison, J. & Bacon-Shone, J. (1984). Log contrast models for experiments with mixtures, *Biometrika* **71**, 323–330.

[15] Snee, R.D. & Marquardt, D.W. (1976). Screening concepts and designs for experiments with mixtures, *Technometrics* **18**, 19–29.

[16] Snee, R.D. (1979). Experimental designs for mixture systems with multicomponent constraints, *Communications in Statistics: Theory and Methods* **A8**, 303–326.

[17] Piepel, G.F. (1988). Programs for generating extreme vertices and centroids of linearly constrained experimental regions, *Journal of Quality Technology* **20**, 125–139.

[18] Piepel, G.F. (1988). A note on models for mixture-amount experiments when the total amount takes a zero value, *Technometrics* **30**, 449–450.

[19] Piepel, G.F. (1985). Models and designs for generalizations of mixture experiments where the response depends on the total amount, Ph.D. Dissertation, University Microfilms International, Ann Arbor. (http://www.umi.com/products_umi/dissertations).

[20] Kowalski, S.M., Cornell, J.A. & Vining, G.G. (2002). Split-plot designs and estimation methods for mixture experiments with process variables, *Technometrics* **44**, 72–79.

[21] Piepel, G.F. (1999). Modeling methods for mixture-of-mixtures experiments applied to a tablet formulation problem, *Pharmaceutical Development and Technology* **4**, 593–606.

[22] Nigam A.K. (1973). Multifactor mixture experiments. *Journal of the Royal Statistical Society* **35**, 51–56.

[23] Piepel, G.F., Anderson, C.M. & Redgate, P.E. (1993) Response surface designs for irregularly-shaped regions (Parts 1, 2, and 3),. *Proceedings of the Section on Physical and Engineering Sciences*, American Statistical Association, Alexandria, pp. 205–227.

Related Articles

Assessment of Experimental Designs; Box–Behnken Designs; Central Composite Designs; Factorial

Experiments; **Fractional Factorial Designs**; **Optimal Design**; **Plackett–Burman Designs**; **Response Surface Methodology**; **Split-Plot Designs**; **Uniform Experimental Designs**.

GREG F. PIEPEL

Mixture Models, Multivariate: Aging and Dependence

Introduction

In this article, we maintain the same language and notation of the article **Aging and Positive Dependence** and consider vectors of lifetimes $(T_1, \ldots, T_n)$ with joint survival function of the specific form

$$\overline{F}(t_1, \ldots, t_n) = \int_{R^m} \prod_{j=1}^{n} \overline{G}_j(t_j|\theta_1, \ldots, \theta_m) \times \mathrm{d}H(\theta_1, \ldots, \theta_m) \quad (1)$$

where, for some $m \geq 1$, H is an m-dimensional probability distribution and, for any vector $(\theta_1, \ldots, \theta_m)$, $\overline{G}_j(\cdot|\, \theta_1, \ldots, \theta_m)$ is a one-dimensional survival function $(j = 1, 2, \ldots, n)$. We can think of H as the joint distribution of a nonobservable random vector $\mathbf{\Theta} = (\Theta_1, \ldots, \Theta_m)$ such that $T_1, \ldots, T_n$ are conditionally independent, given $\mathbf{\Theta}$. Generally, neither $\Theta_1, \ldots, \Theta_m$ nor $T_1, \ldots, T_n$ are independent random variables.

Models of the form in equation (1) are known as *multivariate mixture models*; they arise in several different fields and in a great variety of situations of applied interest. In particular, they can be used to model heterogeneity; for a discussion about the concept of heterogeneity in the field of reliability (see e.g., [1] the references indicated therein).

It is quite natural to analyze multivariate mixture models in the frame of a Bayesian approach. In such a frame, we will discuss here some aspects of aging and positive dependence.

$T_1, \ldots, T_n$ are conditionally independent, generally nonidentically distributed, given $\mathbf{\Theta}$. Special multivariate mixture models of interest are defined by the conditions:

1. $\overline{G}_j(t|\theta_1, \ldots, \theta_m)$, as a function of $(\theta_1, \ldots, \theta_m)$, depends only on θ_j and
2. $\overline{G}_j(t|\theta_1, \ldots, \theta_m)$, as a function of t and θ_j, has a general form that does not depend on j.

Such conditions can be summarized by writing that $\overline{G}_j(t|\theta_1, \ldots, \theta_m)$ is of the form

$$\overline{G}_j(t|\theta_1, \ldots, \theta_m) = \rho(t|\theta_j) \quad (2)$$

and describe special *frailty models*.

In several cases of interest, as an additional condition on these frailty models, we have that $\Theta_1, \ldots, \Theta_m$ are exchangeable. As a remarkable feature, in these cases, we have heterogeneity and similarity at the same time: also $T_1, \ldots, T_n$ turn out to be exchangeable. However, $T_1, \ldots, T_n$ generally do not have an identical conditional distribution, given $\mathbf{\Theta}$. We point out then that we obtain in this way exchangeable variables, for which conditional independence is not, however, paired with identical conditional distributions.

Specially relevant cases of exchangeability are met when $\Theta_1, \ldots, \Theta_m$ are independent and identically distributed, or when $P\{\Theta_1 = \ldots = \Theta_m\} = 1$. The latter case is the basic one encountered in Bayesian statistics: $T_1, \ldots, T_n$ are conditionally independent and identically distributed, given a same scalar parameter Θ; in the former case, instead, $T_1, \ldots, T_n$ are themselves independent and identically distributed. Several other models of interest, however, can also be met in between these two extreme cases.

For these models we also notice that, irrespective of the special form of dependence among $\Theta_1, \ldots, \Theta_m$, the one-dimensional marginal distribution of $T_j (j = 1, 2, \ldots, n)$ is a mixture distribution with a survival function of the form

$$\overline{F}(t) = \int_{R} \rho(t|\theta)\, \mathrm{d}H^{(1)}(\theta) \quad (3)$$

$H^{(1)}$ being the one-dimensional marginal of H.

In a general multivariate mixture model, as in equation (1) (not necessarily exchangeable), the marginal distributions of $T_j (j = 1, 2, \ldots, n)$, are univariate mixture distributions with survival functions given by

$$\overline{F}_j(t) = \int_{R^m} \overline{G}_j(t|\theta_1, \ldots, \theta_m)\, dH(\theta_1, \ldots, \theta_m) \quad (4)$$

In applications, there are different kinds of situations that give rise to univariate mixture distributions. The difference between different situations is not, sometimes, very clear. From a probabilistic point of view, however, one can simply say that these situations are described by distributions obtained as marginals of different multivariate mixture models, just with different types of mixing distributions $H(\theta_1, \ldots, \theta_m)$.

Aspects of Ageing

A fact of recurrent interest in quality and reliability is related with aging properties of (univariate) mixtures as in equation (4) and can be roughly summarized as follows: *whereas negative aging tends to be maintained by mixtures, positive aging tends to be lost under mixtures*. This fact has been made precise by formulating different probabilistic results; at the same time, it has been explained in terms of several heuristic interpretations or empirical analysis. It is, in particular, well-known that the property of decreasing failure rate (DFR) is maintained under mixtures (see e.g., [2]); on the other hand, several interpretations have been given of the fact that increasing failure rate (IFR) may, on the contrary, be lost under mixtures. In a reliability framework, well-known papers on this topic are [3, 4], the latter being in particular based on a Bayesian approach.

Here we want to illustrate the basic idea behind all that, still maintaining a Bayesian approach. To this purpose, we assume simplifying conditions of stochastic monotonicity that more easily allow us to explain the reasons why DFR is maintained by mixture, while IFR can be lost (see also [5–8]). Same type of arguments can also be, respectively, worked out for other notions of positive and negative aging.

Consider a lifetime T and a single positive random variable Θ; T is the (observable) lifetime of a component and Θ is a nonobservable factor that affects the probability distribution of T; Θ can have e.g., a meaning such as the quality or the frailty of the component. For the sake of technical simplicity, we consider cases where the joint probability distribution of (T, Θ) admits a joint probability, described as follows: the marginal distribution of Θ, that is concentrated over $[0, +\infty)$, admits a density $\pi(\cdot)$ and the conditional distributions of T, given $\{\Theta = \theta\}$ admits conditional failure rates $r(t|\theta)$, for $\theta \geq 0$; denote by $R(t|\theta)$ the conditional cumulative failure rate of T given $\{\Theta = \theta\}$:

$$R(t|\theta) = \int_0^t r(u|\theta)\, du \quad (5)$$

The marginal distribution of T, obtained by integrating out θ, has then survival function, density function, and failure rate, respectively given by

$$\overline{F}(t) = \int_0^\infty \exp\{-R(t|\theta)\}\pi(\theta)\, d\theta,$$

$$f(t) = \int_0^\infty r(t|\theta) \exp\{-R(t|\theta)\}\pi(\theta)\, d\theta \quad (6)$$

and

$$\lambda(t) = \frac{f(t)}{\overline{F}(t)} = \int_0^\infty r(t|\theta)\pi_t(\theta)\, d\theta \quad (7)$$

where we set

$$\pi_t(\theta) = \frac{\exp\{-R(t|\theta)\}\pi(\theta)}{\int_0^\infty \exp\{-R(t|\theta)\}\pi(\theta)\, d\theta} \quad (8)$$

As an immediate application of Bayes formula, $\pi_t(\theta)$ can be interpreted as the conditional density of Θ, given the event $\{T > t\}$. In many practical situations, it can be reasonable to assume that $r(t|\theta)$ is, for any fixed t, an increasing function of θ: the bigger the value taken by θ, the stochastically smaller is T, in the hazard order sense; it can be easily shown, in such a case, that the distribution with density π_t is stochastically decreasing in t.

Compare now, for fixed t and for two different densities π', π'', the expressions

$$\int_0^\infty r(t|\theta)\pi'(\theta)\, d\theta, \int_0^\infty r(t|\theta)\pi''(\theta)\, d\theta \quad (9)$$

Since, for fixed t, $r(t|\theta)$ is increasing in θ, we have

$$\int_0^\infty r(t|\theta)\pi'(\theta)\, d\theta \leq \int_0^\infty r(t|\theta)\pi''(\theta) d\theta \quad (10)$$

whenever the distribution with density π' is smaller than the distribution with density π'', in the sense of the ordinary stochastic order.

Put now, for two values $0 < t' < t''$, $\pi'(\theta) = \pi_{t'}(\theta), \pi''(\theta) = \pi_{t''}(\theta)$. For a same fixed value $\tau > 0$, we can then write, since π_t is stochastically decreasing in t and $r(t|\theta)$ is increasing in θ,

$$\int_0^\infty r(\tau|\theta)\pi_{t'}(\theta)\,\mathrm{d}\theta \geq \int_0^\infty r(\tau|\theta)\pi_{t''}(\theta)\,\mathrm{d}\theta \quad (11)$$

Suppose T is conditionally DFR given θ, i.e. $r(t|\theta)$ is decreasing in t. By fixing e.g., $\tau = t'$, we can immediately conclude that T is DFR; in fact

$$\lambda(t') = \int_0^\infty r(t'|\theta)\pi_{t'}(\theta)\,\mathrm{d}\theta \geq \int_0^\infty r(t''|\theta)\pi_{t''}(\theta)\,\mathrm{d}\theta = \lambda(t'') \quad (12)$$

Suppose now that T is IFR in the conditional distribution. Even if, for $t' < t''$, it is then $r(t'|\theta) \leq r(t''|\theta)$, we cannot generally conclude that $\lambda(t') \leq \lambda(t'')$, since in any case $\pi_{t'}$ is stochastically larger than $\pi_{t''}$.

This explains why the distribution of T, which is a mixture of IFR distributions and has the density function given by

$$f(t) = \int_0^\infty r(t|\theta)\exp\{-R(t|\theta)\}\pi(\theta)\,\mathrm{d}\theta \quad (13)$$

may potentially be not IFR!

In passing we have to notice however that, under special conditions, a mixture of IFR distribution can still be IFR (see [9] and [10]).

This specific phenomenon that a mixture of IFR distribution is not generally IFR (and that the mixture can, however, be IFR, under specific conditions), has also been discovered and extensively analyzed in several other fields, related with reliability, such as demography, population dynamics, survival analysis, and medicine, (see e.g., [11–13]); it has also been remarked that the phenomenon is somewhat related to the *Simpson's paradox* (e.g., [13, 14]).

More generally, the aging behavior of univariate distributions obtained by mixtures has often been found to be relevant in different applications and related statistical analysis; the main problem is that many different types of behavior are possible when failure rates, of the distributions which are mixed, are not decreasing. Therefore, studies on the aging properties of mixtures, in an analytical viewpoint, have quite a long history (see e.g., reviews in [15–18]). Specific aspects about this theme concern the asymptotic (or tail) behavior (see e.g., [15, 19]) and conditions that give rise to *bathtub* distributions (see [20] and references therein).

An important problem in reliability is to understand what are the situations under which **burn-in** procedures are justified. From a Bayesian viewpoint, deciding whether it is convenient to adopt some *burn-in* procedure is just a decision problem (see e.g., the introductory discussion in [21]) and establishing the duration of burn-in is just an optimal stopping problem (see [22, 23] and references therein).

Roughly speaking, burn-in is useful in cases of *infant mortality*. For this reason a special interest has been devoted to the theme of negative aging; the issues of concern are understanding what negative aging really means, which analytical conditions imply it, and what real reliability models can manifest such a behavior. In this respect, there are several ways to interpret the fact that, in some practical situations where mixtures appear, burn-in procedures are actually convenient (see also e.g., [6, 23, 24] and references therein). From a mathematical point of view, one can say that infant mortality and univariate negative aging often arise in the case of mixtures.

One can imagine, however, that multivariate mixture models also manifest some kind of negative multivariate aging. Multivariate properties of the Bayesian type (see in particular [18, 23, 25, 26] and the brief sketch in **Aging and Positive Dependence**) have been dealt with in [27] for these models in the exchangeable case; sufficient conditions that guarantee some multivariate negative aging have been proved therein, under suitable assumptions of positive dependence for the vector $\boldsymbol{\Theta}$ and stochastic monotonicity of the conditional distributions of $T_1, \ldots, T_n$, given $\boldsymbol{\Theta}$.

As one conclusion of main interest for this article, we point out that in cases of (univariate) mixture distributions and of multivariate mixture models one can encounter some form of negative aging.

Positive Dependence Properties

Let us now pass to analyze positive dependence for multivariate mixture models. In this analysis, we have to distinguish between the two cases where the unobservable factor $\boldsymbol{\Theta}$, w.r.t. which $T_1, \ldots, T_n$ are conditionally independent, is a random scalar or a random vector. It is a general fact, however, that stochastic monotonicity of the different lifetimes $T_1, \ldots, T_n$ with respect to $\boldsymbol{\Theta}$, combined with suitable positive dependence properties of $\boldsymbol{\Theta}$, gives rise to positive dependence properties for $\mathbf{T} = (T_1, \ldots, T_n)$. This fact admits a rather direct heuristic interpretation in a Bayesian view (*see also* **Aging and Positive Dependence**, the section titled "Notions of Positive Dependence"). When $\boldsymbol{\Theta}$ reduces to a scalar Θ (i.e., if $m = 1$), no assumption is needed for Θ.

A basic result in this direction was given by Jogdeo in [28] (see also [29]).

Theorem 1 *Let $T_1, \ldots, T_n$ be conditionally independent, given Θ and each of them stochastically increasing (or decreasing) in Θ, in the ordinary stochastic order. Then $T_1, \ldots, T_n$ are associated.*

By postulating stronger types of stochastic monotonicity for $T_1, \ldots, T_n$, one can obtain stronger conditions of positive dependence. Examples of that are the following results, proved in [30].

Theorem 2 *Let $T_1, \ldots, T_n$ be conditionally independent, given Θ and each of them stochastically decreasing (or increasing) in Θ, in the hazard rate order. Then $\mathbf{T} = (T_1, \ldots, T_n)$ is weakened by failure.*

For the following result, we need to assume that the support $\mathcal{D}$ of the distribution of $\mathbf{T}$ is a *lattice*, i.e., $\mathbf{u}, \mathbf{v} \in \mathcal{D} \Longrightarrow \mathbf{u} \wedge \mathbf{v} \in \mathcal{D},\ \mathbf{u} \vee \mathbf{v} \in \mathcal{D}$.

Theorem 3 *Let $T_1, \ldots, T_n$ be conditionally independent, given Θ and each of them stochastically decreasing (or increasing) in Θ, in the likelihood ratio order. Then $\mathbf{T}$ is multivariate totally positive of order 2.*

Let us consider now cases in which $T_1, \ldots, T_n$ are conditionally independent, given a vector $\boldsymbol{\Theta}$. In these cases, the space R^m of possible values of $\boldsymbol{\Theta}$ is to be thought of as (partially) ordered in the componentwise sense. Theorem 1 above can be generalized as follows (see again [28]).

Theorem 4 *Let $T_1, \ldots, T_n$ be conditionally independent, given $\boldsymbol{\Theta}$ and each of them stochastically decreasing (or increasing) in $\boldsymbol{\Theta}$, in the ordinary stochastic order; furthermore, let $\boldsymbol{\Theta}$ be associated. Then $T_1, \ldots, T_n$ are associated.*

Theorem 1 is clearly a special case of Theorem 4, in fact, a scalar random variable Θ can be seen as a special case of an associated random vector. Analogously, Theorems 2 and 3 have also been respectively generalized in suitable ways to the case of a vector $\boldsymbol{\Theta}$; in such results, $\boldsymbol{\Theta}$ is to be assumed multivariate totally positive of order 2 (see [30]).

For some other aspects of dependence properties for mixture models, see also [31].

A general concept of positive dependence for a random vector $(X_1, \ldots, X_r)$ is that of *dependence by total positivity with degree* $(k_1, \ldots, k_r)$ (see in particular [32, 33]). This concept is important since several different notions of positive dependence can just be obtained as special cases of it. In the paper [34], Kirmani and Kochar extended the aforementioned results in order to prove *dependence by total positivity with degree* $(k_1, \ldots, k_r)$ for vectors of lifetimes $T_1, \ldots, T_n$, jointly distributed according to multivariate mixture modes.

Conclusions

We mentioned that multivariate mixture models are of remarkable interest for applications and that, in particular, they can be used to model heterogeneity. Some other examples of applications are mentioned, e.g., in [30].

We want to conclude this article, however, by stressing that multivariate mixture models are also relevant for the following aspect of conceptual character: they can likely manifest positive dependence along with some form of negative aging.

References

[1] Arjas, E. & Bhattacharjee, M. (2004). Modelling heterogeneity in repeated failure time data: a hierarchical Bayesian approach, *Mathematical Reliability: An Expository Perspective*, Kluwer Academic Publishers, Boston, pp. 71–86.

[2] Barlow, R.E. & Proschan, F. (1975). *Statistical Theory of Reliability and Life Testing*, Holt, Rinehart and Winston, New York.

[3] Proschan, F. (1963). Theoretical explanation of observed decreasing failure rate, *Technometrics* **5**, 375–383.

[4] Barlow, R.E. (1985). A Bayesian explanation of an apparent failure rate paradox, *IEEE Transactions on Reliability* **R34**, 107–108.

[5] Spizzichino, F. (1992). Reliability decision problems under conditions of ageing, *Bayesian Statistic 4*,

J. Bernardo, J. Berger, A.P. Dawid & A.F.M. Smith, eds, Clarendon Press, Oxford, pp. 803–811.

[6] Block, H.W. & Savits, T.H. (1997). Burn-in, *Statistical Science* **12**, 1–19.

[7] Lynn, N.J. & Singpurwalla, N.D. (1997). "Burn-in" makes us feel good, Comment to [6].

[8] Finkelstein, M.S. & Esaulova, V. (2001). Modeling a failure rate for a mixture of distribution functions, *Probability in the Engineering and Informational Sciences* **15**, 383–400.

[9] Lynch, J.D. (1999). On conditions for mixtures of increasing failure rate distributions to have an increasing failure rate, *Probability in the Engineering and Informational Sciences* **13**, 33–36.

[10] Block, H.W., Li, Y. & Savits, T.H. (2003). Preservation of properties under mixture, *Probability in the Engineering and Informational Sciences* **17**, 205–212.

[11] Vaupel, J.W. & Yashin, A.I. (1985). Heterogeneity's ruses: some surprising effects of selection on population dynamics, *The American Statistician* **39**, 176–185.

[12] Cohen, J.E. (1986). An uncertainty principle in demography and the unisex issue, *The American Statistician* **40**, 32–39.

[13] Gurland, J. & Sethuraman, J. (1995). How pooling failure data may reverse increasing failure rates, *Journal of the American Statistical Association* **90**, 1416–1423.

[14] Scarsini, M. & Spizzichino, F. (1999). Simpson-type paradoxes, dependence and aging, *Journal of Applied Probability* **36**, 119–131.

[15] Shaked, M. & Spizzichino, F. (2001). *Mixtures and Monotonicity of Failure Rate Functions*, *Handbook of Statistics*, N. Balakrishnan & C.R. Rao, eds, Elsevier Science, Amsterdam, Vol. 20, pp. 185–198.

[16] Block, H.W. & Savits, T.H. (2004). The shape of failure rate mixtures, *Mathematical Reliability: An Expository Perspective*, Kluwer Academic Publishers, Boston, pp. 197–206.

[17] Finkelstein, M.S. (2003). A model of aging and a shape of the observed force of mortality, *Lifetime Data Analysis* **9**, 93–109.

[18] Lai, C.-D. & Xie, M. (2006). *Stochastic Ageing and Dependence for Reliability*, Springer-Verlag, New York.

[19] Block, H.W. & Joe, H. (1997). Tail behavior of the failure rate functions of mixtures, *Lifetimes Data Analysis* **3**, 269–288.

[20] Lai, C.D., Xie, M. & Murthy, D.N.P. (2001). *Bathtub-Shaped Failure Rate Life Distributions*, *Handbook of Statistics*, N. Balakrishnan & C.R. Rao, eds, Elsevier Science, Amsterdam, Vol. 20, pp. 69–104.

[21] Clarotti, C.A. & Spizzichino, F. (1990). Bayesian burn-in decision procedures, *Probability in the Engineering and Informational Sciences* **4**, 437–445.

[22] Jensen, U. & Spizzichino, F. (2004). The burn-in problem - a discussion of sequential stop and go strategies *Mathematical Reliability: An Expository Perspective*, Kluwer Academic Publishers, Boston, pp. 207–229.

[23] Spizzichino, F. (2001). *Subjective Probability Models for Lifetimes*, Chapman and Hall/CRC, Boca Raton.

[24] Block, H.W., Mi, J. & Savits, T.H. (1993). Burn-in and mixed populations, *Journal of Applied Probability* **30**, 692–702.

[25] Barlow, R.E. & Mendel, M.B. (1992). de Finetti-type representations for life distributions, *Journal of the American Statistical Association* **87**, 1116–1122.

[26] Bassan, B. & Spizzichino, F. (1999). Stochastic comparison for residual lifetimes and Bayesian notions of multivariate ageing, *Advances in Applied Probability* **31**, 1078–1094.

[27] Spizzichino, F. & Torrisi, G. (2001). Multivariate negative aging in an exchangeable model, *Statistics and Probability Letters* **55**, 71–82.

[28] Jogdeo, K. (1978). On a probability bound of Marshall and Olkin, *Annals of Statistics* **6**, 232–234.

[29] Szekli, R. (1995). *Stochastic Ordering and Dependence in Applied Probability Lecture Notes in Mathematics 97*, Springer-Verlag, New York.

[30] Shaked, M. & Spizzichino, F. (1998). Positive dependence properties of conditionally independent random lifetimes, *Mathematics of Operation Research* **23**, 944–959.

[31] Belzunce, F. & Semeraro, P. (2004). Preservation of positive and negative orthant dependence concepts under mixtures and applications, *Journal of Applied Probability* **41**, 961–974.

[32] Shaked, M. (1977). A concept of positive dependence for exchangeable random variables, *Annals of Statistics* **5**, 505–515.

[33] Lee, M.L.T. (1985). Dependence by total positivity, *Annals of Probability* **13**, 572–582.

[34] Khaledi, B.E. & Kochar, S. (2001). Dependence properties of multivariate mixture distributions and their applications, *Annals of the Institute of Statistical Mathematics* **53**, 620–630.

Related Articles

Aging and Positive Dependence; **Bayesian Reliability Analysis**; **Bayesian Reliability Demonstration**; **Stochastic Orders and Aging**; **Subjective Life Models**.

FABIO SPIZZICHINO

Mixtures *see* Survival Analysis, Nonparametric

Modules and Modular Decomposition

Introduction

In engineering terminology, a *module* is simply a cluster of components that is treated as a single entity in a system. For example, an engine is often treated as a module in the airline industry. If an engine has failed or has signs of serious **degradation**, it is taken off the airplane as a whole and sent to the engine manufacturer for repair and/or rebuild. A self-contained assembly of electronic components and circuitry such as video card, a hard disk, or a motherboard is also referred to as a module. In software engineering, a module is defined as a software entity that groups a set of subprograms and data structures. These software modules are units that can be compiled separately, which makes them reusable and allows multiple programmers to work on different modules simultaneously. A module may also be an integrated unit including both hardware and software.

In system reliability theory, a module represents a group of components that has a single input from and a single output to the rest of the system in the system's reliability block diagram (*see* **Coherent Systems**). The state of the module can be represented by a random variable. The contribution of all components in a module to the performance of the whole system can be represented by the state of the module. Once the state of the module is known, one does not need to know the states of the components within the module to determine the state of the system. If we know the reliabilities of the modules in a system, we can use them to find the reliability of the system.

With the advancement of technology, today's engineering systems are becoming more and more sophisticated. For example, a spacecraft may consist of millions of components. In design optimization of such a system subject to budget, reliability, physical weight, and other constraints, efficient reliability evaluation algorithms are needed. Appropriate identification of modules within a system and use of modular decomposition will enhance the efficiency of system reliability evaluation and, as a result, that of design optimization.

In this article, we will provide the formal definitions of modules and modular decomposition for system reliability analysis, illustrate the technique of modular decomposition, and outline reliability evaluation algorithms for several well-studied modules. References for further reading in this area will also be provided. Unless otherwise stated, we adopt the following assumptions in this article:

1. The systems to be discussed are coherent systems (*see* **Coherent Systems**).
2. The system and its components may be in only two possible states, either working or failed.
3. The states of the components are independent random variables.
4. The mission time of the system is implicitly specified.

On the basis of these assumptions, we deal primarily with system and component reliabilities rather than their reliability functions of time t.

Definitions and General Guidelines

In reliability analysis, a *system* is defined as a set of components that work together to perform a specific function. The reliability of the system is determined by the reliabilities of its components and how the components are logically connected together to form the system. If there exists a subset of components in the system which work together to perform a subfunction of the system, we say that this subset of components forms a subsystem. The reliability of this subsystem is completely determined by its components and how they are logically connected together to form the subsystem. In reliability evaluation of the original system, if we can treat this subsystem as a single "component" with a reliability value of the subsystem and the reliability of the system obtained this way is the same as that obtained when all the components of the subsystem are explicitly considered, then we say that this subsystem of components is actually a "module". This is an intuitive explanation of what a module is. If we use the definition of a coherent system (see Barlow and Proschan [1]), a module can be explained to be a subset of components which form a coherent system structure such that the system as a whole is coherent even if the components in the module are treated as a single "component". The formal definition of a module is given below.

Definition 1 (Barlow and Proschan [1]) The coherent system (A, χ) is a module of the coherent system (C, ϕ) if $\phi(\mathbf{x}) = \psi(\chi(\mathbf{x}^A), \mathbf{x}^{A^c})$, where ψ is a coherent structure function, A is a subset of C, A^c is the subset of C that is complementary to A, and $\mathbf{x}^A$ is a vector representing the states of the components in set A. The set $A \subseteq C$ is called a modular set of (C, ϕ).

On the basis of this definition, each component by itself is a module of the system. The system itself can also be considered to be a module. However, these two extreme cases do not provide any advantage in system reliability evaluation.

If A is a module, A^c must be a module too. If a module A has been identified in a system, we can say that the system has been decomposed into two modules, A and A^c. The system can be considered as consisting of these two modules. Each module has a structure function. The structure function of the system can be expressed as a function of the structure functions of the two modules.

More than two modules may exist in a coherent system. Modular decomposition is a technique that can be used to decompose a coherent system into several disjoint modules. Such a decomposition is useful if the structure function of each module can be easily derived and the structure function relating the system state to the states of these disjoint modules can be easily derived. A formal definition of modular decomposition is given below.

Definition 2 (Barlow and Proschan [1]) A modular decomposition of a coherent system (C, ϕ) is a set of disjoint modules, $\{(A_1, \chi_1), (A_2, \chi_2), \ldots, (A_r, \chi_r)\}$ together with an organizing structure ψ such that

$$C = \bigcup_{i=1}^{r} A_i, \quad \text{and} \quad A_i \cap A_j = \emptyset \quad \text{for } i \neq j \tag{1}$$

$$\phi(\mathbf{x}) = \psi\left[\chi_1(\mathbf{x}^{A_1}), \chi_2(\mathbf{x}^{A_2}), \ldots, \chi_r(\mathbf{x}^{A_r})\right] \tag{2}$$

To apply the modular decomposition technique in the derivation of system structure functions, we need to identify disjoint modules within the system structure. The structure function of each disjoint module must be easy to derive. These modules can then be treated as "supercomponents" or subsystems, and we only need to find the relationship between the state of the system and the states of these modules. Hopefully, the number of such modules is much smaller than the number of components in the system.

We may use the following two approaches individually or in combination to identify modules in a complex system. The first approach defines modules based on a clear understanding of the functions of some components in the system. For example, in a chemical refinery, a primary pump is used to provide the required flow rate and pressure of a certain liquid. A secondary pump is installed as a cold standby unit. These two pumps together with their auxiliary instruments such as the sensing and switching mechanism can be called the *pump module*. The reliability of the pump module can be derived using the standby structure and used in further evaluation of the chemical refinery's reliability. Another example is a power system including generation, transmission, and distribution subsystems. A single power station including several generators may be considered to be a module if all these generators use a single point of delivering power to the transmission network. This module of generators may have a k-out-of-n structure (*see* ***k*-out-of-*n* Systems**).

The second approach assumes that the reliability block diagram of the system has already been developed. From the system reliability block diagram, we then identify modules following Definition 1. We will illustrate the technique of modular decomposition using this second approach in the section titled "Modular Decomposition".

Well-studied modules include the series structure, the parallel structure, the series–parallel structure, the parallel–series structure, the bridge structure, the k-out-of-n structure, the standby structure, and the consecutive-k-out-of-n structure. Efficient reliability evaluation algorithms have been reported for these modules. These modules will be further elaborated in the section titled "Well-Studied Modules".

In subsequent discussions, the following notation will be used:

p_i : reliability of component i

q_i : unreliability of component i, $q_i = 1 - p_i$

R_a : reliability of a where

a is a string of letters or

numerals that may represent

a module or subsystem or the system

$Q_a : 1 - R_a$.

Modular Decomposition

Consider the reliability block diagram depicted in Figure 1(a). The nine components are numbered from 1 to 9. We are interested in finding the reliability of this system.

Examining Figure 1(a), we can identify the following modules:

- Module I consists of components 1 and 2 connected in series.
- Module II consists of components 3 and 4 connected in series.
- Module III consists of components 6, 7, and 8, wherein components 7 and 8 connected in parallel form a submodule which is then connected to component 6 in series.

The three modules listed above are the only meaningful modules in Figure 1(a) that may be used for efficient evaluation of system reliability. They are the only clusters of components that have a single input from and a single output to the rest of the system in the reliability block diagram. The reliabilities of these modules are

$$R_{\text{I}} = p_1 p_2 \tag{3}$$

$$R_{\text{II}} = p_3 p_4 \tag{4}$$

$$R_{\text{III}} = p_6(1 - q_7 q_8) \tag{5}$$

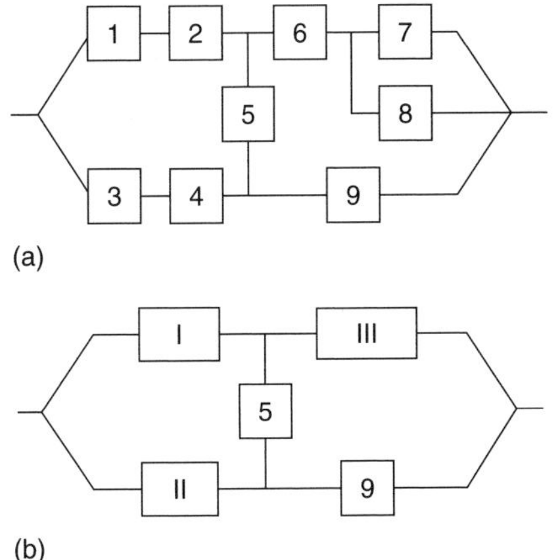

Figure 1 Illustration of modular decomposition

With these identified modules, the system reliability block diagram shown in Figure 1(a) is transformed into the one shown in Figure 1(b). This resulting reliability block diagram is actually called the *bridge structure*.

If direct identification of known modules does not yield a simple organizing structure for system reliability evaluation, there are other techniques one may apply for further identification of modules. These techniques include the so-called pivotal decomposition, Δ–star transformation, and star–Δ transformation. For details on these techniques, readers may refer to Kuo and Zuo [2] and Misra [3].

Well-Studied Modules

Many special system structures have been studied by researchers. Efficient reliability evaluation algorithms have been reported for these structures. In this section, we call these special structures modules and provide their definitions and reliability evaluation algorithms (*see* **Parallel, Series, and Series–Parallel Systems**; ***k*-out-of-*n* Systems**).

Any n components connected in series form the *series module*. The reliability of a series module can be calculated as

$$R_{\text{series}} = p_1 p_2 \cdots p_n \tag{6}$$

Any n components connected in parallel form the *parallel module*. The reliability of a parallel module can be calculated as

$$R_{\text{parallel}} = 1 - q_1 q_2 \cdots q_n \tag{7}$$

A parallel–series system can be considered to consist of m disjoint modules that are connected in parallel and module $i (1 \le i \le m)$ consists of n_i components that are connected in series. If the number of components in each of the m modules is not too large and m is not too large, the whole parallel–series system may be considered to be a module. The reliability of such a *parallel–series module* can be expressed as

$$R_{\text{parallel-series}} = 1 - \prod_{i=1}^{m} \left(1 - \prod_{j=1}^{n_i} p_{ij}\right) \quad (8)$$

where p_{ij} is the reliability of component j in series module i for $1 \le i \le m$ and $1 \le j \le n_i$.

A series–parallel system can be considered to consist of m disjoint modules that are connected in series and module i ($1 \le i \le m$) consists of n_i components that are connected in parallel. If the number of components in each of the m modules is not too large and m is not too large, the whole series–parallel system may be considered to be a module. The reliability of such a *series–parallel module* can be expressed as

$$R_{\text{series-parallel}} = \prod_{i=1}^{m} \left(1 - \prod_{j=1}^{n_i} q_{ij}\right) \quad (9)$$

where q_{ij} is the unreliability of component j in parallel module i for $1 \le i \le m$ and $1 \le j \le n_i$.

A standby structure can also be considered to be a module. Consider a *standby module* with a primary unit and one cold standby unit. There is a sensing and switching mechanism that switches the standby unit into operation when it detects the failure of the primary unit. To evaluate the reliability of such a module, one needs to know the lifetime distribution function of each unit. Under the assumptions that each component being used follows the exponential lifetime distribution with parameter λ, the standby unit is perfect, and the sensing and switching mechanism is perfect, the reliability function of the module can be expressed as

$$R_{\text{standby}}(t) = e^{-\lambda t}(1 + \lambda t) \quad (10)$$

With this equation, one can evaluate the reliability of the module for any specified mission time t. For more advanced analysis of the standby module under more relaxed conditions, readers are referred to Kuo and Zuo [2].

The reliability block diagram of the bridge structure is shown in Figure 2. It can be called the *bridge module*. The reliability of this module can be expressed as

$$R_{\text{bridge}} = p_3(1 - q_1q_4)(1 - q_2q_5) + q_3 \times [1 - (1 - p_1p_2)(1 - p_4p_5)] \quad (11)$$

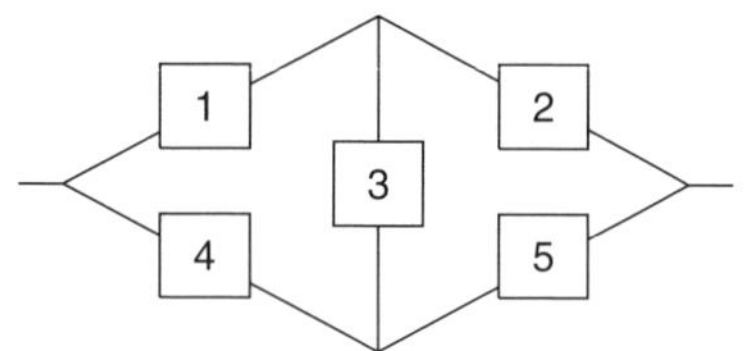

Figure 2 The bridge module

The k-out-of-n structure may also appear in a sophisticated reliability system. A k-out-of-n system works if and only if at least k of the n components work. A recursive formula for reliability evaluation of such a *k-out-of-n module* is given by Rushdi [4] as follows:

$$R(k, n) = p_n R(k - 1, n - 1) + q_n R(k, n - 1) \quad (12)$$

where $R(a, b)$ denotes the probability that at least a components out of the b components work. The boundary conditions for equation (12) are

$$R(0, j) = 1, \quad j \ge 0 \quad (13)$$

$$R(j + 1, j) = 0, \quad j \ge 0 \quad (14)$$

It would be advantageous to treat a k-out-of-n structure as a module in a complex system for its reliability evaluation only if the number of components in the k-out-of-n structure is not too large.

A consecutive-k-out-of-n system consists of linearly arranged components and it is failed if and only if at least k consecutive components are failed. Such a system structure may appear as a module in a sophisticated system. An efficient recursive equation is provided by Hwang [5] for reliability evaluation of such a *consecutive-k-out-of-n module*:

$$R_c(k, n) = R_c(k, n - 1) - R_c(k, n - k - 1) \times p_{n-k} \prod_{j=n-k+1}^{n} q_j \quad (15)$$

where $R_c(a, b)$ is the probability that at least a consecutive components have failed among the first b components. The boundary conditions for equation (15) are

$$R_c(a, b) = 1, \quad b < a \quad (16)$$

$$p_0 \equiv 0 \quad (17)$$

Advanced coverage of the consecutive-k-out-of-n structure and its variations can be found in Chang *et al.* [6].

Summary and Further Reading

A key step in modular decomposition is the definition of modules. Each identified module should be simple enough to enable us to obtain its various reliability characteristics relatively easily. Consider the example reliability block diagram given in Figure 3. There are 10 components in the system. One option is to decompose the system into three modules with module I including components 1 through 4, module II including components 5 through 7, and module III including components 8 through 10. Module I has a parallel structure, module II has a series–parallel structure, while module III has a 2-out-of-3 structure. The organization structure of the system consisting of these three modules is a series structure.

To evaluate a large and complex reliability system, we often have to apply the so-called hierarchical modular decomposition. The system is decomposed into modules (this is called the *level 1 decomposition*). Each module in the system may be further decomposed into submodules (this is called the *level 2 decomposition*). The process continues until the modules or submodules in the lowest level of decomposition are simple enough.

The technique of modular decomposition can be used not only to evaluate the exact system reliability but also to derive bounds on system reliability. This can be achieved through hierarchical modular decomposition. Starting from the lowest decomposition level, we first find the bounds on the reliabilities of all the modules at this level. Once these bounds are obtained, these modules at the lowest level are treated as supercomponents with known reliability bounds. Higher-level modules consisting of these supercomponents are then analyzed. Bounds on these higher-level modules may be obtained. This process is repeated until we reach the system level wherein the upper and/or lower bounds on the system reliability can be obtained from the bounds of its immediate modules. As illustrated by Barlow and Proschan [1], utilization of modular decomposition can improve the bounds obtained.

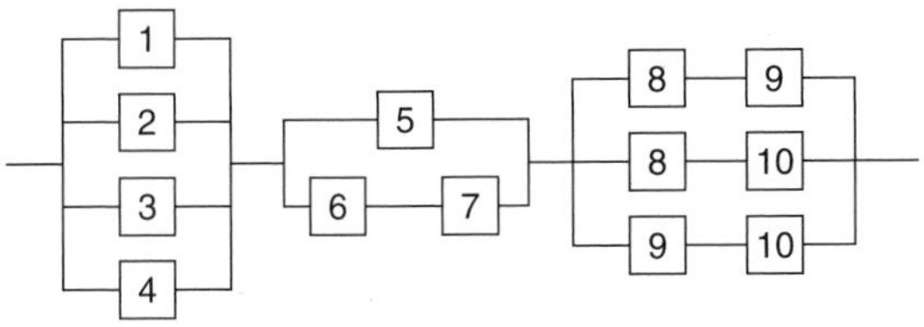

Figure 3 An example system reliability block diagram

The modules summarized in this article are widely seen in engineering systems. The series, parallel, parallel–series, and series–parallel modules are most commonly seen. The bridge module is used in networks. The standby module is often used in mechanical systems. The k-out-of-n module is seen in modeling power generation stations and the triple-modular redundancy (TMR) and N-modular redundancy (NMR) schemes in software engineering and fault tolerance control (see Hecht [7], Xie *et al.* [8]), and N-version programming (NVP) (see Avizienis [9]). The consecutive-k-out-of-n module has been used in modeling pipeline transportation systems (see Chang *et al.* [6]).

Advanced studies of the k-out-of-n and the consecutive-k-out-of-n modules can be found in Chang *et al.* [6], Kuo and Zuo [2], and Lisnianski and Levitin [10]. Both components and the system may assume binary states or multiple states (more than two states). There may be common-cause failures. The system and the components may be repairable or nonrepairable. Other reliability characteristics such as mean time to failure (MTTF), mean time between failure (MTBF), and availability have also been studied. Another topic involving modules that has been well studied is the assembly of components into different modules of a system. This involves allocation of components having different reliabilities to different modules within the system. El-Neweihi *et al.* [11] provided optimal allocations of components to parallel–series and **series–parallel systems**. Du and Hwang [12] provided optimal allocations of components to an s-stage k-out-of-n system, in which the system can be hierarchically decomposed into s stages and each stage has the k-out-of-n system structure.

References

[1] Barlow, R.E. & Proschan, F. (1975). *Statistical Theory of Reliability and Life Testing: Probabilities*, Holt, Rinehart and Winston, New York.

[2] Kuo, W. & Zuo, M.J. (2003). *Optimal Reliability Modeling: Principles and Applications*, John Wiley & Sons, New York.
[3] Misra, K.B. (1992). *Reliability Analysis and Prediction: A Methodology Oriented Treatment*, Elsevier, Amsterdam.
[4] Rushdi, A.M. (1986). Utilization of symmetric switching functions in the computation of k-out-of-n system reliability, *Microelectronics and Reliability* **26**(5), 973–987.
[5] Hwang, F.K. (1982). Fast solutions for consecutive-k-out-of-n:F system, *IEEE Transactions on Reliability* **R-31**(5), 447–448.
[6] Chang, G.J., Cui, L. & Hwang, F.K. (2000). *Reliabilities of Consecutive-k Systems*, Kluwer, Boston.
[7] Hecht, H. (2004). *Systems Reliability and Failure Prevention*, Artech House, Boston.
[8] Xie, M., Dai, Y.S. & Poh, K.L. (2004). *Computing Systems Reliability: Models and Analysis*, Kluwer Academic, New York.
[9] Avizienis, A. (1995). The methodology of N-version programming, in *Software Fault Tolerance*, M.R. Lyu ed, John Wiley & Sons, Chapter 2, pp. 23–46.
[10] Lisnianski, A. & Levitin, G. (2003). *Multi-state System Reliability: Assessment, Optimization and Applications*, World Scientific, Singapore.
[11] El-Neweihi, E., Proschan, F. & Sethuraman, J. (1986). Optimal allocation of components in parallel-series and series-parallel systems, *Journal of Applied Probability* **23**(3), 770–777.
[12] Du, D.Z. & Hwang, F.K. (1990). Optimal assembly of an s-stage k-out-of-n system, *SIAM Journal of Discrete Mathematics* **3**(3), 349–354.

Related Articles

Coherent Systems; ***k*-out-of-*n* Systems**; **Parallel, Series, and Series–Parallel Systems**; **Reliability of Redundant Systems**.

MING J. ZUO

Moments

Introduction

Moments are an important class of *expectations* used to describe probability distributions. Together, the entire set of moments of a random variable will generally determine its probability distribution exactly.

There are three main types of moments:

1. raw moments,
2. central moments, and
3. factorial moments.

Raw Moments

Where a random variable is denoted by the letter X, and k is any positive integer, the kth raw moment of X is defined as $E(X^k)$, the expectation of the random variable X raised to the power k. Raw moments are usually denoted by μ'_k where $\mu'_k = E(X^k)$, if that expectation exists. The first raw moment of X is $\mu'_1 = E(X)$, also referred to as the *mean of* X. The second raw moment of X is $\mu'_2 = E(X^2)$, the third $\mu'_3 = E(X^3)$, and so on. If the kth moment of X exists, then all moments of lower order also exist. Therefore, if the $E(X^2)$ exists, it follows that $E(X)$ exists.

Central Moments

Where X again denotes a random variable and k is any positive integer, the kth central moment of X is defined as $E[(X-c)^k]$, the expectation of X minus a constant, all raised to the power k. Where the constant is the mean of the random variable, this is referred to as the *kth central moment around the mean*. Central moments around the mean are usually denoted by μ_k where $\mu_k = E[(X-\mu_X)^k]$. The first central moment is equal to zero as $\mu_1 = E[(X-\mu_X)] = E(X) - E(\mu_X) = \mu_X - \mu_X = 0$. In fact, if the probability distribution is symmetrical around the mean (e.g., the **normal distribution**) all odd central moments of X around the mean are equal to zero, provided they exist. The most important central moment is the second central moment of X around the mean. This is $\mu_2 = E[(X-\mu_X)^2]$, the *variance* (*see* **Variance**) of X.

The third central moment about the mean, $\mu_3 = E[(X-\mu_X)^3]$, is sometimes used as a measure of asymmetry or **skewness**. As an odd central moment around the mean, μ_3 is equal to zero if the probability distribution is symmetrical. If the distribution is negatively skewed, the third central moment about the mean is negative, and if it is positively skewed,

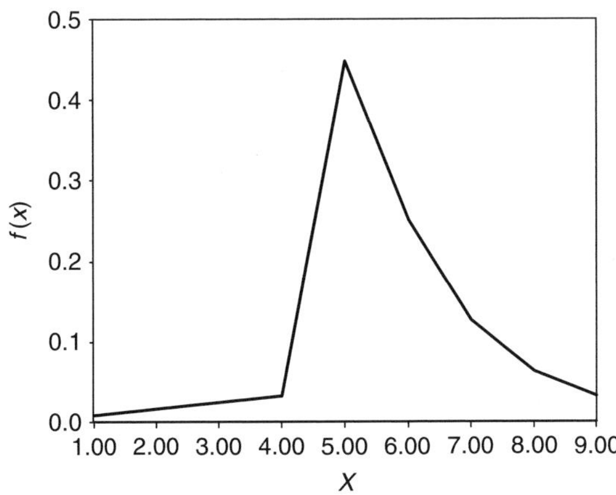

Figure 1 An example of an asymmetrical probability distribution where the third central moment around the mean is equal to zero

the third central moment around the mean is positive. Thus, knowledge of the shape of the distribution provides information about the value of μ_3. Knowledge of μ_3 does not necessarily provide information about the shape of the distribution, however. A value of zero may not indicate that the distribution is symmetrical. As an illustration of this, μ_3 is approximately equal to zero for the distribution depicted in Figure 1, but it is not symmetrical. The third central moment is therefore not used much in practice.

The fourth central moment about the mean, $\mu_4 = E[(X - \mu_X)^4]$, is sometimes used as a measure of excess or **kurtosis**. This is the degree of flatness of the distribution near its center. The coefficient of kurtosis $(\mu_4/\sigma^4 - 3)$ is sometimes used to compare an observed distribution to that of a normal curve. Positive values are thought to be indicative of a distribution that is more peaked around its center than that of a normal curve, and negative values are thought to be indicative of a distribution that is more flat around its center than that of a normal curve. However, as was the case for the third central moment around the mean, the coefficient of kurtosis does not always indicate what it is supposed to.

Factorial Moments

Finally, where X denotes a random variable and k is any positive integer, the kth factorial moment of X is defined as the following expectation $E[X(X-1)\ldots(X-k+1)]$. The first factorial moment of X is therefore $E(X)$, the second factorial moment of X is $E[(X-1+1)(X-2+1)] = E(X^2 - X)$, and so on. Factorial moments are easier to calculate than raw moments for some random variables (usually discrete). As raw moments can be obtained from factorial moments and *vice versa*, it is sometimes easier to obtain the raw moments for a random variable from its factorial moments.

Moment Generating Function

For each type of moment, there is a function that can be used to generate all of the moments of a random variable or probability distribution. This is referred to as the *moment generating function* and denoted by mgf, $m_X(t)$ or $m(t)$. In practice, however, it is often easier to calculate moments directly. The main use of the mgf is therefore in characterizing a distribution and for theoretical purposes. For instance, if a mgf of a random variable exists, then this mgf uniquely determines the corresponding distribution function. As such, it can be shown that if the mgfs of two random variables both exist and are equal for all values of t in an interval around zero, then the two cumulative distribution functions are equal. However, existence of all moments is not equivalent to existence of the mgf.

More information on the topic of moments and mgfs is given in [1-3].

References

[1] Casella, G. & Berger, R.L. (1990). *Statistical Inference*, Duxbury Press, Belmont.

[2] DeGroot, M.H. (1986). *Probability and Statistics*, 2nd Edition, Addison-Wesley, Reading.

[3] Mood, A.M., Graybill, F.A. & Boes, D.C. (1974). *Introduction to the Theory of Statistics*, McGraw-Hill, Singapore.

REBECCA WALWYN

Article originally published in Encyclopedia of Statistics in Behavioral Science *(2005, John Wiley & Sons, Ltd). Minor revisions for this publication by Jeroen de Mast.*

Monitoring of Safety in EU Railways

ERA, the Ambitious Projects on Safety of a Newcomer, under Directive 2004/49/EC

Railway is the safest land transport mode and one of the overall safest transport modes. Directive 2004/49/EC, on railway safety, has been put in place to support the creation of an integrated European railway area by facilitating market opening through the harmonization of safety management and regulation. Directive 2004/49/EC will also contribute to improve safety by: a cultural shift from a deterministic to a risk-based approach, the establishment of safety management systems, the introduction of a system for independent accident investigation, as well as by moving from self-regulation to regulation by independent authorities (National Safety Authorities – NSAs). This aims to create a basis for mutual trust between member states (MSs), mainly through the development of common and transparent methods to monitor safety performance, to set targets and to manage the introduction of significant changes in railway systems with a risk-based systematic approach.

Collection and Publication of Information on Safety

Purpose

Safety information is collected mainly to provide input to the development of an objective tool for setting common safety targets (CSTs) and assessing their achievement. CSTs will be developed in subsequent sets, on the basis of safety performances in MSs and their possible economic impact in terms of costs and benefits. The European Railway Agency (ERA) is in charge of these projects, carried out in conjunction with industry and MSs in working groups of experts.

Development of Common Safety Indicators (CSIs) and Accidents Investigation

The working group on common safety indicators (CSIs) is developing and defining indicators structured as follows:

Safety performances

- type of accidents
- fatalities and injuries classified in categories of people
- precursors to accidents.

Economic aspects of safety performances

- societal benefits and costs of improving and worsening safety performances.

CSIs are expressed in total of events and consequences (fatalities and injuries) per unit of time (1 year), normalized mainly by traffic performances (train*km, passenger*km).

The flow of information and data exchange on safety issues between the concerned parties is completed by the activity of accidents investigation, carried out by National Investigation Bodies (NIBs) and communicated to ERA; the aim of this is to better understand the causes of accidents, which, together with the aforementioned statistics on accidents, precursors to accidents, and economic aspects of safety, provides the necessary information to ERA to set CSTs at EU level (see Figure 1).

Data Sources of CSIs

ERA relies on data from two sources, Eurostat and NSAs. Eurostat has been collecting EU-wide harmonized data on "type of accidents, related consequences, and traffic performances", since 2004; NSAs have been publishing data on CSIs every year, starting from 2006 and according to a common format developed with ERA. Therefore, both Eurostat and NSAs will collect data on accidents and related consequences. In addition, NSAs will also report on precursors to accidents and economic aspects of safety. This double collection of accidents data should be temporary and it is a consequence of the further development, introduced by Directive 2004/49/EC, of the type of accidents classification and categorization of people; ERA and Eurostat are working together to eliminate these differences and to harmonize **data collection** processes.

Consistency between Eurostat and NSAs. ERA has informally collected 2004 and 2005 data on CSIs from NSAs, with the view to improving data reliability through the appraisal of consistency between

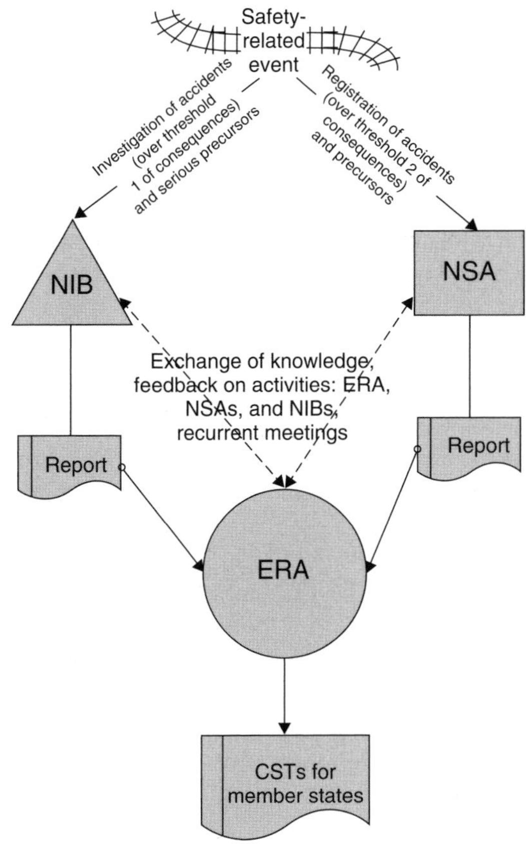

Figure 1 Flow of information to set CSTs

NSAs and Eurostat statistics. The differences identified between NSAs and Eurostat statistics have been registered and sent to both sources, inviting them to further discuss with each other, to eliminate existing inconsistencies.

Collection and Analysis of Time Series of National Data on Railway Safety. In the context of ERA's work to develop CSTs, it has been necessary to complement data delivered by Eurostat and data on CSIs with longer **time series** of national data on railway safety. This in order to provide appropriate input to the analysis aimed at identifying for each MS the current safety performance of national railway systems in terms of risk. The need to look at long time series of data derives from the high degree of oscillation which data on accidents and related consequences (i.e., fatalities and serious injuries) typically show from year to year, making difficult the identification of the current risk levels attributable to national railway systems. Being, in fact, the risk calculated as a product of consequences of accidents and their respective frequencies, the oscillation of data both on frequency and consequences is systematically transmitted to risk estimates. This oscillation is the cumulative effect of rarity and statistical randomness of the events under consideration. Accidents with multiple fatalities and injuries represent rare events, therefore they are considered as high-impact and low-frequency events and generate outliers in time series.

Different techniques have been considered for smoothing down the outliers due to multiple fatality accidents and to calculate the underlying, intrinsic risk levels of national railway systems. These techniques use moving averages or moving **percentiles** (i.e., moving median or upper quartile), which respectively average and cut out isolated outliers in the same moving n-year time window.

The random behavior of accident statistics over time, cumulated with the intrinsic rarity of significant accidents, may also lead to recursive zero values in time series, especially for small, modern railway networks with low traffic volumes. This makes it difficult to estimate the underlying risk levels of these systems through robust statistical methods, especially when adequate time series of data are not available. Empirical solutions are needed in these cases for identifying and targeting risk levels, such as the adoption of empirical tolerance intervals $[0, \sigma]$ for the risk levels of the system, where σ is a predetermined standard deviation from a mean which is assumed to be equal to zero.

ANGELO PIRA, ROBERTO PIAZZA AND JANE RAJAN

Monte Carlo Methods

Monte Carlo methods – definition and common uses

Monte Carlo methods refer to computer-intensive methods which are used to mimic *unknown* statistical distributions of *output* random variables, *via* generation of random numbers from *input* random variables with *known* distributions. Both input and output variables (random or deterministic) are tied together by a stochastic model. Application of **Monte Carlo** methods is conducted in the framework of **Monte Carlo simulation** (namely, simulation that is stochastic in some manner). Major areas of application of Monte Carlo methods include:

- Obtaining numerical solutions to certain mathematical problems which are too complicated to solve analytically (such as the calculation of algebraically intractable integrals *via* Monte Carlo integration).
- Finding optimal solutions for problems too complex to solve mathematically (like NP-complete problems) *via* optimum-search techniques (like simulated annealing and **genetic algorithms**).
- System simulation (like physical, biological, and financial systems) in order to study and predict their behavior under various stochastic scenarios.
- Simulating distributions of statistics with unknown distributions for statistical inference (**bootstrap** and jackknife methods).
- Inverse problems, where a large collection of possible models are pseudorandomly generated *via* Monte Carlo methods and the associated likelihood of model's properties are evaluated.

Some techniques employed in Monte Carlo simulation

A major effort in Monte Carlo simulation relates to generating random numbers from the specified distributions. There are several general methods for random number generation, like rejection methods or the simpler inversion method. The latter refers to the case where random numbers need to be generated from a distribution with a known inverse cumulative distribution function (cdf) (*see* **Cumulative Distribution Function (CDF)**). Since any cdf, expressed as a function of its random variable, is uniformly distributed, introducing a set of numbers, sampled from the uniform distribution on the (0, 1) interval, into the *inverse* cdf would produce a set of random numbers that has the desired distribution. For example, for the inverse cdf of the exponential distribution with parameter λ:

$$x = F^{-1}(P) = -(1/\lambda)\exp(1 - P) \qquad (1)$$

where $P = F(X \leq x)$, X is exponentially distributed with parameter λ. Since P is uniformly distributed, uniformly distributed random numbers introduced into equation (1) would produce exponentially distributed numbers. When the distribution does not have an explicit inverse, like the **normal distribution**, special methods have been devised to generate the required distribution (including approximations for the algebraically intractable inverse cdf). Simulation software packages nowadays have random number generators for most commonly used distributions. These methods actually generate *pseudorandom* numbers since deterministic methods are used for the generation of the random numbers.

A second major effort in application of Monte Carlo methods is variance reduction. Variance reduction techniques aim to generate simulation data with minimal variance in order to increase the precision of estimates derived from the simulated data. There are several available variance reducing techniques, like antithetic **resampling**, the method of control variates, indirect **estimation**, and others.

Monte Carlo methods in quality and reliability engineering

Monte Carlo methods are extensively employed in reliability engineering to compute reliability of systems with complex configurations, or systems with components that are not necessarily statistically independent. Another common application is to calculate algebraically intractable integrals that appear in reliability-related calculations. The effects of various policies of preventive maintenance or various warranty policies are also often examined *via* Monte Carlo methods. In quality engineering these methods are used to assess the impact of a product design change, examine the effect of process improvement, and to optimize operating conditions of a running process.

HAIM SHORE

Monte Carlo Methods, Univariate and Multivariate

Monte Carlo Method

In the world of simulation, a **Monte Carlo** method is a means of solving a deterministic problem using random numbers. The origin of the term comes from a casino in Monaco; the method requires many repetitions of a random process, which is similar to that which occurs in a casino. The term became popular with those using such methods in the 1940s and 1950s, notably Nicholas Metropolis and Stanislaw Ulam [1], among others.

The ideas underpinning the method had been recognized well in advance of this time. For example, the case of Buffon's needle is a classic example of the use of random experiments in order to estimate a deterministic quantity. Since the time of Metropolis and Ulam, the methods have been of direct interest to the statistical community; [2] indeed even in the 1980s it was recognized that such simulation methods were useful, in an age when computers were "powerful" and mathematicians were scarce [3]. Of course, computer power has continued to improve substantially over the past two decades; and such methods are increasingly common [4, 5].

In order to describe the method in more detail, we look at a simple example. Following this, some technical considerations will be highlighted, together with a discussion of how the method can be applied to a more complex situation. We then note how the method relates to that of ensembles. Finally we examine some particular methods.

A Simple Example

In order to demonstrate the idea, we solve a very simple problem using such a method.

Consider the question "What is the area of a circle inscribed within the unit square?" Graphically, the region of interest is highlighted in Figure 1. The total area from which the simulation will be carried out is the white square, with the circle shaded in black.

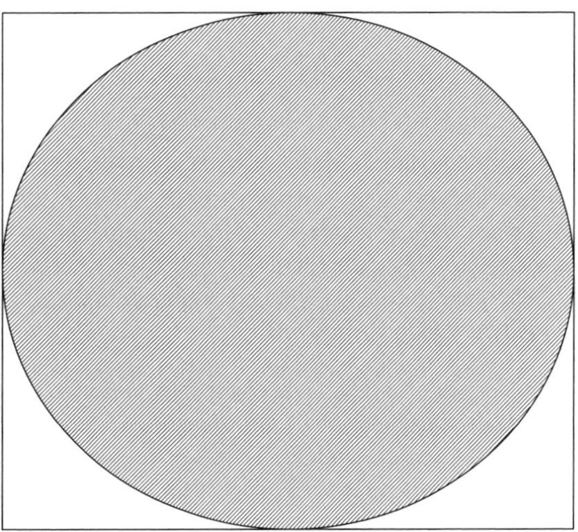

Figure 1 Pictorial representation of circle in unit square

In order to answer this, we consider an "experiment" in which there is an element of chance. The idea is that we "throw" darts uniformly at the unit square. For each of the darts, we then determine whether or not it has fallen within the shaded circle. If it has, then we count it as a success. Otherwise we think of it as a failure. The relative area of the circle to the square is then approximated by the proportion of successes to the total number of throws.

Thus, the formal algorithm, noting that the circle is centered at (0.5,0.5) and has radius 0.5 proceeds as follows;

- Set $s = 0$.
- Repeat N times
  - Generate $x \sim U[0, 1]$.
  - Generate $y \sim U[0, 1]$.
  - If $(x - 0.5)^2 + (y - 0.5)^2 < (0.5)^2$ then $s = s + 1$.

The area is then estimated as the fraction, s divided by the total number of simulations, N.

Characteristics of the Algorithm

We note that the method has the following characteristics;

- The method requires a source of random numbers (the U[0,1]).
- There are many repetitions of a simple process (the throwing of the darts).
- The answer that is obtained is an approximation to that which is required.

Indeed, the first of these conditions is not strictly true; it is sufficient that the source of "random" numbers be pseudorandom – the key requirement is that there is no relationship between what is being estimated and the source we are using.

We (the persons carrying out the experiment) are in charge of the number of such repetitions that we obtain. In practice, since computers are typically used to carry out this simulation process, it is "cheap" to carry out each repetition. Thus the main barrier to the size of the collection of realizations is the time available and the processor power and memory storage available to us.

Even though the answer we obtain is only an approximation to that which is required, we can quantify the size of the error. This is termed *Monte Carlo error*, and by varying the number of simulations we can make this arbitrarily small. Since this last point is important, a little more detail is provided below.

Monte Carlo Error

Where independent simulations have been obtained it is possible to quantify the size of the Monte Carlo error by the following considerations;

Consider the example above. Formally we describe the probability of the "dart" being in the unit square by a parameter, p. The throwing of each "dart" is then a Bernoulli trial; that is the result of the throw is 1 (i.e., success – landing inside the circle) with probability p and 0 (i.e., failure – landing outside the circle) with probability $1 - p$. We denote this random variable as X.

If we consider a large collection (n) of independent realizations of such random variables, we end up with a collection $\{x_1, x_2, \ldots, x_n\}$ of zeroes or ones. The sum of these n realizations may be denoted S_n. It is clear that S_n is then a random variable which has a binomial distribution with parameters (n, p). In this case, the expectation and variance of the random variable may readily be obtained as np and $np(1 - p)$ respectively.

Thus, if we consider the derived random variable $p_{\text{est}}(n)$, given by;

$$p_{\text{est}}(n) = \frac{S_n}{n} \quad (1)$$

then this has expected value p and variance $\frac{p(1-p)}{n}$. Thus, as n gets large, we are increasingly confident that $p_{\text{est}}(n)$ is close to p.

Returning to the question, we wish to estimate the relative area of the circle. By considering the setup, we note that this relative area is exactly the probability of a uniformly thrown "dart" hitting the circle. Thus, it is equal to p, and our task is to estimate p to a given degree of accuracy.

Using the central limit theorem, we note that as n increases there is approximately 95% probability of $p_{\text{est}}(n)$ being contained in a ball that stretches 2 standard deviations from p, and 99.9% probability of being in a ball that extends about 3 standard deviations from p. Thus, since we can construct $p_{\text{est}}(n)$ by "experiment" we have a method of estimating p.

By observing this fact, and noticing that we can increase n to be arbitrarily large, we note that we can obtain estimates of p that are accurate to a given accuracy.

The residual uncertainty that exists because $p_{\text{est}}(n)$ only *approximates* p may be quantified in terms of the standard deviation of the estimator considered as a random variable. This, as outlined above is $\sqrt{(\frac{p(1-p)}{n})}$.

It is this residual uncertainty that is termed the *Monte Carlo error.* In any situation where a Monte Carlo method is used, it is essential that some reference be made to this quantity – either explicitly or even just to note that the relative size of the Monte Carlo error is "small".

In general, the Monte Carlo error may be estimated using the standard deviation of the generated sequence; in the example above, this would be the standard deviation of the sequence of realizations of the Bernoulli random variables.

An Extension of the Example

The example that has been used is a particularly simple one. Indeed, it has the status of a "toy" and is used simply to illustrate the method.

However, we may readily consider situations that are of real practical interest to us.

Consider a system, $\mathcal{S}$, which takes a number of inputs; we may think of the system as being described by $\mathcal{S}(I_1, I_2, \ldots, I_m; 0)$ at time 0. We may be interested in an indicator function, R, which takes the value 1 if the system is functioning and the value 0 if it has failed.

Thus, $R(\mathcal{S}(I_1, I_2, \ldots, I_m; 0)) = 1$ tells us that the system with the given inputs was functioning initially; and if $R(\mathcal{S}(I_1, I_2, \ldots, I_m; t_1)) = 1$ and the quantity at a later time is given by $R(\mathcal{S}(I_1, I_2, \ldots, I_m; t_2)) = 0$ we know that the system was functioning at t_1 but had failed by t_2.

Now, in general, it is possible that the inputs are uncertain – that is that $I_1, I_2, \ldots, I_m$ may be thought of as realizations of a multivariate random variable. It is likely that it is possible to quantify our uncertainty about the inputs in the form of a probability distribution – for example if I_1 is temperature, it may be possible to say it can be thought of as normally distributed centered at room temperature.

Regardless of the fact that the inputs are random variables, we are still concerned with some derived quantity – for example the probability that the system is operational at 100 h, say. This may be obtained by integrating across the probability distribution for the multivariate inputs, and thus calculating the expectation of $R(\mathcal{S}(I_1, I_2, \ldots, I_m; 100))$. That is;

$$\begin{aligned} P(R(\mathcal{S}(\mathbf{I}; 100))) = 1 \\ &= E(R(\mathcal{S}(\mathbf{I}; 100))) \\ &= \int R(\mathcal{S}(\mathbf{I}; 100)) f(\mathbf{I})\, \mathrm{d}\mathbf{I} \qquad (2) \end{aligned}$$

In principle, such a calculation is possible. In practice, it is analytically intractable. Where the input space is of high dimension, it is difficult to do this using quadrature or other standard techniques [6].

Thus, a procedure exactly analogous to that used with the random dart above is employed. In this case it runs as follows;

- Repeat N times
  - Generate $\mathbf{I} \sim f(\mathbf{u})$.
  - Evolve the system to 100 h.
  - If $R(\mathcal{S}(I_1, I_2, \ldots, I_m; 100)) = 1$ then $s = s + 1$

The reliability of the system at 100 h may then be estimated as $\frac{s}{N}$.

Relationship with Method of Ensembles

A closely related idea is the *method of ensembles.* The notion here is to consider a large number (N) of realizations of the system, where for each of the realizations, a new value of the input parameters is simulated. The resulting collection is then termed an *ensemble* and may be thought of as providing the analyst with realizations of the system in N parallel universes.

This method is commonly used in statistical mechanics, and increasingly so in applications where financial risk is the key consideration.

The analyst then examines these multiple realizations to summarize a key quantity of interest. For example, if we are interested in whether the system has failed by 100 h, we examine the ensemble, and the "best guess" at this probability is the fraction of the ensemble for which the system has failed. The method is exactly the same as the Monte Carlo strategy described above, but the way in which we express it is slightly different.

The advantage of expressing our findings in terms of an ensemble is that it is often easier for those not used to the language of probability, marginalization and expectation to think in this fashion. However, the technical considerations are exactly the same; we must concern ourselves with how many members are in the ensemble, and the associated Monte Carlo error.

Particular Examples of Direct Monte Carlo Methods

A number of specific techniques are worthy of exploration. These are briefly outlined in what follows.

Univariate Rejection Method

One method of obtaining deviates from a distribution of known form is the rejection algorithm (sometimes termed the *accept–reject method*). This is a very flexible method, which permits a wide class of probability distributions to be used in Monte Carlo simulation strategies.

The rejection method is a two stage process. Let the probability distribution from which a deviate is required be denoted $f(\cdot)$.

The first step is that a proposed deviate is generated from a proposal probability distribution. This proposal distribution, let us call it $g(\cdot)$ is chosen such that $Mg(x) > f(x)$ for some scalar M and for all x in the domain of interest. Typically the form of $g(\cdot)$ is simple, in that deviates are cheaply available.

The second step involves preferentially accepting proposed deviates from areas of $f(\cdot)$ that have high probability. Explicitly, this is done by accepting a proposed deviate x_{prop} with probability given by;

$$p_{\text{accept}} = \frac{f(x_{\text{prop}})}{Mg(x_{\text{prop}})} \tag{3}$$

If the proposed deviate is not accepted, then a new proposal is generated and the process is repeated.

Thus the algorithm is as follows;

- Generate $x_{\text{prop}} \sim g(x)$.
- Calculate $p_{\text{accept}} = f(x_{\text{prop}})/Mg(x_{\text{prop}})$.
- Generate $u \sim U[0, 1)$.
- If $u < p_{\text{accept}}$ then accept x_{prop};
- Else repeat the process.

The choice of $g(\cdot)$ is of critical importance in determining the efficiency of the algorithm. If $Mg(\cdot)$ does not tightly bound $f(\cdot)$, then a large number of deviates will be generated for each one that is accepted. Depending on the cost of sampling from $g(\cdot)$ the algorithm may then be impractical. It can easily be shown that the expected number of times the sampler will have to draw from $g(\cdot)$ for each successfully accepted deviate is given by the scalar, M.

A pictorial representation of the setup is shown in Figure 2.

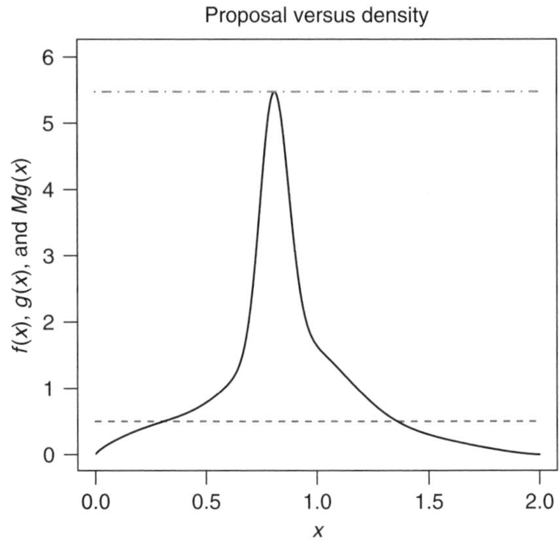

Figure 2 Typical situation where rejection is used

Multivariate Rejection Method

If the density to be simulated from, $f(\mathbf{x}) = f(x_1, x_2, \ldots, x_m)$ which is a multivariate density, then the multivariate version of the algorithm can be used. In order to do this, a multivariate proposal density, $g(\mathbf{x})$ is required, with the same bounding conditions.

In this instance, the scalar, M, typically has to be much bigger, and thus the algorithm is correspondingly less efficient. Thus, in practice, fully multivariate versions of the rejection algorithm are not commonly used.

Importance Sampling

If the cost of generating deviates from the probability distribution of interest is substantial, then it makes sense that we use these as efficiently as possible. One such method, which ensures that we examine the event frequently of interest to us (and consequently reduces the variance of our estimator), is called *importance sampling*.

The idea behind importance sampling is to "adjust" the proposal from which deviates are selected, in order that the Monte Carlo algorithm evaluates the function in the region of interest more frequently. The fact that the proposal density has been adjusted has to subsequently be corrected for; this is done by reweighting the sampled values. An example makes this idea clear.

Consider the use of importance sampling in numerically approximating an integral. The basic principle is as follows:

In wanting to estimate

$$I = \int_{-\infty}^{\infty} g(x) f(x)\,\mathrm{d}x \tag{4}$$

where $f(x)$ is a density function, one could sample n values of x from $f(x)$ and then approximate with

$$\hat{I} = \frac{1}{n}\sum_{i=1}^{n} g(x_i) \tag{5}$$

Alternatively, m values of x could be sampled from another density $h(x)$ and then I could be estimated using

$$\hat{I} = \frac{1}{m}\sum_{i=1}^{m} \frac{g(x_i)f(x_i)}{h(x_i)} \tag{6}$$

We have to be careful about how $h(x)$ is chosen so that the estimator is most efficient. It turns out that the most efficient form for $h(x)$, samples from areas where $g(x)$ is large, provided that $f(x)$ is not small [2].

These results extend easily to the multivariate setting. However, effective choice of $h(x)$ is more difficult in the multidimensional setting. Thus, if we have good intuition about what areas of $g(x)$ are of interest, we can exploit them using this technique. The risk, however, is that if we choose poorly we will obtain a much less efficient **estimation** strategy.

Importance sampling can be particularly useful in reliability modeling where we are concerned with rare or extreme events; for example, we may be interested in extreme loadings to a structure, which have high probability of causing damage. Thus, in order to estimate the overall reliability of the structure, we may wish to ensure we simulate many of these "high-impact events". We then adjust the overall weighting of these according to their true probability, in order to reduce the variability of the estimated survival function in the region where it changes most.

Composition

When a density can be described as a combination of other densities, the method of composition can be used. The simplest example of this is the context in which the density is a mixture distribution.

For example, consider a situation in which we are interested in estimating the probability that a system copes with a random loading event. There may be two possible loading modes, M_1 and M_2; and the damage caused may be described by a random variable for each of the possible modes. Thus, for example, if M_1 occurs, the damage may be $D \sim \text{Gamma}(\alpha_1, \beta_1)$ and if M_2 then $D \sim \text{Gamma}(\alpha_2, \beta_2)$. If we know that the probability of M_1 is p_1 then we can decompose the **pdf** for D into two parts. Specifically;

$$f(D) = p_1 f_1 + p_2 f_2 \tag{7}$$

where f_1 is $\text{Gamma}(\alpha_1, \beta_1)$ and f_2 is $\text{Gamma}(\alpha_2, \beta_2)$.

In order to simulate from such a decomposed distribution, one proceeds using the "broken stick" method as follows;

- Generate $u \sim U[0, 1)$.
- If $u < p_1$ then simulate $D \sim f_1(\cdot)$.
- Else $D \sim f_2(\cdot)$.

Thus, we simulate D from the first component with probability p_1 and from the second with probability p_2. This can be applied to a mixture with any number of components.

It is noted that some distributions can be very well approximated by mixtures; for example, early implementations yielding deviates from a Gaussian used this method, as outlined in [3].

Box–Muller

An example of a transformation method which efficiently generates from the bivariate **normal distribution** is the Box–Muller transform. The method is a simple transformation method, which takes deviates from the uniform unit square and maps them to the bivariate standard normal. It proceeds as follows;

- Generate $u_1 \sim U[0, 1)$.
- Generate $u_2 \sim U[0, 1)$.
- Let $\theta = 2\pi u_1$.
- Let $R = \sqrt{(-2\ln(u_2))}$.
- Return (x, y) by transformation of polar (R, θ).

Then the pair (x, y) are independent standard normal deviates.

The advantage of this method is that it efficiently generates from the required distribution. None of the deviates are "wasted" through rejection. It is easy to implement. Such transformation methods exist for other situations.

Correlated Random Variables

Special concern is needed when the probability distributions of interest to us are correlated. This is critical when considering the tails of distributions. Such issues arise in the context of reliability applications, and applications in **finance**, such as the probability of ruin.

As an example, consider the simple sum of standard normally distributed random variables, $Y = X_1 + X_2$. If X_1 and X_2 are independent of each other, then a Monte Carlo strategy involving Y could proceed with X_1 and X_2 each being simulated without consideration of the other. However, if there is a positive **correlation** between the random variables – that is large values of X_1 are seen more often in combination with large values of X_2 – then extreme values of Y are more likely than we would observe if we simulated them independently. Similarly, if there is a negative correlation between X_1 and X_2, we will see extreme values of the combination less frequently than we might naively expect.

In order to ensure that we correctly account for correlation between the random variables, we could simulate X_1 from the marginal distribution and X_2 from the conditional distribution.

Other Extensions and Considerations

What has been described in this section is the Monte Carlo method in general. It should be clear that much of the work involved in implementing a Monte Carlo strategy involves careful consideration of how to simulate from the density of interest to us. This has not been discussed here as it is the subject of other articles.

In particular, the fact that we use pseudorandom numbers, that is a sequence of deviates generated by a computer which "looks" random, is a fact that has been glossed over. The detail of how such a sequence may be generated, some of the pitfalls, and some of the features of the resulting sequence one may wish to check are all the subject of other articles.

If one wishes to sample from a distribution that is not uniform, then one needs a method of doing this. For example, in considering **system reliability** above, one of the parameters was sampled from a normal distribution. For completeness, it is necessary for us to demonstrate that it is in general possible to do this. It is sufficient to note that, given the cumulative distribution function, this is always possible using the inverse transform sampling method. However, more efficient techniques often exist, as highlighted for particular situations; for example methods for particular parametric distributions are discussed elsewhere [7–10].

Other articles [11, 12], discuss derived techniques such as **Markov chain Monte Carlo**, where the samples that are generated are no longer independent of one another. The extension has substantial advantages in high dimensional problems. However, the consideration of Monte Carlo error is not as straightforward as outlined above. Additionally, care is needed to ensure that the sampler is well behaved.

References

[1] Metropolis, N. & Ulam, S. (1949). The Monte Carlo method, *Journal of the American Statistical Association* **44**(247), 335–341.

[2] Kleijnen, J.P.C. (1974). *Statistical Techniques in Simulation*, M. Dekker, New York.

[3] Ripley, B.D. (1987). *Stochastic Simulation*, John Wiley & Sons, New York.

[4] Fishman, G.S. (1995). *Monte Carlo: Concepts, Algorithms, and Applications*, Springer-Verlag, New York.

[5] Robert, C.P. & Casella, G. (2004). *Monte Carlo Statistical Methods*, 2nd Edition, Springer-Verlag, New York.

[6] Concepcion Ausin, M. (2007). An introduction to quadrature and other numerical integration techniques, *Encyclopedia of Statistics in Quality and Reliability*, John Wiley & Sons.

[7] Rosinski, J. (2007). Simulation of Lévy processes, *Encyclopedia of Statistics in Quality and Reliability*, John Wiley & Sons.

[8] Marseguerra, M. & Zio, E. (2007). Simulation of life distributions, *Encyclopedia of Statistics in Quality and Reliability*, John Wiley & Sons.

[9] De Blasi, P. (2007). Simulation of the beta-Stacy process with application to analysis of censored data, *Encyclopedia of Statistics in Quality and Reliability*, John Wiley & Sons.

[10] Laud, P.W. & Pajewski N.M. (2007). Simulation of the dirichlet process, *Encyclopedia of Statistics in Quality and Reliability*, John Wiley & Sons.

[11] Wiper, M.P. (2007). Introduction to Markov chain Monte Carlo simulation, *Encyclopedia of Statistics in Quality and Reliability*, John Wiley & Sons.

[12] Mira A. (2007). Strategies to improve convergence and mixing in Markov chain Monte Carlo, *Encyclopedia of Statistics in Quality and Reliability*, John Wiley & Sons.

CATHAL D. WALSH

Monte Carlo, Markov Chain *see* Markov Chain Monte Carlo, Introduction

Moving Averages

Introduction

A moving average is a time-dependent weighted average, which is repeatedly recalculated as further observations in a **time series** become available. The name derives from the fact that the average "moves" through the time series. As more observations are collected, they are added to the calculation of the average, and at the same time observations from the most distant past are deleted (or given smaller weights). There are two basic reasons to calculate a **moving average**:

- **Prediction**
  A moving average performs *smoothing* of the various components that comprise the time series (like trend, seasonality, or randomness), so that predicting immediate-future values of these components, based on the appropriate moving average, is less affected by fluctuations that had occurred in these components in the past.
- **Adaptive control**
  To provide predicted values for a monitored process variable that will reflect adaptation, in a controlled manner, of the forecasting method to changes that have recently occurred in the process. The predicted values, based on the moving average, may then serve as baseline to detect out-of-control states for the process (*see* **Exponentially Weighted Moving Average (EWMA)**).

Unlike time-series modeling, a moving average does not attempt to model the data. Rather it attempts to smooth the data so that any *long-term* signal embedded in the observations becomes more visible.

The Standard Moving Average

This average, calculated to predict the observation at time t, and based on the most recent n observations (collected from time $t-n$ until $t-1$), is the weighted average

$$\overline{X}_t = w_1 X_{t-1} + w_2 X_{t-2} + \cdots + w_n X_{t-n} \qquad (1)$$

where the weights sum to 1: $\sum_{i=1}^{n} w_i = 1$. It is customary to assign smaller weights to more ancient data.

The Exponentially Weighted Moving Average

Alternatively, one could use the **exponentially weighted moving average**:

$$\overline{X}_t = \lambda X_t + (1-\lambda)\overline{X}_{t-1} = \overline{X}_{t-1} + \lambda(X_t - \overline{X}_{t-1}), \quad 0 \le \lambda \le 1 \qquad (2)$$

where $\overline{X}_0 = X_0$ (*see* for further details **Exponentially Weighted Moving Average (EWMA)**). A larger λ would provide better (quicker) adaptation to most recent fluctuations, and thus less smoothing of these fluctuations. It is easily shown that the average in equation (2) is in fact an exponentially-weighted average of all previous observations, where weights become smaller the older an observation is.

HAIM SHORE

Moving Range and *R* Charts

Introduction

The quality of a manufactured item is often measured by one or more quality measurements. This measurement(s) varies from item to item following a (joint) distribution. The parameter or parameters of the (joint) distribution of the quality measurement(s) are measures of the quality of the process. As an example consider a production process that fills bottles with a soft drink that are advertised to contain 16 ounces. The fill X is a measure of the quality of a container. The mean fill μ and the standard deviation σ of the distribution of fills X are typical quality measures of the process. Loss of customers can be associated with μ being too low and unnecessary cost of production when μ is too high. Increases in σ are associated with increases in the proportion of unacceptable over and/or under filled bottles and decreases with quality improvement. It is therefore desirable to control both μ and σ.

The meaning of the process being "in control" is most commonly based on what Shewhart [1] referred to as *natural causes of variability*. We assume that these causes of variability are inherent and can only be removed by redesigning the process. Here we use the word *in control* to mean the state of the process when there are no assignable causes of variability present. When our filling process is in control, we will assume that μ and σ have the values μ_0 and σ_0, respectively. If assignable causes of variability are present, then $\mu \neq \mu_0$ or $\sigma \neq \sigma_0$ and the process is said to be "out of control."

Initially, the practitioner must bring the process into a state of statistical control. The Phase I **control chart** is often recommended as an aid to the practitioner for this purpose. The Shewhart Phase I X and $\overline{X}$ charts are the most commonly recommended charts to aid the practitioner in bringing the mean μ into a state of statistical control (see Ryan [2] for a discussion of control charts for the mean). The Phase I Shewhart moving range (MR), the sample range (R), and the sample standard deviation (S) charts are recommended for σ. These charts are designed to look at the data collected from the process in retrospect to answer the question "were these data collected from a process that is in control?" When the process is believed to be in control, it is desirable to monitor the process for a possible change from in control to out of control. The question the practitioner wishes to answer is "has the process changed?" The charts used in the monitoring phase are often referred to as Phase II charts. The Phase II Shewhart MR, R, and S charts are most often used to monitor σ. The R and S (with runs rules) (see Burroughs *et al.* [3], Mitra [4]), the run sum R and S (see Champ and Rigdon [5]), and cumulative sum (**CUSUM**) R and S (see Page [6], Gan [7], Acosta-Mejia [8], and Amin and Wilde [9]), have also been developed to monitor for a change in σ. For a general overview of control charts, see Faltin [10].

Typically, it is recommended that the process standard deviation σ also be controlled when controlling the process mean μ. This is due to the fact that under the normal model the distribution of the plotted statistic for the Shewhart X and $\overline{X}$, CUSUM, and **exponentially weighted moving average** (EWMA) charts depend on σ. Thus, if the chart for monitoring μ indicates a possible change in the process this could be due to a change in σ. It is then argued that a chart for controlling for a change σ be examined first. If there is no indication that σ has changed, then any change in the process indicated by the chart for controlling the mean is interpreted as being due to a change in μ. It can be shown that the MR chart is affected not only by changes in σ but also by changes in μ whereas the R and S charts are only affected by changes in σ.

In this article, we will examine the design and use Phase I and II MR and R charts. The charts are designed under the assumption that the quality measurements are independent each having a **normal distribution**. The relative performance of these charts can be examined under the assumption that both μ_0 and σ_0 are known, one is to be estimated, or both are to be estimated.

Moving Range and *R* Charts

The Shewhart MR and R charts are designed under the assumption that the observed values of X are independent and each has a normal distribution. Also, it is assumed that periodically a random sample of size $n \geq 1$ is taken from the output of the production process. For convenience of discussion, we let $X_{t,j}$ represent the X measurement on the jth item in sample t. The MR chart is often recommend for the case in which $n = 1$. The common scenario in Phase I is that the practitioner will have available X measurements on the items in m independent random samples each of size n that have been taken periodically from the output of the production process. In our example, the practitioner will have available the measurements $\{X_{t,1}, \ldots, X_{t,n}\}$ for $t = 1, \ldots, m$.

When $n = 1$, we define the sequence of moving ranges $MR_2, \ldots, MR_m$ by

$$MR_t = |X_{t-1,1} - X_{t,1}| \qquad (1)$$

It can be shown under our model assumptions that if the process is in control the mean and standard deviation of the distribution MR_t are given, respectively, by $\left(2/\sqrt{\pi}\right)\sigma_0$ and $\sqrt{2(\pi-2)/\pi}\,\sigma_0$. If σ_0 were known, the Shewhart 3σ limits upper control limits (UCL) and lower control limits (LCL) for the Phase I Shewhart *MR* chart are

$$LCL = 0 \quad \text{and}$$
$$UCL = \frac{2+3\sqrt{2(\pi-2)}}{\sqrt{\pi}}\sigma_0 \approx 3.686\sigma_0 \qquad (2)$$

since $\left(2-3\sqrt{2(\pi-2)}\right)/\sqrt{\pi} \approx -1.429 < 0$. Generally, it is not the case that σ_0 is known. The value of σ_0 is then estimated by the average $\overline{MR}$ of the $m-1$ MRs times an unbiasing constant $\sqrt{\pi}/2$, that is, $\overline{MR} = (m-1)^{-1}\sum_{t=2}^{m} MR_t$. The Shewhart 3σ control limits are then given by

$$LCL = 0 \quad \text{and}$$
$$UCL = \frac{2+3\sqrt{2(\pi-2)}}{\sqrt{\pi}}\left(\frac{\sqrt{\pi}}{2}\overline{MR}\right) \doteq 3.2665\overline{MR} \qquad (3)$$

The chart "signals" a potential out-of-control process if $MR_t >$ UCL. If a positive LCL is desired or if it is desired to have an overall probability α of at least one of the MR_t's indicating a potential out-of-control process when the process is in control, the control limits could be adjusted by selecting the control limits as

$$LCL = \left(1 - k_{\mathrm{L}}\sqrt{\pi-1}\right)\overline{MR} = h_{\mathrm{L}}\overline{MR} \quad \text{and}$$
$$UCL = \left(1 + k_{\mathrm{U}}\sqrt{\pi-1}\right)\overline{MR} = h_{\mathrm{U}}\overline{MR} \qquad (4)$$

where k_{L} (h_{L}) and k_{U} (h_{U}) are appropriately selected. It is not difficult to show that as the number of samples m increases the probability α the chart signals a potential out-of-control process for fixed chart parameters k_{L} (h_{L}) and k_{U} (h_{U}) also increases. Some criteria for selecting these chart parameters are such that (a) it is just as likely to observe a value for the plotted statistic less than the LCL as greater than the UCL when the process is in control and (b) it is least likely to observe at least one of the plotted statistics outside the control limits when the process is in control. Using one or both of these criteria for selecting the control limits of the *MR* chart has not been studied in the literature.

The control limits for the Phase II *MR* chart are defined in a similar way to those for the Phase I *MR* chart. The difference is the method for choosing the chart parameters $k_{MR,L}$ ($h_{MR,L}$) and $k_{MR,U}$ ($h_{MR,U}$) and the value of $\overline{MR}$ is based on m independent samples each of size $n = 1$ from a process that is believed to be in control. Typically, the chart parameters of the Phase II *MR* chart are chosen such that the **average run length** (ARL) of the chart is relatively large when the process is in control and as small as possible when the process is out of control. The run length is the number of samples taken in Phase II until the plotted statistic plots either below LCL or above UCL.

For the case of $n > 1$, often the Phase I Shewhart *R* chart is recommended as an aid to the practitioner in bringing the process into a state of statistical control. The sample range *R* is the difference between the maximum and minimum values of the sample. For a discussion of the sample range *R* see Montgomery [11] and Alwan [12]. Under the normal model, the Phase I Shewhart *R* has LCL and UCL are defined by

$$LCL = \left(1 - k_{R,L}\frac{d_3}{d_2}\right)\overline{R} \quad \text{and}$$
$$UCL = \left(1 + k_{R,U}\frac{d_3}{d_2}\right)\overline{R} \qquad (5)$$

where $\mu_R = d_2\sigma_0$ and $\sigma_R = d_3\sigma_0$. Values of d_2 and d_3 are functions of the sample size n and must be determined numerically. Table 1 gives values for d_2 and d_3 for $n = 2, \ldots, 12$. Harter [13], Montgomery [11], and Alwan [12] provide tables of these values.

Table 1 Constants d_2 and d_3^2

n	d_2	d_3^2
2	1.1283791671	0.7267604553
3	1.6925687506	0.7891977107
4	2.0587507460	0.7740624738
5	2.3259289473	0.7466376009
6	2.5344127212	0.7191713092
7	2.7043567512	0.6942311313
8	2.8472006121	0.6721236717
9	2.9700263244	0.6525962151
10	3.0775054617	0.6352897762
11	3.1728727038	0.6198643117
12	3.2584552798	0.6060285277

Often it is recommended that $k_{R,L}$ and $k_{R,L}$ be set to 3 as suggested by Shewhart [1]. As with the MR chart, the chart parameters $k_{R,L}$ and $k_{R,U}$ are functions of the number of samples m for a given value of α. Similar to the Phase II Shewhart MR chart, the control limits for the Phase II Shewhart R charts are defined using the same formulas as in Phase I. The differences are that the value of $\overline{R}$ is based on m independent random samples each of size n that are believed to have been taken from a process that is in control. Further, the chart parameters for these charts are usually selected so that the run length distribution when the process is in control has certain properties such as a given in control ARL.

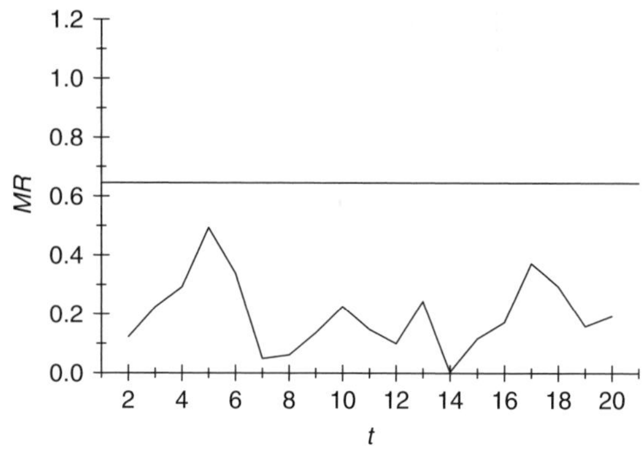

Figure 1 Phase I MR chart with $m = 20$

Example

Periodically, from our bottle filling process, we have selected $m = 20$ independent samples each of size $n = 5$. The X fills of these bottles along with the MR for the first sampled value and the range of each sample are recorded in the Table 2.

To illustrate the MR chart, we will consider that we have available only the measurement $x_{t,1}$ taken on the first item in each sample. For these data, the MR chart has $LCL = 0$ and $UCL = 3.2670 \times 0.1974 = 0.6449$ using equation (3). The Phase I MR is displayed in Figure 1.

The control limits for the Phase I R chart are $LCL = 0$ and $= 2.1145 \times 0.1159 = 0.2451$ using equation (5). There is no indication by the Phase I MR chart that these data are from a process that is out of control.

The Phase I R chart is displayed in Figure 2. We see from Figure 2 that the range of sample taken at sampling stage 5 signals a potential out-of-control

Table 2 Phase I Data

t	$x_{t,1}$	$x_{t,2}$	$x_{t,3}$	$x_{t,4}$	$x_{t,5}$	$MR_{t,1}$	R_t
1	32.0175	31.9947	31.9916	31.9927	32.0093	–	0.0258
2	32.1399	32.0075	31.9786	31.9925	32.0108	0.1224	0.1612
3	31.9159	31.9926	32.0039	31.9993	31.9898	0.2240	0.0880
4	32.2082	31.9892	32.0071	31.9961	31.9990	0.2923	0.2190
5	31.7144	31.9853	31.9933	31.9894	31.9932	0.4938	0.2788
6	32.0526	31.9971	31.9959	32.0160	31.9789	0.3381	0.0737
7	32.0037	31.9977	31.9943	32.0123	32.0011	0.0488	0.0180
8	31.9426	31.9937	31.9842	31.9874	31.9947	0.0611	0.0521
9	32.0792	32.0035	32.0056	32.0029	32.0019	0.1366	0.0774
10	31.8532	31.9883	32.0092	31.9857	31.9892	0.2261	0.1561
11	32.0017	31.9988	31.9972	31.9969	31.9901	0.1485	0.0116
12	31.9017	32.0129	31.9970	31.9925	32.0040	0.1000	0.1112
13	32.1456	31.9828	31.9991	31.9779	31.9872	0.2439	0.1677
14	32.1496	32.0012	31.9983	31.9890	32.0035	0.0039	0.1605
15	32.0326	32.0084	32.0037	32.0028	31.9980	0.1169	0.0346
16	31.8605	32.0242	31.9850	31.9952	31.9961	0.1722	0.1637
17	32.2333	32.0059	31.9956	32.0065	32.0038	0.3728	0.2376
18	31.9382	32.0078	31.9914	31.9930	31.9951	0.2950	0.0696
19	32.0970	32.0186	32.0006	32.0050	32.0087	0.1587	0.0964
20	31.9029	32.0065	32.0165	31.9959	32.0182	0.1940	0.1153

$\overline{x} = 32.0006$, $\overline{MR} = 0.1974$, $\overline{R} = 0.1159$, and $\overline{S} = 0.0485$.

Table 3 Phase II Data

t	$x_{t,1}$	$x_{t,2}$	$x_{t,3}$	$x_{t,4}$	$x_{t,5}$	$\overline{x_t}$	$MR_{t,1}$	R_t
1	31.7649	31.9616	31.9763	31.9691	31.9981	31.9340	–	0.2332
2	32.2237	32.0310	32.0029	32.0056	32.0406	32.0608	0.4588	0.2208
3	31.9838	32.0089	32.0017	31.9927	31.9965	31.9967	0.2399	0.0251
4	31.8351	31.9757	32.0009	31.9754	31.9646	31.9503	0.1487	0.1658
5	31.9435	32.0098	31.9967	32.0081	32.0057	31.9928	0.1084	0.0663
6	32.3005	31.9935	31.9983	31.9894	32.0040	32.0572	0.3570	0.3111

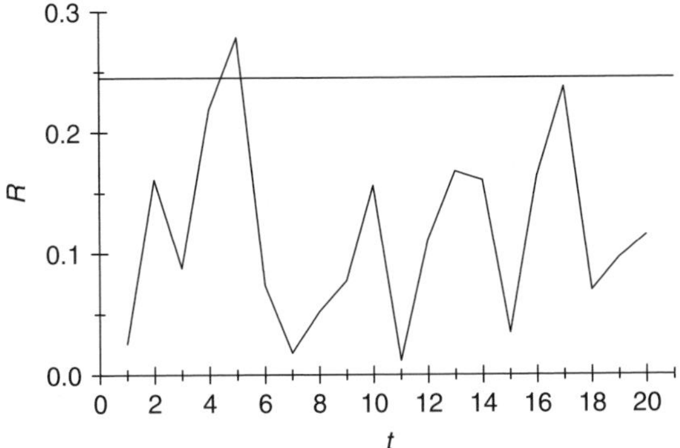

Figure 2 Phase I *R* chart with $m = 20$, $n = 5$

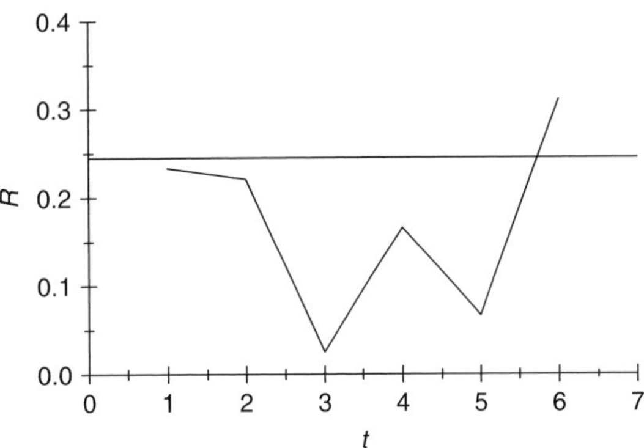

Figure 4 Phase II *R* chart

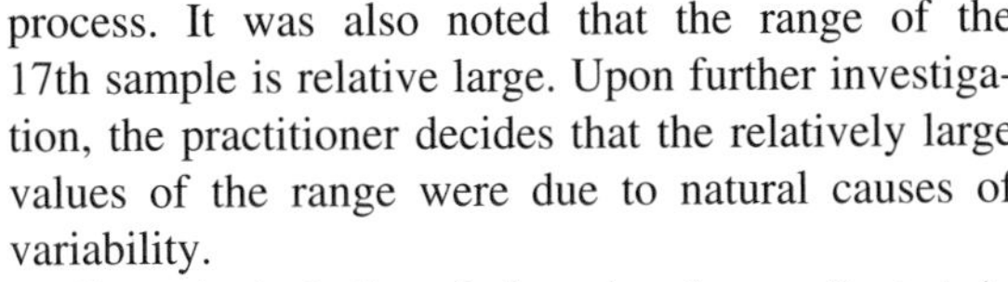
process. It was also noted that the range of the 17th sample is relative large. Upon further investigation, the practitioner decides that the relatively large values of the range were due to natural causes of variability.

Since it is believed that the data collected in Phase I is in control, then the same 3-σ limits are used for the Phase II chart. Table 3 contains the data collected in Phase II for the first six samples.

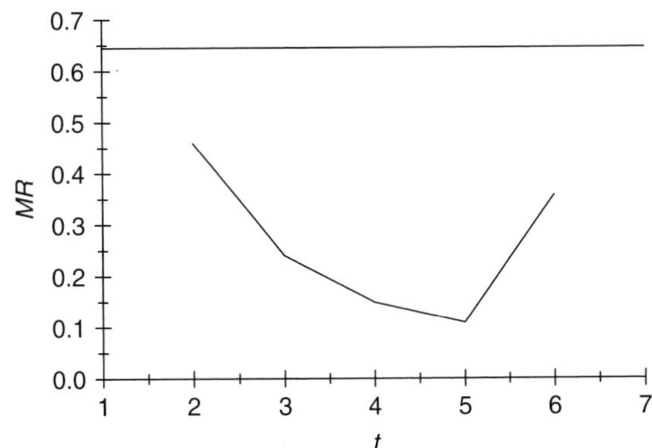

Figure 3 Phase II *MR* chart

The Phase II *MR* and *R* charts are given respectively in Figures 3 and 4.

The Phase II *MR* chart provides no evidence that the process has changed. The sample range R_6 of the sixth sample in Phase II signals a potential out-of-control process suggesting to the practitioner that the process should be examined to see it had changed.

Conclusion

In this article, we have discussed the *MR* and *R* charts. These charts are simple to use and interpret. The *MR* chart unlike the *R* chart is affected by changes in both the mean and the standard deviation of the distribution of the quality measurement. Also, it has been shown by Rigdon *et al.* [14] that the Phase II *X* chart performs about as well as the combined Phase II *X* and *MR* chart. We do not recommend using the *MR* chart.

References

[1] Shewhart, W.A. (1931). *Economic Control of Quality of Manufactured Product*, D. Van Nostrand, New York.

[2] Ryan, T. (2007). Control charts for the mean, *Encyclopedia of Statistics in Quality and Reliability*, John Wiley & Sons, New York.

[3] Burroughs, T.E., Rigdon, S.E. & Champ, C.W. (1995). Analysis of the Shewhart *S*-chart with runs rules when no standards are given, in *Proceedings of the Twenty-Sixth Annual Meeting of the Midwest Decision Sciences Institute*, St. Louis, pp. 268–270, 4–6 May 1995.

[4] Mitra, A. (2007). Standard deviation charts, *Encyclopedia of Statistics in Quality and Reliability*, John Wiley & Sons, New York.

[5] Champ, C.W. & Rigdon, S.E. (1998). Monitoring variability using the run sum chart, in *Proceedings of the Quality and Productivity Section of the American Statistical Association*, Dallas, pp. 24–25, 9–13 August 1998.

[6] Page, E.S. (1963). Controlling the standard deviation by CUSUMs and warning lines, *Technometrics* **5**, 307–315.

[7] Gan, F.F. (2007). CUSUM charts, *Encyclopedia of Statistics in Quality and Reliability*, John Wiley & Sons, New York.

[8] Acosta-Mejia, C.A. (1998). Monitoring reduction in variability with the range, *IIE Transactions* **30**, 515–523.

[9] Amin, R.W. and Wilde, M. (2000). Two-Sided CUSUM Control Charts for Variability, *Allgemeines Statistisches Archiv* **84**, 295–313.

[10] Faltin, F.W. (2007). Control charts, *Encyclopedia of Statistics in Quality and Reliability*, John Wiley & Sons, New York.

[11] Montgomery, D.M. (2005). *Introduction to Statistical Quality Control*, 5th Edition, John Wiley & Sons, New York.

[12] Alwan, L.C. (2001). *Statistical Process Analysis*, Irwin McGraw-Hill, Boston.

[13] Harter, H.L. (1960). Tables of range and studentized range, *Annals of Mathematical Statistics* **31**, 1122–1147.

[14] Rigdon, S.E., Cruthis, E.N. & Champ, C.W. (1994). Design strategies for individual and moving range control charts, *Journal of Quality Technology* **26**, 274–287.

Further Reading

Amin, R.W. & Matthias, W. (2000). Two-sided CUSUM control charts for variability, *Allgemeines Statistisches Archive* **84**, 295–313.

Vander Wiel, S. (2007). Exponentially weighted moving average control charts, *Encyclopedia of Statistics in Quality and Reliability*, John Wiley & Sons, New York.

CHARLES W. CHAMP AND DEBORAH K. SHEPHERD

MSA, Overview *see* Measurement Systems Analysis, Overview

MTBF *see* Mean Time Between Failures

MTTF *see* Mean Time to Failure

MTTR *see* Mean Time to Repair

Multiattribute Warranty Policies

Introduction

Nearly everything sold in the market is covered by a warranty – expressed or implied. Various types of warranties serve a number of purposes, including protection for the buyer against low quality or service and limitation of product liability for the seller. In some instances, warranties are mandated and regulated by state and local governments. Because of this diversity of purpose, warranties have been studied by researchers in many different disciplines, such as engineering, management, marketing, economics, law, and accounting. For more details of warranty applications, *see* **Warranty Servicing**.

A warranty is a seller's guarantee given to the buyer stating that a product is reliable and free from known defects and that the seller will repair or replace the product if it fails to conform to satisfactory performance. Consequently, this obligation generates costs to the producer associated with failures of the product. Thus, one of the decision problems faced by producers in warranty analysis is how to compare various types of warranty plans and estimate the total cost associated with each plan.

A formal development of warranty models can be traced back to a paper by Arrow [1]. Since then, a wide variety of warranty policies has been proposed and various statistical models for determining warranty costs have been developed. Now one can say that warranty analysis constitutes a "field" of study within engineering–statistics–marketing. As further evidenced by a series of review papers by Blischke and Murthy [2] and Murthy and Blischke [3, 4], and a handbook edited by Blischke and Murthy [5], this particular field of study has experienced a rapid growth and extensive applications to various managerial decision problems.

In this note, we first classify various types of warranty plans proposed in the literature, and then review several methods of modeling two-attribute warranty policies. For further development of warranty models and analysis methods, readers are referred to Thomas and Rao [6] and Murthy and Djamaludin [7]. For articles in this encyclopedia that are related to the analysis of warranty policies see the section titled "Related Articles".

Warranty Policies

A warranty can be thought of as a contractual *obligation* or a potential *liability* to the seller. However, the limit of the seller's obligation or liability is usually specified at the time of product sale and the seller is responsible only for the product failures satisfying the prespecified warranty criteria. On the basis of the number of warranty attributes used in determining the warranty eligibility, warranty plans can be classified as (a) *single-attribute* warranty policies and (b) *multiattribute* warranty policies.

In *single-attribute* warranty policies, we use only one of the warranty characteristics, such as the time, the hours of operation, the mileage, or the frequency of product usage. The most popular criterion is the *warranty period* that is measured from the time of sale of the product to a certain point in the product's lifetime. The popularity of the warranty period is twofold. In most instances, the total number of product failures is a function of time. Besides, time is easily measurable and thus tends to minimize potential conflicts between a seller and a buyer on the warranty eligibility of a failed product.

In *multiattribute* warranty policies, on the other hand, more than one warranty characteristics are employed simultaneously as criteria in judging the warranty eligibility of a failed product. Under the 5-year, 50 000-mi protection plan used in the automobile industry, for instance, the warranty eligibility of an automobile is determined by its age and mileage at the time of breakdown. Since its introduction in the late 1960s, the two-attribute warranties have become a critically important segment of the automobile business environment.

It is reported that aircraft manufacturers also offer this type of two-attribute warranty policy, with usage corresponding to the number of flying hours. In the next section, we review various types of single-attribute warranty policies.

Single-Attribute Warranty Policies

Under the typical situation of a single-attribute warranty transaction, a seller provides some kind of repair or replacement service for any product failures occurring during some specified time. For the warranty service, a buyer pays either no cost or a predetermined portion of the full repair or replacement cost. Therefore, two of the major decision variables in a single-attribute warranty policy are the *warranty price* and the *warranty period*: (a) *how much* a buyer should pay to receive the warranty service and (b) *how long* a seller should provide the warranty service.

On the basis of the nature of a buyer's payment of repair or replacement cost upon a product failure, warranty policies can be divided into (a) the free-replacement warranty, (b) *pro rata* warranty, and (c) lump-sum warranty. First, under the *free-replacement* warranty policy shown in Figure 1(a), a producer provides as many replacements or repairs as needed free of charge to a consumer as long as the product failures satisfy the warranty criteria. In many cases, the anticipated cost of honoring the free warranty service has already been included in the selling price of the product or put into reserve to meet future **warranty claims**.

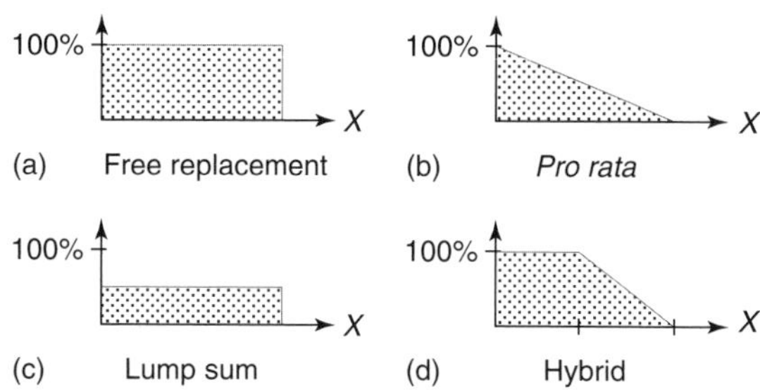

Figure 1 Various types of single-attribute warranty policies

Second, under the *pro rata* warranty (or rebate) policy, a consumer must pay a fraction of the full repair cost. As shown in Figure 1(b), the fraction is usually determined by the percentage of the warranty period in use by the old item prior to the failure. An automobile tire typically carries such a *pro rata* warranty, in which a buyer pays a small portion of the full replacement cost on the basis of the remaining tread depth or the mileage driven.

Third, under the *lump-sum* warranty policy shown in Figure 1(c), a customer receives a fixed or lump-sum rebate for any product failures occurring within the warranty period. Particularly when it is difficult to measure the exact time of product failure, this policy is employed to minimize potential conflicts between a producer and a customer about the amount of compensation due for product failure.

We can also imagine a hybrid form of the aforementioned warranty policies. As shown in Figure 1(d) for example, a total warranty period can be divided into two periods: an initial free-replacement warranty period, followed by a *pro rata* period.

On the basis of how long a producer should provide the warranty service, warranties can be classified as (a) fixed-period and (b) renewable policies. First, under the *fixed-period* warranty policy, the warranty period is specified at the time of product sale and fixed regardless of a producer's repairs or replacements. The warranty terms expire automatically at the end of the warranty period. This type of warranty policy is the most popular in practice.

Second, in the *renewable* warranty policy, the original warranty period is extended whenever a repair or replacement occurs during the current warranty period. When the current warranty period expires without any product failures, the warranty terms expire completely. Renewal theory, described in **Renewal Theory**, can be applied to describe this failure process if we assume that successive failure times form a renewal reward process.

We can imagine different types of multiattribute warranty policies but, in the next section, we restrict our attention to the two-attribute warranty policies, which is the most common in practice. For more technical presentations of warranty models, readers are referred to **Warranty Modeling**.

Multiattribute Warranty Policies

As shown in Figure 2, various types of two-attribute warranty policies have been considered in literature: (a) rectangular, (b) triangular, (c) L-shaped, and (d) parabolic warranty policies. The 5-year, 50 000-mi automobile warranty can be displayed in a *rectangular* form as in Figure 2(a); any breakdowns of an automobile will be eligible for the warranty service as long as its age is less than x^* years and its mileage is less than y^* mi. In terms of set notation, the region covered by the warranty is denoted by $\Omega = \{(X, Y) \text{ for } X < x^* \text{ and } Y < y^*\}$.

In the *triangular* warranty policy, on the other hand, the warranty region is expressed as $\Omega = \{(X, Y) \text{ for } aX + bY < c\}$, where a, b, and c are constants specified by a seller at the time of product sale. We can also imagine the *L-shaped* warranty

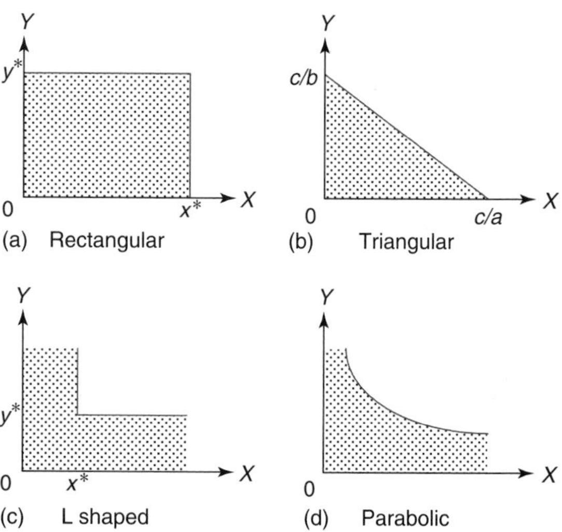

Figure 2 Various types of two-attribute warranty policies

policy as shown in Figure 2(c), in which the product is eligible for the warranty service if X is less than x^* or Y is less than y^*; i.e. $\Omega = \{(X, Y)$ for $X < x^*$ or $Y < y^*\}$.

Since the product failure is a function of X and Y, the ISO **warranty cost** curve can be drawn on the plane for a given cost as shown in Figure 2(d). Moskowitz and Chun [8] proposed such a *parabolic* warranty policy, in which the producer offers a set of warranty plans for the same price and each customer can choose any plan in the set. For example, the producer may offer $\Omega = \{(2$ years, 100 000 mi), (5 years, 50 000 mi), (8 years, 30 000 mi)$\}$. In such a case, a retired grandmother would prefer the 8-year, 30 000-mi warranty policy, whereas the 2-year, 100 000-mi plan is appealing to a traveling salesman. While any other type of two-attribute warranty policy tends to favor one group of consumers over another, the parabolic warranty policy is shown to be more equitable to all consumers.

Chun and Tang [9] analyzed the relationships among the aforementioned two-attribute warranty policies. They also showed that the single-attribute warranty policies are special cases of two-attribute policies.

Modeling Two-Attribute Warranties

Offering warranty coverage leads to additional costs to the producer, associated with claims under warranty. Thus, a producer needs to estimate the expected total cost of offering a two-attribute warranty policy. Various approaches have been proposed in modeling two-attribute warranty policies: (a) two-dimensional, (b) one-dimensional, and (c) Markovian approaches.

First, in two-attribute warranty policies, product failures are dependent on the age X and the usage Y of a product. Thus, the sequence of product failures can be modeled as a counting process on the two-dimensional plane of time and mileage. In the counting process, the time and the usage between successive failures can be described as a bivariate **probability density function**. This method of modeling is called a *two-dimensional approach* to a two-attribute warranty plan. As a bivariate probability density function, the bivariate exponential distribution is a natural choice if we assume that the times and the usages between successive failures are independent and identically distributed random variables. The beta-Stacy distribution and the multivariate Pareto distribution of the second kind have been also considered by other researchers [10].

Second, in many practical situations, the failure process on the two-dimensional plane can be represented by a univariate renewal process. This formulation method is referred to as the *one-dimensional approach*, which takes advantage of the linear relationship between X and Y. For most drivers, for example, the average mileage driven per year is constant; that is, the product usage Y is a linear function of the time X.

Note that not all products of a kind have the same usage rate; a high-intensity user like a traveling salesman for example, has a high usage rate, while a low-intensity user like a retired grandmother presumably has a low rate of use. Thus, the usage rate is assumed to be a nonnegative random variable of the continuous type. The distribution of the usage rate could be modeled satisfactorily by a gamma distribution as in Moskowitz and Chun [8]; by changing its parameter values, we can represent a wide variety of distributions with different location, dispersion, shape, and the like. Other types of distributions, such as the uniform distribution, the β-weighted two-point distribution, and the exponential distribution, have also been considered in literature.

Third, in most practical situations, the age x^* and the usage y^* in two-attribute warranty policies take discrete values; that is, it is unimaginable to offer a 5.2-year, 71 534-mi protection plan, even though those are the optimal values for a given warranty cost. Thus, two-attribute warranty policies can be modeled *via* a Markovian approach [11], in which the usage rate per year is discretized as a *state* and the time is also discretized as a *transition*. Figure 3 for example, shows a transition probability matrix for a 5-year, 50 000-mi warranty plan.

$$P = \begin{array}{c} \\ 0\text{–}10 \\ 10\text{–}20 \\ 20\text{–}30 \\ 30\text{–}40 \\ 40\text{–}50 \\ \infty \end{array} \begin{array}{cccccc} 0\text{–}10 & 10\text{–}20 & 20\text{–}30 & 30\text{–}40 & 40\text{–}50 & \infty \\ \end{array}$$

	0–10	10–20	20–30	30–40	40–50	∞
0–10	0.1	0.2	0.3	0.3	0.1	0
10–20	0	0.1	0.4	0.3	0.1	0.1
P = 20–30	0	0	0.2	0.4	0.2	0.2
30–40	0	0	0	0.2	0.4	0.4
40–50	0	0	0	0	0.3	0.7
∞	0	0	0	0	0	1

Figure 3 An example of the transition probability matrix in the Markovian approach

In the transition probability matrix in Figure 3, the state i is the mileage of an automobile, while the state ∞ represents the automobiles with more than 50 000 mi. The transition probability p_{ij} is the probability that the state of an automobile with the mileage in the range i will be changed to the state j over a period of 1 year. The state ∞ is an *absorbing* state, while other states are *transient*. Thus, two-attribute warranty policies can be easily modeled as an absorbing Markov chain, and we can find many interesting facts about the absorbing chain such as the expected number of times that each state will be entered before we reach an absorbing state. We can also consider the discounting factor per transition and the repair cost for each state.

Concluding Remarks

Warranties have been important elements in business planning and decision making for manufacturers and producers of services. In this note, we have investigated various two-attribute warranty policies and reviewed three different modeling approaches, but no research works have been reported in the open literature that compare the effectiveness of those approaches. There may be some cases where the more mathematically onerous two-dimensional method should be used. There may be other cases in which a simpler method, such as the one-dimensional and Markovian approaches, is sufficient enough to model the aggregate behavior of entire products. But our past experience in decision analysis and on other management science problems suggests that the differences among them will be nominal. An empirical study based on automobile warranty data could answer some questions as to which approaches are more effective in modeling the two-dimensional failure process and in estimating the parameters of warranty models.

Since its appearance in literature in the late 1980s, the two-attribute warranty policy has been extended and generalized in many different directions by releasing some of the basic assumptions in original models and proposing other new policies. But few research models have been successfully applied to the automobile industry. The new research direction in the area of two-attribute warranty policy should be not just to propose another decision model that is based on an artificial fact, but to collect empirical data from the industry to address real problems that are empirically grounded – not only the source that came from the field, but also the solution is of interest to the practitioners.

References

[1] Arrow, K. (1963). Uncertainty and the welfare economics of medical care, *The American Economic Review* **53**, 941–973.

[2] Blischke, W.R. & Murthy, D.N.P. (1992). Product warranty management – I: a taxonomy for warranty policies. *European Journal Operational Research* **62**, 127–148.

[3] Murthy, D.N.P. & Blischke, W.R. (1992). Product warranty management – II: an integrated framework for study, *European Journal of Operational Research* **62**, 261–281.

[4] Murthy, D.N.P. & Blischke, W.R. (1992). Product warranty management – III: a review of mathematical models, *European Journal of Operational Research* **63**, 1–34.

[5] Blischke, W.R. & Murthy, D.N.P. (1995). *Product Warranty Handbook*, Marcel Dekker, New York.

[6] Thomas, M.U. & Rao, S.S. (1999). Warranty economic decision models: a summary and some suggested directions for future research, *Operations Research* **47**, 807–820.

[7] Murthy, D.N.P. & Djamaludin, I. (2002). New product warranty: a literature review, *International Journal of Production Economics* **79**, 231–260.

[8] Moskowitz, H. & Chun, Y.H. (1994). A Poisson regression model for two-attribute warranty policies, *Naval Research Logistics* **41**, 355–376.

[9] Chun, Y.H. & Tang, K. (1999). Cost analysis of two-attribute warranty policies based on the product usage rate. *IEEE Transactions on Engineering Management* **46**, 201–209.

[10] Murthy, D.N.P., Iskandar, B.P. & Wilson, R.J. (1995). Two-dimensional failure-free warranty policies: two-dimensional point process models, *Operations Research* **43**, 356–366.

[11] Balachandran, K.R., Maschmeyer, R.A. & Livingstone, J.L. (1981). Product warranty period: a Markovian approach to estimation and analysis of repair and replacement costs, *The Accounting Review* **56**, 115–124.

Related Articles

Renewal Theory; **Repairable Systems Reliability**; **Warranty Claims and Costs: Statistical Analysis**

of; **Warranty Cost Analysis**; **Warranty Cost Prediction Based on Warranty Data**; **Warranty Modeling**; **Warranty Servicing**; **Warranty: Usage and Wear Process for**.

YOUNG H. CHUN

Multicomponent Maintenance

Introduction

Multicomponent maintenance is the problem of finding optimal **maintenance policies** for a system consisting of several units of machines or many pieces of equipment, which may or may not depend on each other [1]. Owing to the existence of interactions between components in complex systems, applying single-component maintenance policies is often not optimal. Generally, three types of interactions or dependencies can be distinguished: economic dependence, stochastic dependence, and structural dependence. Economic dependence implies that grouping maintenance actions either saves costs (e.g. because to economies of scale) or yields higher costs (e.g. because of machine downtime) compared to multiple individual maintenance actions. Stochastic dependence occurs if the condition of components influences the lifetime distribution of other components. Structural dependence applies if components structurally form a part, so that the maintenance of a failed component implies maintenance of working components. Before we discuss these dependencies in more detail, we will first give a historical perspective on multicomponent maintenance.

Historical Perspective

Multicomponent maintenance is one of the key problems in maintenance management of complex systems. Intense scientific interest in problems related to maintenance management originated only a few decades ago. The basis for the scientific support of maintenance is found in reliability engineering. The book *Mathematical Theory of Reliability* [2] indicates the start of scientific interest in maintenance problems. In the early stages, maintenance was almost exclusively interpreted as replacement of components (*see* **Replacement Strategies**). The development of some simple but insightful models describing aging phenomena of technical components, as well as the choice of appropriate actions to cope with these phenomena, showed that a scientific approach can be useful. In the 1970s and early 1980s, the basic models were extended in several directions. A substantial part of this work was primarily motivated by mathematical curiosity, rather than practical need.

In the 1980s, scientists made a step forward in bridging the gap between analytical models and practice by a systematic analysis of multicomponent systems. Multicomponent systems are much closer to reality than the classical single-component systems. Since maintenance actions in practice quite often come down to the replacement of one or more components in a multicomponent system, the study of these systems was an adequate way to bring reliability studies closer to practical maintenance management. Moreover, similar to many other areas of applied (stochastic) optimization (*see* **Reliability of Redundant Systems**), the major advances in information technology brought the use of quantitative analytic models within computational reach.

During the last 15 years substantial progress has been made in developing quantitative decision support systems for maintenance management of complex systems. Sophisticated decision support systems are in operation nowadays in the oil industry (both for maintenance of refineries as well as offshore installations), road and railways maintenance, and electric power generation. The number of companies that make use of some kind of quantitative tools to support inspection and replacement of equipment is increasing rapidly.

Dependencies in Multicomponent Systems

Thomas [3] defines three different types of interactions between components: economic, stochastic, and structural. Below, we will discuss these dependencies in more detail.

Economic Dependence

In simple terms, economic dependence implies that the cost of joint maintenance of a group of components does not equal the total cost of individual maintenance of these components. The effect of this dependence comes to the fore in the execution of maintenance activities. The joint execution of maintenance activities can save costs in some cases, but can lead to higher costs in other cases or it may not be allowed. So, we can further subdivide the models with economic dependence into positive and negative economic dependence. In many multicomponent systems both forms of economic dependence between components exist. As an example, we will discuss the different dependencies in ***k*-out-of-*n*** systems.

Positive Economic Dependence. We distinguish between the following forms of positive dependence:

- economies of scale
  - general
  - single setup
  - multiple setups
    - hierarchy of setups
- downtime opportunity.

The term *economies of scale* is often used to indicate that joint maintenance activities are cheaper than multiple individual activities. Economies of scale can result from preparatory or setup activities that can be shared when several components are maintained simultaneously. The cost of this setup work is often called *setup cost*. Setup costs can be saved because the execution of a group of activities requires only one setup. In practice, maintenance of multicomponent systems requires different setup activities and in that case there usually is a hierarchy of setups. For instance, consider a system consisting of two components, which both consist of two subcomponents. Maintenance of the subcomponents of the components may require a setup at system level and component level. Firstly, this means that the setup cost at component level is paid only once when the maintenance of two subcomponents of a component is combined. Secondly, the setup cost at system level is paid only once when all subcomponents are maintained at the same time. In multicomponent maintenance models the setup cost structure is incorporated in the objective function of the corresponding optimization problem.

Another form of positive dependence is the downtime opportunity. Component failures can often be regarded as opportunities for **preventive maintenance** of nonfailed components. In a series system (*see* **Parallel, Series, and Series–Parallel Systems**) a component failure results in a nonoperating system. In that case it may be worthwhile to replace other components preventively at the same time (opportunistic maintenance). This way the **system downtime** results in savings in maintenance costs since more components can be replaced at the same time. Moreover, by grouping corrective and preventive maintenance, the downtime can be regulated and in some cases it can even be reduced. In some cases the downtime cost can be included in the setup cost. In general, however, it is difficult to assess the cost associated with the downtime.

Negative Economic Dependence. Negative economic dependence between components occurs when maintaining components simultaneously is more expensive than maintaining components individually. There can be several reasons for this such as:

- manpower restrictions
- safety requirements
- redundancy/production loss.

Firstly, grouping maintenance results in a peak in manpower needs. Manpower restrictions may even be violated and additional labor needs to be hired, which is costly. The problem here is to find the balance between workload fluctuation and grouping maintenance.

Secondly, there are often restrictions on the use of equipment, when executing maintenance activities simultaneously. For instance, use of one equipment may hamper the use of other equipment and cause unsafe operations. Safety requirements often prohibit joint operation.

Thirdly, joint (corrective) maintenance of components in systems in which some kind of redundancy (*see* **Reliability of Redundant Systems**) is available may not be beneficial. Although there may exist economies of scale through simultaneous repair of a number of (identical) components, leaving components in a failed condition for some time increases the risk of costly production losses. Production loss may increase more than linearly with the number of components out of operation. For an example of this type of economic dependence, we refer to Stengos and Thomas [4].

Economic Dependencies in *k-out-of-n Systems*. The k-out-of-n system (*see* **k-out-of-n Systems**) is a typical example of a system with both positive and negative economic dependence between components. A k-out-of-n system functions if at least k components function. If $k = 1$, then it is a parallel system; if $k = n$, then it is a series system. Let us for the moment distinguish between the cases $k = n$ and $k < n$.

In the series system ($k = n$, *see* **Parallel, Series, and Series–Parallel Systems**), positive economic dependence between components is present due to downtime opportunities. The failure of one component results in an expensive downtime of the system and this time can be used to group both preventive and corrective maintenance, i.e. opportunistic grouping. Negative economic dependence is not explicitly present in the series system.

If $k < n$, then there is redundancy in the system and it fails less often than its individual components. In this way a certain reliability can be guaranteed. Typically, the components of this system are identical which allows for economies of scale in the execution of maintenance activities. It is not only possible to obtain savings by grouping preventive maintenance, but also by grouping corrective maintenance. In other words, the redundant components introduce additional positive dependence in the system. Whereas positive economic dependence is present upon failure of a component, negative economic dependence plays a role as long as the system operates. A single failure of a component may not always be an opportunity to combine maintenance activities. Firstly, grouping corrective and preventive maintenance upon the failure of the component increases the probability of system failure and costly production losses. Secondly, leaving components in a failed condition for some time, with the intention to group corrective maintenance at a later stage, has the same effect. So, there is a trade-off between the potential loss resulting from a system failure and the benefit of joint maintenance.

Stochastic Dependence

Stochastic dependence (*see* **Aging and Positive Dependence**) is also referred to as *failure interaction* and *probabilistic/statistical dependence*. It defines a relationship between components upon failure of a component. For example, it may be the case that the failure of one component induces the failure of other components or causes a shock to other components. Generally, the state of components can influence the state of the other components in the system. Here, the state can be given by the age, the failure rate, state of failure, or another condition measure. In the seminal work on stochastic dependence, Murthy and Nguyen [5, 6] introduce two types of failures: natural and induced. Natural failures are modeled by random variables and represent "normal" failures because of, e.g. aging or heavy usage. Induced failures are the result of other failures. Typically, there are two ways of modeling the induced failures. Firstly, a (natural) failure of one component induces a failure of other components with probability p and has no effect with probability $1 - p$. Scarf and O'Deara [7, 8] have compared several age- and block-replacement policies (*see* **Multivariate Age and Multivariate Renewal Replacement**; **Block Replacement**) for a two-component system with both this type of failure dependence and economic dependence. Secondly, a natural failure of one component may act as a shock to another component, resulting in a higher failure rate or condition of this component. This type of failure interaction is often modeled by a nonhomogeneous Poisson process (*see* **Poisson Processes**). Özekici [9] proposes a quite general Markov (*see* **Markov Processes**; **Maintenance and Markov Decision Models**) model for failure interaction. The state of a component at time t, which is given by a continuous random variable, depends on the state of the system up to time t. The state of the system is again given by the vector of component state values.

Finally, we recognize that the maintenance policies for systems with failure interaction are mainly of an opportunistic nature, since the failure of one component is potentially harmful for other components.

Structural Dependence

Structural dependence (*see* **Group Maintenance Policies**; **Total Productive Maintenance**) means that some operating components have to be replaced, or at least dismantled, before failed components can be replaced or repaired. In other words, structural dependence between components indicates that they cannot be maintained independently, but only together. It is not about failure dependence, but about maintenance dependence. Since the failure of a component offers an opportunity to replace other components, opportunistic policies are expected to perform well on systems with structural dependence between components. Obviously, preventive maintenance may also be advantageous, since maintenance of structural dependent components can be grouped.

There may be several reasons for structural dependence. For example, a bicycle chain and a cassette form a union, which should always be replaced together, rather than individually. Another example is from Dekker *et al.* [10], which considers road maintenance. Several deterioration patterns affect roads, e.g., longitudinal and transversal unevenness, cracking, and raveling. For each mechanism one may define a virtual component, but if one applies a maintenance action to such a component it also affects the state with respect to the other failure mechanisms.

The seminal paper in this category is from Sasieni [11]. He considers the production of rubber tires. The machine that produces the tires consists of two "bladders"; one tire is produced on each bladder simultaneously. Upon failure of a bladder, the machine must be stripped down before replacement can be done. This means that the other bladder can be replaced at the same time. Note that immediate replacement is not mandatory, but a failed bladder will produce faulty tires.

Modeling Multicomponent Maintenance

Several models for the maintenance of multicomponent systems have been proposed. Firstly, these models can be classified based on the planning aspect of the model: stationary (long term) and dynamic (short term). Secondly, one can distinguish three types of optimization methods that can be used to select the model: exact, heuristic, or policy optimization. The typical features of the models in these categories are specified below.

Planning Horizon

In stationary models, a long-term stable situation is assumed and mostly these models assume an infinite planning horizon. This assumption facilitates mathematical analysis. Models of this kind provide static rules for maintenance, which do not change over the planning horizon. They generate for example, long-term maintenance frequencies for groups of related activities or control limits for carrying out maintenance, depending on the state of the components. Three main classes of models in this category can be recognized, namely grouping corrective maintenance, grouping preventive maintenance, and opportunistic maintenance. In the first case, components are only correctively maintained and a failed component can be left in the failed state until its corrective maintenance is carried out jointly with that of other failed components. In the second case, preventive maintenance is carried out to prevent failures or to decrease operating costs, and this is planned in advanced and in such a way that setup costs can be saved by simultaneous execution. In the third case, maintenance is not necessarily planned in advance. However, setup savings can be obtained since (corrective or preventive) maintenance of a component yields an opportunity for maintenance of other components.

In dynamic models, short-term information such as a varying deterioration of components or unexpected opportunities can be taken into account. Situations that are not stationary, such as a varying use of components, and unexpected events that may create an opportunity for doing maintenance at lower costs, can now be incorporated. These models generate dynamic decisions that may change over the planning horizon.

The dynamic-grouping models can be further subdivided into two categories: those with a finite horizon and those with a rolling horizon. Finite-horizon models consider the system in this horizon only, and hence assume implicitly that the system is not used afterward, unless a so-called residual value is incorporated to estimate the industrial value of the system at the end of the horizon. Rolling-horizon models also use a finite horizon, but they do so repeatedly and on the basis of a long-term (infinite-horizon) plan. That is, once decisions of the finite horizon are implemented or when new information becomes available, a new horizon is considered, and a tentative plan based on the long-term one is adapted according to short-term circumstances.

Optimization Method

Exact optimization methods are designed to find the global optimal solution to a problem. However, if the computing time of the optimization method increases exponentially with the number of components, then exact methods are desirable only to a certain extent. In that case, solving problems with many components is impossible and heuristics should be used. Heuristics are local optimization methods that do not pretend to find the global optimum, but can be applied to find a solution to the problem in a reasonable time. The quality of such a solution depends on the problem instance. In some cases, it is possible to give an upper bound on the gap between the optimal solution and the solution found by the heuristic.

In many papers, maintenance planning is done by optimizing a certain type of policy. Well-known maintenance policies are the age- and block-replacement policies (*see* **Multivariate Age and Multivariate Renewal Replacement**; **Block Replacement**) and their extensions. The advantage of policy optimization over other optimization methods is that it gives more insight into the solution of the problem. Policy optimization will not always result in the global optimal solution, since there may be another policy that results in a better solution.

Heuristic and exact methods are often used in finite-horizon models, especially when time is discretized and the decision parameters only take integer values. Policy optimization is mainly used in infinite-horizon models. In these models, it is often possible to derive analytical expressions for optimal control parameters and the corresponding optimal costs.

Example

We will now illustrate the preceding review of multicomponent maintenance models with an example. It is based on two studies by Scarf and O'Deara [7, 8], who consider the maintenance of a clutch system in a bus fleet. The system is a series system consisting of the clutch assembly (component 2) and the clutch controller (component 1). In this particular system a (natural) failure of component 1 causes a failure of component 2 with probability 1, whereas a failure of component 2 *never* induces a failure of component 1. Moreover, the system exhibits (positive) economic dependence in that combining (preventive or failure) replacements saves costs. Replacement restores the respective component to a new condition. Below, we highlight two replacement policies for this system: combined failure-based replacement (CFR) and combined age-based replacement (CAR; *see* **Multivariate Age and Multivariate Renewal Replacement**).

Note that because we deal with a series system (*see* **Parallel, Series, and Series–Parallel Systems**), a system failure implies that either both components have failed or just one of the components has failed. It is assumed that failed components are replaced immediately. CFR now prescribes that either (a) both failed components are replaced or (b) the failed component is replaced correctively and the component that has not failed is replaced preventively. CAR dictates that *both* components are replaced at age T or upon failure, whichever occurs first. Observe that both policies prescribe that components that have not failed are replaced preventively. Scarf and O'Deara [7] also consider age- and failure-based replacement policies that do not replace components that have not failed preventively. Since these policies do not take into account the positive economic dependence between components, they are outperformed by their extensions (CAR and CFR, respectively).

Let us introduce some notation. The cost of a preventive replacement of both components is given by C_{PP} and is less than or equal to the independent preventive replacement of both components. The cost of failure replacement of both components C_{FF}, is assumed to be equal to the cost of combined failure and preventive replacements of the components. The lifetime distribution $F_i(x)$ of component i is the Weibull distribution with parameters α_i and β_i, $i = 1, 2$. That is, $F_i(x) = 1 - \exp(-(x/\beta_i)^{\alpha_i})$, $i = 1, 2$. The lifetime is measured in months. Now, it can be shown that the mean cost rates for CFR and CAR are given by

$$C_{\infty}^{\mathrm{CFR}} = \frac{C_{\mathrm{FF}}}{\displaystyle\int_0^{\infty} (1 - F_1(x))(1 - F_2(x))\,\mathrm{d}x} \quad \text{and}$$

$$C_{\infty}^{\mathrm{CAR}}(T) = \frac{C_{\mathrm{FF}} + (C_{\mathrm{PP}} - C_{\mathrm{FF}})(1 - F_1(T))(1 - F_2(T))}{\displaystyle\int_0^{T} (1 - F_1(x))(1 - F_2(x))\,\mathrm{d}x} \tag{1}$$

respectively. Minimizing $C_{\infty}^{\mathrm{CAR}}(T)$ with respect to T yields the optimum replacement age.

Figure 1 shows the (optimal) mean cost rate for CFR and CAR for the lifetime parameters as

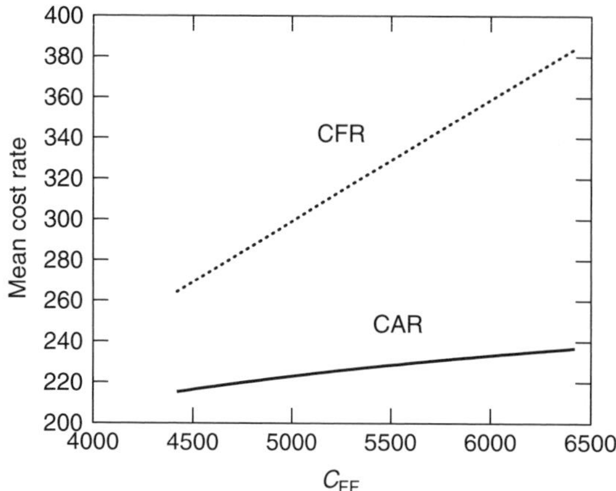

Figure 1 Mean cost rate for replacement policies CFR and CAR *versus* C_{FF} (system failure replacement cost). The Weibull parameters are given by $\alpha_1 = 8$, $\beta_1 = 34.8$, $\alpha_2 = 7.7$, and $\beta_2 = 17.8$. The cost of preventive replacement of both components is $C_{PP} = 2740$. The cost of combined failure and preventive replacements C_{FF} is varied between 4420 and 6420

estimated in [7, 8]. Cost parameter C_{PP} has been set to 2740 and C_{FF} has been varied between 4420 and 6420 (this corresponds with the case studies by Scarf and O'Deara [7], where the unavailability cost is 1600 and the penalty cost of a failure is varied between 0 and 2000). As expected CAR outperforms CFR and the difference becomes more pronounced as the cost of failure increases. Scarf and O'Deara [7] show that the performance (with respect to both mean cost rate and implementation) of CAR is comparable with that of well-performing opportunistic age-replacement policies. Notice that CAR treats the system as a single component and prescribes to replace it at age T. As such it is easy to implement in a decision support system. Moreover, a further study [8] shows that CAR competes with the best (modified) block-replacement policies, which are also relatively easy to manage.

Further Reading

Several overview articles have appeared on multicomponent maintenance models. Cho and Parlar [1] review articles from 1976 up to 1991. It divides the literature into five topical categories: machine-interference/repair models, group/block/cannibalization/opportunistic models, inventory/maintenance models, other maintenance/replacement models, and inspection/maintenance models. Dekker *et al.* [12] exclusively deal with multicomponent maintenance models that are based on economic dependence. The models are classified on the basis of their planning aspect: stationary (long term) or dynamic (short term). Wang [13] gives an overview of maintenance policies of deteriorating systems. The emphasis is on policies for single-component systems. One section is devoted to opportunistic maintenance policies for multicomponent systems. The author primarily considers models with economic dependence. In a recent article, Nicolai and Dekker [14] review articles in which multicomponent maintenance of systems with economic (both positive and negative), stochastic, and structural dependence is modeled (and optimized).

References

[1] Cho, D. & Parlar, M. (1991). A survey of maintenance models for multi-unit systems, *European Journal of Operational Research* **51**, 1–23.

[2] Barlow, R.E. & Proschan, F. (1965). *Mathematical Theory of Reliability*, John Wiley & Sons, New York.

[3] Thomas, L. (1986). A survey of maintenance and replacement models for maintainability and reliability of multi-item systems, *Reliability Engineering* **16**, 297–309.

[4] Stengos, D. & Thomas, L. (1980). The blast furnaces problem, *European Journal of Operational Research* **4**, 330–336.

[5] Murthy, D.N.P. & Nguyen, D. (1985a). Study of a multicomponent system with failure interaction, *European Journal of Operational Research* **21**, 330–338.

[6] Murthy, D.N.P. & Nguyen, D. (1985b). Study of two-component system with failure interaction, *Naval Research Logistics Quarterly* **32**, 239–247.

[7] Scarf, P.A. & O'Deara, M. (1998). On the development and application of maintenance policies for a two-component system with failure dependence, *IMA Journal of Mathematics Applied in Business and Industry* **9**, 91–107.

[8] Scarf, P.A. & O'Deara, M. (2003). Block replacement policies for a two-component system with failure dependence, *Naval Research Logistics* **50**(1), 70–87.

[9] Özekici, S. (1988). Optimal periodic replacement of multicomponent reliability systems, *Operations Research* **36**, 542–552.

[10] Dekker, R., Plasmeijer, R. & Swart, J. (1998). Evaluation of a new maintenance concept for the preservation

of highways, *IMA Journal of Mathematics Applied in Business and Industry* **9**, 109–156.

[11] Sasieni, M. (1956). A Markov chain process in industrial replacement, *Operational Research Quarterly* **7**, 148–155.

[12] Dekker, R., van der Duyn Schouten, F. & Wildeman, R. (1996). A review of multi-component maintenance models with economic dependence, *Mathematical Methods of Operations Research* **45**, 411–435.

[13] Wang, H. (2002). A survey of maintenance policies of deteriorating systems, *European Journal of Operational Research* **139**, 469–489.

[14] Nicolai, R. & Dekker, R. (2007). Optimal maintenance of multi-component systems: a review, in *Maintenance of Complex Systems: Blending Theory with Practice*, K.A.H. Kobbacy & D.N.P. Murthy, eds, Springer-Verlag, Berlin.

Related Articles

Aging and Positive Dependence; **Block Replacement**; **Group Maintenance Policies**; **Inspection Policies for Reliability**; ***k*-out-of-*n* Systems**; **Maintenance and Markov Decision Models**; **Maintenance Optimization**; **Maintenance Optimization in Random Environments**; **Markov Processes**; **Multivariate Age and Multivariate Renewal Replacement**; **Poisson Processes**; **Reliability of Redundant Systems**; **Replacement Strategies**; **Total Productive Maintenance**.

ROBIN P. NICOLAI AND ROMMERT DEKKER

Multiple Modes in Proficiency Test Data

Introduction

Chromium in sediment and biota, chlorobiphenyls in mussels, and aluminum in sediment, are some of the many chemical contaminants in the marine environment. Proficiency Test (PT) schemes are used to provide the external quality assurance for chemical analytical laboratories that measure these types of contaminants [1]. Homogeneous natural test materials are distributed for chemical analysis on a regular basis and the returning data are assessed to determine the extent of the agreement between laboratories. Many scheme providers, such as QUASIMEME[a] [2], allow laboratories to use their own optimized analytical methodology to obtain a single value for the measurement. A consensus value is obtained from these data that forms the basis of the assessment. Ideally, these data should be normally distributed, but frequently they are skewed, bi-, or multimodal. As a consequence the consensus value can be highly influenced by the distribution of these data. Many factors contribute to tailing or bimodality and it is important that we identify the data associated with each mode and, where possible, the root causes of these differences. In some instances, it is due to the relative **bias** of some procedure in the chemical analytical process. In such cases, we need to establish whether the bimodality is true or is an artefact.

We also require qualitative and, where possible, quantitative information on the degree of separation needed between modes before detection is possible and a simple screening test to identify bimodality. In addition, we require information on the population characteristics of each mode, preferably without separating the data and a clear, sensitive, graphical representation of the bimodality. We have described the Cofino model and its applications in **Population Characteristics of Proficiency Test Data** and we describe the specific problem of the evaluation of bimodal data here.

By using the Cofino model (*see* **Population Characteristics of Proficiency Test Data**) [3, 4] to evaluate the data, we can obtain the population characteristics (mean, standard deviation (SD) and the percentage of data attributed to that mean (λ)) for each of the population measurement functions (PMFs). In most cases, over 90% of the data are accounted for by the first two modes, PMF_1 and PMF_2 when the **normal distribution** approximation (NDA) is applied. So, for the bimodality to be significant, the data associated with PMF_2 needs to be distinct from the data associated with PMF_1. Since the data associated with both PMF_1 and PMF_2 have a known mean and SD from the Cofino model, we can use the Student's t test as a simple, preliminary, indicator of bimodality.

The identification of bimodality primarily depends on two factors. The first is the degree of separation between the two modes and the second, the percentage of data in each mode.

Resolution of the Modes

To test the resolving **power** of the model, we randomly generated a series of datasets to give $N(100, 2)$ with 50 values and $N(100 + i, 2)$ with 30 values, where $i = 1 : 10$. The SDs of the two populations s_{pop} are thus equal. Data for each degree of separation $N(100, 2) + N(100 + i, 2)$ were combined to provide bimodal distributions with a separation i between the modes. We then applied the Cofino NDA model to obtain the population characteristics of both modes.

In Figure 1 we have plotted the separation of the two modes expressed as i/s_{pop} (thus normalized with respect to s_{pop}) against (a) their means, (b) the ratio of the percentage of data associated with PMF_1 and PMF_2, and (c) Student's t value.

We detected very little separation up to a $3 \times s_{pop}$ difference in the mode means, as both distributions overlapped and influenced the mean of the two modes, PMF_1 and PMF_2. As we increased the separation, the Cofino model mean of the first mode (PMF_1) decreased to approximately the known value (100), while that of the second mode (PMF_2) increased to its known value. Simultaneously, the ratio of percentage of data in the two modes fell to that of the two independent datasets $50/30 = 1.7$.

Percentage of Data in Each Mode

The second factor affecting the identification of bimodality is the number of observations (NObs) associated with each mode.

In the example we assigned a ratio of 50:30 for the data in the two modes. However, if we change this ratio to 50:50 it becomes more difficult to identify and isolate the two modes due to the way in which the SD, using the NDA model, is constructed.

Again we generated and combined data with $N(100, 2)$ and with NObs from 30 to 50 in. increments of 5 and a second distribution with $N(107, 2)$ and $NObs = 30$. The values for the mean of data associated with PMF_1, and PMF_2 are plotted against the ratio of NObs in the two modes. Superimposed on this plot is the value of the Student's t value for each of the sets of data (Figure 2).

When the ratio of NObs for the two modes is 1 there is no separation. Here the values for the Cofino mean for PMF_1 and PMF_2 are 103.4 and 103.6, respectively. As we increase the ratio of NObs between the two datasets, we begin to detect separation. When the ratio of NObs exceeds 1.7 the Cofino means of the two modes are close to their independent values, the Student's t value is >2, i.e. t_{crit}, and there is a clear separation. Both means of the data associated with PMF_1 and PMF_2 are now

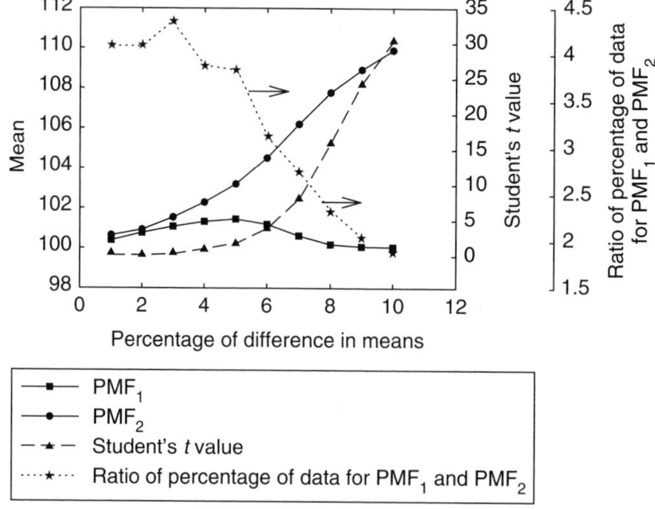

Figure 1 The effect of separation of two overlapping distributions $N(100, 2)$ and $N(100 + i, 2)$ where $i = 1 : 10$ on the mean of data associated with PMF_1 (first mode) and PMF_2 (second mode) obtained using the Cofino model. Both modes are clearly resolved when the difference between the means >6%

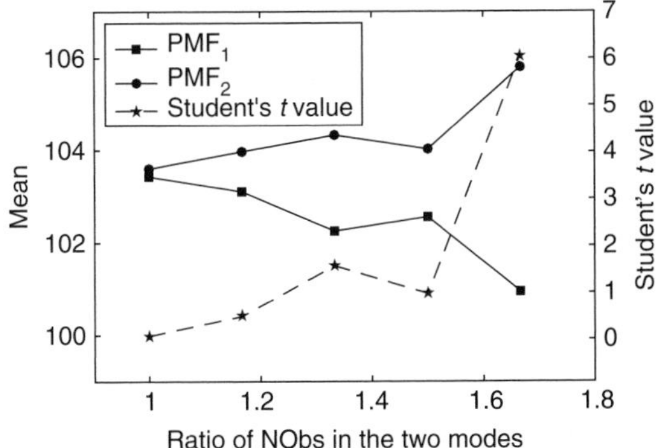

Figure 2 The effect of the ratio of NObs in the two modes $N(100, 2)$ and $N(107, 2)$ on the mean of data associated with PMF_1 and PMF_2. Both modes are resolved when the ratio of NObs in the two modes >1:1.66

close to their independent values 100.2(100) and 106.6(107) respectively.

The breakpoint for separation and obtaining the population characteristics of both modes depends both on the ratio of NObs in the two modes and on the difference between their means and their SDs. In our example, using one and the same s_{pop} of about 2% of population mean, the ratio of NObs had to be greater than 1.7 and the difference between the means greater than $3 \times s_{pop}$. Below these breakpoints it is still possible to detect the bimodality using the matrix overlap, even though the population characteristics of each mode cannot be obtained from the whole dataset (*see* **Population Characteristics of Proficiency Test Data**) (Figure 3b). In such a circumstance, the data in the two modes can be identified and each subset of data treated separately. For many other datasets the assessment can be made in a single calculation. Also by inspection of the distribution of the data in the matrix-overlap plots, it is possible to identify any inhomogeneity that might otherwise go undetected.

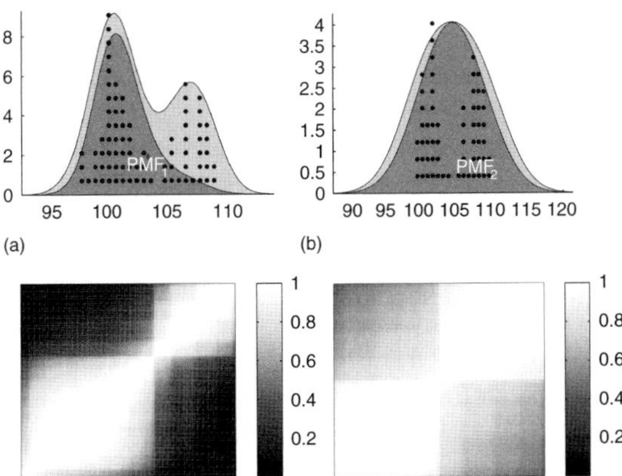

Figure 3 The PMF (upper) for all data (light shade area) and for PMF_1 (dark shades area) and the overlap matrix (lower) for the combined distributions $N(100, 2)$ and $N(107, 2)$ (a) with NObs 50:30 and (b) with NObs 50:50 in each mode. A distinct separation occurs when the means differ by >6% and the ratio of NObs is >1:1.66. Even when there is no evidence of separation of the modes, as in PMF profile (b – upper), the two constituent modes are distinctly evident in the overlap matrix (b – lower)

Evaluation of the Causes of Bimodal Data in PTs

In any PT where the contents of the matrix are not characterized, we need to estimate the population characteristics of each main mode in the data and provide a clear explanation of any bimodal distribution. The Cofino model often allows us to achieve this without separating the data, while the normal arithmetic or robust mean using the Hu [5–7] or Ha [8, 9] estimators will only provide a mean value that may lie somewhere between the two modes.

Chromium in Sediment and Biota

Different laboratories extract the chromium from the sediment using either a nondestructive method or hydrofluoric acid (HF) digest (*total*), or by *partial* digestion methods with aqua regia. For one PT, the robust mean of all data, using the Hu estimator, was $85\,mg\,kg^{-1}$ while the Cofino model identifies two modes at 89 and $67\,mg\,kg^{-1}$ which could be associated with *total* and *partial* digestion methods, respectively. When we subsequently separated the data into two sets according to the digestion method and reapplied the robust statistics, we obtained means of 91.5 and $69.1\,mg\,kg^{-1}$.

The level of chromium in biological tissue is generally low and contamination can easily occur during chemical analysis unless an ultraclean environment is used. Data from a PT for chromium in cod liver were bimodal, with a Cofino mean for the data associated with the main mode (PMF_1) around $110\,\mu g\,kg^{-1}$ (Figure 4). The digestion media included hydrochloric acid, aqua regia, nitric acid, nitric acid with hydrogen peroxide, and five others unspecified. There was no clear preference for either group but most of the laboratories used nitric, with or without hydrogen peroxide. Although there was a range of different digestive media, six laboratories used an open heating (OH) system for digestion and five of these were within the upper 50% of the data, with three being in the second mode. Microwave (M) digestion and the pressure bomb (PB) formed most of the data in the first mode (Figure 4).

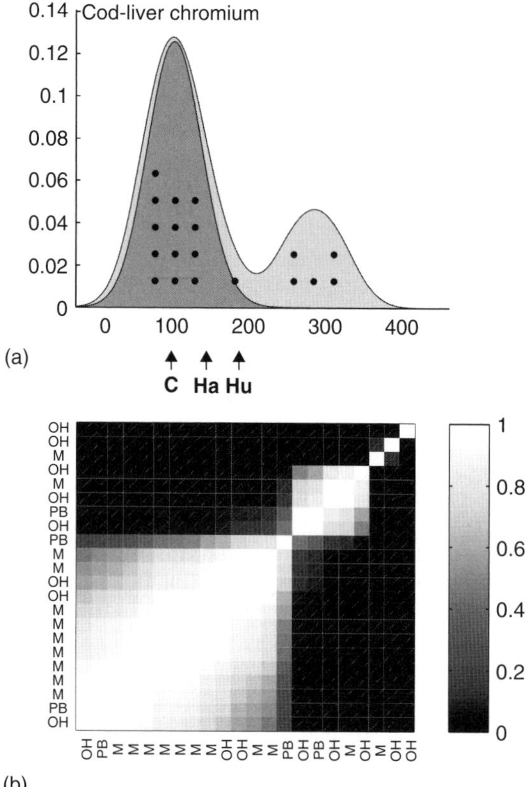

Figure 4 The PMF (a) and the overlap matrix (b) for chromium in cod liver PT. The distribution of returning data is bimodal. The methods of sample preparation used to obtain the data in the main, mode (a) comprise microwave (M) and pressure bomb (PB), both sealed systems, and a few open vessel heating (OH). In the smaller mode at higher concentration, the positively tailing data included five out of eight laboratories that used the OH system, which suggests that this maybe a source of local contamination. C: Cofino mean, Ha: Hampel estimator, Hu: Huber estimator. Our evaluation concluded that the Cofino model provided the best estimates of means for *both* modes

Although there may be other differences between methods, e.g., detection systems, there is a clear bias caused by contamination of the samples digested in open vessels.

A comparison of the data obtained by the Cofino model, the Hu and Ha estimators are given in Table 1. The mean values are also indicated in Figure 4. From the summary statistics in Table 1 and from the matrix-overlap plot in Figure 4 we see that the Cofino model provides a consensus value more closely aligned with the main mode of the data while the robust statistics give a mean between the two modes.

Chlorobiphenyl (CB105) in Mussels

CB105 is amongst the more difficult congeners to completely separate chromatographically from other CBs, in particular, from CB132. Therefore, PT data can be a result of measuring just CB105 or a mixture of CB105 and other, unseparated congeners. Higher values can also be caused by contamination due to insufficient sample cleanup. The matrix-overlap plot (Figure 5a) shows two modes indicated by the lighter colored areas and two very high values (top right) The mode at the lower concentration is more homogeneous with a greater overlap of data. The second mode, at the higher concentration is more diffuse. These higher values in the second mode are most likely to be associated with an insufficiently cleaned up sample, degrading or inappropriate chromatographic column selection, and/or optimization.

The mean of data associated with the first mode at a lower concentration is $0.27\,\mu g\,kg^{-1}$ while the second mode is about 2.5 times higher at $0.64\,\mu g\,kg^{-1}$. In most cases where there is clear information on likely additive interferences, the best estimate for the consensus value is often the mode with the lower value. However, this does not always need to be the case and careful judgment with supporting chemistry is essential. Again, mean using the Hu estimator is *ca* 30% higher than that when using the Cofino model (Table 1).

Aluminum in Sediment

The selection of digestion methods affects the recovery of aluminum from minerals such as feldspar in the sediments. A range of strong acids are normally used, with HF being amongst the most effective at breaking

Table 1 A comparison of the summary statistics from the proficiency test for chromium and chlorobiphenyl, CB105, in mussel and aluminum in sediment. Each of these PT data shows a bimodal distribution due to different methods of analysis (see Figures 4–6). Both robust methods using the Huber (Hu) and Hampel (Ha) to estimate the mean for chromium are affected by the second mode at higher concentrations for chromium and CB105 and by the second mode at lower concentrations for aluminum. In each case, the Cofino mean estimates the mean where there is the greatest overlap of data. The Cofino model can provide the mean and SD for both modes from all data in one computation

	Determinand	Total NObs	LCV NObs	Percentage of data in first mode	Units	Cofino mean	Cofino SD	Huber mean	Huber SD	Hampel mean	Hampel SD	Arithmetic mean	Arithmetic SD
Biota	Chromium	25	3	70	$\mu g\ kg^{-1}$	125.21	80.22	187.39	130.34	150.61	83.50	683.82	2139.42
Mussel	CB105	22	0	75	$\mu g\ kg^{-1}$	0.31	0.16	0.42	0.24	0.39	0.18	1.08	3.12
Sediment	Aluminum	27	0	75	%	4.62	1.46	4.09	2.04	4.18	1.41	4.10	1.84

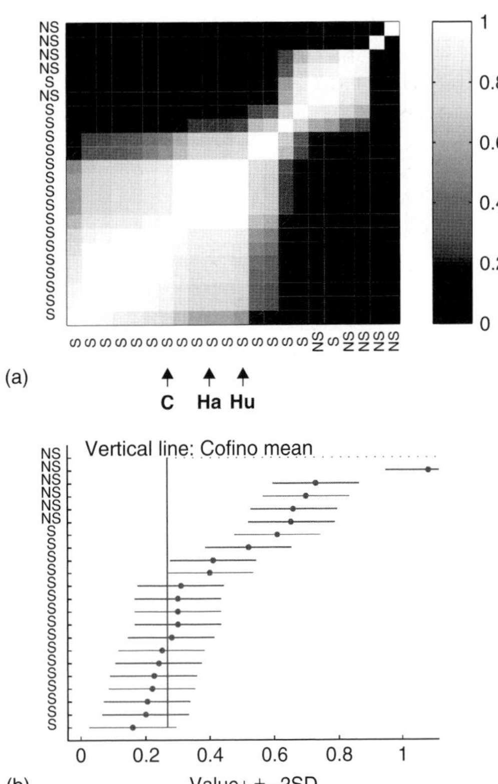

Figure 5 The overlap matrix (a) and the ranked distribution (b) for chlorobiphenyl, CB105 in mussels. The highly tailing data creates a bimodal distribution caused by poor chromatographic resolution, insufficient sample cleanup or contamination (NS) during the chemical analysis which leads to elevated results for these laboratories. The data in the main mode of the distribution (S) have been generated under more optimum conditions. C: Cofino mean, Ha: Hampel estimator, Hu: Huber estimator. The Cofino model provides the best estimates of means for *both* modes

the crystal lattice and releasing the aluminum. Other digestive techniques used are aqua regia, nitric acid as well as nondestructive techniques such as X-ray fluorescence. Most laboratories currently use HF as the preferred digestive "wet" technique.

The results for aluminum in the marine sediment show three obvious modes in the data that are clearly visible in the overlap matrix plot (Figure 6a). The highest mode, around 7.5% Al, is probably a result of two laboratories obtaining a particularly high value. The main mode has a concentration of 4.95% and consists almost exclusively of digestion methods involving HF or nondestructive, *total* techniques. The

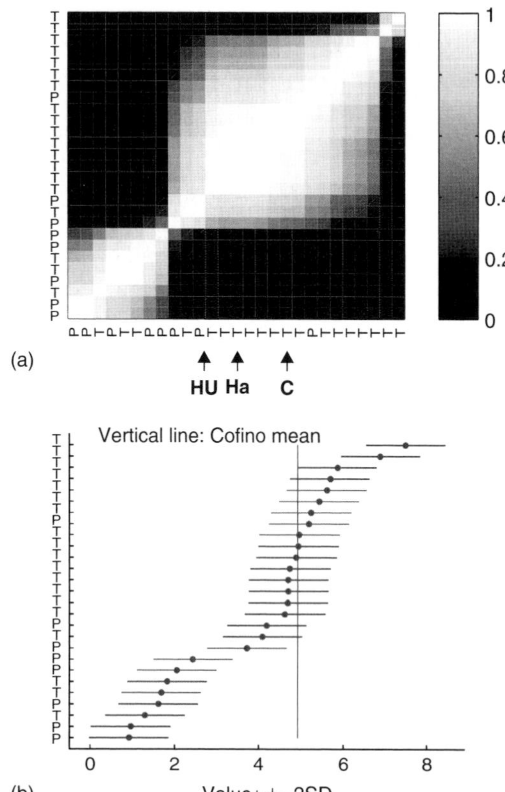

Figure 6 The overlap matrix (a) and the ranked distribution (b) for aluminum in sediment. The bimodal distribution results from different methods of sediment digest. The partial digest using acids other than hydrofluoric acid (HF) form most of the lower mode while the main mode at higher concentrations consists of total digest methods (HF) and nondestructive techniques such as X-ray fluorescence. C: Cofino mean, Ha: Hampel estimator, Hu: Huber estimator. The Cofino model provides the best estimates of means for *both* modes

lower mode at 1.6% Al includes methods using aqua regia (two), nitric acid (two), nitric acid and peroxide (*partial*), as well as three laboratories using HF. The non-HF, wet techniques clearly do not extract all of the aluminum, but the use of HF *per se* does not always guarantee complete recovery of the element. The robust estimates are again affected by the second mode at lower concentration, due mainly to the data obtained by the partial digest of the sediment. In this case, the most homogeneous and reliable consensus value was based on the main mode at the higher concentration.

Benefits of the Cofino Model

Using the Cofino model we are able to estimate the population characteristics of each mode in a bi- or multimodal distribution. In most cases, we can achieve this using the whole dataset, including left censored values (LCVs), without the need to trim the data or remove outliers. The resolution between the modes depends on the ratio of data in the two modes and on the difference between the means in relation to the SDs of the modes. However, even in unfavourable conditions, the matrix overlap plot will still suggest the presence of two modes.

Evaluation using the Hu or Ha robust estimator will provide similar consensus values where the tailing and **outlier** values are *ca* < 10% of all data, but unlike the Cofino model, these techniques are less able to cope effectively with heavily tailing or bimodal data. In some cases, as we have shown here, the robust mean using these estimators can occur between the modes where there are very few data. The combination of the information gained from the Cofino model with scientific expertise will, in many cases as shown above, enhance the reliability of interpretation of PT data.

End Note

[a] QUASIMEME, Quality Assurance of Information for Marine Environmental Monitoring (in Europe), funded initially by the EU (1992–1996) for European Laboratories, is now a worldwide project funded by the subscribing participating laboratories www.quasimeme.org.

Acknowledgments

The authors wish to acknowledge the members of the QUASIMEME team, in particular Judith Scurfield, in testing this model, and the participants of the QUASIMEME scheme for providing invaluable data for evaluation. The authors would also like to acknowledge Ian Davies for his helpful critical review of the manuscript.

References

[1] Thompson, M., Ellison, S.L.R. & Wood, R. (2006). The international harmonized protocol for the proficiency testing of analytical chemistry laboratories, *Pure and Applied Chemistry* **78**, 145–196.

[2] Cofino, W.P. & Wells, D.E. (1994). Design and evaluation of the QUASIMEME inter-laboratory performance studies: a test case for robust statistics, *Marine Pollution Bulletin* **29**, 149–158.

[3] Cofino, W.P., Wells, D.E., Ariese, F., van Stokkum, I.H.M., Wengener, J.W. & Peerboom, R. (2000). A new model for the inference of population characteristics from experimental data using uncertainties, *Journal of Chemometrics and Intelligent Laboratory Systems* **53**, 37–55.

[4] Cofino, W.P., van Stokkum, I.H.M., van Steenwijk, J. & Wells, D.E. (2005). A new model for the inference of population characteristics from experimental data using uncertainties. Part II. Application to censored datasets, *Analytica Chimica Acta* **533**, 31–39.

[5] Huber, P.J. (1972). Robust statistics: a review, *Annals of Mathematical Statistics* **43**, 1041–1067.

[6] Analytical methods committee robust statistics–how not to reject outliers. Part 1. Basic concepts, (1989). *The Analyst* **114**(12), 1693–1697.

[7] Analytical methods committee robust statistics–how not to reject outliers. Part 2. Inter-laboratory trials, (1989). *The Analyst* **114**(12), 1699–1702.

[8] ISO/CD 20612 (2005). *Water Quality-Interlaboratory Comparisons for Proficiency Test of Laboratories*, International Standards Organization, Geneva.

[9] Müller, C.H. & Uhlig, S. (2001). Estimation of variance components with high breakdown point and high efficiency, *Biometrika* **88**(2), 353–366.

David E. Wells and Wim P. Cofino

Multiple Sampling Plans

Introduction

Under a **single sampling** plan (*see* **Single Sampling by Attributes and by Variables**), all items for inspection are sampled in one go, and the decision of acceptance or rejection is based on the evidence from this unique sample. From an intuitive point of view, there seems to be an opportunity to reduce the total inspection amount, if lot quality is either very good or very bad: Sample and inspect a few items; if the sample quality is very good accept, if it is very bad, reject; if the sample quality is doubtful, sample a few further items. Continue with this procedure, until the quality of the cumulative sample clearly indicates acceptance or clearly indicates rejection. If lot quality is on one of the extreme sides, the total sampling amount required for a sufficiently reliable decision should be smaller than under a single sampling plan.

The above idea leads to so-called iterative or multistage sampling plans with the following two special cases: (a) A *multiple m-stage sampling plan*, where an upper limit of m successive samples is imposed. The simplest case is a *double sampling plan* (two-stage sampling plan) with a maximum of two sampling steps. (b) A *sequential sampling plan*, where sampling may continue until 100% inspection of the lot.

The idea of iterative sampling is old and had already appeared in Dodge's technical reports at Bell Telephone Laboratories in the 1920s and 1930s, see [1]. Double and multiple sampling plans are provided by the traditional industrial standards. MIL-STD-105E provides double and seven-stage sampling plans (*see* **Attributes Sampling Schemes in International Standards**). ISO 2859-1 contains double and five-stage plans (*see* **Attributes Sampling Schemes in International Standards**). The Dodge–Romig tables contain double sampling plans (*see* **Rectification Sampling Schemes**). Sequential sampling plans are provided by ISO 8422 and ISO 8423 (*see* **Sampling in Industrial Standards**).

Structure and Operation of Multiple Sampling Plans

In principle, multiple sampling plans can be devised for any of the discrete item quality models of statistical process inspection (SPI) (statistical lot inspection) (*see* **Sampling Inspection of Products** for a survey). As most established schemes of multiple sampling do, we restrict attention to the customary attributes quality models described by Table 1. For a formal exposition of these models *see* **Sampling Inspection of Products**. Sequential sampling procedures under the specification range model are provided by the standard ISO 8423 (*see* **Sampling in Industrial Standards**). We consider multiple sampling plans in a type A framework, where the decision target is a finite lot of items $1, \ldots, N$. The methods may also be used for type B purposes where sampling and decision is directed toward the lot generating process. *See* **Sampling Inspection of Products** for the distinction of type A and type B procedures.

A multiple m-stage sampling plan is defined by a triple $(\boldsymbol{n}, \boldsymbol{a}, \boldsymbol{r})$, where $\boldsymbol{n} = (n_1, \ldots, n_m)$ is a sequence

Table 1 Attributes quality models

	Conforming–nonconforming	Nonconformities
Item quality indicator	Item is conforming (nondefective) or nonconforming (defective)	Number of nonconformities (defects) on the item
Lot quality indicator	Lot proportion nonconforming p	Lot average number of nonconformities k
Sample statistic t	Number of nonconforming items in the sample	Number of nonconformities on items in the sample

of **sample sizes** $n_1, \ldots, n_m \geq 1$, $\boldsymbol{a} = (a_1, \ldots, a_m)$ is a sequence of integer *acceptance numbers* $0 \leq a_1 \leq \cdots \leq a_m$, $\boldsymbol{r} = (r_1, \ldots, r_m)$ is a sequence of integer *rejection numbers* $1 \leq r_1 \leq \cdots \leq r_m$. The m parameters relate to the m stages of the sampling plan as exposed by Table 2. In most **sampling schemes**, e.g., in MIL-STD-105E and ISO 2859, the differences $r_l - a_l$, $l = 1, \ldots, m-1$, of the rejection and acceptance number have few variation in l.

The sampling and decision algorithm of a multiple m-stage sampling plan is explained by Table 3 where the sample statistics t_l conform to the quality models as explained by Table 1. Because of $a_m = r_m - 1$, the algorithm terminates with certainty at stage m, either with acceptance or with rejection.

In most cases, it is not the proper act of sampling in the sense of selecting items from the lot, which involves costs, but rather the act of inspecting the items to calculate the test statistic (number of nonconforming items or total number of nonconformities in the sample. Inspection costs can be reduced by *curtailed inspection.* Two curtailment techniques exist: (a) Total curtailment: At each stage l, the items in the sample are successively inspected, and as soon as the cumulative number $t_1 + \cdots + t_{l-1} + t_{l,j}$ of nonconforming items or the number of nonconformities

Table 2 The parameters of a multiple m-stage sampling plan for the inspection of a lot of size N

Stage	Sample size	Cumulative sample size	Acceptance number	Rejection number
1	n_1	$n_1^{(c)} = n_1$	a_1	$r_1,\ r_1 \geq a_1 + 2$
2	n_2	$n_2^{(c)} = n_1 + n_2$	a_2	$r_2,\ r_2 \geq a_2 + 2$
⋮	⋮	⋮	⋮	⋮
m	n_m	$n_m^{(c)} = n_1 + \cdots + n_m$	a_m	$r_m,\ r_m = a_m + 1$
$0 \leq a_1 \leq a_2 \leq \cdots \leq a_m$, $1 \leq r_1 \leq r_2 \leq \cdots \leq r_m$, $n_1 + \cdots + n_m \leq N$				

Table 3 The sampling and decision algorithm of a multiple m-stage sampling plan with parameters given by Table 2

Stage	Operation	Decision algorithm	
		$t_1 \leq a_1$	$\Longrightarrow$ terminate, accept
1	Sample n_1 items Calculate sample statistic t_1	$t_1 \geq r_1$	$\Longrightarrow$ terminate, reject
		$a_1 + 1 \leq t_1 \leq r_1 - 1$	$\Longrightarrow$ continue with stage 2
		$t_1 + t_2 \leq a_2$	$\Longrightarrow$ terminate, accept
2	Sample n_2 items Calculate sample statistic t_2	$t_1 + t_2 \geq r_2$	$\Longrightarrow$ terminate, reject
		$a_2 + 1 \leq t_1 + t_2 \leq r_2 - 1$	$\Longrightarrow$ continue with stage 3
⋮	⋮	⋮	
m	Sample n_m items Calculate sample statistic t_m	$t_1 + \cdots + t_m \leq a_m$	$\Longrightarrow$ terminate, accept
		$t_1 + \cdots + t_m \geq r_m$	$\Longrightarrow$ terminate, reject

found until the jth inspected item equals r_l, inspection is stopped and the lot is rejected. (b) Second-stage curtailment: Starting with curtailment in stage 2 only, so as to obtain an unbiased estimator of the lot quality indicator from the first stage.

Performance Measures of Multiple Sampling Plans

The OC (operating characteristic) function of a type A m-stage sampling plan $(\boldsymbol{n}, \boldsymbol{a}, \boldsymbol{r})$ is the conditional probability $L_{\boldsymbol{n},\boldsymbol{a},\boldsymbol{r}}(q) = P(\text{accept the lot}|\text{lot quality indicator} = q)$ of accepting the lot under the condition that the lot quality indicator (lot proportion nonconforming or lot average number of defects) adopts a value q (*see* **Sampling Inspection of Products** for a general introduction). Table 4 explains and displays the OC function $L_{\boldsymbol{n},\boldsymbol{a},\boldsymbol{r}}(p)$ as a function of the lot proportion nonconforming p under the conforming–nonconforming quality model, and the OC function $L_{\boldsymbol{n},\boldsymbol{a},\boldsymbol{r}}(k)$ as a function of the lot average number of nonconformities k under the nonconformities quality model. The OC functions and further performance measures, see below, are expressed by means of stage l probabilities $A_{\boldsymbol{n},\boldsymbol{a},\boldsymbol{r},l}(q)$, $R_{\boldsymbol{n},\boldsymbol{a},\boldsymbol{r},l}(q)$, $T_{\boldsymbol{n},\boldsymbol{a},\boldsymbol{r},l}(q)$ as explained by Table 4. From a type A view, the probabilities given by Table 4 for the two attributes quality models are approximations: the hypergeometric distribution is approximated by the binomial distribution, and the negative hypergeometric distribution is approximated by the Poisson distribution, compare the approximations discussed by **Single Sampling by Attributes and by Variables**. The approximations are tolerable for large lot size N. From a type B view, where the lot generating process is the inspection target, the values given by Table 4 are exact. For a detailed analysis, see [2] or [3].

Under multiple sampling, the actually reached cumulative sample size SN is a random variable

Table 4 Stage l probabilities and OC functions of m-stage sampling plans $(\boldsymbol{n}, \boldsymbol{a}, \boldsymbol{r})$

	Generic formulas
$A_{\boldsymbol{n},\boldsymbol{a},\boldsymbol{r},l}(q)$	$A_{\boldsymbol{n},\boldsymbol{a},\boldsymbol{r},l}(q) = P(\text{plan } (\boldsymbol{n},\boldsymbol{a},\boldsymbol{r}) \text{ accepts in stage } l \mid \text{lot quality indicator} = q) = \sum_{\substack{0\le j_1\le n_1,\ldots,0\le j_l\le n_l \\ a_i+1\le j_1+\cdots+j_i\le r_i-1,\ i=1,\ldots,l-1 \\ j_1+\cdots+j_l\le a_l}} \rho_{\boldsymbol{n},\boldsymbol{a},\boldsymbol{r},j_1,\ldots,j_l}(q)$
$R_{\boldsymbol{n},\boldsymbol{a},\boldsymbol{r},l}(q)$	$R_{\boldsymbol{n},\boldsymbol{a},\boldsymbol{r},l}(q) = P(\text{plan } (\boldsymbol{n},\boldsymbol{a},\boldsymbol{r}) \text{ rejects in stage } l \mid \text{lot quality indicator} = q) = \sum_{\substack{0\le j_1\le n_1,\ldots,0\le j_l\le n_l \\ a_i+1\le j_1+\cdots+j_i\le r_i-1,\ i=1,\ldots,l-1 \\ j_1+\cdots+j_l\ge r_l}} \rho_{\boldsymbol{n},\boldsymbol{a},\boldsymbol{r},j_1,\ldots,j_l}(q)$
$T_{\boldsymbol{n},\boldsymbol{a},\boldsymbol{r},l}(q)$	$T_{\boldsymbol{n},\boldsymbol{a},\boldsymbol{r},l}(q) = P(\text{plan } (\boldsymbol{n},\boldsymbol{a},\boldsymbol{r}) \text{ terminates in stage } l \mid \text{lot quality indicator} = q) = A_{\boldsymbol{n},\boldsymbol{a},\boldsymbol{r},l}(q) + R_{\boldsymbol{n},\boldsymbol{a},\boldsymbol{r},l}(q)$
OC function	$L_{\boldsymbol{n},\boldsymbol{a},\boldsymbol{r}}(q) = \text{P(sampling plan } (\boldsymbol{n},\boldsymbol{a},\boldsymbol{r}) \text{ accepts} \mid \text{lot quality indicator} = q) = A_{\boldsymbol{n},\boldsymbol{a},\boldsymbol{r},1}(q) + \cdots + A_{\boldsymbol{n},\boldsymbol{a},\boldsymbol{r},m}(q)$
ASN function	$ASN_{\boldsymbol{n},\boldsymbol{a},\boldsymbol{r}}(q) = E[\text{random cumulative sample size} \mid \text{lot quality indicator} = q] = \sum_{l=1}^{m} n_l \sum_{j=l}^{m} T_{\boldsymbol{n},\boldsymbol{a},\boldsymbol{r},j}(q)$
	m-stage sampling plan $(\boldsymbol{n},\boldsymbol{a},\boldsymbol{r})$ for lot proportion nonconforming p
Hypergeometric	$\rho_{\boldsymbol{n},\boldsymbol{a},\boldsymbol{r},j_1,\ldots,j_l}(p) = \prod_{r=1}^{l} \frac{\binom{Np-(j_1+\cdots+j_{r-1})}{j_r}\binom{N-(n_1+\cdots+n_{r-1})-(Np-(j_1+\cdots+j_{r-1}))}{n_r-j_r}}{\binom{N-(n_1+\cdots+n_{r-1})}{n_r}}$
Binomial	$\rho_{\boldsymbol{n},\boldsymbol{a},\boldsymbol{r},j_1,\ldots,j_l}(p) = \binom{n_1}{j_1}\cdots\binom{n_l}{j_l} p^{j_1+\cdots+j_l}(1-p)^{n_1+\cdots+n_l-j_1-\cdots-j_l}$
Poisson	$\rho_{\boldsymbol{n},\boldsymbol{a},\boldsymbol{r},j_1,\ldots,j_l}(p) = \frac{n_1^{j_1}}{j_1!}\cdots\frac{n_l^{j_l}}{j_l!} p^{j_1+\cdots+j_l} \exp(-(n_1+\cdots+n_l)p)$
	m-stage sampling plan $(\boldsymbol{n},\boldsymbol{a},\boldsymbol{r})$ for lot average number of nonconformities k
Poisson	$\rho_{\boldsymbol{n},\boldsymbol{a},\boldsymbol{r},j_1,\ldots,j_l}(k) = \frac{n_1^{j_1}}{j_1!}\cdots\frac{n_l^{j_l}}{j_l!} k^{j_1+\cdots+j_l} \exp(-(n_1+\cdots+n_l)k)$

Table 5 The sampling and decision algorithm of a sequential sampling plan (n, s, h_1, h_2)

Stage	Operation	Decision algorithm	
1	Sample n items Calculate sample statistic t_1	$t_1 \leq s \cdot n - h_1$ $t_1 \geq s \cdot n + h_2$ $s \cdot n - h_1 < t_1 < s \cdot n + h_2$	$\Longrightarrow$ terminate, accept $\Longrightarrow$ terminate, reject $\Longrightarrow$ continue with stage 2
$\vdots$	$\vdots$	$\vdots$	
l	Sample n items Calculate sample statistic t_l	$t_1 + \cdots t_l \leq sln - h_1$ $t_1 + \cdots t_l \geq sln + h_2$ $-h_1 < t_1 + \cdots + t_l - sln < h_2$	$\Longrightarrow$ terminate, accept $\Longrightarrow$ terminate, reject $\Longrightarrow$ continue with stage $l + 1$
$\vdots$	$\vdots$	$\vdots$	

in which $SN = n_l^{(c)} = n_1 + \cdots + n_l$ iff the procedure terminates at stage l, $1 \leq l \leq m$. A suitable performance measure is the expectation $E[SN]$, the so-called average sample number (*ASN*), which as a function of the sampling plan parameters and the lot quality indicator is denoted as $ASN_{\boldsymbol{n},\boldsymbol{a},\boldsymbol{r}}(q)$. $ASN_{\boldsymbol{n},\boldsymbol{a},\boldsymbol{r}}(q)$ can be calculated by the termination probabilities $T_{\boldsymbol{n},\boldsymbol{a},\boldsymbol{r},1}(q), \ldots, T_{\boldsymbol{n},\boldsymbol{a},\boldsymbol{r},m}(q)$ in the manner described by Table 4.

Under curtailed inspection, the random cumulative sample size is $SN = n_l^{(c)} = n_1 + \cdots + n_l$, if the procedure terminates at stage l with acceptance, and $SN = n_1 + \cdots + n_{l-1} + j$, if the procedure terminates at stage l with rejection, where j is the smallest number such that $t_1 + \cdots + t_{l-1} + j = r_l$. Formulas for the $ASN = E[SN]$ under curtailment are provided by Montgomery [4] and in more technical detail by Uhlmann [2] and Hald [5].

Sequential Sampling

Sequential sampling plans and the corresponding decision algorithm are defined in a way analogous to the definition of multiple sampling plans, except that there is no upper limit m imposed on the number of sampling stages, i.e. the triple $(\boldsymbol{n}, \boldsymbol{a}, \boldsymbol{r})$ consists of infinite sequences. Tables 2 and 3 remain valid for explaining the parameters and the decision algorithm, except that there is no terminating restriction.

Usually, sequential sampling plans are simplified by the following two assumptions: (a) The stage sample sizes are identical, i.e. $n = n_1 = n_2 = \ldots$. (b) The differences of rejection and acceptance numbers remain invariant, i.e., $d = r_l - a_l$ for $l = 1, 2, \ldots$. Under the simplifications (a) and (b), a sequential sampling plan can be defined by a tuple (n, s, h_1, h_2) where n is the invariant sample size, s is a *slope*, h_1, h_2 are *increments*, and where $a_l = sln - h_1$ are the acceptance numbers, $r_l = sln + h_2$ the rejection numbers. The operation of a sequential sampling plan (n, s, h_1, h_2) is illustrated by Table 5. Plans with sample size $n \geq 2$ are called *group sequential plans*. *Item-by-item sequential plans* (*unit sequential plans*) with $n = 1$, i.e. only one item is drawn at each stage, are very common. If there are no economic or administrative restrictions for the sample (group) size n in a group sequential plan, Cowden [6] suggests the use of the smallest group size under which acceptance is possible in the first stage (first sample), i.e. n is the smallest integer exceeding h_1/s.

As for multiple sampling, the essential performance measures are the OC and the ASN function. The OC function of a type A sequential sampling plan (n, s, h_1, h_2) is the conditional probability $L_{n,s,h_1,h_2}(q) = P(\text{accept the lot} \mid \text{lot quality indicator} = q)$ of accepting the lot under the condition that the lot quality indicator (lot proportion nonconforming or lot average number of defects) adopts a value q. The ASN function is $ASN_{n,s,h_1,h_2}(q) = E$ [cumulative sample size | lot quality indicator $= q$). In principle, the formulas given in Table 4 can be adapted to the sequential case by substituting finite by infinite sums. However, the resulting expressions are difficult both for theoretical and for numerical analysis. Better closed expressions and approximations are provided by [5].

Design by Prescribing Two Points of the OC Function

The design of sampling plans based on prescribing two points of the OC function is described for simple sampling plans by **Single Sampling by Attributes and by Variables**. The same approach can be used for designing iterative sampling plans: a good quality level q_1 should be accepted with a high probability $L(q_1) \geq 1 - \alpha$, and a poor quality level q_2 should be accepted with a small probability $L(q_2) \leq \beta$, where $0 < \beta < 1 - \alpha$.

For multiple sampling plans, the two-points design does not determine a unique plan among all plans $(\boldsymbol{n}, \boldsymbol{a}, \boldsymbol{r})$. Additional constraints are necessary to assure uniqueness, e.g. requiring that the sample sizes satisfy a geometric progression $n_{i+1} = cn_i$. See [7] for a survey.

For sequential item-by-item sampling plans $(1, s, h_1, h_2)$, the simple approximate solution developed by [8] is displayed by Table 6. The results obtained by Wald [8] include approximations to the OC and ASN functions of the two-points design plan $(1, s, h_1, h_2)$.

Iterative Sampling with Rectification

Under rectifying sampling for lot proportion nonconforming, a rejected lot is subject to 100% inspection. All nonconforming items discovered in the sample and, in case of rejection, in the lot are replaced by conforming ones. The performance measures of iterative rectification plans are defined analogously to the performance measures of single rectification plans (*see* **Single Sampling by Attributes and by Variables**): as a function of the proportion nonconforming p in the submitted lot, the *average outgoing quality* $AOQ(p)$ is the expected value of the lot proportion nonconforming after application of a rectification plan, and the *average total inspection* $ATI(p)$ is the expected value of inspected items. The *average outgoing quality limit* is the maximum $AOQL = \max_{0 \leq p \leq 1} AOQ(p)$. Table 7 displays exact formulas for multiple sampling plans, and approximate formulas for sequential item-by-item plans designed by prescribing two points of the OC function as shown in Table 6, see [7] for a derivation of the approximations.

Comparison of Single and Iterative Sampling

The intuitive motivation of iterative sampling is a reduction of expected sample size under extreme (very good or very bad) lot quality. For a simple quantitative comparison, one can compare the ASN functions of single and iterative plans which are matched to have approximately the same OC function. An instance of this comparison under inspection

Table 6 Design of an item-by-item sequential sampling $(1, s, h_1, h_2)$ plan by prescribing two points of the OC function

Design rule	$0 < q_1 < q_2 < 1, \quad 0 < \beta < 1 - \alpha, \quad L_{1,s,h_1,h_2}(q_1) \geq 1 - \alpha \geq \beta \geq L_{1,s,h_1,h_2}(q_2)$
	Item-by-item plan $(1, s, h_1, h_2)$ for lot proportion nonconforming p
Increments	$h_1 = \frac{\log(1-\alpha)-\log(\beta)}{\log\left(\frac{p_2}{p_1}\right)+\log\left(\frac{1-p_1}{1-p_2}\right)}, \quad h_2 = \frac{\log(1-\beta)-\log(\alpha)}{\log\left(\frac{p_2}{p_1}\right)+\log\left(\frac{1-p_1}{1-p_2}\right)}$
Slope	$s = \frac{\log(1-p_1)-\log(1-p_2)}{\log\left(\frac{p_2}{p_1}\right)+\log\left(\frac{1-p_1}{1-p_2}\right)}$
	Item-by-item plan $(1, s, h_1, h_2)$ for lot average number of nonconformities k
Increments	$h_1 = \frac{\log(1-\alpha)-\log(\beta)}{\log(k_2)-\log(k_1)}, \; h_2 = \frac{\log(1-\beta)-\log(\alpha)}{\log(k_2)-\log(k_1)}$
Slope	$s = \frac{k_2-k_1}{\log(k_2)-\log(k_1)}$
	Approximate OC and ASN functions
Generic OC	$L_{1,s,h_1,h_2}(q) = \frac{\exp(h_2 t)-1}{\exp(h_2 t)-\exp(-h_1 t)}, \quad t = t(q)$
$q = p$	$t(p)$ defined as the inverse $p(t) = \frac{\exp(st)-1}{\exp(t)-1}$
$q = k$	$t(k)$ defined as the inverse $k(t) = \frac{st}{\exp(t)-1}$
Generic ASN	$\frac{h_2-(h_1+h_2)L_{1,s,h_1,h_2}(q)}{q-s}$, to be used with $q = p$ or $q = k$

Table 7 Performance measures of iterative rectification plans for lot proportion nonconforming p

Multiple m-stage plan $(\boldsymbol{n}, \boldsymbol{a}, \boldsymbol{r})$[(a)]	
$AOQ(p)$	$AOQ(p) = \sum_{l=1}^{m} p\left(1 - \frac{n_1+\cdots+n_l}{N}\right) A_{\boldsymbol{n},\boldsymbol{a},\boldsymbol{r},l}(p)$
$ATI(p)$	$ATI(p) = \sum_{l=1}^{m} (n_1 + \cdots + n_l) A_{\boldsymbol{n},\boldsymbol{a},\boldsymbol{r},l}(p) + N\left(1 - L_{\boldsymbol{n},\boldsymbol{a},\boldsymbol{r},l}(p)\right)$
Item-by-item sequential plan $(1, s, h_1, h_2)$[(b)]	
$AOQ(p)$	$AOQ(p) \approx pL_{1,s,h_1,h_2}(p)$
$ATI(p)$	$ATI(p) \approx \frac{L_{1,s,h_1,h_2}(p)(\log(\beta)-\log(1-\alpha))}{p\log\left(\frac{p_2}{p_1}\right)+(1-p)\log\left(\frac{1-p_2}{1-p_1}\right)} + N\left(1 - L_{1,s,h_1,h_2}(p)\right)$

(a) See Table 4 for the probabilities $A_{\boldsymbol{n},\boldsymbol{a},\boldsymbol{r},l}(p)$ and the OC $L_{\boldsymbol{n},\boldsymbol{a},\boldsymbol{r},l}(p)$
(b) Notation and quantities from Table 6

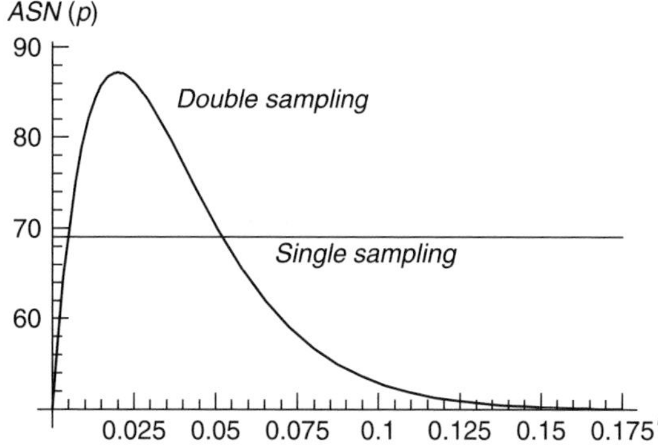

Figure 1 Average sample number $ASN(p)$ for single sampling plan (69, 1) and double sampling plan with $\boldsymbol{n} = (50, 100)$, $\boldsymbol{a} = (0, 3)$, $\boldsymbol{r} = (2, 4)$

for lot proportion nonconforming p is displayed by Figure 1. For the single sampling plan (69, 1) and the double sampling plan with $\boldsymbol{n} = (50, 100)$, $\boldsymbol{a} = (0, 3)$, $\boldsymbol{r} = (2, 4)$, the OC functions are very close. With respect to the ASN, the double plan is superior to the single plan for p smaller than 0.005 and p larger than 0.06. For the large intermediate range from 0.5 to 6% single sampling is better.

Opposed to the advantage of iterative sampling for extreme lot quality are several disadvantages of iterative sampling: administrative difficulties; incompetence of inspectors to execute the algorithm correctly; hidden setup sampling costs from repeated sampling, in particular under sequential plans. Iterative methods should be preferred over single sampling only if the product quality history inspires data-based confidence that lot quality is really on one of the extreme sides.

References

[1] Dodge, H.F. (1969). Notes on the evolution of acceptance sampling plans. Parts I, II, III, *Journal of Quality Technology* **1**, 77–88, 155–162, 225–232.
[2] Uhlmann, W. (1982). *Statistische Qualitätskontrolle*, B.G. Teubner, Stuttgart.
[3] Schilling, E.G. (1983). *Acceptance Sampling in Quality Control*, Marcel Dekker, New York.
[4] Montgomery, D.C. (2005). *Introduction to Statistical Quality Control*, 5th Edition, John Wiley & Sons, New York.
[5] Hald, A. (1981). *Statistical Theory of Sampling Inspection by Attributes*, Academic Press, London.
[6] Cowden, D.J. (1957). *Statistical Methods in Quality Control*, Prentice-Hall, Englewood.
[7] Duncan, A.J. (1986). *Quality Control and Industrial Statistics*, 5th Edition, Irwin Publishing, Homewood.
[8] Wald, A. (1947). *Sequential Analysis*, John Wiley & Sons, New York.

Kodakanallur Krishnaswamy Suresh and Rainer Göb

Multistate Reliability Theory

Introduction

An inherent weakness of traditional reliability theory (*see* **Coherent Systems**) is that the system and the components are always described either as functioning or failed. Early attempts to replace this by a theory of multistate systems for multistate components were made in the late 1970s in [1, 2] and [3]. This was followed up by independent work in [4, 5] and [6] giving proper definitions of a multistate monotone system (MMS) and of multistate **coherent systems** (MCSs) and also of minimal **path and cut** vectors. Furthermore, in [7] upper and lower bounds for the availabilities and unavailabilities, to any level, in a fixed time interval, were obtained for MMSs based on corresponding information on the multistate components. These were assumed to be maintained and interdependent. Such bounds are of great interest when trying to predict the performance process of the system, noting that the exact expressions are obtainable just for trivial systems. Hence, by the mid-1980s the basic multistate reliability theory was established. A review of the early development in this area is given in [8]. Very recently probabilistic modeling of partial monitoring of components with applications to preventive system maintenance has been extended in [9] to MMSs of multistate components.

The theory was applied in [10] to an offshore electrical power generation system for two nearby oilrigs, where the amount of power that may possibly be supplied to the two oilrigs, is considered as system states. This application is also used to illustrate the theory in [9]. In [11] the theory was applied to the Norwegian offshore gas pipeline network in the North Sea, as of the end of the 1980s, transporting gas to Emden in Germany. The system state depends not only on the amount of gas actually delivered, but also to some extent on the amount of gas compressed, mainly by the compressor component closest to Emden. Recently, the first book [12] on multistate system reliability analysis and optimization was printed. The book also contains several examples of application of reliability assessment and optimization methods to real engineering problems.

Basic Concepts and Basic Bounds

Let $S = \{0, 1, \ldots, M\}$ be the set of states of the system; then the $M + 1$ states represent successive levels of performance ranging from the perfect functioning level M down to the complete failure level 0. Furthermore, let $C = \{1, \ldots, n\}$ be the set of components and S_i $(i = 1, \ldots, n)$ the set of states of the ith component. We claim $\{0, M\} \subseteq S_i \subseteq S$. Hence, the states 0 and M are chosen to represent the endpoints of a performance scale that might be used for both the system and its components. Let x_i $(i = 1, \ldots, n)$ denote the state or performance level of the ith component and $\boldsymbol{x} = (x_1, \ldots, x_n)$. It is assumed that the state, ϕ, of the system is given by the structure function $\phi = \phi(\boldsymbol{x})$. For the following type of multistate systems a series of results can be derived.

Definition 1 A system is *an MMS* iff its structure ϕ satisfies:

(i) $\phi(\boldsymbol{x})$ is nondecreasing in each argument
(ii) $\phi(\boldsymbol{0}) = 0$ and $\phi(\boldsymbol{M}) = M(\boldsymbol{0} = (0, \ldots, 0)$, $\boldsymbol{M} = (M, \ldots, M))$.

The first assumption says that improving one of the components cannot harm the system, whereas the second says that if all components are in the complete failure (perfect functioning) state, then the system is in the complete failure (perfect functioning) state.

We now impose some further restrictions on the structure function ϕ. The following notation is required:

$$(\cdot_i, \boldsymbol{x}) = (x_1, \ldots, x_{i-1}, \cdot, x_{i+1}, \ldots, x_n),$$
$$S^0_{i,j} = S_i \cap \{0, \ldots, j-1\} \text{ and}$$
$$S^1_{i,j} = S_i \cap \{j, \ldots, M\} \tag{1}$$

Definition 2 Consider an MMS with structure function ϕ satisfying

(i) $\min_{1\le i\le n} x_i \le \phi(\boldsymbol{x}) \le \max_{1\le i\le n} x_i$.

If in addition $\forall i \in \{1, \ldots, n\}$, $\forall j \in \{1, \ldots, M\}$, $\exists(\cdot_i, \boldsymbol{x})$ such that

(ii) $\phi(k_i, \boldsymbol{x}) \ge j, \phi(\ell_i, \boldsymbol{x}) < j, \forall k \in S^1_{i,j}, \forall \ell \in S^0_{i,j}$, we have a *multistate strongly coherent system (MSCS)*,
(iii) $\phi(k_i, \boldsymbol{x}) > \phi(\ell_i, \boldsymbol{x})$ $\forall k \in S^1_{i,j}$, $\forall \ell \in S^0_{i,j}$, we have an *MCS*,
(iv) $\phi(M_i, \boldsymbol{x}) > \phi(0_i, \boldsymbol{x})$, we have a *multistate weakly coherent system (MWCS)*.

When $M = 1$, all this reduces to the established binary coherent system (BCS) (*see* **Coherent Systems**). The structure function $\min_{1\le i\le n} x_i$ $\left(\max_{1\le i\le n} x_i\right)$ is often denoted as the multistate series (parallel) structure.

Now choose $j \in \{1, \ldots, M\}$ and let the states $S^0_{i,j}$, $(S^1_{i,j})$ correspond to the failure (functioning) state for the ith component, if a binary approach is used. Condition (ii) above, means that for all components i and any level j, there shall exist a combination of the states of the other components, $(\cdot_i, \boldsymbol{x})$, such that if the ith component is in the binary failure (functioning) state, the system itself is in the corresponding binary failure (functioning) state. In general, modifying [6], condition (ii) says that every level of each component is relevant to the same level of the system, condition (iii) says that every level of each component is relevant to the system, whereas condition (iv) simply says that every component is relevant to the system.

For a BCS one can prove the following, practically very useful principle: **redundancy** at the component level is superior to redundancy at the system level except for a parallel system where it makes no difference. Assuming $S_i = S$ $(i = 1, \ldots, n)$ then this is also true for an MCS, but not for an MWCS.

We now discuss a special type of an MSCS. Define the indicators $(j = 1, \ldots, M)$ $I_j(x_i) = 1(0)$ if $x_i \geq j (x_i < j)$ and the indicator vector $(\boldsymbol{I}_j(\boldsymbol{x})) = (I_j(x_1), \ldots, I_j(x_n))$.

Definition 3 An MSCS is said to be a *binary type multistate strongly coherent system (BTMSCS)* iff there exist binary coherent structures ϕ_j, $j = 1, \ldots, M$, such that its structure function ϕ satisfies $\phi(\boldsymbol{x}) \geq j \Leftrightarrow \phi_j(\boldsymbol{I}_j, \boldsymbol{x}) = 1$ for all $j \in \{1, \ldots, M\}$ and all $\boldsymbol{x}$.

Choose again $j \in \{1, \ldots, M\}$ and let the states $S^0_{i,j}(S^1_{i,j})$ correspond to the failure (functioning) state for the ith component, if a binary approach is applied. By the definition above, ϕ_j will, from the binary states of the components, uniquely determine the corresponding binary state of the system.

In what follows $\boldsymbol{y} < \boldsymbol{x}$ means $y_i \leq x_i$ for $i = 1, \ldots, n$, and $y_i < x_i$ for some i.

Definition 4 Let ϕ be the structure function of an MMS and let $j \in \{1, \ldots, M\}$. A vector $\boldsymbol{x}$ is said to be a *minimal path (cut) vector to level* j iff $\phi(\boldsymbol{x}) \geq j$ and $\phi(\boldsymbol{y}) < j$ for all $\boldsymbol{y} < \boldsymbol{x}$ $(\phi(\boldsymbol{x}) < j$ and $\phi(\boldsymbol{y}) \geq j$ for all $\boldsymbol{y} > \boldsymbol{x})$.

Definition 5 The *performance process of the* ith *component* $(i = 1, \ldots, n)$ is a stochastic process $\{X_i(t), t \in [0, \infty)\}$, where for each fixed $t \in [0, \infty)$ $X_i(t)$ is a random variable which takes values in S_i. The *joint performance process for the components* $\{\boldsymbol{X}(t), t \in [0, \infty)\} = \{(X_1(t), \ldots, X_n(t)), t \in [0, \infty)\}$ is the corresponding vector stochastic process. The *performance process of an MMS* with structure function ϕ is a stochastic process $\{\phi(\boldsymbol{X}(t)), t \in [0, \infty)\}$, where for each fixed $t \in [0, \infty)$, $\phi(\boldsymbol{X}(t))$ is a random variable which takes values in S.

Definition 6 The performance processes $\{X_i(t), t \in [0, \infty)\}$, $i = 1, \ldots, n$ are *independent* in the time interval I iff, for any integer m and $\{t_1, \ldots, t_m\} \subset I$ the random vectors $\{X_1(t_1), \ldots, X_1(t_m)\}, \ldots, \{X_n(t_1), \ldots, X_n(t_m)\}$ are independent.

Definition 7 Let $j \in \{1, \ldots, M\}$. The *availability*, $h^{j(I)}_\phi$ and the *unavailability*, $g^{j(I)}_\phi$ to level j in the time interval I for an *MMS* with structure function ϕ are given by

$$h^{j(I)}_\phi = P[\phi(\boldsymbol{X}(s)) \geq j \ \ \forall s \in I],$$

$$g^{j(I)}_\phi = P[\phi(\boldsymbol{X}(s)) < j \ \ \forall s \in I] \tag{2}$$

Note that $h^{j(I)}_\phi + g^{j(I)}_\phi \leq 1$, with equality for the case $I = [t, t]$.

As an example of the bounds for $h^{j(I)}_\phi$ and $g^{j(I)}_\phi$ given in [7], we give the following theorem by first introducing the $n \times M$ matrices

$$\boldsymbol{P}^{(I)}_\phi = \{p^{j(I)}_i\}^{i=1,\ldots,n}_{j=1,\ldots,M}$$

$$\boldsymbol{Q}^{(I)}_\phi = \{q^{j(I)}_i\}^{i=1,\ldots,n}_{j=1,\ldots,M} \tag{3}$$

Note that according to [13] we do not need to assume that each of the performance processes of the components is associated in I.

Theorem 1 *Let (C, ϕ) be an MMS with the marginal performance processes of its components being independent in I. Furthermore, for $j \in \{1, \ldots, M\}$ let $\boldsymbol{y}^j_k = (y^j_{1k}, \ldots, y^j_{nk})$, $k = 1, \ldots, n^j$ $(\boldsymbol{z}^j_k = (z^j_{1k}, \ldots, z^j_{nk})$, $k = 1, \ldots, m^j)$ be its minimal path (cut) vectors to level j. Define*

$$\ell^{j'}_\phi(\boldsymbol{P}^{(I)}_\phi) = \max_{1 \leq k \leq n^j} \prod_{i=1}^n p_i^{y^j_{ik}(I)}$$

$$\bar{\ell}^{j'}_\phi(\boldsymbol{Q}^{(I)}_\phi) = \max_{1 \leq k \leq m^j} \prod_{i=1}^n q_i^{z^j_{ik}+1(I)} \tag{4}$$

$$\ell_{\phi}^{j^*}(\boldsymbol{P}_{\phi}^{(I)}) = \prod_{k=1}^{m^j} \coprod_{i=1}^{n} p_i^{z_{ik}^j+1(I)}$$

$$\bar{\ell}_{\phi}^{j^*}(\boldsymbol{Q}_{\phi}^{(I)}) = \prod_{k=1}^{n^j} \coprod_{i=1}^{n} q_i^{y_{ik}^j(I)} \quad (5)$$

$$B_{\phi}^{j}(\boldsymbol{P}_{\phi}^{(I)}) = \max_{j\le k\le M} \times \left\{\max\left[\ell_{\phi}^{k'}(\boldsymbol{P}_{\phi}^{(I)}), \ell_{\phi}^{k^*}(\boldsymbol{P}_{\phi}^{(I)})\right]\right\} \quad (6)$$

$$\bar{B}_{\phi}^{j}(\boldsymbol{Q}_{\phi}^{(I)}) = \max_{1\le k\le j} \times \left\{\max\left[\bar{\ell}_{\phi}^{k'}(\boldsymbol{Q}_{\phi}^{(I)}), \bar{\ell}_{\phi}^{k^*}(\boldsymbol{Q}_{\phi}^{(I)})\right]\right\} \quad (7)$$

Then

$$B_{\phi}^{j}(\boldsymbol{P}_{\phi}^{(I)}) \le h_{\phi}^{j(I)} \le \inf_{t\in I}\left[1 - \bar{B}_{\phi}^{j}(\boldsymbol{Q}_{\phi}^{([t,t])})\right] \le 1 - \bar{B}_{\phi}^{j}(\boldsymbol{Q}_{\phi}^{(I)}) \quad (8)$$

$$\bar{B}_{\phi}^{j}(\boldsymbol{Q}_{\phi}^{(I)}) \le g_{\phi}^{j(I)} \le \inf_{t\in I}\left[1 - B_{\phi}^{j}(\boldsymbol{P}_{\phi}^{([t,t])})\right] \le 1 - B_{\phi}^{j}(\boldsymbol{P}_{\phi}^{(I)}) \quad (9)$$

Here $\coprod_{i=1}^{n} a_i \stackrel{\text{def}}{=} 1 - \prod_{i=1}^{n}(1 - a_i)$. By specializing $M = 1$ and $I = [t, t]$ the bounds reduce to the familiar ones from binary theory as given in [14].

An Offshore Electrical Power Generation System

The purpose of the offshore electrical power generation system considered in [9] and [10], depicted in Figure 1, is to supply two nearby oilrigs with electrical power. Both oilrigs have their own main generation, represented by equivalent generators A_1 and A_3, each having a capacity of 50 MW. In addition oilrig 1 has a standby generator A_2 that is switched into the network in case of outage of A_1 or A_3. A_2 also has a capacity of 50 MW. The control unit, U, continuously supervises the supply from each of the generators with an automatic control of the switches. If, for instance, the supply from A_3 to oilrig 2 is not sufficient, whereas the supply from A_1 to oilrig 1 is sufficient, U can activate A_2 to supply oilrig 2 with electrical power through the standby subsea cables L.

The components to be considered here are A_1, A_2, A_3, U and L. We let the perfect functioning level M equal 4 and let the set of states of all components be $\{0, 2, 4\}$. For A_1, A_2 and A_3 these states are interpreted as

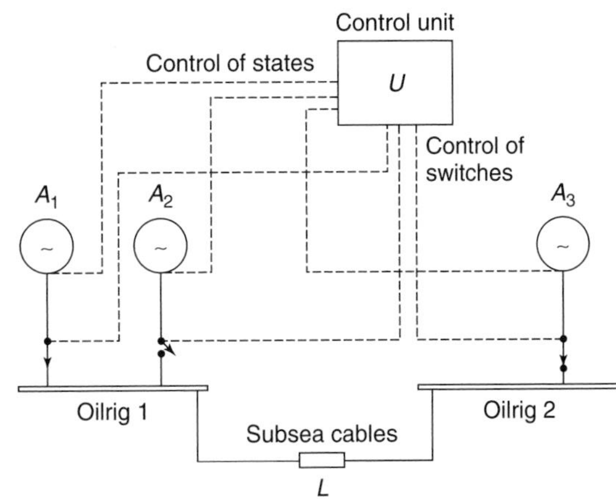

Figure 1 Outline of an offshore electrical power generation system

0: The generator cannot supply any power.
2: The generator can supply maximum 25 MW power.
4: The generator can supply maximum 50 MW power.

Note that as an approximation we have chosen to describe supply capacity of the generators using a discrete scale of three points. The supply capacity is not a measure of the actual amount of power delivered at a fixed point of time.

The control unit U has the states

0: U will by mistake switch off the main generators A_1 and A_3 without switching on A_2;
2: U will not switch on A_2 when needed;
4: U is functioning perfectly.

The subsea cables L are actually assumed to be constructed as double cables transferring half of the power through each cable. This leads to the following states of L

0: No power is transferred.
2: 50% of the power is transferred.
4: 100% of the power is transferred.

Let us now for simplicity assume that the mechanism that distributes the power from A_2 to oilrig 1 or 2 is working perfectly, transferring excess power from A_2 to oilrig 2 if oilrig 1 is ensured a delivery corresponding to state 4. Now let $\phi(A_1, A_2, A_3, U, L)$ = the amount of power that can be supplied to oilrig 2. In addition to the states taken by A_1, A_2, A_3, ϕ can also take the following states

1: The maximum amount of power that can be supplied is 12.5 MW.
3: The maximum amount of power that can be supplied is 37.5 MW.

Number the components $A_1, A_2, A_3, U,$ and L successively as 1, 2, 3, 4, and 5. Then this leads to

$$\phi(\boldsymbol{x}) = I(x_4 > 0)\min(x_3 + \max(x_1 + x_2 I(x_4 = 4) - 4, 0)x_5/4, 4) \quad (10)$$

noting that $\max(x_1 + x_2 I(x_4 = 4) - 4, 0)$ is just the excess power from A_2, which one tries to transfer to oilrig 2. This is obviously a MMS.

References

[1] Barlow, R.E. & Wu, A.S. (1978). Coherent systems with multistate components, *Mathematics of Operations Research* **4**, 275–281.
[2] El-Neweihi, E., Proschan, F. & Sethuraman, J. (1978). Multistate coherent systems, *Journal of Applied Probability* **15**, 675–688.
[3] Ross, S. (1979). Multivalued state component reliability systems, *Annals of Probability* **7**, 379–383.
[4] Griffith, W. (1980). Multistate reliability models, *Journal of Applied Probability* **17**, 735–744.
[5] Natvig, B. (1982). Two suggestions of how to define a multistate coherent system, *Advances in Applied Probability* **14**, 434–455.
[6] Block, H.W. & Savits, T.S. (1982). A decomposition for multistate monotone systems, *Journal of Applied Probability* **19**, 391–402.
[7] Funnemark, E. & Natvig, B. (1985). Bounds for the availabilities in a fixed time interval for multistate monotone systems, *Advances in Applied Probability* **17**, 638–665.
[8] Natvig, B. (1985). Multistate coherent systems, in *Encyclopedia of Statistical Sciences*, N.L. Johnson & S. Kotz, eds, John Wiley & Sons Vol. 5, pp. 732–735.
[9] Gåsemyr, J. & Natvig, B. (2005). Probabilistic modelling of monitoring and maintenance of multistate monotone systems with dependent components, *Methodology and Computing in Applied Probability* **7**, 63–78.
[10] Natvig, B., Sørmo, S., Holen, A.T. & Høgåsen, G.T. (1986). Multistate reliability theory – a case study, *Advances in Applied Probability* **18**, 921–932.
[11] Natvig, B. & Mørch, H.W. (2003). An application of multistate reliability theory to an offshore gas pipeline network, *International Journal of Reliability, Quality and Safety Engineering* **10**, 361–381.
[12] Lisnianski, A. & Levitin, G. (2003). *Multi-State System Reliability*, World Scientific, London.
[13] Natvig, B. (1993). Strict and exact bounds for the availabilities in a fixed time interval for multistate monotone systems, *Scandinavian Journal of Statistics* **20**, 171–175.
[14] Barlow, R.E. & Proschan, F. (1975). *Statistical Theory of Reliability and Life Testing: Probability Models*, Holt, Rinehart and Winston, New York.

Related Articles

Coherent Systems; **Multicomponent Maintenance**; **Path Sets and Cut Sets in System Reliability Modeling**; **Reliability of Redundant Systems**; **System Availability**.

BENT NATVIG

Multi-Vari Charts

What are Multi-Vari Charts?

Multi-vari charts are a graphical form of analysis of variance. You can use them most effectively for problem solving. However, you can also use them to estimate variation in well-behaved processes.

Len Seder

Leonard Seder (1915–2004) invented multi-vari charts in the late 1940s. He published a two-part article "Diagnosis with Diagrams" in Industrial Quality Control in 1950 (January and March). Len gave credit to Dr Joseph M. Juran for the concept of graphing process variability in a manner similar to a stock report with daily high, low, and closing prices represented by a vertical line for the range and a point for the closing. Len described multi-vari charts as a diagnostic tool, and compared them to medical diagnostic tools like X-rays or electrocardiograms. He stressed the power of an effective graphical presentation. "The question is, then, how can these powerful techniques be made more available? One answer is to be found, in the author's opinion, in the use of graphics in place of statistics. Much of the success of the simpler statistical quality control techniques is undeniably attributable to the forcefulness and conciseness of graphic presentation."

Why do Multi-Vari Charts Work?

A well-planned and executed multi-vari chart will separate overall process variation into families. You will clearly see the family causing most of the change and patterns of variation within the largest family, which will provide further clues about the source of variation. The power of a multi-vari chart comes from the application of the **Pareto** principle to the sources of variation in a process.

Dorian Shainin

Dorian Shainin (1914–2000) attended MIT with Len Seder. The two men established a professional relationship with Dr Juran in the 1940s. At that time, Dorian was in charge of quality and reliability at Hamilton Standard (a division of United Aircraft) and Len was with Gillette Safety Razor Company in Boston. Dorian was an early adopter of Len's multi-vari charts. He was also an admirer of Dr Juran's application of the Pareto principle to project selection.

In the 1950s, Dorian concluded that just as the Pareto principle applied to the effect individual problems were having on business performance, it must apply to the effect that individual variables had on changes in product or process performance. If a problem could be expressed in terms of variation, i.e., too much change in the values of product or process outputs, and then the Pareto principle requires that one cause must be responsible for most of the change. Dorian called this dominant cause the *big Red X*. The Red X paradigm does not suggest that the Red X is the only source of variation. In fact, it recognizes that there are thousands of potential causes. The Red X causes most of the variation. Find it and control it and you will make a significant improvement in performance.

Red X Paradigm

The Red X paradigm changes the way problems are attacked. If there is a dominant cause of variation, finding it and controlling it is the only path to improvement. The statistical law underlying analysis of variance states that independent sources of variation combine as the square root of the sum of the squares. If one source causes a range of 5 units and another independent source is causing 1 unit, the combined range will be 5.1 units (the square root of 26), not six (the sum of $5 + 1$). The 5-unit cause is the Red X and it is 25 times stronger than the cause contributing 1 unit.

Once you focus on finding the Red X, the priority becomes finding it quickly, with minimal resources. The best approach is a progressive search, using a process of elimination. Eliminate everything that cannot be the Red X and whatever is left must be the Red X. This is a classic approach for solving criminal cases. Sherlock Holmes expressed it well in Sir Arthur Conan Doyle's "*The Sign of the Four*": "Watson, how often have I told you? When you eliminate the impossible, whatever remains, however improbable must be the truth." Effective methods converge rapidly on the true root cause by eliminating the impossible.

Dictionary Game Thinking

Dorian illustrated the concept with a parlor game called the *dictionary game*. A secret word is chosen. You may only ask questions that can be answered with a "yes" or "no". And you are given a dictionary as a tool to help discover the word. Uninitiated contestants often use a 20-questions approach. The initial common questions are as follows:

- "Is it a person?"
- "Is it a place?"
- "Is it a thing?"

Since we are looking for a word, additional questions might be as follows:

- "Does it have more than five letters?"
- "Is it a noun?"
- "Does it have more than two syllables?"

The keys to the dictionary game are listed below:

1. Use a process of elimination.
2. Take advantage of the dictionary's structure (it is alphabetical).
3. Start with broad questions (eliminate a lot).

4. Narrow the questions as the search progresses.
5. Ask questions that will narrow down the search no matter the answer.

Here is an example. Let us choose "desktop" as the secret word. Start the search by finding the center of the dictionary by page number. In my dictionary, the first "a" word is on page 43 and the last "z" word is on page 1373. The midpoint is page 708. The first word on page 709 is "lucky". Asking, "does the word come before 'lucky' ", will eliminate half the possible words. It does not matter if the answer is "yes" or "no". In our example, the correct answer is "yes". The word "lucky" and all words following it have been eliminated.

The next step is to divide the remaining section in half and repeat the process. The next midpoint is on page 375. The first word on page 375 is "domino". The question is now: "Does your word come before 'domino'?" Again the answer is yes and another quarter of the dictionary has been eliminated.

Within 10 questions, the search will be narrowed to a single page. In my dictionary, "desktop" is on page 344. The pages are divided into two columns. Applying the question to the word atop of column two, "despair", narrows the search to the first column. There are 26 words in the column. Word 13 is "desk". The question is now: "Does your word come before 'desk'?" The answer is no.

Within 17 questions, we would have eliminated all the words in the dictionary, except "desktop". The first question eliminated half the words; the second a quarter; and the last question only eliminated one. Notice that even though the question form stays the same, the questions start broad and become narrower as the search progresses.

Dictionary Game Thinking with Multi-Vari Charts

A multi-vari chart is an important tool for finding a process Red X. By separating the overall process variation into families of variation, all but the largest family can be eliminated. The search for the Red X is quickly narrowed to a small part of the process.

The Case of the Incapable Lathe

In 1966, Dorian published an article in Industrial Quality Control. It was titled "The Case of the Incapable Lathe" and it illustrated the power of multi-vari charts in finding a Red X.

A manufacturer was making jet engine fuel pump controllers. A key component was a rotor shaft. Shaft diameter was an important dimension. The shafts were turned on a lathe. A capability study determined that the process was not capable. Management put pressure on manufacturing to fix the problem. After failing to get tolerance relief from engineering, manufacturing assigned a team of experts to examine the problem and recommend a solution. The team concluded that the lathe was old and worn out and simply not capable of maintaining the tolerance of ± 0.001 in (Figures 1 and 2). They believed the lathe had worn bushings and spindles and needed to be replaced. They recommended that management purchase a new lathe. Management was not convinced and asked Dorian to investigate.

The process produced shafts on the lathe. Finished shafts were deposited in a basket. After cleaning to remove oil and metal chips, the parts were inspected for size. By the time the parts were measured, part orientation on the lathe and order of production were lost.

Planning the Multi-Vari

To start with, identify the families of variation present within the process. Three broad families combine to produce the overall observed variation: within piece, piece-to-piece and time-to-time.

- Within piece describes deviations from the intended shape of the parts.
- Piece-to-piece captures process changes from one machine cycle to the next.
- Time-to-time captures process trends, shifts, or cycles.

Within Piece. Study the physical process the machine uses to create shape. Then consider how the shape might deviate from the design. For example, the lathe will rotate the workpiece about its central axis as the cutting tool moves in from the outside. As the cutting tool meets the workpiece a circular shape is produced. The cutting tool then moves down the workpiece creating a cylinder. A change in the relationship between the workpiece center and the cutting tool during a rotational cycle will create an oval shape (out of round), instead of a circle. A

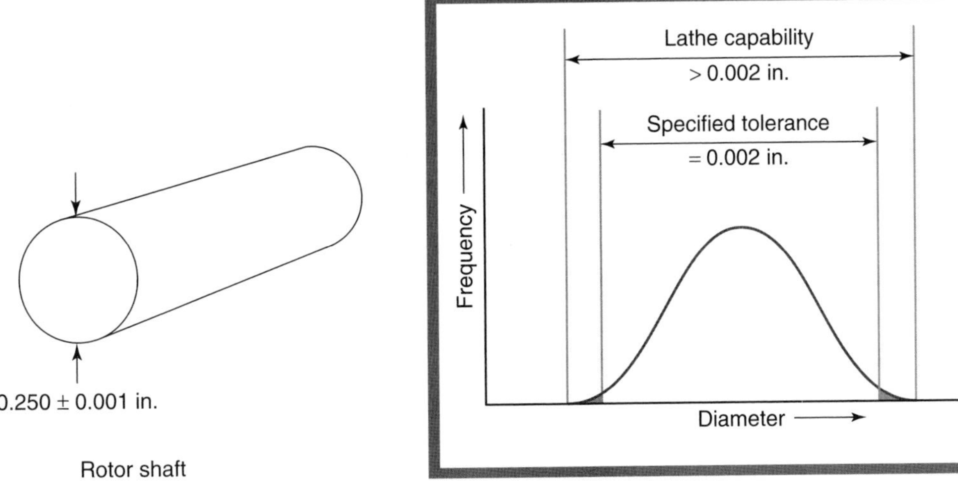

Figure 1 Tolerances of incapable of lathe

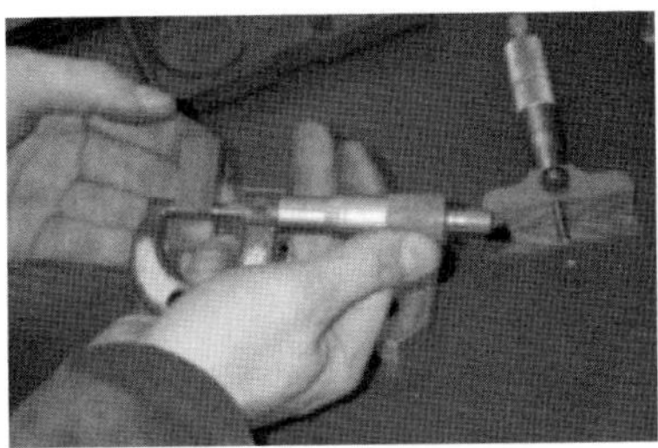

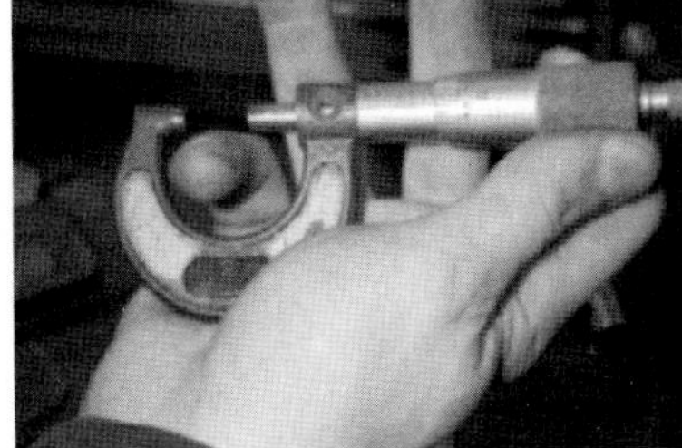

Figure 2 Measuring the diameter of the rotor shaft

change in the relationship between the workpiece center and the cutting tool as the tool travels down the length of the workpiece will produce a tapered shape.

To capture these potential changes, Dorian had the setup man capture parts as they came off the lathe. The setup man marked the end of the part that was held by the lathe chuck for future reference. See Figures 3, 4, and 5. He then found and measured the maximum and minimum diameters at each end of the part (chuck and tail). Four readings represented the range of diameter values within each part.

Piece-to-Piece. Study the organization of the manufacturing process to find piece-to-piece families. If multiple lathes produced the rotor shafts, then piece-to-piece would be consecutive parts from the same lathe, but a new family lathe-to-lathe would have to be captured. If the shafts had been produced on a six-spindle screw machine, then piece-to-piece would be consecutive pieces from the same spindle, not consecutive pieces from the machine. One machine cycle would produce six parts. These broader families come from parallel process paths. Create a detailed process flow diagram to find the parallel paths. Cavity-to-cavity, pallet-to-pallet and line-to-line are common examples.

Only one lathe produced rotor shafts, so there were no parallel paths. Piece-to-piece was captured by sampling consecutive parts from the lathe. Differences in size or shape from one part to the next would indicate instability in the equipment. A sample of three consecutive pieces will reveal trends, shifts, or cycles in the equipment.

Time-to-Time. Processes change over time. Operators make adjustments. They introduce new lots of material. Ambient conditions such as temperature and

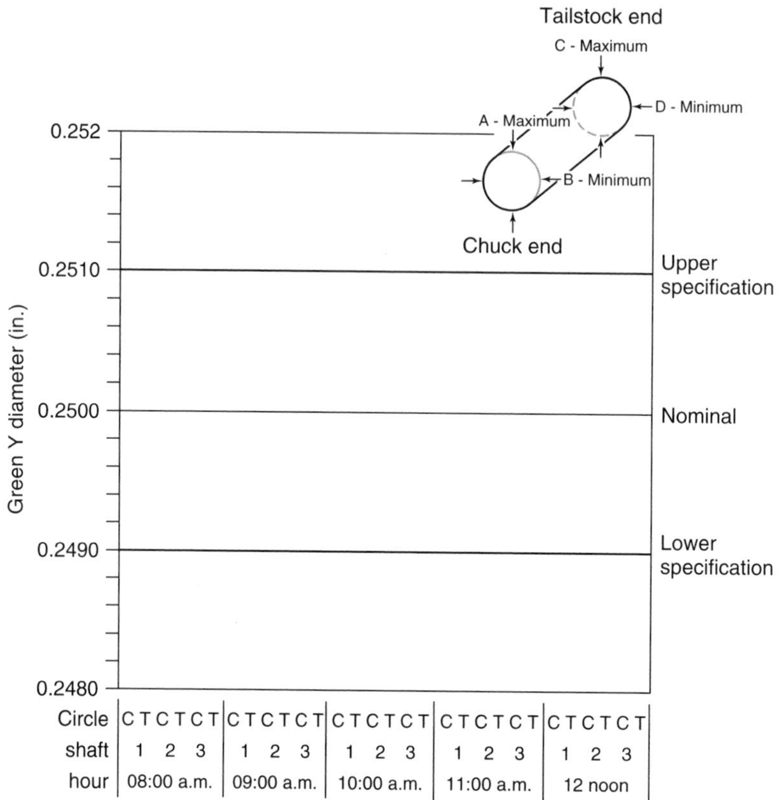

Figure 3 Components of variation for rotor shafts. C: chuck; T: tailstock

humidity vary. These changes may cause changes in process outputs. Select sampling times that are likely to catch important changes. Machine start-up, tool changes, material changes, and operator changes are good candidates for sampling times.

Dorian decided to take samples each hour.

Completing the Plan

Once you identify the families of variation, develop a stratified sampling plan and lay out a graphical format to display the data. A stratified sample means the sample is in the order of production. Stratified samples are not random.

The graphical layout in Figure 3 will capture the data from three consecutive parts taken each hour from 8:00 a.m. to noon.

Executing the Multi-Vari Study

The number of samples required to find the largest family of variation cannot be predetermined. Continue sampling until you capture variation of Red X magnitude (80% of the overall variation should be sufficient). During the sampling, do not try to control or vary specific process parameters. Let the process vary normally. Measure parts immediately and plot the data. Stop sampling when the Red X family is revealed. It is a bit like fishing. Cast your net in a small area. If you do not catch anything, progressively cast the net wider and wider until the catch is made.

Analyzing Multi-Vari Charts

Once the data are plotted, look for the largest family. Calculations are discouraged. The magnitude should be clear from the graph. Also look for patterns in the data. These provide important insights for those who understand the mechanics of the manufacturing process.

Figure 4 shows the multi-vari chart for the rotor shaft after the first three-piece sample at 8:00 a.m.

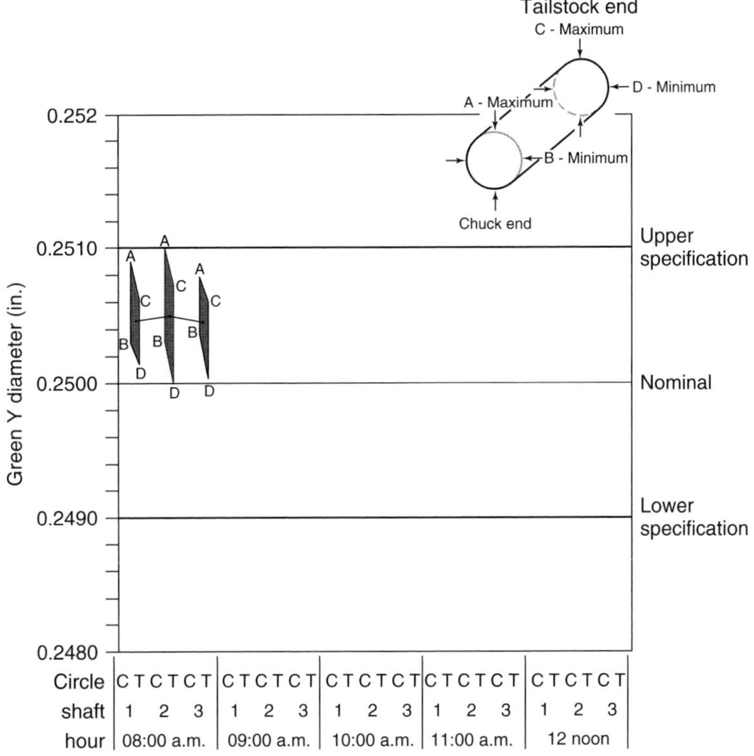

Figure 4 Sources of variation for the rotor shafts. C: chuck; T: tailstock

The vertical axis captures changes in part diameters. The distance from A to B represents out-of-roundness at the chuck end of the part. The distance from C to D is out-of-roundness at the tailstock end. Notice that the first three pieces have the same out-of-roundness at each end. The vertical distance from A to C represents part taper.

The multi-vari chart has revealed considerable diameter variation within each piece. The variation uses half the tolerance. However, we have not seen variation of Red X magnitude. The process capability study revealed variation exceeding the tolerance limits. More data are needed.

Figure 5 shows the multi-vari chart at noon.

Notice that the taper has increased and the out-of-roundness has stayed about the same. More importantly, changes in the position of the hourly subgroups exceed the tolerance and are the largest family. Whatever be the Red X, it is time based. Even though there is considerable within piece variation, that is not where the Red X resides and it must be eliminated for now.

Multi-vari charts reveal the largest family and they also reveal patterns of change. The time-to-time family trends downward from 8:00 to 10:00 a.m.; it shifts between 10:00 and 11:00 a.m.; and it starts another downward trend from 11:00 a.m. to noon.

The statistical portion of the investigation is over. Now process knowledge and an understanding of the physical world take over. The manufacturing supervisor thought excessive tool wear caused the trend and a new tool explained the shift between 10:00 and 11:00 a.m. The tool comes from the outside. If the tool were wearing, the parts would get larger. Our parts are getting smaller. A material lot change could explain a shift, but is not consistent with the trend.

A few questions revealed that the operator took a break at 10:30 a.m. During the break, the process shut down. Temperature is a time-to-time variable that fits the clues. Machining parts involves friction and raises temperature. Metal expands as it heats up. Expanding metal will cause the parts to be smaller after they have cooled. Shutting down the lathe during break

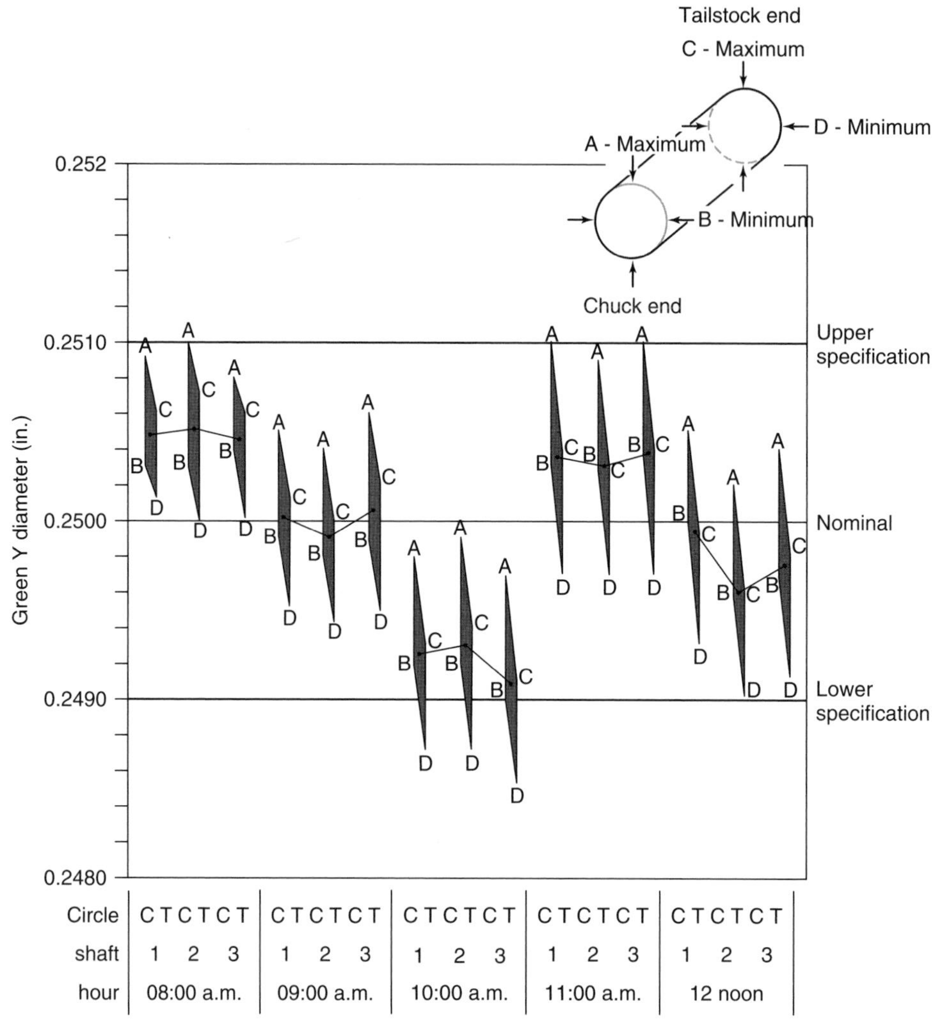

Figure 5 Multi-vari chart for the rotor shafts illustrating that the biggest variation is hour-to-hour. C: chuck; T: tailstock

provides time for cooling, allowing the process to go back to its eight o'clock state. The lathe has a coolant reservoir to control process temperature. An examination of the reservoir revealed that the coolant was well below the prescribed levels. Adding coolant eliminated the time-to-time changes and made the process capable.

Multi-vari charts do not reveal the Red X. They reveal the Red X family. Further investigative steps are always required. Well-designed and well-executed multi-vari charts eliminate broad areas of the process. They allow the investigator to narrow down the search.

Kraft Paper Strength Case

Kraft paper is used for packaging. Paper strength is an important material property. A paper mill was losing market share because their paper strength was inconsistent. The paper process is more complicated than making rotor shafts on a lathe.

Trees are cut and the logs shipped to the mill. A chipper converts the logs to wood chips. A digester cooks the chips in a chemical soup that breaks down the glue that holds the fibers together. After rinsing, the wood fibers are suspended in water, creating a batch called a *furnish*. The furnish is fed into the paper machine where the fibers are distributed across a moving wire. The material dries as it moves down the machine and reels take up the finished paper at the end. The experts blamed the variation in paper strength on the trees. Some trees simply have stronger fiber than others and the paper mill cannot be held responsible.

A multi-vari chart was planned to reveal variation

- across the paper machine;
- in the machine direction; and
- from furnish-to-furnish.

The stratified sampling plan is illustrated in Figure 7. See Figures 6 and 7.

Figure 8 shows the resulting multi-vari chart:

There are several important observations and conclusions:

1. The largest family is furnish-to-furnish.
2. There is a nonrandom pattern in paper strength across the machine.
3. The furnish-to-furnish variation eliminates the paper machine and points to the digester.
4. The nonrandom pattern eliminates the trees. How would the weak fibers know to align on the right side of the paper machine?
5. There is more across machine variation with the weak furnish than with the strong furnish. This signals an interaction.

If every furnish could be made as strong as the best furnish, the across machine variation would not matter and the customers would be delighted. We do not know the name of the Red X, but we know where it resides. It is in the digester.

An investigation into the digesting process uncovered four variables that were supposed to be controlled:

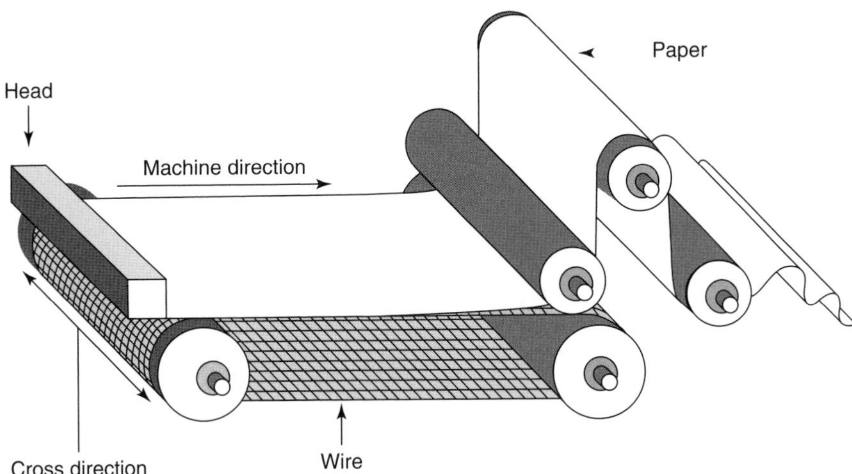

Figure 6 Kraft paper process

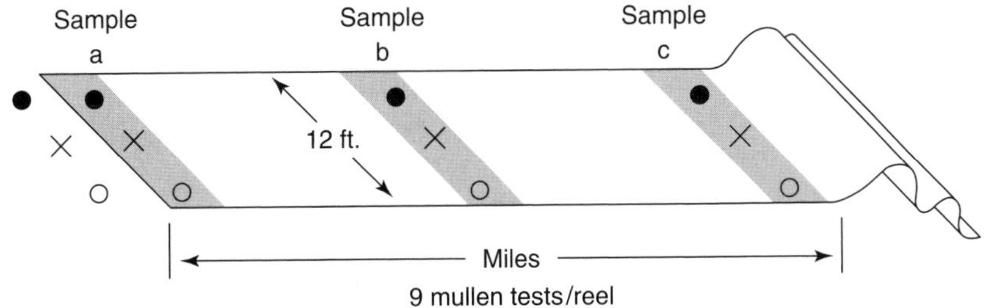

Figure 7 Stratified sampling plan for Kraft paper study

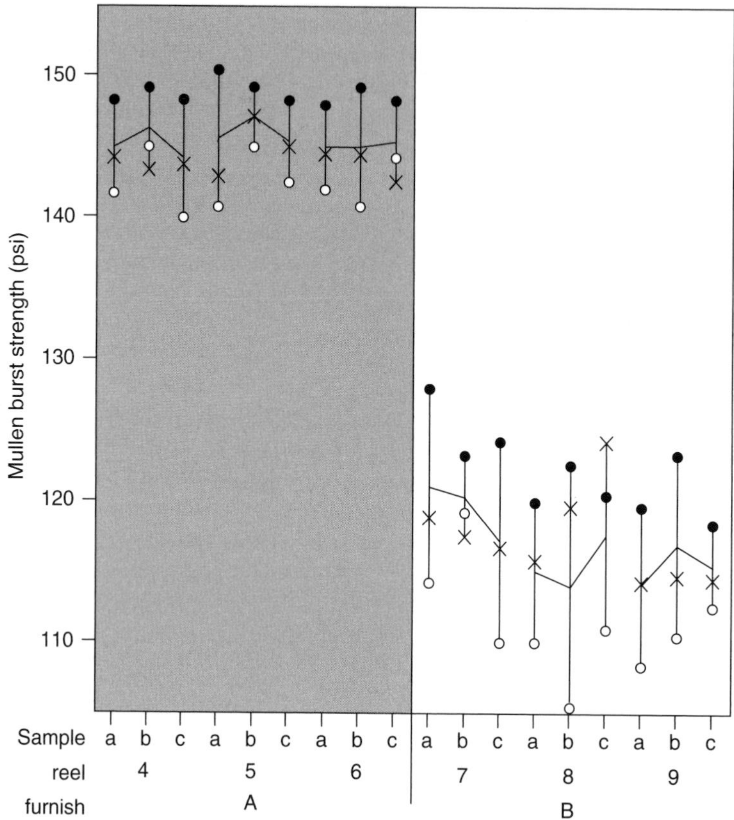

Figure 8 Multi-vari chart for the Kraft paper investigation showing that the biggest source of variation is furnish-to-furnish

1. Cooking temperature
2. Cooking pressure
3. Cooking time
4. Loading ratio (chips per gallon of water).

A full **factorial** designed experiment revealed an interaction between cooking pressure and loading ratio. See Figure 9. Once the interaction was understood and the levels controlled at their optimum, all paper reels were strong, exceeding customer expectations. The lost market share was recovered and the share was raised to new heights.

A molder of plastic parts was having trouble maintaining the proper material packing. Packing is measured as part weight (grams). The families of variation were:

- Cavity-to-cavity
- Shot-to-shot
- Hour-to-hour

The multi-vari chart revealed an interesting pattern (See Figure 10). There is little variation in the shot-to-shot family. There is substantial variation in cavity-to-cavity at 8:00 a.m. and 10:00 a.m., but no cavity-to-cavity variation at nine o'clock. Another interaction has been uncovered.

There is an interaction between the cavity-to-cavity and hour-to-hour families. The investigative team studied the cavities and could not find anything different about cavity C. When looking at the runners, they found C to be slightly smaller. The hour-to-hour variation was by the machine operator. When the operators saw a problem with the parts from cavity C being too small, they increased pack pressure. Notice that all the parts at nine o'clock are heavy. After a while, flash would increase, so they would decrease pack pressure, thus turning on the Red X in cavity C. Increasing the runner size for cavity C solved the problem.

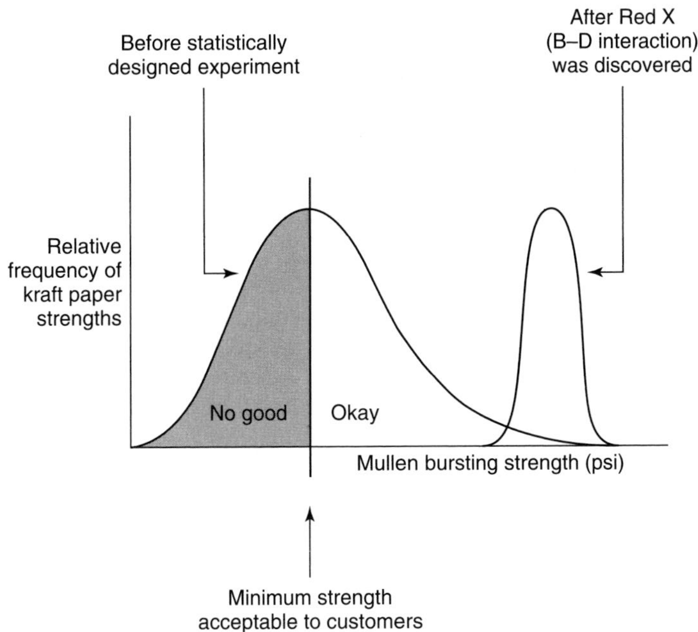

Figure 9 Kraft paper study: Using the interaction of two control variables to reduce variation

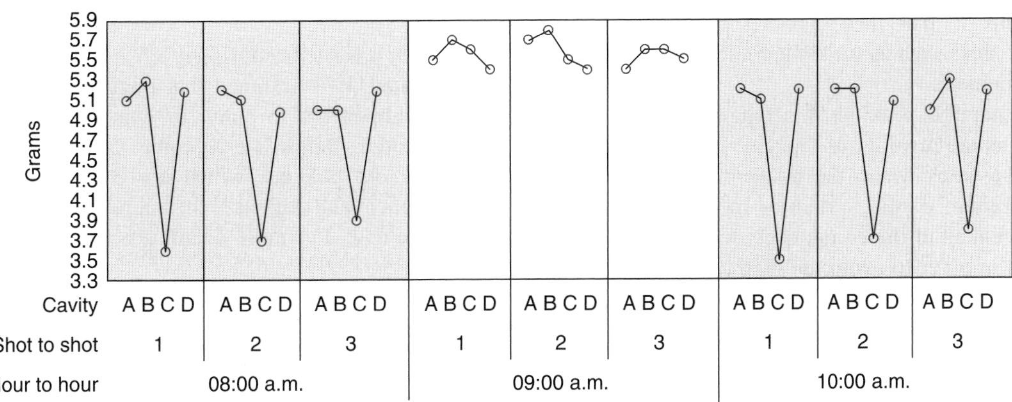

Figure 10 Kraft paper study: Multi-vari chart showing that the biggest source of variation is hour-to-hour. Also notice that cavity-to-cavity variation is the smallest at 9:00 a.m.

Using Multi-Vari Charts Effectively

The multi-vari chart is a powerful statistical tool without the statistical calculations. Properly planned and executed, the chart will reveal the largest family of variation and make a significant, early contribution to the search for the Red X.

Selecting the proper families requires an understanding of how the process creates the shape or property desired. It also requires walking the process to see the parallel paths.

Be sure to take stratified samples and plot the data as soon as they become available. Stop sampling when variation of Red X magnitude is captured.

The chart will reveal the largest family of variation and patterns of change. The patterns will provide guidance on potential next steps.

Remember the multi-vari chart will not identify the Red X. However, as Sherlock Holmes might say, it will eliminate a great number of variables that cannot be the Red X.

RICHARD D. SHAININ

Multivariate Age and Multivariate Renewal Replacement

Introduction

For the past four decades, many researchers have shown interest in the study of **maintenance policies** for systems with stochastic failure. One major reason for this is that these policies can be applied to many areas such as the military, industry, health, and environment.

Generally speaking, in a repairable system, components are repaired or replaced upon failure. There will be cases where the interruption caused in service failure of the system is more costly than the replacement of the components which are in service. Thus, in order to avoid the high cost associated with a failure of the system, an alternative approach is not to wait for components to fail, but to replace or repair them in accordance with a predetermined maintenance schedule. Such preventive maintenance policies reduce unexpected incidents of component or system failure. A large part of reliability theory deals with replacement of single-component systems. Two of the popular **replacement policies** are (a) age- and (b) **block-replacement** policies. Under age replacement a system is replaced only if it has either failed or attained an age, say T. When systems can be replaced in blocks, an example being the replacement of street lamps, a block-replacement policy is used. Here the system is replaced only if either it has failed or the current time is an integer multiple of the block-replacement interval, say T. Many papers have dealt with a stochastic comparison of the number of failures and replacements under these policies. Several authors have also studied different optimal policies under age and block replacement policies. We refer you to Barlow and Proschan [1], Lam [2], Valdez-Flores and Feldman [3] and Shaked and Zhu [4].

As systems become more complicated and require new technologies and methodologies, more sophisticated maintenance policies are needed to solve the maintenance problems. More specifically, it is quite common to find systems consisting of k components which are related. In such cases a legitimate question is: when to replace the components? Obviously, the classical age-and block-replacement policies are unsuitable for such situations, since they only deal with a single component or system. Here, our focus is strictly on the age replacement policy. Also, our survey is confined to multivariate versions of age replacement policies as well as optimal age replacement policies.

Throughout this article "increasing" means "nondecreasing" and "decreasing" means "nonincreasing".

Multivariate Age and Multivariate Renewal Replacements

Consider a system consisting of k components that are related. It is assumed that there are many identical components in stock to every component of this system. Below we describe the following two replacement policies, which are extensions of age replacement and ordinary renewal policies to the multivariate case. For more details see Ebrahimi [5].

Definition 1

1. Under a multivariate age replacement (MAR) policy, the component $i, i = 1, \ldots, k$, of a system is replaced either at age T_i, or upon its failure. We refer to this as the MAR at $(T_1, \ldots, T_k)$;
2. Under a multivariate renewal replacement (MRR), the component $i, i = 1, 2, \ldots, k$, is replaced upon its failure.

To avoid complexity, we confine our attention to the case where $k = 2$. We define the following counting processes:

$N_i(t) =$ the number of component i failures in $(0, t]$ under MRR, $i = 1, 2$; $N_A(t; i, T) =$ the number of component i failures in $(0, t]$ under a MAR $(T, T), i = 1, 2$.

We let X_1 and X_2 be two non-negative continuous random variables representing the times to failure of new components 1 and 2, respectively. We also let $\bar{F}(x_1, x_2) = P(X_1 > x_1, X_2 > x_2)$ be the joint survival function of X_1 and X_2 and $\bar{F}_{X_i}(x_i) = P(X_i > x_i)$ be the marginal survival function of $X_i, i = 1, 2$. Here, $\bar{F}(0, 0) = \bar{F}_{X_1}(0) = \bar{F}_{X_2}(0) = 1$ and $\bar{F}(\infty, \infty) = \bar{F}_{X_1}(\infty) = \bar{F}_{X_2}(\infty) = 0$. It is assumed that as soon as a component of the system fails or it reaches the age T it will be replaced by a new and identical component whose lifetime is independent of the replaced component but is dependent on the lifetimes of the components that are currently in service. It is also assumed that the replacement time (the time it takes to replace either component) is negligible. We denote the conditional survival function of X_1 given $X_2 > y$ and the conditional survival function of X_2 given $X_1 > y$ by

$$\bar{H}_y^{(1)}(x) = P(X_1 > x | X_2 > y) \qquad (1)$$

and

$$\bar{H}_y^{(2)}(x) = P(X_2 > x | X_1 > y) \qquad (2)$$

respectively.

We let $W_i(j)$ and $Y_i(j)$ be the intervals between the $(j-1)$th and jth arrivals of the processes $N_i(t)$ and $N_A(t; i, T), i = 1, 2$ respectively. Now, one can show that under MRR

$$\begin{aligned} P(W_i(j) \geq w_i, i = 1, 2 \big| W_i(\ell) &= a_{i\ell}, \ell = 1, \cdots, j-1, \\ i = 1, 2) &= \bar{F}(w_1, w_2) \end{aligned} \qquad (3)$$

and under MAR(T, T) for $\ell_i T \leq y_i < (\ell_i + 1)T$, $i = 1, 2$

$$\begin{aligned} P(Y_i(j) \geq y_i, i = 1, 2 \big| Y_i(\ell) = a_{i\ell}, \\ \ell = 1, \cdots, j-1, i = 1, 2) \\ &= (\bar{F}(T, T))^{\ell_1} (\bar{H}^{(2)}_{y_1 - \ell_1 T}(T))^{\ell_2 - \ell_1} \\ &\times \bar{F}(y_1 - \ell_1 T, y_2 - \ell_2 T) \\ &\text{if } \ell_1 \leq \ell_2 \text{ and} \\ &= (\bar{F}(T, T))^{\ell_2} (\bar{H}^{(1)}_{y_2 - \ell_2 T}(T))^{\ell_1 - \ell_2} \\ &\times \bar{F}(y_1 - \ell_1 T, y_2 - \ell_2 T) \end{aligned} \qquad (4)$$

if $\ell_1 \geq \ell_2$, see Ebrahimi [5] for more details. It should be noted that the equation (4) can be generalized to the case $k > 2$ by defining conditional survival functions similar to equations (2) and (3) for all permutations of $X_1, X_2, \ldots, X_k$.

Contrast between MAR and MRR

It is clear that the multivariate age planned replacements take into account the ages of the components. It is therefore of interest to compare it with MRR with respect to the number of failures.

The following result provides a stochastic comparison of failures experienced under MAR(T,T) and MRR.

Theorem 1 *If (a)* $\bar{F}(x + t_1, x + t_2) \leq (\geq) \bar{F}(x, x)$ $\bar{F}(t_1, t_2)$ *for all* $x, t_1, t_2 \geq 0$, *and* $\bar{H}_y^{(i)}(t_1 + t_2) \leq (\geq)$ $\bar{H}_y^{(i)}(t_1) \bar{H}_y^{(i)}(t_2), i = 1, 2,$ *for all* $y, t_1, t_2 \geq 0$, *then* $P(N_i(t_i) \leq n_i, i = 1, 2) \leq (\geq) P(N_A(t_i; 1, T) \leq n_i, i = 1, 2)$ *for all* $n_1, n_2, t_1, t_2 \geq 0$.

Theorem 1 states that under the conditions $(a(\leq))$ and $(b(\leq))$ imposed on the joint survival function of X_1 and X_2 the number of failures under MAR is stochastically less than the number of failures under MRR. Next result studies the effect of varying the replacement interval under MAR.

Theorem 2 *If (a)* $\bar{F}(x_1 + t_1, x_2 + t_2) \leq \bar{F}(x_1, x_2)$ $\bar{F}(t_1, t_2)$ *for all* $x_1, x_2, t_1, t_2 \geq 0$ *and (b)* $\bar{H}_y^{(i)}(t_1 + t_2) \leq \bar{H}_y^{(i)}(t_1) \bar{H}_y^{(i)}(t_2), i = 1, 2,$ *for all* $t_1, t_2, y \geq 0$, *then for any* $k \geq 1$, $P(N_A(t_i; i, T) \leq n_i, i = 1, 2) \geq P(N_A(t_i; i, kT) \leq n_i, i = 1, 2)$, *for all* $t_i, n_i, i = 1, 2$.

Finally, our last result gives us information regarding the weakest component (weakest in the sense that it is more likely to fail).

Theorem 3 *If* X_1 *is stochastically larger than* X_2 *(*X_1 *is said to be stochastically larger than* X_2 *if* $\bar{F}_{X_1}(x) > \bar{F}_{X_2}(x)$ *for all* $x \geq 0$, *see Shaked and Shanthikumar [6], then (a)* $P(N_1(t) \leq n) \geq P(N_2(t) \leq n)$ *for all* $n, t \geq 0$ *and (b)* $P(N_A(t; 1, T) \leq n) \geq P(N_A(t; 2, T) \leq n)$, *for all* $n, t \geq 0$.

Intuitively speaking, Theorem 3 says that if component 1 has a larger survival function than component 2, then the number of times that component 1 fails is stochastically less than the number of times that component 2 fails under both MAR(T,T) and MRR.

Optimal Age Replacement under MAR(T,T)

In this section we determine an optimal replacement policy T^* under MAR(T,T) such that

$$C(T) = \frac{\text{the expected cost incurred in a cycle}}{\text{the expected length of a cycle}} \quad (5)$$

is minimized. Here a cycle is the time between two consecutive system failures. We note that $C(T)$ is being used as the criterion for evaluating the replacement policies.

Now, consider the following two cases.

Case I Series system (the system works if both components work). Let c_1 be the constant cost of replacement of one or both components when the system fails and let c_2 be the constant cost of a preventive replacement of both components at age T with $0 < c_2 < c_1 < \infty$. Then, equation (5) reduces to

$$C(T) = \frac{1}{\int_0^T \bar{F}(u,u)\,\mathrm{d}u}(c_1 - \beta \bar{F}(T,T)) \quad (6)$$

where $\beta = c_1 - c_2 > 0$. We can determine the optimal replacement policy T^* by minimizing the equation (6). The minimization procedure can be done by analytical or numerical methods. See Ebrahimi [5] for more details. In fact, one can show that T^* satisfies the following condition.

$$\frac{g(T)}{\bar{F}(T,T)} \int_0^T \bar{F}(u,u)\,\mathrm{d}u + \bar{F}(T,T) = \frac{c_1}{\beta} \quad (7)$$

where $g(T) = -\frac{\mathrm{d}}{\mathrm{d}t}\bar{F}(t,t)|_{t=T}$.

Case II Parallel System (the system works if at least one component works). Under this case equation (5) reduces to

$$C(T) = \frac{1}{\sum_{i=1}^{2} \frac{\int_0^T \bar{F}_{X_i}(u)\,\mathrm{d}u}{1 - \bar{F}_{X_i}(T)} - \frac{\int_0^T \bar{F}(u,u)\,\mathrm{d}u}{1 - \bar{F}(T,T)}} \times \left(c_1 + c_2 \sum_{i=1}^{2} \frac{\bar{F}_{X_i}(T)}{1 - \bar{F}_{X_i}(T)} - \frac{c_2 \bar{F}(T,T)}{1 - \bar{F}(T,T)}\right) \quad (8)$$

We may proceed as in Case 1 and determine the optimal policy T^* by minimizing $C(T)$ in the equation (8).

Applications

This section illustrates applications of our results that are obtained in the previous sections.

Bivariate Gumble Model

Consider X_1 and X_2 with the bivariate Gumble distribution

$$\bar{F}(x_1, x_2) = \exp[-\lambda_1 x_1 - \lambda_2 x_2 - \delta x_1 x_2], \quad x_1, x_2 \geq 0, \lambda_1, \lambda_2, \delta \geq 0 \quad (9)$$

See Gumble [7]. It is clear that $\bar{F}(x_1 + t_1, x_2 + t_2) \leq \bar{F}(x_1, x_2)\bar{F}(t_1, t_2)$. Also, $\bar{H}_y^{(1)}(x) = P(X_1 > x | X_2 > y) = \exp[-\lambda_1 x - \delta x y]$, $\bar{H}_y^{(2)}(x) = P(X_2 > x | X_1 > y) = \exp[-\lambda_2 y - \delta x y]$ and $\bar{H}_y^{(i)}(t_1 + t_2) \leq \bar{H}_y^{(i)}(t_1)$ $\bar{H}_y^{(i)}(t_2)$, $i = 1, 2$, for $t_1, t_2 \geq 0$. Therefore, all the results in Theorems 1 and 2 are true for this family of distributions.

Marshall–Olkin Model

Consider X_1 and X_2 with the joint survival function of Marshall–Olkin bivariate exponential which is given by

$$\bar{F}(x_1, x_2) = \exp(-\lambda_1 x_1 - \lambda_2 x_2 - \delta \max(x_1, x_2)), \quad x_1, x_2 \geq 0, \lambda_1, \lambda_2 > 0 \text{ and } \delta \geq 0 \quad (10)$$

See Marshall and Olkin [8]. It is clear that

$$\bar{F}(x + t_1, x + t_2) \geq \bar{F}(x, x)\bar{F}(t_1, t_2) \quad (11)$$

Also, $\bar{H}_y^{(1)}(x) = \exp(-\lambda_1 x - \delta \max(x, y))$, $\bar{H}_y^{(2)}(x) = \exp(-\lambda_2 x - \delta \max(x, y))$ and $\bar{H}_y^{(i)}(t_1, t_2) \geq \bar{H}_y^{(i)}(t_1)\bar{H}_y^{(i)}(t_2)$, $i = 1, 2$, for $t_1, t_2 \geq 0$. Therefore all the results in Theorem 1 hold.

T^* for Bivariate Gumble and Marshall and Olkin Models

As an application of equation (6), consider the bivariate Gumble family of distributions which was described in the section titled "Bivariate Gumble Model". For this case, equation (7) reduces to

$$\exp[(- (\lambda_1 + \lambda_2)T - \delta T^2)] + (\lambda_1 + \lambda_2 + 2\delta T) \int_0^T \exp(-(\lambda_1 + \lambda_2) \times u - \delta u^2)\,du = \frac{c_1}{\beta} \quad (12)$$

Now, suppose $c_1 = 2, c_2 = 1$ and $\lambda_1 = \lambda_2 = \delta = 1$. Then, we get T^* as approximately 0.27. Since $X_i, i = 1, 2$, has an exponential distribution with mean 1, it is clear that under the univariate age replacement policy $T^* = \infty$. Also under the Marshall and Olkin model for series system $T^* = \infty$. Intuitively speaking, this simply means that age replacement policy is not applicable in univariate case of exponential distribution and under the Marshall and Olkin model.

Concluding Remarks

In this article, we described several maintenance policies for multicomponent systems. Our policies are suitable for revealed failures of components, that is, failures that are detected as soon as they occur. There are also examples of units whose failures are not revealed unless some type of inspection is performed. Berrade [9] proposed a policy for unrevealed failures of components. While the results obtained in section titled "Contrast between MAR and MRR" are appropriate for the case in which components are replaced at the same time, $T_1 = T_2 = T$, it will be interesting to see if similar results are obtained when components are replaced at different times, $T_1 \neq T_2$.

References

[1] Barlow, R. & Proschan, F. (1981). *Statistical Theory of Reliability and Life Testing*, McArdle Press, Silver Spring.

[2] Lam, Y. (1988). A note on the optimal replacement problem, *Advances in Applied Probability* **20**, 479–482.

[3] Valdez-Flores, C. & Feldman, R.M. (1989). A survey of preventive maintenance models for stochastically deteriorating single-unit systems, *Naval Research Logistics* **36**, 419–446.

[4] Shaked, M. & Zhu, H. (1992). Some results on block replacement policies and renewal theory, *Journal of Applied Probability* **29**, 932–946.

[5] Ebrahimi, N. (1997). Multivariate age replacement, *Journal of Applied Probability* **34**, 1032–1040.

[6] Shaked, M. & Shanthikumar, J.G. (1994). *Stochastic Orders and Their Applications*, Academic Press, San Diego.

[7] Gumble, E.J. (1961). Multivariate exponential distributions, *Bulletin of the International Statistical Institute* **39**, 469–475.

[8] Marshall, A.W. & Olkin, I. (1967). A multivariate exponential distribution, *Journal of the American Statistical Association* **62**, 30–44.

[9] Berrade, M.D. (1999). Statistical techniques in reliability: aging properties under mixtures and optimal maintenance policies, Ph.D. thesis, University of Zaragoza, Spain.

Related Articles

Age-Dependent Minimal Repair and Maintenance; **Block Replacement**; **Burn-In and Maintenance Policies**; **Dynamic Programming Methods in Repair and Replacement**; **General Minimal Repair Models**; **Group Maintenance Policies**; **Maintenance and Markov Decision Models**; **Maintenance Optimization**; **Maintenance Optimization in Random Environments**; **Markov Renewal Processes in Reliability Modeling**; **Multicomponent Maintenance**; **Multivariate Imperfect Repair Models**; **Multivariate Stochastic Orders and Aging**; **Replacement Strategies**; **Stationary Replacement Strategies**; **Total Productive Maintenance**.

NADER EBRAHIMI

Multivariate Analysis

Introduction

The body of statistical methodology used to analyze simultaneous measurements on many variables is called *multivariate analysis*. Many multivariate methods are based on an underlying probability model known as the *multivariate normal* (*see* **Normal Distribution**). Other methods are *ad hoc* in nature and are justified by logical or common-sense arguments. Regardless of their origin, multivariate techniques invariably must be implemented on a computer. Current sophisticated statistical software packages make the implementation step easier.

Multivariate analysis is a "mixed bag". It is difficult to establish a classification scheme for multivariate techniques that is both widely accepted and also indicates the appropriateness of the techniques. One classification distinguishes techniques designed to study interdependent relationships from those designed to study dependent relationships. Another classifies techniques according to the number of populations and the number of sets of variables being studied. Yet another classification may distinguish those methods applicable to metric data from those applicable to nonmetric data. This chapter is divided into sections according to inferences about means, inferences about **covariance** structure, and techniques for classification or grouping. This should not, however, be regarded as an attempt to place each method into a slot. Rather, the choice of methods and the types of analyses employed are determined largely by the objectives of the investigation.

The objectives of scientific investigations, for which multivariate methods most naturally lend themselves, include the following:

1. data reduction or structural simplification,
2. sorting and grouping,
3. investigation of the dependence among variables,
4. forecasting and prediction, and
5. hypothesis construction and testing.

A Historical Perspective

Many current multivariate statistical procedures were developed during the first half of the twentieth century. A reasonably complete list of the developers would be voluminous. However, a few individuals can be cited as making important initial contributions to the theory and practice of multivariate analysis. Francis Galton and Karl Pearson did pioneering work in the areas of *correlation* (*see* **Correlation**) and regression analysis. R. A. Fisher's derivation of the exact distribution of the sample correlation coefficient and related quantities provided the impetus for the multivariate distribution theory. C. Spearman and K. Pearson were among the first to work in the area of factor analysis. Significant contributions to multivariate analysis were made during the 1930s by S. S. Wilks (general procedures for testing certain multivariate hypotheses), H. Hotelling (*see* **Hotelling's T^2**), principal component analysis, canonical correlation analysis, R. A. Fisher (discrimination and classification), and P. C. Mahalanobis (generalized distance, **hypothesis testing**). J. Wishart derived an important joint distribution of sample variances and covariances that bears his name. Later, M. S. Bartlett and G. E. P. Box contributed to the large-sample theory associated with certain multivariate test statistics.

Many multivariate methods evolved in concert with the development of electronic computers. Specifically, ingenious graphical methods for displaying multivariate data (e.g., Chernoff faces, Andrews plots) can only be conveniently implemented on a computer. Multidimensional scaling and many clustering procedures were not feasible before the advent of fast computers. Several people have exploited the power of the computer to develop procedures for extracting information from very large data sets, and to create informative lower-dimensional representations of multivariate data (see, for example [1], and the references therein).

Notation

The description of multivariate data and the computations required for their analysis are greatly facilitated by the use of matrix algebra. Consequently, the subsequent discussion will rely heavily on the following notation:

$\mathbf{X}$, a $p \times 1$ random vector,

$\mathbf{x}_j$, a $p \times 1$ multivariate observation on $\mathbf{X}$

$$\mathbf{X} = \begin{bmatrix} x_{11} & x_{12} & \cdots & x_{1p} \\ x_{21} & x_{22} & \cdots & x_{2p} \\ \vdots & \vdots & & \vdots \\ x_{n1} & x_{n2} & \cdots & x_{np} \end{bmatrix} = [\mathbf{x}_1, \mathbf{x}_2, \ldots, \mathbf{x}_p] \tag{1}$$

an $n \times p$ matrix. (Each column of $\mathbf{X}$ represents a multivariate observation.)

$$\bar{\mathbf{x}} = [\bar{x}_1, \bar{x}_2, \ldots, \bar{x}_p]' = \left\{ \bar{x}_i = \frac{1}{n}\sum_{j=1}^{n} x_{ij};\ i = 1, 2, \ldots, p \right\} \tag{2}$$

a $p \times 1$ vector of sample means

$$\mathbf{S} = \left\{ s_{ik} = \frac{1}{n-1} \sum_{j=1}^{n} (x_{ij} - \overline{x}_i)(x_{kj} - \overline{x}_k); \quad i, k = 1, 2, \ldots, p \right\} \tag{3}$$

a $p \times p$ symmetric matrix of sample variances and covariances.

$$\mathbf{R} = \left\{ r_{ik} = \frac{s_{ik}}{\sqrt{s_{ii}}\sqrt{s_{kk}}} \right\} \tag{4}$$

a $p \times p$ symmetric matrix of sample correlation coefficients

$$\boldsymbol{\mu} = \{\mu_i\} = E(\mathbf{X}) \tag{5}$$

a $p \times 1$ vector of population means. ($E(\cdot)$ is the *expectation* (*see* **Expectation**) operator.)

$$\boldsymbol{\Sigma} = \{\sigma_{ij}\} = E(\mathbf{X} - \boldsymbol{\mu})(\mathbf{X} - \boldsymbol{\mu})' \tag{6}$$

a $p \times p$ symmetric matrix of population variances and covariances.

$$\boldsymbol{\rho} = \left\{ \rho_{ij} = \frac{\sigma_{ij}}{\sqrt{\sigma_{ii}}\sqrt{\sigma_{jj}}} \right\} \tag{7}$$

a $p \times p$ symmetric matrix of population correlation coefficients.

$$\mathbf{Z} = \left\{ Z_i = \frac{X_i - \mu_i}{\sqrt{\sigma_{ii}}} \right\} \tag{8}$$

a $p \times 1$ vector of standardized variables.

Multivariate Normal Distribution

A generalization of the familiar bell-shaped normal density to several dimensions plays a fundamental role in multivariate analysis. In fact, many multivariate techniques assume that the data were generated from a *multivariate* normal distribution. Although real data are never *exactly* multivariate normal, the normal density is often a useful approximation to the "true" population distribution. Therefore, the *normal distribution* (*see* **Normal Distribution**) serves as a *bona fide* population model in some instances. Also, the sampling distributions of many multivariate statistics are approximately normal, regardless of the form of the parent population, because of a *central limit effect*.

The p-dimensional normal density for the random vector $\mathbf{X} = [X_1, X_2, \ldots, X_p]'$, evaluated at the point $\mathbf{x} = [x_1, x_2, \ldots, x_p]'$, is given by

$$f(\mathbf{x}) = (2\pi)^{-p/2}|\boldsymbol{\Sigma}|^{-1/2} \times \exp\left\{ -\frac{1}{2}(\mathbf{x} - \boldsymbol{\mu})'\boldsymbol{\Sigma}^{-1}(\mathbf{x} - \boldsymbol{\mu}) \right\} \tag{9}$$

where $-\infty < x_i < \infty, i = 1, 2, \ldots, p$. Here $\boldsymbol{\mu}$ is the population mean vector, $\boldsymbol{\Sigma}$ is the population's variance–covariance matrix, $|\boldsymbol{\Sigma}|$ is the determinant of $\boldsymbol{\Sigma}$, and $\exp(\cdot)$ stands for the exponential function. We denote this p-dimensional normal density by $N_p(\boldsymbol{\mu}, \boldsymbol{\Sigma})$.

Contours of constant density for the p-dimensional normal distribution are ellipsoids defined by $\mathbf{x}$ such that

$$(\mathbf{x} - \boldsymbol{\mu})'\boldsymbol{\Sigma}^{-1}(\mathbf{x} - \boldsymbol{\mu}) = c^2 \tag{10}$$

These ellipsoids are centered at $\boldsymbol{\mu}$ and have axes $\pm c\sqrt{\lambda_i}\,\mathbf{e}_i$, where $\boldsymbol{\Sigma}\,\mathbf{e}_i = \lambda_i\,\mathbf{e}_i, i = 1, 2, \ldots, p$. That is, $\lambda_i, \mathbf{e}_i$ are the eigenvalue (normalized) eigenvector pairs associated with $\boldsymbol{\Sigma}$.

The following are true for a random vector $\mathbf{X}$ having a multivariate normal distribution.

1. Linear combinations of the components of $\mathbf{X}$ are normally distributed.
2. All subsets of the components of $\mathbf{X}$ have a (multivariate) normal distribution.
3. Zero covariance implies that the corresponding components are distributed independently.
4. The conditional distributions of the multivariate components are (multivariate) normal.

(See references [2, Chapter 2] and [3, Section 8a] for more discussion of the multivariate normal distribution.) Because these properties make the normal distribution easy to manipulate, it has been overemphasized as a population model.

To some degree, the *quality* of inferences made by some multivariate methods depend on how closely the true parent population resembles the multivariate normal form. It is imperative, then, that procedures exist for detecting cases in which the data exhibit moderate to extreme departures from what is expected under multivariate normality.

Sometimes nonnormal data can be made to look more normal by considering transformations (*see* **Box and Cox Transformation**) of the data. Normal theory analyses can then be carried out with the suitably transformed data [4, Section 4.8; 5, Section 4.2]. Latest advances in the theory of *discrete* multivariate analysis are contained in reference [6].

Sampling Distributions of $\overline{\mathbf{X}}$ and S

The tentative assumption that the columns of the data matrix, treated as random vectors, constitute a random sample from a normal population, with mean $\boldsymbol{\mu}$ and covariance $\boldsymbol{\Sigma}$, completely determines the sampling distribution of $\overline{\mathbf{X}}$ and **S**. We now summarize the sampling distribution results.

Let $\mathbf{X}_1, \mathbf{X}_2, \ldots, \mathbf{X}_n$ be a random sample of size n from a p-variate *normal* distribution with mean $\boldsymbol{\mu}$ and covariance matrix $\boldsymbol{\Sigma}$. Then (see [2, Sections 3.3 and 7.2; 3, Section 8b]):

1. $\overline{\mathbf{X}}$ is distributed as $N_p(\boldsymbol{\mu}, (1/n)\boldsymbol{\Sigma})$.
2. $(n-1)\mathbf{S}$ is distributed as a Wishart random matrix with $n-1$ **degrees of freedom** (df).
3. $\overline{\mathbf{X}}$ and **S** are independent.

Because $\boldsymbol{\Sigma}$ is unknown, the distribution of $\overline{\mathbf{X}}$ cannot be used directly to make inferences about $\boldsymbol{\mu}$. However, **S** provides independent information about $\boldsymbol{\Sigma}$ and the distribution of **S** does not depend on $\boldsymbol{\mu}$. This allows one to construct a statistic for making inferences about $\boldsymbol{\mu}$.

Selected Problems about Means

Single-Population Mean Vector

An immediate objective in many multivariate studies is to make statistical inferences about population mean vectors. As an initial example, consider the problem of testing whether a multivariate normal population mean vector has a particular value $\boldsymbol{\mu}_0$.

Let the null and alternative hypotheses be $H_0 : \boldsymbol{\mu} = \boldsymbol{\mu}_0$ and $H_1 : \boldsymbol{\mu} \neq \boldsymbol{\mu}_0$, respectively. Once the sample is in hand, a sample mean vector far from $\boldsymbol{\mu}_0$ tends to discredit H_0. A test of H_0 based on the (statistical) distance of $\overline{\mathbf{x}}$ from $\boldsymbol{\mu}_0$ (assuming the population covariance matrix $\boldsymbol{\Sigma}$ is unknown) can be carried out using Hotelling's T^2,

$$T^2 = n(\overline{\mathbf{x}} - \boldsymbol{\mu}_0)'\mathbf{S}^{-1}(\overline{\mathbf{x}} - \boldsymbol{\mu}_0) \tag{11}$$

Hotelling's T^2 is a distance measure that takes account of the joint variability of the p measured variables. Under H_0, T^2 is distributed as

$$[(n-1)p/(n-p)]F_{p,n-p} \tag{12}$$

where n is the **sample size** and $F_{p,n-p}$ denotes an F random variable with p and $n-p$ df. Let $F_{p,n-p}(\alpha)$ be the upper 100αth percentage point of this *F distribution* (*see* **Probability Density Functions**). The hypothesis $H_0 : \boldsymbol{\mu} = \boldsymbol{\mu}_0$ is rejected at the α level of significance if

$$T^2 > [(n-1)p/(n-p)]F_{p,n-p}(\alpha) \tag{13}$$

Example 1 The quality-control department of a microwave oven manufacturer is concerned about the radiation emitted by the ovens. They record radiation measurements with the oven doors opened and closed. Measurements for a random sample of microwave ovens could then be compared with the standards for radiation emission set by the manufacturer.

Other principles of test construction (e.g., likelihood ratio, union–intersection principle) also lead to the use of Hotelling's T^2 in this testing situation.

Hotelling's T^2 statistic has numerous other applications. There are multivariate analogs of the paired and two-sample univariate t statistics. For applications, see references [4, Chapter 5; 7, Chapter 2; 8, Chapter 5].

Several-Population Means, Multivariate Analysis of Variance (MANOVA)*

Multivariate analysis of variance (MANOVA) is concerned with inferences about several population means. It is a direct generalization of the *analysis of variance* (ANOVA) (*see* **Analysis of Variance**) to the case of more than one response variable. In its simplest form, one-way MANOVA, random samples are collected from each of g populations and arranged as

$$\begin{array}{llcl}
\text{Population 1} & \mathbf{X}_{11}, \mathbf{X}_{12}, & \ldots, & \mathbf{X}_{1n_1} \\
\text{Population 2} & \mathbf{X}_{21}, \mathbf{X}_{22}, & \ldots, & \mathbf{X}_{2n_2} \\
\vdots & & \vdots & \\
\text{Population } g & \mathbf{X}_{g1}, \mathbf{X}_{g2}, & \ldots, & \mathbf{X}_{gn_g}
\end{array}$$

MANOVA is used first to investigate whether the population mean vectors are the same, and if not, which mean components differ significantly. It is assumed that

1. $\mathbf{X}_{l1}, \mathbf{X}_{l2}, \ldots, \mathbf{X}_{ln_l}$ is a random sample of size n_l from a population with mean $\boldsymbol{\mu}_l$, $l = 1, 2, \ldots, g$. The random samples from different populations are independent.
2. All populations have a common covariance matrix $\boldsymbol{\Sigma}$.
3. Each population is multivariate normal.

Condition 3 can be relaxed by applying the central limit theorem when the sample sizes n_l are large. The model states that the mean $\boldsymbol{\mu}_l$ consists of a common part $\boldsymbol{\mu}$ plus an amount $\boldsymbol{\tau}_l$ due to the lth treatment. According to the model, *each component* of the observation vector $\mathbf{X}_{lj}$ satisfies the univariate model. The errors for the components of $\mathbf{X}_{lj}$ are correlated, but the covariance matrix $\boldsymbol{\Sigma}$ is the same for all populations (see Table 1).

A vector of observations may be decomposed as suggested by the model. Thus,

$$\mathbf{x}_{lj} = \bar{\mathbf{x}} + (\bar{\mathbf{x}}_l - \bar{\mathbf{x}}) + (\mathbf{x}_{lj} - \bar{\mathbf{x}}_l)$$

$$(\text{observation}) = \begin{bmatrix}\text{overall}\\ \text{sample}\\ \text{mean } \widehat{\boldsymbol{\mu}}\end{bmatrix} + \begin{bmatrix}\text{estimated}\\ \text{treatment}\\ \text{effect } \widehat{\boldsymbol{\tau}}_l\end{bmatrix} + \begin{pmatrix}\text{residual}\\ \widehat{\mathbf{e}}_{lj}\end{pmatrix} \qquad (14)$$

which leads to a decomposition of the **sum of squares** and cross-products matrix

$$\sum_{l=1}^{g}\sum_{j=1}^{n_l}(\mathbf{x}_{lj} - \bar{\mathbf{x}})(\mathbf{x}_{lj} - \bar{\mathbf{x}})' \qquad (15)$$

and the MANOVA table (see Table 2). One test of $H_0 : \boldsymbol{\tau}_1 = \boldsymbol{\tau}_2 = \cdots = \boldsymbol{\tau}_g = \mathbf{0}$ involves *generalized variances*. We reject H_0 if the ratio of generalized variances

$$\Lambda = \frac{|\mathbf{W}|}{|\mathbf{B} + \mathbf{W}|} = \left|\frac{\sum_{l=1}^{g}\sum_{j=1}^{n_l}(\mathbf{x}_{lj} - \bar{\mathbf{x}}_l)(\mathbf{x}_{lj} - \bar{\mathbf{x}}_l)'}{\sum_{l=1}^{g}\sum_{j=1}^{n_l}(\mathbf{x}_{lj} - \bar{\mathbf{x}})(\mathbf{x}_{lj} - \bar{\mathbf{x}})'}\right| \qquad (16)$$

is too small. The quantity $\Lambda = |\mathbf{W}|/|\mathbf{B} + \mathbf{W}|$ is called *Wilks' lambda* after its proposer S. Wilks.

A comparison of the MANOVA table with the familiar $p = 1$ ANOVA table reveals that they are of the same structure. For the multivariate generalization, squares $(\bar{x}_i - \bar{x})^2$ are replaced by sums-of-squares and cross-products matrices $(\bar{\mathbf{x}}_i - \bar{\mathbf{x}})(\bar{\mathbf{x}}_i - \bar{\mathbf{x}})'$. The same type of replacement holds for any fixed ANOVA, so MANOVA tables can be constructed

Table 1 MANOVA model for comparing g population mean vectors

$\mathbf{X}_{lj} = \boldsymbol{\mu} + \boldsymbol{\tau}_l + \mathbf{e}_{lj}$, $j = 1, 2, \ldots, n_l$, and $l = 1, 2, \ldots, g$, where $\mathbf{e}_{lj}$ are independent $N_p(\mathbf{0}, \boldsymbol{\Sigma})$ variables. Here the parameter vector $\boldsymbol{\mu}$ is an overall mean (level) and $\boldsymbol{\tau}_l$ represents the lth treatment effect with, for instance,

$$\sum_{l=1}^{g} n_l\boldsymbol{\tau}_l = \mathbf{0}$$

Table 2 MANOVA table for comparing population mean vectors

Source of variation	Matrix of (SSP)[a]	Degrees of freedom (*df*)
Treatment	$\mathbf{B} = \sum_{l=1}^{g} n_l(\bar{\mathbf{x}}_l - \bar{\mathbf{x}})(\bar{\mathbf{x}}_l - \bar{\mathbf{x}})'$	$g - 1$
Residual (error)	$\mathbf{W} = \sum_{l=1}^{g}\sum_{j=1}^{n_l}(\mathbf{x}_{lj} - \bar{\mathbf{x}}_l)(\mathbf{x}_{lj} - \bar{\mathbf{x}}_l)'$	$\left(\sum_{l=1}^{g} n_l\right) - g$
Total (corrected for the mean)	$\mathbf{B} + \mathbf{W} = \sum_{l=1}^{g}\sum_{j=1}^{n_l}(\mathbf{x}_{lj} - \bar{\mathbf{x}})(\mathbf{x}_{lj} - \bar{\mathbf{x}})'$	$\left(\sum_{l=1}^{g} n_l\right) - 1$

[a] SSP: sum of square and cross product

easily for any of the common designs (see, for example, references [2, Chapter 8; 4, Chapter 6 and 9; 9, Chapter 11]).

Summary Remarks

Multivariate analysis takes into account the joint variation of several responses. One noticeable difference from the univariate situation is that rejection of a null hypothesis $H_0 : \boldsymbol{\mu} = \boldsymbol{\mu}_0$ must be followed by a determination of which component(s) led to the rejection. Technically, it is at least one linear combination $a_1\mu_1 + \cdots + a_p\mu_p = \mathbf{a}'\boldsymbol{\mu}$ that is different from $\mathbf{a}'\boldsymbol{\mu}_0$, but this class typically includes some individual μ_i. As we proceed to several treatments, rejection of the null hypothesis $H_0 : \boldsymbol{\mu}_1 = \boldsymbol{\mu}_2 = \cdots = \boldsymbol{\mu}_g$ must be followed by a comparison of the $\boldsymbol{\mu}_l$ to determine which treatments are different and then to determine which components contribute to the difference.

Multivariate Multiple Regression

Regression analysis is the statistical methodology for predicting values of one or more *response* (dependent) variables from a collection of *predictor* (independent) variable values. It can also be used for assessing the effects of the predictor variables on the responses. In its simplest form, it applies to the fitting of a straight line to data.

The classical linear regression model states that the response Y is composed of a mean, which depends in a linear fashion on the predictor variables z_i and random error ϵ which accounts for measurement error and the effects of other variables not explicitly considered in the model (*see* **Mean Square Error**). The values of the predictor variables recorded from the experiment or set by the investigator are treated as *fixed*. The error (and hence the response) is viewed as a random variable, the behavior of which is characterized by a set of distributional assumptions.

Specifically, the linear regression model with a single response and n measurements on Y and the associated predictors $z_1, z_2, \ldots, z_r$ can be written in matrix notation as

$$\mathbf{Y} = \mathbf{Z}\boldsymbol{\beta} + \boldsymbol{\epsilon} \tag{17}$$

with $E(\boldsymbol{\epsilon}) = \mathbf{0}$ and $\text{Cov}(\boldsymbol{\epsilon}) = \sigma^2\mathbf{I}$.

The *least-squares* (*see* **Least-Squares Estimation**) estimator of $\boldsymbol{\beta}$ is given by $\widehat{\boldsymbol{\beta}} = (\mathbf{Z}'\mathbf{Z})^{-1}\mathbf{Z}'\mathbf{Y}$ and σ^2 is estimated by $(\mathbf{Y} - \mathbf{Z}\widehat{\boldsymbol{\beta}})'(\mathbf{Y} - \mathbf{Z}\widehat{\boldsymbol{\beta}})/(n - r - 1)$. The literature on multiple linear regression is vast; see the numerous books on the subject including references [10, 11].

Multivariate multiple regression is the extension of multiple regression to several response variables. Each response is assumed to follow its own regression model but with the same predictors, so that

$$\begin{aligned}
Y_1 &= \beta_{01} + \beta_{11}z_1 + \cdots + \beta_{r1}z_r + \epsilon_1 \\
Y_2 &= \beta_{02} + \beta_{12}z_1 + \cdots + \beta_{r2}z_r + \epsilon_2 \\
&\;\vdots \qquad\qquad \vdots \\
Y_m &= \beta_{0m} + \beta_{1m}z_1 + \cdots + \beta_{rm}z_r + \epsilon_m
\end{aligned} \tag{18}$$

The error term $\boldsymbol{\epsilon} = [\epsilon_1, \epsilon_2, \ldots, \epsilon_m]'$ has $E(\boldsymbol{\epsilon}) = \mathbf{0}$ and $\text{Var}(\boldsymbol{\epsilon}) = \boldsymbol{\Sigma}$. Thus, the error terms associated with different responses may be correlated.

To establish notation conforming to the classical linear regression model, let $[z_{j0}, z_{j1}, \ldots, z_{jr}]$ denote the values of the predictor variables for the jth trial, $\mathbf{Y}_j = [Y_{j1}, Y_{j2}, \ldots, Y_{jm}]'$ the responses, and $\boldsymbol{\epsilon}_j = [\epsilon_{j1}, \epsilon_{j2}, \ldots, \epsilon_{jm}]'$ the errors. In matrix notation, the design matrix

$$\underset{(n\times(r+1))}{\mathbf{Z}} = \begin{bmatrix} z_{10} & z_{11} & \cdots & z_{1r} \\ z_{20} & z_{21} & \cdots & z_{2r} \\ \vdots & \vdots & & \vdots \\ z_{n0} & z_{n1} & \cdots & z_{nr} \end{bmatrix} \tag{19}$$

is the same as that for the single-response regression model. The other matrix quantities have multivariate counterparts. Set

$$\begin{aligned}
\underset{(n\times m)}{\mathbf{Y}} &= \begin{bmatrix} Y_{11} & Y_{12} & \cdots & Y_{1m} \\ Y_{21} & Y_{22} & \cdots & Y_{2m} \\ \vdots & \vdots & & \vdots \\ Y_{n1} & Y_{n2} & \cdots & Y_{nm} \end{bmatrix} \\
&= [\mathbf{Y}_{(1)} \;\vdots\; \mathbf{Y}_{(2)} \;\vdots\; \cdots \;\vdots\; \mathbf{Y}_{(m)}]
\end{aligned} \tag{20}$$

$$\begin{aligned}
\underset{((r+1)\times m)}{\boldsymbol{\beta}} &= \begin{bmatrix} \beta_{01} & \beta_{02} & \cdots & \beta_{0m} \\ \beta_{11} & \beta_{12} & \cdots & \beta_{1m} \\ \vdots & \vdots & & \vdots \\ \beta_{r1} & \beta_{r2} & \cdots & \beta_{rm} \end{bmatrix} \\
&= [\boldsymbol{\beta}_{(1)} \;\vdots\; \boldsymbol{\beta}_{(2)} \;\vdots\; \cdots \;\vdots\; \boldsymbol{\beta}_{(m)}]
\end{aligned} \tag{21}$$

and

$$\underset{(n\times m)}{\boldsymbol{\epsilon}} = \begin{bmatrix} \epsilon_{11} & \epsilon_{12} & \cdots & \epsilon_{1m} \\ \epsilon_{21} & \epsilon_{22} & \cdots & \epsilon_{2m} \\ \vdots & \vdots & & \vdots \\ \epsilon_{n1} & \epsilon_{n2} & \cdots & \epsilon_{nm} \end{bmatrix} \tag{22}$$
$$= [\boldsymbol{\epsilon}_{(1)} \ \vdots \ \boldsymbol{\epsilon}_{(2)} \ \vdots \ \cdots \ \vdots \ \boldsymbol{\epsilon}_{(m)}]$$

Stated simply, the ith response $Y_{(i)}$ follows the linear regression model (see Table 3)

$$\mathbf{Y}_{(i)} = \mathbf{Z}\boldsymbol{\beta}_{(i)} + \boldsymbol{\epsilon}_{(i)}, \quad i = 1, 2, \ldots, m \tag{23}$$

with $\mathrm{Cov}(\boldsymbol{\epsilon}_{(i)}) = \sigma_{ii}\mathbf{I}$. However, the errors for *different* responses on the *same* trial can be correlated.

The m observations on the jth trial have covariance matrix $\boldsymbol{\Sigma} = \{\sigma_{ik}\}$, but observations from different trials are uncorrelated. Here $\boldsymbol{\beta}$ and $\boldsymbol{\Sigma}$ are matrices of unknown parameters and the design matrix $\mathbf{Z}$ has the jth row $[z_{j0}, \ldots, z_{jr}]$.

Given the outcomes $\mathbf{Y}$ and the values of the predictor variables $\mathbf{Z}$, we determine the least-squares estimates $\widehat{\boldsymbol{\beta}}_{(i)}$ exclusively from the observations, $\mathbf{Y}_{(i)}$, on the ith response. Since

$$\widehat{\boldsymbol{\beta}}_{(i)} = (\mathbf{Z}'\mathbf{Z})^{-1}\mathbf{Z}'\mathbf{Y}_{(i)} \tag{24}$$
$$\widehat{\boldsymbol{\beta}} = \left[\widehat{\boldsymbol{\beta}}_{(1)} \ \vdots \ \widehat{\boldsymbol{\beta}}_{(2)} \ \vdots \ \cdots \ \vdots \ \widehat{\boldsymbol{\beta}}_{(m)}\right]$$
$$= (\mathbf{Z}'\mathbf{Z})^{-1}\mathbf{Z}\mathbf{Y} \tag{25}$$

Using the least-squares estimates $\widehat{\boldsymbol{\beta}}$, we can form the matrices of

$$\text{Predicted values } \widehat{\mathbf{Y}} = \mathbf{Z}\widehat{\boldsymbol{\beta}} = \mathbf{Z}(\mathbf{Z}'\mathbf{Z})^{-1}\mathbf{Z}'\mathbf{Y} \tag{26}$$
$$\text{Residuals } \widehat{\boldsymbol{\epsilon}} = \mathbf{Y} - \widehat{\mathbf{Y}} = \mathbf{I} - \mathbf{Z}(\mathbf{Z}'\mathbf{Z})^{-1}\mathbf{Z}'\mathbf{Y} \tag{27}$$

Table 3 Multivariate linear regression model

$$\underset{(n\times m)}{\mathbf{Y}} = \underset{(n\times(r+1))}{\mathbf{Z}}\ \underset{((r+1)\times m)}{\boldsymbol{\beta}} + \underset{(n\times m)}{\boldsymbol{\epsilon}}$$

with

$$E(\boldsymbol{\epsilon}_{(i)}) = \mathbf{0};$$
$$\mathrm{Cov}(\boldsymbol{\epsilon}_{(i)}, \boldsymbol{\epsilon}_{(k)}) = \sigma_{ik}\mathbf{I}, \quad i, k = 1, 2, \ldots, m.$$

The m observations on the jth trial have covariance matrix $\boldsymbol{\Sigma} = \{\sigma_{ik}\}$, but observations from different trials are uncorrelated. Here $\boldsymbol{\beta}$ and $\boldsymbol{\Sigma}$ are matrices of unknown parameters and the design matrix $\mathbf{Z}$ has the jth row $[z_{j0}, \ldots, z_{jr}]$.

Example 2 Companies considering the purchase of a computer must first assess their future needs in order to determine the proper equipment. Data from several similar company sites can be used to develop a forecast equation of computer hardware requirements for, say, inventory management. The independent variables might include z_1 = customer orders and z_2 = add–delete items. The multivariate responses might include Y_1 = central processing unit (CPU) time and Y_2 = disc input/output capacity [10, Chapter 7].

Summary Remarks

Anderson [2, Section 8.7] derives test statistics and discusses the distribution theory for multivariate regression.

By allowing $\mathbf{Z}$ to have less than full rank, all fixed-effects MANOVA can be incorporated into the multivariate multiple regression framework. This unifying concept is also valuable in connecting ANOVA with the classical multiple linear regression model.

A regression model in which the $((r+1) \times m)$ coefficient matrix $\boldsymbol{\beta}$ is less than full rank is known as a *reduced-rank regression model*. This can arise in a variety of contexts, and can also be linked to the multivariate techniques of principal components and canonical correlation analysis. See reference [12] for a comprehensive discussion of multivariate reduced-rank regression.

Analysis of Covariance Structure

Principal Components

A principal component analysis is concerned with explaining the variance–covariance structure through a few *linear* combinations of the original variables. Its general objectives are (a) data reduction and (b) interpretation.

Although p components are required to reproduce the total system variability, often, much of this variability can be accounted for by a small number k of the principal components. If so, there is (almost) as much information in the k components as there is in the original p variables. The k principal components can then replace the initial p variables, and the original data set is reduced to one consisting of n measurements on k principal components.

Analyses of principal components are more of a means to an end than an end in themselves because they frequently serve as intermediate steps in much larger investigations. For example, principal components may be inputs to a multiple linear regression (as incorporated in the *partial least-squares* approach).

Algebraically, principal components are particular linear combinations of the p random variables $X_1, X_2, \ldots, X_p$. Geometrically, these linear combinations represent the selection of a new coordinate system obtained by rotating the original system with $X_1, X_2, \ldots, X_p$ as the coordinate axes. The new axes represent the directions with maximum variability and provide a simpler and more parsimonious description of the covariance structure.

Principal components depend solely on the covariance matrix $\boldsymbol{\Sigma}$ (or the correlation matrix $\boldsymbol{\rho}$) of $X_1, X_2, \ldots, X_p$. Their development does not require a multivariate normal assumption.

The first principal component is the linear combination with maximum variance. That is, it maximizes $\text{Var}(Y_1) = \mathbf{a}_1' \boldsymbol{\Sigma} \mathbf{a}_1$, where the coefficient vector $\boldsymbol{a}_1$ is restricted to be of unit length. Therefore, we define first principal component = linear combination $\mathbf{a}_1'\mathbf{X}$ that maximizes

$$\text{Var}(\mathbf{a}_1'\mathbf{X}) \quad \text{subject to} \quad \mathbf{a}_1'\mathbf{a}_1 = 1 \tag{28}$$

second principal component = linear combination $\mathbf{a}_2'\mathbf{X}$ that maximizes

$$\text{Var}(\mathbf{a}_2'\mathbf{X}) \quad \text{subject to} \quad \mathbf{a}_2'\mathbf{a}_2 = 1 \tag{29}$$

and

$$\text{Cov}(\mathbf{a}_1'\mathbf{X}, \mathbf{a}_2'\mathbf{X}) = 0 \tag{30}$$

At the ith step, ith principal component = linear combination $\mathbf{a}_i'\mathbf{X}$ that maximizes

$$\text{Var}(\mathbf{a}_i'\mathbf{X}) \quad \text{subject to} \quad \mathbf{a}_i'\mathbf{a}_i = 1 \tag{31}$$

and

$$\text{Cov}(\mathbf{a}_i'\mathbf{X}, \mathbf{a}_k'\mathbf{X}) = 0 \quad \text{for} \quad k < i \tag{32}$$

Let $\boldsymbol{\Sigma}$ be the covariance matrix associated with the random vector $\mathbf{X}' = [X_1, X_2, \ldots, X_p]$. Let $\boldsymbol{\Sigma}$ have the eigenvalue–eigenvector pairs $(\lambda_1, \mathbf{e}_1)$, $(\lambda_2, \mathbf{e}_2), \ldots, (\lambda_p, \mathbf{e}_p)$, where $\lambda_1 \geq \lambda_2 \geq \cdots \geq \lambda_p \geq 0$. The ith *principal component* is given by

$$Y_i = \mathbf{e}_i'\mathbf{X} = e_{i1}X_i + e_{i2}X_2 + \cdots + e_{ip}X_p, \quad i = 1, 2, \ldots, p \tag{33}$$

With these choices,

$$\text{Var}(Y_i) = \mathbf{e}_i' \boldsymbol{\Sigma}\, \mathbf{e}_i = \lambda_i, \quad i = 1, 2, \ldots, p \tag{34}$$

$$\text{Cov}(Y_i, Y_k) = \mathbf{e}_i' \boldsymbol{\Sigma}\, \mathbf{e}_k = 0, \quad i \neq k \tag{35}$$

Therefore, the principal components are uncorrelated and have variances equal to the eigenvalues of $\boldsymbol{\Sigma}$. If some λ_i are equal, the choices of the corresponding coefficient vectors $\mathbf{e}_i$, and hence of $\mathbf{Y}_i$ are not unique.

How do we summarize the sample variation in n measurements on p variables with a few judiciously chosen linear combinations?

Assume the data $\mathbf{x}_1, \mathbf{x}_2, \ldots, \mathbf{x}_n$ represent n independent drawings from some p-dimensional population with mean vector $\boldsymbol{\mu}$ and covariance matrix $\boldsymbol{\Sigma}$. These data yield the sample mean vector $\bar{\mathbf{x}}$, the sample covariance matrix $\mathbf{S}$, and the sample correlation matrix $\mathbf{R}$. These quantities are substituted for the corresponding population quantities above to get *sample principal components*.

Example 3 The weekly rates of return for five stocks (Allied Chemical, DuPont, Union Carbide, Exxon, and Texaco) listed on the New York Stock Exchange were determined for the period January 1975–December 1976. The weekly rates of return are defined as (current Friday closing price – previous Friday closing price)/(previous Friday closing price) adjusted for stock splits and dividends. The observations in 100 successive weeks appear to be distributed independently, but the rates of return *across* stocks are correlated, since, as one expects, stocks tend to move together in response to general economic conditions.

Let $x_1, x_2, \ldots, x_5$ denote observed weekly rates of return for Allied Chemical, DuPont, Union Carbide, Exxon, and Texaco, respectively. Then,

$$\bar{x}' = [0.0054,\ 0.0048,\ 0.0057,\ 0.0063,\ 0.0037] \tag{36}$$

$$\mathbf{R} = \begin{bmatrix} 1.000 & 0.577 & 0.509 & 0.387 & 0.462 \\ 0.577 & 1.000 & 0.599 & 0.389 & 0.322 \\ 0.509 & 0.599 & 1.000 & 0.436 & 0.426 \\ 0.387 & 0.389 & 0.436 & 1.000 & 0.523 \\ 0.462 & 0.322 & 0.426 & 0.523 & 1.000 \end{bmatrix} \tag{37}$$

We note that $\mathbf{R}$ is the covariance matrix of the standardized observations

$$z_1 = \frac{x_1 - \overline{x}_1}{\sqrt{s_{11}}}, \quad z_2 = \frac{x_2 - \overline{x}_2}{\sqrt{s_{22}}}, \quad \ldots,$$
$$z_5 = \frac{x_5 - \overline{x}_5}{\sqrt{s_{55}}} \tag{38}$$

The eigenvalues and corresponding normalized eigenvectors of $\mathbf{R}$ were determined by a computer and are

$$\widehat{\lambda}_1 = 2.857, \widehat{\mathbf{e}'}_1 = [0.464, 0.457, 0.470, 0.421, 0.421] \tag{39}$$

$$\widehat{\lambda}_2 = 0.809, \widehat{\mathbf{e}'}_2 = [0.240, 0.509, 0.260, -0.526, -0.582] \tag{40}$$

$$\widehat{\lambda}_3 = 0.540, \widehat{\mathbf{e}'}_3 = [-0.612, 0.178, 0.335, 0.541, -0.435] \tag{41}$$

$$\widehat{\lambda}_4 = 0.452, \widehat{\mathbf{e}'}_4 = [0.387, 0.206, -0.662, 0.472, -0.382] \tag{42}$$

$$\widehat{\lambda}_5 = 0.343, \widehat{\mathbf{e}'}_5 = [-0.451, 0.676, -0.400, -0.176, 0.385] \tag{43}$$

Using the standardized variables, we obtain the first two sample principal components

$$\widehat{y}_1 = \widehat{\mathbf{e}'}_1\mathbf{z} = 0.464z_1 + 0.457z_2, + 0.470z_3 + 0.421z_4 + 0.421z_5 \tag{44}$$

$$\widehat{y}_2 = \widehat{\mathbf{e}'}_2\mathbf{z} = 0.240z_1 + 0.509z_2 + 0.260z_3 - 0.526z_4 - 0.582z_5 \tag{45}$$

These components, which account for

$$\left(\frac{\widehat{\lambda}_1 + \widehat{\lambda}_2}{p}\right) 100\% = \left(\frac{2.857 + 0.809}{5}\right) 100\% = 73\% \tag{46}$$

of the total (standardized) sample variance, have interesting interpretations. The first component is a (roughly) equally weighted sum or "index", of the five stocks. This component might be called a *general stock-market component* or simply a *market component*. (In fact, at the time the data were collected, these five stocks were included in the Dow–Jones Industrial Average.)

The second component represents a contrast between the chemical stocks (Allied Chemical, DuPont, and Union Carbide) and the oil stocks (Exxon and Texaco). It might be called an *industry component*. Thus, most of the variation in these stock returns is due to market activity and uncorrelated industry activity.

The remaining components are not easy to interpret and, collectively, represent variation that is probably specific to each stock. In any event, they do not explain much of the total sample variance.

Factor Analysis

The essential purpose of factor analysis is to describe, if possible, the covariance relationships among many variables in terms of a few underlying, but unobservable, random quantities called *factors*. Basically, the factor model is motivated by the following argument. Suppose variables can be grouped by their correlations. That is, all variables within a particular group are highly correlated among themselves but have relatively small correlations with variables in a different group. It is conceivable that each group of variables represents a single underlying construct or factor that is responsible for the observed correlations. For example, correlations from the group of test scores in Classics, French, English, Mathematics, and Music collected by Spearman suggested an underlying "intelligence" factor. A second group of variables, perhaps representing physical-fitness scores, might correspond to another factor. It is this type of structure that factor analysis seeks to confirm.

Factor analysis can be considered an extension of principal component analysis. Both can be viewed as attempts to approximate the covariance matrix $\mathbf{\Sigma}$. The approximation based on the factor analysis model is more elaborate; the primary question is whether the data are consistent with a prescribed structure.

The observable random vector $\mathbf{X}$ with p components has mean $\boldsymbol{\mu}$ and covariance matrix $\mathbf{\Sigma}$. The factor model postulates that $\mathbf{X}$ is linearly dependent on a few unobservable random variables $F_1, F_2, \ldots, F_m$, called *common factors* and p additional sources of variation $\epsilon_1, \epsilon_2, \ldots, \epsilon_p$, called *errors* or, sometimes, *specific factors*. In particular, the factor analysis model is

$$
\begin{aligned}
X_1 - \mu_1 &= l_{11}F_1 + l_{12}F_2 + \cdots + l_{1m}F_m + \epsilon_1 \\
X_2 - \mu_2 &= l_{21}F_1 + l_{22}F_2 + \cdots + l_{2m}F_m + \epsilon_2 \\
&\vdots \\
X_p - \mu_p &= l_{p1}F_1 + l_{p2}F_2 + \cdots + l_{pm}F_m + \epsilon_p
\end{aligned} \tag{47}
$$

or, in matrix notation,

$$\underset{(p\times 1)}{\mathbf{X} - \boldsymbol{\mu}} = \underset{(p\times m)}{\mathbf{L}} \underset{(m\times 1)}{\mathbf{F}} + \underset{(p\times 1)}{\boldsymbol{\epsilon}} \tag{48}$$

The coefficient l_{ij} is the *loading* of the ith variable on the jth factor, so the matrix $\mathbf{L}$ is the *matrix of factor loadings*. Note that the ith specific factor ϵ_i is associated only with the ith response X_i. The p deviations $X_1 - \mu_1, X_2 - \mu_2, \ldots, X_p - \mu_p$ are expressed in terms of $p+m$ variables $F_1, F_2, \ldots, F_m, \epsilon_1, \epsilon_2, \ldots, \epsilon_p$, which are unobservable.

With so many unobservable quantities, a direct verification of the factor model from observations on $X_1, X_2, \ldots, X_p$ is hopeless. However, with some additional assumptions about the random vectors $\mathbf{F}$ and $\boldsymbol{\epsilon}$, the preceding model implies certain covariance relationships that can be checked. It follows immediately from the factor model that

1. $\mathrm{Cov}(\mathbf{X}) = \mathbf{LL}' + \boldsymbol{\Psi}$, or

$$\mathrm{Var}(X_i) = l_{i1}^2 + \cdots + l_{im}^2 + \psi_i \text{ and} \tag{49}$$

$$\mathrm{Cov}(X_i, X_k) = l_{i1}l_{k1} + \cdots + l_{im}l_{km} \tag{50}$$

2. $\mathrm{Cov}(\mathbf{X}, \mathbf{F}) = \mathbf{L}$, or $\mathrm{Cov}(X_i, F_j) = l_{ij}$.

That portion of the variance of the ith variable contributed by the m common factors is called the ith *communality*. The portion of $\mathrm{Var}(X_i) = \sigma_{ii}$ due to the specific factor is often called the *uniqueness* or *specific variance*. Denoting the ith communality by h_i^2,

$$\underbrace{\sigma_{ii}}_{\mathrm{Var}(X_i)} = \underbrace{l_{i1}^2 + l_{i2}^2 + \cdots + l_{im}^2}_{\text{communality}} + \underbrace{\psi_i}_{\substack{\text{specific}\\ \text{variance}}} \tag{51}$$

or $h_i^2 = l_{il}^2 + l_{i2}^2 + \cdots + l_{im}^2$, and $\sigma_{ii} = h_i^2 + \psi_i$, $i = 1, 2, \ldots, p$. The ith communality is the sum of squares of the loadings of the ith variable on the m common factors.

Given observations $\mathbf{x}_1, \mathbf{x}_2, \ldots, \mathbf{x}_n$ on p generally correlated variables, factor analysis seeks to answer the question: Does the factor model (see Table 4), with a small number of factors, adequately represent the data? In essence, we tackle this statistical model-building problem by trying to verify covariance relationships 1 and 2.

The sample covariance matrix $\mathbf{S}$ is an estimator of the unknown population covariance matrix $\boldsymbol{\Sigma}$. If the off-diagonal elements of $\mathbf{S}$ are small or those of the sample correlation matrix $\mathbf{R}$ essentially zero, the variables are not related and a factor analysis will not prove useful. In these circumstances, the *specific* factors play the dominant role, whereas the major aim of the factor analysis is to determine a few important *common* factors.

If $\boldsymbol{\Sigma}$ appears to deviate significantly from a diagonal matrix, then a factor model can be entertained. The initial problem is one of estimating the factor loadings l_{ij} and specific variances ψ_i. (Methods of **estimation** are discussed, for example, in references [5, Section 5.4; 7, Section 7.3; 13, Chapter 9].)

Table 4 Orthogonal factor model with m common factors

$$\underset{(p\times l)}{\mathbf{X}} = \underset{(p\times l)}{\boldsymbol{\mu}} + \underset{(p\times m)}{\mathbf{L}}\underset{(m\times l)}{\mathbf{F}} + \underset{(p\times 1)}{\boldsymbol{\epsilon}}$$

$\mu_i =$ *mean* of variable i
$\epsilon_i = i$th *specific factor*
$F_j = j$th *common factor*
$l_{ij} =$ *loading* of the ith variable on the jth factor

The unobservable random vectors $\mathbf{F}$ and $\boldsymbol{\epsilon}$ satisfy

$\mathbf{F}$ and $\boldsymbol{\epsilon}$ are independent.
$E(\mathbf{F}) = \mathbf{0}$, $\mathrm{Cov}(\mathbf{F}) = \mathbf{I}$.
$E(\boldsymbol{\epsilon}) = \mathbf{0}$, $\mathrm{Cov}(\boldsymbol{\epsilon}) = \boldsymbol{\Psi}$, where $\boldsymbol{\Psi}$ is a diagonal matrix.

All factor loadings obtained from the initial loading by an orthogonal transformation have the same ability to reproduce the covariance (or correlation) matrix. From matrix algebra, we know that an orthogonal transformation corresponds to a rigid rotation (or reflection) of the coordinate axes. For this reason, an orthogonal transformation of the factor loadings and the implied orthogonal transformation of the factors is called *factor rotation*.

If $\widehat{\mathbf{L}}$ is the $p \times m$ matrix of estimated factor loadings obtained by any method, then

$$\widehat{\mathbf{L}}^* = \widehat{\mathbf{L}}\mathbf{T}, \qquad \text{where} \quad \mathbf{T}\mathbf{T}' = \mathbf{T}'\mathbf{T} = \mathbf{I} \tag{52}$$

is a $p \times m$ matrix of "rotated" loadings. Moreover, the estimated covariance (or correlation) matrix remains unchanged, since

$$\widehat{\mathbf{L}}\widehat{\mathbf{L}}' + \widehat{\boldsymbol{\Psi}} = \widehat{\mathbf{L}}\mathbf{T}\mathbf{T}'\widehat{\mathbf{L}} + \widehat{\boldsymbol{\Psi}} = \widehat{\mathbf{L}}^*\widehat{\mathbf{L}}^{*'} + \widehat{\boldsymbol{\Psi}} \tag{53}$$

The residual matrix $\mathbf{S} - \widehat{\mathbf{L}}\widehat{\mathbf{L}}' - \widehat{\boldsymbol{\Psi}} = \mathbf{S} - \widehat{\mathbf{L}}^*\widehat{\mathbf{L}}^{*'} - \widehat{\boldsymbol{\Psi}}$ also remains unchanged after rotation. Moreover, the specific variances $\widehat{\psi}_i$ and hence the communalities $\widehat{h}_i^2$ are unaltered. Therefore, from a mathematical viewpoint, it is immaterial whether $\widehat{\mathbf{L}}$ or $\widehat{\mathbf{L}}^*$ is obtained.

Since the original loadings may not be readily interpretable, the usual practice is to rotate them until a "simple structure" is achieved. The rationale is very much akin to sharpening the focus of a microscope in order to see the detail more clearly.

Ideally, we should like to see a pattern of loadings such that each variable loads highly on a single factor and has small-to-moderate loadings on the remaining factors. It is not always possible to get this simple structure.

Kaiser [14] has suggested an analytical measure of simple structure known as the *varimax* (or *normal varimax*) *criterion.* Define $\tilde{l}_{ij}^* = \widehat{l}_{ij}^*/\widehat{h}_i$ to be the final rotated coefficients scaled by the square root of the communalities. The (normal) varimax procedure selects the orthogonal transformation T that makes

$$V = \frac{1}{p}\sum_{j=1}^{m}\left[\sum_{i=1}^{p}\tilde{l}_{ij}^{*4} - \left(\sum_{i=1}^{p}\tilde{l}_{ij}^{*2}\right)^2 \Big/ p\right] \tag{54}$$

as large as possible.

Effectively maximizing V corresponds to "spreading out" the squares of the loadings on each factor as much as possible. Therefore, we hope to find groups of large and negligible coefficients in any *column* of the rotated loadings matrix $\mathbf{L}^*$.

In factor analysis, interest is usually centered on the parameters in the factor model. However, the estimated values of the common factors, called *factor scores*, may also be required. Often, these quantities are used for diagnostic purposes as well as inputs to a subsequent analysis. (See references [4, Section 9.5; 7, Section 7.8; 8, Section 13.6] for further discussion of factor scores.)

Example 4 Beginning with correlations between the scores of the Olympic decathlon events, a factor analysis can be employed to see if the 10 events can be explained in terms of two, three, or four underlying "physical" factors. One interesting study of this kind [15] found that the four factors "explosive arm strength", "explosive leg strength", "running speed", and "running endurance" represented several years of decathlon data quite well.

Canonical Correlations and Variables

Canonical correlation analysis seeks to identify and quantify the associations between two sets of variables. Harold Hotelling [16], who initially developed the technique, provided the example of relating arithmetic speed and arithmetic power to reading speed and reading power. Other examples include relating governmental policy variables to economic goal variables and relating college "performance" variables with precollege "achievement" variables.

Canonical correlation analysis focuses on the correlation between a linear combination of the variables in one set and a linear combination of the variables in another set. When the association between the two sets is expected to be unidirectional – from one set to the other – we label one set the *independent* or *predictor* variables and the other set the *dependent* or *criterion* variables.

The idea of canonical correlation analysis is to first determine the pair of linear combinations having the largest correlation. Next, one determines the pair of predictor set/criterion set linear combinations having the largest correlation among all pairs uncorrelated with the initially selected pair. The process continues by selecting, at each stage, the pair of predictor set/criterion set linear combinations having largest correlation among all pairs that are uncorrelated with the preceding choices. The pairs of linear combinations are the *canonical variables* and their correlations are *canonical correlations*. The following discussion gives the necessary details for obtaining the canonical variables and their correlations. In practice, sample covariance matrices are substituted for the corresponding population quantities, yielding sample canonical variables and sample canonical correlations.

Suppose $p \leq q$, and let the random vectors

$$\underset{(p\times 1)}{\mathbf{X}_1} \quad \text{and} \quad \underset{(q\times 1)}{\mathbf{X}_2}$$

have

$$\text{Cov}(\mathbf{X}_1) = \underset{(p\times p)}{\boldsymbol{\Sigma}_{11}}, \text{Cov}(\mathbf{X}_2) = \underset{(q\times q)}{\boldsymbol{\Sigma}_{22}} \tag{55}$$

and

$$\text{Cov}(\mathbf{X}_1, \mathbf{X}_2) = \underset{(p\times q)}{\boldsymbol{\Sigma}_{12}} \tag{56}$$

For coefficient vectors,

$$\underset{(p\times 1)}{\mathbf{a}} \quad \text{and} \quad \underset{(q\times 1)}{\mathbf{b}}$$

form the linear combinations $U = \mathbf{a}'\mathbf{X}_1$ and $V = \mathbf{b}'\mathbf{X}_2$. Then

$$\underset{\mathbf{a,b}}{\text{Max}}\ \text{Corr}(U, V) = \rho_1^* \tag{57}$$

attained by the linear combination (first canonical variate pair)

$$U_1 = \mathbf{a}'_1\mathbf{X}_1 \quad \text{and} \quad V_1 = \mathbf{b}'_1\mathbf{X}_2 \tag{58}$$

The kth pair of canonical variates, $k = 2, 3, \ldots, p$,

$$U_k = \mathbf{a}'_k\mathbf{X}_1 \quad \text{and} \quad V_k = \mathbf{b}'_k\mathbf{X}_2 \tag{59}$$

maximize

$$\text{Corr}(U_k, V_k) = \rho_k^* \tag{60}$$

among those linear combinations uncorrelated with the preceding $1, 2, \ldots, k-1$ canonical variables.

The canonical variates have the following properties:

$$\text{Var}(U_k) = \text{Var}(V_k) = 1 \tag{61}$$

$$\text{Cov}(U_k, U_l) = \text{Corr}(U_k, U_l) = 0, \quad k \neq l \tag{62}$$

$$\text{Cov}(V_k, V_l) = \text{Corr}(V_k, V_l) = 0, \quad k \neq l \tag{63}$$

$$\text{Cov}(U_k, V_l) = \text{Corr}(U_k, V_l) = 0, \quad k \neq l \tag{64}$$

for $k, l = 1, 2, \ldots, p$.

In general, canonical variables are artificial; they have no physical meaning. If the original variables $\mathbf{X}_1$ and $\mathbf{X}_2$ are used, the canonical coefficients **a** and **b** have units proportional to those of the $\mathbf{X}_1$ and $\mathbf{X}_2$ sets. If the original variables are standardized to have zero means and unit variances, the canonical coefficients have no units of measurement, and they must be interpreted in terms of the standardized variables.

Classification and Grouping Techniques

Discriminant Analysis and Classification

Discriminant analysis and classification are multivariate techniques concerned with *separating* distinct sets of objects (or observations) and with *allocating* new objects (observations) to previously defined groups. Discriminant analysis is rather exploratory in nature. As a separating procedure, it is often employed on a one-time basis in order to investigate observed differences when causal relationships are not well understood. Classification procedures are less exploratory in the sense that they lead to well-defined rules, which can be used for assigning new objects. Classification ordinarily requires more problem structure than discrimination.

Thus, the immediate goals of discrimination and classification, respectively, are

1. To describe either graphically (in three or fewer dimensions) or algebraically, the differential features of objects (observations) from several known collections (populations). We try to find "discriminants" whose numerical values are such that the collections are separated as much as possible.
2. To sort objects (observations) into two or more labeled classes. The emphasis is on deriving a rule that can be used to assign a *new* object to the labeled classes optimally.

To fix ideas, we will list situations in which one may be interested in (a) separating two classes of objects, or (b) assigning a new object to one of the two classes (or both). It is convenient to label the classes π_1 and π_2. The objects are ordinarily separated or classified on the basis of measurements, for instance, on p associated random variables $\mathbf{X}' = [X_1, X_2, \ldots, X_p]$. The observed values of $\mathbf{X}$ differ to some extent from one class to the other. We can

Populations π_1 and π_2	Measured variables **X**
Solvent and distressed property-liability insurance companies	Total assets, cost of stocks and bonds, market value of stocks and bonds, loss expenses, surplus, amount of premiums written
Federalist papers written by James Madison and those written by Alexander Hamilton	Frequencies of different words and length of sentences
Purchasers of a new product and laggards (those "slow" to purchase)	Education, income, family size, amount of previous brand switching
Alcoholics and nonalcoholics	Activity of monoamine oxidase enzyme, activity of adenylate cyclase enzyme

think of the totality of values from the first class as being the population of $\mathbf{x}$ values for π_1 and those from the second class as the population of $\mathbf{x}$ values for π_2. These two populations can then be described by probability density functions $f_1(\mathbf{x})$ and $f_2(\mathbf{x})$, and, consequently, we can talk of assigning observations to populations or objects to classes interchangeably.

You may wonder at this point how it is we *know* some observations belong to a particular population but we are unsure about others. (This, of course, is what makes classification a problem!) There are several conditions that can give rise to this apparent anomaly.

Incomplete knowledge of future performance.
"Perfect" information requires destroying object.
Unavailable or expensive information.

It should be clear that classification rules cannot usually provide an error-free method of assignment. This is because there may not be a clear distinction between the measured characteristics of the populations; that is, the groups may overlap. It is then possible, for example, to incorrectly classify a π_2 object as belonging to π_1 or a π_1 object as belonging to π_2.

For a discussion of discriminant analysis and subsequent classification procedures, see references [2, Chapter 6; 4, Chapter 11; 5, Chapter 6]. Since the literature on this subject is large, we simply display below the "best" allocation rule for two multivariate normal populations with a common covariance matrix $\boldsymbol{\Sigma}$. In practice, sample quantities replace the corresponding population quantities.

Let $\boldsymbol{\mu}_1$ and $\boldsymbol{\mu}_2$ be the two population mean vectors, $c(1|2)$ the cost of incorrectly assigning a population 2 observation to population 1, $c(2|1)$ the cost of incorrectly assigning a population 1 observation to population 2, p_1 the "prior" probability of population 1, and p_2 the "prior" probability of population 2, then we allocate $\mathbf{x}$ to π_1 if

$$(\boldsymbol{\mu}_1 - \boldsymbol{\mu}_2)'\boldsymbol{\Sigma}^{-1}\mathbf{x} - \frac{1}{2}(\boldsymbol{\mu}_1 - \boldsymbol{\mu}_2)'\boldsymbol{\Sigma}^{-1}(\boldsymbol{\mu}_1 + \boldsymbol{\mu}_2) \geq \ln\left[\frac{c(1|2)}{c(2|1)}\left(\frac{p_2}{p_1}\right)\right] \tag{65}$$

Allocate $\mathbf{x}$ to π_2 otherwise.

The first term, $y = (\boldsymbol{\mu}_1 - \boldsymbol{\mu}_2)'\boldsymbol{\Sigma}^{-1}\mathbf{x}$, above is *Fisher's linear discriminant function.* (Fisher actually developed the sample version $\widehat{y} = (\overline{\mathbf{x}}_1 - \overline{\mathbf{x}}_2)'\mathbf{S}_{\text{pooled}}^{-1}\mathbf{x}$, where $\mathbf{S}_{\text{pooled}}$ is the pooled sample covariance matrix.) Assuming a common population covariance matrix, it is the linear function $\mathbf{a}'\mathbf{x}$ with $\mathbf{a} \propto (\boldsymbol{\mu}_1 - \boldsymbol{\mu}_2)'\boldsymbol{\Sigma}^{-1}$ that maximizes the separation between the two populations, as measured by

$$\frac{(\mu_{1y} - \mu_{2y})^2}{\sigma_y^2} = \frac{(\mathbf{a}'\boldsymbol{\mu}_1 - \mathbf{a}'\boldsymbol{\mu}_2)^2}{\mathbf{a}'\boldsymbol{\Sigma}\mathbf{a}} \tag{66}$$

One important way of judging the performance of any classification procedure is to calculate its "error rates", or misclassification probabilities. When the forms of the parent populations are known completely, misclassification probabilities can be calculated with relative ease. Because parent populations are rarely known, one must concentrate on the error rates associated with the sample classification function.

Finally, it should be intuitive that good classification (low error rates) will depend on the separation of the populations. The farther apart the groups, the more likely it is that a *useful* classification rule can be developed.

Fisher also proposed a several-population extension of his discriminant method. The motivation behind Fisher discriminant analysis is the need to obtain a reasonable representation of the populations that involves only a *few* linear combinations of the observations, such as $\mathbf{a}_1'\mathbf{x}$, $\mathbf{a}_2'\mathbf{x}$, and $\mathbf{a}_3'\mathbf{x}$. His approach has several advantages when one is interested in *separating* several populations for visual inspection or graphically descriptive purposes. It allows for the following:

1. Convenient representations of the g populations that reduce the dimension from a very large number of characteristics to a relatively few linear combinations. Of course, some information – needed for optimal classification – may be lost unless the population means lie completely in the lower-dimensional space selected.
2. Plotting of the means of the first two or three linear combinations (discriminants). This helps display the relationships and possible groupings of the populations.
3. Scatterplots of the sample values of the first two discriminants, which can indicate **outliers*** or other abnormalities in the data. (See references [1, Chapter 4; 4, Section 11.7; 17, Chapter 2] for examples of low-dimensional representations.)

Summary Remarks

The linear discriminant functions that we have presented can arise from a multivariate multiple linear regression model with the response variable vector containing binary categorical variables representing the different groups.

Other, typically computer intensive, procedures are available for discrimination and classification. These include logistic regression, classification trees, **neural networks**, and support vector machines. A good reference for these and related methods is [1].

Clustering and Graphical Procedures

Rudimentary, exploratory procedures are often quite helpful in understanding the complex nature of multivariate relationships. Searching the data for a structure of "natural" groupings is an important exploratory technique. Groupings can provide an informal means for assessing dimensionality, identifying outliers, and suggesting interesting hypotheses concerning relationships.

Grouping, or clustering, is distinct from the classification methods discussed earlier. Classification pertains to a *known* number of groups, and the operational objective is to assign new observations to one of these groups. Cluster analysis is a more primitive technique in that no assumptions are made concerning the number of groups or the group structure. Grouping is done on the basis of similarities or distances (dissimilarities). The inputs required are similarity measures or data from which similarities can be computed. Good references on clustering include [1, Section 14.3; 4, Chapter 7; 18, 19].

Clustering and Lower-Dimensional Representations

Even without the precise notion of a natural grouping, we are often able to cluster objects in two- or three-dimensional scatter plots by eye. To take advantage of the mind's ability to group similar objects, several graphical procedures have been developed for depicting high-dimensional observations in two dimensions. Boxes, glyphs, stars, Andrews plots, and Chernoff faces are among the methods that produce two-dimensional "pictures" of multivariate data.

Two-dimensional scatter plots of high-dimensional data can be constructed using principal component axes, or discriminant score axes, or canonical variate axes. In doing so, there is always some information lost. In contrast to this activity, multidimensional scaling seeks to directly "fit" the original data into a low-dimensional space such that any distortion caused by a lack of dimensionality is minimized.

A biplot is a two-dimensional graphical representation of the information in the $n \times p$ data matrix. Correspondence analysis is a graphical procedure for representing association in a two-way table of frequencies or counts.

Extensions of two-dimensional principal component representations of data such as self-organizing maps and principal curves and surfaces are discussed in reference [1, Chapter 14].

Multivariate Methods and Data Mining

Data mining (ref. to Data Mining Review article in CIM section) refers to the process of discovering associations and relationships in very large data sets (perhaps several terabytes of data). Data mining is not possible without appropriate software and fast computers. Many of the multivariate methods discussed here, along with algorithms developed in the machine learning and artificial intelligence fields, play important roles in data mining. Data mining has helped to identify new chemical compounds for prescription drugs, detect fraudulent claims and purchases, create and maintain individual customer relationships, design better engines, improve process control, and develop effective credit scoring rules.

References

[1] Hastie, T., Tibshirani, R. & Friedman, J. (2001). *The Elements of Statistical Learning: Data Mining, Inference and Prediction*, Springer-Verlag, New York.

[2] Anderson, T.W. (2003). *An Introduction to Multivariate Statistical Analysis*, 3rd Edition, John Wiley & Sons, New York.

[3] Rao, C.R. (1973). *Linear Statistical Inference and its Applications*, 2nd Edition, John Wiley & Sons, New York.

[4] Johnson, R.A. & Wichern, D.W. (2002). *Applied Multivariate Statistical Analysis*, 5th Edition, Prentice-Hall, Upper Saddle River.

[5] Seber, G.A.F. (2004). *Multivariate Observations*, John Wiley & Sons, New York.

[6] Agresti, A. (2002). *Categorical Data Analysis*, 2nd Edition, John Wiley & Sons, New York.

[7] Morrison, D.F. (2005). *Multivariate Statistical Methods*, 4th Edition, Brooks/Cole Thompson Learning, Belmont.

[8] Rencher, A.C. (2002). *Methods of Multivariate Analysis*, 2nd Edition, John Wiley & Sons, New York.

[9] Lattin, J.M., Carroll, J.D. & Green, P.E. (2003). *Analyzing Multivariate Data*, Brooks/Cole Thompson Learning, Pacific Grove.

[10] Draper, N.R. & Smith, H. (1998). *Applied Regression Analysis*, 3rd Edition, John Wiley & Sons, New York.

[11] Seber, G.A.F. & Lee, A.J. (2003). *Linear Regression Analysis*, 2nd Edition, John Wiley & Sons, New York.

[12] Reinsel, G.C. & Velu, R.P. (1998). *Multivariate Reduced-Rank Regression: Theory and Applications*, Springer-Verlag, New York.

[13] Mardia, K.V., Kent, J.T. & Bibby, J.M. (1979). *Multivariate Analysis*, Academic Press, New York.

[14] Kaiser, H.F. (1958). The varimax criterion for analytic rotation in factor analysis, *Psychometrika* **23**, 187–200.

[15] Linden, M. (1977). A factor analytic study of Olympic decathlon data, *Res. Quar.* **48**, 562–568.

[16] Hotelling, H. (1936). Relations between two sets of variates, *Biometrika* **28**, 321–377.

[17] Everitt, B. & Dunn, G. (2001). *Applied Multivariate Data Analysis*, 2nd Edition, Oxford University Press, New York.

[18] Everitt, B., Landau, S. & Leese, M. (2001). *Cluster Analysis*, 4th Edition, Oxford University Press, New York.

[19] Hartigan, J.A. (1975). *Clustering Algorithms*, John Wiley & Sons, New York.

Further Reading

Basilevsky, A.T. (1994). *Statistical Factor Analysis and Related Methods: Theory and Applications*, John Wiley & Sons, New York.

Berry, M.J.A. & Linoff, G.S. (2004). *Data Mining Techniques*, 2nd Edition, John Wiley & Sons, New York.

Bishop, C.M. (1995). *Neural Networks for Pattern Recognition*. Oxford University Press, New York.

Bollin, K.A. (1989). *Structural Equation Models with Latent Variables*, John Wiley & Sons, New York.

Breiman, L., Friedman, J., Olshen, R. & Stone, C. (1984). *Classification and Regression Trees*, Wadsworth, Belmont.

Gower, J.C. & Hand, D.J. (1995). *Biplots*, Chapman & Hall, London.

Greenacre, M.J. (1984). *Theory and Applications of Correspondence Analysis*, Academic Press, London.

Hand, D., Mannila, H. & Smyth, P. (2001). *Principles of Data Mining*, MIT Press, Cambridge.

Jolliffe, I.T. (2002). *Principal Component Analysis*, 2nd Edition, Springer-Verlag, New York.

Mason, R.L. & Young, J.C. (2002). *Multivariate Statistical Process Control with Industrial Applications*, ASA-SIAM, Philadelphia.

Reinsel, G.C. (1997). *Elements of Multivariate Time Series Analysis*, 2nd Edition, Springer-Verlag, New York.

RICHARD A. JOHNSON AND DEAN W. WICHERN

Article originally published in Encyclopedia of Statistical Sciences, 2nd Edition *(2005, John Wiley & Sons, Inc.). Minor revisions for this publication by Jeroen de Mast.*

Multivariate Charts for Variability

Introduction

Multivariate control charts (*see* **Multivariate Control Charts Overview**) which are specifically designed to detect changes in a **covariance** matrix are discussed here. The covariance matrix represents the variability of a multivariate process. The scope of the discussion is limited to phase II **control charts** for multivariate normal processes. For a more detailed account of the topic being discussed, readers are referred to the works by Weirda [1], Lowry and Montgomery [2], Montgomery [3], and Yeh *et al.* [4], and the references therein (*see* **Hotelling's T^2 Chart**; **Multivariate Exponentially Weighted Moving Average (MEWMA) Control Chart**; **Multivariate Cumulative Sum (CUSUM) Chart**).

Let $X = (X_1, X_2, \ldots, X_p)'$ denote the random variable that represents the p correlated quality characteristics derived from a process whose quality is to be monitored. When the process is in control, it is assumed that X follows a p-dimensional **normal distribution**, denoted by $N_p(\mu_0, \Sigma_0)$, where μ_0 is the in-control process mean and Σ_0 is the in-control process covariance matrix. On the other hand, when the process is out of control, it is assumed that X follows $N_p(\mu, \Sigma)$, where either $\mu \neq \mu_0$ or $\Sigma \neq \Sigma_0$ or both. The in-control parameters μ_0 and Σ_0 are assumed to be either known or that they can be sufficiently estimated at the end of phase I control such that their values can be assumed known prior to the beginning of the phase II monitoring. Here we discuss two phase II multivariate control charting techniques for detecting changes in the covariance matrix, each dependent on how the samples of observations are collected. In the first case, independent samples of $n(n > p)$ observations each are repeatedly drawn. In the second case, independent individual observations $(n = 1)$ are being drawn from the process.

The $|S|$ Chart

In the case when independent samples each with n observations are available for constructing the control chart, a commonly used Shewhart control chart is known as the $|S|$ *chart* (*see* **Multivariate Control Charts Overview**), or the generalized variance chart. For the tth sample, $t = 1, 2, \ldots,$ the n observations are denoted by $X_{t1}, X_{t2}, \ldots, X_{tn}$, and the corresponding sample mean and covariance matrix are $\bar{X}_t = \sum_{j=1}^{n} X_{tj}/n$ and $S_t = \sum_{j=1}^{n} (X_{tj} - \bar{X}_t)(X_{tj} - \bar{X}_t)'/(n-1)$, respectively. To construct the $|S|$ chart, one calculates $|S_t|, t = 1, 2, \ldots,$ the determinant of the sample covariance matrix of the tth sample and plots $|S_t|$ against the sampling sequence t. For 3σ control limits, the upper control limit (UCL), center line (CL), and lower control limit (LCL) of the $|S|$ chart are defined as

$$UCL = \left(1 + 3\frac{\sqrt{b_2}}{b_1}\right)|\Sigma_0| \qquad (1)$$

$$CL = |\Sigma_0| \qquad (2)$$

$$LCL = \max\left(0, \left(1 - 3\frac{\sqrt{b_2}}{b_1}\right)|\Sigma_0|\right) \qquad (3)$$

where $b_1 = (n-1)^{-p} \prod_{i=1}^{p} (n-i)$ and $b_2 = (n-1)^{-2p} \left[\prod_{i=1}^{p} (n-i)\right] \times \left[\prod_{j=1}^{p} (n-j+2) - \prod_{j=1}^{p} (n-j)\right]$. An out-of-control signal is detected if $|S_t|$ plots outside of the control limits.

We next present the application of the $|S|$ chart using a real-life example. The example from Mitra [5] is related to a component used in the assembly of a transmission mechanism. Two correlated quality characteristics, tensile strength and diameter, are to be monitored. The data set consists of 20 independent samples each with four observations. In this case, $p = 2$ and $n = 4$, which result in $b_1 = \frac{2}{3}$ and $b_2 = \frac{84}{81}$. The centerline $|\Sigma_0|$ was estimated by the average of the sample covariance matrices of all 20 samples which turned out to be 48.4. Therefore the *UCL, CL,* and *LCL* were calculated to be 270.1, 48.4, and 0, respectively. The $|S|$ chart, which was produced using Minitab, is shown in Figure 1. The $|S|$ chart indicated that the covariance matrix was in control.

The MEWMS Chart

In the case when only independent individual observations are available, the $|S|$ chart is no longer applicable since the sample covariance matrix does not exist. The alternative is to find other ways to pool observations together and derive control charts based on the pooled observations. Let $Z_t = \Sigma_0^{-1/2}(X_t - \mu_0), t = 1, 2, \ldots,$ be the transformed variable of the tth observation X_t such that when the process is in control Z_t follows $N_p(0, I_p)$, where I_p is a pxp identity matrix. Let $V_t = \lambda Z_t Z_t' + (1-\lambda)V_{t-1}$, where $0 < \lambda < 1$ is a smoothing constant and $V_0 = Z_1 Z_1'$. For each sampling sequence t, one calculates trace(V_t), the trace of the matrix V_t. The values of trace(V_t) are then plotted on the control chart against the sampling frequencies. For 3σ control limits, the UCL, CL, and LCL are defined as

$$UCL = p + L\sqrt{2p\sum_{i=1}^{t} c_i^2} \qquad (4)$$

$$CL = p \qquad (5)$$

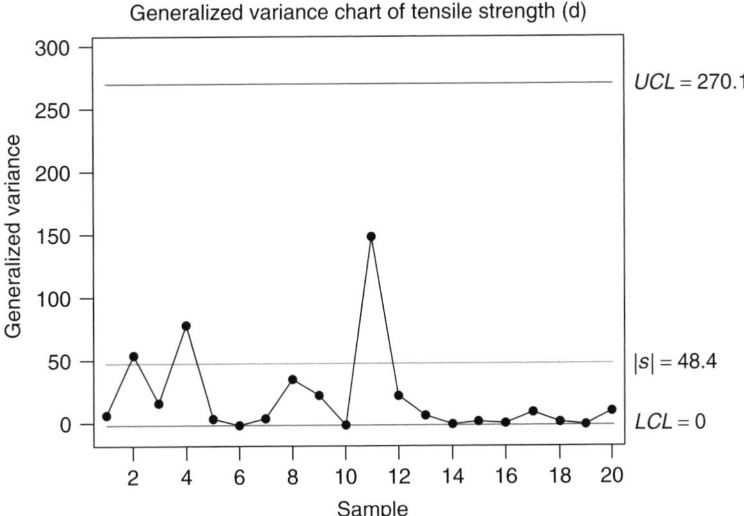

Figure 1 The $|S|$ chart of the transmission assembly example

$$LCL = p - L\sqrt{2p\sum_{i=1}^{t} c_i^2} \quad (6)$$

where L is a multiplier dependent on p, λ and the desired in-control chart performance, and $\sum_{i=1}^{t} c_i^2 = \lambda(2-\lambda)^{-1} + 2(2-\lambda)^{-1}(1-\lambda)^{2t-1}$ which converges to $\lambda/(2-\lambda)$ as $t \to \infty$. Using simulations, Huwang *et al.* [6] provided the values of L for λ ranging from 0.1 to 0.9 and $p = 2$ and 3, such that when the process is in control the control chart will have an in-control **average run length** approximately equal to 370, which is equivalent to a 3σ limits Shewhart chart. Part of the tabulated values of L is provided in Table 1. Note that the chart, called the *multivariate exponentially weighted moving squared-deviation* (MEWMS) chart, is a multivariate extension of the univariate EWMS control chart first proposed by MacGregor and Harris [7]. In general, a smaller value of λ, say $0.1 \leq \lambda \leq 0.3$, gives a better performance of the MEWMS chart – in term of faster detection of out-of-control processes.

Next we present the application of the MEWMS chart. The data were taken from Joshi and Sprague [8] for an application in the semiconductor industry. The observations were related to the quality of wafer manufacturing during photolithography process. There were 74 lots of wafers and one wafer was randomly taken from each lot. Three correlated critical dimension measurements of die were taken on each drawn wafer. Using these samples, except lot 40 which was classified to be out of control, estimates of μ_0 and Σ_0 were obtained and used to establish the control limits for phase II monitoring. The estimates μ_0 and Σ_0 are equal to $\hat{\mu}_0 = \begin{pmatrix} 3.315 \\ 3.108 \\ 3.118 \end{pmatrix}$ and $\hat{\Sigma}_0 = \begin{pmatrix} 0.0093 & 0.0036 & 0.0052 \\ 0.0036 & 0.0085 & 0.0034 \\ 0.0052 & 0.0034 & 0.0088 \end{pmatrix}$, respectively. In phase II, an additional 200 observations were obtained. For every new observation generated, it was transformed using $Z_t = \hat{\Sigma}_0^{-1/2}(X_t - \hat{\mu}_0), t = 1, 2, \ldots, 200$. The matrix $V_t = \lambda Z_t Z_t' + (1-\lambda)V_{t-1}$ was updated and the value of trace(V_t) was calculated and plotted against the sampling sequence. The process was in control for the first 49 observations. Starting from observation 50, a mean shift occurred in the process and the observations starting from sample 50 were generated from a process

Table 1 The values of L of the MEWMS chart for $p = 2$ and 3

λ	$L(p = 2)$	$L(p = 3)$
0.1	2.9013	2.8212
0.2	3.4870	3.3281
0.3	3.8713	3.6621

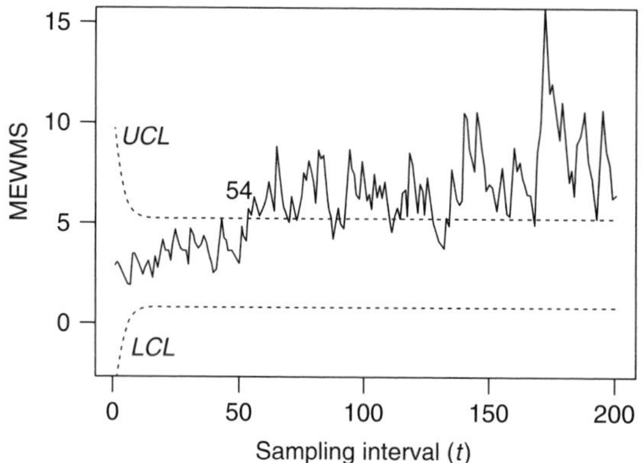

Figure 2 The MEWMS chart of the wafer manufacturing example

having a new mean vector equal to (3.231, 3.200, 3.118) and the same covariance matrix as $\hat{\Sigma}_0$. Beginning at observation 100, an additional change in the covariance matrix also took place. The new covariance matrix starting from sample 100 is equal to $\Sigma = \begin{pmatrix} 0.0186 & 0.0036 & 0.0052 \\ 0.0036 & 0.0170 & 0.0034 \\ 0.0052 & 0.0034 & 0.0088 \end{pmatrix}$. The corresponding phase II MEWMS chart with $p = 3$, $\lambda = 0.2$, and $L = 3.3281$ was produced using S-plus and shown in Figure 2. The MEWMS chart detected out-of-control signals at as early as observation 54, which was primarily due to a mean shift in the process. The chart also showed an out-of-control signal at observation 100 when the change in the covariance matrix actually occurred.

Discussion

The $|S|$ chart is relatively easy to construct and use, and it is available in some of the commonly used commercial software packages such as Minitab. It is also less sensitive to mean shifts in the sense that a shift in the mean is less likely to cause the $|S|$ chart to produce false out-of-control signals. On the other hand, the chart is restricted to the case when $n > p$. Tang and Barnett [9] showed such control limits may lead to considerably larger false alarm rate than the nominal value of 0.0027, especially when n is relatively small. In the same paper, they provided accurate calculations of control limits for the cases when $p = 3$ and 4.

Since the $|S|$ chart is a Shewhart chart, it is generally less effective in detecting small changes in the generalized variance. Yeh *et al.* [4] proposed an exponentially weighted moving averages (EWMA) chart based on the **EWMA** of $\log|S_t|$. Such an EWMA chart is more effective in detecting small changes in the generalized variance. Other existing multivariate control charts for variability derived from the sample generalized variance can be found in works by Guerrero-Cusumano [10], Tang and Barnett [11], Levinson *et al.* [12], Yeh and Lin [13], and Yeh *et al.* [14].

In the case of $n = 1$, none of the aforementioned charts is applicable since the sample covariance matrix cannot be evaluated. The MEWMS chart discussed above is relatively easy to construct. Although most of the commercial software packages currently do not include the MEWMS chart, it was relatively simple for its customized codes. For other charts that are applicable to the case when $n = 1$, readers are referred to the works by Chan and Zhang [15], Khoo and Quah [16], Yeh *et al.* [17], and Reynolds and Cho [18].

Conclusions

We discuss two multivariate control charts for monitoring variability of multivariate normal processes. The $|S|$ chart is applicable when $n > p$, while the MEWMS chart is specifically designed for individual observations ($n = 1$). Examples of other types of charts include the nonparametric procedures designed to detect changes in a covariance matrix as defined by changes of some function of the matrix (Hawkins [19]), the data depth based nonparametric control charts for nonnormal processes (Liu [20]), and the principle component analysis and dissimilarity index based control charts for multivariate time-series data (Kano *et al.* [21]).

The $|S|$ chart is essentially designed to detect changes in the determinant of the covariance matrix. When a point is outside of the control limits on the $|S|$ chart, caution should be taken when interpreting such a signal (*see* **Multivariate Control Charts, Interpretation of**). The out-of-control signal is primarily due to a possible increase or decrease in the determinant of the covariance matrix. It does not necessarily imply that there is an increase or decrease in the process variability. Johnson and Wichern [22] gave three covariance matrices for bivariate data that all have the same determinant and yet have very different correlations.

An important area not discussed here is the diagnostic techniques. This is more complicated than the univariate case, mainly due to the complexity of the covariance matrix. Because so many parameters are contained in the covariance matrix and that changes in one or some of the parameters can trigger an out-of-control signal, it is of eminent importance to be able to further pinpoint which of these parameters are out of control. Some potential diagnostics are discussed in Yeh *et al.* [4], and the references therein. In another recent paper, Apley and Shi [23] proposed a technique based on a fault model which transformed the problem of diagnosing into the problem of estimating the number of faults which contribute to changes in process variability and the linear combinations by which each of the p correlated quality characteristics is affected.

References

[1] Weirda, S.J. (1994). *Multivariate Statistical Process Control*, Wolters-Noordhoff, Groningen.

[2] Lowry, C.A. & Montgomery, D.C. (1995). A review of multivariate control charts, *IIE Transactions in IIE Research* **27**, 800–810.

[3] Montgomery, D.C. (2005). *Introduction to Statistical Quality Control*, 5th Edition, John Wiley & Sons, New York.

[4] Yeh, A.B., Lin, D.K.-J. & McGrath, R.N. (2006). Multivariate control charts for monitoring covariance matrix: a review, *Journal of Quality Technology and Quantitative Management* **3**, 415–436.

[5] Mitra, A. (1993). *Fundamentals of Quality Control and Improvement*, Macmillian, New York.

[6] Huwang, L., Yeh, A.B. & Wu, C.-W. (2007). Monitoring multivariate process variability for individual observations, *Journal of Quality Technology* **39**(3), 258–278.

[7] MacGregor, J.F. & Harris, T.J. (1993). The exponentially weighted moving variance, *Journal of Quality Technology* **25**, 106–118.

[8] Joshi, M. & Sprague, K. (1997). Obtaining and using statistical process control limits in the semiconductor industry, in *Statistical Case Studies for Industry Process Improvement*, V. Czitrom & P.D. Spagon, eds, SIAM, Philadelphia, pp. 337–356.

[9] Tang, P.F. & Barnett, N.S. (1996). Dispersion control for multivariate processes-some comparisons, *The Australian Journal of Statistics* **38**, 253–273.

[10] Guerrero-Cusumano, J.-L. (1995). Testing variability in multivariate quality control: a conditional entropy measure approach, *Information Sciences* **86**, 179–202.

[11] Tang, P.F. & Barnett, N.S. (1996). Dispersion control for multivariate processes, *The Australian Journal of Statistics* **38**, 235–251.

[12] Levinson, W., Holmes, D.S. & Mergen, A.E. (2002). Variation charts for multivariate processes, *Quality Engineering* **14**, 539–545.

[13] Yeh, A.B. & Lin, D.K.-J. (2002). A new variables control chart for simultaneously monitoring multivariate process mean and variability, *International Journal of Reliability, Quality and Safety Engineering* **9**, 41–59.

[14] Yeh, A.B., Lin, D.K.-J., Zhou, H. & Venkataramani, C. (2003). A multivariate exponentially weighted moving average control chart for monitoring process variability, *Journal of Applied Statistics* **30**, 507–536.

[15] Chan, L.K. & Zhang, J. (2001). Cumulative sum control charts for the covariance matrix, *Statistica Sinica* **11**, 767–790.

[16] Khoo, M.B. & Quah, S.H. (2003). Multivariate control chart for process dispersion based on individual observations, *Quality Engineering* **15**, 639–642.

[17] Yeh, A.B., Huwang, L. & Wu, C.-W. (2005). A multivariate EWMA control chart for monitoring process variability with individual observations, *IIE Transactions on Quality and Reliability Engineering* **37**, 1023–1035.

[18] Reynolds Jr, M.R. & Cho, G.-Y. (2006). Multivariate control charts for monitoring the mean vector and covariance matrix, *Journal of Quality Technology* **38**, 230–253.

[19] Hawkins, D.M. (1992). Detecting shifts in functions of multivariate location and covariance parameter, *Journal of Statistical Planning and Inference* **33**, 233–244.

[20] Liu, R.Y. (1995). Control charts for multivariate processes, *Journal of the American Statistical Association* **90**, 1380–1387.

[21] Kano, M., Nagao, K., Hasebe, S., Hashimoto, I., Ohno, H., Strauss, R. & Bakshi, R.R. (2002). Comparison of multivariate statistical process monitoring methods with applications to the Eastman challenge problem, *Computers and Chemical Engineering* **26**, 161–174.

[22] Johnson, R.A. & Wichern, D.W. (2002). *Applied Multivariate Statistical Analysis*, 5th Edition, Prentice Hall.

[23] Apley, D.W. & Shi, J. (2001). A factor-analysis method for diagnosing variability in multivariate manufacturing processes, *Technometrics* **43**, 84–95.

Related Articles

Hotelling's T^2 Chart; Multivariate Control Charts Overview; Multivariate Control Charts, Interpretation of; Multivariate Cumulative Sum (CUSUM) Chart; Multivariate Exponentially Weighted Moving Average (MEWMA) Control Chart.

Dennis K.J. Lin and Arthur B. Yeh

Multivariate Control Charts Overview

Introduction

Multivariate statistical process control (MSPC) is used to simultaneously monitor multiple measurements from a process. Similar to the case for a single variable, the objective is to detect assignable causes of variation. Multiple measurements are quite common in modern processes. Multiple process variables might be measured such as temperatures, pressures, material properties, concentrations, flow rates, valve settings, revolutions, voltages, and so on. Multiple measurements might also be derived from the product (such as numerous physical dimensions, thickness, surface finish, gloss, uniformity, and so on). In general cases there may be very different units of measurement. In special cases of MSPC, there might be multiple measurements in the same units, such as thickness measurements at several sites on a surface, or lengths of multiple vanes on a single part, or an absorption spectrum for a range of wavelengths of light. Common units often occur in **multiple stream processes**, and [1] provides a summary. Profile measurements often provide additional examples (*see* **Profile Monitoring**). Multiple measurements allow one to consider a number of interesting strategies to monitor the process.

In addition to the cases mentioned previously, multiple measurements arise in other ways. A summary statistic might be used to combine data. For example, the mean thickness from several sites on a semiconductor wafer can be used to summarize multiple measurements to one. Consequently, the summary can be monitored with a **control chart** for a single variable. If some sites were near the center of the wafer and some were near the edge, one might also consider additional summaries such as the mean thickness at the edge minus the mean at the center. This leads to another summary, and another variable to be monitored. Consequently, multivariate data occurs again, although the number of metrics might be reduced from every site to fewer summary scores. Still, with more than one summary score to monitor, the application is again within the area of MSPC.

In a batch process one might monitor the temperature over time as the batch starts and then ends. This becomes more interesting when one considers that batches should also be monitored over time. From the view of the process, each batch provides potentially a large number of temperature measurements. Furthermore, a single temperature is probably supplemented with other temperature sensors, and other variable measurements. There may be several variables measured over time for every batch, and then the batches are monitored in sequence. Consequently, a number of alternative designs are feasible for a process monitor, but clearly multiple measurements are involved (*see* **Control Charts for Batch Processes**).

In yet another application for MSPC, the (rpm) revolutions per minute of a pump is often monitored with the pressure change across the pump. One expects an increase in rpm to correspond to an increased pressure differential. Although there are two variables, (the rpm might be considered an input variable and the pressure change an output), the variables do not have the same symmetric relationship that occurs with measurements from different sites on a wafer. Consequently, the nature of a strategy to monitor this data might incorporate the input–output structure. Still, multiple measurements are involved. A regression-type approach for MSPC has been applied, with and without, a direct cause and effect relationship among the variables (*see* **Regression Control Charts**).

As a transaction progresses through a business process, intermediate cycle times become available. The set of cycle times provides multiple measurements for each transaction.

Advantages of MSPC

With multiple measurements, each can be monitored in its own control chart. However, this has two disadvantages. One is that it is difficult to control the number of false alarms. An out-of-control action plan (OCAP) needs to begin whenever any control chart for a process signals. With multiple measurements and a correspondingly large number of control charts, there is much greater opportunity for false alarms. If the measurements plotted on each chart are independent, the increase in false alarms can be computed from basic probability. However, multiple measurements are often correlated and this makes the calculation more difficult. In any case, the control limits on each chart need to be widened to control the false alarms and this may degrade the detection of assignable causes.

The second disadvantage of separate control charts is more important. There are often important relationships between variables that should be considered for MSPC. Multiple measurements from a process are often redundant, and correlated, and reflect data from many fewer true process states. Figure 1 provides a scatter plot of two measurements from a process. Notice the data is not randomly dispersed in the plot. Instead, the data clusters into an ellipse shape (except for a single point denoted as x_0). This is expected from measurements that are correlated. In Figure 1, the single point x_0 is clearly seen as unusual. Figure 2 shows that neither of the separate control charts for x_1 and x_2 indicate any problems. If x_1 and x_2 are measurements from two sites on a wafer, there is an unusual relationship observed for the wafer corresponding to point x_0. As another example, if x_1 is the rpm of a pump and x_2 is the pressure differential, point x_0 suggests a broken blade (increased rpm without a corresponding pressure change). An unusual data point potentially signifies that an **assignable cause** is present. Therefore, MSPC has detected unusual process results that neither separate control chart can identify.

Similar relationships occur between variables in higher dimensions. Although simple graphics cannot be viewed, the shape can be confirmed from numerical calculations. Consequently, MSPC can be used to monitor for unusual points in multiple dimensions and this provides a powerful methodology. Data points that are distant from the typical, in-control data in multiple dimensions signal potential assignable causes.

The data points tend to form an elliptical shape, and the major axis of the ellipse can be described by a line. The data points cluster along this line due to the **correlation**. In higher dimensions, data often clusters into lower-dimensional subspaces such as a plane in three dimensions (or a hyperplane in higher dimensions). An MSPC strategy can identify these subspaces and build monitors that focus on the subspace or deviations from the subspace.

Structure of the Data

MSPC has traditionally assumed that the measurements are available in the form of a table of numbers. Each row represents a time instant, and the numbers in the row are the measurements obtained from the process at that time (temperatures, pressures, etc.). Consequently, categorical measurements such as $\{poor, average, good\}$ need to be quantified. However, additional work might be needed to acquire

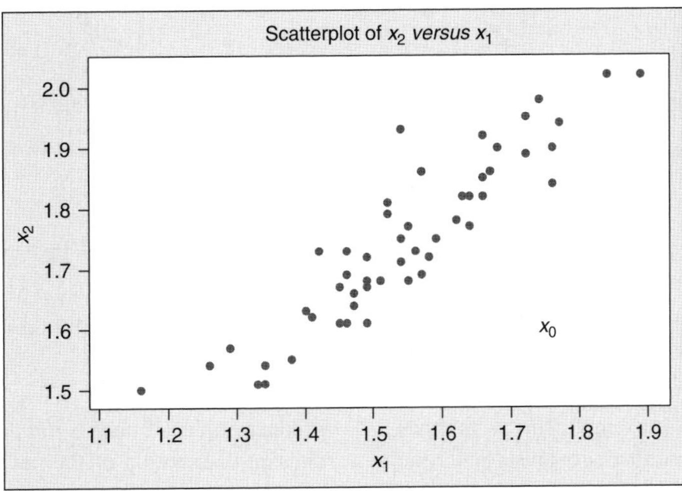

Figure 1 A scatter plot of two correlated measurements x_1 and x_2 displays an elliptical shape, but one point in the lower right does not conform to the pattern

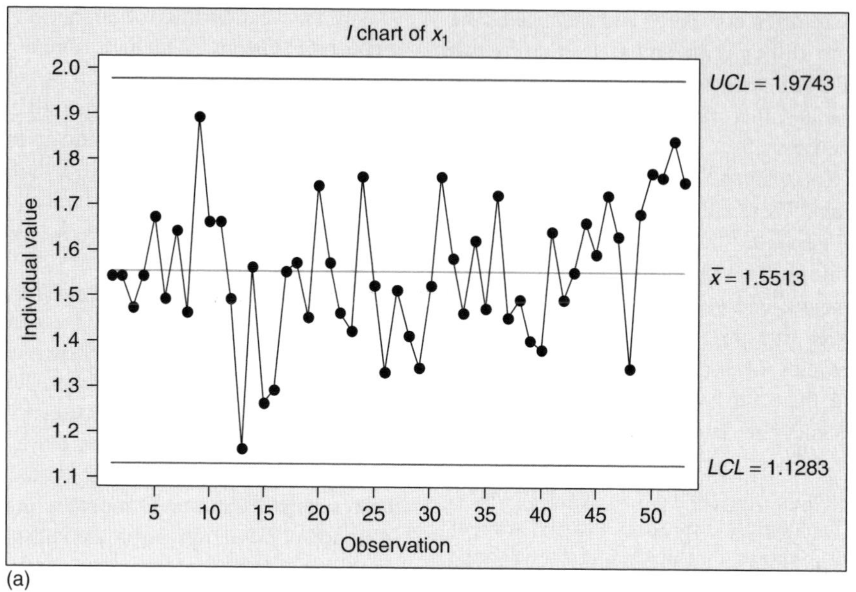

(a)

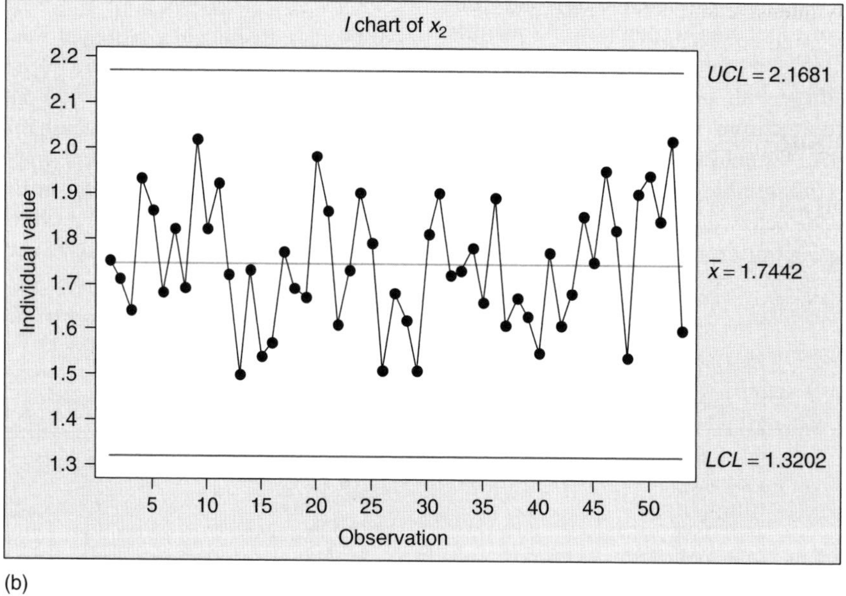

(b)

Figure 2 Individual control charts for each variable x_1 and x_2 do not show any assignable causes. Can you identify the control chart points that correspond to the point in the lower right of Figure 1?

this convenient table. For example, the cycle times in the business process described previously will require some work to fit this table, and several alternatives might be useful (based on the objectives of the analysis). Similarly, the final quality measurement of a product produced now is not dependent on the particle size distribution of the raw material now, but on an earlier, tracked lot of raw material that was used for the current product. Furthermore, many products might have been produced from the same lot of raw material. Laboratory measurements might be taken from only a sample of products. This leads to different frequencies and phases of measurements that probably influence how to construct the table. Also, a particle size distribution may need to be summarized to a few numbers. Consequently, the traditional table might require substantial work as material flows from a start to an end of a process and measurements are derived in between. The batch process described previously presents additional challenges with time of batch, measurement time within the batch, and multiple variables. The design alternatives are driven by the objectives of the analysis.

Also, the numbers in the table are assumed to not be missing. If so, an interpolation or estimation might be needed. The statistical literature has considered several options for measurement **imputation**. The alternative is to eliminate rows that are not complete. However, with different frequencies of measurement this is not an attractive option. Large numbers of rows might be forced to be discarded. More modern, computationally intensive alternatives exist, but these are beyond the scope of this article.

Assignable Causes

MSPC has focused on control procedures to monitor process parameters. The parameters are the natural extensions from the case of a single variable. There is a mean μ_j and variance σ_j^2 for every variable X_j, say for $j = 1, 2, \ldots, p$. Furthermore, MSPC considers the collection of covariances σ_{jk} between variables X_j and X_k. Consequently, there are control charts to monitor means the μ_j's and those for the variances and covariances.

Usually the variances and covariances are organized into a table called the **covariance** *matrix* as

$$\begin{pmatrix} \sigma_1^2 & \sigma_{12} & \cdots & \sigma_{1p} \\ \sigma_{21} & \sigma_2^2 & \cdots & \sigma_{2p} \\ \cdots & \cdots & \cdots & \cdots \\ \sigma_{p1} & \sigma_{p2} & \cdots & \sigma_p^2 \end{pmatrix}$$

and the objective is to monitor for an assignable cause that changes any one of these parameters. The multiple parameters dramatically increase the challenge from the case of a simple range chart. The article **Multivariate Charts for Variability** presents MSPC for such a covariance matrix.

The means of the variables are also assembled into a mean vector,

$$\mu = (\mu_1, \mu_2, \ldots, \mu_p)$$

and monitors for a change to *any* element of this vector are of substantial interest. As shown in the previous figures an integrated strategy that uses the data from all variables can detect assignable causes not visible to separate control charts.

MSPC has focused on a change to any or all elements of the mean vector. Such an omnibus strategy can simultaneously monitor the means of all process variables. However, the chapters here for MSPC show that the omnibus strategy suffers as the number of process variables (denoted p) increases. Consequently, there is also interest in monitors for more restricted assignable causes.

As mentioned previously, high-dimensional data often cluster into lower-dimensional subspaces because of the correlations between the variables. Consequently, assignable causes that preserve the correlation are expected to shift the mean vector within this subspace. Alternatively, an assignable cause that specifically changes the relationships between the variables, might be expected to shift the mean vector from the subspace. When a pump blade breaks the relation between rpm and pressure differential is no longer valid. In either case, these are restrictions on the mean vector under an assignable cause, and such knowledge can be used to improve the performance of MSPC. Principal components analysis (PCA) has been widely used in this context. Much of the PCA work was summarized by Jackson [2], and further development of dimensional reduction currently continues. A more general subspace was considered by Runger [3]. For related examples, *see* **Control Charts for Batch Processes**.

One-sided control charts are common for a single variable. Therefore, the same motivation applies to multivariate data. Interest might be only in an increase in the mean of one or more variables. It is surprisingly more difficult to accommodate such simple information into a multivariate control chart, but solutions have been developed [4].

Nonparametric approaches have also been applied to the MSPC problem. For an introduction, *see* **Control Charts, Nonparametric**. Nonparametric density estimates and monitors to detect changes in these have been considered. Computationally intensive alternatives such as neural networks (*see* **Neural**

Networks: Construction and Evaluation) and other methods have also been considered, but these extensions are beyond the scope of this discussion.

Multivariate Control Charts

Common control charts for MSPC parallel the charts for a single variable. Consequently, there are multivariate extensions to Shewhart, **exponentially weighted moving average** (EWMA), and cumulative sum (**CUSUM**) control charts. The chapters **Hotelling's T^2 Chart**, **Multivariate Exponentially Weighted Moving Average (MEWMA) Control Chart**, and **Multivariate Cumulative Sum (CUSUM) Chart** describe each of these charts in more detail. The charts apply to a data row, or vector, of arbitrary length p. Consequently, the charts can be applied either before or after a dimensional reduction is used with the data. Suitably modified charts can monitor the mean vector or covariance matrix of the process. The charts are developed with a phase I and phase II strategy identical to the one used for a single variable (*see* **Control Charts, Overview**). Trial control limits might need to be modified on the basis of assignable causes. Rational subgrouping principles apply, although individual observations are typically used.

MSPC charts are designed for numerical data, and for processes that are characterized in terms of the mean vector and covariance matrix (called *second-order* **moments**). Although the charts are robust to nonnormality, the process measurements are expected to form elliptical regions, with greater density at the centers of the ellipse. The distance from an observed data vector x_i (or a appropriate sum or average) to the historical process mean vector μ is an important component of these charts. Rather than Euclidean distance, Mahalanobis distance is used

$$(x_i - \mu)'\Sigma^{-1}(x_i - \mu)$$

where Σ denoted the covariance matrix of the variables. Of course, the unknown parameters in the distance calculation need to be estimated in the phase I analysis. From basic algebra it can be shown that points with equal Mahalanobis distance from μ fall on the surface of an ellipse centered at μ. Consequently, a point with large Mahalanobis distance does not follow the expected elliptical pattern as described in Figure 1.

These charts have been effective for measurements from numerous processes in diverse industries. A further assumption is that the measurement vectors from different sample times are independent. If violated, this leads to the case of autocorrelated data, and suitably modified charts may be needed (*see* **Autocorrelated Data**; **Large Autoregressive Integrated Moving Average (ARIMA) Modeling**).

Signal Interpretation

One might presume that a signal from a control chart for a single variable is easier to analyze than MSPC chart. However, it is the process, not a variable, that must be interrogated as part of an OCAP. For example, if the length of an injection-molded part exceeds its control limit, there is not a single dial for part length. Instead, there is a combination of settings for temperatures, pressures, holding times, and so on that might need to be adjusted. The same is true for a multivariate control chart signal – interpretation of the signal and corrective action is not any different. Furthermore, in the previous pump example, a signal is not "caused" by either rpm or pressure differential. It might be caused by a broken blade, or a clog in a pipe. Consequently, the cause is in the process, not from a variable.

Still, in a multivariate control chart, one might eliminate some variables and focus on others that contribute to the signal. This can help corrective actions. If rpm and pressure differential were only two of several variable monitored by MSPC, then the change in the relationship between these two would no doubt benefit diagnosis and OCAP. Methods for identifying contributors to an MSPC signal have received substantial attention (*see* **Multivariate Control Charts, Interpretation of**).

Conclusions

With modern measurement automation, the importance of MSPC increases. It is now unusual to be concerned with a single measurement. And the ubiquitous computational resources have eliminated the traditional burden of onerous calculations. MSPC is an important tool for modern processes and a rich area for further innovations. Other chapters provide details on many of the topics mentioned here.

Acknowledgments

This material is based upon work supported by the National Science Foundation under Grant No. 0355575.

References

[1] Mortell, R. & Runger, G.C. (1995). Process control of multiple stream processes, *Journal of Quality Technology* **27**(1), 1–12.

[2] Jackson, J.E. (1991). *A User's Guide to Principal Components*, John Wiley & Sons, New York.

[3] Runger, G.C. (1996). Projections and the U^2 control chart for multivariate statistical process control, *Journal of Quality Technology* **28**(3), 313–319.

[4] Testik, M. & Runger, G.C. (2005). One-sided multivariate control charts, *IIE Transactions* **38**(8), 635–645.

GEORGE C. RUNGER

Multivariate Control Charts, Interpretation of

Introduction

Statistical process control procedures are designed to work well in a variety of conditions. The general idea is that a stable process containing only common-cause variation is easiest to control, but when the process variation is mixed with special-cause components, statistical input is needed to help monitor the process and detect the changes. One of the goals of a statistical process control (SPC) procedure is to detect the presence of special causes of variation quickly and reliably, and to locate the source of the variation in terms of the process variables so that corrective action may be taken (*see* **Control Charts, Overview**). This is achieved by using a charting statistic and interpreting the output when it signals that a process change has occurred.

When only one process variable is used to measure process performance, changes in the mean and/or changes in the variation of the variable can be detected using single-variable control charts. When several variables are used to monitor a process, the interpretation of signals in a multivariate control chart is not so simple. In addition to changes in location and variation, there may also be changes in the relationships between the many variables. This occurs when there is a **covariance** or **correlation** change between two variables or among a group of variables. Any of these changes can indicate a signal on the control chart for the multivariate process. For example, a change in the correlation between temperature and pressure might indicate that the observed pressure is either too high or too low for the given observed value of the temperature, relative to the historical process data.

Multivariate process control (*see* **Multivariate Control Charts Overview**) can be based on observing an individual observation vector, or on observing a statistic, such as a mean vector or covariance matrix estimate, computed from a sample of observations taken at each time period. In general, process industries such as the chemical industry favor monitoring individual observations, while other industries such as the manufacturing industry often monitor statistics based on the mean of a sample of observations. We will examine signal interpretation for the case where control is based on an individual multivariate observation vector taken in a monitoring stage (labeled phase II). However, the procedures we present can be adapted for control based on the sample mean as well as for phase I operations where a baseline or historical data set (HDS) is being constructed.

Signal Interpretation

One of the more popular multivariate control procedures is based on Hotelling's T^2 statistic (*see* **Hotelling's T^2 Chart**). Consider a p-dimensional vector of observations, $x' = (x_1, x_2, \ldots, x_p)$, that is made on a process at a specified time period. We assume that, when the process is in control, the $\mathbf{x}$ vectors are independent and follow a multivariate **normal distribution** with an unknown mean vector $\boldsymbol{\mu}$ and an unknown covariance matrix $\boldsymbol{\Sigma}$. The form and distribution of the T^2 statistic for the observation vector $\mathbf{x}$

used in phase II is given by

$$T^2 = (\mathbf{x} - \overline{\mathbf{x}})' \mathbf{S}^{-1} (\mathbf{x} - \overline{\mathbf{x}}) \sim \left[\frac{p(n+1)(n-1)}{n(n-p)}\right] F_{(p,n-p)} \qquad (1)$$

The term $F_{(p,n-p)}$ denotes an F distribution with p and $n-p$ **degrees of freedom**, and the estimates, $\overline{\mathbf{x}}$ and $\mathbf{S}$, are obtained from a phase I operation where the created HDS is of size n.

The popularity of the T^2 statistic is due in part to its ease of calculation and also because of the availability of procedures that can be used to interpret its signals. Two well-known procedures are based on using transformations to decompose a signaling T^2 statistic into the sum of p orthogonal components. These orthogonal components can be related to the contributions of individual process variables or groups of variables to the signal. Thus, we are quickly able to determine if the signaling observation vector agrees with the measures of location or variation/covariation of the in-control multivariate distribution.

The oldest of the decomposition procedures is obtained by transforming the signaling T^2 value into a function of the principal components of the estimated covariance matrix, $\mathbf{S}$, or the corresponding estimated correlation matrix. Using the estimated covariance matrix, we define the ith principal component of $\mathbf{S}$ as the linear combination of the p process variables given by

$$z_i = \mathbf{u}_i'(\mathbf{x} - \overline{\mathbf{x}}) \qquad (2)$$

for $i = 1, 2, \ldots, p$, where $\mathbf{u}_i$ is the ith eigenvector of $\mathbf{S}$ and λ_i is the corresponding eigenvalue [1]. Using this transformation, the T^2 statistic in equation (1) becomes

$$T^2 = \frac{z_1^2}{\lambda_1} + \frac{z_2^2}{\lambda_2} + \cdots + \frac{z_p^2}{\lambda_p} \qquad (3)$$

Since the principal component in equation (2) is a linear combination of the p process variables, it can be used to identify the groups of variables contributing to the signal. However, it has an inherent weakness in that it cannot provide a straightforward interpretation in terms of specific process variables.

A more efficient procedure, known as the MYT decomposition, is described below. The MYT decomposition [2] is defined by using the following transformation of the p process variables

$$\mathbf{y} = \mathbf{L}^{-1}(\mathbf{x} - \overline{\mathbf{x}}) \qquad (4)$$

where $\mathbf{L}$ is a lower triangular invertible matrix with the property that $\mathbf{S} = \mathbf{LL}'$. Using this transformation, the T^2 statistic in equation (1) becomes

$$T^2 = y_1^2 + y_2^2 + \cdots + y_p^2 = T_1^2 + T_{2.1}^2 + \cdots + T_{p.1,2,\ldots,p-1}^2 \qquad (5)$$

The T_1^2 term in equation (5) is labeled an unconditional term and the $T_{j.1,2,\ldots,j-1}^2$ terms are labeled conditional terms.

Using equation (4), the general form of the unconditional term contained in any phase II decomposition is given as

$$T_j^2 = \frac{(x_j - \overline{x}_j)^2}{s_j^2} \qquad (6)$$

where x_j is the jth component of $\mathbf{x}$, and $\overline{x}_j$ and s_j^2 are its corresponding mean and variance estimates as determined using the HDS. Under the assumption of no signal, the T_j^2 statistic is described by an F distribution given as

$$T_j^2 \sim \left(\frac{n+1}{n}\right) F_{(1,n-1)} \qquad (7)$$

We use this term to check if the value of the jth variable is outside of its operational range.

Similarly, the general form of the conditional term in any phase II decomposition is given as

$$T_{j.1,2,\ldots,j-1}^2 = \frac{(x_j - \overline{x}_{j.1,2,\ldots,j-1})^2}{s_{j.1,2,\ldots,j-1}^2} \qquad (8)$$

This is the square of the value of the jth variable adjusted by the estimates of the mean and variance of the conditional distribution of x_j adjusted for the variables $x_1, x_2, \ldots, x_{j-1}$. The estimates, $\overline{x}_{j.1,2,\ldots,j-1}$ and $s_{j.1,2,\ldots,j-1}^2$, are obtained from the HDS [3]. With no signal present, the distribution of a conditional term is given as

$$T_{j.1,2,\ldots,j-1}^2 \sim \left(\frac{(n+1)(n-1)}{n(n-k-1)}\right) F_{(1,n-k-1)} \qquad (9)$$

where $k = j - 1$. We use this term to determine if the actual value of the jth variable is close to the value predicted using the values of the remaining variables.

Since the subscripts $\{1, 2, \ldots, p\}$ in the T^2 terms on the right-hand side of equation (5) can be replaced by any permutation of $\{1, 2, \ldots, p\}$, the corresponding distributions of the T^2 statistics in equations (7) and (9) remain invariant with respect to the permutation.

Computer algorithms based on equation (5) are readily available for determining which of the p process variables are contributing to the signaling T^2 value [4]. Generally, these systems work by examining all possible decompositions of a signaling T^2 value. A total decomposition includes computing all the possible distinct MYT components: the p unconditional terms, T_j^2; all two-way conditional terms $T_{i.j}^2$; all three-way terms, $T_{i.jk}^2$; etc. There are $2^{(p-1)} \times (p)$ such terms, but several efficient computational shortcuts exist for obtaining the relevant values.

For example, a useful stepwise process involves first using the p unconditional terms described in equation (6) to check the tolerance of each individual variable. Any variable with a significant F value as given by equation (7) is declared to be out of tolerance. This variable would then be considered as part of the signal and removed from further consideration. Next the two-way terms are checked using equation (9). Any signaling $T_{i.j}^2$ component implies that the relationship between x_i and x_j differs from that given in **S**. If there is a signal, both variables are removed from further consideration and the three-way terms are checked and so on.

Data Example

The MYT decomposition procedure is illustrated in the following data example. More details on how to use it and how to interpret its orthogonal components is given in [5]. A steam turbine system is used to generate electricity for industrial use. Our purpose is to show how signal interpretation can be used in locating the source of a signal in this process. To simplify the problem, we consider only two of numerous variables that are used in controlling the turbine system. These variables include the amount of fuel (F) being used and the amount of megawatts (MW) that is being produced by the turbine. Summary statistics for the HDS for the two variables are contained in Table 1.

A typical T^2 control chart used in monitoring the turbine system is given in Figure 1. All T^2 values below the upper control limit (UCL), obtained using an $\alpha = 0.0027$, indicate that performance is

Table 1 Summary statistics for the historical data set

	MW	F
Mean	53.72	692.45
Standard deviation	3.41	30.41
Correlation matrix		
	MW	**F**
MW	1.000	0.7469
F	0.7469	1.000

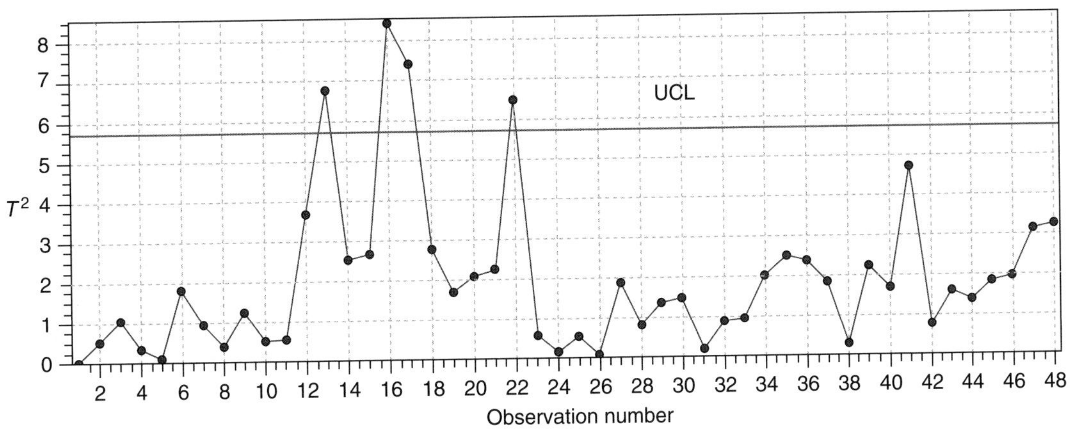

Figure 1 T^2 chart for monitoring the turbine system

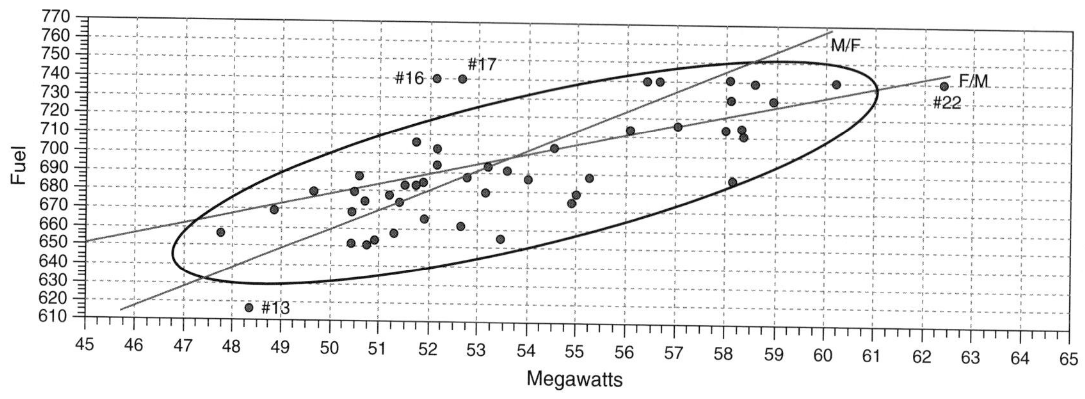

Figure 2 Scatter plot of data

in agreement with the baseline conditions. The T^2 values above the UCL indicate signals.

A scatter plot of the turbine data (fuel *vs* megawatts) along with the corresponding T^2 control ellipse using the formula in equation (1) is presented in Figure 2. Note that the signaling T^2 points identified in Figure 1 are located outside the control ellipse. The two lines in the plot represent the regression of fuel on megawatts (F/M) and the regression of megawatts on fuel (M/F).

In order to illustrate signal interpretation, we apply the MYT decomposition to the four signaling T^2 values corresponding to observations #13, #16, #17, and #22. A total decomposition giving all components and their values for the signaling T^2 values are presented in Table 2. The subscript 1 refers to the megawatt variable while the subscript 2 refers to the fuel variable.

Discussion of Results

Observation #13, located at the lower end of the control ellipse, has a significant unconditional component T_1^2 and a significant conditional component $T_{1.2}^2$. The signaling unconditional term, T_1^2, indicates that the megawatt value is outside the tolerance limits (i.e., range) established using the HDS. The signaling conditional term, $T_{1.2}^2$, indicates that, given the value of fuel for this observation, the value of megawatts is not where it should be relative to the HDS. This is further denoted by the large vertical distance (residual) between the observation and the regression line of megawatts on fuel (M/F) given in Figure 2.

Both of the T^2 conditional terms for observations #16 and #17 are significant. These signals indicate that there is a fouled linear relationship between the megawatts being produced and the fuel being used, and that the two variables are countercorrelated relative to the correlations observed using the HDS. From the plot given in Figure 2, it appears that for these two observations either the megawatts value is too low for the given value of fuel, or the fuel value is too high for the given value of megawatts.

Observation #22, located at the upper end of the control region given in Figure 2, has a significant unconditional term T_2^2. This indicates that the fuel

Table 2 Decomposition components of T^2 signals

Observation	#13	#16	#17	#22
T_1^2	6.4940[(a)]	0.2094	0.0967	2.4639
$T_{2.1}^2$	0.2580	8.2104[(a)]	7.2936[(a)]	4.0189
T_2^2	2.4515	2.4441	2.4441	6.2770[(a)]
$T_{1.2}^2$	4.3005[(a)]	5.9758[(a)]	4.9462[(a)]	0.2059

[(a)]Indicates significant using $\alpha = 0.0027$

usage for this observation is outside the tolerance limits (i.e., range) established using the HDS.

When only two variables are being considered, making it possible to plot the control ellipse, we usually find that points located outside the ends of the ellipse, such as observation #22 in our example, have out-of-tolerance variables. Those observations located around the middle, but outside of the control ellipse, such as observations #16 and #17 in our example, will have fouled relationships between the two variables. Similarly, those observations located between the middle and the end, but outside of the ellipse, such as observation #13 in our example, may have both problems.

Summary

The MYT decomposition has many good features. It provides an orthogonal decomposition of a T^2 value, and the corresponding components indicate which variables in the signaling observation vector disagree with the historical data. The cause of the signal can be directly attributed to a variable or set of variables being out of tolerance, and/or to the correlation between a set of variables being counter to the correlation observed in the baseline data. What it does not answer is whether the cause of the problem is due to a change in the mean vector or to a change in the covariance matrix, since either could produce variables that are out of tolerance and/or countercorrelated. This problem also exists when one uses principal components analysis. In these types of decompositions, we are basically identifying the change and not necessarily isolating the mechanism leading to the change.

Other approaches exist for use in interpreting T^2 signals. Several of these have been shown to be related to the components of the MYT decomposition (see [2]), though none are as comprehensive. These include the method in [6, 7] based on using regression-adjusted variables to improve diagnostics as well as the step-down method in [8, 9] that sequentially tests subsets of the variables assuming there is an *a priori* ordering of the variables. Another related procedure [10] is based on using a univariate t statistic to rank the components of an observation vector according to their contribution to the signal. And in [11] a method is given that is a function of a sum of a subset of the conditional T^2 components.

Signal interpretation can also be accomplished using other techniques. Fuchs and Benjamin [12] present a graphical method based on multivariate profile plots, while Sparks *et al.* [13] use the Gabriel biplot. Kouti and MacGregor [14] provide an approach based on usage of normalized principal components and contribution plots. Maravelakis *et al.* [15] propose a modified approach based on principal components. Nedumaran and Pignatiello [16] describe an approach based on Scheffe-type intervals and usage of univariate diagnostic charts for the individual variables as well as univariate charts for the principal components. Although the list of alternative methods for interpreting signals in a multivariate control chart continues to expand, orthogonal decomposition techniques, particularly those based on the MYT decomposition or principal components, remain the most useful and informative.

References

[1] Jackson, J.E. (1991). *A User's Guide to Principal Components*, John Wiley & Sons, New York.

[2] Mason, R.L., Tracy, N.D. & Young, J.C. (1995). Decomposition of T^2 for multivariate control chart interpretation, *Journal of Quality Technology* **27**, 99–108.

[3] Mason, R.L. & Young, J.C. (2002). *Multivariate Statistical Process Control with Industrial Applications*, ASA-SIAM, Philadelphia.

[4] QualStat V5.0 (1996). *InControl Technologies, Inc.*, Houston.

[5] Mason, R.L., Tracy, N.D. & Young, J.C. (1997). A practical approach for interpreting multivariate T^2 control chart signals, *Journal of Quality Technology* **29**, 396–406.

[6] Hawkins, D.M. (1991). Multivariate quality control based on regression-adjusted variables, *Technometrics* **33**, 61–75.

[7] Hawkins, D.M. (1993). Regression adjustment for variables in multivariate quality control, *Journal of Quality Technology* **25**, 170–182.

[8] Roy, J. (1958). Step-down procedure in multivariate analysis, *Annals of Mathematical Statistics* **29**, 1177–1187.

[9] Wierda, S.J. (1994). *Multivariate Statistical Process Control: Groningen Theses in Economics*, Wolters-Noordhoff, Groningen.

[10] Doganaksoy, N., Faltin, F.W. & Tucker, W.T. (1991). Identification of out-of-control quality characteristics in a multivariate manufacturing environment, *Communications in Statistics: Theory and Methods* **20**, 2775–2790.

[11] Murphy, B.J. (1987). Selecting out of control variables with the T^2 multivariate quality control procedure, *The Statistician* **36**, 571–583.

[12] Fuchs, C. & Benjamini, Y. (1994). Multivariate profile charts for statistical process control, *Technometrics* **36**, 182–195.

[13] Sparks, R.S., Adolphson, A. & Phatak, A. (1997). Multivariate process monitoring using dynamic biplot, *International Statistical Review* **65**, 325–349.

[14] Kourti, T. & MacGregor, J.F. (1996). Multivariate SPC methods for process and product monitoring, *Journal of Quality Technology* **28**, 409–428.

[15] Maravelakis, P.E., Bersimis, S., Panaretos, J. & Psarakis, S. (2002). Identifying the out of control variable in a multivariate control chart, *Communications in Statistics: Theory and Methods* **31**, 2391–2408.

[16] Nedumaran, G. & Pignatiello, J.J. (1998). Diagnosing signals from T^2 and χ^2 multivariate control charts, *Quality Engineering* **10**, 657–667.

Related Articles

Hotelling's T^2 Chart; **Multivariate Control Charts Overview**; **Multivariate Exponentially Weighted Moving Average (MEWMA) Control Chart**; **Regression Control Charts**.

ROBERT L. MASON AND JOHN C. YOUNG

Multivariate Cumulative Sum (CUSUM) Chart

Introduction

Slight deviations of process variables from their target values are inevitable even when the production process is running normally. Under normal process operating conditions, these deviations in the variables are the interplay of multiple small causes, which are practically unavoidable and generally called the *common causes* or *chance causes of variation*. A process operating under common causes of variation is often considered to be statistically *in control* or *stable*. By statistically analyzing the common variation or the in-control process, process monitoring can be established for early detection of shifts to an out-of-control process. By assigning causes of these shifts, corrective actions can be taken and quality can be improved.

Multivariate control charts simultaneously utilize several related measures of process properties for collectively monitoring the statistical state of a process. The intention is rapid detection of abnormal situations and occurrences of faults, which are assumed to shift the process to an out-of-control state and are referred to as *assignable causes of variation*. Although routine practice of multivariate procedures may require computer computations, over the past decade multivariate control charts have been widely utilized in industry and new multivariate control charting techniques have been developed. This is due, at least in part, to the increased use of computers on the production floor and the amplified ability of computers and sensors to capture data on several process properties simultaneously. Various multivariate control charts (*see* **Multivariate Control Charts Overview**) exist in the literature. Among these multivariate control charts, the Hotelling's T^2 (*see* **Hotelling's T^2 Chart**; [1]), the multivariate **exponentially weighted moving average** (MEWMA) (*see* **Multivariate Exponentially Weighted Moving Average (MEWMA) Control Chart**; [2]), and the multivariate cumulative sum (MCUSUM) are the familiar ones.

In this article, the state of the art for the MCUSUM control charts is reviewed. The univariate cumulative sum (CUSUM) (*see* **Cumulative Sum (CUSUM) Chart**) procedure is described first since it is the basis of the multivariate extensions. Next, some multivariate extensions to the univariate CUSUM control chart are explained with their descriptions and basic formulas. References are provided for the other MCUSUM procedures. Advantages and disadvantages of each multivariate procedure are discussed.

The Cumulative Sum Procedure

CUSUM control charts are important tools for the real-time, on-line monitoring of *univariate* (with one variable of interest) processes. These charts are especially effective in detecting small to moderate sized shifts in a process mean.

The classical one-sided CUSUM control chart can be derived from a sequential probability ratio test (SPRT) for the mean of a **normal distribution** or from a likelihood-ratio test for a change point in a normal distributed mean. Page [3] used SPRT to develop a one-sided CUSUM control for detecting shifts in the mean of a normal distributed variable. The null and the alternative hypotheses are

$$H_0 : \mu = \mu_0 \text{ and } H_1 : \mu = \mu_1 \quad (1)$$

respectively. Here, μ represents a process mean whose true value is unknown, μ_0 is the known on-target (in-control process) mean value and μ_1 is a specified off-target (out-of-control process) mean value to be detected quickly.

To generate a procedure for monitoring two-sided mean shifts, i.e., the mean μ increases to some value $\mu_1 > \mu_0$ or decreases to some value $\mu_1 < \mu_0$, Page proposed to simultaneously operate two one-sided CUSUMs. These two one-sided CUSUMs are designed typically with the same shift magnitude value δ for the increasing and decreasing off-target means. That is, $\mu_1 = \mu_0 \pm \delta$ and $\delta = |\mu_1 - \mu_0|$. The monitoring problem is easily accomplished with the recursive calculations that signal if either one-sided CUSUM

$$s_t^+ = \max\left(s_{t-1}^+ + \sigma^{-1}\left(y_t - \mu_0 - \frac{\delta}{2}\right), 0\right) \quad \text{or}$$

$$s_t^- = \max\left(s_{t-1}^- + \sigma^{-1}\left(\mu_0 - y_t - \frac{\delta}{2}\right), 0\right) \quad (2)$$

exceeds a predetermined decision interval, $h (s_t^+ > h$ or $s_t^- > h)$. Here, t denotes the time or sequence index, y denotes the observation on the variable of interest distributed as normal with mean μ and constant variance σ^2, and the statistics s_t^+ and s_t^- (generally with, $s_0^+ = s_0^- = 0$) are the one-sided upper and lower CUSUMs, respectively. It can be seen from equation (2) that sums are accumulated as the name implies, but an observation is accumulated only if it differs from the on-target mean value by more than one-half the specified shift magnitude δ. Such accumulation of information across successive observations is what makes the CUSUM procedure powerful in detecting small to medium sized shifts in the mean. In fact, this chart is the optimal diagnostic for a shift of the mean from the on-target mean value μ_0 to the off-target mean value μ_1. Since the development of CUSUM by Page, a large number of studies have evaluated their theoretical and practical properties. A good recent review can be found in Hawkins and Olwell [4].

Multivariate Cumulative Sum Procedures

Several authors have proposed multivariate (multiple variables of interest) extensions to the univariate CUSUM procedure for monitoring processes with multiple related variables. Many of the key steps in the development of these MCUSUM procedures also appear in the construction of the CUSUM control charts for two-sided shifts (Runger and Testik [5]).

There are mainly two general approaches for the multivariate process monitoring problems. One approach comes from the univariate perspective where multiple univariate control charts are used simultaneously as a collection. Each of the univariate charts in the collection aims in some direction and these charts may be based on measured variables, principal component variables, regression adjusted variables, etc. A detailed discussion is provided in Runger and Montgomery [6]. An out-of-control alarm can be triggered individually by any of the univariate control charts of the collection. This approach may be ineffective in detecting some types of assignable causes since the multivariate characteristic of the data is no longer present directly. The other approach comes from the multivariate perspective, which utilizes the relationships between the variables. Multivariate observation vectors or a statistic of multivariate observation vectors are reduced to a scalar statistic by utilizing the relationships among the variables or the process knowledge before being plotted on a single control chart. Then this single control chart is used for monitoring the variables of the process jointly. In the literature, MCUSUM procedures have been developed using one of these two approaches. However, before going into detail there is one more important characteristic of multivariate control charts that should be explained. Multivariate control charts are one of two types; directionally invariant and direction specific.

Direction-Specific versus Directionally Invariant Control Charts

Let us consider a univariate monitoring problem first. Univariate control charts for monitoring a process mean are specifically designed for the purpose of detecting shifts in the mean along its axis. Since the mean is a scalar value, it does not have a direction but has magnitude. If an **assignable cause** results in a shift of the mean, the shifted mean value relative to the on-target mean value would be either an increase or a decrease with some magnitude on the monitored variable's axis. Therefore, a two-sided control chart is used when both increases and decreases in the mean are important to detect. Consider the two-sided CUSUM chart in equation (2) where the δ term used in the upper CUSUM s_t^+ and the lower CUSUM s_t^- is the same. Because the observations are assumed to be distributed as normal, which is symmetric, performance of the two-sided CUSUM chart would be the same, on average, for detecting a mean shift having some magnitude λ, either an increase to $\mu = \mu_0 + \lambda$ or a decrease to $\mu = \mu_0 - \lambda$. Generally, the metric used for performance evaluation of control charts is the **average run length** (ARL), which is the average number of points plotted on a control chart before the control chart signals an alarm the first time. Consequently, ARL performance of the two-sided CUSUM chart is invariant to an increase or decrease of the process mean relative to the on-target mean but depends on the magnitude or equivalently the distance $\lambda(\mu) = |\mu - \mu_0|$ of the shifted mean from the on-target mean.

Now consider multivariate monitoring problems. Suppose that the variables are arranged in a vector. Unlike the scalar process mean values in a univariate setting, multivariate process mean vectors have both magnitude and direction. Due to correlations among the monitored variables, an assignable cause would often result in a shift of the mean vector with some or all of the on-target mean values being changed. The shifted mean vector may have infinite number of possible shift directions that need to be monitored jointly, if not restricted. One approach to multivariate process mean monitoring is to sensitize the control chart to some known or anticipated shift directions when such process knowledge is available. Therefore, the control chart is expected to perform well in detecting mean shifts in these directions but not necessarily be as effective in other directions. The other approach is to design the control chart to simultaneously look at all shift directions such that the ARL performance is the same, on average, in detecting all mean vectors $\boldsymbol{\mu}$ which are equidistant from $\boldsymbol{\mu}_0$. The distance metric used here as a measure of shift magnitude is the statistical (Mahalanobis) distance, which is the square root of the noncentrality parameter

$$\lambda^2(\boldsymbol{\mu}) = (\boldsymbol{\mu} - \boldsymbol{\mu}_0)'\boldsymbol{\Sigma}^{-1}(\boldsymbol{\mu} - \boldsymbol{\mu}_0) \qquad (3)$$

Consequently, a multivariate control chart is called *directionally invariant* if its ARL performance in detecting a shift of the mean from the on-target mean is determined solely by the statistical distance of the shifted mean from the on-target mean. Note that the univariate symmetric two-sided CUSUM in equation (2) is also directionally invariant. On the other hand, a multivariate control chart is direction specific if its ARL performance in detecting a shift of the mean from the on-target mean is a function of both the statistical distance and the particular direction of the shifted mean relative to the on-target mean. Multiple univariate control charts aim in specified directions and hence are direction specific.

When selecting a multivariate control chart it is important to consider the types of assignable causes that may affect the process. MCUSUM procedures in the literature are either directionally invariant or direction specific. In the following, considered MCUSUM procedures are classified based on their directional characteristic. Advantages and disadvantages of each are discussed.

The Process Model

Suppose that we have observations on p variables $y_1, y_2, \ldots, y_p$ and these are arranged in a $p \times 1$ vector $\mathbf{y}$. Assume a sequence $\mathbf{y}_1, \mathbf{y}_2, \ldots$ of the p-dimensional observation vectors and let these successive observation vectors be independent and identically distributed as p-variate normal with known on-target mean vector $\boldsymbol{\mu}$ and known and constant $p \times p$ **covariance** matrix $\boldsymbol{\Sigma}$, i.e., $\mathbf{y}_t \sim N_p(\boldsymbol{\mu}, \boldsymbol{\Sigma})$ at time t. In the following, the vector $\boldsymbol{\mu}$ will represent a multivariate normal process mean whose true value is unknown. Several cases when testing a shift in the mean vector from a known on-target value $\boldsymbol{\mu}_0$ will be considered by using the characteristics of the off-target mean vector.

Direction-Specific MCUSUM Procedures

Let us first consider the simplest case where the off-target mean vector is known in terms of both magnitude and direction. Denote this anticipated off-target mean vector by $\boldsymbol{\mu}_1$. Note that the on-target and off-target process distributions are completely specified as $\mathbf{y}_t \sim N_p(\boldsymbol{\mu}_0, \boldsymbol{\Sigma})$ and $\mathbf{y}_t \sim N_p(\boldsymbol{\mu}_1, \boldsymbol{\Sigma})$, respectively. The null and alternative hypotheses are

$$H_0 : \boldsymbol{\mu} = \boldsymbol{\mu}_0 \text{ and } H_1 : \boldsymbol{\mu} = \boldsymbol{\mu}_1 \quad (4)$$

respectively. To detect a mean shift in one particular direction, a univariate CUSUM chart aimed in that one direction will give the best ARL performance. The solution introduced by Healy [7] is to apply a univariate CUSUM based on a linear combination of the observed variables. For two-sided shifts along the direction of the off-target mean vector, the chart signals if either

$$s_t^+ = \max\left(s_{t-1}^+ + \mathbf{a}'(\mathbf{y}_t - \boldsymbol{\mu}_0) - \frac{\delta}{2}, 0\right) > h \text{ or}$$
$$s_t^- = \max\left(s_{t-1}^- + \mathbf{a}'(\boldsymbol{\mu}_0 - \mathbf{y}_t) - \frac{\delta}{2}, 0\right) > h \quad (5)$$

Here, δ is equal to the Mahalanobis distance of $\boldsymbol{\mu}_1$ to $\boldsymbol{\mu}_0$, i.e., $\lambda(\boldsymbol{\mu}_1)$ and the vector of the linear combination coefficients is $\mathbf{a} = \boldsymbol{\Sigma}^{-1}(\boldsymbol{\mu}_1 - \boldsymbol{\mu}_0)/\delta$. The terms $\mathbf{a}'(\mathbf{y}_t - \boldsymbol{\mu}_0)$ and $\mathbf{a}'(\boldsymbol{\mu}_0 - \mathbf{y}_t)$ both have a standard univariate normal distribution when $\boldsymbol{\mu} = \boldsymbol{\mu}_0$. All of the available theory for designing univariate CUSUM charts and ARL performance analysis can be used for this MCUSUM chart.

Nevertheless, the solution is not very general since the shifted mean needs to be known exactly in terms of magnitude and direction. If the mean shifts some other direction, this procedure may be anything from fairly effective to ineffective.

The problem is more interesting when it is only assumed that the process mean vector $\boldsymbol{\mu}$ has a known shift magnitude δ (in terms of the Mahalanobis distance of $\boldsymbol{\mu}$ to $\boldsymbol{\mu}_0$) but unknown direction when the process experiences a mean shift. Consider the following null and alternative hypothesis, respectively

$$H_0 : \boldsymbol{\mu} = \boldsymbol{\mu}_0 \text{ and } H_1 : \{\boldsymbol{\mu}|\lambda(\boldsymbol{\mu}) = \delta\} \quad (6)$$

It is convenient to standardize the observation vectors as

$$\mathbf{z}_t = \boldsymbol{\Sigma}^{-1/2}(\mathbf{y}_t - \boldsymbol{\mu}_0) \quad (7)$$

where $\boldsymbol{\Sigma}^{-1/2}$ is a square root of $\boldsymbol{\Sigma}$. In terms of $\mathbf{z}$, we can write

$$H_0 : \boldsymbol{\mu} = \mathbf{0} \text{ and } H_1 : \{\boldsymbol{\mu}|\|\boldsymbol{\mu}\| = \delta > 0\} \quad (8)$$

where $\|.\|$ denotes the norm of a vector. One obvious solution to this problem is to apply a set of p two-sided CUSUM charts, one for each element of the vector $\mathbf{y}$ or one for each element of the vector $\mathbf{z}$. Woodall and Ncube [8] essentially considered this approach.

Charts based on $\mathbf{y}$ have the drawback that they do not directly use the **correlation** between the variables. Charts based on $\mathbf{z}$ have the advantages accrued from a principal components transformation and this is a rather natural choice. This is well known to be easily accomplished with the recursive calculations that signal if either

$$s_t^+(j) = \max\left[s_{t-1}^+(j) + \mathbf{z}_t(j) - \frac{\delta(j)}{2}, 0\right] > h \text{ or}$$
$$s_t^-(j) = \max\left[s_{t-1}^-(j) - \mathbf{z}_t(j) - \frac{\delta(j)}{2}, 0\right] > h \quad (9)$$

Here, the index $j (j = 1, 2, \ldots, p)$ is used to indicate the parameters of the CUSUM chart for the jth component $\mathbf{z}(j)$ of the standardized observation vector $\mathbf{z}$. Runger and Testik [5] proposed to name this MCUSUM procedure as MPYRAMID due to its geometric characteristic. Because the collection of CUSUM charts is considered as a single monitoring procedure, it is necessary to adjust the decision interval h of the charts to keep the total false alarm probability low. A method for selecting h and determining the total false alarm probability is given in Woodall and Ncube [8].

The ARL performance of the MPYRAMID chart depends on the direction of the shift vector and this dependency is lessened, but not removed, by the use of principal component variables $\mathbf{z}(j)$ rather than original variables. It is important to consider the type of mean shift that is important to detect. Charts based on $\mathbf{y}$ would aim along the axes of the original variables and would generally do a good job in detecting shifts along these axes. However, they may not be as effective in detecting shifts in a different direction such as a principal components axis. On the other hand, the MPYRAMID chart based on $\mathbf{z}$ would aim along the axes of the principal component variables and would generally do a good job in detecting shifts along these axes. Similarly, these charts may not be as effective in detecting shifts in different directions such as the axes of original variables.

Directionally Invariant MCUSUM Procedures

For a known off-target magnitude δ but unknown direction another approach is to apply a generalized likelihood-ratio (GLR) procedure. This procedure signals if

$$g_t = \max_{1\le\tau\le t}\left[\delta\left\|\sum_{i=\tau}^{t}\mathbf{z}_i\right\| - \frac{\delta^2}{2}(t-\tau+1)\right] > c \quad (10)$$

Here, g is the control statistic, τ is an unknown shift time ($1 \le \tau \le t$), and c is a decision interval. The derivation was provided by Basseville and Nikiforov [9]. This MCUSUM procedure is referred to as *MCONE* by Runger and Testik [5] because of its geometric characteristic.

Unfortunately, simplification of the procedure by recursive calculations is not available. However, an advantage of MCONE is that it is equally sensitive to shifts of the same magnitude (Mahalanobis distance) in all directions from the on-target mean. Consequently, it can be thought of as a collection of two-sided CUSUM charts operated in all directions simultaneously.

An example MCONE chart is shown in Figure 1 for a multivariate normal process monitoring problem with seven variables. Here the mean shifts to an off-target value at time 15 and the chart triggers an alarm at time 18. To illustrate how the calculations are performed, the control statistic g plotted on the control chart is calculated for the first three observation vectors as given in Table 1. Note that the value for δ is selected to be 1 in

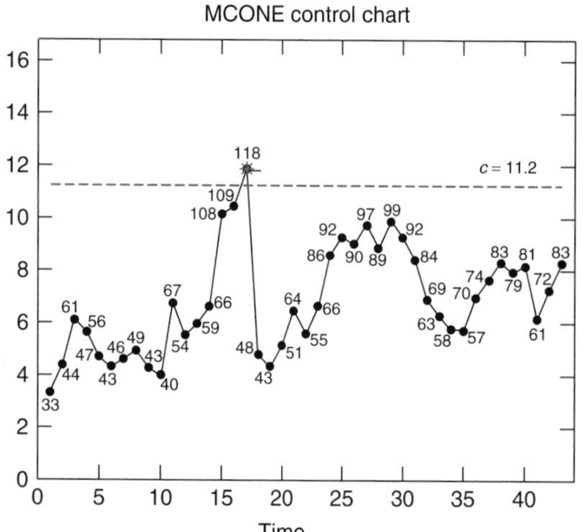

Figure 1 An example of the MCONE control chart

Table 1 An example of the MCONE calculations ($\delta = 1$)

t	τ	$\left\|\sum_{i=\tau}^{t}\mathbf{z}_i\right\| - \frac{1}{2}(t-\tau+1)$	g_t
1	1	3.83 − 0.5 = 3.33	3.33
2	2	2.37 − 0.5 = 1.87	
2	1	5.40 − 1.0 = 4.40	max(1.87, 4.40) = 4.40
3	3	2.96 − 0.5 = 2.46	
3	2	4.46 − 1.0 = 3.46	
3	1	7.59 − 1.5 = 6.09	max(2.46, 3.46, 6.09) = 6.09

the calculations and the first three observation vectors are

$$\mathbf{z}_1 = [-2.28 \quad 0.06 \quad 1.86 \quad -1.45 \quad 1.46 \quad -0.52 \quad 1.23]'$$

$$\mathbf{z}_2 = [0.04 \quad -1.42 \quad 0.63 \quad -1.03 \quad 1.44 \quad -0.19 \quad -0.22]'$$

$$\mathbf{z}_3 = [-2.26 \quad 0.08 \quad 0.03 \quad -0.56 \quad 1.43 \quad -0.45 \quad -1.04]'$$

Other MCUSUM Procedures

Another direction-specific procedure for multivariate observations is the CUSUM of regression adjusted variables proposed by Hawkins [10]. Hawkins extended the approach of a known shift direction to the case in which several shift directions of interest are specified. Some other directionally invariant MCUSUM procedures in the literature include the COT and MCUSUM of Crosier [11] and MC1 and MC2 of Pignatiello and Runger [12]. Of these procedures, MCUSUM and MC1 are the ones with better performance. Although these two MCUSUM procedures have no known optimality properties, they have good practical performance. However, for these two procedures worst-case scenarios exist such that their steady-state performance is impacted by an accumulation of data in one direction before the process shifts another direction, namely the inertia problem. This is because one distinctive characteristic of univariate CUSUM procedure, that is that the steady-state performance is superior to the initial performance, does not appear in these charts. A detailed discussion of the MCUSUM procedures considered in this article as well as some others with performance comparisons can be found in Runger and Testik [5].

Summary

There are several multivariate extensions to the univariate CUSUM control chart. This work reviewed some of these multivariate extensions and compared their advantages and disadvantages. Basic formulas were provided and references were given to allow the interested reader to investigate further.

References

[1] Hotelling, H.H. (1947). Multivariate quality control-illustrated by the air testing of sample bombsights, in *Techniques of Statistical Analysis*, C. Eisenhart, M.W. Hastay & W.A. Wallis, eds, McGraw-Hill, New York, pp. 111–184.

[2] Lowry, C.A., Woodall, W.H., Champ, C.W. & Rigdon, S.E. (1992). A multivariate exponentially weighted moving average chart, *Technometrics* **34**, 46–53.

[3] Page, E.S. (1954). Continuous inspection schemes, *Biometrika* **41**, 100–114.

[4] Hawkins, D.M. & Olwell, D.H. (1998). *Cumulative Sum Charts and Charting for Quality Improvement*, Springer-Verlag, New York.

[5] Runger, G.C. & Testik, M.C. (2004). Multivariate extensions to cumulative sum control charts, *Quality and Reliability Engineering International* **20**, 587–606.

[6] Runger, G.C. & Montgomery, D.C. (1997). Multivariate and univariate process control: geometry and shift directions, *Quality and Reliability Engineering International* **13**, 153–158.

[7] Healy, J.D. (1987). A note on multivariate CUSUM procedures, *Technometrics* **29**, 409–412.

[8] Woodall, W.H. & Ncube, M.M. (1985). Multivariate CUSUM quality control procedures, *Technometrics* **27**, 285–292.

[9] Basseville, M. & Nikiforov, I.V. (1993). *Detection of Abrupt Changes: Theory and Application*, Prentice Hall, Englewood Cliffs.

[10] Hawkins, D.M. (1991). Multivariate quality control based on regression-adjusted variables, *Technometrics* **33**, 61–75.

[11] Crosier, R.B. (1988). Multivariate generalizations of cumulative sum quality control schemes, *Technometrics* **30**, 291–303.

[12] Pignatiello, J.J. & Runger, G.C. (1990). Comparisons of multivariate *CUSUM* charts, *Journal of Quality Technology* **22**, 173–186.

MURAT C. TESTIK AND GEORGE C. RUNGER

Multivariate Exponentially Weighted Moving Average (MEWMA) Control Chart

Introduction

The multivariate **exponentially weighted moving average** (MEWMA) **control chart** is a tool for simultaneously monitoring several correlated process variables. Historically, univariate charts have been used to determine if a process is in a state of statistical control (*see* **Control Charts, Overview**; **Control Charts for the Mean**). If multiple variables need to be considered, then separate charts have been used for monitoring each individual variable. However, many variables that affect the quality of a process are often correlated. Using separate univariate charts assumes that the variables are independent and ignores the **correlation** among them. Multivariate control charts are designed to incorporate the correlation structure of the variables into the chart when determining the state of statistical control of a process (*see* **Multivariate Control Charts Overview**). The exponentially weighted moving average (EWMA) statistic computes a weighted average over time in an attempt to detect mean changes more quickly.

MEWMA Statistic

To detect small changes in a process mean vector, a commonly used chart is the MEWMA control chart. In the univariate case, the EWMA chart (*see* **Exponentially Weighted Moving Average (EWMA) Control Chart**) has been shown to provide quicker detection of small changes in the process mean than the traditional $\overline{x}$ chart (*see* **Control Charts for the Mean**). Extending this to the multivariate situation was proposed by Lowry *et al.* [1]. Given a set of vectors $X_1, X_2, \ldots, X_n$ that follow a multivariate **normal distribution** of dimension p with mean vector μ and variance–**covariance** matrix Σ, the MEWMA statistic can be defined as in equation (1)

$$Z_i = R[X_i - \mu_0] + (I - R)Z_{i-1} \tag{1}$$

for $i = 1, 2, \ldots, n$ where R is a diagonal matrix, $R = \text{diag}(r_1, \ldots, r_p)$ with $0 < r_i \le 1$, and Z_0 is usually initialized to the target mean vector, μ_0. The elements of the matrix R may be chosen to reflect the relative importance of the recorded data. If all variables are equally important, then $r_1 = r_2 = \ldots = r_p$. Using this choice for the matrix R yields $Z_i \sim N_p(0, \Sigma_z)$ where the variance–covariance matrix is given in equation (2).

$$\Sigma_{z_i} = \frac{r[1-(1-r)^{2i}]}{2-r}\Sigma \qquad (2)$$

Since the variance depends on the weights of the previous time periods it increases over time to its limiting value of $\frac{r}{2-r}\Sigma$. For values of $r \ge 0.5$, this limiting value is achieved rapidly. However, for values of $0 < r < 0.5$, it may take more than 25 time periods to get close to this limiting value. Using the exact variances in the series ensures that the probability of a false alarm is constant for all plotted statistics.

Plotted Statistic

The form of the statistic plotted on the control chart depends on whether Σ is known or not. If Σ is known then a χ^2 statistic may be computed

Table 1 Data used for example MEWMA control chart

Index	X_1	X_2	X_3	X_4	X_5	Z_1	Z_2	Z_3	Z_4	Z_5	T^2
1	5.358	4.924	4.431	3.093	4.487	0.0716	−0.015	−0.114	−0.381	−0.1026	4.34248
2	2.148	4.156	4.001	5.865	4.371	−0.513	−0.181	−0.291	−0.132	−0.208	6.389396
3	4.161	3.618	3.836	4.505	6.543	−0.578	−0.421	−0.465	−0.205	0.142	8.366432
4	4.879	6.672	5.742	4.68	5.275	−0.487	−0.003	−0.224	−0.228	0.169	3.286738
5	2.612	7.032	4.242	5.184	6.012	−0.867	0.404	−0.331	−0.145	0.337	9.219914
6	5.208	5.251	5.575	5.477	5.847	−0.652	0.374	−0.150	−0.021	0.439	5.762943
7	5.875	5.32	6.257	3.191	3.045	−0.347	0.363	0.132	−0.379	−0.039	4.006224
8	3.884	6.257	4.894	3.95	5.082	−0.501	0.542	0.084	−0.513	−0.015	7.274752
9	5.753	5.027	5.112	6.295	5.67	−0.250	0.439	0.090	−0.151	0.122	2.379272
10	6.138	5.312	6.101	5.402	3.569	0.028	0.413	0.292	−0.041	−0.189	2.694534
11	4.728	4.547	5.035	5.266	6.13	−0.032	0.240	0.241	0.021	0.075	1.153995
12	6.336	4.227	6.07	4.775	5.962	0.241	0.038	0.406	−0.028	0.252	2.406579
13	4.035	4.256	5.876	5.387	5.51	0.000	−0.119	0.500	0.055	0.304	3.436191
14	4.707	5.394	5.214	5.427	3.586	−0.058	−0.016	0.443	0.129	−0.040	2.698012
15	4.78	6.347	5.311	5.011	5.448	−0.091	0.256	0.417	0.105	0.058	2.817496
16	3.1	5.306	3.938	4.102	4.949	−0.453	0.266	0.121	−0.095	0.036	2.737835
17	3.512	5.559	5.688	3.657	4.912	−0.660	0.325	0.234	−0.345	0.011	7.085626
18	3.636	3.778	3.853	5.002	3.468	−0.801	0.015	−0.042	−0.275	−0.297	8.250673
19	4.532	3.322	6.813	3.897	7.156	−0.734	−0.323	0.329	−0.441	0.193	9.862655
20	5.59	4.301	3.737	5.855	4.56	−0.469	−0.398	0.011	−0.182	0.067	3.65062
21	6.596	4.47	5.165	7.168	4.18	−0.056	−0.425	0.042	0.288	−0.111	2.063864
22	5.162	5.873	5.245	3.02	6.678	−0.013	−0.165	0.082	−0.165	0.247	0.897306
23	5.666	3.901	2.89	4.596	4.194	0.123	−0.352	−0.356	−0.213	0.036	3.229715
24	4.105	4.256	6.375	4.951	6.004	−0.080	−0.430	−0.010	−0.180	0.230	2.192427
25	5.486	4.584	4.564	2.917	4.481	0.033	−0.427	−0.095	−0.561	0.080	4.631125
26	4.12	4.309	5.689	5.217	5.058	−0.150	−0.480	0.062	−0.405	0.076	3.970696
27	5.257	6.448	4.698	3.678	6.807	−0.068	−0.095	−0.011	−0.589	0.422	3.986599
28	3.976	5.837	3.75	4.88	5.332	−0.260	0.092	−0.259	−0.495	0.404	4.263655
29	4.705	5.076	5.318	4.384	5.471	−0.267	0.089	−0.143	−0.519	0.417	3.960487
30	4.928	3.942	4.491	4.616	4.719	−0.228	−0.141	−0.217	−0.492	0.278	3.586314
31	3.667	2.771	5.078	5.735	4.877	−0.449	−0.558	−0.158	−0.247	0.198	5.02378
32	5.895	5.502	4.25	3.95	6.476	−0.180	−0.346	−0.276	−0.407	0.453	4.912568
33	4.839	7.25	4.816	6.68	3.481	−0.176	0.173	−0.258	0.010	0.059	1.176238
34	6.808	5.034	4.715	3.373	3.204	0.221	0.145	−0.263	−0.317	−0.312	2.815775
35	5.610	6.081	6.005	3.898	4.175	0.299	0.332	−0.010	−0.474	−0.415	4.350111
36	4.777	5.939	5.243	11.000	11.000	0.194	0.454	0.041	0.821	0.868	17.82526

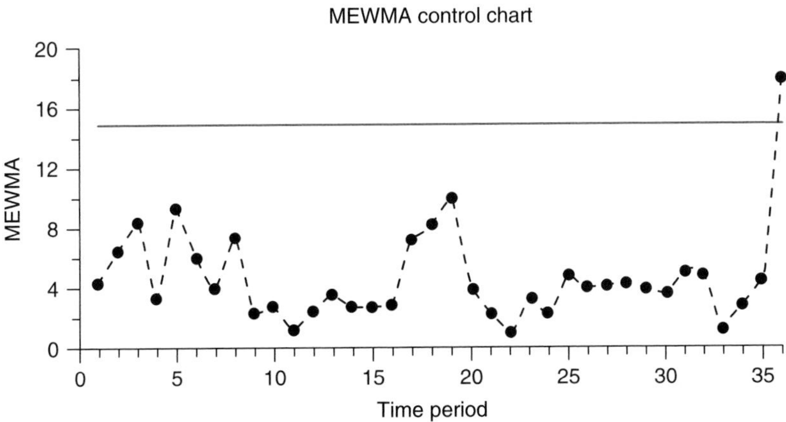

Figure 1 MEWMA control chart is based on the computed T2 statistics

as $Q_i^2 = Z_i'\Sigma_{Z_i}^{-1}Z_i$ with an upper control limit based on the χ_p^2 distribution with a false alarm rate of α. If Σ is not known, one can use the sample variance–covariance matrix (S) as an estimate of Σ (*see* **Multivariate Analysis**) and compute the Hotelling's T^2 value for each successive element of the series, Z_i. The Hotelling's T^2 value is computed as follows: $T_i^2 = Z_i'S_{Z_i}^{-1}Z_i$, where $S_{Z_i} = \frac{r[1-(1-r)^{2i}]}{2-r}S$. The T^2 value can be plotted on a control chart with the upper control limit defined by $\frac{(n-1)p}{n-p}F^{-1}_{(\alpha,p,n-p)}$ where $F^{-1}_{(\alpha,p,n-p)}$ denotes the upper $100(1-\alpha)$th percentile of the F distribution with p and $n-p$ **degrees of freedom**. This is similar to the T^2 control chart (*see* **Hotelling's T^2 Chart**), but replaces observations at time t with the EWMA value computed for time t. In practice, one would prefer to use the χ^2 statistic if Σ is known, since it will result in a smaller **average run length** (ARL). For situations where there is not a great deal of historical data, the Hotelling's T^2 form should be used.

Example

A simulated example is constructed with five variables ($p = 5$) and 35 observations. The simulated data represent measurements for five variables ($X_1 - X_5$). The data set contains 36 observations where the last observation, which is out of control, yields a signal. The first 35 data points were generated from an in-control process and used to estimate the sample variance–covariance matrix. Fixing the target mean to be $\mu_0' = (5, 5, 5, 5, 5)$ the MEWMA vectors were calculated using $r = 0.2$ for all variables. Table 1 contains the original data (X_1 – X_5), the MEWMA values (Z_1 – Z_5) and the resulting T^2 statistic for each vector. These T^2 statistics are also plotted in Figure 1. The control limit is computed as $\frac{(n-1)\times p}{n-p}F^{-1}_{(0.05,p,n-p)} = \frac{34\times 5}{30}2.534 = 14.357$. In this example the first 35 observations are inside the control limit, but the 36th observation falls above the upper control limit. For this demonstration the false alarm rate was 0.05, which may yield too many false alarms. It is customary to use false alarm rates of 0.01 or less.

Related Topics

When a value is plotted above the upper control limit (UCL), it is called a *signal*. While a signal indicates a potential problem, it does not indicate which variable or set of variables are problematic. See **Multivariate Control Charts, Interpretation of** for techniques to determine which variable(s) are problematic in such a signal.

MEWMA charts are useful for detecting small shifts in the overall mean. Typically one uses values such as $0 < r \leq 0.2$ for these situations. In practice, some authors have suggested using two simultaneous MEWMA charts, one with a small value for r and another with $r \geq 0.5$. This latter chart will be able to detect large shifts better. Controlling the overall false alarm rate for these combined charts is somewhat problematic and has not been thoroughly addressed in the literature.

There are some difficulties with the MEWMA chart that also need to be addressed. First, there are some computational issues to deal with. However, all of the calculations in the example above were done using Microsoft Excel. Secondly, there is an "inertia" problem that arises when a process that has been in-control shifts out of control. It may take several time periods before the control chart signals. This problem can be diminished to some extent with the combination chart mentioned above. Calculating the false alarm rate for the MEWMA chart can be challenging. Rigdon [2] and [3] has developed an algorithm for computing false alarm rates. Molnau *et al.* [4] have developed a **Markov chain** based algorithm for calculating the false alarm rates. Linderman and Love [5] and Prabhu and Runger [6] discuss the design of MEWMA control charts.

Despite the computation and interpretation challenges of the MEWMA control chart, it is a good choice when monitoring the mean vector of several correlated process variables. This chart is especially useful when one is interested in detecting relatively small shifts in the mean vector. Competing techniques include the Hotelling's T^2 control chart (*see* **Hotelling's T^2 Chart**) and the multivariate **CUSUM** control chart (*see* **Multivariate Cumulative Sum (CUSUM) Chart**). A general discussion of multivariate control chart may be found in **Multivariate Control Charts Overview**.

References

[1] Lowry, C.A., Woodall, W.H., Champ, C.W. & Rigdon, S.E. (1992). A multivariate exponentially weighted moving average control chart, *Technometrics* **34**, 46–53.

[2] Rigdon, S.E. (1995). An integral equation for the in-control average run length of a multivariate exponentially weighted moving average control chart, *Journal of Statistical Computation and Simulation* **52**, 351–365.

[3] Rigdon, S.E. (1995). A double-integral equation for the average run length of a multivariate exponentially weighted moving average control chart, *Statistics and Probability Letters* **24**, 365–373.

[4] Molnau, W.E., Runger, G.C., Montgomery, D.C., Skinner, K.R., Loredo, E.N. & Prabhu, S.S. (2001). A program for ARL calculation for multivariate EWMA charts, *Journal of Quality Technology* **33**, 515–521.

[5] Linderman, K. & Love, T.E. (2000). Economic and economic statistical designs for MEWMA control charts, *Journal of Quality Technology* **32**, 410–417.

[6] Prabhu, S.S. & Runger, G.C. (1997). Designing a multivariate EWMA control chart, *Journal of Quality Technology* **29**, 8–15.

MICHAEL D. CONERLY

Multivariate Imperfect Repair Models

Introduction

In the study of **maintenance policies** for simple **repairable systems**, i.e. a one-component repairable system with a single repairman, a common assumption is that the system after repair is "as good as new". This constitutes a perfect repair model. However, because of aging and cumulative wear effects, it is known in practice that repair of a failed system may not yield a functioning system that is as good as new. Barlow and Hunter [1] presented a **minimal repair** model in which repair does not change the age of the system. Brown and Proschan [2] and Berg and Cleroux [3] examined a maintenance action, called the **imperfect repair** *model*, in which a system is repaired at failure, with probability α, and that it is returned to the as good as new state (perfect repair) and with probability $(1 - \alpha)$ that it is returned to the functioning state, but is only as good as a system of its age at failure (minimal repair). This model has been generalized by Block *et al.* [4] to the case in which α depends on the age of the system at failure. Much work has been performed in this direction since then. See for example, Balaban and Singpurwalla [5], Uematsu and Nishida [6], Langberg [7], Kijima [8], Ebrahimi [9], Last and Szekli [10], Li and Shaked [11], Belzunce and Semeraro [12], Wang and Zhang [13], and many references cited there.

As systems become more complicated and require new technologies and methodologies, more sophisticated repair policies are needed. More specifically, it is quite common to find systems consisting of k components which are related. In such cases, a legitimate question is how to generalize an imperfect repair model to a system with several components. This article deals with multivariate versions of imperfect repair models that may be used for exploring effects of imperfect repairs of some components on the residual lives of components that are still functioning. Throughout this article "increasing" means "nondecreasing" and "decreasing" means "nonincreasing".

Models, Notations, and Terminology

Consider a system consisting of k components that are related. Let $T = (T_1, \ldots, T_k)$ be the vector of lifetimes of components when the components do not undergo any kind of repair. Also, let $T^* = (T_1^*, \ldots, T_k^*)$ be the waiting times of components until the first perfect repair starting with new components when they are subjected to imperfect repair. Suppose $\bar{F}$ and $\bar{G}$ are joint survival functions of T and T^*, respectively. That is, $\bar{F}(t_1, \ldots, t_k) = P(T_1 > t_1, \ldots, T_k > t_k)$ and $\bar{G}(t_1, \ldots, t_k) = P(T_1^* > t_1, \ldots, T_k^* > t_k)$. We use F and G for joint distribution functions of T and T^*, respectively. For $k = 1$, Brown and Proschan [2] observed that if $\lambda(t) = \frac{f(t)}{\bar{F}(t)}$ is the original hazard function of T_1, then the resulting hazard function of T_1^*, $\lambda^*(t) = \frac{g(t)}{\bar{G}(t)}$ is $\lambda^*(t) = \alpha\lambda(t)$. Hence $\bar{G}(t) = (\bar{F}(t))^{\alpha}$. Here f and g are the probability density functions of T_1 and T_1^*, respectively.

Consider models in which, if at some time t_0, component i_0 fails and is imperfectly repaired, then i_0 is as good as an identical functioning component of age t_0 (minimal repair), and the remaining functioning components "do not know" about the component i_0's failure and imperfect repair. Specifically, suppose that components $j_1, \ldots, j_\ell$ have been perfectly repaired at times $t_{j_1}, \ldots, t_{j_\ell}$ respectively, and $t_{j_i} < t_0$, for $i = 1, \ldots, \ell$. Also, components $i_1, \ldots, i_m$ are still functioning at time t_0 and the component i_0 has just been imperfectly repaired. Note that $\{j_1, \ldots, j_\ell, i_1, \ldots, i_m, i_0\} = \{1, \ldots, k\}$. Then, it is clear that the distribution of the times to next failure of components $i_0, i_1, \ldots, i_m$ is

$$P\Big\{T_{i0} \le t_{i0}, T_{i_1} \le t_{i_1}, \ldots, T_{i_m} \le t_{i_m} | T_{i0} > t_0, T_{i1} > t_0, \ldots, T_{im} > t_0; T_{j_1} = t_{j_1}, \ldots, T_{j\ell} = t_{j\ell}\Big\} \quad (1)$$

On the other hand, if the component i_0 at time t_0 has been perfectly repaired, then

$$P\Big\{T_{i1} \le t_{i1}, \ldots, T_{im} \le t_{i_m} | T_{i1} > t_0, \ldots, T_{i_m} > t_0; T_{j1} = t_{j1}, \ldots, T_{j\ell} = t_{j\ell}; T_{i_0} = t_0\Big\} \quad (2)$$

Assuming that $\bar{F}(t_1, \ldots, t_k)$ is absolutely continuous, i.e., no more than one component can fail at a time, below we describe two multivariate imperfect repair models:

Model 1 k components start to function at (the same) time zero. Upon failure, a component undergoes a repair. With probability α the repair is unsuccessful and the component is scrapped (perfect repair). With probability $(1 - \alpha)$, the repair is imperfect repair. It is clear that the joint distribution of the times to next failure of the functioning components after an imperfect (perfect) repair is given by equations (1) and (2).

Model 2 k components start to function at time zero. Upon failure, a component undergoes a repair. If n components $(n = 0, 1, \ldots, k - 1)$ have already been perfectly repaired, with probability α_{n+1} the repair of the component is perfect and with probability $1 - \alpha_{n+1}$ the repair is imperfect repair. It is clear that the joint distribution of the times to next failure of functioning components after an imperfect (perfect) repair is given by equations (1) and (2). For more details about these models, see Shaked and Shanthikumar [14] and Natvig [15].

If we consider the case in which there is a positive probability that at least two of the original components can fail at the same time, that is, $\bar{F}(t_1, \ldots, t_k)$ has singularities on sets of the form $\{(t_1, \ldots, t_k); t_{i_1} = \cdots = t_{i_\ell}\}$ for some $1 \le i_1 < i_2 < \cdots < i_\ell \le k$. Then, the following two interpretations of model 1 are possible.

Interpretation 1 If two or more components fail at the same time, all the failed components are scrapped with probability α or all the failed components are minimally repaired with probability $(1 - \alpha)$.

Interpretation 2 If two or more components fail at the same time, then each of them, independently of the others, is either successfully minimally repaired with probability $(1 - \alpha)$ or scrapped with probability α.

The next model gives a very general repair policy when $\bar{F}(t_1, \ldots, t_k)$ has singularities. To describe this model, let S denote the set of vectors $(s_1, s_2, \ldots, s_k)$, where $s_j = 0$ or 1 for $j = 1, \ldots, k$. Here $s_j = 0$ means that the component j is minimally repaired and $s_j = 1$ means that the component j is scrapped.

Model 3 k components start at time zero. Upon failure, a component undergoes a repair. There are several sources which cause the failures of components $1, 2, \ldots,$ and k. Source $\ell(j)$ causes the failure of component j; with probability $\alpha(j)$ the repair of this component is perfect and with probability $(1-\alpha(j))$ the repair is imperfect, $j = 1, \ldots, k$. For given i and $j, i \neq j, i, j = 1, \ldots, k$, source $\ell(i, j)$ causes the failures of both components i and j, with probability $\alpha(i, j; s_i, s_j)$ the component i (component j) is minimally repaired if $s_i = 0 (s_j = 0)$. Otherwise it is scrapped. For example, if $s_i = s_j = 1$, then $\alpha(i, j, 0, 0)$ stands for the probability that both components are minimally repaired. If $s_i = 0, s_j = 1$, then $\alpha(i, j, 1, 0)$ stands for the probability that component j is scrapped and component i is minimally repaired. Here $\alpha(i, j, 1, 1) + \alpha(i, j, 0, 0) + \alpha(i, j, 1, 0) + \alpha(i, j, 0, 1) = 1$. For a given $i, j, m; i \neq j \neq m, i, j, m = 1, \ldots, k$, the source $\ell(i, j, m)$ causes the failure of three components i, j, and m; with probability $\alpha(i, j, m; s_i, s_j, s_m)$ either one or two or three are minimally repaired depending on whether s_i, s_j, or s_m are zero or one. Again here, $\sum_{s_i}\sum_{s_j}\sum_{s_m} \alpha(i, j, m, s_i, s_j, s_m) = 1$. Continuing this, suppose source $\ell(1, 1, \ldots, 1)$ causes the failures of all k components. Then $\alpha(1, 1, \ldots, 1, s_1, \ldots, s_k)$ is the probability that either one, two, ... or all k components are minimally repaired depending on whether $s_1, s_2, \ldots,$ or s_k are zero or one. Note that $\sum_S \alpha(1, 1, \ldots, 1, s_1, s_2, \ldots, s_k) = 1$. It is clear that $k = 2$, then $\alpha(1)$ and $\alpha(2)$ are the probabilities of perfect repairs of components 1 and 2, respectively. Also $\alpha(1, 1, 0, 0)$, $\alpha(1, 1, 0, 1), \alpha(1, 1, 1, 0)$, and $\alpha(1, 1, 1, 1)$ are the probabilities that both components are minimally repaired, component 1 is minimally repaired, component 2 is minimally repaired, and both components are scrapped. This coincides with the model (3.5) of Shen and Griffith [16].

Main Results

In this section, we obtain the distribution of G in terms of F. We treat the bivariate case ($k = 2$) to keep the notations manageable. The results, however, can be easily extended to the k-dimensional case.

The following theorem gives the distribution of G in terms of F under Models 1 and 2. See Shaked and Shanthikumar [14] for more details.

Theorem 1 *Suppose that (T_1, T_2) has the joint density function f and (T_1^*, T_2^*) has the joint density function g:*

(a) Under Model 1,

$$g(t_1, t_2) = f(t_1, t_2)\alpha^2 \left(\bar{F}(t_1, t_2)\right)^{\alpha-1} \times \left[\frac{\bar{F}_2(t_2|t_1)}{\bar{F}_2(t_1|t_1)}\right]^{\alpha-1} \quad \textit{if } 0 \le t_1 \le t_2 \tag{3}$$

$$= f(t_1, t_2)\alpha^2 \left(\bar{F}(t_1, t_2)\right)^{\alpha-1} \times \left[\frac{\bar{F}_1(t_1|t_2)}{\bar{F}_1(t_2|t_2)}\right]^{\alpha-1} \quad \textit{if } 0 \le t_2 \le t_1 \tag{4}$$

where $\bar{F}_1(x|y) = P(T_1 > x|T_2 = y)$ and $\bar{F}_2(x|y) = P(T_2 > x|T_1 = y)$.

(b) Under Model 2,

$$g(t_1, t_2) = f(t_1, t_2)\alpha_1\alpha_2 \left(\bar{F}(t_1, t_2)\right)^{\alpha_1-1} \times \left[\frac{\bar{F}_2(t_2|t_1)}{\bar{F}_2(t_1|t_1)}\right]^{\alpha_2-1} \quad \textit{if } 0 \le t_1 \le t_2$$

$$= f(t_1, t_2)\alpha_1\alpha_2 \left(\bar{F}(t_1, t_2)\right)^{\alpha_1-1} \times \left[\frac{\bar{F}_1(t_1|t_2)}{\bar{F}_1(t_1|t_1)}\right]^{\alpha_2-1} \quad \textit{if } 0 \le t_2 \le t_1 \tag{5}$$

From Theorem 1, under Model 2, one can easily show that

$$P(T_1^* \le T_2^*) = \int_0^\infty \alpha_1 \left(\bar{F}(t, t)\right)^{\alpha_1-1} \times \bar{F}_2(t|t) f_1(t)\,dt \tag{6}$$

where f_1 is the marginal density of T_1. Also, the marginal density function of T_1^* is

$$g_1(t) = \alpha_1 \left(\bar{F}(t, t)\right)^{\alpha_1-1} f_1(t)\bar{F}_2(t|t) + \alpha_1\alpha_2 \int_0^t \left\{ f(t, x) \left(\bar{F}(x, x)\right)^{\alpha_1-1} \times \left[\frac{\bar{F}_1(t|x)}{\bar{F}_1(x|x)}\right]^{\alpha_2-1} \right\} dx \tag{7}$$

The density of T_2^* can be written in a similar manner.

Suppose that

$$\bar{F}(t_1, t_2) = P(U_1 > t_1,\ U_2 > t_2,\ U_{12} > \max(t_1, t_2)) = \bar{F}_1(t_1)\bar{F}_2(t_2)\bar{F}_{12}(\max(t_1, t_2)) \tag{8}$$

where U_1, U_2, and U_{12} are independent random variables having cumulative distribution functions F_1, F_2, and F_{12}, respectively.

Given equation (8), under Model 3, the following theorem gives the distribution of G in terms of F_1, F_2, and F_{12}. See Shen and Griffith [16] for more details.

Theorem 2 *Under Model 3,*

$$\bar{G}(t_1, t_2) = \left(\bar{F}_1(t_1)\right)^{\alpha(1)} \left(\bar{F}_{12}(\max(t_1, t_2))\right)^{\alpha(1,1,1,0)} \times \left(\bar{F}_2(t_2)\right)^{\alpha(2)} \left(\bar{F}_{12}(\max(t_1, t_2))\right)^{\alpha(1,1,0,1)} \times \left(\bar{F}_{12}(\max(t_1, t_2))\right)^{\alpha(1,1,1,1)} \quad (9)$$

Examples

In this section, we apply our results to two families of distributions.

Marshall–Olkin Distribution

The random vector (T_1, T_2) is said to have the bivariate Marshall–Olkin exponential distribution, see Marshall and Olkin [17], if

$$\bar{F}(t_1, t_2) = \exp(-\lambda_1 t_1 - \lambda_2 t_2 - \delta \ \max(t_1, t_2)), \quad t_1, t_2 \geq 0, \ \lambda_1, \lambda_2 > 0, \delta \geq 0 \quad (10)$$

Given that $\alpha(1) = \alpha(2) = \frac{1}{2}$ and $\alpha(1, 1, 1, 0) = \alpha(1, 1, 1, 1) = \alpha(1, 1, 0, 1) = \alpha(1, 1, 0, 0) = \frac{1}{4}$. From Theorem 2,

$$\bar{G}(t_1, t_2) = \exp\left(-\frac{\lambda_1}{2} t_1 - \frac{\lambda_2}{2} t_2 - \frac{3}{4}\delta \max(t_1, t_2)\right) \quad (11)$$

Bivariate Gumble Model

Consider T_1 and T_2 with the bivariate Gumble distribution, see Gumble [18],

$$\bar{F}(t_1, t_2) = \exp(-\lambda_1 t_1 - \lambda_2 t_2 - \delta t_1 t_2), t_1, t_2 \geq 0, \quad \lambda_1, \lambda_2 > 0, \ \delta \geq 0 \quad (12)$$

It is clear that

$$f(t_1, t_2) = [(\lambda_1 + \delta t_2)(\lambda_2 + \delta t_1) - \delta] \times \exp(-\lambda_1 t_1 - \lambda_2 t_2 - \delta t_1 t_2) \quad (13)$$

$$\bar{F}_2(x|y) = P(T_2 > x|T_1 = y) = \frac{(\lambda_1 + \delta x)}{\lambda_1} \exp(-\lambda_2 x - \delta x y) \quad (14)$$

and

$$\bar{F}_1(x|y) = P(T_1 > x|T_2 = y) = \frac{(\lambda_2 + \delta x)}{\lambda_2} \exp(-\lambda_1 x - \delta x y) \quad (15)$$

Now, suppose that $\alpha_1 = \alpha_2 = 1/2$, then from Theorem 1

$$g(t_1, t_2) = \frac{1}{4}[(\lambda_1 + \delta t_2)(\lambda_2 + \delta t_1) - \delta] \times \exp\left(-\frac{1}{2}(\lambda_1 t_1 + \lambda_2 t_2 + \delta t_1 t_2)\right) \times \frac{\lambda_1 + \delta t_2}{\lambda_1 + \delta t_1}\left[\frac{\exp(-\lambda_2 t_2 - \delta t_1 t_2)}{\exp(-\lambda_2 t_1 - \delta t_1^2)}\right], \quad 0 \leq t_1 \leq t_2 \quad (16)$$

and

$$g(t_1, t_2) = \frac{1}{4}[(\lambda_1 + \delta t_2)(\lambda_2 + \delta t_1) - \delta] \times \exp\left(-\frac{1}{2}(\lambda_1 t_1 + \lambda_2 t_2 + \delta t_1 t_2)\right) \times \frac{\lambda_2 + \delta t_1}{\lambda_2 + \delta t_2}\left[\frac{\exp\ (-\lambda_1 t_1 - \delta t_1 t_2)}{\exp(-\lambda_1 t_2 - \delta t_2^2)}\right], \quad 0 \leq t_2 \leq t_1 \quad (17)$$

Concluding Remarks

In this article, we describe several extensions of univariate imperfect repair models to multivariate imperfect models. These models are applicable for situations where a system has k components that are related. It should be noted that proposed models in this article do not incorporate any maintenance schedule or cost. For such models, we refer you to

Li and Xu [19], Wang and Zhang [13], and many references cited there.

References

[1] Barlow, R.E. & Hunter, L.C. (1960). Optimum preventive maintenance policy, *Operations Research* **8**, 851–859.

[2] Brown, M & Proschan, F. (1983). Imperfect repairs, *Journal of Applied Probability* **20**, 851–859.

[3] Berg, M. & Cleroux, R. (1982). A marginal cost analysis for an age replacement policy with minimal repair, *Infor* **20**, 256–263.

[4] Block, H., Borges, W. & Savits, W.S. (1985). Age dependent minimal repair, *Journal of Applied Probability* **22**, 370–385.

[5] Balaban, H.S. & Singpurwalla, N.D. (1984). Stochastic properties of a sequence of interfailure times under minimal repair and survival, in *Reliability Theory and Models*, M.A. Abdel-Hameed, E. Cinlar & J. Quinn, eds, Academic press, New York, Vol. 65–80.

[6] Uematsu, K. & Nishida, T. (1987). One unit system with a failure rate depending upon the degree of repair, *Mathematica Japonica* **32**, 139–147.

[7] Langberg, N. (1988). Comparison of replacement policies, *Journal of Applied Probability* **25**, 780–788.

[8] Kijima, M. (1989). Some results for repairable systems, *Journal of Applied Probability* **26**, 89–102.

[9] Ebrahimi, N. (1993). Modeling repairable systems, *Advances in Applied Probability* **25**, 926–938.

[10] Last, G. & Szekli, R. (1998). Stochastic comparison of repairable systems by coupling, *Journal of Applied Probability* **35**, 348–370.

[11] Li, H. & Shaked, M. (2003). Imperfect repair model with preventive maintenance, *Journal of Applied Probability* **40**, 1043–1059.

[12] Belzunce, F. & Semeraro, P. (2004). Preservation of positive and negative orthant dependence concepts under mixture applications, *Journal of Applied Probability* **41**, 961–974.

[13] Wang, G.J. & Zhang, Y.L. (2006). Optimal periodic preventive repair and replacement policy assuming geometric process repair, *IEEE Transactions on Reliability* **55**, 118–122.

[14] Shaked, M. & Shanthikumar, J.G. (1986). Multivariate imperfect repair, *Operations Research* **34**, 437–448.

[15] Natvig, B. (1990). On Information based minimal repair and the reduction in remaining system lifetime due to the failure of specific module, *Journal of Applied Probability* **27**, 365–375.

[16] Shen, S.H. & Griffith, W.S. (1992). Multivariate imperfect repair, *Journal of Applied Probability* **29**, 947–956.

[17] Marshall, A.W. & Olkin, I. (1967). A multivariate exponential distribution, *Journal of the American Statistical Association* **62**, 30–44.

[18] Gumble, E.J. (1960). Bivariate exponential distribution, *Journal of the American Statistical Association* **55**, 698–707.

[19] Li, H. & Xu, S. (2006). A Multivariate Cumulative Damage Shock Model with Block Preventive Maintenance, *Stochasic models* **22**, 341–360..

Related Articles

Block Replacement; General Minimal Repair Models; Markov Renewal Processes in Reliability Modeling; Multivariate Age and Multivariate Renewal Replacement; Multivariate Stochastic Orders and Aging; Replacement Strategies.

NADER EBRAHIMI

Multivariate Mixture Models *see* Mixture Models, Multivariate: Aging and Dependence

Multivariate Process Capability Indices, Comparison of

Introduction

Quality measures can be used to evaluate a process's performance. In general, there are several issues that need to be discussed in assessing product quality. One important issue is the process capability index (PCI). A PCI is a numerical summary that compares the behavior of a product or process characteristic to engineering specifications. The capability indices, C_p, C_{pk}, and C_{pm}, are widely used to evaluate process performance based on a single engineering specification. However, products usually have two or more quality characteristics in processes. Thus, these univariate capability indices do not satisfy this situation, and multivariate methods for assessing process capability are needed. Process capability and related analysis have been studied for well over 20 years. Kotz and Johnson [1] provided a compact survey and brief interpretations and comments on some 170 publications on PCIs from 1992 to 2000.

In general, multivariate PCIs can be divided into two categories: the ratio of the tolerance region to the process region and the **process yield** index. In the following section, several multivariate PCIs which were proposed by Chan *et al.* [2], Pearn *et al.* [3], Taam *et al.* [4], Chen [5], Shahriari *et al.* [6], Wang and Du [7], Walter *et al.* [8], and Pearn *et al.* [9] will be reviewed. In the third section, an example is used to compare the performance of different multivariate PCIs. The conclusion is made in the final section.

Multiple Characteristics Processes and Multivariate Process Capability Indices

Multiple Characteristics Processes

In most processes, products usually have multiple quality characteristics. Each of these characteristics must satisfy its specification. So, the assessed quality of a product depends on the combined effects of these characteristics, rather than on their individual values. With respect to the tolerance region of multiple characteristics, it can be a rectangular tolerance region for two-dimensional cases. In higher dimensions, they form a hypercube. For more complex engineering specifications, the tolerance region will be more complicated.

In most **multivariate capability indices**, it is usually assumed that the observations X have a multivariate **normal distribution**, $N_v(\mu, \sum)$, where v is the number of variables, μ is the mean vector, and Σ represents the variance–**covariance** matrix of X. Also, T is the target vector, upper specification limits (USL) and lower specification limits (LSL) are the specification limits, $\overline{X}$ is the sample mean vector, and S is the sample covariance matrix. For convenience, the assumption and notations are used in the following subsection.

Multivariate Process Capability Indices

Chan *et al.* [2] introduced a new version of the multivariate index C_{pm}, which is defined by

$$C_{pm} = \sqrt{\frac{nv}{\sum_{i=1}^{n}(X_i - T)'A^{-1}(X_i - T)}} \tag{1}$$

This value indicates the **degrees of freedom** associated, while the denominator represents the sum of the observed Mahalanobis distances from the target [10]. Smaller values of C_{pm} suggest that the process is unable to meet the specifications, while larger values of C_{pm} suggest that the process is indeed capable of meeting its specifications.

Pearn *et al.* [3] introduced two multivariate PCIs ${}_vC_p$ and ${}_vC_{pm}$, which are a generalization of C_p and C_{pm}, respectively. They are defined as

$${}_vC_p = \frac{K}{\theta\chi_{v,0.9973}} \tag{2}$$

and

$${}_vC_{pm} = \frac{{}_vC_p}{\left(1 + \dfrac{(\mu - T)'A^{-1}(\mu - T)}{v}\right)^{1/2}} \tag{3}$$

where the region R is defined by $(X - T)'A^{-1}(X - T) \leq K^2$ and $V_0 = \theta^2 A$.

Taam *et al.* [4] proposed a multivariate capability index defined as a ratio of two volumes: $MC_{pm} = \dfrac{\text{vol}(R_1)}{\text{vol}(R_2)} = \dfrac{\text{vol(modified tolerance region)}}{\text{vol}((X-\mu)'\Sigma_T^{-1}(X-\mu) \leq k(q))}$, where R_1 is a modified tolerance region and R_2 is a scaled 99.73% process region from a multivariate normal assumption. The modified tolerance region is the largest ellipsoid that is centered at the target completely within the original tolerance region. The estimate for MC_{pm} can be expressed as

$$\widehat{MC}_{pm} = \frac{\widehat{C}_p}{\widehat{D}} = \frac{\dfrac{\text{vol(modified tolerance region)}}{|S|^{1/2}(\pi k)^{v/2}\left[\Gamma(v/2+1)^{-1}\right]}}{\left[1 + \dfrac{n}{n-1}(\overline{X} - T)'S^{-1}(\overline{X} - T)\right]^{1/2}} \tag{4}$$

where K is the 99.73% percent quantile of a χ^2 distribution and $|\cdot|$ denotes the determinant. The numerator $\widehat{C}_p$ is analogous to C_p, that is, a value greater than 1 implies that the process has a smaller variation than allowed by specification limits with a certain confidence level; a value less than 1 implies more variation. Similarly, $0 < 1/\widehat{D} < 1$ measures the closeness between the process mean and the target; a larger $1/\widehat{D}$ implies the mean is closer to target.

Chen [5] proposed a multivariate process index over a tolerance zone. A tolerance zone is expressed as $V = \{X \in R^v : h(X - \mu_0) \le r_0\}$, where r_0 is a positive number and $h(X - \mu_0)$ is a cumulative distribution function. A process is capable if $P(X \in V) \ge 1 - \alpha$, where α is the allowable expected proportion of nonconforming production from a process. Let $r = \min\{c : P(h(X - \mu_0) \le c) \ge 1 - \alpha\}$. If the cumulative distribution function of $h(X - \mu_0)$ is increasing in a neighborhood of r, then r is simply the unique root of the equation $P(h(X - \mu_0) \le r) = 1 - \alpha$. The process is deemed capable if $r \le r_0$, that is, $(r_0/r) \ge 1$. The multivariate process capability is defined as

$$MC_p = \frac{r_0}{r} \tag{5}$$

Shahriari *et al.* [6] proposed a multivariate capability vector, which consists of three components (C_{pM}, PV, LI). The first component of the vector is a ratio of areas or volumes, analogous to the ratio of lengths of the univariate C_p index. The denominator is the area (two-dimensional case) or the volume (three or more dimensions) defined by the engineering tolerance region, which is defined as $C_{pM} = \left[\frac{\text{volume of engineering tolerance}}{\text{volume of modified process region}}\right]^{1/v}$. The approach forms a "modified process region" by drawing the smallest rectangle around the ellipse. The edges of the rectangle are determined by solving the equations of the first derivatives, with respect to each x_i, of the quadratic form $(X - \mu_0)'\Sigma(X - \mu_0) = \chi^2_{v,\alpha}$. Two solutions to this equation are given in the following:

$$UPL_i = \mu_i + \sqrt{\frac{\chi^2_{v,\alpha}\det(\Sigma_i^{-1})}{\det(\Sigma^{-1})}},$$

$$LPL_i = \mu_i - \sqrt{\frac{\chi^2_{v,\alpha}\det(\Sigma_i^{-1})}{\det(\Sigma^{-1})}}, \quad \forall \cdots i = 1, 2, \ldots, v \tag{6}$$

where $\chi^2_{v,\alpha}$ is the upper $100\alpha\%$ of a χ^2 distribution with v degrees of freedom associated with the probability contour, and $\det(\Sigma_i^{-1})$ is the determinant of Σ_i^{-1}, which is a matrix obtained from Σ^{-1} by deleting the ith row and column. As a result, the equation can be expressed as

$$C_{pM} = \left[\frac{\prod_{i=1}^{v}(USL_i - LSL_i)}{\prod_{i=1}^{v}(UPL_i - LPL_i)}\right]^{1/v} \tag{7}$$

If the value is greater than 1, then it shows that the circumscribed modified process region is smaller than the engineering specified region, that it is "acceptable". Necessarily, the modified process region is influenced by the shape of the elliptical contour and the size of the contour. The second component is the **significance level** of the Hotelling's T^2 statistic ($T^2 = n(\overline{X} - \mu)'S^{-1}(\overline{X} - \mu)$) which is expressed as

$$PV = P_r\left(F_{v,n-v} > \frac{n - v}{v(n - 1)}T^2\right) \tag{8}$$

where $F_{v,n-v}$ denotes a variable having the F distribution with $v, n - v$ degrees of freedom. The PV value will never exceed 1, and it indicates that the center of process is far from the engineering target value when the value is close to 0. The third component summarizes a comparison of the location of the modified process region and the tolerance region. It has a value 1 if the entire modified process region is contained within the tolerance region and, otherwise, a value of 0. That is,

$$LI = \begin{cases} 1 & \text{if modified process region is contained} \\ & \text{within the tolerance} \\ 0 & \text{otherwise} \end{cases} \tag{9}$$

Wang and Du [7] proposed a very useful method using principal component analysis (PCA) in process performance for multivariate data. By using the spectral decomposition, we can get a matrix $D = U^{\mathrm{T}}SU$, where D is a diagonal matrix. The elements of D, $\lambda_1, \lambda_2, \ldots, \lambda_v$, are the eigenvalues of S and the columns of U, $u_1, u_2, \ldots, u_v$ are the eigenvectors of S. So the ith principal component is expressed as $PC_i = u^{\mathrm{T}}x, i = 1, 2, \ldots, v$, where xs are $v \times 1$ vectors of original observations on variables. The engineering specifications and target values of PC_is are $\begin{cases} LSL_{PC_i} = u_i^{\mathrm{T}}LSL \\ USL_{PC_i} = u_i^{\mathrm{T}}USL \ \forall \ i = 1, 2, \ldots, v. \\ T_{PC_i} = u_i^{\mathrm{T}}T \end{cases}$ Similarly, the relevant sample estimators, S^2 and $\overline{X}$ of PC_is can defined as $\begin{cases} S^2_{PC_i} = \lambda_i \\ \overline{X}_{PC_i} = u_i^T \overline{X} \end{cases} \forall\ i = 1, 2, \ldots, v$. Now, only few principal components can explain most of the total variability (about 70 ~ 90%). Jackson [11] proposed a χ^2 test for identifying the significant components. Referring to this method, we can choose the suitable number of $PC_i s$ easily. Now, the related formulas of multivariate PCIs for the multivariate normal data and non-multivariate normal data can be displayed for the following: (a) two multivariate PCIs for multivariate normal data are defined as

$$\widehat{MC}_p = \left(\prod_{i=1}^{v} \widehat{C}_{P;PC_i}\right)^{1/v} \quad (10)$$

and

$$\widehat{MC}_{pk} = \left(\prod_{i=1}^{v} \widehat{C}_{pk;PC_i}\right)^{1/v} \quad (11)$$

where

$$\widehat{C}_{p;PC_i} = \frac{(USL_{PC_i} - LSL_{PC_i})}{6\sqrt{S^2_{PC_i}}} \quad (12)$$

and

$$\widehat{C}_{pk;PC_i} = \frac{\min\left(USL_{PC_i} - \overline{X}_{PC_i}, \overline{X}_{PC_i} - LSL_{PC_i}\right)}{3\sqrt{S^2_{PC_i}}} \quad (13)$$

The approximate $100(1-\alpha)\%$ **confidence intervals** for MC_p and MC_{pk} are

$$\left[\left(\prod_{i=1}^{v} \widehat{C}_{p;PC_i}\sqrt{\frac{\chi^2_{1-\alpha/2,n-1}}{n-1}}\right)^{1/v}, \left(\prod_{i=1}^{v} \widehat{C}_{p;PC_i}\sqrt{\frac{\chi^2_{\alpha/2,n-1}}{n-1}}\right)^{1/v}\right] \quad (14)$$

and

$$\left[\left(\prod_{i=1}^{v} \widehat{C}_{pk;PC_i}\left[1 - Z_{\alpha/2}\sqrt{\frac{1}{9n\widehat{C}_{pk,PC_i}} + \frac{1}{2(n-1)}}\right]\right)^{1/v}, \left(\prod_{i=1}^{v} \widehat{C}_{pk;PC_i}\left[1 + Z_{\alpha/2}\sqrt{\frac{1}{9n\widehat{C}_{pk;PC_i}} + \frac{1}{2(n-1)}}\right]\right)^{1/v}\right] \quad (15)$$

where $Z \sim N(0, 1)$, respectively. (b) The multivariate PCI for non-multivariate normal data is based on a PCI by Luceño [12], which is designed to consider both the process location and spread, as well as to provide for confidence bounds that are insensitive to departures from normality. The multivariate PCI is defined as

$$\widehat{MC}_{pc} = \left(\prod_{i=1}^{v} \widehat{C}_{pc;PC_i}\right)^{1/v} \quad (16)$$

where $\widehat{C}_{pc;PC_i} = (USL_{PC_i} - LSL_{PC_i})/6\sqrt{\pi/2\overline{c}}$. The approximate $100(1-\alpha)\%$ confidence interval for MC_{pc} is

$$\left[\left(\prod_{i=1}^{v} \frac{\widehat{C}_{pc;PC_i}}{1 + t_{\alpha,n-1}S_{C_i}/(\overline{c}_i\sqrt{n})}\right)^{1/v}, \left(\prod_{i=1}^{v} \frac{\widehat{C}_{pc;PC_i}}{1 - t_{\alpha,n-1}S_{C_i}/(\overline{c}_i\sqrt{n})}\right)^{1/v}\right] \quad (17)$$

where

$$\overline{c_i} = \frac{1}{n}\sum_{j=1}^{n}\left|PC_{ij} - \frac{(USL_{PC_i} + LSL_{PC_i})}{2}\right| \quad (18)$$

$$S_{C_i} = \frac{1}{n-1}\left(\sum_{j=1}^{n}\left|PC_{ij} - \frac{(USL_{PC_i} + LSL_{PC_i})}{2}\right|^2 - n\overline{c}_i^2\right) \quad (19)$$

and t is the Student's t distribution.

Walter *et al.* [8] proposed the multivariate PCIs based on the eigenvalue problem and quadratic form. First, we can generalize the formula of the univariate PCI $C_p = (USL - LSL)/6s$ using the expression diag $((USL-LSL)/6)\, S_{X'X}^{-1}\text{diag}\,((USL-LSL)/6)$. Then, we can define the uncorrected multivariate PCI by

$$MC_p^{(B1)} = \min_{j=1,\ldots,v}\left\{\text{eigenvalues} \times \left[\text{diag}\left(\frac{USL-LSL}{6}\right) S_{X'X}^{-1} \times \text{diag}\left(\frac{USL-LSL}{6}\right)\right]\right\}^{1/2} \quad (20)$$

This formula compares the diagonal of a standardized tolerance region with the length of the longest space diagonal of the standardized variance matrix. Furthermore, a quadratic correction of the equation (20) is given by

$$MC_{pk}^{(B1)} = \frac{MC_p^{(B1)}}{D_B} \quad (21)$$

where $D_B = \sqrt{1 + (\overline{X} - T)^{\mathrm{T}} \left(\sum \text{th}\right)^{-1} (\overline{X} - T)}$ and $\sum \text{th} = \text{diag}((USL_j - LSL_j)/6) \times ((USL_j - LSL_j)/6)$.

The quadratic correction above does not consider the dependence between the product variables. That is, the tolerance region can be a circle or a sphere. The interpretation of this index is similar to the univariate index C_{pk}. They also proposed the multivariate PCIs based on the eigenvalue problem (see the details in Walter *et al.* [8]).

Pearn *et al.* [9] provided the probability of producing a good product satisfying all its specifications. The process yield is defined as

$$p = \int_{[LSL,USL]} N_v(X|\mu, \Sigma)\, \mathrm{d}X \quad (22)$$

Furthermore, they proposed a generalized process yield index $TS_{pk,PC}$, based on the index S_{pk} [13] and the PCA method which is similar to Wang and Du [7]. They also provided a lower confidence bound for the true process yield.

Discussion

These methods proposed by Chan *et al.* [2], Pearn *et al.* [3], Taam *et al.* [4], Shahriari *et al.* [6], Walter *et al.* [8], and Pearn *et al.* [9], respectively, must use the assumption of multivariate normality, while those proposed by Chen [5] and Wang and Du [7] do not require this assumption of multivariate normality. The tolerance regions of these methods using the assumption of multivariate normality are ellipsoidal except for Shahriari *et al.* [6]. On one hand, Chen [5] and Wang and Du [7] provided more flexible methods in assessing the capability for multivariate data. Wang *et al.* [14] presented a comparison of three methods by Tamm *et al.* [4], Shahriari *et al.* [6], and Chen [5]. In general, these proposed approaches all have their own reasonable points of view, but we still cannot obtain a consistent method of assessing the multivariate PCIs. Apparently, we should make an effort to study the relevant issues in the multivariate PCIs. Here, we give a simple summary for the multivariate PCIs (see Table 1).

Example

Let us consider an example of a device on an integrated circuit. The width (x_1) and thickness (x_2) are two important characteristics for a device on an integrated circuit to ensure good conductivity. The specifications for these two variables are (5, 6) (μm) and (0.6, 1.2) (μm), respectively. The targets of the

Table 1 The summary of multivariate process capability indices

Authors	PCI	Distribution application	Tolerance form	Category
Chan *et al.* [2]	C_{pm}	Multivariate Normal	Elliptical	Ratio value
Pearn *et al.* [3]	${}_vC_{pm}, {}_vC_p$	Multivariate Normal	Elliptical	Ratio value
Taam *et al.* [4]	MC_{pm}, MC_p	Multivariate Normal	Elliptical	Ratio value
Chen [5]	MC_p	No specific	No specific	Process yield
Shahriari *et al.* [6]	$[C_{pM}, PV, LI]$	Multivariate Normal	Elliptical	Ratio value
Wang and Du [7]	MC_{pm}, MC_{pk}, MC_p	No specific	No specific	Ratio value
Walter *et al.* [8]	$MC_{pk}^{(B1)}, MC_p^{(B1)}$	Multivariate Normal	Elliptical	Ratio value
Pearn *et al.* [9]	$2\Phi(3TS_{pk;PC}) - 1$	Multivariate Normal	Elliptical	Process yield

Table 2 The results for example

Authors	Estimated values
Chan *et al.* [2]	$C_{pm} = 1.0157$
Taam *et al.* [4]	$MC_p = 2.9099$, $MC_{pm} = 2.9099/1.0166 = 2.8765$
Shahriari *et al.* [6]	$(C_{pM}, PV, LI) = (1.1899, 0.7250, 1)$
Wang and Du [7]	$MC_p = 1.1133$, $MC_{pk} = 1.091^{(a)}$, $MC_{pm} = 1.1110$
Walter *et al.* [8]	$MC_p^{(B1)} = 0.8733$, $MC_{pk}^{(B1)} = 0.8733/1.0063 = 0.8678$
Pearn *et al.* [9]	The estimated process yield is 0.9984

[(a)] 95% lower confidence bound: $MC_p = 0.8699$, $MC_{pk} = 0.8388$

specifications are (5.5, 0.9). From a sample of 30 observations, we have the sample mean vector and the sample covariance matrix which are $\overline{X} = \begin{bmatrix} 5.512 \\ 0.901 \end{bmatrix}$ and $S = \begin{bmatrix} 0.0257 & 0.0097 \\ 0.0097 & 0.0044 \end{bmatrix}$, respectively. Now we calculate the capability values for this case depending on several methods we introduce and we give a summary in Table 2. From Table 2, all capability indices are larger than 1 except the indices of Walter *et al.* [8]. This means that the variation of the process is small, the mean of the process is close to the center of specification and the mean of the process is around the target. Based on this result, we can conclude that this process is capable. By observing the spread of the 30 observations in the above figures, it is clear that this conclusion is reasonable. Here we get the estimated process yield 0.9984 by using the maple program. This means that the process produces high percentage of conforming products and implies the process is capable. Hence, the process should have high process capability values. According to the above process capability values calculated, we can infer that the estimated process yield is rational.

Conclusion

Often, a manufactured product is described in multiple characteristics. That is, manufactured items require values of several different characteristics for adequate description of their quality. Each of those characteristics must satisfy certain specifications. The assessed quality of a product depends on the combined effects of these characteristics, rather than on their individual values.

Currently, the research of multivariate PCIs is still very sparse in comparison to the research of the univariate PCIs. Also, there is no consistency regarding a methodology for evaluating the process capability with multiple quality characteristics. In addition, it is very difficult to obtain the relevant statistical properties to make a more detailed inference for the multivariate PCIs. Thus, we realize that there are still critical difficulties in trying to assess the value of multivariate systems in terms of a single value. Clearly, further investigations in this field are needed.

References

[1] Kotz, S. & Johnson, N.L. (2002). Process capability indices – a review, 1992–2000, *Journal of Quality Technology* **34**, 2–19.

[2] Chan, L.K., Cheng, S.W. & Spiring, F.A. (1991). A multivariate measure of process capability, *Journal of Modeling and Simulation* **11**, 1–6.

[3] Pearn, W.L., Kotz, S. & Johnson, N.L. (1992). Distributional and inferential properties of process capability indices, *Journal of Quality Technology* **24**, 216–231.

[4] Taam, W., Subbaiah, P. & Liddy, J.W. (1993). A note on multivariate capability indices, *Journal of Applied Statistics* **20**, 339–351.

[5] Chen, H. (1994). A multivariate process capability index over a rectangular solid tolerance zone, *Statistica Sinica* **4**, 749–758.

[6] Shahriari, H., Hubele, N.F. & Lawrence, F.P. (1995). A multivariate process capability vector, *Proceedings of the 4th Industrial Engineering Research Conference*, Institute of Industrial Engineers, Nashville, pp. 304–309.

[7] Wang, F.K. & Du, T.C.T. (2000). Using principal component analysis in process performance for multivariate data, *Omega* **28**, 185–194.

[8] Walter, J., Anghel, C. & Braun, L. (2004). Multivariate process capability indices: new approaches and comparisons with known indices, working paper.

[9] Pearn, W.L., Wang, F.K. & Yen, C.H. (2006). Measuring production yield for processes with multiple quality characteristics, *International Journal of Production Research* **44**, 4649–4661.

[10] Mahalanobis, P.C. (1936). On the generalized distance in statistics, *Proceedings of the National Institute of Science of India* **12**, 49–55.

[11] Jackson, J.E. (1980). Principal component and factor analysis: part I – principal components, *Journal of Quality Technology* **12**, 201–213.

[12] Luceño, A. (1996). A process capability index with reliable confidence intervals, *Communications in Statistics – Simulation and Computation* **25**, 235–245.

[13] Boyles, R.A. (1994). Process capability with asymmetric tolerances, *Communications in Statistics – Simulation* **23**, 615–643.

[14] Wang, F.K., Hubele, N.F., Lawrence, F.P., Miskulin, J.D. & Shahriari, H. (2000). Comparison of three multivariate process capability indices, *Journal of Quality Technology* **32**, 263–275.

FU-KWUN WANG

Multivariate Process Control *see* Hotelling's T^2

Multivariate Renewal Replacement *see* Multivariate Age and Multivariate Renewal Replacement

Multivariate Stochastic Orders and Aging

Dynamic Multivariate IFR Notions

A random vector $\boldsymbol{X} = (X_1, X_2, \ldots, X_n)$ is said to be smaller than the random vector $\boldsymbol{Y} = (Y_1, Y_2, \ldots, Y_n)$ in the *multivariate ordinary stochastic order* (denoted by $\boldsymbol{X} \le_{\text{st}} \boldsymbol{Y}$) if $E[\phi(\boldsymbol{X})] \le E[\phi(\boldsymbol{Y})]$ for every function ϕ that is increasing with respect to the coordinate-wise order in $\mathbb{R}^n$, such that the expectations above are well defined. See Shaked and Shanthikumar [1] or Müller and Stoyan [2] for detailed studies of this order.

Let $\boldsymbol{X} = (X_1, X_2, \ldots, X_n)$ be a nonnegative random vector with an absolutely continuous distribution function. Here it is helpful to think about $X_1, X_2, \ldots, X_n$ as the lifetimes of n components $1, 2, \ldots, n$ that make up some system. Suppose that an observer observes the system continuously in time and records the failure times and the identities of the components that fail as time passes. Thus, a typical "history" that the observer has observed by time $t \ge 0$ is of the form

$$h_t = \{\boldsymbol{X}_I = \boldsymbol{x}_I, \boldsymbol{X}_{\overline{I}} > t\boldsymbol{e}\}, \quad 0\boldsymbol{e} \le \boldsymbol{x}_I \le t\boldsymbol{e}, \quad I \subseteq \{1, 2, \ldots, n\} \tag{1}$$

(Here, for every n-dimensional vector $\boldsymbol{x}$ and for any set $I \subseteq \{1, 2, \ldots, n\}$, $\boldsymbol{x}_I$ denotes the vector of the x_i's such that $i \in I$. Also, here $\overline{I}$ denotes the complement of I in $\{1, 2, \ldots, n\}$, and $\boldsymbol{e}$ denotes the vector of 1's.) In equation (1) I is the set of components that have already failed by time t (with failure times $\boldsymbol{x}_I$) and $\overline{I}$ is the set of components that are still alive at time t.

Let

$$h'_s = \{\boldsymbol{X}_J = \boldsymbol{y}_J, \boldsymbol{X}_{\overline{J}} > s\boldsymbol{e}\}, \quad 0\boldsymbol{e} \le \boldsymbol{y}_J \le s\boldsymbol{e}, \quad J \subseteq \{1, 2, \ldots, n\} \tag{2}$$

be another history. If $t \le s$ and the histories h_t and h'_s are such that each component that failed in h_t also failed in h'_s, and, for components that failed in both histories, the failures in h'_s are earlier than the failures in h_t, then we say that the history h_t is *less severe* or *more pleasant* than the history h'_s and we denote it by $h_t \le h'_s$. Note that if h_t and h'_s are as in equations (1) and (2), then $h_t \le h'_s$ if, and only if, $I \subseteq J$ and $\boldsymbol{y}_I \le \boldsymbol{x}_I$.

Recall that a univariate nonnegative random variable X has the increasing failure rate (IFR) aging property if, and only if, $[X - t | X > t] \ge_{\text{st}} [X - t' | X > t']$ whenever $t \le t'$ (*see* **Stochastic Orders and Aging**).

For every vector $\boldsymbol{a} = (a_1, a_2, \ldots, a_n)$ denote by $\boldsymbol{a}_+$ the vector $(\max\{a_1, 0\}, \max\{a_2, 0\}, \ldots, \max\{a_n, 0\})$. Generalizing the univariate IFR characterization mentioned above, we can define a nonnegative random vector $\boldsymbol{X}$ as multivariate increasing failure rate (MIFR) if for $t \leq s$ we have

$$[(\boldsymbol{X} - t\boldsymbol{e})_+ | h_t] \geq_{\text{st}} [(\boldsymbol{X} - s\boldsymbol{e})_+ | h'_s] \quad \text{whenever } h_t \leq h'_s \tag{3}$$

Another possibility is to call the nonnegative random vector $\boldsymbol{X}$ MIFR if for $t \leq s$ we have

$$[(\boldsymbol{X} - t\boldsymbol{e})_+ | h_t] \geq_{\text{st}} [(\boldsymbol{X} - s\boldsymbol{e})_+ | h'_s] \quad \text{whenever } h_t \text{ and } h'_s \text{ coincide on } [0, t) \tag{4}$$

The latter definition is the MIFR$|\mathcal{F}_t$ definition of Arjas [3] where $\mathcal{F}_t$ is the minimal σ field generated by $\boldsymbol{X}$; see Arjas [3] for more details. Obviously the two MIFR notions, defined by equations (3) and (4), satisfy

$$\text{MIFR(3)} \Longrightarrow \text{MIFR(4)}$$

The two *different* definitions of MIFR, given in equations (3) and (4), have some desirable properties. For example, a vector consisting of independent IFR random variables is MIFR according to either one of these two definitions. However, perhaps the most important feature of these kinds of definitions is their intuitive interpretation. In the univariate case these two definitions coincide with the univariate definition of IFR. More properties of these MIFR notions can be found in Arjas [3] and in Shaked and Shanthikumar [4].

Let us now recall another characterization of the univariate IFR aging notion, that is, a nonnegative random variable X has the IFR aging property if, and only if, one of the following equivalent conditions hold:

$$[X - t | X > t] \geq_{\text{hr}} [X - t' | X > t'] \quad \text{whenever } t \leq t' \tag{5}$$

or

$$X \geq_{\text{hr}} [X - t | X > t] \quad \text{for all} \quad t \geq 0 \tag{6}$$

where $\leq_{\text{hr}}$ denotes the univariate hazard rate stochastic order (*see* **Stochastic Orders and Aging**). Generalizing these univariate characterizations to the multivariate case we obtain other MIFR notions as described next.

Consider a typical "history" of $\boldsymbol{X}$ as in equation (1). Let $i \in \overline{I}$ be a component that is still alive at time t. Its multivariate conditional hazard rate, at time t, is defined as follows:

$$\lambda_{i|I}(t|\boldsymbol{x}_I) = \lim_{\Delta t \downarrow 0} \frac{1}{\Delta t} P\left\{t < X_i \leq t + \Delta t | \boldsymbol{X}_I = \boldsymbol{x}_I, \boldsymbol{X}_{\overline{I}} > t\boldsymbol{e}\right\} \tag{7}$$

where, of course, $0\boldsymbol{e} \leq \boldsymbol{x}_I \leq t\boldsymbol{e}$, and $I \subseteq \{1, 2, \ldots, n\}$. Let $\boldsymbol{Y} = (Y_1, Y_2, \ldots, Y_n)$ be another nonnegative random vector with an absolutely continuous distribution function. Denote its multivariate conditional hazard rate functions by $\eta_{\cdot|\cdot}(\cdot|\cdot)$, where the η's are defined analogously to the λ's in equation (7). Suppose that

$$\eta_{i|I \cup J}(u|\boldsymbol{y}_I, \boldsymbol{y}_J) \geq \lambda_{i|I}(u|\boldsymbol{x}_I) \quad \text{whenever } J \cap I = \emptyset, \quad \boldsymbol{y}_I \leq \boldsymbol{x}_I \leq u\boldsymbol{e}, \text{ and } \boldsymbol{y}_J \leq u\boldsymbol{e}$$

where $i \in \overline{I \cup J}$. Then $\boldsymbol{Y}$ is said to be *smaller than* $\boldsymbol{X}$ *in the dynamic multivariate hazard rate order* (denoted as $\boldsymbol{Y} \leq_{\text{dyn-hr}} \boldsymbol{X}$); see Shaked and Shanthikumar [1] for more details about this multivariate stochastic order. Recalling the univariate characterization (equation (5)) of the IFR aging notion, one may define a multivariate aging property of $\boldsymbol{X}$ by requiring $\boldsymbol{X}$ to satisfy, for $t \leq s$ and histories h_t and h'_s,

$$[(\boldsymbol{X} - t\boldsymbol{e})_+ | h_t] \geq_{\text{dyn-hr}} [(\boldsymbol{X} - s\boldsymbol{e})_+ | h'_s] \quad \text{whenever } h_t \leq h'_s \tag{8}$$

Yet another possible multivariate analog of the characterization (equation (5)) is to require $\boldsymbol{X}$ to satisfy, for $t \leq s$,

$$[(\boldsymbol{X} - t\boldsymbol{e})_+ | h_t] \geq_{\text{dyn-hr}} [(\boldsymbol{X} - t\boldsymbol{e})_+ | h'_s] \quad \text{whenever } h_t \text{ and } h'_s \text{ coincide on } [0, t) \tag{9}$$

An analog of the characterization (equation (6)) is to require $\boldsymbol{X}$ to satisfy equations (8) or (9) with $t = 0$; that is,

$$\boldsymbol{X} \geq_{\text{dyn-hr}} [(\boldsymbol{X} - s\boldsymbol{e})_+ | h'_s] \quad \text{for any history } h'_s, \ s \geq 0 \tag{10}$$

It turns out that the three conditions, defined in equations (8), (9), and (10), are equivalent:

$$\text{MIFR(8)} \Longleftrightarrow \text{MIFR(9)} \Longleftrightarrow \text{MIFR(10)}$$

and as such they define a notion of MIFR property. Since the order $\leq_{\text{dyn-hr}}$ is stronger than the order $\leq_{\text{st}}$ (see Shaked and Shanthikumar [1]), this notion is stronger than the MIFR notions defined by equations (3) and (4), that is,

$$\text{MIFR}(8) \Longrightarrow \text{MIFR}(3)$$

See more details in Shaked and Shanthikumar [4].

Arjas [3] also defined a multivariate notion of decreasing failure rate (DFR). This definition is a version of equation (4) in which the inequality is reversed, but some care is needed for such a definition – only the lifetimes of components that are alive at time s are compared.

A Dynamic Multivariate NBU Notion

Recall that a univariate nonnegative random variable X has the new better than used (NBU) aging property if, and only if, $X \geq_{\text{st}} [X - t | X > t]$ whenever $t \geq 0$ (*see* **Stochastic Orders and Aging**). Following the idea that led to the MIFR aging notion given in equation (4), one can define a multivariate new better than used (MNBU) aging notion. Explicitly, one can call the nonnegative random vector $\boldsymbol{X}$ MNBU if

$$\boldsymbol{X} \geq_{\text{st}} [(\boldsymbol{X} - t\boldsymbol{e})_+ | h_t] \quad \text{for all} \quad t \geq 0 \tag{11}$$

This is the MNBU$|\mathcal{F}_t$ definition of Arjas [3] where $\mathcal{F}_t$ is the minimal σ field generated by $\boldsymbol{X}$; see Arjas [3] for more details. Obviously the MIFR notion in equation (4) implies this NBU notion.

Arjas [3] also defined a multivariate notion of new worse than used (NWU). This definition is a version of equation (11) in which the inequality is reversed, but some care is needed for such a definition – only the lifetimes of components that are alive at time t are compared.

Multivariate Aging Notions Based on Comparisons with Multivariate Exponential Distributions

The univariate aging notions of IFR, IFRA (increasing failure rate average), and NBU, can be characterized, respectively, by the univariate convex transform order $\leq_{\text{c}}$, by the univariate star order $\leq_*$, and by the univariate superadditive order $\leq_{\text{su}}$ (*see* **Stochastic Orders and Aging**). Explicitly, if X is a nonnegative random variable, and Exp denotes an exponential random variable (no matter what the mean is) then

$$X \text{ is IFR} \Longleftrightarrow X \leq_{\text{c}} \text{Exp} \tag{12}$$

$$X \text{ is IFRA} \Longleftrightarrow X \leq_* \text{Exp} \tag{13}$$

$$X \text{ is NBU} \Longleftrightarrow X \leq_{\text{su}} \text{Exp} \tag{14}$$

Motivated by equations (12–14), one may define MIFR, multivariate increasing failure rate average (MIFRA), and MNBU aging notions. In order to do that one needs multivariate analogs of the univariate stochastic orders $\leq_{\text{c}}$, $\leq_*$, and $\leq_{\text{su}}$, and one needs to choose a suitable multivariate exponential distribution to replace Exp in equations (12–14). Below we first describe multivariate analogs of the orders $\leq_{\text{c}}$, $\leq_*$, and $\leq_{\text{su}}$.

Roy [5] considered the following multivariate extensions of the univariate stochastic orders $\leq_{\text{c}}$, $\leq_*$, and $\leq_{\text{su}}$. Let $\boldsymbol{X} = (X_1, X_2, \ldots, X_n)$ and $\boldsymbol{Y} = (Y_1, Y_2, \ldots, Y_n)$ be two nonnegative random vectors, with survival functions $\overline{F}$ and $\overline{G}$, respectively; that is, $\overline{F}(\boldsymbol{x}) = P\{\boldsymbol{X} > \boldsymbol{x}\}$ and $\overline{G}(\boldsymbol{x}) = P\{\boldsymbol{Y} > \boldsymbol{x}\}$ for $\boldsymbol{x} = (x_1, x_2, \ldots, x_n) \in \mathbb{R}_+^n$. For $i = 1, 2, \ldots, n$, and $\boldsymbol{x} \in \mathbb{R}_+^n$, denote by $\overline{G}_i^{-1}$ the inverse of $\overline{G}(\boldsymbol{x})/\overline{G}(x_1, \ldots, x_{i-1}, 0, x_{i+1}, \ldots, x_n)$ as a function of x_i, and also denote $\overline{F}_i(\boldsymbol{x}) = \overline{F}(\boldsymbol{x})/\overline{F}(x_1, \ldots, x_{i-1}, 0, x_{i+1}, \ldots, x_n)$. Then $\boldsymbol{X}$ is said to be less than $\boldsymbol{Y}$ in the multivariate convex transform (star, superadditive) order (denoted by $\boldsymbol{X} \leq_{\text{c}} (\leq_*, \leq_{\text{su}})\ \boldsymbol{Y}$) if for each $i = 1, 2, \ldots, n$, it holds that $\overline{G}_i^{-1}(\overline{F}_i(x_1, x_2, \ldots, x_n))$ is convex (star-shaped, superadditive) in x_i for any given $x_1, \ldots, x_{i-1}, x_{i+1}, \ldots, x_n$.

Next, let **Exp** denote any random vector that has a multivariate Gumbel exponential distribution with the survival function

$$\overline{G}(x_1, x_2, \ldots, x_n) = \exp\left[-\sum_i \lambda_i x_i - \sum_{i<j} \lambda_{ij} x_i x_j \cdots - \lambda_{1,2,\ldots,n} x_1 x_2 \cdots x_n\right], \quad (x_1, x_2, \ldots, x_n) \in \mathbb{R}_+^n \tag{15}$$

where $\lambda_{i_1, i_2, \ldots, i_k} > 0$ for each $\{i_1, i_2, \ldots, i_k\} \subseteq \{1, 2, \ldots, n\}$; see Kotz *et al.* [6, p. 406].

Now one can define MIFR, MIFRA, and MNBU aging notions for a random vector $\boldsymbol{X}$ by replacing Exp in equations (12–14) by **Exp**, using the multivariate stochastic orders $\le_{\rm c}$, $\le_*$, and $\le_{\rm su}$ described above. However, historically, proper definitions of MIFR, MIFRA, and MNBU aging notions were introduced first, and then analogs of equations (12–14) were obtained. So this is the way we will proceed here.

Roy [7] proposed multivariate extensions of the IFR, IFRA, and NBU notions as described below. Let $\boldsymbol{X} = (X_1, X_2, \ldots, X_n)$ be an absolutely continuous nonnegative random vector with survival function $\overline{F}$. For $i = 1, 2, \ldots, n$, denote by $r_i(\boldsymbol{x})$ and $A_i(\boldsymbol{x})$ the failure rate and the failure rate average of the random variable $\left[X_i \middle| \bigcap_{j\neq i} X_j > x_j\right]$ at the point x_i, where $\boldsymbol{x} = (x_1, x_2, \ldots, x_n) \in \mathbb{R}_+^n$; that is, writing $R(\boldsymbol{x}) = -\log \overline{F}(\boldsymbol{x})$, we have

$$r_i(\boldsymbol{x}) = \frac{\partial}{\partial x_i} R(\boldsymbol{x})$$

and

$$A_i(\boldsymbol{x}) = \frac{1}{x_i}\int_0^{x_i} r_i(x_1, \ldots, x_{i-1}, y, x_{i+1}, \ldots, x_n)\,\mathrm{d}y$$

According to Roy [7], $\boldsymbol{X}$ has the

1. MIFR property if $r_i(\boldsymbol{x})$ is increasing in x_i for all $(x_1, \ldots, x_{i-1}, x_{i+1}, \ldots, x_n) \in \mathbb{R}_+^{n-1}$ and each i.
2. MIFRA property if $A_i(\boldsymbol{x})$ is increasing in x_i for all $(x_1, \ldots, x_{i-1}, x_{i+1}, \ldots, x_n) \in \mathbb{R}_+^{n-1}$ and each i,
3. MNBU property if

$$\begin{aligned}&\overline{F}(x_1, \ldots, x_{i-1}, x_i + y, x_{i+1}, \ldots, x_n)\\ &\quad\times \overline{F}(x_1, \ldots, x_{i-1}, 0, x_{i+1}, \ldots, x_n)\\ &\quad\le \overline{F}(x_1, \ldots, x_{i-1}, x_i, x_{i+1}, \ldots, x_n)\\ &\quad\times \overline{F}(x_1, \ldots, x_{i-1}, y, x_{i+1}, \ldots, x_n),\end{aligned}$$

for all $y \ge 0$ and for all $\boldsymbol{x} \in \mathbb{R}_+^n$ and each i.

Roy [5] provided the following characterizations

$$\boldsymbol{X} \text{ is MIFR} \Longleftrightarrow \boldsymbol{X} \le_{\rm c} \mathbf{Exp} \quad (16)$$

$$\boldsymbol{X} \text{ is MIFRA} \Longleftrightarrow \boldsymbol{X} \le_* \mathbf{Exp} \quad (17)$$

$$\boldsymbol{X} \text{ is MNBU} \Longleftrightarrow \boldsymbol{X} \le_{\rm su} \mathbf{Exp} \quad (18)$$

where **Exp** denotes any random vector with survival function given in equation (15). Roy [5] also noted that equations (16–18) hold if **Exp** is replaced by a vector of independent exponential random variables.

References

[1] Shaked, M. & Shanthikumar, J.G. (2007). *Stochastic Orders*, Springer, New York.

[2] Müller, A. & Stoyan, D. (2002). *Comparison Methods for Stochastic Models and Risks*, John Wiley & Sons, New York.

[3] Arjas, E. (1981). A stochastic process approach to multivariate reliability systems: notions based on conditional stochastic order, *Mathematics of Operations Research* **6**, 263–276.

[4] Shaked, M. & Shanthikumar, J.G. (1991). Dynamic multivariate aging notions in reliability theory, *Stochastic Processes and Their Applications* **38**, 85–97.

[5] Roy, D. (2002). Classification of multivariate life distributions based on partial ordering, *Probability in the Engineering and Informational Sciences* **16**, 129–137.

[6] Kotz, S., Balakrishnan, N. & Johnson, N.L. (2000). *Continuous Multivariate Distributions, Volume 1: Models and Applications*, 2nd Edition, John Wiley & Sons, New York.

[7] Roy, D. (1994). Classifications of life distribution under multivariate model, *IEEE Transaction on Reliability* **43**, 224–229.

Related Article

Aging and Positive Dependence.

FÉLIX BELZUNCE AND MOSHE SHAKED

Network Reliability *see* Computational Issues in Network Reliability

Neural Networks in Statistical Process Control

Introduction

The resurgence of research interests in neural computing came about in the early 1980s. Enormous interest exists in studying and applying neural computing in almost all fields of science and technology (engineering, business, and medical science, to name a few). The application of **neural networks** in industrial, production engineering or related domains started gaining popularity in the latter 1980s. The momentum of the growing research interest peaked around the mid-1990s and is still sustaining a high interest level as of 2006. The tremendous interest and wide applicability of neural computing theories and technology motivate researchers to explore how this class of modeling tools can be effectively applied to statistical process control (SPC).

The fundamental issue at the core of classical statistical quality control has remained relatively unchanged since the introduction of **control charts** by Walter Shewhart in the 1920s (*see* **Statistical Quality Control**). The basic principle for a control scheme is to signal an out-of-control state when the behavior of critical quality variables in a process is considered statistically abnormal, but also to leave the process intact when no evidence of statistical abnormality is present. However, as a result of modern computer and data acquisition technology, the manifestation of an out-of-control state in a process may take various forms, e.g., characteristically patterned or highly autocorrelated. Under the circumstances, the data collected often render the traditional statistical control charts ineffective or inapplicable. From the process improvement point of view, practitioners desire to have the knowledge about a faulty process beyond just "a violation of a certain set of control limits, e.g., 3σ" (*see* **Control Charts and Process Capability**; **Quality Management, Overview**; **Six Sigma**). The aforementioned concerns have been recognized and the need to identify nonrandom or unnatural patterns was well documented long ago, e.g., in the widely referenced *Statistical Quality Control Handbook* [1]. They also have drawn keen attention from the statistical quality control research community in the last couple of decades (*see* **Sampling Inspection of Products and Statistical Process Control**; **Control Charts, Overview**; **Control Charts for the Mean**; **Exponentially Weighted Moving Average (EWMA) Control Chart**; **Cumulative Sum (CUSUM) Chart**; **Multivariate Control Charts Overview**; **Hotelling's T^2 Chart**; **Multivariate Exponentially Weighted Moving Average (MEWMA) Control Chart**; **Multivariate Control Charts, Interpretation of**; **Control**

Charts for Short Production Runs; Control Charts for Batch Processes; Control Charts, Selection of; Statistical Process Control in Clinical Medicine).

Two primary categories of problems are the detection of shifts and the recognition of unnatural patterns from the process monitored. The former is concerned with process mean shift, which may occur when the process variable is independently and identically distributed (i.i.d.), or under some underlying structure such as **autocorrelation** (*see* **Autocorrelated Data**). The latter focuses on patterns exhibited within the typical control limits, which cannot be recognized using conventional decision criteria. The neural network–based schemes are flexible and adaptable enough to deal with both categories of problems. The expanded capabilities to distinguish between mean shift and structural change or patterns are of great value for the purpose of process control. The article focuses on various models appropriate in the SPC context and the general modeling approach employing various neural network algorithms.

Process Models

Modeling Unnatural Patterns and Shifts

The realization of a process variable over time, X_t, can be considered comprising of the process mean and two noise components, i.e., one from the random cause and the other is a special disturbance from assignable causes, as expressed in equation (1).

$$X_t = \mu + x_t + d_t \tag{1}$$

where

- $\mathbf{X}_t$: quality measurement at time t,
- $\boldsymbol{\mu}$: process mean when the process is in control,
- $\boldsymbol{\sigma}$: process standard deviation when the process is in control,
- $\mathbf{x}_t$: random normal variate at time $t, x_t \sim N(0, k\sigma)$, $0 < k \leq 1$,
- $\mathbf{d}_t$: special disturbance at time t.

Note that d_t may take in various forms to model unnatural patterns such as trends, cycles, systematic variables, mixtures, stratification, and sudden shift in level, and so on [2]. A shift in process mean can be expressed as a type of unnatural pattern.

Modeling Shifts under Autocorrelation

Let Z_t represent the generic form of an autoregressive and moving average (ARMA) model (*see* **Large Autoregressive Integrated Moving Average (ARIMA) Modeling**). Suppose the process under consideration is one that has an additive, deterministic step shift of magnitude, δ, at some point of time, ψ. Then the process can be represented by equation (2).

$$X_t = Z_t + \delta h_t \tag{2}$$

where

$$h_t = \begin{cases} 0 & if \quad t < \psi \\ 1 & if \quad t \geq \psi \end{cases} \tag{3}$$

Modeling Shifts in Multivariate Processes

Suppose there are p process variables which are jointly distributed following a p-variate **normal distribution** with known mean vector $\boldsymbol{\mu}$, and known **covariance** matrix $\boldsymbol{\Sigma}$. Equation (4) represents a multivariate process.

$$X_t = \mu + x_t + \Delta_t \tag{4}$$

where

- $\boldsymbol{X}_t$: a vector of p quality measurements at time t,
- $\boldsymbol{\mu}$: a vector of process means when the process is in control,
- $\boldsymbol{x}_t$: a vector of random normal variate at time t, x_t follows a p-variate normal distribution with mean vector of zeros and known covariance matrix $\boldsymbol{\Sigma}$,
- $\boldsymbol{\Delta}_t$: step shift vector at time t.

Modeling Shifts in Multivariate Autocorrelated Processes

A multivariate autocorrelated process of p variables can be expressed as a vector autoregressive model. Equation (5) represents a Vector Autoregressive model of order p, or VAR(p).

$$\begin{aligned} Y_t - \mu_t &= \Phi_1(Y_{t-1} - \mu_{t-1}) + \Phi_2(Y_{t-2} - \mu_{t-2}) \\ &\quad + \cdots + \Phi_p(Y_{t-p} - \mu_{t-p}) + \varepsilon_t \end{aligned} \tag{5}$$

where $\boldsymbol{\mu}_t$ is the vector of mean values at time t, ε_t is an independent multivariate normal random vector with the mean vector of zeros and covariance matrix $\boldsymbol{\Sigma}$, and $\boldsymbol{\Phi}_i (i = 1, 2, \ldots, p)$ is a matrix of autocorrelation parameters.

Let $\boldsymbol{Z}_t = \boldsymbol{Y}_t - \boldsymbol{\mu}_t$. Equation (6) describes a process that has an additive, deterministic step shift vector, $\boldsymbol{\Delta}$, beginning at some point of time, ψ.

$$X_t = Z_t + \Delta h_t \tag{6}$$

where

$$h_t = \begin{cases} 0 & if \quad t < \psi \\ I & if \quad t \geq \psi \end{cases} \tag{7}$$

The above models represent some of the widely characterized processes. Other forms of process models are possible.

Neural Network Modeling

Algorithms

Some of the neural network algorithms that have been adopted include two-layer, multilayer perceptrons with the widely used and proven learning algorithm backpropagation [2, 3], radial basis function (RBF) [4], Boltzmann machines [5, 6], adaptive resonance theory (ART) [7, 8] and its variants ARTMAP and fuzzy ARTMAP [9], and learning vector quantization (LVQ) network [10]. The data types handled can be coded binary or raw analog data series. The network output is in the form of pattern class, predicted magnitude of the shift, or value of the observation in the series. The network training regimes include supervised, e.g., backpropagation, and unsupervised, e.g., ART. Supervised training requires the desired or target output to be presented during the learning, while unsupervised training does not. In terms of the functionality of the system developed, it can be a general-purpose or special-purpose system.

Beyond doubt, the backpropagation algorithm is the most widely adopted in building neural networks for a wide range of applications including both pattern recognition and shift detection. The standard backpropagation uses a generalized Delta rule [3] that updates network connection weights according to the gradient descent method using equation (8).

$$\Delta w_{ij} = \eta \delta_i o_j \tag{8}$$

where o_j is the actual output of the jth unit, η is the learning constant and δ_i is defined as follows:

$$\delta_i = (t_i - o_i) f'(net_i) \tag{9}$$

where $f'(net_i)$ is the derivative of the activation function f for the ith unit evaluated at the net total input net_i and t_i is the desired output of the ith unit. Two of the most widely used activation functions are the sigmoid (equation 10) and the hyperbolic tangent (equation 11).

$$f(z) = (1 + e^{-z})^{-1} \tag{10}$$

$$f(z) = \frac{e^z - e^{-z}}{e^z + e^{-z}} \tag{11}$$

For more discussions about neural networks, *see* [11, 12] and **Neural Networks: Construction and Evaluation**. For a review of applying neural networks to SPC, see [13].

The Modeling Process

The modeling process typically consists of two stages, namely, network development and performance evaluation. In the network development stage, one has to determine the following key parameters: the window size of input data, the number of hidden layers and hidden nodes, the number of output nodes, the size of the training file, the size of the testing file, the learning rule, the **transfer function**, the training termination or convergence criterion, and so on. These decisions have direct impact on the capabilities of the trained network. Depending on detecting shifts or recognizing patterns, the requirement of the window size varies. The window size should be as small as possible, say five or lesser, and yet sufficient to capture the nature of the data series. When the underlying structure of the process data is more complex, such as autocorrelation, or the patterns of interest are more complicated, such as concurrent presence of multiple patterns or features, the window size has to be sufficiently large, say 50 or more. See [14–16] for some guiding principles, e.g., avoid ambiguity, adequate representation, and balanced representation, in designing the training data sets. The goal at the end of this stage is to develop a network that has not only learned all the training examples well but can also be applied to new, unseen data series with reasonable generalization capabilities. Root mean squared

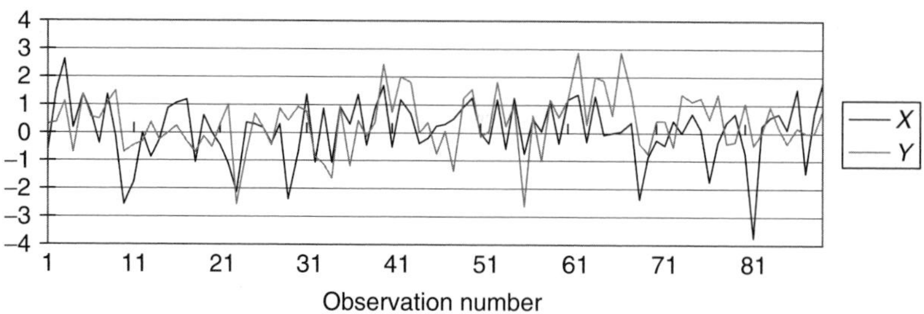

Figure 1 VAR(1): The standardized order quantities

(RMS) errors are frequently used as the measure to determine the stability of learning or if a satisfactory level of training has been reached.

Performance Evaluation

To evaluate the capabilities of the trained network, one needs to measure the probabilities of type I and type II errors. In the case of shift detection, the primary concern is how quickly the trained network can detect the shift when it actually occurs without generating too many false alarms. Similar to the conventional SPC measures, average run length (ARL) is a useful performance measure (*see* **Average Run Lengths and Operating Characteristic Curves**). Given a prespecified in-control ARL, the trained network is "tuned" to a corresponding level of discriminating **power**. For instance, this level will be determined by choosing an appropriate output activation threshold value in a backpropagation neural network.

In the case of pattern recognition, one is concerned with not only the correct classification rates but also the timeliness of the recognition. Run length measures similar to the conventional ARL should be considered. However, depending on the number of pattern classes to be recognized, the possible outcomes for each classification are not dichotomous. Therefore, the run length must be based only on the correct classifications. Target-pattern-based measures and ARL indices, which combine both the correct classification rate and the ARL, should be used [2].

An Illustrated Example

Consider a situation whereby two vendors order the same type of goods from the same supplier. The order quantities placed by each vendor form a **time series** with autocorrelation. Owing to the common demand population and the common source of supply, the two vendors' order quantities are also cross-correlated, thus, the order quantities follow a VAR(1) model. At some point in time, one of the vendors embarked on a promotional drive to stimulate the demand. Over a 3-month period, order quantities were collected and shown as in Figure 1. The data are preprocessed as moving windows of data (see Figure 2) and presented to a backpropagation neural network previously trained for similar data. The network has a configuration of 100 input nodes and 5 output nodes, with no hidden layers. The 100 input nodes receive 50 pairs of (X, Y) observations. The five output nodes represent shift in X, shift in Y, autocorrelation in X, autocorrelation in Y, and cross-**correlation** between X and Y, respectively.

Figure 3 presents the neural network output chart for the first two output nodes, i.e., shift in X and shift in Y. The chart clearly indicates the progression of the two time series. The upward shift in order quantities from vendor Y can be clearly observed. In fact, the mean increases by an amount equivalent to 0.5 times the standard deviation of the existing ordering process from observation no. 60 onward. With proper tuning of in-control ARL, the cutoff threshold is

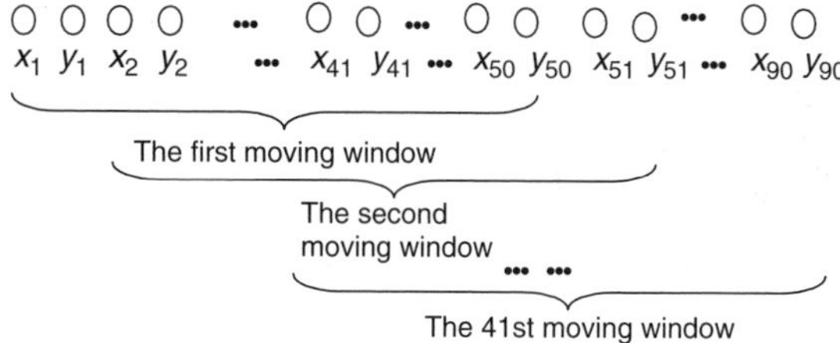

Figure 2 Moving windows for neural network input

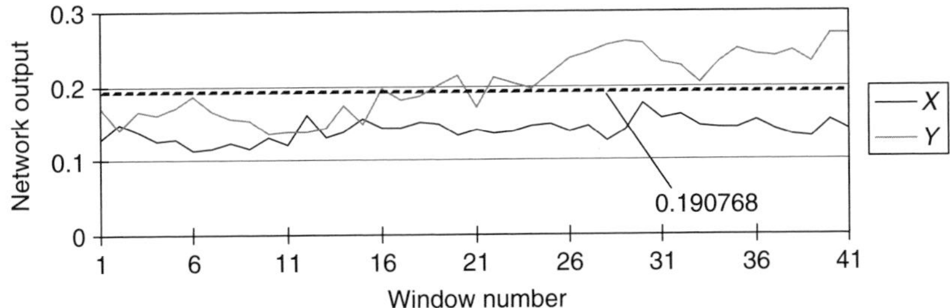

Figure 3 The neural network output chart

set at 0.190768. The neural network–based control scheme detected the shift and identified the source at observation no. 65 (or in window no. 16).

Strengths

Being flexible, adaptable, and capable of learning are the three major strengths. A neural network can be trained for a wide range of data. Unlike traditional statistical approaches, to apply a trained network does not require assumptions about the data distribution, the time series model, or the prediction errors of the new data. The approach is relieved of the requirement of the validity of typical statistical assumptions. Neural networks are more flexible and can be applied to a wider range of data.

The capabilities of a network are closely dependent on the data upon which it was trained. By carefully designing and selecting the training data set, a network can be trained to adapt to any particular data set or a set of features. For example, one may adapt a network to focus more on small to moderate shifts by providing more training data of such features. Neural networks are more adaptable and ideal for developing special-purpose systems.

By presenting the training data to a neural network, the network learns to capture the features or form clusters from the data. In addition to learning from the given desired outputs, so-called supervised learning, neural networks can also learn to form unknown clusters or pattern classes. The progression of learning can be in stages at different times and cumulative. In other words, what is learned in the past can become part of the memory and the basis upon which new learning may take place.

Weaknesses

A lack of analytical robustness and the need for intensive data are the two major weaknesses. The production of the data sets used in training and/or testing involves some degree of randomness. The process of iterative training itself also involves randomness. Despite the efforts in producing pseudorandom data and in training for stability, the resultant trained network is not an outcome of some analytical modeling. As a result, the outcome is often data-dependent or training-dependent making it difficult to offer a robust analysis or explanation of the results. In terms of the amount of data required, the neural network approach suffers from the scalability problem, e.g., when dealing with processes of high dimension.

Summary

The pattern classification capability of neural networks enables one to develop pattern recognizers while the adaptive learning capability allows networks to be "equipped" for different types of patterns or different purposes for process control. As the nature of modern industry or business transactions evolves and the mode of data acquisition advances, the need for dealing with complex and enormous amounts of data becomes more pressing. As demonstrated in this article, it is promising to explore further the utility of neural networks in dealing with data collected from autocorrelated processes in a multivariate setting. The advantages of the neural network approach over the traditional statistical approaches in such applications will depend highly on the ability to interpret the neural network outputs and the diagnosis of the assignable causes associated with each of the variables. These two domains must be integrated to develop a highly automated and intelligent neural network–based SPC system. Muck work remains to be done.

References

[1] Western Electric Company. (1956). *Statistical Quality Control Handbook*, Western Electric Company, New York.

[2] Hwarng, H. & Hubele, N. (1993). Back-propagation pattern recognizers for $\overline{X}$ control charts: methodology and performance, *Computers and Industrial Engineering* **24**, 219–235.

[3] Rumelhart, D., Hinton, G. & Williams, R. (1986). Learning internal representation by error propagation, in *Parallel Distributed Processing: Explorations in the Microstructure of Cognition, Vol. 1: Foundations*, D. Rumelhart & J. McClelland, eds, MIT Press, Cambridge, pp. 318–362.

[4] Moody, J. & Darken, C. (1989). Fast learning in networks of locally-tuned processing units, *Neural Computing* **1**, 281–294.

[5] Hinton, G. & Sejnowski, T. (1986). Learning and relearning in Boltzmann machines, in *Parallel Distributed Processing: Explorations in the Microstructure of Cognition, Vol. 1: Foundations*, D. Rumelhart & J. McClelland, eds, MIT Press, Cambridge, pp. 282–317.

[6] Hwarng, H. & Hubele, N. (1992). Boltzmann machines that learn to recognize patterns on control charts, *Statistics and Computing* **2**, 191–202.

[7] Carpenter, G. & Grossberg, S. (1987). A massively parallel architecture for a self-organizing neural pattern recognition machine, *Computer Vision, Graphics, and Image Processing* **37**, 54–115.

[8] Hwarng, H. & Chong, C. (1995). Detecting process non-randomness through a fast and cumulative learning ART-based pattern recognizer, *International Journal of Production Research* **33**, 1817–1833.

[9] Hwarng, H. (1997). A neural network approach to identifying cyclic behavior on control charts: a comparative study, *International Journal of Systems Science* **28**, 99–112.

[10] Kohonen, T. (1989). *Self-Organization and Associative Memory*, Springer-Verlag, Berlin.

[11] Fu, L. (1994). *Neural Networks in Computer Intelligence*, McGraw-Hill, New York.

[12] Hertz, J., Krogh, A. & Palmer, R. (1991). *Introduction to the Theory of Neural Computation*, Addison-Wesley, Redwood.

[13] Zorriassatine, F. & Tannock, J. (1998). A review of neural networks for statistical process control, *Journal of Intelligent Manufacturing* **9**, 209–224.

[14] Hwarng, H. & Hubele, N. (1993). $\overline{X}$ control chart pattern identification through efficient off-line neural network training, *IIE Transactions* **25**, 27–40.

[15] Hwarng, H. (1995). Proper and effective training of a pattern recognizer for cyclic data, *IIE Transactions* **27**, 746–756.

[16] Hwarng, H. (2004). Detecting process mean shift in the presence of autocorrelation: a neural network based monitoring scheme, *International Journal of Production Research* **42**, 573–595.

H. BRIAN HWARNG

Neural Networks: Construction and Evaluation

Introduction

Neural network (NN) stems from the thought of trying to mimic the ability of the human brain in both structure and function. The human brain contains many billions of nerve cells, or neurons. Each neuron is a cell, which connects to other neurons through synapses. Each neuron also uses biochemical reactions to receive, process, and transmit information. Following this idea, NNs have been designed to achieve human brain-like performance and have been applied in many fields of study including statistics.

A NN model (*see* **Neural Networks in Statistical Process Control**) is usually demonstrated as layers of functional units, called nodes or neurons or perceptrons. There are at least two layers, input and output. However, a NN normally has an additional layer referred to as a "*hidden layer*". Nodes of the NN are connected by weights. The nodes and weights of the NN are comparable to neurons and synapses, respectively. Hence, the NN model solves a problem by finding optimal weights. Typically, NNs are utilized in three problem areas: (a) classification and diagnosis, (b) function approximation and prediction, and (c) optimization.

The type of NN is defined the NN architecture or topology. Lippmann [1] provides an introduction to six important NNs that can be used for pattern classification: (a) Hopfield net, (b) Hamming net, (c) Carpenter–Grosberg classifier, (d) perceptron, (e) multi-layer perceptron (MLP), and

(f) Kohonen self-organizing feature maps. Existing NN architectures currently used are: MLP, radial basis function (RBF), learning vector quantization (LVQ), adaptive resonance theory (ART), auto-associative NNs, and Kohonen self-organizing maps (SOM).

To construct a NN model, there are five important factors for consideration.

1. **NN architecture**

Different NN architectures have different rules for solving modeling problems. Learning or training is the term used to explain the process of finding the weights. There are two types of NN training. Supervised learning is known as learning with a teacher. It happens when there is a known target value associated with the input nodes. One of the most popular algorithms used in supervised learning is back-propagation. Unsupervised learning is known as learning without a teacher. It occurs when data lack target output values relating to the pattern of input nodes. The algorithms of unsupervised learning can be found in Lippmann [1].

2. **Number of hidden layers**

Since the nodes in the input layer only pass data to other layers, computation occurs in both hidden layers and the output layer. Adding additional hidden layers to solve a problem may provide a better result than use of only two-layer, input and output, networks. However, there is no rule of thumb for the suitable number of hidden layers. The more the hidden layers, the larger the NN size, which leads to overfitting (overtraining) problems. On the other hand, too few hidden layers cause underfitting (undertraining) problems.

3. **Number of nodes in each layer**

Generally, input layer nodes are directly related to the number of observations needed to represent each set of input variables. No guidelines exist for the number of hidden layer nodes as with the number of hidden layers. Overfitting problems occur when there are too many hidden nodes, and underfitting problems when there are too few hidden nodes. Output layer nodes correspond to the anticipated number of distinct patterns or classifications that are to be decided by the network.

4. **Type of connection**

In a full-connection network every single node in a layer connects to every node in the next layer. Otherwise, it is a partial-connection network. In addition, there is a feed-forward network in which nodes in each layer can only be connected to other nodes in the next layer, while in the recurrent network nodes in each layer can be connected to the nodes in the previous layer, next layer and even nodes in the same layer.

5. **Activation function or transfer function**

This function occurs at every computational node, both hidden and output. It can be either linear or nonlinear, symmetrical or nonsymmetrical. Two common functions widely used are the sigmoid function (logistic function) and hyperbolic tangent function (*see* **Neural Networks in Statistical Process Control**).

An Example: NNs for Multivariate Quality Control

Traditional **multivariate control charts**, such as the chi-square (*see* **Hotelling's T^2 Chart**) and the multivariate exponentially weighted moving average (MEWMA) charts (*see* **Multivariate Exponentially Weighted Moving Average (MEWMA) Control Chart**), are commonly used for monitoring a process and detecting process shifts when there are several correlated quality characteristics. These two traditional techniques make three important assumptions:

1. the data follow a multivariate **normal distribution**,
2. the **covariance** matrix is constant over time, and
3. the observations are independent over time.

The performance of both charts is seriously affected if any of these assumptions is not met. The NN then becomes an alternative for consideration because it does not require such strict assumptions. The NN approach finds the relationships between input variables and output variables from existing data. Therefore, an appropriately designed NN may be easier for practitioners to use and more robust to traditional assumptions, while providing equal or improved performance relative to traditional multivariate statistical process control (SPC) methods.

The bivariate case is illustrated using simulated data written by the interactive matrix language procedure (Proc IML) of Statistical Analysis System (SAS) for generating multivariate observations and assessing the performance of the three different methods based on the average run length (ARLs). For each method, ARLs were calculated for both the in-control and out-of-control state, over the total number of observations, to compare the performance of all three methods. As described briefly in the previous section, there are many NN architectures that may be applied to classification problems, which are focused on a change in structure for mean shifts, assuming constant variance. The MLP is one feasible NN architecture. Advantages of the MLP are its simplicity and flexibility.

Construction of a Neural Network for SPC

1. **NN architecture**

The feed-forward MLP with back-propagation is a feasible and suitable NN because of its simplicity.

2. **Type of connection**

A full connection is used following the recommendation of Pugh [2] and Chang and Aw [3] as this approach speeds network convergence.

3. **Number of hidden layers**

Only one hidden layer is used following the result of Pugh [2, 4] as no significant performance difference is reported for the MLP with one-hidden layer or two-hidden layers.

4. **Number of nodes**

(a) Input layer nodes

There are two issues that affect the appropriate number of nodes for a NN.

First, if the problem does not address pattern recognition, the input layer contains only the current observation, regardless of preceding and succeeding data. Hence, each input node might be the vector of current quality characteristic mean of each observation sized n. There are then two input nodes. Each of them represents the sample mean of each ith observation for the first and second quality variable, respectively, as the vector of $(\overline{\mathbf{X}}_1, \overline{\mathbf{X}}_2)$, where $\overline{\mathbf{X}}_j = \frac{\sum_{k=1}^{n} x_{ijk}}{n}$. Summarizing a multivariate sample by using the sample mean vector NN inputs may be considered as preprocessing of the data. Note that this procedure is based on the assumption of a constant variance matrix of two correlated variables. One should also note that the training phase for NN modeling with SPC requires data from both in-control and out-of-control process states. This is quite different from the traditional methods which require phase I data to be from an in-control process.

The other situation is that when pattern recognition is addressed, preceding data are supplied to the NN through the input nodes, in addition to current data. Three consecutive observations then are considered at time periods t, $t-1$, and $t-2$, respectively. In each time period, there is the vector of sample means for the first and second quality variable. Hence, there are six input nodes expressed by $(\overline{\mathbf{X}}_{(t,1)}, \overline{\mathbf{X}}_{(t,2)}, \overline{\mathbf{X}}_{(t-1,1)}, \overline{\mathbf{X}}_{(t-1,2)}, \overline{\mathbf{X}}_{(t-2,1)}, \overline{\mathbf{X}}_{(t-2,2)})$, where $\overline{\mathbf{X}}_{(i,j)}$ is the sample mean for the quality characteristic j, $j = 1, 2$, at time period i for $i = t, t-1, t-2$.

(b) Hidden layer nodes

As Guo and Dooley [5] reported, there is no standard rule for deciding the optimum number of hidden layer nodes. The primary concern is the overtraining or undertraining problem. Thus the performance of NN is illustrated with three and five hidden nodes.

(c) Output layer nodes

Only one output node is required since we are only interested in a change in structure from an in-control state to an out-of-control state. One might consider multiple outputs to signal various changes such as changes in the variance–covariance structures or mean shifts in specific directions.

Four NN architectures are illustrated. They are graphically displayed in Figures 1–4.

5. **Cutoff point estimation**

At the final step of NN process monitoring, the NN output must be checked against its appropriate cutoff

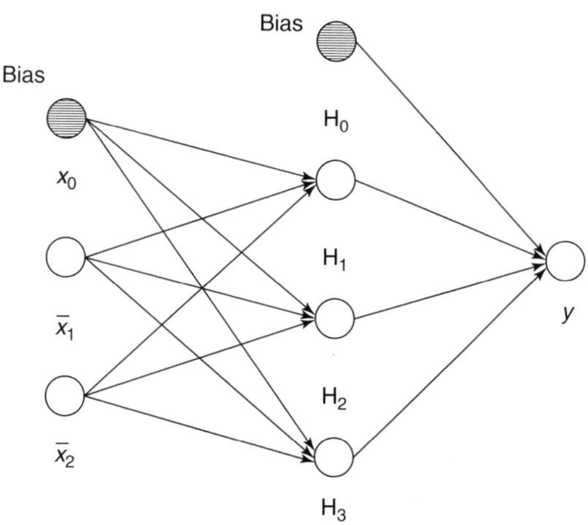

Figure 1 MLP with two input nodes, and three hidden nodes in a hidden layer (NN2(3))

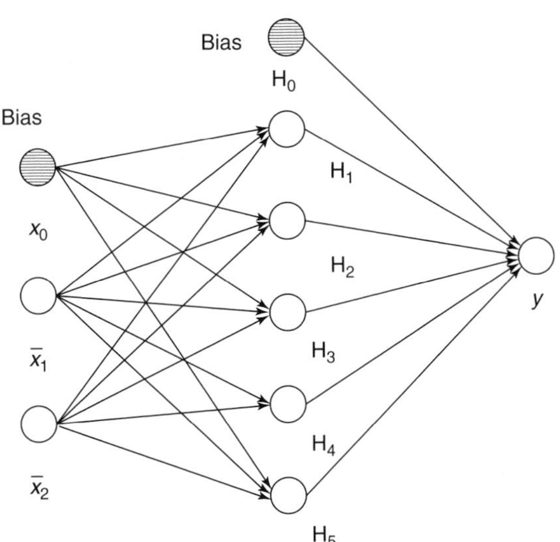

Figure 2 MLP with two input nodes, and five hidden nodes in a hidden layer (NN2(5))

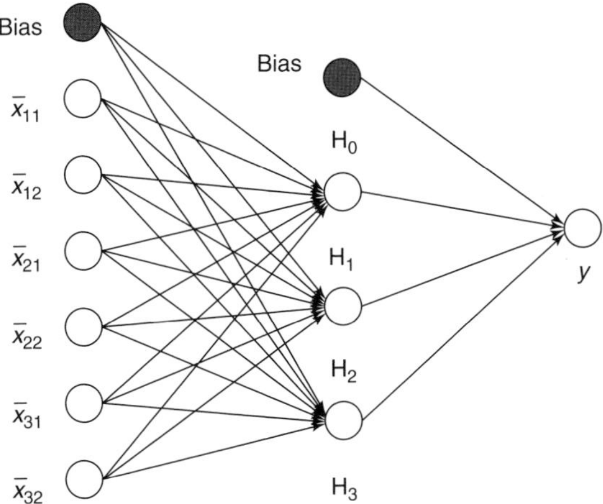

Figure 3 MLP with six input nodes, and three hidden nodes in a hidden layer (NN6(3))

point (control limit) to determine whether the process is in control. Based on the idea of a false alarm rate, α, the cutoff point is determined from the **quantiles** of the predicted values for in-control data. For example, if the false alarm rate, α, equals 0.005 and there are 10 000 and 15 000 observations of in-control data, so the cutoff point is at the 0.005 quantile of the network output distribution. Yi *et al.* [6] used this method to compare **ARL** performances between

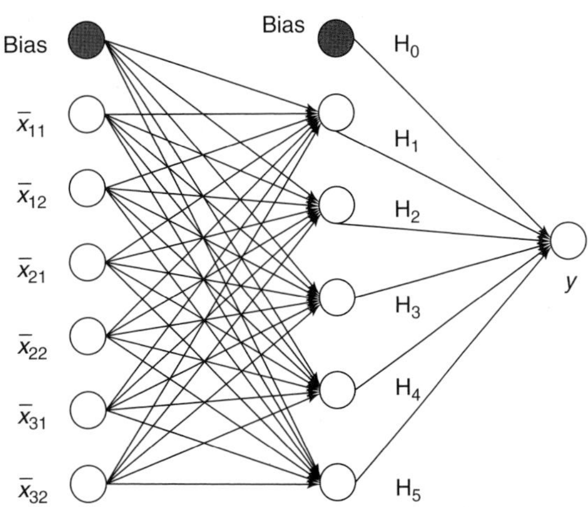

Figure 4 MLP with six input nodes, and five hidden nodes in a hidden layer (NN6(5))

MLP NNs and $\overline{X}$ control charts for investigating nonnormally distributed quality characteristics.

Since the distribution of the cutoff point is also not known, using a smoothed estimate rather than the sample quantile may be better. The smoothed estimate of the 99.995th percentile, known as the Harrell-Davis [7] estimator, might be provided a starting point for determining more accurate cutoff points through simulation.

Performance of the Illustrated NN Systems

All competing monitoring schemes are designed to provide equal in-control ARLs of 200. The out-of-control ARL of each method is used to compare performances under two different shift sizes, small and large shifts, as measured by the noncentrality parameter term, δ, as

$$\delta^2(\boldsymbol{\mu}) = (\boldsymbol{\mu} - \boldsymbol{\mu}_0)' \sum^{-1} (\boldsymbol{\mu} - \boldsymbol{\mu}_0)$$

where $\boldsymbol{\mu}_0$ is the in-control mean vector and $\boldsymbol{\mu}$ is the current mean vector. A large value of δ indicates that there are big shifts in the mean. If the value of $\delta = 0$ ($\boldsymbol{\mu} = \boldsymbol{\mu}_0$), the process is in control. In addition to the size of the shift, the performances of the NN methods are evaluated for one-variable shifts along the x-axis and two-variable shifts along the major and minor axes of the bivariate distribution. A major axis shift is defined in the eigenvector direction associated with the largest eigenvalue of the inverse of covariance (correlation) matrix and a minor axis shift is in the eigenvector direction associated with the smallest eigenvalue of the inverse of covariance (correlation) matrix. NNs are trained using two different sizes of data sets, small and large sizes, consisting of 20 000 and 30 000 observations. Monitoring schemes exhibiting best ARL performances across the various illustrated conditions are provided in Figures 5 and 6.

As for training NN with smaller training set, the NN6(3) provides the best performance at the large shift. For small shifts, the MEWMA method provides the best performance if the variables are not highly correlated. However, if the variables are highly correlated, the NN6(3) provides the best performance. Similar results are obtained for training NN with larger training sets.

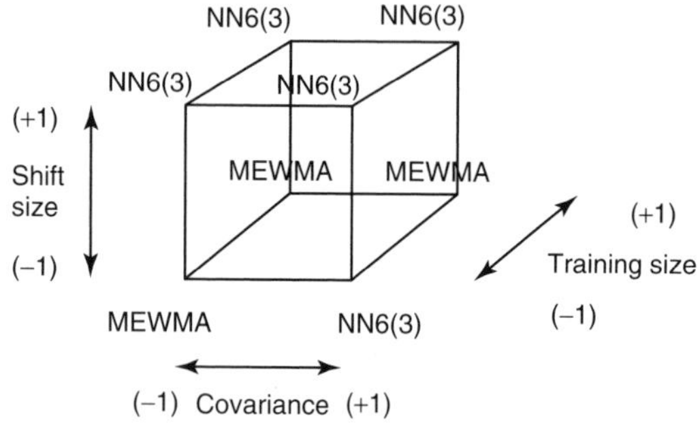

Figure 5 A graphical display of optimal methods for a one-variable shift

For two-variable shifts, four factors, covariance (A), shift size (B), training size (E), and direction of shift (F), each at two levels, are of interest. The graphical display of the optimal method at each condition is shown in Figure 6.

The interested reader can find additional information regarding the use of NN models in multivariate quality in Saithanu [8].

Strengths

The practitioner will find many advantages to using NN-based monitoring schemes. These advantages include good performance and flexibility. The flexibility of this method refers not only to relaxation of distributional assumptions associated with traditional methods, but also includes flexibility in applications. Univariate, multivariate, and regression-based applications (*see* **Regression Control Charts**), as well as special purpose systems, are legitimate candidates for NN modeling.

Weaknesses

NNs come with some disadvantages or weaknesses. Some of these disadvantages are associated with the lack of experience most practitioners have with NN models. The selection of an appropriate architecture as well as training issues may require steep learning curves for the inexperienced. There is also the matter of model interpretation that is problematic for NN models. Finally, NN models and

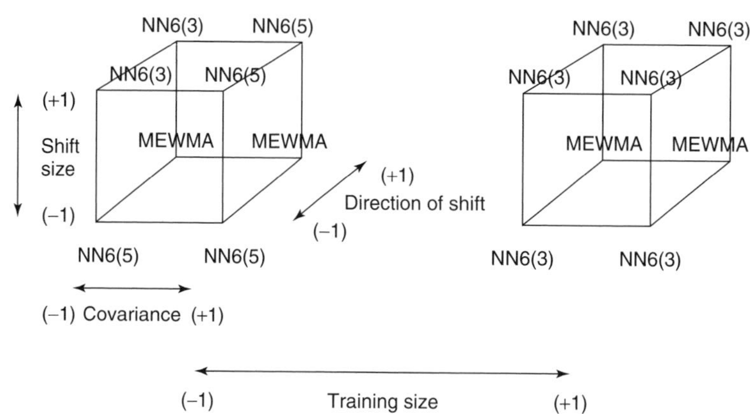

Figure 6 A graphical display of optimal methods for two-variable shifts

their development are computational intensive. These computational requirements come in several forms. Two of the most immediate issues are the need for specialized software and the need for large training datasets.

Conclusions

The NN is a very flexible tool that provides competitive performances relative to more traditional process monitoring schemes. The development of these monitoring systems has requirements that are somewhat unique to this method. The need for both in-control and out-of-control data for the training phase, large training datasets and specialized software have been mentioned in this article. The software required to design and develop NN models continues to become more mainstream and available to practitioners. In complex high volume production environments, where large data sets are common, the practitioner may be well served by NN-based monitoring systems.

References

[1] Lippmann, P.R. (1987). An introduction to computing with neural nets, *IEEE ASSP Magazine* **4**, 4–22.

[2] Pugh, G.A. (1991). A comparison of neural networks to SPC charts, *Computers and Industrial Engineering* **21**, 253–255.

[3] Chang, S.I. & Aw, C.A. (1996). A neural fuzzy control chart for detecting and classifying process mean shifts, *International Journal of Production Research* **34**, 2265–2278.

[4] Pugh, G.A. (1989). Synthetic neural networks for process control, *Computers and Industrial Engineering* **17**, 24–26.

[5] Guo, Y. & Dooley, K.J. (1992). Identification of change structure in statistical process control, *International Journal of Production Research* **30**, 1655–1669.

[6] Yi, J., Prybutok, R.V. & Clayton, R.H. (2001). ARL comparisons between neural network models and $\overline{X}$ control charts for quality characteristics that are nonnormally distributed, *Economic Quality Control* **16**, 5–15.

[7] Harrell, F.E. & Davis, C.E. (1982). A new distribution-free quantile estimator, *Biometrika* **69**, 635–640.

[8] Saithanu, K. (2006). *Neural networks and multivariate quality control*, Ph.D. Dissertation, University of Alabama, Tuscaloosa.

KIDAKAN SAITHANU

Nonhomogeneous Poisson Process

see Poisson Processes

Nonparametric and Semiparametric Bayesian Reliability Analysis

Introduction

Suppose that T is a nonnegative random variable denoting lifetime. We define the survival function at time t to be $S(t) = P(T > t) = 1 - F(t)$, where F is the cumulative distribution function (**cdf**). If T has density $f(t)$, the hazard rate function $r(t)$ at time t is given by $r(t) = \frac{f(t)}{S(t)}$. This gives the instantaneous failure rate at time t given survival just prior to time t. The corresponding cumulative hazard at time t is $R(t) = \int_0^t r(s)\,\mathrm{d}s$. It is well known that $S(t) = \exp(-R(t))$. Hence, knowing any one of density, cdf, survival function, hazard rate function, or cumulative hazard function is equivalent for inference about the properties of the life distribution. Sometimes, the phrase "reliability function" is used to denote the survival function of a component or a system of components connected in some structure.

In reliability and survival analysis, it is often of interest to estimate the reliability of the system/component from the observed lifetime data. Suppose n components are put to test and $T_1, \ldots, T_n$ are the corresponding lifetimes. If all $T_1, \ldots, T_n$ are actually observed, we have complete data. Usually, however, some of the T_i's are censored. Suppose that T_i and C_i are the lifetime and (right) censoring times, respectively, of the ith component. One observes $X_i = \min(T_i, C_i)$ and $\delta_i = I_{T_i \le C_i}$. On the basis of observed data $(X_i, \delta_i), i = 1, \ldots, n$, one would like to infer about the lifetime distribution as a whole.

No matter whether one has complete or censored data, a host of techniques have been developed for their analysis. They can be broadly classified as parametric and nonparametric approaches. The advantages of parametric methods are their simplicity in the sense that only a handful of parameters can be used to explain the behavior. For example, Kvam and Samaniego [1] used a parametric approach for analyzing the life-history data from an r-out-of-k system whereby the component lifetimes were assumed to be exponentially distributed. The unknown parameter was estimated using the maximum-likelihood method. However, as is well known, parametric models impose certain structural restrictions and are less than optimal if one is unable to verify/justify the model assumptions. An alternative is to use nonparametric methods such as the ones used by Kvam and Samaniego [1] or Chen [2] in estimating **component reliability**.

In this article, we first review some nonparametric and semiparametric Bayesian methods available for the analysis of life-history data. The advantage of using Bayesian methods is the ability to incorporate prior information in the inferential procedure, and semiparametric models allow for partial specification of model structure through parameters. We then present a semiparametric Bayesian approach for estimating the component lifetime distribution on the basis of the system lifetime data, along with additional information (such as the censoring indicator and the number of failed components) for multiple k-out-of-n systems [3].

In the section titled "Nonparametric Methods", we briefly review past work on the nonparametric Bayesian analysis for survival data. In the section titled "Semiparametric Method", we present a theoretical development of a semiparametric procedure. Under "Sampling Procedure", we briefly discuss the sampling procedure that will be used to implement the model. In the section titled "Illustration", we present the results of a simulation study based on data from Chen [2]. Finally, in the last section, we close with some concluding remarks.

Nonparametric Methods

In Bayesian analysis, it is very important to have as close to correct a likelihood function as possible. Often, the standard parametric models are not rich enough to capture the uncertainty in the observed data. In such situations, nonparametric Bayesian methods provide a more flexible alternative. Nonparametric Bayesian methods in reliability can be broadly classified into three groups, depending on the quantity on which one places a prior distribution: prior on the class of all distributions, prior on the class of all hazard rates, and prior on the class of all cumulative hazards. In each case, the underlying distribution is assumed to be free of any parameters and the prior information is combined with life-history data to obtain the corresponding posterior. Owing to the nonparametric nature, the **prior distributions** are taken to be stochastic processes. We provide a brief review of the three methods in the following subsections. For a comprehensive review of various nonparametric methods of estimating the survival function, see Ferguson *et al.* [4]. Also see Sinha and Dey [5] and Singpurwalla [6] for a detailed discussion on nonparametric approaches to reliability **estimation**.

Prior on Distributions

Dirichlet processes, introduced by Ferguson [7] are one of the fundamental concepts in the nonparametric Bayesian analysis literature. A Dirichlet process can be used to provide a nonparametric prior for a distribution function.

Suppose that F is a cdf. We say that F has a Dirichlet process prior, i.e. $F \sim \mathcal{D}(M, F_0)$ if the following happens. For any partition $A_1 \cup A_2 \cup \cdots \cup A_k = \Re$,

$$\begin{aligned}(F(A_1), F(A_2), \ldots, F(A_k)) \sim \text{Dirichlet}(&MF_0(A_1), \\ &MF_0(A_2), \ldots, \\ &MF_0(A_k)) \qquad (1)\end{aligned}$$

where Dirichlet$(\alpha_1, \ldots, \alpha_k)$ denotes a Dirichlet distribution with parameters $(\alpha_1, \ldots, \alpha_k)$. See Wilks [8] for more on Dirichlet distribution. $M > 0$ is called the *precision parameter* and F_0 is called the *baseline distribution*. F_0 can be thought of as the "average value" of F and M as the amount of "concentration" of F around F_0. A high value of M signifies that F is very close to F_0 and a low value of M signifies large dispersion around F_0. However, see Sethuraman and Tiwari [9] for another interpretation of small values of M.

Ferguson [7] showed that if $\theta | F \sim F$ and $F \sim \mathcal{D}(M, F_0)$, then $F|\theta \sim \frac{1}{M+1}[MF_0 + \delta_\theta]$. Here, δ_θ denotes the degenerate part of the distribution at a specific value of θ. Repeated application of this result shows that if $T_1, \ldots, T_n$ are observations generated by F, and F has a Dirichlet process prior $\mathcal{D}(M, F_0)$, the Bayes estimator of F is

$$\hat{F}(t) = \frac{M}{M+n} F_0(t) + \frac{1}{M+n} F_n(t) \qquad (2)$$

where $F_n(t)$ is the empirical distribution function of the sample.

The above requires complete information about the lifetimes (i.e., no censoring). Susarla and Van-Ryzin [10] developed the nonparametric Bayes estimator of the cdf in the presence of right censoring and assuming a Dirichlet process prior on the underlying distribution. Let $(X_i, \delta_i), i = 1, \ldots, n$ be the observed data with $X_i = \min(T_i, C_i)$ and $\delta_i = I_{T_i \le C_i}$. Let $u_1 < \cdots < u_k$ be the distinct values among $X_1, \ldots, X_n$ and λ_j be the number of censored observations at u_j. Let $k(t)$ be the number of u_j's that are less than or equal to u_k and h_k be the number greater than u_k. Then, the Bayes estimator of the survival function under squared error loss is given by

$$\frac{M(1 - F_0(t)) + h_k(t)}{M+n} \prod_{j=1}^{k(t)} \frac{M(1 - F_0(u_j)) + h_j + \lambda_j}{M(1 - F_0(u_j)) + h_j} \qquad (3)$$

This estimator reduces to the Kaplan–Meier estimator as $M \to 0$, and under no censoring, reduces to the estimator given earlier.

Dirichlet processes give rise to distributions as priors that are discrete with probability one. In addition, the corresponding Bayes estimator gets complicated for right-censored data. To avoid these difficulties, one may use neutral-to-the-right (NTR) priors. A random distribution function F on the real line is said to be NTR if for every m and $t_1 < t_2 < \cdots < t_m$, there exist independent random variables $V_1, \ldots, V_m$ such that $(1 - F(t_1), \ldots, 1 - F(t_m))$ has the same distribution as $(V_1, V_1V_2, \ldots, \prod_{i=1}^m V_i)$. Doksum [11] and Ferguson and Phadia [12] show that if $X_1, \ldots, X_n$ is a sample from F and F is NTR, then the posterior distribution of F given $X_1, \ldots, X_n$ is also NTR. It turns out that the censored case is simpler to treat than the uncensored case. The implementation of the results for practical applications gets cumbersome. Damien *et al.* [13] and Walker and Damien [14] have proposed simulation-based approaches for a full Bayesian analysis involving NTR processes.

Prior on Hazard Functions

An alternative approach is to put a nonparametric prior on the class of all hazard functions. Suppose we partition the time axis into $(k+1)$ intervals $0 < t_1 < \cdots < t_k < \infty$. Let $p_1 = F(t_1)$, $p_i = F(t_i) - F(t_{i-1}), i = 2, \ldots, k$. Note that $F(0) = 0$ and $F(\infty) = 1$. Define $Z_1 = p_1$, $Z_i = p_i/(1 - p_1 - \cdots - p_{i-1})$ for $i = 1, 2, \ldots, k$. Note that Z_i is the failure rate over the interval $[t_{i-1}, t_i)$ and $(Z_1, \ldots, Z_k)$ gives the piecewise constant hazard function over $[0, t_k)$. Let n_i denote the number of failures in the interval $[t_{i-1}, t_i)$, $\boldsymbol{p} = (p_1, \ldots, p_k)$ and $\boldsymbol{d} = (n_1, \ldots, n_{k+1})$. The observed data is given by $\boldsymbol{d}$.

In such a scenario, one may assume that the prior distribution of the Z_i's is independent, $\beta(\nu_{1i}, \nu_{2i})$. This results in a generalized Dirichlet prior for $\boldsymbol{p}$ given by the probability function

$$f(p_1, \ldots, p_k) = \prod_{i=1}^{k} \frac{\Gamma(\nu_{1i} + \nu_{2i})}{\Gamma(\nu_{1i})\Gamma(\nu_{2i})} p_i^{\nu_{1i}-1} \times (1 - p_1 - \cdots - p_k)_i^{\gamma} \qquad (4)$$

where $\gamma_i = \nu_{2i} - \nu_{1,i+1} - \nu_{2,i+1}$ for $i = 1, \ldots, k-1$ and $\gamma_k = \nu_{2k} - 1$. See Basu and Tiwari [15] for the special case when the generalized Dirichlet prior for $\boldsymbol{p}$ reduces to the Dirichlet prior. Given the observed count data $\boldsymbol{d}$, the posterior of the piecewise hazards turns out also to be generalized Dirichlet. Lochner [16] and Tiwari and Rao [17] further discuss the use of this approach to estimate F. This method has several disadvantages. First, it uses count data, leading to loss of information by not using the actual failure times. Second, the inferential conclusions depend on whether the cells are combined at the prior or at the posterior stage. In addition, one needs an excessive number of parameters to specify the generalized Dirichlet distribution. See also Wilks [8] for more on generalized Dirichlet distributions.

The above procedure cannot account for any structural pattern in the hazard rate function. Padgett and Wei [18] and Arjas and Gasbarra [19] take the prior on the hazard rate function to be a **Poisson process** with constant jump size. Mazzuchi and Singpurwalla [20] take the prior on hazard rate function to be ordered Dirichlet. The former ensures that the hazard rate function is nondecreasing and the latter ensures that it is monotone.

Dykstra and Laud [21] introduce an extended gamma process and use it to model the prior distribution of a nondecreasing hazard rate process. They show that when $t_1, \ldots, t_n$ are the right-censoring times of n observations and $\Gamma(a(s), \beta(s))$ is the prior on the hazard rate process; the posterior on the hazard rate process is also an extended gamma process $\Gamma(a(s), \hat{\beta}(s))$ where

$$\hat{\beta}(s) = \frac{\beta(s)}{1 + \beta(s)\sum_{i=1}^{n}(t_i - s)^+} \tag{5}$$

with $a^+ = \max(a, 0)$. When one observes the actual failure times $t_1, \ldots, t_n$, the resulting posterior hazard rate is a mixture of extended gamma processes, which is complicated to calculate. Laud *et al.* [22] approximate this posterior by approximating its random independent increments *via* a Gibbs sampler. Once the posterior hazard rate process is obtained, the corresponding survival function is given by

$$P(T \geq t) = \exp\left[-\int_0^t \log(1 + \hat{\beta}(s)(t - s))\,\mathrm{d}a(s)\right] \tag{6}$$

Other methods of specifying the prior on hazard rates include use of Markov beta and Markov gamma processes considered by Nieto-Barajas and Walker [23]. See also Tiwari and Rao [17], Tiwari and Jammalamadaka [24], and Tiwari and Kumar [25].

Prior on Cumulative Hazard

An alternative to putting a prior on the hazard rate function is to put a prior on the cumulative hazard function. This is particularly attractive especially when there is no density, since the cumulative hazard still exists in such situations. Kalbfleisch [26] proposed using a gamma process as a prior for the cumulative hazard function $R(t)$. Take any partition $0 \equiv t_0 < t_1 < t_2 < \cdots < t_{k-1} < t_k \equiv \infty$, of the time points and define $r_i = \log(1 - Z_i)$ where $Z_i = P(T \in [t_{i-1}, t_i) | T \geq t_{i-1})$ is the hazard rate over the interval $[t_{i-1}, t_i)$. Assume that r_i's are independent gamma random variables with shape $\alpha_i - \alpha_{i-1}$ and scale c, where $\alpha_i = cR^*(t_i)$ and $R^*(t)$ is interpreted as the best guess of $R(t)$. c is the measure of strength of conviction about the guess and large values indicate a strong conviction.

Given n failure times $\tau_1, \ldots, \tau_n$, the posterior cumulative hazard will be a process with independent increments. Kalbfleisch [26] has shown that the posterior cumulative hazard has an increment at τ_i which is given by a density $A(c + A_i, c + A_{i+1})$ at u, where $A_i = n - i$ and $A(a, b)$ is of the form

$$\frac{\exp(-bu) - \exp(au)}{u \log(a/b)}$$

Between τ_{i-i} and τ_i, the increments are prescribed by a gamma process with shape function $cR^*(\cdot)$ and scale $c + A_i$. The survival function is recovered either by simulation or by approximation using expected value of the process.

The above formulation suffers from some difficulties. First, there is a lack of intuition regarding the assumption of gamma distribution on r_i. Second, the independent increments property of $R(t)$ may not be meaningful, since under aging and wear, the successive Z_i's would be judged to be increasing. Finally, presence of ties in the failure time data presents problems in the model fitting.

A different model was proposed by Hjort [27] to model the randomness in the cumulative hazard rate. Suppose R has a beta process prior with parameters $c(\cdot)$ and $R_0(\cdot)$. The posterior of R given life-history data $(X_1, \delta_1), \ldots, (X_n, \delta_n)$ is also a beta process of the form

$$R|\text{data} \sim \beta\left[c(\cdot) + Y(\cdot), \int_0^{(\cdot)} \frac{c\,\mathrm{d}R_0 + \mathrm{d}N}{c + Y}\right] \tag{7}$$

where

$$N(t) = \sum_{i=1}^{n} I(X_i \leq t, \delta_i = 1) \tag{8}$$

and

$$Y(t) = \sum_{i=1}^{n} I(X_i \geq t) \tag{9}$$

are two counting processes derived from the data. The Bayes estimator of $R(t)$ under squared error loss is

$$\hat{R}(t) = \int_0^t \frac{c\,\mathrm{d}R_0 + \mathrm{d}N}{c + Y} \tag{10}$$

and the corresponding estimator of F is

$$\hat{F}(t) = 1 - \prod_{[0,t]}\left[1 - \frac{c\,\mathrm{d}R_0 + \mathrm{d}N}{c + Y}\right] \tag{11}$$

As $c(\cdot)$ decreases to zero, $\hat{F}$ tends to the Kaplan–Meier estimator.

Semiparametric Method

Often, it may be necessary to impose certain structural restrictions on the underlying survival/reliability model to aid in the physical understanding and interpretation, all the while maintaining considerable generality. In such situations, one may decide to use semiparametric models, whereby parts of the model are parametric (reflecting the desired structural restriction) and the remainder nonparametric. An example of this is the **Cox** model, introduced by Cox [28]. To perform Bayesian analysis, the nonparametric part is assumed to be a realization of a stochastic process. As before in the nonparametric Bayesian analysis case, one can put nonparametric priors on the hazard rates. Different methods have been discussed in Sinha and Dey [5] for dealing with these models.

Recently, Merrick *et al.* [29] developed a semiparametric Bayesian proportional hazards model for reliability and maintenance of machine tools. Their proposed model uses a mixture of Dirichlet process (MDP) prior for the baseline failure rate in the proportional hazards model. Such priors were introduced by Antoniak [30] and have been popularized by various authors such as MacEachern [31] and West *et al.* [32]. Another application of Bayesian semiparametric models in reliability is presented in Merrick *et al.* [33], where the authors use a proportional intensities model and present a semiparametric Bayesian approach for nonhomogeneous Poisson processes. Apart from the above, we were unable to find Bayesian semiparametric models in the context of reliability analysis. This emphasizes the fact that while Bayesian semiparametric models have been popular in survival analysis, they have not been as popular in (engineering) reliability analysis. Below, we present a new model to estimate component reliability using system reliability data from multiple r-out-of-k systems.

In industrial and biological problems, one often has a multicomponent system (consisting of, say, k components) and observes the time of system failure. Owing to the system architecture, failure of the system occurs if and only if at least a certain number of components (say, r) fail. For example, a liquid crystal display (LCD) may be said to correctly function when at least 80% of its pixels function properly. Such a system is called an *r-out-of-k system*, special cases of which are a series system ($r=1$) and a parallel system ($r=k$). r-out-of-k systems have been well studied in the reliability context and are a favorite way of increasing system **redundancy**. See Hoyland and Rausand [34] for more details on r-out-of-k systems and associated examples.

Suppose that we have an r-out-of-k system that consists of k components with identical life distributions that act independent of each other. Assume that the component life distribution is given by $F(\cdot|\theta)$ where θ is an unknown p-dimensional parameter. Suppose that T is the system failure time and C is the censoring time. We observe $X = T \wedge C$ and $\delta = I(X = T)$, the censoring indicator. When $\delta = 1$, X is distributed as the rth-order statistic based on a sample of size k from $F(\cdot|\theta)$. When $\delta = 0$ (i.e., the system was alive and censored at X), we may or may not have information on the number of components $s(< r)$ in the system that have failed. Let $\gamma = 1$ indicate that s is observed and $\gamma = 0$ indicate that s is unobserved. Assuming that $F(\cdot|\theta)$ is absolutely continuous with respect to the Lebesgue measure, the likelihood contribution of the system is

$$L(\theta) = \left[r\binom{k}{r}F^{r-1}(x|\theta)f(x|\theta)\{1-F(x|\theta)\}^{k-r}\right]^{\delta} \left[\binom{k}{s}F^{s}(x|\theta)\{1-F(x|\theta)\}^{k-s}\right]^{(1-\delta)\gamma} \left[\sum_{s=0}^{r-1}\binom{k}{s}F^{s}(x|\theta)\{1-F(x|\theta)\}^{(k-s)}\right]^{(1-\delta)(1-\gamma)} \quad (12)$$

When $\delta = 1$, we take $\gamma = 1$ and $s = r$.

Suppose that for $i = 1, \ldots, n$, we have m_i copies of an r_i-out-of-k_i system. The m_i copies are assumed to be independently distributed with the same component distribution $F(\cdot|\theta_i)$. For $i \neq j$, the systems are also assumed to be conditionally independent of each other and the parameters θ_i and θ_j may or may not be equal.

The resulting data $\boldsymbol{X}$ is presented in Table 1.

Table 1 Failure time data from several r-out-of-k systems

System	r	k	Observations		
1	r_1	k_1	$(X_{11}, \delta_{11}, \gamma_{11}, s_{11})$	...	$(X_{1m_1}, \delta_{1m_1}, \gamma_{1m_1}, s_{1m_1})$
2	r_2	k_2	$(X_{21}, \delta_{21}, \gamma_{21}, s_{21})$	...	$(X_{2m_2}, \delta_{2m_2}, \gamma_{2m_2}, s_{2m_2})$
...	...	...	...	...	...
n	r_n	k_n	$(X_{n1}, \delta_{n1}, \gamma_{n1}, s_{n1})$	...	$(X_{nm_n}, \delta_{nm_n}, \gamma_{nm_n}, s_{nm_n})$

Denoting $\boldsymbol{\theta}_n = (\theta_1, \ldots, \theta_n)$, the likelihood is given by

$$L(\boldsymbol{\theta}_n|\boldsymbol{X}) = \prod_{i=1}^{n}\prod_{j=1}^{m_i}\left[r_i\binom{k_i}{r_i}\{F(x_{ij}|\theta_i)\}^{r_i-1}f(x_{ij}|\theta_i) \times \{1 - F(x_{ij}|\theta_i)\}^{k_i-r_i}\right]^{\delta_{ij}}$$
$$\times \left[\binom{k_i}{s_{ij}}\{F(x_{ij}|\theta_i)\}^{s_{ij}}\{1 - F(x_{ij}|\theta_i)\}^{k_i-s_{ij}}\right]^{(1-\delta_{ij})\gamma_{ij}}$$
$$\times \left[\sum_{s=0}^{r_i-1}\binom{k_i}{s}\{F(x_{ij}|\theta_i)\}^{s}\{1-F(x_{ij}|\theta_i)\}^{k_i-s}\right]^{(1-\delta_{ij})(1-\gamma_{ij})} \quad (13)$$

We assume that the θ_i's are independent and identically distributed from a distribution G with a Dirichlet process prior having baseline distribution G_0 and precision M. Hence the prior distribution of $\theta_1, \ldots, \theta_n$ assuming M and G_0 are known is given by (see Antoniak [30], Blackwell and MacQueen [35]):

$$\pi(\boldsymbol{\theta}_n) = \prod_{i=1}^{n}\left[\frac{MG_0(\mathrm{d}\theta_i) + \sum_{j<i}\delta_{\theta_j}(\mathrm{d}\theta_i)}{M + i - 1}\right] \quad (14)$$

As mentioned earlier, under the Dirichlet process setup, some of the system parameters θ_i may be identical. This is a reflection of the fact that some of the systems may be built using components from the same manufacturer and thus would have similar behavior.

Our goal will be to estimate the reliability of a component using the predictive approach. Assuming that a future system has a parameter θ_{n+1} and that X_{n+1} is the lifetime of a component of the system, we want

$$\begin{aligned} S(x) &= P(X_{n+1} > x|\boldsymbol{X}) \\ &= \int P(X_{n+1} > x|\theta_{n+1}, \boldsymbol{X}, \boldsymbol{\theta}_n) \times f(\theta_{n+1}|\boldsymbol{\theta}_n, \boldsymbol{X})f(\boldsymbol{\theta}_n|\boldsymbol{X})\,\mathrm{d}\boldsymbol{\theta}_{n+1} \\ &= \int \overline{F}_{n+1}(x|\theta_{n+1})\frac{1}{M+n}[MG_0(\mathrm{d}\theta_{n+1}) + \sum_{j=1}^{n}\delta_{\theta_j}(\mathrm{d}\theta_{n+1})]f(\boldsymbol{\theta}_n|\boldsymbol{X}_n)\,\mathrm{d}\boldsymbol{\theta}_n \\ &= \frac{1}{M+n}\int[M\int\overline{F}(x|\theta)G_0(\mathrm{d}\theta) + \sum_{j=1}^{n}\overline{F}(x|\theta_j)]f(\boldsymbol{\theta}_n|\boldsymbol{X})\,\mathrm{d}\boldsymbol{\theta}_n \end{aligned} \quad (15)$$

Note that the complicated nature of the likelihood precludes any conjugate choice of baseline prior to simplify the posterior distribution calculations.

Ferguson [36], Kuo [37], and Tiwari and Kumar [25] have used this type of mixture approach for estimating parameters such as density function and reliability. We will use the Gibbs sampler to sample from the posterior and draw inferences. The Gibbs sampler is difficult to implement as is, since the likelihood contributions involve calculations of the cdf, the **probability density function**, and/or sums involving them. While several algorithms have been proposed to deal with such nonconjugate setups for sampling from the posterior in an MDP setting [38], they are computationally intensive.

The problem arises because we have several unobserved lifetimes, essentially making data incomplete. Here, we introduce a data-augmentation technique (see Tanner and Wong [39], Tanner [40]) whereby the observed data are augmented to get the "complete" data $\widetilde{\boldsymbol{X}} = \{X_{ijl}\}$, i.e. the exact failure times of all the components in a system. The likelihood based on the augmented data can then be written as

$$\widetilde{L}(\boldsymbol{\theta}_n|\widetilde{X}) = \prod_{i=1}^{n}\prod_{j=1}^{m_i}\prod_{l=1}^{k_i} f(x_{ijl}|\theta_i) \qquad (16)$$

Similar techniques were also used in Kim and Arnold [41]. This simplifies calculations by taking advantage of the conjugate structure of the likelihood and the baseline prior. Note that the posterior distribution of the number of failed components s_{ij} for systems in which such data are missing arises naturally and can be interpreted as the posterior based on a uniform prior on $\{0, \ldots, k_i\}$.

Sampling Procedure

1. Generate a set of starting values of $\theta_1, \ldots, \theta_n$.
2. If $\delta_{ij} = \gamma_{ij} = 0$, sample s_{ij} from binomial(k_i, $F(X_{ij}|\theta_i)$) truncated to $\{0, \ldots, r_i - 1\}$.
3. If $\delta_{ij} = 1$, generate $\{X_{ij(l)}\}_{l=1}^{r-1}$ as order statistics from

$$g_{\mathrm{L}}(x|X_{ij}, \theta_i) = \frac{f(x|\theta_i)}{F(X_{ij}|\theta_i)}, \quad 0 < x < X_{ij} \qquad (17)$$

Also generate $\{X_{ij(l)}\}_{l=r+1}^{k_i}$ as order statistics from

$$g_{\mathrm{U}}(x|X_{ij}, \theta_i) = \frac{f(x|\theta_i)}{1 - F(X_{ij}|\theta_i)}, \quad X_{ij} < x < \infty \qquad (18)$$

Also, set $X_{ij(r)} \equiv X_{ij}$.

Note that $g_{\mathrm{L}}(\cdot|X_{ij}, \theta_i)$ is the density of observations from $F(\cdot|\theta_i)$ truncated above at X_{ij}. Similarly, $g_{\mathrm{U}}(\cdot|X_{ij}, \theta_i)$ is the density for observations that are truncated below.

4. If $\delta_{ij} = 0$, generate $\{X_{ij(l)}\}_{l=1}^{s_{ij}}$ as order statistics from $g_{\mathrm{L}}(\cdot|X_{ij}, \theta_i)$ and also generate $\{X_{ij(l)}\}_{l=s_{ij}+1}^{k_i}$ as order statistics from $g_{\mathrm{U}}(\cdot|X_{ij}, \theta_i)$.
5. Having generated the "complete data" $\{X_{ijl}\}$, we update the θ's conditionally as

$$\theta_i|\text{ rest} \sim \prod_{j=1}^{m_i}\prod_{l=1}^{k_i} f(X_{ijl}|\theta_i) \times \left[\frac{MG_0(\mathrm{d}\theta_i) + \sum_{j\neq i}\delta_{\theta_j}(\mathrm{d}\theta_i)}{M + n - 1}\right] \qquad (19)$$

This is done using the Gibbs sampler. Note that in the above expression, we need the exact failure times X_{ijl}, not the ordered values $X_{ij(l)}$. If the sufficient statistic does not depend on the ordering of the observations, one can replace the exact values by the ordered values. Such is the case, for example, when one is interested in inferring the mean of a normal or the scale of a Weibull distribution.

6. Go back to 2 and repeat until convergence to the steady state.

Below, we give a special case for illustration:

Lognormal Distribution

If the component failure times are distributed as lognormal(μ, σ^2) $(LN(\mu, \sigma^2))$, we can generate samples from the truncated distributions given by $g_{\mathrm{L}}(\cdot)$ and $g_{\mathrm{U}}(\cdot)$ using

$$X_{\mathrm{L}} = \exp\left[\mu + \sigma\Phi^{-1}\left(U\Phi\left(\frac{\log X - \mu}{\sigma}\right)\right)\right] \qquad (20)$$

and

$$X_{\mathrm{U}} = \exp\left[\mu + \sigma\Phi^{-1}\left(U + (1 - U)\Phi \times \left(\frac{\log X - \mu}{\sigma}\right)\right)\right] \qquad (21)$$

where $U \sim U(0, 1)$. Note that the random variables X_{L} and X_{U} above are generated by inverting the cdf's corresponding to g_{L} and g_{U}, respectively.

In this case,

$$\prod_{j=1}^{m_i}\prod_{l=1}^{k_i} f(x_{ij(l)}|\theta_i) = \frac{1}{(2\pi\sigma^2)^{m_i k_i/2}}\frac{1}{\prod_{j,l} x_{ij(l)}} \times \exp\left[-\frac{1}{2\sigma^2}\sum_{j,l}\left(\log x_{ij(l)} - \mu_i\right)^2\right] \qquad (22)$$

Note that $\sum_{j=1}^{m_i}\sum_{l=1}^{k_i}\log x_{ij(l)} = \sum_{j=1}^{m_i}\sum_{l=1}^{k_i}\log x_{ijl}$ is the sufficient statistic and is not order dependent.

Illustration

We used the system configuration given in Table 2 to generate our data [see 2].

Table 2 Configuration of several r-out-of-k systems

i	1	2	3	4	5	6
r_i	1	3	5	1	4	7
k_i	5	5	5	7	7	7
m_i	10	9	11	12	10	13

To generate an observation on an r_i-out-of-k_i system, a simple random sample of size k_i is generated from F and another simple random sample of size k_i is generated from G. Let $u_{i(r_i)}$ and $v_{i(r_i)}$ be the r_ith-order statistics from the two samples, respectively. If $u_{i(r_i)} \leq v_{i(r_i)}$, the generated observation is taken as $(X_i, \delta_i, \gamma_i, s_i) = (u_{i(r_i)}, 1, 1, r_i)$. Otherwise, let r be the rank of the largest-order statistic from F that is smaller than $v_{i(r_i)}$. With 90% probability, the generated observation is taken as $(X_i, \delta_i, \gamma_i, s_i) = (v_{i(r_i)}, 0, 1, r)$ and with 10% probability, the generated observation is taken as $(X_i, \delta_i, \gamma_i, s_i) = (v_{i(r_i)}, 0, 0, 0)$. We used lognormal distribution with $\mu_{01} = 1, \sigma_{01} = 1$ as the F and a lognormal distribution with $\mu_{02} = \{0.8, 2.8\}, \sigma_{02} = 1$ as G. This procedure was seen to give rise to 68% censoring when $\mu_{02} = 0.8$ and about 0% censoring when $\mu_{02} = 2.8$. As indicated earlier, for a censored lifetime, the number of failed components was noted with probability 0.9 in each of the censoring schemes.

Once the dataset was generated according to the above procedure, we estimated the reliability of a single component $S(x)$ as outlined in equation (15). We assumed that the underlying population is $LN(\mu, 1)$ with $\mu \sim \mathcal{D}(M, G_0)$ and $G_0 \sim N(1, 1)$. We also assumed $M \sim \gamma(0.1, 0.1)$ and ran a full Bayesian approach. The updating of M was done along the lines of Escobar and West [42] and for improved mixing, the μ_i's were updated using "Algorithm 3" in Neal [38]. The predictive reliability is based on 5000 runs of the Gibbs sampler with a thinning of 10, obtained after a **burn-in** of 5000 iterations. Convergence was ascertained using CODA [43]. The results are presented in Figure 1. The precision of the estimate is measured by RMSE (root mean square error), which is the RMSE of the estimate at selected points and is given in Table 3.

We see that as the percentage of censoring increases, the estimated values get farther away from the true values. In the case where $\mu_{02} = 0.8$, we also kept track of the true (but unobserved) number of component failures at the data generation stage and the estimated values at the simulation stage. The average number of failures turned out to be (0, 0, 2.18, 2.77, 5.70, 5.97), while the true values are (0, 0, 2, 1, 6, 6).

Conclusion

We have presented a brief overview of nonparametric and semiparametric Bayes methods in lifetime data analysis. In the semiparametric method outlined here,

Table 3 Accuracy of the estimates based on data from various censoring schemes

μ_{02}	*RMSE*	Censoring proportion (%)
0.8	0.055	67.69
2.8	0.048	0

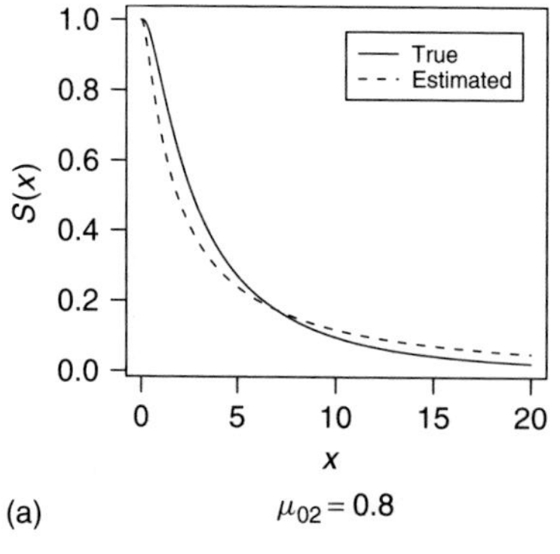

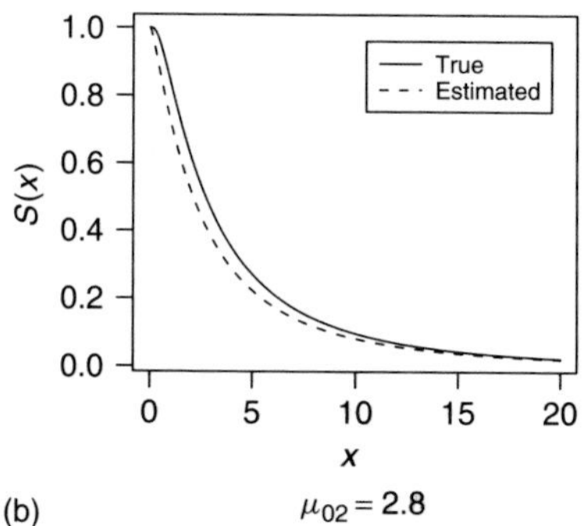

Figure 1 Estimated reliability of a component from the system configuration in Table 2 using different censoring distributions

one can, as a by-product, infer the number of failed components for a censored system lifetime when it is unobserved. The method of data augmentation can be used when the sufficient statistic is not order dependent–otherwise one can always use the original likelihood and use one of the nonconjugate sampling methods outlined in Neal [38]. This discussion is not exhaustive but is intended to give a flavor of the current state of the art. Owing to advances in computing, realistic models, which were once avoided because of difficulty in implementation, will be becoming more popular.

References

[1] Kvam, P.H. & Samaniego, F.J. (1993). On maximum likelihood estimation based on ranked set samples with applications to reliability, in *Advances in Reliability*, Elsevier Science B. V., The Netherlands, pp. 215–229.

[2] Chen, Z. (2003). Component reliability analysis of k-out-of-n systems with censored data, *Journal of Statistical Planning and Inference* **116**, 305–315.

[3] Boyles, R.A. & Samaniego, F.J. (1987). On estimating component reliability for systems with random redundancy levels, *IEEE Transactions in Reliability* **R-36**, 403–407.

[4] Ferguson, T.S., Phadia, E.G. & Tiwari, R.C. (1992). Bayesian nonparametric inference, in *Current Issues in Statistical Inference: Essays in Honor of D. Basu, No. 17 in IMS Lecture Notes Monograph Series*, Institute of Mathematical Statistics, Hayward, pp. 127–150.

[5] Sinha, D. & Dey, D.K. (1997). Semiparametric Bayesian analysis of survival data, *Journal of the American Statistical Association* **92**, 1195–1212.

[6] Singpurwalla, N.D. (2006). *Reliability and Risk: A Bayesian Perspective*, John Wiley & Sons, New York.

[7] Ferguson, T.S. (1973). A Bayesian analysis of some nonparametric problems, *The Annals of Statistics* **1**, 209–230.

[8] Wilks, S.S. (1962). *Mathematical Statistics*, John Wiley & Sons, New York.

[9] Sethuraman, J. & Tiwari, R.C. (1982). Convergence of Dirichlet measures and the interpretation of their parameter, in *Statistical Decision Theory and Related Topics III*, Academic Press, Vol. 2, pp. 305–315.

[10] Susarla, V. & VanRyzin, J. (1976). Nonparametric Bayesian estimation of survival curves from incomplete observations, *Journal of the American Statistical Association* **71**, 740–754.

[11] Doksum, K.A. (1974). Tailfree and neutral random probabilities and their posterior distributions, *The Annals of Probability* **2**, 183–201.

[12] Ferguson, T.S. & Phadia, E.G. (1979). Bayesian nonparametric estimation based on censored data, *The Annals of Statistics* **7**, 163–176.

[13] Damien, P., Laud, P.W. & Smith, A.F.M. (1995). Approximate random variate generation from infinitely divisible distributions with applications to Bayesian inference, *Journal of the Royal Statistical Society. Series B* **57**, 547–563.

[14] Walker, S. & Damien, P. (1998). A full Bayesian non-parametric analysis involving neutral to the right process, *Scandinavian Journal of Statistics* **25**, 669–680.

[15] Basu, D. & Tiwari, R.C. (1982). A note on the Dirichlet process, in *Statistics and Probability: Essays in Honor of C. R. Rao*, North-Holland, New York, pp. 89–103.

[16] Lochner, R.H. (1975). A generalized Dirichlet distribution in Bayesian life testing, *Journal of the Royal Statistical Society. Series B* **37**, 103–113.

[17] Tiwari, R.C. & Rao, J.S. (1983). Bayesian nonparametric estimation of failure rates with censored data, *Calcutta Statistical Association Bulletin* **32**, 79–90.

[18] Padgett, W.J. & Wei, L.J. (1981). A Bayesian nonparametric estimator of survival probability assuming increasing failure rate, *Communications in Statistics: Theory and Methods* **A10**, 49–63.

[19] Arjas, E. & Gasbarra, D. (1994). Nonparametric Bayesian inference from right-censored survival data using the Gibbs sampler, *Statistica Sinica* **4**, 505–524.

[20] Mazzuchi, T.A. & Singpurwalla, N.D. (1985). A Bayesian approach to inference for monotone failure rates, *Statistics and Probability Letters* **3**, 135–142.

[21] Dykstra, R.L. & Laud, P. (1981). A Bayesian approach to reliability, *The Annals of Statistics* **9**, 356–367.

[22] Laud, P.W., Smith, A.F.M. & Damien, P. (1996). Monte Carlo methods for approximating a posterior hazard rate process, *Statistics and Computing* **6**, 77–83.

[23] Nieto-Barajas, L.E. & Walker, S.G. (2002). Markov beta and gamma processes for modelling hazard rates, *Scandinavian Journal of Statistics* **29**, 413–424.

[24] Tiwari, R.C. & Jammalamadaka, S.R. (1985). Estimation of survival function and failure rates with censored data, *Statistics* **16**, 535–540.

[25] Tiwari, R.C. & Kumar, S. (1989). Bayes reliability estimation under a random environment governed by a Dirichlet prior, *IEEE Transactions on Reliability* **R-37**, 218–223.

[26] Kalbfleisch, J.D. (1978). Non-parametric Bayesian analysis of survival time data, *Journal of the Royal Statistical Society. Series B* **40**, 214–221.

[27] Hjort, N.L. (1990). Nonparametric Bayes estimators based on beta processes in models for life history data, *The Annals of Statistics* **18**, 1259–1294.

[28] Cox, D.R. (1972). Regression models and life tables, *Journal of the Royal Statistical Society. Series B* **34**, 187–220.

[29] Merrick, J.R.W., Soyer, R. & Mazzuchi, T.A. (2003). A Bayesian semiparametric analysis of the reliability and maintenance of machine tools, *Technometrics* **45**, 58–69.

[30] Antoniak, C.E. (1974). Mixtures of Dirichlet processes with applications to Bayesian nonparametric problems, *The Annals of Statistics* **2**, 1152–1174.

[31] MacEachern, S.N. (1994). Estimating normal means with a conjugate-style Dirichlet process prior, *Communications in Statistics: Simulation and Computation* **23**, 727–741.

[32] West, M., Muller, P. & Escobar, M.D. (1994). Hierarchical priors and mixture models with application in regression density estimation, in *Aspects of Uncertainty: A Tribute to D. V. Lindley*, John Wiley & Sons, London, pp. 363–368.

[33] Merrick, J.R., Soyer, R. & Mazzuchi, T.A. (2005). Are maintenance practices for railroad tracks effective? *Journal of the American Statistical Association* **100**, 17–25.

[34] Hoyland, A. & Rausand, M. (1994). *System Reliability Theory: Models and Statistical Methods*, John Wiley & Sons, New York.

[35] Blackwell, D. & MacQueen, J.B. (1973). Ferguson distribution via Polya urn schemes, *The Annals of Statistics* **1**, 353–355.

[36] Ferguson, T.S. (1983). Bayesian density estimation by mixtures of normal distributions, in *Recent Advances in Statistics: Papers in Honor of Herman Chernoff on his Sixtieth Birthday*, Academic Press, New York; London, pp. 287–302.

[37] Kuo, L. (1986). Computations of mixtures of Dirichlet processes, *SIAM Journal of Scientific Statistical Computing* **7**, 60–71.

[38] Neal, R.M. (2000). Markov chain sampling methods for Dirichlet process mixture models, *Journal of Computational and Graphical Statistics* **9**, 249–265.

[39] Tanner, M.A. & Wong, W.H. (1987). The calculation of posterior distributions by data augmentation, *Journal of the American Statistical Association* **82**, 528–540.

[40] Tanner, M.A. (1993). *Tools for Statistical Inference*, Springer, New York.

[41] Kim, Y. & Arnold, B.C. (1999). Parameter estimation under generalized ranked set sampling, *Statistics and Probability Letters* **42**, 353–360.

[42] Escobar, M.D. & West, M. (1995). Bayesian density estimation and inference using mixtures, *Journal of the American Statistical Association* **90**, 577–588.

[43] Best, N.G., Cowles, M.K. & Vines, K. (1995). *CODA: Convergence Diagnosis and Output Analysis Software for Gibbs Sampling Output, Version 0.30*, Technical Report, MRC Biostatistics Unit, University of Cambridge.

Related Articles

Cumulative Damage Models Based on Gamma Processes; Dirichlet Process, Simulation of; *k*-out-of-*n* Systems; Markov Chain Monte Carlo, Introduction; Nonparametric Methods for Analysis of Repair Data; Poisson Processes; Survival Analysis, Nonparametric.

Kaushik Ghosh and Ram C. Tiwari

Nonparametric Methods for Analysis of Repair Data

Introduction

Engineering systems and human systems are subject to failure. Rather than replacing the failed system by a new system, repair is implemented because substituting a new item for the failed item is impractical, not feasible, or too expensive. Thus we are led to the situation where a system, undergoing continued maintenance by having some type of repair applied after each failure, yields a data set consisting of the failure times and the type of repair administered at each failure.

A repair model is a stochastic formulation of the joint distribution of the failure times, or equivalently, the joint distribution of the interfailure times. We let $\{S_j\}$ denote the failure times of the system and $\{T_j\}$, the interfailure times, where $T_j = S_j - S_{j-1}$ and $S_0 \stackrel{\text{def}}{=} 0$. Let F be the distribution of the time to first failure of the system. It is of interest to estimate F (or equivalently the survival function $\overline{F} = 1 - F$) nonparametrically from the failure times and knowledge of the type of repairs made after each failure. Other parameters, such as the expected values of the interfailure times and other features of the repair process, can be expressed in terms of F. **Estimation** of F is not straightforward because data that arise from repair processes will typically have dependencies that are induced, for example, by the nature of the repair, other environmental factors, and the specification of the period of observation.

In the section titled "Repair Models" we discuss several of the prominent repair models in use. In the section titled "A General Repair Model" we address nonparametric estimation of F and describe a nonparametric asymptotic simultaneous confidence band for F. The section titled "Additional Nonparametric Methods" considers related nonparametric estimation and testing procedures for **repair data**.

Repair Models

In the *perfect repair model*, a failed system is replaced by a new one having the same stochastic properties as the original. That is, $\{T_j\}$ are independent and identically distributed (i.i.d.) according to F. The imperfect repair model known as the **minimal repair** *model*, due to Ascher [1] (*see* **Imperfect Repair**), specifies that the repaired system has the same distribution as an operational new system with the age of the recently failed system. In symbols, the survival distribution function of the minimally repaired system is given by $\overline{F}_t$, where

$$\overline{F}_t(s) = \frac{1 - F(s+t)}{1 - F(t)}, \quad s > 0 \tag{1}$$

and t denotes the time of the previous failure. The Brown and Proschan (BP) [2] repair model is a generalization of the minimal repair model. The BP model is an imperfect repair model where at the time of each repair, a perfect repair is performed with probability p and a minimal repair is performed with probability $1 - p$. The Block, Borges, and Savits (BBS) [3] model generalizes the BP model by allowing the probability of a perfect repair to depend on the age of the failed system. Kijima [4] introduced two models, models I and II, that allow for repairs to be better than minimal but not as good as perfect repairs. In the Kijima models, the effective age to which the system is restored after repair depends on the effective age just before failure and the degree of repair random variables. Kijima's models I and II are special cases of a repair model considered by Uematsu and Nishida [5]. Dorado, Hollander, and Sethuraman (DHS) [6] defined a general repair that contains many popular repair models and introduces many others. Because of its generality, we focus on it in the section titled "A General Repair Model".

A General Repair Model

The DHS [6] model of imperfect repair is based on the family of survival distributions

$$\overline{F}_{a,\theta}(x) \stackrel{\text{def}}{=} \frac{\overline{F}(\theta x + a)}{\overline{F}(a)} \tag{2}$$

where θ is a positive parameter. This family enjoys a stochastic ordering property, namely, $\theta < \theta'$ implies $F_{a,\theta} \stackrel{\text{st}}{\geq} F_{a,\theta'}$ for each a. Thus lower values of θ imply longer remaining life. The DHS model is based on two sequences $\{A_j\}$, $\{\theta_j\}$ where the A's are called *effective ages* and the θ's are known as *life supplements*. The A's and θ's are assumed to satisfy

$$A_1 = 0,\ \theta_1 = 1,\ A_j \geq 0,\ \theta_j \in (0, 1] \text{ and}$$
$$A_j \leq A_{j-1} + \theta_{j-1} T_{j-1},\ j \geq 2 \tag{3}$$

The DHS model is completely described by the joint distribution of the interfailure times given by

$$P(T_j \leq t | A_1, \ldots, A_j, \theta_1, \ldots, \theta_j, T_1, \ldots, T_{j-1}) = F_{A_j,\theta_j}(t) \tag{4}$$

Note that, for $j \geq 1$, the effective age A_{j+1} of the system after the jth repair is less than the effective age $X_j \stackrel{\text{def}}{=} A_j + \theta_j T_j$ just before the jth failure, and because $\theta_j \leq 1$, X_j is less than the actual age S_j. The DHS model also encompasses an i.i.d. renewal model studied by Peña *et al.* [7] and Hollander and Peña [8].

All of the repair models mentioned in the section titled "Repair Models" are special cases of the DHS model. For example, setting $\theta_j = 1$, $A_j = S_{j-1}$ yields the minimal repair model. For the specifications of the A's and θ's that yield the other models of the section titled "Repair Models" and also many new repair models, see Dorado *et al.* [6].

DHS develops asymptotic inferential procedures for model (4) by invoking an analogy to a survival model with censored data. We refer the reader to Dorado *et al.* [6] for details and only summarize the results here.

DHS fixes a $T > 0$ and defines two processes $N(\cdot)$ and $Y(\cdot)$ by

$$N(t) = \sum_j I\{X_j \leq t, S_j \leq t\} \tag{5}$$

and

$$Y(t) = \sum_j I\{A_j < t \leq (X_j \wedge [A_j + \theta_j (T - S_{j-1})])\} \tag{6}$$

where $X_j = A_j + \theta_j T_j$ is the effective age just before the jth failure. DHS assumes that n independent copies of the processes N and Y are observed on a finite interval. Let N_n and Y_n denote the sum of the n independent copies of N and Y, respectively. Let

Λ denote the cumulative hazard function of F. The Nelson–Aalen estimator of Λ of is

$$\hat{\Lambda}_n(t) = \int_0^t \frac{J_n}{Y_n} \, \mathrm{d}N_n \tag{7}$$

where $J_n(t) = I(Y_n(t) > 0)$. The DHS estimator of $\overline{F}$ is

$$\hat{\overline{F}}_n(t) = \prod_{s \le t} (1 - \mathrm{d}\hat{\Lambda}(s)) \tag{8}$$

where $\prod_{s \le t}(1 - \mathrm{d}\hat{\Lambda}(s))$ denotes the product integral (see Gill and Johansen [9]). The estimate $\hat{F}_n(t)$ can be shown to be the solution of the Volterra integral equation

$$\hat{F}_n(t) = \int_0^t (1 - \hat{F}_n(s-)) \, \mathrm{d}\hat{\Lambda}_n(s) \tag{9}$$

DHS obtained weak convergence results for $\hat{F}_n(t)$ and also derived a simultaneous confidence band for F. For $t \in [0, T]$, let $L_n = I(\overline{F}_n(t) < 1)$ and set

$$\hat{C}_n(t) = \int_0^t \frac{J_n L_n}{(Y_n/n)(1 - \hat{F}_n)} \, \mathrm{d}\hat{F}_n,$$

$$\hat{K}_n(t) = \frac{\hat{C}_n(t)}{1 + \hat{C}_n(t)} \tag{10}$$

For t such $\hat{F}_n(t) = 1$, set $\hat{K}_n(t) = 1$. A nonparametric asymptotic simultaneous confidence band for F with confidence coefficient at least $100(1 - \alpha)$ is

$$\hat{F}_n \pm n^{-1/2} \frac{\lambda_\alpha (1 - \hat{F}_n)}{(1 - \hat{K}_n)} \tag{11}$$

where λ_α is such that

$$P\left(\sup_{t \in [0,1]} |B^0(t)| \le \lambda_\alpha \right) = 1 - \alpha \tag{12}$$

and B^0 denotes a Brownian bridge on [0, 1].

The following formulas are computational simplifications. Let $X_{(1)}, X_{(2)}, \ldots, X_{(r)}$ be the distinct ordered values of the X's whose corresponding failure times are within $[0, T]$. Let δ_j be the number of observations with value $X_{(j)}$. Then

$$\hat{\overline{F}}_n(t) = \prod_{X_{(j)} \le t} \left(1 - \frac{\delta_j}{Y_n(X_{(j)})} \right) \tag{13}$$

and

$$\hat{C}_n(t) = n \sum_{X_{(j)} \le t} \frac{\hat{F}_n(X_{(j)}) - \hat{F}_n(X_{(j-1)})}{Y_n(X_{(j)})(1 - \hat{F}_n(X_{(j)}))} \tag{14}$$

Some data sets will lead to $\hat{F}_n(t_0) = 1$ for some $0 < t_0 < T$. For such data sets, the band obtained is a confidence band only on the interval $(0, \sigma)$ where $\sigma = \inf\{t \in [0, T] : \hat{F}_n(t) = 1\}$.

The restriction $\theta_j \in (0, 1]$ imposed by DHS (see equation (3)) does not allow for destructive repair. Last and Szekli (LS) [10] do permit deterioration due to repair by extending Kijima's model II. LS show that their repair model contains many models proposed in the literature including those of Stadje and Zuckerman [11] and Baxter *et al.* [12].

Additional Nonparametric Methods

Let P denote the probability measure associated with the life distribution F. The restriction P_A of a probability measure P to a set A is defined by

$$P_A(B) = \frac{P(A \cap B)}{P(A)} \text{ if } P(A) > 0 \tag{15}$$

To complete the formal definition when $P(A) = 0$, define $P_A = \mu$ where μ is some conveniently chosen probability measure. Another generalization of the BBS model can be formed by postulating that T_1 has distribution P and that for $n \ge 1$, the distribution of T_{n+1}, given the failure times $T_1, \ldots, T_n$ and other environmental factors $E_1, \ldots, E_n$ observed at those failure times, is

$$P(T_{n+1} \in B) = P_{A_n}(B) \tag{16}$$

where A_n is a set that depends (measurably) on $T_1, \ldots, T_n, E_1, \ldots, E_n$. Appropriate choices of the restriction sets $A_1, A_2, \ldots$ generate different repair models. Bayesian nonparametrics for repair model (16) puts a prior distribution on P. For integrated squared-error loss, the Bayes estimator of P is the expected value of the posterior distribution of P, given the data. Sethuraman and Hollander (SH) [13, 14] introduced a new class of priors for Bayesian nonparametrics called *partition-based (PB) priors* and a subclass called *partition-based Dirichlet (PBD) priors*. They utilize these priors to show how to obtain posterior distributions under such priors, and obtain Bayes estimators of F in model (16). When specialized to the BBS model, the SH nonparametric Bayes estimators are competitors to the Whitaker and Samaniego (WS) [15] nonparametric estimator of F (*see* **Imperfect Repair**).

Kvam *et al.* [16] use isotonic regression techniques to find maximum-likelihood estimators of F and its failure rate when F is known to have an increasing failure rate. Suppose that we have data from a BBS repair consisting of n systems, each observed till the time of a perfect repair. Presnell, Hollander and Sethuraman (PHS) [17] proposed **nonparametric tests** of the minimal repair assumption in the BBS model. Their Mann–Whitney–Wilcoxon-type statistic is of the form $\int \hat{F}_e \, d\hat{F}$ where $\hat{F}$ is the WS estimator and $\hat{F}_e$ is the empirical distribution based on the initial failure times of n systems. They also studied a Kolmogorov–Smirnov (KS)-type test based on the maximum absolute difference between $\hat{F}$ and $\hat{F}_e$. We briefly describe the KS-type test. A consistent estimator of F is provided by $\hat{F}_e$ and, under the BBS model, $\hat{F}$ also consistently estimates F. If, however, the minimal repair assumption does not hold, $\hat{F}$ may likely diverge from F. Let $\hat{H}$ be the empirical distribution of the time to perfect repair of the n systems. Let

$$D(t) = \int_0^t \frac{dF(s)}{(1 - H(s))(1 - F(s))}$$

$$\hat{D}(t) = \int_0^t \frac{d\hat{F}(s)}{(1 - \hat{H}(s))(1 - \hat{F}(s))}$$

$$L(t) = \frac{1}{(1 - F(t))} - 1 - D(t)$$

$$\hat{L}(t) = \frac{1}{(1 - \hat{F}(t))} - 1 - \hat{D}(t)$$

$$G = \frac{L}{1 + L}, \text{ and } \hat{G} = \frac{\hat{L}}{1 + \hat{L}} \qquad (17)$$

Corollary 2.2 of Presnell *et al.* [17] shows that a test of the minimal repair assumption can be performed by referring

$$S_\tau = \sup_{0 \le t \le \tau} \sqrt{n} \frac{(1 - \hat{G}(t))}{(1 - \hat{F}(t))} |\hat{F}(t) - \hat{F}_e(t)| \qquad (18)$$

to a table of the supremum of the absolute value of the Brownian bridge over the interval $[0, \hat{G}(\tau)]$ (*cf.* Hall and Wellner [18] and Koziol and Byar [19]).

Hollander *et al.* [20] considered the case where two BBS processes are observed, each until its first perfect repair. They test $H_0 : F_1 = F_2$, where F_i is the distribution of the time to first failure for process $i, i = 1, 2$. Their test statistic is $W = \int_0^\infty \hat{F}_1 \, d\hat{F}_2$ where $\hat{F}_i$ is the WS estimator for the ith process. For the BBS model, Augustin and Peña [21, 22] derived goodness-of-fit tests of $H_0 : \Lambda = \Lambda_0$ where Λ_0 is completely specified. Augustin and Peña [23] developed tests for the composite case where the functional form of the failure rate $\lambda_0(\cdot, \xi)$ is known except for the value of ξ. Peña and Hollander (PH) [24] consider a general model for recurrent events that allows for interventions and incorporates covariates. Their general model contains the DHS and LS models as special cases. González *et al.* [25] apply the PH model for modeling intervention effects after cancer relapses.

References

[1] Ascher, H. (1968). Evaluation of repairable systems using the "bad as old" concept, *IEEE Transactions on Reliability* **17**, 105–110.

[2] Brown, M. & Proschan, F. (1983). Imperfect repair, *Journal of Applied Probability* **20**, 851–859.

[3] Block, H., Borges, W. & Savits, T. (1985). Age-dependent minimal repair, *Journal of Applied Probability* **22**, 370–385.

[4] Kijima, M. (1989). Some results for repairable systems with general repair, *Journal of Applied Probability* **26**, 89–102.

[5] Uematsu, K. & Nishida, T. (1987). One unit system with a failure rate depending on the degree of repair, *Mathematica Japonica* **32**, 139–147.

[6] Dorado, C., Hollander, M. & Sethuraman, J. (1997). Nonparametric estimation for a general repair model, *Annals of Statistics* **25**, 1140–1160.

[7] Peña, E.A., Strawderman, R.L. & Hollander, M. (2001). Nonparametric estimation with recurrent event data, *Journal of the American Statistical Association* **96**, 1299–1315.

[8] Hollander, M. & Peña, E.A. (2004). Nonparametric methods in reliability, *Statistical Science* **19**, 644–651.

[9] Gill, R.D. & Johansen, S. (1990). A survey of product-integration with a view toward application in survival analysis, *Annals of Statistics* **18**, 1501–1555.

[10] Last, G. & Szekli, R. (1998). Asymptotic and monotonicity properties of some repairable systems, *Advances in Applied Probability* **30**, 1089–1110.

[11] Stadje, W. & Zuckerman, D. (1991). Optimal maintenance strategies for repairable systems with general

degree of repair, *Journal of Applied Probability* **28**, 384–396.

[12] Baxter, L., Kijima, M. & Tortella, M. (1996). A point process model for the reliability of a maintained system subject to general repair, *Communications in Statistics: Stochastic Models* **12**, 37–65.

[13] Sethuraman, J. & Hollander, M. (2006a). Bayesian methods in repair models, in *Proceedings of the International Conference on Degradation, Damage, Fatigue and Accelerated Life Models in Reliability Testing*, Angers, France, pp. 217–221.

[14] Sethuraman, J. & Hollander, M. (2006b). Nonparametric Bayes Estimation in Repair Models, Submitted.

[15] Whitaker, L.R. & Samaniego, F.J. (1989). Estimating the reliability of systems subject to imperfect repair, *Journal of the American Statistical Association* **84**, 301–309.

[16] Kvam, P.H., Singh, H. & Whitaker, L.R. (2002). Estimating distributions with increasing failure rate in an imperfect repair model, *Lifetime Data Analysis* **8**, 53–67.

[17] Presnell, B., Hollander, M. & Sethuraman, J. (1994). Testing the minimal repair assumption in an imperfect repair model, *Journal of the American Statistical Association* **89**, 289–297.

[18] Hall, W.J. & Wellner, J. (1980). Confidence bands for a survival curve from censored data, *Biometrika* **67**, 133–143.

[19] Koziol, J.A. & Byar, D.P. (1975). Percentage points of the asymptotic distributions of one- and two-sample KS statistics for truncated or censored data, *Technometrics* **17**, 507–510.

[20] Hollander, M., Presnell, B. & Sethuraman, J. (1992). Nonparametric methods for imperfect repair models, *Annals of Statistics* **20**, 879–896.

[21] Augustin, Z. & Peña, E.A. (1999). Order statistic properties, random generation, and goodness-of-fit testing for a minimal repair model, *Journal of the American Statistical Association* **94**, 266–272.

[22] Augustin, Z. & Peña, E.A. (2001). Goodness-of-fit of the distribution of time-to-first-occurrence in recurrent event models, *Lifetime Data Analysis* **7**, 289–306.

[23] Augustin, Z. & Peña, E.A. (2000). Testing whether the distribution of time to first event occurrence is in a specified family, in *Abstracts Book of the MMR' 2000 Second International Conference on the Mathematical Methods in Reliability: Methodology, Practice and Inference*, Bordeaux, Vol. 1, pp. 47–50.

[24] Peña, E.A. & Hollander, M. (2004). Models for recurrent phenomena in survival analysis and reliability, in *Mathematical Reliability: An Expository Perspective*, T. Mazzuchi, N. Singpurwalla & R. Soyer, eds, Kluwer, Dordrecht, pp. 105–123.

[25] González, J.R., Peña, E.A. & Slate, E.H. (2005). Modelling intervention effects after cancer relapses, *Statistics in Medicine* **24**, 3959–3975.

Related Articles

General Minimal Repair Models; **Imperfect Repair**; **Imperfect Repair, Counting Processes**; **Multivariate Imperfect Repair Models**; **Repair Data, Sets of: How to Graph, Analyze, and Compare**.

MYLES HOLLANDER, FRANCISCO J. SAMANIEGO AND JAYARAM SETHURAMAN

Nonparametric Tests

Introduction

Nonparametric tests are the counterparts to *parametric tests* (*see* **Parametric Tests**). Typically, parametric tests assume that the data are drawn from a distribution, which has a finite number of unknown parameters, often the **normal distribution** with unknown mean μ and σ^2. In contrast, nonparametric tests do not rely on such rather restrictive distributional assumptions, and hence are also known as *distribution-free tests*.

The parameter space is the set of all possible values of the unknown parameters. In mathematical statistics, a statistical method is called *parametric* when the parameter space is finite dimensional, and *nonparametric* when the parameter space is infinite dimensional, such as the space of all possible distribution functions. To complicate matters, there are also semiparametric models in which the parameter vector consists of a finite-dimensional part and an infinite-dimensional part. For instance, if we assume that we are dealing with data from a distribution that is symmetric around some unknown median η, then we are in essence dealing with a semiparametric model, as the parameter vector consists of a finite-dimensional part η and an infinite-dimensional part, a distribution that is symmetric around zero. Below, we shall ignore the distinction between nonparametric and semiparametric methods.

Permutation Tests

Nonparametric tests have been around since the early days of statistical inference. Perhaps the oldest form of nonparametric test is the permutation test, see [1, 2]. If data have been obtained from experiments that involve **randomization**, then we may assess the statistical significance of a permutation test by recalculating the test statistic for every permutation of the data consistent with the randomization. Obviously, permutation tests are computer intensive, which has impeded practical use for a long time. Recently, there is renewed interest in permutation tests.

Rank Tests

The computational demands of a permutation test may be drastically reduced by replacing the observations with their (transformed) ranks before actually performing the test. The resulting test is called a *rank test*.

An illustrative example is the signed rank test, see [3]. This test assumes that n observations are drawn from a distribution that is symmetric around an unknown location θ. The signed rank test may also be applied to paired data.

The null hypothesis to be tested is $H_0 : \theta = \theta_0$. The test assigns ranks to the observations according to their distance to θ_0. Therefore, the observation closest to θ_0 is assigned rank 1. The test statistic T is obtained by summing the ranks belonging to the observations smaller than θ_0, and is asymptotically normal.

To assess statistical significance (*see* **Significance Level**), we recalculate this rank sum for every permutation of the data. For example, if $n = 4$, then under the null hypothesis there are 2^4 equally likely outcomes for the ranks assigned to the observations smaller than θ_0 see Table 1.

Table 1 Derivation of the null distribution of the rank sum test statistic for $n = 4$. The rank sum test statistic is the sum of the ranks assigned to observations smaller than θ_0

Ranks assigned	Rank sum T	Ranks assigned	Rank sum T
–	0	2, 3	5
1	1	2, 4	6
2	2	1, 2, 3	6
3	3	3, 4	7
1, 2	3	1, 2, 4	7
4	4	1, 3, 4	8
1, 3	4	2, 3, 4	9
1, 4	5	1, 2, 3, 4	10

Thus, the test statistic T takes values 0, 1, 2, 8, 9, and 10 with probability 1/16, and values 3, 4, 5, 6, and 7 with probability 1/8.

It is known that the test statistic T has an approximately normal distribution for $n \geq 20$. Under the null hypothesis, the test statistic T has expected value $n(n+1)/4$ and variance $n(n+1)(2n+1)/24$, and hence approximate critical values of the standardized test statistic

$$\frac{T - \dfrac{n(n+1)}{4}}{\sqrt{\dfrac{n(n+1)(2n+1)}{24}}} \tag{1}$$

may be determined by means of the standard normal distribution for $n \geq 20$. For $n < 20$, tabulated critical values have to be used.

Another example is the two-sample rank sum test, see [3, 4]. It involves two independent samples. Suppose that the first sample has size n_1 and is drawn from a distribution with population median η_1 and that the second sample has size n_2 and is drawn from a distribution with population median η_2, the null hypothesis to be tested is $H_0 : \eta_1 = \eta_2$.

The combined sample is ranked from the least to the greatest. The Wilcoxon rank sum test statistic W is the sum of the ranks assigned to the first sample. It is known that the test statistic W has an approximately normal distribution when both sample sizes n_1 and n_2 are larger than 10. Under the null hypothesis, the test statistic W has expected value $n_1(n_1 + n_2 + 1)/2$ and variance $n_1 n_2(n_1 + n_2 + 1)/12$. Therefore, approximate critical values of the standardized test statistic

$$\frac{W - \dfrac{n_1(n_1+n_2+1)}{2}}{\sqrt{\dfrac{n_1 n_2(n_1+n_2+1)}{12}}} \tag{2}$$

may be determined by means of the standard normal distribution for $n_1, n_2 \geq 10$. The Mann–Whitney test statistic U is related to W, and is, in fact, given by

$$U = W - \frac{n_1(n_1+1)}{2} \tag{3}$$

It follows that the Wilcoxon rank sum test and the Mann–Whitney test are equivalent.

A further example is the Kruskal–Wallis test, which compares the effects of $k > 2$ treatments. To each treatment, there belongs a sample, and these samples are independent. The Kruskal–Wallis test assumes that

$$X_{ij} = \eta + \tau_j + \epsilon_{ij} \quad (4)$$

where X_{ij} is the jth observation in the ith sample, η is an unknown constant, τ_j is the unknown effect of treatment j, and the error variables, ϵ_{ij}'s, are a random sample from a distribution with median 0. The null hypothesis to be tested is

$$H_0 : \tau_1 = \tau_2 = \cdots = \tau_k \quad (5)$$

Rank the combined sample from the least to the greatest, and let $\overline{R}_j$ denote the mean of the ranks assigned to sample j. The Kruskal–Wallis test statistic is defined as

$$H = \frac{12}{N(N+1)} \sum_{j=1}^{k} n_j \left(\overline{R}_j - \overline{\overline{R}}\right)^2 \quad (6)$$

where n_j is the size of sample j, $N = n_1 + \cdots + n_k$ is the size of the combined sample, and $\overline{\overline{R}} = (N+1)/2$ is the average rank in the combined sample. The null hypothesis is rejected for large values of H. Under the null hypothesis, H approximately follows a χ^2 distribution with $k-1$ **degrees of freedom** for $\min(n_1, \ldots, n_k) \geq 5$.

Instead of rank sums $\sum R_i$, one may use sums of transformed ranks $\sum a(R_i)$. For instance, the two-sample Savage test employs the transformed ranks

$$a(R_i) = \sum_{j=N-R_i+1}^{N} 1/j \quad (7)$$

see [5].

Rank tests may be generalized so as to accommodate censored observations. Generalizations of the two-sample rank sum test and the Savage test are found in [6] and [7], respectively. The latter test is commonly known as the *logrank test* and is related to the Cox regression model.

For an extensive overview of rank tests, refer to [8, 9].

Resampling Methods

The advent of high-speed computing has had a beneficial effect on nonparametric statistics. As computation became fast and cheap, ideas that had been around for some time suddenly became applicable. In particular, this holds true for permutation tests and resampling methods, see [10].

Resampling methods are related to permutation tests. However, instead of permuting the data, subsamples are drawn. The most prominent resampling method is the **bootstrap**, see [11]. In its simplest form, the bootstrap constructs bootstrap samples by sampling with replacement from the initial sample. Often, the bootstrap is used to construct confidence intervals. However, the bootstrap may also be used to obtain critical values for some test statistic.

Let $X_1, \ldots, X_n$ be a sample drawn from P_θ, where θ is a (possibly infinite-dimensional) parameter that belongs to some parameter space. We intend using the test statistic $T_n = T_n(X_1, \ldots, X_n)$ to test the null hypothesis $H_0 : \theta \in \Theta_0$ *versus* $H_1 : \theta \in \Theta \backslash \Theta_0$, where $\Theta_0 \subset \Theta$. Let $\hat{\theta} = \hat{\theta}(X_1, \ldots, X_n)$ be a consistent estimator of θ under the null hypothesis, and let $T_n^* = T_n(X_1^*, \ldots, X_n^*)$, where $X_1^*, \ldots, X_n^*$ is a bootstrap sample drawn from $P_{\hat{\theta}}$. The conditional distribution of T_n^* given $X_1, \ldots, X_n$ is the bootstrap estimate for the null distribution of T_n. **Quantiles** of the bootstrap null distribution may act as bootstrap critical values for T_n.

The error in the probability of a type I error is of smaller order for a bootstrap test than for the corresponding asymptotical theory test if and only if T_n is an asymptotic pivotal quantity, that is, the asymptotic null distribution of the test statistic T_n does not depend on θ, see [12]. Therefore, a bootstrap test should preferably be based on an asymptotic pivotal quantity.

For instance, let Θ be the collection of distributions on the real line whose first 10 **moments** are finite, and let $X_1, \ldots, X_n$ be a random sample drawn from a distribution belonging to Θ. Let Θ_0 be the set of elements of Θ that have mean μ_0. Then, the bootstrap samples should be resampled from $X_1^0, \ldots, X_n^0$, where $X_i^0 = X_i + \mu_0 - \overline{X}_n$. That is, X_i^0 is obtained by shifting X_i over a distance $\mu_0 - \overline{X}_n$. Note that the mean of the shifted sample $X_1^0, \ldots, X_n^0$ is equal to μ_0, and hence the empirical distribution of $X_1^0, \ldots, X_n^0$ belongs to Θ_0. Moreover, the bootstrap test may be based on the studentized statistic $T_n = \sqrt{n}(\overline{X}_n - \mu_0)/S_n$, which is an asymptotic pivotal quantity.

References

[1] Fisher, R.A. (1935). *The Design of Experiments*, Oliver & Boyd, London.

[2] Pitman, E.J.G. (1937). Significance tests which may be applied to samples from any populations, *Supplement to the Journal of the Royal Statistical Society* **4**, 119–130.

[3] Wilcoxon, F. (1945). Individual comparisons by ranking methods, *Biometrics Bulletin* **1**, 80–83.

[4] Mann, H.B. & Whitney, D.R. (1947). On a test of whether one of two random variables is stochastically larger than the other, *Annals of Mathematical Statistics* **18**, 50–60.

[5] Savage, R.I. (1956). Contributions to the theory of rank order statistics – the two-sample case, *Annals of Mathematical Statistics* **27**, 590–615.

[6] Gehan, E.A. (1965). A generalized Wilcoxon test for comparing arbitrarily singly-censored samples, *Biometrika* **52**, 203–223.

[7] Mantel, N. (1966). Evaluation of survival data and two new rank order statistics arising in its consideration, *Cancer Chemotherapy Reports* **50**, 163–170.

[8] Hollander, M. & Wolfe, D.A. (1999). *Nonparametric Statistical Methods*, John Wiley & Sons, New York.

[9] Hettmansperger, T.P. & McKean, J.W. (1998). *Robust Nonparametric Statistical Methods*, Edward Arnold, London.

[10] Efron, B. (1979). Computers and the theory of statistics: thinking the unthinkable, *SIAM Review* **21**(4), 460–480.

[11] Efron, B. (1979). Bootstrap methods: another look at the jackknife, *Annals of Statistics* **7**, 1–26.

[12] Beran, R. (1988). Prepivoting test statistics: a bootstrap view of asymptotic refinements, *Journal of the American Statistical Association* **83**, 687–697.

ALEX J. KONING

Normal Distribution

Introduction

This article describes the univariate normal distribution, with brief references also to the bivariate normal distribution and the multivariate normal. For each of these, there are brief historical remarks, and discussions of distributional properties, sampling distributions, and related applications.

The Univariate Normal Distribution

Historical Remarks

The univariate normal distribution is probably *the* most important distribution in classical statistical theory and methods. This distribution has a "bell-shaped" continuous *probability density function* (*see* **Probability Density Function (PDF)**). In the history of statistics, it was first discovered by the German mathematician Carl F. Gauss in the early nineteenth century while he was studying certain problems in physics and astronomy. As a result, this distribution is also known as the *Gaussian distribution*.

Distributional Properties

A continuous univariate random variable X is said to follow a normal distribution with parameters μ and σ^2, if its probability density function $f(x)$ is of the form

$$f(x) = \left[\frac{1}{(2\pi)^{1/2}\sigma}\right]\exp\left[\frac{-(x-\mu)^2}{2\sigma^2}\right]$$
$$-\infty < x, \mu < \infty, \text{ and } \sigma > 0 \quad (1)$$

in symbols, $X \sim N(\mu, \sigma^2)$. It can be shown by elementary calculus that the mean and the variance of this distribution are μ and σ^2, respectively.

A random variable Z is said to follow a *standard* normal distribution if $Z \sim N(0, 1)$. The distribution function of Z, $\Phi(z)$, is then given by

$$\Phi(z) = \int_{-\infty}^{z}\left[\frac{1}{(2\pi)^{1/2}}\right]\exp\left(\frac{-u^2}{2}\right)\mathrm{d}u \quad (2)$$

Since the function $\exp(-u^2/2)$ is symmetric about 0, $\Phi(z)$ satisfies $\Phi(z) = 1 - \Phi(-z)$ for all z. The $(1-\alpha)$th *quantile* (*see* **Quantiles**) (or $100(1-\alpha)$th percentile) of a standard normal distribution, often denoted z_α in most textbooks, is the quantity that satisfies $\Phi(z_\alpha) = 1 - \alpha, \alpha \in (0, 1)$. Since the function $\Phi(z)$ cannot be expressed in a closed form, tables for the numerical values of $\Phi(z)$ and z_α are needed. Such tables can be found in most statistics books.

The following theorem, which can be proved by calculus, shows how a normal distribution with mean μ and variance σ^2 is related to the standard normal distribution.

Theorem 1 *If $X \sim N(\mu, \sigma^2)$, then the random variable $Z = (X - \mu)/\sigma$ is distributed according to the standard normal distribution.*

As a simple application of Theorem 1, it follows that

Fact 1. If $X \sim N(\mu, \sigma^2)$, then,

1. $\Pr[a < X \le b] = \Phi((b - \mu)/\sigma) - \Phi((a - \mu)/\sigma)$ for all $a < b$;
2. the $(1 - \alpha)$th quantile of the distribution of X is $\mu + z_\alpha \sigma$.

Thus, tables for the standard normal distribution can be used for any univariate normal distribution.

More detailed distribution properties of the univariate normal distribution can be found in Patel and Read [1] and other related sources.

Sampling Distributions

The χ^2 distribution, Student's t distribution, and the F distribution (*see* **Probability Density Functions**), generally considered to be the cornerstones of classical statistical analysis, are closely related to the normal distribution. Specifically, let $X_1, X_2, \ldots, X_N$ be a random sample of size N from an $N(\mu, \sigma^2)$ distribution; let

$$\overline{X} = \frac{1}{N} \sum_{i=1}^{N} X_i \quad \text{and}$$

$$S^2 = \frac{1}{N-1} \sum_{i}^{N} = 1(X_i - \overline{X})^2 \qquad (3)$$

denote the sample mean and the sample variance, respectively. Then,

1. $\overline{X}$ has an $N(\mu, \sigma^2/N)$ distribution and $(N - 1) S^2/\sigma^2$ has a χ^2 distribution with $N - 1$ **degrees of freedom**. Furthermore, $\overline{X}$ and S^2 are independent.
2. $t = N^{1/2}(\overline{X} - \mu)/S$ has a Student's t distribution with $N - 1$ degrees of freedom, where $S = \sqrt{S^2}$ is the sample standard deviation.
3. t^2 has an F distribution with degrees of freedom $(1, N - 1)$.

Another important sampling distribution result related to the normal distribution is a fundamental theorem in **probability theory**, called the *central limit theorem*.

Theorem 2 *Let $X_1, X_2, \ldots, X_N$ be a random sample of size N from any population with mean μ and finite variance σ^2. Then, for every fixed z,*

$$\lim_{N\to\infty} \Pr[\{N^{1/2}(\overline{X} - \mu)/\sigma\} \le z] = \Phi(z) \qquad (4)$$

This theorem provides an approximation for the distribution of $\overline{X}$ when N is large. In most applications,

$$\Pr[a < \overline{X} \le b] \doteq \Phi(N^{1/2}(b - \mu)/\sigma) - \Phi(N^{1/2}(a - \mu)/\sigma) \qquad (5)$$

when $N \ge 30$.

Related Applications

In various applications of statistical inference problems – including *estimation* (*see* **Estimation**) of the parameters and *hypothesis testing* (*see* **Hypothesis Testing**) – the above sampling distribution results are applied. These include the following:

1. For inference on μ when σ^2 is known, the results for the distribution of $\overline{X}$ and Theorem 2 may be applied.
2. For inference on μ when σ^2 is unknown under the assumption of normality, Student's t distribution may be used.
3. When making statistical inference on σ^2 under the assumption of normality, the χ^2 distribution may be used.
4. The normal distribution may be applied for inference on the difference of two normal means when their variances are known. Similarly, the Student's t distribution may be applied for the same purpose when the variances are unknown but equal; in the hypotheses-testing problem, this is known as the *two-sample t test* (*see* **Parametric Tests**).
5. The F distribution may be applied for testing the equality of $k \ge 2$ normal means when the variances are assumed to be equal but unknown.

This method is known as the *one-way analysis of variance (ANOVA) method* (*see* **Analysis of Variance**). When $k = 2$, it reduces to the two-sample t test as a special case.

6. Other results related to Theorem 2 are useful in large-sample inference problems. For example, it is known that under regularity conditions the *maximum-likelihood* (*see* **Maximum Likelihood**) estimator has an asymptotically normal distribution, and that the asymptotic null distribution of $-2\ln\lambda$ is χ^2, where λ is the likelihood-ratio function in a likelihood-ratio test.

The Bivariate Normal Distribution

Historical Remarks

Studies of the bivariate normal distribution seem to begin in the middle of the nineteenth century, and moved forward dramatically when Galton published his work [2] on the applications of **correlation** analysis in genetics. As Karl Pearson noted in his 1920 *Biometrika* paper [3], "In 1885 Galton had completed the theory of bivariate normal correlation" but, because he "was very modest and throughout his life underrated his own mathematical powers, he did not at once write down the equation" of the bivariate normal density function. Consequently, it was Pearson himself who gave a definitive mathematical formulation of the bivariate normal distribution in his 1896 paper [4] on regression and heredity.

Distributional Properties

A two-dimensional random vector (X_1, X_2) is said to have a bivariate normal distribution if their joint density function $f(x_1, x_2)$ is of the form

$$f(x_1, x_2) = [2\pi\sigma_1\sigma_2(1-\rho^2)^{1/2}]^{-1} \exp\left[-\tfrac{1}{2}Q_2(x_1, x_2; \boldsymbol{\mu}, \boldsymbol{\Sigma})\right], \quad -\infty < x_1, x_2 < \infty \tag{6}$$

where

$$Q_2(x_1, x_2; \boldsymbol{\mu}, \boldsymbol{\Sigma}) = (x_1 - \mu_1, x_2 - \mu_2)\boldsymbol{\Sigma}^{-1} \times \begin{pmatrix} x_1 - \mu_1 \\ x_2 - \mu_2 \end{pmatrix} \tag{7}$$

defines an ellipse centered at $(\mu_1, \mu_2) \equiv \boldsymbol{\mu}$ (which is the mean vector), the 2×2 matrix

$$\boldsymbol{\Sigma} = \begin{pmatrix} \sigma_1^2 & \rho\sigma_1\sigma_2 \\ \rho\sigma_1\sigma_2 & \sigma_2^2 \end{pmatrix} \tag{8}$$

is the **covariance** matrix, and $\rho \in (-1, 1)$ is the *correlation* (*see* **Correlation**) coefficient.

The marginal and conditional distributions of a bivariate normal random vector are univariate normal. For details, see Tong [5, Section 2.1].

Sampling Distributions and Related Applications

The sampling distribution results involve the distributions of the sample mean vector $\overline{\mathbf{X}}$, the sample covariance matrix $\mathbf{S}$, the independence property of $\overline{\mathbf{X}}$ and $\mathbf{S}$, and the distribution of the sample correlation coefficient. Those results may be applied for estimation and hypothesis testing purposes. For details, see Anderson [6, Section 2.3] and Tong [5, Sections 2.1 and 2.2].

The Multivariate Normal Distribution

Historical Remarks

The development of the multivariate normal distribution theory, which originated mainly from the studies of regression analysis and multiple and partial correlation analysis, was treated comprehensively for the first time by Edgeworth in his 1892 paper [7]. The development of the sampling distribution theory under the assumption of normality (such as Fisher's work on the distributions of sample correlation coefficients and *Hotelling's* T^2 distribution; *see* **Hotelling's** $\boldsymbol{T^2}$) then followed.

Distributional Properties

An n-dimensional random vector $\mathbf{X} = (X_1, \ldots, X_n)$ is said to follow a multivariate normal distribution with mean vector $\boldsymbol{\mu} = (\mu_1, \ldots, \mu_n)$ and covariance matrix $\boldsymbol{\Sigma}_{n\times n} = (\sigma_{ij})$, in symbols $\mathbf{X} \sim N_n(\boldsymbol{\mu}, \boldsymbol{\Sigma})$, if its joint probability density function is of the form

$$f(\mathbf{x}) = [(2\pi)^{n/2}|\boldsymbol{\Sigma}|^{1/2}]^{-1} \exp\left[-\tfrac{1}{2}Q_n(\mathbf{x}; \boldsymbol{\mu}, \boldsymbol{\Sigma})\right], \quad \mathbf{x} \in \mathbb{R}^n \tag{9}$$

where

$$Q_n(\mathbf{x}; \boldsymbol{\mu}, \boldsymbol{\Sigma}) = (\mathbf{x} - \boldsymbol{\mu})\boldsymbol{\Sigma}^{-1}(\mathbf{x} - \boldsymbol{\mu})' \tag{10}$$

The marginal and conditional distributions of a multivariate normal random vector are also normal. For details, see Anderson [6, Sections 2.3, 2.4, and 2.5] and Tong [5, Section 3.3].

Sampling Distributions and Related Applications

The sampling distribution results also involve $\overline{\mathbf{X}}$, $\mathbf{S}$, and their independence property; in particular, the results are related to Hotelling's T^2 distribution and the Wishart distribution (generalizations of the Student's t distribution and χ^2 distribution, respectively). There also exist distributional results on the sample regression equations and various types of sample correlation coefficients. Such results have been found useful for the purposes of prediction and correlation analysis. For details on sampling distributions, see Anderson [6, Chapters 4, 5, and 7], Tong [5, Sections 3.4 and 3.5], and other related sources. For related applications in inference, a classical reference is Anderson [6].

References

[1] Patel, J.K. & Read, C.B. (1982). *Handbook of the Normal Distribution*, Revised Edition 1996, Marcel Dekker, New York.

[2] Galton, F. (1888). Co-relations and their measurements, chiefly from anthropometric data, *Proceedings of the Royal Society of London* **45**, 135–145.

[3] Pearson, K. (1920). Notes on the history of correlation, *Biometrika* **13**, 25–45.

[4] Pearson, K. (1896). Mathematical contributions to the theory of evolution – III. Regression, heredity and panmixia, *Philosophical Transactions of the Royal Society of London. Series A* **187**, 253–318.

[5] Tong, Y.L. (1990). *The Multivariate Normal Distribution*, Springer-Verlag, New York.

[6] Anderson, T.W. (1984). *An Introduction to Multivariate Statistical Analysis*, 2nd Edition, John Wiley & Sons, New York.

[7] Edgeworth, F.Y. (1892). Correlated averages, *Philosophical Magazine, Series 5* **34**, 190–204.

Related Articles

Normality Tests

Y.L. TONG

Article originally published in Encyclopedia of Biostatistics, 2nd Edition *(2005, John Wiley & Sons, Ltd). Minor revisions for this publication by Jeroen de Mast.*

Normality Tests

Introduction

In testing normality, one should distinguish between the simple null hypothesis (parameters μ and σ^2 known) and the composite null hypothesis (parameters μ and σ^2 unknown). The simple null hypothesis is of little practical value.

Under the composite null hypothesis, the unknown parameters μ and σ^2 should be estimated. By using plug-in estimators, we may readily extend tests developed for the simple null hypothesis to the composite null hypothesis. However, plugging in estimators has a complicating effect on the null distribution of the test statistic. This makes it difficult to determine critical values and/or *p values* (*see* ***P*** **Values**).

Distance Methods

Distance methods are based on measuring the distance between a parametric and a nonparametric estimator of some function describing the distribution, such as the cumulative distribution function, the cumulative hazard function, or the moment generating function. In testing normality, the estimated normal cumulative distribution function and the empirical distribution function are typically used.

The asymptotic distribution of the composite null hypothesis versions of the Kolmogorov, the Cramér–von Mises, or the Anderson–Darling tests cannot be captured by a mathematical expression, and hence we have to resort to tabulation. The composite null hypothesis version of the Kolmogorov test is generally known as the *Lilliefors test*; [1] tabulates the null distribution of the test statistic.

Skewness and Kurtosis Tests

Instead of measuring the distance between the empirical distribution function and the estimated cumulative normal probability function, we may focus on certain characteristics of the **normal distribution**, such as the higher **moments**. For instance, it is known that the *skewness* (*see* **Skewness**) and *kurtosis* (*see* **Kurtosis**) of any normal distribution are equal to 0 and 3, respectively, and which suggests the use of the third and fourth sample *moments* (*see* **Moments**) for testing normality. These tests are referred to as *directional*, as they lack the omnibus property of the distance-based tests. In [2] percentage points for various tests derived from skewness and kurtosis are given.

In econometrics, the Jarque–Bera test, see [3], is commonly used to test normality. This test is also derived from skewness and kurtosis.

Shapiro–Wilk-Type Tests

The normal *probability plot* (*see* **Probability Plots**) is a popular technique for exploring normality, and is a plot of the observations *versus* their corresponding ideal values, the so-called normal scores. If the data follow a normal distribution, the normal probability plot should be approximately linear. This suggests that we may test normality by quantifying the degree of nonlinearity in the normal probability plot.

The first test along these lines is the Shapiro–Wilk test, see [4]. Unfortunately, this test is computationally demanding. In particular, the Shapiro–Wilk test requires the inversion of a certain $n \times n$ **covariance** matrix V, where n denotes the sample size. In [5] it is argued that for large sample sizes, the covariance matrix V may be replaced by the identity matrix. This yields the Shapiro–Francia test, a simplified version of the Shapiro–Wilk test, which may be used for sample sizes larger than 50. In [6] it is recognized that the Shapiro–Francia test statistic is in fact the squared normal probability plot **correlation**, that is, the squared correlation between the observations and their normal scores; moreover, critical values for the normality probability plot correlation are given; see Figure 1. The Ryan–Joiner test in Minitab, see [7], only differs from the Shapiro–Francia test in the way the normal scores are obtained.

The close relation to the normal probability plot is an attractive aspect of the Shapiro–Wilk-type tests.

Comparison

In [8] results of a comparative study of various normality tests are given. Their findings include the following:

- the Shapiro–Wilk test provides a generally superior omnibus measure of nonnormality;
- the distance-based tests are typically very insensitive;
- a combination of both sample skewness and sample kurtosis usually provides a sensitive judgment, but even their combined performance is usually dominated by the Shapiro–Wilk test.

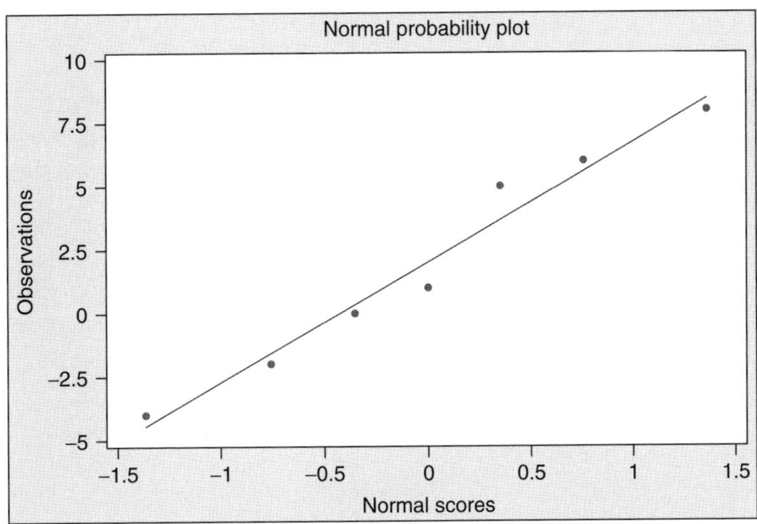

Figure 1 An application of the Shapiro –Francia test. The correlation between the observations 6, 1, −4, 8, −2, 5, 0 and their normal scores is 0.984. The corresponding critical value at the 5% **significance level** is 0.899 [see 6 Table 1]. As $0.984 > 0.899$, the normality hypothesis is not rejected

In [9], the superiority of the Shapiro–Wilk test is confirmed. However, it is pointed out that the findings in [8] are misleading with respect to the distance-based methods: the Cramér–von Mises and Anderson–Darling tests with plug-in estimators perform slightly worse than the Shapiro–Wilk test.

References

[1] Lilliefors, H.W. (1967). On the Kolmogorov-Smirnov test for normality with mean and variance unknown, *Journal of the American Statistical Association* **62**, 399–402.

[2] Bowman, K.O. & Shenton, L.R. (1975). Omnibus test contours for departures from normality based on $\sqrt{b_1}$ and b_2, *Biometrika* **62**, 243–250.

[3] Jarque, C.M. & Bera, A.K. (1980). Efficient tests for normality, homoscedasticity and serial independence of regression residuals, *Economics Letters* **6**, 255–259.

[4] Shapiro, S.S. & Wilk, M.B. (1965). An analysis of variance test for normality: complete samples, *Biometrika* **52**, 591–611.

[5] Shapiro, S.S. & Francia, R.S. (1972). Approximate analysis of variance test for normality, *Journal of the American Statistical Association* **67**, 215–216.

[6] Filliben, J.J. (1975). The probability plot correlation coefficient test for normality, *Technometrics* **17**, 111–117.

[7] Ryan, T.A. & Joiner, B.L. (1976). Normal probability plots and tests for normality, Technical report, The Pennsylvania State University, State College. URL http://www.minitab.com/resources/articles/normprob.aspx.

[8] Shapiro, S.S., Wilk, M.B. & Chen, H.J. (1968). A comparative study of various tests for normality, *Journal of the American Statistical Association* **63**, 1343–1372.

[9] Stephens, M.A. (1974). EDF statistics for goodness of fit and some comparisons, *Journal of the American Statistical Association* **69**, 730–737.

Alex J. Koning

Numerical Integration *see* Quadrature and Numerical Integration

Observational Studies

An observational study is an empirical investigation that attempts to estimate the effects caused by a factor when it is not possible to perform an experiment. Whereas in an experiment, factor settings are assigned to runs by the inquirer, in observational studies, the inquirer does not manipulate the factors, but merely records the values that they happen to have. Contrary to experimental data, observational data demonstrate **correlation** rather than causation (*see* **Causality**).

Operating Characteristic Curves *see* Average Run Lengths and Operating Characteristic Curves

Optimal Decision Search *see* Decision Search via Simulation

Optimal Design

Introduction

With luck, and perhaps after data transformation, the analysis of a well-designed experiment may require little more than a few simple plots that reveal how the responses depend on the factors. Subsequent **estimation** of the parameters modeling this dependence usually requires the use of **least squares**. In a good experiment the variances and covariances of the estimated parameters will be small. Optimal experimental designs minimize functions of these variances and so provide good estimates of the parameters.

This article is mainly concerned with D-optimality in which the generalized variance of the parameter estimates is minimized. Many of the standard designs, such as the 2^m factorials and the series of 2^{m-f} fractional factorials, as well as standard designs for **mixture experiments** and some designs for second-order response surfaces can be justified because they are D-optimal, in addition to other justifications. However, a major feature of the methods of optimal experimental design is the ability to find good designs in nonstandard situations. Here are some advantages of the approach:

1. the availability of algorithms for the construction of designs;
2. the calculation of good designs for a specified number of trials;
3. the provision of simple, but incisive, methods for the comparison of designs;

4. the ability to divide response-surface and other designs into blocks of a specified size;
5. the determination of good response-surface designs over nonregular regions;
6. the construction of designs for mixture models, perhaps over nonregular regions;
7. the ability to find designs for nonstandard models, such as nonlinear models.

The first six of these seven points are illustrated in the sections that follow. Chapter 17 of Atkinson *et al.* [1] gives a thorough coverage of optimal designs for nonlinear models.

The article starts in the section titled "Simple Regression" with linear and quadratic regression in one variable. Several criteria of optimality are introduced in the third section. Examples of numerically constructed D-optimal designs are given in the section titled "Second-Order **Response Surface** in Two Factors". These include designs for an arbitrary number of trials, blocked response-surface designs and designs over nonregular design regions. The distinction between exact and continuous designs is made in the following section in "Exact and Continuous Designs" and used in the section titled "The General Equivalence Theorem and Design Efficiency" in the description of the general equivalence theorem for D- and G-optimal designs. The theorem is also used to demonstrate the optimality of designs and calculate efficiencies. Algorithms for design construction are in the section titled "Design Algorithms". Extensions, including designs for detecting **lack of fit**, and references to the literature on optimal experimental design are in the section titled "Discussion and Literature".

Simple Regression

Since optimal designs focus on the variances of the estimated parameters, we need to specify a model. We start with the simplest interesting case, linear regression in one variable, which serves to introduce some important ideas. There is a single response y, the expected value of which depends linearly on the value of the factor or process variable x through the relationship

$$E(y) = \beta_0 + \beta_1 x \tag{1}$$

The design problem is to choose N values of x at which measurements are to be taken. Although there are N observations, the number of distinct values of x used in the experiment, which we call n, may be much less than N. These n values of x must lie in the experimental region $\mathcal{X}$. For a single variable we can scale the values of x such that $-1 \leq x \leq 1$, where ± 1 correspond to the minimum and maximum values of the variable, which might typically be time, temperature, pressure, stirring speed, or catalyst concentration.

The model is to be fitted to the data by least squares. For this method to be appropriate, the variance of the observations should be constant and we write $\text{var}(y) = \sigma^2$. The least-squares estimators of the parameters in equation (1) are called $\hat{\beta}_0$ and $\hat{\beta}_1$ with

$$\text{var}(\hat{\beta}_1) = \frac{\sigma^2}{\sum (x_i - \bar{x})^2} \tag{2}$$

where the sample average $\bar{x} = \sum x_i / N$. Here and in equation (2) the summations are over all N values x_i. From equation (2) the variance of $\hat{\beta}_1$ is minimized when $\sum (x_i - \bar{x})^2$ is maximized. For a fixed number of trials N this occurs when all trials are at $+1$ or -1. If N is even, exactly half the trials will be at $x = +1$ and the other half at $x = -1$. This design has $\bar{x} = 0$, so that $\text{var}(\hat{\beta}_0) = \sigma^2/N$ which does not depend any further on the design. This design is therefore optimal for both parameters.

This design has many of the features of an optimal design. The number of support points of the design n, here two, may be appreciably less than the number of trials N. The fine structure of the design depends on the values of N, here whether it is even or odd, in which case one more trial is put at one end of the design region; it does not matter which. A third feature is that the design covers the whole of the design region, using the most extreme values of x available. In any application of the design, the order in which the two treatments $x = -1$ and $x = 1$ are applied should be randomized (*see* **Randomization in Experimental Designs**), to avoid any confounding with omitted variables. For example, if experiments are performed over time, gradual fouling of catalysts may mean that yields slowly decline during the experiment. If all trials at one level of x were performed in the second half of the experiment, the decline in yield would be confounded with any effect of the factor. Joiner [2] gives further examples of such "lurking variables".

Fitted models are often used for prediction. The prediction at the point x, not necessarily included in the data from which the parameters were estimated, is

$$\hat{y}(x) = \hat{\beta}_0 + \hat{\beta}_1 x = \bar{y} + \hat{\beta}_1(x - \bar{x}) \tag{3}$$

with variance

$$\text{var}\{\hat{y}(x)\} = \sigma^2 \left\{ \frac{1}{N} + \frac{(x-\bar{x})^2}{\sum(x_i - \bar{x})^2} \right\} \tag{4}$$

The optimal design found in this section for parameter estimation also has a minimax property for prediction: it minimizes the maximum variance of the prediction $\hat{y}(x)$ over $\mathcal{X}$. We discuss design for prediction further in the section titled "Criteria of Optimality".

Optimal designs are tailored to the model that is assumed; particularly when there is one factor, the designs often have the same number of design points as there are parameters. An example is the design at two points for the two-parameter regression model. An objection to such a design is that it does not provide for checking the model, for which trials at three or more values of x are needed. This suggests that the design should also be efficient in case the relationship is a second-order model

$$E(y) = \beta_0 + \beta_1 x + \beta_2 x^2 \tag{5}$$

It is always possible that an even-higher-order model is required, but empirical evidence from the results of decades of experimentation has shown that third- or higher-order models are rarely needed. If such models seem to be required, data transformations, such as the power transformations analyzed by **Box** and Cox [3], are often called for.

Polynomial models such as equations (1) and (5) can often be thought of as Taylor series approximations of some underlying more complicated model. In such cases, relatively simple polynomial models may provide good predictions of the response, the **bias** of the approximation being offset by a low variance due to a model with few estimated parameters. The trade-off between bias and variance and its effect on the design of experiments is one of the central topics of response-surface methodology (*see* **Response Surface Methodology**).

The design for the three-parameter second-order model, equation (5), introduces further aspects of the dependence of the optimal design on the precise purpose of experimentation. Unlike the two-point design for the first-order model, there is now no overall optimal design for all parameters. Testing whether the first-order model is adequate is equivalent to testing whether $\beta_2 = 0$. The variance of $\hat{\beta}_2$ is minimized by a design that puts half the trials at $x = 0$ and divides the remaining half between $x = -1$ and $x = 1$. But the design that provided the best estimates of all three parameters, in the sense of D-optimality described in the next section, puts one-third of the trials at each of these three points. In a satisfactory manner, the optimal design depends on the precise purpose of the experiment.

Criteria of Optimality

This section mainly describes the criterion of D-optimality which provides designs minimizing the generalized variance of the estimated parameters.

We consider a general experiment in which there are m factors and write the linear model as

$$E(y) = F\beta \tag{6}$$

Here y is the $N \times 1$ vector of responses, β is a vector of p unknown parameters, and F is the $N \times p$ *extended* design matrix. The ith row of F is $f^{\mathrm{T}}(x_i)$, a known function of the m explanatory variables. For the quadratic model equation (5), for which $m = 1$ and $p = 3$,

$$f^{\mathrm{T}}(x_i) = (1 \quad x_i \quad x_i^2) \tag{7}$$

The design is determined by the experimental values of x. The *extended* design matrix reflects not only the design but in addition the model to be fitted.

We assumed in the section titled "Simple Regression" that the observational errors were independent with constant variance σ^2. This value usually does not depend on the design. However, failure to block the experiment (*see* **Blocking**), when there are block effects, can lead to an increase in the estimate of σ^2 and a loss of power of tests about parameter values.

The least-squares estimator of the parameters is

$$\hat{\beta} = (F^{\mathrm{T}}F)^{-1}F^{\mathrm{T}}y \tag{8}$$

where the $p \times p$ matrix $F^{\mathrm{T}}F$ is the information matrix for β. The larger $F^{\mathrm{T}}F$, the greater is the information in the experiment.

With σ^2 constant, the **covariance** matrix of the least-squares estimator is

$$\text{var}\,\hat{\beta} = \sigma^2(F^{\mathrm{T}}F)^{-1} \tag{9}$$

The variance of $\hat{\beta}_j$ is proportional to the jth diagonal element of $(F^{\mathrm{T}}F)^{-1}$ with the covariance of $\hat{\beta}_j$ and $\hat{\beta}_k$ proportional to the (j, k)th off-diagonal element. If we are interested in the comparison of experimental designs, the value of σ^2 is not relevant, since the value is the same for all proposed designs for a particular experiment.

Confidence regions for all p elements of β are of the form

$$(\beta - \hat{\beta})^{\mathrm{T}}F^{\mathrm{T}}F(\beta - \hat{\beta}) \le k \tag{10}$$

In the p-dimensional space of the parameters, equation (10) defines an ellipsoid, the boundary of which is a contour of constant residual **sum of squares**. The volume of the ellipsoid is inversely proportional to the square root of the determinant $|F^{\mathrm{T}}F|$. From the expression for $\text{var}\,\hat{\beta}$ in equation (9), $\sigma^2|(F^{\mathrm{T}}F)^{-1}| = \sigma^2/|F^{\mathrm{T}}F|$ is called the *generalized variance of* $\hat{\beta}$. Designs which maximize $|F^{\mathrm{T}}F|$ minimize this generalized variance and are called *D-optimal* (for *D*eterminant).

Interest in the fitted model may be not only in the parameters, but also in the quality of the predictions. The predicted value of the response for simple regression is given by equation (3). For the general model with p parameters, equation (6), the prediction can be written

$$\hat{y}(x) = \hat{\beta}^{\mathrm{T}}f(x) \tag{11}$$

with variance

$$\text{var}\{\hat{y}(x)\} = \sigma^2 f^{\mathrm{T}}(x)(F^{\mathrm{T}}F)^{-1}f(x) \tag{12}$$

Designs which minimize the maximum over $\mathcal{X}$ of $\text{var}\,\hat{y}(x)$ are called *G-optimal*. As the theorem of the section titled "The General Equivalence Theorem and Design Efficiency" indicates, as N increases, D-optimality and G-optimality give increasingly similar designs.

An alternative to this minimax approach to $\text{var}\,\hat{y}(x)$ is to find designs that minimize the average value of the variance over $\mathcal{X}$. Such designs are variously called *I-optimal* or *V-optimal* from *I*ntegrated and *V*ariance. Another criterion of importance in applications, particularly for nonlinear models, is that of *c-optimality* in which the variance of the linear combination $c^{\mathrm{T}}\hat{\beta}$ is minimized, where c is a $p \times 1$ vector of constants. We do not exhibit such designs in this article, but focus on D- and G-optimality. Details of these and other design criteria are in Chapter 10 of Atkinson *et al.* [1].

Second-Order Response Surface in Two Factors

Many standard designs which have been derived over the years by a variety of methods share the property that they are D- and G-optimal. One example is the series of 2^m factorial designs (*see* **Factorial Experiments**) and their symmetric fractions forming the 2^{m-f} fractional factorial designs (*see* **Fractional Factorial Designs**), which can also be used to construct D-optimally blocked 2^m designs. Some properties of the designs are discussed as Example 6. But these excellent properties do depend on the number of trials N. If this is not a power of 2 then only some of the 2^m points of the full factorial will be selected as design points or to be replicated, depending on whether N is less than or greater than 2^m; the methods of optimal design construction will then be needed to determine these points. Similarly, the designs depend on the design region $\mathcal{X}$. If not all 2^m combinations of the high and low (+1 and −1) values of the factors are available, designs will need to be calculated over this irregular region. This section presents examples of the construction of designs for arbitrary N, for nonregular design regions and also for **blocking**. In the case of the blocked 2^m factorial, we might be forced by, for example, the size of batches of raw materials, to have blocks which are not all of size 2^{m-f}.

We now use the second-order polynomial in two factors to demonstrate some of the properties of D-optimal designs and of the versatility of the approach based on the theory of design optimality. The model is

$$E(y) = \beta_0 + \beta_1x_1 + \beta_2x_2 + \beta_{11}x_1^2 + \beta_{22}x_2^2 + \beta_{12}x_1x_2 \tag{13}$$

Although optimal designs depend on the model, the designs we find also have high efficiency for models in which some of the terms in equation (13) are absent.

Initially we take $\mathcal{X}$ as the square region $[-1, 1]^2$. Numerical searches for the design maximizing $|F^T F|$ will be confined to a grid of N_C candidate points, rather than to searches over the continuous region. Very little is lost in terms of the value of $|F^T F|$, the maximization is simplified and a grid of values corresponds to much experimental practice where settings are rounded to convenient values, such as 5 °C intervals of temperature. The algorithms used to find the designs are outlined in the section titled "Design Algorithms".

Example 1: Dependence of D-Optimal Design on N When $N = 9$ the D-optimal design maximizing $|F^T F|$ is the 3^2 factorial with trials at all combinations of the values $-1, 0$, and 1 for x_1 and x_2. This design is plotted in Figure 1(a). For $N = 13$, the 3^2 design is augmented by the 2^2 factorial at ± 1, that is by replicating the corner points of the 3^2 factorial, as shown in Figure 1(b). This is distinct from the conventional central composite design (*see* **Central Composite Designs**) in which the center point, that is the point $x_1 = x_2 = 0$ is replicated (*see* **Center Points**). In either design the replicated observations can be used to provide a model free estimate of the error variance.

For this second-order model, p, the number of parameters is six, which is therefore the minimum value of N. The D-optimal design for $N = 6$ has three design points that belong to the 3^2 factorial and three that do not. This design is only slightly better than the best design supported at six points of the 3^2. As N increases, the number of points of the optimal design that are not factorial points decreases and the efficiency of fractions of the factorial increases. Some details, together with the definition of design efficiency, are in the section titled "The General Equivalence Theorem and Design Efficiency".

Example 2: Blocking We now find designs for the second-order model (13) in which the trials are divided into two blocks of predetermined size. We assume additive blocking, so the model in equation (13) is augmented by an indicator variable for blocks at two levels. As an example, Figure 2 shows the division of 13 trials into two blocks, one of size 6 and the other of size 7.

This design is one of several with the same value of $|F^T F|$, including some designs from searching over the points of the 5^2 factorial in which the 3^2 factorial is augmented by points with at least one coordinate at ± 0.5. The design of Figure 2 is symmetrical in x_1 and x_2. But, in general, the labeling of the variables can be interchanged, as can the order of the blocks. This design has the advantage that its projection is the design for $N = 13$ in Figure 1(b). Thus the design is D-optimal whether or not blocking is required.

Example 3: Nonregular Design Region We consider a nonregular design region in which some

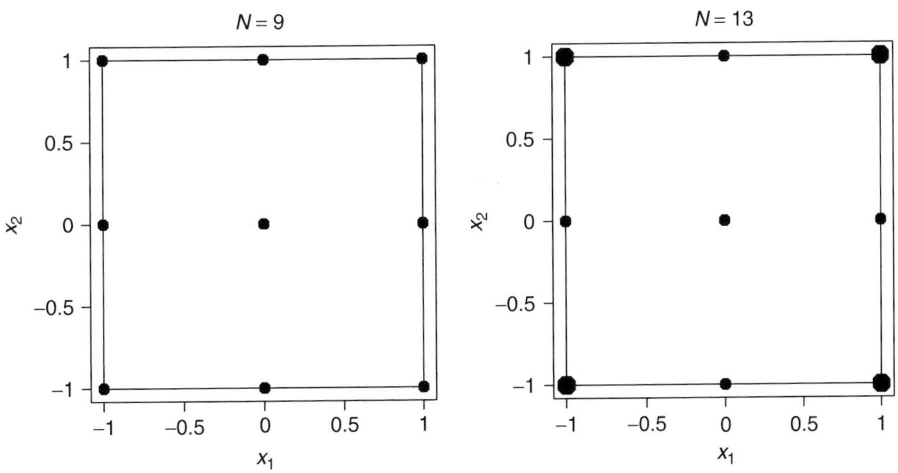

Figure 1 Second-order response surface in two factors: D-optimal designs for number of trials $N = 9$ and 13; heavy dots indicate support points with two observations

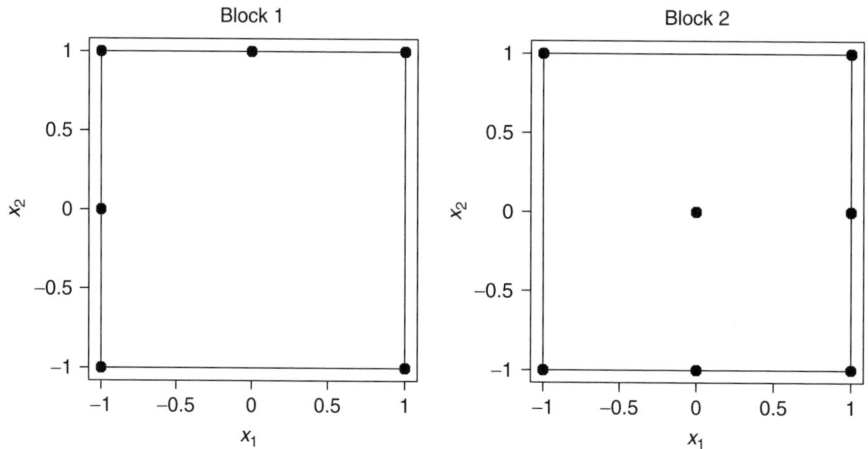

Figure 2 Second-order response surface in two factors. Optimal design for blocks of 6 and 7 trials, a total of $N = 13$ trials

extreme values of x_1 and x_2 are excluded; in a chemical process these might correspond to conditions so severe that the product is degraded, or so dilute or cold that the desired reactions do not occur at a significant rate. Such a design region is shown in Figure 3. Optimal designs were found by searching over the 63 points of the 11^2 factorial that are included in the design region.

Figure 3 gives designs for $N = 9$ and 13. Although there are some other designs that are not far from these designs in efficiency, it is hard to summon up any intuition that leads to efficient designs; the algorithmic methods of optimal experimental design seem essential for the construction of efficient designs.

Finally, Figure 4 shows the D-optimal design when the 13 trials are divided into two blocks of sizes 4 and 9. In this case the projection of the two designs does not quite give the optimal unblocked design for $N = 13$ in Figure 3.

Example 4: Mixture Experiments In mixture experiments (*see* **Mixture Experiments**) the response depends on the proportion of the various

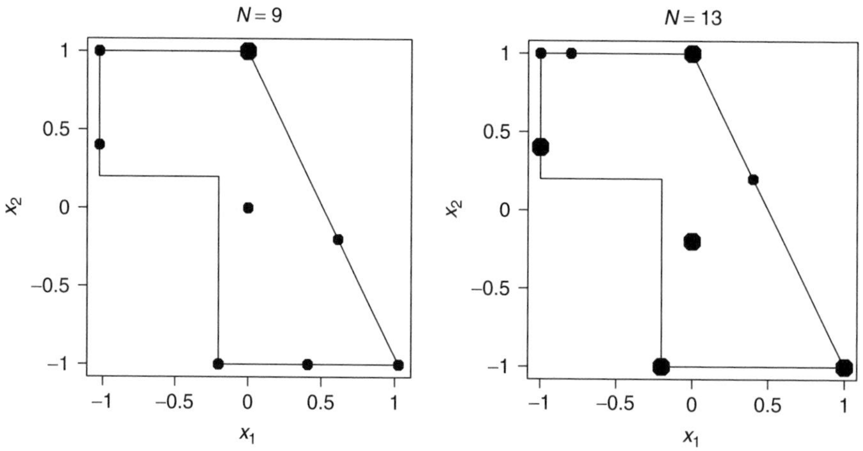

Figure 3 Second-order response surface in two factors with a nonregular design region: D-optimal designs for number of trials $N = 9$ and 13; heavy dots indicate support points with two observations

components, but not on their amount. This imposes constraints on the values of the m factors and the resulting experimental region is a simplex in $m-1$ dimensions.

The statistical study of mixture experiments was introduced by Scheffé [4]. Cornell [5] gives a book-length treatment. Because of the constraints on the factors, not all terms in standard polynomial models are estimable. However, optimal design methods provide D-optimal designs for Scheffé's linear models. Chapter 16 of Atkinson *et al.* [1] gives the details of the designs, including blocked designs similar to those of Figure 4.

It is often necessary to introduce further constraints on the values of the factors to ensure that all mixtures in the experimental region are of interest; Martin *et al.* [6] give examples in glass manufacture. The resulting design regions can be hard to visualize. However, once the design region has been given a mathematical description, optimal designs can readily be found by methods similar to those illustrated above for experiments involving response-surface models.

Exact and Continuous Designs

The D-optimal designs found so far have been for a specified model, design region $\mathcal{X}$ and number of trials N. As Figures 1 and 3 show, the structure of the designs can depend on the value of N. However, as N increases, the dependence decreases and the series of *exact* designs for specific values of N tends to a limiting *continuous* design independent of N. This section describes some uses of continuous optimal designs: they are often easier to find than an exact design for one specific N and they can readily be approximated to provide good exact designs for many values of N. The next section shows how continuous designs can be checked for optimality. Implications for algorithms for finding exact designs are given in the section titled "Design Algorithms".

The theory was introduced in Kiefer [7] and developed in a series of further papers. Kiefer's work on optimal design is collected in Brown *et al.* [8]. A basic idea is to write designs as probability measures.

Suppose an exact design has n points of support with r_i the integer number of trials at experimental conditions x_i. This design can be written as

$$\xi_N = \left\{ \begin{matrix} x_1 & x_2 \ldots x_n \\ r_1/N & r_2/N \ldots r_n/N \end{matrix} \right\} \quad (14)$$

where $\sum r_i = N$. The measure ξ_N specifies the experimental conditions and the proportion of experimental effort at each condition. For example, for a response-surface design in m factors, the vector x_i gives the values of all m factors for the r_i replicate observations at the ith set of conditions; often for an exact design r_i will be 1.

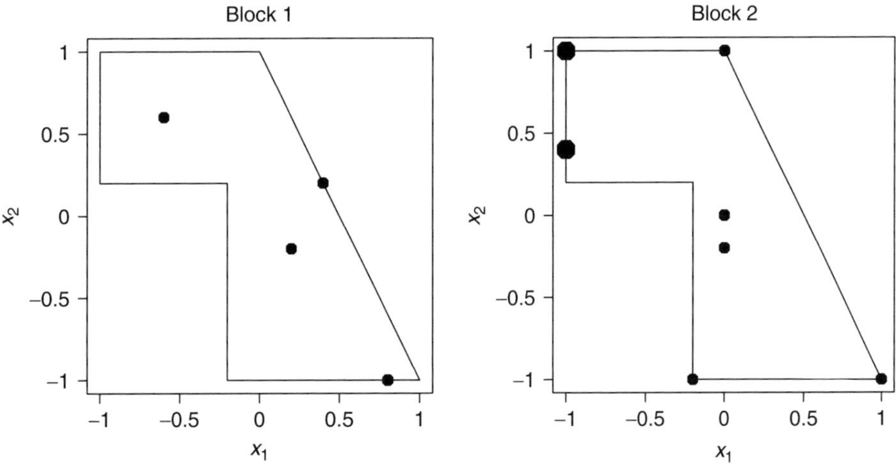

Figure 4 Second-order response surface in two factors with a nonregular design region: D-optimal design for blocks of 4 and 9 trials, a total of $N = 13$ trials; heavy dots indicate support points with two observations

The information matrix introduced in equation (8) can be written in terms of this measure as

$$F^{\mathrm{T}}F = NM(\xi_N) \tag{15}$$

Likewise, the variance of the predicted response, introduced in equation (12), can be written in the standardized form

$$d(x, \xi_N) = \frac{N\mathrm{var}\{\hat{y}(x)\}}{\sigma^2} = f^{\mathrm{T}}(x)M^{-1}(\xi_N)f(x) \tag{16}$$

The continuous design ξ is found by replacing the fractions r_i/N in ξ_N by weights w_i. At the end of the section titled "Simple Regression" it was stated that the D-optimal continuous design for the quadratic model in one variable with $\mathcal{X} = [-1, 1]$ was

$$\xi^* = \begin{Bmatrix} -1 & 0 & 1 \\ 1/3 & 1/3 & 1/3 \end{Bmatrix} \tag{17}$$

where ξ^*, rather than ξ, indicates optimality. This design puts one-third of the trials at each of the three levels of x; $-1, 0$, and 1. For any N that is a multiple of three we can find the exact optimal design by putting $N/3$ trials at each of these sets of experimental conditions. For other values of N a good design is to distribute the trials as evenly as possible over these three points.

The General Equivalence Theorem and Design Efficiency

Exact D-optimal designs such as those of the section titled "Second-Order Response Surface in Two Factors", maximize $|F^{\mathrm{T}}F|$. From equation (15) D-optimal continuous designs therefore maximize $|M(\xi)|$. Kiefer and Wolfowitz [9] provide an equivalence theorem relating D- and G-optimal designs that provides a method of checking the optimality of any continuous design.

If ξ^* is a D-optimal design measure, then ξ^* also minimizes the maximum over $\mathcal{X}$ of the standardized variance of prediction $d(x, \xi)$ given by equation (16); that is ξ^* is also G-optimal. If we write this maximum as

$$\bar{d}(\xi) = \max_{x \in \mathcal{X}} d(x, \xi) \tag{18}$$

then $\bar{d}(\xi^*) = p$, the number of parameters of the linear model. For any nonoptimal design ξ the maximum over $\mathcal{X}$ of $d(x, \xi)$ is greater than p, which provides a method for checking for D-optimality.

The theorem holds for continuous optimal designs. Exact optimal designs such as that of equation (16) for the quadratic model when N is a multiple of 3 will clearly also satisfy it. But, for example, the exact D-optimal design when $N = 9$ of Figure 1 will not satisfy the theorem.

The equivalence theorem provides a **benchmark** for the comparison of designs. To compare a continuous design ξ with the optimal design ξ^* we define the D-efficiency as

$$D_{\mathrm{eff}} = \left\{ \frac{|M(\xi)|}{|M(\xi^*)|} \right\}^{1/p} \tag{19}$$

The comparison of information matrices for designs that are measures removes the effect of the number of observations. Taking the pth root of the ratio of the determinants in equation (19) results in an efficiency measure which has the dimensions of a variance, irrespective of the dimension of the model. Since the variance of estimated regression coefficients is inversely proportional to the number of observations, two replicates of a design measure for which $D_{\mathrm{eff}} = 50\%$ would be as efficient as one replicate of the optimal measure.

Example 5: Quadratic Model in One Variable The D-optimal continuous design for the quadratic model in one variable with $\mathcal{X} = [-1, 1]$ is given in equation (16). The claim that this design is D-optimal can be checked using the general equivalence theorem through calculation of the variance of the predicted response.

For this design $d(x, \xi^*)$ is the quartic given by

$$d(x, \xi^*) = 3 - \frac{9x^2}{2} + \frac{9x^4}{2} \tag{20}$$

which has a maximum over $\mathcal{X}$ of 3 at $x = -1, 0$, or 1, the three design points. Further, this maximum value is equal to the number of parameters p. The plot in Figure 5 exhibits the optimality of the design since the values of three at the design points are indeed the optima over $\mathcal{X}$.

Example 6: 2^m Factorial and Fractions The first-order model in m factors with main effects and all

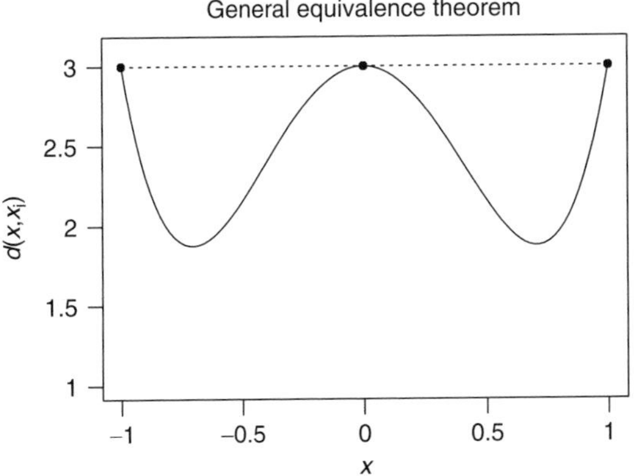

Figure 5 Quadratic model in one variable; $d(x, \xi^*)$ over $\mathcal{X}$ for the design of equation (16) which puts equal weights at $x = -1, 0$, and 1. Since the maximum value is three, which is the number of parameters in the model, the design is D-optimal

possible interactions may be written as

$$E(y) = \beta_0 + \sum_{i=1}^{m} \beta_i x_i + \sum_{i=1}^{m-1} \sum_{j=1+1}^{m} \beta_{ij} x_i x_j + \cdots + \beta_{ijk\ldots} x_i x_j x_k \cdots \quad (21)$$

When $\mathcal{X}$ is cuboidal, the D-optimal continuous design is the equireplicated 2^m factorial, which has $n = 2^m$ points of support at all combinations of values of ± 1 for each factor. If the full model (21) is fitted $p = N$ and $M(\xi^*)$ is the identity matrix. Then $d(x, \xi^*) = p$ at the design points and the equivalence theorem for continuous D-optimality is satisfied. Because of the orthogonality of the design, reflected in the diagonal nature of $M(\xi^*)$, the same design is optimal if terms are omitted from equation (21) or if the design is projected by the omission of one or more factors. Provided all the terms in the model can be estimated, the regular fractions of the design, that is the series of 2^{m-f} fractional factorials, are also the continuous D-optimal designs.

It is clear when analyzing the properties of this design that the maximum values of $d(x, \xi^*)$ occur at the points $x_i = \pm 1$. However, with more complicated examples, it is prudent to search $\mathcal{X}$ more carefully. Suppose the design used was a shrunken 2^m factorial with $x_i = \pm c, c < 1$. Then $d(x, \xi)$ would again equal one at the points of support of the design. However, this would not be the optimal design, since these would not be the maximum values of $d(x, \xi)$ over $\mathcal{X}$.

Example 7: Second-Order Response Surface in *m* Factors The second-order polynomial in m factors is

$$E(y) = \beta_0 + \sum_{j=1}^{m} \beta_j x_j + \sum_{j=1}^{m-1} \sum_{k=j+1}^{m} \beta_{jk} x_j x_k + \sum_{j=1}^{m} \beta_{jj} x_j^2 \quad (22)$$

Farrell *et al.* [10] show that the D-optimal continuous design for a cuboidal experimental region is supported on subsets of the points of the 3^m factorial, with the members of each subset having the same number of nonzero coordinates. Only three subsets are required, over each of which a specified design weight is uniformly distributed. For $m \leq 5$ the subsets have support ([0], $[m-1]$, $[m]$), that is the center point, the midpoints of edges, and the corner points of the 3^m factorial. For larger values of m, other subsets also support optimal designs. It is interesting to note that the central composite designs ([11]; *see* **Central Composite Designs**), which belong to the family ([0], [1], $[m]$), cannot provide the support for a D-optimal continuous design when $m > 2$.

The number of support points of these continuous D-optimal designs increases rapidly with m, being 113 for $m = 5$. Exact designs with drastically reduced support are therefore required. Since the D-optimal continuous designs have support on the points of the 3^m factorial, it is reasonable to expect that good exact designs for small N can be found by searching over the points of these factorials. Table 11.7 of Atkinson *et al.* [1] summarizes designs found by such searches.

The minimal design has $N = p$. For $2 \leq m \leq 5$, these design all have efficiencies above 85% but designs with appreciably higher efficiencies can be found at the cost of only a few more design points. For $m = 3$, when $p = 10$, one design with $N = 14$ has an efficiency of 97.59%. When $m = 4$, $p = 15$ and a design with $N = 18$ has an efficiency of 93.11%. Finally, for $m = 5$ and $p = 21$, the best design with $N = 26$ is 95.19% efficient. The addition of one or a few trials above $N = p$ causes an appreciable increase in the efficiency of the design, in addition to the reduced variance of parameter estimates coming from a larger design with larger N.

Section 11.5 of Atkinson *et al.* [1] tabulates both continuous and exact D-optimal response-surface designs including designs for spheroidal regions. Chapter 15 of the same book discusses the blocking of response-surface designs.

Example 8: Second-Order Response Surface in Two Factors The designs in Figure 1 for the two-factor response surface model when $N = 9$ and 13 illustrate the dependence of exact optimal designs on the value of N. The continuous D-optimal design, like these two designs, is supported on the points of the 3^2 factorial with weight 0.1458 at each corner point (± 1, ± 1), weight 0.0802 at each center points of the edges (0, ± 1; ± 1, 0) and weight 0.0960 at the center point of the design. Relative to this continuous design the exact design for $N = 9$ has a D-efficiency of 97.40%, whereas the value for $N = 13$ is 99.77%.

The smallest design, the exact D-optimal design for $N = 6$, has three support points that are not those of the 3^2 factorial and a D-efficiency of 89.15% relative to the continuous optimal design. The best design on six points of the 3^2 factorial has a very slightly lower efficiency of 88.49%. The best designs of both types for $N = 7$ and 8 already have efficiencies close to, or above, 95%. The conclusion is that, even for these very small designs, fractions of the 3^2 factorial provide efficient designs, as they do for higher values of m in Example 7. Application of the methods of optimal experimental design allows us to quantify the loss involved in experimenting at only these three levels.

Design Algorithms

Algorithms for the construction of continuous designs can either use relatively straightforward function maximization or be based on the general equivalence theorem through the properties of the variance $d(x, \xi)$. These show that the optimal design contains only points with the highest possible value of $d(x, \xi)$, namely p.

In the sequential construction of continuous D-optimal designs, observations are added at the value of x in $\mathcal{X}$ for which this variance is a maximum; $\mathcal{X}$ can either be treated as a continuous region or discretized to give a set of N_C candidate points. For an N-trial design written as the measure ξ_N, let the measure $\bar{\xi}_N$ put unit mass at the point where this maximum occurs. Then, provided $\bar{d}(\xi) > p$, we take

$$\xi_{N+1} = \frac{(N\xi_N + \bar{\xi}_N)}{(N+1)} \tag{23}$$

If $\bar{d}(\xi) = p$, the equivalence theorem of the section titled "The General Equivalence Theorem and Design Efficiency" shows that the design is D-optimal and the algorithm stops.

Convergence is not monotonic. As an instance, starting from the nine-trial design of Figure 1 for the two-factor response-surface model, we find that the variance is maximum at the corners of the region and one trial would be added at one of these corners. Continuing the procedure with $N = 10$ would lead to the addition of a trial at another corner point, and so on, until the design for $N = 13$ is obtained when $\bar{d}(\xi_{13})$ is very close to p. Addition of one further trial must destroy the symmetry of this design and there will be a slight decrease in D-efficiency and an increase in the value of $\bar{d}(\xi)$, with $\bar{d}(\xi_{14}) > \bar{d}(\xi_{13})$. However, continuing with the sequential addition of trials in this way will eventually lead to a design with the weights given in the first paragraph of Example 8.

The convergence of such algorithms can be slow, but ultimate convergence is usually not required. Starting from an arbitrary design, the algorithm typically adds mass at a restricted number of conditions in $\mathcal{X}$. These conditions suggest the support of the optimal design. Once this pattern becomes clear, either the arbitrary starting design can be rejected and the algorithm restarted or the information gained can be used to guide maximization of the function $|M(\xi)|$. The advantage of the sequential algorithm is that it reduces the search for an optimal design to a sequence of optimizations in m dimensions.

Exact designs for N trials are usually found using exchange algorithms. It now follows from the equivalence theorem that the n support points of the design should have high values of $d(x, \xi)$ and the $N_\mathrm{C} - n$ unused candidate points should have low values. Exchange algorithms explore the replacement of one of the N trials of the design by any of the other $N_\mathrm{C} - 1$ candidate points, thereby allowing for a possible increase in replication of points already in the design. In the original algorithm [12] all $(N_\mathrm{C} - 1) \times n$ exchanges were considered and the one leading to the largest increase in $|M(\xi_N)|$ made. At this point n may change, although N is constant. The process was then repeated until no further improvement was found. More recent algorithms [13, 14] independently order the values of $d(x, \xi_N)$ for the candidate and support points and make the first exchanges in the ordered lists that increases $|M(\xi_N)|$. Again the process is repeated until no further improvement is found. Whatever algorithm is used, the search is repeated several times from random starting points.

Computer procedures for finding D-optimal designs of the kind given here, such as Example 3, accordingly require the ability to specify a model and a list of candidate points. Commercial packages with this flexibility include Design Expert, JMP, SAS, and WebDOE. The R package AlgDesign is downloadable from `http://cran.r-project.org` and the Gosset package of Hardin and Sloane [15] is at `http://www.research.att.com/~njas/gosset/`.

Discussion and Literature

The motivation for optimal designs in the introduction to this article was that they provide a means of minimizing functions of the variances of the estimated parameters. Box [16, Chapter 9] argues that it is doubtful if single criterion optimal designs are useful in locating designs satisfying the compromises between objectives required in a practical experiment. Although the focus of this article is on D-optimality, the theory and algorithms extend straightforwardly to compound designs in which a weighted combination of criteria can be employed to find a design that is, for example, simultaneously good for parameter estimation, establishing a response transformation, and for prediction in a specified region. A general approach to compound designs is given in Chapter 21 of Atkinson *et al.* [1].

The list of 14 characteristics of a good experimental design in Box and Draper [17] includes the ability to check models for lack of fit. The early stages of an experimental program often use 2^m factorials and their fractions to screen factors. To check whether a second-order model such as equation (13) is required, one or more trials are included at the center of the experimental region. An informal extension of this procedure is to augment the optimal design by inclusion of a few extra trials in unused regions of $\mathcal{X}$. A more formal approach to designs for model checking, using the methods of optimal design, is given by DuMouchel and Jones [18], who specify both a primary model, for which good estimation of the parameters is required, and secondary terms, which need to be detected if they are nonzero. It is hoped that the primary model will provide an adequate fit. Examples of designs and the exploration of the relationship with compound D-optimality are given in Chapter 20 of Atkinson *et al.* [1].

The first book in English on optimal experimental design was Fedorov [12]. Pukelsheim [19] focuses on theory, as does the shorter book of Fedorov and Hackl [20]. Atkinson *et al.* [1] is more concerned with applications and provides SAS code for the construction of many designs. In addition to the linear models covered in this article, methods are given for the optimal design of experiments for correlated data, generalized linear models and, in particular, in Chapter 17, for the design of experiments for nonlinear models. The papers in Berger and Wong [21] focus on recent applications of optimal design, including some in the pharmaceutical industry. As with any topic, a large amount of information can be obtained by searching the Web. About 75% of the references on Google are to "optimal design" while 25% use the term *optimum design*.

References

[1] Atkinson, A.C., Donev, A.N. & Tobias, R.D. (2007). *Optimum Experimental Designs, with SAS*, Oxford University Press, Oxford.

[2] Joiner, B.L. (1981). Lurking variables: some examples, *American Statistician* **35**, 227–233.

[3] Box, G.E.P. & Cox, D.R. (1964). An analysis of transformations (with discussion), *Journal of the Royal Statistical Society. Series B* **26**, 211–246.

[4] Scheffé, H. (1958). Experiments with mixtures, *Journal of the Royal Statistical Society. Series B* **20**, 344–360.

[5] Cornell, J.A. (1990). *Experiments with Mixtures*, 2nd Edition, John Wiley & Sons, New York.

[6] Martin, R.J., Bursnall, M.C. & Stillman, E.C. (2001). Further results on optimal and efficient designs for constrained mixture experiments, in *Optimal Design 2000*, A.C. Atkinson, B. Bogacka & A. Zhigljavsky, eds, Kluwer Academic Publishers, Dordrecht, pp. 225–239.

[7] Kiefer, J. (1959). Optimum experimental designs (with discussion), *Journal of the Royal Statistical Society. Series B* **21**, 272–319.

[8] Brown, L.D., Olkin, I., Sacks J. & Wynn, H.P. (eds) (1985). *Jack Carl Kiefer Collected Papers III*, John Wiley & Sons, New York.

[9] Kiefer, J. & Wolfowitz, J. (1959). Optimum designs in regression problems, *Annals of Mathematical Statistics* **30**, 271–294.

[10] Farrell, R.H., Kiefer, J. & Walbran, A. (1968). Optimum multivariate designs, in *Proceedings of the 5th Berkeley Symposium*, University of California Press, Berkeley, Vol. 1, pp. 113–138.

[11] Box, G.E.P. & Draper, N.R. (1963). The choice of a second order rotatable design, *Biometrika* **50**, 335–352.

[12] Fedorov, V.V. (1972). *Theory of Optimal Experiments*, Academic Press, New York.

[13] Cook, R.D. & Nachtsheim, C.J. (1980). A comparison of algorithms for constructing exact optimal designs, *Technometrics* **22**, 315–324.

[14] Atkinson, A.C. & Donev, A.N. (1989). The construction of exact D–optimum experimental designs with application to blocking response surface designs, *Biometrika* **76**, 515–526.

[15] Hardin, R.H. & Sloane, N.J.A. (1993). A new approach to the construction of optimal designs, *Journal of Statistical Planning and Inference* **37**, 339–369.

[16] Box, G.E.P. (2006). *Improving Almost Anything*, 2nd Edition, John Wiley & Sons, New York.

[17] Box, G.E.P. & Draper, N.R. (1975). Robust designs, *Biometrika* **62**, 347–352.

[18] DuMouchel, W. & Jones, B. (1994). A simple Bayesian modification of D–optimal designs to reduce dependence on an assumed model, *Technometrics* **36**, 37–47.

[19] Pukelsheim, F. (1993). *Optimal Design of Experiments*, John Wiley & Sons, New York.

[20] Fedorov, V.V. and Hackl, P. (1997). *Model-Oriented Design of Experiments, Lecture Notes in Statistics, Vol. 125*, Springer-Verlag, New York.

[21] Berger, M. & Wong, W.-K. (eds) (2005). *Applied Optimal Designs*, John Wiley & Sons, New York.

ANTHONY C. ATKINSON

Optimal Reliability Design– Algorithms and Comparisons

Introduction

Optimal reliability design problems are known to be NP hard [1]. Finding efficient optimization algorithms is always a hot spot in this field. Classification of the literature by **reliability optimization** techniques is summarized in Table 1. This paper reviews and compares the recent developments of heuristic algorithms, metaheuristic algorithms, exact methods, and other optimization techniques in optimal reliability design. Owing to their robustness and feasibility, metaheuristic algorithms, especially genetic algorithms (GAs), have been widely and successfully applied (*see* **Genetic Algorithms in Reliability**). To improve computation efficiency or to avoid premature convergence, an important part of this work has been devoted in recent years to developing hybrid GAs, which usually combine a GA with heuristic algorithms, simulation annealing methods, **neural network** techniques, or other local search methods. Though more computation effort is involved, exact methods are particularly advantageous for small problems, and their solutions can be used to measure the performance of the heuristic or metaheuristic methods [2]. No obviously superior heuristic method has been proposed, but several of them have been well combined with exact or metaheuristic methods to improve their computational efficiency.

Metaheuristic Methods

Metaheuristic methods inspired by natural phenomena usually include the GAs, tabu search, simulated annealing algorithm, and ant colony optimization method. ant colony optimization (ACO) has been recently introduced into optimal reliability design, and it is proving to be a very promising general

Table 1 Reference classification by reliability optimization methods

	Metaheuristic algorithms
ACO	[3–8]
GA	[9–39]
HGA	[40–50]
TS	[51]
SA	[10, 52–55]
IA	[56]
GDA	[57]
CEA	[58]
	Exact methods
	[59–67]
	Max–min approaches
	[68, 69]
	Heuristic methods
	[70–72]
	Dynamic programming
	[73, 74]

method in this field. Altiparmak *et al.* [10] provide a comparison of metaheuristics for the optimal design of computer networks.

Ant Colony Optimization Method

ACO is one of the adaptive metaheuristic optimization methods developed by M. Dorigo for traveling salesman problems in [75]. It is inspired by the behavior of real-life ants that consistently establish the shortest path from their nest to food. The essential trait of the ACO algorithm is the combination of *a priori* information about the structure of a promising solution with *posteriori* information about the structure of previously obtained good solutions [76].

Liang and Smith [3] first develop an ant colony metaheuristic optimization method to solve the reliability–**redundancy** allocation problem for a ***k*-out-of-*n***: G series system. The proposed ACO approach includes four stages:

- Construction stage: construct an initial solution by selecting component j for subsystem i according to its specific heuristic η_{ij} and pheromone trail intensity τ_{ij}, which also sets up the transition probability mass function P_{ij}.
- Evaluation stage: evaluate the corresponding system reliability and penalized system reliability providing the specified penalized parameter.
- Improvement stage: improve the constructed solutions through local search.
- Updating stage: update the pheromone value online and offline given the corresponding penalized system reliability and the controlling parameter for pheromone persistence.

Nahas and Nourelfath [5] present an application of the ant system in a reliability optimization problem to maximize the system reliability subject to the system budget, with multichoice constraints incorporated at each subsystem. It also combines a local search algorithm and a specific improvement algorithm that uses the remaining budget to improve the quality of a solution.

The ACO algorithm has also been applied to a multiobjective reliability optimization problem [6] and to the optimal design of multistate series–parallel power systems [4].

Hybrid Genetic Algorithm

GA is a population-based directed random search technique inspired by the principles of evolution. Although it provides only heuristic solutions, it can be effectively applied to almost all complex combinatorial problems, and, thereby, it has been employed in a large number of references as shown in Table 1. Gen and Kim [77] provide a state-of-the-art survey of GA-based reliability design.

To improve computational efficiency, or to avoid premature convergence, numerous researchers have been inspired to seek effective combinations of GAs with heuristic algorithms, simulation annealing methods, neural network techniques, steepest-descent methods, or other local search methods. The combinations are generally called hybrid GAs, and they represent one of the most promising developmental directions in optimization techniques.

Considering a complex system with a known system structure function, Zhao and Liu [49] provide a unified modeling idea for both active and cold-standby redundancy optimization problems. The model prohibits any mixture of component types within subsystems. Both the lifetime and the cost of redundancy components are considered as random variables, so stochastic simulation is used to estimate the system performance, including the mean lifetime, percentile lifetime, and reliability. To speed up the solution process, these simulation results become the training data for training a neural network to approximate the system performance. The trained neural network is finally embedded into a GA to form a hybrid intelligent algorithm for solving the proposed model. Later, Wattanapongsakorn and Levitan [54] used random fuzzy lifetimes as the basic parameters and employed a random fuzzy simulation to generate the training data.

Lee *et al.* [42] develop a two-phase NN-hGA in which NN is used as a rough search technique to devise the initial solutions for a GA. By bounding the broad continuous search space with the NN technique, the NN-hGA derives the optimum robustly. However, in some cases, this algorithm may require too much computational time to be practical. To improve the computation efficiency, Lee *et al.* [43] present an NN-flcGA to effectively control the balance between exploitation and exploration, which characterizes the behavior of GAs. The essential features of the NN-flcGA include the following:

- combination with an NN technique to devise initial values for the GA;

- application of a fuzzy logic controller when tuning strategy GA parameters dynamically; and
- incorporation of the revised simplex search method.

  Later, Lee *et al.* [44] proposed a similar hybrid GA called f-hGA for the redundancy allocation problem of a series–parallel system. It is based on
- application of a fuzzy logic controller to automatically regulate the GA parameters;
- incorporation of the iterative hill climbing method to perform local exploitation around the near-optimum solution.

Hsieh and Hsieh [40] considered the optimal task allocation strategy and hardware redundancy level for a cycle-free distributed computing system so that the system cost during the period of task execution is minimized. The proposed hybrid heuristic combines the GA and the steepest-descent method. Later, Hsieh [41] seeks similar optimal solutions to minimize system cost under constraints on the hardware redundancy levels.

Tabu Search

Although Kuo [2] described the promise of tabu search, Kulturel-Konak *et al.* [51] first developed an tabu search (TS) approach with the application of near feasible threshold (NFT) [78] for reliability optimization problems. This method uses a subsystem-based tabu entry and dynamic length tabu list to reduce the sensitivity of the algorithm to selection of the tabu list length. The definition of the moves in this approach offers an advantage in efficiency, since it does not require recalculating the entire system reliability, but only the reliability of the changed subsystem. The results of several examples demonstrate the superior performance of this TS approach in terms of efficiency and solution superiority when compared to that of a GA.

Other Metaheuristic Methods

Some other adaptive metaheuristic optimization methods inspired by activities in nature have also been proposed and applied in optimal reliability design. Chen and You [56] developed an approach based on immune algorithms inspired by the natural immune system of all animals. It analogizes antibodies and antigens as the solutions and objection functions, respectively. Rocco *et al.* [58] proposed a cellular evolutionary approach combining the multimember evolution strategy with concepts from cellular automata [79] for the selection step. In this approach, the selection of the parents is performed only in the neighborhood in contrast to the general evolutionary strategy that searches for parents in the whole population. And a great deluge algorithm is extended and applied to optimize the reliability of complex systems in [57]. When both accuracy and speed are considered simultaneously, it is proved to be an efficient alternative to ACO and other existing optimization techniques.

Exact Methods

Unlike metaheuristic algorithms, exact methods provide exact optimal solutions though much more computation complexity is involved. The development of exact methods, such as the branch-and-bound approach and lexicographic search, has recently been concentrated on techniques to reduce the search space of discrete optimization methods.

Sung and Cho [66] consider a reliability–redundancy allocation problem in which multiple choice and resource constraints are incorporated. The problem is first transformed into a bicriteria nonlinear integer programming problem by introducing 0–1 variables. Given a good feasible solution, the lower reliability bound of a subsystem is determined by the product of the maximal component reliabilities of all the other subsystems in the solution, while the upper bound is determined by the maximal amount of available sources of this subsystem. A branch-and-bound procedure, based on this reduced solution space, is then derived to search for the global optimal solution. Later, Sung and Cho [67] even combine the lower and upper bounds of the system reliability, which are obtained by variable relaxation and Lagrangian relaxation techniques, to further reduce the search space.

Also, with a branch-and-bound algorithm, Djerdjour and Rekab [59] obtain the upper bound of series–parallel system reliability from its continuous relaxation problem. The relaxed problem is efficiently solved by the greedy procedure described in [80], combining heuristic methods to make use of some slack in the constraints obtained from rounding down. This technique assumes that the objective and constraint functions are monotonically increasing. Ha and Kuo [61] present an efficient branch-and-bound approach for **coherent systems** based on a one-neighborhood local maximum obtained from the steepest-ascent heuristic method. Numerical examples of a bridge system and a hierarchical series–parallel system demonstrate the advantages of this proposed algorithm in flexibility and efficiency.

Apart from the branch-and-bound approach, Prasad and Kuo [63] present a partial enumeration method for a wide range of complex optimization problems based on a lexicographic search. The proposed upper bound of system reliability is very useful in eliminating several inferior feasible or infeasible solutions as shown in either big or small numerical examples. It also shows that the search process described in [84] does not necessarily give an exact optimal solution due to its logical flows.

Lin and Kuo [62] develop a strong Lin and Kuo heuristic to search for an ideal allocation through the application of the reliability importance (*see* **Component Reliability Importance**). It concludes that, if there exists an invariant optimal allocation for a system, the optimal allocation is to assign component reliabilities according to B-importance ordering. This Lin and Kuo heuristic can provide an exact optimal allocation.

Assuming the existence of a convex and differential reliability cost function $C_i(y_{ij})$, $y_{ij} = \log(1 - r_{ij})$ for all components j in any subsystem i, Elegbede *et al.* [60] prove that the components in each subsystem of a series–parallel system must have identical reliability for the purpose of cost minimization. The solution of the corresponding unconstrained problem provides the upper bound of the cost, while a doubly minimization problem gives its lower bound. With these results, the algorithm ECAY, which can provide either exact or approximate solutions depending on different stop criteria, is proposed for series–**parallel systems**.

Other Optimization Techniques

For series–parallel systems, Ramirez-Marquez *et al.* [69] formulate the reliability–redundancy optimizat ion problem with the objective of maximizing the minimal subsystem reliability. Assuming linear constraints, this problem is equivalent to a linear formulation through an easy logarithm transformation [81], which can be solved by readily available commercial software. It can also serve as a surrogate for traditional reliability optimization problems by sequentially solving a series of max–min subproblems. Lee *et al.* [68] present a comparison between the Nakagawa and Nakashima method [82] and the max–min approach used by Ramirez-Marquez from the standpoint of solution quality and computational complexity. The experimental results show that the max–min approach is superior to the Nakagawa and Nakashima method in terms of solution quality in small-scale problems, but the analysis of its computational complexity demonstrates that the max–min approach is inferior to other greedy heuristics.

You and Chen [72] develop a heuristic approach inspired by the greedy method and a GA. The structure of this algorithm includes (a) randomly generating a specified population size number of minimum workable solutions, (b) assigning components either according to the greedy method or to the random selection method, and (c) improving solutions through an inner-system and intersystem solution revision process.

Kevin and Sancho [73] apply a hybrid dynamic programming/depth-first search algorithm to redundancy allocation problems with more than one constraint. Given the tightest upper bound, the knapsack relaxation problem is formulated with only one constraint, and its solution $f_1(b)$ is obtained by a dynamic programming method. After choosing a small specified parameter e, the depth-first search technique is used to find all near-optimal solutions with objectives between $f_1(b)$ and $f_1(b) - e$. The optimal solution is given by the best feasible solution among all of the near-optimal solutions. Also with dynamic programming, Yalaoui *et al.* [74] consider a reliability–redundancy allocation problem in series–parallel systems with components chosen among a finite set. This pseudopolynomial YCC algorithm is composed of two steps: the solution of the subproblems, one for each subsystem, and the global resolution using previous results. It shows that the solutions converge quickly toward the optimum as a function of the required precision.

Comparisons and Discussions of Algorithms Reported in the Literature

In this section, we provide a comparison of several heuristic or metaheuristic algorithms reported in the literature. The compared numerical results are from the GA in Coit and Smith [78], the ACO in Liang and Smith [3], TS in Kulturel-Konak *et al.* [51], linear approximation in Hsieh [81], the immune algorithm (IA) in Chen and You [56] and the heuristic method in You and Chen [72]. The 33 variations of the Fyffe *et al.* problem, as devised by Nakagawa and Miyazaki [83], are used to test their performance, where different types are allowed to reside in parallel. In this problem set, the cost constraint is maintained at 130 and the weight constraint varies from 191 to 159.

As shown in literature, ACO [3], TS [51], IA [56], and heuristic methods [72] generally yield solutions with a higher reliability. When compared to GA [78],

- ACO [3] is reported to consistently perform well over different problem sizes and parameters and improve on GA's random behavior.
- TS [51] results in a superior performance in terms of best solutions found and reduced variability and greater efficiency based on the number of objective function evaluations required.
- IA [56] finds better or equally good solutions for all 33 test problems, but the performance of this IA-based approach is sensitive to value-combinations of the parameters, whose best values are case-dependent and only based upon the experience from preliminary runs and more CPU time is taken by IAs.
- The best solutions found by heuristic methods [72] are all better than, or as good as, the well-known best solutions from other approaches. With this method, the average CPU time for each problem is within 8 s.
- In terms of solution quality, the linear approximation approach in [81] is inferior, but very efficient, and the CPU time for all of the test problems is within 1 s.
- If a decision-maker is considering the max–min approach as a surrogate for system reliability maximization, the max–min approach [69] is shown to be capable of obtaining a close solution (within 0.22%), but it is unknown whether this performance will continue as problem sizes become larger.

For all the optimization techniques mentioned above, it might be hard to discuss about which tool is superior because in different design problems or even in the same problem with different parameters, these tools will perform varyingly.

Generally, if computational efficiency is of most concern to designers, linear approximation or heuristic methods can obtain competitive feasible solutions within a very short time (few seconds), as in [72, 81]. The proposed linear approximation [81] is also easy to implement with any LP software. However, the main limitation of those reported approaches is that the constraints must be linear and separable.

Owing to their robustness and feasibility, metaheuristic methods such as GA and the recently developed TS and ACO could be successfully applied to almost all NP-hard reliability optimization problems. However, they cannot guarantee the optimality and sometimes can suffer from premature convergence because they have many unknown parameters and neither use a prior knowledge nor exploit local search information. Compared to traditional metaheuristic methods, a set of promising algorithms, hybrid GAs [40–44, 49, 50], are attractive since they retain the advantages of GAs in robustness and feasibility, but significantly improve their computational efficiency and searching ability in finding a global optimum by combining heuristic algorithms, neural network techniques, steepest-descent methods, or other local search methods.

For reliability optimization problems, exact solutions are not necessarily desirable because it is generally difficult to develop exact methods for reliability optimization problems that are equivalent to methods used for nonlinear integer programming problems [2]. However, exact methods may be particularly advantageous when the problem is not large. More importantly, such methods can be used to measure the performance of heuristic or metaheuristic methods.

Conclusions and Discussions

Heuristic, metaheuristic, and exact methods are significantly applied in optimal reliability design. Recently, many advances in metaheuristics and exact methods have been reported. Particularly, a new metaheuristic method called *ant colony optimization* has been introduced and demonstrated to be a very promising general method in this field. Hybrid GAs may be the most important recent development among the optimization techniques since they retain the advantages of GAs in robustness and feasibility but significantly improve their computational efficiency.

Optimal reliability design has attracted many researchers, who have produced hundreds of publications since 1960. Owing to the increasing complexity of practical engineering systems and the critical importance of reliability in these complex systems, this still seems to be a very fruitful area for future research. From the view of optimization techniques, there are opportunities for improved effectiveness and efficiency of reported ACO, TS, IA, and great deluge algorithm (GDA), while some new metaheuristic algorithms such as harmony search algorithm and particle swarm optimization may offer excellent solutions for reliability optimization problems. Hybrid optimization techniques are another very promising general developmental direction in this field. They may combine heuristic methods, NN or some local search methods with all kinds of metaheuristics to improve computational efficiency or with exact methods to reduce search space. We may even be able to combine two metaheuristic algorithms such as GA and SA or ACO in future research on optimal reliability design.

Acknowledgments

This work was partly supported by NSF projects DMI-0429176. It is republished with the permission of IEEE *Transactions on Systems, Man, and Cybernetics, Part A: Systems and Humans*.

References

[1] Chen, M.S. (1987). On the computational complexity of reliability redundancy allocation in a series system, *Operations Research Letters* **SE-13**, 582–592.

[2] Kuo, W. & Prasad, V.R. (2000). An annotated overview of system-reliability optimization, *IEEE Transactions on Reliability* **49**, 487–493.

[3] Liang, Y.C. & Smith, A.E. (2004). An ant colony optimization algorithm for the redundancy allocation problem, *IEEE Transactions on Reliability* **53**, 417–423.

[4] Massim, Y., Zeblah, A., Meziane, R., Benguediab, M. & Ghouraf, A. (2005). Optimal design and reliability evaluation of multi-state series-parallel power systems, *Nonlinear Dynamics* **40**, 309–321.

[5] Nahas, N. & Nourelfath, M. (2005). Ant system for reliability optimization of a series system with multiple-choice and budget constraints, *Reliability Engineering and System Safety* **87**, 1–12.

[6] Shelokar, P.S., Jayaraman, V.K. & Kulkarni, B.D. (2002). Ant algorithm for single and multiobjective reliability optimization problems, *Quality and Reliability Engineering International* **18**, 497–514.

[7] Yin, P. & Wang, J. (2006). Ant colony optimization for the nonlinear resource allocation problem, *Applied Mathematics and Computation* **174**, 1438–1453.

[8] Zhao, J., Liu, Z. & Dao, M. (2007). Reliability optimization using multiobjective ant colony system approaches, *Reliability Engineering and System Safety* **92**, 109–120.

[9] Agarwal, M. & Gupta, R. (2006). Genetic search for redundancy optimization in complex systems, *Journal of Quality in Maintenance Engineering* **12**, 338–353.

[10] Altiparmak, F., Dengiz, B. & Smith, A.E. (2003). Optimal design of reliable computer networks: a comparison of meta heuristics, *Journal of Heuristics* **9**, 471–487.

[11] Busacca, P.G., Marseguerra, M. & Zio, E. (2001). Multiobjective optimization by genetic algorithms: application to safety systems, *Reliability Engineering and System Safety* **72**, 59–74.

[12] Coit, D.W. & Smith, A.E. (1998). Redundancy allocation to maximize a lower percentile of the system time-to-failure distribution, *IEEE Transactions on Reliability* **47**, 79–87.

[13] Coit, D.W. & Smith, A.E. (2002). Genetic algorithm to maximize a lower-bound for system time-to-failure with uncertain component Weibull parameters, *Computers and Industrial Engineering* **41**, 423–440.

[14] Elegbede, C. & Adjallah, K. (2003). Availability allocation to repairable systems with genetic algorithms: a multi-objective formulation, *Reliability Engineering and System Safety*, **82**, 319–330.

[15] Hsieh, Y.C., Chen, T.C. & Bricker, D.L. (1998). Genetic algorithm for reliability design problems, *Microelectronics Reliability* **38**, 1599–1605.

[16] Knoak, A., Coit, D.W. & Smith, A.E. (2006). Multiobjective optimization using genetic algorithms: a tutorial, *Reliability Engineering and System Safety* **91**, 992–1007.

[17] Levitin, G., Lisnianski, A., Ben-Haim, H. & Elmakis, D. (1998). Redundancy optimization for series-parallel multi-state systems, *IEEE Transactions on Reliability* **47**, 165–172.

[18] Levitin, G. & Lisnianski, A. (1998). Structure optimization of power system with bridge topology, *Electric Power Systems Research* **45**, 201–208.

[19] Levitin, G. & Lisnianski, A. (1999). Joint redundancy and maintenance optimization for multistate series-parallel systems, *Reliability Engineering and System Safety* **64**, 33–42.

[20] Levitin, G. & Lisnianski, A. (2000). Survivability maximization for vulnerable multi-state systems with bridge topology, *Reliability Engineering and System Safety* **70**, 125–140.

[21] Levitin, G. (2001). Redundancy optimization for multistate system with fixed resource-requirements and unreliable sources, *IEEE Transactions on Reliability* **50**, 52–59.

[22] Levitin, G. & Lisnianski, A. (2001). Reliability optimization for weighted voting system, *Reliability Engineering and System Safety* **71**, 131–138.

[23] Levitin, G. & Lisnianski, A. (2001). Structure optimization of multi-state system with two failure modes, *Reliability Engineering and System Safety* **72**, 75–89.

[24] Levitin, G. & Lisnianski, A. (2001). Optimal separation of elements in vulnerable multi-state systems, *Reliability Engineering and System Safety* **73**, 55–66.

[25] Levitin, G. & Lisnianski, A. (2001). A new approach to solving problems of multi-state system reliability optimization, *Quality and Reliability Engineering International*, **17**, 93–104.

[26] Levitin, G. (2002). Optimal allocation of multi-state retransmitters in acyclic transmission networks, *Reliability Engineering and System Safety* **75**, 73–82.

[27] Levitin, G. (2002). Optimal allocation of elements in a linear multi-state sliding window system, *Reliability Engineering and System Safety* **76**, 245–254.

[28] Levitin, G. (2002). Optimal series-parallel topology of multi-state system with two failure modes, *Reliability Engineering and System Safety* **77**, 93–107.

[29] Levitin, G. (2003). Optimal allocation of multi-state elements in linear consecutively connected systems with vulnerable nodes, *European Journal of Operational Research* **150**, 406–419.

[30] Levitin, G. (2003). Optimal allocation of multistate elements in a linear consecutively-connected system, *IEEE Transactions on Reliability* **52**, 192–199.

[31] Levitin, G. & Lisnianski, A. (2003). Optimizing survivability of vulnerable series-parallel multi-state systems, *Reliability Engineering and System Safety* **79**, 319–331.

[32] Levitin, G. (2003). Optimal multilevel protection in series-parallel systems, *Reliability Engineering and System Safety* **81**, 93–102.

[33] Levitin, G., Dai, Y., Xie, M. & Poh, K.L. (2003). Optimizing survivability of multi-state systems with multilevel protection by multi-processor genetic algorithm, *Reliability Engineering and System Safety* **82**, 93–104.

[34] Levitin, G. (2005). Uneven allocation of elements in linear multi-state sliding window system, *European Journal of Operational Research* **163**, 418–433.

[35] Levitin, G. (2005). Optimal structure of fault-tolerant software systems, *Reliability Engineering and System Safety* **89**, 286–295.

[36] Lisnianski, A., Levitin, G. & Ben-Haim, H. (2000). Structure optimization of multi-state system with time redundancy, *Reliability Engineering and System Safety* **67**, 103–112.

[37] Marseguerra, M., Zio, E., Podofillini, L. & Coit, D.C. (2005). Optimal design of reliable network systems in presence of uncertainty, *IEEE Transactions on Reliability* **54**, 243–253.

[38] Yamachi H., Yamachi, H., Tsujimura, Y., Kambayashi, Y. & Yamamoto, H. (2006). Multi-objective genetic algorithm for solving N-version program design problem. *Reliability Engineering and System Safety* **91**, 1083–1094.

[39] Yang, J.E., Hwang, M.J., Sung, T.Y. & Jin, Y. (1999). Application of genetic algorithm for reliability allocation in nuclear power plants, *Reliability Engineering and System Safety* **65**, 229–238.

[40] Hsieh, C.C. & Hsieh, Y.C. (2003). Reliability and cost optimization in distributed computing systems, *Computers and Operations Research* **30**, 1103–1119.

[41] Hsieh, C.C. (2003). Optimal task allocation and hardware redundancy policies in distributed computing system, *European Journal of Operational Research* **147**, 430–447.

[42] Lee, C.Y., Gen, M. & Kuo, W. (2002). Reliability optimization design using hybridized genetic algorithm with a neural network technique, *IEICE Transactions on Fundamentals of Electronics, Communications and Computer Sciences* **E84-A**, 627–637.

[43] Lee, C.Y., Yun, Y. & Gen, M. (2002). Reliability optimization design using hybrid NN-GA with fuzzy logic controller, *IEICE Transactions on Fundamentals of Electronics, Communications and Computer Sciences* **E85-A**, 432–447.

[44] Lee, C.Y., Gen, M. & Tsujimura, Y. (2002). Reliability optimization design for coherent systems by hybrid GA with fuzzy logic controller and local search, *IEICE Transactions on Fundamentals of Electronics, Communications and Computer Sciences* **E85-A**, 880–891.

[45] Sasaki, M. & Gen, M. (2003). A method of fuzzy multiobjective nonlinear programming with GUB structure by hybrid genetic algorithm, *International Journal of Smart Engineering Design* **5**, 281–288.

[46] Sasaki, M. & Gen, M. (2003). Fuzzy multiple objective optimal system design by hybrid genetic algorithm, *Applied Soft Computing* **2**, 189–196.

[47] Taguchi, T., Yokota, T. & Gen, M. (1998). Reliability optimal design problem with interval coefficients using hybrid genetic algorithms, *Computers and Industrial Engineering* **35**, 373–376.

[48] Taguchi, T. & Yokota, T. (1999). Optimal design problem of system reliability with interval coefficient using improved genetic algorithms, *Computers and Industrial Engineering* **37**, 145–149.

[49] Zhao, R. & Liu, B. (2003). Stochastic programming models for general redundancy-optimization problems, *IEEE Transactions on Reliability* **52**, 181–191.

[50] Zhao, R. & Liu, B. (2004). Redundancy optimization problems with uncertainty of combining randomness and fuzziness, *European Journal of Operation Research* **157**, 716–735.

[51] Kulturel-Konak, S., Smith, A.E. & Coit, D.W. (2003). Efficiently solving the redundancy allocation problem using tabu search, *IIE Transactions* **35**, 515–526.

[52] Kim, H., Bae, C. & Park, D. (2006). Reliability-redundancy optimization using simulated annealing algorithms, *Journal of Quality in Maintenance Engineering* **12**, 354–363.

[53] Suman, B. (2003). Simulated annealing-based multiobjective algorithm and their application for system reliability, *Engineering Optimization* **35**, 391–416.

[54] Wattanapongsakorn, N. & Levitan, S.P. (2004). Reliability optimization models for embedded systems with multiple applications, *IEEE Transactions on Reliability* **53**, 406–416.

[55] Zafiropoulos, E.P. & Dialynas, E.N. (2004). Reliability and cost optimization of electronic devices considering the component failure rate uncertainty, *Reliability Engineering and System Safety* **84**, 271–284.

[56] Chen, T.C. & You, P.S. (2005). Immune algorithms-based approach for redundant reliability problems with multiple component choices, *Computers in Industry* **56**, 195–205.

[57] Ravi, V. (2004). Optimization of complex system reliability by a modified great deluge algorithm, *Asia-Pacific Journal of Operational Research* **21**, 487–497.

[58] Rocco, C.M., Moreno, J.A. & Carrasquero, N. (2000). A cellular evolution approach applied to reliability optimization of complex systems, in Proceedings Annual Reliability and Maintainability Symposium, Los Angeles, California. pp. 210–215.

[59] Djerdjour, M. & Rekab, K. (2001). A branch and bound algorithm for designing reliable systems at a minimum cost, *Applied Mathematics and Computation* **118**, 247–259.

[60] Elegbede, C., Chu, C., Adjallah, K. & Yalaoui, F. (2003). Reliability allocation through cost minimization, *IEEE Transactions on Reliability* **52**, 106–111.

[61] Ha, C. & Kuo, W. (2006). Reliability redundancy allocation: an improved realization for nonconvex nonlinear programming problems, *European Journal of Operational Research* **171**, 24–38.

[62] Lin, F. & Kuo, W. (2002). Reliability importance and invariant optimal allocation, *Journal of Heuristics* **8**, 155–172.

[63] Prasad, V.R. & Kuo, W. (2000). Reliability optimization of coherent systems, *IEEE Transactions on Reliability* **49**, 323–330.

[64] Prasad, V.R., Kuo, W. & Kim, K.O. (2001). Maximization of a percentile life of a series system through component redundancy allocation, *IIE Transactions* **33**, 1071–1079.

[65] Ruan, N. & Sun, X. (2006). An exact algorithm for cost minimization in series reliability systems with multiple component choices, *Applied Mathematics and Computation* **181**, 732–741.

[66] Sung, C.S. & Cho, Y.K. (1999). Branch-and-bound redundancy optimization for a series system with multiple-choice constraints, *IEEE Transactions on Reliability* **48**, 108–117.

[67] Sung, C.S. & Cho, Y.K. (2000). Reliability optimization of a series system with multiple-choice and budget constraints, *European Journal of Operational Research* **127**, 158–171.

[68] Lee, H., Kuo, W. & Ha, C. (2003). Comparison of max-min approach and NN method for reliability optimization of series-parallel system, *Journal of System Science and Systems Engineering* **12**, 39–48.

[69] Ramirez-Marquez, J.E., Coit, D.W. & Konak, A. (2004). Redundancy allocation for series-parallel systems using a max-min approach, *IIE Transactions* **36**, 891–898.

[70] Coit, D.W. & Konak, A. (2006). Multiple weighted objective heuristic for the redundancy allocation problem, *IEEE Transactions on Reliability* **55**, 551–558.

[71] Ramirez-Marquez, J.E. & Coit, D.W. (2004). A heuristic for solving the redundancy allocation problem for multi-state series-parallel system, *Reliability Engineering and System Safety* **83**, 341–349.

[72] You, P.S. & Chen, T.C. (2005). An efficient heuristic for series-parallel redundant reliability problems, *Computer and Operations Research* **32**, 2117–2127.

[73] Kevin, Y.K.N.G. & Sancho, N.G.F. (2001). A hybrid dynamic programming/depth-first search algorithm with an application to redundancy allocation, *IIE Transactions* **33**, 1047–1058.

[74] Yalaoui, A., Châtelet, E. & Chu, C. (2005). A new dynamic programming method for reliability & redundancy allocation in a parallel-series system, *IEEE Transactions on Reliability* **54**, 254–261.

[75] Dorigo, M. (1992). *Optimization learning and natural algorithm*, Ph.D. Thesis, Politecnico di Milano, Italy.

[76] Maniezzo, V., Gambardella, L.M. & De Luigi, F. (2004). Ant colony optimization, in *New Optimization Techniques in Engineering*, G.C. Onwubolu & B.V. Babu, eds, Springer-Verlag Berlin Heidelberg, pp. 101–117.

[77] Gen, M. & Kim, J.R. (1999). GA-based reliability design: state-of-the-art survey, *Computers and Industrial Engineering* **37**, 151–155.

[78] Coit, D.W. & Smith, A.E. (1996). Reliability optimization of series-parallel systems using a genetic algorithm, *IEEE Transactions on Reliability* **45**, 254–260.

[79] Wolfram, S. (1984). Cellular automata as models of complexity, *Nature* **311**, 419–424.

[80] Djerdjour, M. (1997). An enumerative algorithm framework for a class of nonlinear programming problems, *European Journal of Operation Research* **101**, 101–121.

[81] Hsieh, Y.C. (2002). A linear approximation for redundant reliability problems with multiple component choices, *Computers and Industrial Engineering* **44**, 91–103.

[82] Kuo, W., Hwang, C.L. & Tillman, F.A. (1978). A note on heuristic method for in optimal system reliability, *IEEE Transactions on Reliability* **27**, 320–324.

[83] Nakagawa, Y. & Miyazaki, S. (1981). Surrogate constraints algorithm for reliability optimization problems with two constraints, *IEEE Transactions on Reliability* **R-30**, 175–180.

[84] Misra, K.B. (1991). An algorithm to solve integer programming problems: an efficient tool for reliability design, *Microelectronics and Reliability* **31**, 285–294.

Further Reading

Gen, M. & Yun, Y. (2006). Soft computing approach for reliability optimization: state-of-art survey, *Reliability Engineering and System Safety* **91**, 1008–1026.

Ha, C. & Kuo, W. (2005). Multi-path approach for reliability-redundancy allocation using a scaling method, *Journal of Heuristics* **11**, 201–237.

Ha, C. & Kuo, W. (2006). Multi-paths iterative heuristic for redundancy allocation: the tree heuristic, *IEEE Transactions on Reliability* **55**, 37–43.

Kim, K.O. & Kuo, W. (2003). Percentile life and reliability as a performance measures in optimal system design, *IIE Transactions* **35**, 1133–1142.

Kumar, U.D. & Knezevic, J. (1998). Availability based spare optimization using renewal process, *Reliability Engineering and System Safety* **59**, 217–223.

Kuo, W., Chien, K. & Kim, T. (1998). *Reliability, Yield and Stress Burn-in*, Kluwer.

Kuo, W., Prasad, V.R., Tillman, F.A. & Hwang, C.L. (2001). *Optimal Reliability Design: Fundamentals and Applications*, Cambridge University Press.

Kuo, W. & Zuo, M.J. (2003). *Optimal Reliability Modeling: Principles and Applications*, John Wiley & Sons.

Levitin, G. & Lisnianski, A. (1999). Importance and sensitivity analysis of multi-state systems using the universal generating function method, *Reliability Engineering and System Safety* **65**, 271–282.

Lin, F., Kuo, W. & Hwang, F. (1999). Structure importance of consecutive-k-out-of-n systems, *Operations Research Letters* **25**, 101–107.

Marseguerra, M. & Zio, E. (2004). System design optimization by genetic algorithm, *IEEE Transactions on Reliability* **53**, 424–434.

Misra, K.B. (1986). On optimal reliability design: a review, *System Science* **12**, 5–30.

Tillman, F.A., Hwang, C.L. & Kuo, W. (1977). Optimization techniques for system reliability with redundancy- a review, *IEEE Transactions on Reliability* **R-26**, 148–152.

WAY KUO AND RUI WAN

Optimal Reliability Design-Modeling

Introduction

Reliability has become an even greater concern in recent years because high-tech industrial processes, with increasing levels of sophistication, comprise most engineering systems today. Based on enhancing **component reliability** and providing **redundancy**, while considering the trade-off between system performance and resources, optimal reliability design that aims to determine an optimal system-level configuration has long been an important topic in reliability engineering (*see* **Reliability Optimization**). Since 1960, many publications have addressed this problem using different system structures, performance measures, optimization techniques, and options for reliability improvement.

References 1, 2 and 3 provide good overview of the early work in system reliability optimization. Tillman *et al.* [3] were the first to classify articles by system structure, problem type, and solution methods. Also described and analyzed in [4] are the advantages and shortcomings of various optimization techniques. It was during the 1970s that various heuristics were developed to solve complex system reliability problems in cases where the traditional parametric optimization techniques were insufficient. In their 2000 report, Kuo and Prasad [1] summarize the developments in optimization techniques, along with recent optimization methods such as metaheuristics. This chapter discusses the contributions made to the literature since the publication of [1]. Most of the recent work in this area is devoted to:

- multistate system (MSS) optimization;
- percentile life as a system performance measure;
- multiobjective optimization;
- active and cold-standby redundancy;
- optimization techniques.

Based on their system performance, reliability systems can be classified as binary-state systems and MSS. A binary-state system and its components may exist in only two possible states–either working or failed. Binary-state system reliability models have played very important roles in practical reliability engineering. To satisfactorily describe the performance of a complex system we may need more than two levels of satisfaction, for example, excellent, average, and poor [5]. For this reason, MSS reliability models were proposed in the 1970s and since 1998 a large part of the work devoted to MSS optimal design has emerged (*see* **Multistate Reliability Theory**). The primary task of MSS optimization is to define the relationship between component states and system states.

Measures of system performance are basically of four kinds: reliability, availability, mean time-to-failure (TTF), and percentile life (*see* **System Availability**). Reliability has been widely used and thoroughly studied as the primary performance measure for nonmaintained systems. For a maintained system, however, availability, which describes the percentage of time the system really functions, should be considered instead of reliability. Availability is most commonly employed as the performance measure for renewable MSS. Meanwhile, percentile life is preferred to reliability and mean TTF when the system mission time is indeterminate, as in most practical cases.

Some important design principles for improving system performance are summarized in [1]. This chapter primarily reviews articles that address either the provision of redundant components in parallel or the combination of structural redundancy with the enhancement of component reliability (*see* **Reliability of Redundant Systems**; **Reliability Allocation**). These are called *redundancy allocation problems* and *reliability-redundancy allocation problems*, respectively. Redundancy allocation problems are well documented in [1] and [5], which employ a special case of reliability-redundancy allocation problems without exploring the alternatives of component-combined improvement. Recently, a lot of the effort in optimal reliability design has been placed on general reliability-redundancy allocation problems, rather than on redundancy allocation problems.

In practice, two redundancy schemes are available: active and cold-standby. Cold-standby redundancy provides higher reliability, but it is hard to implement because of the difficulty of failure detection. Reliability design problems have generally been formulated considering active redundancy; however, an actual optimal design may include active redundancy or cold-standby redundancy or both.

However, any effort for improvement usually requires resources. Quite often it is hard for a single-objective system to adequately describe a real optimal design problem. For this reason, multiobjective system design problem always deserves a lot of attention.

This chapter describes the state-of-the-art of optimal reliability design. Emphasizing the foci mentioned above, it includes four main problem formulations in optimal **reliability allocation**; describes advances related to those four types of optimization problems; and also provides conclusions and a discussion of future challenges related to reliability optimization problems.

Problem Formulations

Among the diversified problems in optimal reliability design, the following four basic formulations are widely covered.

Problem 1 (P_1):

$$\begin{aligned}&\max R_S = f(\mathrm{x})\\&\text{s.t.}\\&g_i(\mathrm{x}) \le b_i, \quad \text{for} \quad i = 1, \ldots, m\\&\mathrm{x} \in \mathrm{X}\end{aligned}$$

or

$$\begin{aligned}&\min C_S = f(\mathrm{x})\\&\text{s.t.}\\&Rs \ge R_0\\&g_i(\mathrm{x}) \le b_i, \quad \text{for} \quad i = 1, \ldots, m\\&\mathrm{x} \in \mathrm{X}\end{aligned}$$

Problem 1 formulates the traditional reliability-redundancy allocation problem with either reliability or cost as the objective function. Its solution includes two parts: the component choices and their corresponding optimal redundancy levels.

Problem 2 (P_2):

$$\begin{aligned}&\max t_{\alpha,\mathrm{x}}\\&\text{s.t.}\\&g_i(t_{\alpha,\mathrm{x}}; \mathrm{x}) \le b_i, \quad \text{for} \quad i = 1, \ldots, m\\&\mathrm{x} \in \mathrm{X}\end{aligned}$$

Problem 2 uses percentile life as the system performance measure instead of reliability. Percentile life is preferred especially when the system mission time is indeterminate. However, it is hard to find a closed analytical form of percentile life in decision variables.

Problem 3 (P_3):

$$\begin{aligned}&\max E(\mathrm{x},\mathrm{T},\mathrm{W}^*)\\&\text{s.t.}\\&g_i(\mathrm{x}) \le b_i, \quad \text{for} \quad i = 1, \ldots, m\\&\mathrm{x} \in \mathrm{X}\end{aligned}$$

or

$$\begin{aligned}&\min C_s(\mathrm{x})\\&\text{s.t.}\\&E(\mathrm{x},\mathrm{T},\mathrm{W}^*) \ge E_0\\&g_i(\mathrm{x}) \le b_i, \quad for \quad i = 1, \ldots, m\\&\mathrm{x} \in \mathrm{X}\end{aligned}$$

Problem 3 represents MSS optimization problems. Here, E is used as a measure of the entire **system availability** to satisfy the custom demand represented by a cumulative demand curve with a known **T** and **W***.

Problem 4 (P_4):

$$\begin{aligned}&\max z = [f_1(\mathrm{x}), f_2(\mathrm{x}), \ldots, f_S(\mathrm{x})]\\&\text{s.t.}\\&g_i(\mathrm{x}) \leq b_i, \quad \text{for} \quad i = 1, \ldots, m\\&\mathrm{x} \in \mathrm{X}\end{aligned}$$

For multiobjective optimization, as formulated by Problem 4, a Pareto optimal set, which includes all of the best possible trade-offs between given objectives, rather than a single optimal solution, is usually identified.

In the above formulations, the resource constraints may be linear or nonlinear or both.

The articles, classified by problem formulations, is summarized in Table 1.

Brief Reviews of Advances in P_1–P_4

Active and Cold-Standby Redundancy (P_1)

P_1 is generally limited to active redundancy. A new optimal system configuration is obtained when active and cold-standby redundancies are both involved in the design. A cold-standby redundant component does not fail before it is put into operation by the action of switching, whereas the failure pattern of an active redundant component does not depend on whether the component is idle or in operation. Cold-standby redundancy can provide higher reliability, but it is hard to implement because of the difficulties involved in failure detection and switching.

In reference 10, optimal solutions to reliability-redundancy allocation problems are determined for nonrepairable systems designed with multiple ***k*-out-of-*n*** subsystems in series. The individual subsystems may use either active or cold-standby redundancy, or they may require no redundancy. Assuming an exponentially distributed component TTF with rate λ_{ij}, the failure process of subsystem i with cold-standby redundancy can be described by a **Poisson process** with rate $\lambda_{ij}k_i$, while the subsystem reliability with active redundancy is computed by standard binominal techniques. For series-**parallel systems** with only cold-standby redundancy, reference 11 employs the more flexible and realistic Erlang distributed component TTF and $\rho_i(t)$ is introduced to describe the reliability of the imperfect detection/switching mechanism for each subsystem. Reference 12 directly extends this earlier work by introducing the choice of redundancy strategies as an additional decision variable. For each of these methods, however, no mixture of component types or redundancy strategies is allowed within any of the subsystems.

In addition, reference 74 investigates the problem of where to allocate a spare in a k-out-of-n: F system of dependent components through minimal standby redundancy; and reference 75 studies the allocation of one active redundancy when it differs based on the component with which it is to be allocated. Reference 28 considers the problem of optimally allocating a fixed number of s-identical multifunctional spares for a deterministic or stochastic mission time. In spite of some sufficiency conditions for optimality, the proposed algorithm can be easily implemented even for large systems.

Percentile Life Optimization (P_2)

Many diversified models and solution methods, where reliability is used as the system performance measure, have been proposed and developed since the 1960s. However, this is not an appropriate choice when mission time cannot be clearly specified or a system

Table 1 Reference classification by problem formulations

P_1	[6], [7], [8], [9], [10], [11], [12], [13], [14], [15], [16], [17], [18], [19], [20], [21], [22], [23], [24], [25], [26], [27], [28], [29], [30], [31], [32], [33], [4], [34], [35], [36], [37], [38]
P_2	[39], [40], [41], [42], [37], [38]
P_3	[43], [44], [45], [46], [47], [48], [49], [50], [51], [52], [53], [54], [55], [56], [57], [58], [59], [60], [22], [61], [62], [63], [64]
P_4	[65], [66], [67], [68], [69], [70], [71], [72], [73], [37], [38]

is intended for use as long as it functions. Average life is also not reliable, especially when the implications of failure are critical or the variance in the system life is high. Percentile life is considered to be a more appropriate measure, since it incorporates system designer and user risk. When using percentile life as the objective function, the main difficulty is its mathematical inconvenience, because it is hard to find a closed analytical form of percentile life in the decision variables.

Reference 39 solves redundancy allocation problems for series-parallel systems where the objective is to maximize a lower percentile of the system TTF distribution. Component TTF has a Weibull distribution with known deterministic parameters. Later in the literature, reference 40 addresses similar problems where Weibull shape parameters are accurately estimated but scale parameters are random variables following a uniform distribution.

Reference 42 develops a lexicographic search methodology that is, the first to provide exact optimal redundancy allocations for percentile life optimization problems. This algorithm is general for any continuous increasing lifetime distribution.

Three important results are presented in [41], which describe the general relationships between reliability and percentile life maximizing problems.

- S_2 equals S_1 given $\alpha_t = 1 - R_s(t, \mathrm{x}_t^*)$, where $\mathrm{x}_t^* \in S_1$;
- S_1 equals S_2 given $t_{\alpha,\mathbf{x}_\alpha^*}$, where $\mathbf{x}_\alpha^* \in S_2$;
- let $\psi(t)$ be the optimal objective value of P_1. For a fixed α, $t_\alpha = \inf\{t \geq 0 : \Psi(t) \leq 1 - \alpha\}$ is the optimal objective value of P_2.

MSS Optimization (P_3)

MSS is defined as a system that can unambiguously perform its task at different performance levels, depending on the state of its components, which can be characterized by nominal performance rate, availability and cost. Based on their physical nature, MSS can be classified into two important types: Type I MSS (e.g., power systems) and Type II MSS (e.g., computer systems), which use capacity and operation time as their performance measures, respectively.

In the article, an important optimization strategy, combining universal generating function (UGF) and genetic algorithm (GA), has been well developed and widely applied to reliability optimization problems of renewable MSS. In this strategy, there are two main tasks:

- According to the system structure and the system physical nature, obtain the system UGF from the component UGFs.
- Find an effective decoding and encoding technique to improve the efficiency of the GA.

Reference 43 first uses a UGF approach to evaluate the availability of a series-parallel MSS with relatively small computational resources. The essential property of the U-transform enables the total U-function for a MSS to be obtained by simple algebraic operations involving individual component U-functions. Later, reference 76 combines importance and sensitivity analysis and reference 77 extends this UGF approach to MSS with dependent elements. Table 2 summarizes the application of UGF to some typical MSS structures in optimal reliability design.

With this UGF and GA strategy, reference 61 solves the structure optimization of a MSS with time redundancy; reference 47 considers a MSS consisting of two parts: RGS including a number of resource generating subsystems and MPS including elements that consume a fixed amount of resources to perform their tasks; references 49, 54, 78 considers multistate series-parallel systems with two failure modes – open mode and closed mode, and references 46, 50, 55, 57–59 use survivability, instead of reliability to describe the ability of a MSS to tolerate both internal failures and external attacks.

In addition to a GA, reference 62 presents an ant colony method that combines with a UGF technique

Table 2 Application of UGF approach

Series-parallel system	[43], [45], [76], [49], [51], [54], [57] [58], [59], [62]
Bridge system	[44], [46], [78], [61]
LMSSWS	[53], [79], [60]
WVS	[48], [80]
ATN	[52], [81]
LMCCS	[55], [56]
ACCN	[82]

LMSSWS: linear multi-state sliding-window system; WVS: weighted voting system; ATN: acyclic transmission network; LMCCS: linear multi-state consecutively connected system; ACCN: acyclic consecutively connected network to find an optimal series-parallel power-structure configuration.

Besides this primary UGF approach, a few other methods have been proposed for MSS reliability optimization problems. Reference 64 develops a heuristic algorithm RAMC for a Type I multistate series-parallel system, while reference 83 presents a novel continuous-state system mode approximating the objective reliability function by a **neural network** approach.

Multiobjective Optimization (P_4)

In the previous discussion, all problems were single-objective problems. Rarely does a single-objective problem with several hard constraints adequately represent a real problem for which an optimal design is required. When designing a reliable system, as formulated by P_4, it is always desirable to simultaneously optimize several opposing design objectives such as reliability, cost, even volume, and weight. For this reason, a recently proposed multiobjective system design problem deserves a lot of attention. The objectives of this problem are to maximize the system reliability estimates and minimize their associated variance while considering the uncertainty of the component reliability estimations. A Pareto optimal set, which includes all of the best possible trade-offs between the given objectives, rather than a single optimal solution, is usually identified for multiobjective optimization problems.

When considering complex systems, the reliability optimization problem has been modeled as a fuzzy multiobjective optimization problem in reference 69, considering the influence of various kinds of aggregators.

With a weighting technique, references 70, 71 solve multiobjective reliability-redundancy allocation problems using similar linear membership functions for both objectives and constraints; reference 67 transfers P_4 into a single-objective optimization problem and proposes a GA-based approach whose parameters can be adjusted with the experimental plan technique; and reference 65 develops a multiobjective GA to obtain an optimal system configuration and **inspection policy** by considering every target as a separate objective.

P_4 is considered for series-parallel systems, *RB*/1/1, and bridge systems in [66] with multiple objectives to maximize the system reliability while minimizing its associated variance when the component reliability estimates are treated as random variables. For series-parallel systems, component reliabilities of the same type are considered to be dependent since they usually share the same reliability estimate from a pooled data set. The system variance is straightforwardly expressed as a function in the higher **moments** of the component unreliability estimates [84]. For *RB*/1/1, the hardware components are considered identical and statistically independent, while even independently developed software versions are found to have related faults as presented by the parameters *Prv* and *Pall*. Pareto optimal solutions are found by solving a series of weighted objective problems with incrementally varied weights. It is worth noting that significantly different designs are obtained when the formulation incorporates estimation uncertainty or when the component reliability estimates are treated as statistically dependent. Similarly, [68] uses a multiobjective GA to select an optimal network design that balances the dual objectives of high reliability and low uncertainty in its estimation. But the latter exploits **Monte Carlo simulation** as the objective function evaluation engine (*see also* **Monte Carlo Methods, Univariate and Multivariate**).

Besides GA, reference 72 illustrates the application of the ant colony optimization algorithm to solve both continuous function and combinatorial optimization problems, while reference 73 tests five simulated annealing-based multiobjective algorithms–suppapitnarm multi-objective simulated annealing (SMOSA), ulungu multi-objective simulated annealing (UMOSA), pareto simulated annealing (PSA), pareto domination based multi-objective simulated annealing (PDMOSA) and weight based multi-objective simulated annealing (WMOSA).

Conclusions and Discussions

We have reviewed the recent research on optimal reliability design. Many publications have addressed this problem using different system structures, performance measures, problem formulations and optimization techniques.

The systems considered here mainly include series-parallel systems, k-out-of-n: G systems, bridge networks, n-version programming architecture, recovery block architecture and other unspecified coherent systems (*see* **Coherent Systems**; **Parallel, Series, and Series–Parallel Systems**; ***k*-out-of-*n* Systems**). The recently introduced *N-version programming* (*NVP*) and *recovery block* (*RB*) belong to the category of fault tolerant architecture, which usually considers both software and hardware.

Reliability is still employed as a system performance measure in most cases, but percentile life does provide a new perspective on optimal design without the requirement of a specified mission time. Availability is primarily used as the performance measure of renewable MSS whose optimal design has been emphasized and well developed in the past 10 years. Optimal design problems are generally formulated to maximize an appropriate system performance measure under resource constraints, and more realistic problems involving multiobjective programming are also being considered.

Optimal reliability design has attracted many researchers, who have produced hundreds of publications since 1960. Because of the increasing complexity of practical engineering systems and the critical importance of reliability in these complex systems, this still seems to be a very fruitful area for future research.

Compared to traditional binary-state systems, there are still many unsolved topics in MSS optimal design, which include the following:

- using percentile life as a system performance measure;
- involving cold-standby redundancy;
- nonrenewable MSS optimal design.

The research dealing with the understanding and application of reliability at the nano-level has also demonstrated its attraction and vitality in recent years. Optimal system design that considers reliability within the uniqueness of nano-systems has seldom been reported in the literature. It deserves a lot more attention in the future. In addition, uncertainty and component dependency will be critical areas to consider in future research on optimal reliability design.

Acknowledgment

This work was partly supported by NSF Projects DMI-0429176. It is republished with the permission of *IEEE Transactions on Systems, Man, Cybernetics, Part A: Systems and Humans*.

Notations

x_j	number of components at subsystem j
r_j	component reliability at subsystem j
n	number of subsystems in the system
m	number of resources
x	$(\mathrm{x}_1, \ldots, \mathrm{x}_n, \mathrm{r}_1, \ldots, \mathrm{r}_n)$
$g_i(\mathrm{x})$	total amount of resource i required for **x**
R_S	system reliability
C_S	total system cost
R_0	a specified minimum R_S
C_0	a specified minimum C_S
α	system user's risk level, $0 < \alpha < 1$
$t_{\alpha,\mathrm{x}}$	system percentile life, $t_{\alpha,\mathrm{x}} = \inf\{t \geq 0 : R_S \leq 1 - \alpha\}$
E_S	generalized MSS availability index
E_0	a specified minimum E_S

References

[1] Kuo, W. & Prasad, V.R. (2000). An annotated overview of system-reliability optimization, *IEEE Transactions on Reliability* **49**, 487–493.

[2] Misra, K.B. (1986). On optimal reliability design: a review, *System Science* **12**, 5–30.

[3] Tillman, F.A., Hwang, C.L. & Kuo, W. (1977). Optimization techniques for system reliability with redundancy- a review, *IEEE Transactions on Reliability* **R-26**, 148–152.

[4] Taguchi, T. & Yokota, T. (1999). Optimal design problem of system reliability with interval coefficient using improved genetic algorithms, *Computers and Industrial Engineering* **37**, 145–149.

[5] Kuo, W., Prasad, V.R., Tillman, F.A. & Hwang, C.L. (2001). *Optimal Reliability Design: Fundamentals and Applications*, Cambridge University Press.

[6] Altiparmak, F., Dengiz, B. & Smith, A.E. (2003). Optimal design of reliable computer networks: a comparison of meta heuristics, *Journal of Heuristics* **9**, 471–487.

[7] Amari, S.V., Dugan, J.B. & Misra, R.B. (1999). Optimal reliability of systems subject to imperfect fault-coverage, *IEEE Transactions on Reliability* **48**, 275–284.

[8] Amari, S.V., Pham, H. & Dill, G. (2004). Optimal design of k-out-of-n: G subsystems subjected to imperfect fault-coverage, *IEEE Transactions on Reliability* **53**, 567–575.

[9] Chen, T.C. & You, P.S. (2005). Immune algorithms-based approach for redundant reliability problems with multiple component choices, *Computers in Industry* **56**, 195–205.

[10] Coit, D.W. & Liu, J. (2000). System reliability optimization with k-out-of-n subsystems, *International Journal of Reliability, Quality and Safety Engineering* **7**, 129–142.

[11] Coit, D.W. (2001). Cold-standby redundancy optimization for nonrepairable systems, *IIE Transactions* **33**, 471–478.

[12] Coit, D.W. (2003). Maximization of system reliability with a choice of redundancy strategies, *IIE Transactions* **35**, 535–543.

[13] Djerdjour, M. & Rekab, K. (2001). A branch and bound algorithm for designing reliable systems at a minimum cost, *Applied Mathematics and Computation* **118**, 247–259.

[14] Elegbede, C., Chu, C., Adjallah, K. & Yalaoui, F. (2003). Reliability allocation through cost minimization, *IEEE Transactions on Reliability* **52**, 106–111.

[15] Ha, C. & Kuo, W. (2005). Multi-path approach for reliability-redundancy allocation using a scaling method, *Journal of Heuristics* **11**, 201–237.

[16] Hsieh, Y.C., Chen, T.C. & Bricker, D.L. (1998). Genetic algorithm for reliability design problems, *Microelectronics Reliability* **38**, 1599–1605.

[17] Hsieh, Y.C. (2002). A linear approximation for redundant reliability problems with multiple component choices, *Computers and Industrial Engineering* **44**, 91–103.

[18] Kulturel-Konak, S., Smith, A.E. & Coit, D.W. (2003). Efficiently solving the redundancy allocation problem using tabu search, *IIE Transactions* **35**, 515–526.

[19] Lee, C.Y., Gen, M. & Kuo, W. (2002). Reliability optimization design using hybridized genetic algorithm with a neural network technique, *IEICE Transactions on Fundamentals of Electronics, Communications and Computer Sciences* **E84-A**, 627–637.

[20] Lee, C.Y., Yun, Y. & Gen, M. (2002). Reliability optimization design using hybrid NN-GA with fuzzy logic controller, *IEICE Transactions on Fundamentals of Electronics, Communications and Computer Sciences* **E85-A**, 432–447.

[21] Lee, C.Y., Gen, M. & Tsujimura, Y. (2002). Reliability optimization design for coherent systems by hybrid GA with fuzzy logic controller and local search, *IEICE Transactions on Fundamentals of Electronics, Communications and Computer Sciences* **E85-A**, 880–891.

[22] Levitin, G. (2005). Optimal structure of fault-tolerant software systems, *Reliability Engineering and System Safety* **89**, 286–295.

[23] Liang, Y.C. & Smith, A.E. (2004). An ant colony optimization algorithm for the redundancy allocation problem, *IEEE Transactions on Reliability* **53**, 417–423.

[24] Lin, F. & Kuo, W. (2002). Reliability importance and invariant optimal allocation, *Journal of Heuristics* **8**, 155–172.

[25] Nahas, N. & Nourelfath, M. (2005). Ant system for reliability optimization of a series system with multiple-choice and budget constraints, *Reliability Engineering and System Safety* **87**, 1–12.

[26] Kevin, Y.K.N.G. & Sancho, N.G.F. (2001). A hybrid dynamic programming/depth-first search algorithm with an application to redundancy allocation, *IIE Transactions* **33**, 1047–1058.

[27] Prasad, V.R. & Raghavachari, M. (1998). Optimal allocation of interchangeable components in a series-parallel system, *IEEE Transactions on Reliability* **47**, 255–260.

[28] Prasad, V.R., Kuo, W. & Kim, K.O. (2000). Optimal allocation of s-identical multi-functional spares in a series system, *IEEE Transactions on Reliability* **48**, 118–126.

[29] Prasad, V.R. & Kuo, W. (2000). Reliability optimization of coherent systems, *IEEE Transactions on Reliability* **49**, 323–330.

[30] Rocco, C.M., Moreno, J.A. & Carrasquero, N. (2000). A cellular evolution approach applied to reliability optimization of complex systems, in Proceedings Annual Reliability and Maintainability Symposium, Los Angeles, pp. 210–215.

[31] Sung, C.S. & Cho, Y.K. (1999). Branch-and-bound redundancy optimization for a series system with multiple-choice constraints, *IEEE Transactions on Reliability* **48**, 108–117.

[32] Sung, C.S. & Cho, Y.K. (2000). Reliability optimization of a series system with multiple-choice and budget constraints, *European Journal of Operational Research* **127**, 158–171.

[33] Taguchi, T., Yokota, T. & Gen, M. (1998). Reliability optimal design problem with interval coefficients using hybrid genetic algorithms, *Computers and Industrial Engineering* **35**, 373–376.

[34] Wattanapongsakorn, N. & Levitan, S.P. (2004). Reliability optimization models for embedded systems with multiple applications, *IEEE Transactions on Reliability* **53**, 406–416.

[35] Yalaoui, A., Châtelet, E. & Chu, C. (2005). A new dynamic programming method for reliability & redundancy allocation in a parallel-series system, *IEEE Transactions on Reliability* **54**, 254–261.

[36] You, P.S. & Chen, T.C. (2005). An efficient heuristic for series-parallel redundant reliability problems, *Computers and Operations Research* **32**, 2117–2127.

[37] Zhao, R. & Liu, B. (2003). Stochastic programming models for general redundancy-optimization problems, *IEEE Transactions on Reliability* **52**, 181–191.

[38] Zhao, R. & Liu, B. (2004). Redundancy optimization problems with uncertainty of combining randomness and fuzziness, *European Journal of Operation Research* **157**, 716–735.

[39] Coit, D.W. & Smith, A.E. (1998). Redundancy allocation to maximize a lower percentile of the system time-to-failure distribution, *IEEE Transactions on Reliability* **47**, 79–87.

[40] Coit, D.W. & Smith, A.E. (2002). Genetic algorithm to maximize a lower-bound for system time-to-failure with uncertain component Weibull parameters, *Computers and Industrial Engineering* **41**, 423–440.

[41] Kim, K.O. & Kuo, W. (2003). Percentile life and reliability as a performance measures in optimal system design, *IIE Transactions* **35**, 1133–1142.

[42] Prasad, V.R., Kuo, W. & Kim, K.O. (2001). Maximization of a percentile life of a series system through component redundancy allocation, *IIE Transactions*, **33**, 1071–1079.

[43] Levitin, G., Lisnianski, A., Ben-Haim, H. & Elmakis, D. (1998). Redundancy optimization for series-parallel multi-state systems, *IEEE Transactions on Reliability* **47**, 165–172.

[44] Levitin, G. & Lisnianski, A. (1998). Structure optimization of power system with bridge topology, *Electric Power Systems Research* **45**, 201–208.

[45] Levitin, G. & Lisnianski, A. (1999). Joint redundancy and maintenance optimization for multistate series-parallel systems, *Reliability Engineering and System Safety* **64**, 33–42.

[46] Levitin, G. & Lisnianski, A. (2000). Survivability maximization for vulnerable multi-state systems with bridge topology, *Reliability Engineering and System Safety* **70**, 125–140.

[47] Levitin, G. (2001). Redundancy optimization for multi-state system with fixed resource-requirements and unreliable sources, *IEEE Transactions on Reliability* **50**, 52–59.

[48] Levitin, G. & Lisnianski, A. (2001). Reliability optimization for weighted voting system, *Reliability Engineering and System Safety* **71**, 131–138.

[49] Levitin, G. & Lisnianski, A. (2001). Structure optimization of multi-state system with two failure modes, *Reliability Engineering and System Safety* **72**, 75–89.

[50] Levitin, G. & Lisnianski, A. (2001). Optimal separation of elements in vulnerable multi-state systems, *Reliability Engineering and System Safety* **73**, 55–66.

[51] Levitin, G. & Lisnianski, A. (2001). A new approach to solving problems of multi-state system reliability optimization, *Quality and Reliability Engineering International* **17**, 93–104.

[52] Levitin, G. (2002). Optimal allocation of multi-state retransmitters in acyclic transmission networks, *Reliability Engineering and System Safety* **75**, 73–82.

[53] Levitin, G. (2002). Optimal allocation of elements in a linear multi-state sliding window system, *Reliability Engineering and System Safety* **76**, 245–254.

[54] Levitin, G. (2002). Optimal series-parallel topology of multi-state system with two failure modes, *Reliability Engineering and System Safety* **77**, 93–107.

[55] Levitin, G. (2003). Optimal allocation of multi-state elements in linear consecutively connected systems with vulnerable nodes, *European Journal of Operational Research* **150**, 406–419.

[56] Levitin, G. (2003). Optimal allocation of multistate elements in a linear consecutively-connected system, *IEEE Transactions on Reliability* **52**, 192–199.

[57] Levitin, G. & Lisnianski, A. (2003). Optimizing survivability of vulnerable series-parallel multi-state systems, *Reliability Engineering and System Safety* **79**, 319–331.

[58] Levitin, G. (2003). Optimal multilevel protection in series-parallel systems, *Reliability Engineering and System Safety* **81**, 93–102.

[59] Levitin, G., Dai, Y., Xie, M. & Poh, K.L. (2003). Optimizing survivability of multi-state systems with multi-level protection by multi-processor genetic algorithm, *Reliability Engineering and System Safety* **82**, 93–104.

[60] Levitin, G. (2005). Uneven allocation of elements in linear multi-state sliding window system, *European Journal of Operational Research* **163**, 418–433.

[61] Lisnianski, A., Levitin, G. & Ben-Haim, H. (2000). Structure optimization of multi-state system with time redundancy, *Reliability Engineering and System Safety* **67**, 103–112.

[62] Massim, Y., Zeblah, A., Meziane, R., Benguediab, M. & Ghouraf A. (2005). Optimal design and reliability evaluation of multi-state series-parallel power systems. *Nonlinear Dynamics* **40**, 309–321.

[63] Nourelfath, M. & Dutuit, Y. (2004). A combined approach to solve the redundancy optimization problem for multi-state systems under repair policies, *Reliability Engineering and System Safety* **84**, 205–213.

[64] Ramirez-Marquez, J.E. & Coit, D.W. (2004). A heuristic for solving the redundancy allocation problem for multi-state series-parallel system, *Reliability Engineering and System Safety* **83**, 341–349.

[65] Busacca, P.G., Marseguerra, M. & Zio, E. (2001). Multi-objective optimization by genetic algorithms: application to safety systems, *Reliability Engineering and System Safety* **72**, 59–74.

[66] Coit, D.W., Jin, T. & Wattanapongsakorn, N. (2004). System optimization with component reliability estimation uncertainty: a multi-criteria approach, *IEEE Transactions on Reliability* **53**, 369–380.

[67] Elegbede, C. & Adjallah, K. (2003). Availability allocation to repairable systems with genetic algorithms: a multi-objective formulation, *Reliability Engineering and System Safety*, **82**, 319–330.

[68] Marseguerra, M., Zio, E., Podofillini, L. & Coit, D.W. (2005). Optimal design of reliable network systems in presence of uncertainty. *IEEE Transactions on Reliability* **54**, 243–253.

[69] Ravi, V., Reddy, P.J. & Zimmermann, H.J. (2000). Fuzzy global optimization of complex system reliability, *IEEE Transactions on Fuzzy Systems* **8**, 241–248.

[70] Sasaki, M. & Gen, M. (2003). A method of fuzzy multi-objective nonlinear programming with GUB structure by hybrid genetic algorithm, *International Journal of Smart Engineering Design* **5**, 281–288.

[71] Sasaki, M. & Gen, M. (2003). Fuzzy multiple objective optimal system design by hybrid genetic algorithm, *Applied Soft Computing* **2**, 189–196.

[72] Shelokar, P.S., Jayaraman, V.K. & Kulkarni, B.D. (2002). Ant algorithm for single and multiobjective reliability optimization problems, *Quality and Reliability Engineering International* **18**, 497–514.

[73] Suman, B. (2003). Simulated annealing-based multi-objective algorithm and their application for system reliability, *Engineering Optimization* **35**, 391–416.

[74] Bueno, V.C. (2005). Minimal standby redundancy allocation in a k-out-of-n: F system of dependent components, *European Journal of Operation Research* **165**, 786–793.

[75] Romera, R., Valdés, J.E. & Zequeira, R.I. (2004). Active-redundancy allocation in systems, *IEEE Transactions on Reliability* **53**, 313–318.

[76] Levitin, G. & Lisnianski, A. (1999). Importance and sensitivity analysis of multi-state systems using the universal generating function method, *Reliability Engineering and System Safety* **65**, 271–282.

[77] Levitin, G. (2004). A universal generating function approach for the analysis of multi-state systems with dependent elements, *Reliability Engineering and System Safety* **84**, 285–292.

[78] Levitin, G. (2003). Reliability of multi-state systems with two failure-modes, *IEEE Transactions on Reliability* **52**, 340–348.

[79] Levitin, G. (2003). Linear multi-state sliding-window systems, *IEEE Transactions on Reliability* **52**, 263–269.

[80] Levitin, G. (2002). Evaluating correct classification probability for weighted voting classifiers with plurality voting, *European Journal of Operational Research* **141**, 596–607.

[81] Levitin, G. (2003). Reliability evaluation for acyclic transmission networks of multi-state elements with delays, *IEEE Transactions on Reliability* **52**, 231–237.

[82] Levitin, G. (2001). Reliability evaluation for acyclic consecutively connected network with multistate elements, *Reliability Engineering and System Safety* **73**, 137–143.

[83] Liu, P.X., Zuo, M.J. & Meng, M.Q. (2003). Using neural network function approximation for optimal design of continuous-state parallel-series systems, *Computers and Operations Research* **30**, 339–352.

[84] Jin, T. & Coit, D.W. (2001). Variance of system-reliability estimates with arbitrarily repeated components, *IEEE Transactions on Reliability* **50**, 409–413.

Further Reading

Cui, L., Kuo, W., Loh, H.T. & Xie, M. (2004). Optimal allocation of minimal & perfect repairs under resource constraints, *IEEE Transactions on Reliability* **53**, 193–199.

Gupta, R. & Agarwal, M. (2006). Penalty guided genetic search for redundancy optimization in multi-state series-parallel power system, *Journal of Combinational Optimization* **12**, 257–277.

Ha, C. & Kuo, W. (2006). Reliability redundancy allocation: an improved realization for nonconvex nonlinear programming problems, *European Journal of Operational Research* **171**, 24–38.

Ha, C. & Kuo, W. (2006). Multi-paths iterative heuristic for redundancy allocation: the tree heuristic, *IEEE Transactions on Reliability* **55**, 37–43.

Konak, A., Coit, D.W. & Smith, A.E. (2006). Multi-objective optimization using genetic algorithms: a tutorial, *Reliability Engineering and System Safety* **91**, 992–1007.

Kuo, W., Chien, K. & Kim, T. (1998). *Reliability, Yield and Stress Burn-in*, Kluwer.

Kuo, W. & Zuo, M.J. (2003). *Optimal Reliability Modeling: Principles and Applications*, John Wiley & Sons.

Levitin, G. (2004). Consecutive k-out-of-r-from-n system with multiple failure criteria, *IEEE Transactions on Reliability* **53**, 394–400.

Levitin, G. (2004). Reliability optimization models for embedded systems with multiple applications, *IEEE Transactions on Reliability* **53**, 406–416.

Lin, F., Kuo, W. & Hwang, F. (1999). Structure importance of consecutive-k-out-of-n systems, *Operations Research Letters* **25**, 101–107.

Lisianski, A. & Levitin, G. (2003). *Multi-State System Reliability, Assessment, Optimization and Applications*, World Scientific Publishing.

Mahapatra, G.S. (2006). Fuzzy multi-objective mathematical programming on reliability optimization model, *Applied Mathematic and Computation* **174**, 643–659.

Marseguerra, M. & Zio, E. (2000). Optimal reliability/availability of uncertain systems via multi-objective genetic algorithms, Proceedings Annual Reliability and Maintainability Symposium, Los Angels, pp. 222–227.

Salazar, D., Rocco, C.M. & Galvan, B.J. (2006). Optimization of constrained multiple-objective reliability problems using evolutionary algorithms, *Reliability Engineering and System Safety* **91**, 1057–1070.

Tian, Z. & Zuo, M.J. (2006). Redundancy allocation for multi-state systems using physical programming and genetic algorithms, *Reliability Engineering and System Safety* **91**, 1049–1056.

Yamachi, H., Tsujimuraa, Y., Kambayashia, Y. & Yamamotob, H. (2006). Multi-objective algorithm for solving N-version program design problem, *Reliability Engineering and System Safety* **91**, 1083–1094.

Yang, J.-E., Hwang, M.-J., Sung, T.-Y. & Jin, Y. (1999). Application of genetic algorithm for reliability allocation in nuclear power plants, *Reliability Engineering and System Safety* **65**, 229–238.

Yu H., Yalaoui, F., Châtelet, E. & Chu, C. (2007). Optimal design of a maintainable cold-standby system. *Reliability Engineering and System Safety* **92**, 85–91.

Zafiropoulos, E.P. & Dialynas, E.N. (2004). Reliability and cost optimization of electronic devices considering the component failure rate uncertainty, *Reliability Engineering and System Safety* **84**, 271–284.

WAY KUO AND RUI WAN

Optimization, Steepest Ascent *see* Response Surface Methodology

Organizational Assessment Models

The Quality Assurance Models

The seeds of organizational assessment models can be traced back to the period between the 1930s and 50s, when leading companies operating in strategic sectors started to elaborate the quality assurance (QA) concept. The concept was based on the definition of system requirements (QA standards) and on independent assessments (audits) aimed at checking compliance of the involved organization with such requirements. The purpose of QA was to give executives confidence that the audited organization was capable to meet the quality targets and all the relevant activities (planning, execution, measurement, review) had been put in place. The first structured QA standards appeared in the 1950s/1960s, in the wake of the progresses made in the definition of the "product development cycle".

QA extended rapidly to all sectors where defect prevention was a strategic issue (defense, aerospace, nuclear), and then to commercial sectors. Geographic diffusion was also rapid. But companies using QA soon realized that the next logical step was to extend the concept to their suppliers and to assess them for compliance (second-party audits). *External* QA standards were created. Proprietary at the beginning – being derived by the customer internal standards – external standards could not obviously enter into the same details as internal standards. External QA standards soon took the lead over internal standards. In B2B and B2A (Business to Business and Business to Administration) relations, earlier system assessments were seen as an important means to save time and money with respect to the traditional product testing procedures [1].

The need to reduce the burden, for both the suppliers and customers, coming from multiple standards and audits, fostered joint definition of sector standards (e.g., automotive). Increased awareness of the commonalities among production systems – whatever the product – led some supply-intensive administrations (typically defense) to create a wide scope of general standards. The next step was the creation of national QA standards, and finally, in 1987, of the ISO 9000 quality system standards [2]. Such international standards marked the decisive turn from second- to third-party audits, conducted by independent, recognized bodies (certifications).

Total Quality Management Models and Awards

The 1980s showed the limits of a standard-based quality vision. The new customer- focused quality approach inaugurated by the Japanese was winning out over the inward- focused approach induced by standards. The Western "quality prophets" (Deming, Juran, Feigenbaum), unheard at home in the past, were rediscovered. New quality models – called *total* **quality management** (TQM) *models* – proliferated, rotating around the concepts of customer focus, process-based management, employee involvement, and continuous improvement (*see* **Total Quality Management (TQM)**). Proliferation dwindled sharply when some highly respected quality organizations thought of creating national and international quality awards, as a means to diffuse the new quality culture. To that purpose, experts were gathered to review and consolidate the findings of the period and define, through a consensus process, the relevant award assessment model and process. The American Malcolm Baldrige (MB) National Quality Award [3] was issued in 1987, followed, in 1991, by the European Foundation for Quality Management's (EFQM) European Quality Award [4]. Figures 1 and 2 represent the block diagram of the two models.

The two models brought forth significant changes with respect to QA standards. First, they do not look for compliance with minimum requirements but define methods to measure performance and capabilities, based on a number of factors (the blocks in the figures) and subfactors that are considered critical in relation to excellence. The award assessment is based on the analysis of an application report presented by the candidate and a site visit. Scores are assigned according to defined measurement rules. Second change: results appear on the assessment scene – *perceived results* (or outcome) in particular. That means recognition that *compliance* with a model is not sufficient for a quality judgment; *effectiveness*, measured by results, completes the assessment scenario. Finally, the subdivision between *results* and *enablers*, introduced by the EFQM model, starts to shape a real organizational model, where management can identify – and test – the links between

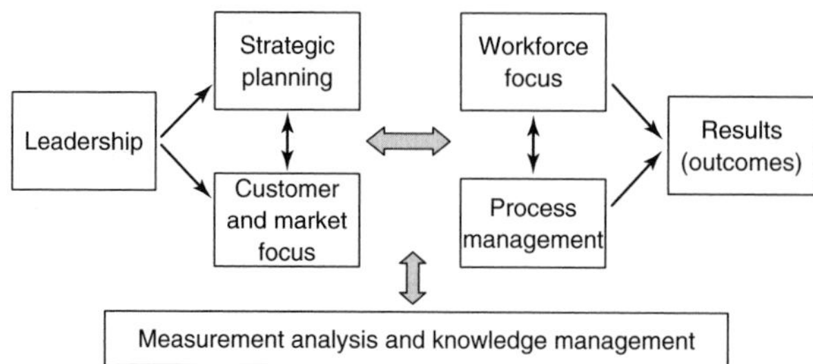

Figure 1 The Malcolm Baldrige model [Source: The National Institute for Standards and Technology – `http://www.quality.nist.gov`.]

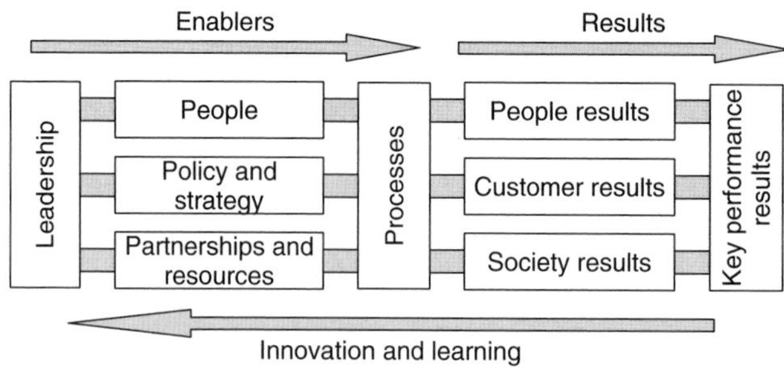

Figure 2 The EFQM excellence model, 2007 [Source: European Foundation for Quality Management – `http://www.efqm.org`. © EFQM. 1999–2003.]

organizational causes and their effects. That helps to understand the underlying dynamics of TQM models: continuous improvement.

Models for Continuous Improvement

The narrower the scope of a model, the better it works. Award models are conceived to measure, compare, and publicly recognize organizational excellence, and to that purpose they need rigorously defined standards. Is it correct to use the same standardized models and processes for performance improvement also? Arbitrarily enlarging the scope of a model is risky; better to reshape it. Unfortunately, the award models and processes have become, in the last 20 years, the normal choice for self-assessment. An acceptable choice when seeking for comparable quantitative measurements of performance; less acceptable when the aim is organizational diagnosis, to identify areas that need improvement (we use this term here both for incremental and quantum leap improvement). Experience shows that focus on scoring is often at odds with focus on diagnosis.

The dominance of award models in TQM – combined with the dominance of ISO 9000 in QA – created a kind of conformism that somehow hampered the development of free, improvement-oriented models. In fact, diagnostic effectiveness is enhanced by the use of *contingency* models (that is, models that fit the organization's specific characters). Standardized models and procedures, necessary for the awards, turn out to be a burden.

The Present Situation

In the area of QA and certification, ISO 9000 is the reference worldwide. In the area of TQM, the MB and EFQM are the most popular models in the West, the Deming Prize Model [5] in Japan. Recognition that the more the model fits the organizations' characters the better it performs, has led to the development of sector TQM models (e.g., education, public administration).

Several TQM/excellence models have been developed by companies, consultant organizations, and experts, often merging TQM concepts with other successful approaches, such as the *balanced scorecards* [6]. They can be divided into two categories.

1. **Models aimed at measuring organizational performance and capabilities**
Similar in principle to the award models, they are however – usually – more inward than outward directed. The aim is to endow management with a set of high-level indicators to measure the state of the organization from a plurality of perspectives. Indicators are chosen from among those factors that are considered critical for excellence (critical success factors, CSF). The literature provides many examples of CSF [7–10]. Albeit experts agree on comprehensive lists of CSF, differences arise when a short list (typically, 7–10 elements) is pursued to get to a manageable model. CSFs are normally high-level, abstract constructs (e.g., leadership, organizational learning) that require careful declination into assessable elements. Separation of results and enablers helps, since it confines the more intangible evaluation to the enabler side. Inconsistencies between enablers and results turn out to be important symptoms for further investigation. For these models, tests and validation of the chosen CSFs, taking into account interdependence, is important, as well as measurement reliability. Structural equation modeling is often used for the purpose.

2. **Models aimed at organizational diagnosis and improvement**
These too imply performance and capability measurements. But the role of measurement is secondary, in relation to results. Measurement is mainly used to identify performance gaps (the starting point for diagnosis); in relation to enablers, it serves to quantify weaknesses and then define priorities of intervention. Possible errors are controllable, since the normally used heuristic cycle "plan/execute/test" evidences them. Diagnostic models tend to overcome the use of abstract constructs, like the critical success factors, as basic elements of the model. Given their purpose, the use of elements that are more directly linked to the reality of the organization is recommended. They can be called *critical systemic factors*, since they are normally subsystems or important elements in the systems perspective. The critical systemic factor constructs are normally deployed to lower level elements and questions, to the point that the critical organizational causes of problems (very often interdependent) can be addressed. Diagnostic models, albeit focused on assessment, have shown the potential to be also used in planning and execution, that is, in all phases of the organization PDCA (plan/do/check/act) cycle [11].

Development Trends: The Systems View

The evolution from organizational assessment models to organizational improvement models seems the most promising trend. However, progress is slow. What is the reason? Why do not managers generally recognize TQM models as management tools, useful to keep their organizations fit. Probably one of the answers (possibly the most relevant) is lack of systems perspective. One could object that all TQM/excellence models claim the systems view. Declarations of interdependence between the components of the system are made, but little is said on how to enhance value generation capabilities through relations – and how to assess consistency of relation maps with strategies. In fact, by developing appropriate relations within a system, unique properties can emerge (the so-called emergent properties) – and organizational excellence is a unique property, that cannot be reached through technology or individual excellence alone.

Separation of enablers from results is a first step toward integration of the systems concepts into the model. Enablers in fact represent the components of the systems (elements, subsystems, and throughput processes) that are critical in relation to the purpose (Figure 3). The purpose is to be found mainly outside the system, in the environment, where customers and stakeholders are located. In the figure, internal stakeholders (like managers and employees) are placed across the border, since they are independent as persons but part of the system in relation to their role. The environment (that represents the suprasystems where the examined system lives and operates) is divided into two parts: the transactional environment (that the organization can influence) and the independent environment, that the organization is bound to understand as well as possible.

A challenge today seems to be developing organizational models that merge quality thinking and

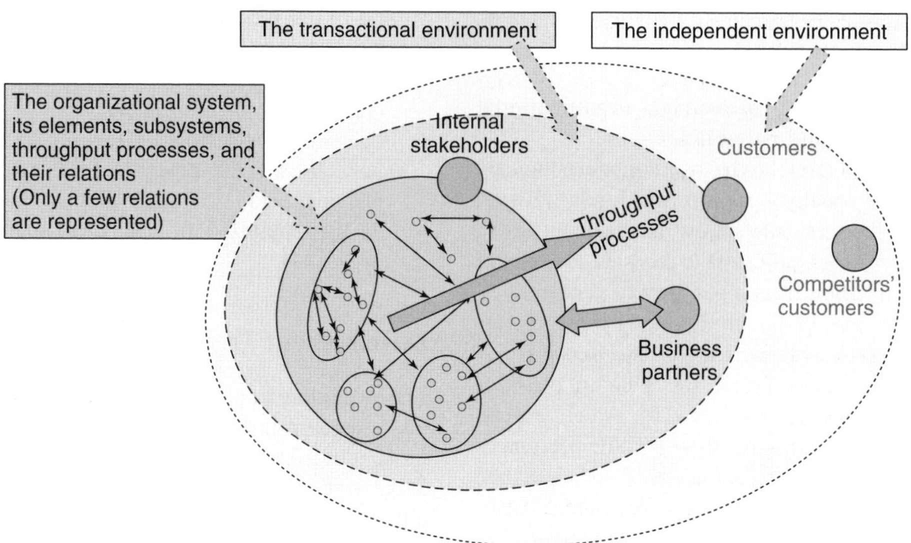

Figure 3 A systemic model of the organization highlighting the main relations within the system and between the system and its environment. It can be used as the basis for building TQM models [12]

systems thinking – and making them acceptable to managers [12].

References

[1] Juran, J.M. & Godfrey, A.B. (1999). *Juran's Quality Handbook*, 5th Edition, McGraw-Hill, New York, pp. 2.13–2.14, 11.1–11.7.

[2] Juran, J.M. & Godfrey, A.B. (1999). *Juran's Quality Handbook*, 5th Edition, McGraw-Hill, New York, Section 11.

[3] Baldrige National Quality Program, http://www.quality.nist.gov.

[4] EFQM, European Foundation for Quality Management, http://www.efqm.org.

[5] The W. Edward Deming Institute, Deming Prize, http://www.deming.org/demingprize.

[6] Kaplan, R.S. & Norton, D.P. (1996). *The Balanced Scorecards*, Harvard Business School Press, Boston.

[7] Black, S. & Porter, L. (1996). Identification of the critical factors of TQM, *Decision Sciences* **27**(1), 1–20.

[8] Bittici, U.S., Suwignjo, P. & Carrie, A.S. (2001). Strategic management through quantitative modeling of performance measurement systems, *International Journal of Production Economics* **69**, 15–22.

[9] Anderson, J., Rungtusanatham, M. & Schroeder, R. (1994). A theory of quality management underlying the Deming management method, *The Academy of Management Review* **19**(3), 472–509.

[10] Curcovic, S., Melnyk, S., Calantone, R. & Hardfield, R. (2000). Validating the Malcolm Baldrige National Quality Award framework through structural equation modeling, *International Journal of Production Research* **38**(4), 765–791.

[11] Conti, T. (1999). *Organizational Self-Assessment*, Kluwer Academic Publishers, Dordrecht, Chapter 3.

[12] Schnauber, H., ed. (2006). Conti T: integration of quality concepts into systems thinking, *Kreative un Konsequent*, Carl Hanser Verlag, Munchen, Chapter 4.

Tito A. Conti

Orthogonal Arrays

Introduction and Motivating Example

We wish to conduct a simple experiment to determine the effects of two factors on a quantitative response Y. An experiment to investigate the effects of several factors on a response is referred to as a *factorial experiment* (*see* **Factorial Experiments**). Assume that each factor is to be observed at only two levels (two values), which we denote -1 and $+1$. If a factor is quantitative, but we decide to observe it at only two of its possible values, −1 is used to denote the smaller of the two levels.

Suppose that we decide to take a total of four observations (runs). At what combinations of values (called the *treatments*) should we observe the factors? We will represent the treatments that we observe for each run by a design matrix or array. Columns represent factors and rows runs. The entry in a given column and row indicates the level of the factor represented by the column that is observed for the run represented by the row. Table 1 gives three of several possibilities. Notice that, in all three designs, the first run consists of the treatment with factor 1 at level 1 and factor 2 at level 1.

A minimal requirement of any design is that it should allow us to estimate all the factor effects (main effects and interactions) that we believe are present. In this experiment, suppose that only main effects are present. Then a design should allow us to estimate the main effects of both factors. To investigate this, for each design we generated observations Y with value equal to the sum of the levels of the two factors plus random error (taken to be normal with mean 0 and standard deviation 0.1). We analyzed the data with statistical software using a main-effects-only analysis of variance model (*see* **Analysis of Variance**) and the type-III sums of squares (**SS**). For more about this type of analysis, see [1, 2] or [3]. The results are summarized in Table 2.

For the first design, the software does not provide estimates of the sum of squares due to the main effects of factors 1 and 2. These sums of squares are estimates of the variation in the response due to these main effects. On reflection, it is not surprising that we do not obtain estimates. The effect of a factor is observed by changing its level and seeing how the response changes. For design 1, whenever we change factor 1 we also change factor 2. Hence, we cannot determine if observed changes in the response are due to factor 1, factor 2, or the sum of their effects. In this case, we say factors 1 and 2 are *completely aliased* (*see* **Aliasing in Fractional Designs**). In general, this occurs when the columns in a design corresponding to two different factors are perfectly correlated (a **correlation** of either 1 or −1). For design 1, the correlation between the four pairs of values is 1.

Table 1 The design matrices for three designs for studying two factors with four runs

	Design matrix 1 factors		Design matrix 2 factors		Design matrix 3 factors	
Run	1	2	1	2	1	2
1	1	1	1	1	1	1
2	1	1	1	1	1	−1
3	−1	−1	−1	1	−1	1
4	−1	−1	−1	−1	−1	−1

Table 2 **Degrees of freedom** and type-III sums of squares for simulated data for the three designs

	Design matrix 1		Design matrix 2		Design matrix 3	
Source	*df*	*SS*	*df*	*SS*	*df*	*SS*
Factor 1	0	–	1	2.1236	1	4.8323
Factor 2	0	–	1	2.2881	1	3.8384
Error	2	0.0148	1	0.0015	1	0.000027
Total	3	16.3689	3	10.4370	3	8.6706

In the second and third designs, we obtain estimates of the main effects of factors 1 and 2. However, additional analysis reveals a problem with the second design. Suppose we had assumed that there were no main effects of factor 2 and fit a main-effects model that included only factor 1. The results (for both designs, 2 and 3) are summarized in Table 3.

As should be the case, for both designs the sum of squares for factor 2 is incorporated (added) into the error sum of squares and the total sum of squares is unchanged. However, for design 2 the effect of factor 1 has changed (increased). For design 3 the effect of factor 1 is unchanged.

The problem is that for design 2 the levels of factors 1 and 2 are correlated (the correlation is 0.5774). This is known as *collinearity* in regression and as *partial* **aliasing** in factorial experiments. When partial aliasing is present the effects of the two factors cannot be completely separated. It is manifested when the sum of squares for one effect (factor 1 in the

Table 3 Degrees of freedom and type-III sums of squares for simulated data for designs 2 and 3 when only factor 1 is included

	Design matrix 2		Design matrix 3	
Source	*df*	*SS*	*df*	*SS*
Factor 1	1	8.1473	1	4.8323
Error	2	2.2896	2	3.8384
Total	3	10.4370	3	8.6706

example) changes when a second effect (factor 2 in the example) is removed from the model.

In design 3, the levels of factors 1 and 2 are uncorrelated (the correlation is 0). In this case, factor effects are independent of each other and the sum of squares for the **main effect** of factor 1 is the same whether or not the main effect of factor 2 is included in the model. When the levels of two factors are uncorrelated, the factors have a property known as *orthogonality*. Orthogonality is formally defined in the next section. Orthogonality is also equivalent (provided the levels are coded properly) to perpendicularity of vectors. When viewed as 4×1 column vectors, the two columns of design 3 are perpendicular.

This simple experiment illustrates the fact that to avoid aliasing (complete or partial), and thus obtain independent estimates of the effects of factors, it is desirable that every pair of factors in a design be orthogonal. This is formalized in the notion of an orthogonal array.

Definitions and Examples

In a factorial experiment, if all the level combinations of two factors appear in the same number of runs, the two factors are said to be *orthogonal*. A design is called *orthogonal* if all pairs of factors in the design are orthogonal. More generally, in a matrix two columns are said to be orthogonal if all possible combinations of symbols in the two columns appear equally often together. If all pairs of columns of a matrix are orthogonal, the matrix is said to be orthogonal. For the case where all factors have s levels, we now define a rich class of designs that have this feature and even stronger properties.

Definition 1 Let S be a set of s symbols or "levels" denoted $0, 1, \ldots, s-1$. An $N \times k$ matrix A with entries from S is called an *orthogonal array* with s levels, strength $t (0 \le t \le k)$, and index λ if every $N \times t$ submatrix of A contains each t-tuple of symbols from S exactly λ times as a row. We use the notation OA(N, k, s, t) to denote such an array.

Because there are s^t possible t-tuples of elements from S, an immediate consequence of this definition is that

$$N = \lambda s^t \tag{1}$$

In particular, equation (1) shows that λ is a function of N, s, and t so the parameters N, k, s, and t are sufficient to define an orthogonal array.

Orthogonal arrays as defined above are sometimes referred to as *symmetrical orthogonal arrays* (see for example [4]). Orthogonal arrays were first introduced by Rao [5, 6]. In the context of two-level factorial experiments, it is not uncommon to denote the elements of S -1, $+1$ rather than 0, 1.

Example 1 An OA(8, 4, 2, 3) is given in Table 4. Rows represent experimental runs and columns factors. Notice that every 8×3 submatrix (every selection of three columns of the matrix) has the property that all eight, three-tuples of elements of $S = \{0, 1\}$ occur exactly $\lambda = 1$ times as a row.

Example 2 An OA(12, 11, 2, 2) with elements of S denoted by -1 and $+1$ is given in Table 5. Notice that every 12×2 submatrix has the property that all four two-tuples of elements of $S = \{-1, 1\}$ occur exactly $\lambda = 3$ times as a row.

In some experiments it may not be possible or desirable to use the same number of levels for each factor. An extension of orthogonal arrays to the case

Table 4 An OA(8, 4, 2, 3)

0	0	0	0
0	0	1	1
0	1	0	1
0	1	1	0
1	0	0	1
1	0	1	0
1	1	0	0
1	1	1	1

Table 5 An OA(12, 11, 2, 2)

+1	+1	−1	+1	+1	+1	−1	−1	−1	+1	−1
−1	+1	+1	−1	+1	+1	+1	−1	−1	−1	+1
+1	−1	+1	+1	−1	+1	+1	+1	−1	−1	−1
−1	+1	−1	+1	+1	−1	+1	+1	+1	−1	−1
−1	−1	+1	−1	+1	+1	−1	+1	+1	+1	−1
−1	−1	−1	+1	−1	+1	+1	−1	+1	+1	+1
+1	−1	−1	−1	+1	−1	+1	+1	−1	+1	+1
+1	+1	−1	−1	−1	+1	−1	+1	+1	−1	+1
+1	+1	+1	−1	−1	−1	+1	−1	+1	+1	−1
−1	+1	+1	+1	−1	−1	−1	+1	−1	+1	+1
+1	−1	+1	+1	+1	−1	−1	−1	+1	−1	+1
−1	−1	−1	−1	−1	−1	−1	−1	−1	−1	−1

where factors may have different numbers of levels is the following.

Definition 2 A *mixed orthogonal array*, denoted $\mathrm{OA}(N, s_1^{k_1} s_2^{k_2}, \ldots, s_v^{k_v}, t)$ of strength t is an $N \times k$ matrix where $k = k_1 + k_2 + \cdots + k_v$ is the total number of factors, in which the first k_1 columns have symbols from $\{0, 1, \ldots, s_1 - 1\}$, the next k_2 columns have symbols from $\{0, 1, \ldots, s_2 - 1\}$, and so on, with the property that in any $N \times t$ submatrix every possible t-tuple occurs an equal number of times as a row.

Example 3 An OA(12, 2^4, 3, 2) is given in Table 6.

Mixed orthogonal arrays are sometimes referred to as *mixed-level orthogonal arrays* or *asymmetrical orthogonal arrays*. Also, an $\mathrm{OA}(N, s_1^{k_1}, s_2^{k_2}, \ldots, s_v^{k_v}, t)$ is sometimes denoted by $L_N(s_1^{k_1} s_2^{k_2}, \ldots, s_v^{k_v})$, when the strength is assumed to be two. Mixed orthogonal arrays were studied as early as 1961 by Addelman and Kempthorne [7] and were called *asymmetrical orthogonal arrays* by Rao [8].

Construction and Tables of Orthogonal Arrays

A variety of techniques are available for constructing orthogonal arrays. For example, an s^{k-p} **fractional factorial** design of resolution $k - p + 1$ is an $\mathrm{OA}(s^{k-p}, k, s, k - p)$. The N-run Plackett–Burman designs (*see* **Plackett–Burman Designs**), which are designs for experiments with all factors at two levels, can be generated cyclically given the first row. The second row is obtained from the first by shifting all entries to the right by one position and then placing the last entry in the first position. The third row is generated from the second row in the same way and this process continues until $N - 1$ rows are generated. All entries in the last row are -1. You can verify that the 12-run Plackett–Burman design given previously is generated in this way. Table 7.5 of [4] lists the first row for 12-, 20-, 24-, 36-, and 44-run Plackett–Burman designs.

Most techniques involve the use of certain combinatorial objects (linear codes, mutually orthogonal Latin squares, Hadamard matrices) or abstract algebra (Galois fields, projective geometry). Consult [9] for details, as well as a wealth of information about orthogonal arrays. Extensive tables of orthogonal arrays exist (see for example [9, 10]). N. J. A. Sloane maintains on on-line library of orthogonal arrays at the web site http://www.research.att.com/~njas/oadir/index.html. One can also find Plackett–Burman designs at this web site.

Table 6 An OA(12, 2^4, 3, 2)

0	0	0	0	0
0	1	0	1	0
1	0	1	1	0
1	1	1	0	0
0	0	1	1	1
0	1	1	0	1
1	0	0	1	1
1	1	0	0	1
0	0	1	0	2
0	1	0	1	2
1	0	0	0	2
1	1	1	1	2

Applications

Fractional Factorial Experiments

Some General Comments. Orthogonal arrays with two levels provide useful designs for two-level fractional factorial experiments with k factors (*see* **Fractional Factorial Designs**). In particular, an $\mathrm{OA}(N, k, 2, t)$ yields a design matrix for an N-run experiment in which no factorial effect of order $w \le t$ is aliased with any other factorial effect of order $t + 1 - w$ or less. A design with this property is said to have *resolution* $t + 1$ (*see* **Factorial Designs, Resolution of**). While it is true that a standard 2^{k-p} fractional factorial design of resolution $k - p + 1$ is an $\mathrm{OA}(2^{k-p}, k, 2, k - p)$, it is not true that every $\mathrm{OA}(2^{k-p}, k, 2, k - p)$ necessarily be a standard 2^{k-p} fractional factorial design. In other words, an arbitrary $\mathrm{OA}(2^{k-p}, k, 2, k - p)$ may exhibit complex aliasing, i.e., some higher-order interactions may be partially aliased (rather than fully aliased) with lower-order effects.

Data from an $\mathrm{OA}(N, k, 2, t)$ can be analyzed as follows. Use a model with k factors that includes only main effects and all t and fewer factor interactions of the k factors. Fit the model using standard statistical software. When the number of runs is equal to the number of effects in the model use half-normal plots (*see* **Half-Normal Plot**) or Lenth's method (*see* **Lenth's Method for the Analysis of Unreplicated Experiments**). Consult [1, 2] or [3] for more on the analysis of data from a fractional factorial experiment.

More generally, symmetric orthogonal arrays with more than two levels provide useful designs for higher-level factorial experiments. Also, asymmetric orthogonal arrays are useful for constructing designs in mixed-level factorial experiments. See [4] for further discussion.

Main-Effects Plans. When two-factor and higher-order interactions are assumed to be negligible, a model that includes only main effects would be considered adequate for describing the relationship between the response and the factors (*see* **Main Effect Designs**). Designs in which all pairs of factors are orthogonal have desirable properties in that no main effects are aliased with each other. In 1946, Plackett and Burman [11] constructed a large collection of orthogonal designs that are OA($4m$, $4m - 1$, 2, 2), where $4m$ is not a power of 2. The design in Table 5 is an example. Their designs are known as *Plackett–Burman designs* (*see* **Plackett–Burman Designs**) in the literature and it has become common practice to refer to all OA($4m, 4m - 1, 2, 2$) as Plackett–Burman designs. Selecting $F \leq 4m - 1$ columns of an OA($4m, 4m - 1, 2, 2$) yields a design with $4m$ runs that allows one to estimate the main effects of F factors without aliasing.

Screening Designs. Screening experiments (*see* **Screening Designs**) are used when one is interested in investigating a large number of factors simultaneously, but only a few are thought to be *active* (have nonneglible effects). In screening experiments it is usually only possible to run small fractions of the full factorial. As a consequence, such designs are likely to have low resolution. However, if one discovers that only a few factors are active it is desirable that the subdesign, called a *projection*, consisting only of the active factors be a full factorial. We typically do not know in advance which factors are likely to be active, so designs with the property that the projections onto all sufficiently small subsets of factors are full factorial designs are attractive. When the projection onto the active factors is a full factorial, a reanalysis of the data using only the active factors can be run as a full factorial and all main effects and interactions estimated.

Cheng [12] shows that certain orthogonal arrays on two symbols have this desirable projection property, making these orthogonal arrays useful for screening experiments. A design is said to be of ***projectivity*** p if, for every subset of p factors, a complete factorial design (possibly with some combinations replicated) is produced. If the two symbols in the orthogonal array are -1 and $+1$, the orthogonal array is said to have a *defining word of length* r if there exist r columns of the array with the property that the column whose ith row is the product of all the entries in the ith row of the r columns is either the vector in which all entries are 1 or the vector in which all entries are -1. Cheng [12] shows that an OA($N, k, 2, t$) with $k \geq t + 1$ has projectivity $t + 1$ if and only if it has no defining word of length $t + 1$. For example, one can verify that the 12-run Plackett–Burman design (the OA(12, 11, 2, 2) given in Table 5) has no defining word of length three and hence has projectivity three.

As an extension, one might also be interested in designs with the property that all projections onto a small number of factors be of resolution III (but not necessarily a complete factorial design). Such designs are said to have a hidden projection property. See [4] for more on this.

Optimal Designs for Response Surface Modeling

Suppose N units are observed in an experiment. On the ith, unit a response y_i and p explanatory variables (also called *predictors, regressors*, or *covariates*) $x_{i1}, x_{i2}, \ldots, x_{ip}$ are measured, $1 \leq i \leq N$. Suppose also that the response and explanatory variables are related by the following equation or *response surface model* (*see* **Response Surface Methodology**);

$$y_i = \beta_0 + \beta_1 x_{i1} + \cdots + \beta_p x_{ip} + \varepsilon_i \qquad (2)$$

where the ε_i represent independent random errors that are assumed to be normally distributed with mean 0 and variance σ^2. $\beta_0, \beta_1, \ldots, \beta_p$ are unknown constants or parameters to be estimated. The model in equation (2) is sometimes referred to as a *first-order response surface model.* After appropriately scaling each x_{ij}, it may be plausible to assume all x_{ij} satisfy $-1 \leq x_{ij} \leq +1$. It is customary to arrange the values of the x_{ij} in an $N \times (p+1)$ model matrix

$$\mathbf{X} = \begin{pmatrix} 1 & x_{11} & \cdots & x_{1p} \\ \vdots & \vdots & \ddots & \vdots \\ 1 & x_{N1} & \cdots & x_{Np} \end{pmatrix} \quad (3)$$

If the last p columns of $\mathbf{X}$ in equation (3) are an OA$(N, p, 2, t)$, using -1 and $+1$ to denote the levels, and if $t \geq 2$, then the resulting design is optimal (*see* **Optimal Design**) under a wide range of standard criteria such as A-, D-, and E-optimality. For example, the OA(12, 11, 2, 2) given previously is a D-optimal 12-run design for a first-order response surface model with $p = 11$.

When equation (2) includes not only all linear terms (the $x_{i1}, x_{i2}, \ldots, x_{ip}$) but also all cross-products of subsets of size q or less of these terms ($q \leq p/2$), then a design for which

$$\begin{pmatrix} x_{11} & \cdots & x_{1p} \\ \vdots & \ddots & \vdots \\ x_{N1} & \cdots & x_{Np} \end{pmatrix} \quad (4)$$

is an OA$(N, p, 2, t)$, using -1 and $+1$ to denote the levels, and for which $t \geq 2q$, is again optimal under a wide range of standard criteria such as A-, D-, and E-optimality. For $p/2 < q \leq p$, an OA$(N, p, 2, p)$ is needed in equation (4) for optimality. For example, an OA$(N, p, 2, t)$ with $t \geq 4$ is a D-**optimal design** for the model

$$y_i = \beta_0 + \sum_{j=1}^{p} \beta_j x_{ij} + \sum_{1 \leq j < \ell \leq p} \beta_{j\ell} x_{ij} x_{i\ell} + \varepsilon_i \quad (5)$$

with main effects and two-factor interactions.

Robust Parameter Design

In robust parameter design (*see* **Robust Design**), factors are divided into *control factors* (factors that are fixed once they are selected by a product designer) and *noise factors* (factors that are hard to control or may vary in normal use of a product). A typical objective is to find values of the control factors that make the product insensitive to variation in the noise factors. This can be accomplished by conducting experiments in which the noise factors are varied systematically in order to represent their variation in normal use. One strategy for designing such experiments is to choose separate designs for the control and noise factors. The design for the control factors is called the *control array* and the design for the noise factors the *noise array* (*see* **Product Array Designs**). Then, for each setting of the control factors in the control array, all settings in the noise array are observed. Such a design is called a *cross array*. The total number of runs in the cross array is the product of the number of factor settings in the control array and the number of factor settings in the noise array. Orthogonal arrays (symmetric or asymmetric) are often used for both the control and noise arrays. A name closely associated with robust parameter design is Genichi Taguchi. For more on robust parameter design, see [4].

Computer Experiments

There are situations in which a physical experiment is not feasible. In such cases it may be possible to describe the physical process of interest by a mathematical model implemented with code on a computer. One can then experiment with the code by varying the inputs. This is called a *computer experiment* (*see* **Computer Experiments**). For complex code that runs slowly (for example, taking 1 day for a single run), the number of runs is limited and one must design the computer experiment carefully. One principle is to select runs that are, in some sense, spread evenly over the design space. Orthogonal arrays have been used as the basis for designs in **computer experiments**. For k predictor variables each constrained to be between 0 and 1, and for some positive integer N, let $S = \{1/(N+1), 2/(N+1), \ldots, N/(N+1)\}$. For this choice of symbols, an OA(N^t, k, N, t) is an N^t-run design $(1 \leq t \leq k)$ with the property that for any t columns, if the rows are taken as points, we have a set of evenly spaced points on the t-dimensional unit cube $[0, 1]^t$. Thus, an OA(N^t, k, N, t) provides a design with the property that when projected onto any t, or smaller, dimensional subspace (any subset of t or fewer columns), the points are spread evenly. When $t = 1$, this produces what is known as a *Latin hypercube design* (*see* **Latin Hypercube Designs**). Koehler and Owen [13] provide additional examples of the use of orthogonal arrays in computer experiments.

An Example

Hunter *et al.* [14] ran a screening experiment to study the effects of seven factors; initial structure (A), bead size (B), pressure treatment (C), heat treatment (D), cooling rate (E), polish (F), and final treatment (G), on the fatigue life of weld-repaired castings. Possible designs included an 8-run, resolution-III fractional factorial and a 16-run, resolution-IV fractional factorial. The eight-run design allows one to fit only a model with main effects but no interactions, and no separate estimate for error is possible. The 16-run design allows one to add some two-factor interactions (assuming others are negligible) and allows an estimate of the error. To reduce the number of runs, avoid aliasing main effects with each other, and still accommodate some two-factor interactions (assuming others are negligible – see [4, Chapter 8] for discussion of the actual aliasing present in the design), a 12-run OA(12, 7, 2, 2) can be used. This was the design actually employed. The design and response (log lifetime) are given in Table 7.

The objective was to identify the factors that affect casting lifetime. A main-effects-only model can be fit to the data. Using a normal plot or Lenth's method, only factor F is identified as significant. A traditional analysis of variance gives a P value of 0.0556 and R^2 of 0.75. A follow-up analysis can be conducted by adding various subsets of all two-factor interactions with F to a model including all main effects. This process further identifies the FG interaction as significant (the P value for F is now 0.0144 and 0.042 for FG), with a marked improvement in the model fit (R^2 increases to 0.95).

Table 7 The OA(12, 7, 2, 2) for the 12-run, seven-factor experiment, with the log of the observed lifetimes

A	B	C	D	E	F	G	log lifetime
+1	+1	−1	+1	+1	+1	−1	6.058
+1	−1	+1	+1	+1	−1	−1	4.733
−1	+1	+1	+1	−1	−1	−1	4.625
+1	+1	+1	−1	−1	−1	+1	5.899
+1	+1	−1	−1	−1	+1	−1	7.000
+1	−1	−1	−1	+1	−1	+1	5.752
−1	−1	−1	+1	−1	+1	+1	5.683
−1	−1	+1	−1	+1	+1	−1	6.607
−1	+1	−1	+1	+1	−1	+1	5.818
+1	−1	+1	+1	−1	+1	+1	5.917
−1	+1	+1	−1	+1	+1	+1	5.863
−1	−1	−1	−1	−1	−1	−1	4.809

The conclusion is that polish and its interaction with the final treatment affect lifetime. Examining the factor effects indicates that setting F at its highest level and G at its lowest level produces the longest lifetime. This is consistent with what one observes in the data in Table 7.

References

[1] Box, G.E.P., Hunter, W.G. & Hunter, J.S. (2005). *Statistics for Experimenters. An Introduction to Design, Data Analysis, and Model Building*, 2nd Edition, John Wiley & Sons, New York, p. 633.

[2] Dean, A. & Voss, D. (1999). *Design and Analysis of Experiments*, Springer-Verlag, New York, p. 740.

[3] Montgomery, D.C. (2004). *Design and Analysis of Experiments*, 5th Edition, John Wiley & Sons, New York, p. 660.

[4] Wu, C.F.J. & Hamada, M. (2000). *Experiments*, John Wiley & Sons, New York, p. 630.

[5] Rao, C.R. (1946). Hypercubes of strength "d" leading to confounded designs in factorial experiments, *Bulletin of the Calcutta Mathematical Society* **38**, 67–78.

[6] Rao, C.R. (1947). Factorial experiments derivable from combinatorial arrangements of arrays, *Journal of the Royal Statistical Society, Supplement* **9**, 128–139.

[7] Addelman, S. & Kempthorne, O. (1961). Orthogonal main-effect plans, Technical Report ARL 79, Aeronautical Research Lab, Wright-Patterson Air Force Base, November 1961.

[8] Rao, C.R. (1973). Some combinatorial problems of arrays and applications to the design of experiments, in *A Survey of Combinatorial Theory*, J.N. Srivastava, ed, North-Holland, Amsterdam, pp. 349–359.

[9] Hedayat, A.S., Sloane, N.J.A. & Stufken, J. (1999). *Orthogonal Arrays*, Springer, New York, p. 416.

[10] Raghavarao, D. (1971). *Construction and Combinatorial Problems in Design of Experiments*, John Wiley & Sons, New York, p. 386.

[11] Plackett, R.L. & Burman, J.P. (1946). The design of optimum multifactorial experiments, *Biometrika* **33**, 305–325.

[12] Cheng, C.-S. (2006). Projection properties of factorial designs for screening experiments, in *Screening. Methods for Experimentation in Industry, Drug Discovery, and Genetics*, A. Dean & S. Lewis, eds, Springer Science+Business Media, New York, pp. 156–168.

[13] Koehler, J.R. & Owen, A.B. (1996). Computer experiments, in *Handbook of Statistics*, S. Ghosh & C. Rao, eds, Elsevier Science B. V., Amsterdam, Vol. 13, pp. 261–308.

[14] Hunter, G.B., Hodi, F.S. & Eager, T.W. (1982). High-cycle fatigue of weld repaired cast Ti-6A1-4V, *Metallurgical Transactions* **13A**, 1589–1594.

Related Articles

Fractional Factorial Designs; **Latin Hypercube Designs**; **Plackett–Burman Designs**; **Product Array Designs**; **Screening Designs**.

WILLIAM I. NOTZ

Outliers

Outliers are observations that stand apart from the majority of the observations. Outliers may be caused by reading or typing errors, contamination (data comes from more than one source), faults of the **measurement system**, or incidents in the process. It is important to detect outliers, because statistical analyses may be heavily affected (and even become invalid) when outliers are present. Basically, there are two ways to deal with outliers. One approach is to discard or reject outliers using tests of discordancy (preferably after additional checking of the circumstances under which they were obtained). Another approach is accommodation, i.e., to adjust analyses so that they become robust against possible outliers. The former approach comes down to deletion of the data point or replacement with the missing data value, while the latter approach involves the use of robust procedures such the median or the **trimmed mean** instead of the sample mean.

The box plot is a simple graphical tool to detect outliers. As a rule of thumb, observations that are more than 1.5 or 2 times the interquartile range (IQR, *see* **Quartiles**) away from the median of the data are marked in box plots as outliers by plotting them as crosses. Formal tests for outliers include the Dixon's tests and the Grubbs tests. Both tests assume normality (*see* **Normality Tests**). Dixon's test statistic for detecting one outlier on the right is given by $(x_{(n)} - x_{(n-1)})/(x_{(n)} - x_{(1)})$, where $x_{(i)}$ denotes the ith order statistic, i.e., the ith smallest observation. Similar test statistics exist for one or two outliers on the left or on both sides. Thus $x_{(n)}$ is the largest observation and possibly an outlier. Critical values of the Dixon's test must be looked up in a statistical table as there is no closed-form formula for them.

The Grubbs tests are based on a transformation of the maximum studentized value of the sample, i.e., the largest among the values $(x_i - \bar{x})/s$, where $\bar{x}$ denotes the sample mean and s the sample standard deviation. The Grubbs test statistic for detecting one outlier on the right is given by $(x_{(n)} - \bar{x})/s$, with critical value $(n-1)t_{n-2,\alpha/n}/\sqrt{n(n-2+t^2_{n-2,\alpha/n})}$ where $t_{n-2,\alpha/n}$ is the value such that a Student's t-distribution with $n-2$ **degrees of freedom** has probability α/n of exceeding it.

Caution should be exercised when using such tests, because they may behave poorly for other alternative hypotheses. For example, Dixon's test for one outlier on the right may have poor ***power*** when there are two outliers on the right. We refer to [1] for a comprehensive overview on this topic and to [2] for a shorter, practical guide.

References

[1] Barnett, V. & Lewis, T. (1994). *Outliers in Statistical Data*, John Wiley & Sons, New York.

[2] Iglewicz, B. & Hoaglin, D.C. (1993). *How to Detect and Handle Outliers, ASQC Basic References in Quality Control*, ASQ Quality Press, Milwaukee, Vol. 16.

Related Article

Normality Tests.

ALESSANDRO DI BUCCHIANICO

Overview of Statistics

Introduction

The term statistics is used in different ways. "Accident statistics" or "sales statistics" refer to numerical data in these areas. (The corresponding theoretical terminology defines statistics to be any functions of observable random variables.) Common problems encountered in the work with such data and those collected by scientists, engineers, government officials, lawyers, doctors, and so on have led to the development of general methods and principles concerning the collection, presentation, and analysis of data. The term statistics also denotes the discipline concerned with such methods.

This article considers statistics broadly as the field comprising all of the above concerns. It might be described as the enterprise dealing with the collection of data sets, and extraction and presentation of the information they contain.

New methods and formulations are constantly being developed, and new types of application come into view and in turn give rise to new problems. This flow of research is disseminated through a multitude of journals being published around the world. A listing of annual contributions is available in the *Current Index to Statistics: Applications, Methods, and Theory*, published since 1975. Much of the material prior to 1970 is listed in the five-volume *Index to Statistics and Probability* (Ross and Tukey, eds. R and D Press).

What establishes statistics as a discipline is that the same kind of data requiring the same kind of concepts and methods turn up in many different fields. On the other hand, special areas of application also may require some specialization and adaptation to particular needs. The *Encyclopedia of Statistics* gives a comprehensive overview in nine volumes and a number of updates.

Data Interpretation

Statistical Methodology

It is a central fact, underlying essentially all statistical thinking, that an actual data set typically is only one of many possible such sets that might have been obtained under the given circumstances. (For a possible exception, see Diaconis [1].) Measurements vary when they are repeated. A store inventory, besides depending on the day on which it is taken, will be affected by bookkeeping and counting errors. In a survey, data will be obtained from different households if a new sample is to be drawn; even if the same households are visited on another occasion, different members may be at home and provide different answers; and, finally, even the same family member may answer the same questions differently on another day. As a consequence, interpretation of a data set depends not only on the actual data, but also on what (if anything) is assumed about the possible alternative observations that might have been obtained instead. The following sections consider three categories of such assumptions, and briefly indicate the kinds of statistical procedures that can be based on each.

Data Analysis

It is rare that a data set is studied without any preconceived notions. Consider, however, the idealized approach of pure data analysis in which the data are considered on their own terms. The statistical methods developed on this basis have as their primary aim (a) exploration of the data to uncover the features of principal interest and (b) presentation of the data in a manner that will bring out and highlight these features. The set of techniques dealing with (a) and (b) are called *exploratory data analysis* (*see* **Exploratory Data Analysis**) and *descriptive statistics* (*see* **Descriptive Statistics**), respectively. Both employ a great variety of numerical and *graphical* (*see* **Graphical Representation of Data**) techniques.

The simplest, most basic data-analytic methods concern a single batch of numbers, for example, the first-year sales of 12 novels of a successful author, 40 measurements of corrosion taken at different locations on a copper plate, or the ages of 350 000 cancer patients listed in a tumor registry. Histograms, stem and leaf displays, and one-dimensional scatter plots (*see* **Graphical Representation of Data**) are some of the many ways of presenting the numbers of a batch. From these one can get an impression of where the data are centered (the general level of their values), and how spread out they are. Observations that lie far from the bulk of the data, the so-called *outliers* (*see* **Outliers**), may correspond to errors or exceptional cases and may deserve special attention. The display may exhibit unusual features such as bimodality, suggesting perhaps the possibility of a mixture of two batches, each unimodal but with different *modes*. If there is marked asymmetry, the possibility of making a transformation (*see* **Box and Cox Transformation**) (such as taking logarithms) to obtain a more symmetric data set arises.

Instead of a fairly detailed display of the data, authors frequently present only one or two summary statistics; for example, the mean or median to indicate the general level of the numbers (*see* **Measures of Location**), and a measure of their variability, such as the standard deviation (SD), median absolute deviation, or interquartile range (*see* **Measures of Scale**). A compromise is the *five-number summary*, consisting of the smallest, largest, and median observation, and the first and third *quartiles* (*see* **Quartiles**). These may be displayed graphically as a box *plot* (*see* **Graphical Representation of Data**).

Additional information concerning the numbers of a batch may be available and are important. If the order in which the 12 novels appeared is known, the data may show that the sales steadily increased, or that they increased up to a certain point and then leveled off, and so on, thus providing an indication of the author's changing reputation and success.

Example 1: Corrosion Data The exploratory use to which even very simple data can be put is illustrated by 40 corrosion readings taken at random over a metal plate (Campbell [2], Wolfowitz [3]). No particular pattern emerged when the observations were plotted according to their position on the plate. However, when they were plotted in the order in which they were taken, a bunching together in *runs* (*see* **Runs, Runs Tests**) of high and low observations suggested (as it turned out, correctly) a malfunctioning of the delicate measuring device.

Much of data analysis deals with batches of more complex units, such as pairs of numbers, more general vectors, matrices, or curves, and it need not be restricted to a single batch. With multivariate data, one may be interested in the separate features of the different variables, in relationships among the variables or sets of variables, or in aspects of the overall pattern of the multidimensional data set. The development of graphical (including, in particular, computer-displayed) and numerical methods for these purposes is a very active area of statistical research (*see* **Multivariate Analysis**). It includes, among others, approaches such as cluster analysis, and multivariate classification, multidimensional scaling, pattern recognition, and factor analysis. When several batches are being considered simultaneously, comparisons of the batches will tend to be of primary interest. Given today's large volumes of data collected from industrial processes, much recent research focuses on the adoption of multivariate techniques for studying high-dimensional process data (see [4] and [5]).

Statistical Analysis Based on Probability Models

The data-analytic approach indicated in the preceding section can provide clarification of the phenomena represented by both simple and very complex data sets, and can lead to important new insights and hypotheses. In its simplest forms such as numerical summaries and histograms, it is the statistical presentation that is most frequently encountered by the general public in newspaper articles and magazine reports. (Through misleading use, it also lends itself to much mischief. See for example, Huff [6].) However, this approach lacks an often essential requirement of the resulting inferences: because of the fact mentioned earlier, that the same phenomena might have led to different observations, it is impossible to assess the reliability of the conclusions.

Such an assessment requires knowledge concerning the alternative data sets that might have been observed in the given situation, instead of the set that was observed. The crucial step underlying modern theories of statistical inference and decision making is to consider the observed data as realizations of random variables. The possible values of these variables are governed by probability distributions specifying the probabilities of observing the various possible data sets. These distributions are not assumed to be known, but only to belong to a postulated family of possible distributions. The lack of complete specification represents the unknown aspect of the situation for the clarification of which the data were collected.

For a single batch $X_1, \ldots, X_n$, for example, a set of n measurements of some quantity, investigators often assume that the n observations are independent, and that each has the same probability distribution. If this common distribution is denoted by F, the model is completed by specifying a family $\mathbb{F}$ to which F is assumed to belong. This may for example be the family of all possible distributions F (i.e., no further assumptions are made) or the family of all distributions that are symmetric with respect to a specified point of symmetry. Such broad families are called *nonparametric* in distinction to parametric families where F is known except for the values of some parameters, for example, the family of all normal, Poisson, or Weibull distributions (*see* **Probability Density Functions**). Once the model is specified, one can now ask, and answer, the type of question concerning a single batch considered in the preceding section, with more precision. In particular, the conclusions no longer refer to this particular data set, but to the underlying process that produced it.

Suppose for example that the aspect of interest is the overall level previously represented by the average of the observations. Suppose that $\mathbb{F}$ is the family of normal distributions with mean ξ and unit variance. Then ξ is the average value, not of these particular n measurements but rather of all potential measurements, each weighted according to its probability. The average $\overline{X} = (X_1 + \cdots + X_n)/n$, which was a descriptive measure of the general level of the batch earlier, now becomes an estimator of the unknown ξ: for example, the true value of the quantity being measured or the average value of the characteristic (e.g., height, age, or income) in the population from which the X's were obtained as a sample.

The model assumptions make it possible to get an idea of the accuracy of an estimator such as $\overline{X}$, for example in terms of the expected closeness to the true value ξ. One commonly used measure of this closeness is the expected squared error, which for the case of n independent measurements with variance 1 equals

$$E(\overline{X} - \xi)^2 = \frac{1}{n} \quad (1)$$

This formula shows for instance how the accuracy improves with n, and enables one to determine the **sample size** n required to achieve a given accuracy.

This approach also provides a basis for comparing the accuracy of competing estimators. For the median $\tilde{X}$ of the X's for example, one finds that approximately (if n is not too small) $E(\tilde{X} - \xi)^2 = 1.57/n$ – more than 50% larger than the corresponding value $1/n$ for $\overline{X}$. The median is thus considerably less accurate than the mean. This conclusion depends strongly on the assumption that the X's are normally distributed. For other distributions the result may be just the reverse (*see* **Estimation**).

Another way of describing the accuracy of $\overline{X}$ is obtained by noting that

$$P\left[|\overline{X} - \xi| \le \frac{1.96}{\sqrt{n}}\right] = 0.95 \quad (2)$$

so that with probability 0.95 the estimator $\overline{X}$ will differ from the true value ξ by less than $1.96/\sqrt{n}$. The statement (1) can be paraphrased by saying that the random interval $(\overline{X} - 1.96/\sqrt{n}, \overline{X} + 1.96/\sqrt{n})$ will contain the unknown ξ with probability 0.95. Random intervals that cover an unknown parameter ξ with probability greater than or equal to some prescribed value γ are called *confidence intervals* (*see* **Confidence Intervals**) at confidence level γ.

Point **estimation** and estimation by confidence intervals provide two of the classical approaches to statistical inference. The third is *hypothesis testing* (*see* **Hypothesis Testing**).

Example 2: Extrasensory Perception Suppose the claim of a subject A to have extrasensory perception (ESP) is to be tested by tossing a coin 100 times at a location invisible to A, and recording A's perception (heads or tails) for each toss. Suppose that A obtains the correct result on 54 of the tosses. This clearly is not an indication of a strong ability at ESP. However, even a very slight ability would be of extraordinary interest. Is there support for such a finding, or is the result compatible with purely random guessing? Under the null hypothesis of pure guessing, the probability of calling a toss correctly is 1/2 for each toss, and the probability of getting 54 or more right is then 1/4. The result is therefore not particularly surprising under H, and a case for A's ability to do better than chance has not been made. Had A correctly identified 65 tosses rather than 54, the conclusion would have been quite different. The probability of 65 or more correct calls is only 0.002 when H is true. In the light of so extreme a result, one would have to give serious attention to A's claim.

Hypothesis testing, and point and interval estimation are all used extensively in applications, with each being more useful and popular in some areas than in others. The theory of these three methodologies is concerned with the performance of proposed procedures (including the determination of sample size to achieve a desired performance), the comparison of different procedures, and the determination of optimal ones. An important consideration is the robustness of a given procedure under violation of the model assumptions. If the procedures are very sensitive to these assumptions, one may want to study the problem in a nonparametric setting of the kind described for a single batch at the beginning of this subsection. Nonparametric (and semiparametric) models are particularly important for the analysis of large data sets, which has become more practicable as a result of increased computer capabilities and availability.

A unified framework for the three areas is provided by Wald's decision theory. This very general theoretical approach deals with the choice of one of a set of possible decisions d on the basis of observations x_i. Let the chosen d be denoted by $\delta(\mathbf{x})$. The observed value $\mathbf{x}$ is assumed to be the realization of a random quantity $\mathbf{X}$ with probability density $p_\theta(\mathbf{x})$, θ unknown. A loss function $L(\theta, d)$ measures the loss resulting from decision d when θ is the true parameter value, and the performance of a decision procedure δ is measured by its *risk function*

$$R(\theta, \delta) = E_\theta\,[L(\theta, \delta(\mathbf{X})] \qquad (3)$$

that is, the average loss incurred by δ when θ is true. A principal concern of the theory is the determination of a δ for which the risk function is as small as possible in some suitable sense. Another problem is the characterization of all admissible procedures, i.e., all procedures whose risk cannot be uniformly improved.

Decision theory can also be made to encompass sequential analysis, the **design of experiments** (by letting the loss function take account of the cost of observations), and the choice of model (by imposing a penalty that increases with the complexity of the model). However, it is an abstract approach that has been useful primarily for exhibiting general relationships rather than for its impact on specific methods. An important consequence of Wald's theory has been its liberating effect on the formal consideration of new types of situations, among them multiple comparisons and other simultaneous inference procedures.

Exploration versus Verification

To illustrate both the relation and the difference between the approaches described in the preceding two sections, consider once more the ESP example.

Example 2: Extrasensory Perception (Continued) When looking at the results of the 100 tosses, suppose the experimenter notices that of the 54 successes, 33 occurred during the last and only 21 during the first 50 tosses. The probability of 33 or more successes in 50 tosses with success probability $p = 1/2$ is only about 0.01. One might be tempted to explain away the poor performance in the first half of the experiment by the theory that it requires some warming up before the ability hits its stride, and to declare the success of the second half significant. However, such a conclusion would not be justified on the basis of this analysis since the calculation does not take account of the fact that the particular test (restricting oneself to the second half) was not originally planned but was suggested by the data.

Suppose that the situation had been reversed, that there had been 33 successes in the first and 21 in the second half. A possible explanation is that the exercise of ESP requires great concentration, and after 50 attempts the subject is likely to get tired. Similar explanations could have been found if success had been unusually high in the middle 50, or on the last 25, or the first 25, and so on. Instead of high concentration of successes in a particular segment of the sequence, other patterns might have struck an observer: for example, a gradual rise in the frequency of success, or a cyclic pattern of successes and failures, and so on. For each, an explanation could have been found.

This is the problem of *multiplicity*. Every set of observations – even a completely random one – will show some special features, and explanations can usually be found *post facto* to account for them. Unfettered examination of many different aspects of a data set is legitimate, and in fact a primary purpose, of exploratory data analysis. However, the results will then tend to look more impressive than they really are since attention is likely to focus on the extremes of a large number of possibilities. To legitimize a theory suggested by the data, one must test it, for example, on a separate part of the data not used at the exploratory stage or from new data specially obtained for this purpose. (A somewhat more limiting alternative to such a two-stage procedure is to formulate a number of possible theories to be considered before any observations are taken. The simultaneous testing of such a number of possibilities provides a legitimate calculation for the probability of the most striking of the associated results.)

The two stages, exploration of the data leading to the formulation of a hypothesis, followed by an independent test of this hypothesis, constitute the basic pattern of scientific progress as described by scientists (see for example, Feynman [7, Chapter 7]) and discussed by philosophers of science. Before considering a third aspect in the next section, let us briefly mention another distinction, which relates to the purpose of a statistical investigation. This is the difference between inference and decision making. It may be illustrated by the problem faced by a doctor who wants to arrive at a diagnosis (inference) but must also select a treatment (decision).

Example 3: Medical Diagnosis Suppose there are k possible conditions (diagnoses) $\theta_1, \ldots, \theta_k$ that might have led to the observed symptoms and test results **X** (the data), including for example measurements of temperature, blood pressure, and so on. Under condition θ, the observations **X** have a distribution $p_\theta(\mathbf{x})$. The problem of diagnosis is thus the statistical problem of using the observed value of x to determine the correct value of θ. (A standard procedure is to select the value $\hat{\theta}$ of θ that maximizes $p_\theta(\mathbf{x})$ for the given observation **x**, the so-called maximum-likelihood estimate (*see* **Maximum Likelihood**) of θ.) The associated decision procedure might be to select the treatment that would be most appropriate for the chosen θ. The situation is however more complicated since the choice of treatment must also take account of the severity of the consequences in case of an incorrect diagnosis resulting in a nonoptimal treatment, the so-called loss function. The distinction between inference and decision making is reviewed in Barnett [8]. For a discussion of computer-based medical diagnosis and treatment choice, see Shortliffe [9] and Spiegelhalter and Knill-Jones [10].

Bayesian Inference and Decision Making

Example 3: Medical Diagnosis (Continued) Suppose a patient P being tested for the conditions $\theta_1, \ldots, \theta_k$ of the last example is told that the tests point to $\hat{\theta}(= \theta_1$, say) as the most likely cause of the symptoms. Naturally, P wants to know just how likely it is that θ_1 is in fact the true cause. The doctor has to admit that the term *most likely* was used imprecisely; that $\hat{\theta} = \theta_1$ is the condition that assigns the highest probability to the observed test results, not necessarily the most likely of the conditions $\theta_1, \ldots, \theta_k$, and that in fact no probability can be assigned to these conditions.

Actually, in this example it may be possible to make such an assignment. Suppose that π_i is the incidence of condition θ_i in the population of sufferers from the given symptoms, and hence is the probability that a patient drawn at random from the population of such sufferers has condition θ_i. The condition of such a patient is then a random variable Θ, with prior probability $P(\Theta = \theta_i) = \pi_i$ before the tests are taken, and posterior probability $P(\Theta = \theta_i|\mathbf{x})$ in the light of the test results **x**. The latter probability can be calculated from the π_i and the $p_{\theta_i}(\mathbf{x})$ by Bayes' theorem.

On being presented with this probability, P may however still not be satisfied but complain to the doctor: "You have treated me for 20 years, you know my complete medical history, lifestyle, and habits. In the light of all this additional information, what is the probability of suffering from θ_i not for a random patient but for me personally?" Unfortunately, the interpretation of probability as frequency, which was tacitly assumed up to now, and which in particular applied to the prior probabilities π_i of the preceding paragraph, precludes assigning probabilities (other than 0 or 1) to unique events such as this particular patient's suffering from condition θ_i.

This difficulty is met head-on by the Bayesian approach according to which π_i can be chosen to represent the physician's probability that condition θ_i obtains for this particular patient. This meaning of probability is, however, different from the earlier one. Probability is no longer a frequency but the degree of belief attached to the event in question. (If the event is repetitive, this probability typically draws close to the observed frequency as the number of cases gets large.)

In general, the Bayesian approach assigns a *prior probability distribution* to a parameter θ before the observations **X** are taken. Once the values **x** of **X** are available, the prior distribution of θ is updated to the posterior (i.e., conditional) distribution of θ given $\mathbf{X} = \mathbf{x}$, which shows how the prior beliefs regarding the chances of different θ values have changed in the light of the data.

In the decision theoretic terminology introduced earlier, the relevant assessment of the performance of a procedure δ from a Bayesian point of view is not the risk function $R(\Theta, \delta)$ but rather the posterior expected loss

$$r(\mathbf{x}, \delta) = E\,[L(\theta, \delta(\mathbf{x})|\mathbf{x}] \qquad (4)$$

calculated according to the conditional distribution of θ given **x**.

Ideally a Bayesian has a comprehensive, consistent view of the world with a probability attached to every unknown fact. These probabilities must satisfy an appealing set of axioms (the axioms of *coherence*), and must be updated as new information becomes available. For an individual's response to the world (or even a specific problem), a chief difficulty in implementing this program is the determination of the prior distribution. A considerable literature deals with methods for eliciting a person's degrees of belief with respect to a given situation, but there is also an *objective Bayesian school* in which the prior distribution represents "total lack of information".

Ideally, each person has a single correct personal probability regarding any unknown event. However, in practice, these probabilities "can never be quantified or elicited exactly (i.e., without error) especially in a finite amount of time" (Berger [11, p. 64]). This has led to the suggestion of a robust Bayesian viewpoint according to which one "should strive for Bayesian behavior which is satisfactory for all **prior distributions** which remain plausible after the prior elicitation process has been terminated" (Berger [11]).

A second difficulty arises in situations in which the decision or opinion concerns not a single person, but represents a joint problem for a group or occurs in the public domain, as for example in the publication of the analysis of a scientific investigation. Some aspects of this problem will be considered in the next section.

The Bayesian/Frequentist Controversy

The mutual criticism of Bayesians and frequentists has given rise to a lively (and sometimes acrimonious) debate, which has helped to clarify a number of basic statistical issues. (There are many different variants of Bayesians and frequentists, not all of which will agree with the positions ascribed to these approaches here.) One of the central concerns is that of subjective *versus* objective data evaluation in scientific inference and reporting. Fisher, Neyman, and Pearson developed their frequentist theories in a deliberate effort to free statistics from the Bayesian dependence on a prior distribution, and this aspect has continued as the central frequentist objection. The Bayesian response to this criticism is twofold. On the one hand, it is pointed out that frequentist analysis involves similar types of specification. There is the choice of model and loss function, both of which must be chosen in the light of previous experience and involve judgments that are likely to vary from one person to another. In addition, there is the problem of selecting a frame of reference that forms the basis of the frequency calculations. In assessing the incidence of the conditions $\theta_1, \ldots, \theta_k$ of the preceding section, for example, for what population should this be calculated: the population of the world, or the country, or the city in which the person lives; should the comparison be restricted to patients of the same age, sex, and so on? This is the problem of conditional inference, which so far has found no satisfactory frequentist solution. (For the Bayesian, the problem does not arise in this form since the probabilities will always refer to this particular patient, but the same issue arises when one must decide how to weigh the experience with other patients in forming an opinion about this one.)

As a more positive response, there have been Bayesian suggestions (for example, Dickey [12], Smith [13]), that in scientific inference the analysis should be reported under a variety of different priors that – it is hoped – will include the opinions of the readers. The view that "any approach to scientific inference which seeks to legitimize *an* answer to complex uncertainty is a totalitarian parody of a would-be rational human learning process" (Smith [13]) considerably narrows the gulf between the two approaches.

The reason for this narrowing can be found in the combination of two facts. The first is a basic result of Wald's (frequentist) decision theory to the effect that every admissible procedure is a Bayes solution or a limit of Bayes solutions. Secondly, frequentists tend not to believe in a unique correct approach, and may therefore try a number of different solutions corresponding to different optimality principles, robustness properties, and so on. If these lead to similar conclusions, any one of them can be adopted. Otherwise, a careful examination of the differences may clarify the reason for the discrepancies and point to one as the most appropriate. Lacking such a resolution, one may instead prefer to report a number of different procedures.

The Bayesian and frequency approaches lead to different ways of assessing the performance of a decision procedure. From a strict Bayesian point of view, only the posterior distribution of θ given $\mathbf{x}$, and the posterior expected loss $r(\mathbf{x}, \delta)$, are relevant, while frequentists measure the performance of a procedure by its risk function. However, on this issue also, an accommodation to statistical practice and the need for communication has narrowed the gap by generating a Bayesian interest in risk functions. Thus Rubin [14, p. 116] writes, "Frequency calculations that investigate the operating characteristic of Bayesian procedures are relevant and justifiable for a Bayesian when investigating or recommending procedures for general consumption." Similar considerations can be found in Berger [15].

In the other direction, the frequentist approach has been strongly influenced by Bayesian ideas, in particular, by recognizing that it is natural and useful to consider the prior distribution leading to a proposed admissible procedure. An example in which such a Bayesian interpretation has made an important contribution to a theory developed in a decision theoretic framework is that of Stein estimation.

Data Acquisition

Measuring Single Units

The first part of this entry dealt with the interpretation of data once they have been collected. In this and the following sections, we consider some of the processes that produce data, and the statistical problems arising at this earlier stage. (An extensive discussion of various types of data is provided in Hoaglin *et al.* [16].)

The basic data units are the numbers, symbols, words, or other entries making up a data set; for example, measurements of the height of a person or plant, or of the weight of a wagonload of fruit or a minute amount of some chemical; barometer or temperature readings; or the scores of a student in an intelligence or aptitude test. Alternatively, the observations could be the information provided by a person answering a questionnaire or an interviewer, such as family size, last year's income, or religious and political affiliation.

A concern for **data quality**, for their reliability and validity, is an important task preceding the collection of data (*see* **Measurement Systems Analysis, Overview**). Efforts must be made to eliminate **bias**, reduce variability, and eliminate sources of error. The repeatability and reproducibility of measuring gauges should be assessed, and if needed improved. In constructing a questionnaire, great care is required to avoid ambiguities. Checks on the reliability of responses can often be built into the data set, for example by asking for the same information in a number of different ways or in different contexts.

Another aspect of data improvement arises after the data have been obtained. Data editing (or "cleaning") involves the deletion or modification of entries that do not appear to be in consonance with the rest of the data and that sometimes represent obvious errors (for example, in a series of monthly measurements of a baby's head circumference when one month's measurement is smaller than the preceding ones). A variety of statistical methods have been developed for this purpose. In addition, robust statistical procedures are available that satisfactorily control the effect of outlying observations (see for example Hoaglin *et al.* [17] and Hampel [18]). On the other hand, data cleaning by inspection, without clearly stated rules, runs the risk of introducing a subjective element. (For example, just which observations are to be singled out for such treatment?) Even with good rules, it may remove valuable evidence and destroy the basis for probability calculations. For this reason, it is typically better to consider the editing of data as part of the statistical analysis and, while perhaps indicating definite or suspected errors, not to change the original data.

The improvement of data quality mentioned above is not the only aspect of **data collection** to be considered before obtaining the observations for an investigation. Two questions, in particular, that are of crucial importance and that will be discussed in the next sections are how many and what kind of observations are required.

Assessing Population Characteristics

The target of most investigations is not a single unit but a population of such units (or the process generating the units). A set of professors, apples, days, mice, or light bulbs is typically examined not because of interest in these particular specimens but in order to reach conclusions about the populations they represent. (It is this aspect and the associated quantification that has earned statistics its reputation of being antihumanistic.)

Efforts to obtain information about every member of the target population through a census go back to at least the third millennium B.C. in Babylonia, China, and Egypt. (For a nontechnical history of the census, see Alterman [19].) The purpose was usually to provide a basis for taxation or proscription.

Today, population censuses seeking a great variety of information are carried out, many of them at regular intervals, in nearly all countries of the world.

However, taking a complete inventory is not the only way to obtain accurate statistical information about a population (and in the case of an infinite population, such as the one consisting of all products that a process manufactures, it is not even possible). The same end can usually be achieved more economically by collecting information only from the members of a sample, taken from the population by a suitable sampling method. The much smaller size of the sampling operation tends to make this procedure not only cheaper but also more accurate since it permits better control of the whole process and hence the quality of the resulting data. On the other hand, a census has the advantage – not shared by any sample – of providing information even for very small subpopulations.

Suppose a population Π consists of N units, each of which has a value v of some characteristic of interest, such as the age or income in the case of a human population, the number or weight of the apples on each of the N trees in an orchard, or the length of life or brightness of the lightbulbs in a shipment received from the factory. To obtain an estimate of the average v value $\bar{v} = (v_1 + \cdots + v_N)/N$, a sample is taken from the population. In the early days of sampling, investigators usually relied on judgment samples, in which judgment is used to obtain a sample "representative" of the population. It is now realized that such sampling tends to lead to biases, and does not provide a basis for calculating accuracy or the sample size needed to achieve a desired accuracy. The methods used instead are probability **sampling schemes** that select the units to be included in the sample according to stated probabilities. The simplest such scheme is simple random sampling, according to which n units are chosen in such a way that every possible sample of size n has the same probability of being drawn. The average of the sample v values is then the natural estimate of $\bar{v}$.

Better accuracy can often be attained, and the needed sample size and resulting cost therefore reduced, by dividing the population to be sampled into strata within which the v values are more homogeneous than in the population as a whole but which differ widely among each other. (The population of school children of a city might for example be stratified by school, grade, and gender.) A stratified sample of size n is then obtained by drawing a simple random sample of size n_i from the ith stratum for each i, where $\sum n_i = n$.

In a different direction, the cost of sampling can often be reduced by combining units into clusters (for example, all the apartments in an apartment house, all houses in a city block, or all patients in a hospital ward), and obtaining the required information for each member of a sampled cluster. The two approaches can be combined into stratified cluster sampling, and many other designs are possible.

Sample surveys to obtain information about a population are in widespread use and have become familiar through election polls, market surveys, and surveys of television viewers to establish ratings. However, they are not very well understood. In particular, it appears puzzling how it is possible to obtain an accurate estimate of the opinions or intentions of many millions from information concerning just a few thousands.

Example 4: Election Poll To get some insight into this question, let us simplify the situation, and suppose that a population Π consists of a large number N of voters, each of whom supports either candidate A or B. A simple random sample of n voters is drawn from Π, and the preference of each member of the sample is ascertained. If X is the number of voters in the sample favoring A, then X/n, the proportion of A supporters in the sample, is the natural estimate of the proportion p of A supporters in the population.

The question at issue is how large a sample is required for X/n to achieve a prescribed accuracy as an estimate of p, for example, for the standard deviation (SD) of X/n to satisfy

$$\mathrm{SD}\left(\frac{X}{n}\right) \leq 0.01 \tag{5}$$

It is often felt intuitively that the required sample size n should be roughly proportional to the population size N. However, it turns out that this intuition is misleading and that in fact, for large populations, the required n is essentially independent of the value of N.

To see this, suppose for a moment that the sampling is done "with replacement", i.e., that the members are drawn successively at random, with each – after giving the required information – being put back into the population before the next member is drawn, again at random. This method is slightly less efficient than the original simple random sampling (and therefore requires a larger sample size), because it allows the same unit to be drawn more than once. It is introduced here because it is particularly simple to analyze. Sampling with replacement is characterized by two properties: (a) on each draw, the probability of obtaining an A supporter is p; (b) the results of the n draws are independent in the sense that the probability of getting an A supporter on any given draw does not depend on the results of the earlier draws.

Let us now compare this situation with a quite different one. Suppose a coin with a probability p of falling heads when spun on its edge (this probability may be far from 1/2), is spun n times and that X is the number of times it falls heads. Then (a) the probability of heads is p on each spin, and (b) the results of the n spins are independent. The standard deviation of X/n is therefore the same for this coin problem as in the election sampling with replacement. The number n required to reduce this standard deviation to 0.01 is therefore also the same in both cases. Note however that the coin problem involves only n and p; no population is involved. Therefore the required n cannot depend on N.

This argument depends of course crucially on the assumption that the sampling was random. It is the randomness that insures that with high probability the sample contains approximately the same proportion of A supporters as the population. In addition, it was tacitly assumed that the preference of each voter can be ascertained without error. In practice, the possibility of "response error" can rarely be ruled out.

Data from Experiments

A study is called an *experiment* if its data are produced by an intervention (i.e., do not occur naturally) for the purpose of gathering information. It is a *comparative experiment* if its purpose is to compare several ways of doing something (e.g., different machines, procedures, medical treatments, fertilizers, and so on) rather than to determine some absolute value.

Example 5: Weather Modification Consider a company's claim to be able to increase precipitation by seeding the clouds of suitable storms. Here the comparison is between seeding and not seeding. How can one obtain data to test the claim and to provide an estimate of the amount of increase?

As one possibility, the company might seed all suitable storms in the given location, and compare the resulting rainfall with that during the corresponding period in the preceding years. Of course, if the results are very striking (for example, if an enormous downpour occurs immediately following each seeding) this may settle the issue. However, typically the results are less clear. Suppose, for example, that the total rainfall matches, or even slightly exceeds, that of the wettest of the last five years. This may be the result of the seeding; or it may just be the consequence of an exceptionally wet year.

This difficulty is inherent in studies in which there is no randomness in the assignment to the experimental units of the conditions or treatments being compared. A better basis for the establishment of a causal relationship – in this case that the increased rainfall is due to the seeding – is obtained if storms are compared within the same season, and if they are assigned to the two treatments (seeding and not seeding) according to a random mechanism. This can be done in a variety of ways, corresponding to different experimental designs.

As a simple possibility, suppose that the experiment is to extend to the first 20 storms that are suitable for seeding. Of these, 10 will be seeded and 10 not, the latter providing the *controls*. According to one design (*complete* **randomization**), 10 of the numbers $1, \ldots, 20$ are selected at random; the storms bearing these numbers will be seeded, the remaining 10 will not. An alternative design (*paired comparisons*) pairs the storms $(1, 2), (3, 4), \ldots, (19, 20)$, and within each pair assigns at random (e.g., by tossing a coin) one storm to seeding, the other to control. This second design will be particularly effective if storms occurring close together in time are more likely to be similar in strength than storms separated by longer intervals.

It is interesting to note the close relationship of these designs to the sampling schemes of the preceding section. In the first design, the storms assigned to seeding are a simple random sample of the 20 available storms; in the second case, they constitute a stratified sample, with samples of size 1 from each of 10 strata of size 2.

Unfortunately, random assignment of the conditions being compared is not always possible. Consider for example a study that reports that married men tend to live longer than unmarried ones. It is tempting to draw the conclusion that marriage prolongs life, perhaps by providing a more regulated lifestyle. However, such a conclusion is not justified. Married and unmarried men constitute groups that differ in many ways. The latter for instance includes men with health problems that preclude marriage and that also tend to shorten life. It is thus not clear whether the observed effect is the result of the difference in marital status or of conditions leading to this difference. All that can be safely concluded is that marriage is associated with longer life expectancy. This may be enough for an insurance company but does not answer the sociological or public health question regarding the effect of marriage. Quite generally, *observational studies* (*see* **Observational Studies**) in which subjects have not been assigned to the conditions being compared (marital status, smoking habits, religious beliefs, and so on) can establish associations (such as that between marital status and longevity) but have great difficulty in validating causal relationships. To establish *causation* (*see* **Causality**) is a cherished goal of statistical methodology but tends to be rather elusive. (For some further discussion of this point, see for example Mosteller and Tukey [20, pp. 260–261] and Holland [21].)

Example 6: Headache Remedy Suppose that you wake up with a headache, take a headache remedy, and an hour later find that the pain is gone. On the basis of this isolated instance (which, in words attributed to R. A. Fisher, is an experience rather than an experiment), it is clearly not possible to conclude that the result is due to the remedy. The cause might instead have been another intervening event such as breakfast, or possibly the headache had run its course and would have dissipated in any case.

The attribution of the cure to the medication becomes much stronger if the incident is not isolated. If you have suffered from headaches before, took the medicine sometimes soon after onset, at other times only after having waited in vain for the pain to subside on its own; if on different occasions you took the tablets at different times of day, and if under all these different circumstances the headache disappeared shortly after treatment and never (or only very rarely) before, the robustness of the effect over many different conditions would tend to carry conviction where a single instance would not. The finding could be strengthened further if your own experience could be merged with that of others. (However, even if the evidence for a treatment effect is convincing, observations of this kind cannot determine whether the ingredients of the medication are responsible for the improvement, or whether it is a *placebo effect*, i.e. the patient's belief in the effectiveness of the treatment, which relieves the pain.)

To see how to systematize the informal reasoning of the preceding paragraph, consider another example.

Example 7: New Traffic Signs Suppose that four-way stop signs have been installed at a dangerous intersection, and that four accidents had occurred in the month preceding the installment of the signs but only one in the month following it. These facts by themselves provide little basis for an inference. Suppose for example that the monthly accidental frequencies constitute a purely random sequence of which the value 4 was by chance high, which however caused the installment of the signs. Then even without this intervention, a decrease could have been expected in the succeeding month.

A more meaningful comparison can be obtained if the accident statistics are available not only for one month before and after, but for several months in both directions. One is then dealing with an interrupted times series, for which it is possible to compare the observations before with those after the "interruption" (the installment of the signs). Even the nonrandom choice of the time of intervention would then have only a relatively minor effect.

While the interrupted **time series** can establish that the accident rate is lower after installment than it was before, it cannot establish the new signs as the cause of the decrease since other changes might have occurred at the same time. To mention only one possibility, the community may have been affected by editorials published at the time of the third and fourth accidents. Some control – although not as firm as that resulting from randomization – can be obtained by studying the accident rates during the same months at some other intersections also, a multiple time series design. If the other series do not show a corresponding decrease, this will clearly strengthen the causal argument.

Quasi-experiments (a better term might be "quasi-controlled experiments") such as the interrupted time series and multiple time series described above, which try to identify and control the most plausible alternatives to the treatment being the cause of the change, are treated by Cook and Campbell [22] and Cochran [23] (for an elementary discussion see Campbell [24]). They can greatly strengthen the causal attribution although they cannot be expected to be as convincing as a controlled randomized experiment.

Serial Data

Most of the data considered so far were assumed to be collected on a one-shot basis. Often, they are obtained instead consecutively over a period of time. Such data are particularly useful in assessing changes over time. (Are winters getting colder? Is the birth rate in the United States falling? Is a process producing products of stable quality?)

An important application is provided by statistical **quality control**. An established production process is monitored by taking observations of the quality of the product at regular intervals. The process is said to be in control if the successive observations are independent and have the same distribution. A *control chart* (*see* **Control Charts, Overview**) on which the successive observations are plotted provides signals when the process appears to be going out of control. A similar approach is used in monitoring the cardiogram of a heart patient in **intensive care**, where the data consist of a continuous graph instead of a discrete sequence of points. Analogously, seismographs provide continuous data for the monitoring of seismic activity while persons watching their weight through regular weighings provide another illustration of the discrete case.

In these applications, a single process is followed to check that no changes are occurring and to alert us when they occur. A different situation arises when one is interested in assessing the changes occurring in a population as part of a developing pattern or as the result of some intervention. For example, to study the dynamic behavior of a chemical process, or stock prices, one collects a sequence of observations. To study the pattern of growth (the *growth curve*) of children, plants, or institutions, one observes each member of a sample from the population over a period of time. In such time series data and *longitudinal studies* the observations at different times often are dependent. Probability models for sequences of dependent observations are considerably more complicated than those for sequences of independent and identically distributed (i.i.d.) variables. Such models and their statistical analysis are treated in the theory of *time series* (*see* **Time Series Analysis**) or more generally of stochastic processes.

Observations arising serially, for instance on patients coming to a clinic, successive books by an author, or stock market performance in successive time periods, provide an opportunity to economize by letting the size of the study be determined by the data (according to a clearly specified rule) instead of fixing it in advance. Suppose for example that a shipment of goods is being sampled to determine whether its overall quality meets certain specifications. Items are being drawn at random and examined one by one. It may then be possible to stop early when the quality of the initial observations is either very satisfactory or very unsatisfactory, while one may wish to take a larger sample when the initial observations are mixed. The working out of economical stopping rules providing the desired statistical information, and the analysis of the resulting data, is the problem treated in *sequential analysis*.

Designing Experiments

The first part of this article was concerned with the interpretation of data once they have been obtained, and the preceding sections with various processes used to acquire the data. However, preceding even this stage, it is necessary to plan how many and what kind of data will be needed to answer the questions under consideration.

As an illustration, consider once more the ESP study of Example 2, based on 100 tosses of a coin. Suppose the investigator had decided, before the experiment was carried out, to give serious attention to the possibility of ESP, provided the number of subject A's correct calls would be at least 65. As pointed out in Example 2, under the hypothesis of pure guessing, the probability of 65 or more answers is 0.002. There is thus little danger of paying attention to the claim if it has no validity.

However, is this study giving subject A a fair shake? What is the probability of getting 65 or more of the tosses right if A really does possess the claimed ability? The answer depends of course on the extent of the ability, which can be measured by the probability of calling a toss correctly. Here are some values, computed under the simplifying assumptions that the 100 calls are independent, and that the probability p of a correct call is the same for each toss.

p	0.55	0.6	0.7	0.75
P(number of correct calls ≥ 65) $=$ power	0.027	0.179	0.884	0.990

The probability in this table measures the **power** (*see* **Power**) of the test of the hypothesis $H : p = 1/2$, i.e., the probability of following up A's claim against various alternative values of p. It shows that the power is quite satisfactory when p is 0.7 or more, but not, for example, when $p = 0.6$. In this case, despite the large discrepancy between pure guessing and an ability corresponding to $p = 0.6$, the probability is less than 20% that serious attention would be given to the claim. To get higher power would require a larger number of tosses. For example, to increase the power against $p = 0.6$ to 0.9, while keeping the probability of following up the claim when $p = 0.5$ at 0.002, would require a sample of about 425 tosses instead of the previous 100. A statement of the goals to be achieved, e.g., the required power of a test or accuracy of an estimate, and determination of the sample size needed for this purpose, are crucial aspects of planning a study.

Taking a fixed number of observations (100 or 425, etc.) is not necessarily the best design. Suppose for example that A calls all of the first 25 tosses correctly. This may be enough to provide the desired evidence of A's ability (or suggest that something is wrong with the experiment). Thus, a sequential stopping rule of the type indicated at the end of the previous section might be more efficient.

Consider next the problem of determining an **optimal design** when different types of observations are involved. Suppose that we are concerned with the comparison of two treatments, and that the total number of observations is fixed at $N = 2k$. Then it will typically be best to assign k subjects to each treatment, since this will tend to maximize the power of the tests and minimize the variance of the estimate of the difference. This may of course not be the case if observations on one of the treatments have smaller variance than on the other.

Finding the optimum assignment becomes much more difficult when observations are taken sequentially, and a decision is required at each stage as to which treatment to apply next. What is the best (or even a good) rule depends strongly on the purpose of the procedure. If the only concern is to decide whether the treatments are equally effective and, if not, which is better and by how much, the treatments should be assigned in a balanced way, i.e., so that each is received by the same number of subjects. However, if one is comparing two treatments of a serious medical condition, an additional consideration arises: a desire to minimize the number of patients in the study who receive the inferior treatment. A sequential procedure that has been suggested for such a case is the "play the winner rule" in which the next patient receives the treatment that at that point looks better. (For details, see for example, Flehinger and Louis [25] and Siegmund [26].) It should be noted however, that such an unbalanced procedure will require more observations than the best balanced rule, and hence will take longer to lead to termination of the study and hence to a recommendation. Thus, the inferior treatment will be assigned to fewer patients within the experimental group but to a larger number outside of it.

Comparative experiments typically involve more than the study of just one difference. Consider an experiment to investigate the effect of a number of factors (to which experimental units can be assigned at random) on some observable response such as rainfall, crop yield, or length of life. Two possible designs are one-at-a-time experiments, in which each factor is studied in a separate experiment with all other factors held constant, and *factorial experiments* (*see* **Factorial Experiments**), which provide a joint study of the effects of all factors simultaneously. The latter type of design has two important advantages.

Two or more effects may interact in the sense that the effect of one factor depends on the level of the other. In a study of the effect of textbooks and instructors on the performance of students, for example, it may turn out that a textbook that works very well for one instructor is quite uncongenial to another. The existence and size of such interactions are most easily investigated by studying the various factors simultaneously. When no interactions are present, joint experimentation has the advantage of requiring far fewer observations to estimate or test the various effects than would be needed if each factor were studied separately. (For further discussion of these and related issues see for example, Cox [27] and Box *et al.* [28].)

Much of the theory of factorial designs was originated by R. A. Fisher in the 1920s, in the context of agricultural field trials, where it continues to play a central role; in addition, its uses have since expanded to many other areas of application. The paper on **response surface** analysis by Box and Wilson [29] marked the starting point of their adoption in industry.

Conclusion

Statistics has been discussed in this article as dealing with the collection and interpretation of *data* to obtain information. We have been particularly concerned with the uncertainty caused by the fact that the observations might have taken on other values, and with the resulting variability of the data. An alternative view of statistics is obtained by noting that uncertainty attaches not only to data, but also to many other "chancy" events such as length of life, the occurrence of accidents, the quality of a manufactured article, or the fall of a die. Correspondingly, statistics has also been defined as the subject concerned with understanding, controlling, and reducing *uncertainty*. Actually, there is little difference between these two descriptions: data involve uncertainty, and the study of any particular uncertainty requires the collection of appropriate data. It is a matter of emphasis.

Earlier sections have given a general indication of the pervasiveness and impact of statistical considerations relating to both data and uncertainty. To illustrate the power and wide range of the statistical approach, we shall in this concluding section take a brief look at three specific studies.

Example 8: Polio Vaccine Trial To test whether a proposed polio vaccine (the Salk vaccine) was effective, a large scale study was carried out in 1954. The most useful part of the study was a completely randomized trial in which about half of approximately 400 000 school children were assigned at random to receive the vaccine, with the other half receiving a placebo (injection of an ineffective salt solution). The study was double blind, i.e., neither the children nor the physicians making the diagnosis knew to which of the two groups any given child belonged. The results in Table 1 show that vaccination has cut the rate of polio by about 60%. The probability of that marked a decrease under the hypothesis that the vaccine has no effect is about $1/10^7$, much too small to reasonably attribute the effect to pure chance. (Note that this example is of the same type as the hypothetical Example 2 (ESP).)

Table 1 [(a)]

	Total number	Number with confirmed polio
Vaccinated	200 745	57
Placebo	201 229	142

[(a)] Adapted from Meier

Example 9: Medical Progress In the preceding example we were concerned with the effectiveness of a single medical innovation – the Salk vaccine. The study reported in the present example (Gilbert *et al.* [30]) addresses a much broader question: What can be said about the effectiveness of present-day medical (or rather more specifically, surgical and anesthetic) innovations as a whole? Such an investigation of the combined implications of a whole area of studies is called a *meta-analysis*. (For an account of meta-analysis with many references to the literature, see Hedges and Olkin [31].)

A first step toward such an analysis is to decide what data to collect, in particular, which studies to include in the investigation. Ideally, one would like to have a list of all relevant studies for the period in question. For medical research, an approximation to this ideal is available in National Library of Medicine's MEDLARS (MEDical Literature And Retrieval System), which provides exhaustive coverage of the world's medical literature since 1964. From MEDLARS, the authors obtained a set of 107 papers ("the sample") evaluating the success of specific innovative surgical and anesthetic treatments, and used these to assess the effectiveness of such treatments as a whole. The authors view this set of papers as a sample from the flow of such research studies, and therefore believe that the results should give a realistic idea of what to expect from future innovations, at least for the short term.

The sample contained a total of 48 comparisons of a new treatment with a standard (control). In 36 of the cases, the assignment to treatment and control was randomized, but not in the remaining 12. The results are summarized in Table 2.

A striking feature of the results for the randomized trials is that only 5 out of 36 or about 14% of the

Table 2 [(a)]

	Innovation highly preferred	Innovation about standard			Standard highly preferred	Total
		Preferred	Equal	Preferred		
Randomized	5	7	14	6	4	36
Nonrandomized	5	2	3	1	1	12

innovations were highly preferred. This suggests, the authors point out, that medical science is sufficiently well established so that substantial improvements over the standard treatments are difficult to achieve yet that it is not so settled that only a major theoretical advance will lead to any substantial further improvements.

A more sanguine assessment is obtained from the nonrandomized studies, where 5 out of 12 or about 42% of the innovations were highly preferred. Since randomization provides a surer foundation, the results in the first row may be deemed to be more reliable than those in the second. However, such a judgment overlooks the fact that the comparison of randomized with nonrandomized studies is itself not randomized, i.e., the assignment of studies to these two types was not, and was in fact far from, random. As a result, the two types of studies may not be directly comparable. For example, a surgeon who is strongly convinced of the great superiority of the new treatment, might for ethical or other reasons not be willing to apply the standard treatment, and would therefore have to look for controls from an earlier period or from other surgeons, while no such qualms might arise for a less dramatic innovation where a randomized study would then be acceptable. Many other explanations could be imagined, and much more information would be needed before one could attempt to assign a cause.

The authors go beyond the frequencies displayed in Table 2 to estimate the size of the treatment effects. For this purpose, they utilize a sophisticated technique, **empirical Bayes**, which is particularly suited for meta-analysis.

Example 10: Literary Detection The Federalist papers are a historically important set of 85 short political essays published mostly anonymously in 1787–1788 by Alexander Hamilton, James Madison, and John Jay. The authorship of most of these papers was eventually established but that of 12 of them has remained in dispute between Madison (M) and Hamilton (H), with a recent historical research leaning toward, but not clearly deciding in favor of, Madison.

An effort to resolve the doubt by statistical methods was undertaken by Mosteller and Wallace [34]. The basic idea of such literary detection is to find aspects of the writing styles of the two (or more) authors in question that have good ability to discriminate between the various possibilities. This was particularly difficult in the present case because the two authors have very similar styles. However, by comparing known texts of M and H, Mosteller and Wallace were able to identify a number of words that were used with much higher frequency by one of the authors than the other. As an illustration, Table 3 shows the frequency of occurrence of the word "on" in blocks of about 200 words from texts by the two authors. (For example, the number of blocks in which the word "on" occurred exactly twice was 27 for the

Table 3 [(a)]

	Word frequency of "on" in 200-word blocks							Total number of blocks
	0	1	2	3	4	5	6	
H	145	67	27	7	1	–	–	247
M	63	80	55	32	20	8	4	262

[(a)]Adapted from Mosteller and Wallace [34]

247 H blocks, but 55 for the 262 M blocks.) The rate of occurrence of the word "on" per thousand words of text was 3.38 for H and 7.57 for M.

The present problem is similar to the diagnostic problem of Example 3, with θ taking on the two values H and M. (The statistical methodology dealing with the attribution of an item to one of two or more classes, a patient to a disease, or a piece of writing to an author, is called *classification* or *discrimination*.) Mosteller and Wallace's first task was to decide on observations X that might provide them with the evidence needed to settle the question. They selected a total of 30 words (including the word "on") with high discriminating ability, and used as their observations X the frequencies with which these words occurred in a given text of disputed authorship.

Next the distribution $p_\theta(x)$ for the frequency of a given word had to be specified for each author with the help of the known H and M texts. Bayesian analyses for each of the 30 words based on the chosen family of distributions were combined into overall odds for M and H. For all but two of the papers, these overwhelmingly (several hundred million to one) favored Madison, given any reasonable prior odds for M and H. For each of these papers, the analysis leaves little room for doubt. In the remaining two cases Madison is also favored, but the odds are more modest, and the statistical attribution therefore less certain.

Statistical considerations arise in nearly all fields of human endeavor. Many of the associated activities are carried out by large numbers of full- or part-time professional statisticians. The basic training in statistics, as in most other fields, occurs at universities and colleges where departments of statistics offer both undergraduate and graduate degrees in statistics. (At some institutions, the principal statistics courses are instead provided by the mathematics or economics department.) In addition, there may be degrees in biostatistics. Statistics courses and quantitative programs with a strong statistical component may also be offered in operations research, business schools, demography, economics, education, psychology, and sociology. In addition to courses and programs preparing for a profession in statistics, statistics departments also provide "service courses" both at introductory and advanced levels to students in other fields who may need to use statistical methods in their work.

Another need of statistical education is filled by courses in statistical concepts as a part of general education. Acquiring the ability to think in statistical terms is of great importance, even for persons without a quantitative bent, in view of the pervasive occurrence of statistical ideas in newspapers and magazines, and in the terminology we use to describe and discuss the world around us. What do we mean by saying that women tend to live longer than men, or that cancer patients survive longer today than they did 10 years ago? Does it mean they live longer on the average, that the median length of their life is longer, or that they have a better chance to survive to any given age? And is the longer survival of persons diagnosed to have cancer primarily due to the availability of more effective treatments, including earlier diagnosis? In fact, is there even an increase in length of survival, or is the apparent increase just a statistical consequence of earlier diagnosis? If the disease is diagnosed a year earlier, the survival after diagnosis has increased by a year even if nothing else has changed.

To consider another example, what is meant by the "rate of unemployment"? Roughly speaking, it is the proportion among the people wishing to be employed who are not. But do the official rates include those past seekers for jobs who have given up in despair? A change in the figure for unemployment may be the result of a small change in the definition, or of some sociological change such as the increased entry of women into the workforce.

Many of the considerations involved in such issues are statistical. Fortunately, the basic ideas needed for general discussion and comprehension can be communicated at a fairly nontechnical level, as is done, for example, in Tanur *et al.* [35], Mosteller *et al.* [33], and Freedman *et al.* [32]. Books such as these, and the increasing availability of first courses in probability and statistics in high schools, should have the effect of gradually raising the general level of statistical literacy.

Acknowledgments

I would like to thank David Cox, Persi Diaconis, David Freedman, William Kruskal, Frederick Mosteller, Juliet Shaffer, and the editors for reading a draft of this article, and providing me with comments and suggestions that resulted in many improvements.

References

[1] Diaconis, P. (1985). In *Exploring Data Tables, Trends, and Shapes*, D.C. Hoaglin, F. Mosteller and J.W. Tukey, eds, John Wiley & Sons, New York.

[2] Campbell, W.E. (1942). Monograph No. B-1350, Bell Telephone System Technical Publication.

[3] Wolfowitz, J. (1943). On the theory of runs with some applications to quality control, *Annals of Mathematical Statistics* **14**, 280–288.

[4] Nomikos, P. & MacGregor, J.F. (1995). Multivariate SPC charts for monitoring batch processes, *Technometrics* **37**(1), 41–59.

[5] Nomikos, P. & MacGregor, J.F. (1995). Multiway partial least squares in monitoring batch processes, *Chemometrics and Intelligent Laboratory Systems* **30**, 97–108.

[6] Huff, D. (1954). *How to Lie with Statistics*, Norton, New York.

[7] Feynman, R. (1965). *The Character of Physical Law*, The British Broadcasting Corporation, London, England.

[8] Barnett, V. (1982). *Comparative Statistical Inference*, 2nd Edition, John Wiley & Sons, New York.

[9] Shortliffe, E.H. (1976). *Computer-Based Medical Consultations, MYCIN*, Elsevier, New York.

[10] Spiegelhalter, D.J. & Knill-Jones, R.P. (1984). Statistical and knowledge-based approaches to clinical decision-support systems, *Journal of the Royal Statistical Society. Series A* **147**, 35–77.

[11] Berger, J. (1984). In *Robustness of Bayesian Analysis*, J. Kadane, ed, North-Holland, Amsterdam, The Netherlands.

[12] Dickey, J.M. (1973). Scientific reporting and personal probabilities: student's hypothesis, *Journal of the Royal Statistical Society. Series B* **35**, 285–305.

[13] Smith, A.F.M. (1984). Present position and potential developments: some personal views. Bayesian statistics, *Journal of the Royal Statistical Society. Series A* **147**, 245–259.

[14] Rubin, D.B. (1984). Bayesianly justifiable and relevant frequency calculations for the applied statistician, *Annals of Statistics* **12**, 1151–1172.

[15] Berger, J. (1985). *Statistical Decision Theory and Bayesian Analysis*, 2nd Edition, Springer, New York.

[16] Hoaglin, D.C., Light, R., McPeek, B., Mosteller, F. & Stoto, M.A. (1982). *Data for Decisions*, Abt Associates, Cambridge.

[17] Hoaglin, D.C., Mosteller, F. & Tukey, J.W. (1983). *Understanding Robust and Exploratory Data Analysis*, John Wiley & Sons, New York.

[18] Hampel, F.R., Ronchetti, E.M., Rousseeuw, P.J. & Stahel, W.A. (1986). In *Robust Statistics*, John Wiley & Sons, New York.

[19] Alterman, H. (1969). *Counting People*, Harcourt, Brace and World, New York.

[20] Mosteller, F. & Tukey, J.W. (1977). *Data Analysis and Regression*, Addison-Wesley, Reading, MA.

[21] Holland, P.W. (1986). Statistics and causal inference, *Journal of the American Statistical Association* **81**(396), 945–960.

[22] Cook, T.D. & Campbell, D.T. (1979). *Quasi-Experimentation*, Rand McNally, Chicago, IL.

[23] Cochran, W.G. (1983). *Planning and Analysis of Observational Studies*, John Wiley & Sons, New York.

[24] Campbell, D.T. (1978). In *Statistics: A Guide to the Unknown*, 2nd Edition, J. Tanur *et al.*, eds. Wadsworth, Belmont, CA.

[25] Flehinger, B.J. & Louis, T.A. (1972). Sequential treatment allocation in clinical trials, *Biometrika* **58**, 419–426.

[26] Siegmund, D. (1985). *Sequential Analysis*, Springer, New York.

[27] Cox, D.R. (1958). *Planning of Experiments*, John Wiley & Sons, New York.

[28] Box, G.E.P., Hunter, W.G. & Hunter, J.S. (1978). *Statistics for Experimenters*, John Wiley & Sons, New York.

[29] Box, G.E.P. & Wilson, K.B. (1951). On the experimental attainment of optimum conditions, *Journal of the Royal Statistical Society. Series B* **8**(11), 1–45.

[30] Gilbert, J.P., McPeek, B. & Mosteller, F. (1977). In *Costs, Risks and Benefits of Surgery*, J.P. Bunker, B.A. Barnes, F. Mosteller, eds, Oxford University Press, New York.

[31] Hedges, L.V. & Olkin, I. (1985). *Statistical Methods for Meta-Analysis*, Academic, Orlando.

[32] Freedman, D., Pisani, R. & Purves, R. (1978). *Statistics*, Norton, New York.

[33] Mosteller, F., Kruskal, W.H., Link, R.F., Pieters, R.S. & Rising, G.R. (eds) (1973). *Statistics by Example*. Addison-Wesley, Reading, MA, 4 Vols.

[34] Mosteller, F. & Wallace, D.L. (1984). *Applied Bayesian and Classical Inference*, Springer, New York.

[35] Tanur, J., Mosteller, F., Kruskal, W.H., Link, R.F., Pieters, R.S., Rising, G.R. & Lehmann, E.L. (eds) (1978). *Statistics: A Guide to the Unknown*, 2nd ed, Wordsworth, Belmont, CA.

Further Reading

Healy, M.J.R. (1973). The varieties of statistician, *Journal of the Royal Statistical Society. Series A* **136**, 71–74.

Hoaglin, D.C., Mosteller, F. & Tukey, J.W. (1985). *Exploring Data Tables, Trends, and Shapes*, John Wiley & Sons, New York.

Meier, P. (1978). In *Statistics: A Guide to the Unknown*, 2nd Edition, J. Tanur *et al.*, eds, Wordsworth, Belmont.

Moser, C.A. (1973). *Journal of the Royal Statistical Society. Series A* **136**, 75–88.

ERICH L. LEHMANN

Article originally published in Encyclopedia of Statistical Sciences, 2nd Edition *(2005, John Wiley & Sons, Inc.). Minor revisions for this publication by Jeroen de Mast.*

P Values

A *P* value is frequently used to report the result of a *hypothesis test* (*see* **Hypothesis Testing**). A *P* value is defined as the probability, calculated under the null hypothesis, of obtaining a test result as extreme as that observed in the sample. Let the null hypothesis H_0 state that $\theta = \theta_0$, and let T be a sample statistic, then $P = P_{\theta_0}(T \geq t)$, or $P = P_{\theta_0}(T \leq t)$, or $P = P_{\theta_0}(|T| \geq t)$ (depending on the direction specified by the alternative hypothesis). In the classical approach, one rejects H_0 if the *P* value is smaller than the desired *significance level* (*see* **Significance Level**) α of the test. Common choices for the significance level are 0.1, 0.05, and 0.01. *See also* **Power**.

Parallel, Series, and Series–Parallel Systems

Introduction

In the area of reliability, often the objective is to determine the probability that a system will function for a certain period. Such systems consist of multiple components connected according to a certain structure. Whether a system functions depends on how the components are connected as well as the state of each component. Each component will have a state that can be classified as either "functioning" or "failed". In most cases, there is an underlying probability distribution that is used to determine if the component is in the functioning state or the failed state at some time t.

Three important system configurations that are widely used in reliability are series systems, parallel systems, and series–parallel systems. In a series system, components are connected in such a way that the failure of a single component results in system failure. A series system is also referred to as a **competing risks** *system* since the failure of an individual (system) can be classified as one of n possible risks (component) that competes for the individual's failure. An example of a series system is illustrated in Agustin and Peña [1] where a series system of softwares is considered with the failure of a software (component) resulting in the failure of the system. On the other hand, a system where the failure of all components will result in the failure of the system is called a *parallel system*. Some papers that have considered parallel systems include Kvam and Peña [2] on an equal load-sharing dynamic model and Sinkovic *et al.* [3] in the area of communication. A common interest in the aforementioned papers is in modeling how the load can be shared by the components in parallel whenever critical jobs or tasks are given to the system to complete. In addition, standby or redundant systems can also be associated with parallel systems (see Osaki and Nakagawa [4]). For such systems, the operational unit can be considered to be connected in parallel with all the standby units. As soon the unit fails, one of the standby units goes into operation. Finally, a series–parallel system is a system in which m subsystems are connected in series and each subsystem consists of k components connected in parallel (see Baxter and Harche [5]). For such a system, failure is observed if a subsystem has failed. Note that the failure of a subsystem occurs when all of the k components have failed, hence the name *series–parallel system.*

One of the goals in reliability is to ensure that at any given instant, we observe a functioning system with a high probability. The probability that the system is functioning is defined as the system reliability. In this article, we will obtain the system reliability for the series, parallel, and series–parallel systems. The system reliability will depend on the reliability of each component as well as on how the components are connected. Moreover, we will assume that the reliability of each component is a function of how long it is observed. The time to failure of a component will depend on a known probability distribution.

System Reliability

Structure Function

Consider a system of n components in which a component is either in a "functioning" state or in a "failed" state. For $i = 1, \ldots, n$, let X_i be the binary random variable defined to be

$$X_i = \begin{cases} 1, & \text{component } i \text{ is functioning} \\ 0, & \text{otherwise} \end{cases} \quad (1)$$

The random variables $X_1, \ldots, X_n$ are assumed to be independent. Thus, the components function independent of one another. We will define $\boldsymbol{X} = (X_1, \ldots, X_n)$ to be the state vector of the system. The state of the system depends on a function that considers how the components are connected as well as the corresponding state of each component. This is called the *structure function* of the system and is denoted by $\phi(\boldsymbol{X})$. The function $\phi(\boldsymbol{X})$ is a binary random variable where

$$\phi(\boldsymbol{X}) = \begin{cases} 1, & \text{system is functioning} \\ 0, & \text{otherwise} \end{cases} \quad (2)$$

To describe the systems of interest in this article, we need to define the following: vector inequality, relevant components, and a nondecreasing structure function.

Definition Let $\boldsymbol{X} \in \mathbf{R}^n$ and $\boldsymbol{Y} \in \mathbf{R}^n$ be two state vectors.

1. We say that $\boldsymbol{X} \leq \boldsymbol{Y}$ if $X_i \leq Y_i$ for all $i = 1, \ldots, n$. We have $\boldsymbol{X} < \boldsymbol{Y}$ if $X_j < Y_j$ for at least one j.
2. Component i is irrelevant if for all values of X_j, where $j = 1, \ldots, n$ and $j \neq i$,

$$\begin{aligned} &\phi(X_1, \ldots, X_{i-1}, 0, X_{i+1}, \ldots, X_n) \\ &\quad = \phi(X_1, \ldots, X_{i-1}, 1, X_{i+1}, \ldots, X_n) \end{aligned} \quad (3)$$

3. A structure function is a nondecreasing function if $\boldsymbol{X} \leq \boldsymbol{Y}$ implies that $\phi(\boldsymbol{X}) \leq \phi(\boldsymbol{Y})$.

The primary interest in reliability are systems where replacing a failed component by a functioning component does not result in the system being "worse off". In other words, the structure function of the system is nondecreasing. Such a system is called a *coherent system* (*see* **Coherent Systems**).

Definition *A system of n components where all components are relevant and whose structure function is nondecreasing is called a coherent system.*

In what follows, we will show that a series system, a parallel system, and a series–parallel system are all examples of coherent systems.

Series System. A series system is defined to be a system that functions if and only if all components are functioning. One can also consider such a system as having failed as soon as one component has failed. Thus, the structure function of a series system is

$$\phi(\boldsymbol{X}) = X_1 X_2 \ldots X_n = \prod_{i=1}^{n} X_i \quad (4)$$

Since the random variables $X_1, \ldots, X_n$ are all binary, the structure function of a series system can be written as $\phi(\boldsymbol{X}) = \min(X_1, \ldots, X_n)$. Figure 1 illustrates a series system with three components.

As an illustration, consider a series system where component 1 has failed while the remaining $n - 1$ components are functioning. Thus, the state vector is $\boldsymbol{X} = (X_1, X_2, \ldots, X_n) = (0, 1, \ldots, 1)$, which results

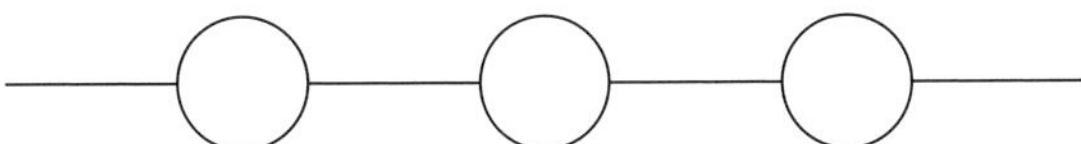

Figure 1 Diagram for a series system with three components

in $\phi(\boldsymbol{X}) = \min(0, 1, \ldots, 1) = 0$. On the other hand, if all components are functioning, then $\phi(\boldsymbol{X}) = \min(1, \ldots, 1) = 1$. Clearly, $\phi(\boldsymbol{X}) = 0$ whenever $X_i = 0$ for some i. Thus, to show that a series system is a coherent system, we need to consider two state vectors $\boldsymbol{X}$ and $\boldsymbol{Y}$ such that $X_i = 0$ for some $i = 1, \ldots, n$ while $Y_j = 1$ for all $j = 1, \ldots, n$. Thus, $\boldsymbol{X} < \boldsymbol{Y}$ implies that $\phi(\boldsymbol{X}) \leq \phi(\boldsymbol{Y})$, which means that ϕ is nondecreasing. In addition, it is easily seen that a series system with state vector $\boldsymbol{Y} = (1, 1, \ldots, 1)$ will fail as soon as one component fails. Hence, each component in a series system is relevant. It follows that a series system is coherent.

Parallel System. A parallel system is a system that functions if and only if at least one component is functioning. Since $1 - X_j = 0$ whenever component j has failed, the structure function of a parallel system is

$$\begin{aligned}\phi(\boldsymbol{X}) &= 1 - (1 - X_1) \cdots (1 - X_n) \\ &= 1 - \prod_{i=1}^{n} (1 - X_i)\end{aligned} \tag{5}$$

It should be noted that an alternative way of expressing the structure function of a parallel system is $\phi(\boldsymbol{X}) = \max(X_1, \ldots, X_n)$. Figure 2 presents a diagram for a parallel system with two components.

To illustrate, suppose that all components have failed except for component 1. Thus, $X_1 = 1$ while $X_2 = 0, \ldots X_n = 0$. It follows that $\phi(\boldsymbol{X}) = \max(1, 0, \ldots, 0) = 1$. On the other hand, if $X_1 = 0, \ldots, X_n = 0$, then $\phi(\boldsymbol{X}) = \max(0, \ldots, 0) = 0$. To show that a parallel system is coherent, it suffices to consider two state vectors $\boldsymbol{X}$ and $\boldsymbol{Y}$ such that $X_j = 0$ for all $j = 1, \ldots, n$ and $Y_i = 1$ for some $i = 1, \ldots, n$. Clearly, $\boldsymbol{X} < \boldsymbol{Y}$ and it follows that $\phi(\boldsymbol{X}) < \phi(\boldsymbol{Y})$. Moreover, every component is relevant since the structure function is equal to zero whenever all components have failed whereas the structure function is one if the state of one component changes its state

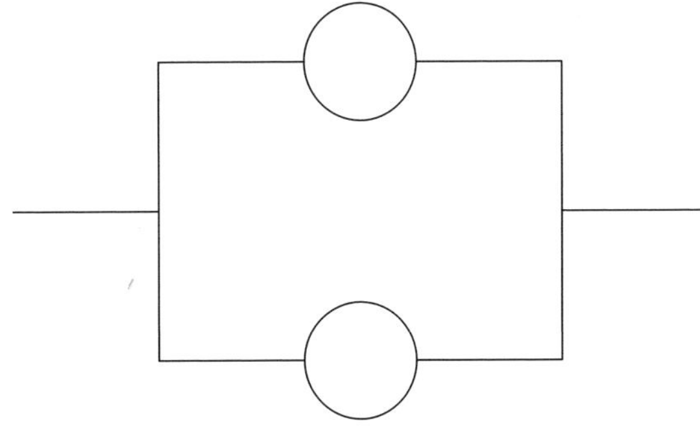

Figure 2 Diagram for a parallel system with two components

from failed to functioning. Hence, a parallel system is coherent.

Series–Parallel System. We define a series–parallel system with mk components as a system comprised of m subsystems connected in series. Each subsystem consists of k components connected in parallel. Thus, if at least one component in each subsystem is functioning, then the system is functioning. On the other hand, if all components in one subsystem have failed, then the system has failed regardless of the state of the components in the other subsystems. To obtain the structure function of a series–parallel system, let us define the structure function of the jth subsystem to be $\phi_j(\boldsymbol{X}_j)$, for $j = 1, \ldots, m$. The vector $\boldsymbol{X}_j = (X_{j1}, \ldots, X_{jk})$ represents the state vector of the jth subsystem, where X_{jl} is the random variable that represents the state of the lth component in the subsystem, for $l = 1, \ldots, k$. One can obtain

$$\phi_j(\boldsymbol{X}_j) = 1 - \prod_{l=1}^{k} (1 - X_{jl}) = \max(X_{j1}, \ldots, X_{jk}) \tag{6}$$

Hence, by setting $\boldsymbol{X} = (\boldsymbol{X}_1, \ldots, \boldsymbol{X}_m)$, the structure function of the series–parallel system is

$$\phi(\boldsymbol{X}) = \prod_{j=1}^{m} \phi_j(\boldsymbol{X}_j) = \prod_{j=1}^{m} \left[1 - \prod_{l=1}^{k} (1 - X_{jl}) \right] \tag{7}$$

An alternative form for the structure function of the series–parallel system is

$$\phi(\boldsymbol{X}) = \min_{1 \leq j \leq m} \left[\max_{1 \leq l \leq k} X_{jl} \right] \tag{8}$$

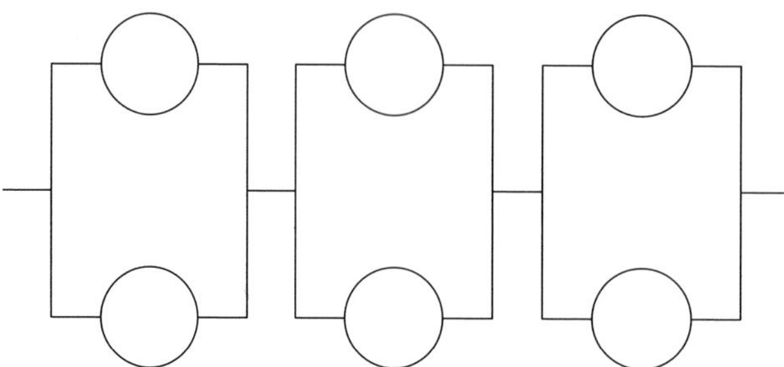

Figure 3 Diagram for a series–parallel system with $m = 3$ subsystems and $k = 2$ components per subsystem

To provide an illustration of a series–parallel system, consider a series–parallel system with $m = 3$ subsystems with each subsystem containing $k = 2$ components. Hence, the system has a total of six components. An illustration for this system is given in Figure 3. For example, suppose that $X_{11} = 0$, $X_{12} = 0$, and $X_{22} = 0$ while the remaining X_{jl} are all equal to 1. One can easily see that subsystem 1 has failed while the remaining subsystems are still functioning. It follows that

$$\phi_1(X_{11}, X_{12}) = \max(0, 0) = 0 \tag{9}$$

$$\phi_2(X_{21}, X_{22}) = \max(1, 0) = 1 \tag{10}$$

$$\phi_3(X_{31}, X_{32}) = \max(1, 1) = 1 \tag{11}$$

which results in $\phi(\boldsymbol{X}) = \min(0, 1, 1) = 0$. To show that a series–parallel system is coherent, consider a state vector $\boldsymbol{Y}$ such that all subsystems are functioning. In other words, for all $j = 1, \ldots, m$, there exists a component, denoted by l, such that $Y_{jl} = 1$. In addition, consider a state vector $\boldsymbol{X}$ such that there exists a subsystem j where $X_{j1} = \cdots = X_{jk} = 0$. It follows that $\boldsymbol{X} < \boldsymbol{Y}$ and $\phi(X) < \phi(Y)$. Moreover, the relevance of each component follows from the fact that components in series systems and parallel systems are all relevant.

System Reliability

Let $p_i = \mathrm{P}\{X_i = 1\}$ be the probability that the ith component is functioning and suppose $\boldsymbol{p} = (p_1, \ldots, p_n)$. We define

$$h(\boldsymbol{p}) = \mathrm{E}[\phi(\boldsymbol{X})] = \mathrm{P}\{\phi(\boldsymbol{X}) = 1\} \tag{12}$$

to be the reliability of the system. Recall that $X_1, \ldots, X_n$ are independent random variables. Thus, for a series system, the reliability of the system is

$$\begin{aligned} h(\boldsymbol{p}) &= \mathrm{P}\{X_1 = 1, \ldots, X_n = 1\} \\ &= \mathrm{P}\{X_1 = 1\} \ldots \mathrm{P}\{X_n = 1\} = \prod_{i=1}^{n} p_i \end{aligned} \tag{13}$$

On the other hand, to find the reliability of a parallel system, it is more convenient to write the reliability as

$$h(\boldsymbol{p}) = 1 - \mathrm{P}\{\phi(\boldsymbol{X}) = 0\} \tag{14}$$

By the independence of the X_i's, we get

$$\begin{aligned} h(\boldsymbol{p}) &= 1 - \mathrm{P}\{X_1 = 0, \ldots, X_n = 0\} \\ &= 1 - \mathrm{P}\{X_1 = 0\} \ldots \mathrm{P}\{X_n = 0\} \end{aligned} \tag{15}$$

It follows that the reliability of a parallel system is

$$\begin{aligned} h(\boldsymbol{p}) &= 1 - \left[1 - \mathrm{P}\{X_1 = 1\}\right] \ldots \left[1 - \mathrm{P}\{X_n = 1\}\right] \\ &= 1 - \prod_{i=1}^{n} [1 - p_i] \end{aligned} \tag{16}$$

For a series–parallel system, we first define for $j = 1, \ldots, m$ and $l = 1, \ldots, k$, $p_{jl} = \mathrm{P}\{X_{jl} = 1\}$ to be the probability that the lth component in the jth subsystem is functioning. We will set $\boldsymbol{p} = (\boldsymbol{p}_1, \ldots, \boldsymbol{p}_m)$ where $\boldsymbol{p}_j = (p_{j1}, \ldots, p_{jk})$. Since the subsystems are connected in series, it follows from equation (13) that

$$\begin{aligned} h(\boldsymbol{p}) &= \mathrm{P}\{\phi_1(\boldsymbol{X}_1) = 1, \ldots, \phi_m(\boldsymbol{X}_m) = 1\} \\ &= \prod_{j=1}^{m} \mathrm{P}\{\phi_j(\boldsymbol{X}_j) = 1\} \end{aligned} \tag{17}$$

Since the components in a subsystem are connected in parallel, we use equation (16) to show that

$$P\left\{\phi_j(\boldsymbol{X}_j) = 1\right\} = 1 - \prod_{l=1}^{m}\left(1 - p_{jl}\right) \tag{18}$$

Thus, the reliability of a series–parallel system is

$$\begin{aligned} h(\boldsymbol{p}) &= \left[1 - \prod_{l=1}^{k}(1 - p_{1l})\right] \cdots \left[1 - \prod_{l=1}^{k}(1 - p_{ml})\right] \\ &= \prod_{j=1}^{m}\left[1 - \prod_{l=1}^{k}\left(1 - p_{jl}\right)\right] \end{aligned} \tag{19}$$

Time-Dependent System Reliability

In this section, we will consider a system of components where the lifetime of the ith component is a random variable Y_i. For example, a car (system) will have parts (components) such as the wheel, engine, or air conditioning system where the time to failure of each part is a random variable that depends on a number of factors (e.g., driving conditions, manufacturing process of the part, etc.) We will assume that the components operate independently of one another. In other words, the component lifetime variables $Y_1, Y_2, \ldots, Y_n$ are independent. Our main interest is to obtain the system reliability on the basis of the underlying assumption on the distribution of the component lifetimes. In doing so, we need to provide the appropriate representation of the distribution of the lifetime of a system. Leemis [6] provides an extensive discussion on five functions that are typically used in reliability. We will focus on three of these functions, namely, the probability density function, the survivor function, and the hazard function.

Let Y_i be a random variable with **probability density function** $f_i(y;\theta)$ and cumulative distribution function $F_i(y;\theta) = P\{Y_i \le y\}$, for $i = 1, \ldots, n$ and some parameter θ. It follows that the reliability of component i is

$$\begin{aligned} S_i(y;\theta) \equiv P\{Y_i > y\} &= \int_y^{\infty} f_i(u;\theta)\,\mathrm{d}u \\ &= 1 - F_i(y;\theta) \end{aligned} \tag{20}$$

The reliability of component i given in equation (20) is also known as the survivor function of Y_i. Since $F_i(y;\theta) = \int_0^y f_i(t;\theta)\,\mathrm{d}t$, the density function of the ith component, in terms of the reliability of component i, is

$$f_i(y;\theta) = -\frac{\mathrm{d}}{\mathrm{d}y}S_i(y;\theta) = -S_i'(y;\theta) \tag{21}$$

Another representation for the lifetime of a component is *via* the hazard function $h_i(y;\theta)$. This is especially important due to its interpretation as the amount of risk associated with a component. In terms of the **component reliability**, the hazard function can be expressed as

$$\begin{aligned} h_i(y;\theta) &= \lim_{\Delta y \searrow 0} \frac{S_i(y;\theta) - S_i(y+\Delta y;\theta)}{S_i(y;\theta)\Delta y} \\ &= -\frac{S_i'(y;\theta)}{S_i(y;\theta)} = \frac{f_i(y;\theta)}{S_i(y;\theta)} \end{aligned} \tag{22}$$

Let Y be the random variable associated with the system lifetime and $S(y;\theta)$ be the system reliability. The corresponding density function, cumulative distribution function, and hazard function of the system lifetime are $f(y;\theta)$, $F(y;\theta)$, and $h(y;\theta)$, respectively. In terms of the density function and cumulative distribution function, the hazard function is of the form

$$h(y;\theta) = \frac{f(y;\theta)}{S(y;\theta)} \tag{23}$$

Our main interest is to obtain the appropriate expressions for the series, parallel, and series–parallel systems.

For a series system, note that the system lifetime can be expressed as $Y = \min(Y_1, \ldots, Y_n)$. This results in a system reliability equal to

$$\begin{aligned} S(y;\theta) = P\{Y > y\} &= P\{\min(Y_1, \ldots, Y_n) > y\} \\ &= P\{Y_1 > y, \ldots, Y_n > y\} \end{aligned} \tag{24}$$

Since the component lifetimes are assumed to be independent, the system reliability simplifies to

$$S(y;\theta) = \prod_{i=1}^{n} P\{Y_i > y\} = \prod_{i=1}^{n} S_i(y;\theta) \tag{25}$$

Following equation (21), the density function of Y is obtained as

$$f(y;\theta) = -\sum_{i=1}^{n} S_i'(y;\theta) \prod_{\substack{j=1 \\ j \ne i}}^{n} S_j(y;\theta) \tag{26}$$

The hazard function for a series system is obtained by substituting the system reliability given in equation (25) and the density function given in equation (26) into equation (23).

For a parallel system, the random variable Y can be expressed as $Y = \max(Y_1, \ldots, Y_n)$. It follows that the system reliability for a parallel system is

$$\begin{aligned} S(y;\theta) &= \mathrm{P}\{\max(Y_1,\ldots,Y_n) > y\} \\ &= 1 - \mathrm{P}\{\max(Y_1,\ldots,Y_n) \le y\} \\ &= 1 - \mathrm{P}\{Y_1 \le y, \ldots, Y_n \le y\} \end{aligned} \quad (27)$$

By the independence of the component lifetimes, the resulting system reliability is

$$S(y;\theta) = 1 - \prod_{i=1}^{n} \mathrm{P}\{Y_i \le y\} = 1 - \prod_{i=1}^{n} [1 - S_i(y;\theta)] \quad (28)$$

The corresponding density function for the parallel system is

$$f(y;\theta) = -\sum_{i=1}^{n} S_i'(y;\theta) \prod_{\substack{j=1 \\ j\ne i}}^{n} [1 - S_j(y;\theta)] \quad (29)$$

For a series–parallel system with m subsystems connected in series with each subsystem containing k components connected in parallel, let Y_j be the lifetime of the jth subsystem. The random variable Y_{jl} denotes the lifetime of the lth component in the jth subsystem, for $j = 1, \ldots, m$ and $l = 1, \ldots, k$. Since the components are all independent, it follows that the subsystem lifetimes are also independent. By following the same arguments used to obtain equation (25), it follows that

$$S(y;\theta) = \mathrm{P}\{Y > y\} = \prod_{j=1}^{m} \mathrm{P}\{Y_j > y\} \quad (30)$$

Since $\mathrm{P}\{Y_j \le y\} = \mathrm{P}\{\max(Y_{j1}, \ldots, Y_{jk}) \le y\}$, we can apply equation (28) for each subsystem to obtain the system reliability. Thus, after straightforward computations, equation (30) simplifies to

$$\begin{aligned} S(y;\theta) &= \prod_{j=1}^{m} \left[1 - \prod_{l=1}^{k} \mathrm{P}\{Y_{jl} \le y\}\right] \\ &= \prod_{j=1}^{m} \left[1 - \prod_{l=1}^{k} [1 - S_{jl}(y;\theta)]\right] \end{aligned} \quad (31)$$

From equation (31), the density function of Y for the series–parallel system is

$$\begin{aligned} f(y;\theta) = &-\sum_{j=1}^{m} \left[\sum_{l=1}^{k} S_{jl}'(y;\theta) \prod_{\substack{l^*=1 \\ l^*\ne l}}^{k} [1 - S_{jl^*}(y;\theta)]\right] \\ &\times \left[\prod_{\substack{j^*=1 \\ j^*\ne j}}^{m} \left(1 - \prod_{l=1}^{k} (1 - S_{j^*l}(y;\theta))\right)\right] \end{aligned} \quad (32)$$

Example As an illustration, consider the case when each component lifetime follows an exponential distribution with mean $\theta = \lambda_i$, that is, $Y_i \sim \exp(\lambda_i)$. In other words, the random variable Y_i has density function

$$f_i(y;\lambda_i) = \frac{1}{\lambda_i} \exp\left(-\frac{y}{\lambda_i}\right), \quad y > 0 \quad (33)$$

Using equation (20), the survivor function of Y_i is

$$\begin{aligned} S_i(y;\lambda_i) &= \int_y^{\infty} \frac{1}{\lambda_i} \exp\left(-\frac{u}{\lambda_i}\right) \mathrm{d}u \\ &= \exp\left(-\frac{y}{\lambda_i}\right), \quad y > 0 \end{aligned} \quad (34)$$

while the hazard function is

$$h_i(y;\lambda_i) = \frac{f_i(y;\lambda_i)}{S_i(y;\lambda_i)} = \frac{1}{\lambda_i}, \quad y > 0 \quad (35)$$

Let $\boldsymbol{\lambda} = (\lambda_1, \ldots, \lambda_n)$ be the parameter vector of interest. For a series system with n components, equation (25) simplifies to

$$S(y;\boldsymbol{\lambda}) = \prod_{i=1}^{n} \exp\left(-\frac{y}{\lambda_i}\right) = \exp\left(-y \sum_{i=1}^{n} \frac{1}{\lambda_i}\right) \quad (36)$$

Since the density function of Y is

$$\begin{aligned} f(y;\boldsymbol{\lambda}) = -\frac{\mathrm{d}}{\mathrm{d}y} S(y;\boldsymbol{\lambda}) &= \left(\sum_{i=1}^{n} \frac{1}{\lambda_i}\right) \\ &\times \exp\left(-y \sum_{j=1}^{n} \frac{1}{\lambda_j}\right) \end{aligned} \quad (37)$$

it follows that for a series system, the random variable Y is exponentially distributed with mean $[\sum_{i=1}^{n}(\lambda_i)^{-1}]^{-1}$. It can be easily seen that by using equation (23), the hazard function is

$$h(y;\boldsymbol{\lambda}) = \sum_{i=1}^{n}\frac{1}{\lambda_i} \tag{38}$$

On the other hand, for a parallel system, using the survivor function of an exponential random variable given in equation (34) in equation (28) results in a system reliability equal to

$$S(y;\boldsymbol{\lambda}) = 1 - \prod_{i=1}^{n}\left[1 - \exp\left(-\frac{y}{\lambda_i}\right)\right], \quad y > 0 \tag{39}$$

The resulting density function of the system lifetime for a parallel system is

$$\begin{aligned} f(y;\boldsymbol{\lambda}) &= -\frac{\mathrm{d}}{\mathrm{d}y}S(y;\boldsymbol{\lambda}) \\ &= \frac{\mathrm{d}}{\mathrm{d}y}\left[\prod_{i=1}^{n}\left(1 - \exp\left(-\frac{y}{\lambda_i}\right)\right)\right] \\ &= \sum_{i=1}^{n}\left(\frac{1}{\lambda_i}\right)\exp\left(-\frac{y}{\lambda_i}\right) \\ &\quad\times\left[\prod_{\substack{j=1\\ j\neq i}}^{n}\left(1 - \exp\left(-\frac{y}{\lambda_j}\right)\right)\right] \end{aligned} \tag{40}$$

This can also be obtained by replacing the appropriate expressions in equation (29) by the survivor function of an exponential random variable given in equation (34).

Finally, for a series–parallel system, let λ_{jl} be the mean lifetime of the lth component in the jth subsystem, for $j = 1, \ldots, m$ and $l = 1, \ldots, k$. The parameter vector of interest is $\boldsymbol{\lambda} = (\lambda_{11}, \ldots, \lambda_{1k}, \lambda_{21}, \ldots, \lambda_{2k}, \ldots, \lambda_{m1}, \ldots, \lambda_{mk})$. Equation (31) simplifies to

$$S(y;\boldsymbol{\lambda}) = \prod_{j=1}^{m}\left[1 - \prod_{l=1}^{k}\left(1 - \exp\left(-\frac{y}{\lambda_{jl}}\right)\right)\right], \quad y > 0 \tag{41}$$

Using equations (32) and (34), the density function of Y for the series–parallel system is

$$\begin{aligned} f(y;\boldsymbol{\lambda}) &= -\frac{\mathrm{d}}{\mathrm{d}y}S(y;\boldsymbol{\lambda}) \\ &= \sum_{j=1}^{m}\left\{\left[\prod_{\substack{j^*=1\\ j^*\neq j}}^{m}\left(1 - \prod_{l=1}^{k}\left(1 - \exp\left(-\frac{y}{\lambda_{j^*l}}\right)\right)\right)\right]\right. \\ &\quad\times\left[\sum_{l=1}^{k}\left(\prod_{\substack{l^*=1\\ l^*\neq l}}^{k}\left(1 - \exp\left(-\frac{y}{\lambda_{jl^*}}\right)\right)\right)\right. \\ &\quad\left.\left.\times\left(\frac{1}{\lambda_{jl}}\right)\exp\left(-\frac{y}{\lambda_{jl}}\right)\right]\right\} \end{aligned} \tag{42}$$

As a special case, suppose $Y_i \sim \exp(\alpha)$ for all $i = 1, \ldots, n$. For ease of notation, we will let α be the parameter of interest. The system reliability of a series system simplifies to

$$S(y;\alpha) = \exp\left(-\frac{n}{\alpha}y\right) = \exp\left(-\frac{y}{\alpha/n}\right), \quad y > 0 \tag{43}$$

which shows that the system lifetime Y is exponentially distributed with mean α/n. On the other hand, the system reliability of a parallel system can be written as

$$S(y;\alpha) = 1 - \left[1 - \exp\left(-\frac{y}{\alpha}\right)\right]^n, \quad y > 0 \tag{44}$$

while the system reliability of a series–parallel system simplifies to

$$S(y;\alpha) = \left[1 - \left(1 - \exp\left(-\frac{y}{\alpha}\right)\right)^k\right]^m, \quad y > 0 \tag{45}$$

The appropriate density function for each system can then be obtained by differentiating the corresponding system reliability.

Example Whenever the risk of failure rapidly increases with time, a failure time distribution that is commonly used is the Weibull distribution. For instance, Ferdous *et al.* [7] assumed **software failure** times that follow the Weibull distribution. Their goal was to predict failure rates and mean time to failure of the software.

In this example, we consider the case where the lifetime for component i follows a Weibull distribution with parameters α_i and β_i. Hence, the random variable Y_i has density function

$$f_i(y;\alpha_i,\beta_i) = \frac{\beta_i}{\alpha_i}\left(\frac{y}{\alpha_i}\right)^{\beta_i-1}\exp\left[-\left(\frac{y}{\alpha_i}\right)^{\beta_i}\right],\quad y>0 \tag{46}$$

It is easily seen that $\beta_i = 1$ will result in $Y_i \sim \exp(\alpha_i)$. The survivor function of Y_i is

$$S_i(y;\alpha_i,\beta_i) = \int_y^{\infty}\frac{\beta_i}{\alpha_i}\left(\frac{u}{\alpha_i}\right)^{\beta_i-1}\exp\left[-\left(\frac{u}{\alpha_i}\right)^{\beta_i}\right]\mathrm{d}u$$

$$= \exp\left[-\left(\frac{y}{\alpha_i}\right)^{\beta_i}\right],\quad y>0 \tag{47}$$

while the hazard function is

$$h_i(y;\alpha_i,\beta_i) = \frac{\beta_i}{\alpha_i}\left(\frac{y}{\alpha_i}\right)^{\beta_i-1},\quad y>0 \tag{48}$$

Let $\boldsymbol{\alpha} = (\alpha_1,\ldots,\alpha_n)$ and $\boldsymbol{\beta} = (\beta_1,\ldots,\beta_n)$ denote the two parameter vectors of interest. For a series system, the system reliability is

$$S(y;\boldsymbol{\alpha},\boldsymbol{\beta}) = \prod_{i=1}^{n}\exp\left[-\left(\frac{y}{\alpha_i}\right)^{\beta_i}\right]$$

$$= \exp\left[-\sum_{i=1}^{n}\left(\frac{y}{\alpha_i}\right)^{\beta_i}\right],\quad y>0 \tag{49}$$

In the special case where $\alpha_i = \alpha$ and $\beta_i = \beta$ for $i = 1,\ldots,n$, equation (49) simplifies to

$$S(y;\alpha,\beta) = \exp\left[-\sum_{i=1}^{n}\left(\frac{y}{\alpha}\right)^{\beta}\right] = \exp\left[-n\left(\frac{y}{\alpha}\right)^{\beta}\right]$$

$$= \exp\left[-\left(\frac{n^{\frac{1}{\beta}}y}{\alpha}\right)^{\beta}\right],\quad y>0 \tag{50}$$

which implies that Y is a Weibull random variable with parameters $\alpha/(n^{1/\beta})$ and β. On the other hand, by using equations (28) and (47), the system reliability for a parallel system is

$$S(y;\boldsymbol{\alpha},\boldsymbol{\beta}) = 1-\prod_{i=1}^{n}\left[1-\exp\left[-\left(\frac{y}{\alpha_i}\right)^{\beta_i}\right]\right],\quad y>0 \tag{51}$$

Finally, for a series–parallel system with k components in each of the m subsystems, the system reliability is

$$S(y;\boldsymbol{\alpha},\boldsymbol{\beta}) = \prod_{j=1}^{m}\left[1-\prod_{l=1}^{k}\left(1-\exp\left[-\left(\frac{y}{\alpha_{jl}}\right)^{\beta_{jl}}\right]\right)\right],\quad y>0 \tag{52}$$

References

[1] Agustin, M. & Peña, E. (1999). A dynamic competing risks model, *Probability in the Engineering and Informational Sciences* **13**, 333–358.

[2] Kvam, P. & Peña, E. (2005). Estimating load-sharing properties in a dynamic reliability system, *Journal of the American Statistical Association* **100**, 262–272.

[3] Sinkovic, V., Lovrek, I. & Nemeth, G. (1999). Load balancing in distributed parallel systems for telecommunications, *Computing* **63**, 201–218.

[4] Osaki, S. & Nakagawa, T. (1971). On a two-unit standby redundant system with standby failure, *Operations Research* **19**, 510–523.

[5] Baxter, L. & Harche, F. (1992). On the optimal assembly of series-parallel systems, *Operations Research Letters* **11**, 153–157.

[6] Leemis, L. (1995). *Reliability: Probabilistic Models and Statistical Methods*, Prentice Hall, Englewood Cliffs.

[7] Ferdous, J., Borhan Uddin, M. & Pandey, M. (1995). Reliability estimation with Weibull inter failure times, *Reliability Engineering and System Safety* **50**, 285–296.

MARCUS A. AGUSTIN

Parametric Tests

Introduction

One way to look at the definition of a parametric test is to note that a parameter is a measure taken from a population. Thus, the mean and standard deviation of data for a whole population are both parameters. Parametric tests make certain assumptions about the nature of parameters, including the appropriate probability distribution, which can be used to decide whether the result of a statistical test would be significant. In addition, they can be used to find a **confidence interval** for a given parameter.

The types of tests given here have been categorized into those that compare means, those that compare variances, those that relate to **correlation** and regression, and those that deal with frequencies and proportions.

Comparing Means

z-Tests

z-tests compare a statistic (or single score) from a sample with the expected value of that statistic in the population under the null hypothesis. The expected value of the statistic under H_0 is subtracted from the observed value of the statistic and the result is divided by its standard deviation. When the sample statistic is based on more than a single score, then the standard deviation for that statistic is its standard error (*see* **Standard Error**). Thus, this type of test can only be used when the expected value of the statistic and its standard deviation are known. The probability of a z-value is found from the standardized **normal distribution**, which has a mean of zero and a standard deviation of one (*see* **Probability Density Functions**).

One-Group z-Test for a Single Score. In this version of the test, the equation is

$$z = \frac{x - \mu}{\sigma} \tag{1}$$

where

- x is the single score
- μ is the mean for the scores in the population
- σ is the standard deviation of the population.

Example 1

$$\text{Single score} = 10$$
$$\mu = 5$$
$$\sigma = 2$$
$$z = \frac{10 - 5}{2} = 2.5.$$

Critical value for z at $\alpha = 0.05$ with a two-tailed test is 1.96.
Decision: reject H_0.

One-Group z-Test for a Single Mean. This version of the test is a modification of the previous one because the distribution of means is the standard error of the mean: $\sigma/\sqrt{n}$, where n is the sample size.

The equation for this z-test is

$$z = \frac{m - \mu}{\left(\frac{\sigma}{\sqrt{n}}\right)} \tag{2}$$

where

- m is the mean of the sample
- μ is the mean of the population
- σ is the standard deviation of the population
- n is the **sample size**.

Example 2

$$m = 5.5$$
$$\mu = 5$$
$$\sigma = 2$$
$$n = 20$$
$$z = \frac{5.5 - 5}{\left(\frac{2}{\sqrt{20}}\right)} = 1.12.$$

The critical value for z with $\alpha = 0.05$ and a one-tailed test is 1.64.
Decision: fail to reject H_0.

t-Tests

t-tests form a family of tests that derive their probability from Student's t distribution (*see* **Probability Density Functions**). The shape of a particular t distribution is a function of the **degrees of freedom**. t-tests differ from z-tests in that they estimate the population standard deviation from the standard deviation(s) of the sample(s). Different versions of the t-test have different ways in which the degrees of freedom are calculated.

One-Sample t-Test. This test is used to compare a mean from a sample with that of a population or a hypothesized value from the population, when the standard deviation for the population is not known. The null hypothesis is that the sample is from the (hypothesized) population (that is, $\mu = \mu_h$, where μ is the mean of the population from which the sample came and μ_h is the mean of the population with which it is being compared).
The equation for this version of the t-test is

$$t = \frac{m - \mu_h}{\left(\frac{s}{\sqrt{n}}\right)} \tag{3}$$

where

m is the mean of the sample
μ_h is the mean or assumed mean for the population
s is the standard deviation of the sample
n is the sample size.

Degrees of freedom
In this version of the t-test, df $= n - 1$.

Example 3

$$m = 9, \mu_h = 7$$
$$s = 3.2, n = 10$$
$$\text{df} = 9$$
$$t_{(9)} = 1.98.$$

The critical t with df $= 9$ at $\alpha = 0.05$ for a two-tailed test is 2.26.
Decision: fail to reject H_0.

Two-Samples t-Test. This test is used to compare the means of two different samples. The null hypothesis is $\mu_1 = \mu_2$ (i.e., $\mu_1 - \mu_2 = 0$, where μ_1 and μ_2 are the means of the two populations from which the samples come).

There are two versions of the equation for this test – one when the variances of the two populations are homogeneous (*see* **Homogeneity of Variances**) and the variances are pooled, and one when the variances are heterogeneous and are entered separately into the equation.
Homogeneous variance

$$t = \frac{m_1 - m_2}{\sqrt{pv \times \left(\frac{1}{n_1} + \frac{1}{n_2}\right)}} \tag{4}$$

where m_1 and m_2 are the sample means of groups 1 and 2 respectively, n_1 and n_2 are the sample sizes of the two groups, and pv is the **pooled variance**

$$pv = \frac{(n_1 - 1) \times s_1^2 + (n_2 - 1) \times s_2^2}{n_1 + n_2 - 2} \tag{5}$$

where s_1^2 & s_2^2 are the variances of groups 1 and 2. When the two group sizes are the same, this simplifies to

$$t = \frac{m_1 - m_2}{\sqrt{\frac{s_1^2 + s_2^2}{n}}} \tag{6}$$

where n is the sample size of each group.
Heterogeneous variances

$$t = \frac{m_1 - m_2}{\sqrt{\frac{s_1^2}{n_1} + \frac{s_2^2}{n_2}}} \tag{7}$$

Degrees of freedom
Homogeneous variance

$$df = n_1 + n_2 - 2 \tag{8}$$

Heterogeneous variance

$$df = \frac{\left(\frac{s_1^2}{n_1} + \frac{s_2^2}{n_2}\right)^2}{\frac{\left(\frac{s_1^2}{n_1}\right)^2}{n_1 - 1} + \frac{\left(\frac{s_2^2}{n_2}\right)^2}{n_2 - 1}} \tag{9}$$

Example 4 Groups with homogeneous variance

$$m_1 = 5.3, m_2 = 4.1$$
$$n_1 = n_2 = 20$$
$$s_1^2 = 1.3, s_2^2 = 1.5$$
$$df = 38$$
$$t_{(38)} = 3.21.$$

The critical t for a two-tailed probability with $df = 38$, at $\alpha = 0.05$ is 2.02.
Decision: reject H_0.

Paired t-Test. This version of the t-test is used to compare two means that have been gained from the same sample, for example, quality of products

before and after a corrective action is applied to them, or from two matched samples. For each product, a difference score is found between the two scores for that product (quality after the correction minus the quality before). The null hypothesis is that the mean of the difference scores is zero ($\mu_d = 0$, where μ_d is the mean of the difference scores in the population).
The equation for this test is

$$t = \frac{m_d}{\left(\frac{s_d}{\sqrt{n}}\right)} \tag{10}$$

where

- m_d is the mean of the difference scores
- s_d is the standard deviation of the difference scores
- n is the number of difference scores.

Degrees of freedom
In this version of the t-test, the df are $n - 1$.

Example 5

$$m_1 = 153.2, m_2 = 145.1\, m_d = 8.1$$
$$s_d = 14.6, n = 20$$
$$df = 19$$
$$t_{(19)} = 2.48.$$

The critical t for $df = 19$ at $\alpha = 0.05$ for a one-tailed probability is 1.729.
Decision: reject H_0.

ANOVA

Analysis of variance (ANOVA) (*see* **Analysis of Variance**) allows the comparison of the means of more than two different conditions to be compared at the same time in a single omnibus test. As an example, researchers might wish to study differences in productivity of shifts. The null hypothesis, which is tested, is that the mean productivity is equal for all shifts. Thus, the null hypothesis would be $\mu_1 = \mu_2 = \mu_3$, where the μ_i denote the mean daily productivity of the morning, evening, and night shifts. ANOVA partitions the overall variance in a set of data into different parts.

The statistic that is created from an ANOVA is the F-ratio. It is the ratio of an estimate of the variance between groups and an estimate of the variance, which is not explicable in terms of differences between groups, that which is due to individual differences (sometimes referred to as *error*).

$$F = \frac{\text{variance between groups}}{\text{variance due to individual differences}} \tag{11}$$

If the null hypothesis is true, then these two variance estimates will both be due to individual differences and F will be close to 1. If the values from the different groups do differ, then F will tend to be larger than 1. The probability of an F-value is found from the F distribution (*see* **Probability Density Functions**). The value of F, which is statistically significant, depends on the degrees of freedom. In this test, there are two different degrees of freedom that determine the shape of the F distribution: the df for the variance between groups and the df for the error.

The variance estimates are usually termed the *mean squares (MS)*. These are formed by dividing a sum of squared deviations from a mean (usually referred to as the **sum of squares**) by the appropriate degrees of freedom.

The F-ratio is formed in different ways, depending on aspects of the design. In addition, the F-value will be calculated on a different basis if the independent variables (IVs) are fixed or random. The examples given here are for fixed IVs. For variations on the calculations, see [1]. The methods of calculation shown will be ones designed to explicate what the equation is doing rather than the computationally simplest version.

One-Way ANOVA

This version of the test partitions the total variance into two components: between groups (attributed to the IV) and within groups (the error).

Sums of squares

The sum of squares between the groups (SS_{bg}) is formed from

$$SS_{bg} = \sum \left[n_i \times (m_i - m)^2\right] \tag{12}$$

where

- n_i is the sample size in group i
- m_i is the mean of group i
- m is the overall mean.

The sum of squares within the groups (SS_{wg}) is formed from

$$SS_{wg} = \sum\sum (x_{ij} - m_i)^2 \quad (13)$$

where

x_{ij} is the jth data point in group i
m_i is the mean of group i.

Degrees of freedom
The df for between groups is one fewer than the number of groups:

$$df_{bg} = k - 1 \quad (14)$$

where k is the number of groups.

The degrees of freedom for within groups is the total sample size minus the number of groups:

$$df_{wg} = N - k \quad (15)$$

where

N is the total sample size
k is the number of groups

Mean squares
The MS are formed by dividing the sum of squares by the appropriate degrees of freedom:

$$MS_{bg} = \frac{SS_{bg}}{df_{bg}} \quad (16)$$

$$MS_{wg} = \frac{SS_{wg}}{df_{wg}} \quad (17)$$

F-ratio
The F-ratio is formed by

$$F = \frac{MS_{bg}}{MS_{wg}} \quad (18)$$

Example 6 Three groups each with six observations are compared (Table 1).

$$Overall mean\,(m) = 4.278$$

$$SS_{bg} = \sum [6 \times (m_i - 4.278)^2] = 40.444$$

$$SS_{wg} = 19.167$$

$$df_{bg} = 3 - 1 = 2$$

$$df_{wg} = 18 - 3 = 15$$

$$MS_{bg} = \frac{40.444}{2} = 20.222$$

$$MS_{wg} = \frac{19.167}{15} = 1.278$$

$$F_{(2,15)} = \frac{20.222}{1.278} = 15.826.$$

The critical F-value for $\alpha = 0.05$, with df of 2 and 15 is 3.68.
Decision: reject H_0.

Multiway ANOVA

When there is more than one IV, the way in which these variables work together can be investigated to see whether some act as moderators for others. An example of a design with two IVs would be if researchers wanted to test whether the effects of different types of music (jazz, classical, or pop) on blood pressure might vary, depending on the age of the listeners. The moderating effects of age on the effects of music might be indicated by an interaction between age and music type. In other words, the pattern of the link between blood pressure and music type differed between the two age groups. Therefore, an ANOVA with two IVs will have three F-ratios, each of which is testing a different null hypothesis. The first will ignore the presence of the second IV and test the main effect of the first IV, such that if there were two conditions in the first IV, then the null hypothesis would be $\mu_1 = \mu_2$, where μ_1 and μ_2 are the means in the two populations for the first IV. The second F-ratio would test the second null hypothesis, which would refer to the **main effect** of the second IV with the existence of the first being ignored. Thus, if there were two conditions in the second IV, then the second H_0 would be $\mu_a = \mu_b$ where μ_a and μ_b are the means in the population

Table 1 Observations and group means in a one-way design

	Group		
	A	B	C
	2	5	7
	3	7	5
	3	5	6
	3	5	4
	1	5	4
	1	4	7
Mean (m_i)	2.167	5.167	5.500

for the second IV. The third F-ratio would address the third H_0, which would relate to the interaction between the two IVs. When each IV has two levels, the null hypothesis would be $\mu_{a1} - \mu_{a2} = \mu_{b1} - \mu_{b2}$, where the μ_{a1} denotes the mean for the combination of the first condition of the first IV and the first condition of the second IV.

Examples are only given here of ANOVAs with two IVs. For more complex designs, there will be higher-order interactions as well. When there are k IVs, there will be 2-, 3-, 4-, ..., k-way interactions, which can be tested. For details of such designs, see [1].

This version of ANOVA partitions the overall variance into four parts: the main effect of the first IV, the main effect of the second IV, the interaction between the two IVs, and the error term, which is used in all three F-ratios.

Sums of squares

The total sum of squares (SS_{Total}) is calculated from

$$SS_{\text{Total}} = \sum\sum\sum (x_{ijp} - m)^2 \qquad (19)$$

where x_{ijp} is the pth observation where IV_1 is on level i, and IV_2 is on level j. A simpler description is that it is the sum of the squared deviations of each observation from the overall mean.

The sum of squares for the first IV (SS_A) is calculated from

$$SS_A = \sum [n_i \times (m_i - m)^2] \qquad (20)$$

where

n_i is the number of observations having level i for IV_1

m_i is the mean of the observations having level i for IV_1

m is the overall mean.

The sum of squares for the second IV (SS_B) is calculated from

$$SS_B = \sum [n_j \times (m_j - m)^2] \qquad (21)$$

where

n_i is the number of observations having level j for IV_2

m_j is the mean of the observations having level j for IV_2

m is the overall mean.

The interaction sum of squares (SS_{AB}) can be found by finding the between-cells sum of squares ($SS_{\text{b.cells}}$), where a cell refers to the combination of levels in the two IVs: for example, first level of IV_1 and first level of IV_2. $SS_{\text{b.cells}}$ is found from

$$SS_{\text{b.cells}} = \sum\sum [n_{ij} \times (m_{ij} - m)^2] \qquad (22)$$

where

n_{ij} is the number of observations for which IV_1 is on level i and IV_2 is on level j,

m_{ij} is the mean of the observations for which IV_1 is on level i and IV_2 is on level j,

m is the overall mean.

$$SS_{AB} = SS_{\text{b.cells}} - (SS_A + SS_B) \qquad (23)$$

The sum of squares for the residual (SS_{res}) can be found from

$$SS_{\text{res}} = SS_{\text{Total}} - (SS_A + SS_B + SS_{AB}) \qquad (24)$$

Degrees of freedom

The total degrees of freedom (df_{Total}) are found from

$$df_{\text{Total}} = N - 1 \qquad (25)$$

where N is the total sample size.

The degrees of freedom for each main effect (for example, df_A) are found from

$$df_A = k - 1 \qquad (26)$$

where k is the number of levels in that IV.

The interactiondegrees of freedom (df_{AB}) are found from

$$df_{AB} = df_A \times df_B \qquad (27)$$

where df_A and df_B are the degrees of freedom of the two IVs.

The degrees of freedom for the residual (df_{res}) are calculated from

$$df_{\text{res}} = df_{\text{Total}} - (df_A + df_B + df_{AB}) \qquad (28)$$

Mean squares

Each mean square is found by dividing the sum of squares by the appropriate df. For example, the mean square for the interaction (MS_{AB}) is found from

$$MS_{AB} = \frac{SS_{AB}}{df_{AB}} \quad (29)$$

F-ratios

Each F-ratio is found by dividing a given mean square by the mean square for the residual. For example, the F-ratio for the interaction is found from

$$F = \frac{MS_{AB}}{MS_{res}} \quad (30)$$

Example 7 Twenty observations are collected, equally distributed over the four combinations of two factors, each having two levels (Table 2).
Overall mean = 8.2

Sums of squares

$$SS_{Total} = 139.2$$
$$SS_A = 51.2$$
$$SS_B = 5.0$$
$$SS_{b.cells} = 56.4$$
$$SS_{AB} = 56.4 - (51.2 + 5.0) = 0.2$$
$$SS_{res} = 139.2 - (51.2 + 5.0 + 0.2) = 82.8$$

Degrees of freedom

$$df_{Total} = 20 - 1 = 19$$
$$df_A = 2 - 1 = 1$$
$$df_B = 2 - 1 = 1$$
$$df_{AB} = 1 \times 1 = 1$$
$$df_{res} = 19 - (1 + 1 + 1) = 16$$

Mean squares

$$MS_A = \frac{51.2}{1} = 51.2$$
$$MS_B = \frac{5}{1} = 5$$
$$MS_{AB} = \frac{0.2}{1} = 0.2$$
$$MS_{res} = \frac{82.8}{16} = 5.175$$

F-ratios

$$F_{A(1,16)} = \frac{51.2}{5.175} = 9.89$$

Table 2 Observations in a two-way design

IV_1 (A)	1		2	
IV_2 (B)	1	2	1	2
	9	7	6	4
	11	10	4	4
	10	14	9	10
	10	10	5	9
	7	10	6	9
Means	9.4	10.2	6	7.2

The critical value for F at $\alpha = 0.05$ with df of 1 and 16 is 4.49.

Decision: reject H_0

$$F_{B(1,16)} = \frac{5}{5.175} = 0.97.$$

Decision: fail to reject H_0

$$F_{AB(1,16)} = \frac{0.2}{5.175} = 0.04.$$

Decision: fail to reject H_0.

Extensions of ANOVA. ANOVA can be extended in a number of ways, including multiple factors, random and fixed factors, crossed and nested designs, and multivariate responses. See [1] for a general description.

Comparing Variances

F-Test for Difference between Variances

Two Independent Variances. This test compares two variances from different samples to see whether they are significantly different. An example of its use could be where we want to see whether the variability in quality of one production line differs from the variability at another line.
The equation for the F-test is

$$F = \frac{s_1^2}{s_2^2} \quad (31)$$

where the sample variance in one sample is divided by the variance in the other sample.

If the research hypothesis is that one particular group will have the larger variance, then that should be treated as group 1 in this equation. As usual, an F-ratio close to 1 would suggest no difference in the variances of the two groups. A large

F-ratio would suggest that group 1 has a larger variance than group 2, but it is worth noting that a particularly small F-ratio, and therefore a probability close to 1, would suggest that group 2 has the larger variance.

Degrees of freedom
The degrees of freedom for each variance are one fewer than the sample size in that group.

Example 8

Group 1
Variance: 16
Sample size: 100

Group 2
Variance: 11
Sample size: 150
$F = \frac{16}{11} = 1.455$

Degrees of freedom
Group 1 $df = 100 - 1 = 99$; group 2 $\mathrm{df} = 150 - 1 = 149$
The critical value of F with df of 99 and 149 for $\alpha = 0.05$ is 1.346.
Decision: reject H_0.

***k* Independent Variances.** The following procedure was devised by Bartlett [2] to test differences between the variances from more than two independent groups $i = 1, \ldots, k$.

The finding of the statistic B (which can be tested with the χ^2 distribution) involves a number of stages. The first stage is to find an estimate of the overall variance (S^2). This is achieved by multiplying each sample variance by its related df, which is one fewer than the size of that sample, summing the results and dividing that sum by the sum of all the df:

$$S^2 = \frac{\sum[(n_i - 1) \times s_i^2]}{N - k} \tag{32}$$

where

n_i is the sample size for the ith group
s_i^2 is the sample variance in group i
N is the total sample size
k is the number of groups.

Next, we need to calculate a statistic known as C, using

$$C = 1 + \frac{1}{3 \times (k - 1)} \times \left[\sum\left(\frac{1}{n_i - 1}\right) - \frac{1}{N - k}\right] \tag{33}$$

We are now in a position to calculate B:

$$B = \frac{2.3026}{C} \times \left[(N - k) \times \log(S^2) - \sum\{(n_i - 1) \times \log(s_i^2)\}\right] \tag{34}$$

where log means taking logarithm to the base 10.

Degrees of freedom

$$df = k - 1, \quad \text{where } k \text{ is the number of groups} \tag{35}$$

Kanji [3] cautions against using the chi-square distribution when the sample sizes are smaller than 6 and provides a table of critical values for a statistic derived from B when this is the case.

Example 9 We wish to compare the variances of three groups: 2.62, 3.66, and 2.49, with each group having the same sample size of 10.

$N = 30, k = 3, N - k = 27, C = 0.994$
$S^2 = 2.923, \quad \log(s_1^2) = 0.418, \quad \log(s_2^2) = 0.563,$
$\log(s_3^2) = 0.396$
$\log(S^2) = 0.466,$
$(N - k) \times \log(S^2) = 12.579$
$\sum[(n_i - 1) \times \log(s_i^2)] = 12.402$
$B = 0.4098$
$df = 3 - 1 = 2$

The critical value of χ^2 for $\alpha = 0.05$ for $df = 2$ is 5.99.
Decision: fail to reject H_0.

Correlation and Regression

***t*-Test for a Single Correlation Coefficient.** This test can be used to test the statistical significance of a correlation (*see* **Correlation**) between two variables. It makes the assumption under the null hypothesis that there is no correlation between these variables in the population. Therefore, the null hypothesis is $\rho = 0$, where ρ is the correlation in the population.

The equation for this test is

$$t = \frac{r \times \sqrt{n-2}}{\sqrt{1-r^2}} \tag{36}$$

where

r is the correlation in the sample
n is the sample size

Degrees of freedom
In this version of the t-Test, $df = n - 2$, where n is the sample size.

Example 10

$$r = 0.4, n = 15$$
$$df = 13$$
$$t_{(13)} = 1.57.$$

Critical t for a two-tailed test with $\alpha = 0.05$ and $df = 13$ is 2.16.
Decision: fail to reject H_0.

***t*-Test for Regression Coefficient.** This tests the size of an unstandardized regression coefficient. In the case of simple regression, where there is only one predictor variable, the null hypothesis is that the regression in the population is 0; that is, that the variance in the predictor variable does not account for any of the variance in the criterion variable. In multiple linear regression, the null hypothesis is that the predictor variable does not account for any variance in the criterion variable, which is not accounted for by the other predictor variables.

The version of the t-test is

$$t = \frac{b}{SE} \tag{37}$$

where b is the unstandardized regression coefficient
SE is the standard error for the regression coefficient.
In simple regression, the standard error is found from

$$SE = \sqrt{\frac{MS_{\text{res}}}{SS_p}} \tag{38}$$

where MS_{res} is the MS of the residual for the regression

SS_p is the sum of squares for the predictor variable.

For multiple regression, the standard error takes into account the interrelationship between the predictor variable for which the SE is being calculated and the other predictor variables in the regression (see [4]).

Degrees of freedom
The degrees of freedom for this version of the t-test are based on p (the number of predictor variables) and n (the sample size): $df = n - p - 1$.

Example 11 In a simple regression ($p = 1$), $b = 1.3$, and the standard error of the regression coefficient is 0.5

$$n = 30$$
$$df = 28$$
$$t_{(28)} = 2.6$$

The critical value for t with $df = 28$, for a two-tailed test with $\alpha = 0.05$ is 2.048.
Decision: reject H_0.

In multiple regression, it can be argued that a correction to α should be made to allow for multiple testing. This could be achieved by using a Bonferroni adjustment.

***F*-Test for a Single R^2.** This is the equivalent of a one-way ANOVA. In this case, the overall variance within the variable to be predicted (the response variable, for example, moisture percentage) is separated into two sources: that which can be explained by the relationship between the predictor variable(s) (such as various machine settings) and the response variable, and that which cannot be explained by this relationship, the residual.

The equation for this F-test is

$$F = \frac{(N - p - 1) \times R^2}{(1 - R^2) \times p} \tag{39}$$

where N is the sample size
p is the number of predictor variables
R^2 is the squared multiple correlation coefficient (*see* **Coefficient of Determination (R^2)**).

Degrees of freedom
The regression degrees of freedom are p (the number of predictor variables). The residual degrees of freedom are $N - p - 1$.

Example 12 Number of predictor variables: 3
Sample size: 336
Regression df: 3
Residual df: 332

$$R^2 = 0.03343$$
$$F_{(3,332)} = \frac{(336 - 3 - 1) \times 0.03343}{(1 - 0.03343) \times 3} = 3.83$$

Critical F-value with df of 3 and 332 for $\alpha = 0.05$ is 2.63.
Decision: reject H_0.

***F*-Test for Comparison of Two R^2.** This tests whether the addition of predictor variables to an existing regression model adds significantly to the amount of variance that the model explains. If only one variable is being added to an existing model, then the information about whether it adds significantly is already supplied by the t-test for the regression coefficient of the newly added variable.

The equation for this F-test is

$$F = \frac{(N - p_1 - 1) \times \left(R_1^2 - R_2^2\right)}{(p_1 - p_2) \times \left(1 - R_1^2\right)} \tag{40}$$

where N is the sample size.

The subscript 1 refers to the regression with more predictor variables and subscript 2, the regression with fewer predictor variables.
p is the number of predictor variables.
R^2 is the squared multiple correlation coefficient from a regression.

Degrees of freedom
The regression $df = p_1 - p_2$, while the residual $df = N - p_1 - 1$.

Example 13 $R_1^2 = 0.047504$

Number of predictor variables (p_1): 5

$R_2^2 = 0.03343$

Number of predictor variables (p_2): 3
Total sample size: 336

df for regression $= 5 - 3 = 2$
df for residual $= 336 - 5 - 1 = 330$

$$F_{(2,330)} = 2.238.$$

The critical value of F with df of 2 and 330 for $\alpha = 0.05$ is 3.023.
Decision: fail to reject H_0.

Frequencies and Proportions

Chi-Square

χ^2 is a distribution against which the results of a number of statistical tests are compared. The two most frequently used tests that use this distribution are themselves called χ^2 *tests* and are used when the data take the form of frequencies. Both types involve the comparison of frequencies that have been found (observed) to fall into particular categories with the frequencies that could be expected if the null hypothesis were correct. The categories have to be mutually exclusive; that is, a case cannot appear in more than one category.

The first type of test involves comparing the frequencies in each category for a single variable, for example, the number of smokers and nonsmokers in a sample. It has two variants, which have a different way of viewing the null **hypothesis testing** process. One version is like the example given above, and might have the null hypothesis that the number of smokers in the population is equal to the number of nonsmokers. However, the null hypothesis does not have to be that the frequencies are equal; it could be that they divide the population into particular proportions, for example, 0.4 of the population are smokers and 0.6 are nonsmokers.

The second way in which this test can be used is to test whether a set of data is distributed in a particular way, for example, that they form a normal distribution. Here, the expected proportions in different intervals are derived from the proportions of a normal curve, with the observed mean and standard deviation, that would lie in each interval, and H_0 is that the data are normally distributed.

The second most frequent χ^2 test is for contingency tables where two variables are involved, for example, the number of smokers and nonsmokers in two different socioeconomic groups.

All the tests described here are calculated in the following way:

$$\chi^2 = \sum \frac{(f_o - f_e)^2}{f_e} \tag{41}$$

where f_o and f_e are the observed and expected frequencies, respectively.

The degrees of freedom in these tests are based on the number of categories and not on the sample size.

One assumption of this test is over the size of the expected frequencies. When the degrees of freedom are 1, the assumption is that all the expected frequencies will be at least 5. When the df is greater than 1, the assumption is that at least 20% of the expected frequencies will be 5.

Yates [5] devised a correction for χ^2 when the degrees of freedom are 1 to allow for the fact that the χ^2 distribution is continuous and yet when df = 1, there are so few categories that the χ^2 values from such a test will be far from continuous; hence, the Yates test is referred to as a *correction for continuity*. However, it is considered that this variant on the χ^2 test is only appropriate when the marginal totals are fixed, that is, that they have been chosen in advance [6]. In most uses of χ^2, this would not be true. If we were looking at gender and smoking status, it would make little sense to set, in advance, how many males and females you were going to sample, as well as how many smokers and nonsmokers.

χ^2 corrected for continuity is found from

$$\chi^2_{(1)} = \sum \frac{(|f_o - f_e| - 0.5)^2}{f_e} \tag{42}$$

where $|f_o - f_e|$ means taking the absolute value, in other words, if the result is negative, treat it as positive.

One-Group Chi-Square/Goodness of Fit

Equal Proportions. In this version of the test, the observed frequencies that occur in each category of a single variable are compared with the expected frequencies, which are such that the same proportion will occur in each category.

Degrees of freedom

The df are based on the number of categories (k); $df = k - 1$.

Example 14 A sample of 45 products are placed into three categories, with 25 in category A, 15 in B, and 5 in C.

The expected frequencies (given the null hypothesis of equal proportions) are calculated by dividing the total sample by the number of categories. Therefore, each category would be expected to have 15 people in it.

$\chi^2 = 13.33$

$df = 2.$

The critical value of the χ^2 distribution with $df = 2$ and $\alpha = 0.05$ is 5.99.

Decision: reject H_0.

Other procedures for rates and proportions are discussed in **Statistical Methods for Counts, Rates, and Proportions**.

Test of Distribution. This test is another use of the previous test, but the example will show how it is possible to test whether a set of data is distributed according to a particular pattern. The distribution could be uniform, as in the previous example, or nonuniform.

Example 15 One hundred scores have been obtained with a mean of 1.67 and a standard deviation of 0.51. In order to test whether the distribution of the scores deviates from being normally distributed, the scores can be converted into z-scores by subtracting the mean from each and dividing the result by the standard deviation. The z-scores can be put into ranges. Given the sample size and the need to maintain at least 80% of the expected frequencies at 5 or more, the width of the ranges can be approximately half a standard deviation except for the two outer ranges, where the expected frequencies get smaller, the further they go from the mean. At the bottom of the range, as the lowest possible score is 0, the equivalent z-score will be −3.27. At the top end of the range, there is no limit set on the scale.

By referring to standard tables of probabilities for a normal distribution, we can find out what the expected frequency would be within a given range of

Table 3 The expected (under the assumption of normal distribution) and observed frequencies of a sample of 100 values

From z	To z	f_e	f_o
−3.275	−1.500	6.620	9
−1.499	−1.000	9.172	7
−0.999	−0.500	14.964	15
−0.499	0.000	19.111	22
0.001	0.500	19.106	14
0.501	1.000	14.953	15
1.001	1.500	9.161	11
1.501	5.000	6.661	7

z-scores. Table 3 shows the expected and observed frequencies in each range.

$\chi^2 = 3.56$
$df = 8 - 1 = 7.$

The critical value for a χ^2 distribution with $df = 7$ at $\alpha = 0.05$ is 14.07.
Decision: fail to reject H_0.
See **Normality Tests** for other tests on normality.

Chi-Square Contingency Test

This version of the χ^2 test investigates the way in which the frequencies in the levels of one variable differ across the other variable. Once again it is for categorical data. An example could be a sample of blind people and a sample of sighted people; both groups are aged over 80 years. Each person is asked whether they go out of their house in a normal day. Therefore, we have the variable visual condition with the levels blind and sighted, and another variable whether the person goes out with the levels yes and no. The null hypothesis of this test would be that the proportions of sighted and blind people going out would be the same (which is the same as saying that the proportions staying in would be the same in each group). This can be rephrased to say that the two variables are independent of each other: the likelihood of a person going out is not linked to that person's visual condition.

The expected frequencies are based on the marginal probabilities. In this example, that would be the number of blind people, the number of sighted people, the number of the people in the whole sample who go out, and the number of people in the whole sample who do not go out. Thus, if 25% of the entire sample went out, then the expected frequencies would be based on 25% of each group going out and, therefore, 75% of each not going out.

The degrees of freedom for this version of the test are calculated from the number of rows and columns in the contingency table: $df = (r - 1) \times (c - 1)$, where r is the number of rows and c, the number of columns in the table.

Example 16 Two variables A and B each have two levels. Level 1 of variable A has 27 people and 13 are in level 2. Level 1 of variable B has 21 people and 19 are in level 2 (Table 4).

If variables A and B are independent, then the expected frequency for the number who are in the first level of both variables will be based on the fact that 27 out of 40 (or 0.675) were in level 1 of variable A and 21 out of 40 (or 0.525) were in level 1 of variable B. Therefore, the proportion who would be expected to be in level 1 of both variables would be $0.675 \times 0.525 = 0.354375$, and the expected frequency would be $0.354375 \times 40 = 14.175$.

$\chi^2 = 2.16$
$df = (2 - 1) \times (2 - 1) = 1.$

The critical value of χ^2 with $df = 1$ and $\alpha = 0.05$ is 3.84.
Decision: fail to reject H_0.

z-Test for Proportions

Comparison between a Sample and a Population Proportion. This test compares the proportion in a sample with that in a population (or that which might be assumed to exist in a population).

Table 4 A contingency table showing the cell and marginal frequencies of 40 participants

		Variable A		
		1	2	Total
Variable B	1	12	9	21
	2	15	4	19
	Total	27	13	40

The standard error for this test is $\sqrt{\dfrac{\pi \times (1-\pi)}{n}}$ where

π is the proportion in the population
n is the sample size.

The equation for the z-test is

$$\frac{p-\pi}{\sqrt{\dfrac{\pi \times (1-\pi)}{n}}} \tag{43}$$

where p is the proportion in the sample.

Example 17 In a sample of 25, the proportion to be tested is 0.7

Under the null hypothesis, the proportion in the population is 0.5

$$z = \frac{0.7-0.5}{\sqrt{\dfrac{0.5 \times (1-0.5)}{25}}} = 2.$$

The critical value for z with $\alpha = 0.05$ for a two-tailed test is 1.96.
Decision: reject H_0.

Comparison of Two Independent Samples. Given two populations, the null hypothesis to be tested is that the proportion having a certain characteristic is equal in both populations. Given a sample from each population, the standard error for this test is

$$\sqrt{\frac{\pi_1 \times (1-\pi_1)}{n_1} + \frac{\pi_2 \times (1-\pi_2)}{n_2}} \tag{44}$$

where

π_1 and π_2 are the proportions in each population
n_1 and n_2 are the sizes of the two samples.

When the population proportions are not known, they are estimated from the sample proportions. The equation for the z-test is

$$z = \frac{p_1 - p_2}{\sqrt{\dfrac{p_1 \times (1-p_1)}{n_1} + \dfrac{p_2 \times (1-p_2)}{n_2}}} \tag{45}$$

where p_1 and p_2 are the proportions in the two samples.

Example 18

Sample 1: $n = 30,\ p = 0.7$
Sample 2: $n = 25,\ p = 0.6$

$$z = \frac{0.7-0.6}{\sqrt{\dfrac{0.7 \times (1-0.7)}{30} + \dfrac{0.6 \times (1-0.6)}{25}}} = 0.776.$$

Critical value for z at $\alpha = 0.05$ with a two-tailed test is 1.96.
Decision: fail to reject H_0.

Comparison of Two Correlated Samples. The main use for this test is to judge whether there has been change across two occasions when a measure was taken from a sample. For example, researchers might be interested in whether people's attitudes to a ban on smoking in public places had changed after seeing a video on the dangers of passive smoking compared with attitudes held before seeing the video. A complication with this version of the test is over the estimate of the standard error. A number of versions exist, which produce slightly different results. However, one version will be presented here from Agresti [7]. This will be followed by a simplification, which is found in a commonly used test.

Table 5 shows that originally 40 people were in category A and 44 in category B, and on a second occasion, this had changed to 50 people being in category A and 34 in category B. This test is only interested in the cells where change has occurred: the 25 who were in A originally but changed to *B* and the 35 who were in *B* originally but changed to A. By converting each of these to proportions of the entire sample, 25/84 = 0.297619 and 35/84 = 0.416667, we have the two proportions we wish to compare. The standard error for the

Table 5 The frequencies of 84 participants placed in two categories and noted at two different times

		After		
		A	B	Total
Before	A	15	25	40
	B	35	9	44
	Total	50	34	84

test is

$$\sqrt{\frac{(p_1 + p_2) - (p_1 - p_2)^2}{n}}$$

where n is the total sample size.

The equation for the z-test is

$$z = \frac{p_1 - p_2}{\sqrt{\frac{(p_1 + p_2) - (p_1 - p_2)^2}{n}}} \quad (46)$$

Example 19 Using the data in Table 5,

$$p_1 = 0.297619$$
$$p_2 = 0.416667$$
$$n = 84$$
$$z = \frac{0.297619 - 0.416667}{\sqrt{\frac{(0.297619 + 0.416667) - (0.297619 - 0.416667)^2}{84}}}$$
$$= -1.304$$

The critical z with $\alpha = 0.05$ for a two-tailed test is -1.96.
Decision: fail to reject H_0.

When the z from this version of the test is squared, this produces the Wald test statistic.

A simplified version of the standard error allows another test to be derived from the resulting z-test: McNemar's test of change.

In this version, the equation for the z-test is

$$z = \frac{p_1 - p_2}{\sqrt{\frac{(p_1 + p_2)}{n}}} \quad (47)$$

Example 20 Once again using the same data as that in the previous example,

$$z = -1.291.$$

If this z-value is squared, then the statistic is McNemar's test of change, which is often presented in a further simplified version of the calculations, which produces the same result.

References

[1] Montgomery, D.C. (1997). *Design and Analysis of Experiments*, 4th Edition, John Wiley & Sons.

[2] Bartlett, M.S. (1937). Some examples of statistical methods of research in agriculture and applied biology, *Supplement to the Journal of the Royal Statistical Society* **4**, 137–170.

[3] Kanji, G.K. (1993). *100 Statistical Tests*, Sage Publications, London.

[4] Draper, N.R. & Smith, H. (1998). *Applied Regression Analysis*, 3rd Edition, John Wiley & Sons.

[5] Yates, F. (1934). Contingency tables involving small numbers and the $\chi 2$ test, *Supplement to the Journal of the Royal Statistical Society* **1**, 217–235.

[6] Neave, H.R. & Worthington, P.L. (1988). *Distribution-Free Tests*, Routledge, London.

[7] Agresti, A. (2002). *Categorical Data Analysis*, 2nd Edition, John Wiley & Sons, Hoboken.

DAVID CLARK-CARTER

Article originally published in Encyclopedia of Statistics in Behavioral Science *(2005, John Wiley & Sons, Ltd). Minor revisions for this publication by Jeroen de Mast.*

Pareto Chart

Pareto Analysis

Pareto analysis is a problem-solving technique for prioritizing which root causes for a given problem should be handled first. The main idea behind the technique is the empirical observation that 80% of the problems come from 20% of the causes. This empirical observation was named by Juran [1] after the Italian econometrician Pareto (1848–1923), who made this observation during his studies of the distribution of wealth. The 80–20 rule has later been shown to apply in many other contexts. Pareto analysis is one of Ishikawa's seven basic quality tools.

The Pareto chart is a bar chart where the root causes or categories are shown ordered from high importance or frequency to lower importance or frequency. Pareto analyses may only be performed on data collected in a stable situation, so no trends may be present. Depending on the context, causes may be weighted by frequency, costs, impact. If causes are aggregated into categories, then care must be taken to ascertain that categories allow fair comparisons.

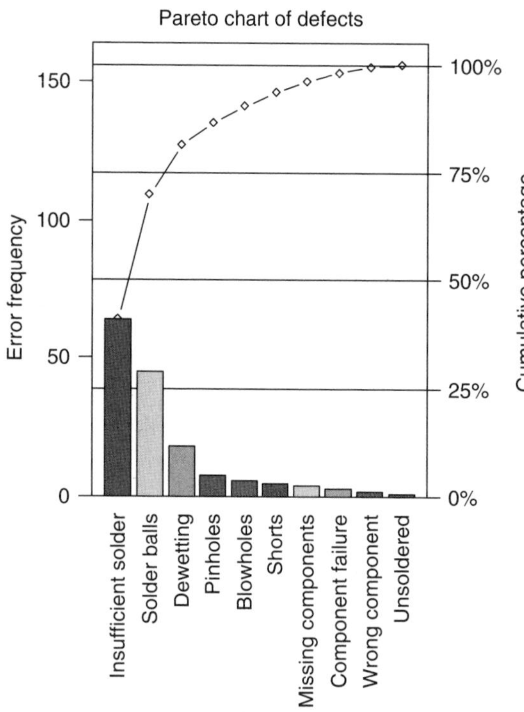

Figure 1 Pareto chart of printed circuit defects

An example adapted from [2] shows defects found in printed circuit boards. The bars indicate the frequency of the errors ordered from left (most frequent) to right (less frequent). As can be seen from the scale on the right-hand side of Figure 1, the three most frequent defects account for slightly over 80% of the defects.

References

[1] Juran, J.M. (1951). *Juran's Quality Control Handbook*, 1st Edition, McGraw-Hill.

[2] Montgomery, D.C. & Runger, G.C. (2006). *Applied Statistics and Probability for Engineers*, 4th Edition, John Wiley & Sons.

Related Articles

Cause-and-Effect Diagrams; **Failure Modes and Effects Analysis**; **Graphical Representation of Data**.

ALESSANDRO DI BUCCHIANICO

Path Sets and Cut Sets in System Reliability Modeling

Background and Preliminaries

Graphs

Definitions. A *graph* is an ordered pair of sets $(\mathcal{N}, \mathcal{L}) = \mathcal{G}$ with $\mathcal{L} \subset \mathcal{N} \times \mathcal{N}$. $\mathcal{N}$ is called the *set of nodes* of the graph and $\mathcal{L}$ is called the *set of links* of the graph. Typically, a graph is identified with a drawing in which the nodes are represented as points in the plane and the links are represented as lines drawn to join two nodes. In other terminology in common use, the nodes may be called *vertices* and the links may be called *arcs* or *edges*.

A *labeled graph* is a graph in which the nodes and/or links have names. That is, there is a one-to-one correspondence between the nodes of the graph and a set of $|\mathcal{N}|$ objects (the node labels) and/or between the links of the graph and a set of $|\mathcal{L}|$ objects (the link labels).

A *directed graph* is a graph in which each link is assigned an orientation or direction. In a directed graph, the links (i, j) and (j, i) are different, whereas in an ordinary (undirected) graph, they are identical. The concept of "a link from i to j" makes sense in a directed graph; in an undirected graph it would be proper to say, rather, "a link between i and j".

Examples. The cities in a state, together with the roads joining them, may be described as a graph. In this example, the cities form the nodes of the graph and the roads joining them form the links. A natural labeling of the nodes and links exists. A natural gas pipeline may be modeled as a graph with the terminals and transfer stations as the nodes and the pipes as the links. Graphs find application in the social sciences also. A graph may be constructed from a set of individuals (who form the nodes) in which two individuals are joined by a link if they are known to each other. Indeed, any binary reflexive relation between discrete entities gives rise to a graph in which the entities are the nodes and a link joins two nodes if the relation holds for those two nodes. If the relation is not reflexive, then the result is a directed graph. If there is a notion of "strength" of the relationship and we wish to incorporate this notion into the model, a more appropriate approach is to model the graph as a capacitated network; *see* the section titled "Paths, Cuts, and Demand Satisfaction in Capacitated Networks".

References. Other useful graph concepts, such as degree, are not covered in this article because they are not relevant to the reliability modeling application. For a review of the basic facts of graph theory, see [1–3].

Paths and Connectedness

Definitions. Two nodes i and j are *adjacent* if $(i, j) \in \mathcal{L}$. The *adjacency matrix* of $\mathcal{G}$ is an $|\mathcal{N}| \times |\mathcal{N}|$ matrix whose ij entry is one if $(i, j) \in \mathcal{L}$ and is zero otherwise. In other terminology in common use, the adjacency matrix may be called the *incidence matrix*. A *path* in a graph is a sequence of alternating adjacent nodes and the links joining them, beginning and ending with a node. Two nodes i and j are said to be *connected* if there is a path having i as its initial node and j as its terminal node. That is, the path takes the form $\{i, (i, \nu_1), \nu_1, (\nu_1, \nu_2), \ldots, \nu_k, (\nu_k, j), j\}$ for some $\nu_1, \ldots, \nu_k \in \mathcal{N}$ and $(i, \nu_1), (\nu_1, \nu_2), \ldots, (\nu_k, j) \in \mathcal{L}$. There is no loss in abbreviating this to $\{(i, \nu_1), (\nu_1, \nu_2), \ldots, (\nu_k, j)\}$. When it is necessary or desirable to explicitly indicate the nodes being connected, the path will be called an (i, j) *path*. Clearly, adjacent nodes are connected but connected nodes need not be adjacent. If the graph is directed, the links in the path must be considered with the proper orientation.

A *cut* for two given nodes is a set of nodes and/or links whose removal from the graph disconnects the two nodes. When it is necessary or desirable to explicitly indicate the nodes being disconnected, the cut will be called an (i, j) *cut*.

A path connecting two given nodes is called *minimal* if it contains no proper subset that is also a path connecting the same two nodes. A cut for two given nodes is called *minimal* if it contains no proper subset that is also a cut disconnecting those nodes.

Examples. The Washington, DC, Metro subway system http://www.wmata.com/metrorail/systemmap.cfm may be modeled as a graph with the stations as the nodes. In this graph, the Pentagon and College Park – University of Maryland are connected but not adjacent. DuPont Circle and Farragut North are both connected and adjacent. The Manhattan (NY) city streets may be modeled as a graph with the intersections as the nodes. For vehicle traffic, this graph is directed because many streets in Manhattan are one-way. For pedestrian traffic, the graph need not be directed.

Random Graphs

Definitions. A *random graph* is a graph in which a stochastic indicator variable is attached to each node and link. When the variable is zero, it indicates that, that node or link is not present in the graph. When it is one, it indicates that, that node or link is present in the graph. Each choice of values for these indicator variables, by whatever random mechanism is at play, produces a different graph (the choice is not completely unrestricted; if the indicator of a node is zero, the indicators of all the links emanating from that node must be zero also). The adjacency matrix of a random graph is a random matrix.

In a reliability engineering application, the indicator variable for a link or node describes the functioning or nonfunctioning of the link or node. The usual convention is that the indicator variable is one when the link or node functions and zero if it does not function. There is no fundamental reason why the opposite assignment could not be used, and it is sometimes seen.

References. For purposes of the reliability modeling described in this article, this definition is as much of the subject of random graphs as we will use. Additional material on random graphs may be found in [4] and [5].

Paths, Cuts, and System Reliability Modeling

Introduction: Structure Functions and Reliability Block Diagrams

For the purposes of this article, reliability is considered an all-or-nothing condition: either a unit or system functions properly and completely at a stated time, or it does not function at all at that time. Thus, we may imagine the functioning-or-failed condition of a unit or system as a zero–one, or indicator, random variable that, in the general case, will be allowed to depend on time. The system *structure function* is a Boolean function that maps $\{0, 1\}^c$ into $\{0, 1\}$, where the number of components in the system is c. The structure function is the indicator variable of the system functioning when the arguments in the structure function are replaced by the indicator variables of the components' functioning. Let X_k indicate the functioning of component k and $x_k \in \{0, 1\}(k = 1, \ldots, c)$. The system structure function is $\varphi(x_1, \ldots, x_c)$ and the system reliability is $P\{\varphi(X_1, \ldots, X_c) = 1\}$. For example, the structure function for a series system of c components is $\varphi(x_1, \ldots, x_c) = x_1 x_2 \cdots x_c$.

A structure function is called *coherent* if it is nondecreasing in each variable separately and each component is relevant (that is, the state of the system depends on the states of all the components in the structure function).

The system reliability block diagram is a labeled random graph whose links represent the components or subsystems whose reliability description is known or provided (for analysis of a system reliability block diagram, it is sufficient to allow the nodes to be perfectly reliable). The system reliability block diagram expresses the reliability logic of a system in the sense that it shows how the system fails when constituent components and subsystems fail. It is a pictorial representation of the system structure function. Two special nodes are called out: a source or origin node and a terminal or destination node. The system functions if and only if in the random graph there is a path connecting the source node to the terminal node.

In many cases, the system reliability block diagram is a *series–parallel* structure. In such cases, the probability that the system functions is easily concluded from nesting of the standard formulas for the reliability of series systems and parallel systems (*see* **Parallel, Series, and Series–Parallel Systems**). Other structures, such as the ***k*-out-of-*n*** hot standby and k-out-of-n cold standby structures, are also amenable to similar probabilistic analysis. Some other structures, such as the bridge structure shown in Figure 1, lend themselves less readily to this type of analysis. In such cases, it may be convenient to use the path set or the cut set methods that are the subject of this article.

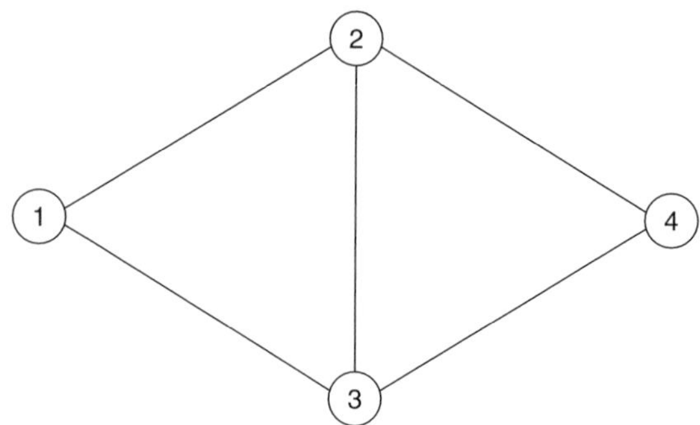

Figure 1 Bridge network

Path Sets and Cut Sets

Given two nodes in a graph, the *path set* for those two nodes is the union of all paths connecting those two nodes. In other terminology in common use, a path set may be called a *tie set*. The *cut set* for those two nodes is the union of all cuts for those two nodes. The *minimal path set* is the union of all minimal paths. The *minimal cut set* is the union of all minimal cuts.

Some authors use the phrases "path set" for what we call a *path* and "cut set" for what we call a *cut*. This makes the naming of the path set or cut set (as defined here) less natural and we avoid that difficulty by using this scheme.

The key concept is that the system functions if and only if there is at least one minimal path whose components are all in a functioning condition. Similarly, the system does not function if and only if there is as least one minimal cut whose components are all in a failed (nonfunctioning) condition. The random graph model provides a framework for computing probabilities of system functioning and failure (nonfunctioning) based on these concepts.

Examples

Consider the reliability block diagram depicted in Figure 2. The small circles represent the nodes of this random graph and these are assumed not to fail. The links representing subsystems that can fail individually in this system have been labeled for convenience by the letters A, B, C, D, and E. The two unlabeled links in the center of the diagram are not associated with any subsystems and are assumed not to fail. Note that, like the bridge structure of Figure 1, this is not a series–parallel graph. In this model, the

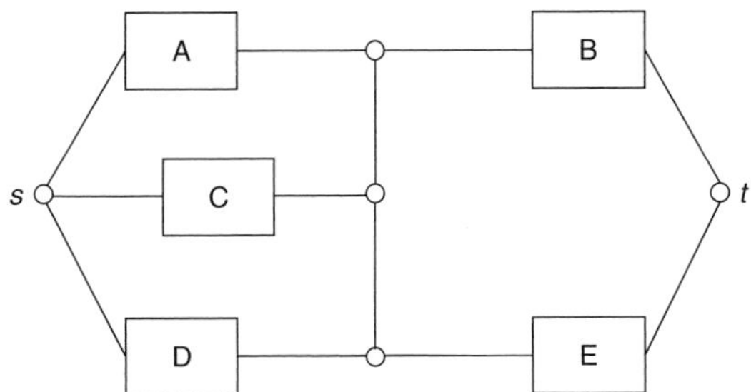

Figure 2 Example of a reliability block diagram

system functions if the node s at the left-hand edge of the diagram and the node t at the right-hand edge of the diagram are connected. This representation indicates that the system functions if any one of the sets {A, B}, {D, E}, {C, B}, or {C, E} of units consists entirely of functioning units. Each of these is an (s, t) path. The union of these five paths constitutes the path set for the node pair (s, t). Note that all these paths are minimal, and these are the only minimal paths, so the path set {A, B} ∪ {D, E} ∪ {C, B} ∪ {C, E} is the minimal path set.

Similarly, the system fails to function if any of the sets {B, E}, {A, C, D}, {C, B, D}, or {C, A, E} consists entirely of nonfunctioning, or failed, units. Any one of these is a cut for (s, t), or an (s, t) cut. There are other (s, t) cuts in this graph, such as {A, B, E}, but the four cuts enumerated above are the only minimal cuts. The minimal cut set for (s, t) is their union {B, E} ∪ {A, C, D} ∪ {C, B, D} ∪ {C, A, E}.

References

See [4, 6, 7] for additional discussion of path sets and cut sets as models for reliability of complex systems.

System Reliability Modeling and Computation Using Path Sets and Cut Sets

Path Set Method

Because the path set contains all paths connecting s to t, for the system to function it suffices that at least one path be made up entirely of functioning units. Therefore, the probability that the system functions is given by the probability of the path set in the labeled random graph representing the system reliability block diagram. However, only minimal paths need be considered because if a path is not a minimal path, then it has a proper subset that is still a path and is a member of the minimal path set. In other words, the union of all (s, t) paths is equal to the union of all (s, t) minimal paths. Consequently, we have the following result:

Proposition 1. The probability that the system functions is given by the probability of the system's minimal path set.

Example. Consider again the system shown in Figure 2. Letting $p_A = P\ \{A = 1\}$ (where we have abused notation slightly by identifying the indicator random variable's letter with the unit's label) and similarly for B, C, D, and E, the probability that the system functions is given by

$$P\big(\{A=1, B=1\} \cup \{D=1, E=1\} \cup \{C=1, B=1\} \cup \{C=1, E=1\}\big) \quad (1)$$

This equation shows the strength and weaknesses of the path set method. Its strength is that it is completely straightforward and mechanical to write the expression for the probability that the system functions once the path sets are known. Its weaknesses are that (a) enumerating the paths connecting s and t is tedious for all but the simplest of graphs and (b) the expression that results is the probability of a large union of events that are not, in general, disjoint. However, these weaknesses pertain mainly to manual execution; the algorithmic nature of the procedure means that software for path set reliability analysis is within reach, and indeed has been available for some time [8].

General Representation of System Reliability *via* the Minimal Path Set. As usual, let $\boldsymbol{x} = (x_1, \ldots, x_c)$ denote the vector if indicators of the functioning of the c components of the system. Enumerate the minimal paths of the system; suppose there are m of them called $\pi_1, \ldots, \pi_m$. The structure function for the series system represented by the minimal path π_k is

$$\varphi_k(x) = \prod_{i \in \pi_k} x_i \quad (2)$$

for $k = 1, \ldots, m$. The system functions if and only if at least one of the minimal paths consists entirely of functioning units, so it follows that the structure function for the system may be written as

$$\varphi(\mathbf{x}) = 1 - \prod_{k=1}^{m} [1 - \varphi_k(\mathbf{x})] = 1 - \prod_{k=1}^{m} [1 - \prod_{i \in \pi_k} x_i] \quad (3)$$

This equation shows how the system structure function may be represented in terms of the structure functions of the system's minimal paths.

Cut Set Method

Because the cut set contains all (s, t) cuts, for the system to fail it suffices that at least one cut be made up entirely of nonfunctioning units. However, only minimal cuts need be considered because if a cut is not a minimal cut, then it has a proper subset that is still a cut and is a member of the minimal cut set. In other words, the union of all (s, t) cuts is equal to the union of all (s, t) minimal cuts. Therefore, the probability that the system fails to function is given by the probability of the minimal cut set in the labeled random graph representing the system reliability block diagram. Consequently, we have the following result:

Proposition 2. The probability that the system fails is given by the probability of the system's minimal cut set.

Example. Consider again the system shown in Figure 2. The probability that the system fails to function is given by

$$P(\{\mathrm{B}=0,\mathrm{E}=0\}\cup\{\mathrm{A}=0,\mathrm{C}=0,\mathrm{D}=0\}\cup\{\mathrm{C}=0,\mathrm{B}=0,\mathrm{D}=0\}\cup\{\mathrm{C}=0,\mathrm{A}=0,\mathrm{E}=0\}) \quad (4)$$

Again, equation (4) represents the probability of a union of events that are not, in general, disjoint so manual computation can become cumbersome. An algorithm for system reliability evaluation using cut sets may be found in [3].

General Representation of System Reliability *via* the Minimal Cut Set. Enumerate the minimal cuts of the system; suppose there are n of them called $\chi_1, \ldots, \chi_n$. The structure function for the series system represented by the minimal cut χ_k is

$$\varphi_k(x) = 1 - \prod_{i\in\chi_k}(1 - x_i) \quad (5)$$

for $k = 1, \ldots, n$. The system fails if and only if at least one of the minimal cuts consists entirely of nonfunctioning units, so it follows that the structure function for the system may be written as

$$\varphi(\mathbf{x}) = \prod_{k=1}^{n}\varphi_k(\mathbf{x}) = \prod_{k=1}^{n}\left[1 - \prod_{i\in\chi_k}(1 - x_i)\right] \quad (6)$$

This equation shows how the system structure function may be represented in terms of the structure functions of the system's minimal cuts.

References. Additional information on the use of path sets and cuts sets for system reliability modeling and computation may be found in [6] and [9].

Bounds

The minimal path set and minimal cut set representations for the system reliability lend themselves readily to the development for bounds on the system reliability. The first such bounds were developed by Esary and Proschan [3]. Letting C (resp., W) denote the minimal cut (resp., path) set for the system, i.e., for the nodes (s, t), Esary and Proschan's lower bound for the system reliability is

$$\prod_{c\in C}\left(1 - \prod_{i\in c}P\{X_i = 0\}\right) \quad (7)$$

and their upper bound is

$$1 - \prod_{w\in W}\left(1 - \prod_{i\in w}P\{X_i = 1\}\right) \quad (8)$$

The lower bound gives good approximations for highly reliable systems, while the upper bound works better for systems whose components have low reliability. Numerous improvements have been developed; we refer especially to [10] and [11] for later developments.

Paths, Cuts, and Demand Satisfaction in Capacitated Networks

A *network* is a directed graph containing two distinguished subsets $\mathcal{O}$ and $\mathcal{D}$ of $\mathcal{N}$, the sets of origin nodes and destination nodes, respectively. A network is *capacitated* if there is an assignment of nonnegative real numbers to each link and node that represents the *capacity* of that link or node. Origin nodes and destination nodes are called so because the network is required to transport given quantities of some commodity, discrete (package shipments, telecom/datacom packets) or continuum (oil, gas), tangible (postal letters) or intangible (sensor readings) from the origin node(s) to the destination node(s). The matrix of required quantities is the *exogenous demand*. In other terminology in common use, the origin nodes may be called *sources* and the destination nodes may be called *sinks*.

A *flow* in a network is a map that assigns to each link a nonnegative real number that represents the quantity of the commodity present on that link; see [10]. Sometimes, we speak of the flow on a link (or at a node) to mean the real number in the flow that is attached to that link or node. In a capacitated network, this is required to be no greater than the capacity of the link or node. Determining the flow from the origin-to-destination demands usually requires solution of some optimization problem (such as minimum cost) and is beyond the scope of this article. Nonetheless, in seeking the probability that the demand is satisfied, that is, that the required number of units is transported to the destination nodes, the concepts of path and cut may be generalized to yield useful methods.

The flow into a node is equal to the sum of the flows on the links terminating on that node. Similarly, the flow out of a node is equal to the sum of the flows on the links originating from that node. To say that a demand is satisfied is to say that the flows into the destination nodes are all at least as large as the amount of the commodity required to be delivered to the destinations (in the exogenous demand). Demand satisfaction is a generalization of connectivity. Connectivity is necessary for demand satisfaction: if two nodes are not connected by any paths, then the flow into the terminal node will be zero. Ramirez-Marquez *et al.* [12] developed a generalization of the cut set method described above that is appropriate for demand satisfaction in capacitated networks.

References

[1] Harary, F. (1971). Graph Theory, *Addison-Wesley*, Reading.

[2] Bollobas, B. (1998). *Modern Graph Theory*, Springer-Verlag, New York.

[3] Esary, J.D. & Proschan, F. (1963). Coherent structures of non-identical components, *Technometrics* **5**, 191–209.

[4] Bollobas, B. (2001). *Random Graphs*, 2nd Edition, Cambridge University Press, New York.

[5] Spencer, J. (2001). *The Strange Logic of Random Graphs*, Springer-Verlag, New York.

[6] Barlow, R.E. & Proschan, F. (1981). *Statistical Theory of Reliability and Life Testing: Probability Models*, To Begin With, Silver Spring.

[7] Birolini, A. (2004). *Reliability Engineering Theory and Practice*, Springer-Verlag, Berlin.

[8] Shen, Y. (1995). A New Simple Algorithm for Enumerating All Minimal Paths and Cuts of a graph, *Microelectronics & Reliability*, **35**(6), 973–976.

[9] Pages, A. & Gondran, M. (1986). *System Reliability Evaluation and Prediction in Engineering*, Springer-Verlag, New York.

[10] Ford, L.R. & Fulkerson, D.R. (1962). *Flows in Networks*, Princeton University Press, Princeton.

[11] Koutras, M.V. & Papastavridis, S.G. (1993). Application of the Stein-Chen method for bounds and limit theorems in the reliability of coherent structures, *Naval Research Logistics* **40**, 617–631.

[12] Ramirez-Marquez, J.E., Coit, D. & Tortorella, M. (2005). Multistate two-terminal reliability: a generalized cut-set approach, *IEEE Transactions on Reliability* (under review).

Further Reading

Billinton, R. & Allan, R.N. (1983). *Reliability Evaluation of Engineering Systems*, Plenum Press, New York.

Elsayed, E.A. (1996). *Reliability Engineering*, Addison-Wesley, Reading.

Fu, J.C. & Koutras, M.V. (1995). Reliability bounds for coherent structures with independent components, *Statistics and Probability Letters* **22**, 137–148.

Harary, F. (1969). *Graph Theory*, Addison-Wesley, Reading.

Kolchin, V.F. (1999). *Random Graphs*, Cambridge University Press, New York.

Rabinowitz, L. (1968). *Reachability and connectedness in random graphs and digraphs*, Ph. D. thesis, Rutgers University, New Brunswick.

Michael Tortorella

Patient Opinion Measures

Introduction

In healthcare a wide range of patient opinion measures has been developed. Most of these measures are questionnaires, which may be provided to patients by (e-)mail, in face-to-face or telephone contacts, or through the Internet. Many measures have focused on either patient reported symptoms and health-related quality of life, or on patient opinions on healthcare – the latter is the focus of this contribution. Patient opinion measures have been used in scientific research for a long time, but only in recent decades they are increasingly used for quality improvement, accreditation of healthcare providers, and public reporting on quality of care providers. In these applications the users are not necessarily researchers, but citizens, patients, healthcare providers, health authorities, and policy makers. In many countries a market of patient survey research has developed in recent years and users of patient opinion measures have to choose from a range of competing measures.

From a methodological and statistical perspective, there is nothing unique in the validation or application of patient opinion measures compared to other measures. More recent developments in questionnaire methodology, such as computer-assisted interviewing and item–response theory, have not yet reached the field of patient opinion measures. In fact, many patient opinion measures, which are currently used in applications, have hardly been validated. A few measures have been validated reasonably, but it can be observed that methodological issues related to their application of the instrument tend to be ignored [1]. This can seriously **bias** the results, also if the measure itself was carefully developed and validated. This contribution aims to provide an overview of some statistical issues in the application of patient opinion measures. It uses the Europep instrument to illustrate the points made [2] (Box 1).

Methods

Sampling Participants

An appropriate sampling procedure is crucial for the validity of any population-based study. This requires, among others, a clear definition of the sampling frame (from which cases are sampled) and a sampling procedure that avoids bias (ideally, random sampling). Bias caused by the sampling procedure cannot be compensated for by larger sample sizes. A small but appropriately sampled study population may be preferable to a large but biased sample. New statistical methods for handling biased samples (e.g., sampling participants through the Internet) may widen the opportunities in the future. The Europep instrument was not made for one particular sampling procedure. Many users have consecutively recruited patients among visitors of a general practice, but others have randomly sampled patients from practice registers or population registers.

Response rate can be defined as the number of completed questionnaires divided by the total of questionnaires that reached the target population. A low response rate could induce selection bias. Many studies with the Europep instrument have achieved response rates of 70% or higher. It is recommended to strive for a response rate of at least 60% in using the instrument. Sending reminders increases the response rate by about 10% and repeated mailing of the questionnaire was only marginally more effective than a simple postcard reminder [3]. In many situations it is necessary to use reminders to achieve acceptable response rates. It is recommended to interpret results carefully if the response rate is low (e.g., lower than 60%). While the results may still be valuable for educational feedback and scientific research, they are probably less useful for accreditation and public reporting.

It is difficult to provide recommendations on absolute figures for the **sample size** in studies, which use patient opinion measures. Inclusion of more patients increases the **power** of the study and results in more accurate estimations (smaller **confidence intervals**). If figures are presented for specific practices or practitioners, it should be taken into account that the data are clustered. Statistical advice on appropriate power calculations may be needed. Some studies have suggested that a minimum of 60 respondents are needed per practice/practitioner to allow for reliable figures at the level of practice/practitioner [4]. Lower numbers may be acceptable in educational feedback or scientific research, depending on the research question.

Handling of Items

Health services research is a multidisciplinary research domain, which borrows from clinical epidemiology and social science methodology. The two traditions do not always take the same approach, which is certainly the case in the handling of questionnaire items. The psychometrics tradition regards items as repeated measurements for an

Box 1 Construction and revision of the Europep questionnaire

The Europep instrument was developed by an international consortium of researchers and family physicians in the years 1995–1998. It was developed from the beginning as an international instrument for patient evaluations of general practice, using rigorous translation and validation procedures. We aimed at use for educational purposes in general practices and regions as well as nationwide surveys and international comparisons. A series of studies were performed for its development, including an international study on patient priorities and studies to examine protoversions of the questionnaire. The questionnaire is focused on evaluations of specific aspects of general practice (not: priorities, wishes, reports, experiences, satisfaction, utilities, etc.). It comprises 23 items, which use a five-point scale (poor–excellent). Explicit criteria were used for the final selection of items, which focused on coverage of domains of general practice, importance to patients, item response, language problems, and discrimination (specific quantitative criteria were formulated). Providing effective feedback on Europep data was not explicitly addressed, but some countries have developed elaborated feedback procedures. More information can be found on www.swisspep.ch/pages/EUROPEP.html.
In 2006, the users of the Europep instrument felt a need to revise and possibly extend the Europep questionnaire. The questionnaire had been used in more than 20 European countries, in some cases on a very large scale (e.g., in Denmark and Switzerland). We sought a balance between improving the questionnaire and maintaining a core set of items – needed for comparisons over time. We planned to exclude and revise items on the basis of a systematic process with a group of users of the Europep instrument. The criteria were specified *a priori*: (a) An item would be omitted or revised if there was a serious ambiguity or translation problem in any country, leading to misinterpretations (item interpretation). (b) An item would be omitted or revised if there would be an unexpected response of lower than 30% in more than half of the countries (item response). (c) An item would be omitted or revised if more than 80% of the respondents used only one of the five answering categories in most (>80%) countries (item discrimination). The EPA project data on Europep data (collected in the period 2000–2005) provided data on item response and discrimination (www.ru.nl/topas-europe). Eight rounds of proposals and feedback were done in the group of users over the email, using a cumulative list of interpretation problems and revision proposals. The criteria regarding item response and discrimination did not lead to omission of items, but a range of interpretation problems was identified which led to a number of small revisions. The result was the prototype of the Europep 2006 questionnaire:

What is your assessment of the general practitioner over the last 12 months with respect to:
1. Making you feel you have time during consultation*
2. Showing interest in your personal situation*
3. Making it easy for you to tell him or her about your problem
4. Involving you in decisions about your medical care
5. Listening to you
6. Keeping your records and data confidential
7. Providing quick relief of your symptoms*
8. Helping you to feel well so that you can perform your normal daily activities
9. Thoroughness of the approach to your problems*
10. Physical examination of you
11. Offering you services for preventing diseases (e.g., screening, health checks, immunizations)
12. Explaining the purpose of examinations, tests, and treatments*
13. Telling you enough about your symptoms and/or illness*
14. Helping you deal with emotions related to your health status*
15. Helping understand why it is important to follow the GP's advice*
16. Knowing what has been done or told during previous contacts in the practice*
17. Preparing you for what to expect from specialists, hospital care, or other care providers*

Box 1 continued

What is your assessment of the general practice over the last 12 months with respect to:
18. The helpfulness of the practice staff (other than the doctor) to you*
19. Getting an appointment to suit you?
20. Getting through to the practice on telephone?
21. Being able to talk to the general practitioner on the telephone*
22. Waiting time in the waiting room?
23. Providing quick services for urgent health problems?

*Revised items

underlying concept, so the scores of correlated items should be aggregated to compose a measure of the underlying concept. It is this aggregated measure, which is most relevant for the study. The clinimetrics tradition, on the other hand, tends to regard items either as independent measures (each item reflects one concept) or as aspects of a known biological process (e.g., symptoms of a disease). In the latter case, items are simply added up without a need for an observed high **correlation** between the item scores. In health services research, often a balance has to be found between the two approaches.

In the Europep instrument each item has been developed to reflect a specific concept, such as "availability of the GP on the telephone" or "time in the consultation". In other words, different items were not primarily designed as repeated measures for underlying concepts. Therefore, items were presented and interpreted separately. Scales are aggregations of different items, which reflect an underlying concept. While the precise meaning of the individual items gets lost, the advantage of scales is that these are usually more accurate than individual items (smaller confidence intervals). Explorative psychometric analyses of Europep data (principal factor analysis, orthogonal rotation, accepting factors with eigenvalue higher than 1) showed that the Europep instrument contains two scales: an evaluation of clinical management and an evaluation of practice management.

The following procedure was used in the analysis of Europep data. Reliability analysis was used to verify the internal consistency (Cronbach's α) on each of the two scales in the Europep instrument. Raw scores were used for analysis; cases with missing values were excluded by list-wise deletion. It was felt that Cronbach's α should be 0.70 or higher (an arbitrary value); in reality, α's were much higher. While different procedures for calculation of scale scores may be acceptable, the following procedure was recommended as a standard for Europep users. Items are dichotomized according to 5 *versus* 1–4 ("excellent *vs* less than excellent"). The category "excellent" is coded 1, while the other valid values (excluding the missing values) are coded 0. Then the mean value of the items in the scale is determined. For this purpose a new variable is defined, which is the mean value of the selected items. The result is a score for each individual, which can be interpreted as the proportion of items, which were assessed as "excellent".

Missing values are a problem for any study, which cannot be fully solved satisfactorily. Many users of the Europep instrument have regarded the category "do not know/not applicable" as a missing value. This implies that cases with missing values were not included in the presentation of scores on the items. It is recommended to Europep users to use this approach as the standard procedure. In the calculation of scales of the Europep instrument, missing values were handled as follows. Patients are excluded if more than one-third of the items in the scale were missing values (no answers or not applicable/not relevant). **Imputation** of missing values in scales (e.g., by mean on the other items in the scale) is not recommended, because this artificially increases the consistency of the scale scores.

Presentation of Findings

For educational purposes little more than presentation of absolute numbers and percentages in each answering category for each item may be needed.

For comparison with other providers or patient groups potential confounders should be considered. The most obvious confounder in studies of patient views is patient age, although the absolute size of effect of age is small. It has been found fairly consistently that older patients have more positive views of the care received (for which no evidence-based explanation is available). The impact of other patient characteristics is less consistent and probably small, after adjustment for patient age. In the application of the Europep instrument it is recommended to control for age in comparisons, particularly if they are used for accreditation or public reporting.

Furthermore, it is recommended to present all figures with confidence intervals, certainly if they are used for accreditation or public reporting. In many, if not all, applications data are clustered within practices or practitioners (different patients from the same practice/practitioner are included). This implies that random effects models should be used, for which expert statistical advice is often required. Inappropriate statistics are most likely to overestimate differences and changes, because confidence intervals tend to be wider after correction for clustering. This implies that differences between care providers or changes over time are observed, while these are in fact not significant.

In practical applications, the ultimate goal is to improve quality of care (which may be operationalized in different ways). Despite the enormous number of patient opinion studies, the evidence on its impact on quality of care or patient outcomes is very scarce. A randomized trial showed that providing educational feedback on patient opinions of general practice was positively received by physicians, but did not lead to observable changes in professional behavior or practice management [5]. The use of patient opinion measures in accreditation and public reporting may be more effective, but research evidence to support this claim is currently not available.

Conclusions

Patient opinion measures need to be validated and used sensibly, like any measure. Some statistical aspects of sampling participants, handling of items, and presentation of findings were discussed in this contribution. The Europep instrument was used throughout the paper to illustrate the points made. Statistical aspects of applying patient opinion measures should be addressed to avoid bias, also if the measure itself was carefully developed and validated.

References

[1] Wensing, M. & Elwyn, G. (2003). Methods for incorporating patients' views in health care: validity, effectiveness and implementation, *British Medical Journal* **326**, 877–879.

[2] Grol, R., Wensing, M., Mainz, J., Jung, H.P., Ferreira, P., Hearnshaw, H., Hjortdahl, P., Olesen, F., Reis, S., Ribacke, M. & Szecsenyi, J. (2000). Patients in Europe evaluate general practice care: an international comparison, *British Journal of General Practice* **50**, 882–887.

[3] Wensing, M. & Schattenberg, G. (2005). Response rates after three follow-up procedures in a survey among elderly adults: a randomised trial, *Journal of Clinical Epidemiology* **58**, 959–961.

[4] Wensing, M., Van der Vleuten, C., Grol, R. & Felling, A. (1997). The reliability of patients' judgements of general practice care: how many questions and patients are needed? *Quality in Health Care* **6**, 80–85.

[5] Vingerhoets, E., Wensing, M. & Grol, R. (2001). Educational feedback on patients' evaluations of care: a randomised trial, *Quality in Health Care* **10**, 224–228.

MICHEL WENSING

Perfect Sampling

Introduction

A common requirement of many problems in a variety of disciplines is for a sample to be drawn from some probability distribution π on a space $\mathcal{X}$. Often this distribution cannot be specified completely: for example, it may only be known up to a normalization constant. One widely used method for doing this is Markov chain **Monte Carlo** (MCMC). Such an approach involves designing an ergodic Markov chain X, which has π as its stationary distribution, and then running a simulation of X until it is near equilibrium (*see* **Markov Chain Monte Carlo, Introduction**; **Convergence and Mixing in Markov Chain Monte Carlo**).

However, an obvious problem with MCMC is the following: for how long should we run the chain before sampling (in other words, what is an appropriate burn-in time)? If we sample from X when it is not 'close enough' to equilibrium then the algorithm output may be from a distribution that is far from π.

An attractive alternative to estimating a sufficient burn-in period is to adapt the MCMC algorithm to form a *perfect simulation* algorithm. Such a procedure has two desirable features:

1. The algorithm determines for itself when it should stop.
2. If the algorithm is successful, then it returns a sample drawn *exactly* from π.

In [1], an algorithm known as *coupling from the past* (CFTP) was introduced and used to sample from the exact equilibrium distribution of the critical Ising model on a finite lattice. Although there do exist other perfect simulation algorithms (such as the FMMR algorithm [2]), in this article we shall principally present the CFTP algorithm, and a variant of this method named *dominated CFTP* (also called *CIAFTP*).

A friendly warning: perfect simulation is a technical area that has mainly evolved in the stochastic process literature, and which involves a lot of stochastic process theory. Unfortunately, there's no escaping from this fact, but hopefully the simple examples below will help newcomers to this field gain some insight into the beauty of the approach.

Coupling from the Past

As the name suggests, CFTP is based on the idea of coupling Markov chains together (see [3] for a comprehensive introduction). For our purpose, a coupling of two chains X and X' (with common transition kernel) is a prescription for specifying their joint transition kernel such that if the two chains meet, then they stay together thereafter.

A general method of producing such a coupling is to define the updates of X and X' using a common update function. More specifically, it is possible to describe the transitions of X using a deterministic function $f : \mathcal{X} \times [0, 1] \to \mathcal{X}$ and a sequence of *Uniform*[0, 1] random variables $\{\xi_n\}$, such that

$$X_{n+1} = f(X_n, \xi_n) \tag{1}$$

(This construction is known as a *stochastic recursive sequence*.) Using this representation we can couple X and X' such that they stay together once they agree: define their updates by

$$X_{n+1} = f(X_n, \xi_n), \quad X'_{n+1} = f(X'_n, \xi_n) \tag{2}$$

The beauty of this representation is that it enables us to couple *any number* of chains in this way, using the same f and the same randomness $\{\xi_n\}$ for all the chains.

The basic concept behind CFTP is the following: Consider a hypothetical copy of our chain of interest, say $\tilde{X}$, which has been running since time $-\infty$. At time zero this chain will be in equilibrium, i.e., $\tilde{X}_0 \sim \pi$. This would be of obvious use if we could determine $\tilde{X}_0$, but clearly we cannot run a chain from time $-\infty$ in practice! However, it may be possible to determine $\tilde{X}_0$ by looking back only a *finite number of steps* into the past of $\tilde{X}$.

To describe the algorithm, suppose that we have available to us an independent and identically distributed (i.i.d.) sequence $\{\xi_n\}_{-\infty}^{0}$ of *Uniform*[0, 1] random variables and a deterministic update function f. For $v \geq -u$, define the random composite *input–output* map $F_{(-u,v]}(x) : \mathcal{X} \to \mathcal{X}$ by

$$F_{(-u,v]}(x) = f(f(\ldots f(f(x, \xi_{-u}), \xi_{-u+1}) \ldots, \xi_{v-2}), \xi_{v-1}) \tag{3}$$

Thus if X is begun at time $-u$ with the value $X_{-u} = x$, then we may set $X_v = F_{(-u,v]}(x)$. We write $X_v^{x,-u}$ for the value of the chain X at time v, when X is started at time $-u$ from state x.

Definition 1 For a given update function f, the *backwards coalescence time* T^* of the chain X is the random variable defined by

$$T^* = \min\{n \geq 0 \;:\; F_{(-t,0]}(x) = F_{(-s,0]}(y), \text{ for all } x, y \in \mathcal{X} \text{ and for all } s, t \geq n\} \tag{4}$$

Now consider starting chains $X^{x,-n}$ from *all* states $x \in \mathcal{X}$ at time $-n$. Using f we can couple all of these chains, as well as the hypothetical $\tilde{X}$, using the same source of randomness $\{\xi_n\}_{-\infty}^{0}$. If $n \geq T^*$ then $X_0^{x,-n} =: X_0$ is the same for all x, and so $X_0 = \tilde{X}_0 \sim \pi$. That is, if we start our chains $X^{x,-n}$ far enough back into the past that they have all coalesced by time zero, then their common value at zero is an exact draw from π, as required! Of course, we do not know the value of T^*, and so the CFTP algorithm simply repeats the above procedure for larger and larger n until $n \geq T^*$ is achieved:

CFTP Algorithm

```
set n ← 1
while F_(−n,0](X) not constant
    n ← 2n
return F_(−n,0](X)
```

The most important aspect of the algorithm is that the random sequence $\{\xi_n\}$ is *reused* in each execution of the `while` loop. This implies a possible issue with computer memory in practice, but there is a trick that avoids this, known as *read-once CFTP* [4]. The use of binary search for T^* contained within the `while` loop is also noteworthy: we are free to increase n in any way we like, but the binary search means that the total number of simulated Markov chain steps is linear in T^*, whereas if we used $n \leftarrow n+1$ (for example) it would grow quadratically. Furthermore, this strategy means that the number of simulation steps comes within a factor of 4 of the true value of T^* [1, 4].

Theorem 1 *If the backward coalescence time T^* is almost surely finite, then CFTP samples from equilibrium.*

The proof of Theorem 1 is very simple [1, 5]. Note that if $\mathcal{X}$ is finite, then T^* is almost surely finite (but very large) for the trivial independence coupling, where chains evolve independently until they meet, after which they agree forever. In fact, $\mathbb{P}(T^* < \infty)$ is always either zero or one, whatever the choice of f [6]. A good CFTP algorithm uses a function f that has a high probability of making target chains coalesce quickly.

A Simple Example

The following example was considered in [7], with variants appearing in [1, 8]. Consider a symmetric reflecting random walk X on $\mathcal{X} = \{1, 2, 3, 4\}$, satisfying $\pi(i) = 1/4$ for all $i \in \mathcal{X}$ (see Figure 1).

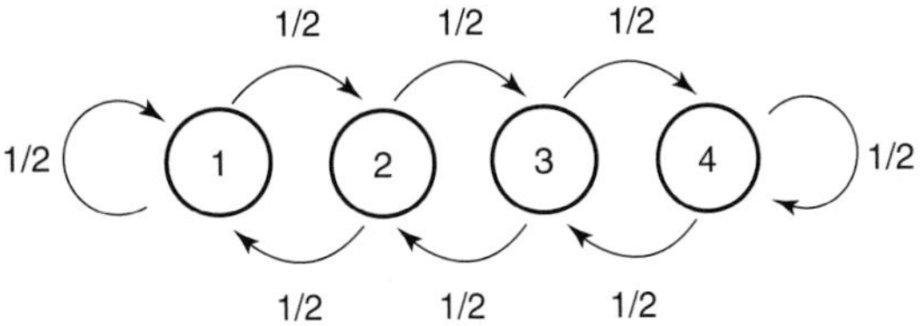

Figure 1 Simple symmetric reflecting random walk

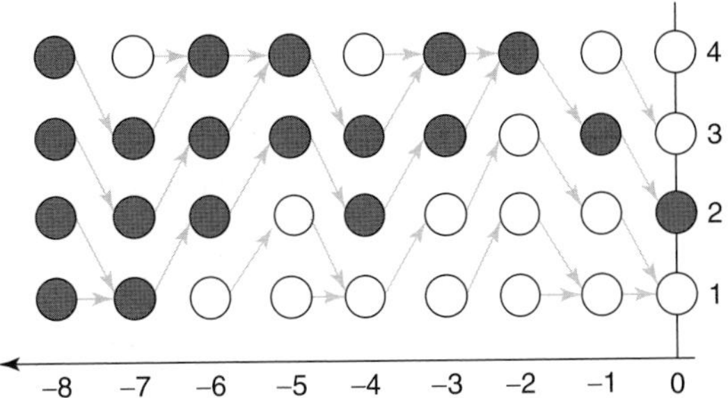

Figure 2 CFTP for a simple symmetric reflecting random walk

For our update function f we may choose the following:

$$f(x, u) = \begin{cases} \min(x+1, 4) & \text{if } u \leq 1/2 \\ \max(x-1, 1) & \text{if } u > 1/2 \end{cases} \tag{5}$$

Figure 2 demonstrates the CFTP algorithm for this example. The gray arrows indicate the transitions determined by f and the sequence $\{\xi_n\}$. The algorithm runs chains $X^{x,-n}$ started from all states x, with $n = 1, 2, 4, \ldots$. When $n = 8$ is reached, all of the target chains have the same value at time zero: $X_0^{x,-8} = 2$ for this realization (for which $T^* = 7$). The shaded circles highlight the possible values of the chains $X^{x,-8}$ at each step: coalescence occurs the first time that $X^{1,-n}$ hits four or $X^{4,-n}$ hits one.

Implementation of CFTP for this example is trivial, and Figure 3(a) shows a histogram of the results of 10 000 runs of the algorithm. A χ^2 test to compare this output to π yields a p value of 0.79: it is clear that the algorithm is drawing from the correct distribution. In contrast, Figure 3(b) shows the output of a wrongly implemented algorithm, in which the random sequence $\{\xi_n\}$ is not reused when the chains are restarted further into the past. Here a definite departure from uniformity is visible, and indeed a χ^2 test gives a tiny p value of 2×10^{-16}. The reader may

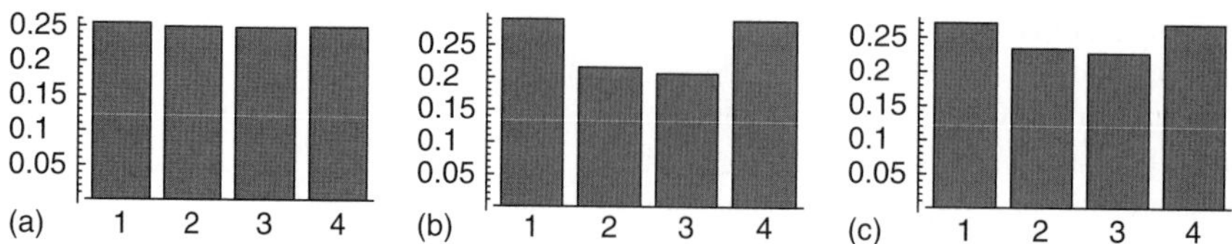

Figure 3 Output of 10 000 runs of the CFTP algorithm for the random walk described above: (a) a proper implementation, reusing randomness; (b) not reusing randomness; (c) reusing randomness, but with simulations with "long"run times interrupted and discarded

like to see if his/her intuition can explain this bias toward a distribution with more mass at the extremal states! For more examples of such bias see [1, 4, 8]. Finally, Figure 3(c) shows another possible source of bias: that which is introduced by *user impatience*. This happens when simulations with long run times are interrupted and discarded. Since the output of the CFTP algorithm is in general not independent of the algorithm run time, this introduces a bias. In this implementation, any simulation that had not coalesced when starting chains from time -8 were discarded. The p value for this sample is again very small: $p = 2.6 \times 10^{-11}$. It should be noted that the FMMR algorithm (under suitable implementation) does not suffer from this bias [9].

Although this is clearly a toy example, it does highlight the ease with which CFTP may be applied to state spaces with a partial order. More specifically, suppose that $\mathcal{X}$ admits a partial order $\preceq$, which is respected by the update function f. That is,

$$x \preceq y \quad \Rightarrow \quad f(x,u) \preceq f(y,u) \quad \text{for all } x, y \in \mathcal{X} \tag{6}$$

Furthermore, suppose that $\mathcal{X}$ contains maximal and minimal elements, x^{max} and x^{min}, satisfying $x^{min} \preceq x \preceq x^{max}$ for all $x \in \mathcal{X}$. With this setup it becomes simple to check for coalescence in the CFTP algorithm, since the monotonicity of f guarantees that

$$F_{(-n,0]}(x^{\min}) = F_{(-n,0]}(x^{\max}) \implies F_{(-n,0]}(\mathcal{X}) \quad \text{is constant} \tag{7}$$

Thus, instead of running target chains, $X^{x,-n}$, starting from all states $x \in \mathcal{X}$ at time $-n$, we now only need to simulate chains started from $x^{\min}$ and $x^{\max}$. This is exactly the case in our example above, where f is monotonic (using the normal ordering on the integers): coalescence occurs exactly when the two chains $X^{1,-n}$ and $X^{4,-n}$ meet.

Evading Monotonicity

Monotonicity, while useful for CFTP, is by no means essential. Consider modifying the random walk above by setting $\mathbb{P}(X_{n+1} = 2 \mid X_n = 1) = 1$, and leaving all other transitions the same. This chain, say $\hat{X}$, is no longer reversible, and there is no monotonic update function under the usual ordering on the integers (although a different ordering, or the use of subsampling, may reintroduce monotonicity). However, a CFTP algorithm for this chain can be constructed using the *crossover trick* [7]: this first appeared in [10] to deal with *antimonotone chains*, and is also used in [11].

A variant of this approach is to use the original monotonic random walk X as an *envelope process* for $\hat{X}$. Suppose we define an update function $\hat{f}$ as follows:

$$\hat{f}(1,u) = 2, \quad \text{and}$$

$$\hat{f}(x,u) = \begin{cases} \min(x+1, 4) & \text{if } u \le 1/2 \\ \max(x-1, 1) & \text{if } u > 1/2 \end{cases} \quad \text{for } x = 2, 3, 4 \tag{8}$$

Note that $\hat{f}$ is a valid update function for $\hat{X}$, and that $f(x,u) \le \hat{f}(x,u)$ for all $x \in \mathcal{X}$ and $u \in [0,1]$ (where f is defined in equation 5). This suggests the following perfect simulation algorithm:

- run a target chain $X^{1,-n}$ using the update function f and random sequence $\{\xi_i\}$, until the first time $S \le 0$ that state 4 is hit (if $S \not\le 0$, increase n and repeat, reusing $\{\xi_i\}$);
- due to the ordering of f and $\hat{f}$, $X_S^{1,-n} = \hat{X}_S^{x,-n} = 4$ for all x, and so all target chains $\hat{X}^{x,-n}$ have coalesced by time S;
- run the chain $\hat{X}^{4,S}$ up to time zero, using $\hat{f}$ and the same $\{\xi_i\}$. Return $\hat{X}_0^{4,S}$.

The idea of using an envelope process was studied in more depth, under the name *bounding chains*, in [12]: the technique can be used in situations which are neither monotonic nor antimonotonic, and can also give bounds on the expected run time of CFTP. This trick, along with the crossover trick mentioned above, increases the possibility of efficient implementation of CFTP for a variety of chains. Furthermore, there are other variants of CFTP, which make the technique applicable to even more general chains, even those on infinite state spaces. For example, the method of *small set CFTP* [13] uses the idea of small set regeneration to determine when target chains have coalesced.

Dominated CFTP

The classic CFTP algorithm of [1] has a major drawback: a successful CFTP algorithm for X exists if and only if X is *uniformly ergodic* [6]. However, there does exist a major extension of CFTP, known as *dominated CFTP* [14, 15], or *coupling into and from the past* (CIAFTP), which can be applied to chains not satisfying this restriction. Indeed, dominated coupling from the past (domCFTP) can (in principle) be implemented for any *geometrically ergodic* chain [16], and a class of *subgeometrically ergodic* chains for which a domCFTP algorithm exists is introduced in [17], greatly extending the possible applicability of perfect simulation.

Whereas CFTP works by checking for *vertical* coalescence (target chains started from all states have coalesced by time zero), domCFTP checks for *horizontal* coalescence (all sufficiently early starts from a specific state lead to the same result at time zero). A second chain Y is used to identify how far into the past one has to go to determine that this coalescence has occurred. To describe the algorithm in the simplest possible way, we consider here only a monotonic chain X on $\mathcal{X} = [0, \infty)$: see [18] for a much more general formulation.

Suppose that copies of X can be coupled such that, for each $x, t, u \geq 0$, and $s \geq -t$, we can construct $X^{x,-t}$ (begun at state x at time $-t$) satisfying

$$X_s^{x,-t} \leq X_s^{x,-u} \Longrightarrow X_{s+1}^{x,-t} \leq X_{s+1}^{x,-u} \qquad (9)$$

Suppose too that we can construct a dominating process Y on $\mathcal{X}$ which is stationary, defined for all time, and may be coupled to the target chains X such that

$$X_s^{x,-t} \leq Y_s \Longrightarrow X_{s+1}^{x,-t} \leq Y_{s+1} \qquad (10)$$

The domCFTP algorithm then proceeds as follows:

1. Draw Y_0 from its stationary distribution.
2. Simulate Y backwards to time $-n$.
3. Set $y = Y_{-n}$. Simulate $X^{y,-n}$ and $X^{0,-n}$ forwards to time zero (coupled to each other and to Y so that equations (9) and (10) are satisfied).
4. If $X_0^{y,-n} = X_0^{0,-n}$, then return this value as a perfect draw from π. If not, extend the realization of Y back to time $-2n$, set $n \leftarrow 2n$, and go to step 3.

A proof that this algorithm returns a perfect draw from π (so long as it terminates almost surely) may be found in [5, 15]. A consequence of the coupling in equation (9) is that the target processes are *funneled*: the earlier the two target chains (X^0 and X^y) are started, the closer they will be at time zero. As a simple example of the algorithm in practice, consider a birth–death process X with transitions $x \to x+1$ at rate $\alpha_x \leq \alpha < \infty$, and $x \to x-1$ at rate μx. This chain is clearly monotonic, and may be dominated by the birth–death chain Y, which has births at rate α and deaths at rate μ. It is easy to see that Y is reversible and has a *Poisson*(α/μ) equilibrium distribution. A realization of the domCFTP algorithm for this chain when $\alpha_x = 10 - \log(x+1)/(x+1)$ and $\mu = 1$ is shown in Figure 4. This algorithm can also be made to work if multiple simultaneous deaths are allowed to occur, or if the birth rate μ varies with x, as is the case with many problems in stochastic geometry [10, 15].

Summary

Perfect simulation algorithms are both a useful practical tool and an exciting area of research. Unlike with MCMC, there is no issue of estimating a burn-in time, and the algorithm output is an exact draw from the required distribution. We have seen above how to apply CFTP and domCFTP to a couple of toy examples: generalization to more interesting chains will often require little additional theory. However, for complicated Markov chains such algorithms are typically harder to implement than MCMC, mainly due to the potential

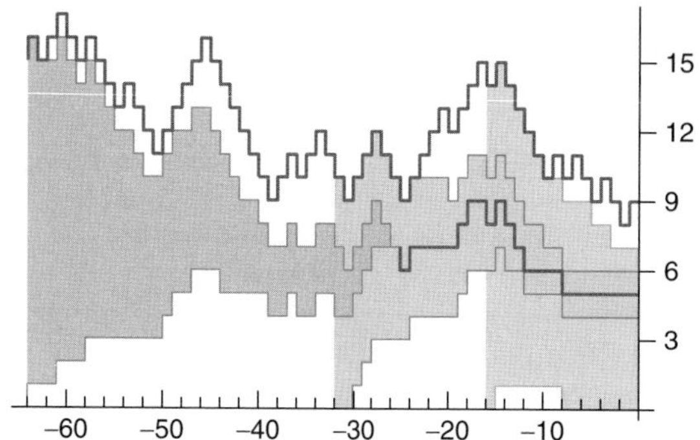

Figure 4 Implementation of domCFTP for a birth–death process. The topmost line shows the evolution of the dominating process into the past, and the shaded regions demonstrate the funneling of target processes. In this realization, all target chains started beneath the dominating process at time −64 have coalesced by time zero

difficulty in finding an efficient method of coupling target chains together. That said, since the arrival of CFTP, more and more variants on this and other techniques mean that there is now a wide range of perfect simulation algorithms available. A comprehensive resource for further investigation of perfect simulation is the on-line bibliography of David Wilson: http://research.microsoft.com/˜dbwilson/exact/.

References

[1] Propp, J.G. & Wilson, D.B. (1996). Exact sampling with coupled Markov chains and applications to statistical mechanics, *Random Structures and Algorithms* **9**, 223–252.

[2] Fill, J.A., Machida, M., Murdoch, D.J. & Rosenthal, J.S. (2000). Extension of Fill's perfect rejection sampling algorithm to general chains, *Random Structures and Algorithms* **17**(3–4), 290–316.

[3] Lindvall, T. (2002). *Lectures on the Coupling Method*, Dover Publications.

[4] Wilson, D.B. (2000). How to couple from the past using a read-once source of randomness, *Random Structures and Algorithms* **16**(1), 85–113.

[5] Kendall, W.S. (2005). Notes on perfect simulation, in *Markov Chain Monte Carlo: Innovations and Applications, Vol. 7 of Lecture Notes Series*, W.S. Kendall, F. Liang & J.-S. Wang, eds, Institute for Mathematical Sciences, National University of Singapore, World Scientific.

[6] Foss, S.G. & Tweedie, R.L. (1998). Perfect simulation and backward coupling, *Stochastic Models* **14**, 187–203.

[7] Thönnes E. (2000). A primer on perfect simulation, *Statistical Physics and Spatial Statistics (Wuppertal, 1999), Vol. 554 of Lecture Notes in Physics*, Springer, Berlin, pp. 349–378.

[8] Häggström, O. (2002). *Finite Markov Chains and Algorithmic Applications, Vol. 52 of London Mathematical Society Student Texts*, Cambridge University Press.

[9] Thönnes, E. (1999). Perfect simulation of some point processes for the impatient user, *Advances in Applied Probability* **31**(1), 69–87.

[10] Kendall, W.S. (1998). Perfect simulation for the area-interaction point process, in *Probability Towards 2000*, L. Accardi & C. Heyde, eds, Springer-Verlag, pp. 218–234.

[11] Häggström, O. & Nelander, K. (1998). Exact sampling from anti-monotone systems, *Statistica Neerlandica* **52**(3), 360–380.

[12] Huber, M. (2004). Perfect sampling using bounding chains, *Annals of Applied Probability* **14**(2), 734–753.

[13] Murdoch, D.J. & Green, P.J. (1998). Exact sampling from a continuous state space, *Scandinavian Journal of Statistics* **25**(3), 483–502.

[14] Kendall, W.S. (1997). Perfect simulation for spatial point processes, in *Proceedings of the 51st Session of the ISI*, Istanbul, August 1997, Vol. 3, pp. 163–166.

[15] Kendall, W.S. & Møller, J. (2000). Perfect simulation using dominating processes on ordered spaces, with application to locally stable point processes, *Advances in Applied Probability* **32**, 844–865.

[16] Kendall, W.S. (2004). Geometric ergodicity and perfect simulation, *Electronic Communications in Probability* **9**, 140–151 (electronic).

[17] Connor, S.B. & Kendall, W.S. Perfect simulation for a class of positive recurrent Markov chains, *Annals of Applied Probability* **17**(3), 781–808.

[18] Cai, Y. & Kendall, W.S. (2002). Perfect simulation for correlated Poisson random variables conditioned to be positive, *Statistics and Computing* **12**(3), 229–243.

STEPHEN B. CONNOR

Performance Measures for Robust Design

Introducing Taguchi's Signal-to-Noise Ratio and Two-Step Procedures

The objective of robust design (*see* **Robust Design**) is to find the vector of design variable settings that minimizes the quadratic loss function popularized by Taguchi and Wu [2],

$$L(Y, t) = A_0(Y - t)^2 \tag{1}$$

where A_0 is a constant, Y is the process response, and t the desired target value for Y. The average loss may be decomposed as follows:

$$\begin{aligned} E_{L(Y,t)} &= R(\mathbf{C}) \\ &= A_0 E_{\mathbf{N},\epsilon}(Y - t)^2 \\ &= A_0 \left[\mathrm{Var}_{\mathbf{N},\epsilon}(Y) + (E_{\mathbf{N},\epsilon}(Y) - t)^2\right] \end{aligned} \tag{2}$$

where **C** is a vector of design variables that can be partitioned into two mutually exclusive parts corresponding to "nonadjustment" factors, that is, $\mathbf{d} = (d_1, \ldots, d_p)$, and "adjustment" factors, that is, $\mathbf{a} = (a_1, \ldots, a_q)$. Note that this partitioning of design variables is a theoretical ideal, and mutually exclusive groupings may not exist in practice. $\mathrm{Var}_{\mathbf{N},\epsilon}(Y)$ and $E_{\mathbf{N},\epsilon}(Y)$ are the variance and mean of the process response over the random sources of variability represented by **N** and ϵ. **N** is a vector of uncontrollable "noise" variables (*see* **Factorial Experiments**) and ϵ represents purely random variation. For ease of notation we will simply write Var(Y) and $E(Y)$ from here onward.

The adjustment factors making up **a** are those that affect the mean, that is, $E(Y)$, independently of Var(Y), whereas the nonadjustment variables that make up **d** are assumed to affect only the variance Var(Y). Equation (2) indicates that the minimization of average quadratic loss is mathematically equivalent to the minimization of the sum of process variance and the square of the deviation of process mean from the target.

Taguchi [3] introduced a family of performance measures called ***signal-to-noise*** *ratios* (SNRs; *see* **Signal-to-Noise Ratios for Robust Design**) whose specific form depends on the desired response outcome. The case where the response has a fixed nonzero target is called the *nominal-the-best* (NTB) case and is defined as follows:

$$SNR = 10\log_{10}\left[\frac{\overline{Y}^2}{S_Y^2}\right] \tag{3}$$

where $\overline{Y}$ and S_Y^2 are respectively the sample mean and variance of the response estimated at each test combination of design variables selected for experimentation. Responses having unique smaller-the-better or larger-the-better targets are respectively called the *STB* and *LTB* cases. We have only presented the formula of the SNR for the NTB case because it most transparently conveys the concept of a "signal", that is, $\overline{Y}$, being compared to systemic background variability or "noise", that is, S_Y^2.

To accomplish the objective of minimal expected squared-error loss for the NTB case, Taguchi [3] proposed the following two-step optimization procedure:

1. Calculate and model the SNRs and find the values of the nonadjustment factor settings, that is, **d**, which maximize the SNR.
2. Shift mean response to the target by changing the values of the adjustment factor(s) **a**.

Although he provided a performance measure, that is, the SNR, and a related optimization scheme, that is, the two-step procedure, it was not explained how these two entities collectively achieve the goal of minimizing average quadratic loss.

Relating Performance Measure to Optimization: The Performance Measure Independent of Adjustment (PerMIA)

León *et al.* [1] showed that under certain circumstances Taguchi's use of the SNR, within the context of his two-step optimization procedure, does in fact minimize expected quadratic loss. Specifically this happens when the response conforms to a **transfer function** of the following form

$$Y = E_Y(\mathbf{a}, \mathbf{d})\epsilon(\mathbf{N}, \mathbf{d}) \tag{4}$$

where **N** are the noncontrollable noise variables and $(\mathbf{a}, \mathbf{d})$ are the adjustment and nonadjustment factors. They observed that for this class of models, the SNR of equation (3) can be minimized independently of the mean. This was a rigorous articulation of when the use of Taguchi's SNR within his two-step optimization procedure accomplished the stated goal of minimizing expected quadratic loss. They caution that when a response is not compliant with the functional form of equation (4), the use of the SNR in the two-step procedure of Taguchi does not necessarily minimize expected quadratic loss.

In order to provide a consistent framework within which to employ Taguchi's two-step procedure, they defined a new entity called the *performance measure independent of adjustment*, that is, the PerMIA. The PerMIA is a performance measure, that is, a function, which can be minimized independently of the mean and is specifically defined for the particular class of models being considered. Assuming that a PerMIA can be defined for the situation under consideration, use of a PerMIA within the framework of the two-step procedure guarantees the minimization of expected quadratic loss. This is because, in the first step, settings of the nonadjustment factors are chosen to minimize the PerMIA and, in the second step, the mean is shifted as closely as possible to its target value by choosing corresponding values for the adjustment factors. Because of the defining properties of the PerMIA, the adjustment of the mean in the second step does not affect the value of the PerMIA. When the process under consideration follows the model of equation (4), the SNR of equation (3) is a PerMIA but for other models it is not. This explained why use of the SNR in all situations does not always yield results that minimize the value of expected quadratic loss.

In particular León *et al.* [1] indicate that for an additive transfer function, that is,

$$Y = E_Y(\mathbf{a}, \mathbf{d}) + \epsilon(\mathbf{N}, \mathbf{d}) \tag{5}$$

that $\text{Var}(Y)$ is a PerMIA and should be used in place of the SNR. León *et al.* [1] conclude by proposing a modification of Taguchi's two-step procedure which first finds values of the nonadjustment factors $\mathbf{d}$ which minimize the PerMIA and, conditional on those values of $\mathbf{d}$, finds the values of the adjustment factors $\mathbf{a}$ that minimize expected quadratic loss.

Nair and Pregibon [4] advocate a three-phase process for robust design consisting of exploration, modeling, and optimization. Their performance measure is process response which they assume is a function of the design and noise variables, that is, $\mathbf{C}$ and $\mathbf{N}$ referred to in equation (2). They reduce the robust design problem to minimization of Var_Y subject to the constraint that E_Y is held as close as possible to the target level t.

They suggest that process variance can typically be modeled as a function of adjustment factors ($\mathbf{a}$) and nonadjustment factors ($\mathbf{d}$) as follows:

$$\text{Var}(Y)(\mathbf{C}) = f_1(E(Y)(\mathbf{a}, \mathbf{d}))\epsilon(\mathbf{N}, \mathbf{d}) \tag{6}$$

This model implies that minimization of variance relies primarily on the nonadjustment factors ($\mathbf{d}$) with residual variability determined by the distance between the mean and the target. They recommend a transformation-based approach which allows the separation of the **dispersion effects**, that is, $\epsilon(\mathbf{d})$, and the location effects, that is, $f_1(E(Y)(\mathbf{a}, \mathbf{d}))$. The specific transformation is identified in the exploratory phase with plots of the response data at each treatment combination supplemented with mean–variance plots. Their graphical examination plots the means and variances on a log–log scale which is approximately linear with slope k for the common situation where

$$f_1(E_Y(\mathbf{a}, \mathbf{d})) = [E_Y(\mathbf{a}, \mathbf{d})]^k \tag{7}$$

which represents an important specific case of equation (6). The values of k suggest the appropriate transformation with a value of $k = 0$ indicating a unity transformation due to the lack of a strong relationship between mean and variance.

eqr231(5)Box [5] introduced a generalized transformation of response approach that applies to commonly found relations between location and dispersion. We remind the reader that León *et al.* [1] demonstrated that Taguchi's SNR is a PerMIA when Var_Y is proportional to E_Y. Box defines a PerMIA as

$$P(\mathbf{d}) = \frac{\text{Var}_Y(\mathbf{a}, \mathbf{d})}{f_2[(E_Y(\mathbf{a}, \mathbf{d})]} \tag{8}$$

under the assumption that the function $f_2[(E_Y(\mathbf{a}, \mathbf{d})]$ can be found for which the resulting PerMIA is a function of only the nonadjustment design variables $\mathbf{d}$. Because $f_2[(E_Y(\mathbf{a}, \mathbf{d})]$ is not known *a priori*, Box provides a data analytic technique called the λ *plot* to empirically find it. The λ plot assumes a transformation of the data, that is, $Y^{(\lambda)}$, such that a value of λ that achieves maximum separation of location and dispersion effects of the process response can be graphically identified.

On a strategic level, Box [5] asserted that the two crucial aspects of robust design are the choice of an appropriate performance criterion and its **estimation**. He states that experimentation, use of simple graphical techniques and sequential testing are the best ways to pursue robust design rather than the use of preordained performance criteria such as the SNR. Box also argued that the SNRs for the STB and LTB cases are inadequate summaries of data and extremely inefficient measures of location.

León and Wu [6] extend the work of León *et al.* [1]. The first thing they explain is how a two-step procedure converts a constrained robust design problem into an unconstrained one. They continue by articulating when a simple adjustment of mean to target can be substituted for solving for values of the adjustment variables **a** that minimize expected loss conditional on the values of the nonadjustment variables **d** that minimize the PerMIA. They introduce a corresponding procedure called the *constrained two-step procedure* (C2P) to simplify the identification of a PerMIA and solve the constrained minimization problem with a second step that simply adjusts the process mean.

Lastly, León and Wu [6] cite the fallacy of assuming a quadratic loss function and mention asymmetric losses around a target as one instance where the quadratic loss function is inappropriate. For nonquadratic loss functions they introduce general dispersion, location, and off-target measures for use in a two-step process. This generalization of the loss function and related two-step procedure includes identification of adjustment functions, that is, functions of design factors used to make adjustments of process location. Though Taguchi used the process mean as an adjustment function, they point to the median as a possible alternative. They apply these new techniques in a number of examples featuring additive and multiplicative models with nonquadratic loss functions.

In an extension of León and Wu's discussion of nonquadratic loss functions [6], Moorhead and Wu [7] develop modeling and analysis strategies for a general loss function where the quality characteristic follows a location-scale model. Their procedure adds a third step to the traditional two-step process, an adjustment step which moves the mean to the side of the target with lower cost. Although limited by its need for a location-scale model, this procedure builds upon the familiar two-step procedures used for quadratic functions. An alternative approach for asymmetric loss that considers both univariate and multivariate cases is presented in Joseph [8].

Concluding Remarks

In defining the PerMIA, León *et al.* [1] provided a unifying framework in which a performance measure explicitly achieves the optimization goal of minimizing expected quadratic loss via the two-step procedures promoted by Taguchi. León and Wu [6] showed how the PerMIA represents a family of performance measures appropriate for nonquadratic and nonsymmetric loss functions as well. Box [5] and others, including Nair and Pregibon [4], Vining and Myers [9], Shoemaker *et al.* [10], Tsui [11–14], Tsui and Li [15], Lunani *et al.* [16], and Bérubé and Wu [17] posit that using process response or its transformation as a performance measure provides a more flexible approach to robust design than the SNR-based techniques of Taguchi.

References

[1] León, R.V., Shoemaker, A.C. & Kackar, R.N. (1987). Performance measure independent of adjustment: an explanation and extension of Taguchi's signal to noise ratio, *Technometrics* **29**, 253–285.

[2] Taguchi, G. & Wu, Y. (1980). *Introduction to Off-Line Quality Control*, Central Japan Quality Control Association, Nagoya.

[3] Taguchi, G. (1986). *Introduction to Quality Engineering: Designing Quality into Products and Processes*, Kraus International Publications, White Plains.

[4] Nair, V.N. & Pregibon, D. (1986). A data analysis strategy for quality engineering experiments, *AT&T Technical Journal* **65**, 73–84.

[5] Box, G.E.P. (1988). Signal to noise ratios, performance criteria and transformations, *Technometrics* **30**, 1–31.

[6] León, R.V. & Wu, C.F.J. (1992). Theory of performance measures in parameter design, *Statistica Sinica* **2**, 335–358.

[7] Moorhead, P.R. & Wu, C.F.J. (1998). Cost-driven parameter design, *Technometrics* **40**, 111–119.

[8] Joseph, V.R. (2004). Quality Loss functions for nonnegative variables and their applications, *Journal of Quality Technology* **36**, 129–138.

[9] Vining, G.G. & Myers, R.H. (1990). Combining Taguchi and response surface philosophies: a dual response approach, *Journal of Quality Technology* **22**, 38–45.

[10] Shoemaker, A.C., Tsui, K.-L. & Wu, C.F.J. (1991). Economical experimentation methods for robust parameter design, *Technometrics* **33**, 415–428.

[11] Tsui, K.-L. (1992). An overview of Taguchi method and newly developed statistical methods for robust design, *IIE Transactions on Quality and Reliability* **24**(5), 44–57.

[12] Tsui, K.-L. (1996). A multi-step analysis procedure for robust design, *Statistica Sinica* **6**, 631–648.

[13] Tsui, K.-L. (1999a). Robust design optimization with multiple characteristics, *International Journal of Production Research* **37**, 433–445.

[14] Tsui, K.-L. (1999b). Response model analysis of dynamic robust design experiments, *IIE Transactions on Quality and Reliability* **31**, 1113–1122.

[15] Tsui, K.-L. & Li, A. (1994). Analysis of smaller and larger the better robust design experiments, *International Journal of Industrial Engineering* **1**, 193–202.

[16] Lunani, M., Nair, V.N. & Wasserman, G.S. (1997). Graphical methods for robust design with dynamic characteristics, *Journal of Quality Technology* **29**, 327–338.

[17] Bérubé, J. & Wu, C.F.J. (1998). Signal-to-noise ratio and related measures parameter design optimization, Technical Report 321, University of Michigan.

Related Articles

Robust Design; **Signal-to-Noise Ratios for Robust Design**.

TERRENCE E. MURPHY AND
KWOK-LEUNG TSUI

Physical Degradation Models

Introduction

The focus within the reliability area has traditionally been on the collection and analysis of time-to-failure (TTF) data. High reliability implies few failures, so assessing and improving product/process reliability based on TTF data can be difficult. Accelerated life testing (ALT) can shorten the product life during test intervals by stressing the product beyond its normal use. However, ALT involves considerable amount of extrapolation. One should incorporate as much knowledge about the product as possible in designing the acceleration experiments.

Many physical systems degrade over time. Recent advances in sensing and measurement technologies are making it feasible to collect extensive amounts of data on physical **degradation** associated with components, systems, and manufacturing equipment. Davies [1] discusses a variety of engineering applications, types of degradation data, and recent developments in the area of condition monitoring and system maintenance. These cover data from vibration and acoustical monitoring, thermography, lubricant and wear debris analysis, and so on. Meeker and Escobar ([2], Chapters 14 and 21 and references therein) describe applications in luminosity of light bulbs, corrosion of batteries, semiconductor metal-oxide semiconductor (MOS) devices, and so on.

In practice, a number of subjects are under study. Degradation measurements of each subject are taken over time. The number of observation times and observation times themselves are allowed to vary across the subjects. These types of data are referred to as regular *degradation data*. For some applications, only one meaningful measurement can be taken on each test unit. The degradation measurement process destroys or changes the physical or mechanical characteristics of test units. This is known as *destructive degradation*. For example, the data from fatigue crack growth subject to loading cycles [3, 4] and bridge beams subject to erosion of chloride ion ingression [5] belong to regular degradation data. Data from the strength of adhesive bonds subject to environmental effects and internal erosion [6] are from destructive degradation experiments since the test is destructive and strength can be measured only once on each unit.

When the amount of degradation reaches a prespecified critical level D, failure occurs. Let T denote the TTF and $X(t)$ be the degradation process, then $T = \inf\{t : X(t) \geq D\}$. In many tests, the failure data is supplemented by degradation data. Degradation data are a very rich source of reliability information and offer many advantages over TTF data. The goal of this article is to present some common models for physical degradation. Specifically, we shall mainly discuss random-effect models and some Lévy process models. There is extensive literature on nonlinear random-effect models (e.g., [7, 8]), and it is flexible and has been found quite useful in reliability applications. The Lévy processes such as Wiener process and gamma process have also been found useful as **degradation models**. These models lead to tractable forms for TTF distributions and thus we can make reliability inferences based on a combination of degradation and TTF data.

The paper is organized as follows. The section titled "Physical Degradation" gives some examples of physical degradation that provide useful motivation for degradation models. The section titled "Degradation Models" discusses two common physical degradation models and the reliability inferences drawn from them. In the section titled "An Illustrative Example", we use a simulation example to illustrate that there are tremendous gains to be realized from degradation data. Finally we summarize the paper and describe some additional research topics.

Physical Degradation

In any given application, the problem context and appropriate subject matter knowledge must be used to model the underlying degradation mechanisms ([9] and [10] are good examples). For well-understood failure mechanisms, one may have a model based on physical/chemical theories that describe the underlying degradation process. Usually degradation models start with a deterministic description of the degradation process, often in the form of a differential equation. Then randomness can be incorporated into the model to describe the material parameters, component-to-component variability, initial conditions, and operating and environmental conditions.

Example 1: Automobile tire wear. If $\Lambda(t)$ is the amount of automobile tire tread wear at time t. This can be modeled simply by linear rate of degradation and wear rate is $\mathrm{d}\Lambda(t)/\mathrm{d}t = C$, then $\Lambda(t) = \Lambda(0) + Ct$, where $\Lambda(0)$ and C could be taken as constant for individual units but random from component-to-component.

Example 2: Fatigue crack growth. The prediction of crack growth accumulation is crucial for the reliability analysis of various materials in manufacturing. Let $\Lambda(t)$ be the crack size at time t and $\Delta K(\Lambda)$ be the stress intensity range, which is a function of the current crack size. A common used crack growth rate model is the Paris-rule model (e.g., [11]),

$$\frac{\mathrm{d}\Lambda(t)}{\mathrm{d}t} = C[\Delta K(\Lambda)]^m \tag{1}$$

where C and m are material properties. For example, when modeling a two-dimensional edge crack in a plate, $\Delta K(\Lambda) \propto \sqrt{\Lambda}$. Then we have

$$\Lambda(t) = \begin{cases} (C_0 + C_1 t)^{2/(2-m)} & m \neq 2 \\ C_3 \mathrm{e}^{C_4 t} & m = 2 \end{cases} \tag{2}$$

where $C_i, i = 1, \ldots, 4$ are constants which are functions of m. Note that different choices of m lead to different shapes of degradation paths.

Example 3: Current-gain degradation of bipolar transistors. In self-aligned polysilicon emitter transistors, a large electric field existing at the periphery of the emitter–base junction under reverse bias can create hot-carrier-induced degradation. Let Λ be the current gain, Burnett and Hu [12] studied a hot-electron degradation model,

$$\frac{\mathrm{d}\Lambda(t)}{\mathrm{d}t} = CG(\Lambda) I_R \mathrm{e}^{-\phi/kT_\mathrm{e}} \tag{3}$$

where ϕ is the critical energy needed for device damage, T_e is electron temperature, I_R represents the density of electrons, and $G(\Lambda)$ represents the dependence of degradation rate on existing damage. Assuming $G(\Lambda) \propto \Lambda^{-h}$, we have

$$\Lambda(t) \propto [t I_R \mathrm{e}^{-\phi/kT_\mathrm{e}}]^{1/(h+1)} \tag{4}$$

This relationship can be used to predict MOS device degradation dynamics.

Example 4: Photodegradation of polymeric materials. Excessive exposure of polymeric materials leads to photodegradation failure. Let Λ be the total number of absorbed photons that effectively contribute to the photodegradation of a material during an exposure period. It has the form [13]

$$\Lambda(t) = \int_0^t \int_{\lambda_0}^{\lambda_1} E_0(\lambda, s)(1 - 10^{-A(\lambda,s)})\phi(\lambda)\, \mathrm{d}\lambda \mathrm{d}s \tag{5}$$

where λ_0 and λ_1 are the minimum and maximum photolytically effective UV–visible wavelengths, respectively, $A(\lambda, s)$ is the absorbance of sample at specified UV–visible wavelength and at time s, $E(\lambda, s)$ is the incident spectral UV–visible radiation dose to which a polymeric material is exposed at time s, and $\phi(\lambda)$ is the damage at wavelength λ relative to a reference wavelength. The overall photolytic damage to a polymeric material can be a linear, power, or exponential response of Λ.

Degradation Models

There are many factors that cause variability in the degradation curves and in the failure times. There are several approaches to modeling random degradation process in the literature. These include random-effect degradation models [4, 9, 10], stochastic process models such as Wiener process models [14–17], gamma process models [18–20], and compound **Poisson process** models [21], random shock process [22, 23], cumulative **damage models** [24], and multistage models [25]. For a general discussion of different degradation models, see Singpurwalla [26] and Meeker and Escobar [2]. We will mainly focus on the random-effect models and some nonhomogeneous Lévy process models.

Let $X_i(t)$ be the observed degradation path for unit i. Consider the nonlinear random-effect model

$$X_i(t) = \Lambda(t; \theta, \omega_i) + \epsilon_i(t) \tag{6}$$

where $\Lambda(t)$ is a smooth monotone function of t, θ is the population-level parameter characterizing degradation, ω_i is the random effect associated with subject i, and $\epsilon_i(t)$ is measurement error. In general, the degradation path $\Lambda(t;\theta,\omega_i)$ will be nonlinear and could be chosen by physical mechanism theory as discussed in the section titled "Physical Degradation". For example, data from a linear random-effect model is given by

$$X_i(t_j) = \omega_i t_j + \epsilon_i(t_j),\ i = 1,\dots,n,\ j = 1,\dots,m \quad (7)$$

Suppose we assume that ω_i's are i.i.d. lognormal and $\epsilon_i(t_j)$'s are i.i.d. normally. TTF is defined as the first-passage time of the true degradation path $\omega_i t$ to the prespecified threshold D. The distribution function of TTF $T_i = D/\omega_i$ is $F_T(t) = P(\omega_j t > D) = 1 - F_{\mathrm{LN}}(D/t)$, where F_{LN} is the lognormal distribution function. The TTF distribution function can be obtained analytically for simple models. For general nonlinear random-effect model in equation (6), Meeker and Escobar ([2], Chapter 21) also describe a simulation-based approach and the use of **bootstrap** methods for **estimation** and inference. Inference under the random-effect model (6) has been discussed in the literature, see, for example, Lu and Meeker [4] and Meeker *et al.* [27]. The common approach is to maximize the likelihood function directly. When the likelihood does not have a closed form, various approximations have been proposed [28] and are implemented in S-plus functions `lme` and `nmle`.

Next, we shall discuss two common nonhomogeneous Lévy processes models. First consider a nonhomogeneous Gaussian process $\{X(t): t \ge 0\}$. Here, $X(t)$ represents the degradation measurement for an individual unit at time t. Let $\Lambda(t)$ be a continuous increasing function. The nonhomogeneous Gaussian process has the following properties: (a) the increments $\Delta X(t) = X(t+\Delta t) - X(t)$ are independent and (b) $\Delta X(t)$ has a **normal distribution** with mean $\Delta\Lambda(t) = \Lambda(t+\Delta t) - \Lambda(t)$ and variance $\sigma^2\Delta\Lambda(t)$. Let $Z(t)$ be the basic Wiener process to have linear drift with $\nu = 1$. Then, we get the nonhomogeneous Gaussian process through time transformation as $Y(t) = Z(\Lambda(t))$. Thus, we can rewrite the model as

$$X(t) = \Lambda(t) + \sigma W(\Lambda(t)) \quad (8)$$

Note that this process $X(t)$ has continuous sample path and independent increments. The level crossing of the cumulative degradation threshold D by the process $X(t)$ can be obtained in terms of the inverse Gaussian (IG) distribution as $T_\Lambda = \Lambda^{-1}(T_{\mathrm{IG}})$, where T_{IG} is $\mathrm{IG}(D, D^2/\sigma^2)$. There is a huge literature on IG distribution and its applications ([29–31] and references therein). Assume that Λ belongs to a known parametric family from the knowledge of the degradation mechanism. For example, the family of power law curves $\Lambda(t;\nu,\beta) = \nu t^\beta$ can be used to model increasing, constant, and decreasing degradation rates. An example with a finite asymptote is $\Lambda(t;\nu,\beta) = C(1-\exp(-\nu t^\beta))$. These two families have been considered by Whitmore and Schenkelberg [17] in the context of accelerated degradation models. It is straightforward to get maximum-likelihood estimates of the parameters based on regular degradation data, since the differences of consecutive degradation measurements are independent normal random variables. In most cases this has to be done numerically. The Fisher information can be used, as usual, to get **confidence intervals** and test hypotheses.

Inference under nonhomogeneous gamma process is similar. The gamma process, $Y(t)$, has an independent increment $\Delta Y(t)$ which has gamma distribution $\mathrm{Gam}(\Delta\Lambda(t),\alpha)$ with mean $\alpha\Delta\Lambda(t)$ and variance $\alpha^2\Delta\Lambda(t)$. Unlike the Wiener process that has a nonmonotone and continuous path, the gamma process has a monotone path and evolves on each time interval by means of an infinity of jumps. The TTF distribution can be obtained directly,

$$\begin{aligned} P(T < t) &= P(Y(t) > d) = 1 - G(d;\Lambda(t),\alpha) \\ &= I\left(\Lambda(t), \frac{d}{\alpha}\right) \end{aligned} \quad (9)$$

where $G(\cdot;\Lambda(t),\alpha)$ is the gamma distribution function with parameters $\Lambda(t)$ and α and $I(t,\alpha)$ is the upper incomplete gamma function, $I(t,\alpha) = \int_\alpha^\infty x^{t-1}\mathrm{e}^{-x}\mathrm{d}x/\Gamma(t)$. The unknown parameters are again estimated by maximizing the likelihood function.

An Illustrative Example

We will use simulated data to illustrate the parametric analysis and compare the estimators from degradation data and TTF data. We use a power law model $X(t) = \nu t^\beta + \sigma W(t^\beta)$ with $\nu = 1$, $\beta = 2$, and $\tau = \sigma^2 = 4$ be the true values of the parameters. Degradation data are simulated for $n = 40$ specimens at 10 time points $t_1 = 1, \dots, t_{10} = 10$. Figure 1(a) is the plot of this simulated data. Failure is defined when the degradation path crosses $D = 100$.

The maximum-likelihood estimators are given as: $[\hat{\nu}, \hat{\beta}, \hat{\tau}] = [1.14, 1.96, 4.28]$ and the estimated standard errors are: 0.040, 0.096, and 0.345, respectively.

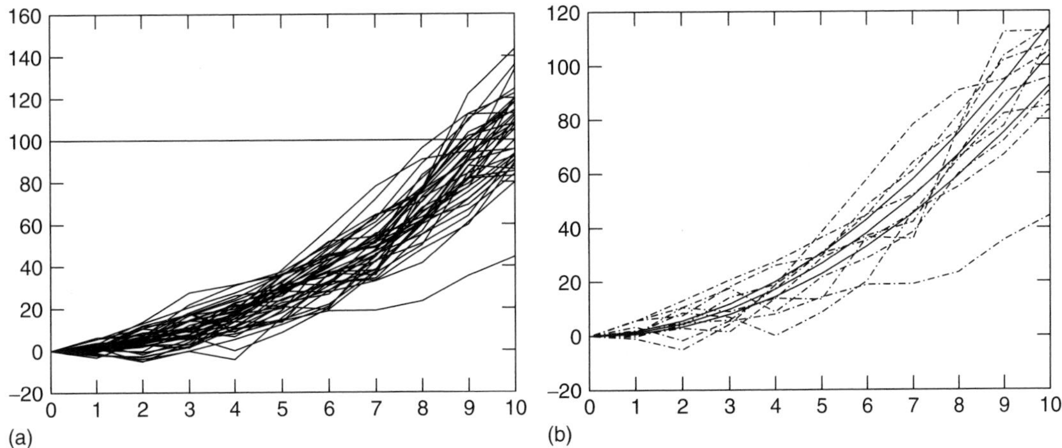

Figure 1 Simulated degradation data and estimated Λ and its pointwise 95% confidence intervals

Figure 1(b) shows the estimated $\Lambda(t)$ superimposed on the data (in dash-dotted line) and the pointwise 95% confidence intervals. As we can see, the estimated $\hat{\Lambda}$ follows the data very well.

For the simulated data in Figure 1, suppose we observed only TTF data corresponding to the first-passage times above the threshold D, that is, the interval censored data. On the basis of this data, the MLEs are given by Table 1: $[\hat{\nu}, \hat{\beta}, \hat{\tau}] = [0.24, 2.62, 6.41]$, which are not good estimates compared to the estimates from degradation data. For the total 40 units, 3 units failed at interval $[t_8, t_9]$, 22 failed at $[t_9, t_{10}]$, and the other 11 failed after t_{10}. We do not have enough information to give good estimates of the parameters. We simulate the data 1000 times and the standard errors of these estimators from TTF are given in Table 1, which are very large. It is not uncommon to see situations where reliability engineers collect degradation data but convert them to "TTF" data and then do a standard Weibull or lognormal analysis with the pseudodata. Such converting is quite inefficient and there is tremendous gain to be realized from degradation data.

Table 1 Maximum-likelihood estimators and their standard errors for degradation data and TTF data

Parameter	ν	β	τ
True	1.00	2.00	4.00
TTF data	0.24 (0.182)	2.62 (8.719)	6.41 (30.498)
Degradation data	1.14 (0.040)	1.96 (0.096)	4.28 (0.345)

Discussion

An interesting extension for current degradation models is when we have information on covariates. For example, in accelerated degradation experiment, the stress level such as temperature is a **covariate**. Boulanger and Escobar [32] studied the design of accelerated degradation tests under the nonlinear random-effect model

$$X_{ij}(t) = \alpha_{ij} F(t, \xi_{ij}) + \epsilon_{ij}(t) \tag{10}$$

where $X_{ij}(t)$ is the degradation measurement at time t for the jth unit under stress condition i. F is a known continuous monotone increasing function of t with $F(0, \xi_{ij}) = 0$ and $F(\infty, \xi_{ij}) = 1$. Hence, α_{ij} measures the asymptote of the degradation level. Also, (α_{ij}, ξ_{ij}) are random coefficients with $E[\log \alpha_{ij}] = A + Bs_i$, where s_i is the ith stress level. If we have data from accelerated degradation tests, we may use model (10) to fit the data and estimate the unknown parameters. We can also use model (10) to develop the optimal degradation planning. The D-optimal plan is to maximize the determinant of Fisher information of the maximum-likelihood estimator of unknown parameters.

For nonhomogeneous Lévy processes models, Bagdonavicius and Nikulin [19] studied nonhomogeneous gamma process with covariates via an **accelerated life model** by replacing $\Lambda(t)$ by $\Lambda(te^{x^{\mathrm{T}}\beta})$, where x is the covariate vector. Lawless and Crowder [20] treated scale parameter α in a gamma process as a function of x to accommodate the covariate. Similar to **Cox** model, we may study the proportional mean model replacing $\Lambda(t)$ by $\Lambda(t)e^{x^T\beta}$ to incorporate the covariate information.

Another interesting issue to study is the optimal maintenance strategies for physical degradation systems. Good degradation models can be used as a tool to optimize maintenance decisions and to minimize the total maintenance cost of the system. Grall *et al.* [33] studied a predictive maintenance structure for a degradation system. They assumed the degradation process is a homogeneous gamma process and provided the maintenance policy to (a) determine whether the system should be replaced preventively or correctively, or whether the system should be left as is; (b) determine the time to the next inspection. There is a need for additional research to develop maintenance policy for other more complicated models.

References

[1] Davies, A. (1998). *Handbook of Condition Monitoring: Techniques and Methodology*, Chapman & Hall, London.

[2] Meeker, W.Q. & Escobar, L.A. (1998). *Statistical Methods for Reliability Data*, John Wiley & Sons, New York.

[3] Hudak Jr, S.J., Saxena, A., Bucci, R.J. & Malcolm, R.C. (1978). Development of standard methods of testing and analyzing fatigue crack growth rate data, Technical Report AFML-TR-78-40, Westinghouse R& D Center, Westinghouse Electric Corporation, Pittsburgh.

[4] Lu, C.J. & Meeker, W. (1993). Using degradation measures to estimate a time-to-failure distribution, *Technometrics* **35**, 161–174.

[5] Elsayed, E.A. & Liao, H.T. (2004). A geometric Brownian motion model for field degradation data, *International Journal of Materials and Product Technology* **20**, 51–72.

[6] Escobar, L.A., Meeker, W., Kugler, D. & Kramer, L. (2004). Accelerated destructive degradation tests: data, models and analysis, *Mathematical and Statistical Methods in Reliability*, B. H.Lindqvist & K. A. Doksum, eds, World Scientific Publishing Company.

[7] Longford, N.T. (1993). *Random Coefficient Models*, Clarendon Press, Oxford.

[8] Lindsey, J.K. (1993). *Models for Repeat Measures*, Clarendon Press, Oxford.

[9] LuValle, M.J., Welsher, T.L. & Svoboda, K. (1988). Acceleration transforms and statistical kinetic models, *Journal of Statistical Physics* **52**, 311–320.

[10] Meeker, W.Q. & LuValle, M.J. (1995). An Accelerated life test model based on reliability kinetics, *Technometrics* **37**, 133–146.

[11] Dowling, N.E. (1993). *Mechanical Behavior of Materials*, Prentice Hall, Englewood Cliffs.

[12] Burnett, J. & Hu, C. (1988). Modeling hot-carrier effects in polysilicon emitter bipolar transistors, *IEEE Transactions on Electron Devices* **35**, 2238–2244.

[13] Signor, A.W., VanLandingham, M.R. & Chin, J.W. (2003). Effects of ultraviolet radiation exposure on vinyl ester resins: characterization of chemical, physical and mechanical damage *Polymer Degradation and Stability* **79**(2), 359–368.

[14] Doksum, K. (1991). Degradation rate models for failure time and survival data, *CWI Quarterly* **4**, 195–203.

[15] Doksum, K. & Hoyland, A. (1992). Models for variable stress accelerated life testing experiments based on wiener process and the inverse Gaussian distribution, *Technometrics* **34**, 74–82.

[16] Whitmore, G.A. (1995). Estimation degradation by a Wiener diffusion process subject to measurement error. *Lifetime Data Analysis* **1**, 307–319.

[17] Whitmore, G.A. & Schenkelberg, F. (1997). Modelling accelerated degradation data using wiener diffusion with a time scale transformation. *Lifetime Data Analysis* **3**, 27–43.

[18] Singpurwalla, N.D. & Youngren, M.A. (1998). Multivariate distributions induced by dynamic environments. *Scandinavian Journal of Statistics* **20**, 251–261.

[19] Bagdonavicius, V. & Nikulin, M. (2000). Estimation in degradation models with explanatory variables, *Lifetime Data Analysis* **7**, 85–103.

[20] Lawless, J. & Crowder, M. (2004). Covariate and random effects in a Gamma process model with application to degradation and failure, *Lifetime Data Analysis* **10**, 213–227.

[21] Zacks, S. (2004). Distributions of failure times associated with non-homogeneous compound poisson damage processes, in *A Festchrift for Herman Rubin, Lecture Notes-Monograph Series Vol. 45*, A. DasGupta, ed, Institute of Mathematical Statistics, Beachwood, pp. 396–407.

[22] Gertsbakh, I.B. (1989). *Statistical Reliability Theory*, Marcel Dekker, New York.

[23] Esary, J.D., Marshall, A.W. & Proschan, F. (1973). Shock models and wear processes, *Annals of Probability* **1**, 627–649.

[24] Bogdanoff, J.L. & Kozin, F. (1985). *Probabilistic Models of Cumulative Damage*, John Wiley & Sons, New York.

[25] Hougaard, P. (1999). Multi-state models: a review, *Lifetime Data Analysis* **5**, 239–246.

[26] Singpurwalla, N.D. (1995). Survival in dynamic environments, *Statistical Science* **1**, 86–103.

[27] Meeker, W.Q., Escobar, L.A. & Lu, C.J. (1998). Accelerated degradation tests: modeling and analysis, *Technometrics* **40**, 89–99.

[28] Pinheiro, J.C. & Bates, D.M. (2000). *Mixed Effects Models in S and S-PLUS*, Springer-Verlag, New York.

[29] Chikkara, R.S. & Folks, J.L. (1989). *The Inverse Gaussian Distribution*, Marcell Dekker, New York.

[30] Sheshadri, V. (1999). *The Inverse Gaussian Distribution*, Springer, New York.

[31] Jorgensen, B. (1982). *Statistical Properties of the Generalized Inverse Gaussian Distribution*, Springer-Verlag, New York.

[32] Boulanger, M. & Escobar, L.A. (1994). Experimental design for a class of degradation tests, *Technometrics* **36**, 260–272.

[33] Grall, A., Dieulle, L., Berenguer, C. & Roussignol, M. (2002). Continuous-time predictive-maintenance scheduling for a deteriorating system, *IEEE Transactions on Reliability* **51**(2), 141–150.

Further Reading

Dirkse, J.P. (1975). An absorption probability for the Ornstein-Uhlenbeck process, *Journal of Applied Probability* **12**, 595–599.

Related Articles

Degradation and Failure; **Degradation Models**; **Degradation Processes**.

Xiao Wang

Plackett–Burman Designs

Construction of Plackett–Burman Designs

Plackett and Burman (PB) [1] provided us with a class of two-level orthogonal designs which now just are referred to as *PB designs*. If we restrict the name to only those designs introduced by Plackett and Burman, they exist for the number of runs $n = 4m$, $m = 1, 2, \ldots, 25$ (except 23 which was given by Baumert *et al.* [2]).

Plackett and Burman used three ways of constructing these designs. One is cyclic generation. For a cyclic generated orthogonal array (*see* **Orthogonal Arrays**), it is possible to write down all the $n - 1$ contrast columns knowing only the sequence of ± 1's in the first row. For instance for the 12-run PB design this string omitting the 1's and only writing down the signs is given by: $+-+---+++-+$. Shifting this row cyclically one place 10 times and adding a row of minus signs produces a design matrix for the 12-run PB design. Adding an initial column of plus signs gives us the 12×12 Hadamard matrix (*see* **Projectivity in Experimental Designs**), shown in Table 1 where the 11 factors are denoted $x_A, x_B, \ldots, x_K$.

Plackett and Burman [1] used this way of constructing orthogonal arrays for $n = 8$, 12, 16, 20, 24, 32, 36, 44, 48, 60, 68, 72, 80, and 84. For $n = 28$, 52, 76, and 100 they used block cycling and for $n = 40$, 56, 64, 88, and 96 their way of obtaining orthogonal arrays was by doubling. The technique of doubling can be explained as follows. Let H_n be a $n \times n$ Hadamard matrix with all entries in the first column equal to $+1$. The matrix $\begin{bmatrix} H_n & H_n \\ H_n & -H_n \end{bmatrix}$ is then a $2n \times 2n$ Hadamard matrix whose last $2n - 1$ columns constitute the orthogonal two-level array.

For $n = 2^k$ $n = 1$, 2, 3, 4, 5, and 6, PB designs coincide with the well known fractional factorial two-level designs (*see* **Fractional Factorial Designs**), which also are called *geometric designs* or sometimes *regular designs*. The rest of the PB designs are called *nongeometric designs* or sometimes *nonregular designs*. The signs for cyclic generation of the 20-, 24-, and 36-run PB designs are:

$$n = 20: ++--++++-+-+-- \\ --++-$$

$$n = 24: +++++-+-++--++ \\ --+-+----$$

$$n = 36: -+-+++---+++++ \\ -+++--+----+-+- \\ ++--+-$$

How to obtain the other PB designs is described in great detail in [1]. PB designs are also available on Neil Sloane's web site: `www.research.att.com/~njas/index.html#TABLES`.

The rest of this article focuses on **projectivity** of PB design, methods for analyzing them, an example to show their efficiency for experimentation and some techniques that can be used to construct designs with efficient run sizes and good projective properties from PB designs.

Projectivity of Plackett–Burman Designs

PB designs were introduced as main-effects plans (*see* **Main Effect Designs**), and nongeometric PB

Table 1 A 12 × 12 Hadamard matrix where the last 11 columns constitute a 12-run PB design

	x_A	x_B	x_C	x_D	x_E	x_F	x_G	x_H	x_I	x_J	x_K
1	1	−1	1	−1	−1	−1	1	1	1	−1	1
1	1	1	−1	1	−1	−1	−1	1	1	1	−1
1	−1	1	1	−1	1	−1	−1	−1	1	1	1
1	1	−1	1	1	−1	1	−1	−1	−1	1	1
1	1	1	−1	1	1	−1	1	−1	−1	−1	1
1	1	1	1	−1	1	1	−1	1	−1	−1	−1
1	−1	1	1	1	−1	1	1	−1	1	−1	−1
1	−1	−1	1	1	1	−1	1	1	−1	1	−1
1	−1	−1	−1	1	1	1	−1	1	1	−1	1
1	1	−1	−1	−1	1	1	1	−1	1	1	−1
1	−1	1	−1	−1	−1	1	1	1	−1	1	1
1	−1	−1	−1	−1	−1	−1	−1	−1	−1	−1	−1

designs have for a long time been considered to be important **screening designs** in situations where only main effects were expected to show up. Later (Lin and Draper [3], Box and Bisgaard [4], Cheng [5], Box and Tyssedal [6], and Samset and Tyssedal [7]) it was discovered that the nongeometric PB designs have quite remarkable projective properties compared to the geometric fractional **factorial** designs. The concept of design projectivity for two-level designs was defined by Box and Tyssedal [6]. A $n \times k$ design with n runs and k factors each at two levels is said to be of projectivity P if the design contains a complete 2^P factorial in every possible subset of P out of the k factors, possibly with some points replicated. Box and Tyssedal [6] describe such designs as (n, k, P) screens.

Box and Tyssedal [6] were able to prove three propositions for orthogonal two-level arrays that can be used for classifying PB designs with respect to being of projectivity $P = 2$ and only two or of projectivity at least $P = 3$ (*see* **Projectivity in Experimental Designs**). Later Samset and Tyssedal [7] discovered by a computer search that PB designs with $n = 68$, 72, 80, and 84 also are $(n, n - 1, 4)$ screens. An orthogonal design with $k = n - 1$ is called *saturated.* A complete listing of the projectivity of saturated PB designs is given in Table 2.

We observe that all saturated geometric fractional factorial designs are of projectivity $P = 2$ and only 2. However for all of them it is possible to remove $n/2 - 1$ columns and obtain a $(n, n/2, 3)$ screen (Box and Tyssedal [6]).

In their computer search Samset and Tyssedal [7] also discovered that all PB designs with projectivity $P = 3$ have the hidden projection properties (Lin and Draper [3] and Wang and Wu [8]) that all main-effects and two-factor interactions for any four factors are estimable under the assumption that higher-order interactions are negligible.

Methods for Analyzing Plackett–Burman Designs

Nongeometric designs have more complex aliasing (*see* **Aliasing in Fractional Designs**), than the geometric designs (Lin and Draper [9]). For instance in a 12-run PB design every **main effect** may be partially aliased with 45 two-factor interactions and a single two-factor interaction will appear in the alias pattern of all main effects not involved with this two-factor interaction. As a result standard methods often used for analyzing unreplicated two-level designs such as normal and half-normal plots (Daniel [10]), or more quantitative methods such as **Lenth**'s method (Lenth [11]) may be of little value since these methods are based on being able to separate active contrasts from contrasts estimating only noise.

Table 2 The projectivity of Plackett–Burman designs

Screens	Number of runs, n
$P = 2$	4, 8, 16, 32, 40, 56, 64, 88, 96
$P = 3$	12, 20, 24, 28, 36, 44, 48, 52, 60, 76, 100
$P = 4$	68, 72, 80, 84

However, several methods are available also for analyzing these designs. These can mainly be classified as effect based or factor based. Effect-based methods aim at identifying significant effects. A model consisting of main effects and interactions (*see* **Interactions**), is assumed to give an adequate approximation of the response. The principle of effect heredity; i.e., an interaction is excluded from the model unless at least one (weak heredity) or both (strong heredity) of the main effects associated with the interaction also are included, is often a precept. Examples of such methods can be found in Hamada and Wu [12] and Chipman *et al.* [13]. An effect-based method that doesn't depend on the heredity principle is given in Tyssedal and Kulahci [14]. Factor-based methods aim at identifying active factors and they are less dependent on model assumptions, heredity included (*see* **Projectivity in Experimental Designs**). Their drawback may be that if the factor effects are small relative to the level of noise, it may be difficult to discriminate among several candidate sets of active factors. Examples of such methods are given in Box and Meyer [15], Kulahci and Box [16] and Tyssedal and Samset [17]. For both methods it is crucial that sparsity of either effects or factors respectively can be assumed. In addition projective properties of the design used are conclusive in order for a factor-based search to work well (*see* **Projectivity in Experimental Designs**).

An Example

Montgomery *et al.* [18] published a paper in quality engineering with the title: *Some Cautions in the Use of Plackett–Burman designs*. They used a 12-run PB designs, see Table 1, and data were simulated from the following model

$$y = 200 + 8x_A + 10x_D + 12x_E - 12x_Ax_D + 9x_Ax_E + \varepsilon \tag{1}$$

where ε is a normally distributed random error term with mean 0 and variance 9. The columns for the factors x_A, x_D, and x_E correspond to the ones in Table 1. We note that the model obeys both the weak and the strong heredity principle. Montgomery *et al.* [18] used normal **probability plots** in combination with the method suggested by Hamada and Wu [11] for the analysis. Due to large two-factor interactions these two methods are not very suitable for analyzing this response and their conclusion was that a sequential strategy, starting with a 2_{III}^{11-7} fractional factorial and adding two additional follow-up runs in this situation outperformed a PB design.

Now suppose we instead chose to use a factor-based method as proposed in Tyssedal and Samset [17] which we will call a projective-based search. This method exploits the projective properties of the design (Lin and Draper [3] and Box and Tyssedal [6]). In particular, the 12-run PB design projects into a 2^1 design replicated six times in any one factor. It projects into a 2^2 design replicated three times in any two factors and it projects into a 2^3 design with four runs replicated in any three factors.

Thereby assuming at most three active factors, we can investigate how well every set of k factors explain the data $1 \le k \le 3$, by examining the fit for replicated runs, calculating

$$\widehat{\sigma}^2 = \frac{1}{\sum_{i=1}^{s}(r_i - 1)} \sum_{i:\text{replicated}} \sum_{j=1}^{r_i} (y_{ij} - \bar{y}_i)^2 \tag{2}$$

where s is the number of runs replicated and r_i is the number of replications for replicated run number i. Sets of factors that give small values of $\widehat{\sigma}^2$ are candidates for being declared active.

We simulated a new set of data from the model given in equation (1). A candidate list showing the five set of factors minimizing $\widehat{\sigma}$ for $k = 1$, 2, and 3 is given in Table 3. Clearly assuming at most three active factors one would in this case end up with x_A, x_D, and x_E after 12 runs and no heredity assumptions need to be imposed on the model.

In cases where several set of factors have approximately the same $\widehat{\sigma}$, backward regression on a model fully expanded with main effects and interactions will sometimes eliminate the ambiguity. For more on this method we refer to Tyssedal and Samset [17] and

Table 3 Results from a projective-based search on the response given in equation (1)

1 active		2 active		3 active	
$\widehat{\sigma}$	Factor	$\widehat{\sigma}$	Factors	$\widehat{\sigma}$	Factors
18.15	x_E	16.78	x_A, x_E	4.26	x_A, x_D, x_E
24.62	x_F	16.80	x_C, x_E	8.05	x_B, x_E, x_I
24.75	x_A	17.40	x_E, x_I	8.29	x_C, x_E, x_J
24.87	x_D	17.66	x_B, x_J	8.34	x_C, x_E, x_H
24.87	x_G	17.98	x_E, x_F	9.25	x_D, x_E, x_F

Tyssedal *et al.* [19] where also a successful application of the 12-run PB design is given. Finally we remark that in this case a search through all possible models consisting of main-effects and two-factor interactions for up to three factors would work equally well. The success of such an effect-based search, however, depends on the underlying model (*see* **Projectivity in Experimental Designs**).

Construction of Designs from PB Designs

The **foldover** of a $n \times k$ two-level design is the design obtained when all mirror images of the original design's runs are added in addition to a column where the first n entries are $+1$ and the last n entries are -1 (*see* **Projectivity in Experimental Designs**). If $\mathbf{X}$ is a nongeometric PB design, Samset and Tyssedal [7] found by computer search that the foldover is always of projectivity $P = 4$ or a $(2n, n+1, 4)$ screen except for $n = 40$, 56, 88, and 96. Designs constructed in this way from PB designs will always have the hidden projection property that all main-effects and two-factor interactions of any five factors can be estimated (Samset and Tyssedal [7]).

A two-level design with more than $n - 1$ factors is called *supersaturated* (*see* **Supersaturated Designs**). Lin [20] introduced a new class of supersaturated designs using half-fractions of orthogonal arrays, i.e., PB designs. For any arbitrary column the $n/2$ rows corresponding to either $+1$ or -1 in that column will constitute a design allowing $n - 2$ factors in $n/2$ runs. If the PB designs used for construction is of projectivity $P = 4$, the supersaturated designs will have projectivity at least $P = 3$, Samset and Tyssedal [7].

PB designs have also been suggested for use in **split-plot** experimentation (*see* **Split-Plot Designs**), both as whole-plot and subplot arrays in order to reduce the number of individual tests (Kulahci and Bisgaard [21] and Tyssedal and Kulahci [14]).

Concluding Remarks

Due to their complex alias pattern practitioners have often been warned against using nongeometric PB designs. This warning seems to be exaggerated and may in some situations act as a barrier against efficient experimentation. Nongeometric PB designs provide us with a lot more design alternatives. If factor sparsity can be assumed their excellent projective properties compared to the geometric fractional factorials allow us to identify active factor spaces [22] under weak assumptions on the underlying model, and their hidden projection properties are very attractive if the model is adequately approximated in terms of main-effects and two-factor interactions.

One of the reasons why these designs are not used as often as they should is the lack of good available software to do the analysis. In the future more effort is needed to overcome this unnecessary obstacle.

References

[1] Plackett, R.L. & Burman, J.P. (1946). The design of optimum multifactorial experiments, *Biometrika* **33**, 305–325.

[2] Baumert, L.D., Golomb, S.W. & Hall, M. (1962). Discovery of a Hadamard matrix of order 92, *Bulletin of the American Mathematical Society* **68**, 237–238.

[3] Lin, D.K.J. & Draper, N.R. (1992). Projection properties of Plackett and Burman designs, *Technometrics* **34**, 423–428.

[4] Box, G.E.P. & Bisgaard, S. (1993). What can you find out from 12 experimental runs, *Quality Engineering* **5**, 663–668.

[5] Cheng, C.S. (1995). Some projection properties of orthogonal arrays, *The Annals of Statistics* **23**, 1223–1233.

[6] Box, G.E.P. & Tyssedal, J. (1996). Projective properties of certain orthogonal arrays, *Biometrika* **83**(4), 950–955.

[7] Tyssedal, J. & Samset, O. (1999). *Two-Level Designs with Good Projection Properties*, Preprint, Statistics No. 12/1999, Norwegian University of Science and Technology, Norway.

[8] Wang, J.C. & Wu, C.F.J. (1995). A hidden projection property of Plackett-Burman and related designs, *Statistica Sinica* **5**, 235–250.

[9] Lin, D.K.J. & Draper, N.R. (1993). Generating alias relationships for two-level Plackett and Burman designs, *Computational Statistics and Data Analysis* **15**, 147–157.

[10] Daniels, C. (1959). Use of half-normal plots in interpreting factorial two-level experiments, *Technometrics* **1**, 311–340.

[11] Lenth, R.V. (1989). Quick and easy analysis of unreplicated factorials, *Technometrics* **31**(4), 469–473.

[12] Hamada, H. & Wu, C.F.J. (1992). Analysis of designed experiments with complex aliasing. *Journal of Quality Technology* **24**(3), 130–137.

[13] Chipman, H., Hamada, H. & Wu, C.F.J. (1997). A Bayesian variable-selection approach for analyzing designed experiments, *Technometrics* **39**, 372–381.

[14] Tyssedal, J. & Kulahci, M. (2005). Analysis of split-plot designs with mirror image pairs as sub-plots. *Quality and Reliability Engineering* **21**(5), 539–551.

[15] Box, G.E.P. & Meyer, R.D. (1993). Finding the active factors in fractionated screening experiments, *Journal of Quality Technology* **25**, 94–105.

[16] Kulahci, M. & Box, G.E.P. (2003). Catalysis of discovery and development in engineering and industry, *Quality Engineering* **15**(3), 509–513.

[17] Tyssedal, J. & Samset, O. (1997). *Analysis of the 12 Run Plackett-Burman Design*, Preprint, Statistics No. 8/1997, Norwegian University of Science and Technology, Norway.

[18] Montgomery, D.C., Borror, C.M. & Stanley, J.D. (1997). Some cautions in the use of Plackett-Burman designs. *Quality Engineering* **10**(2), 371–381.

[19] Tyssedal, J., Grinde, H. & Røstad, C.C. (2006). The use of a 12 run Plackett-Burman design in the injection moulding of a technical plastic component, *Quality and Reliability Engineering* **22**, 651–657.

[20] Lin, D.K.J. (1993). A new class of supersaturated designs, *Technometrics* **35**, 28–31.

[21] Kulahci, M. & Bisgaard, S. (2005). The use of Plackett-Burman designs to construct split-plot designs, *Technometrics* **47**(4), 495–501.

[22] Box, G.E.P. & Tyssedal, J. (2001). Sixteen run designs of high projectivity for screening, *Communications in Statistics: Simulation and Computation* **30**(2), 217–228.

JOHN TYSSEDAL

Poisson Processes

Introduction

Many processes in everyday life that "count" events up to a particular point in time can be accurately described by the so-called Poisson process, named after the French scientist Siméon Poisson (1781–1840; appointed as full professor at the Ecole Polytechnique, Paris, in 1806 as a successor of Fourier). An (ordinary) Poisson process is a special Markov process (*see* **Markov Processes**), in continuous time, in which the only possible jumps are to the next higher state. A Poisson process may also be viewed as a counting process that has particular, desirable, properties. In this article we shall first give a few equivalent definitions of the Poisson process. Subsequently, we describe the relation between the Poisson process and the (negative) exponential distribution; we then show relations between the Poisson process and the uniform distribution, and between the Poisson process and the binomial distribution. Next we discuss "Poisson conservation" results that are extremely useful in, for example, analyzing queuing networks in which customers arrive according to a Poisson process. The next section is devoted to the uniformization principle, a useful principle in studying the transient behavior of **Markov processes**. Finally, a generalization of the Poisson process, *viz.*, the compound Poisson process is considered.

Definitions of the Poisson Process

A counting process $\{C(t), t \geq 0\}$ is a stochastic process that keeps count of the number of events that have occurred up to time t. Obviously, $C(t)$ is nonnegative and integer valued for all $t \geq 0$. Furthermore, $C(t)$ is nondecreasing in t. $C(t) - C(s)$ equals the number of events in the time interval $(s, t]$, $s < t$.

$C(t)$ could, for example, denote the number of arrivals of customers at a railway station in $(0, t]$, or the number of accidents on a particular highway in that time interval, or the number of births of animals in a particular zoo in $(0, t]$, or the number of calls to a telephone call center during that period. A Poisson process is a counting process that has the desirable additional properties that the numbers of events in disjoint intervals are independent (*independent increments*) and that the number of events in any given interval depends only on the length of that interval, and not on its particular position in time (*stationary increments*). In the case of the arrivals at the railway station, the stationarity assumption is clearly not fulfilled; there will be many more arrivals between 5 p.m. and 6 p.m. than between, say, 5 a.m. and 6 a.m. Still, one might wish to study the arrival process at the railway station during the rush hour. Restricting oneself to subsequent working days between 5 p.m. and 6 p.m. does allow one to use the stationary increments assumption.

Similarly, the independent increments assumption may be violated in the zoo example, but it will be a reasonably accurate representation of reality in many cases. From the viewpoint of mathematical tractability, these two properties are extremely important. We refer to Feller [1] for an extensive and lucid discussion of stochastic processes with stationary and independent increments. In the sequel we make an additional assumption, which reduces counting processes with stationary and independent increments to Poisson processes.

Definition 1 A Poisson process $\{N(t), t \geq 0\}$ is a counting process with the following additional properties:

1. $N(0) = 0$.
2. The process has stationary and independent increments.
3. $\mathbb{P}(N(h) = 1) = \lambda h + o(h)$ and $\mathbb{P}(N(h) \geq 2) = o(h)$, $h \downarrow 0$, for some $\lambda > 0$.

Above, the $o(h)$ symbol indicates that the ratio $\dfrac{\mathbb{P}(N(h) \geq 2)}{h}$ tends to zero for $h \downarrow 0$.

An equivalent definition is as follows:

Definition 2 A Poisson process $\{N(t), t \geq 0\}$ is a counting process with the following additional properties:

1. $N(0) = 0$.
2. The process has stationary and independent increments.
3. $\mathbb{P}(N(t) = n) = e^{-\lambda t} \dfrac{(\lambda t)^n}{n!}, \quad n = 0, 1, 2, \ldots$.

Of course, the last property states that the number of events in any interval of length t is Poisson distributed with mean λt. λ is called the *rate* of the Poisson process. It readily follows that the probability generating function of $N(t)$ is given by $\mathbb{E}[z^{N(t)}] = \sum_{n=0}^{\infty} z^n \mathbb{P}(N(t) = n) = e^{-\lambda(1-z)t}$. Differentiation yields $\mathbb{E}[N(t)] = \lambda t$, $\mathbb{E}[N(t)(N(t) - 1)] = (\lambda t)^2$, and hence $\mathrm{Var}(N(t)) = \lambda t$.

The last property of Definition 1 may look awkward at first sight, but is insightful. It states that having two or more events in a small time interval is extremely unlikely, while the probability of a single event is approximately proportional to the length of that small interval.

Remark 1 A Poisson process arises naturally in large populations. Consider such a population of size n, and observe the number of events of a certain type, occurring during a unit time interval. For example, if the probability of an individual calling a call center between, say, 9:00 and 9:01 is p, then the total number of calls during that minute is binomially distributed with parameters n and p. If n becomes large and p gets small such that $np \to \lambda > 0$, the result is a Poisson process with rate λ of calls per minute.

We close this section by giving yet another, equivalent, definition of the Poisson process.

Definition 3 A Poisson process $\{N(t), t \geq 0\}$ is a counting process with the following additional properties:

1. $N(0) = 0$.
2. The only changes in the process are unit jumps upward. The intervals between jumps are independent, exponentially distributed random variables with mean $1/\lambda$, $\lambda > 0$.

Notice that part (3) of Definition 2 indeed implies that $\mathbb{P}(N(t) = 0) = e^{-\lambda t}$, that is, that the interval until the first event is exponentially distributed with mean $1/\lambda$, denoted $\exp(\lambda)$. The exponential distribution has the memoryless property (uniquely among all continuous distributions): If $T \sim \exp(\lambda)$, then $\mathbb{P}(T > s + t | T > s) = \mathbb{P}(T > t)$. It should also be noted that the memoryless property of the exponential distribution and the independence of successive jump intervals indeed imply that increments are stationary and independent.

Each of the above equivalent definitions has features that make it useful for deriving particular properties, as we shall see in the sequel.

Relation between the Poisson Process and the Exponential Distribution

There is an intimate relation between the Poisson process and the exponential distribution, as already revealed by Definition 3. In this section we go somewhat deeper into this relation.

Let T_1 denote the time of the first event of a Poisson process and let T_n denote the time between the $(n-1)$th and the nth event, $n = 2, 3, \ldots$. Let $S_n = \sum_{i=1}^{n} T_i$ denote the instant of the nth event. An important observation is that

$$N(t) \geq n \quad \leftrightarrow \quad S_n \leq t \qquad (1)$$

That is, the number of events during $(0, t]$ is at least n if the time until the nth event is no larger than t. Hence

$$\mathbb{P}(N(t) \geq n) = \mathbb{P}(S_n \leq t), \quad n = 1, 2, \ldots, t \geq 0 \qquad (2)$$

Definition 2 implies that $\mathbb{P}(N(t) \geq n) = \sum_{j=n}^{\infty} \mathrm{e}^{-\lambda t} \frac{(\lambda t)^j}{j!}$, and hence

$$\mathbb{P}(S_n \leq t) = 1 - \sum_{j=0}^{n-1} \mathrm{e}^{-\lambda t} \frac{(\lambda t)^j}{j!}, \quad n = 1, 2, \ldots, (t \geq 0) \qquad (3)$$

This is the well-known result that the sum of n independent, $\exp(\lambda)$ distributed, random variables is $\text{Erlang}(n; \lambda)$ distributed. In particular, $T_1 = S_1$ is $\exp(\lambda)$. The density of this Erlang distribution equals $\lambda \mathrm{e}^{-\lambda t} \frac{(\lambda t)^{n-1}}{(n-1)!}$, $n = 1, 2, \ldots, (t > 0)$, which also follows from the reasoning below:

$$\begin{aligned}
&\mathbb{P}(t < S_n \leq t + h) \\
&\quad = \mathbb{P}(N(t) = n - 1, \text{ one event in } (t, t + h]) \\
&\qquad + \mathrm{o}(h) \\
&\quad = \mathbb{P}(N(t) = n - 1)\mathbb{P}(\text{one event in } (t, t + h]) \\
&\qquad + \mathrm{o}(h) \\
&\quad = \mathrm{e}^{-\lambda t} \frac{(\lambda t)^{n-1}}{(n-1)!} \lambda h + \mathrm{o}(h)
\end{aligned} \qquad (4)$$

Now divide by h and let $h \downarrow 0$ to get the density.

A well-known property of the exponential distribution is that it has a constant failure rate equal to λ: if T is $\exp(\lambda)$ distributed, then

$$\mathbb{P}(T \leq t + h | T \geq t) = \lambda h + \mathrm{o}(h), \quad h \downarrow 0 \qquad (5)$$

implying that $\lim_{h \downarrow 0} \mathbb{P}(T \leq t + h | T \geq t)/h = \lambda$. We have also seen this property in Definition 1 of the Poisson process.

To illustrate the concept of failure rate, consider a probability distribution function $F(\cdot)$ describing the lifetime of a device and denote its density by $f(\cdot)$. Then the failure rate of $F(\cdot)$ is defined as $r(t) = f(t)/[1 - F(t)]$; $r(t)\,\mathrm{d}t$ gives the probability that the device fails during $(t, t + \mathrm{d}t]$, given that it is "alive" at time t.

Relations between the Poisson Process and the Uniform Distribution

In this section we discuss a property of the Poisson process that is often very useful in applications. If exactly one event of a Poisson process has occurred in $(0, t]$, then the time of that occurrence is uniformly distributed on $(0, t)$. The informal explanation is that, because of the stationary and independent increments, each subinterval of equal length in $(0, t)$ has the same probability to contain that event. The formal derivation is

$$\begin{aligned}
&\mathbb{P}(T_1 \leq s | N(t) = 1) \\
&\quad = \frac{\mathbb{P}(\text{one event in } (0, s], \text{ no event in } (s, t])}{\mathbb{P}(N(t) = 1)} \\
&\quad = \frac{\mathbb{P}(N(s) = 1)\mathbb{P}(N(t - s) = 0)}{\mathbb{P}(N(t) = 1)} = \frac{s}{t}, \\
&\qquad 0 \leq s \leq t
\end{aligned} \qquad (6)$$

More generally, the following can be proved for a Poisson process: If $N(t) = n$, then the event times $S_1, \ldots, S_n$ are distributed like the order statistics of n independent random variables that are uniformly distributed on $(0, t)$.

The property that a Poisson arrival "is just as likely to occur in any interval" has proved to be extremely useful in, for example, queuing theory. Firstly, it forms the basis of the well-known PASTA (*Poisson Arrivals See Time Averages*) property. In queuing terms this property states that an outside observer, arriving to a queue according to a Poisson process, sees the system as if it were in steady state. Consider, for example, an $M/G/s$ queue. This is a queuing model in which arrivals occur according to a Poisson process (M), requiring a generally distributed (G) service time at one of s servers. The PASTA property implies for the $M/G/s$ queue that the number of customers seen by an arriving customer has the same distribution as the steady-state number of customers. The PASTA property was first rigorously proved by Wolff [2].

Another well-known application of the property that Poisson arrivals are uniformly distributed over an interval is provided by the $M/G/\infty$ queue [3]. This is an $M/G/s$ queue with an ample ($s = \infty$) number of servers. The $M/G/\infty$ system may be used, for example, to model certain production or transportation systems; the customers might then be pallets moving on a conveyor belt, or cars on a road. Given that n arrivals have occurred in the $M/G/\infty$ system up to t, the above property specifies the distribution of the arrival epochs. One can then easily determine the probability that such a customer, who has arrived in $(0, t]$, is still present at t. Thus one can obtain the distribution of the number of customers $L(t)$ in the $M/G/\infty$ system at t, starting from an empty system at 0 (*cf.* Takács [3, Chapter 3]). It turns out that $L(t)$ has a Poisson distribution with time-dependent rate $\lambda \int_0^t \mathbb{P}(B > y)\,\mathrm{d}y$, B denoting service time of an individual customer. When $t \to \infty$, one gets the so-called stationary distribution of $L(t)$, which is Poissonian with rate $\lambda \mathbb{E} B$. Furthermore, the number of customers served in $(0, t]$ is also Poisson distributed, with rate $\lambda \int_0^t \mathbb{P}(B \le y)\,\mathrm{d}y$, and it is statistically independent of the process $\{L(t), t \ge 0\}$.

Relations between the Poisson Process and the Binomial Distribution

Two important consequences of the stationary and independent increments properties are the following:

1. For any $t > u \ge 0$ and integers $n \ge k$,
$$\mathbb{P}(N(u) = k | N(t) = n) = \binom{n}{k} \left(\frac{u}{t}\right)^k \times \left(1 - \frac{u}{t}\right)^{n-k}, \quad k = 0, 1, \ldots, n \tag{7}$$
That is, given that n events occurred in $(0, t]$, the probability that k of them occurred in $(0, u]$ is given by the binomial distribution with parameters n and "success" probability $p = u/t$.
2. For two independent Poisson processes $N_1(t)$ and $N_2(t)$, having rates λ_1 and λ_2, respectively,
$$\mathbb{P}(N_1(t) = k | N_1(t) + N_2(t) = n) = \binom{n}{k} \times \left(\frac{\lambda_1}{\lambda_1 + \lambda_2}\right)^k \left(\frac{\lambda_2}{\lambda_1 + \lambda_2}\right)^{n-k}, \quad k = 0, 1, \ldots, n \tag{8}$$
This result implies, in particular, that if two such processes "compete" on which one will be the first to occur, the probability that $N_i(t)$ is "quicker" is given by $\dfrac{\lambda_i}{\lambda_1 + \lambda_2}$, $i = 1, 2$.

See again Remark 1 for yet another relation between the Poisson process and the binomial distribution.

Conservation Properties

The Poisson process satisfies some conservation properties that greatly enhance its applicability. In particular, random splitting of a Poisson process results in independent Poisson processes; similarly, merging independent Poisson processes again produces a Poisson process. We now formalize these statements, without proof.

Proposition 1 *Suppose that events occur according to a Poisson process $\{N(t), t \ge 0\}$ of rate λ. Further, suppose that each event is classified as a type i event with probability $p_i \in (0, 1)$, $i = 1, \ldots, K$. Let $N_i(t)$ denote the number of type-i events in $(0, t]$. Then $\{N_1(t), t \ge 0\}, \ldots, \{N_K(t), t \ge 0\}$, are independent Poisson processes with corresponding rates $\lambda p_1, \ldots, \lambda p_K$.*

Indeed, it is easy to see that an arbitrary interval of length h will contain a type-i event with probability $\lambda p_i h + \mathrm{o}(h)$, $h \downarrow 0$. The independence of the various Poisson processes is less obvious. But remember that the knowledge that, say, $N_1(t) = j$ just implies that j out of the t/h intervals of length h contain a type-1 event; it does not really imply anything about the occurrence of events of other types.

Now consider the dual situation, in which there are K independent Poisson processes $\{N_1(t), t \ge 0\}, \ldots, \{N_K(t), t \ge 0\}$ with corresponding rates $\lambda_1, \ldots, \lambda_K$, and in which we count together all events of all these processes.

Proposition 2 *The sum process $\{N(t), t \ge 0\}$ of K independent Poisson processes $\{N_i(t), t \ge 0\}$, $i = 1, \ldots, K$, with $N(t) = \sum_{i=1}^K N_i(t)$, $t \ge 0$, is a Poisson process with rate equal to the sum $\sum_{i=1}^K \lambda_i$ of the rates of the individual processes.*

This proposition can be easily proved using any of the three Definitions 1, 2, or 3. For example, it is straightforward (and intuitive) to verify the following property of Definition 1: $\mathbb{P}(N(h) = 1) = \lambda h + \mathrm{o}(h)$, from the similar property of the individual Poisson processes. The proposition also follows using the multiplication property of probability generating function (PGF) of independent random variables and using the fact that the PGF of $N_i(t)$ equals $\mathrm{e}^{-\lambda_i(1-z)t}$. Alternatively, Definition 3 also works well here; use the fact that the minimum of K independent $\exp(\lambda_i)$ distributed random variables is $\exp\left(\sum_{i=1}^K \lambda_i\right)$ distributed, combined with the memoryless property of the exponential distribution.

Networks of Queues

In 1956 Paul Burke [4] proved the output theorem, a result that plays a crucial role in the study of networks of **queues**. This theorem states the following: Consider an $M/M/s$ queue, *viz.*, a queuing system with s servers, first-come-first-served service operation, a Poisson(λ) arrival process of customers, and independent exp(μ) distributed service times. Assume that $\lambda < s\mu$, implying that the process of number of customers in the system reaches an equilibrium distribution. Then the output process of the $M/M/s$ queue, counting the number of departures in $(0, t]$, is a Poisson process with rate λ. It can be shown that, when $s < \infty$, the above "conservation of Poisson flows" property only holds for the multiserver queue when service times are exponential. However, when $s = \infty$, it holds for general service times.

In 1957 Reich [5] provided an extremely elegant proof of the output theorem. He observed that the process constituted by the number of customers in the $M/M/s$ system is a reversible stochastic process. Viewed backward in time, this process is statistically indistinguishable from the original process. The output process of the reversed-in-time $M/M/s$ queue coincides with the Poisson input process of the original $M/M/s$ queue, and hence is also Poissonian. Using once more the reversibility property leads to the conclusion that the output process of the original $M/M/s$ queue is a Poisson process.

The previously discussed properties of conservation of Poisson flows under merging, splitting, and passing through an $M/M/s$ system allow one to analyze the queue-length process in a network of queues $Q_1, \ldots, Q_K$, all having external Poisson arrival processes, exponential service times, possibly multiple servers, and with Markovian routing of customers from queue to queue. Indeed, the output of an $M/M/s_i$ queue Q_i is Poisson. If a fraction p_{ij} is routed to Q_j, then the resulting flow is again Poisson; and if it merges with another, independent, Poisson process, the resulting process is also Poisson. This is only part of the explanation why such networks of exponential multiserver queues have a joint queue-length distribution that exhibits a product form, the ith term of the product corresponding to an $M/M/s_i$ queue Q_i in isolation. For an excellent exposition of the theory of product-form queuing networks we refer to Kelly [6].

Uniformization

In many application areas it is important to determine the *transient* behavior of a Markov process. Consider a Markov process $\{X(t), t \geq 0\}$, with transition probabilities P_{ij} and visit times to all states being exponentially distributed with the *same* mean $1/\lambda$. Hence the number of transitions in $(0, t]$ is Poisson distributed with mean λt. Denoting the corresponding Poisson process by $\{N(t), t \geq 0\}$ we have the following:

$$\mathbb{P}(X(t) = j | X(0) = i) = \sum_{n=0}^{\infty} P_{ij}^{(n)} \mathrm{e}^{-\lambda t} \frac{(\lambda t)^n}{n!} \tag{9}$$

where $P_{ij}^{(n)}$ denote the n-step transition probabilities between the states (with $P_{(i,i)}^{(0)} = 1$); put differently, these are the n-step transition probabilities of the discrete-time Markov chain underlying the Markov process. The above formula is derived by conditioning on the number of transitions in $(0, t]$.

Equation (9) is computationally advantageous, as the infinite sum can be truncated while the $P_{ij}^{(n)}$ can be evaluated efficiently. An interesting idea, apparently due to Jensen [7], allows extension of this principle to the case of *unequal* mean visit times. Let us assume that the mean visit time to state i is $1/\lambda_i$. Let λ be such that $\lambda \geq \lambda_i$ for all i. Consider a Poisson process with rate λ. Now consider a Markov process that spends exp(λ) in any state i and then jumps with probability $\hat{P}_{ij} = \frac{\lambda_i}{\lambda} P_{ij}$ to j and with probability $\hat{P}_{ii} = 1 - \frac{\lambda_i}{\lambda}$ back to i; one might call this a fictitious transition. Remember that a sum of a geometrically distributed number of independent exponentially distributed random variables is again exponentially distributed; hence the sum of consecutive visit times to state i, before another state is visited, is exp(λ_i). A little thought now shows that this new Markov process is really the same as the original Markov process. The transient behavior of the original Markov process can hence be inferred from that of the new Markov process, by using equation (9), with $P_{ij}^{(n)}$ being replaced by the n-step transition probabilities of the new discrete-time Markov chain that has transition probabilities $\hat{P}_{ij}$.

As an application of this uniformization technique, consider a system that alternates between up and down periods, up (down) periods being exponentially distributed with parameter μ_u (μ_d). We leave it to the reader to verify that by taking $\lambda = \mu_u + \mu_d$ as uniformization parameter, one gets new transition probabilities $\hat{P}_{uu} = \hat{P}_{du} = \frac{\mu_d}{\mu_u + \mu_d}$, and hence

$\hat{P}_{uu}^{(n)} = \hat{P}_{du}^{(n)} = \dfrac{\mu_d}{\mu_u + \mu_d}$, $n = 1, 2, \ldots$, leading to

$$P(X(t) = u | X(0) = u) = \frac{\mu_d}{\mu_u + \mu_d} + \frac{\mu_u}{\mu_u + \mu_d} e^{-(\mu_u + \mu_d)t} \quad (10)$$

and by symmetry,

$$P(X(t) = d | X(0) = d) = \frac{\mu_u}{\mu_u + \mu_d} + \frac{\mu_d}{\mu_u + \mu_d} e^{-(\mu_u + \mu_d)t} \quad (11)$$

There is a large variety of other applications of the uniformization property in stochastic analysis and simulation, that may be found in various textbooks; the present example is taken from Ross [8, Section 5.8].

The Compound Poisson Process

A limitation of the Poisson process is that the jumps are always of unit size. A stochastic process $\{X(t), t \geq 0\}$ is called a *compound Poisson process* if it can be represented by

$$X(t) = \sum_{i=0}^{N(t)} Y_i, \quad t \geq 0 \quad (12)$$

where $\{N(t), t \geq 0\}$ is a Poisson process and Y_1, Y_2, ... are independent, identically distributed random variables that are also independent of $\{N(t), t \geq 0\}$. $X(t)$ could, for example, represent the accumulated workload input into a queuing system in $(0, t]$: Customers arrive according to a Poisson process $\{N(t), t \geq 0\}$, and the ith customer requires a service time of length Y_i. Alternatively, the Poisson arrival process might represent the number of insurance claims in $(0, t]$, while the Y_i represent independent claim sizes. $X(t)$ is then the total amount of monetary claims up to time t.

As another example one can envision a device subject to a series of independent random shocks. If the damage caused by the ith shock is Y_i and the number of shocks in $(0, t]$ is $N(t)$, then the total accumulated damage up to time t is given by $X(t)$.

It is easily seen that, if the rate of the Poisson process equals λ and the Y_i have a common Laplace–Stieltjes transform $\beta(s) = \mathbb{E}[e^{-sY_i}]$, then

$$\mathbb{E}[e^{-sX(t)}] = e^{-\lambda(1-\beta(s))t} \quad (13)$$

Differentiation then readily yields that $\mathbb{E}[X(t)] = \lambda t \mathbb{E} Y_1$ and $\mathrm{Var}[X(t)] = \lambda t \mathbb{E}[Y_1^2]$.

Compound Poisson processes are an important subclass of Lévy processes (*see* **Lévy Processes**). We refer to Bertoin [9] for a detailed account of the theory of Lévy processes.

Epilogue

The Poisson process is a stochastic counting process that arises naturally in daily life situations in which there is a large population of individuals who, more or less independently of each other, have a small probability of contributing to the count in the next small time interval. The (compound) Poisson process has beautiful mathematical properties, which make it a very powerful tool for stochastic modeling and analysis. We refer the interested reader to the monograph of Kingman [10]. Excellent textbook treatments can be found in the books of Ross [8] and Tijms [11].

References

[1] Feller, W. (1966). *An Introduction to Probability Theory and its Applications*, John Wiley & Sons, New York, Vol. II.

[2] Wolff, R.W. (1982). Poisson arrivals see time averages, *Operations Research* **30**, 223–231.

[3] Takács, L. (1962). *Introduction to the Theory of Queues*, Oxford University Press, Oxford.

[4] Burke, P.J. (1956). The output of a queuing system, *Operations Research* **4**, 699–704.

[5] Reich, E. (1957). Waiting times when queues are in tandem, *Annals of Mathematical Statistics* **28**, 768–773.

[6] Kelly, F.P. (1979). *Reversibility and Stochastic Networks*, John Wiley & Sons, New York.

[7] Jensen, A. (1953). Markoff chains as an aid in the study of Markoff processes, *Skandinavsk Aktuarietidskrift* **36**, 87–91.

[8] Ross, S.M. (1983). *Stochastic Processes*, John Wiley & Sons, New York.

[9] Bertoin, J. (1996). *Lévy Processes*, Cambridge University Press.

[10] Kingman, J.F.C. (1993). *Poisson Processes*, Oxford Studies in Probability, 3, The Clarendon Press, Oxford University Press, New York.

[11] Tijms, H.C. (1994). *Stochastic Models. An Algorithmic Approach*, John Wiley & Sons, Chichester.

Related Articles

Intensity Function; **Intensity Functions for Nonhomogeneous Poisson Processes**.

ONNO J. BOXMA AND URI YECHIALI

Policy Deployment for Performance Improvement

Introduction

Organizations that win in the long term "plan their work and work their plan". Realization of strategy – the long-term vision of an organization is achieved by a disciplined approach to setting direction and then executing that direction through the effective use of the organization's resources. In Japan this method is called *policy deployment* – which has also been called the *secret weapon* in the Japanese management system. Policy deployment is the strategic direction-setting methodology used to identify business goals, as well as to formulate and deploy major change management projects throughout an organization. It describes how strategy cascades from vision to execution in the workplace through a collaborative engagement process that also includes implementation details like performance, self-assessment, and management review. It describes a systematic relationship between strategy development and the organization's daily imperative to measure and manage its operations using a system that aligns the actions of its people to produce collaboration among the various business functions and processes to produce requirements for customers.

Historical Development of Policy Deployment

What were the circumstances under which policy deployment originated? Interest in strategy, market focus, and long-term, balanced planning were generated by the visits of Dr. Peter F. Drucker to Japan in the early 1950s [1]. As a result of his teaching, "policy and planning" was added to the Deming Prize checklist in 1958. Bridgestone Tire Corporation first used *hoshin kanri*, the Japanese term for policy deployment, in 1965. In 1976, Dr. Yoji Akao and Dr. Shigeru Mizuno were involved in the implementation of *hoshin kanri* in Yokagawa Hewlett–Packard (YHP) as part of its pursuit of the Deming Prize. By 1982, YHP had used *hoshin* to manage a strategic change that moved it from the least profitable Hewlett–Packard (HP) division to the most profitable. In 1985, this *hoshin* methodology was introduced to the rest of HP as a lesson learned from the YHP Deming Prize journey. From HP this methodology was transferred to other leading companies including Proctor & Gamble, Ford, Xerox, and Florida Power & Light, involving several advisors and councilors of the Union of Japanese Scientists and Engineers (JUSE). The work of the GOAL/QPC research committee also extended the managerial technology of policy deployment and was a key ingredient in introducing policy deployment across North America and through multinational companies, to the world [2].

Foundations of Policy Deployment

Mizuno defined *hoshin kanri* as the process for "deploying and sharing the direction, goals, and approaches of corporate management from top management to employees, and for each unit of the organization to conduct work according to the plan". *Hoshin kanri* is a comprehensive, closed-loop management planning, objectives deployment, and operational review process that coordinates activities to achieve desired strategic objectives. The word *hoshin* refers to the long-range strategic direction that anticipates competitive developments while the word *kanri* refers to a control system for managing the process [2].

This management system does not encourage random business improvement, but focuses the organization on projects that move it toward its strategic direction. It builds strength from its relationship with the daily management system that is focused on *kaizen* – continuous improvement. *Hoshin* seeks breakthrough improvement in business processes by allocating strategic business resources (both financial and human resources) to projects that balance short-term business performance to sustain improvement toward its long-term objectives. In a policy deployment management system this two-pronged approach integrates operational excellence in the daily management system with architectural design of its long-term future. This planning process contains two objectives: *hoshin* – the long-range planning objectives for strategic change that allow an organization to achieve its vision and *nichijo kanri* – the daily, routine management control system (or daily management system) that translates the strategic objectives into the work that must be accomplished for an organization to fulfill its mission. The blending of these two elements into a consensus management process to achieve a shared purpose is the key to success for the policy deployment process. In a *hoshin* planning system, strategy is observed through the persistence of its vision – how it is deployed across cycles of learning in project improvement projects that move the performance of the organization's daily management system toward its direction of desired progress.

The fundamental premise in policy deployment is that the best way to obtain desired results for an organization is for all employees to understand the long-range direction and participate in designing the practical steps to achieve the results. This form of participative management evolved and was influenced by the Japanese refinement of Drucker's management by objectives (MBO) through the birth and growth of the quality circle movement. To manage their workplace effectively, workers must have measures of their processes and monitor these measures to assure that they are contributing to continuous improvement as well as closing the gap toward the strategic targets. Policy deployment became the tool that Japanese business leaders used to engage their workers in a strategic dialog and align their work with the consensus strategic direction of their firm. When HP first implemented *hoshin* planning, many of its business leaders explained how it worked by calling it "turbo-MBO".

Policy deployment links breakthrough projects that deliver the long-term strategic direction to achieve sustainable business strength while, at the same time, delivering an operating plan to achieve short-term performance. The methods of policy deployment anticipate long-term requirements by focusing on annual plans and actions that must be met each year to accumulate into long-term strength. Policy deployment processes begin when senior management identifies the key issues or statements of vulnerability, where improvement will have its greatest impact on business performance. This perspective is an essential starting point for policy deployment and allows management to focus a strategic dialog to solicit ideas from frontline workers regarding the opportunities for improvement that exist in their workplace. As Dr. Noriaki Kano of the Tokyo Science University points out, without such direction "the ship would be rudderless". The communication of the focus area or theme for improvement provides a cohesive direction to assure alignment of the entire organization and to build consensus between the management team and the workers on business priorities.

Hoshin helps to create the type of organization that William McKnight, former CEO of 3M, expressed as his desire: "an *organization* that would continually self-mutate from within impelled forward by employees exercising their individual initiative" [3]. In this type of organization, creativity is managed through a combination of self-initiated improvement projects, which engage teams to combine individual capabilities for achieving strategic projects that make a difference to the whole organization. How does this change management process work at the frontline where these strategic *hoshin* projects interface with the organization's routine work processes?

Perhaps the reason *hoshin kanri* took hold within HP is that this methodology demonstrated its ability to translate qualitative, directional, or strategic goals of an organization into quantitative, achievable actions that focus on fundamental business priorities achieving significant competitive breakthroughs – in short, the leaders at HP recognized *hoshin kanri* as MBO done right![a] The extension of this methodology beyond HP to other leading firms came about because HP was recognized as possessing the best practice for linking its strategic direction with its operational management systems.

Daily Management System

Policy deployment uses a systems approach to manage organization-wide improvement of key business processes. It combines the efforts of focused teams on breakthrough projects with the efforts of intact work groups who continuously improve the performance of their work processes. All strategic change occurs in projects that accomplish those activities that are necessary to achieve the stretch business objectives that assure sustained success for the organization. Policy deployment systematically plans ways to link strategic direction with those business fundamentals that are required to run the business routine successfully. Policy deployment allows management to commission change projects for implementation and to review the implementation of a system of projects and thereby to manage change. It seeks opportunities disguised as problems – and elevates those high-priority changes required for the improvement of the daily management system and work processes into business change objectives that are accomplished as *hoshin* projects.

Routine operation of the daily management system requires a foundation in management by fact, or the combination of business measurement with statistical analysis and graphical reports that illustrates the current state of performance, historical trends, and is able to extrapolate trends through statistical inference. A key ingredient is the business fundamentals measurement system that includes the set of basic process results measures that are monitored at control points within the organization where the flow of its throughput can be managed based on the requirements that are driven (using a pull system) by the customer requirements. This measurement system should include both predictive and diagnostic capabilities.

HP embedded its daily management system into a work process measurement system that they initially called *business fundamentals tables*. Other companies refer to the set of measures that translate strategic goals into operational measures of work (in units such as quality, cost, and time) as either a customer dashboard or a balanced scorecard. These systems are used to monitor the daily operations of a business and to report to the management on the progress in the process for developing and delivering value to customers. This measurement process must operate in close-to-real time to permit process owners to take appropriate corrective action that will limit the "escapes" of defects to external customers by catching and correcting errors before they are released from the organization, and finding and fixing mistakes as they occur at the source. Such measures of core work processes are called *business fundamentals* because they must operate under control for the business to achieve its fundamental performance objectives.

These measures must also be captured at the point where control may be exercised by process operators to adjust the real-time operation of the process and assure meeting the customer's performance requirements on a continuing basis. As the great Dutch architect Miles van der Rohe once observed "God is in the details" and it is in these details that business must effectively operate. A daily management system defines the details of an organization's operations. Thus, the measurement and the point at which it is both monitored and controlled are parts of the daily management system and at this point they must be related to their contribution to deliver organizational performance objectives. In the language of Six Sigma, a "Business Y" (such as "profitable growth") that must be achieved is the strategic goal, while a "Process X" (such as "creditworthy customers") delivers this result by process-level performance through the **transfer function** $Y = f(X)$ and the "X" is therefore a fundamental business measure of the organization's daily management system.

Collins and Porras describe how leading companies stimulate improvement by *evolutionary progress*. "Evolutionary" describes progress that resembles the organic growth or the way that species evolve and adapt to their natural environments. Evolutionary progress differs from the big hairy audacious goals (BHAG) of strategic progress in two ways. First, whereas BHAG progress involves clear and unambiguous goals ("We are going to climb *that* mountain"), evolutionary progress involves ambiguity ("By trying lots of different approaches, we're bound to stumble onto something that works; we just don't know ahead of time what it will be"). Second, whereas a BHAG involves bold discontinuous leaps, evolutionary progress begins with small *incremental* steps or mutations, often in the form of quickly seizing unexpected opportunities that eventually grow into major – and often unanticipated – strategic shifts. Evolutionary progress represents a means to take advantage of unplanned opportunities for improvement that are observed at the point of application – the daily management system. The accumulation of many evolutionary improvements results in what looks like part of a brilliant overall strategic plan [4]. Both types of change are needed to stimulate the organic growth of an enterprise. If an organization can make improvements in the "right X's" then it will improve its performance on the critical Business Y.

Choosing Strategic Direction

Hoshin kanri begins with a process for choosing strategic change. In most firms, this process is called strategic planning. Proposed changes are usually identified to either increase the competitive performance of a process or to create the competitive "attractiveness" of a product to its targeted market. Strategic choice in both dimensions is essential to have a globally competitive organization. As pointed out by Dr. Hiroshi Osada, many Japanese companies have not paid enough attention to the critical aspects of strategy formulation as they have to the deployment of their strategy using *hoshin kanri*. This leads to an error of effectively deploying a poorly chosen strategy. When management confuses the mechanistic aspects of policy deployment with its own crucial obligation to establish strategic direction, then they create a grievous error that is truly an abrogation of leadership. An organization may effectively deploy management's strategic choice, however, if the choice of strategy is not carefully directed it will not lead to improvement [5].

Osada notes that in traditional Japanese management systems, ideas flow "bottom-up" from the workplace to the management. However, in policy deployment, there is also a top-down approach to planning change. As Osada comments, policy deployment "is a simple tool for effectively deploying a given policy, and has therefore been broadly adopted by Japanese industry, it does not aid in policy formulation. Even when employing management by policy (MBP), therefore, the question of whether or not a given policy is appropriate will remain. It is thus possible for an inappropriate policy to be effectively deployed – to a counterproductive effect" [6]. Strategic direction must be determined by discovering the alternatives for achieving the organization's vision and choosing the direction that will accomplish it. This direction is modified through the power of the incremental change to act as the "rudder" that steers the ship by making "finely tuned" changes to the general direction of the strategy.

What are the essential ingredients in choosing strategic direction? This process of management integrates strategic planning, change management, and project management with the performance management methods that focus on delivering results. Some specific work activities in designing and implementing a policy deployment system include the following:

- identifying critical business assumptions and areas of vulnerability,
- identifying specific opportunities for improvement,
- establishing business objectives to address the most imperative issues,
- setting performance improvement goals for the organization,
- developing change management strategies for addressing business objectives,
- defining goals project charters for implementing each change strategy,
- creating operational definitions of performance measures for key business processes, and
- defining business fundamental measures for all subprocesses to the working level.

Once a strategy has been set, the next challenge is to align the strategic direction with the work that is being performed in the daily management system.

Aligning Operations with Strategies

A critical challenge for organizations is to align their strategic direction with their daily work systems so that they work in concert to achieve the desired state. Alignment must include linking cultural practices, strategies, tactics, organization systems, structure, pay and incentive systems, building layout, accounting systems, job design, and measurement systems – *everything*. In short, alignment means that all elements of the company work together much like an orchestra leader integrates the various instruments to conduct a coordinated symphony. Organizations that apply the most mature aspects of policy deployment do not put in place any random mechanisms or processes, but they make careful, reasoned strategic choices that reinforce each other and achieve synergy. These organizations will "obliterate misalignments". If you evaluate your company's systems, you can probably identify some specific items that have misaligned with its vision and impede its progress. These "inappropriate" practices have been maintained over time and have not been abandoned when they no longer align with the organizational purpose. "Does the incentive system reward behaviors inconsistent with your core values? Does the organization's structure get in the way of progress? Do goals and strategies drive the company away from its basic purpose? Do corporate policies inhibit change and improvement? Does the office and building layout stifle progress? Attaining alignment is not just a process of adding new things; it is also a never-ending process of identifying and doggedly correcting misalignments that push a company away from its core ideology or impede progress" [7].

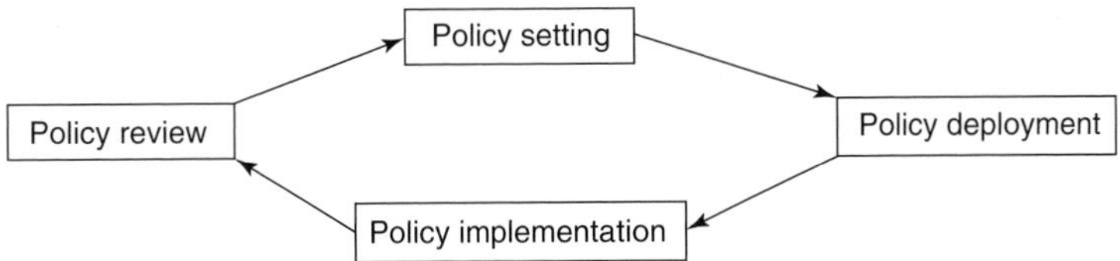

Figure 1 The system of hoshin kanri of managing policy

The System for Policy Management

Policy deployment combines both the "target and the means to achieve the target" into a consensus-generating, management decision-making process. Improvement targets are described using four elements: a performance measure to be improved, direction and rate of improvement desired, targeted improvement magnitude, and timeframe for achievement of the target. A means to achieve the target describes a set of specific actions that will be taken to deliver the desired results. These means may differ across the organization, based upon the initial, local management self-assessment, or "current state analysis" that is conducted to assess the business area's starting point for change and determine the magnitude and nature of the performance gap to be closed by the change management or *hoshin* project to deliver the desired state.

Peter Drucker once commented "for full effectiveness all the work needs to be integrated into a unified *program for performance*" [8]. The program for performance is designed by the top management team to provide a specific, effective course of action to achieve its desired results. To achieve these results, all the dimensions of the business must be consistent with each other. This is the job of the policy deployment system.

This system for managing policy consists of *kanri* or control mechanisms that deploy business policy through four essential steps in order to execute management's program for the business direction using a systematic sequence of steps that achieve *hoshin* project objectives within the constraints of assigned resources. These four steps define policy setting (or establishment of *hoshin* projects), deployment (or propagation of these projects throughout the organization), implementation (or integration of the results of change into the daily management system), and review (or assessment of the results achieved from the process) (see Figure 1). These four steps will be described in the next four sections of this chapter.

Policy Setting

Policy setting is a top-down "catchball" whereby management conducts "strategic dialog" with employees to collect ideas and opinions about chronic major problems and their aspirations regarding the business future, and then processes this information in conjunction with environmental data analysis and scenario analysis to formulate the annual business change objectives (which some organization call their *hoshin* projects): strategic change projects (identified by both targets to be achieved and means for achievement). In this phase, organizations recognize the most critical projects that must be accomplished to eliminate vulnerabilities or capture the benefits from potential change initiatives or newly emerging improvement opportunities. For organizations to succeed they must undergo a rigorous analysis of both their fundamental work processes to identify business imperatives (things that must change) and their current strategic direction to determine potential business vulnerabilities from competitive, economic, or technological changes.

This system structures application of continuous improvement into strategic and operational dimensions. David Packard incessantly used the term *continuous improvement* beginning in the early 1940s – it is not a new term – but as Collins and Porras describe its adaptation in leading companies that have adapted to change, they observe that it is an essential structural ingredient in those companies that have been *Built to Last*: "Visionary companies apply the concept of self-improvement in a much broader sense than just process improvement. It means long-term investments for the future, it means investment in the development of employees; it means adoption of new ideas and technologies. In short, it means doing *everything* possible to make the company stronger tomorrow than it is today" [9].

Most organizations operate on three levels of managerial thinking: enterprise thinking assures their long-term viability; strategic thinking focuses on products, markets, and customers; and operational thinking focuses on the daily work that delivers the organization's results. Strategies align to these three areas of focus: "Management strategies can be classified into three types – corporate strategy, business strategy, as well as functional and cross-functional strategy – depending on the level of the corporate organization to which they apply. The corporate strategy, which delineates the fundamental direction of the whole company, is certainly very important for realizing a management vision; but it would be no exaggeration to say that the success or failure of the corporate strategy is determined by the particular business strategies, since it is through these business strategies that the aims of the corporate strategy are actually implemented" [10]. The portfolio of specific strategies that any organization pursues must be managed to deliver the risk – benefit performance desired by the organization in order to achieve its desired results – whether for breakthrough results or just for incremental improvement of a specific business area.

How is this approach to planning conducted? The corporate planning process should deliver increasing business brand value to balance financial risk and reward. This planning process consists of three elements: strategic planning, business planning, and functional planning that must all fit together in an integrated planning system. The strategic planning process is conducted at the enterprise level of the business thinking to identify which business opportunities to exploit and how to sustain the ability of the organization to meet or exceed its annual performance objectives. The business planning process is conducted at the business level of thinking and its objectives are to drive market share to accelerate financial payback, build customer loyalty, and decrease market risk. At the operational level of thinking, the functional planning process improves all process performance to reduce cost, cycle time, and defects while enhancing responsiveness to customers and delivering customer satisfaction.

A business strategy should deliver "visionary" performance: strategy is the persistence of a vision over the long term – and it requires both vision setting and vision deployment to assure alignment in strategic direction. Policy deployment provides direction to guide these plans and assures that the organization moves in a coherent direction. The more robust a plan for required change, the more effective the organization's ability is to accomplish this plan. Robustness is a function of the management team's ability to see beyond its operating horizon and understand what may occur in its planning horizon that requires its focus and attention today.

What is a planning horizon? It is the distance that an organization "sees" into the future to study and understand the potential impacts of events on its policies and prepare it for evolving situations that may impact its current or future performance. Organizations tend to have four distinct planning horizons.

- **Business foresight**
  Managing for the long term to assure that the organization is not surprised by changes in the assumptions that it has made in the design of its business model and product line strategy (focusing on a 3–10-year business outlook).
- **Strategic direction**
  Managing for the intermediate-term changes in technology and competitive dimensions to assure that vulnerabilities in the business model are not exploited and to bridge the chasm that may exist between product line introductions (focusing on the next 3–5 years of business operation, depending on the degree of change that is anticipated in the business environment).
- **Business plans**
  Managing the short-term fluctuations of the market – a planning horizon that delivers against short-term fluctuations in demand or supply (focus on quarterly and annual operating plans).
- **Business controls**
  Managing the current state of a business – a planning horizon that delivers today's performance and assures rapid responses for corrective actions required to sustain advertised service levels (focus on the daily–weekly–monthly–quarterly operating plans to deliver the targeted results in the annual plan).

How is strategic policy formulated? Strategic direction is best established using cross-functional dialog to capture all the inputs of the organization and to build a common direction, based on the consensus of how to boost organizational strengths and overcome organizational weaknesses in the face of critical business threats to capture the most important market opportunities. Most organizations have just two kinds of strategic decisions: those that may be executed within the areas of their direct oversight of top management (e.g., personnel decisions, budgeting, merger, capital budgeting, etc.) and those that require cross-organizational collaboration for implementation. These cross-functional projects require special attention and project management to realize the objectives of the change initiative. Such change strategies that require mutual consent and collaboration are ideal for a policy deployment system. In addition to planned continuous improvement that is a result of problem solving, continuous improvement may also result from process management, whenever a process is consciously enhanced over time.

Osada encourages strategic engagement of all employees through the following ways:

- recognizing product life stage (product life cycle analysis)
- analyzing business and product position objectively (positioning analysis)
- analyzing competitiveness (competitive analysis)
- perceiving strengths and weaknesses of products (SWOT analysis)
- forecasting future competitiveness using **time series** data (time series analysis)
- maintaining transparency through visualization (visual method); involving all employees [11].

Osada encourages using seven strategic tools (the S-7 tools) in policy setting:

1. environment analysis
2. product analysis
3. market analysis
4. product–market analysis
5. product portfolio analysis (product portfolio management, PPM)
6. strategic elements analysis
7. resource allocation analysis [12].

But, all these tools and methods are employed as staff-directed preconditions for strategic planning in the planning approaches of many Western organizations where they link the three planning systems (strategic, business, and functional) with the business and environmental assessment analyses that precede strategic decision-making. While this linkage may be a bit weaker in Japanese firms, in Western firms, such issues do not appear to be a critical shortfall. However, without complete integration of these planning processes it is difficult to obtain the degree of effectiveness in deployment of shared resources that permits breakthrough achievement to occur. What is breakthrough achievement in management? Breakthroughs represent at least an order-of-magnitude change in performance that is accomplished over a relatively short period of time. Breakthrough is achieved by developing a capability to choose the right objectives for planned change and then aggressively executing these objectives. This requires two factors: identification of what to change and the timing of when to change it. The job of top management is to decide which lever of change must be pulled in order to accomplish the desired result.

Other success factors that are significant in achieving breakthrough plans include the right action to achieve the desired state of change. Right does not mean comprehensive or exhaustive, but it implies a budgeting of energy that focuses an organization on catalytic actions that stimulate organizational response in the desired direction – applying a limited capital budget and the best people to accomplish those important objectives that they have been personally developed to concentrate on. A second success factor in the management of breakthrough projects is the capacity of an organization to convert objectives into results. Excellence comes from execution of plans, not just from planning. To execute, an organization must be mobilized to consolidate their energy and coordinate their actions to achieve shared objectives for the common good of the organization as an organism – a living entity that requires appropriate nourishment and execution of all its bodily functions. A third critical success factor for breakthrough management is the capacity of an organization to integrate specific improvements into standard operating practices that are consolidated across the entire organization for maximum leverage effect. This success factor is based on the existence of a business control management system that holds the gains from improvement projects and is able to assure that performance degradation does not occur – people do not slip back into their "old way of doing things", but that they embrace the new methods as their routine way of working. This success factor is addressed in the policy deployment and policy implementation steps.

It should be noted that breakthrough change projects can only be accomplished if the daily work processes are operating under reasonable control. If a business is not operating under control, then the "breakthrough" activity should focus on bringing itself under control before making a significant investment in strategic change. When an organization's daily management system is operating under control, then more time is available for strategic change because management is not "fighting fires" or "expediting execution" of routine work. This requires that the management start this journey by evaluating its readiness to make strategic changes in its operating system.

To maximize the effectiveness of policy deployment, the strategic direction setting and implementation processes should identify the highest priority business process improvement projects and the requirements for accomplishing them. For instance, only some of these projects will require the degree of diagnostic sophistication that is available from a Six Sigma Black Belt while others will require intensive capital investment or software development to accomplish their objectives. Only when management chooses change projects that improve the infrastructure of its business processes – work processes whose performance contributes to the common cause variation of the business performance – can the most significant gains be realized from a comprehensive business improvement effort. How does management achieve this focus? The short answer is that management must put in place the methods to "recognize" their priority business improvement needs by linking their choice of change management projects to the strategic direction of their business so that the portfolio of change management projects drives the policy changes that are necessary to achieve the long-term performance objectives of the organization. Some of the questions that must be addressed during policy setting include the following:

- What is our business and what results do we expect to achieve? How will we know that we have achieved these results? Is this the best we could do?
- What are our assumptions about society, the economy, market and customers, technology, and knowledge? Are they still valid?
- Has anything happened that would change the dynamics of our industry or markets?
- What would change mean for our business position?
- Are there any opportunities that we should anticipate and capitalize upon to our long-term advantage?
- Where should we choose to excel? Can we take action on this opportunity?

Policy Deployment

While policy setting sets organizational policy, the policy deployment step deploys policy change to the organization by changing the way that effective work is accomplished in the daily management system.

How are the *hoshin* or strategic change objectives deployed? Policy deployment is the heart of this policy management system and has received much attention because of the "catchball" approach that aligns the objectives of the organization and then balances work by resource leveling and prioritization of improvement activities. An implementation plan for a change project is a living document – it acts like a compass to guide an organization while allowing employees to take ownership by participating in choices that define the reasoning behind the project as well as the steps in the project's execution. Change projects can identify two types of improvement effort – either by a breakthrough project or by a continuous improvement project – to change the way work process activities function. Breakthrough activities are strategic change projects that make a significant shift in the organization's capability to perform routine operational work processes or deliver products (either goods or services) to the marketplace. Work process continuous improvement (*kaizen*) activities are part of a daily management system that defines how work is accomplished. The *kaizen* change activities are the responsibility of all work process owners. This planning approach focuses on guidelines for major improvement projects (while small incremental or continuous improvements are made through the regular course of continuous improvement of the daily routine work). The distinction between these activities is twofold: first, more of the organization's resources are focused on breakthrough projects, and second, accomplishment of a breakthrough project usually occurs over a multiple-year period (or as a series of coordinated improvement projects). One important management consideration in choosing breakthrough projects is that the combination of all the annual breakthrough projects (also called the *portfolio of change projects*) will define the steps that an organization chooses for accomplishing strategic change in the range of its midterm planning horizon (1–3 years).

To achieve "saturation" of policy (which consists of both targets and the means for their achievement) in the organization – or complete deployment of change projects within the whole organization that is affected by the defined policy change – and assure collaboration of all the affected work groups, the objectives cascade of an action plan for a particular improvement project must involve not only functional deployment of policy but also engage its cross-functional aspects. It is across the functional seams of an organization where most significant difficulties are encountered and these boundaries represent focus areas for management to assure continuous collaboration in the execution of change projects and consensus among the various functional organizations that engage all the decision-making managers in the areas where the change will have a direct effect. To understand the difficulty that the boundary condition dynamics have, consider what happens as change is managed when organizations shift work activities from internal to external units (e.g., from internal manufacturing to an external contract manufacture). At such boundary conditions, conflicting objectives and political issues of the organizations often can interfere with performance improvement work and it is the job of the management team to eliminate any such barriers to the success of their project team. This also happens within organizations at their functional boundaries.

Catchball is the process that is used to build a consensus through dialog about the targets and means to achieve the change. This process is data driven and uses tools that permit management by facts. Catchball links annual change projects to midrange and long-term plans – deployment prior to annual fiscal year commencement, incorporated into target setting and annual employee objectives cascade; coordinated both vertically within functions and horizontally across processes and negotiated across the processes to allocate resources (competence, funding, and equipment) to achieve the shared and agreed objectives. The catchball process includes four activities:

- building alignment through linked cascade of means
- setting business performance targets and objectives
- cascading business objectives to the workplace (*gemba*)
- achieving alignment of improvement and effective resource allocation.

Policy deployment is a structured, systematic, and standardized process and it has an ability to empower organizations to achieve strategic change. Policy deployment includes a few key elements that assure that an organization is properly and fully engaged in change projects:

- *Policy* – a general rule or operating principle that describes a management-approved process to approach a business condition or situation based on how it chooses to control its work and manage risk. Once the right policy has been determined, then the organization can handle similar situations with a pragmatic response by adapting its policies to the concrete situation that it faces. Truly unique business situations that run counter to the critical business assumptions require the full attention of the senior management team to evaluate how these situations challenge the boundary conditions of the business model and threaten its policies of operations with change that is imposed from externalities. Policies consist of targets and means.
- *Target* – the measurable results that are to be achieved within a specific timeframe for performance. Targets have checkpoints.
- *Checkpoints* – a measurement point that is used to evaluate an intermediate state in the policy deployment process to demonstrate that progress is being made. The data collected at a checkpoint can be reported to management in interim project status reports. The checkpoint of one process is the control point of the next process – the checkpoints and control points work together to formulate a "waterfall" that cascades across the implementation plan flow and is part of the business measurement system.
- *Check items* – check items and process or project variables that are evaluated to enable organizations to understand the causes that contribute to the outcome of a particular policy.
- *Means* – the sequence of actions that an organization will take to implement a policy or choice of the management team that is an outcome of the strategic direction-setting process. Means have control points.

- *Control points* – a point in a sequence of work activities where corrective action may be taken or countermeasures are put in place to resolve a concern or issue that has been identified at a checkpoint.
- *Control items* – control items verify whether results agree with the established goals – does the work demonstrate progress in accomplishments that will enable the final achievement of targets?
- *Deployment* – the process of engaging the entire organization in an appropriate participation in the strategic direction both vertically (within functional areas) and horizontally (across the functional areas) by creating shared ownership of the implementation actions. The entire system of deployment is "connected" from the long-term vision to the daily management activities. The plans are progressively more detailed as they are refined in deployment from the top levels of the organization to the frontline employees and teams. The plan is deployed through an organization by negotiating the means between management layers (levels) as well as across the functional departments. Targets are not negotiated.
- *Catchball* – the joint analysis process that encourages a strategic dialog between levels of organizational deployment is called a *catchball.* The means at one level become the desired outcomes of the subsequent level. This cascading of targets and means establishes the linkage and alignment of objectives across organizational levels. Mutual discussion between the parties – a two-way communication that is both top-down in general direction and bottom-up in adaptation to the workplace using the existing hierarchical management structure and matrix process structure to engage all parts of the organization in the dialog. This dialog is a negotiation process (see *nemawashi* and *sureawashi* below) that arrives at a collective wisdom to develop and refine the implementation plan.
- *Nemawashi* – negotiation – prior consultation to achieve consensus; careful preparation of the roots of a plant for transplantation; and seeking to achieve *wa* or harmony, consensus, and absence of conflict.
- *Sureawashi* – the sharpening of a sword requires four ingredients: a blade, a template for the angle to be produced, oil to ease the process, and a sharpening stone to remove the unwanted material. This analogy is used for policy deployment. The use of data makes the objectives cascade a fact-based process, not just a subjective negotiation process. Mutual consultation occurs between the organizational levels in order to test the feasibility of planned process improvements and to refine any conflicts between the objectives of the organizational layers. This process opens communication channels and establishes both agreement and alignment in the way people work. This dialog is necessary to obtain buy-in and define achievable plans that middle managers are committed to implement.
- *Shibui* – a state of uncluttered, beautifully efficient austerity – the perfect balance or harmony (*wa*) between not enough and too much – used to describe the desired state or vision of a business system.

Peter Drucker quotes the Roman law in order to focus management on the things that are most important: "*De minimis non curat praetor*" (The magistrate does not consider trifles) [13]. This warning to management against what has been called "micro-management" is a reason for senior executives to focus on the vital few issues that are critical in the business that they manage. If they do not take the time to manage these important things, then no one else will. If they choose to spend their time focused at the detail level of project execution, then they will squander a more effective use of their time on those vital activities that engage the higher thinking levels of the organization that cannot be reasonably delegated to others for effective action. Management must work on a long-term planning horizon in order to deliver sustained organizational strength. It must also review current actions to assure that short-term profitability is being achieved. But, whenever management spends more time on the short-term issues than it does on the long-term ones, it sacrifices future strength in favor of current results – and displays to the entire organization its lack of trust in the ability of the organization to perform its daily work. This behavior signals to the entire organization that a crisis exists and reinforces stagnation as the workers wait for the top management to intervene and make the decisions that they should more properly make. A very important benefit of an effective policy deployment system is delegation of appropriate decision rights to the proper place in the organization where the best information exists and where action will be taken to implement that decision.

Some of the questions that are addressed during policy deployment include the following:

- What are the consequences of not doing this project?
- What are the risks inherent in this project?
- What will happen to the business if this project does not succeed?
- What will success in this project commit the business to?
- How does this project add to the total economic results of performance?
- Have we assigned our best people to work on breakthrough opportunities?
- Have we communicated clearly and taken into consideration all objections before chartering the project?

Policy Implementation

Policy implementation consists of the execution of the project plan – both the actions taken by the team involved in the change and the in-process management reviews. All change is implemented on a project-by-project basis according to management's priorities and the logical sequence for attacking each opportunity for improvement. The project plan typically will use a Gantt chart to assign clear responsibility for each improvement item in the implementation plan and to record its activity progress in accomplishing the project subtasks. Senior managers should also conduct regular progress reviews of each change project to monitor team progress in improvement, assure that the projects advance, and eliminate any possible barriers, roadblocks, or bottlenecks that restrict the project's advancement. Senior management should also monitor the execution of the change projects that they sponsor to assure that these projects will deliver the desired gain in the daily management system. If the project review indicates that insufficient progress is being made, then they can develop some countermeasures or reallocate resources so that appropriate corrective actions are taken to assure continued progress.

Another activity that occurs during the policy implementation is publication of information about the ongoing change projects so that the entire organization is informed of the actions being taken to improve performance. This communication can help the organization to align other activities with progress being made on these strategically focused change projects. As a guideline for communication, the management should inform all involved parties of any changes to the change project team's mission, vision of the outcome, guiding principles, or objectives. If the management team communicates effectively and often, then it will translate the planning rhetoric into action realities. Peter F. Drucker observed that "The most time consuming step in the process is not making the decision, but putting it into effect. Unless a decision has degenerated into work it is not a decision; it is at best a good intention" [14].

Some of the questions addressed during policy implementation include the following:

- Have we placed the right people in the right jobs to give the project the best opportunity to succeed?
- Does this project team have everything that it needs to get the job done?
- Are all the people who need to know about this project being informed?
- Are all the right actions across the organization being taken to assure success?
- Is this project implementation the best utilization of the knowledge and ability of the organization's people?
- Does this project implementation make the best overall contribution from use of the organization's limited resources (people, time, and money)?

Policy Review

As the annual cycle of change projects nears completion, results of project implementation efforts should be evaluated to determine performance against targets, shortfalls from expected performance, completion rate, and causes for both under- and overachievement. Specific action must be identified to compensate for performance deficiencies and prevent recurrence of such problems in future change management projects. Diagnosis of the performance of the policy planning process is conducted to drive improvements in planning systems. "Feedback has to be built into the decision to provide a continuous testing, against actual events, of the expectations that underlie a decision" [15].

Policy review is accomplished in two ways: through management self-assessment (by senior managers as well as by local managers evaluating their activities to determine where they have opportunities for improvement: either performance enhancements or problem resolution) and through operating reviews of the results produced by the local organization where senior managers identify areas where results are not aligned with expectations for performance. Policy review applies two subprocesses to perform these duties: performance review and business measurement.

Aligning Objectives through Performance Review. The review process seeks to identify conformance to plans (e.g., is there any shortfall or overachievement in targets?). Once nonconformity is identified, then the root cause of the deviation is discerned to determine an appropriate response to the out-of-control type of condition. Both corrective actions and countermeasures are identified to realign the process and assure that process integrity and stability are achieved in the business control system. The following actions may be taken in response to an out-of-control condition:

- emergency countermeasures to alleviate the immediate issue, concern, or problem;
- short-term corrective action to prevent the specific problem from recurring; or
- long-term preventive action to remove problem root causes and mistake-proof the process making a permanent solution and preventing the problem from recurring.

The review step facilitates organizational learning by examining problem areas and critical success factors to discover what directional shifts need to be accomplished in order to achieve the desired end state or vision of the business. Strategy is the persistence of the vision – achieved one project at a time through exercising constancy of purpose in the business planning process. These project reviews are conducted to assess achievement relative to the following planning elements:

- change project objectives
- business planning objectives and corporate commitments
- business improvement plans
- economic plans and projections
- customer requirements and expectations
- competitive performance analysis
- business excellence self-assessment.

Questions addressed during this policy review include the following:

- What results have been demonstrated from this project?
- Which results were expected and which results were unexpected?
- What will this project's outcome do for customers?
- What have we done well that our competitors have done poorly?
- What have we done poorly that our competitors seem to have no problem with?

Business Control and Management Responsibility. The ultimate objective of MBP is to establish a reliable organization – one that creates predictable results through the effective coordination of value-adding work that customers perceive as meeting their needs. In this environment, all employees are aware of their personal contribution to the objectives of the entire organization and are able to make local choices that are aligned with the strategic direction because they understand how the strategy affects their work and *vice versa*. To assure that these local decisions are aligned with strategic direction, it is the responsibility of the management team to develop a measurement system that provides employees with the visible line of sight from their work activities to its contribution to strategic direction. In this measurement system, it is essential that causal linkages (e.g., built using a Six Sigma $Y = f(X)$ transfer function) be established so that effective control can be executed at the local operating level.

Benefits of Policy Deployment

A policy deployment system orchestrates continuous improvements with breakthroughs to assure that the organization attains its long-range goals. Those elements of long-range plans that can be achieved in a one-year period are identified and become the focus or "vital few" goals to be achieved during that year. Policy deployment plans the way that change is implemented in an organization's daily management process. Accomplishments of such a planning system include

- communicating the vision required for sustained success,
- identifying and choosing breakthrough activities or projects required for the vision,
- orchestrating the direction of an organization's change,
- developing plans and projects that support the business objectives,
- aligning the organization's change efforts both vertically and horizontally,
- ensuring that the plan is effectively and efficiently executed,
- reviewing the progress in executing plans,
- changing plans when it is proved necessary to achieve targets,
- learning from the experience of planning and executing.

"If you can think of new methods to preserve the core, then by all means put them in place. If you can invent powerful new methods to stimulate progress, then give them a try. Use the proven methods *and* create new methods. You must do both" [16]. The imperative for organizations that endure is to do both breakthrough improvement and evolutionary improvement – both change management and routine management – at the same time. This is what Collins and Porras call "the genius of AND" – an inclusive approach to planning and executing change that requires organizations to embrace both aspects of change simultaneously.

The most important thing about priority decisions that face a business is that they are made and communicated deliberately and conscientiously. In a system of MBP, all the important decisions are visible and there is an opportunity for dialog to guide these decisions into the direction that the organization, as a whole, is influenced by the knowledge of all its members. In such a system the key decisions that drive toward its common goals are not made haphazardly, but with the full awareness of the organization. Such open decision processes elicit cooperation of the entire organization in the implementation and review of its activities to assure that it will be able to meet its desired outcomes. The responsibility of the management is to put in place a system of decision- making that generates this degree of collaborative work toward the common end.

Criticisms of Policy Deployment

Despite their application in many leading companies, policy deployment systems have been criticized for their mechanistic use of forms and templates that some see as restricting individual creativity. Some also believe that these planning systems lack strategic emphasis and do not engage the full organization as participants in strategy formulation. Osada summarizes the shortcomings of policy deployment as observed in some Japanese companies.

1. It is difficult for those at the middle management level and below to understand the process of formulating strategic policy. Compared with [editor note: the process for] policy deployment, the process of policy formulation is unclear [editor note: poorly understood and communicated] an indication of management's view that such form of communication is of little value.
2. Strategic policy is ostensibly based on the long-term interests of the firm, but there is no way to judge whether a policy is appropriate, or even truly 'strategic' [editor note: in the essential nature of the policy itself].
3. Several problems in formulating a long-term strategic plan are not addressed, for instance:
   - 3a. Changes in operating environment and other uncertainties are not adequately accounted for; possible difficulties are therefore not foreseen.
   - 3b. Positioning of business is not perceived objectively. The question of whether business aims are optimum and clear is not addressed.
   - 3c. Only one part of the staff, at the top level, participates in strategic policy formulation; it is therefore difficult to judge whether a policy reflects the reality at the "front line" of operations [10].

It must be observed that not all of these objections are strongly negative. Organizations must ask if they must really involve frontline employees actively in formulating strategy. Nokia Mobile Phones uses a current state analysis for self-assessment of frontline operations and then rolls this data into their strategy-setting process. They also create a "strategic dialog" that builds participation of midlevel managers in conversations about strategy. Other organizations open communication lines through email forums and internal surveys. In such instances, the objection is not critical to the total impact of policy deployment implementation. Additionally, any argument that says "Every employee should have an *interest* in matters of strategic policy" is a very different argument than saying that every employee should be actively involved in *formulation* of business strategy. Satisfying employee interest in strategy can be addressed by improving communications. Also, a broad involvement of employees in formulation of strategy increases risk of inappropriate public statements or inadvertent disclosure of the company's strategy in venues where competitors may discover sensitive information that can be used against the organization. Whenever this occurs, a company loses its competitive advantage. The challenge for management is to build a strong consensus, without risking disclosure of their strategic direction to competition. It is another challenge of the management to balance these concerns in a way that is appropriate to their way of working and business culture.

Summary

Policy deployment, when coupled with a statistically based business measurement system, has been observed in several leading companies to create a robust management process that engages an entire organization in the strategic planning process. It assures line of sight from the strategic goals of the organization to the operational tasks that workers perform at the front line as they do the work that produces the organization's goods or services. The nature of this process can be described using the term "robustness" – a statistical state in which a process is able to accept variation in its inputs, without influencing the variation of its outputs. Such a process is capable of performing consistently – delivering consistent results according to its design intent. Because policy deployment engages the workforce in achieving its common goal of sustained success, this methodology has become a strategic tool for assuring sustained competitive advantage over current and potential business rivals.

Sustained success must be "dynamic" to achieve its enduring state. That is, it must provide continuous advantage despite changes in the environment, regulatory shifts, technological breakthroughs, or competitive market. Anticipating potential actions by rivals is critical to delivering sustained success. To enjoy such sustained success, an organization must master the skills of priority setting and project management to assure that they effectively define and deploy the right initiatives that result in sustained success. Advantage means staying ahead of rivals and this requires that organizations not only make continuous improvements but also use "breakthrough" opportunities to distinguish themselves in their marketplace as a differentiated provider of products and services. This type of management requires managerial competence in three areas: business vulnerability analysis, action planning administration, and operational excellence. The best implementations of policy deployment will engage its employees in the strategy-setting processes as well as the organization's change management process.

End Note

[a] It must be noted that Peter F. Drucker initially discussed MBO in Japan in the mid-1950s. Drucker taught management concepts to the Japanese along with Dr. Joseph M. Juran and Dr. W. Edwards Deming. At that time Dr. Juran and Dr. Deming worked in the Graduate Management School of New York University under the supervision of Dr. Drucker.

References

[1] Drucker, P.F. (2002). Keynote Address, *56th Annual Quality Congress*, 20 May 2002.

[2] Akao, Y. (ed) (1991). *Hoshin Kanri: Policy Deployment for Successful TQM*, Productivity Press, Portland, p. xxx.

[3] Collins, J.C. & Porras, J.I. (1994). *Built to Last: Successful Habits of Visionary Companies*, Harper Business, New York, p. 156.

[4] Collins, J.C. & Porras, J.I. (1994). *Built to Last: Successful Habits of Visionary Companies*, Harper Business, New York, p. 146.

[5] Osada, H. (1998). Strategic management by policy in total quality management, *Strategic Change* **7**, 277–287.

[6] Osada, H. (1998). Strategic management by policy in total quality management, *Strategic Change* **7**, 277.

[7] Collins, J.C. & Porras, J.I. (1994). *Built to Last: Successful Habits of Visionary Companies*, Harper Business, New York, p. 215.

[8] Drucker, P.F. (1964). *Managing for Results*, Harper & Row, New York, p. 193.

[9] Collins, J.C. & Porras, J.I. (1994). *Built to Last: Successful Habits of Visionary Companies*, Harper Business, New York, p. 186.

[10] Osada, H. (1998). Strategic management by policy in total quality management, *Strategic Change* **7**, 278.

[11] Osada, H. (1998). Strategic management by policy in total quality management, *Strategic Change* **7**, 279.

[12] Osada, H. (1998). Strategic management by policy in total quality management, *Strategic Change* **7**, 281.

[13] Drucker, P.F. (1985). *The Effective Executive*, Harper & Row, New York, p. 156.

[14] Drucker, P.F. (1985). *The Effective Executive*, Harper & Row, New York, p. 114.

[15] Drucker, P.F. (1985). *The Effective Executive*, Harper & Row, New York, p. 139.

[16] Collins, J.C. & Porras, J.I. (1994). *Built to Last: Successful Habits of Visionary Companies*, Harper Business, New York, p. 216.

GREGORY H. WATSON

Polya Trees and Their Use in Reliability and Survival Analysis

A Polya Tree Prior Centered at the Weibull Family

Polya trees generalize and add flexibility to standard parametric models such normal, Weibull, lognormal, Pareto, log-logistic, gamma, and others used for the statistical analysis of time to event data. A simple Weibull model for survival times $\mathbf{T} = (T_1, \ldots, T_n)'$ assumes a distribution function G with a particular form, i.e.

$$T_1, \ldots, T_n | G \overset{\text{i.i.d.}}{\sim} G,\ G(t|\alpha, \lambda) = 1 - \exp(-t^{\alpha}/\lambda) \quad \text{for } t \geq 0 \quad (1)$$

This type of model is referred to as *parametric* because the class of possible density shapes is indexed by a small number of parameters, here $\boldsymbol{\theta} = (\alpha, \lambda)$. Parametric survival models are straightforward to fit and draw inferences from, but the trade-off for this ease of implementation is lack of flexibility to capture more complex features (e.g., multimodality, **skewness**, and clustering) that may be present in time to event data. This motivates expanding the class of distributions to include not only the Weibull family (or more generally any parametric family) but also deviations from it supported by the data. A Bayesian nonparametric Polya tree prior for G achieves this flexibility. Other approaches include the Dirichlet process [1], Dirichlet process mixtures, beta, and gamma processes (*see* **Dirichlet Process, Simulation of**; **Survival Analysis, Nonparametric**; **Lévy Processes, Simulation of**).

A Polya tree adds $2^{J+1} - 2$ additional parameters, $\mathcal{X}_J$, to equation (1) that collectively "adjust" the overall shape of G to add more probability mass where data are clumped and remove mass where data are sparse relative to the Weibull distribution $G(t|\alpha, \lambda) = 1 - \exp(-t^{\alpha}/\lambda)$. The parameters $\boldsymbol{\theta} = (\alpha, \lambda)$ serve to fix the overall shape, location, and spread of the distribution G; the parameters $\mathcal{X}_J$ refine its shape to match the observed data more closely. In fact, the parameters $\mathcal{X}_J$ adjust G on intervals derived from **quantiles** of the Weibull(α, λ) distribution, obtained from the quantile function $G^{-1}(p|\boldsymbol{\theta}) = (-\lambda \log(1-p))^{1/\alpha}$. These intervals partition the sample space and are constructed as follows.

Initially, $(0, \infty)$ is split into two pieces, $B_{\boldsymbol{\theta}}(0) = [0, G^{-1}(0.5|\boldsymbol{\theta}))$ and $B_{\boldsymbol{\theta}}(1) = [G^{-1}(0.5|\boldsymbol{\theta}), \infty)$. These two sets partition $[0, \infty)$ and have equal probability, 0.5, under the parametric Weibull(α, λ) model. The set $B_{\boldsymbol{\theta}}(0)$ is further split into two subsets $B_{\boldsymbol{\theta}}(00) = [0, G^{-1}(0.25|\boldsymbol{\theta}))$ and $B_{\boldsymbol{\theta}}(01) = [G^{-1}(0.25|\boldsymbol{\theta}), G^{-1}(0.5|\boldsymbol{\theta}))$, each having equal probability, 0.25, under the Weibull(α, λ) model. Similarly $B_{\boldsymbol{\theta}}(1)$ is split into two subsets $B_{\boldsymbol{\theta}}(10) = [G^{-1}(0.5|\boldsymbol{\theta}), G^{-1}(0.75|\boldsymbol{\theta}))$ and $B_{\boldsymbol{\theta}}(11) = [G^{-1}(0.75|\boldsymbol{\theta}), \infty)$, each with Weibull probability 0.25. This defines the first and second partitions of the Polya tree: $B_{\boldsymbol{\theta}}(0), B_{\boldsymbol{\theta}}(1) \in \Pi_{\boldsymbol{\theta}1}$, and $B_{\boldsymbol{\theta}}(00), B_{\boldsymbol{\theta}}(01), B_{\boldsymbol{\theta}}(10), B_{\boldsymbol{\theta}}(11) \in \Pi_{\boldsymbol{\theta}2}$. For $\alpha = 2$ and $\lambda = 10$, $B_{(2,10)}(0) = [0, 2.63)$ and $B_{(2,10)}(1) = [2.63, \infty)$ at the first level and $B_{(2,10)}(00) = [0, 1.70)$, $B_{(2,10)}(01) = [1.70, 2.63)$, $B_{(2,10)}(10) = [2.63, 3.72)$, and $B_{(2,10)}(11) = [3.72, \infty)$ at the second.

Nested partitions are sequentially defined in this manner by splitting each set in $\Pi_{\boldsymbol{\theta}j}$, having probability 2^{-j} under Weibull(α, λ), into two sets in $\Pi_{\boldsymbol{\theta},j+1}$, each having probability $2^{-(j+1)}$. Let $\epsilon_1\epsilon_2 \cdots \epsilon_j$ be a binary expansion that indexes a set $B_{\boldsymbol{\theta}}(\epsilon_1 \cdots \epsilon_j) \in \Pi_{\boldsymbol{\theta}j}$. Then $B_{\boldsymbol{\theta}}(\epsilon_1 \cdots \epsilon_j) = B_{\boldsymbol{\theta}}(\epsilon_1 \cdots \epsilon_j 0) \cup B_{\boldsymbol{\theta}}(\epsilon_1 \cdots \epsilon_j 1)$. Here, $B_{\boldsymbol{\theta}}(\epsilon_1 \cdots \epsilon_j) = [G^{-1}(N/2^j), G^{-1}((N+1)/2^j))$, where N is the decimal representation of $\epsilon_1 \cdots \epsilon_j$ and ranges over the integers from 0 to $2^j - 1$.

The probability of an observation T_i falling into $B_{\boldsymbol{\theta}}(0)$ is denoted $X(0) = G(B_{\boldsymbol{\theta}}(0))$. Then the probability of T_i falling into the companion set $B_{\boldsymbol{\theta}}(1)$

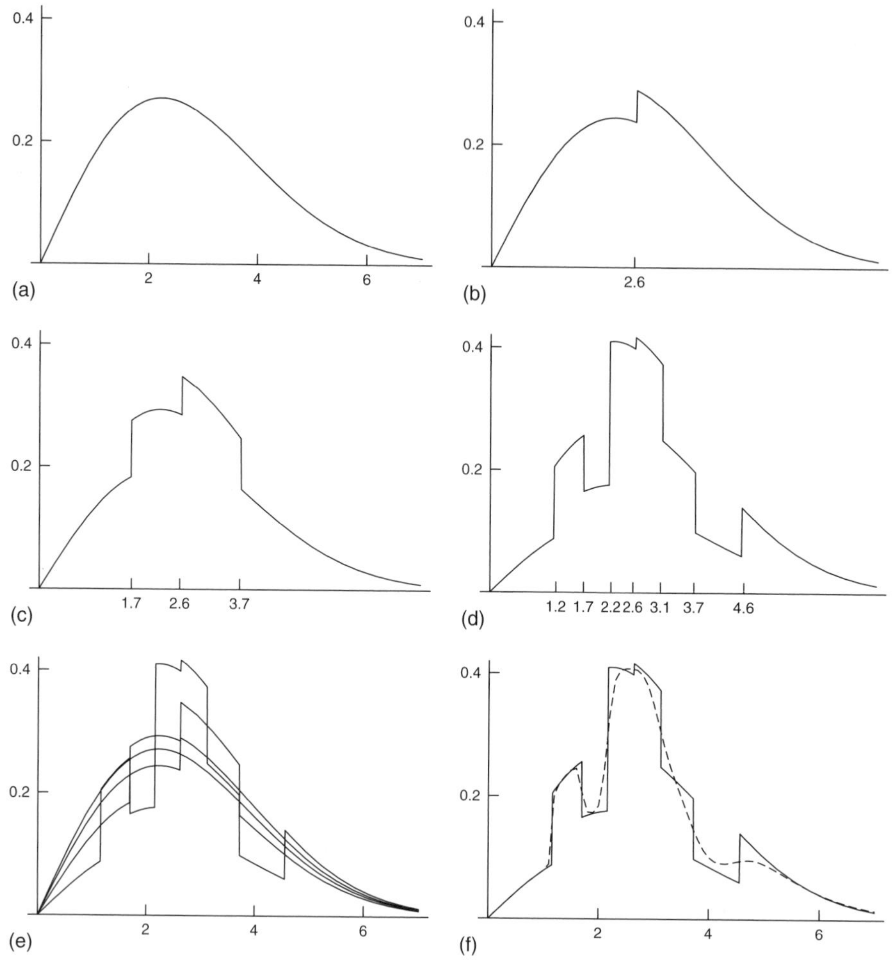

Figure 1 (a) Weibull(2, 10) ($J = 0$) density, (b–d) $J = 1, 2, 3$ respectively, (e) all four levels on same plot, (f) PT and MPT densities

is 1 minus this, $X(1) = G(B_\theta(1)) = 1 - G(B_\theta(0))$. Given that $T_i \in B_\theta(\epsilon_1 \cdots \epsilon_j)$, the *conditional probability* of T_i falling into $B_\theta(\epsilon_1 \cdots \epsilon_j 0)$ *versus* its companion set $B_\theta(\epsilon_1 \cdots \epsilon_j 1)$ is given by $X(\epsilon_1 \cdots \epsilon_j 0)$ *versus* $X(\epsilon_1 \cdots \epsilon_j 1)$. That is, for $\epsilon_1\epsilon_2 \cdots \epsilon_j \in \{0, 1\}^j$, $X(\epsilon_1 \cdots \epsilon_j 0) = P(T_i \in B_\theta(\epsilon_1 \cdots \epsilon_j 0) | T_i \in B_\theta(\epsilon_1 \cdots \epsilon_j))$ and $X(\epsilon_1 \cdots \epsilon_j 1) = 1 - X(\epsilon_1 \cdots \epsilon_j 0) = P(T_i \in B_\theta(\epsilon_1 \cdots \epsilon_j 1) | T_i \in B_\theta(\epsilon_1 \cdots \epsilon_j))$.

Let $E_J = \bigcup_{j=1}^{J} \{0, 1\}^j$ denote all binary numbers from 0 to $2^J - 1$ and let $\mathcal{X}_J = \{X(\boldsymbol{\varepsilon}) : \boldsymbol{\varepsilon} \in E_J\}$ be the set of all conditional probabilities up to level J that define G. A fully specified Polya tree has an infinite number of levels ($J \to \infty$). A finite Polya tree has J levels. If J is chosen to be large, one can further assume G to be flat on intervals $B_\theta(\boldsymbol{\varepsilon})$ of finite length at level J [2]. Alternatively, one can assume that G follows the Weibull (α, λ) on sets at level J [3] – the approach taken here. When $X(\boldsymbol{\varepsilon}) = 0.5$ for all $\boldsymbol{\varepsilon} \in E_J$, G is simply the Weibull(α, λ) density we started with.

Let us see how particular choices of conditional probabilities affect the density. Figure 1(a) shows a Weibull(2, 10) density. This is also the density of G governed by a Polya tree but with all conditional probabilities $X(\epsilon_1 \cdots \epsilon_j) = 0.5$. In Figure 1(b), fixing $X(0) = 0.45$ serves to adjust the Weibull density at the first level of the tree on the sets $B_{(2,10)}(0)$ and $B_{(2,10)}(1)$. Figure 1(c) shows the density further refined at the second level from fixing the conditional probabilities $X(00) = 0.4$ and $X(10) = 0.6$. Figure 1(d) shows a finite Polya tree with three levels obtained by further fixing $X(000) = 0.3$, $X(010) = 0.3$, $X(100) = 0.6$, and $X(110) = 0.3$. Figure 1(e) shows the original density along with the three refinements. Figure 1(f) is obtained by keeping the fixed values in $\mathcal{X}_3$ above, but by considering a mixture of Polya trees (MPT) from taking $\alpha \sim N(2, 0.1^2)$ independent of $\lambda \sim N(10, 0.5^2)$. In the mixture, $E(\alpha) = 2$ and $E(\lambda) = 10$ so the prior on G is centered at the same Weibull(2, 10) density, but by allowing the mixing parameters (α, λ) to be random the partitioning effects (e.g., jagged edges in the density at partition interval endpoints) apparent in the simple Polya tree are smoothed over. This is because the partition sets $\Pi_{\theta j}$ are random under the mixture.

Instead of treating the conditional probabilities as fixed, unknown parameters, the Bayesian considers them random. The probability of an observation T_i falling into $B_\theta(0)$ is modeled as $X(0) = G(B_\theta(0)) \sim \beta(\alpha(0), \alpha(1))$. That is,

$$\begin{bmatrix} X(0) \\ X(1) \end{bmatrix} = \begin{bmatrix} G(B_\theta(0)) \\ G(B_\theta(1)) \end{bmatrix} \sim \text{Dirichlet}\left(\begin{bmatrix} \alpha(0) \\ \alpha(1) \end{bmatrix}\right) \tag{2}$$

Recall that, in general, for $\epsilon_1\epsilon_2\cdots\epsilon_j \in \{0,1\}^j$, $X(\epsilon_1\cdots\epsilon_j 0) = P(T_i \in B_\theta(\epsilon_1\cdots\epsilon_j 0)|T_i \in B_\theta(\epsilon_1\cdots\epsilon_j))$ and $X(\epsilon_1\cdots\epsilon_j 1) = 1 - X(\epsilon_1\cdots\epsilon_j 0) = P(T_i \in B_\theta(\epsilon_1\cdots\epsilon_j 1)|T_i \in B_\theta(\epsilon_1\cdots\epsilon_j))$. Rewritten in compact notation, for $\boldsymbol{\varepsilon} \in \{0,1\}^j$, $X(\boldsymbol{\varepsilon}0) = P(T_i \in B_\theta(\boldsymbol{\varepsilon}0)|T_i \in B_\theta(\boldsymbol{\varepsilon}))$ and $X(\boldsymbol{\varepsilon}1) = 1 - X(\boldsymbol{\varepsilon}0) = P(T_i \in B_\theta(\boldsymbol{\varepsilon}1)|T_i \in B_\theta(\boldsymbol{\varepsilon}))$. These pairs of conditional probabilities are also Dirichlet:

$$\begin{bmatrix} X(\boldsymbol{\varepsilon}0) \\ X(\boldsymbol{\varepsilon}1) \end{bmatrix} \sim \text{Dirichlet}\left(\begin{bmatrix} \alpha(\boldsymbol{\varepsilon}0) \\ \alpha(\boldsymbol{\varepsilon}1) \end{bmatrix}\right) \tag{3}$$

So the prior on G is in fact a tree of pairs of conditional probabilities of falling into one set *versus* an adjoining set. These pairs of probabilities are assumed to be independent. Recall that fixing $X(\boldsymbol{\varepsilon}) = 0.5$ for all $\boldsymbol{\varepsilon} \in E_J$ ensures that G is Weibull(α, λ). By setting $\alpha(\boldsymbol{\varepsilon}0) = \alpha(\boldsymbol{\varepsilon}1)$ for each $\boldsymbol{\varepsilon} = \epsilon_1\cdots\epsilon_j \in E_J$, random G is instead *centered* at the Weibull(α, λ) distribution on all partition sets:

$$E\{G(B_\theta(\boldsymbol{\varepsilon}))\} = \int_{B_\theta(\boldsymbol{\varepsilon})} \mathrm{d}G(t|\alpha, \lambda) \tag{4}$$

from the fact that $E(X(\boldsymbol{\varepsilon})) = 0.5$ for all $\boldsymbol{\varepsilon} \in E_J$. The parameters $\alpha(\epsilon_1\cdots\epsilon_j)$ are often set to $\alpha(\epsilon_1\cdots\epsilon_j) = cj^2$ for $c > 0$ in applications. This ensures that the conditional probabilities get more highly concentrated around 0.5 as the level J increases, giving larger sets at lower levels more flexibility to adapt to large trends or clumps in the observed data.

The tree structure and independence among probability pairs in $\mathcal{X}_J$ gives the measure of any set

$$G\{B_\theta(\epsilon_1\cdots\epsilon_j)\} = \prod_{i=1}^{j} X(\epsilon_1\cdots\epsilon_i) \tag{5}$$

A key result for Polya trees is that the conditional probability pairs in $\mathcal{X}_J$ given data $\mathbf{T} = (T_1, \ldots, T_n)$ are again independent Dirichlet vectors, and are thus easily sampled for a given (α, λ). This conjugacy result has played a part in several papers that involve Polya tree priors (e.g., [2, 4, 5]) but in many models conjugacy is lost and more sophisticated methods for obtaining posterior inference, such as Markov chain **Monte Carlo** (MCMC), are employed (*see* **Markov Chain Monte Carlo, Introduction**).

The definition of a Polya tree can be more complicated than what is outlined so far, but this gives a particular development that has been used with great success in a number of papers. More general Polya trees involve definition of elements $\mathcal{A} = \{\alpha(\boldsymbol{\varepsilon})\}$ one at a time and consideration of different partitions $\{\Pi_1, \ldots, \Pi_J\}$. An MPT prior on G assumes $\boldsymbol{\theta} = (\alpha, \lambda)$ is random in addition to the elements in $\mathcal{X}_J$. See [6] and [7] for more details and definitions. Figure 1 illustrates that the MPT prior smooths over partitioning effects apparent in a simple Polya tree; see [6, 8].

Polya Trees and Survival Modeling

The literature on survival modeling with Polya trees can be sorted into two categories: the no-covariates case focusing on the estimation of survival, hazard, and density functions; and semiparametric models that extend these models to include **covariate** information.

Ferguson [9] codifies and fleshes out various results published in the 1960s on Polya trees. This paper also contrasts Polya trees to the special case of the Dirichlet process [1]. Lavine [6, 7] carefully develops and catalogs many additional aspects of Polya tree and MPT theory and methodology; these papers were foundational, and provided core material from which many further developments grew. Lavine [6] analyzes reliability data on the lifetimes of spherical pressure vessels from four different MPT models, each with different sets $\mathcal{A} = \{\alpha(\boldsymbol{\varepsilon})\}$, centered at the exponential family $G(t|\alpha) = 1 - \exp(-\alpha t)$. The data considered in [6] were observed exactly; censored survival data is considered by Muliere and Walker [10] through the construction of Polya trees with some partition points coinciding with censored observations. The resulting estimator reduces to that of Kaplan and Meier [11] as a limiting case. When the partition is fixed ahead of time, i.e. not picked to coincide with censored values, Neath [12] shows that the posterior distribution for G is an MPT, derives the mixing distribution, and provides a reanalysis of data considered in [10, 11].

These approaches all consider survival data without covariates. Often in reliability settings such as accelerated testing, covariates for the ith experimental unit $\mathbf{x}_i$ are fixed at predetermined values and the resultant lifetime T_i, or possibly a censoring time, is recorded. In survival analysis settings, patient information such as age, treatment, gender, etc., are collected into $\mathbf{x}_i$ and a time to event T_i observed. In either case, the covariates $\mathbf{x}_i$ are linked to T_i through a probability model. In what follows, let $S_i(t) = P(T_i > t)$ be the survival function of subject i with covariates $\mathbf{x}_i$ given all model parameters.

Walker and Mallick [2] develop an accelerated failure time model

$$\log T_i = \mu + \mathbf{x}_i'\boldsymbol{\beta} + \sigma\epsilon_i \quad (6)$$

where the ϵ_i are independent and identically distributed (i.i.d.) from a distribution G assigned a Polya tree prior; this model is also considered in [13]. In applications, $\sigma = 1$ is fixed and the Polya tree is centered at a median-zero **normal distribution** with a large variance. Hanson [8] considers proportional hazards

$$S_i(t) = S_0(t)^{\exp(\mathbf{x}_i'\boldsymbol{\beta})} \quad (7)$$

proportional odds

$$\frac{1 - S_i(t)}{S_i(t)} = \exp(\mathbf{x}_i'\boldsymbol{\beta})\frac{1 - S_0(t)}{S_0(t)} \quad (8)$$

and a reparameterized version of equation (6),

$$S_i(t) = S_0\{\exp(\mathbf{x}_i'\boldsymbol{\beta})t\} \quad (9)$$

where equations (7–9) assume the same MPT prior centered at the Weibull(α, λ) family for baseline survival S_0. Mallick and Walker [5] develop a semiparametric transformation model

$$h(T_i) = \mu + \mathbf{x}_i'\boldsymbol{\beta} + \epsilon_i \quad (10)$$

where $h(\cdot)$ is a monotone function modeled as a transformation of a small mixture of beta cumulative distribution function (**cdf**'s). As in equation (6), the ϵ_i are i.i.d. from a distribution G assigned a Polya tree prior. This model reduces to the proportional odds model (8) when G is fixed to be the standard logistic distribution, reduces to the proportional hazards model (7) when G is extreme value, and reduces to equation (6) when $h(\cdot) = \log(\cdot)$ but with a fixed scale $\sigma = 1$. All of these papers analyzed data on $n = 121$ patients diagnosed with small cell lung cancer, considered in the next section. The accelerated failure time model was found to fit these data better than the proportional odds and proportional hazards models [8, 5].

Hanson and Johnson [3] allow for random σ in equation (6), rather than a simple Polya tree, smoothing out partitioning effects. A frailty version of this model has been used to analyze data on fetal lifetime in cows with parametric frailties accounting for different herd effects [14]. Damien *et al.* [15] considered the analysis of reliability data through equation (9). The proportional odds model (8) was extended by Hanson and Yang [16] to include parametric frailties. Survival data on the lifetimes of $n = 97$ men with advanced, inoperable lung cancer were analyzed by Mallick and Walker [5] and Hanson and Yang [16]. The proportional odds model was picked to be superior in both papers.

Walker and Mallick [4] consider a Polya tree prior on the random effects distribution in generalized linear mixed models, and, in particular, use a Polya tree to model frailties in the Cox model [17]. The use of Polya trees to model random effects in hierarchical models was further developed in [8]. These models are in contrast to the models considered in [5, 14, 16], which considered parametric frailty models coupled with semiparametric survival models. Although common, parametric frailty assumptions are typically made for computational convenience; the MPT model provides a powerful method to relax these strict parametric assumptions in survival and reliability modeling.

In related work, Karabatsos and Walker [18] consider estimation of distributions from several populations that are subject to a stochastic order constraint, and apply their methodology to data comprising the reaction times of schizophrenics and nonschizophrenics.

An Example: Survival Data on Lung Cancer Patients

Data on the treatment of limited-stage small cell lung cancer in $n = 121$ patients were analyzed by Walker and Mallick [2], Mallick and Walker [5], Hanson [8], and Walker *et al.* [13] among others. In the study, it was of interest to determine which sequencing of the drugs cisplatin and etoposide increased the lifetime from time of diagnosis, measured in days, of those with limited-stage small cell lung cancer. Treatment A applied cisplatin followed by etoposide, while treatment B applied etoposide followed by cisplatin; $x_{i1} = 0, 1$ for treatments A, B respectively. The patients' ages in years at entry into the study, x_{i2}, were also included so $\mathbf{x}_i = (x_{i1}, x_{i2})'$. Treatment A was administered to 62 patients, while treatment B was administered to 59 patients; 23 patients were administratively right censored.

Here we consider model (9) with an MPT baseline S_0 centered at the Weibull(α, λ) family of distributions. The flat prior $p(\alpha, \lambda, \beta_1, \beta_2) \propto 1$ was assumed. The remaining parameters were fixed at $c = 1$ and $J = 4$ and inference obtained *via* MCMC [8]. The estimated age effect $\hat{\beta}_2$ and 95% credible interval is 0.004 (−0.005, 0.034); for treatment the estimated effect $\hat{\beta}_1$ and 95% credible interval is 0.36 (0.15, 0.71). Treatment A is estimated to prolong mean and median life by a factor of $e^{0.36} \approx 1.4$ times, or about 40%, relative to B.

The MPT approach models the baseline survival distribution S_0 in tandem with the regression coefficients $\boldsymbol{\beta}$, so, arbitrary posterior functionals of $(S_0, \boldsymbol{\beta})$ are easily obtained from the MCMC output. For example, one can obtain the hazard ratio for treatment B *versus* A at any time point t, the predictive density at t, arbitrary survival quantiles, etc. Figure 2(b) is a plot of the estimated posterior median hazard ratio for treatment B *versus* A at the mean entry age of 61,

$$hr(t|\text{data}) = \frac{h_{(1,61)}(t)}{h_{(0,61)}(t)}|\text{data} \qquad (11)$$

versus time with pointwise 95% credible intervals. Here $[1 - S_x(t)]$ by just $S_x(t)$, the hazard function at t. We see that those under regimen B are typically about 1.5 times more likely to die at any instant *versus* those under A, but there is considerable variability of the hazard ratio over time. For example, we see peaks under 500 days where those under B are estimated to be almost three times as likely to instantaneously expire. In Figure 2(a) we have the two predictive densities at baseline age 61 for treatments A and B. Note that unlike the Weibull model, the MPT generalization allows for multimodality in these predictive distributions which may have biological interpretations in some applications.

Discussion

This paper introduces the Polya tree prior as a generalization of a parametric family of distributions, and provides a review of literature on the use of Polya trees in reliability and survival analysis. Details of model development and fitting are left to the references. Good references on Polya tree and MPT modeling include [6–9, 13].

One aspect of Polya tree distributions that can be exploited in the modeling of survival data is that the tails of an MPT G retain the parametric flavor of the centering family, essential in some cure-rate models and reliability settings where extrapolation is common. Also, MPT survival models can accommodate arbitrarily censored and truncated data. For example, interval-censored and right-truncated data are easily

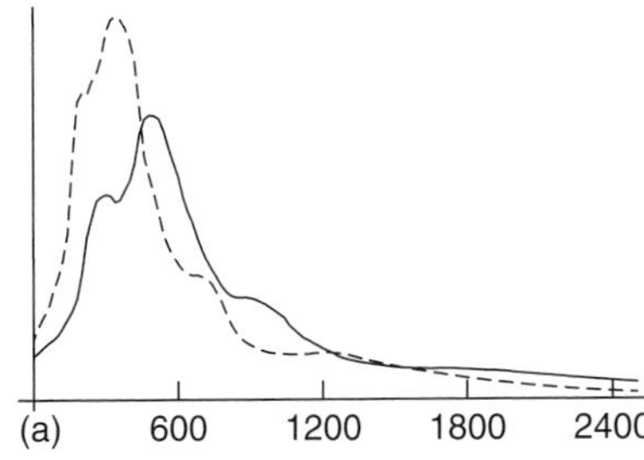

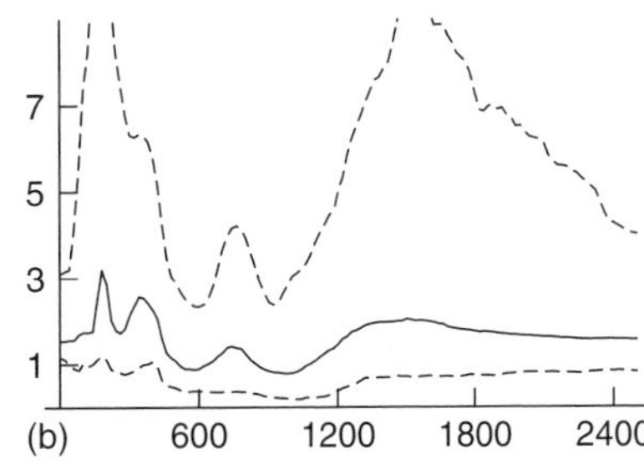

Figure 2 (a) Survival density estimates for treatments A (solid) and B (dashed) for 61-year-old patients. (b) Hazard ratio (B *vs* A) and pointwise 95% credible intervals

handled in the proportional hazards model. This is not typically the case with other nonparametric Bayesian survival models.

References

[1] Ferguson, T. (1973). A Bayesian analysis of some nonparametric problems, *Annals of Statistics* **1**, 209–230.

[2] Walker, S. & Mallick, B. (1999). Semiparametric accelerated life time model, *Biometrics* **55**, 477–483.

[3] Hanson, T. & Johnson, W. (2002). Modeling regression error with a mixture of Polya trees, *Journal of the American Statistical Association* **97**, 1020–1033.

[4] Walker, S. & Mallick, B. (1997). Hierarchical generalized linear models and frailty models with Bayesian nonparametric mixing, *Journal of the Royal Statistical Society. Series B* **59**, 845–860.

[5] Mallick, B. & Walker, S. (2003). A Bayesian semiparametric transformation model incorporating frailties, *Journal of Statistical Planning and Inference* **112**, 159–174.

[6] Lavine, M. (1992). Some aspects of Polya tree distributions for statistical modeling, *Annals of Statistics* **20**, 1222–1235.

[7] Lavine, M. (1994). More aspects of Polya tree distributions for statistical modeling, *Annals of Statistics* **22**, 1161–1176.

[8] Hanson, T. (2006). Inference for mixtures of finite Polya tree models, *Journal of the American Statistical Association* **101**, 1548–1565.

[9] Ferguson, T. (1974). Prior distributions on spaces of probability measures, *Annals of Statistics* **2**, 615–629.

[10] Muliere, P. & Walker, S. (1997). A Bayesian nonparametric approach to survival analysis using Polya trees, *Scandinavian Journal of Statistics* **24**, 331–340.

[11] Kaplan, E. & Meier, P. (1958). Nonparametric estimation from incomplete observations, *Journal of the American Statistical Association* **53**, 457–481.

[12] Neath, A. (2003). Polya tree distributions for statistical modeling of censored data, *Journal of Applied Mathematics and Decision Sciences* **7**, 175–186.

[13] Walker, S., Damien, P., Laud, P. & Smith, A. (1999). Bayesian nonparametric inference for random distributions and related functions, *Journal of the Royal Statistical Society. Series B* **61**, 485–527.

[14] Hanson, T., Bedrick, E., Johnson, W. & Thurmond, M. (2003). A mixture model for bovine abortion and fetal survival, *Statistics in Medicine* **22**, 1725–1739.

[15] Damien, P., Galenko, A., Popova, E. & Hanson, T. (2007). Bayesian semiparametric analysis for a single item maintenance optimization, *European Journal of Operational Research* **182**, 794–805.

[16] Hanson, T. and Yang, M. (2007). Bayesian semiparametric proportional odds models, *Biometrics* **63**, 88–95.

[17] Cox, D. (1972). Regression models and life-tables (with discussion), *Journal of the Royal Statistical Society. Series B* **34**, 187–220.

[18] Karabatsos, G. & Walker, S. (2007). Bayesian nonparametric inference of stochastically ordered distributions, with Polya trees and Bernstein polynomials, *Statistics and Probability Letters* **77**, 907–913.

Related Article

Accelerated Life Models; **Bayesian Reliability Analysis**.

TIMOTHY E. HANSON

Pooled Variance, Pooled Estimate

When estimators are based on more than one sample, then pooling is the name for obtaining an estimator for a characteristic of the combined samples by combining estimates based on the individual samples. The reason for doing this is that a pooled estimator is more accurate because it is based on more observations. The basic example is the pooled *variance* that appears in **confidence intervals** and tests for the difference of the mean of two samples from *normal distributions* with the same variance. In this case the sample *variance* of the combined sample may be a biased estimator (*see* **Bias of an Estimator**) of the common variance, because it is a measure of deviation from the overall mean (which may be different from the means of the individual samples). The correct way is to use the pooled variance $((n_1 - 1)S_1^2 + (n_2 - 1)S_2^2)/(n_1 + n_2 - 2)$, which is an unbiased estimator of the common *variance* of the two samples. Note that pooling means in this case is not a problem, since the pooled mean $((n_1 - 1)\bar{X}_1 + (n_2 - 1)\bar{X}_2)/(n_1 + n_2)$ equals the overall mean of the combined sample. However, see [1] for a very readable, practical discussion whether to pool or not. More generally, estimators are pooled in *analysis of variance*.

In case one is not certain whether it is allowed to pool the data, statistical tests may be performed to decide whether to pool or not. Such estimators are known under the name sometimes-pool estimator (see [2, 3] for further details).

References

[1] Mosteller, F. (1948). On pooling data, *Journal of American Statistical Association* **43**, 231–242.

[2] Bancroft, T.A. (1944). On biases in estimation due to the use of preliminary tests of significance, *Annals of Mathematical Statistics* **15**, 190–204.

[3] Han, C.P. & Bancroft, T.A. (1968). On pooling means when variance is unknown, *Journal of American Statistical Association* **63**, 1333–1342.

Related Articles

Bias of an Estimator; **Confidence Intervals**.

ALESSANDRO DI BUCCHIANICO

Population Characteristics of Proficiency Test Data

Introduction

Silicate in seawater, acenaphthylene in sediment, and DDT in fish muscle, are some of the many chemicals found in the sea. Interpretation of information on the concentrations of such chemical contaminants in the environment is highly dependent on the reliability of the analytical measurement. The quality of these measurements is frequently assessed through a laboratory's participation in an appropriate proficiency test (PT) scheme [1, 2]. Homogeneous environmental test materials are provided for analysis and the resultant data from the chemical measurements are compared by the scheme organizer to determine the level of agreement between participants.

A statistical evaluation of the PT should provide a robust and reliable estimate of the true value and the variance of the data, without being influenced by extreme values, many of which are random in origin. The assessment of the individual laboratory performance can then be based on these estimates.

The true value of the concentration of a determinant in a natural PT is usually unknown. Therefore we need to estimate a consensus value from the data provided by the laboratories. When data are generated under statistical control they will generally be normally distributed. We can then use the arithmetic mean and standard deviation (SD) as summary statistics. In reality, however, we find that overlying any **normal distribution** are positively or negatively skewed values that can create bi- or multimodal distributions due to outliers, biased measurements, and left-censored values (LCVs). **Outlier** tests have been used with some caution to eliminate extreme values, but we need alternative methods to evaluate the nonnormal distributions generated in PTs. Robust methods using the Huber [3–5] or Hampel [6, 7] estimators effectively weigh down the effect of skewed and outlying values on the mean where they comprise $<10\%$ of the data. These methods, however, cannot cope with highly skewed data, bimodality, or LCVs.

The Cofino model [8, 9] overcomes many of these limitations by taking an alternative approach. Our model identifies clusters of values within a dataset that exhibit a high level of agreement and, for each cluster, calculates the mean, variance, and the percentage of the whole dataset. This model, therefore, minimizes the contribution of the skewed or outlying data to the mean of the main mode of the data.

Theoretical Background

The Cofino model [8, 9] uses a **probability density function (pdf)** for each observation as a starting point. We obtain a pdf representing all data in a PT by summing each pdf associated with the individual measurement. The overall pdf is the starting point for our model. Instead of calculating the mean of the data, our model establishes the most probable value, given the overall pdf. As an analog to the wave functions in quantum mechanics, observation measurement functions (OMFs, φ_i) are defined as the square root of the pdf attributed to the individual observation, which is the key to calculating the population characteristics. The set of OMFs forms a space, or a basis set, in which we construct the so-called *population measurement functions* (*PMFs*). The PMF ψ_i is constructed as a linear combination of OMFs: $\Psi_i = \sum c_{ij}\varphi_j$.

We obtain the coefficients by locating the, unnormalized, PMF_1, with the highest probability in the basis set. This probability is given by the integral $\int \psi_1^2\, \mathrm{d}c$. We use Lagrange multipliers and impose the additional constraint, that the sum of the squared coefficients is equal to one. The mathematical elaboration requires a solution to the eigenvector–eigenvalue equation $Sc = \lambda c$. In this equation, S represents the matrix of overlap integrals, with S_{ij} given by $\int \phi_i \phi_j \, dc$. The overlap integral S_{ij} can range between 0 (no overlap) and 1 (100% overlap) when the observations have identical pdfs.

From a set of n basis vectors, OMFs, the model creates a total of n eigenvectors, c, with eigenvalues, λ. The eigenvalue λ_i gives the probability, in the basis set, of the corresponding eigenfunction i. The highest probability and thus maximum value for λ is equal to the number of data n, which is obtained when all data have exactly the same pdf, and each OMF has a coefficient equal to $1/\sqrt{n}$. The eigenvector with the highest eigenvalue λ is the PMF_1. The remaining n-1 linear combinations are ranked according to probability (i.e., eigenvalue) and are denoted as $PMF_2 \ldots PMF_n$. PMF_2 and higher PMFs may sometimes be additional modes, but are frequently only clusters of data ordered according to their degree of overlap. When the squared OMFs of all PMFs are summed together over the entire concentration range, the pdf of the entire dataset is reconstructed.

For each PMF ψ_i the expectation value and variance can be calculated as follows:

$$\bar{x}_i = \frac{\int c \times \psi_i^2 \, dc}{\int \psi_i^2 \, dc} \quad (1)$$

$$s_i^2 = \frac{\int c^2 \times \psi_i^2 \, dc}{\int \psi_i^2 \, dc} - \bar{x}_i^2 \quad (2)$$

Also, the eigenvalues λ enable a quantitative assessment of the degree of comparability and the character (unimodal, bimodal) of the dataset. So we can convert the eigenvalue of the mode, or cluster proper into a percentage of the overall pdf. This percentage, therefore, quantitatively describes the fraction of the dataset that is accounted for by the PMF in question.

We have extended the model to include LCVs by defining the appropriate pdfs for these observations. The simplest approach is to assume that any concentration between zero and the limit of quantization (LOQ) has an equal probability. We can then use the square root of a rectangular pdf as basis function. Explicitly, when an LCV is reported, the basis function is equal to $\sqrt{1/LOQ}$ in the interval between zero and LOQ, or otherwise zero. These basis functions have an expectation value $\int \phi_i^2 x \, dx = LOQ/2$ and a variance $\int \phi_i^2 x^2 \, dx - \bar{x}^2 = LOQ^2/12$. Other pdf profiles, such as triangular, could be used, but we have concluded that in our studies we have no justification to use pdfs that would require more assumptions.

Inputs for the Model

In our studies we use normal distributions to define the basis functions. Therefore, in addition to the measured values, an estimate of the *within-laboratory* SD is required to define the pdfs and thus the OMFs. The *within-laboratory SD* may be calculated, for instance, when laboratories submit replicate data. The main constraint here is the ability to assess the robustness of the methods used by the laboratories to replicate the measurements.

We have specifically developed the Cofino model using a normal distribution approximation (NDA) to apply to situations where the reliability of the *within-laboratory SD* cannot be assessed, or when *within-laboratory SD*s are not available at all [8, 9]. The NDA approach is based on the premise that the population characteristics of a normal distribution can be reproduced when each data point is given a normal distribution with the reported concentration as mean and a *within-laboratory SD* constructed as

$$SD_{\mathrm{NDA}} = 1.168 \times \text{median}\ (\text{abs}(x_i - \text{median}(x)) \quad (3)$$

The NDA assumption has been fully tested and verified [8]. We have developed the model for normally distributed populations, but it will also handle nonnormally distributed populations as equally well as the robust techniques developed by Huber and Hampel [8].

Output of the Model

In this application, we have used the Cofino NDA model. We obtain both numerical **descriptive statistics** and a graphical output to aid interpretation of PT data. We have also compared similar information using the Huber and Hampel estimators. The Cofino mean, SD of each mode in the data, and the percentage of data associated with each mode were calculated. The main mode (PMF_1) accounts for >75% of data when the distribution is close to normal. In such cases, the mean and SD we obtained by the different methods of calculation are very comparable. The measurement of silicate in seawater is given as an example (Table 1 and Figure 1).

Table 1 A comparison of the summary statistics from the proficiency test for silicate in seawater and acenaphthylene in sediment

Matrix	Determinant	Total NObs	LCV NObs	Percentage of data in first mode	Units	Cofino mean	Cofino SD	Huber mean	Huber SD	Hampel mean	Hampel SD	Arithmetic mean	Arithmetic SD
Seawater	Silicate	47	0	79	$\mu mol\, l^{-1}$	9.79	0.70	9.91	0.73	9.88	0.72	9.95	1.72
Sediment	Acenaphthylene	14	1	69	$mg\, kg^{-1}$	12.16	14.79	23.72	25.10	20.94	17.35	24.76	24.29

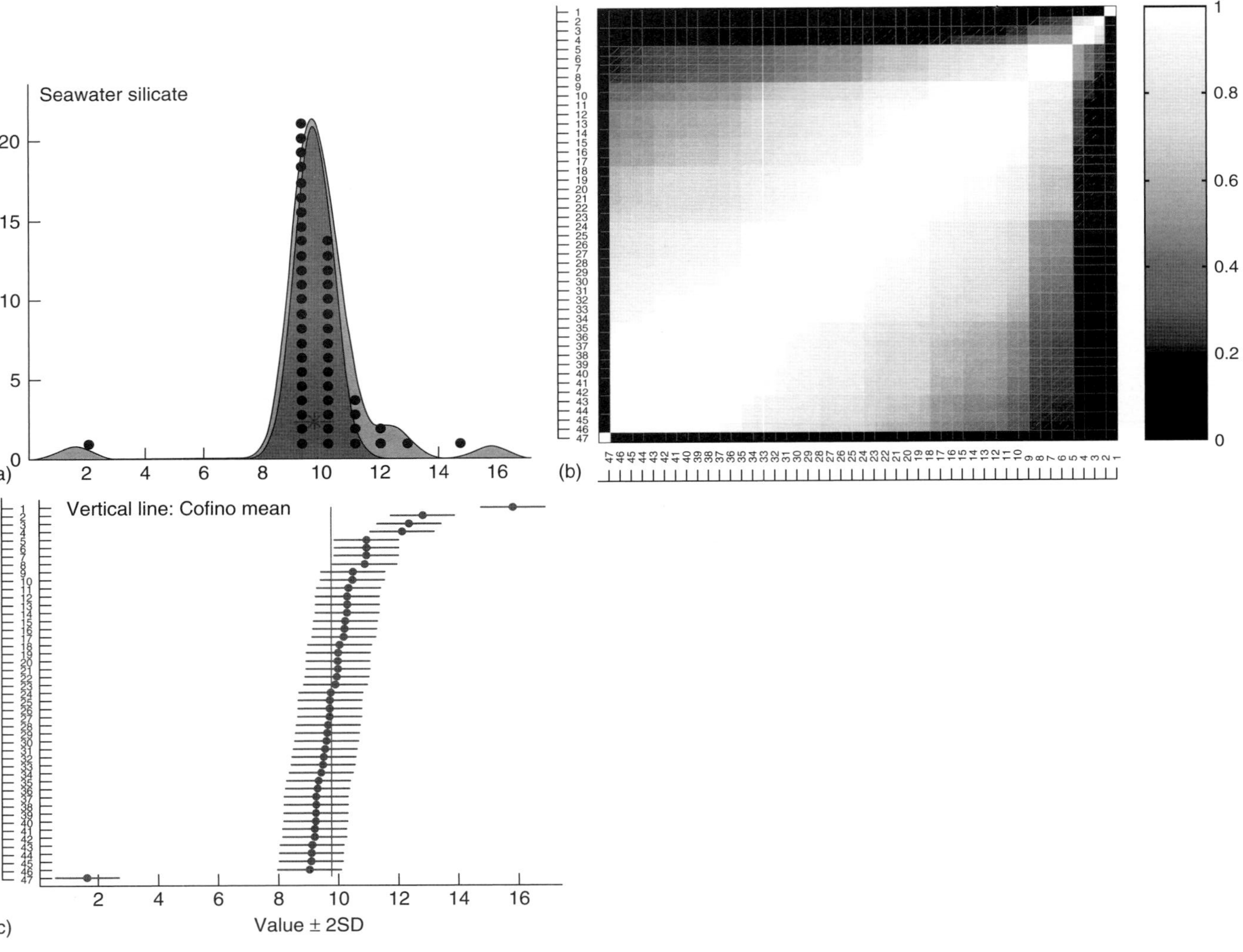

Figure 1 Proficiency test for silicate in seawater showing the Cofino model graphical output. (a) The population measurement function for all data (light area) and for the main mode PMF_1 (dark area). Histogram of data. (b) The overlap matrix. Scale 0 (no overlap of data) to 1 (complete overlap). The extensive, central light area shows a high level of agreement and homogeneity of the data. (c) Ranked overview of data $\pm 2SD_{NDA}$

The three main graphical outputs from the Cofino statistical calculations provide an insight into the structure of the distribution (Figure 1). The pdf conforming for all data (light area) and for data in the main mode (PMF_1) (dark area), allows us to compare the measurements associated with the consensus value with that of the whole distribution. The overlap matrix of each value with all other values is obtained during the calculation which provides us with a map of the homogeneity of the data. The white area (1) denotes overlap (agreement) and the black area (0) denotes no overlap (no agreement). Modes (square areas of similar color) that can be seen within the overlap matrix are not always immediately obvious in the PMF plots (*see* **Multiple Modes in Proficiency Test Data**). The ranked overview provides us with the individual data (or mean of replicates) with the SD reconstructed from the NDA (equation 3) and also gives the location of the consensus value in relation to the distribution. The range of the LCVs is also included.

We have provided two examples for comparison. The mean and SD for silicate in seawater determined by the Huber and Hampel robust estimators differ from our model by <2%. So where data closely reflect a normal distribution, different methods of calculation can produce similar results. Larger differences between these methods of calculation arise with skewed and tailing data. The example of data from the analysis of acenaphthylene in sediment (Table 1 and Figure 2) shows almost 100% difference in the mean using the Huber estimator ($23\,\mathrm{mg\,kg^{-1}}$) and the Cofino model ($12.2\,\mathrm{mg\,kg^{-1}}$). The positive tailing values are *ca* 50% of all numerical values (Figure 2) and have a clear influence on the robust estimators compared with the Cofino model. Why is this so? In Figure 2(b) we can see that there is no overlap of the tailing data with the values associated with the first, main mode, PMF_1 and so effectively do not contribute to the construction of the mean of this mode. The Cofino model, therefore, provides a more meaningful consensus value for the purposes of this PT.

The evaluation of the LCVs by the Cofino model allows us to include all data and provide a more complete assessment. In some cases, where the concentration of the determinant is very low, the LCVs may constitute most of the data. Ignoring these LCVs can lead to a biased interpretation. We have, as an example, measurements of *op′*-DDT in fish muscle (Figure 3) where there are 6 numerical values and 11 LCVs. The LCVs and the two lowest numerical values are in good agreement while the remaining four numerical values are at considerably higher concentration, possibly due to contamination during chemical analysis. The Cofino model *without* LCVs gives a consensus value of $13.9\,\mathrm{\mu g\,kg^{-1}}$ while the robust mean (Huber estimator) is $18.9\,\mathrm{\mu g\,kg^{-1}}$. Using the full Cofino model *with* the LCVs, the consensus value reduces to *ca* $1.5\,\mathrm{\mu g\,kg^{-1}}$, which is more in line with the range of the majority of the data. We should emphasize that in this case the consensus value would most likely be indicative, due to the high uncertainty of the consensus value.

Benefits of the Model

The Cofino model has been developed and tested for use in the determination of population characteristics specifically, but not exclusively, for PTs. The current algorithm used in the NDA implementation of the model assumes an underlying normal distribution, but has a same level of robustness as the Huber and Hampel approaches. The model can be applied to a wider range of data types.

Currently, this model provides comprehensive summary statistics for each mode in a dataset combined with an extensive graphical description of the structure of the data. These outputs are obtained in a single computation without any manipulation of the data, trimming or subjective elimination, including LCVs and missing values.

The model can be applied directly to tailing and skewed data, including extreme values (outliers) and bimodal (multimodal) distributions. In the case of bimodal data, the model may identify each mode and provide an estimate of the mean, SD, and percentage of data associated with each mode (*see* **Multiple Modes in Proficiency Test Data**). The model can include LCVs and has been tested on datasets with up to 60% LCVs. The model works well without the need to have the *within-laboratory* SD by using the NDA developed for this purpose. However, where these values are both available and reliable then they may provide a more sensitive replacement for the NDA.

The detailed, graphical information from the model provide a plot of the PMFs and/or of the population density functions. The color density plot of the

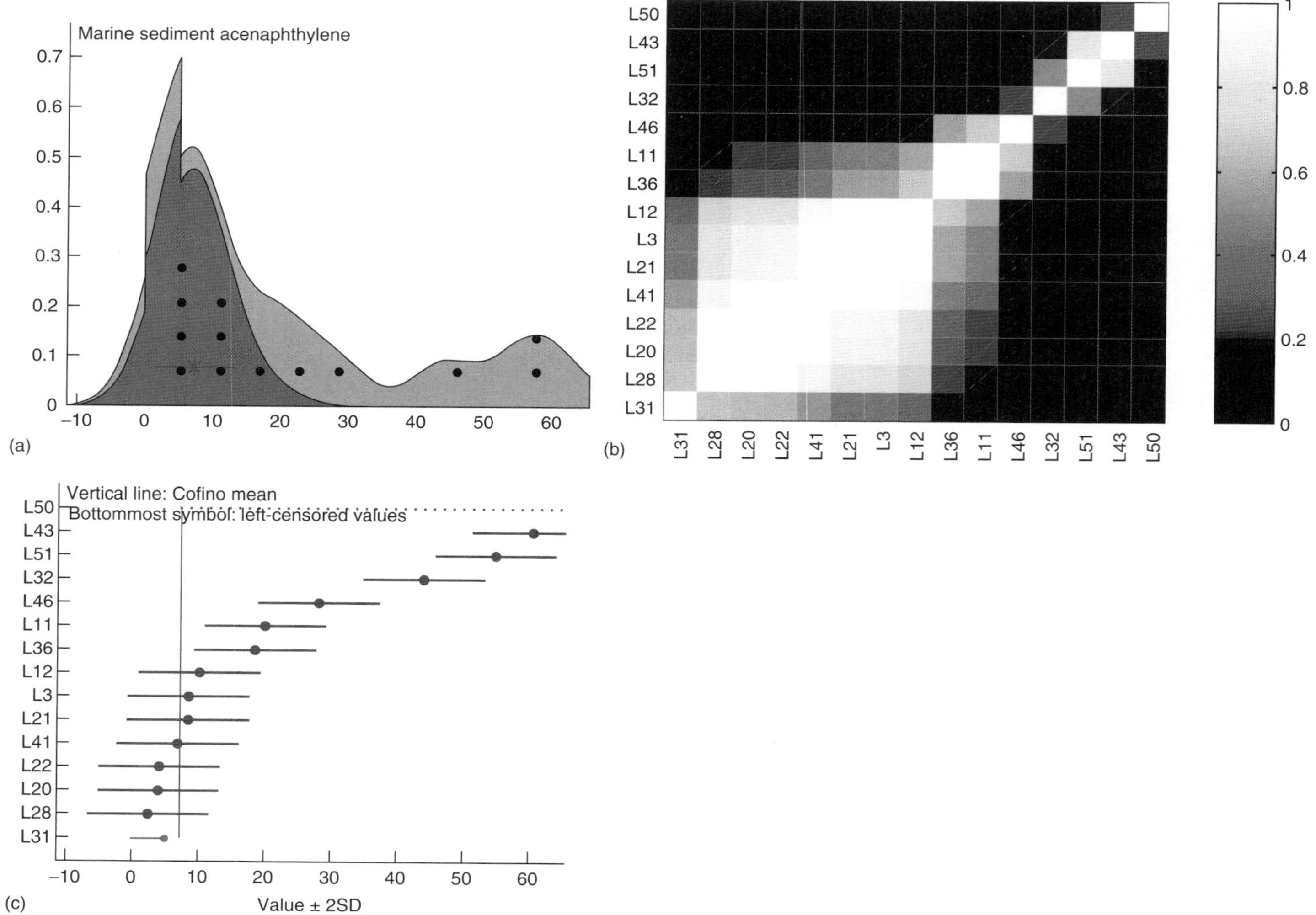

Figure 2 Proficiency test for acenaphthylene in sediment showing the Cofino model output. (a) The population measurement function for all data (light area) showing extensive tailing, and for the main mode PMF_1 (dark area). Histogram of data. (b) Matrix overlap. Scale 0 (no overlap of data) to 1 (complete overlap). The light square shows data below L36 in agreement with the positive tailing values. (C) Ranked overview of data $\pm 2SD_{NDA}$ dark values denote LVCs

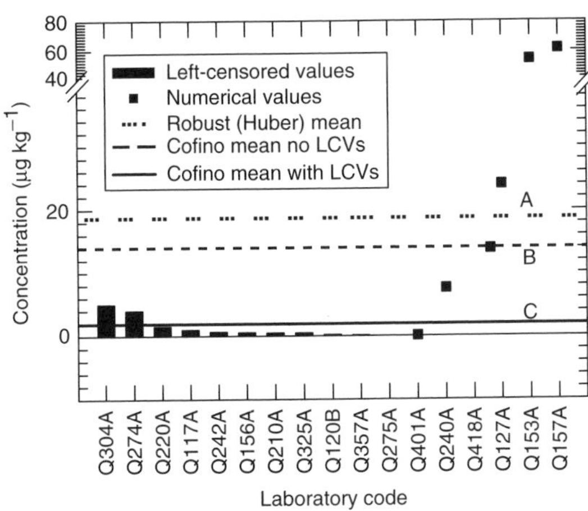

Figure 3 The ranked distribution of data for *op*′-DDT in fish muscle. Of the 17 data, 11 are LCVs. Only the output from the Cofino model with the LCVs included give an estimate that reflects the values found by most laboratories. i.e., $<2\,\mu g\,kg^{-1}$. A: Robust (Huber) mean $18.9\,\mu g\,kg^{-1}$ also only includes numerical data. B: Cofino mean $13.9\,\mu g\,kg^{-1}$ without LCVs. C: Cofino mean $1.5\,\mu g\,kg^{-1}$ with LCVs included

matrix overlap is very sensitive to identifying structure of data, especially modality, where it is not so clear in other plots. Bimodality can be detected with the overlap matrix plot, even when both modes have equal density (*see* **Multiple Modes in Proficiency Test Data**). A ranked overview shows the means, the SD of each laboratory, and the consensus value that includes both numeric and LCVs.

The model has been fully tested with randomly generated normal distributions with the number of observations (NObs) 10–200, relative SD 2–30%, to compare performance with conventional statistics. We have also tested normal distributions overlaid to construct bimodal data. The sensitivity to bimodality, with respect to the distance, and the relative amplitude of the modes has also been tested. An evaluation of over 2000 datasets taken from the QUASIMEME Laboratory Performance Studies has been made using this model.[a]

End Note

[a] The Cofino model has been developed using the MATLAB programming language. Copies of the code and associate information are available from the authors. A compiled version is available for use without the need to obtain the MATLAB programs. The compiled versions cannot be amended.

Acknowledgments

The authors acknowledge the members of the QUASIMEME team, in particular Judith Scurfield, in testing this model, and the participants of the QUASIMEME scheme for providing invaluable data for evaluation. The authors also acknowledge Ian Davies for his helpful critical review of the manuscript.

References

[1] Cofino, W.P. & Wells, D.E. (1994). Design and evaluation of the QUASIMEME inter-laboratory performance studies: a test case for robust statistics, *Marine Pollution Bulletin* **29**, 149–158.

[2] Thompson, M., Ellison, S.L.R. & Wood, R. (2006). The international harmonized protocol for the proficiency testing of analytical chemistry laboratories, *Pure and Applied Chemistry* **78**, 145–196.

[3] Huber, P.J. (1972). Robust statistics: a review, *Annals of Mathematical Statistics* **43**, 1041–1067.

[4] Analytical methods committee robust statistics–how not to reject outliers. Part 1. Basic concepts, *The Analyst* **114**(12) 1693–1697 (1989).

[5] Analytical methods committee robust statistics–how not to reject outliers. Part 2. Inter-laboratory trials, *The Analyst* **114**(12), 1699–1702 (1989).

[6] ISO/CD 20612 (2005). *Water Quality-Interlaboratory Comparisons for Proficiency Test of Laboratories*, International Standards Organization, Geneva.

[7] Müller, C.H. & Uhlig, S. (2001). Estimation of variance components with high breakdown point and high efficiency, *Biometrika* **88**(2), 353–366.

[8] Cofino, W.P., Wells, D.E., Ariese, F., van Stokkum, I.H.M., Wengener, J.W. & Peerboom, R. (2000). A new model for the inference of population characteristics from experimental data using uncertainties, *Journal of Chemometrics and Intelligent Laboratory Systems* **53**, 37–55.

[9] Cofino, W.P., van Stokkum, I.H.M., van Steenwijk, J. & Wells, D.E. (2005). A new model for the inference of population characteristics from experimental data using uncertainties. Part II. Application to censored datasets, *Analytica Chimica Acta* **533**, 31–39.

David E. Wells and Wim P. Cofino

Positive Dependence *see* Aging and Positive Dependence

Power

The power of a *hypothesis test* (*see* **Hypothesis Testing**) is the probability that the null hypothesis H_0 is rejected when some other hypothesis H is true. Suppose the hypotheses are about a parameter θ, and suppose H_0 specifies $\theta = \theta_0$, then the power function is defined as $\beta(\theta) = P_\theta(H_0 \text{ rejected})$. The **significance level** of the test is $\alpha = \beta(\theta_0)$. For $\theta \neq \theta_0$, $\beta(\theta)$ gives the probability that H_0 is correctly rejected given the effect size $\theta - \theta_0$.

Related Article

Sample-Size Determination.

P_p, P_{pk}

P_p and P_{pk} are indices used to express the capability to meet specifications for a product or process characteristic. Higher values denote better capability. P_p and P_{pk} are calculated for stable, normally distributed data according to the equations:

$$P_p = \frac{(\text{Upper specification limit} - \text{Lower specification limit})}{6 \times (\text{Long-term process standard deviation})} \quad (1)$$

$$P_{pk} = \frac{\text{Min(Upper specification limit} - \text{Mean, Mean} - \text{Lower specification limit)}}{3 \times (\text{Long-term process standard deviation})} \quad (2)$$

Note that P_{pk} accounts for the location of the data (through its average or mean value), whereas P_p does not. P_{pk} is defined for either one- or two-sided specification limits, whereas P_p is defined only for two-sided limits. For these reasons, P_{pk} is usually the preferred metric in practice. Because these indices are calculated using a measure of long-term variability, they will generally be lower than the corresponding short-term capability measures, C_p and C_{pk} (*see* $\boldsymbol{C_p}$, $\boldsymbol{C_{pk}}$).

Precedence Tests

Review on Precedence-Type Procedures

In measuring reliability, we observe lifetimes of products or systems of interest. In order to gain a sound knowledge about product or system failure-time distributions, *life-testing and reliability experiments* are carried out before (and while) products are put on the market or systems are used. Life-testing experimentation is a practical way to determine the product reliability and product quality. During the life test, successive times to failure are noted and lifetime data are thus collected. These lifetime data are often used to make decisions on accepting a new design over an existing design. Life testing is useful in many industrial environments including the automobile, materials, chemical, telecommunications, electronics, and pharmaceutical industries. For instance, a manufacturer of electronic components may wish to compare two designs A and B in terms of life. Specifically, he/she may want to abandon design A if there is evidence at the 0.05 level that it has shorter life in comparison to design B.

Assume that a random sample of n_1 units is from F_X (existing design), another independent sample of n_2 units is from F_Y (new design), and that all these sample units are placed simultaneously on a life-testing experiment. We use $X_1, X_2, \ldots, X_{n_1}$ to denote the sample from distribution F_X, and $Y_1, Y_2, \ldots, Y_{n_2}$ to denote the sample from distribution F_Y. A natural null hypothesis is that the two distributions are equal (i.e., there is no difference between the existing design and the new design) and we are generally concerned with the alternative models where one distribution is stochastically larger than the other; for example, the alternative that F_Y is stochastically larger than F_X (the new design produces units with longer life as compared to the existing design) which can be expressed as

$$H_0 : F_X = F_Y \text{ against } H_1 : F_X > F_Y \quad (1)$$

There has been a considerable amount of research work carried out on this problem from a nonparametric point of view; for more details, see the books [1–5].

In order to save on test time and/or to save on the number of test items, it is common that not all the items under test will be observed until failure in an industrial setting. In other words, some of the test items will be withdrawn or removed from the life test. This situation is known as *censoring*. For example, in the situation mentioned above with regard to making a decision on accepting a new design over an existing design, usually the cost of producing items from the new design will be much higher than the cost of producing items from the existing design; therefore, we may decide to observe only a few failures (say, the first r failures) from the new design in order to save a number of items that could potentially be used for some other tests. In this scenario, *precedence test* is useful to test the hypotheses in equation (1). The precedence test is a distribution-free two-sample life test based on the order of early failures, and was first proposed by Nelson [6]. The precedence test allows a simple and robust comparison of two distribution functions. Precedence test, based on the number of X failures that precede the rth Y failure, will be useful (a) when life tests involve expensive units since the units that had not failed could be used for some other testing purposes, and (b) to make quick and reliable decisions early on in the life-testing experiment.

We denote the order statistics from the X sample and the Y sample by $X_{1:n_1} \leq X_{2:n_1} \leq \cdots \leq X_{n_1:n_1}$ and $Y_{1:n_2} \leq Y_{2:n_2} \leq \cdots \leq Y_{n_2:n_2}$, respectively. Without loss of generality, we assume that $n_1 \leq n_2$. Moreover, we let M_1 be the number of X failures before $Y_{1:n_2}$ and M_i be the number of X failures between $Y_{i-1:n_2}$ and $Y_{i:n_2}$, $i = 2, 3, \ldots, r$. We denote the observed value of M_i by m_i. It is of interest to mention here that these M_i's are related to the so-called exceedance statistics whose distributional properties have been discussed, for example, by Fligner and Wolfe [7] and Randles and Wolfe [2].

Nelson [6] provided tables of critical values, which cover all combinations of sample sizes up to 20 for one-sided (two-sided) significance levels of 0.05 (0.10), 0.025 (0.05), and 0.005 (0.01). After Nelson [6] first introduced the precedence life test, many authors have studied the **power** properties of the precedence test and have also proposed some alternatives. For example, Eilbott and Nadler [8] investigated the properties of precedence tests under the assumption that both underlying distributions are exponential. They obtained closed-form expressions for the small-sample and asymptotic power under the exponential distribution. Then, Shorack [9] actually showed that these expressions are valid for a large class of distributions. Subsequently, Young [10] presented some asymptotic results for the precedence test. He also applied the precedence sampling to a population selection problem concerning population **quantiles**. Katzenbeisser [11] derived the distribution as well as the first two **moments** of precedence (he used the term *exceedance* instead) test statistics under Lehmann alternatives. Moreover, Katzenbeisser [12] studied the exact power of two-sample location tests based on precedence test statistics against shifts in exponential, logistic, and rectangular distributions. Liu [13] investigated the properties of the precedence probabilities and their applications, while Lin and Sukhatme [14] considered the best precedence test and compared the power of the best precedence test with other nonparametric and **parametric tests**.

Nelson [15] examined the power of the precedence test when the underlying distributions were normal. After Lin and Sukhatme [14] discussed the best precedence test, van der Laan and Chakraborti [16] used their results to derive the best precedence test under Lehmann alternatives. Recently, a book by Balakrishnan and Ng [17] provides a comprehensive overview of theoretical and applied results relating to a wide variety of problems in which precedence test and its alternatives have been used. They have demonstrated the effectiveness of these tests in life-testing situations designed for making quick and reliable decisions in the early stages of a life-testing experiment.

Precedence Test

Following the aforementioned setting and notation, the precedence test statistic $P_{(r)}$ is defined as the number of failures from the X sample that precede the rth failure from the Y sample; that is,

$$P_{(r)} = \sum_{i=1}^{r} M_i \tag{2}$$

For example, from Figure 1, with $r = 4$, the precedence test statistic takes on the value $P_{(4)} = 0 + 3 + 4 + 1 = 8$.

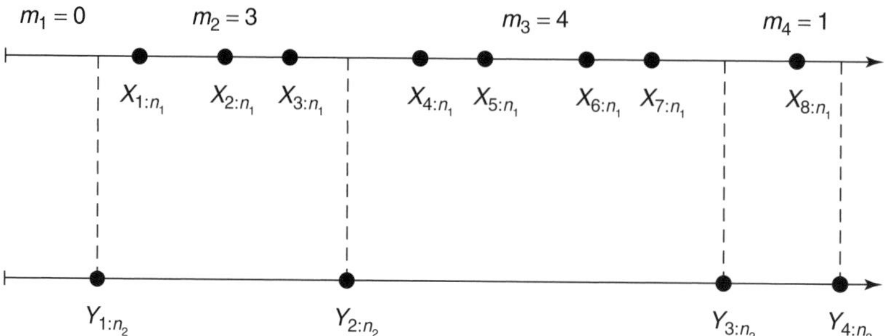

Figure 1 Schematic representation of a precedence life test

It is evident that large values of $P_{(r)}$ lead to the rejection of H_0 in equation (1). For a fixed level of significance α, the critical region will be $\{s, s+1, \ldots, n_1\}$, where

$$\alpha = \mathbf{P}(P_{(r)} \geq s | F_X = F_Y) \qquad (3)$$

For specified values of n_1, n_2, s, and r, an expression for α in equation (3) is given by

$$\alpha = \frac{\sum_{j=s}^{n_1} \binom{s+r-1}{j} \binom{n_1+n_2-s-r+1}{n_1-j}}{\binom{n_1+n_2}{n_2}} \qquad (4)$$

with the summation terminating as soon as any of the factorials involve negative arguments. The critical value s and the exact level of significance α, as close as possible to 5 and 10%, are presented in Balakrishnan and Ng [17] for different choices of the sample sizes n_1 and n_2 and r.

Maximal Precedence Test

Since the precedence test is based on the sum of the frequencies of failures from the X sample between the first r failures of the Y sample, it does not take into consideration the distribution of these frequencies of failures. As a result, there will be a *masking effect* present. Balakrishnan and Frattina [18] noted that such a "masking effect" affects the performance of the precedence test and, therefore, proposed a maximal precedence test which is based on the maximum of the numbers of X failures occurring before the first and between the first and second Y-failures. Balakrishnan and Ng [19] generalized this procedure for up to r Y failures which is called *general maximal precedence test*. It is a test procedure based on the maximum number of failures occurring from the X sample before the first, between the first and the second,..., between the $(r-1)$th and the rth failures, from the Y sample. It is a test procedure useful for testing the hypotheses in equation (1) and it avoids the masking effect.

With the same notation as above, the general maximal precedence test statistic is defined as

$$M_{(r)} = \max(M_1, M_2, \ldots, M_r) \qquad (5)$$

For example, if we refer to Figure 1, with $r = 4$, the maximal precedence test statistic is $M_{(4)} = \max(0, 3, 4, 1) = 4$. Large values of $M_{(r)}$ will lead to the rejection of H_0 and in favor of H_1 in equation (1). The null distribution of $M_{(r)}$, for the special case when $r = 2$, was derived by Balakrishnan and Frattina [18]. Subsequently, Balakrishnan and Ng [19] derived the joint distribution of $M_1, \ldots, M_r$, under $H_0: F_X = F_Y$, as

$$\mathbf{P}(M_1 = m_1, \ldots, M_r = m_r \mid F_X = F_Y) = \frac{\binom{n_1+n_2-\sum_{i=1}^{r} m_i - r}{n_2 - r}}{\binom{n_1+n_2}{n_2}} \qquad (6)$$

and they used equation (6) to derive the null distribution of the general maximal precedence test statistic $M_{(r)}$ for $r \geq 2$. They examined the power properties of the maximal precedence test and compared them with those of the precedence test and Wilcoxon's rank-sum test (based on complete samples). Through their simulation work, they demonstrated that the maximal precedence test does avoid the masking effect affecting the precedence test.

Wilcoxon-Type Precedence Test

We can see that even though an alternative to the precedence test has been proposed in the form of maximal precedence test, it too is based on frequencies of failures preceding the rth Y failure. Although Balakrishnan and Ng [19] demonstrated that the maximal precedence test does avoid the masking effect, it still suffers from a loss of power when compared to the classical Wilcoxon's rank-sum test (based on complete samples). For this reason, Ng and Balakrishnan [20, 21] extended the precedence test in another way by constructing Wilcoxon-type rank based precedence test that takes into account the magnitude of the failure times.

The Wilcoxon rank-sum test is a well-known nonparametric testing procedure for testing the hypothesis in equation (1) based on complete samples. For testing the hypotheses in equation (1), if complete samples of sizes n_1 and n_2 were available, then one can use the standard Wilcoxon's rank-sum statistic, proposed by Wilcoxon [22], which is simply the sum of ranks of X observations in the combined sample.

Ng and Balakrishnan [20] proposed the Wilcoxon-type rank-sum precedence tests for testing the hypotheses in equation (1) when the Y sample is Type-II censored. This test is a variation of the precedence test and a generalization of the Wilcoxon rank-sum test. Three Wilcoxon-type rank-sum precedence test statistics – the minimal, maximal, and expected rank-sum statistics – have been proposed by these authors. In order to test the hypotheses in equation (1), instead of using the maximum of the frequencies of failures from the X sample between the first r failures of the Y sample, one could use the sum of the ranks of those failures. More specifically, suppose $m_1, m_2, \ldots, m_r$ denote the number of X failures that occurred before the first, between the first and the second,..., between the $(r-1)$th and the rth Y failures, respectively; see Figure 1. Let W be the rank sum of the X failures that occurred before the rth Y failure. The Wilcoxon's test statistic will be smallest when all the remaining $\left(n_1 - \sum_{i=1}^{r} m_i\right)$ X failures occur between the rth and $(r+1)$th Y failures. The test statistic in this case would be

$$
\begin{aligned}
W_{\min,r} &= W + \left[\left(\sum_{i=1}^{r} m_i + r + 1\right)\right. \\
&\quad \left. + \left(\sum_{i=1}^{r} m_i + r + 2\right) + \cdots + (n_1 + r)\right] \\
&= \frac{n_1(n_1 + 2r + 1)}{2} - (r+1)\sum_{i=1}^{r} m_i + \sum_{i=1}^{r} i m_i
\end{aligned}
\tag{7}
$$

This is called the *minimal rank-sum statistic.*

The Wilcoxon's test statistic will be the largest when all the remaining $\left(n_1 - \sum_{i=1}^{r} m_i\right)$ X failures occur after the n_2th Y failure. Such a test statistic is called the *maximal rank-sum statistic* and is given by

$$
\begin{aligned}
W_{\max,r} &= \frac{n_1(n_1 + 2n_2 + 1)}{2} \\
&\quad - (n_2 + 1)\sum_{i=1}^{r} m_i + \sum_{i=1}^{r} i m_i
\end{aligned}
\tag{8}
$$

We could similarly propose a rank-sum statistic using the expected rank sums of failures from the first sample between the rth and the $(r+1)$th,..., after the n_2th failures of the second sample, denoted by $W_{E,r}$. It can be shown that $W_{E,r}$ is simply the average of $W_{\min,r}$ and $W_{\max,r}$, and is given by

$$
\begin{aligned}
W_{E,r} &= \frac{n_1(n_1 + n_2 + r + 1)}{2} \\
&\quad - \left(\frac{n_2 + r}{2} + 1\right)\sum_{i=1}^{r} m_i + \sum_{i=1}^{r} i m_i
\end{aligned}
\tag{9}
$$

For example, from Figure 1, when $n_1 = n_2 = 10$ with $r = 4$, we have

$$
\begin{aligned}
W_{\min,4} &= 2 + 3 + 4 + 6 + 7 + 8 + 9 \\
&\quad + 11 + 13 + 14 = 77, \\
W_{\max,4} &= 2 + 3 + 4 + 6 + 7 + 8 + 9 \\
&\quad + 11 + 19 + 20 = 89, \\
W_{E,4} &= \frac{77 + 89}{2} = 83
\end{aligned}
\tag{10}
$$

It is evident that small values of $W_{\min,r}$, $W_{\max,r}$, and $W_{E,r}$ lead to the rejection of H_0 and in favor of H_1 in equation (1). Moreover, in the special case of

$r = n_2$ (i.e., when we observe all the failures from the Y sample), we have $W_{\min,n_2} = W_{\max,n_2} = W_{E,n_2}$ and in this case they are all equivalent to the classical Wilcoxon's rank-sum statistic mentioned earlier.

Ng and Balakrishnan [21] derived the null distributions of these three Wilcoxon-type rank-sum precedence test statistics based on equation (6), while Ng and Balakrishnan [20] observed that the large-sample normal approximation for the null distribution is not satisfactory in the case of small or moderate sample sizes. For this reason, they developed an Edgeworth expansion to approximate the significance probabilities. They also derived the exact power function under the Lehmann alternative and examined the power properties of the Wilcoxon-type rank-sum precedence tests under a location-shift alternative through extensive **Monte Carlo** simulations.

Weighted Precedence and Maximal Precedence Tests

Weighted precedence and maximal precedence tests for testing the hypotheses in equation (1) is another logical extension of the precedence and maximal precedence tests. The motivation for this is explained in the following two cases based on the example mentioned earlier.

Example 1 A manufacturer of electronic components wishes to compare two designs A and B in terms of their life lengths. Specifically, he/she wishes to abandon design A if there is enough evidence at 5% level of significance that it has shorter life in comparison to design B. $n_1 = 10$ components from design A and $n_2 = 10$ components from design B are placed simultaneously on a life test, and the test is to be terminated when the 5th failure from design B occurs. Figures 2 and 3 show two possible outcomes of the life-testing experiment in this example.

The precedence and maximal precedence test statistics are equal in both cases, and they are $P_{(5)} = 8$ and $M_{(5)} = 5$. We have the critical values with $n_1 = n_2 = 10$ and $r = 5$ as 9 (with level of significance 0.02864) and 6 (with level of significance 0.02709) for the precedence and maximal precedence tests, respectively. Therefore, we will not reject the null hypothesis that two distributions are equal in both

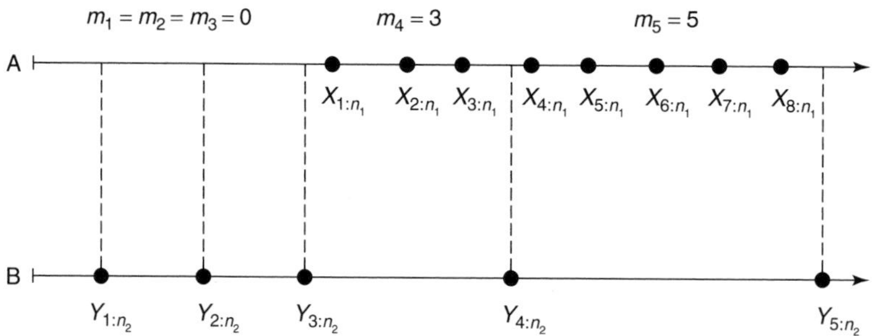

Figure 2 Case 1 of the precedence life test for Example 1

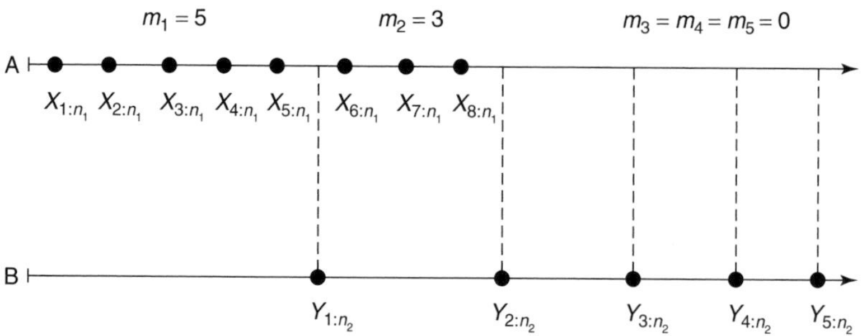

Figure 3 Case 2 of the precedence life test for Example 1

cases at the same level of significance. However, we feel that Case 2 provides much more evidence that design B is better than design A. This suggests that we should develop a test procedure that distinguishes Cases 1 and 2. On this basis, Ng and Balakrishnan [23] proposed the weighted precedence and maximal precedence tests.

The weighted precedence and maximal precedence tests are defined by giving decreasing weights to m_i for increasing i. For example, the weighted precedence test statistic $P^*_{(r)}$ is defined as

$$P^*_{(r)} = \sum_{i=1}^{r} (n_2 - i + 1)\, m_i \tag{11}$$

and the weighted precedence test statistic $M^*_{(r)}$ is defined as

$$M^*_{(r)} = \max_{1 \le i \le r} \{(n_2 - i + 1)\, m_i\} \tag{12}$$

For example, from Figure 1, when $n_1 = n_2 = 10$ with $r = 4$, we have

$$\begin{aligned} P^*_{(4)} &= (9 \times 3) + (8 \times 4) + (7 \times 1) = 66, \\ M^*_{(4)} &= \max\{(9 \times 3), (8 \times 4), (7 \times 1)\} = 32 \end{aligned} \tag{13}$$

Once again, it is clear that large values of $P^*_{(r)}$ or $M^*_{(r)}$ would lead to the rejection of H_0 and in favor of H_1 in equation (1).

Ng and Balakrishnan [23] derived the null distributions of these test statistics and the exact power functions under the Lehmann alternative. They also compared the power (under location shift) of the weighted precedence and maximal precedence tests with those of the original precedence and maximal precedence tests.

We note here that in a precedence test, since the life testing continues until the occurrence of the rth Y failure, it may be viewed as a test based on a Type-II right censored Y sample. Therefore, if we want to save the testing units at the early stage of the life test, a Type-II progressive censoring scheme may instead be employed on the Y sample. For a comprehensive treatment on progressive censoring, one may refer to Balakrishnan and Aggarwala [24]. This will then be a logical extension of the precedence and maximal precedence tests.

Suppose a Type-II progressive censoring scheme is to be adopted on the Y sample which means that a number $r < n_2$ is prefixed for the number of complete failures to be observed and the censoring scheme $(R_1, R_2, \ldots, R_r)$ with $R_j \ge 0$ and $\sum_{j=1}^{r} R_j + r = n_2$ is to be employed at the failure times. During the life-testing experiment, at the time of the jth failure, R_j functioning items are randomly removed from the test. We denote such an observed ordered Y sample by $Y_{1:r:n_2} \le Y_{2:r:n_2} \le \ldots \le Y_{r:r:n_2}$. Moreover, we denote by M_1 the number of X failures before $Y_{1:r:n_2}$ and by M_i the number of X failures between $Y_{i-1:r:n_2}$ and $Y_{i:r:n_2}$, $i = 2, 3, \ldots, r$.

An extension of the weighted precedence and weighted maximal precedence tests to Type-II progressive censoring has been discussed by Ng and Balakrishnan [23] and Balakrishnan and Ng [17]. Then, based on $M_1, M_2, \ldots, M_r$ obtained under Type-II progressive censoring on the Y sample with censoring scheme $(R_1, R_2, \ldots, R_r)$, we propose the weighted precedence test statistic $P^*_{(r)}$ (analogous to equation (11)) as

$$P^*_{(r)} = \sum_{i=1}^{r} \left[n_2 - \left(\sum_{j=1}^{i-1} R_j \right) - i + 1 \right] m_i \tag{14}$$

and the weighted maximal precedence test statistic $M^*_{(r)}$ (analogous to equation (12)) as

$$M^*_{(r)} = \max_{1 \le i \le r} \left\{ \left[n_2 - \left(\sum_{j=1}^{i-1} R_j \right) - i + 1 \right] m_i \right\} \tag{15}$$

For example, from Figure 4, with $n_2 = 10, r = 4$ and censoring scheme (2, 1, 1, 2), the weighted precedence test statistic becomes $P^*_{(4)} = (10 \times 0) + (7 \times 3) + (5 \times 4) + (3 \times 1) = 44$, while the weighted maximal precedence test statistic becomes $M^*_{(4)} = \max\{(10 \times 0), (7 \times 3), (5 \times 4), (3 \times 1)\} = 21$. It is clear that large values of $P^*_{(r)}$ or $M^*_{(r)}$ would lead to the rejection of H_0 and in favor of H_1 in equation (1).

It is important to mention here that the weighted precedence and maximal precedence test statistics defined earlier in equations (11) and (12) become special cases when we set $R_1 = R_2 = \cdots = R_{r-1} = 0$ and $R_r = n_2 - r$.

Exact Power under Lehmann Alternative

The Lehmann alternative $H_1 : [F_X]^\gamma = F_Y$ for some γ is a subclass of the alternative $H_1 : F_X > F_Y$ when $\gamma > 1$; see Lehmann [1] and Gibbons and

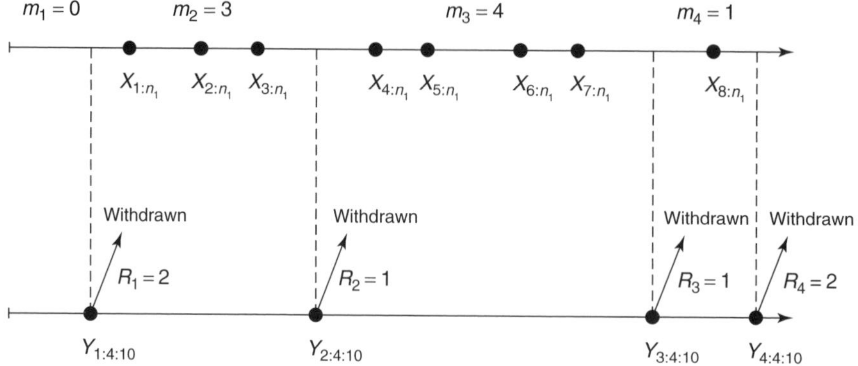

Figure 4 Schematic representation of a precedence life test with progressive censoring

Chakraborti [5]. The joint probability mass function of $M_1, \ldots, M_r$ under the Lehmann alternative is

$$\mathbf{P}\{M_1 = m_1, \ldots, M_r = m_r \mid [F_X]^\gamma = F_Y\}$$
$$= \frac{n_1!n_2!\gamma^r}{m_1!(n_2-r)!}\left\{\prod_{j=1}^{r-1}\frac{\Gamma\left(\sum_{i=1}^{j} m_i + j\gamma\right)}{\Gamma\left(\sum_{i=1}^{j+1} m_i + j\gamma + 1\right)}\right\}$$
$$\times \sum_{k=0}^{n_2-r}\binom{n_2-r}{k}(-1)^k$$
$$\times \frac{\Gamma\left(\sum_{i=1}^{r} m_i + (r+k)\gamma\right)}{\Gamma\left(n_1 + (r+k)\gamma + 1\right)} \tag{16}$$

The power of a test is the probability of rejecting the null hypothesis when the alternative hypothesis is indeed true. So under the Lehmann alternative, the power function for the precedence test and it's alternatives can be determined. Note that the null distribution of the test procedures are obtained simply by putting $\gamma = 1$ in the above formulas. We can see that $H_1 : [F_X]^\gamma = F_Y$ is a subclass of the alternative $H_1 : F_X > F_Y$ when $\gamma > 1$.

The power values of the precedence, maximal precedence, weighted precedence, weighted maximal precedence, and Wilcoxon-type rank-sum precedence tests under the Lehmann alternative have all being computed and presented in Balakrishnan and Ng [17]. From those values, it is seen that the Wilcoxon-type rank-sum precedence tests discussed here give higher power values under the Lehmann alternative for all the cases considered.

With a Type-II progressive censoring scheme $(R_1, R_2, \ldots, R_r)$ on the Y sample, under the Lehmann alternative $[F_X]^\gamma = F_Y, \gamma > 1$, the joint probability mass function of $M_1, M_2, \ldots, M_r$ is given by

$$\mathbf{P}(M_1 = m_1, M_2 = m_2, \ldots, M_r = m_r \mid [F_X]^\gamma = F_Y)$$
$$= C\gamma^r \sum_{j_i (i=1,2,\ldots,r)=0}^{R_i}\binom{R_1}{j_1}\binom{R_2}{j_2}\cdots\binom{R_r}{j_r}$$
$$\times (-1)^{\left(\sum_{i=1}^{r} j_i\right)}$$
$$\times \left\{\prod_{k=1}^{r-1} B\left(\sum_{i=1}^{k} m_i + \gamma\left(\sum_{i=1}^{k} j_i\right) + k\gamma, m_{k+1} + 1\right)\right\}$$
$$\times B\left(\sum_{i=1}^{r} m_i + \gamma\left(\sum_{i=1}^{r} j_i\right) + r\gamma, n_1 - \sum_{i=1}^{r} m_i + 1\right) \tag{17}$$

Note that when $R_1 = R_2 = \cdots = R_{r-1} = 0$ and $R_r = n_2 - r$, the joint probability mass function in equation (17) reduces to the expression in equation (16). The exact power values for different choices of n_1, n_2,

r, and γ for several progressive censoring schemes have been computed and presented in Balakrishnan and Ng [17].

Selection of the Best Population

When we have two or more competing populations, a problem of interest often is to make a decision as to which is the best if we reject the null (homogeneity) hypothesis in equation (1). In the literature of ranking and selection methodology, parametric and nonparametric procedures for selecting the best among k populations based on samples from these populations have been discussed in great detail; see, for example, Bechhofer *et al.* [25], Gibbons *et al.* [26], and Gupta and Panchapakesan [27]. In order to select the best population based on observing only a few failures from two or more samples under life testing, test procedures based on the nonparametric precedence-type statistics with selecting the best population as an objective in the alternative hypotheses will be useful, especially when budget and/or facility constraints are in place in an industrial setting.

Recently, Balakrishnan *et al.* [28] proposed a nonparametric test procedure based on the precedence statistic to test the equality of two distributions against alternatives that a specified population is better than the other. They also studied the k-sample situation and showed that the procedure works well for different underlying distributions by means of Monte Carlo simulations. Similarly, Ng *et al.* [29] used the minimal Wilcoxon rank-sum precedence statistic instead of the precedence statistic to select the best population based on a test for equality. They also compared the probabilities of correct selection of the minimal Wilcoxon rank-sum precedence statistic with those of the precedence statistic and showed that the performance of minimal Wilcoxon rank-sum precedence statistic is better for selecting the best population. For more details about the use of precedence-type statistics in ranking and selection, one may refer to the recent book by Balakrishnan and Ng [17].

References

[1] Lehmann, E.L. (1975). *Nonparametrics: Statistical Methods Based on Ranks*, McGraw-Hill, New York.

[2] Randles, R.H. & Wolfe, D.A. (1979). *Introduction to the Theory of Nonparametric Statistics*, John Wiley & Sons, New York.

[3] Hettmansperger, T.P. & McKean, J.W. (1998). *Robust Nonparametric Statistical Methods*, Edward Arnold, London.

[4] Hollander, M. & Wolfe, D.A. (1999). *Nonparametric Statistical Methods*, 2nd Edition, John Wiley & Sons, New York.

[5] Gibbons, J.D. & Chakraborti, S. (2003). *Nonparametric Statistical Inference*, 4th Edition, Marcel Dekker, New York.

[6] Nelson, L.S. (1963). Tables of a precedence life test, *Technometrics* **5**, 491–499.

[7] Fligner, M.A. & Wolfe, D.A. (1976). Some applications of sample analogues to the probability integral transformation and a coverage property, *The American Statistician* **30**, 78–85.

[8] Eilbott, J. & Nadler, J. (1965). On precedence life testing, *Technometrics* **7**, 359–377.

[9] Shorack, R.A. (1967). On the power of precedence life tests, *Technometrics* **9**, 154–158.

[10] Young, D.H. (1973). A note on some asymptotic properties of the precedence test and applications to a selection problem concerning quantiles, *Sankhyā Series B* **35**, 35–44.

[11] Katzenbeisser, W. (1985). The distribution of two-sample location exceedance test statistics under Lehmann alternatives, *Statistical Papers* **26**, 131–138.

[12] Katzenbeisser, W. (1989). The exact power of two-sample location tests based on exceedance statistics against shift alternatives, *Mathematische Operationsforschung und Statistik* **20**, 47–54.

[13] Liu, J. (1992). Precedence probabilities and their applications, *Communications in Statistics: Theory and Methods* **21**, 1667–1682.

[14] Lin, C.H. & Sukhatme, S. (1992). On the choice of precedence tests, *Communications in Statistics: Theory and Methods* **21**, 2949–2968.

[15] Nelson, L.S. (1993). Tests on early failures – the precedence life test, *Journal of Quality Technology* **25**, 140–143.

[16] van der Laan, P. & Chakraborti, S. (2001). Precedence tests and Lehmann alternatives, *Statistical Papers* **42**, 301–312.

[17] Balakrishnan, N. & Ng, H.K.T. (2006). *Precedence-Type Tests and Applications*, John Wiley & Sons, Hoboken.

[18] Balakrishnan, N. & Frattina, R. (2000). Precedence test and maximal precedence test, in *Recent Advances in Reliability Theory: Methodology, Practice, and Inference*, N. Limnios & M. Nikulin, eds, Birkhäuser, Boston, pp. 355–378.

[19] Balakrishnan, N. & Ng, H.K.T. (2001). A general maximal precedence test, in *System and Bayesian Reliability*, Y. Hayakawa, T. Irony & M. Xie, eds, World Scientific Publishing, Singarpore, pp. 105–122.

[20] Ng, H.K.T. & Balakrishnan, N. (2002). Wilcoxon-type rank-sum precedence tests: large-sample approximation and evaluation, *Applied Stochastic Models in Business and Industry* **18**, 271–286.

[21] Ng, H.K.T. & Balakrishnan, N. (2004). Wilcoxon-type rank-sum precedence tests, *Australia and New Zealand Journal of Statistics* **46**, 631–648.
[22] Wilcoxon, F. (1943). Individual comparisons by ranking methods, *Biometrics Bulletin* **1**, 80–83.
[23] Ng, H.K.T. & Balakrishnan, N. (2002). Weighted precedence and maximal precedence tests and an extension to progressive censoring, *Journal of Statistical Planning and Inference* **135**, 197–221.
[24] Balakrishnan, N. & Aggarwala, R. (2000). *Progressive Censoring: Theory, Methods and Applications*, Birkhäuser, Boston.
[25] Bechhofer, R.E., Santner, T.J. & Goldsman, D.M. (1995). *Design and Analysis of Experiments for Statistical Selection, Screening, and Multiple Comparisons*, John Wiley & Sons, New York.
[26] Gibbons, J.D., Olkin, I. & Sobel, M. (1999). *Selecting and Ordering Populations: A New Statistical Methodology, Classics in Applied Mathematics, Vol. 26*, Society for Industrial and Applied Mathematics, Philadelphia. Unabridged reproduction of the same title, John Wiley & Sons: New York, 1977.
[27] Gupta, S.S. & Panchapakesan, S. (2002). *Multiple Decision Procedures: Theory and Methodology of Selecting and Ranking Populations, Classics in Applied Mathematics, Vol. 44*, Society for Industrial and Applied Mathematics, Philadelphia. Unabridged reproduction of the same title, John Wiley & Sons: New York, 1979.
[28] Balakrishnan, N., Ng, H.K.T. & Panchapakesan, S. (2006). A nonparametric procedure based on early failures for selecting the best population using a test for equality, *Journal of Statistical Planning and Inference* **136**, 2087–2111.
[29] Ng, H.K.T., Balakrishnan, N. & Panchapakesan, S. (2007). Selecting the Best Population Using a Test for Equality Based on Minimal Wilcoxon Rank-sum Precedence Statistic, *Methodology and Computing in Applied Probability* **9** 263–305.

NARAYANASWAMY BALAKRISHNAN AND
HON KEUNG TONY NG

Precontrol

Description of the Method

Precontrol monitors the acceptability of a process by comparing individual process values against their specification limits. Its goal is the prevention of nonconforming units. Precontrol, also referred to as *stoplight control*, classifies measurement data into several groups and monitors the process on the basis of the resulting attribute data. In the case of two-sided tolerance, precontrol lines are drawn at the midpoints between the target and the lower (upper) specification limits. The area between the precontrol limits is referred to as the *green zone*, the areas between the precontrol limits and the specification limits are called the *yellow zones*, and the areas outside the specification limits are referred to as the *red zones*.

Initial process acceptability (capability) is established through a simple setup approval procedure. Five consecutive units are taken from the process, and all five measurements must fall within the green zone for the production to commence. If a single yellow result is observed, additional observations are taken and the count of units falling into the green zone is started anew. If two consecutive yellows or a red are encountered, the process is adjusted (see below) and the approval process is restarted.

Following setup approval, two consecutive units are sampled from the process at periodic intervals. If both units fall into the green zone, or if only one unit falls into a yellow zone, production continues. The process is stopped if both units fall into yellow zones, or if one or more unit falls into a red zone. After stopping the process, one must find the cause of the problem and take action. If both units fall into the same yellow zone, it is likely that the process is off-target. In this case, level adjustments are made using feedback control on the deviations from the target. If consecutive yellows (or reds) are on opposite sides of the target, it is likely that the variability has increased. Causes for the extra variability need to be identified and their occurrence must be prevented in the future. The setup approval stage is repeated after each stop. Five consecutive units must be in the green zone before production can resume.

It is suggested that the frequency of the sampling be set such that the average number of sample pairs between stoppages is about six. For example, if samples are taken every hour and stoppages occur on average every 6 h, the sampling interval is set correctly. The sampling interval should be cut in half if stoppages occur on average every 3 h, and it should be doubled if stoppages occur on average every 12 h.

Several modifications of the basic precontrol technique are discussed in the literature. A two-stage sampling approach that takes an additional sample if the initial sample yields ambiguous results is discussed by Salvia [1]. A modification that replaces the specification limits with control limits (which use the standard deviation of process measurements) is proposed by Gurska and Heaphy [2].

Historical Note

Precontrol was first described by Satterthwaite [3] who attributes the development of the technique to himself and C.W. Carter, W.R. Purcell, and D. Shainin. The method is described in Bhote [4, 5], Juran and Gryna [6], Shainin [7], and Shainin and Shainin [8].

Example

Bhote [4, p. 228] describes an example that deals with the thickness of certain chrome and gold deposits. For chrome the target thickness is 9 units, with specification limits of 6 and 12, resulting in precontrol limits of 7.5 and 10.5. For gold the target thickness is 32, with specification limits of 23 and 41, resulting in precontrol limits of 27.5 and 36.5, respectively. The data in Table 1 indicate no problems with the capability of the thickness of gold deposits. However, problems are indicated for the thickness of chrome deposits. Two consecutive yellows on the same side of the target are observed on April 4. The first sample on April 5 leads to two consecutive units on opposite sides, while in the second sample one of the observations is in the red zone. The process must be investigated and reasons for the unacceptable results must be found. Often this is easier said than done, as considerable detective work may be required. Bhote makes no mention of how the process was adjusted in this particular example.

Differences between Precontrol and Control Charts

Both precontrol and control charts take periodic observations from a process. Despite their similarities, the two techniques differ with respect to their objectives. The goal of precontrol is the prevention of nonconforming units. It is an algorithm for controlling a process on its tolerances (without explicit consideration of stability), while control charts monitor the stability of the process over time (without specifically looking at its capability). Control charts (*see* **Control Charts, Overview**), process capability, and links between control charts and capability (*see* **Control Charts and Process Capability**) are described in more detail elsewhere.

For very capable processes, a control chart may signal lack of control even though the shifted process is still well within its specifications. It has been argued that in this case control charts are of limited value, as operators would question the wisdom of disturbing a process that produces good items.

Strengths of Precontrol

In situations where processes are highly capable, precontrol offers a simple useful technique for checking the capability of the process. Precontrol reduces the temptation of interfering with processes that are still capable. The procedure will not flag unstable processes as long as they remain capable. The approach has the virtue of drawing management attention only to priority problems.

Table 1 Specification limits. Chrome: 6 and 12. Gold: 25 and 41[(a)]

	Unit 1		Unit 2		
	Chrome	Gold	Chrome	Gold	
April 3	9.6	32.0	10.2	30.0	
April 4	10.2	33.0	9.6	33.5	
April 4	**10.7**	33.5	**11.2**	35.0	Two yellows on same side (chrome)
April 4	7.6	30.5	8.6	25.4	
April 5	**10.5**	34.0	**6.5**	34.5	Two yellows on opposite sides (chrome)
April 5	**12.2**	30.5	10.2	32.0	One red (chrome)
April 6	7.6	33.5	9.6	35.0	

[(a)]Values outside the green zone (7.5 and 10.5 for chrome, and 27.5 and 36.5 for gold) are in boldface

Precontrol was originally developed with machining operations in mind. There, the operator is faced with the problem of first setting up the machine, and deciding whether the setup has been done properly and whether the machine is ready for full production. In many machining operations the output variable will drift owing to factors such as tool wear. Simple feedback algorithms that bring the process back to its target are usually possible, and operators with considerable machining experience know how to make these adjustments.

Precontrol (just like control charts) forces the operator to take periodic measurements on a process. Quite often it is not the limits or the zones on these charts that are important, but the fact that process information is obtained. Sudden shifts, trends, cycles, and the presence of overcompensation become visible.

Weaknesses of Precontrol

Precontrol is easy to implement, but its *ad hoc* decision rules on grouped data have led to criticism. In particular, it has been questioned whether the small setup inspection sample size of five consecutive items is able to detect processes with low capability. Furthermore, grouping discards information, and there is a real danger that poor processes slip through the certification stage (Ledolter and Swersey [9]).

Also, there are important reasons why one should worry about process instability even though products are still acceptable. (a) Specifications may change (what is good today may not be good tomorrow); (b) unstable processes may not stay capable for long (within specifications today, but not tomorrow); (c) any deviation from a target, even if the unit is within the specification limits, leads to a loss; and (4) it is difficult to learn about root causes if a process is not in statistical control.

Precontrol is poorly suited for processes with low capability, as its frequent signals will lead to corrective actions that in the absence of profound process understanding amount to little more than process tampering. Processes with low capability must be improved. This is best done by first making sure that assignable (special) causes of variation have been identified and removed. Typically, this requires a thorough improvement cycle that utilizes control charts as well as other improvement techniques. It would be optimistic to expect that the *ad hoc* adjustments of precontrol can fix the problem. Precontrol alone provides little guidance for process improvement.

Further Discussion

Detailed discussion of the strengths and weaknesses of precontrol can be found in Traver [10], Logothetis [11], Mackertick [12], Ermer and Roepke [13], Steiner [14], Ledolter and Swersey [9], and Ledolter and Burrill [15].

References

[1] Salvia, A. (1988). Stoplight control, *Quality Progress* **21**, 39–42.

[2] Gurska, G. & Heaphy, M. (1991). Stop light control – revisited, *ASQC Statistics Division Newsletter Fall* **11**(4), 10–11.

[3] Satterthwaite, F. (1954). A simple effective process control method, Report 41-1, Rath and Strong, Boston.

[4] Bhote, K. (1988). *Strategic Supply Management*, American Management Association, New York.

[5] Bhote, K. (1991). *World Class Quality*, American Management Association, New York.

[6] Juran, J. & Gryna, F. (1988). *Quality Control Handbook*, 4th Edition, McGraw-Hill, New York.

[7] Shainin, D. (1984). Better than good old Xbar and R charts asked by vendees, in *ASQC Quality Congress Transactions*, American Society for Quality Control, Milwaukee, pp. 302–307.

[8] Shainin, D. & Shainin, P. (1989). Pre-control versus Xbar and R charting: continuous or immediate quality improvement, *Quality Engineering* **1**, 419–429.

[9] Ledolter, J. & Swersey, A. (1997). An evaluation of pre-control, *Journal of Quality Technology* **29**, 163–171.

[10] Traver, R. (1985). Pre-control: a good alternative to Xbar-R charts, *Quality Progress* **18**(9), 11–14.

[11] Logothetis, N. (1990). The theory of pre-control. A serious method or a colorful naivity? *International Journal of Total Quality Management* **1**, 207–220.

[12] Mackertich, N. (1990). Precontrol versus control charting. A critical comparison [with discussion by Dorian Shainin], *Quality Engineering* **2**, 253–268.

[13] Ermer, D. & Roepke, J. (1991). An analytical analysis of pre-control, in *ASQC Quality Congress Transactions*, American Society for Quality Control, Milwaukee, pp. 522–527.

[14] Steiner, S. (1997). Pre-control and some simple alternatives, *Quality Engineering* **10**, 65–74.

[15] Ledolter, J. & Burrill, C. (1999). *Statistical Quality Control: Strategies and Tools for Continual Improvement*, John Wiley & Sons, New York.

Johannes Ledolter

Prediction of Expected Fatigue Lives of Fiber Reinforced Plastic Joints

Introduction

Analysis of fatigue is typically based on the concept of a function D that represents the accumulated damage due to stresses and strains occurring at any given time. This function is presumed to increase monotonically, and failure is expected when the accumulated damage reaches some critical level. Usually the damage function is normalized so that it reaches unity at failure $-D = 1$ at the time of failure. This reasonable concept is very compatible with stochastic modeling, with $\{D\}$ becoming a stochastic process when stresses are stochastic. The problem of predicting time of failure then becomes one of studying the $\dot{D}$ rate of growth of $\{D\}$. The practical difficulty in implementing the accumulated **damage model** is that D is not observable by currently available means. D is known at only two points; it is presumed to be at least approximately zero for a specimen that has not yet been subjected to loads and it is unity when failure occurs [1].

Background

Current fatigue analysis methods are based on various approximations and assumptions about the function D. The goal in formulating such approximations is to achieve compatibility with the results of experiments. These experiments are sometimes performed with quite complicated time histories of loading, but more typically they involve simple periodic (possibly harmonic) loads in so-called *constant-amplitude tests*. It is presumed that each period of the motion contains only one peak and one valley. In this situation, the number of cycles until fatigue failure is usually found to depend primarily on the amplitude of the cycle, although this is usually characterized with the alternative nomenclature of *stress range*, which is essentially the double amplitude, being equal to a peak value minus a valley value. There is a secondary dependence on the mean stress level, which can be taken as the average of the peak and valley stresses [1].

Constant-Amplitude Fatigue

The notation S_r denotes the stress range of a cycle and N_f designate the number of cycles until failure in a constant-amplitude test. A typical experimental investigation of constant-amplitude fatigue for specimens of a given configuration and material involves performing a large number of tests including a number of values of S_r, and then plotting the (S_r, N_f) results. This is called an *S/N curve*, and it forms the basis for most information and assumptions about D. One can emphasize the dependence of fatigue life on stress range by writing $N_f\ (S_r)$ for the fatigue life observed for a given value of the stress range. In principle, the S/N curve of $N_f\ (S_r)$ *versus* S_r could be any nonincreasing curve, but experimental data commonly show that a large portion of that curve is well approximated by an equation of the form

$$N_f(S_r) = K S_r^{-m} \tag{1}$$

in which K and m are positive constants whose values depend on both the material and the geometry of the specimen. This S/N curve becomes a straight line on log–log paper. If S_r is given either by small or very large values, then the form of equation (1) will generally no longer be appropriate, but fatigue analysis usually consists of predicting failure due to moderately large loads, so equation (1) is often quite useful.

Because of the considerable scatter in S/N data, it is necessary to use a statistical procedure to account for specimen variability and choose the best $N_f(S_r)$ curve. For the common power-law form of equation (1), the value of K and m are usually found from linear regression, which minimizes the mean-squared error in fitting the data points. It should be noted that this linear regression of data does not involve the use of any probability model.

As noted, the fatigue life also has a secondary dependence on the mean stress. Although this effect is often neglected in fatigue prediction, there are simple empirical formulas that can be used to account for it in an approximate way. One such approach is called the *Goodman correction*, and it postulates that the damage done by a loading $x(t)$ with mean stress x_m and stress range S_r is the same as would be done by an "equivalent" mean-zero loading with an increased stress range of S_e, given by $S_e = S_r/(1 - x_m/x_u)$, in which x_u denotes the ultimate stress capacity of the material. The *Gerber correction* is similar, $S_e = S_r/[1 - (x_m/x_u)^2]$, and has often been determined to be in better agreement with empirical data. Note that use of the Goodman or Gerber correction allows one to include both mean stress and stress range in fatigue prediction, while maintaining the simplicity of a one-dimensional S/N relationship [1].

Another power-law relationship established in metal fatigue is the Manson–Coffin relationship, usually expressed as [2]

$$\varepsilon_{ap} = \varepsilon_f(N)^c \tag{2}$$

in which ε_{ap} is the applied plastic strain amplitude, and ε_f and c are fatigue coefficients. The Manson–Coffin equation is usually employed in the context of local approaches to fatigue damage accumulation; only in such a situation can the plastic strain magnitude be quantified from local material response due to stress concentration and/or residual stresses.

Because it is difficult to define plastic strain for composite materials, the Manson–Coffin relation has not found any application. However, owing to its universality, the power-law S/N curve is often used to characterize fatigue failures, especially when little information on the fatigue behavior of a particular material exists. In composite fatigue, this relationship is usually associated with the names of Hahn and Kim [3] who preferred to express it as

$$sN^d = c \tag{3}$$

In this equation d and c are model (material) parameters, and s is the ratio of the applied stress, S, and the material's ultimate strength, S_{ult}, i.e.,

$$s = \frac{S}{S_{ult}} \tag{4}$$

This ratio is frequently applied as a measure of loading severity in fatigue studies.

Hahn and Kim also consider the semilogarithmic S/N curve model in which the applied stress level is related to $\log(N)$, expressing it as

$$s = k\log(N) + b \tag{5}$$

where k and b are material parameters and s is given by equation (4) [2].

Stochastic Fatigue

As previously noted, the fundamental problem in stochastic analysis of fatigue is to define a stochastic process $\{D\}$ to model the accumulation of damage caused by a stochastic stress time history $\{X\}$. It seems that most predictions of fatigue life in practical problems are based on simpler models in which information about the accumulated damage is obtained from the constant-amplitude S/N curve.

Hwang and Han [4] suggested that a composite material D can be most generally expressed as

$$D = D(D_0, n, S, f, T, M, \ldots) \tag{6}$$

where D_0 is the initial damage existing prior to the application of the considered loading segment, n is the number of fatigue cycles, S is the applied stress level, f is the loading frequency, T is the temperature, and M is the moisture content. The effect of the last three parameters on fatigue damage will not be considered in this section.

The combined influence of the other three parameters, D_0, n, and S, on the fatigue damage accumulation can be considered in the assumption that when damage is caused by a multilevel stress history, not only does the damage contribution from each loading cycle need to be added up, but the influence of the loading of each cycle on the damage caused by subsequent cycles needs to be identified. This is quite a complex situation that is rarely employed in practice. The first simplification is done by neglecting the effect of D_0. That is, by assuming that damage caused by the sequence of n cycles with magnitude S is always the same regardless of where this sequence occurs in the loading history. Such damage accumulation rules have been considered in the context of fatigue of fiber reinforced plastic (FRP) laminates and are discussed below.

The simplest damage accumulation law, which is commonly known as *Palmgren–Miner's rule*, assumes that damage contribution from each cycle of the loading history is independent of the other cycles. Therefore, the damage inflicted by n stress cycles with magnitude S is simply

$$D = \frac{n}{N} \tag{7}$$

where N denotes the cycles to failure at S from the constant-amplitude S/N curve. For all stress levels this damage rule yields

$$D = \sum_{i=1}^{m} \frac{n_i}{N_i} \tag{8}$$

where n_i is the number of cycles having amplitude S_i and N_i is the cycles to failure at S_i.

In many situations, the rate of fatigue damage accumulation in composites may deviate significantly from that assumed in equation (7). For example, Howe and Owen [5] reported that

$$D = B\left(\frac{n}{N}\right) - C\left(\frac{n}{N}\right)^2 \quad B, C > 0 \tag{9}$$

fits much better with constant-amplitude data that they obtained from chopped strand mat (CSM) composites. Such an equation, however, has no particular benefit over the linear damage accumulation rule, equation (7). Making damage a function of the ratio of n/N only implies that it is always the same regardless of the cycle's position in the loading sequence. Therefore, there is always a measure of damage that accumulates linearly under constant-amplitude loadings.

In connection with equation (1), Palmgren–Miner's rule equation (8) is used almost exclusively in metal fatigue. By treating a multilevel stress time history as a random process, equations (1) and (8) sometimes allow quite a simple stochastic approach to be used to estimate cumulative damage. This approach is applicable when the loading can be described as a zero-mean, Gaussian, and narrowband stochastic process. In such a case, the amplitudes of the fatigue cycles possess a Rayleigh probability distribution, which can be expressed as

$$p_S(S) = \frac{S}{S_{\text{rms}}^2} \exp\left(-\frac{S^2}{2S_{\text{rms}}^2}\right) \tag{10}$$

In the above formula, S_{rms} denotes the root-mean-square (RMS) value of the stochastic stress history. By using the Palmgren–Miner hypothesis equation (7) and the power-law S/N curve equation (1), the expected fatigue damage $E[D]$ and the expected cycles to failure $E[N]$ are expressed as

$$E[D] = \frac{1}{E[N]} = \frac{1}{K} E[S^m] = \frac{1}{K} \int_0^\infty \frac{S^{m+1}}{S_{\text{rms}}^2} \exp\left(-\frac{S^2}{2S_{\text{rms}}^2}\right) \mathrm{d}S \tag{11}$$

The integration of the above equation yields

$$E[D] = \frac{1}{E[N]} = \frac{1}{K} 2^{\frac{m}{2}} S_{\text{rms}}^m \Gamma\left(1 + \frac{m}{2}\right) \tag{12}$$

where $\Gamma(\cdot)$ is the gamma function. Equation (12) is known as the *Rayleigh approximation.*

An expression similar to equation (12) can be derived if the Palmgren–Miner rule is applied in conjunction with the semilogarithmic S/N curve model equation (5). To this end, assume first that the semilogarithmic model is expressed as

$$\log(N) = b - aS \tag{13}$$

in which b and a are the parameters of the model obtained from a linear regression of the experimental data. This equation can be further rearranged as

$$N = 10^{b-aS} = \mathrm{e}^{\overline{b}-\overline{a}S} \tag{14}$$

where

$$\overline{b} = b\ln(10)\overline{a} = a\ln(10) \tag{15}$$

The expected damage can then be expressed as

$$E[D] = \frac{1}{E[N]} = \mathrm{e}^{-\overline{b}} \int_0^\infty \frac{S}{S_{\text{rms}}^2} \times \exp\left(-\frac{S^2}{2S_{\text{rms}}^2} + \overline{a}S\right) \mathrm{d}S \tag{16}$$

Performing the integration in equation (16) gives

$$E[D] = \frac{1}{E[N]} = \mathrm{e}^{-\overline{b}} \left[1 + \overline{a}S_{\text{rms}} \exp\left(\frac{\overline{a}^2 S_{\text{rms}}^2}{2}\right) \times \Phi(\overline{a}S_{\text{rms}})\sqrt{2\pi}\right] \tag{17}$$

in which $\Phi(\cdot)$ denotes the Gaussian cumulative function [6].

Experimental Approach

The composite test program discussed in this paper was set up to provide a more realistic insight into the static and fatigue behavior and damage accumulation of FRP joints. To study the fatigue damage accumulation in FRP specimens, several different specimen configurations were designed and fabricated for characterization under constant- and variable-amplitude fatigue loadings.

The base laminate consisted of an alternating 0/90° stacking sequence of 30 plies made of 0.025 in. thick Woven Roving (WR) glass fabric and 510-A vinyl ester resin. This layup was fabricated by Hardcore DuPont Composites, L.L.C., into several panels measuring 2 ft by 3 ft using the Seeman composite resin infusion molding process (SCRIMP) process. The postcure cycle for the panels was 2 h at 160 °F. The panels were then cut into specimens.

The behavior of composite joints under static, constant-amplitude, and variable-amplitude fatigue was studied. Each configuration was realized in a single geometry butt-strap joint. Specimens were made of the same laminate as previously discussed with the addition of 15 ply laminates of the same layup; joint adherents were oriented along the 0/90° material direction. Figure 1 shows the typical geometries of joint specimens [6].

Static Tests

The static tests consisted of ultimate tension and compression strength tests, which were designed to characterize the static behavior of the specimens and to aid in selecting the appropriate stress levels for the fatigue tests. In all configurations, some strain gages were mounted on the surface of the joints in order to monitor the evolution of the state at the points of interest. All bonded specimens failed in an adhesive mode. The bolted joint exhibited a combination of cleavage-tension and shear-out failure and the bolted-bonded joint, after demonstrating first a bond failure, had finally failed in a shear-out mode. In the compressive regime, the bonded joint had delaminated in both laps and one of the adherents owing to the combined effect of contraction and buckling. Similarly, the bolted joint had shortened and buckled antisymmetrically (mode 2) and all joint components had delaminated as a consequence. The bonded-bolted joint showed an early bond failure and a final failure due to a symmetric buckling and a consequent delamination in all components. The bolted-bonded joint is the most advantageous configuration, as far as the tensile strength is concerned, since it has a higher initial stiffness, a higher failure load and a very good ductility compared to the others. On the other hand, in the compressive regime, although it has a higher initial stiffness, its strength and ductility are lower than those of the bolted configuration, which may be due to the difference in the buckling mode, since a mode 2 buckling requires a higher load to occur [6].

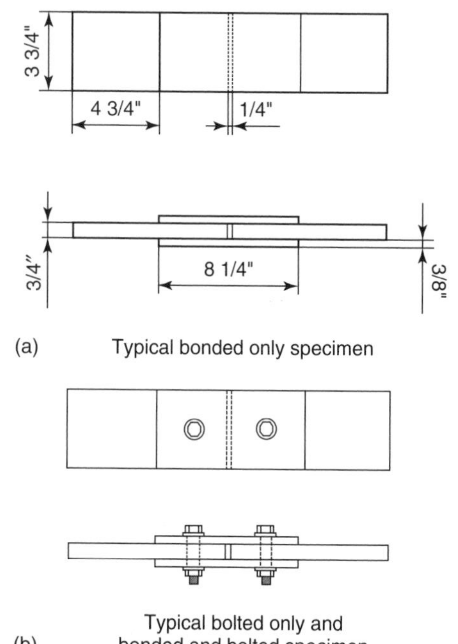

Figure 1 Geometry of different types of composite joints

Fatigue Tests

The fatigue tests typically consisted of two to three specimens tested at each stress level, or RMS level for constant or variable amplitude, respectively. All tests were performed at constant frequency. The variable-amplitude loading represents a segment of a sinusoidal time history composed of 10 000 end points, which with the assumption of stationary and ergodicity can be shown to be a narrowband Gaussian process. This segment is repeated until failure of the specimen occurs [6].

Tables 1 and 2 show the constant-amplitude and variable-amplitude test results for different types of composite joints.

The failure of the bonded joints is an adhesive failure of the bond, while that of the bolted joints is a bearing failure followed by a partial shear-out failure. The failure of the bonded-bolted joints is a two-stage one. First, and very early, comes the adhesive failure of the bonds and then, typically after a long time, the bearing and partial shear-out failures of the bolts occurs.

The experimental fatigue lives from the constant-amplitude tests of all joint configurations are presented in Figure 2 on a log(S)–log(N) scale and in Figure 3 on a S-log(N) scale. Both logarithmic and semilogarithmic S/N curve models were fit to these data by linear regression. The curves indicate a definite deviation from the logarithmic S/N curve model for all joint configurations. Neither S/N curve model appears to represent the data well.

Variable-Amplitude Fatigue Tests Analysis

Experimental variable-amplitude fatigue lives are obtained by applying the aforementioned time

Table 1 Constant-amplitude fatigue tests of FRP joints

Joint type	ID name	Maximum stress level (ksi)	Maximum load level (lbs)	Cycles to failure
Bonded only	JA-9	2	5044	2 411 800
	JA-10	2	5042	3 369 600
	JA-20	2	5035	2 748 520
	JA-7	3	7735	51 900
	JA-8	3	7591	255 400
	JA-19	3	7735	1 000 300
	JA-17	4	10 134	98 600
	JA-18	4	10 170	86 200
	JA-5	5	12 806	3170
	JA-6	5	12 846	4220
	JA-3	7.5	18 737	980
	JA-4	7.5	18 764	1060
Bolted only	JB-7	3	7607	589 400
	JB-8	3	7019	20 000 000[(a)]
	JB-13	3	7307	23 200
	JB-14	3	7319	899 100
	JB-11	3.5	8617	12 600
	JB-12	3.5	8409	8720
	JB-9	4	9623	1810
	JB-10	4	9464	2530
	JB-5	5	11 795	890
	JB-6	5	11 695	800
	JB-3	7.5	17 774	305
	JB-4	7.5	17 793	320
Bonded and bolted	JAB-9	4	9511	6 964 000
	JAB-10	4	9667	6 578 900
	JAB-11	4.5	10 958	517 800
	JAB-12	4.5	10 765	2 733 500
	JAB-5	5	11 878	475 900
	JAB-6	5	11 775	333 500
	JAB-7	6	14 463	13 000
	JAB-8	6	14 191	16 200
	JAB-3	7.5	17 850	6560
	JAB-4	7.5	17 692	5420
	JAB-13	9	21 331	1480
	JAB-14	9	21 331	2780

[(a)] Stopped

history to the specimens at different RMSs; and the analytic random fatigue lives are calculated by following the fatigue damage accumulation models (equations (12) and (17)). Comparison between the variable-amplitude fatigue results and the Rayleigh approximation, based on both logarithmic and semilogarithmic S/N curve model is provided in Table 3. It can be seen that for the lower RMS levels considered, the classical Rayleigh approximation underestimates the actual fatigue life from 37 to 67%. The same trend is observed for Gaussian approximation method and the expected fatigue lives are underestimated from 56 to 66%.

The fatigue lives at the higher RMS levels are significantly overestimated by using both Rayleigh and Gaussian methods. For both bonded only and bolted only specimen types, the predictions at the higher RMS levels (1.75 ksi) are somewhat improved by the use of Rayleigh approximation method, but this improvement is quite marginal. At the same time, Rayleigh approximation leads to further deviation from the experimental mean for the two lower stress RMS levels of bonded only and bonded and bolted

Table 2 Variable-amplitude fatigue tests of FRP joints

Joint type	ID number	RMS (ksi)	Maximum–Minimum stress (ksi)	Maximum load level (lbs)	Cycles to failure
Bonded only	JA-14	1.25	±5.426	13 812	3 654 000
	JA-15	1.25	±5.426	13 912	1 441 900
	JA-16	1.25	±5.426	13 839	4 419 200
	JA-11	1.75	±7.597	19 949	62 500
	JA-12	1.75	±7.597	19 103	22 500
	JA-13	1.75	±7.597	19 306	26 700
Bolted only	JB-18	1.25	±5.426	13 116	253 900
	JB-19	1.25	±5.426	13 060	492 500
	JB-20	1.25	±5.426	13 161	356 700
	JB-15	1.75	±7.597	18 671	7520
	JB-16	1.75	±7.597	18 407	9390
	JB-17	1.75	±7.597	18 357	8220
Bolted and bonded	JAB-18	1.75	±7.597	19 279	5 343 100
	JAB-19	1.75	±7.597	18 070	11 934 700
	JAB-20	1.75	±7.597	17 986	9 977 000
	JAB-15	2.25	±9.765	23 169	203 100
	JAB-16	2.25	±9.765	22 900	305 700
	JAB-17	2.25	±9.765	24 664	67 400

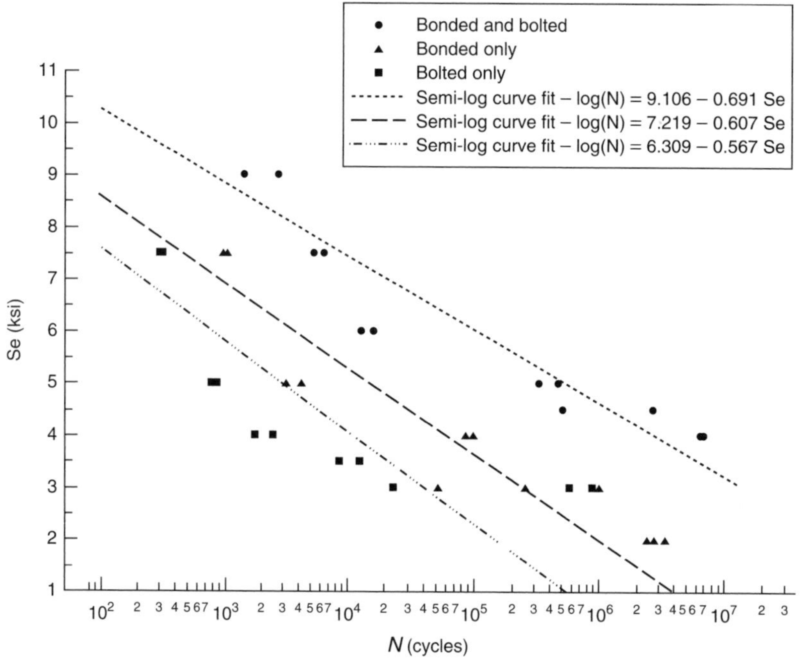

Figure 2 S/N curves of FRP joints (log–log curve fit)

specimens (1.25 and 1.75, respectively). These large deviations mainly come from the basic assumptions built in the Rayleigh approximation.

As pointed out in the previous section, the classical Rayleigh and Gaussian approximation formulas (12) and (17) is based on both equation (1) and on the

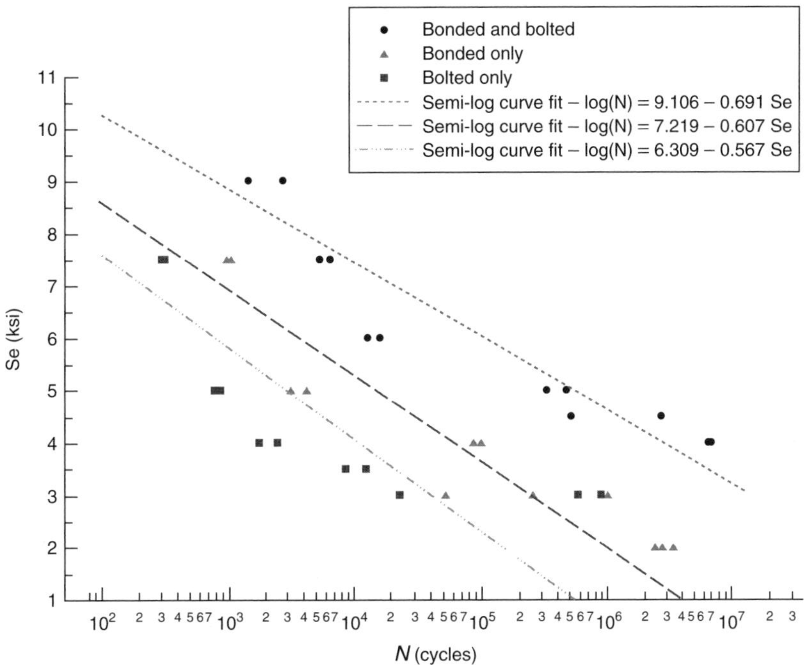

Figure 3 S/N curves of FRP joints (semi-log curve fit)

Table 3 Comparison of Rayleigh and Gaussian fatigue lives with test results

Joint type	RMS level (ksi)	Experimental (geometrical mean)	Rayleigh approximation	Gaussian method
Bonded only	1.25	2 855 600	925 300	979 200
	1.75	33 500	113 600	148 300
Bolted only	1.25	354 600	225 500	129 800
	1.75	8340	22 100	26 100
Bonded and bolted	1.75	8 600 700	3 510 200	3 789 200
	2.25	161 100	275 400	234 500

Palmgren–Miner damage accumulation rule equation (8). Therefore, misrepresentation of the actual stress-life relationship caused by a poor S/N curve function or insufficient S/N data, or disregarding the sequence effects by using the Miner rule, are main reasons for the shortcomings of the Rayleigh approximation in estimating the random fatigue lives.

Conclusion

Fatigue damage in FRP laminates occurs in the form of diffuse zones containing microcracks in the matrix, fiber–matrix debonding sites, ply delamination, and broken fibers. This damage leads to **degradation** of the elastic moduli of the laminate, and is therefore best modeled by stiffness degradation approaches. Various S/N curve models for constant-amplitude fatigue damage characterization can be described in terms of the fatigue modulus concept of Hwang and Han. In contrast, fatigue damage accumulation techniques constitute a much less investigated area; most data available deal with two-stage loading cases only [6]. The constant- and variable-amplitude fatigue studies on the laminate joints tested show a definite trend toward deviation from the power-law S/N curve model. Variable-amplitude fatigue tests with narrowband stress histories demonstrate significant deviation of the observed fatigue lives from those predicted with the Rayleigh approximation and Gaussian cumulative function. Such a deviation can be attributed to the inadequacy of the Palmgren–Miner's damage accumulation rule for composite materials and to poor representation or insufficiency of the constant-amplitude data as well as load frequency and sequence effects, which have not yet been quantified and incorporated into a fatigue damage accumulation model.

References

[1] Lutes, L.D. & Sarkani, S. (2004). *Random Vibration, Elsevier, Butterworth-Heinemann*, pp. 520–523.

[2] Momenkhani, K. & Sarkani, S. (2006). A new method for predicting the fatigue life of fiber-reinforced plastic laminates, *Journal of Composite Materials* **40**(21), 1971–1982.

[3] Hahn, H.T. & Kim, R.Y. (1976). Fatigue behavior of composite laminates, *Journal of Composite Materials* **10**, 156–180.

[4] Hwang, W. & Han, K.S. (1986). Cumulative damage models and multi-stress fatigue life prediction, *Journal of Composite Materials* **20**, 125–153.

[5] Howe, R.J. & Owen, M.J. (1972). Cumulative damage in chopped strand mat polyester resin laminates, *Proceedings of the 8th International Reinforced Plastics Conference*, BPF, London, pp. 137–148.

[6] Michaelov, G., Kharrazi, M.,Momenkhani, K., Sarkani, S. & Kihl, D.P. (1997). *Fatigue Behavior and Damage Accumulation of Fiber-Reinforced Plastic Laminates and Joints*, NSWCCD-TR-65-97/36, Naval Surface Warfare Center, Carderock Division.

KOUROSH MOMENKHANI AND SHAHRAM SARKANI

Prior Distribution Elicitation

Introduction

Expert engineering judgment plays a vital role in assessing and modeling the quality and reliability of systems [1, 2], although presenting subjective judgments as formal probability distributions is not without controversy [3]. Like any **data collection** process, a sound methodology for elicitation is a prerequisite for acquiring reliable subjective expert judgment, which will be summarized in the form of a prior distribution.

A prior distribution is a subjective probability distribution describing some unknown quantity of interest that has been elicited from a relevant person(s) prior to observing data. Using Bayes theorem, the prior distribution is updated in the light of data to produce a posterior distribution. Updation requires a conditional probability distribution, known as the *likelihood function*, which describes the probability of observing the data conditional on the state of the unknown quantity of interest.

A subjective probability distribution is one that describes a person's belief in the value of an unknown quantity. Winkler [4] believes that there is no single correct subjective probability distribution for an unknown quantity because each person forms his/her beliefs from personal experience. Cooke [5] points out that there is no way of measuring the correctness of a subjective distribution. In contrast, O'Hagan [6] distinguishes between true and stated probabilities. The former would result if an expert were capable of perfectly accurate assessments, while the latter would result from attempts to specify the underlying true probabilities. The authors agree that a good prior distribution will conform to the laws of probability, although the differing perspective has implications for validation.

The Bayes coherency principal states that uncertainties are described by probabilities and subjective probabilities should be such as to ensure self-consistent betting behavior [7]. This means that a person assesses subjective probabilities in such a way that they would avoid gambles involving certain losses (a so-called Dutch book). In principle, subjective probabilities can be measured through preference behavior, when a subject declares indifference between the event under discussion and a lottery with a known probability (assuming of course that the "prizes" are comparable).

Even if a prior distribution conforms to the laws of probability, some elicited distributions provide more useful inference than others. Research in experimental psychology has demonstrated that accurate subjective probabilities are unobtainable by simply asking someone to provide a probability number; instead a structured elicitation process is required

[8, 9]. These processes consist of an "expert" (or set of experts) and an "analyst". Following the definition of Ferrell [10], an expert is "a person with substantive knowledge about the events whose uncertainty is to be assessed", while an analyst should be a suitably qualified individual who will collect the data from the expert in order to formulate the prior distribution for use in modeling. The aim of an elicitation process is to minimize the impact of biases inherent in surfacing and capturing subjective expert judgment.

After examining aspects of modeling relevant to the elicitation of a prior distribution in the section titled "Model Parameterization and Uncertainties", the section titled "Heuristics and Biases" describes the typical biases encountered during elicitation. The section titled "Assessing Prior Distributions" further considers characteristics of good prior distributions, while the section titled "Improving Probability Assessors" highlights issues associated with improving the ability of assessors to provide useful inference. The section titled "Generic Processes" outlines a generic process for supporting the elicitation of a prior distribution and provides references to specific processes used in reliability. Finally, the section titled "Summary and Conclusions" reflects upon advanced theoretical and practical issues concerning the elicitation of prior distributions.

Model Parameterization and Uncertainties

It is useful to distinguish between different types of uncertainty in modeling: aleatory uncertainty, which corresponds to inherent randomness in the process generating observations; and epistemic uncertainty, which corresponds to state of knowledge of an expert. Within a Bayesian statistical model, the aleatory uncertainty is described by the likelihood function, while the prior distribution describes the epistemic uncertainty. Typically, as more relevant data are acquired, the epistemic uncertainty is reduced although the aleatory uncertainty is not. As an example, consider the uncertainty associated with determining the length of time a component will operate without failing. The uncertainty associated with the mean time to failure of a component from a particular class is epistemic and will reduce through observing the times to failure for such components. However, even if the mean time to failure of components within this class is known, the aleatory uncertainty would still remain with respect to the actual time to failure for any one component.

Many of the favored model parameters, such as a failure rate and mean time to failure, are not actually observable quantities, but are simply parameters of a model used to make predictions about the future. It has been strongly argued on foundational grounds, for example the discussion in Chapter 2 of Bedford and Cooke [11], that an expert should only be asked for probability assessments on observable quantities. This raises an interesting issue concerning the degree to which the model construct relates to an expert's perception and hence the coverage of the elicitation. For example, Bedford *et al.* [2] propose that elicitation should include qualitative structuring of the model as well as the quantitative assessment of the prior distribution for the quantity of interest to facilitate appropriate parameterization.

Heuristics and Biases

Research in psychology is relevant to the elicitation of a prior distribution. The identification of the major sources of bias within expert assessments motivates the need for supportive elicitation processes and validation of data.

Sunstein [12] discusses two alternative cognitive systems to describe how a person makes decisions: System 1 uses intuition, and results in quick decisions, which can be subject to error; System 2 is more deliberative, calculative, and less likely to produce errors. System 1 is based on heuristics, whereby a person develops shortcut approaches to making decisions. While heuristics can provide insight into a problem, careful deliberation should result in an improved assessment. This implies that the elicitation of a prior distribution requires an analyst to engage the expert in System-2 thinking. Hammond (see [13]) describes a person's thinking as a continuum between the aforementioned dichotomy.

Tversky and Kahneman [14] were the first authors to make substantial contributions to the identification of heuristics used within probability assessments and to discuss the implications for bias. Three key heuristics – availability, representative, anchoring – are typically used when assessing probabilities.

The *availability heuristic* refers to a person's judgment being influenced by how easily historical events are recalled. As an example, an expert may believe the likelihood of a particular system failing is high because there was a recent failure event. This can lead to a biased assessment as the expert is not considering a full failure history of the system.

The *representative heuristic* refers to the similarity between an outcome and a population. For example, on assessing the probability that a new component design will have a mean lifetime in excess of 1000 operating hours, one may consider how well the existing component population represents the design being evaluated. Care is required in using such a heuristic because an expert may assign an intuitive measure of similarity rather than considering how the probability will change with respect to that of the existing population.

The *anchoring and adjusting heuristic* refers to the situation when an expert provides an initial assessment and all future assessments are derived from an adjustment of the initial assessment. For example, consider the situation where a prior distribution for the mean time to failure for a component is required from an expert. An initial assessment may correspond to a best guess, from which the expert may anchor upon the initial assessment to adjust for upper and lower bounds from which the distribution is determined. Research shows that the adjustments made are typically insufficient [4, 15].

Assessing Prior Distributions

If an expert provides subjective probabilities that are coherent, then these are not incorrect insofar as they reflect the beliefs of the expert. However, it is clear that different experts give different assessments, and so this raises the question of how one might assess the "quality" or "usefulness" of such distributions. As Cooke [5] pointed out, there is no absolute measure of the correctness of a subjective assessment. However two measures are often used either quantitatively or qualitatively: calibration and information.

To assess the degree to which an expert's assessments are well calibrated requires the subjective probabilities to be compared with a collection of observed realizations. For example, two experts could assess the same event with one expert providing a probability of 0.9 and the other a probability of 0.2. This event will either be realized or not. Since neither expert assigned a probability of 1 or 0, regardless of the outcome neither will be incorrect. The first expert would be calibrated if 90% of all events assessed by that expert at the 90% level are realized and 10% are not. While for the second expert, 20% of all events assessed by that expert at the 20% level are realized and 80% are not. For this example, both experts could be perfectly calibrated but provide very different probabilities.

Direct assessment of calibration is only possible if we are able to make observations directly on the quantity of interest. In many cases, for example the uncertainty of a failure rate, this is not the case and so the distribution of any directly observable quantities (such as the number of units failing up to a given time) would be a predictive distribution obtained as a mixture of the sampling distribution weighted by the prior. Hence calibration measurements would naturally confound the sampling distribution and the prior.

A second consideration for the usefulness of an expert's prior distribution is information, which can be assessed through **mean squared error** between the observed realizations and the subjective probabilities, information scores, or a similar such formula. Stronger conclusions can be drawn from an informative prior distribution than from an uninformative one. It is ideal to have an informative, well-calibrated expert. Note that it is possible to have an uninformative, calibrated expert. For example, if upon assessing a mean time to failure an expert provides the entire range of positive real numbers below the median 50% of the time and above the median 50% of the time, then the expert is calibrated but not informative.

Improving Probability Assessors

The quality of subjective probabilities from an expert is dependent upon both the expert's experience and the method of elicitation. If the expert lacks experience, the prior distribution will be uninformative or misleading, regardless of the elicitation approach employed. Poorly designed elicitation techniques may degrade the quality of the information provided by experts. Fischhoff [16] proposes four necessary conditions to support improved judgment skills:

1. abundant practice with a set of reasonably homogeneous tasks;

2. clear-cut criterion events for outcome feedback;
3. task-specific reinforcement;
4. explicit admission of the need for learning.

There is extensive evidence that these are often not achieved in practice [10, 17].

Standard techniques for eliciting prior distributions that reflect the uncertainty of the expert are reported in the literature. See, for example, Cooke [5] and Meyer and Booker [18]. The choice of technique depends upon the quantity of interest, the selection of which has been given little attention in the literature.

One important consideration when selecting the quantity of interest is feedback to the expert. This is considered crucial for calibrating the expert and should be event specific [10, 16, 17]. In other words, the feedback must be with respect to assigning probabilities to particular classes of relevant events and not only feedback on the ability of the expert to assign probabilities to any situation. To increase the effectiveness of feedback in terms of learning, conditions that influence the event should reoccur as often as possible [16, 19]. Therefore, the factors on which the measure is conditioned should be as few and as general as possible.

Generic Processes

A necessary first step toward developing a method for eliciting expert judgment is to start with a general approach that possesses the required key features for any elicitation process and to tailor it to the desired situation.

Phillips [20], Clemen and Reilly [21], and Garthwaite *et al.* [22], for example, all describe generic processes. Another generic process, first developed by the Stanford Research Institute (SRI) for eliciting expert judgment is discussed in Spetzler *et al.* [23], Ferrell [24], and Merkhofer [9]. The issues addressed within the SRI process are common considerations within the alternatives. There are seven stages to the SRI approach: motivate, structure, condition, encode, verify, aggregate, and discretize.

Motivating the expert can be achieved by explaining the elicitation process and how the results will be used. This phase can be used to enhance the comprehension of an expert regarding the process, encourage an expert to provide an accurate assessment of his/her understanding, and determine potential biases, such as motivational biases. These include expert bias where an expert becomes overconfident merely because he/she has been given the title "expert" and management bias where an expert provides goals as opposed to judgment. For example, an expert states the aspiration that there will be no faults within this system by time of manufacture, rather than an assessment of his beliefs.

Structuring involves defining the specific event being considered to ensure there is no ambiguity in the questions and to explore how an expert thinks about the quantity for which probability judgment is to be elicited. The aim is to manage cognitive bias by simplifying the complex task of assigning probabilities by disaggregating the quantity of interest into more elementary variables [25]. However, the unpacking principle [13] may be the consequence. This refers to the situation where the more detailed the description of the event, the greater the likelihood assigned to it. For example, an expert may provide an assessment for the probability of a component failing and subsequently during the elicitation process provide probabilities associated with causes of failure that may result in a system probability exceeding the initial assessment.

Sometimes experts maybe unclear about whether they are being asked for conditional events or conjoined events (P(A|B) or P(A and B)), as these can be difficult to distinguish in natural language. Judgments can be conditioned upon all relevant information necessary to ensure that an expert's awareness has been stimulated and to manage biases such as anchoring and availability.

Encoding refers to an expert providing a numerical expression that reflects his/her uncertainty regarding an outcome. There are a number of techniques available for encoding subjective probabilities. Common approaches range from direct **estimation** whereby a person is simply asked for the probability to inferring probabilities from elicited preferences with specified lotteries. A few devices have been used to support such elicitation, such as probability wheels or visual analog scales. An interesting discussion of alternative methods is given by O'Hagan *et al.* [13].

Encoding a probability distribution is more challenging because additional information is required than for an individual probability. One could partition the distribution into discrete intervals and elicit probabilities for the value of interest belonging to each of the intervals. Evidence exists to suggest that people are better at estimating the median and mode rather than the mean; moreover people are poor at assessing or interpreting the variance [13].

A popular alternative encoding procedure for distributions is the fractile method [5], in which the expert assesses the median value of his/her subjective probability distribution along with the (25th, 75th) and the (5th, 95th) **percentiles**. Once these values have been elicited, a parametric distribution can be sought to maximize fit. The main drawback of this technique is that the elicited data often reflect a central bias [26]. However, after percentiles of the prior distribution have been assessed, graphical techniques can be applied to enhance the quality of the prior distribution [27].

Verification is required to ensure that an expert has provided a reflection of his/her true beliefs. If problems are encountered then the previous stages are to be repeated.

If multiple experts are assessing the same quantity of interest, then their individual probability distributions should be aggregated. There are two approaches to aggregation – mathematical and behavioral. The former implies the experts should not influence each others' decisions [24]. The latter requires experts to share their judgment and reassess their distributions and includes techniques such as Delphi [24] and nominal group technique [28]. There are several mathematical approaches to aggregation, most of which aim to evaluate a weighted average across the experts. See Cooke [5] for a fuller discussion.

It is often necessary to treat continuous random variables as discrete during the encoding stage. *Discretizing* refers to techniques for fitting continuous distributions to the elicited data while preserving important **moments**. This is accomplished by dividing the range of all possible values for the uncertain variable into intervals, selecting a representative point from each interval, and assigning that point the probability that the actual value will fall within the corresponding interval. The moments can be preserved through Gaussian quadrature techniques [29].

Two elicitation processes specific to reliability include PREDICT (Kerscher [30, 31] and REMM [32–34]). Both provide modeling frameworks designed to estimate reliability throughout system development beginning with a problem structuring phase, which consists of eliciting a graphical representation of the relationship between the potential failure modes and the system reliability. The graphs form the basis for the elicitation.

Summary and Conclusions

An overview of the key issues associated with the elicitation of a prior distribution has been presented; in particular, the shortcomings due to the prevalence of heuristics employed by an expert when assessing probabilities have been highlighted. Evidence indicates that an expert can improve his/her assessments given feedback about meaningful quantities of interest and this learning can be accelerated under the right conditions. Moreover, good subjective probability assessment requires a supportive process for the experts, whereby key stages are sequenced for managing specific biases. For further details about these issues see O'Hagan *et al.* [13], Meyer and Booker [18], and Cooke [5].

Practically, the design and implementation of an elicitation process requires considered **project management** to ensure that, for example, the correct experts are selected, the choice of elicitation method is relevant to the model requirements and parameterization, the integrity of the probabilities assessed are maintained, and that sufficient resources are obtained to support an effective, and efficient, process. See Meyer and Booker [18] for further discussion.

Theoretically, there is a need to infer from the assessments which probability distributions on model parameters are consistent. This approach has been developed by Cooke [35] with more algorithms and underlying theory for probabilistic inversion in Kraan and Bedford [36]. Taking a more standard Bayesian perspective, Percy [37] discusses the indirect assessment of parameters from generic prior distributions through the direct assessment of quantiles of the predictive distribution which will have observable outcomes. Gutierrez-Pulido *et al.* [38] take a similar line, considering both moments and quantiles of the time to failure for a system as sources of information from which prior distributions can be fitted. Such methods could also be applicable to other Bayesian contexts where prior distributions on lifetime distribution parameters are to be assessed – for example in **Bayesian accelerated** or proportional hazards life modeling [39].

Elicitation of a prior distribution places a heavy cognitive burden on the subjective probability assessments by an expert. Yet observed data can exist for earlier generations of a system, component or process, although it is not clear how these data can be adapted to inform the assessment of a new design. **Empirical Bayes** provides a framework for integrating relevant historical data and hence can reduce the cognitive burden on an expert. See for example Quigley *et al.* [40] and Bedford *et al.* [41].

References

[1] Keeney, R.L. & von Winterfeldt, D. (1991). Eliciting probabilities from experts in complex technical problems, *IEEE Transactions on Engineering Management* **38**, 191–201.

[2] Bedford, T., Quigley, J. & Walls, L. (2006). Expert elicitation for reliable system design, *Statistical Science* **21**(4), 428–450.

[3] Evans, R.A. (1989). Bayes is for the Birds, *IEEE Transactions in Reliability* **38**, 401.

[4] Winkler, R. (1967). The assessment of prior distribution in Bayesian analysis, *Journal of the American Statistical Association* **62**, 776–800.

[5] Cooke, R.M. (1991). *Experts in Uncertainty*, Oxford University Press.

[6] O'Hagan, A. (1988). *Probability: Methods and Measurement*, Chapman and Hall, London.

[7] De Finette, B. (1974). *Theory of Probability*, John Wiley & Sons, New York.

[8] Kahneman, D., Slovic, P. & Tversky, A. (1982). *Judgement Under Uncertainty: Heuristics and Biases*, Cambridge University Press.

[9] Merkhofer, M. (1987). Quantifying judgemental uncertainty: methodology, experiences, and insights, *IEEE Transactions on Systems, Man, and Cybernetics* **17**, 741–752.

[10] Ferrell, W. (1994). Discrete subjective probabilities and decision analysis: elicitation, calibration and combination, in *Subjective Probability*, G. Wright & P. Ayton, ed, John Wiley & Sons.

[11] Bedford, T. & Cooke, R. (2001). *Probabilistic Risk Analysis: Foundations and Methods*, Cambridge University Press, Cambridge.

[12] Sunstein, C.R. (2005). *Laws of Fear*, Cambridge University Press.

[13] O'Hagan, A., Buck, C.E., Daneshkhah, A., Eiser, J.R., Garthwaite, P.H., Jenkinson, D.J., Oakley, J.E. & Rakow, T. (2006). *Uncertain Judgements Eliciting Experts' Probabilities*, John Wiley & Sons.

[14] Tversky, A. & Kahneman, D. (1974). Judgement under uncertainty: heuristics and biases, *Science* **185**, 257–263.

[15] Alpert, M. & Raiffa, H. (1982). A progress report on the training of probability assessors, in *Judgement Under Uncertainty: Heuristics and Biases*, D. Kahneman, P. Slovic & A. Tversky, eds, Cambridge University Press.

[16] Fischhoff, B. (1989). Eliciting knowledge for analytical representation, *IEEE Transactions on Systems, Man, and Cybernetics* **19**, 448–461.

[17] Wright, G. & Bolger, F., eds. (1992). *Expertise and Decision Support*, Plenum Press.

[18] Meyer, M.A. & Booker, J.D. (2001). *Eliciting and Analysing Expert Judgement: A Practical Guide, ASA-SIAM Series on Statistics and Applied Probability*, 2nd Edition.

[19] Kadane, J. & Wolfson, L. (1997). Experiences in elicitation, *The Statistician* **47**, 3–20.

[20] Phillips, L.D. (1999). Group elicitation of probability distributions: are many heads better than one? in *Decision Science and Technology: Reflections on the Contributions of Ward Edwards*, J. Shanteau, B. Mellors & D. Schum, eds, Kluwer Academic Publishers.

[21] Clemen, R.T. & Reilly, T. (2001). *Making Hard Decisions with Decision Tools*, Duxbury Press.

[22] Gartwaite, P.H., Kadane, J.B. & O'Hagan, A. (2005). Statistical methods for eliciting probability distributions, *Journal of the American Statistical Association* **100**, 680–701.

[23] Spetzler, C. & Stael von Holstein, C.A. (1975). Probability encoding in decision analysis, *TIMS* **22**, 340–358.

[24] Ferrell, W. (1985). In *Combining Individual Judgements, Behaviour Decision Making*, G. Wright, ed, Plenum Press.

[25] Armstrong, J.S., Denniston, W.B. & Gordon, M.M. (1975). The use of decomposition principle in making judgements, *Organizational Behavior and Human Performance* **14**, 257–263.

[26] Seaver, D., von Winterfeldt, D. & Edwards, W. (1978). Eliciting subjective probability distributions on continuous variables, *Organizational Behavior and Human Performance* **21**, 379–391.

[27] Chaloner, K., Church, T., Lois, T. & Mattis, J. (1993). Graphical elicitation of a prior distribution for a clinical trial, *The Statistician* **42**, 341–353.

[28] Moore, C.M. (1987). *Group Techniques for Idea Building*, Sage.

[29] Miller, A.C. & Rice, T.R. (1983). Discrete approximations of probability distributions, *Management Science* **29**, 352–362.

[30] Kerscher, W.J.I., Booker, J.M., Bement, T.R. & Meyer, M.A. (1998). *Characterizing Reliability in a Product/Process Design-Assurance Program. Annual Reliability and Maintainability Symposium 1998 Proceedings. International Symposium on Product Quality and Integrity (Cat. No.98CH36161)*, IEEE, Anaheim, pp. 105–112.

[31] Kerscher, W., Booker, J. & Meyer, M. (2003). Predict: a case study, using fuzzy logic, in *Proceedings of the Reliability and Maintainability Symposium*, Tampa, FL, USA.

[32] Walls, L., Quigley, J. & Marshall, J. (2006). Modeling to support reliability enhancement during product development with applications in the UK Aerospace Industry, *IEEE Transactions on Engineering Management* **53**(2), 263–274.

[33] Walls, L. & Quigley, J. (2001). Building prior distributions to support Bayesian reliability growth modelling using expert judgement, *Reliability Engineering and System Safety* **74**, 117–128.

[34] Hodge, R., Evans, M., Marshall, J., Quigley, J. & Walls, L. (2001). Eliciting engineering knowledge about reliability during design - lessons learnt from implementation, *Quality and Reliability Engineering International* **17**, 169–179.

[35] Cooke, R.M. (1994). Parameter fitting for uncertain models: modelling uncertainty in small models, *Reliability Engineering and System Safety* **44**, 89–102.

[36] Kraan, B. & Bedford, T. (2005). Probabilistic inversion of expert judgments in the quantification of model uncertainty, *Management Science* **51**(6), 995–1006.

[37] Percy, D.F. (2002). Bayesian enhanced strategic decision making for reliability, *European Journal of Operational Research* **139**(1), 133–145.

[38] Gutierrez-Pulido, H., Aguirre-Torres, V. & Christen, J.A. (2005). A practical method for obtaining prior distributions in reliability, *IEEE Transactions on Reliability* **54**(2), 262–269.

[39] Bunea, C. & Mazzuchi, T.A. (2005). Bayesian accelerated life testing under competing failure modes. In *Annual Reliability and Maintainability Symposium, 2005 Proceedings. Proceedings Annual Reliability and Maintainability Symposium*, Alexandria, VA, USA, 152–157.

[40] Quigley, J., Bedford, T. & Walls, L. (2007). Estimating rate of occurrence of rare events with empirical Bayes: a railway application, *Reliability Engineering and System Safety* **92**(5), 619–627.

[41] Bedford, T., Quigley, J. & Walls, L. (2006). Fault tree inference through combining Bayes and empirical Bayes methods, in *Proceedings of the ESREL Conference*, Estoril, pp. 859–865.

JOHN QUIGLEY, TIM BEDFORD AND LESLEY WALLS

Related Articles

Accelerated Life Tests: Bayesian Models; **Bayesian Robustness: Theory and Computation**; **Bayesian Reliability Analysis**; **Bayesian Reliability Demonstration**; **Empirical Bayes Estimation of Reliability**; **Expert Opinion in Reliability**; **Masked Failure Data: Bayesian Modeling**; **Repairable Systems: Bayesian Analysis**; **Software Reliability: Bayesian Analysis**.

Prior Information in Sampling Schemes

Introduction

The choice of a sampling scheme and the amount of inspection is guided, at least informally, by prior information on quality. Formal rules for adapting the rigor of inspection to past inspection records are, for example, the switching rules contained in many standard sampling schemes like MIL-STD-105E or ISO 2859, *see* **Attributes Sampling Schemes in International Standards** and **Variables Sampling Schemes in International Standards**. These are often based on heuristic arguments, but there are also rules derived from a strict mathematical treatment. An interesting recent example is an article by Baillie and Klaassen [1]. However, these authors do not model prior information explicitly either. Here we are concerned with the formal models of prior information. Together with a loss function, in a very broad sense, these may be used to design optimal sampling strategies or to analyze and compare the performance of existing sampling schemes.

In the "Bayesian approach", prior knowledge is modeled in terms of a prior distribution. According to Hald [2], a complete statistical model of sampling inspection should contain the expected distribution of submitted lots according to quality (prior distribution) and the cost of sampling inspection, acceptance, and rejection. The best sampling plan is defined as the one within the class minimizing the average costs. This is essentially the paradigm of Bayesian decision theory, see [3]. However, the latter is not restricted to costs in a monetary sense. It only requires a numerical goal function that measures the consequences of possible decisions. Bayesian sampling plans will be introduced in the first section of this chapter. In practice, basic assumptions are often not satisfied. Therefore, modifications may be necessary and these will be addressed in the second section.

Bayesian Sampling Plans

A useful guide to practitioners on how and when to use Bayesian sampling plans has been described by Calvin [4]. The basic concepts are introduced in the following example. We use the framework established by chapters **Sampling Inspection of Products** and **Single Sampling by Attributes and by Variables**: A lot of items $1, \ldots, N$ is described by the sequence $z_1, \ldots, z_N$ of item quality indicators. Lot quality q is a function of the item quality indicators. Let $z_{i1}, \ldots, z_{in}$ be the item quality indicators in a sample of size n from the lot. A sample-based decision rule is represented by a function $\Delta(z_{i1}, \ldots, z_{in}) = d$. A **single sampling** plan is given by (n, Δ) where the values d of Δ range over $d = a$ "accept" and $d = r$ "reject". There are essentially two types of cost models: "type A" means that the costs of items in the sample do not depend on the decision about the remainder of the lot, whereas in "type B" the costs of defective items in the sample and those in the reminder of the lot depend in the same way on the final decision.

Introductory Example

In single sampling by attributes, the item quality indicators are binary variables where $z_i = 1$ means that the i-th item is defective (nonconforming) and $z_i = 0$ means that it is nondefective (conforming). The lot quality indicator is the number $m = z_1 + \ldots + z_N$ of defective items in the lot, or the fraction defective $p = mN^{-1}$.

Linear Cost Model of Type B. As a simple example, we introduce a linear cost model of type B. For the interpretation of the parameters, see [2]. The cost of decision d per item with characteristic z is given by $k_d(z) = a_d + b_d\, z$, i.e., it is a_d if the item is conforming and $a_d + b_d$ if the item is nonconforming. The cost K_d of acceptance and rejection of the whole lot is the sum of the individual costs, it depends obviously on p:

$$K_d = \sum_{i=1}^{N} k_d(z_i) = N a_d + m b_d = N(a_d + b_d p)$$
$$= N\, k_d(p) = K_d(p) \qquad (1)$$

Moreover, there are sampling costs $n \cdot s$, where s is the sampling cost per item.

Prior Distribution and Bayesian Decision Rule. The decision d is based on the number X of defective items in a random sample without replacement. For fixed m, its probability distribution $P[X = x|m]$ is hypergeometric. In addition, there is a prior distribution $\Psi(m)$ for m (prior to sampling), i.e., the number of nonconforming items in the lot is a random variable M with distribution Ψ. In a frequency interpretation, m varies from lot to lot according to Ψ.

Given a particular observation $X = x$, Bayes theorem allows us to calculate the posterior probability of $M = m$ as $P[M = m|X = x] = P[X = x|m]\Psi(m)\left(\sum_{l=0}^{N} P[X = x|l]\Psi(l)\right)^{-1}$. From this, the expected posterior loss of decision d can be derived as

$$E(K_d|X = x) = \sum_{m=0}^{N} K_d\left(\frac{m}{N}\right) P[M = m|X = x] \qquad (2)$$

One should accept if $E(K_a|X = x) \leq E(K_r|X = x)$ and reject otherwise. It can be shown that there exists c such that $E(K_a|X = x) \leq E(K_r|X = x)$ if $x \leq c$, where $c = -1$ is admitted with the interpretation "rejection regardless of the outcome of the sampling". This justifies using a sampling plan (n, c). This is the first way of looking at Bayes: we have several possible decisions; given an observation, a decision is chosen, which has the smallest posterior loss. However, this does not answer the question which **sample size** n should be used. For this purpose, one has to compare the expected costs for different sampling plans, including sampling costs. For fixed m, the average loss of (n, c) is given by

$$R\left(n, c|\frac{m}{N}\right) = n \cdot s + k_a\left(\frac{m}{N}\right) P[X \leq c|m]$$
$$+ k_r\left(\frac{m}{N}\right) P[X > c|m] \qquad (3)$$

Considered as a function of p, $R(n, c|p)$ is called the risk function of $(n,\ c)$. Its expected value with respect to Ψ,

$$R_\Psi(n, c) = E_\Psi\left[R\left(n, c|\frac{m}{N}\right)\right]$$
$$= \sum_{m=0}^{N} \Psi(m) R\left(n, c|\frac{m}{N}\right) \qquad (4)$$

is called the Bayes risk of using (n, c). A Bayesian sampling plan (n_Ψ, c_Ψ) minimizes $R_\Psi(n, c)$. This constitutes the second way of looking at Bayes: a decision rule tells us which action to take depending on the observation. We choose the "Bayesian rule" which minimizes the expected costs.

Priors Related to the Production Process, Additional Examples

Frequently, the distribution of the quality of the lot is related to the distribution of quality in the production process: if the items of lots of size N are drawn independently at random from a production process with constant probability π of producing a defective item, the number m of defective items in the lot varies according to the binomial frequency function $b(m|N,\pi)$. If π itself varies at random from lot to lot according to a distribution $W(\pi)$, the distribution of lot quality becomes a mixed binomial distribution

$$\Psi(m) = \int_0^1 b(m|N,\pi)\,\mathrm{d}W(\pi) \tag{5}$$

This model is frequently mentioned in the literature: it allows analysis of various situations where sampling inspection is advantageous, and derivation of adequate strategies, see [2]. For example, W may represent a production process in statistical control with occasional out-of-control periods. The lots produced during these periods are considered as outliers, and if the product in the in-control state is satisfactory, the problem of sampling inspection is to detect the outliers. Formally, W can be taken as a beta distribution with outliers.

Cost Model of Type A under Single Sampling. In such environments, cost models of type A have a simple structure. Thyregod [5] provides a very elegant treatment, which is restricted to neither the attributes sampling nor the linear cost models. The item quality indicators $Z_1 \ldots Z_N$ are assumed to be independent and distributed according to a density $f(z|\omega)$, where ω is the process parameter. This includes various models, in particular: (a) attributes sampling where $Z \in \{0,1\}$, $\omega = P[Z=1]$ and (b) sampling by variables where Z is normally distributed with expectation μ and standard deviation σ and $\omega = (\mu, \sigma)$. The parameter ω varies in the production process from lot to lot according to a distribution $W(\omega)$. The decision $d \in \{a, r\}$ is based on a sample of size n, which can be identified with $Z_1 \ldots Z_n$. The cost of the decision depends only on the remaining $N - n$ item characteristics, it is the sum of the individual costs $k_d(z_j)$ of the items $j = n+1, \ldots, N$. The function $k_d(z)$ is not restricted to a special type; it includes the above cost model, slight generalizations (e.g., sampling by variables with upper specification limit U : $k_d(z) = a_d$, if $z \leq U$ (item good) and $= a_d + b_d$, if $z > U$ (item defective)) as well as nonlinear cost functions. The cost of sampling is the sum of individual sampling costs $k_s(z_i)$ of the items in the sample.

For i = a,r,s, let $k_i^*(\omega) = \int k_i(z)\,f(z|\omega)\,\mathrm{d}z$ be the expectations of the costs with respect to $f(z|\omega)$. Moreover, let $k_s^* = \int k_s^*(\omega)\,\mathrm{d}W(\omega)$ be the overall expected sampling costs. Since, for given ω, the n observations in the sample are independent from the remaining $N - n$ variables, the expected cost when using the sampling plan (n, Δ) is $R_W(n, \Delta) = n \cdot k_s^* + (N-n)\,r_W(n, \Delta)$, where

$$r_W(n,\Delta) = \int [P\,[\Delta = a|\omega]\,k_a^*(\omega) + P\,[\Delta = r|\omega]\,k_r^*(\omega)]\,\mathrm{d}W(\omega) \tag{6}$$

Equation (6) has the formal structure of equations (3) and (4), with Ψ replaced by W and $R(n, c|p)$ replaced by the integrand in brackets in equation (3). This means that the original decision problem regarding the lot is formally equivalent to a decision problem regarding the process parameter ω with prior distribution W. A Bayesian sampling plan minimizes $R_W(n, \Delta)$. Thyregod [5] shows that it has usually a familiar structure: there is a test statistic t, and the lot is accepted if $t \leq c$. For example, in the case of a normally distributed quality characteristic with parameters μ and σ and upper specification limit U,

$$t = \sqrt{n}(\bar{X} - U)\,S^{-1} \tag{7}$$

where S is the sample standard deviation.

Extensions and More Recent Results

The Bayesian approach is neither restricted to decision rules of this simple structure nor to monetary costs. For example, Bayesian multistage or **sequential sampling** plans may be designed to minimize the average sample size, which now plays the role of a risk function. Here the class of decision rules is much richer than above. A considerable amount of literature exists on Bayesian sampling, and with increasing computer **power** more sophisticated models can be studied. For example, Gonzalez *et al.* [6] consider the number of defects per unit of product, whereas Chen *et al.* [7] introduce a Bayesian variable **acceptance sampling** scheme for life testing with mixed censoring. In both the articles, different types of nonlinear loss functions are used.

Modifications of Full Bayes

Robustness with Respect to Prior and Economic Model

Correct assignment of the prior distribution and a monetary cost model, as well as the stability of the prior are problematic issues of the Bayesian approach. General concepts of **Bayesian robustness** are discussed by Berger [3]. Bayesian sampling plans often have a poor discriminating power, which is a problem if the assumed prior distribution is misspecified or unstable. Therefore, Hald [2] recommends using restricted Bayesian sampling plans, where the Bayes risk is minimized within a smaller class of sampling plans with sufficient discriminating power. If full information about costs is not available, a partial cost function or the expected amount of inspection can be minimized under conventional constraints on the consumer's or the producer's risk, see [2]. The latter approach avoids the problem of estimating cost parameters completely. Bayes theorem also allows us to calculate posterior risks. In this spirit, Schafer [8] suggests to minimize the sample size under constraints on the posterior risk to the producer and the consumer risk.

Γ-Minimax

A concept which automatically yields more robust strategies is partial prior information, also called Γ-minimax. For example, Krumbholz and Pflaumer [9] derive sampling strategies which minimize expected costs subject to prior information on the probability that the fraction nonconforming does not exceed a certain upper bound. In a similar spirit, Collani [10] considers the so-called α-optimal sampling schemes, *see* **Economic Sampling Schemes**. These conditions are examples of "generalized moment conditions" of the prior. Other generalized or ordinary **moments** can be estimated more easily from past inspections (see below), but the calculation of a corresponding sampling strategy is usually a complex "Γ-minimax" problem. This can be solved in a numerically efficient way by semi-infinite programming, see [11].

Availability of Information on the Prior

If the fraction nonconforming p or the process parameter ω varies from lot to lot according to a prior Q, it is possible to gain information about Q from past inspections. However, one does not observe p or ω directly. The observable distribution is the mixture of the conditional distribution of the test statistic, given p or ω, with mixing distribution Q. Methods which use information about the observable distribution to derive a Bayesian decision rule are called **empirical Bayes**; for a general outline we refer again to [3], in particular, the section titled "Nonparametric Empirical Bayes Analysis". Empirical Bayes has been applied to sampling inspection several times. Another way is to estimate ordinary or generalized moments of the prior indirectly from the observable distribution [12], and to use a Γ-minimax sampling plan. This method is able to incorporate information about the uncertainty of the estimator, for example, a confidence band around the empirical estimate of the marginal distribution.

Adaptive Sampling Strategies

A final issue is stability of the prior. Rendtel and Lenz [13] have developed a sampling system based on sequential tests, which is able to adapt to changing priors. Here it is not necessary to model these changes explicitly. There have also been attempts to model prior information by stochastic processes more general than independent and identically distributed (i.i.d.), but it seems that these have not gained much popularity.

References

[1] Baillie, D. & Klaassen, C.A.J. (2006). Credit to and in acceptance sampling, *Statistica Neerlandica* **60**, 283–291.

[2] Hald, A. (1981). *Statistical Theory of Sampling Inspection by Attributes*, Academic Press, London.

[3] Berger, J.O. (1993). *Statistical Decision Theory and Bayesian Analysis*, 3rd Edition, Springer, New York.

[4] Calvin, T.M. (1990). *How and When to Perform Bayesian Sampling*, Revised Edition, American Society for Quality.

[5] Thyregod, P. (1974). Bayesian single sampling acceptance plans for finite lot sizes, *Journal of the Royal Statistical Society. Series B* **36**, 305–319.

[6] González, C. & Palomo, G. (2003). Bayesian acceptance sampling plans following economic criteria: an application to paper pulp manufacturing, *Journal of Applied Statistics* **30**, 319–334.

[7] Chen, J., Chou, W., Wu, H. & Zhou, H. (2004). Designing acceptance sampling schemes for life testing with mixed censoring, *Naval Research Logistics* **51**, 597–612.

[8] Schafer, R.E. (1967). Bayes single sampling plans by attributes based on the posterior risk, *Naval Research Logistics Quarterly* **14**, 81–88.

[9] Krumbholz, W. & Pflaumer, P. (1982). Möglichkeiten der Kosteneinsparung bei der Qualitätskontrolle durch Berücksichtigung von unvollständigen Vorinformationen, *Zeitschrift für Betriebswirtschaftslehre* **52**, 1088–1102.

[10] von Collani, E. (1986). The α-optimal sampling scheme, *Journal of Quality Technology* **18**, 63–66.

[11] Fandom Noubiap, R. & Seidel, W. (2001). An algorithm for calculating Γ-minimax decision rules under generalized moment conditions, *Annals of Statistics* **29**, 1094–1116.

[12] Seidel, W. (1997). *Unbiased Estimation of Generalized Moments of Process Curves*, In *Frontiers in Statistical Quality Control, Vol. 5*, H.-J. Lenz & P.-Th. Wilrich, eds, Physica-Verlag, Heidelberg, pp. 21–37.

[13] Rendtel, U. & Lenz, H.-J. (1990). *Adaptive Bayes'sche Stichprobensysteme für die Gut-Schlecht-Prüfung*, Physica-Verlag, Heidelberg.

WILFRIED SEIDEL

Probability Density Function (PDF)

Probability density functions (PDFs) are mathematical formulas that provide information about how the values of random variables are distributed. For a discrete random variable X assuming values in a sample space Ω, the density is usually named the *frequency distribution*, defined as $p_X(x) = P(X = x)$, for each x in Ω. If X is a continuous random variable, its density function f_X determines the probability that X is in an interval $[a, b]$, namely,

$$P(a \leq X \leq b) = \int_a^b f_X(x)\,\mathrm{d}x \qquad (1)$$

In particular, the **PDF** determines the probability that X is smaller than x, namely, $F_X(x) = P(X \leq x) = \int_{-\infty}^{x} f_X(x)\,\mathrm{d}x$. The function F_X is called the *(cumulative) distribution function of X* (*see* **Cumulative Distribution Function (CDF)**). The best-known density function is the Gauss curve, which defines the **normal distribution**. (*See* **Probability Density Functions**.)

Probability Density Functions

Introduction

Probability density functions (PDFs) (*see* **Probability Density Function (PDF)**) are mathematical formulas that provide information about how the values of random variables are distributed. For a discrete random variable (one taking only particular values within some interval), the formula gives the probability of each value of the variable. For a continuous random variable (one that can take any value within some interval), the formula specifies the probability that the variable falls within a particular range. This is given by the area under the curve defined by the **PDF**. Here a list of the most commonly encountered PDFs and their most important properties is provided. More comprehensive accounts of density functions can be found in [1, 2].

Probability Density Functions for Discrete Random Variables

Bernoulli

A simple PDF for a random variable, X, that can take only two values, for example, 0 or 1. Tossing a single coin provides a simple example. Explicitly the density is defined as

$$P(X = x_1) = 1 - P(X = x_0) = p \qquad (1)$$

where x_1 and x_0 are the two values that the random variable can take (often labeled as "success" and "failure") and P denotes probability. The single parameter of the Bernoulli density function is the probability of a "success", p. The expected value of X is p and its variance is $p(1 - p)$.

Binomial

A PDF for a random variable, X, that is the number of "successes" in a series of n independent Bernoulli variables. The number of heads in n tosses of a coin provides a simple practical example. The binomial density function is given by

$$P(X = x) = \frac{n!}{x!(n-x)!} p^x (1-p)^{n-x}, \quad x = 0, 1, 2, \ldots, n \qquad (2)$$

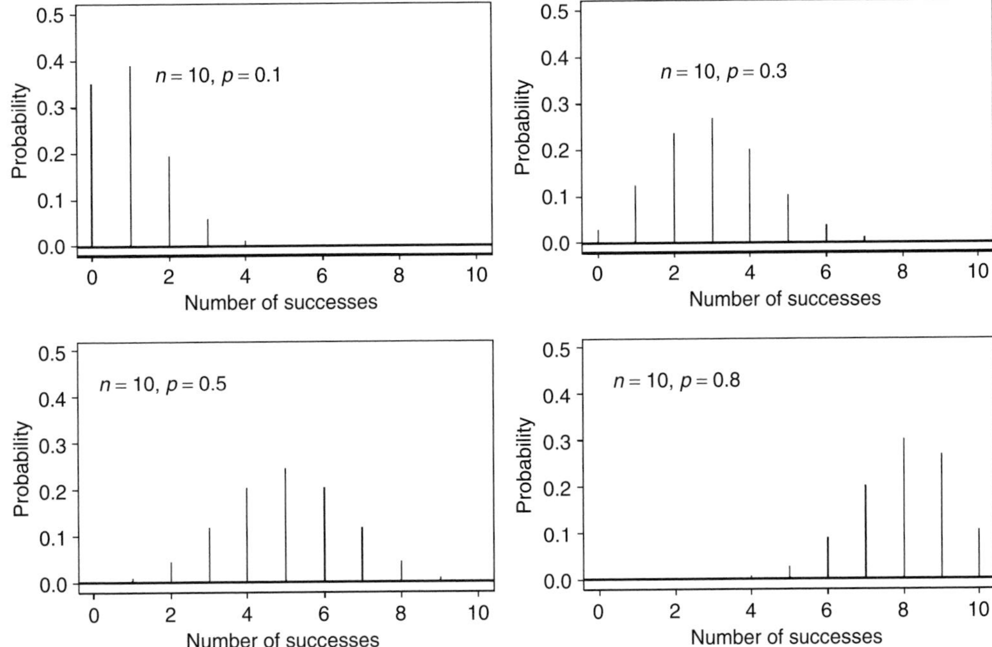

Figure 1 Examples of binomial density functions

The probability of the random variable taking a value in some range of values is found by simply summing the PDF over the required range. The expected value of X is np and its variance is $np(1-p)$. Some examples of binomial density functions are shown in Figure 1. The binomial is important in assigning probabilities to simple chance events such as the probability of getting three or more sixes in 10 throws of a fair die, in the development of simple statistical significance tests such as the sign test (*see* **Nonparametric Tests**) and as the error distribution used in logistic regression. See also the multinomial distribution defined later in this article.

Geometric

The PDF of a random variable, X, that is the number of "failures" in a series of independent Bernoulli variables before the first "success". An example is provided by the number of tails before the first head, when a coin is tossed a number of times. The density function is given by

$$P(X = x) = p(1-p)^{x-1}, \quad x = 1, 2, \ldots \tag{3}$$

The geometric density function possesses a *lack of memory property*, by which we mean that in a series of Bernoulli variables the probability of the next n trials being "failures" followed immediately by a "success" remains the same geometric PDF regardless of what the previous trials were. The mean of X is $1/p$ and its variance is $(1-p)/p^2$. Some examples of geometric density functions are shown in Figure 2.

Negative Binomial

The PDF of a random variable, X, that is the number of "failures" in a series of independent Bernoulli variables before the kth "success". For example, the number of tails before the 10th head in a series of coin tosses. The density function is given by

$$P(X = x) = \frac{(k+x-1)!}{(x-1)!k!}p^k(1-p)^x, \quad x = 0, 1, 2 \ldots \tag{4}$$

The mean of X is $k(1-p)/p$ and its variance is $k(1-p)/p^2$. Some examples of the negative binomial distribution are given in Figure 3. The density is important in discussions of overdispersion in the context of **Generalized Linear Models**.

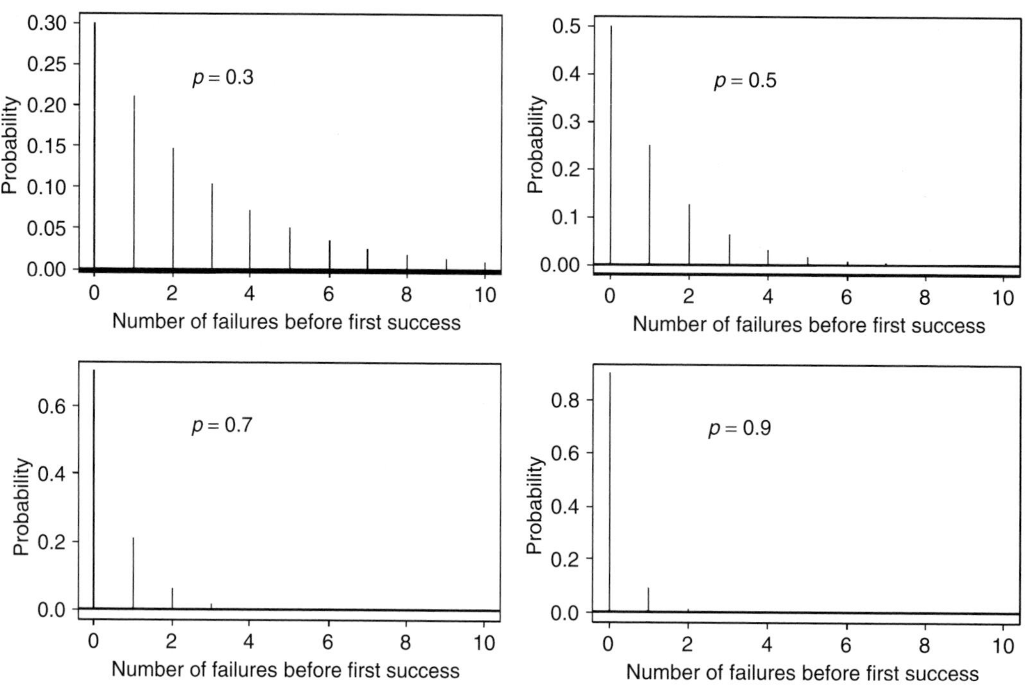

Figure 2 Examples of geometric density functions

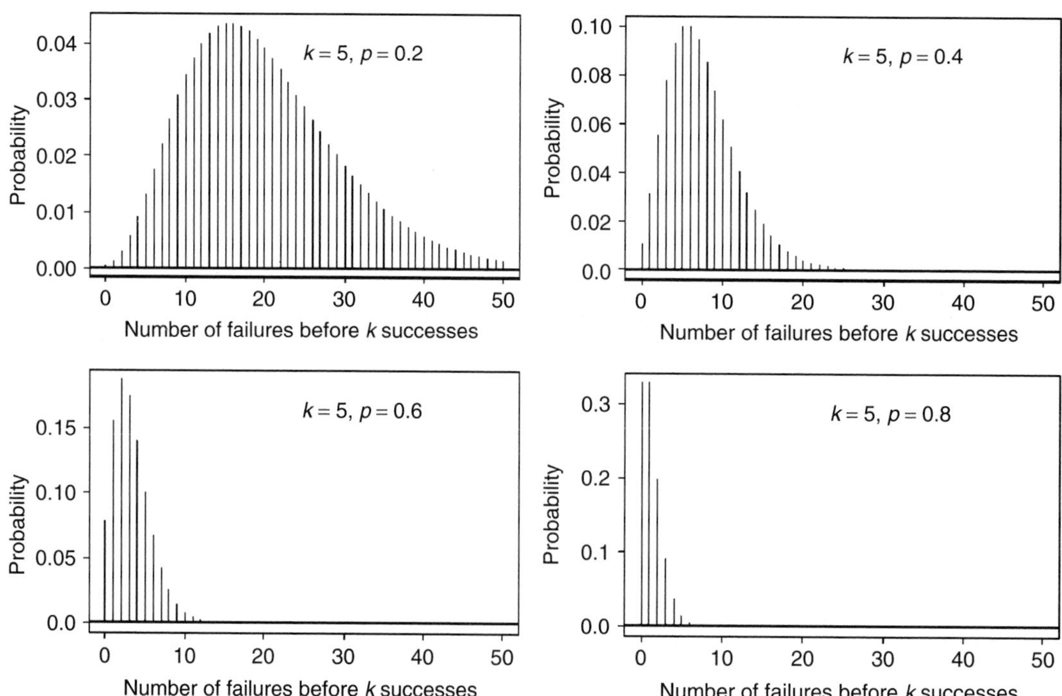

Figure 3 Examples of negative binomial density functions

Hypergeometric

A PDF associated with *sampling without replacement* from a population of finite size. If the population consists of r elements of one kind and $N-r$ of another, then the hypergeometric is the PDF of the random variable X defined as the number of elements of the first kind when a random sample of n is drawn. The density function is given by

$$P(X = x) = \frac{\dfrac{r!(N-r)!}{x!(r-x)!(n-x)!(N-r-n+x)!}}{\dfrac{N!}{n!(N-n)!}} \quad (5)$$

The mean of X is nr/N and its variance is $(nr/N)(1 - r/n)[(N - n)/(N - 1)]$. The hypergeometric density function is the basis of Fisher's exact test used in the analysis of sparse contingency tables.

Poisson

A PDF that arises naturally in many instances, particularly as a probability model for the occurrence of rare events, for example, the emission of radioactive particles. In addition, the Poisson density function is the limiting distribution of the binomial when p is small and n is large. The Poisson density function for a random variable X taking integer values from 0 to ∞ is defined as

$$P(X = x) = \frac{e^{-\lambda}\lambda^x}{x!}, \quad x = 0, 1, 2 \ldots \infty \quad (6)$$

The single parameter of the Poisson density function, λ, is both the expected value and variance, that is, the mean and variance of a Poisson random variable are equal. Some examples of Poisson density functions are given in Figure 4.

Probability Density Functions for Continuous Random Variables

Normal

The PDF, $f(x)$, of a continuous random variable X defined as follows

$$f(x) = \frac{1}{\sigma\sqrt{2\pi}} \exp\left[-\frac{1}{2}\frac{(x-\mu)^2}{\sigma^2}\right] \quad -\infty < x < \infty \quad (7)$$

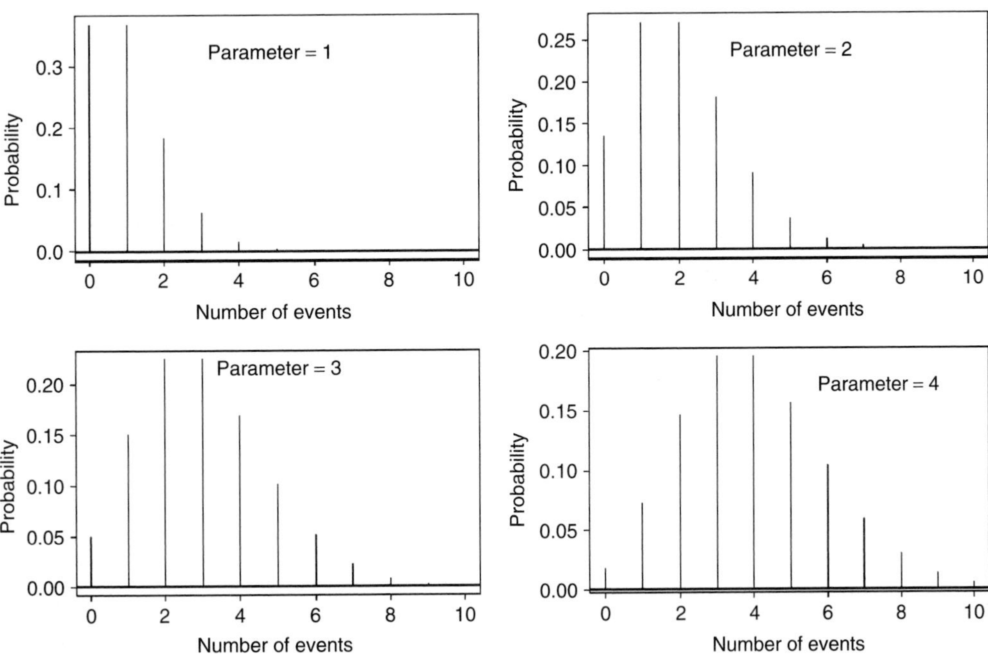

Figure 4 Examples of Poisson density functions

where μ and σ^2 are, respectively the mean and variance of X. When the mean is zero and the variance one, the resulting density is labeled as the *standard normal*. The normal density function is bell-shaped as is seen in Figure 5, where a number of normal densities are shown.

In the case of continuous random variables, the probability that the random variable takes a particular value is strictly zero; nonzero probabilities can only be assigned to some range of values of the variable. So, for example, we say that $f(x)\,\mathrm{d}x$ gives the probability of X falling in the very small interval, $\mathrm{d}x$, centered on x, and the probability that X falls in some interval, [A, B] say, is given by integrating $f(x)\,\mathrm{d}x$ from A to B.

The normal density function is ubiquitous in statistics. The vast majority of statistical methods are based on the assumption of a normal density for the observed data or for the error terms in models for the data. In part this can be justified by an appeal to the central limit theorem. The density function first appeared in the papers of de Moivre at the beginning of the eighteenth century and some decades later was given by Gauss and Laplace in the theory of errors and the least squares method. For this reason, the normal is also often referred to as the *Gaussian* or *Gaussian–Laplace*.

Uniform

The mean of such a random variable is having constant probability over an interval. The density function is given by

$$f(x) = \frac{1}{\beta - \alpha}, \quad \alpha < x < \beta \tag{8}$$

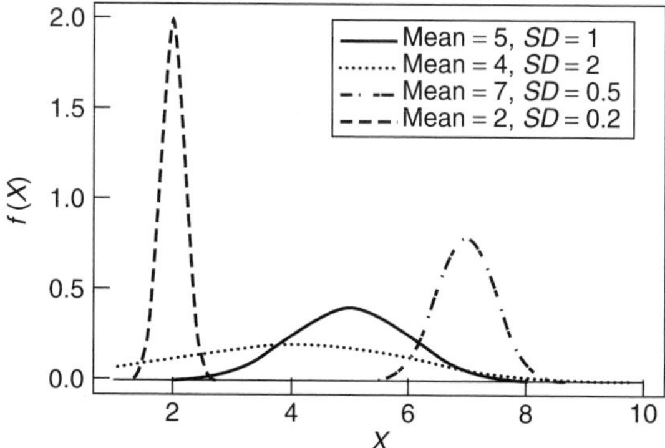

Figure 5 Normal density functions

The mean of the density function is $(\alpha + \beta)/2$ and the variance is $(\beta - \alpha)^2/12$. The most commonly encountered version of this density function is one in which the parameters α and β take the values 0 and 1 respectively, and is used in generating quasi-random numbers.

Exponential

The PDF of a continuous random variable, X, taking only positive values. The density function is given by

$$f(x) = \lambda \mathrm{e}^{-\lambda x}, \quad x > 0 \tag{9}$$

The single parameter of the exponential density function, λ, determines the shape of the density function, as we see in Figure 6, where a number of different exponential density functions are shown. The mean of an exponential variable is $1/\lambda$ and the variance is $1/\lambda^2$.

The exponential density plays an important role in some aspects of reliability engineering and also gives the distribution of the time intervals between independent consecutive random events such as particles emitted by radioactive material.

Beta

A PDF for a continuous random variable, X, taking values between 0 and 1. The density function is defined as

$$f(x) = \frac{\Gamma(p+q)}{\Gamma(p)\Gamma(q)} x^{p-1}(1-x)^{q-1}, \quad 0 < x < 1 \tag{10}$$

where $p > 0$ and $q > 0$ are parameters that define particular aspects of the shape of the beta density and Γ is the gamma function (see [3]). The mean

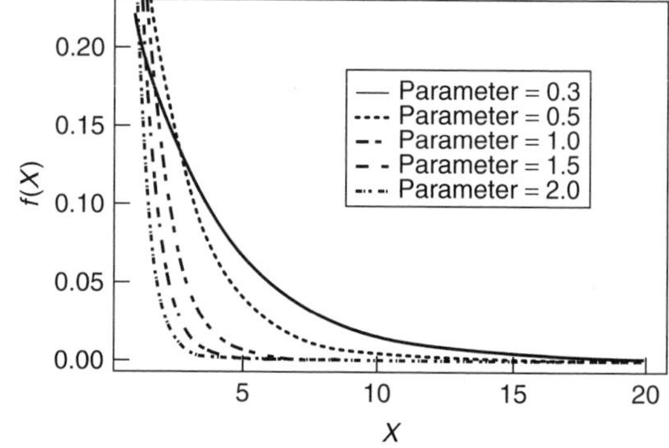

Figure 6 Exponential density functions

of a beta variable is $p/(p+q)$ and the variance is $pq/(p+q)^2(p+q+1)$.

Gamma

A PDF for a random variable X that can take only positive values. The density function is defined as

$$f(x) = \frac{1}{\Gamma(\alpha)} x^{\alpha-1} e^{-x}, \quad x > 0 \qquad (11)$$

where the single parameter, α, is both the mean of X and also its variance.

The gamma density function can often act as a useful model for a variable that cannot reasonably be assumed to have a normal density function because of its positive **Skewness**.

Chi-Squared

The PDF of the **sum of squares** of a number (ν) of independent standard normal variables given by

$$f(x) = \frac{1}{2^{\nu/2}\Gamma(\nu/2)} x^{(\nu-2)/2} e^{-x/2}, \quad x > 0 \qquad (12)$$

that is, a gamma density with $\alpha = \nu/2$. The parameter ν is usually known as the ***degrees of freedom*** of the density. The mean and variance of a χ^2 distribution are ν and 2ν respectively. This density function arises in many areas of statistics, most commonly as the null distribution of the chi-squared goodness of fit statistics, or as the distribution associated with estimators of variance.

Student's T

The PDF of a variable defined as

$$t = \frac{\bar{X} - \mu}{s/\sqrt{n}} \qquad (13)$$

where $\bar{X}$ is the arithmetic mean of n observations from a normal density with mean μ and s is the sample standard deviation. The density function is given by

$$f(t) = \frac{\Gamma\left\{\frac{1}{2}(\nu+1)\right\}}{(\nu\pi)^{1/2}\Gamma\left(\frac{1}{2}\nu\right)} \left(1 + \frac{t^2}{\nu}\right)^{-\frac{1}{2}(\nu+1)},$$

$$-\infty < t < \infty \qquad (14)$$

where $\nu = n - 1$. The shape of the density function varies with ν, and as ν gets larger the shape of the t-density function approaches that of a standard normal. This density function is the null distribution of Student's t-statistic used for testing hypotheses about population means (*see* **Parametric Tests**).

Fisher's F

The PDF of the ratio of two independent random variables each having a χ^2 distribution, divided by their respective degrees of freedom. The form of the density is that of a beta density with $p = \nu_1/2$ and $q = \nu_2/2$, where ν_1 and ν_2 are, respectively, the degrees of freedom of the numerator and denominator χ^2 variables. Fisher's F density is used to assess the equality of variances in, for example, the F-test and the analysis of variance (*see* **Parametric Tests**).

Weibull distribution

Named after the Swedish Professor Waloddi Weibull, the class of Weibull distributions is sometimes thought of as a generalization of the exponential distribution (which it includes as a special case). The PDF is

$$f(x) = \frac{c}{b}\left(\frac{x-a}{b}\right)^{c-1} \exp\left\{-\left[\frac{(x-a)}{b}\right]^c\right\}, x > a \qquad (15)$$

The parameters a and b are location and scale parameters (a is sometimes called the *threshold*, and is often assumed to be 0), while c is a shape parameter. For $a = 0$ and $c = 1$ we obtain the density of the exponential distribution. Weibull distributions are used often in lifetime and time-to-failure models.

Lognormal distribution

If for some value of a, $\ln(X - a)$ has a **normal distribution** with mean μ and variance σ^2, then X has a lognormal distribution with parameters a, μ and σ. Reparameterizing as $b = \exp(\mu)$, a, b, and c are location, scale, and shape parameters. The PDF is

$$f(x) = \frac{1}{\sigma(x-a)\sqrt{2\pi}} \exp\left[-\frac{1}{2\sigma^2}\left\{\log\left(\frac{x-a}{b}\right)\right\}^2\right],$$

$$x > a \qquad (16)$$

The mean and variance are $E(X) = a + b\exp(\sigma^2/2)$ and $Var(X) = b^2\exp(\sigma^2)[\exp(\sigma^2) - 1]$. The lognormal distribution is a possible model for variables that have a skewed distribution.

Multivariate Density Functions

Multivariate density functions play the same role for *vector random variables* as the density functions described earlier play in the univariate situation. Here we shall look at two such density functions, the multinomial and the multivariate normal.

Multinomial

The multinomial PDF is a multivariate generalization of the binomial density function described earlier. The density function is associated with a vector of random variables $X' = [X_1, X_2, \ldots, X_k]$ which arises from a sequence of n independent and identical trials each of which can result in one of k possible mutually exclusive and collectively exhaustive events, with probabilities $p_1, p_2, \ldots, p_k$. The density function is defined as follows:

$$P(X_1 = x_1, X_2 = x_2, \ldots, X_k = x_k)$$

$$= \frac{n!}{x_1!x_2!\ldots x_k!} p_1^{x_1} p_2^{x_2} \ldots p_k^{x_k} \quad (17)$$

where x_i is the number of trials with outcome i. The expected value of X_i is np_i and its variance is $np_i(1 - p_i)$. The **covariance** of X_i and X_j is $-np_i p_j$.

Multivariate Normal

The PDF of a vector of continuous random variables, $X' = [X_1, X_2, \ldots, X_p]$ defined as follows

$$f(\mathbf{x}_1, \mathbf{x}_2, \ldots, \mathbf{x}_p) = (2\pi)^{-p/2}|\boldsymbol{\Sigma}| \times \exp\left[-\frac{1}{2}(\mathbf{x} - \boldsymbol{\mu})'\boldsymbol{\Sigma}^{-1}(\mathbf{x} - \boldsymbol{\mu})\right] \quad (18)$$

where μ is the mean vector of the variables and Σ is their covariance matrix.

The simplest version of the multivariate normal density function is that for two random variables

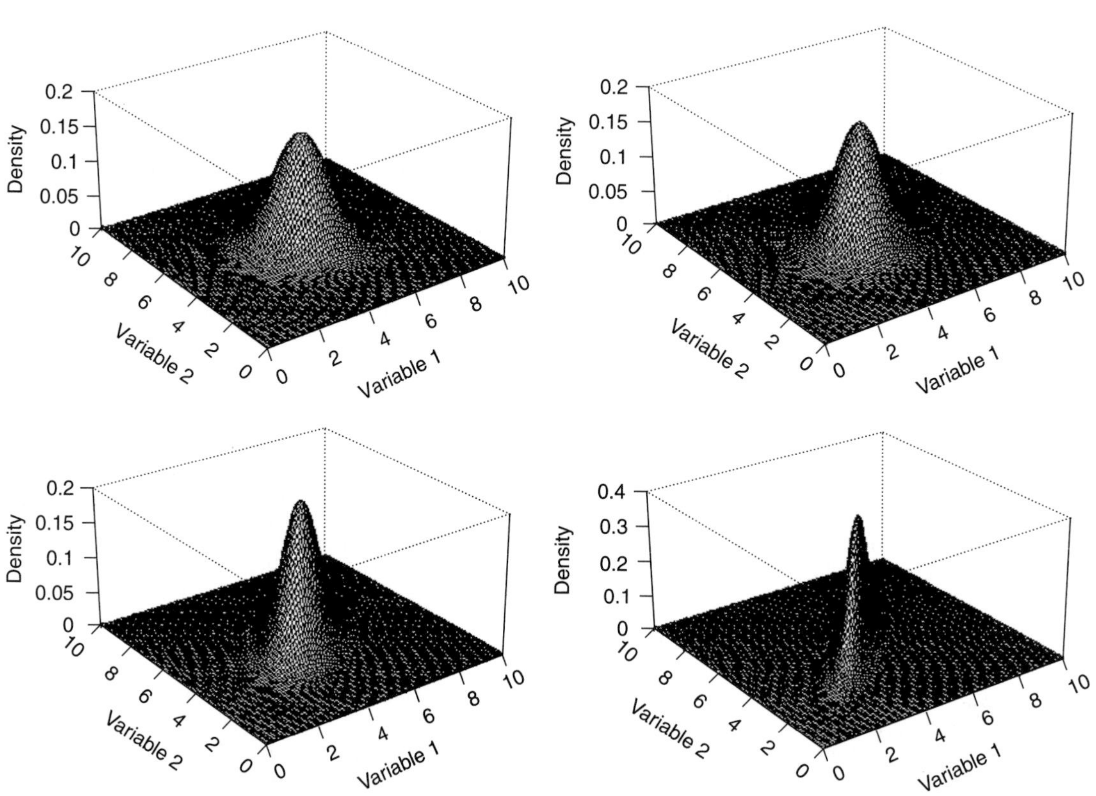

Figure 7 Four bivariate normal density functions with means 5 and standard deviations 1, for variables 1 and 2, and with correlations equal to 0.0, 0.3, 0.6, and 0.9

and known as the *bivariate normal density function*. The density function formula given above now reduces to

$$f(x_1, x_2) = \frac{1}{2\pi\sigma_1\sigma_2\sqrt{1-\rho^2}} \exp\left\{-\frac{1}{1-\rho^2}\right. \times \left[\frac{(x_1-\mu_1)^2}{\sigma_1^2} - 2\rho\frac{(x_1-\mu_1)}{\sigma_1} \times \frac{(x_2-\mu_2)}{\sigma_2} + \frac{(x_2-\mu_2)^2}{\sigma_2^2}\right]\left.\right\} \quad (19)$$

where $\mu_1, \mu_2, \sigma_1, \sigma_2, \rho$ are, respectively, the means, standard deviations, and **correlation** of the two variables. Perspective plots of a number of such density functions are shown in Figure 7.

References

[1] Balakrishnan, N. & Nevzorov, V.B. (2003). *A Primer on Statistical Distributions*, John Wiley & Sons, Hoboken.

[2] Evans, M., Hastings, N. & Peacock, B. (2000). *Statistical Distributions*, 3rd Edition, John Wiley & Sons, New York.

[3] Everitt, B.S. (2002). *Cambridge Dictionary of Statistics*, 2nd Edition, Cambridge University Press, Cambridge.

BRIAN S. EVERITT

Article originally published in Encyclopedia of Statistics in Behavioral Science *(2005, John Wiley & Sons, Ltd). Minor revisions for this publication by Jeroen de Mast.*

Probability Plots

Introduction

Probability plots are often used to compare two empirical distributions (*see* **Estimation**), but they are particularly useful to evaluate the shape of an empirical distribution against a theoretical distribution. For example, we can use a probability plot to examine the degree to which an empirical distribution differs from a **normal distribution**.

There are two major forms of probability plots, known as *probability–probability (P–P) plots and quantile–quantile (Q–Q) plots*. They both serve essentially the same purpose, but plot different functions of the data.

An Example

It is easiest to see the difference between P–P and Q–Q plots with a simple example. The data displayed in Table 1 are derived from data collected by Compas (personal communication) on the effects of stress on cancer patients. Each of the 41 patients completed the externalizing behavior scale of the Brief Symptom Inventory. We wish to judge the normality of the sample data. A histogram for these data is presented in Figure 1 with a normal distribution superimposed. The sample mean is 12.10, and the sample standard deviation is 10.32.

P–P Plots

A P–P plot displays the cumulative probability of the obtained data on the x axis and the corresponding expected cumulative probability for the reference distribution (in this case, the normal distribution) on the y axis. For example, a raw score of 15 for our data had a cumulative probability of 0.756. It also had a z score, given the mean and standard deviation of the sample, of 0.28. For a normal distribution, we expect to find 61% of the distribution falling at or below $z = 0.28$. So, for this observation, we had considerably more scores (75.6%) falling at or below 15 than we would have expected if the distribution were normal (61%). From Table 1, you can see the results of similar calculations for the full data set. These are displayed in columns 4 and 6.

If we plot the obtained cumulative percentage on the x axis and the expected cumulative percentage on the y axis, we obtain the results shown in Figure 2. Here, a line drawn at 45° is superimposed. If the data had been exactly normally distributed, the points would have fallen on that line. It is clear from the figure that this was not the case. The points deviate noticeably from the line, and thus, from normality.

Q–Q Plots

A Q–Q plot resembles a P–P plot in many ways, but it displays the observed values of the data on the x axis and the corresponding expected values from a normal distribution on the y axis. The expected values are based on the *quantiles* of the

Table 1 Frequency distribution of externalizing scores with normal probabilities and **quantiles**

Score	Frequency	Percent	Cumulative percent	z	CDF normal	Normal quantile
0	3	7.3	7.3	−1.17	0.12	−3.01
1	2	4.9	12.2	−1.07	0.14	−0.03
2	2	4.9	17.1	−0.98	0.16	2.19
4	1	2.4	19.5	−0.78	0.22	3.13
5	3	7.3	26.8	−0.69	0.25	5.61
6	2	4.9	31.7	−0.59	0.28	7.08
7	1	2.4	34.1	−0.49	0.31	7.77
8	1	2.4	36.6	−0.40	0.35	8.46
9	6	14.6	51.2	−0.30	0.38	12.31
10	2	4.9	56.1	−0.20	0.42	13.58
11	3	7.3	63.4	−0.1	0.46	15.54
12	2	4.9	68.3	−0.01	0.50	16.92
14	2	4.9	73.2	0.18	0.57	18.39
15	1	2.4	75.6	0.28	0.61	19.16
16	2	4.9	80.5	0.38	0.65	20.87
19	1	2.4	82.9	0.67	0.75	21.81
24	1	2.4	85.4	1.15	0.88	22.88
25	1	2.4	87.8	1.25	0.89	24.03
28	1	2.4	90.2	1.54	0.94	25.35
29	1	2.4	92.7	1.64	0.95	27.01
31	1	2.4	95.1	1.83	0.97	29.08
40	1	2.4	97.6	2.70	0.99	32.41
42	1	2.4	100.0	2.90	1.00	36.02

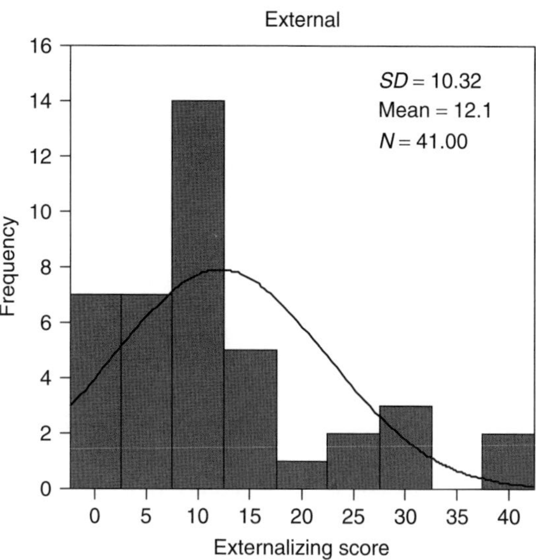

Figure 1 Histogram of externalizing scores with normal curve superimposed

distribution. To take our example of a score of 15 again, we know that 75.6% of the observations fell at or below 15, placing 15 at the 75.6 percentile. If we had a normal distribution with a mean of 12.10 and a standard deviation of 10.32, we would expect the 75.6 percentile to fall at a score of 19.16. We can make similar calculations for the remaining data, and these are shown in column 7 of Table 1.

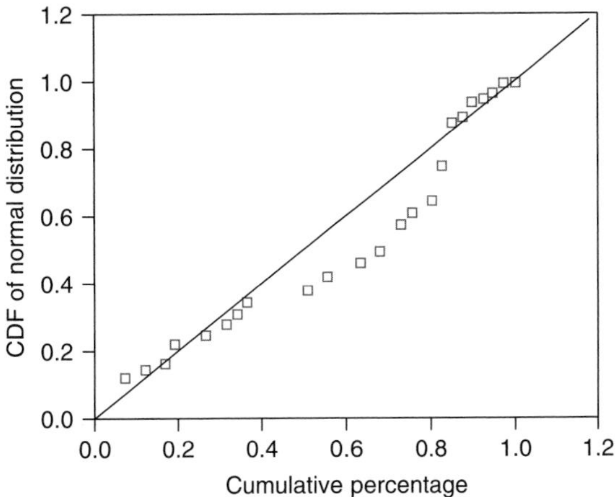

Figure 2 P–P plot of obtained distribution against the normal distribution

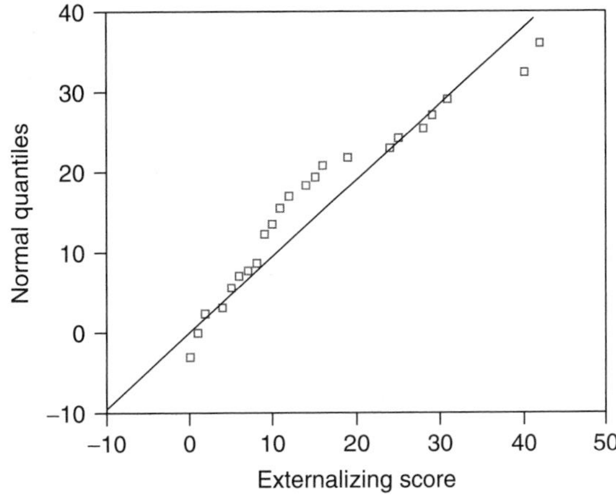

Figure 3 Q–Q plot of obtained distribution against the normal distribution

(The final cumulative percentage has been treated as 99.9 instead of 100 because the normal distribution runs to infinity and the quantile for 100% would be undefined.)

The plot of these values can be seen in Figure 3, where, again, a straight line is superimposed representing the results that we would have if the data were perfectly normally distributed. Again, we can see significant departure from normality.

Related Article

Half-Normal Plot.

DAVID C. HOWELL

Article originally published in Encyclopedia of Statistics in Behavioral Science *(2005, John Wiley & Sons, Ltd). Minor revisions for this publication by Jeroen de Mast.*

Probability Theory

Introduction

Probability theory provides a mathematical framework for the description of phenomena whose outcomes are not deterministic but rather are subject to chance. This framework, historically motivated by a frequency interpretation of probability, is provided by a formal set of three innocent-looking axioms and the consequences thereof. These are far-reaching, and a rich and diverse theory has been developed. It provides a flexible and effective model of physical reality and furnishes, in particular, the mathematical theory needed for the study of the discipline of statistics.

The Formal Framework

Elementary (nonmeasure theoretic) treatments of probability can readily deal with situations in which the set of distinguishable outcomes of an experiment subject to chance (called *elementary events*) is finite. This encompasses most simple gambling situations, for example.

Let the set of all elementary events be Ω (called the *sample space*) and let $\mathcal{F}$ be the Boolean field of subsets of Ω (meaning just that (a) $A \in \mathcal{F}$ implies that $\overline{A} \in \mathcal{F}$, the bar denoting complementation with respect to Ω and (b) $A_1, A_2 \in \mathcal{F}$ implies that $A_1 \cup A_2 \in \mathcal{F}$). Then the probability P is a set function on $\mathcal{F}$ satisfying the axioms:

Axiom 1 $P(A) \geq 0, A \in \mathcal{F}$.

Axiom 2 If $A_1, A_2 \in \mathcal{F}$ and $A_1 \cap A_2 = \emptyset$ (the empty set), then

$$P(A_1 \cup A_2) = P(A_1) + P(A_2) \qquad (1)$$

Axiom 3 $P(\Omega) = 1$.

These axioms enable probabilities to be assigned to events in simple games of chance. For example, in the toss of an ordinary six-sided die, Ω consists of six equiprobable elementary events and for $A \in \mathcal{F}$, $(A) = n(A)/6$, where $n(A)$ is the number of elementary events in A.

The formulation above can be extended straightforwardly to the case where the number of elementary events in the sample space is at most countable, by extending Axiom 2 to deal with countable unions of disjoint sets. However, a full measure-theoretic formulation is necessary to deal with general sample spaces Ω and to provide a framework which is rich enough for a comprehensive study of limit results on combinations of events (sets).

The measure-theoretic formulation has a very similar appearance to the elementary version. We start from a general sample space Ω and let $\mathcal{F}$ be the Borel σ-field of subsets of Ω (meaning that (a) $A \in \mathcal{F}$ implies that $\overline{A} \in \mathcal{F}$ and (b) $A_i \in \mathcal{F}$, $1 \leq i < \infty$, implies that $\bigcup_{i=1}^{\infty} A_i \in \mathcal{F}$). Then as axioms for the probability measure P we have Axioms 1 and 3 as before, while Axiom 2 is replaced by:

Axiom 2′ If $A_i \in \mathcal{F}$, $1 \leq i < \infty$, and $A_i \cap A_j = \emptyset$, $i \neq j$, then

$$P\left(\bigcup_{i=1}^{\infty} A_i\right) = \sum_{i=1}^{\infty} P(A_i) \qquad (2)$$

The triplet $(\Omega, \mathcal{F}, P)$ is called a *probability space*. This framework is due to Kolmogorov [1] in 1933 and is very widely accepted, although some authors reject the countable additivity Axiom 2′ and rely on the finite additivity version (Axiom 2) (e.g., Dubins and Savage [2]). An extension of the Kolmogorov theory in which unbounded measures are allowed and conditional probabilities are taken as the fundamental concept has been provided by Rényi [3, Chapter 2]. For $A, B \in \mathcal{F}$ with $P(B) > 0$, the conditional probability of A given B is defined as $P(A|B) = P(A \cap B)/P(B)$.

Suppose that the sample space Ω has points ω. A real-valued function $X(\omega)$ defined on the probability space $(\Omega, \mathcal{F}, P)$ is called a *random variable* (or, in the language of measure theory, $X(\omega)$ is a *measurable function*) if the set $\{\omega : X(\omega) \leq x\}$ belongs to $\mathcal{F}$ for every $x \in \mathbb{R}$ (the real line). Sets of the form $\{\omega : X(\omega) \leq x\}$ are usually written as $\{X \leq x\}$ for convenience and we shall follow this practice. Random variables and their relationships are the principal objects of study in much of probability and statistics.

Important Concepts

Each random variable X has an associated *distribution function* defined for each real x by

$$F(x) = P(X \leq x) \qquad (3)$$

The function F is nondecreasing, right continuous, and has at most a countable set of points of discontinuity. It also uniquely specifies the probability measure that X induces on the real line.

Random variables are categorized as having a discrete distribution if the corresponding distribution function is a step function and as having a continuous distribution if it is continuous.

Most continuous distributions of practical interest are absolutely continuous, meaning that the distribution function $F(x)$ has a representation of the form

$$F(x) = \int_{-\infty}^{x} f(u)\,\mathrm{d}u \tag{4}$$

for some $f \geq 0$; f is called the **probability density function**.

A relatively small number of families of distributions are widely used in applications of probability theory (*see* **Probability Density Functions**). Among the most important are the binomial, which is discrete and for which

$$P(X = r) = \binom{n}{r} p^r (1-p)^{n-r}, \quad 0 \leq r \leq n, \quad 0 < p < 1 \tag{5}$$

and the (unit) normal, which is absolutely continuous and for which

$$P(X \leq x) = (2\pi)^{-1/2} \int_{-\infty}^{x} \mathrm{e}^{-(1/2)u^2} \mathrm{d}u \quad -\infty < x < \infty \tag{6}$$

The binomial random variable represents the numbers of successful outcomes in n trials, repeated under the same conditions, where the probability of success in each individual trial is p. Normal distributions, or close approximations thereto, occur widely in practice, for example, in such measurements as heights or weights or examination scores, where a large number of individuals are involved.

Two other particularly important distributions are the Poisson, for which

$$P(X = r) = \mathrm{e}^{-\lambda} \lambda^r / r! \quad r = 0, 1, 2, \ldots, \quad \lambda > 0 \tag{7}$$

and the exponential, for which

$$P(X \leq x) = \lambda \int_{0}^{x} \mathrm{e}^{-\lambda u}\,\mathrm{d}u, \qquad x > 0, \quad \lambda > 0$$
$$= 0 \text{ otherwise} \tag{8}$$

These commonly occur in situations of rare events and as waiting times between phenomena, respectively.

A sequence of random variables $\{X_1, X_2, \ldots\}$ defined on the same probability space is called a *stochastic process*. For a finite collection of such random variables $\{X_1, X_2, \ldots, X_m\}$ we can define the joint distribution function

$$F(x_1, x_2, \ldots, x_m) = P(X_1 \leq x_1, X_2 \leq x_2, \ldots, X_m \leq x_m) \tag{9}$$

for $x_i \in \mathbb{R}$, $1 \leq i \leq m$, and the random variables are said to be *independent* if

$$P(X_1 \leq x_1, \ldots, X_m \leq x_m) = P(X_1 \leq x_1) \cdots P(X_m \leq x_m) \tag{10}$$

for all $x_1, \ldots, x_m$ (*see* **Dependence**). Independence is often an essential basic property and much of probability and statistics is concerned with random sampling, namely independent observations (random variables) with identical distributions.

One step beyond independence is the Markov chain $\{X_1, X_2, \ldots\}$, in which case

$$P(X_m \leq x_m | X_1 = i_1, \ldots, X_n = i_n) = P(X_m \leq x_m | X_n = i_n) \tag{11}$$

for any $n < m$ and all $i_1, \ldots, i_n$ and x_m. That is, dependence on a past history involves only the most recent of the observations. Such random sequences with a natural timescale belong to the domain of stochastic processes; *see* **Markov Processes**.

Considerable summary information about random variables and their distributions is contained in quantities called *moments* (*see* **Moments**). For a random variable X with distribution function F, the rth *moment* ($r > 0$) is defined by

$$EX^r = \int_{-\infty}^{\infty} x^r\,\mathrm{d}F(x) \tag{12}$$

provided that $\int_{-\infty}^{\infty} |x|^r\,\mathrm{d}F(x) < \infty$. $E|X|^r$ is called the rth *absolute moment*. The operator E is called *expectation* (*see* **Expectation**) and the first moment ($r = 1$) is termed the *mean*. Moments about the mean $E(X - EX)^r$, $r = 2, 3, \ldots$ are widely used and $E(X - EX)^2$ is termed the *variance* (*see* **Variance**) of X, while $(E(X - EX)^2)^{1/2}$ is its *standard deviation*. The mean and variance of X are, respectively, measures of the location and spread of its distribution.

Convergence

Much of probability theory is concerned with the limit behavior of sequences of random variables or their distributions. The great diversity of possible probabilistic behavior when the **sample size** is small is often replaced by unique and well-defined probabilistic behavior for large sample sizes. In this lies the great virtue of the limit theory: statistical regularity emerging out of statistical chaos as the sample size increases.

A sequence of random variables $\{X_n\}$ is said to converge *in distribution* (or *in law*) to that of a random variable X (written $X_n \xrightarrow{d} X$) if the sequence of distribution functions $\{F_n(x) = P(X_n \le x)\}$ converges to the distribution function $F(x) = P(X \le x)$ at all points x of continuity of F.

Convergence in distribution does not involve the random variables explicitly, only their distributions. Other commonly used modes of convergence are the following:

1. $\{X_n\}$ converges *in probability* to X (written $X_n \xrightarrow{p} X$) if, for every $\epsilon > 0$,
$$\lim_{n\to\infty} P(|X_n - X| > \epsilon) = 0 \tag{13}$$
2. For $r > 0$, $\{X_n\}$ converges *in the mean of order* r to X (written $X_n \xrightarrow{r} X$) if
$$\lim_{n\to\infty} E|X_n - X|^r = 0 \tag{14}$$
3. $\{X_n\}$ converges *almost surely* to X (written $X_n \overset{\text{a.s.}}{\to} X$) if for every $\epsilon > 0$,
$$\lim_{n\to\infty} P\left(\sup_{k\ge n}|X_k - X| > \epsilon\right) = 0 \tag{15}$$

It is not difficult to show that 2 implies 1, which in turn implies convergence in distribution while 3 implies 1. None of the converse implications hold in general.

Convergence in distribution is frequently established by the method of characteristic functions. The *characteristic function* of a random variable X with distribution function F is defined by

$$f(t) = E\mathrm{e}^{itX} = \int_{-\infty}^{\infty} \mathrm{e}^{itx}\,\mathrm{d}F(x) \tag{16}$$

where $i = \sqrt{-1}$. If $f_n(t)$ and $f(t)$ are, respectively, the characteristic functions of X_n and X, then $X_n \xrightarrow{d} X$ is equivalent to $f_n(t) \to f(t)$. Characteristic functions are particularly convenient for dealing with sums of independent random variables as a consequence of the multiplicative property

$$E\mathrm{e}^{it(X+Y)} = E\mathrm{e}^{itX}E\mathrm{e}^{itY} \tag{17}$$

for independent X and Y. They have been extensively used in most treatments of convergence in distribution for sums of independent random variables (the classical central limit problem). The basic central limit theorem (of which there are many generalizations) deals with independent and identically distributed random variables X_i with mean μ and variance σ^2. It gives, upon writing $S_n = \sum_{i=1}^n X_i$, that

$$\lim_{n\to\infty} P(\sigma^{-1}n^{-1/2}(S_n - n\mu) \le x) = (2\pi)^{-1/2}\int_{-\infty}^{x} \mathrm{e}^{-(1/2)u^2}\,\mathrm{d}u \tag{18}$$

the limit being the distribution function of the unit normal law.

Convergence in probability results are often obtained with the aid of Markov's inequality

$$P(|X_n| > c) \le c^{-r}E|X_n|^r \tag{19}$$

for $r > 0$ and a corresponding convergence in the rth mean result. The useful particular case $r = 2$ of Markov's inequality is ordinarily called *Chebyshev's inequality*. For example, in the notation of equation (20),

$$P(|n^{-1}S_n - \mu| > \epsilon) \le \epsilon^{-2}n^{-2}E(S_n - n\mu)^2 \to 0 \tag{20}$$

as $n \to \infty$, which is a version of the weak (*see* **Laws of Large Numbers**), $n^{-1}S_n \xrightarrow{p} \mu$.

Almost sure convergence results rely heavily on the Borel–Cantelli lemmas. These deal with a sequence of events (sets) $\{E_n\}$ and $E = \bigcap_{r=1}^{\infty}\bigcup_{n=r}^{\infty} E_n$, the set of elements common to infinitely many of the E_n. Then

1. $\sum P(E_n) < \infty$ implies that $P(E) = 0$.
2. $\sum P(E_n) = \infty$ and the events E_n are independent implies that $P(E) = 1$.

With the aid of 1, equation (21) can be strengthened to the strong *law of large numbers* (*see* **Laws of Large Numbers**),

$$n^{-1}S_n \xrightarrow{\text{a.s.}} \mu \tag{21}$$

as $n \to \infty$.

The strong law of large numbers embodies the idea of a probability as a strong limit of relative frequencies; if we set $X_i = I_i(A)$, the indicator function of the set A at the ith trial, then $\mu = P(A)$. This property is an important requirement of a physically realistic theory.

The law of large numbers, together with the central limit theorem, provides a basis for a fundamental piece of statistical theory. The former yields an estimator of the location parameter μ and the latter enables approximate *confidence intervals* (*see* **Confidence Intervals**) for μ to be prescribed and tests of hypotheses (*see* **Hypothesis Testing**) constructed.

Other Approaches

The discussion above is based on a frequency interpretation of probability, and this provides the richest theory. There are, however, significant limitations to a frequency interpretation in describing uncertainty in some contexts, and various other approaches have been suggested. Examples are the development of subjective probability by de Finetti and Savage (e.g., references [4] and [5]) and the so-called logical probability of Carnap [6]. The former is the concept of probability used in Bayesian inference and the latter is concerned with objective assessment of the degree to which evidence supports a hypothesis. A detailed comparative discussion of the principal theories that have been proposed is given in Fine [7].

The emergence of probability theory as a scientific discipline dates from around 1650.

The Literature

Systematic accounts of probability theory have been provided in book form by many authors. These are necessarily at two distinct levels since a complete treatment of the subject has a measure-theory prerequisite. Elementary discussions, not requiring measure theory, are given for example, in Chung [8], Feller [9], Gnedenko [10], and Parzen [11]. A full treatment can be found in Billingsley [12], Breiman [13], Chow and Teicher [14], Chung [15], Feller [16], Hennequin and Tortrat [17], Loéve [18], and Rényi [19].

References

[1] Kolmogorov, A. (1933). *Grundbegriffe der Wahrscheinlichkeitsrechnung*, Springer-Verlag, Berlin. (English trans.: N. Morrison, in Foundations of the Theory of Probability. Chelsea, New York, 1956.)

[2] Dubins, L.E. & Savage, L.J. (1965). *How to Gamble if you Must*, McGraw-Hill, New York.

[3] Rényi, A. (1970). *Foundations of Probability*, Holden-Day, San Francisco.

[4] de Finetti, B. (1974). *Theory of Probability*, John Wiley & Sons, New York, Vols. 1–2.

[5] Savage, L.J. (1962). *The Foundations of Statistical Inference: A Discussion*, John Wiley & Sons, New York.

[6] Carnap, R. (1962). *The Logical Foundations of Probability*, 2nd Edition, University of Chicago Press, Chicago.

[7] Fine, T.L. (1973). *Theories of Probability. An Examination of the Foundations*, Academic Press, New York.

[8] Chung, K.L. (1974). *Elementary Probability Theory with Stochastic Processes*, Springer-Verlag, Berlin.

[9] Feller, W. (1968). *An Introduction to Probability Theory and its Applications*, 3rd Edition, John Wiley & Sons, New York, Vol. 1.

[10] Gnedenko, B.V. (1967). *The Theory of Probability*, 4th Edition, Chelsea Publishing, New York. (Translated from the Russian by B.D. Seckler.)

[11] Parzen, E. (1960). *Modern Probability Theory and its Applications*, John Wiley & Sons, New York.

[12] Billingsley, P. (1979). *Probability and Measure*, John Wiley & Sons, New York.

[13] Breiman, L. (1968). *Probability*, Addison-Wesley, Reading.

[14] Chow, Y.S. & Teicher, H. (1978). *Probability Theory. Independence, Interchangeability, Martingales*, Springer-Verlag, New York.

[15] Chung, K.L. (1974). *A Course in Probability Theory*, 2nd Edition, Academic Press, New York.

[16] Feller, W. (1971). *An Introduction to Probability Theory and its Applications*, 2nd Edition, John Wiley & Sons, New York, Vol. 2.

[17] Hennequin, P.L. & Tortrat, A. (1965). *Théorie des Probabilités et Quelques Applications*, Masson, Paris.

[18] Loéve, M. (1977). *Probability Theory*, 4th Edition, Springer-Verlag, Berlin, Vols. 1–2.

[19] Rényi, A. (1970). *Probability Theory*, North-Holland, Amsterdam. (Earlier versions of this book appeared as Wahrscheinlichkeitsrechnung, VEB Deutscher Verlag, Berlin, 1962; and Calcul des probabilités, Dunod, Paris, 1966.)

CHRISTOPHER. C. HEYDE

Article originally published in Encyclopedia of Statistical Sciences, 2nd Edition *(2005, John Wiley & Sons, Inc.). Minor revisions for this publication by Jeroen de Mast.*

Process Capability Indices for Multiple Stream Processes

Introduction

Separate process streams are created in manufacturing when the flow of parts is divided into several presumably identical paths. This split in production occurs when an operation has different molds, spindles, lines, stations, lines, fixtures, or pallets. Presumably, the quality of parts produced by each stream is identical but, in practice, there are often substantial differences between streams. The capability index introduced in this article may be applied when a measure of capability is desired for the combined output of all streams.

The C_{pk} Index

As in all capability studies, the first prerequisite is to verify that each process stream is in a reasonably good state of statistical control (*see* **Control Charts and Process Capability**). The output from each stream may have a different average and/or standard deviation, but every stream must have a stable, predictable output.

For example, an injection-molding operation utilizes a four-cavity die to produce a plastic part. Through the use of **control charts**, each cavity demonstrates good stability for the inner diameter (i.d.) of the part being molded, but the output of hole size varies considerably from cavity to cavity, as is illustrated in Figure 1.

After the parts are molded, they fall down a chute into a common container, which is then shipped to a customer. What should the customer who receives this shipment – which contains a mixture of parts from all four cavities – expect in terms of parts meeting the specification requirements for hole size?

One popular metric for assessing capability is called the C_{pk} *index* [1], which is defined in equations (1–3). Here, μ is the process mean, σ is the process standard deviation, LSL is the lower specification limit for the feature being studied, and USL is its upper specification limit (*see* **Control Charts for the Mean**).

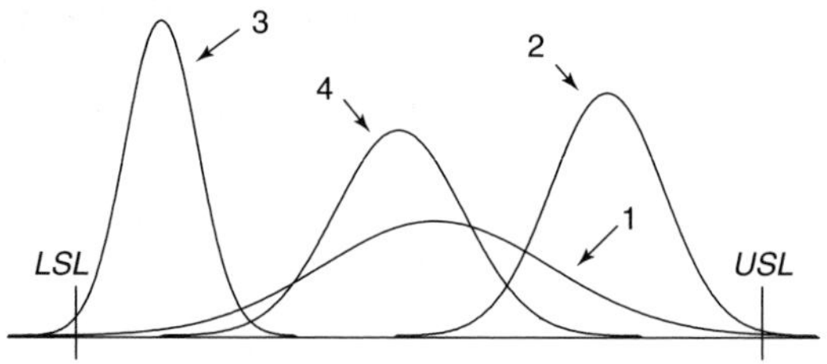

Figure 1 Distributions of hole size for each of four cavities (labeled 1–4)

$$C_{pk} = \text{minimum}(C_{pl}, C_{pu}) \quad (1)$$

$$C_{pl} = \frac{\mu - LSL}{3\sigma} \quad (2)$$

$$C_{pu} = \frac{USL - \mu}{3\sigma} \quad (3)$$

If only an LSL is given for the feature being studied, then C_{pk} equals C_{pl}. Conversely, when just a USL is provided, C_{pk} equals C_{pu}.

When the process output is close to a **normal distribution**, the percentage of parts below the LSL, p_{LSL}, is determined with equation (4), where $\Phi(\cdot)$ is the cumulative distribution function of the standard normal distribution (with $\mu = 0$ and $\sigma = 1$) [2]. For example, $\Phi(0)$ equals 0.50000 (50%), $\Phi(2)$ equals 0.97725, and $\Phi(3)$ equals 0.99865. In the spreadsheet program Excel, $\Phi(2)$ corresponds to the statistical function NORMSDIST(2), which returns the value of 0.97725.

$$p_{LSL} = 1 - \Phi(3C_{pl}) \quad (4)$$

Likewise, the percentage of parts above the USL, p_{USL}, is found with equation (5).

$$p_{USL} = 1 - \Phi(3C_{pu}) \quad (5)$$

The total percentage of nonconforming parts, p_T, is then computed as demonstrated in equation (6).

$$p_T = p_{LSL} + p_{USL} \quad (6)$$

Typically, because the output for each process stream will have a different mean and standard deviation, each stream will have different C_{pk}, p_{LSL}, and p_{USL} values.

Past Attempts

When faced with differences between process streams that cannot be easily corrected, some practitioners compute the C_{pk} index for all streams and then report just the highest value to customers. This practice is deceptive as it would cause customers to believe they are purchasing a higher level of quality than what they will actually receive.

To avoid misleading customers, other practitioners go to the opposite extreme and report just the lowest C_{pk} index of all the individual streams. Doing so means the customer will receive at least this quality level because the percentages of conforming parts are greater for the outputs from all the other streams. However, if there are large differences between the streams, reporting results for only the worst stream will make the manufacturer's overall quality level look much poorer than it really is.

A third approach has been to report an overall capability index by averaging the C_{pk} indices for the individual streams. To demonstrate why this technique is invalid, suppose an operation has just two process streams; the output of the first stream has a C_{pk} of 2.00 while the output of the second has a C_{pk} of 0.00. The average of these two C_{pk} indices is 1.00 $[(2.00 + 0.00)/2 = 1.00]$.

From equation (1), a C_{pk} index of 1.00 means that both C_{pl} and C_{pu} must also be at least 1.00. Then, from equations (4) and (5), neither p_{LSL} nor p_{USL} can be greater than 0.135%. Thus, a C_{pk} index of 1.00 implies that p_T for this shipment is at most 0.27%.

However, the process stream with a C_{pk} index of 2.00 is producing essentially 0.0% nonconforming parts (because C_{pu} and C_{pl} are both equal to, or greater than, 2.00, both p_{LSL} and p_{LSL} are very close to 0). On the other hand, the second stream is generating at least 50% nonconforming parts (with either C_{pu} or C_{pl} equal to 0.00, either p_{LSL} or p_{LSL} equals 0.5000 while the other is equal to, or less than, 0.5000).

Assuming both streams produce an equal number of parts in a given shipment, there will be an overall average of at least 25% nonconforming parts $(0 + 50\%)/2$, which is much larger than the 0.27% maximum expected from a shipment with a C_{pk} rating of 1.00. Thus, reporting a C_{pk} index of 1.00 for a shipment containing a mixture of the outputs from these two process streams would falsely imply a much higher quality level than will be received by the customer.

Computing the Average C_{pk} Index

To correctly calculate the "average" C_{pk} of the combined output from several process streams, the

Table 1 Percentage nonconforming for each cavity

Cavity, i	$p_{i,LSL}$	$p_{i,USL}$
1	0.001	0.004
2	0.000	0.008
3	0.015	0.000
4	0.000	0.000
Sum for all four cavities	0.016	0.012
Average percentage nonconforming	0.004	0.003

percentage of out-of-specification parts produced by each stream should first be computed. For the four process streams of the plastic injection-molding operation mentioned earlier, these calculations generate the results displayed in Table 1. Here, i represents the cavity number, $p_{i,LSL}$ is the expected percentage of parts with hole diameters below the LSL for the ith cavity, while $p_{i,USL}$ is the expected percentage of parts with hole diameters above the USL for the ith cavity.

The individual percentages of parts with hole sizes below the LSL for each of these four cavities sum to a total of 0.016, or 1.6%, as is seen in the second-last row of Table 1. This total is then divided by four – in general, q, the number of process streams – to come up with 0.004 (0.016/4). (This calculation assumes each of the four cavities contributes 1/4 of the total production of parts.) This last value is labeled $\overline{p}_{LSL}$, which represents the expected average percentage of parts with hole diameters below the LSL for the combined output of the four cavities. Equation (7) demonstrates this calculation for the injection-molding example.

$$\overline{p}_{LSL} = \frac{\sum_{i=1}^{q} p_{i,LSL}}{q} = \frac{\sum_{i=1}^{4} p_{i,LSL}}{4} = \frac{0.016}{4} = 0.004 \tag{7}$$

The above average is an estimate of the percentage of parts below the LSL that the customer will receive, and is thus an indication of the process's ability to mold parts with hole sizes above the lower specification limit.

In a similar manner, an estimate of the expected average percentage of parts produced by this process with hole diameters above the USL is 0.3%, as is shown by equation (8).

$$\overline{p}_{USL} = \frac{\sum_{i=1}^{4} p_{i,USL}}{4} = \frac{0.012}{4} = 0.003 \tag{8}$$

The C_{pl} value, labeled $\overline{C}_{PL}$, corresponding to a given $\overline{p}_{LSL}$ is found by rearranging equation (4) to derive equation (9). The $\Phi^{-1}(\cdot)$ operator is the inverse of the $\Phi(\cdot)$ operator previously defined for equation (4). In Excel, $\Phi^{-1}(0.97725)$ corresponds to the statistical function NORMSINV(0.97725), which returns the value of 2.0000.

$$\overline{C}_{PL} = \frac{\Phi^{-1}(1 - \overline{p}_{LSL})}{3} \tag{9}$$

In the four-cavity die example, the $\overline{p}_{LSL}$ of 0.004 is associated with a $\overline{C}_{PL}$ value of 0.88, as is shown in equation (10).

$$\overline{C}_{PL} = \frac{\Phi^{-1}(1 - 0.004)}{3} = \frac{2.652}{3} = 0.88 \tag{10}$$

In a similar manner, equation (11) demonstrates how the $\overline{C}_{PU}$ value corresponding to a $\overline{p}_{USL}$ of 0.003 is calculated to be 0.92.

$$\overline{C}_{PU} = \frac{\Phi^{-1}(1 - \overline{p}_{USL})}{3} = \frac{\Phi^{-1}(1 - 0.003)}{3} = \frac{2.748}{3} = 0.92 \tag{11}$$

The average C_{pk} index is then defined as shown in equation (12).

$$\text{Average } C_{pk} = \text{minimum}(\overline{C}_{PL}, \overline{C}_{PU}) = \text{minimum}\left[\frac{\Phi^{-1}(1 - \overline{p}_{LSL})}{3}, \frac{\Phi^{-1}(1 - \overline{p}_{USL})}{3}\right] \tag{12}$$

Using the numbers from the previous example, an estimate of the average C_{pk} index for the hole diameters of parts produced by this four-cavity die molding process is computed in equation (13) to be 0.88.

$$\text{Average}C_{pk} = \text{Minimum}(0.88, 0.92) = 0.88 \tag{13}$$

Because this index is quite a bit less than 1.00, the combined output of the four cavities is *not* capable of consistently producing at least 99.73% conforming parts as far as hole size is concerned. In fact, it is producing 0.4% below the LSL and 0.3% above the USL, for a total of 0.7% nonconforming parts, which equates to 99.3% conforming parts. This is the quality level a customer would expect from a process having a traditional C_{pk} rating of 0.88.

Although the average C_{pk} index would be reported to customers, the individual cavity results are used to target process improvement efforts. In the four-cavity die example, cavity three has the largest percentage of parts (1.5%) with hole diameters below the LSL. Because it has no parts with holes above the USL, the initial capability improvement effort for this process should focus on increasing the average output of cavity three (refer back to Figure 1).

Average C_{pk} for Attribute Data

The average C_{pk} index can also be used when assessing the combined capability of several process streams that are generating attribute-type data (*see* **Measurement Systems Analysis, Attribute**), as this next example demonstrates.

Dough for frozen pizzas is arranged in six rows on a conveyor. The conveyor then moves under several dispensing stations that deposit various vegetables, meats, sauces, and cheeses on the dough. At the end of this line, the pizzas coming off each row are checked to see if the ingredients are evenly distributed over the face of the pizza. Those with an unacceptable ingredient distribution are rejected. After achieving stability for each of the six rows, their rejection rates are estimated and summarized in Table 2.

The average C_{pk} index for the combined output of these six process streams is computed by first determining the average percentage of pizzas that do not satisfy the appearance requirement, as seen in equation (14).

$$\overline{p} = \frac{\sum_{i=1}^{6} p_i}{6} = \frac{0.000072}{6} = 0.000012 \tag{14}$$

Table 2 Rejection rate for each row

Row, i	p_i
1	0.000016
2	0.000003
3	0.000022
4	0.000012
5	0.000005
6	0.000014
Sum	0.000072
Average	0.000012

The average C_{pk} index is then computed as displayed in equation (15).

$$\text{Average } C_{pk} = \frac{\Phi^{-1}(1-\overline{p})}{3} = \frac{4.224}{3} = 1.41 \quad (15)$$

Given a C_{pk} goal of 1.33, the combined output of these six rows demonstrates very good capability for meeting the appearance requirement.

Summary

The average C_{pk} index provides a method for assessing the outgoing quality of a mixture of outputs from similar process streams. This metric does not require the outputs from all streams to be centered at the same value nor have the same process spread, which is important because, in most cases, they will not. In addition to part features (with either unilateral or bilateral specifications) whose measurements generate variable-type data, this index applies equally well to features characterized by attribute-type data.

References

[1] Kane, V. (1986). Process capability indices, *Journal of Quality Technology* **18**, 41–52.

[2] Montgomery, D. (1991). *Introduction to Statistical Quality Control*, 2nd Edition, John Wiley & Sons, New York, p. 43.

Further Reading

Bothe, D. (1977). *Measuring Process Capability*, Landmark Publishing, Cedarburg, pp. 721–728.

DAVIS ROSS BOTHE

Process Capability Indices, Alternatives to

Introduction

The entire issue of the journal of quality technology (JQT) in October 1992 JQT was devoted to the topic of process capability indices (PCIs). The five articles covered recent developments of PCIs, confidence bounds on PCIs, relationships to squared error loss, and the distributional and inferential properties of PCIs [4–8]. Additionally, the editor, Peter Nelson discussed the various problems with PCIs and further stated the reason for the publication was "to give those unfortunate enough to have to use capability indices a clear understanding of the assumptions involved and of what the indices are actually measuring" [9]. Finally, with the publication of that issue of JQT, *everything* you ever wanted to know about PCIs had been discussed and PCIs would no longer be abused. Unfortunately, the articles were probably too heavily involved mathematically and the messages contained in the JQT issue were not fully appreciated. Many companies still rely on PCIs to measure and compare the quality of their vendors' processes without understanding the underlying assumptions of PCIs and the properties of the *estimators* of these indices.

Preceding the JQT issue, the topic of capability indices had been a source of much interest and debate from their popularization in the early 1980s. The indices C_p and C_{pk} were regularly taught in most introductory statistical process control (SPC) classes and usually without caveats about their shortcomings. Efforts to improve PCIs continue as evidenced by the fact that in a recent issue of JQT [10], an article by Wang *et al.* compared three multivariate PCIs.

A multitude of indices exist and $C_p, C_{pk}, C_{pm}, C_{pmk}, C_{psk}, C_{pw}, P_p, P_{pk}, C_pM, MC_{pM}$, and so on, are a few examples. Many improvements have been proposed such as the use of **confidence intervals** to show the uncertainties in the calculated estimates of the indices. Unfortunately, these are not practical for large sets of indices and the widths of these intervals increase as the estimate increases. It is not clear how to demonstrate that a required index has been met. Must the required value exceed the lower confidence limit or is it sufficient to be contained in the interval? Another suggestion has been to use tolerance regions (see Wang [10]), but how can these be used in a practical way for large sets of indices? In addition, there is some difficulty in interpretation of these regions.

The purpose of this paper is to introduce a basic approach to meet most of the goals of PCIs that are presently being used without some of the associated problems that have been addressed in the previous literature. The intention is to create a foundation and direction for future expansion rather than to formulate a total solution. In cases where simple assumptions (normality and symmetric specifications) are not justified, the usual modifications can easily be introduced to meet the requirements of the application.

Specification Limits and Process Capability Indices

PCIs are often used in two specific phases. In phase I, the capability of a process to meet required specifications is determined. Three approaches are often used and are explained below.

The first approach is to determine the process specifications by needs (product performance, process optimization, customer requirements, etc.). A technology development group then determines the process distribution. The PCIs are calculated and if they are not met, the process is returned to the development group for the needed improvement to meet the required PCIs.

The second alternative is to determine the process specifications by needs as in the initial approach. A set of PCIs is specified and the technology group determines the process distribution. If the process does not meet the PCIs, a new set of indices is calculated that can be met.

The third technique always works the first time. The technology group determines the process distribution. A set of PCIs are specified and then the process specifications are determined from the combination of the process distribution and the PCIs. Unfortunately, this method is used more often than expected particularly when the specifications are on intermediate process steps and not on final product parameters.

In phase II, the goal is to measure changes in the process (either degradations or improvements) over time.

Before continuing, perhaps a few comments on how specification limits might best be determined are worthwhile. When the limits are determined for process and product development, the following ideas are often utilized. The design requirements of the product are determined, simulation models are run, engineering judgment is factored into the picture, and last, but not least, customer agreement is needed. These limits are then validated by data from product or process window characterizations and/or by historical data if it exists. Specifications used in manufacturing are often obtained by this data validation approach.

An Alternative Approach to Process Capability Indices in Phase I

The alternative is to measure whether the process distribution meets the required specifications by matching the mean (or a measure of the center) and the variance (or a measure of the spread) against those values assumed by the specifications. The criteria for the matching of the mean and standard deviation will be based on ratios that encompass both the sample size of the data set as well as a level of confidence associated with the ratio. Consequently, it would be feasible, if desired, to make comparisons between requirements, time periods, processes, and so on, in a statistically meaningful procedure. The usual caveats applied to PCIs must still hold. The process must be stable and the specifications must be technically based. In addition, a set of initial assumptions will be made. For this discussion it will be assumed that the process is normally distributed with mean μ and variance σ^2. The specifications will be assumed to be symmetric about the mean and the specification width reflects $k\sigma$. These assumptions can be modified for future work. The matching of the variance will be addressed first.

Let $x_1, x_2, \ldots, x_n$ represent a sample of n data points from the normal distributed process under study with $\overline{X}$ and S^2 the unbiased estimators of the mean and variance. Thus,

$$\frac{(n-1)S^2}{\sigma^2} \sim (\chi^2_{n-1}) \text{ so that} \tag{1}$$

$$\Pr\left[\frac{(n-1)S^2}{\sigma^2} < \chi^2_{1-\alpha,n-1}\right] = 1-\alpha \text{ and} \tag{2}$$

$$\Pr\left[S^2 < \frac{(\chi^2_{1-\alpha,n-1})\sigma^2}{n-1}\right] = 1-\alpha \tag{3}$$

Hence,

$$\Pr\left[S < \sigma\sqrt{\frac{(\chi^2_{1-\alpha,n-1})}{n-1}}\right] = 1-\alpha \tag{4}$$

Similarly, at the lower tail of the distribution,

$$\Pr\left[\sigma\sqrt{\frac{(\chi^2_{\alpha,n-1})}{n-1}} < S\right] = 1-\alpha \tag{5}$$

Denote σ_{upper} and σ_{lower} as follows:

$$\sigma_{\text{upper}} = \sigma\sqrt{\frac{(\chi^2_{1-\alpha,n-1})}{n-1}} \quad (6)$$

and

$$\sigma_{\text{lower}} = \sigma\sqrt{\frac{(\chi^2_{\alpha,n-1})}{n-1}} \quad (7)$$

Using the sample size of the data set, a predetermined probability level α, and the value of σ reflected by the specification, the values of σ_{lower} and σ_{upper} can be determined. From process data, an unbiased estimate of σ^2, s^2, may be obtained.

Now define the ratios:

$$S_{\text{ru}} = -\frac{s}{\sigma_{\text{upper}}} \quad (8)$$

and

$$S_{\text{rl}} = \frac{\sigma_{\text{lower}}}{s} \quad (9)$$

With these definitions,

$$\Pr[S_{\text{ru}} < -1] = \alpha \text{ and } \Pr[S_{\text{rl}} > 1] = \alpha \quad (10)$$

If $S_{\text{ru}} < -1, s > \sigma$ with α significance (based on the sample size) then the process needs improvement with respect to σ.

If $S_{\text{rl}} > 1, s < \sigma$ with α significance (based on the sample size) then the process is performing better than expected and is clearly capable with respect to σ.

If neither condition exists, the difference between s and σ is not statistically significant and no action is necessary.

A similar approach may be used for matching the mean of the process distribution. However, with large sample sizes, a statistical significance may not reflect the practical or engineering significance. Therefore, it is necessary to define a "minimum engineering significant difference" (MESD) that is of real importance.

As stated earlier, $\overline{X}$ and S^2 are the unbiased estimators of the mean and variance. Thus,

$$\sqrt{n}\frac{\overline{X}-\mu}{S} \sim (t_{n-1}) \quad (11)$$

so that

$$\Pr\left[-t_{\alpha/2,n-1} < \sqrt{n}\frac{\overline{X}-\mu}{S} < t_{\alpha/2,n-1}\right] = 1-\alpha \quad (12)$$

and

$$\Pr\left[-1 < \sqrt{n}\frac{\overline{X}-\mu}{t_{\alpha/2,n-1}S} < 1\right] = 1-\alpha \quad (13)$$

Define the mean ratio (MR) as

$$\sqrt{n}\frac{\overline{X}-\mu}{t_{\alpha/2,n-1}S} \quad (14)$$

Hence,

If $MR > 1$ and $(\overline{X}-\mu) > MESD$, then $\overline{X} > \mu$, with significance α.

If $MR < -1$ and $(\overline{X}-\mu) < -MESD$, then $\overline{X} < \mu$, with significance α.

Otherwise, no significant difference exists.

Phase II – Measure Changes in the Process Over Time

The approach to measure changes over time uses the two-sample analogs of phase I, which are to test the significance of the differences in the variance (spread) and the mean (center) from the previous time period. The initial assumptions are similar to those of the one-sample case with respect to symmetry, spec width, and the underlying distributions for the present (new) and the past (old) time periods. Additionally, assume no change has taken place between the period so that $\sigma_{\text{old}} = \sigma_{\text{new}}$ and $\mu_{\text{old}} = \mu_{\text{new}}$. As in the earlier case the comparison of variances will be addressed first.

Now,

$$\left(\frac{S^2_{\text{new}}}{S^2_{\text{old}}}\right) \sim (Fn_{\text{new}}-1, n_{\text{old}}-1) \quad (15)$$

so that

$$\Pr\left[\left(\frac{S^2_{\text{new}}}{S^2_{\text{old}}}\right) < (F_{1-\alpha}, n_{\text{new}}-1, n_{\text{old}}-1)\right] = 1-\alpha \quad (16)$$

This leads to

$$\Pr[S^2_{\text{new}} > (S^2_{\text{old}})(F_{1-\alpha}, n_{\text{new}}-1, n_{\text{old}}-1)] = \alpha \quad (17)$$

and

$$\Pr[S^2_{\text{new}} < (S^2_{\text{old}})(F_{\alpha}, n_{\text{new}}-1, n_{\text{old}}-1)] = \alpha \quad (18)$$

As with the one-sample case, denote S_{upper} and S_{lower} as follows:

$$S_{\text{upper}} = (S_{\text{old}})\sqrt{(F_{1-\alpha}, n_{\text{new}}-1, n_{\text{old}}-1)} \quad (19)$$

and

$$S_{\text{lower}} = (S_{\text{old}})\sqrt{(F_{\alpha}, n_{\text{new}}-1, n_{\text{old}}-1)} \quad (20)$$

From the past and present process data calculate s^2_{old} and s^2_{new}. Using the sample sizes of the two sets of data and a predetermined probability level, α, S_{lower}, and S_{upper} can be determined.

Now define the ratios:

$$S_{\text{cru}} = -\frac{s_{\text{new}}}{s_{\text{upper}}} \quad (21)$$

and

$$S_{\text{crl}} = \frac{s_{\text{lower}}}{s_{\text{new}}} \quad (22)$$

With these definitions,

$$\Pr[S_{\text{cru}} < -1] = \alpha \text{ and } \Pr[S_{\text{crl}} > 1] = \alpha \quad (23)$$

If $S_{\text{cru}} < -1$, $s_{\text{new}} > s_{\text{old}}$ with α significance based on the sample sizes then the process needs improvement with respect to s_{old}.

If $S_{\text{crl}} > 1$, $s_{\text{new}} < s_{\text{old}}$ with α significance based on the sample sizes then the process is performing better than expected and is clearly capable with respect to s_{old}.

If neither condition exists, the difference between s_{old} and s_{new} is not statistically significant and no action is necessary.

Continuing with the same assumptions previously stated and using an analog to the one-sample case, define the comparative mean ratio (CMR) as follows:

$$\frac{(\overline{X}_{\text{new}} - \overline{X}_{\text{old}})}{(s^*)(t_{\alpha/2,\nu})} \quad (24)$$

where s^* is the appropriate **standard error** of the estimate determined after the variances have been tested, using a pooled estimate of σ if no change is indicated and the calculated values of s^2_{old} and s^2_{new} if a change is indicated, and ν is the **degrees of freedom** for the combined sample.

Hence,

If $CMR > 1$ and $(\overline{X}_{\text{new}} - \overline{X}_{\text{old}}) > MESD$, then $\overline{X}_{\text{new}} > \overline{X}_{\text{old}}$ with significance α.

If $CMR < -1$ and $(\overline{X}_{\text{new}} - \overline{X}_{\text{old}}) < -MESD$, then $\overline{X}_{\text{new}} < \overline{X}_{\text{old}}$ with significance α.

Otherwise, no significant difference exists.

Example

The example is discussed in Spiring's article in JQT 29 [11]. The process is a blow-molding procedure for plastic bottles where the outside lip diameter is the measure under study. He states that the upper specification limit (USL) is 0.846, lower specification limit (LSL) is 0.814, and the target is 0.830. A sample of 100 observations were collected with $\overline{X} = 0.8254$ and $\hat{\sigma}$ (m.l.e.) = 0.005. The author then calculates the following estimates and 95% confidence intervals:

$$\hat{C}\text{p} = 1.067\ (0.914, 1.209) \quad (25)$$

$$\hat{C}\text{pk}^* = 0.760\ (0.688, 0.881) \quad (26)$$

$$\hat{C}\text{pm} = 0.785\ (0.686, 0.836) \quad (27)$$

What has changed? Does a problem exist with this process? The first index confirms that the process spread is not a problem, but does not address the possibility that it is better than the specification suggests. The second and third indices and their confidence intervals point to a problem in the mean of the process.

Assuming that the process is normally distributed and that the specification is defined as $\mu \pm 3\sigma$, then $\mu = 0.830$ and $\sigma = 0.016/3 = 0.0053$.

Since $\hat{\sigma}$(m.l.e.) = 0.005, then $s = 0.005025$ which is less than the hypothesized σ. Therefore, $S_{\text{rl}} = \frac{\sigma_{\text{lower}}}{s}$ is the appropriate statistic to evaluate. For this example,

$$\sigma_{\text{lower}} = \sigma\sqrt{\frac{(\chi 2_{\alpha,n-1})}{n-1}} = (0.0053)\sqrt{\frac{77.046}{99}} = 0.0047 \quad (28)$$

and $S_{\text{rl}} = 0.9353 < 1$. Consequently, based on the sample size of 100 and the one-sided 95% confidence interval, there is no significant difference in the standard deviation of the process.

Looking at the $MR = \sqrt{n}\frac{\overline{X}-\mu}{t_{\alpha/2,n-1}S}$ and assuming the MESD is approximately one σ, the

$$MR = \sqrt{100}\left(\frac{0.8254 - 0.830}{(1.98422)(0.005025)}\right) = -4.614 \quad (29)$$

Now, MR $= -4.614 < -1$, but since $(\overline{X} - \mu) = -0.0046$ and the MESD was assumed to be $\sigma = 0.0053$, no action would be taken despite the statistical significance of the result. However, if the *MESD* was less than 0.0046, the statistical significance would also be of engineering significance.

Other Issues

This paper has addressed only the relatively simple case in which the specifications are two-sided symmetrical and the process is assumed to be normally distributed. However, because the technique uses the distribution parameters in the metric, it may easily be adapted to applications involving other assumed distributions and their parameters such as discrete and nonsymmetric distributions without an intermediate transformation or fitting to a **normal distribution**.

In cases where these conditions are not true, the use of medians, interquartile ranges, and other standard robust procedures may be used in a similar manner.

Comments and Conclusions

The advantages of this technique include the following:

1. Since the sample size and confidence levels are built directly into the ratios, they may be sorted to prioritize improvement efforts unlike the present PCIs.
2. The ratios can be used to measure changes over time.
3. The concepts can be expanded to other distributions in a direct manner.
4. The ratios add statistical thinking to the measurement processes so that meaningful changes are recognized.
5. Many of the problems associated with the traditional PCIs have been eliminated.

Although the advantages are clear, the implementation of this procedure to the areas where PCIs are presently used will be difficult to accomplish. A great deal of effort will be required to change the mindset with respect to the *status quo* and this must occur at all levels within a company from top management to the worker on the factory floor.

References

[1] Gunter, B. (1989). The use and abuse of Cpk: parts 1–4, *Quality Progress* (January): **22**(1), 72–73; (March): **22**(3), 108–109; (May): **22**(5), 79–80; (July): **22**(7), 86–87.

[2] Benson, E., Kotz, S. & Alt, F.B. (1994). The next generation of PCI's, in *Proceedings Joint Statistical Meetings*, August 18, Toronto.

[3] Process Capability Indices Issue, *Journal of Quality Technology* **24**, (October).

[4] Rodriguez, R.N. (1992). Recent developments in process capability analysis, *Journal of Quality Technology* **24**, 176–187. Also, an extensive reference list appears in this article.

[5] Kushler, R.H. & Hurley, P. (1992). Confidence bounds for capability indices, *Journal of Quality Technology* **24**, 188–195.

[6] Franklin, L.A. & Wasserman, G.S. (1992). Bootstrap lower confidence limits for capability indices, *Journal of Quality Technology* **24**, 196–210.

[7] Johnson, T. (1992). The relationship of Cpm to squared error loss, *Journal of Quality Technology* **24**, 211–215.

[8] Pearn, W.L., Kotz, S. & Johnson, N.L. (1992). Distributional and inferential properties of process capability indices, *Journal of Quality Technology* **24**, 216–231.

[9] Nelson, P. (1992). Editorial, *Journal of Quality Technology* **24**, 175.

[10] Wang, F.K., Hubele, N.F., Lawrence, F.P., Miskulin, J.D. & Shahriari, H. (2000). Comparison of three multivariate process capability indices, *Journal of Quality Technology* **32**, 263–275.

[11] Spiring, F.A. (1997). A unifying approach to process capability indices, *Journal of Quality Technology* **29**, 49–58.

Further Reading

Dovich, R.A. (1991). Statistical terrorists, *Quality in Manufacturing* 14–15.

Johnson, N.L. & Kotz, S. (1993). *Process Capability Indices*, Chapman & Hall, London.

Potter, R.W. & Pantula, S.G. (1991). *Confidence Intervals for Capability Indices in Nested Experiments*, SEMATECH, Austin.

RICHARD I. POST

Process Capability Indices, Bayesian Estimation of

Process Capability Indices

The process capability index, *Cp*, relates process variation to customer requirements in the form of a ratio

$$Cp = \frac{USL - LSL}{6\sigma} \quad (1)$$

where the difference between the upper specification limit (USL) and the lower specification limit (LSL) provides a measure of allowable process spread and 6σ, σ^2 being the process variance, a measure of actual process spread.

The failure of *Cp* to consider a target value resulted in the development of several indices. Those indices that assume the target (T) to be the midpoint of the upper and lower specification limits include

$$Cpu = \frac{USL - \mu}{3\sigma} \quad (2)$$

$$Cpl = \frac{\mu - LSL}{3\sigma} \quad (3)$$

$$Cpk = \min(Cpl, Cpu) \quad (4)$$

$$Cpm = \frac{USL - LSL}{6\sqrt{\sigma^2 + (\mu - T)^2}} \quad (5)$$

and

$$Cpk = (1 - k)Cp \quad (6)$$

where μ denotes the process mean, $k = 2|T - \mu|/(USL - LSL)$, $0 \le k \le 1$ and $LSL < \mu < USL$. The two definitions of Cpk are equivalent when $0 \le k \le 1$.

Generalized versions of these indices, that do not assume T to be the midpoint of the specification limits, include

$$Cpu^* = \frac{USL - T}{3\sigma}\left(1 - \frac{|T - \mu|}{USL - T}\right) \quad (7)$$

$$Cpl^* = \frac{T - LSL}{3\sigma}\left(1 - \frac{|T - \mu|}{T - LSL}\right) \quad (8)$$

$$Cpk^* = \min(Cpl^*, Cpu^*) \quad (9)$$

and

$$Cpm^* = \frac{\min[USL - T, T - LSL]}{3\sqrt{\sigma^2 + (\mu - T)^2}} \quad (10)$$

A hybrid of Cpk^* and Cpm^* is defined to be

$$Cpmk = \frac{\min[USL - \mu, \mu - LSL]}{3\sqrt{\sigma^2 + (\mu - T)^2}} \quad (11)$$

As all of these indices use a function of σ as a measure of actual process spread in their determination of process capability and, in practice, are translated into parts per million nonconforming, none should be used when the characteristic under investigation is not normally distributed. To illustrate, suppose that precisely 99.73% of the process measurements fall within the specification limits (i.e., a **process yield** of 2700 ppm) and the process is centered at the target. The values of *Cp* (as well as *Cpl, Cpu, Cpk, Cpm,* and *Cpmk*) are 0.5766, 0.7954, 1.0000, 1.2210, and 1.4030 when the measurements arise from a uniform, triangular, normal, logistic, and double exponential distribution respectively, yet the process yields are identical. Under these circumstances, the assumption that the underlying process is normally distributed is critical to both sampling results and parameter values.

A Bayesian Approach

The general Bayesian approach assumes the parameter(s) to be stochastic and uses sampling results to adjust an assumed prior distribution to develop a posterior distribution for the associated parameter(s). In the case of process capability indices, and following the assumption that the observations must arise from a **normal distribution**, the likelihood function for a sample of size n is

$$L(\mu, \sigma^2) = (2\pi\sigma^2)^{-n/2}\, \mathrm{e}^{-\sum_{i=1}^{n}(x_i - \mu)^2/(2\sigma^2)} \quad (12)$$

The noninformative prior density function

$$h(\mu, \sigma^2) = 1/\sigma \qquad -\infty < \mu < \infty, \quad 0 < \sigma < \infty \quad (13)$$

results in the posterior distribution

$$f(\sigma|x_1, x_2, x_3, \ldots, x_n) = \frac{2}{\Gamma\left[(n-1)/2\right]}\left(\frac{(n-1)s^2}{2}\right)^{\frac{n-1}{2}} \sigma^{-n}\mathrm{e}^{\frac{-(n-1)s^2}{2\sigma^2}} \quad (14)$$

where $s^2 = \sum_{i=1}^{n}(x_i - \overline{x})^2/(n-1)$ and $\overline{x} = \sum_{i=1}^{n} x_i/n$. As each of the process capability indices is a function of the process parameters μ and σ, and judged based upon a value (e.g., $Cp > 1$) this allows equation (14) to be written as

$$\Pr(Cp > c|x_1, x_2, x_3, \ldots, x_n) \quad (15)$$

In the case of *Cp*, Cheng and Spiring [1] rewrote equation (15)

$$\begin{aligned}\Pr(Cp > c|x_1, x_2, x_3, \ldots, x_n) &= \Pr\left(\sigma < \left(\frac{USL - LSL}{6c}\right)|x_1, x_2, x_3, \ldots, x_n\right)\\ &= \int_0^{(USL-LSL)/(6c)} f(\sigma|x_1, x_2, \ldots, x_n)\, d\sigma\\ &= \int_0^{(USL-LSL)/(6c)} \frac{2}{\Gamma\left[(n-1)/2\right]}\\ &\quad \times \left(\frac{(n-1)s^2}{2}\right)^{\frac{n-1}{2}} \sigma^{-n}\, \mathrm{e}^{\frac{-(n-1)s^2}{2\sigma^2}}\, d\sigma \quad (16)\end{aligned}$$

resulting in a credible interval that depends upon n, s^2, *USL, LSL* and c.

Program *Cp: Mathematica* v5.2 Code for determining Pr(*Cp* > *c*|$x_1, x_2, x_3, \ldots, x_n$)

In[1]:

```
n:= ;
s2:= ;
specdiff:= ;
c:= ;
N[Integrate[((2/Evaluate[Gamma
[(n−1)/2]]((n−1)s2/2)^((n−1)/2))(x^(−n))
Exp[−((n−1)s2/2)/(x^2)]), {x,0, (specd-
iff/(6c))}]]
```

The user must enter: **n** the sample size, **s2** the sample variance, **specdiff** = (*USL* – *LSL*), and **c** the desired level of *Cp*.

Pearn and Wu [2] extended the credible interval approach for *Cp* to the case where $\widehat{C}p^*$ is essentially a pooled estimate of *Cp* based upon a series (m) of repeated samples (n). Minimum values for $\widehat{C}p^*$ have been developed for $m = 2(2)10$, $n = 10(5)30$, and $p = 0.95,\ 0.975,\ 0.99$ providing a Bayesian alternative to the capability chart developed in [3].

In the case of *Cpm* and the noninformative prior distribution defined in equation (13), the conditional distribution of equation (14) becomes

$$f(\sigma'|x_1, x_2, x_3, \ldots, x_n) = \frac{2\sigma'}{\Gamma[n/2]}\left(\frac{(n-1)s^2 + n(\overline{x}-\mu)^2}{2}\right)^{\frac{n}{2}} \times (\sigma'^2 - (\mu - T)^2)^{-\left(\frac{n}{2}+1\right)} \times \mathrm{e}^{\frac{(n-1)s^2+n(\overline{x}-\mu)^2}{-2(\sigma'^2-(\mu-T)^2)}} \quad (17)$$

where $\sigma'^2 = \sigma^2 + (\mu - T)^2$ and $(\mu - T)^2 < \sigma'^2 < \infty$. The resulting credible interval depends upon n, x, s^2, *USL, LSL*, T, μ and c and is of the form [4]

$$\Pr(Cpm > c|x_1, x_2, x_3, \ldots, x_n) = \int_0^{(USL-LSL)/(6c)} f(\sigma'|x_1, x_2, \ldots, x_n)\, \mathrm{d}\sigma' = \int_0^{(USL-LSL)/(6c)} \frac{2\sigma'}{\Gamma[n/2]} \times \left(\frac{(n-1)s^2 + n(\overline{x}-\mu)^2}{2}\right)^{\frac{n}{2}} \times (\sigma'^2 - (\mu - T)^2)^{-\left(\frac{n}{2}+1\right)}\, \mathrm{e}^{\frac{(n-1)s^2+n(\overline{x}-\mu)^2}{-2(\sigma'^2-(\mu-T)^2)}}\, \mathrm{d}\sigma' \quad (18)$$

Program *Cpm: Mathematica* v5.2 Code for determining Pr(*Cpm*>*c*|$x_1, x_2, x_3, \ldots, x_n$)

In[1]:

```
n:= ;
s2:= ;
mu:= ;
T:= ;
xbar:= ;
specdiff:= ;
c:= ;
Integrate[(2x/Evaluate[Gamma[n/2]])
(((n−1)s2+n(xbar−mu)^2)/2)^(n/2)(x^2−
(mu-T)^2)^(−(n/2 + 1)) Exp[((n−1)s2+n
(xbar−mu)^2)/(−2((x^2)−(mu-T)^2))],
{x,0,(specdiff/(6c))}]
```

The user must enter: **n** the sample size, **s2** the sample variance, **mu** the population average, **T** the target value, **xbar** = the sample average, **specdiff** = (*USL* – *LSL*), and **c** the desired level of *Cpm*.

In the case of *Cpk*, Shiau *et al.* [5] used the same approach to develop a credible interval restricted to the case where $m = (USL + LSL)/2$ by considering

$$\Pr(Cpk > c|x_1, x_2, x_3, \ldots, x_n) = \Pr\left(\sigma < \left(\frac{USL - LSL}{6c}\right) |x_1, x_2, x_3, \ldots, x_n\right) \quad (19)$$

which for the posterior distribution defined in equation (14) becomes

$$\Pr(Cpk > c|x_1, x_2, x_3, \ldots, x_n) = \int_0^{(USL-LSL)/(6c)} f(\sigma|x_1, x_2, \ldots, x_n)\, \mathrm{d}\sigma = \int_0^{(USL-LSL)/(6c)} \frac{2}{\Gamma[n/2]}\left(\frac{ns'^2}{2}\right)^{\frac{n}{2}} \times \sigma^{-(n+1)}\mathrm{e}^{\frac{-ns'^2}{2\sigma^2}}\, \mathrm{d}\sigma \quad (20)$$

where $s'^2 = \sum_{i=1}^{n}((x_i - m)^2/n)$, resulting in a credible interval that depends upon n, s'^2, *USL, LSL* and c. The authors further simplify equation (20) to get

$$\Pr(Cpk > c|x_1, x_2, x_3, \ldots, x_n) = \int_b^{\infty} \frac{1}{\Gamma[n/2]} y^{(n/2)-1}\mathrm{e}^{-y}\mathrm{d}y \quad (21)$$

for $b = n/2\left(c/\widehat{C}pk\right)^2$.

Program ***Cpk: Mathematica*** **v5.2 Code for determining Pr(*Cpk* $> c|x_1, x_2, x_3, \ldots, x_n$)**

In[1]:
```
n:= ;
cpkh:= ;
c:= ;
1-Integrate[(1/Evaluate[Gamma[n/2]])
(y^((n/2)-1))
Exp[-y],{y,0,(n/2)(c/cpkh)^2}]
```

The user must enter: **n** the sample size, **cpkh** = $(USL - LSL)/(6s')$, and **c** the desired level of *Cpk*.

In the case of Cpm^*, Lin *et al.* [6] developed a credible interval, $\mathbf{P}(Cpm^* > c|x_1, x_2, x_3, \ldots, x_n) = p$, of the form

$$p = \begin{cases} \left\{ \displaystyle\int_0^{t^*} \frac{\Phi(b_1(y)+b_U^*(y)) - \Phi(b_1(y)-b_L^*(y))}{\gamma^\alpha y^{\alpha+1}\Gamma(\alpha)} \exp\left(-\frac{1}{\gamma y}\right) dy \right\} & \text{for } \overline{x} < T \\ \left\{ \displaystyle\int_0^{t^*} \frac{\Phi(b_1(y)+b_L^*(y)) - \Phi(b_1(y)-b_U^*(y))}{\gamma^\alpha y^{\alpha+1}\Gamma(\alpha)} \exp\left(-\frac{1}{\gamma y}\right) dy \right\} & \text{for } \overline{x} > T \end{cases} \tag{22}$$

where $t^* = 2[\min[USL - T, T - LSL])/(2c)] \times [(n-1)s^2 + n(\overline{x} - T)^2]^{-1}$,

Φ is the cumulative distribution function of the $N(0, 1)$,

$$b_1(y) = \left[(2/y)(1 - (1/(1 + ((n(\overline{x} - T)^2)/((n-1)s^2)))))\right]^{0.5},$$

$$b_U^*(y) = \left[2(USL - T)/(USL - LSL)\right] \times \left[n((2(\min[USL - T, T - LSL]/(3c))^2 \times ((n-1)s^2 + n(\overline{x} - T)^2)^{-1})/y - 1)\right]^{0.5},$$

$$\gamma^2 = (\overline{x} - T)^2/s^2,$$

$$b_2(y) = \left[n((2\widehat{C}pm^2/nyc^2) - 1)\right]^{0.5}, \text{ and}$$

$$b_L^*(y) = \left[2(T - LSL)/(USL - LSL)\right] \times \left[n((2(\min[USL - T, T - LSL]/(3c))^2 \times ((n-1)s^2 + n(\overline{x} - T)^2)^{-1})/y - 1)\right]^{0.5},$$

resulting in an interval that depends upon n, s^2, γ, *USL*, *LSL* and c.

The authors indicate that although equation (22) reduces to

$$p = \left\{ \int_0^t \frac{\Phi(b_1(y) + b_2(y)) - \Phi(b_1(y) - b_2(y))}{\gamma^\alpha y^{\alpha+1}\Gamma(\alpha)} \times \exp\left(-\frac{1}{\gamma y}\right) dy \right\} \tag{23}$$

where $t = 2\widehat{C}pm^2/(nc^2)$ when T is the midpoint of the specification limits, and the equivalent of equation (18) when $m = (USL + LSL)/2$, equation (22), is "complicated and computationally inefficient".

Extension of the credible interval approach to *Cpmk* proves to be a difficult problem mathematically. Pearn and Lin [7] have examined some properties of *Cpmk* from a Bayesian estimation perspective.

Example

Braverman [8] discussed an example from a **quality control** department that wanted to assess the capability of a shaft manufacturing process. The shaft specifications were $USL = 2.15$ and $LSL = 2.01$ with $T = 2.08$. A random sample of 50 shafts was selected and measured with the following results (see Table 1)

Using the above ***Mathematica*** programs for *Cp*, *Cpm*, and *Cpk*, the associated Bayesian credible interval probabilities are listed in Table 2 suggesting that the probability that the process is deemed capable given the sampling results is approximately 0.7 for each of *Cp*, *Cpm*, and *Cpk*.

Comments

The amount of material associated with the Bayesian approach for assessing process capability represents a small portion of the work in the area of drawing inferences regarding a process capability. A list of additional related manuscripts that deal with specific situations has been included. Much like the general development, the evolution of Bayesian inference in the area of process capability assessment appears to be dominated by academic interests and has had little practitioner input.

Table 1 Process specifications and summaries

Sampling summaries			Specifications			Process capability estimates			
n	$\overline{x}$	s	T	USL	LSL	$\widehat{Cp}$	$\widehat{Cpm}$	$\widehat{Cpk}$	$\widehat{Cpmk}$
50	2.08	0.022	2.08	2.15	2.01	1.06	1.06	1.06	1.06

Table 2 Associated credible intervals

Cp		Cpm		Cpk	
$\widehat{Cp}$	$\Pr(Cp > 1\vert\widehat{C}p)^{(a)}$	$\widehat{Cpm}$	$\Pr(Cpm > 1\vert\widehat{C}pm)^{(b)}$	$\widehat{Cpk}$	$\Pr(Cpk > 1\vert\widehat{C}pk)^{(c)}$
1.06	0.6926	1.06	0.7279	1.06	0.6929

(a)From **Program *Cp***
(b)From **Program *Cpm*** for $\mu = T$
(c)From **Program *Cpk*** for $m = (USL + LSL)/2$

References

[1] Cheng, S.W. & Spiring, F.A. (1989). Assessing process capability: a Bayesian approach, *IIE Transactions* **21**, 97–98.

[2] Pearn, W.L. & Wu, C.-W. (2005). A Bayesian approach for assessing process precision based on multiple samples, *European Journal of Operational Research* **165**(3), 685–695.

[3] Spiring, F.A. (1995). Process capability: a total quality management tool, *Total Quality Management* **6**, 21–33.

[4] Fan, S.-K. & Kao, C.-K. (2004). Lower Bayesian confidence limits on the process capability index *Cpm*: a comparative study, *Quality and Reliability Engineering International* **5**, 444–452.

[5] Shiau, J.H., Chiang, C. & Hung, H. (1999). A Bayesian procedure for process capability assessment, *Quality and Reliability Engineering International* **15**, 369–378.

[6] Lin, G.H., Pearn, W.L. & Yang, Y.S. (2005). A Bayesian approach to obtain a lower bound for the *Cpm* capability index, *Quality and Reliability Engineering International* **21**, 655–668.

[7] Pearn, W.L. & Lin, G.H. (2003). A Bayesian-like estimator of the process capability index *Cpmk*, *Metrika* **57**, 303–312.

[8] Braverman, J.D. (1981). *Fundamentals of Statistical Quality Control*, Prentice-Hall, Reston.

Further Reading

Bernardo, J.M. & Irony, T.Z. (1996). A general multivariate Bayesian process capability index, *The Statistician: Journal of the Institute of Statisticians* **45**, 487–502.

Lin, G.H., Pearn, W.L. & Chen, S. (2002). A note on the distributions of the estimated *Cpk* with asymmetric tolerances, *Far East Journal of Theoretical Statistics* **6**, 25–37.

van der Merwe, A.J. & Chikobvu, D. (2004). Bayesian estimation of the process capability index *Cpk*, *South African Statistical Journal* **38**, 139–158.

Niverthi, M. & Dey, D.K. (2000). Multivariate process capability: a Bayesian perspective, *Communications in Statistics: Simulation and Computation* **29**, 667–687.

Pearn, W.L. & Chen, K.S. (1996). A Bayesian-like estimator of *Cpk*, *Communications in Statistics: Simulation and Computation* **25**, 321–329.

Pearn, W.L. & Liao, M.-Y. (2005). One-sided process capability assessment in the presence of measurement errors, *Quality and Reliability Engineering International* **22**(7), 771–785.

Shiau, J.H., Hung, H. & Chiang, C. (1999). A note on Bayesian estimation of process capability indices, *Statistics and Probability Letters* **45**, 215–224.

Singh, H.P. & Saxena, S. (2005). Bayesian shrinkage estimation of process capability index *Cp*, *Communications in Statistics: Theory and Methods* **34**, 205–228.

de Souza Borges, W. & Ho, L.L. (2001). A fraction defective based capability index, *Quality and Reliability Engineering International* **17**, 447–458.

FRED SPIRING AND SMILEY CHENG

Process Capability Indices, Comparison of

Measuring Process Capability

Historically process capability was synonymous with process variation and expressed as either 6σ or the population range. Neither of these measures consider customer requirements nor permit general comparisons among processes as both measures are unit dependent. Juran [1] suggests that Japanese companies have initiated the use of process capability indices by relating process variation to customer requirements in the form of the ratio

$$C_p = \frac{USL - LSL}{6\sigma} \quad (1)$$

where the difference between the upper specification limit (USL) and the lower specification limit (LSL) provides a measure of allowable process spread (i.e., customer requirements) and 6σ, σ^2 being the process variance, a measure of actual process spread (i.e., process performance). Bissell [2] suggests that the British Standards Institution proposed an analogous measure referred to as the relative precision index, which was withdrawn and replaced by C_p in 1942.

Incorporating customer specification limits into the assessment of process capability results in a more meaningful measure that fosters comparisons across all types of processes. However Cp uses only the customer's USL and LSL in its assessment and fails to consider a target (or nominal) value. The three processes depicted in Figure 1 have identical values of σ^2 and therefore identical values of C_p. However, processes 2 and 3 deviate from the target (T), and additional work and/or costs due to these departures may be incurred. As a result processes 2 and 3 are considered less capable of meeting customer requirements than process 1.

Processes with small variability, but poor proximity to the target, have sparked the derivation of several indices that incorporate targets into their assessment. The most common of these measures assume the target (T) to be the midpoint of the specification limits and include

$$C_{pu} = \frac{USL - \mu}{3\sigma} \quad (2)$$

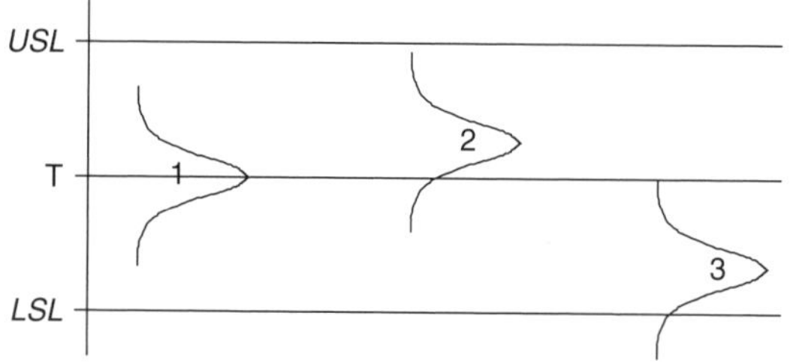

Figure 1 Three processes with identical values of C_p

$$C_{pl} = \frac{\mu - LSL}{3\sigma} \quad (3)$$

$$C_{pk} = \min(C_{pl}, C_{pu}) \quad (4)$$

$$C_{pm} = \frac{USL - LSL}{6\sqrt{\sigma^2 + (\mu - T)^2}} \quad (5)$$

and

$$C_{pk} = (1 - k)C_p \quad (6)$$

where μ is the process mean, $k = \frac{2|T-\mu|}{USL-LSL}$, $0 \leq k \leq 1$ and $LSL < \mu < USL$. The two definitions of C_{pk} are presented interchangeably and are equivalent when $0 \leq k \leq 1$.

The generalized analogs of these measures do not assume T to be the midpoint of the specifications and are

$$C^*_{pu} = \frac{USL - T}{3\sigma}\left(1 - \frac{|T - \mu|}{USL - T}\right) \quad (7)$$

$$C^*_{pl} = \frac{T - LSL}{3\sigma}\left(1 - \frac{|T - \mu|}{T - LSL}\right) \quad (8)$$

$$C^*_{pk} = \min(C^*_{pl}, C^*_{pu}) \quad (9)$$

and

$$C^*_{pm} = \frac{\min[USL - T, T - LSL]}{3\sqrt{\sigma^2 + (\mu - T)^2}} \quad (10)$$

A hybrid of these measures, C_{pmk}, is defined to be

$$C_{pmk} = \frac{\min[USL - \mu, \mu - LSL]}{3\sqrt{\sigma^2 + (\mu - T)^2}} \quad (11)$$

Individually C_{pu}, C^*_{pu}, C_{pl}, and C^*_{pl} consider only unilateral tolerances (i.e., USL or LSL respectively) in assessing process capability. Each uses 3σ as a measure of actual process spread, while the distance from where the process is centered (μ) to the USL (for C_{pu}) or LSL (for C_{pl}) is used as a measure of allowable process spread. Both C_{pu} and C_{pl} compare the length of one tail of the **normal distribution** (3σ) with the distance between the process mean and the respective specification limit. In the case of bilateral tolerances, C_{pu} and C_{pl} have an inverse relationship and individually do not provide a complete assessment of process capability. However, conservatively taking the minimum of C_{pu} and C_{pl} results in the bilateral tolerance measure defined as C_{pk}. Similarly C^*_{pu} and C^*_{pl} use adjusted distances between their respective specification limits and μ that incorporate the target in assessing allowable spread.

C_{pm} incorporates a measure of proximity to the target by replacing the process variance in the definition of C_p, with the process mean square error around the target. C_p and C_{pm} are identical when the process is centered at the target (i.e., $\mu = T$). But since C_p is not a valid measure of process capability when the process is not centered at the target, C_{pm} dominates C_p as a measure of process capability.

$C^*_{pl}, C^*_{pu}, C^*_{pk}, C^*_{pm}$, and C_{pmk} form the basis for a large number of process capability indices, sometimes referred to as third generation. Effectively these indices extend the class of allowable/customer process requirements versus actual process spread/performance by allowing the target to be other than the midpoint of the specification limits.

By definition, each of $C_p, C_{pl}, C_{pu}, C_{pk}, C_{pm}$, their generalized analogues, and C_{pmk} are unitless, thereby fostering comparisons among and within processes, regardless of the underlying mechanics of the product or service being monitored. In all cases, as process performance improves, either through reductions in variation and/or moving closer to the target, these indices increase in magnitude. As process performance improves relative to customer requirements, customer satisfaction increases as the process has a greater ability to be near the target. In all cases larger index values indicate a more capable process. Using the indices to assess the ability of the process to meet customer expectations, with larger values of the indices indicating a higher customer satisfaction and lower values, a poorer customer satisfaction is consistent with the concept of "fitness for use".

A unified index, C_{pw} [3] of the form

$$C_{pw} = \frac{USL - LSL}{6\sqrt{\sigma^2 + w(\mu - T)^2}} \tag{12}$$

can, by varying w, be used to represent a wide spectrum of capability indices.

Allowing w to take on various values permits C_{pw} to assume equivalent computational algorithms for a variety of indices, including C_p, C_{pm}, C_{pk}, and C_{pmk}. For example, setting $w = 0$, results in

$$C_{pw} = \frac{USL - LSL}{6\sqrt{\sigma^2}} = C_p \tag{13}$$

while for $w = 1$,

$$C_{pw} = \frac{USL - LSL}{6\sqrt{\sigma^2 + (\mu - T)^2}} = C_{pm} \tag{14}$$

Defining $d = \frac{USL-LSL}{2}$, $a = \mu - \frac{USL+LSL}{2}$ and $p = \frac{|\mu - T|}{\sigma}$ then

$$w = \begin{cases} \left(\dfrac{d^2}{(d-|a|)^2} - 1\right)\dfrac{1}{p^2} & 0 < p \\ 0 & \text{elsewhere} \end{cases} \tag{15}$$

allows $C_{pw} = C_{pk}$.

Similarly,

$$w = \begin{cases} \dfrac{k(2-k)}{(1-k)^2 p^2} & 0 < k < 1 \\ 0 & \text{elsewhere} \end{cases} \tag{16}$$

where $k = \dfrac{2|T - \mu|}{USL - LSL}$, allows $C_{pw} = C^*_{pk}$, while

$$w = \begin{cases} \left(\dfrac{d}{d-|a|}\right)^2\left(\dfrac{1}{p^2} + 1\right) - \dfrac{1}{p^2}, & 0 < p \\ 0 & \end{cases} \tag{17}$$

elsewhere results in $C_{pw} = C_{pmk}$. Table 1 includes several weights including the appropriate weights associated with $C_p(u, v)$ [4].

It is important to note that the universal assumptions associated with the unified index C_{pw} include that the underlying process is stable and the process measurements are normally distributed.

Interpreting Process Capability Measures

Process capability measures have traditionally been used to provide insights into the number (or proportion) of a nonconforming product (i.e., yield). Practitioners cite a C_p value of 1 as representing 2700 ppm nonconforming, while 1.33 represents 63 ppm; 1.66 corresponds to 0.6 ppm; and 2 indicates <0.1 ppm. C_{pk} has similar connotations with a C_{pk} of 1.33 representing a maximum of 63 ppm nonconforming. Practitioners use the value of the process capability index and its associated number, nonconforming, to identify capable processes. A process with a C_p greater than or equal to 1 has traditionally been deemed capable, while a C_p of less than 1 indicates that the process is producing more than 2700 ppm nonconforming and used as an indication that the process is not capable of meeting customer requirements. In the case of C_{pk}, the auto industry frequently uses 1.33 as a benchmark in assessing the capability of a process.

Inherent in any discussion of yield as a measure of process capability is the assumption that product produced just inside the specification limit is of

Table 1 Other C_{pw} weights and the associated index

w	Resulting index
$\begin{cases} \dfrac{6C_p - p}{(3C_p - p)^2 p}, & 0 < \dfrac{p}{3} < C_p \\ 0, & \text{elsewhere} \end{cases}$	$C^*_{pk} = (1-k)\dfrac{USL - LSL}{6\sigma}$
$\begin{cases} \left(\dfrac{d}{d-\lvert a\rvert}\right)^2 \left(\dfrac{1}{p^2} + 1\right) - \dfrac{1}{p^2}, & 0 < p \\ 0, & \text{elsewhere} \end{cases}$	$C_{pmk} = \dfrac{\min(USL - \mu, \mu - LSL)}{3\sqrt{\sigma^2 + (\mu - T)^2}} = \dfrac{d - \lvert\mu - M\rvert}{3\sqrt{\sigma^2 + (\mu - T)^2}}$
$\begin{cases} \left(\dfrac{d}{d-u\lvert a\rvert}\right)^2 \left(\dfrac{1}{p^2} + v\right) - \dfrac{1}{p^2}, & 0 < p \\ 0, & \text{elsewhere} \end{cases}$	$C_p(u, v) = \dfrac{d - u\lvert\mu - M\rvert}{3\sqrt{\sigma^2 + v(\mu - T)^2}}$

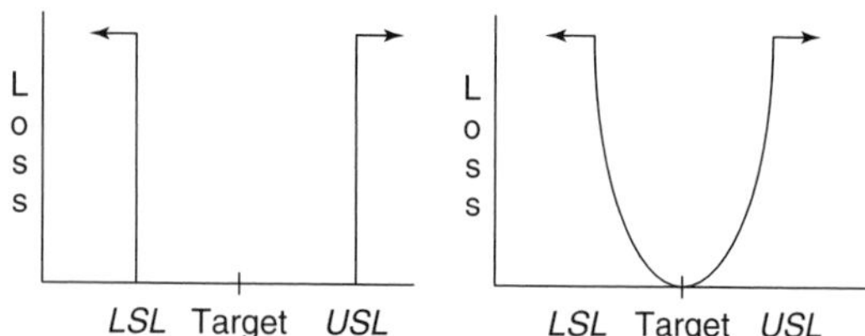

Figure 2 Square-well and modified quadratic loss functions

equal quality to that produced at the target. This is equivalent to assuming a square-well loss function [5] for the quality variable (see Figure 2).

In practice the magnitudes of $C_p, C_{pl}, C_{pu}, C_{pk}$, their generalized analogues, and C_{pmk} are interpreted as a measure of nonconforming and therefore are represented by the square-well loss function. Any change in the magnitude of these indices (holding the customer requirements constant) is due to changes in the distance between the specification limits and the process mean. As Boyles [6] emphatically points out "C_{pk} does not in itself say anything about the distance between μ and T" and "is essentially a measure of **process yield** only." By design $C_p, C_{pl}, C_{pu}, C_{pk}, C^*_{pl}, C^*_{pu}, C^*_{pk}$, and C_{pmk} are used to identify changes in the amount of product beyond the specification limits (not proximity to the target) and therefore consistent with the square-well loss function.

Taguchi used the quadratic loss function to motivate the idea that a product imparts no loss, only if that product is produced at its target. He maintains that even small deviations from the target result in a loss of quality and that as the product increasingly deviates from its target, there are larger and larger losses in quality. This approach to quality and quality assessment is different from the traditional approach, where no loss in quality is assumed until the product deviates beyond its USL and LSL (i.e., square-well loss function). Taguchi's philosophy highlights the need to have small variability around the target. Clearly in this context, the most capable process will be one that produces its entire product at the target, with the next best being the process with the smallest variability around the target.

The motivation for C_{pm} and C^*_{pm} does not arise from examining yield, but from looking at the ability of the process to be in the neighborhood of the target. This motivation has little to do with the number of nonconforming although upper bounds on the number of nonconforming can be determined for numerical values of C_{pm} [7]. Several authors [3, 6, 8] have discussed the relationship between C_{pm} and the quadratic loss function and its affinity with the philosophies that support a loss in quality for any departure from the target. The loss associated with the quality variable under investigation is then depicted using a modified quadratic loss function (see Figure 2).

C_{pk}, C_{pm}, and C_{pmk} have different functional forms, are represented by different loss functions, and have different relationships with C_p as the process drifts from the target. C_{pk} and C_{pm} are often called second-generation measures of process capability whose motivations arise directly from the inability of C_p to consider the target value. The differences in their associated loss functions demarcate the measures, while the magnitudinal relationship between C_p and C_{pk}, C_{pm}, C_{pmk} are also different. C_{pk}, C_{pm}, and C_{pmk} are functions of C_p that attempt to penalize the process for not being centered on the target

(T). Expressing C_{pm}, C_{pk}, and C_{pmk} in the following manner

$$C_{pm} = \frac{1}{\sqrt{1+\left(\frac{\mu-T}{\sigma}\right)^2}} C_p \tag{18}$$

$$C_{pk} = \left(1 - \frac{2|\mu - T|}{USL - LSL}\right) C_p \tag{19}$$

and

$$C_{pmk} = \left(1 - \frac{2|\mu - T|}{USL - LSL}\right)\left(\frac{1}{\sqrt{1+p^2}}\right) C_p \tag{20}$$

illustrates the "penalizing" relationship between C_p and C_{pm}, C_{pk}, C_{pmk} respectively. As the process mean drifts from the target (measured by $p = \frac{|\mu-T|}{\sigma}$), C_{pm}, C_{pk}, and C_{pmk} decline as a percentage of C_p (see Figure 3). In the case of C_{pm}, this relationship is independent of the magnitude of C_p. On the other hand, the decline in C_{pk} and C_{pmk} as a percentage of C_p depends on the magnitude of C_p. For example, in Figure 3, $C_{pk}(1)$, $C_{pk}(2)$, and $C_{pk}(3)$ represent the relationships between C_{pk} and C_p for $C_p = 1$, $C_p = 2$ and $C_p = 3$, respectively. Similarly $C_{pmk}(1)$, $C_{pmk}(2)$, and $C_{pmk}(3)$ represent the relationships between C_{pmk} and C_p for $C_p = 1$, $C_p = 2$, and $C_p = 3$.

C_{pk}, C_{pmk}, and C_{pm} have different functional forms, are represented by different loss functions, and have different relationships with C_p as the process drifts from the target. Hence, although C_{pm} and C_{pk} are lumped together as second-generation measures and C_{pmk} as a hybrid of the two, all are different in their development and assessment of process capability. A discussion on the relationship of capability indices (C_p, C_{pk}, and C_{pm}) among themselves, linked to specifications at the supplier level and specifications to the assembly level is cited in [9]. Similarly C^*_{pk} and C^*_{pmk} differ from C^*_{pm} in their assessment of process capability.

Effects of Nonnormality

If the process measurements do not arise from a normal distribution, none of the indices discussed provides valid measures of yield. Each index uses a function of σ as a measure of actual process spread in its determination of process capability. But as many authors [10–13] have pointed out, although the standard deviation has become synonymous with the term dispersion, its physical meaning need not be the same for different families of distributions, or for that matter, within a family of distributions. Therefore the actual process spread (a function of 6σ) does not provide a consistent meaning over various distributions. To illustrate, suppose that precisely 99.73%

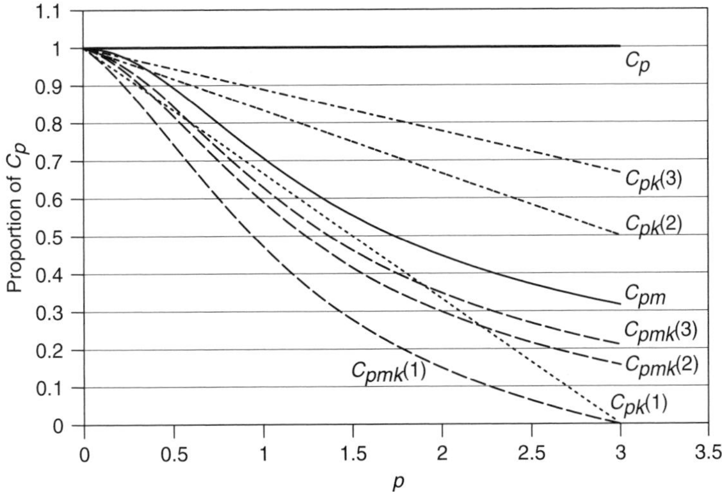

Figure 3 Comparison of C_p, C_{pk}, C_{pm}, and C_{pmk}

of the process measurements fall within the specification limits (i.e., a yield of 2700 ppm). The values of $C_p, C_{pl}, C_{pu}, C_{pk}, C_{pm}$, and C_{pmk} are 0.5766, 0.7954, 1.0000, 1.2210, and 1.4030 respectively when the measurements arise from a uniform, triangular, normal, logistic, and double exponential distribution. As long as 6σ carries a yield interpretation when assessing process capability (i.e., is translated into ppm nonconforming), none of the indices should be used if the underlying process distribution is not normal.

If we assume process capability assessments to be studies of the ability of the process to produce product around the target, then C^*_{pm} and C_{pmk} provide practitioners with an assessment of capability regardless of the distribution associated with the measurements. Clustering around the target, rather than a measure of nonconforming releases the physical meaning attached to 6σ. The denominators of C^*_{pm} and C_{pmk} then provide a measure of the clustering around the target and compare this with customer tolerance.

Eliminating the physical meaning of ppm and investigating Euclidean distances around the target allows C_{pm} and C_{pmk} to compare the capability of various processes, or processes over time, regardless of the underlying distribution. The underlying distribution will impact the inferences that can be made from samples gathered from the population; however, the population parameter is no longer distributionally sensitive.

References

[1] Juran, J.M. (1979). *Quality Control Handbook*, McGraw-Hill, New York.

[2] Bissell, A.F. (1990). How reliable is your capability index? *Applied Statistics* **39**, 331–340.

[3] Spiring, F.A. (1997). A unifying approach to process capability indices, *Journal of Quality Technology* **29**, 49–58.

[4] Vannman, K. (1995). A unified approach to capability indices, *Statistica Sinica* **5**, 805–820.

[5] Tribus, M. & Szonyi, G. (1989). An alternate view of the Taguchi approach, *Quality Progress* **22**(5), 46–52.

[6] Boyles, R.A. (1991). The Taguchi capability index & corrigenda, *Journal of Quality Technology* **23**, 17–26.

[7] Spiring, F.A. (1991). A new measure of process capability, *Quality Progress* **24**(2), 57–61.

[8] Johnson, T. (1992). The relationship of cpm to squared error loss, *Journal of Quality Technology* **24**, 211–215.

[9] Parlar, M. & Wesolowsky, G.O. (1999). Specification limits, capability indices, and process centering in assembly manufacturing, *Journal of Quality Technology* **31**, 317–325.

[10] Hoaglin, D.C., Mosteller, F. & Tukey, J.W. (1983). *Understanding Robust and Exploratory Data Analysis*, John Wiley & Sons, New York.

[11] Mosteller, F. & Tukey, J.W. (1977). *Data Analysis and Regression*, Addison-Wesley, Reading.

[12] Tukey, J.W. (1970). A survey of sampling from contaminated distributions, *Contributions to Probability and Statistics*, Stanford University Press, Stanford, pp. 448–485.

[13] Huber, P.J. (1977). *Robust Statistical Procedures*, Society for Industrial and Applied Mathematics, Philadelphia.

FRED SPIRING

Process Capability Indices, Multivariate

Basic Capability Indices

Process capability indices are becoming a common approach to report process performance, both internally and in managing customer–supplier relationships. In the univariate case, *capability indices* are defined as ratios of spread between the process specification limits to variability of the process. For the univariate case, C_p, C_{pk}, C_{pm}, and C_{pmk} are the most commonly known *process capability indices* [1–3]. Most process capability indices assume that the process performance data are normally distributed. Let μ and σ be the mean and standard deviation, respectively, of a process performance characteristic such as weight, assay, particle size, or voltage levels. Furthermore let USL, LSL, and T be the upper and lower specification limits and the target value, respectively. The above mentioned population capability indices are defined as follows:

$$C_p = \frac{USL - LSL}{6\sigma} \tag{1}$$

$$C_{pk} = \min\left[\frac{USL - \mu}{3\sigma}, \frac{\mu - LSL}{3\sigma}\right] \tag{2}$$

$$C_{pm} = \frac{USL - LSL}{6\sqrt{\sigma^2 + (\mu - T)^2}} \tag{3}$$

$$C_{pmk} = \min\left[\frac{USL-\mu}{3\sqrt{\sigma^2+(\mu-T)^2}}, \frac{\mu-LSL}{3\sqrt{\sigma^2+(\mu-T)^2}}\right] \quad (4)$$

with the sample estimates being obtained by replacing the population μ and σ by their sample estimates, $\bar{x}$ and s, respectively.

General Multivariate Capability Indices

However, many industrial processes have more than one performance or quality characteristic so that multivariate evaluation of process performance becomes more and more important. Although univariate process capability indices have been extensively studied in the literature, their multivariate counterparts have received relatively little attention.

Taam *et al.* [4] and later Pal [5] proposed the index $C_{pb} = S_R/A_P$, where S_R represents the surface defined by the rectangle of the upper and lower tolerances and A_P is the process surface including 99.73% of the process data. Wang and Chen [6] proposed multivariate equivalents for C_p, C_{pk}, C_{pm}, and C_{pmk} based on principal component analysis decomposition. Wang and Du [7] proposed the same indices with extension to the nonnormal multivariate case. Wang *et al.* [8] proposed a process capability multivariate vector to evaluate the process performance. Such a vector has the following components: a multivariate C_p index, the p value of a T^2 Hotelling test, and a location index. Chen [9] also investigated a method to estimate the multivariate C_p using a nonconforming proportion approach. Castagliola and Castellanos [10] propose a method for estimating both C_p and C_{pk} indices for two quality variables labeled BC_p. The proposed bivariate C_{pk} index is calculated by first finding four probabilities corresponding to the theoretical fraction of values that exceed the surface formed by the process tolerances (nonconforming proportion). They then use a numerical quadrature method based on Green's theorem for computing the probabilities associated with each convex polygon. The capability index BC_p is found by maximizing the capability index BC_{pk}, i.e. by maximizing the theoretical proportion of conforming items.

A general formulation of multivariate capability indices is based on the region $L \le g(\mathbf{X}) \le U$ where $g(\mathbf{X})$ is a monotonic function of the joint **probability density function** of $\mathbf{X}$ and often, L is zero. If $\mathbf{X}$ is assumed to have a multivariate normal $N_\nu\ (\boldsymbol{\mu}, \boldsymbol{\Sigma})$ distribution a natural $g(\mathbf{X})$ function is $g(\mathbf{X}) = (\mathbf{X}-\boldsymbol{\mu})\boldsymbol{\Sigma}^{-1}(\mathbf{X}-\boldsymbol{\mu})$ and one may regard an item as nonconforming if $(\mathbf{X}-\boldsymbol{\mu})\boldsymbol{\Sigma}^{-1}(\mathbf{X}-\boldsymbol{\mu}) > U$. We thus obtain the ellipsoidal specification region $(\mathbf{X}-\boldsymbol{\mu})\boldsymbol{\Sigma}^{-1}(\mathbf{X}-\boldsymbol{\mu}) \le \mathrm{U}$.

An analo of C_p is

$$\frac{\text{Volume of }\left\{(X-\mu)\Sigma^{-1}(X-\mu) \le U\right\}}{\text{Volume of }\left\{(X-\mu)\Sigma^{-1}(X-\mu) \le R\right\}} \quad (5)$$

where $\Pr[(\mathbf{X}-\boldsymbol{\mu})\boldsymbol{\Sigma}^{-1}(\mathbf{X}-\boldsymbol{\mu}) \le \mathrm{R}] = 1-p$.

Again, if $\mathbf{X}$ has a multivariate **normal distribution** $(\mathbf{X}-\boldsymbol{\mu})\boldsymbol{\Sigma}^{-1}(\mathbf{X}-\boldsymbol{\mu})$ has a χ^2 distribution with ν **degrees of freedom** and $R = \chi^2_{\nu,1-p}$ is the corresponding upper $100(1-p)\%$ point of the χ^2 distribution with ν degrees of freedom. For more examples of such indices see Wierda [11], Chan *et al.* [12], Kotz and Johnson [2], Foster *et al.* [13], Pearn *et al.* [14], Pearn [15], and Pearn and Kotz [16].

Capability Indices Based on Tolerance Regions

Tolerance regions provide an alternative to the $L \le g(\mathbf{X}) \le U$ region, that is based on estimates of **percentiles** from a multivariate distribution with parameters either known or estimated from the data.

Fuchs and Kenett [17] suggest the establishment of a process control schema based on tolerance regions by estimating the level set $\{f \ge c\}$ with a prespecified probability content $1-\alpha$, of the density f which generates the data. A reasonable criterion would be to reject an additional observation $\mathbf{X}_{n+1}$ whenever $\mathbf{X}_{n+1} \in \{f < c\}$, which would give an exact false alarm probability α. Since f is usually unknown, the population tolerance region $\{f \ge c\}$ can be estimated by plugging in an estimator function of f. For the multivariate normal case, Krishnamoorthy and Mathew [18] provide various estimates and clarify an inconsistency in notation adopted by John [19] and Fuchs and Kenett [17]. Baillo and Cuevas [20] develop a nonparametric solution of the procedure suggested by Fuchs and Kenett [17]. Di Bucchianico *et al.* [21] develop a nonparametric procedure for constructing multivariate tolerance regions that relies on the choice of a class of sets to which the tolerance region belongs (for example, ellipsoids or hyperrectangles). Polansky [22] also uses the tolerance regions approach to evaluate the capability of a manufacturing process.

Thus, a natural extension of process capability measures to the multivariate case builds on the concept of tolerance regions. This area is an active research area and will probably remain so in the next few years. Applying tolerance regions to the general formulation of **multivariate process capability indices** offers new definitions of process capability indicators.

References

[1] Kotz, S. & Johnson, N.L. (1993). *Process Capability Indices*, Chapman & Hall, London.

[2] Kotz, S. & Johnson, N.L. (2002). Process capability indices – a review 1992–2000 with discussion, *Journal of Quality Technology* **34**, 2–53.

[3] Kenett, R. & Zacks, S. (1998). *Modern Industrial Statistics: Design and Control of Quality and Reliability*, 2nd Edition, Chinese edition 2004, Duxbury Press, San Francisco, 2003.

[4] Taam, W., Subbaiah, P. & Liddy, W.L. (1993). A note on multivariate capability indices, *Journal of Applied Statistics* **20**(3), 339–351.

[5] Pal, S. (1999). Performance evaluation of a bivariate normal process, *Quality Engineering* **11**(3), 379–386.

[6] Wang, F.K. & Chen, J.C. (1998). Capability index using principal components analysis, *Quality Engineering* **11**(1), 21–27.

[7] Wang, F.K. & Du, T.C.T. (2000). Using principal component analysis in process performance for multivariate data, *Omega: The International Journal of Management Science* **28**, 185–194.

[8] Wang, F.K., Hubele, N.F., Lawrence, F.P., Miskulin, J.D. & Shahriari, H. (2000). Comparison of three multivariate process capability indices, *Journal of Quality Technology* **32**(3), 263–275.

[9] Chen, H. (1994). A multivariate process capability index over a rectangular solid zone, *Statistica Sinica* **4**, 749–758.

[10] Castagliola, P. & Castellanos, J.V. (2005). Capability indices dedicated to the two quality characteristics case, *Quality Technology and Quantitative Management* **2**(2), 201–220.

[11] Wierda, S. (1994). *Multivariate Statistical Process Control*, Groningen Theses in Economics, Management and Organization, Wolters-Noordhoff, Groningen.

[12] Chan, L.K., Cheng, S.W. & Spiring, F.A. (1991). A multivariate measure of process capability, *International Journal of Modelling and Simulation* **11**, 1–6.

[13] Foster, E.J., Barton, R.R., Gautam, N., Truss, L.T. & Tew, J.D. (2005). The process-oriented multivariate capability index, *International Journal of Production Research* **43**(10), 2135–2148.

[14] Pearn, W.L., Kotz, S. & Johnson, K.L. (1992). Distribution and inferential properties of process capability indices, *Journal of Quality Technology* **24**, 216–231.

[15] Pearn, W.L. (1998). New generalization of process capability index Cpk, *Journal of Applied Statistics* **25**(6), 801–810.

[16] Pearn, W.L. & Kotz, S. (2006). *Encyclopedia and Handbook of Process Capability Indices: A Comprehensive Exposition of Quality Control Measures, Series on Quality, Reliability and Engineering Statistics*, World Scientific Publishing Company.

[17] Fuchs, C. & Kenett, R.S. (1987). Multivariate tolerance regions and F-tests, *Journal of Quality Technology* **19**, 122–131.

[18] Krishnamoorthy, K. & Mathew, T. (1999). Comparison of approximation methods for computing tolerance factors for a multivariate normal population, *Technometrics* **41**(3), 234–249.

[19] John, S. (1963). A tolerance region for multivariate normal distributions, *Sankhya, Series A* **25**, 363–368.

[20] Baillo, A. & Cuevas, A. (2006). Parametric versus nonparametric tolerance regions in detection problems, *Computational Statistics* **21**, 523–536.

[21] Di Bucchianico, A., Einmahl, J. & Mushkudiani, N. (2001). Smallest nonparametric tolerance regions, *Annals of Statistics* **29**, 1320–1343.

[22] Polansky, A.M. (2001). A smooth nonparametric approach to multivariate process capability, *Technometrics* **43**, 199–211.

Further Reading

Fuchs, C. and Kenett, R. (1998). *Multivariate Quality Control: Theory and Application, Quality and Reliability Series, Vol. 54*, Marcel Dekker, New York.

Hubele, N.F., Shahriari, H. & Cheng, C.S. (1991). A bivariate process capability vector, in *Statistical Process Control in Manufacturing*, J.B. Keats & D.C. Montgomery, eds, Marcel Dekker, New York, pp. 299–310.

Kane, V.E. (1986). Process capability indices, *Journal of Quality Technology* **18**(1), 41–52.

Kotz, S. & Lovelace, C.R. (1998). *Introduction to Process Capability Indices: Theory and Practice*, Arnold, London.

Mason, R.L. & Young, J.C. (2002). *Multivariate Statistical Process Control with Industrial Applications, ASA-SIAM Series on Statistics and Applied Probability*, ASA-SIAM Philadelphia.

Palmer, K. & Tsui, K.-L. (1999). A review and interpretations of process capability indices. *Annals of Operations Research* **87**(1), 31–47.

Vannman, K. & Kotz, S. (1995). A superstructure of capability indices-distributional properties and implications, *Scandinavian Journal of Statistics* **22**, 477–491.

Yeh, A.B. & Chen, H. (2001). A nonparametric multivariate process capability index, *International Journal of Modelling and Simulation* **21**(3), 218–223.

Camil Fuchs and Ron S. Kenett

Process Capability Indices, Nonnormal

Introduction

Process capability indices measure the ability of a process to manufacture product that meets specifications when the process is in statistical control. Most process capability studies assume that the process quality characteristic under study, X, is normally distributed. However, nonnormal process distributions are common in semiconductor and other manufacturing industries. For example, the particle counts on wafers and chemical impurity levels in semiconductor manufacturing often have nonnormal distributions. A one-sided specification is often an indication that the underlying distribution may be nonnormal. Normal distributions generally do not fit constrained data. Process capability indices for nonnormal distributions have received increasing attention in recent years. Books that devote some chapters to covering process capability indices for nonnormal distributions include Kotz and Johnson [1], Kotz and Lovelace [2], Bothe [3], and Pearn and Kotz [4]. A recent bibliography of process capability papers is given in Spiring *et al.* [5].

Many authors study the effects of nonnormality on process capability indices and conclude that using normal-based process capability indices such as C_p, C_{pu}, C_{pl}, and C_{pk}, or the equivalent superstructure $C_p(u, v)$ of Vännman [6], to analyze nonnormal data may lead to erroneous results. Somerville and Montgomery [7] investigate the errors incurred in using normal-based analysis to make inference about expected process fallout rate based on nonnormal data. Discussions of the effects of nonnormality are in Kotz and Johnson [8], Lu and Rudy [9], and Spiring *et al.* [10].

Some approaches to deal with nonnormality in process capability analysis include: (a) transforming nonnormal data to normality so that the normal-based process capability indices can be applied to the transformed data, (b) fitting data by a distribution so that quantile-related process capability indices can be evaluated, and (c) developing other non-quantile-based indices that are applicable to nonnormal distributions. Discussions of various approaches include [2] and [8].

Data Transformation

One approach to the nonnormality problem is to transform the data using some well-known transformations such as the **Box**–Cox power transformations and the Johnson system of transformations (see Box and Cox [11], Page [12], Farnum [13], Chou *et al.* [14] and Polansky *et al.* [15, 16]). A test of normality must be performed to check the normality of the transformed data. If the test supports normality, the normal-based process capability indices are appropriate for the transformed data. A powerful test of normality is given by Chen and Shapiro [17].

In a comparative study, Tang and Than [18] use the normal-based index C_{pu} as the standard and compare the performance of seven nonnormal-based indices when applied to lognormal and Weibull data. Their results show that the Box–Cox and Johnson transformation methods perform very well.

Quantile-Based Process Capability Indices

The second approach is to fit the data by an empirical distribution function or by a three or four-parameter distribution in order to obtain the **quantiles** of the distribution of X and define quantile-based process capability indices. The pth quantile of the distribution of X is defined to be a value x_p such that $\Pr(X \le x_p) = p$. If X has a normal $N(\mu, \sigma^2)$ distribution, then $\mu = x_{0.5}$, $\mu - 3\sigma = x_{0.00135}$, and $\mu + 3\sigma = x_{0.99865}$.

The idea is to modify the usual (μ, σ)-related indices so that they are applicable to both normal and nonnormal distributions. In this approach, normal-based process capability indices are modified in order to increase their robustness (see [2]). For example, the modified index of C_p is given by

$$C'_p = \frac{USL - LSL}{x_{0.99865} - x_{0.00135}} \tag{1}$$

where LSL and USL are the lower and upper specification limits, respectively. Veevers [19] introduces the viability index which is equal to $1 - \frac{1}{C'_p}$ for any symmetric unimodal distribution.

Clements [20] suggests using a Pearson system distribution to fit the underlying distribution in order to determine the quantiles for calculating the process capability indices. The Pearson system has been used by Pearn and Kotz [21] and Bittanti *et al.* [22].

Castagliola [23] fits nonnormal data by a Burr distribution. Polansky applies a kernel smoothing method to fit data (*see* **Process Capability Indices, Nonparametric**). Krishnamoorthi and Khatwani [24] use a Weibull distribution to fit data. Ding [25] uses only the first four sample **moments** to fit the underlying distribution by a Chebyshev–Hermite polynomial up to the 10th order, where the original process data need not be available. Pal [26] uses the generalized lambda distribution to fit data. In fitting data by a distribution, a goodness-of-fit test must be performed to check if the distribution fits the data well.

Pearn and Chen [27] generalize the superstructure $C_p(u, v)$ and introduce quantile-based indices, $C_{Np}(u, v)$ and $C'_{Np}(u, v)$, for nonnormal distributions. The index $C_{Np}(u, v)$ is defined as

$$C_{Np}(u, v) = \frac{d - u|M - m|}{3\sqrt{\left(\frac{x_{0.99865} - x_{0.00135}}{6}\right)^2 + v(M - T)^2}} \tag{2}$$

where $M = x_{0.5}, m = (USL + LSL)/2, d = (USL - LSL)/2$, T is the target, and $u, v \geq 0$. $C'_{Np}(u, v)$ has exactly the same form as $C_{Np}(u, v)$ except that the median M is replaced by μ. Their study shows that the generalizations $C_{Np}(u, v)$ are superior to $C'_{Np}(u, v)$ in assessing process capability. Both indices are applicable to normal distributions since they reduce to $C_p(u, v)$. More details about the indices are in [4].

McCormack *et al.* [28] find that the empirical distribution function fits their process data very well for samples of size 100 and conclude that their index $\hat{C}_{Npk}$ is a viable estimator of process capability for nonnormal distributions, where

$$\hat{C}_{Npk} = \min\left(\frac{USL - x_{0.5}}{x_{0.995} - x_{0.5}}, \frac{x_{0.5} - LSL}{x_{0.5} - x_{0.005}}\right) \tag{3}$$

Other Process Capability Indices

The third approach to deal with nonnormality is to develop indices that are not quantile-based. The process capability index, C_{pm}, developed by Chan *et al.* [29], can be applied to assess process capability for both normal and nonnormal distributions since C_{pm} is not distribution sensitive (see [10]).

Luceno [30] introduces the index C_{pc} to handle the nonnormal data and estimate the index by an approximate **confidence interval**, where C_{pc} is defined as

$$C_{pc} = \frac{USL - LSL}{6\sqrt{\frac{\pi}{2}}E|X - T|} \tag{4}$$

A point estimator of C_{pc} is given by

$$\hat{C}_{pc} = \frac{USL - LSL}{6\sqrt{\frac{\pi}{2n}\sum_{i=1}^{n}|X_i - T|}} \tag{5}$$

Wright [31] proposes an index, C_s, to account for skewness (*see* **Process Capability Indices, Skewed**). C_s is defined as

$$C_s = \frac{\min(USL - \mu, \mu - LSL)}{3\sqrt{\sigma^2 + (\mu - T)^2 + |\mu_3/\sigma|}} \tag{6}$$

Bootstrap confidence intervals for C_s are given in Han *et al.* [32]. Nahar *et al.* [33] assess the performance of Wright's index C_s and a modified index C_{sm}, for positively skewed distribution, where C_{sm} is obtained from C_s by replacing $|\mu_3/\sigma|$ by $(-\mu_3/\sigma)$. They find that the modified Wright's index C_{sm} has some useful attributes in reflecting process performance with respect to the percent nonconforming and the accuracy for relatively large samples.

Wu and Swain [34] propose process capability indices based on the weighted variance method. Chang *et al.* [35] propose indices based on the weighted standard deviation method. The idea for both methods is to divide a skewed or asymmetric distribution into two normal distributions that have the same mean but different variances or standard deviations. Simulation studies indicate these methods are fairly consistent in estimating process fallout rate for some nonnormal distributions.

Perakis and Xekalaki [36, 37] propose an index $(1 - p_0)/(1 - p)$, where p and p_0 denote the proportion of conformance (yield) and the minimum allowable proportion of conformance of the process under study. They state that the index can be used for both continuous and discrete process distributions.

Chao and Lin [38] believe a process capability index must represent the true yield and propose a universal index, C_y, defined as

$$Cy = \frac{1}{3}\Phi^{-1}\left[\frac{1}{2}F(USL) - F(LSL) + 1\right] \tag{7}$$

where Φ is the standard **normal distribution** function and F is the distribution function of X. The index can be extended to a multivariate process capability index (*see* **Process Capability Indices, Multivariate**) when F is a multivariate distribution.

Future Research

Many process capability indices have been proposed for nonnormal distributions or data. Various methods of estimating a process capability index can lead to different estimates of the rate of nonconforming product. Kaminsky *et al.* [39] show that it is difficult to demonstrate that a good process is indeed a good process by using a point estimator. Information about confidence interval estimation of the process capability indices is generally lacking.

As pointed out in [10], there has been little research in the area of loss and loss functions as methods for assessing process capability. The potential problem of nonnormality leads researchers to develop new indices or investigate the properties of particular process capability indices by evaluating their estimators using a variety of nonnormal process data. A comprehensive study of the indices and situations where they are applicable seems necessary.

References

[1] Kotz, S. & Johnson, N.L. (1993). *Process Capability Indices*, Chapman & Hall, London.

[2] Kotz, S. & Lovelace, C.R. (1998). *Process Capability Indices in Theory and Practice*, Arnold, London.

[3] Bothe, D.R. (2001). *Measuring Process Capability*, Landmark Publishing, Cedarburg.

[4] Pearn, W.L. & Kotz, S. (2006). *Encyclopedia and Handbook of Process Capability Indices*, World Scientific Publishing.

[5] Spiring, F., Leung, B., Cheng, S.W. & Yeung, A. (2003). A bibliography of process capability papers, *Quality and Reliability Engineering International* **19**, 445–460.

[6] Vännman, K. (1995). A unified approach to capability indices, *Statistica Sinica* **5**, 805–820.

[7] Somerville, S.E. & Montgomery, D.C. (1996–1997). Process capability indices and non-normal distributions, *Quality Engineering* **9**, 305–316.

[8] Kotz, S. & Johnson, N.L. (2002). Process capability indices – a review, 1992–2000, *Journal of Quality Technology* **34**, 2–19.

[9] Lu, M.W. & Rudy, R.J. (2002). "Discussion on process capability indices – a review, 1992–2000", *Journal of Quality Technology* **34**, 38–39.

[10] Spiring, F., Cheng, S.W., Yeung, A. & Leung, B. (2002). "Discussion on Process capability indices – a review, 1992–2000", *Journal of Quality Technology* **34**, 23–27.

[11] Box, G.E.P. & Cox, D.R. (1964). An analysis of transformations, *Journal of the Royal Statistical Society, Series B* **26**(2), 211–252.

[12] Page, M. (1994). Analysis of non-normal process distributions, *Semiconductor International* **17**, 88–96.

[13] Farnum, N.R. (1996–1997). Using Johnson curves to describe non-normal process data, *Quality Engineering* **9**(2), 329–336.

[14] Chou, Y.M., Polansky, A.M. & Mason, R.L. (1998). Transforming non-normal data to normality in statistical process control, *Journal of Quality Technology* **30**, 133–141.

[15] Polansky, A.M., Chou, Y.M. & Mason, R.L. (1999). An algorithm for fitting Johnson transformations to non-normal data, *Journal of Quality Technology* **31**, 345–350.

[16] Polansky, A.M., Chou, Y.M. & Mason, R.L. (1998–1999). Estimating process capability indices for a truncated normal distribution, *Quality Engineering* **11**(2), 257–265.

[17] Chen, L. & Shapiro, S.S. (1995). An alternative test for normality based on normalized spacings, *Journal of Statistical Computation and Simulation* **53**, 269–287.

[18] Tang, L.C. & Than, S.E. (1999). Computing process capability indices for non-normal data: a review and comparative study, *Quality and Reliability Engineering International* **15**, 339–353.

[19] Veevers, A. (1998). Viability and capability indices for multi-response processes, *Journal of Applied Statistics* **25**, 545–558.

[20] Clements, J.A. (1989). Process capability calculations for non-normal distributions, *Quality Progress* **22**(9), 95–100.

[21] Pearn, W.L. & Kotz, S. (1994–1995). Application of Clements' method for calculating second- and third-generation process capability indices for non-normal Pearsonian populations, *Quality Engineering* **7**(1), 139–145.

[22] Bittanti, S., Lovera, M. & Moiraghi, L. (1998). Application of non-normal process capability indices to semiconductor quality control, *IEEE Transactions on Semiconductor Manufacturing* **11**(2), 296–303.

[23] Castagliola, P. (1996). Evaluation of non-normal process capability indices using Burr's distributions, *Quality Engineering* **8**, 587–593.

[24] Krishnamoorthi, K.S. & Khatwani, S. (2000). A capability index for all occasions, *Annual Quality Congress*, Indianapolis, Vol. 54, pp. 77–81.

[25] Ding, J. (2004). A method of estimating the process capability index from the first four moments of non-normal data, *Quality and Reliability Engineering International* **20**, 787–805.

[26] Pal, S. (2005). Evaluation of nonnormal process capability indices using generalized lambda distribution, *Quality Engineering* **17**(1), 77–85.

[27] Pearn, W.L. & Chen, K.S. (1997). Capability indices for non-normal distributions with an application in electrolytic capacitor manufacturing, *Microelectronics and Reliability* **37**(12), 1853–1858.

[28] McCormack, D.W., Harris, I.R., Hurwitz, A.M. & Spagon, P.D. (2000). Capability indices for non-normal data, *Quality Engineering* **12**(4), 489–495.

[29] Chan, L.K., Cheng, S.W. & Spiring, F.A. (1988). A new measure of process capability: Cpm, *Journal of Quality Technology* **20**(3), 162–175.

[30] Luceno, A. (1996). A process capability index with reliable confidence intervals, *Communications in Statistics: Simulation and Computation* **25**, 235–245.

[31] Wright, P.A. (1995). A process capability index sensitive to skewness, *Journal of Statistical Computation and Simulation* **52**, 195–203.

[32] Han, J., Cho, J. & Leem, C.S. (2000). Bootstrap confidence limits for Wright's Cs, *Communications in Statistics: Theory and Methods* **29**(3), 485–505.

[33] Nahar, P.C., Hubele, N.F. & Zimmer, L.S. (2001). Assessment of a capability index sensitive to skewness, *Quality and Reliability Engineering International* **17**, 233–241.

[34] Wu, H.H. & Swain, J.J. (2001). A Monte Carlo comparison of capability indices when processes are nonnormally distributed, *Quality and Reliability Engineering International* **17**, 219–231.

[35] Chang, Y.S., Choi, I.S. & Bai, D.S. (2002). Process capability indices for skewed populations, *Quality and Reliability Engineering International* **18**, 383–393.

[36] Perakis, M. & Xekalaki, E. (2002). A process capability index that is based on the proportion of conformance, *Journal of Statistical Computation and Simulation* **72**(9), 707–718.

[37] Perakis, M. & Xekalaki, E. (2005). A process capability index for discrete processes, *Journal of Statistical Computation and Simulation* **75**(3), 175–187.

[38] Chao, M.T. & Lin, D.K.J. (2006). Another look at the process capability index, *Quality and Reliability Engineering International* **22**(2), 153–163.

[39] Kaminsky, F.C., Dovich, R.A. & Burke, R.J. (1998). Process capability indices: now and in the future, *Quality Engineering* **10**(3), 445–453.

Related Articles

Process Capability Indices, Multivariate; **Process Capability Indices, Nonparametric**; **Process Capability Indices, Skewed**.

YOUN-MIN CHOU AND ALAN M. POLANSKY

Process Capability Indices, Nonparametric

Introduction

A manufacturing process is considered to be capable if it consistently produces items within specifications. The consistency of a process, usually assessed using control chart techniques, is typically the first concern when studying a process (*see* **Control Charts, Overview**). Once consistency has been established, a capability study then focuses on the ability of the process to produce items within specifications. While the primary focus of such a study is usually on the **process yield**, process capability indices are generally the measure used to study the capability of a process. These indices are typically unitless ratios that compare the size of the specification set to the natural variability of the process. Interpretation of the indices and their relationship to the process yield are usually reliant on specific assumptions about the distribution of the quality characteristic. The most common assumption is that the distribution of the quality characteristic follows a **normal distribution**, though indices related to other parametric distributions have been developed.

Many researchers have carefully studied examples of process data and have generally concluded that normal process data may be rare in practice [1–5]. It has also been shown that violations of the normal assumption can have a large effect on the ability of the process capability indices to properly indicate the actual capability of the process [6–8]. To address this problem, many indices have been developed to measure the capability of processes whose quality characteristics follow specific nonnormal distributions, such as the exponential and Weibull. Other indices have been developed that use more flexible families of distributions such as the Burr, Johnson, and Pearson systems (*see* **Process Capability Indices, Nonnormal**). These indices are not strictly nonparametric indices in the sense that one is tacitly assuming that the distribution of the quality characteristic is within the family of distributions. Still other indices have implemented penalties for **skewness** (*see* **Process Capability Indices, Skewed**).

A nonparametric process capability index is one that properly indicates the capability of a process without making specific assumptions about the form of the underlying distribution of the quality characteristic. Some relatively minor assumptions are usually made. For example, it is usually assumed that the distribution of the quality characteristic has some smoothness properties and has a certain number of finite **moments**. This article will focus on these types of methods for assessing the capability of a process. The first part of the article discusses nonparametric process capability indices. The second part of the article discusses nonparametric methods for evaluating the process yield. Alternative types of process capability indices that are a function of the process yield have been suggested by several researchers. The final part of this article addresses issues related to formal statistical inference using these methods.

Capability Indices

Consider a manufacturing process that produces units with a univariate quality characteristic X that follows an unknown distribution F with $\mu = E(X)$ and $\sigma^2 = V(X)$. Assume that the specification set for X is an interval of the form $\mathcal{S} = (L, U) \subset \mathbb{R}$. In the usual case, where F is unknown, we consider observing a sample $X_1, \ldots, X_n$ from F by randomly sampling items from the process. A typical process capability index compares the allowable variability of the process, given by the size of the specification set, to the actual process variability as a ratio. A simple measure of the allowable process variability is given by $U - L$, regardless of the form of F. The actual process variability is related to the variability of F, which is usually measured as $\theta\sigma$, where θ is chosen so that $P(X - \theta\sigma < X < \mu + \theta\sigma) \simeq 1$ [3]. Therefore, the general form of the process capability index is $C_\theta = (U - L)/\theta\sigma$, where σ is often estimated using the sample standard deviation s. When F is normal, the usual choice is $\theta = 6$. The more universal choice of $\theta = 5.15$ has also been suggested by [3]. A nonparametric index can be formed by replacing $\theta\sigma$ with the width of a 99.73% distribution-free tolerance interval for F [9]. Note that these approaches are direct analogs to the C_p (process capability index), which assumes that the process mean is centered in the specification set. Otherwise, these indices merely indicate the best possible capability of the process. See Section 2.2 of [10].

The standard deviation may not be an appropriate measure of variability for all process distributions. A measure based on the **percentiles** of F has been suggested by [11]. The analog of the C_p process capability index they propose has the form

$$\hat{C}_{np} = \frac{U - L}{\hat{\xi}_{99.5} - \hat{\xi}_{0.5}} \tag{1}$$

where $\hat{\xi}_\eta$ is the ηth sample percentile of the observed process data. The same authors suggest an analog of the C_{pk} index, which accounts for the location of the process distribution. This index is given by $\hat{C}_{npk} = \min\{\hat{C}_{npl}, \hat{C}_{npu}\}$, where

$$\hat{C}_{npl} = \frac{\hat{\xi}_{50.0} - L}{\hat{\xi}_{50.0} - \hat{\xi}_{0.5}} \tag{2}$$

and

$$\hat{C}_{npu} = \frac{U - \hat{\xi}_{50.0}}{\hat{\xi}_{99.5} - \hat{\xi}_{50.0}} \tag{3}$$

An empirical study presented in [11] compared the performance of the usual $\hat{C}_{pk}$ process capability index and the proposed $\hat{C}_{npk}$ index. These simulations showed that the **bias** of the $\hat{C}_{npk}$ index was lower than that of the $\hat{C}_{pk}$ index for nonnormal distributions. Similar proposals have been suggested by [12–14].

Another modification of the C_p index is suggested by [15]. The sample form of this index is

$$\hat{C}_{pc} = \frac{U - L}{6\bar{c}\sqrt{\pi/2}} \tag{4}$$

where

$$\bar{c} = n^{-1} \sum_{i=1}^{n} |X_i - (U + L)/2| \tag{5}$$

The choice of the denominator of this index takes advantage of the quick convergence properties of the central limit theorem so that reliable **confidence intervals** for this index are simple to compute, regardless of the form of F. While the process yield associated with this index does depend on F, it is shown by [15] that the relationship is more stable than the relationship between the C_p process capability index and the process yield.

Indices Based on the Process Yield

The process yield is defined to be $p = P(L \leq X \leq U)$, the expected proportion of items in a sample that are within specification. Several authors have suggested that process capability indices that are based on estimates of the process yield may be more useful, particularly for nonnormal process data [16–20]. The advantages of this approach are that the interpretation of the process yield does not depend on the process distribution, the indices easily generalize to the case of a multivariate quality characteristic, and that the index can be applied to discrete processes [21]. For example [20] suggests a process capability index of the form

$$C_{pc} = \frac{1 - p_0}{1 - p} \tag{6}$$

where p_0 is the minimum allowable process yield. The process fallout $q = 1 - p$ can be used as an equivalent basis for these indices. Let $q_L = P(X < L)$ and $q_U = P(X > U)$. Then [22] suggests a process capability index of the form

$$C_f = \min\left\{\frac{q_L^0}{q_L}, \frac{q_U^0}{q_U}\right\} \tag{7}$$

where q_L^0 and q_U^0 are the maximum allowable process fallout rates below and above the specifications, respectively.

In order to implement these indices in the nonparametric setting, a reliable nonparametric method for estimating the process yield p must be used. A simple nonparametric approach consists of estimating the process yield with the proportion of items in the sample that are within specification, that is,

$$\hat{p} = n^{-1} \sum_{i=1}^{n} \delta(L \leq X_i \leq U) \tag{8}$$

where δ is the indicator function. Confidence limits for $\hat{p}$ are easily computed using the binomial distribution [23]. Such an interval is usually quite wide unless the sample size is very large, as the method does not take advantage of information in the sample about the form of F. Methods for computing upper bounds on p based on a modification of Tchebysheff's theorem have been considered by [24]. The use of extreme value theory has been suggested by [22]. Note that if F is continuous then $p = F(U) - F(L)$. This implies that the process yield can also be estimated by replacing F with a nonparametric estimate, such as a kernel estimate [25, 26]. Bandwidth selection for this estimate can be achieved using the plug-in method of [27] or the cross validation technique of [28]. Another approach suggested by [29] takes advantage of the nearly pivotal transformation $(x - \bar{x})/s$ and the **bootstrap** to develop a nonparametric approach to estimating the process fallout rate. Using the fact that a density can be written as a series of Chebyshev–Hermite polynomials [30] derive a method for estimating process yield based only on the first four sample moments.

The Multivariate Case

The case of a multivariate quality characteristic presents challenges for any measure of process capability. This is particularly true for nonparametric process capability indices. Let $\mathbf{X}$ be a d-dimensional quality characteristic and let $\mathcal{S} \subset \mathbb{R}^d$ be the specification set for the characteristic. It will be assumed that $\mathbf{X}$ follows a d-dimensional distribution F. In the usual sample case, $\mathbf{X}_1, \ldots, \mathbf{X}_n$ will denote the quality characteristics observed from items sampled from the process.

A multivariate process capability index has been developed by [31] that is directly related to the process fallout rate. This approach requires a specification set of the form

$$\mathcal{S} = \{\mathbf{x} \in \mathbb{R}^d : h(\mathbf{x} - \mathbf{T}) \leq r\} \tag{9}$$

where $h : \mathbb{R}^d \rightarrow \mathbb{R}^+$, $\mathbf{T}$ is a target vector, and $r > 0$ is a constant. Such a specification set is general enough to include ellipsoidal and rectangular specification sets. In this case, the process fallout rate is $q = P[h(\mathbf{X} - \mathbf{T}) > r]$, and [31] defines the multivariate process capability index as

$$MC_p = \frac{q_0}{q} \tag{10}$$

where q_0 is the maximum acceptable fallout rate. This index is generalized to the nonparametric setting by [32], who use extreme value theory to develop a nonparametric method for estimating q based on $\mathbf{X}_1, \ldots, \mathbf{X}_n$. The process fallout rate can also be estimated in the nonparametric setting using a multivariate kernel density estimate. With this method, one must assume that F has an absolutely continuous d-dimensional density f so that the process fallout rate can be written as

$$q = 1 - \int_{\mathcal{S}} f(\mathbf{x})\,d\mathbf{x} \qquad (11)$$

The fallout rate is then estimated by replacing f with a multivariate kernel density estimate. The bandwidth for this estimate must be specifically tailored to the specific shape of $\mathcal{S}$. A bootstrap method for selecting the bandwidth matrix for the kernel estimate based on observations $\mathbf{X}_1, \ldots, \mathbf{X}_n$ is given by [33].

Statistical Inference

A complete plan for studying the capability of a process should include methods for reliable statistical inference, including confidence intervals and tests of hypotheses. Most of the nonparametric methods developed so far have dealt exclusively with **estimation**. The form of the estimates and the lack of assumptions of the distribution of the quality characteristic generally exclude the possibility of simple closed form inferential methods for the case of finite sample sizes. Asymptotic inference for the indices may also be problematic due to the complexity of the indices. One exception is the index proposed by [15] which was specifically designed for simple inference. In general, most of the methods discussed here may require the application of the nonparametric approaches to inference afforded by such methods as the jackknife and the bootstrap.

References

[1] Dovich, R.A. (1987). CPI, C_{pk}, capability ratio – measures of performance or contributors to confusion? *Machine and Tool Bluebook*, Hitchcock Publishing Company, Wheaton, Illinois, USA, pp. 10–14.

[2] Gunter, B.H. (1989). The use and abuse of C_{pk}, *Quality Progress* **22**(1), 72–73.

[3] Gunter, B.H. (1989). The use and abuse of C_{pk}, part 2, *Quality Progress* **22**(3), 108–109.

[4] McCoy, P.F. (1991). Using performance indices to monitor production processes, *Quality Progress* **24**(2), 49–55.

[5] Pyzdek, T. (1995). Why normal distributions aren't [all that normal], *Quality Engineering* **7**, 769–777.

[6] Chan, L.K., Cheng, S.W. & Spiring, F.A. (1988). The robustness of the process capability index, C_p, to departures from normality, *Statistical Theory and Data Analysis II*, Elsevier Science Publishers.

[7] Somerville, S.E. & Montgomery, D.C. (1996). Process capability indices and non-normal distributions, *Quality Engineering* **9**, 305–316.

[8] Pearn, W.L., Kotz, S. & Johnson, N.L. (1992). Distributional and inferential properties of process capability indices, *Journal of Quality Technology* **24**, 216–231.

[9] Chan, L.K., Cheng, S.W. & Spiring, F.A. (1988). A graphical technique for process capability, *ASQC Quality Congress Transactions - Dallas* **42**, 268–278.

[10] Kotz, S. & Johnson, N.L. (1993). *Process Capability Indices*, Chapman & Hall, London.

[11] McCormack, D.W., Harris, I.R., Hurwitz, A.M. & Spagon, P.D. (2000). Capability indices for non-normal data, *Quality Engineering* **12**, 489–495.

[12] Schneider, H.J., Pruett, J. & Lagrange, C. (1995). Uses of process capability indices in the supplier certification process, *Quality Engineering* **8**, 225–235.

[13] Pearn, W.L. & Chen, K.S. (1997). Capability indices for non-normal distributions with an application in electrolytic capacitor manufacturing, *Microelectronics and Reliability* **37**(12), 1853–1858.

[14] Tang, L.C. & Than, S.E. (1999). Computing process capability indices for non-normal data: a review and comparative study, *Quality and Reliability Engineering International* **15**, 339–353.

[15] Luceño, A. (1996). A process capability index with reliable confidence intervals, *Communications in Statistics: Simulation and Computation* **25**, 235–245.

[16] Carr, W.E. (1991). A new process capability index: parts per million, *Quality Progress* **24**, 152.

[17] Chao, M.T. & Lin, D.K.J. (2006). Another look at the process capability index, *Quality and Reliability Engineering International* **22**, 153–163.

[18] Constable, G.K. & Hobbs, J.R. (1992). Small samples and non-normal capability, *ASQC Quality Congress Transactions - Nashville* **46**, 37–43.

[19] Kaminsky, F.C., Dovich, R.A. & Burke, R.J. (1998). Process capability indices: now and in the future, *Quality Engineering* **10**, 445–453.

[20] Perakis, M. & Xekalaki, E. (2002). A process capability index that is based on the proportion of conformance, *Journal of Statistical Computation and Simulation* **72**, 707–718.

[21] Perakis, M. & Xekalaki, E. (2005). A process capability index for discrete processes, *Journal of Statistical Computation and Simulation* **75**, 175–187.

[22] Yeh, A.B. & Bhattacharya, S. (1998). A robust process capability index, *Communications in Statistics: Simulation and Computation* **27**, 565–589.

[23] Hollander, M. & Wolfe, D.A. (1999). *Nonparametric Statistical Methods*, 2nd Edition, John Wiley & Sons, New York, pp. 31–33.

[24] Flaig, J.J. (1996). A new approach to process capability analysis, *Quality Engineering* **9**, 205–211.

[25] Polansky, A.M. (1998). A smooth nonparametric approach to process capability, *Quality and Reliability Engineering International* **14**, 38–43.

[26] Rodriguez, R.N. (1992). Recent developments in process capability analysis, *Journal of Quality Technology* **24**, 176–187.

[27] Polansky, A.M. & Baker, E.R. (2000). Multistage plug-in bandwidth selection for kernel distribution function estimates, *Journal of Statistical Computation and Simulation* **65**, 63–80.

[28] Bowman, A., Hall, P. & Prvan, T. (1998). Bandwidth selection for the smoothing of distribution functions, *Biometrika* **85**, 799–808.

[29] Ciarlini, P., Gigli, A. & Regoliosi, G. (1999). The computation of accuracy of quality parameters by means of Monte Carlo simulation, *Communications in Statistics: Simulation and Computation* **28**, 821–848.

[30] Ding, J. (2004). A method of estimating the process capability index from the first four moments of non-normal data, *Quality and Reliability Engineering International* **20**, 787–805.

[31] Chen, H. (1994). A multivariate process capability index over a rectangular solid tolerance zone, *Statistica Sinica* **4**, 749–758.

[32] Yeh, A.B. & Chen, H. (2001). A non-parametric multivariate process capability index, *International Journal of Modelling and Simulation* **21**, 218–224.

[33] Polansky, A.M. (2001). A smooth nonparametric approach to multivariate process capability, *Technometrics* **43**, 199–211.

Related Articles

Process Capability Indices, Comparison of; **Process Capability Indices, Multivariate**; **Process Capability Indices, Nonnormal**; **Process Capability Indices, Skewed**; **Process Yield**.

ALAN M. POLANSKY

Process Capability Indices, Skewed

Introduction

The conventional process capability indices (PCIs) such as C_p, C_{pk}, C_{pm}, or C_{pmk} are usually determined under the assumption that quality characteristics follow a **normal distribution**. However, the normality assumption is usually difficult to justify and is often not appropriate in practice. For example, the measurements from drilling processes, coating processes, chemical processes, and semiconductor processes are often skewed.

In general, a process is considered to be capable if the process consistently produces items within given specifications. Therefore, many practitioners have utilized the PCIs, especially C_p or C_{pk}, in connection with the **process yield**. If a process does not follow a normal distribution, however, none of the conventional PCIs provide valid measures of the number of parts nonconforming. Table 1 gives the values of C_p as skewness (*see* **Skewness**) α_3 increases for Weibull, lognormal, and gamma distributions. The table also gives expected number of nonconforming items per million (NPM) and "equivalent $C_p(EC_p)$" calculated by $-(1/3)\Phi^{-1}((NPM/2) \times 10^{-6})$ since $NPM \times 10^{-6} = 2\Phi(-3C_p)$ under normality, where $\Phi(\cdot)$ is the standard normal distribution function. If the value of a PCI is close to that of the equivalent C_p, the PCI describes the process capability very well. In each case in the table, it is assumed that the lower and upper specification limits (LSL and USL) are -3 and 3, respectively, and the distribution is shifted and scaled to produce the same value of the process mean $\mu = 0$ and the standard deviation $\sigma = 1$, so that $C_p = 1$ for all cases. The

Table 1 C_p, NPM, and EC_p *versus* α_3

Distribution	α_3	NPM	EC_p	C_p
Normal	0.0	2700	1.00	1.00
Weibull	0.5	4227	0.95	1.00
	1.0	9870	0.86	1.00
	1.5	14 915	0.81	1.00
	2.0	18 316	0.79	1.00
	2.5	20 280	0.77	1.00
	3.0	21 256	0.76	1.00
Lognormal	0.5	5639	0.92	1.00
	1.0	10 461	0.85	1.00
	1.5	14 087	0.82	1.00
	2.0	16 358	0.80	1.00
	2.5	17 653	0.79	1.00
	3.0	18 325	0.79	1.00
Gamma	0.5	5431	0.93	1.00
	1.0	10 336	0.86	1.00
	1.5	14 782	0.81	1.00
	2.0	18 316	0.79	1.00
	2.5	20 856	0.77	1.00
	3.0	22 528	0.76	1.00

table shows that the NPM increases and the EC_p decreases as skewness increases. The C_p, however, does not reflect this phenomenon and overestimates the process capability. Therefore, the index could not be used to evaluate process capability if the distribution is skewed. For more detailed discussions on the effects of nonnormality (see Somerville and Montgomery [1] and Kotz and Lovelace [2]).

Skewed PCIs to remedy the shortcoming of the conventional PCIs can be divided into five main lines: (a) To use normalizing transformations, (b) to replace the unknown distribution by an empirical distribution or by a known three- or four-parameter distribution, (c) to modify the standard definition of PCIs in order to increase their robustness, (d) to construct PCIs with the estimate of the process yield, and (e) to develop heuristic methods adequate for skewed populations.

Normalizing Transformation Method

The normalizing transformation method is the simplest way to deal with nonnormal data, which transforms original process data to normal or close to normal with some mathematical functions so that the conventional PCIs can be used. For example, if a skewed distribution becomes normal by a square root transformation, the transformed data and specification limits can be used to estimate capability indices. Such transformations include Box–Cox power transformation (*see* **Box and Cox Transformation**) [1] or Johnson transformation [3]. The method is easy to understand, but it may be difficult to translate the results with transformed data back to the original scale and it is likely to require considerable work due to its trial and error nature for finding a proper transformation function.

Quantile Estimation Method

Since $\mu = X_{0.5}$, $\mu - 3\sigma = X_{0.00135}$, and $\mu + 3\sigma = X_{0.99865}$ for a normal distribution, where X_p is the pth quantile (*see* **Quantiles**), C_p and C_{pk} can be redefined as

$$C_p = \frac{USL - LSL}{X_{0.99865} - X_{0.00135}} \tag{1}$$

$$C_{pk} = \min\left\{\frac{USL - X_{0.5}}{X_{0.99865} - X_{0.5}}, \frac{X_{0.5} - LSL}{X_{0.5} - X_{0.00135}}\right\} \tag{2}$$

If a proper distribution can be specified, accurate measures of the three quantiles can be obtained. This involves modeling the process data with a probability model such as Weibull, lognormal, or gamma. If the variability pattern of the data cannot be described by a particular distribution, a flexible distribution approach or a nonparametric approach can be considered. The flexible distribution approach is to fit a family of distributions such as Burr or Pearson distribution to the data and use the quantiles of the fitted distribution. The nonparametric approach is to perform a nonparametric **estimation** of the quantiles. See Clements [4], Castagliola [5], and Ding [6] for the flexible distribution approach and Chang and Lu [7], Polansky [8], and **Process Capability Indices, Nonparametric**.

Clements [4] proposed to use Pearson distributions in estimating the quantiles, $X_{0.5}$, $X_{0.00135}$, and $X_{0.99865}$. These quantiles are calculated by

$$X_{0.5} = \overline{x} + z_{0.5}{\cdot}s \tag{3}$$

$$X_{0.00135} = \overline{x} + z_{0.00135}{\cdot}s \tag{4}$$

$$X_{0.99865} = \overline{x} + z_{0.99865}{\cdot}s \tag{5}$$

where z_p is the standardized quantile of the Pearson curve expressed as a function of skewness and **kurtosis** and can be obtained from the tables in Clements [4] and $\overline{x}$ and s are the sample mean and the sample standard deviation, respectively. The process capability can then be obtained by equations (1) and (2). Pearn and Kotz [9] applied the Clements' method to calculating second and third generation PCIs, C_{pm} and C_{pmk}, for nonnormal populations. The Clements' PCIs can be easily calculated since no complicated distribution fitting is required. Clements' approach, however, requires estimates of the third and fourth **moments** of the data and it is not possible to establish a functional relationship between Clements' C_p and process yield that can be applied to various distributions.

Shore [10] proposed a PCI with the family of distributions introduced by Shore [11] which can be fitted with the estimated first and second moments. Note that the third and fourth moments are needed to obtain Clements' PCIs. Based on the distribution, Shore [10] suggested to calculate the three quantiles by

$$X_{0.5} = M \tag{6}$$

$$X_{0.00135} = A_1 \cdot 0.001352^{B_1} \tag{7}$$

$$X_{0.99865} = A_2 \cdot \log(739.7) + B_2 \tag{8}$$

where M is the sample median and $\{A_i, B_i\}(i = 1, 2)$ are parameters of the assumed distribution determined by the modified two-moment procedure. The **standard error** of the estimator of the Shore's PCI is fairly small, but it does not perform as well as the Clements' PCI, partly because Shore uses only the first and second moments for distribution fitting.

The reliability of the capability estimates can only be assured when the data follow at least approximately the parametric distribution being used even if a flexible distribution is assumed. To avoid this weakness of the flexible distribution method, Polansky [8] used a nonparametric approach based on the kernel technique.

The quantile estimation method can be applied to a broad range of distributions, but the method is somewhat complicated and needs considerably large samples.

Robust PCI

Pearn *et al.* [12] introduced C_θ defined as

$$C_\theta = \frac{USL - LSL}{\theta\sigma} \tag{9}$$

where θ is chosen so that the probability

$$\Pr\left\{\mu - \frac{1}{2}\theta\sigma < X < \mu + \frac{1}{2}\theta\sigma\right\} \tag{10}$$

would be close to one and as independent as possible of the distribution. They recommended using $\theta = 5.15$ because the approximate relation

$$\Pr\{\mu - 2.575\sigma \le X \le \mu + 2.575\sigma\} \simeq 0.99 \tag{11}$$

varies only slightly for a number of distributions including χ^2 distributions with various **degrees of freedom**. This robustness is accomplished with the trade-off giving up the 2700 NPM for the normal population and replacing it by 10 000.

Rodriguez [13] proposed a PCI using the robust estimators of the process mean and dispersion as follows.

$$\tilde{C}_{pk} = \min\left\{\frac{USL - \tilde{X}}{4.45MAD}, \frac{\tilde{X} - LSL}{4.45MAD}\right\} \tag{12}$$

where $\tilde{X}$ is the median of a process and MAD is the median of $(|X_1 - \bar{X}|, \ldots, |X_n - \bar{X}|)$.

The robust method may be useful if there are insufficient data to provide accurate distribution fitting. The value of robust PCIs, however, may not coincide with that of conventional PCIs with "6σ" even for a normal distribution.

Process Yield Estimation Method

If $USL = \mu + k\sigma$ and $LSL = \mu - k\sigma$, a simple way to assess the process capability, is to obtain a nonconforming proportion like $p_{NC} = \Pr\{X < \mu - k\sigma \text{ or } X > \mu + k\sigma\}$, and transform the probability into the PCI using $C_p = -(1/3)\Phi^{-1}(p_{NC}/2)$. To obtain the limit of the probability for an arbitrary distribution, the Chebyshev inequality could be used, however, the limit is so broad that it is of little practical value. Flaig [14] proposed to tighten the probability bounds and estimate PCIs using an extension of the Camp–Meidell theorem. The Camp–Meidell inequality

$$\Pr\{X < \mu - k\sigma \text{ or } X > \mu + k\sigma\} \le \frac{1}{2.25k^2} \tag{13}$$

is applicable only if the underlying distribution is unimodal and symmetrical, but he found a way to overcome this limitation. The Camp–Meidell inequality leads to the calculation of the probability that the process produces individuals outside the specification limits. It cannot be viewed as a PCI in the strict sense [2].

Chao and Lin [15] proposed to construct a PCI with an estimate of the process yield using the relation between the process yield and the C_p used under normality;

$$C_p = \frac{1}{3}\Phi^{-1}\left[\frac{1}{2}\{F(USL) - F(LSL) + 1\}\right] \tag{14}$$

where $F(\cdot)$ is the distribution function of the process. They also suggested use of a nonparametric method for estimating $F(USL)$ and $F(LSL)$ like

$$\widehat{F}(x) = \frac{1}{n}\sum_{i=1}^{n}\Phi\left(\frac{x - X_i}{1.06 \cdot s \cdot n^{-1/5}}\right) \tag{15}$$

The nonparametric method for estimating the tail probabilities, however, does not provide accurate estimates when the sample size is small.

Heuristic Method

Wright [16] proposed a modification of C_{pmk} which incorporates a skewness term in the denominator to reduce the index value when nonsymmetry is present. Wright's index C_s is defined as

$$C_s = \frac{\min\{USL - \mu, \mu - LSL\}}{3\sqrt{\sigma^2 + (\mu - T)^2 + |\mu_3/\sigma|}} = \frac{(d - |\mu - T|)\sigma}{3\sqrt{1 + [(\mu - T)/\sigma]^2 + |\alpha_3|}} \quad (16)$$

where T is the target, $d = (USL - LSL)/2$, and α_3 is skewness. C_s decreases as skewness increases. The index, however, cannot be related to process yield.

Chang *et al.* [17] proposed PCIs for skewed populations based on a weighted standard deviation (WSD) approach, which improves the performance of the weighted variance (WV) PCIs suggested by Choi and Bai [18]. The WSD method utilizes different divisors at the upper and lower limits of the specification interval, and the PCIs are defined as

$$C_p^{\mathrm{WSD}} = \min\left\{\frac{USL - LSL}{6\sigma{\cdot}2P}, \frac{USL - LSL}{6\sigma{\cdot}2(1 - P)}\right\} \quad (17)$$

$$C_{pk}^{\mathrm{WSD}} = \min\left\{\frac{USL - \mu}{3\sigma{\cdot}2P}, \frac{\mu - LSL}{3\sigma{\cdot}2(1 - P)}\right\} \quad (18)$$

where $P = \Pr\{X \le \mu\}$. Wu *et al.* [19] proposed a new WV-based PCI calculated as

$$C_p^{\mathrm{WV}} = \frac{USL - LSL}{3(S_1 + S_2)} \quad (19)$$

where $S_j^2 = 2\sum_{i=1}^{n_j}(X_i - \bar{X})^2/(2n_j - 1)$, n_1 is the number of observations less than or equal to $\bar{X}$, and n_2 is that greater than $\bar{X}$. The WV and WSD methods are based on the idea that the asymmetric distribution can be divided into upper and lower parts from the process mean and the standard deviations above and below the mean can be calculated separately with the divided distributions.

The heuristic approach may not describe the capability of a process accurately. Therefore, if the type of a distribution is known or sufficient process data are available, experienced practitioners can estimate the capability fairly accurately with the data transformation or the curve fitting method. Earlier in the product life cycle, however, quality data are not sufficient. In such a case, the heuristic method may be adequate.

Multivariate PCI

Capability analyses involving more than one quality characteristic are sometimes of interests, and multivariate statistical techniques can be used to analyze several quality characteristics simultaneously. A difficulty of defining multivariate PCIs (*see* **Process Capability Indices, Multivariate**) in that there is no consensus on the methodology for assessing capability, which arises since the multivariate relationship among quality characteristics may or may not be reflected in the engineering specifications. In general, the upper and lower specification limits may be given for each quality characteristic and these specification ranges, taken together, form a rectangle or hypercube. Only a few methods applied to the multivariate skewed populations as well as multivariate normal populations have been suggested. For example, Polansky [20] proposed assessing the capability of a process using a nonparametric estimator based on a kernel estimate, and Chang and Bai [21] applied the WSD method to the modified region approach by Hubele *et al.* [22]. Further studies on multivariate PCIs for nonnormal populations are needed.

References

[1] Somerville, S.E. & Montgomery, D.C. (1996–1997). Process capability indices and non-normal distributions, *Quality Engineering* **9**, 305–316.

[2] Kotz, S. & Lovelace, C.R. (1998). *Process Capability Indices in Theory and Practice*, Arnold, New York.

[3] Polansky, A.M., Chou, Y.M. & Mason, R.L. (1999). An algorithm for fitting Johnson transformations to non-normal data, *Journal of Quality Technology* **31**, 345–350.

[4] Clements, J.A. (1989). Process capability calculations for non-normal distributions, *Quality Progress* **22**, 95–100.

[5] Castagliola, P. (1996). Evaluation of non-normal process capability indices using Burr's distribution, *Quality Engineering* **8**, 587–593.

[6] Ding, J. (2004). A method of estimating the process capability index from the first four moments of non-normal data, *Quality and Reliability Engineering International* **20**, 787–805.

[7] Chang, P. & Lu, K. (1994). PCI calculations for any shape of distribution with percentile, *Quality World, Technical Supplement* **20**, 110–114.

[8] Polansky, A.M. (1998). A smooth nonparametric approach to process capability, *Quality and Reliability Engineering International* **14**, 43–48.

[9] Pearn, W.L. & Kotz, S. (1994). Application of Clements' method for calculating second and third generation process capability indices for non-normal Pearsonian populations, *Quality Engineering* **7**, 139–145.

[10] Shore, H. (1998). A new approach to analyzing non-normal quality data with application to process capability analysis, *International Journal of Production Research* **36**, 1917–1933.

[11] Shore, H. (1995). Fitting a distribution by the first two sample moments (partial and complete), *Computational Statistics and Data Analysis* **19**, 563–577.

[12] Pearn, W.L., Kotz, S. & Johnson, N.L. (1992). Distributional and inferential properties of process capability indices, *Journal of Quality Technology* **24**, 216–231.

[13] Rodriguez, R.N. (1992). Recent developments in process capability analysis, *Journal of Quality Technology* **24**, 176–187.

[14] Flaig, J.J. (1996). A new approach to process capability analysis, *Quality Engineering* **9**, 205–211.

[15] Chao, M.T. & Lin, D.K.J. (2006). Another look at the process capability index, *Quality and Reliability Engineering International* **22**, 153–163.

[16] Wright, P.A. (1995). A process capability index sensitive to skewness, *Journal of Statistical Computation and Simulation* **52**, 195–203.

[17] Chang, Y.S., Choi, I.S. & Bai, D.S. (2002). Process capability indices for skewed populations, *Quality and Reliability Engineering International* **18**, 383–393.

[18] Choi, I.S. & Bai, D.S. (1996). Process capability indices for skewed populations, *Proceedings 20th International Conference on Computer and Industrial Engineering*, Kyongju, Korea, pp. 1211–1214.

[19] Wu, H.H., Swain, J.J., Farrington, P.A. & Messimer, S.L. (1999). A weighted variance capability index for general non-normal processes, *Quality and Reliability Engineering International* **15**, 397–402.

[20] Polansky, A.M. (2001). A smooth nonparametric approach to multivariate process capability, *Technometrics* **43**, 199–211.

[21] Chang, Y.S. & Bai, D.S. (2003). Multivariate process capability indices for skewed populations with weighted standard deviations, *Journal of the Korean Institute of Industrial Engineers* **29**, 114–125.

[22] Hubele, N.F., Shahriari, H. & Cheng, C.S. (1992). A bivariate process capability vector, in *Statistical Process Control in Manufacturing*, J.B. Keats & D.C. Montgomery, eds, Marcel Dekker, New York, pp. 299–310.

Young Soon Chang and Do Sun Bai

Process Capability Plots

Introduction

When measuring the capability of a manufacturing process, some form of process capability index is often used (*see* **Process Capability Indices, Comparison of**). Vännman [1] unified the most common indices by a family of capability indices, defined as

$$C_p(u,v) = \frac{d - u|\mu - T|}{3\sqrt{\sigma^2 + v(\mu - T)^2}} \tag{1}$$

where $u \geq 0$, $v \geq 0$, μ is the process mean and σ is the process standard deviation of the in-control process. In equation (1) the target value T equals the midpoint of the specification interval [*LSL, USL*]. Furthermore, $d = (USL - LSL)/2$, i.e. half the length of the specification interval. Setting $u = 0$ or 1 and $v = 0$ or 1 in equation (1) we obtain $C_p(0,0) = C_p$, $C_p(1,0) = C_{pk}$, $C_p(0,1) = C_{pm}$, and $C_p(1,1) = C_{pmk}$. When using process capability indices, a process is defined to be capable if the process capability index exceeds a certain threshold value $k > 0$. Some commonly used values are $k = 1$, 4/3, 5/3.

Assume that we use an index defined by the family $C_p(u,v)$ in equation (1) and define the process to be capable if $C_p(u,v) > k$, for given values of u, v, and k. Different choices of u, v, and k impose different restrictions on the process parameters (μ, σ). This is easily seen in a process capability plot, which is a contour plot of $C_p(u,v) = k$ as a function of μ and σ, or as a function of μ_t and σ_t, where

$$\mu_t = \frac{\mu - T}{d} \quad \text{and} \quad \sigma_t = \frac{\sigma}{d} \tag{2}$$

The contour curve is obtained by rewriting the index in equation (1) as a function of μ_t and σ_t, solving the equation $C_p(u,v) = k$ with respect to σ_t and then plotting σ_t as a function of μ_t. We easily find that $C_p(u,v) = k$ is equivalent to

$$\sigma_t = \sqrt{\frac{(1 - u|\mu_t|)^2}{9k^2} - v\mu_t^2},$$

$$|\mu_t| \leq \frac{1}{u + 3k\sqrt{v}}, \quad (u,v) \neq (0,0) \tag{3}$$

When $u = v = 0$, i.e. when considering the index $C_p = k$, we have

$$\sigma_t = \frac{1}{3k}, \quad |\mu_t| \le 1 \tag{4}$$

The reason for making the contour plot a function of (μ_t, σ_t) instead of a function of (μ, σ) is to obtain a plot where the scale is invariable, irrespective of the values of the specification limits. This is useful when considering several different characteristics (see the section titled "Several Characteristics in the Same Plot").

To obtain the expression for σ_t in the case of C_{pk}, we let $u = 1$ and $v = 0$ in equation (3) and find that the contour curve is composed of two straight lines. To obtain the expression for σ_t in the case of C_{pm}, we let $u = 0$ and $v = 1$ in equation (3) and find that the contour curve is a semicircle (see Figure 1).

Values of the process parameters μ and σ which give (μ_t, σ_t) values inside the region bounded by the contour curve $C_p(u, v) = k$ and the μ_t axis will give rise to a $C_p(u, v)$ value larger than k, i.e. a capable process. We call this region the capability region. Furthermore, values of μ and σ which give (μ_t, σ_t) values outside this region will give a $C_p(u, v)$ value smaller than k, i.e. a non-capable process.

Note that the process capability plot is based on the process parameters, which are usually unknown in practice. Hence, this plot is a theoretical process capability plot useful to understand the restrictions the index puts on the process parameters. But it cannot immediately be used to deem if a process is capable, since it does not take the randomness of the estimators of the process parameter into account. The following sections discuss how to deal with the case in which the process parameters μ and σ are unknown.

Estimated Process Capability Plots

The process capability plot can now be generalized to an estimated process capability plot, which can be used to judge if a process is capable, when the process parameters μ and σ are unknown and need to be estimated. We assume that the studied characteristic of the process is normally distributed. Let $X_1, X_2, \ldots, X_n$ be a random sample from a normal distribution with mean μ and variance σ^2 as the parametric process quality indicators.

To obtain an appropriate decision rule we consider a hypothesis test with the null hypothesis H_0: $C_p(u, v) \le k_0$ and the alternative hypothesis H_1: $C_p(u, v) > k_0$. As a test statistic we use the estimator $\widehat{C}_p(u, v)$, which is obtained by estimating μ and σ^2 with their maximum-likelihood estimators, i.e.

$$\hat{\mu} = \overline{X} = \frac{1}{n}\sum_{i=1}^{n} X_i \quad \text{and} \quad \hat{\sigma}^2 = \frac{1}{n}\sum_{i=1}^{n}(X_i - \overline{X})^2 \tag{5}$$

The null hypothesis is rejected, and the process is deemed capable, whenever $\widehat{C}_p(u, v) > c_\alpha$, where the constant $c_\alpha > k_0$ is determined so that the **significance level** of the test is α. Details of how to obtain the critical value c_α can be found, e.g. in Vännman [2, 3]. Hubele and Vännman [4] showed that, when

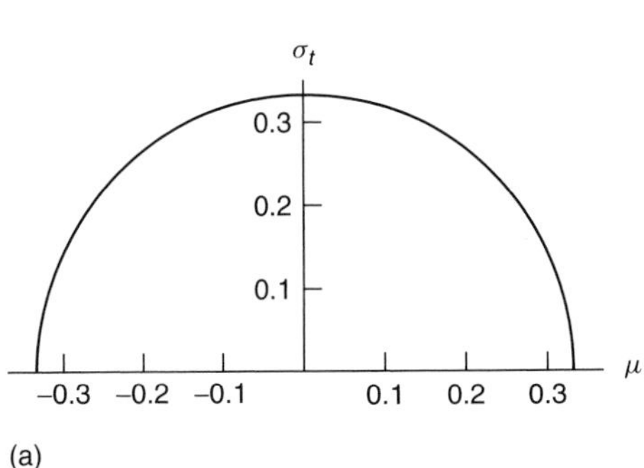

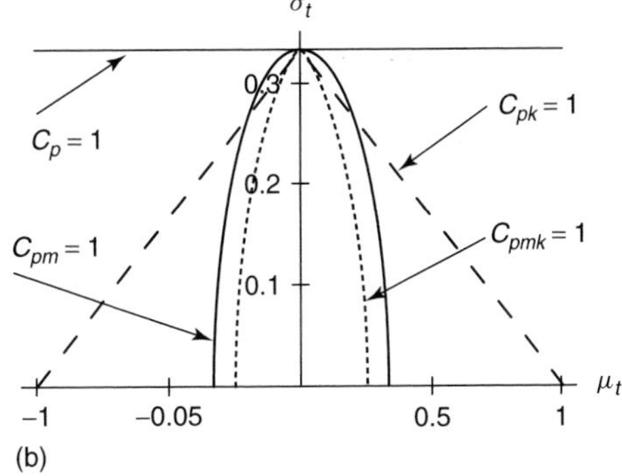

Figure 1 (a) The process capability plot for $C_{pm} = 1$. (b) The contour curves for the process capability indices C_p, C_{pk}, C_{pm}, and C_{pmk} when $k = 1$. The region bounded by the contour curve and the μ_t axis is the corresponding capability region

using the capability index C_{pm}, the critical value at significance level α is obtained as

$$c_\alpha = k_0\sqrt{n/\chi^2_{\alpha,n}} \tag{6}$$

where $\chi^2_{\alpha,n}$ is the αth quantile from a χ^2 distribution with n **degrees of freedom**. When using the capability index C_{pk} it can be shown that the critical value at significance level α is obtained from the equation

$$\int_{-\infty}^{3k_0\sqrt{n}} F_\xi\left(\frac{(3k_0\sqrt{n}-z)^2}{9c_\alpha^2}\right) f_\eta(z)\,\mathrm{d}z = \alpha \tag{7}$$

where F_ξ denotes the cumulative distribution function of a central χ^2 distribution with $n-1$ degrees of freedom and f_η denotes the probability density function of the standardized normal distribution $N(0,1)$. For arbitrary values of $u \geq 0$ or $v \geq 0$ the critical value c_α is determined so that the probability $P(\widehat{C}_p(u,v) > c_\alpha)$ is at most α for all values of (μ_t, σ_t) such that $C_p(u,v) = k_0$, where

$$P(\widehat{C}_p(u,v) > c_\alpha) = \int_{\frac{-\sqrt{n}/\sigma_t}{u+3c_\alpha\sqrt{v}}}^{\frac{\sqrt{n}/\sigma_t}{u+3c_\alpha\sqrt{v}}} F_\xi\left(\frac{(\sqrt{n}/\sigma_t - u|t|)^2}{9c_\alpha^2} - vt^2\right) \times f_\eta\left(t - \frac{\mu_t\sqrt{n}}{\sigma_t}\right)\mathrm{d}t \tag{8}$$

MATLAB codes for solving equation (7), as well as for determining the critical value c_α for arbitrary values of u and v using equation (8), can be obtained from the author.

The estimated process capability plot is then obtained by replacing μ, σ, and k in equations (2–4) with $\hat{\mu}$, $\hat{\sigma}$, and c_α, respectively, and plotting the contour curve $\widehat{C}_p(u,v) = c_\alpha$ as a function of $\hat{\mu}_t$ and $\hat{\sigma}_t$. We can now use the estimated process capability plot to define an estimated capability region as the region bounded by the contour curve $\widehat{C}_p(u,v) = c_\alpha$ and the $\hat{\mu}_t$ axis. To determine if a process can be deemed capable we first calculate

$$\hat{\mu}_t = \frac{\hat{\mu}-T}{d} \quad \text{and} \quad \hat{\sigma}_t = \frac{\hat{\sigma}}{d} \tag{9}$$

Then the point $(\hat{\mu}_t, \hat{\sigma}_t)$ is plotted in the estimated process capability plot. If this point is inside the estimated capability region, the process will be considered capable at significance level α.

To illustrate the estimated process capability plot we use an example from Vännman [2], where the process is defined capable if $C_{pm} > 1$ and the significance level is $\alpha = 5\%$. The example consists of two datasets, each of size 80, from a process where $T = 8.70$ and $d = 0.24$. When $n = 80$ and $k_0 = 1$ we get the critical value $c_{0.05} = 1.151$ from equation (6). First a random sample was taken from a process, giving rise to estimates of μ and σ being equal to $\hat{\mu} = 8.6234$ and $\hat{\sigma} = 0.0519$, respectively. Later, a second random sample was taken from the same process, after it had been adjusted. Then, from the second sample, the obtained estimates of μ and σ were equal to $\hat{\mu} = 8.6628$ and $\hat{\sigma} = 0.0516$, respectively.

On the basis of the sample data, we find the estimated values of μ_t and σ_t to be $\hat{\mu}_t = (\hat{\mu} - T)/d = -0.319$ and $\hat{\sigma}_t = \hat{\sigma}/d = 0.216$ for the first sample, and $\hat{\mu}_t = -0.155$ and $\hat{\sigma}_t = 0.215$ for the second sample. We then plot the contour curve $\widehat{C}_p(u,v) = c_\alpha = 1.151$ together with the points with coordinates $(-0.319, 0.216)$ and $(-0.155, 0.215)$, for the first and second sample, respectively (see Figure 2).

We can see that for the first sample, the point $(-0.319, 0.216)$ is outside the estimated capability region and hence the process cannot be considered capable. For the second sample, the point $(-0.155, 0.215)$ is now inside the estimated capability region and hence the process can be deemed capable at the 5% significance level.

What is obvious from the graphical method used here is that the non-capability in Figure 2(a) is, to a large extent, due to the deviation of the process mean from the target value. By looking at the estimated process capability plot we get information instantly about both the deviation from the target value and the process spread as well as information about the capability. We can also see how much the deviation from target must be reduced in order to achieve a capable process.

Safety Regions

As an alternative to the estimated process capability plot, we can use the theoretical process capability plot together with a safety region. A safety region is

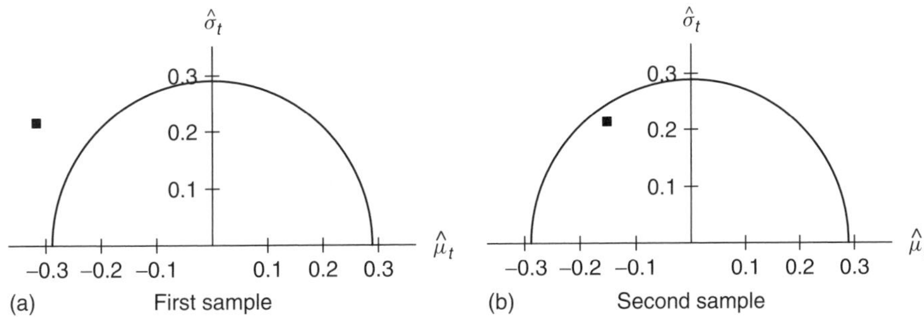

Figure 2 The estimated process capability plot defined by $\widehat{C}_{pm} = c_{0.05} = 1.151$. In (a) the first sample with the observed value of $(\hat{\mu}_t, \hat{\sigma}_t)$ outside the estimated capability region, illustrates a non-capable process. In (b) the second sample, with observed value of $(\hat{\mu}_t, \hat{\sigma}_t)$ inside the estimated process capability region, illustrates a capable process

a region plotted in the theoretical process capability plot for drawing conclusions about process capability at a given significance level. It is built around the estimated point $(\hat{\mu}_t, \hat{\sigma}_t)$ to take the randomness of the estimator $(\hat{\mu}_t, \hat{\sigma}_t)$ into account. The size of the region is determined in such a way that it corresponds to significance level α.

Vännman [3] suggests putting a circular safety region around the estimated point $(\hat{\mu}_t, \hat{\sigma}_t)$. The circular safety region is defined as the region bounded by the circle

$$(\delta - \hat{\mu}_t)^2 + (\gamma - \hat{\sigma}_t)^2 = R^2\hat{\sigma}_t^2 \tag{10}$$

The constant R depends on the sample size n and the significance level α. The process is deemed capable if the whole circular safety region in equation (11) is inside the capability region in the theoretical process capability plot defined by $C_p(u, v) = k_0$. The constant R is chosen so that the probability that the circular safety region is inside the capability region, given that $C_p(u, v) = k_0$, is at most equal to α. When using the index C_{pm} to define a capable process, the probability $\Pi(\mu_t, k_0)$ that the circular region is inside the capability region defined by $C_{pm} = k_0$, given that $C_{pm} = k_0$, can be expressed as follows.

$$\Pi(\mu_t, k_0) = \int_0^{Q(\mu_t,k_0)} F_\tau((S(\mu_t, k_0) - R\sqrt{y})^2 - y) f_\xi(y)\,\mathrm{d}y \tag{11}$$

where

$$S(\mu_t, k_0) = \sqrt{\frac{n}{1-9k_0^2\mu_t^2}} \quad \text{and}$$

$$Q(\mu_t, k_0) = \frac{S(\mu_t, k_0)^2}{(R+1)^2} \tag{12}$$

F_τ denotes the cumulative distribution function of a noncentral χ^2 distribution with 1 degree of freedom and noncentrality parameter λ, where λ is

$$\lambda = \frac{(\mu - T)^2 n}{\sigma^2} = \frac{\mu_t^2 n}{\sigma_t^2} \tag{13}$$

f_ξ denotes the probability density function of a central χ^2 distribution with n -1 degrees of freedom. The probability in equation (11) is defined for $|\mu_t| < 1/(3k)$. The constant R is determined so that the probability $\Pi(\mu_t, k_0)$ in equation (11) is at most α for all values of μ_t such that $C_{pm} = k_0$. MATLAB codes for determining the R-value using equation (11) can be obtained from the author.

If we apply this circular safety region to the previous example, where $n = 80$ and $\alpha = 0.05$ and the process is defined as capable when $C_{pm} > 1$, we find $R = 0.1903$. In Figure 3 we see the circular regions plotted. The conclusions about the process capability are the same as before.

Deleryd and Vännman [5] suggest a rectangular safety region. This region is, however, less powerful than the circular safety region.

Several Characteristics in the Same Plot

Several characteristics of a process can be monitored in the same plot and at the same time the information on the location and spread of the process is retained. As an example consider the following situation.

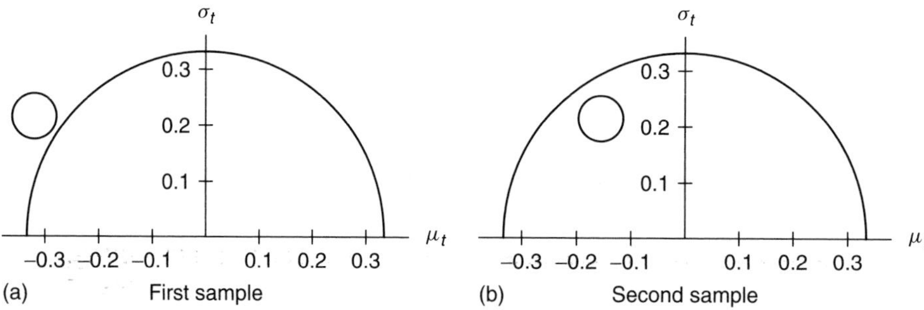

Figure 3 The process capability plot defined by $C_{pm} = 1$ together with the circular region. In (a) the process cannot be considered capable, but in (b) the whole circle is inside the capability region and hence the process is considered capable at 5% significance level

Three different quality characteristics A, B, and C are of interest. They have the following different specification intervals: A: [200, 210]; B: [12.0, 16.0]; C: [1.40, 2.20], with target values equal to the midpoints of each interval. They are assumed to be normally distributed and in statistical control. The process will be considered capable when all three characteristics are considered capable according to the definition that $C_{pm} > 1$.

A sample of size 50 is taken from each of the three characteristics and the point $(\hat{\mu}_t, \hat{\sigma}_t)$ is calculated for each of them. The decision rule is that the process will be considered capable if all three points $(\hat{\mu}_t, \hat{\sigma}_t)$ fall inside the estimated capability region, where the significance level for the estimated capability curve is 1%. Then, according to Bonferroni's inequality, the significance level for the decision rule will be at most 3%. The critical value, when $n = 50$ and $\alpha = 0.01$, is $c_{0.01} = 1.199$ according to equation (6). In Figure 4 we can see the results having the following observed values of $(\hat{\mu}_t, \hat{\sigma}_t)$: A: (0.05, 0.20); B: (0.25, 0.24); C: (−0.15, 0.17).

In Figure 4 we cannot consider the process to be capable, since characteristic B falls outside the semicircle. Figure 4 shows that from one single plot we get, for each of the three characteristics A, B, and C, information about the deviation from target and the standard deviation, as well as information on how to adjust the process to make it capable, if it is not capable.

When monitoring several characteristics in the estimated process capability plot, the same sample size is needed for each characteristic. With different sample sizes, the safety regions can be used instead. Figure 5 shows an example of this situation. In Figure 5(a) the sample size for characteristic A is 50, for B is 30, and for C is 70. Figure 5(b) illustrates the same example as in Figure 4.

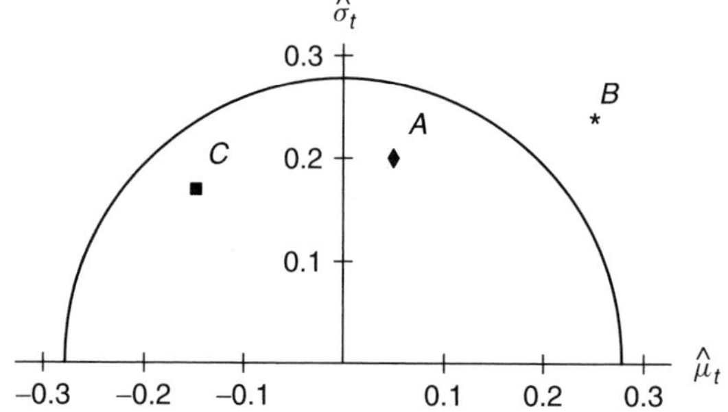

Figure 4 The estimated process capability plot defined by $\widehat{C}_{pm} = c_{0.01} = 1.199$. A, B, and C indicate the observed value of $(\hat{\mu}_t, \hat{\sigma}_t)$ for the three characteristics A, B, and C, respectively

Conclusions

The graphical methods discussed here are efficient in the respect that in one single plot we get visual information simultaneously about the location and spread of the studied characteristic, as well as information about the capability of the process at a given significance level. We can at a glance relate the deviation from target and the spread to each other and to the capability index in such a way that we are able to see whether the non-capability is caused by the fact that the process output is off target, or that the process spread is too large, or if the result is a combination of these two factors. Furthermore, we can easily see

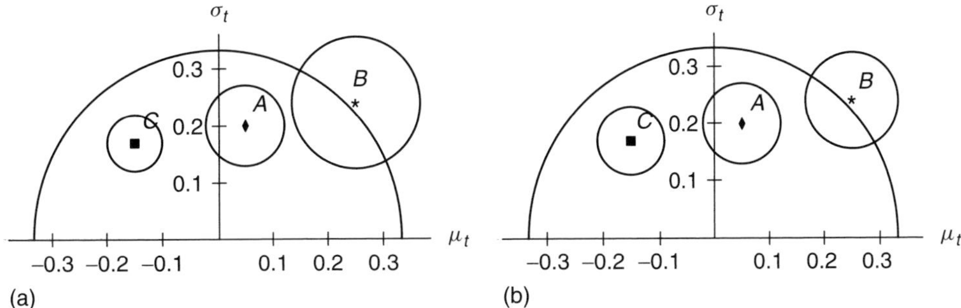

Figure 5 Process capability plots with circular regions for each of characteristics A, B, and C, each at 1% significance level. In (a) the sample sizes differ. In (b) all have sample size 50. The observed values of $(\hat{\mu}_t, \hat{\sigma}_t)$ are the same in (a) and (b)

how large a change is needed to obtain a capable process. For more details regarding the different plots, see Vännman [2, 3] and Deleryd and Vännman [5].

References

[1] Vännman, K. (1995). A unified approach to capability indices, *Statistica Sinica* **5**, 805–820.

[2] Vännman, K. (2001). *A Graphical Method to Control Process Capability, Frontiers in Statistical Quality Control, No 6*, H.-J. Lenz & P.-T.H. Wilrich, eds, Physica-Verlag, Heidelberg, pp. 290–311.

[3] Vännman, K. (2006). Safety regions in process capability plots, *Quality Technology and Quantitative Management* **3**, 227–246.

[4] Hubele, N. & Vännman, K. (2004). The effect of pooled and un-pooled variance estimators on C_{pm} when using subsamples, *Journal of Quality Technology* **36**, 207–222.

[5] Deleryd, M. & Vännman, K. (1999). Process capability plots–a quality improvement tool, *Quality and Reliability Engineering International* **15**, 213–227.

KERSTIN VÄNNMAN

Process Capability, Error in Estimation of

Introduction

The inevitable variations in process measurements come from two sources: the manufacturing process and the gauge. Gauge capability reflects the gauge's precision, or lack of variation, but is not the same as **calibration**, which assures the gauge's accuracy. Process capability measures the ability of a manufacturing process to meet preassigned specifications. Nowadays, many customers use process capability to judge a supplier's ability to deliver quality products. Suppliers ought to be aware of how gauges affect various process capability estimates.

The gauge capability includes two components of **measurement systems** variability: repeatability and reproducibility. Repeatability represents the variability from the gauge or measurement instrument when it is used to measure the same specimen (with the same operator or setup or in the same time period). Reproducibility reflects the variability arising from different operators, setups, time periods, or in general, different conditions. These studies are often referred to as *gauge repeatability and reproducibility* (GR&R) studies. To summarize we have

$$\begin{aligned}\sigma^2_{\text{measurementerror}} &= \sigma^2_{\text{gauge}} \\ &= \sigma^2_{\text{repeatability}} + \sigma^2_{\text{reproducibility}} \qquad (1)\end{aligned}$$

Estimates for $\sigma^2_{\text{repeatability}}$ and $\sigma^2_{\text{reproducibility}}$ come from a gauge study, or a GR&R study. Barrentine [1], Levinson [2], and Montgomery [3, 4], among others, provided procedures for GR&R studies.

Gauge capability is a gauge's ability to repeat and reproduce measurements. Its measurement is the percentage of tolerance consumed by (gauge) capability. Montgomery [4] referred to it as the *precision-to-tolerance* (or P/T) ratio. It is the ratio of the gauge's variation to the specification width; its smaller numbers are of course preferable. Denoting the gauge's standard deviation by σ_{gauge}, we have

$$P/T = \frac{6\sigma_{\text{gauge}}}{USL - LSL} \times 100\% \quad (2)$$

where *USL* and *LSL* represent the upper and lower specification limits, respectively. Some authors and practitioners use the coefficient 5.15 instead of 6 (see, e.g. Barrentine [1] and Levinson [2, 5]). This formula uses 6σ as the natural tolerance width for the gauge, based on and motivated by the **normal distribution** assumptions. Values of the estimated ratio P/T of 0.1 or less often are taken to imply adequate gauge capability. This is based on the generally used rule that requires a measurement device to be calibrated in units one-tenth as large as the accuracy required in the final measurement.

GR&R studies focus on quantifying the measurement errors. Common approaches to GR&R studies, such as the range method [6] and the **ANOVA** (analysis of variance) method [7, 8], assume that the distribution of the measurement errors is normal with a mean error of zero. Let the measurement errors be described by a random variable $M \sim N(0, \sigma_M^2)$, Montgomery and Runger [6] presented the gauge capability λ by means of the formula:

$$\lambda = \frac{6\sigma_M}{USL - LSL} \times 100\% \quad (3)$$

For the measurement system to be deemed acceptable, the variability in the measurements due to the measurement system ought to be less than a predetermined percentage of the engineering tolerance. The recommended guidelines for gauge acceptance are presented in the Automotive Industry Action Group (AIAG) Measurement Systems Analysis (**MSA**) manual. The purpose of a GR&R study is to determine if the variability of the measurement system is small relative to the variability of the monitored process. Several commonly reported ratios (functions of parameters of interest) in GR&R studies are discussed in Burdick *et al.* [9]. Burdick *et al.* [9] also provided a review of methods for constructing measurement systems capability studies with emphasis on the ANOVA approach to analyze the results. More recently, Burdick *et al.* [10] applied the generalized inference methods to construct **confidence intervals** for assessing the ability of a measurement system to correctly classify parts in a GR&R study. The reported simulation suggests that, for appropriately sized experiments, the proposed method provides two-sided intervals and upper bounds that generally maintain the stated confidence level.

Process capability indices (PCIs) are convenient and powerful tools for measuring process performance. In recent years, PCIs have received considerable research attention in the quality assurance and statistical literature. Those indices quantify process performance by taking into consideration process location, process variation, and manufacturing specifications, which reflect process consistency, process accuracy, **process yield**, and process loss. A substantial majority of capability research works that appeared in the literature do not take into account gauge measurement errors. However, the gauge measurement errors have a significant effect on estimating and assessing process capability since measurement systems are not 100% precise. An inaccurate measurement system can thwart all the benefits of improvement endeavors resulting in poor quality. Analyzing process capability without considering gauge capability may often lead to unreliable decisions. It could result in a serious loss to producers if gauge capability is not being considered in process capability **estimation** and testing. On the other hand, improving the gauge measurements and having properly trained operators can reduce the measurement errors. Since measurement errors unfortunately cannot be avoided, using appropriate confidence coefficients and power becomes an essential task. However, the reality is that no measurement is free from error or uncertainty even if it is carried out with the aid of highly sophisticated and precise measuring instruments. Montgomery and Runger [6] pointed out that the quality of the data related to the process characteristics relies substantially on the gauge. Any variation in the measurement process has a direct impact on the ability to execute sound judgment about the manufacturing process. Conclusions about capability of a process based only on the single numerical value of the index are not reliable. Analyzing the effects of measurement errors on PCIs, Mittag [11, 12] and Levinson [2] developed very definitive techniques for quantifying the percentage error in PCIs estimation in the pence of measurement errors. Bordignon and Scagliarini [13] studied the inferential properties of the estimators of C_p and C_{pk} when observations are contaminated by measurement errors, while Scagliarini [14] analyzed the properties of the estimator of C_p for autocorrelated data in the presence of measurement errors. Bordignon and Scagliarini [15] studied the properties of the estimator of C_{pm} when observations are affected by measurement errors. They compared the performances of the estimator on the error case with those of the estimator in the error-free case. The results indicated that the presence of measurement errors in the data leads to different behavior of the estimator according to the entity of the error variability.

Estimating Process Capability with Measurement Errors

Most research works related to PCIs are carried out under the assumption of no gauge measurement errors. Unfortunately, such an assumption does not adequately accommodate real-world situations even with modern, highly sophisticated measuring instruments and devices. In this section, we have emphasized constructing confidence intervals and testing procedures on the most widely used index C_{pk} in presence of gauge measurement errors.

Let $X \sim N(\mu, \sigma^2)$ represent the relevant quality characteristic of a manufacturing process and C_{pk} measure the "true" process capability. However, in practice, we deal with the observed variable Y rather than with the "true" variable X. Assume that X and M (the measurement error) are stochastically independent. In this case, we have $Y \sim N(\mu_Y = \mu, \sigma_Y^2 = \sigma^2 + \sigma_M^2)$, and the empirical process capability index C_{pk}^Y will be obtained after substituting σ_Y for σ. The relationship between the true process capability $C_{pk} = \min\{(USL - \mu)/3\sigma, (\mu - LSL)/3\sigma\} = d - |\mu - m|/(3\sigma)$ and the empirical process capability C_{pk}^Y can be expressed as follows:

$$\frac{C_{pk}^Y}{C_{pk}} = \frac{1}{\sqrt{1 + \lambda^2 C_p^2}} \tag{4}$$

where $d = (USL - LSL)/2$ refers to the half-length of the allowable tolerance of the process, $m = (USL + LSL)/2$ is the midpoint of the specification interval, $C_p = d/3\sigma$, and λ is the gauge capability index.

Since the variation of the observed data is larger than the variation of the original one, the denominator of the index C_{pk} becomes larger, and the true capability of the process will be understated if calculations of PCIs are based on the empirical data Y (the observed measurement contaminated with errors). Suppose that the empirical data $\{Y_i, i = 1, 2, \ldots, n\}$ is available, then the natural estimator

$$\hat{C}_{pk}^Y = \frac{d - |\overline{Y} - m|}{3S_Y} \tag{5}$$

obtained by replacing the process mean μ and the process standard deviation σ by their conventional estimators $\overline{Y} = \sum_{i=1}^n Y_i/n$ and $S_Y = \left[\frac{\sum_{i=1}^n (Y_i - \overline{Y})}{(n-1)}\right]^{1/2}$, from a stable process. Accordingly, applying the technique used in Kotz and Johnson [16] and Pearn and Lin [17], the cumulative distribution function (**CDF**) of $\hat{C}_{pk}^Y$ is obtained as follows:

$$F_{\hat{C}_{pk}^Y}(x) = 1 - \int_0^{3C_p^Y\sqrt{n}} G\left(\frac{(n-1)(3C_p^Y\sqrt{n} - t)^2}{9nx^2}\right) \times \phi[t + 3(C_p^Y - C_{pk}^Y)\sqrt{n}] + \phi[t - 3(C_p^Y - C_{pk}^Y)\sqrt{n}]\,\mathrm{d}t \tag{6}$$

where $C_p^Y = \frac{C_p}{\sqrt{1+\lambda^2 C_p^2}}$, $C_{pk}^Y = \frac{C_{pk}}{\sqrt{1+\lambda^2 C_p^2}}$, $\phi(\cdot)$ is the probability density function (PDF) of the standard normal distribution, and, as above, $\lambda = 6\sigma_M/(USL - LSL) \times 100\%$.

To determine whether a given process meets the preset capability requirement, we could consider the statistical testing with the null hypothesis $H_0 : C_{pk} \le c$ (the process is not capable) and the alternative hypothesis $H_1 : C_{pk} > c$ (the process is capable), where c is the required capability level. If the calculated process capability is greater than the corresponding critical value, we reject the null hypothesis and conclude that the process is capable.

Discussions in Pearn and Liao [18] indicated that the true process capability would be severely underestimated if $\hat{C}_{pk}^Y$ is used. The probability that $\hat{C}_{pk}^Y$ is greater than c_0 would be less than when using $\hat{C}_{pk}$. Thus, the α risk of using $\hat{C}_{pk}^Y$ is less than the α risk of using $\hat{C}_{pk}$ when estimating C_{pk}. The power of the test based on $\hat{C}_{pk}^Y$ is also less than that of the one based on $\hat{C}_{pk}$; that is, the α risk and the power of the test decrease when the gauge measurement error increases. Since the lower confidence bound is severely underestimated and the power becomes small, producers cannot firmly state that their processes meet the capability requirement even if their processes are indeed sufficiently capable. Adequate and even superior product units could be incorrectly rejected in this case. To remedy the situation, Pearn and Liao [18] considered the adjustment of the confidence bounds and of the critical values to provide a superior capability assessment. Suppose that the desired confidence coefficient is θ, the adjusted confidence interval of C_{pk} with the lower confidence bound L^*, can be shown to be

$$\begin{aligned}\theta &= P(C_{pk} > L^*)\\ &= 1 - \int_0^{b^*\sqrt{n}} G\left(\frac{(n-1)(b^*\sqrt{n}-t)^2}{9n(\hat{C}^Y_{pk})^2}\right)\\ &\quad \times [\phi(t+\xi^Y\sqrt{n})+\Phi(t-\xi^Y\sqrt{n})]\,\mathrm{d}t \qquad (7)\end{aligned}$$

where $b^* = \frac{3L^*}{\sqrt{1+\lambda^2 C_p^2}} + |\xi^Y|$ and $\xi^Y = 3(C^Y_p - C^Y_{pk})$. To eliminate the need for estimating ξ^Y, Pearn and Liao [18] followed the technique of Pearn and Shu [19] (by setting $\xi^Y = 1.00$) to find the adjusted lower confidence bound L^*, where C_p can be obtained from the equation $\frac{3(C_p - L^*)}{\sqrt{1+\lambda^2 C_p^2}} = 1.00$ as

$$\begin{aligned}C_p &= \frac{18L^* + \sqrt{324L^{*^2} - 4(9-\lambda^2)(9L^{*^2}-1)}}{2(9-\lambda^2)}\\ &= C_{p1} \qquad (8)\end{aligned}$$

To improve the power of the test, revised critical values c^*_0 can be determined from

$$\begin{aligned}\alpha^* &= P(\hat{C}^Y_{pk} \geq c^*_0 | C_{pk} = c)\\ &= \int_0^{3C^Y_p\sqrt{n}} G\left(\frac{(n-1)(3C^Y_p\sqrt{n}-t)^2}{9n(c^*_o)^2}\right)\\ &\quad \times \phi[t+3(C^Y_p - C^Y_{pk})\sqrt{n}]\\ &\quad + \phi[t-3(C^Y_p - C^Y_{pk})\sqrt{n}]\,\mathrm{d}t \qquad (9)\end{aligned}$$

where α^* is the α risk corresponding to the test using $\hat{C}^Y_{pk}$ and c^*_0, $C^Y_p = \frac{C_p}{\sqrt{1+\lambda^2 C_p^2}}$, $C^Y_{pk} = \frac{c}{\sqrt{1+\lambda^2 C_p^2}}$, and c is the capability requirement. To eliminate the need for further estimation of the characteristic parameter C_p, we set $C^Y_p = C^Y_{pk} + 1/3$ as suggested by Pearn and Lin [17], and find a reliable adjusted critical value c^*_0, where C_p can be obtained using the equation $\frac{C_p}{\sqrt{1+\lambda^2 C_p^2}} = \frac{c}{\sqrt{1+\lambda^2 C_p^2}} + 1/3$ as

$$\begin{aligned}C_p &= \frac{18c + \sqrt{324c^2 - 4(9-\lambda^2)(9c^2-1)}}{2(9-\lambda^2)}\\ &= C_{p2} \qquad (10)\end{aligned}$$

To ensure that the α risk is within the preset magnitude, let $\alpha^* = \alpha$ and solve the above equation to obtain c^*_0. The revised power of the test (denoted by π^*) can be calculated as

$$\begin{aligned}\pi^*(C_{pk}) &= P(\hat{C}^Y_{pk} \geq c_0 | C_{pk}, C_p = C_{p2})\\ &= \int_0^{3C^Y_p\sqrt{n}} G\left(\frac{(n-1)(3C^Y_p\sqrt{n}-t)^2}{9n(c^*_o)^2}\right)\\ &\quad \times [\phi(t+\sqrt{n}) + \Phi(t-\sqrt{n})]\,\mathrm{d}t \qquad (11)\end{aligned}$$

where $C^Y_p = C^Y_{pk} + 1/3$ and $C^Y_{pk} = \frac{C_{pk}}{\sqrt{1+\lambda^2 C_{p2}^2}}$. Pearn and Liao [18] adjusted the confidence intervals and the critical values in order to ensure that the intervals possess the desired confidence coefficients and will improve the power of the test with an appropriate α risk. When estimating the capability, the estimator $\hat{C}^Y_{pk}$ severely underestimates the true capability in the presence of measurement errors. Consequently, if a statistical test is used to determine whether the process meets the capability requirement, the power of the test would decrease substantially. Consequently, lower confidence bounds and critical values ought to be adjusted to improve the accuracy of capability assessment.

An Application Example

Consider the following case taken from a manufacturing factory that makes a type of low-power, fast warm-up, and excellent stability precision voltage reference (PVR) of 15 V. The output voltage is extremely insensitive to both line and load variations and can be externally adjusted with minimal effect on drift and stability. They are ideal for communications equipment, data acquisition systems, instrumentation and process control, high-precision power supplies, and portable battery-powered equipment. Portable battery-powered equipment (Notebook computers, PDAs, DVMs, GPS, etc.). Initial accuracy is one critical quality characteristic of this PVR which has a significant impact on the PVR quality/reliability. This characteristic is usually for room temperature only, and it provides a starting point for most of the other specifications. The output voltage tolerance of a reference is measured without a load applied after the device is turned on and warmed up. Manufacturers specify a reference with a tight initial error so that they do not have to perform room-temperature systems calibration after assembly.

For this particular model of PVR product, the specification limits are $T = m = 15\,\text{V}$, $USL = 15.025\,\text{V}$, and $LSL = 14.975\,\text{V}$. A total of 80 observations are collected from the manufacturing process in the factory. Histograms and normal **probability plots** show that the collected data follows the normal distribution. The Shapiro–Wilk test is applied to further justify the assumption. The knowledge of the gauge capability λ is important in order to determine the effects of measurement errors on the performance of the estimator for C_{pk}. In practice, we do not know the value of λ, however, it is possible to obtain a point estimate and confidence interval through a GR&R study performed with the ANOVA method. Burdick and Larsen [20] reported the upper and lower bounds for approximate $100(1-\alpha)\%$ confidence intervals for the parameters σ_Y^2, σ^2, and σ_M^2 based on modified large-sample (MLS) methods. To determine whether the process is "excellent" ($C_{pk} > 1.33$) with unavoidable measurement errors $\lambda = 0.20$ (provided by the gauge manufacturing factory), we first determine that $c = 1.33$ and $\alpha = 0.05$. Then, based on the sample data of 80 observations, we obtain the sample mean $\overline{y} = 15.0049$, the sample standard deviation $s_y = 0.0045$ ($\overline{y}$ and s_y are the realized sample values for $\overline{Y}$ and S_Y), and the point estimator $\hat{C}_{pk}^Y = 1.489$. Thus, the 95% lower confidence bound of the true process capability can be obtained as $L^* = 1.356$. Moreover, similar to the adjusted lower confidence bound, we obtain the adjusted critical value 1.464 based on $\alpha = 0.05$, $\lambda = 0.20$, and $n = 80$. Since $\hat{C}_{pk}^Y > 1.464$, we conclude that the process is "excellent". We also see that if we ignore the measurement errors and evaluate the critical value without any correction, the critical value may be calculated as $c_0 = 1.543$. In this case we would reject that the process is "excellent" since $\hat{C}_{pk}^Y$ is not greater than the uncorrected critical value 1.543.

Conclusions

To summarize our discussion: when estimating capability the confidence bounds are substantially underestimated in the presence of measurement errors. When we use statistical testing to determine if the process meets the capability requirements, we find that the power of the test decreases with the measurement errors. Since measurement errors are, unfortunately, unavoidable in many branches of the manufacturing industry, to obtain a more accurate confidence bound and to improve the power with an appropriate α risk, adjusted confidence bounds and the critical values should definitely be applied. If the producers do not take into account the effects of the gauge capability on process capability estimation and testing, it may cause substantial and serious loss. In that case, producers are unable to affirm anymore that their processes will meet the capability requirements even if they are indeed sufficiently capable. The producers may incur substantial costs as (large) quantities of acceptable product units will be incorrectly rejected. Improving the gauge measurement accuracy and training the operators to properly execute the production procedures at each step are essential for reducing the measurement errors.

References

[1] Barrentine, L.B. (1991). *Concepts for R&R Studies*, ASQC Quality Press, Milwaukee.

[2] Levinson, W.A. (1995). How good is your gage? *Semiconductor International*, **October**, 165–168.

[3] Montgomery, D.C. (2001). *Introduction to Statistical Quality Control*, 4th Edition, John Wiley & Sons, New York.

[4] Montgomery, D.C. (2005). *Introduction to Statistical Quality Control*, 5th Edition, John Wiley & Sons, New York.

[5] Levinson, W.A. (1996). Do you need a new gage? *Semiconductor International*, **February**, 113–116.

[6] Montgomery, D.C. & Runger, G.C. (1993). Gauge capability and designed experiments. Part I: basic methods, *Quality Engineering* **6**(1), 115–135.

[7] Mandel, J. (1972). Repeatability and reproducibility, *Journal of Quality Technology* **4**(2), 74–85.
[8] Montgomery, D.C. & Runger, G.C. (1993). Gauge capability analysis and designed experiments. Part II: experimental design models and variance component estimation, *Quality Engineering* **6**(2), 289–305.
[9] Burdick, R.K., Borror, C.M. & Montgomery, D.C. (2003). A review of methods for measurement systems capability analysis, *Journal of Quality Technology* **35**(4), 342–354.
[10] Burdick, R.K., Park, Y.J., Montgomery, D.C. & Borror, C.M. (2005). Confidence intervals for misclassification rates in a gauge R&R study, *Journal of Quality Technology* **37**(4), 294–303.
[11] Mittag, H.J. (1994). Measurement error effects on the performance of process capability indices. in Paper presented at the *5th International Workshop on Intelligent Statistical Quality Control*, Osaka.
[12] Mittag, H.J. (1997). Measurement error effects on the performance of process capability indices, *Frontiers in Statistical Quality Control* **5**, 195–206.
[13] Bordignon, S. & Scagliarini, M. (2002). Statistical analysis of process capability indices with measurement errors, *Quality and Reliability Engineering International* **18**, 321–332.
[14] Scagliarini, M. (2002). Estimation of C_p for autocorrelated data and measurement errors, *Communications in Statistics: Theory and Methods* **31**, 1647–1664.
[15] Bordignon, S. & Scagliarini, M. (2006). Estimation of C_{pm} when measurement error is present, *Quality and Reliability Engineering International*, **22**(7), 787–801.
[16] Kotz, S. & Johnson, N.L. (1993). *Process Capability Indices*, Chapman & Hall, London.
[17] Pearn, W.L. & Lin, P.C. (2004). Testing process performance based on the capability index C_{pk} with critical values, *Computers and Industrial Engineering* **47**, 351–369.
[18] Pearn, W.L. & Liao, M.Y. (2005). Measuring process capability based on C_{pk} with gauge measurement errors, *Microelectronics Reliability* **45**, 739–751.
[19] Pearn, W.L. & Shu, M.H. (2003). Manufacturing capability control for multiple power distribution switch processes based on modified C_{pk} MPPAC, *Microelectronics Reliability* **43**, 963–975.
[20] Burdick, R.K. & Larsen, G.A. (1997). Confidence intervals for a ratio in a R&R study, *Journal of Quality Technology* **29**(3), 261–273.

Further Reading

Hamada, M. & Weerahandi, S. (2000). Measurement system assessment via generalized inference, *Journal of Quality Technology* **32**(3), 241–253.

WEN LEA PEARN AND CHIEN-WEI WU

Process Maps and Statistics

Introduction

A process map is an invaluable tool for directing the statistical study of a process. When we use statistical methods to gain knowledge of a process with the intention of improving that process, we are working with a process that is already in current operation. Consequently, the methods, materials, work practices, machine settings, and other operating conditions are already in place. At issue then is selecting what set of methods, practices, or parameters to change to improve operation. Our statistical study of the process can only be effective when it is coupled with knowledge of current process behavior. The process map is designed to capture, in the most effective manner, this kind of process knowledge.

In particular, the process map should include knowledge on the process that will guide process study and improvement. We should look to the process map to address the following required information:

1. What process output characteristics are critical for providing customer value?
2. What has been the past experience in maintaining appropriate variation and levels of critical characteristics?
3. What processing factors affect critical characteristics and are they or can they be managed?

The answer to the above questions will clearly require a mixture of organization, engineering, and statistical knowledge to answer. Organizing this information into a manner useful for process study is the aim of the process map.

How to Draw a Process Map

First, it is necessary to determine the scope of the process map. This will be a function of the traditional view of the process, the process owner's sphere of influence, and an expectation regarding the work necessary to improve the process. As a general comment, individuals constructing maps tend to make the scope of the process too narrow [1]. Before finally deciding on the scope, we should consider those factors, which we are considering as initial inputs to the process. Are some of these inputs internal to the organization? If so, we should consider expanding the process map to include the activities that generated those inputs.

The second step in drawing a process map is to draw a flowchart or other graphical representation of the process. Instructions on flowcharting can be found in the literature [2]. Then, categorized inputs and outputs of the process are added. These inputs and outputs are similar to the kind of information shown on a **cause and effect** (C/E) diagram. The advantage of placing the inputs and outputs on a process map is that, where they effect the process is displayed. Consequently, the nature of how they impact the process can be better understood and **data collection** plans devised to evaluate their anticipated effects. Constructing the process map as described requires that the process be observed several times before the map can approach comprehensiveness and as work is done to improve and maintain the process, the process map will need to be revised to reflect these changes.

Further Characterizing the Inputs and Outputs

The critical outputs of a process (usually referred to as *the big Y's* in **Six Sigma** terminology) can be used to determine how the process is performing. Big *Y*'s relate to the overall process and are typically those characteristics of prime importance to the customer. "Little *y*'s", again in the parlance of Six Sigma, are intermediate outputs of a process, often measured at the end of process stages. The process map will capture our knowledge of what these intermediate outputs are and indicate whether these "little *y*'s" are currently measured or not.

On a process map the inputs, or process "*X*'s" (which include measurements made on incoming materials to the process, process methods, parameter settings, environmental factors, etc.) are categorized as controllable (*C*) or noise (*N*). Choosing to control an input and thereby labeling it with a "*C*" indicates that, given the current technology, resources, and process knowledge, these are the variables that need to be managed. Typically, these variables will have target values and requirements on their limits (possibly communicated as specifications). "Noise" refers to variables that either cannot be controlled (without excessive resources and technology) or are thought to have little influence on the output, so are not controlled. Noise could be intermittent disturbances or common cause, consistent, sources of variation.

Example

A better understanding of a process map is gained by looking at an actual map of a process. The process map shown in Figure 1 was used to describe a crimp operation. It was very helpful in thinking about how the variability in the crimp could be set to a target with minimal variation. A successful crimp required considering the incoming raw material properties, standardizing the operator procedures, and determining correct machine settings. Examples of other project work supported by a process map can be found in Leitnaker and Cooper [1].

Summary Comments

As previously discussed, a process map is an invaluable tool for directing the statistical study of a process. Some specific examples of uses of the process map are listed, and discussed, below.

1. The process map exists to document current process knowledge. As such it should serve the following purposes.
   (a) Capture current knowledge as to which inputs need to be controlled.
   (b) Day-to-day management of processes often focuses on a few, conveniently measured variables, but effective process management requires retaining a more comprehensive picture.
   (c) A common failure in process improvement work is to make changes to a process without regard to what currently exists and why it is there. The process map helps direct attention to important process characteristics that must be considered when process changes are contemplated.
2. The process map will also serve to enhance improvement activities. Because of the wealth of information captured on a process map, it can be used to generate and support actionable, manageable projects.

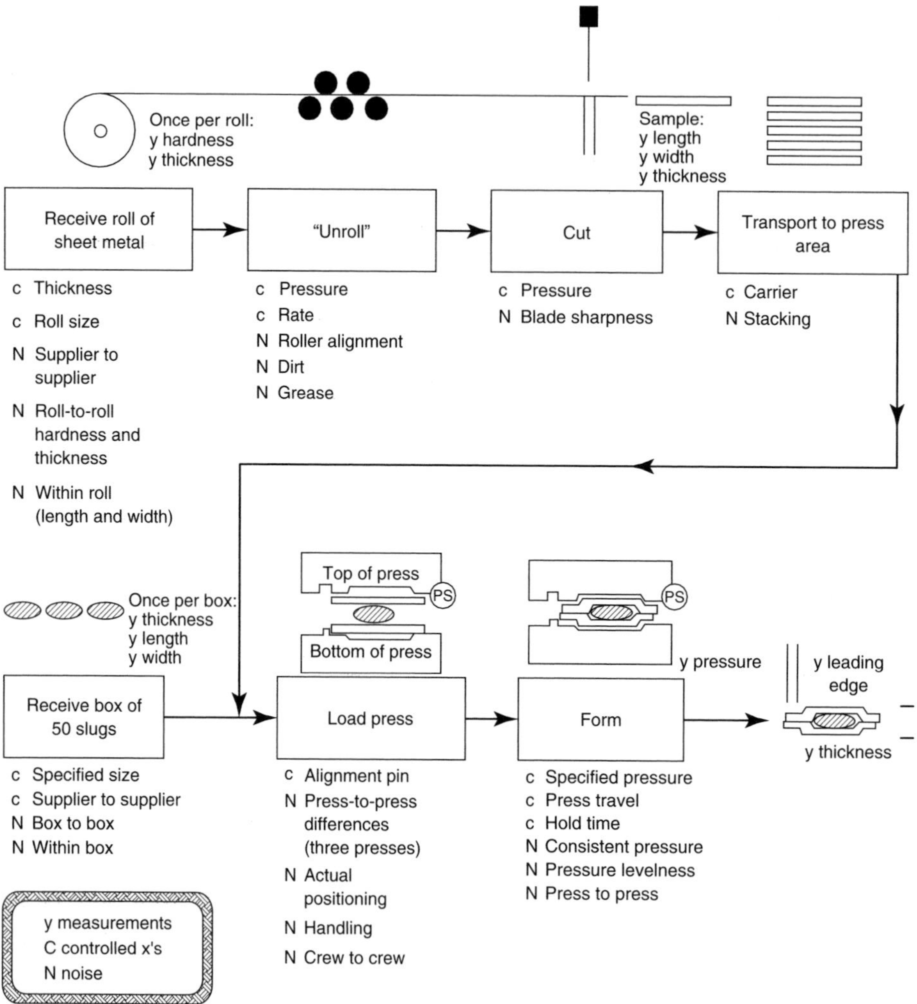

Figure 1 Process map of a crimp operation

3. Most importantly, the process map should be used to provide guidance in planning statistical studies.
   (a) When the objective of our process study is to evaluate the consistency of process outputs, either intermediate or final, the map provides specific information about the processing conditions across which the study should occur. For example, if an incoming raw material is known to be critical to a process output, a study evaluating the behavior of that output will need to cover a sufficient time across which that material will change.
   (b) If our process study is to include a statistically designed experiment then the process map provides immediate candidates for possible factors to be included in the designed experiment.
   (c) A more difficult task in designing an experiment is to understand the effect that process variation will have on experimental error. The process map provides not only a list of sources of variation that affect the process, but also provides guidance about where and at what frequency this variation will occur in the process.

(d) By necessity, the conditions seen during any statistical study will be a subset of the conditions that will occur in the future. In the same way that it is necessary to identify the frame and population in statistical surveys, an understanding of the process being studied is necessary for making inferences based on a statistical study. The inference space of a statistical study can be better understood by the use of a process map.

References

[1] Leitnaker, M.G. & Cooper, A. (2005). Using statistical thinking and designed experiments to understand process operation, *Quality Engineering* **17**, 279–289.

[2] Scholtes, P.R., Joiner, B.L. & Streibel, B.J. (1996). *The Team Handbook*, 2nd Edition, Joiner/Oriel, June 1.

Further Reading

Sanders, D., Ross, B. & Coleman, J. (1999). The process map, *Quality Engineering* **114**(4), 555–561.

Mary Leitnaker and Tony Cooper

Process Maps, Construction of

What is Process Mapping

A process is a series of steps that converts an input(s) to an output(s). This series of steps is usually done repeatedly. As the familiar proverb says "A picture is worth a thousand words", so a good way to explain these steps is pictorially. Therefore, process mapping is a key component in any quality effort to control, improve, document, and streamline a process. Process mapping can take on many forms but the commonality is that the specific steps or events that are required to provide a service or product to a customer are listed. The customer could be an external paying customer or an internal customer.

Process mapping can be called by several names as well as have several appearances. A generic process map may be called a *process flowchart, flowchart diagram*, or a *business map*. Specific maps can take on special names like SIPOC (suppliers–inputs–process–outputs–customer), functional deployment map (FDM), or opportunity map.

Reasons for Creating a Process Map

There are many reasons to create a process map. Four major reasons are to document the current state, to find opportunities for improvement in the current process, to document a change in the process, or to create a new process.

Many quality systems, such as ISO 9000, require documentation of key processes in the organization. Many companies also have their own internal requirements for process documentation. This documentation can be helpful to training employees or to get a common understanding.

One of the most common uses for maps is to identify changes to the process that would increase efficiencies, reduce cycle times, reduce costs, or decrease defects. Improvement teams often find steps that are redundant, unnecessary, and do not add value to the product or service. In the **Six Sigma** methodology define–measure–analyze–improve–control (DMAIC), mapping is used in the define phase to look for quick wins. Quick wins are improvement ideas that can be implemented easily and quickly. In the measure phase, maps are used to identify points in the process to collect data to look for more complicated or less obvious improvements. In the improve phase, they are used to demonstrate the proposed new, improved process.

Process mapping has also been used as part of the groundwork before beginning to develop software programs to support the process. Software developers have long used mapping as a tool to help define the requirements for code. In addition, mapping is a precursor to the use of process simulation. Process simulation is the technique of creating a computer model of the process.

Process Map Formats

Process maps may take on many different looks. Some of the common ones are SIPOC, detailed process map, an FDM, and an opportunity map. Each map has it own purpose and therefore an appropriate time to use it.

SIPOC Map

A process map that has gained popularity with the increase of Six Sigma training is the SIPOC map. SIPOC is an acronym for suppliers–inputs–process–outputs–customers. The SIPOC is a very high-level map that does not use traditional mapping symbols. The first column of the SIPOC lists the key suppliers to the process. These may be internal or external suppliers. The second column lists the key inputs to the process. The inputs are usually those components that are transformed during the process to create the output. The third column is the process column where the major process steps are cataloged. Since this is a top-level map, there usually are only a few major steps listed. The fourth column is where the outputs of the process are listed. The last column would list the main customers or customer types of the outputs. Again, the customer could be an external or an internal customer.

Although the SIPOC is listed and displayed in this order, this does not imply that the SIPOC would be created in this order. Since there has been an increase in the focus on the process from the customers and their requirements, it makes sense to start mapping the process from the customer's perspective. So, the mapping may start with a focus on who the customers are and what are the outputs of interest. To place this emphasis on the customer, there is a similar process map called *COPIS*. This is the SIPOC listed in the opposite way.

The SIPOC is good as a starting point for the team before adding additional detail. It is also good as a display to upper management to get a general sense of the process. It can help define who should be on the team for the more detailed mapping to follow.

An example of a SIPOC of purchasing a plane ticket is shown in Table 1. In this example, the customer could be either the traveler or a travel agent. The traveler is also a supplier in that she/he must supply information about travel requirements. The airlines also are supplier since they provide schedule information. Usually, the outputs and inputs listed are the few key ones. In this case, the key output would be the purchased ticket. The major steps of the process, listed without worrying about decisions or feedback loops, are as follows: entering information in computer system, review of flight schedules, flight selection, entering credit information, and then purchasing the ticket.

Simple Detailed Process Map

A simple detailed process map is usually done to a deeper level than a SIPOC. The detailed process map utilizes symbols to explain what happens in the process. The detailed process map is usually displayed to show the easiest flow going right to left. Many different symbols are used but there are some basic symbols that are traditionally used. The first symbol is the rectangle. The rectangle is used to indicate an activity or operation. The diamond is used to illustrate a decision point. Arrows indicate the flow of process. Ovals indicate the start and stop of the process. Circles are connectors between pages. Table 2 shows these symbols, the use of the symbols, and a simple example of what the symbols might represent.

The detailed process map is useful for finding areas for improvement, understanding redundancies in the process, to find waste, and to identify areas for collecting data to monitor the process.

An example of a simple detailed map of a customer service representative answering a customer's question is illustrated in Figure 1.

Besides indicating activities, decisions, and process flow, detailed process maps may include a myriad of other information. A partial list of this additional information is as follows:

Table 1 SIPOC of purchasing a plane ticket

Suppliers	Inputs	Process	Outputs	Customers
Traveler Airlines	Traveler requirements Flight schedules	Enter information Review flight schedules Select flight Take credit information Purchase ticket	Ticket issued	Traveler Travel agent

Table 2 Traditionally used symbols for detailed process mapping including usage and a simple example

Symbol	Usage	Example
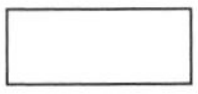 (rectangle)	Activity or operation	Assemble parts
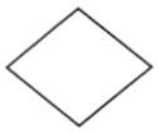 (diamond)	Decision	Request fulfilled?
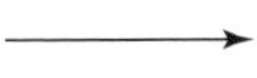 (arrow)	Flow	Go to next step
 (oval)	Start or stop of process to be mapped	Receive customer request
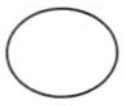 (circle)	Connector to another page	Go to page 2

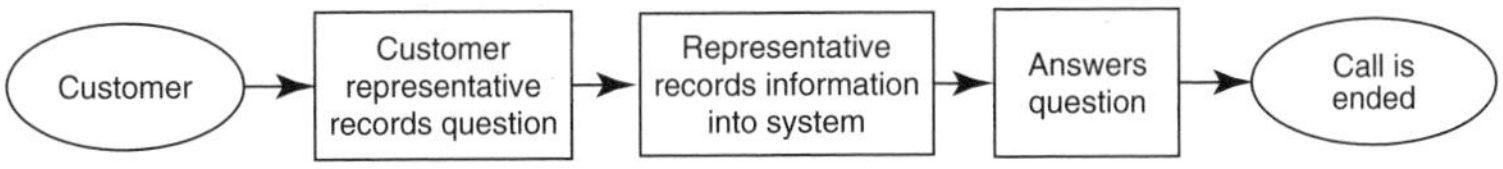

Figure 1 Detailed process map for the process of a customer service representative answering a customer's question

- the number of people required for each step
- the machinery required at each step
- the tooling required at each step
- special inputs required
- information required
- the time to conduct the step
- the cost of performing the step
- the defects associated with specific points in the process
- delays
- queues
- setup times required at each step.

Beyond this list, there are two other types of information that can be added to a map that will make two specialized maps: a functional deployment map, also called a *swimming lanes map*, and an *opportunity map*.

Functional Deployment Map

An FDM takes a traditional detailed process map and adds an extra dimension by listing the functions that actually perform the process step. In an FDM, you would establish columns for each function, department, or person involved in the process. Then each activity or decision is placed in the associated column of the function that performs that step. The purpose of the map would be to identify the key communication or transfer points. These points are often where queues or defects occur. These maps are most useful in environments where the process involves many groups.

A very simple FDM for the process of a customer representative handling a customer call is shown in Figure 2. Three different groups are involved in this process: the customer, the customer service representative, and the engineer who handles the more difficult calls.

Opportunity Map

An opportunity map is a simple detailed process map with two additional columns. The two columns indicate whether a step in the process is either value added or non-value added. The purpose of the opportunity map is to identify opportunities to eliminate or reduce waste in the process. A value-added step is considered to be a step that is inherent to converting the inputs into the product or service. Non-value-added steps are associated with waste and can include, but are not limited to inspection, audits, queues, transportation of product, checking work, and

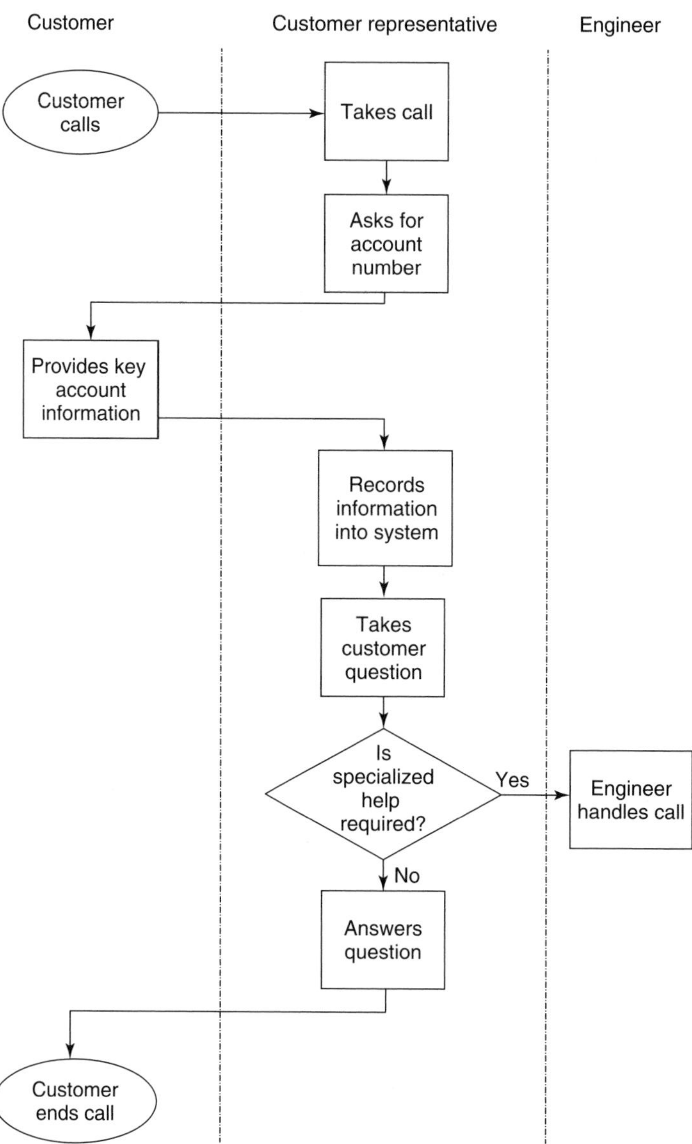

Figure 2 Functional deployment map for the process of a customer service representative handling a customer's call

so on. These activities just increase costs and should be primary targets of improvement.

An opportunity map for a very simple assembly process is shown in Figure 3. In this process, any assembly operation would be considered value added. These steps are turning the product into what the customer expects. The inspection and transportation steps are considered non-value added, or wasteful.

Creating the Process Map

Most process mapping sessions involve the participation of a team of people. Often, if the mapping session is designed to capture the current state of the process, the team finds differing views about what the real steps are. This is often a significant output from process mapping, getting the team to see the process from all points of view.

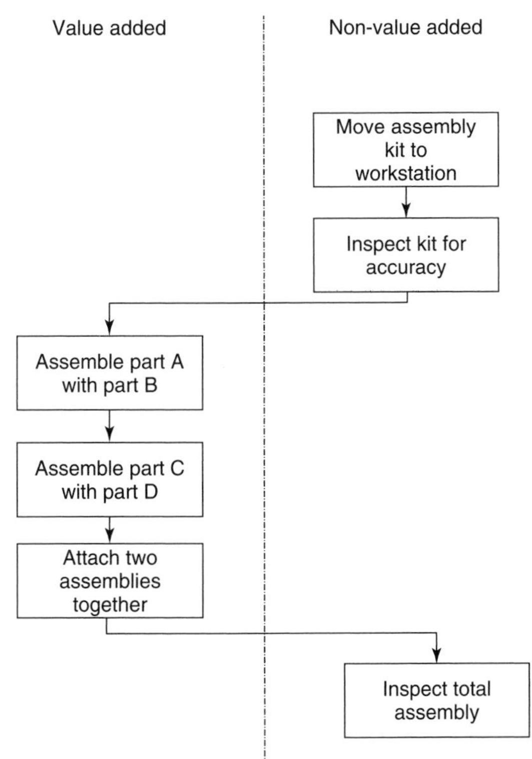

Figure 3 Opportunity map for an assembly process

When doing mapping, there are some basic steps that the team should follow. These steps are as follows:

1. Establish the process to be mapped. Give the process a name.
2. Determine the objective of the mapping exercise – how will the map be used. This will influence your choice of the map that you create.
3. Invite the correct people to participate. Often it is helpful to include people who do not know the process. These people may ask questions that people close to the process may not think to ask.
4. Define the boundaries to be mapped. These can change as the team finds that issues of concern may be influenced by upstream or downstream activities. The people involved in the mapping exercise may change as the boundaries are altered.
5. Observe the process as it is operating. Make sure the team captures the information while observing.
6. Agree on the major steps and the order that they occur.
7. Add detail to the major steps as necessary.
8. Document in the appropriate format.
9. Validate the map with other people in the process.

If process mapping was part of a continuous improvement effort, after this "as-is" map is established, the team would start to question the status quo. Each step in the process should be reviewed to look for opportunities to eliminate unnecessary or redundant steps, combine steps to improve efficiencies, or rearrange steps to facilitate a cleaner, simpler flow. The team would then create the "to-be" process map based on the changes agreed upon.

Best Practices in Process Mapping

There are some key practices that will facilitate the mapping process: involving the right people, making the map understandable, and being clear in the purpose of the map.

If a map is created by one person, the perspective may not really reflect reality. Often, there is someone who originally defined the process. Their perspective may be the wished-for state. However, the people in the process may have intentionally or unintentionally changed the defined steps. One reason may be that the process users have a misunderstanding of what the process should look like. They might not have received proper training or the steps were not properly communicated. Another reason might be that the process users discovered a better or easier way for the process to work. The original steps may not have covered all possible contingencies.

The process may also have been a change from previous steps and the process users did not see the benefit of the change, so they reverted back to the original steps.

For all these reasons, it is important to include the process users in the creation of the process map, to capture the true state of the process. Other people who might be included in the formation of the map are the suppliers, customers, and process owners.

Since there are many different kinds of process maps, and different levels of detail, it is imperative that the creator(s) specify up front what the purpose of the map is, so that the proper format is used. Often, it makes sense to start at a very top level and add complexity as required. Knowing the purpose will also help clarify the types of information to be added.

Difficulties in Process Mapping

One of the most difficult aspects of mapping is the level of detail required. Too much detail can waste time and will not be productive. Too little detail can prevent a deep enough understanding of the process to see how to improve or to provide complete training. It is important to move from less detail to more detail. If the team tries to jump into too much detail in the beginning, they can get very bogged down in things that may be of little consequence. More detail can be added as deemed necessary.

Another difficulty of process mapping can be that the team members are in different locations. Most mapping sessions are done on a wall with a group of people using sticky notes. However, in today's global environment, it is becoming more frequent for the mapping session to take place in an electronic environment. Electronic conference tools can help facilitate the mapping session but it can still add a level of difficulty.

To capture all the information about a process on a document can create an overly complicated document. Since one of the primary functions of process maps is as a communication tool, too much text can be daunting. It is important that this additional information be displayed in as easy-to-read format as possible.

Summary

Process maps are a key part of any quality program for any kind of organization. People tend to understand pictures better than words and understanding the process is the best way to maintain and improve it. Although a process map might take on several looks or formats, the primary goal should be to keep it simple enough to be appealing to read and understandable to all people involved in the process.

LORRAINE DANIELS

Process Reengineering

Background and Definition

Business process reengineering (BPR) is the *fundamental* rethinking and *radical* redesign of *business processes* [1, 2] – that is, the flow of work and information within and between organizations. Its purpose is to achieve *drastic* improvement over existing performance in quality, cost, service, and cycle time. It is also known as business process *redesign*, business process *transformation*, or business process *innovation*.

Rather than pursuing continuous, but incremental improvement like some of its predecessor methodologies such as **Kaizen**, and total quality management (**TQM**), BPR seeks radical improvement in a single effort.

BPR reached the peak of its popularity in the mid-to-late 1990s after Michael Hammer and James Champy published their best-selling book *Reengineering the Corporation* in 1993.

Use of the Methodology

BPR was initially intended to significantly improve, in current terms, the business performance and competitiveness of the enterprise. The extensive reengineering of administrative, information systems, procurement, and manufacturing processes in the early 1990s is the well-known result of this approach. More recently, management has discovered that business performance can further be improved by the systematic reengineering of its *core* processes: the aggregation of work that creates customer value.

Accordingly, the selection of business processes for reengineering predominantly targets those core processes that directly affect the customers of the firm: manufacturing, distribution, customer service, invoicing, and order entry.

Process reengineering is not intended to replace or substitute for *automation, software reengineering*, or traditional *reorganization.* Its original proponents (Hammer [2] and Davenport [3]) and other lead practitioners also emphasized that reengineering should not lead to *downsizing* – its basic concept is to "do more with the same".

Although it is difficult to offer general guidelines on the use of BPR, certain conditions are conducive to its application:

- There are serious quality and customer satisfaction concerns.
- Continuous improvement methods are not producing satisfactory results.
- Competition is consistently outperforming the enterprise.
- The organization exhibits considerable internal organizational conflict.
- The company is experiencing excessive product and administrative cost overruns.

Most of the companies with one or more of these conditions have clearly differentiated functions: vertical organizational "silos" – structures built around the concept of the division of labor and hierarchical management. These functional silos comprise only narrow pieces of a multitude of processes that in totality make up the workings of the enterprise. Managers and executives are in charge of increasingly large segments of the *function*, but no one is in charge of the *processes*. This is known as *process fragmentation*. Because of process fragmentation, handoff problems between functions proliferate throughout the organization.

As a result, processes become highly error prone as each handoff requires coordination and communication and entails wait times and mismatches in schedules. And while these processes have always existed – they are, in essence, the way work is done by the enterprise – only a few people are aware of them, and they are not managed in any consistent or rational way.

Traditional BPR is about the reintegration of work and the elimination of process fragmentation and the related handoff problem.

Basic Concepts and Methods

In many cases, radical redesigning and reorganization of an enterprise is necessary to lower costs and improve the quality of products and services. Information technology (IT) is the key enabler of such radical change. The nine principles of successful reengineering to achieve quantum performance improvement include:

1. Develop the business vision and process goals.
2. Identify all the processes in an organization and prioritize them in order of redesign need (based on the importance of the process, its level of dysfunctionality, and the likelihood of success).
3. Understand and measure the performance of existing processes.
4. Organize around results and customers, not tasks or departments.
5. Place decision points where the work is performed by building appropriate controls into the process.
6. Treat geographically dispersed resources as if they were concentrated.
7. Capture information only once and do this at its source.
8. Identify IT resources and capabilities; integrate IT into the material workflow for maximum leverage.
9. Design and build a prototype of the new process as the base for successive design iterations and tests.

IT plays a critical role in implementing the BPR concept. It is viewed as an important enabler of new ways of communicating and collaborating within an organization and across organizational boundaries.

Generally, the application of IT to process reengineering requires *inductive* thinking in that its power of disrupting existing rules and paradigms is first recognized, and then process situations are sought out where this provides a radically new solution. As part of the BPR approach, *workflow management* systems are considered as significant contributors to process improvement.

Where an enterprise resource planning (ERP) solution is an integral part of the BPR effort, the project planning and management function of the ERP system may be utilized. Major ERP systems offer these capabilities as part of their system packages.

BPR treats business processes as proceduralized workflows. It links functional expertise with process management teams to reintegrate the existing, fractionated workflow into contiguous, cross-functional work processes that are driven by *customer needs*.

Infrastructure

Reengineering is done by trained and qualified people. Selecting and organizing those who actually implement reengineering is critical to success. Certain people have unique roles that focus either on the individual process, or on aligning multiple processes at the enterprise level; others play dual roles as they carry out the tasks of bridging and integration at both levels.

Although there is no one-size-fits-all list of roles, many years of reengineering experience have helped identify the following generic reengineering infrastructure:

- *Enterprise transformation executive*–the senior executive who has the authority and resources to decide and motivate the overall reengineering endeavor.
- *Enterprise transformation council*–a group of influential policy makers who develop and champion the overall process reengineering strategy of the enterprise, commit the necessary resources, and oversee its progress. They are responsible for achieving synergy across the separate, individual reengineering projects of the enterprise and for creating optimal leverage on overall business performance.
- *Process management executive* – a senior manager who is the owner of the enterprise methodology of business process management, and is responsible for developing and maintaining methods, tools, and techniques – including the latest IT and computer software – related to BPR.
- *Business process owner* – an executive or middle manager with the responsibility and accountability for a specific, single business process, its reengineering, and subsequent operational implementation. There are usually a number of business process owners within an enterprise.
- *Business process management team* – a group of selected individuals (usually managers or senior professionals) working with the process owner and committed to the reengineering and subsequent operation of a specific, single business process.
- *Business process management team leader* – a member of the business process management team selected by the process owner to lead and administer the day-to-day work of the process team.

These roles do not represent a hierarchical set. Rather, they are part of a two-level team structure where the transformation council is the strategic entity, and the business process management teams represent the operational part.

Budget

The process implementation budget contains both expense and capital items. It is always preferable to establish a consolidated process reengineering budget, as opposed to distributing the cost implications among the affected functional operating budgets.

Costing methods used in establishing the implementation budget should be consistent with those employed to plan and measure the continuing process costs once the reengineered business process has become operational. Since most process costing methods are activity or task based, it is recommended that *activity-based costing* (ABC) be used in building the process implementation budget.

It is important to point out that all implementation activities in the project plan must contain, beyond people and technology, all the required resources: materials, energy, plant and facilities, intellectual capital, and other assets needed to ensure the successful plan implementation.

It may happen at times that the deployment of a newly reengineered business process may take more time and effort than its design. This will more likely be the case where a massive upgrading of IT, such as the implementation of an ERP system, is part of the reengineering effort.

Strengths

If implemented properly and with the committed and continuing support of senior management, BPR can yield very significant returns. It helped corporate giants like Procter & Gamble and General Motors recover after financial drawbacks due to competitive pressures; Southwest Airlines reengineered the company using IT; Michael Dell understood how and where to apply the concept of BPR to reengineer Dell Computers; Ford reengineered its manufacturing processes where *job #1* is product quality.

Through a smoothly functioning reengineering infrastructure, accountable process ownership, and effective teamwork, reengineering provides the solution to the two known dilemmas of process management:

- variation and dispersion and
- scale and complexity.

Variation and dispersion is the apparent conflict between the need for process consistency across the enterprise and the variation caused by geographic dispersion and differences in market requirements. Reengineering tries to solve this dilemma by creating different versions of the same basic process flow, each under the leadership of a working owner, and assigning an operations director to each instantiation of each version, to deal with variation in market (customer) requirements.

Scale and complexity are the combined effects of process size (amount of resources, number of operating locations) and structure (number of constituent subprocesses and activities, handoff points, multiple process versions). Scale and complexity can make a single process owner's job extremely challenging. Again, the suggested approach is subdivision of the process and the assignment of additional process management resources. Subprocess owners operate under the overall management umbrella provided by the executive process owner.

The idea of reengineering was rapidly adopted by a huge number of firms around the world, but most notably in North America. They were striving for renewed competitiveness, which they had lost due to:

- the market entrance of new foreign competitors (e.g., Japanese manufacturers in the United States);
- their inability to satisfy changing customer needs (e.g., IBM in the mid-1980s); or
- their flawed cost structure.

Reengineering was implemented at an accelerating pace and by the late 1990s more than half of the Fortune 500 companies claimed to either have actually initiated reengineering efforts, or to have firmly planned to do so.

Weaknesses

To this day, a number of critics claim that BPR dehumanizes the workplace. By the late 1990s, BPR gained the somewhat unfair reputation of being an outright excuse for "downsizing", e.g. major and irreversible reductions of the workforce.

A number of factors can cause BPR initiatives to fail to reach their initial objectives. The reasons for such failure are many; however, most can be avoided or corrected:

- lack of sustained management commitment and leadership;
- unrealistic scope and expectations, or prior constraints on problem definition or project scope;
- underestimation of the cultural resistance to change;
- overconfidence in technological solutions;
- poor **project management**;
- concentration on redesign as the only alternative;
- neglecting people's values and beliefs;
- settling for partial or minor results, or quitting too early;
- making reengineering happen as a grassroots effort;
- dissipating energy and resources across an excessive number of reengineering projects; and
- failure to distinguish reengineering from other improvement programs.

The traditional process reengineering methodology evolved around the concept of physical and material workflow, its improvement and redesign. This is because the underlying methodologies have their roots in manufacturing and, in particular, in IT, whose purpose is to automate defined procedures.

This approach to BPR has led to the development of skillful ways of redesigning business processes to measurably improve performance, and to use IT for the streamlining and coordination of complex workflows. However, as companies embarked on using this approach to performance improvement, it eventually became clear that some important things were being overlooked.

The traditional process reengineering approach did not adequately address several fundamental areas [4]:

- the human dimension, including the flow of people-to-people communications in the process;
- the planned adaptability of redesigned processes to deal with further unpredictable change, and
- overall economic and business performance of the enterprise.

Overlooking the Human Dimension

Overlooking the human dimension – mobilizing and motivating people – has been viewed as a major shortcoming of classic workflow-based process reengineering. The downsizing, done without concern for the people, has made the term *reengineering* imply that people do not matter because the approach taken is mechanical and impersonal.

Another limitation of traditional reengineering is its inherent difficulty in harnessing the great variety of intellectual, managerial, and professional communications where there are no easily definable physical work products – e.g., most work results are agreements, decisions, commitments, and the transfer of information and knowledge.

Problem with Unpredictable Change

One of the greatest challenges to BPR is that it must engender thinking that helps anticipate and deal with *unpredictable change* – the ability to quickly identify and abandon old paradigms that underlie current operations.

The Process Paradox

Success at the business process level, but no improvement or even negative overall impact at the enterprise level, is the essence of what has become known as the *process paradox*. After months or even years of careful process redesign, many companies have achieved impressive improvements in individual business processes – only to watch overall results still decline. By now, paradoxical outcomes of this kind have become almost proverbial.

- IBM's small business division won the Malcolm Baldrige Award in 1992 in recognition of the successful redesign of the development and production processes that produced one of IBM's all-time success products: the AS400. At the same time, the IBM Corporation was facing the worst crisis of its history including the downsizing of half of its workforce worldwide.
- In 1995, Corning's telecommunications products division earned the Malcolm Baldrige Award. Six years later the company was experiencing what many believed was its fight for corporate survival.

Summary

Today, defining business processes as the starting point for business analysis and redesign is a widely accepted practice and standard part of the change management portfolio. Change, however, is sought and implemented in a less radical way than what was originally proposed by BPR.

The key task of twentieth century executives and managers was *performance improvement.* The solutions of TQM and BPR were *process solutions* – focusing primarily on key, individual business processes.

Now, management faces the compelling requirement of doing all this continuously, faster, and better than competition – and in the face of accelerating and unpredictable change. The double challenge of the twenty-first century is to be successful while continuously *adapting* and responding to a highly competitive, fast moving, information-driven, and near-turbulent environment. In fact, "continuous adaptation" is the motto for the new century.

References

[1] Hammer, M. (1990). Reengineering work: don't automate, obliterate, *Harvard Business Review* July-August 1990, 104–112.

[2] Hammer, M. & Champy, J. (1993). *Reengineering the Corporation: A Manifesto for Business Revolution*, Harper Business, New York.

[3] Davenport, T. (1993). *Process Innovation: Reengineering Work Through Information Technology*, Harvard Business School Press, Boston.

[4] Pall, G.A. (2000). *The Process-Centered Enterprise: The Power of Commitments*, CRC Press, Boca Raton.

GABRIEL A. PALL

Process Sigma (Level) *see* Sigma Level of a Process

Process Yield

Process Yield

Process yield has, for a long time, been the most common and standard criterion used in various manufacturing industries to characterize the process performance. Process yield is currently defined as the percentage of the processed product units passing inspection. That is, the product characteristic, X, must fall within the preset specification interval.

For processes involving two-sided manufacturing specifications, the process yield can be calculated as: $Yield = \Pr(LSL < X < USL)$, where USL and LSL are the upper and the lower specification limits, respectively. For the-smaller-the-better processes whose preset specifications require only the upper specification limit USL, the process yield is: $Yield = \Pr(X < USL)$. For the-larger-the-better processes whose preset specifications require only the lower specification limit LSL, the process yield is: $Yield = \Pr(X > LSL)$.

Suppose the production process is operating in a stable manner such that the fraction of the nonconformities (NC), the probability that any unit will not conform to specifications, is $P(\text{NC})$, and that successive units produced are independent. Note that $P(\text{NC}) = 1 - Yield$. If a random sample of n units of a product is selected and D is the number of units of the product that are nonconforming, then D follows $b(n, p)$ with $p = P(\text{NC})$, a binomial distribution with parameters n and p. Generally, the actual fraction of nonconformities, p, is unknown. Traditionally, the parameter p is estimated by the sample fraction nonconforming, the estimated p, which is defined as: $\hat{p} = D/n$. The actual *Yield* of a process is also unknown, which can be estimated by $1 - \hat{p} = (n - D)/n$.

The above approach is applicable only for processes with large $P(\text{NC})$, say, $p \geq 0.01$. For processes with small p it is inappropriate, since $\hat{p}$ will be most likely zero in most cases.

Process Capability Indices

Process capability indices (PCIs) are intended to provide single-number assessments of the ability to meet specification limits on quality characteristic(s) of interest. Existing PCIs, $C_p, C_a, C_{PU}, C_{PL}, C_{pk}$, C_{pm}, and C_{pmk} have been proposed to the manufacturing industry as capability measures based on various criteria including process variation, departure, yield, and loss. The process yield is one of the important factors, though not the only one, considered when accessing or improving a process. In this article, PCIs dedicated to the measurement of the yield of normally distributed processes with two-sided specification limits as well as one-sided specification limit are discussed. These indices are explicitly defined as follows [1–4, **Process Capability Indices, Comparison of**]:

$$C_p = \frac{USL - LSL}{6\sigma}, \quad C_a = 1 - \frac{|\mu - m|}{d},$$

$$C_{PU} = \frac{USL - \mu}{3\sigma}, \quad C_{PL} = \frac{\mu - LSL}{3\sigma},$$

$$C_{pk} = \min\left\{\frac{USL - \mu}{3\sigma}, \frac{\mu - LSL}{3\sigma}\right\},$$

$$C_{pm} = \frac{USL - LSL}{6\sqrt{\sigma^2 + (\mu - T)^2}},$$

$$C_{pmk} = \min\left\{\frac{USL - \mu}{3\sqrt{\sigma^2 + (\mu - T)^2}}, \frac{\mu - LSL}{3\sqrt{\sigma^2 + (\mu - T)^2}}\right\} \quad (1)$$

where μ is the process mean, σ is the process standard deviation, $m = (USL + LSL)/2$ is the midpoint of the specification interval, $d = (USL - LSL)/2$ is the half-length of the specification tolerance, and T is the target value preset by the product designer or manufacturing engineer. This paper will focus on the situation in which the target value T is set to the specification center, i.e. $T = m$, the most common case in practical applications.

The index C_p, a function of the process standard deviation σ and the specification limits, has been referred to as the *precision index* which is defined to measure the consistency of the process quality characteristic relative to the manufacturing tolerance. The index C_a, a function of the process mean and the specification limits, has been referred to as the *accuracy index*, which is defined to measure the degree of process centering relative to the manufacturing tolerance. Note that $0 \leq C_a \leq 1$ for $LSL \leq \mu \leq USL$, and $C_a < 0$ for $\mu < LSL$ or $\mu > USL$. Especially, $C_a = 1$ for $\mu = m$.

PCIs for Two-Sided Processes

There are many capability indices including C_p, C_a, C_{pk}, C_{pm}, and C_{pmk} designed for assessing processes with two-sided specification limits (which require

USL and *LSL*). The three indices C_{pk}, C_{pm}, and C_{pmk} may be rewritten respectively as follows

$$C_{pk} = C_p C_a,\ C_{pm} = \frac{C_p}{\sqrt{1+[3C_p(1-C_a)]^2}},\ \text{and}$$

$$C_{pmk} = \frac{C_p C_a}{\sqrt{1+[3C_p(1-C_a)]^2}} \tag{2}$$

which are all functions of the two basic indices C_p and C_a.

For a normally distributed process, the probability fraction of nonconforming $P(\text{NC})$ is

$$P(\text{NC}) = \Phi[-3C_p(2-C_a)] + \Phi[-3C_p C_a] \tag{3}$$

which is also a function of the two basic indices C_p and C_a. Especially, $P(\text{NC}) = 2700$ ppm when $C_p = 1$ and $C_a = 1$, i.e. when $USL - LSL = 6\sigma$ and $\mu = $ m.

Pearn and Lin [5] obtained a yield measure formula based on C_{pk} as:

$$P(\text{NC}) = \Phi[-3C_{pk}] + \Phi\left[-\frac{3C_{pk}(2-C_a)}{C_a}\right] \tag{4}$$

for $C_{pk} > 0$ and $C_a > 0$. Based on C_{pk}, Boyles [6] noted a bound for the yield as:

$$2\Phi[3C_{pk}] - 1 \le Yield \le \Phi[3C_{pk}] \tag{5}$$

or

$$\Phi[-3C_{pk}] \le P(\text{NC}) \le 2\Phi[-3C_{pk}] \tag{6}$$

for $C_{pk} \ge 0$.

Pearn and Lin [5] also obtained a yield measure formula based on C_{pm} as

$$P(\text{NC}) = \Phi\left[-\frac{2-C_a}{\sqrt{\dfrac{1}{(3\,C_{pm})^2} - (1-C_a)^2}}\right] + \Phi\left[-\frac{C_a}{\sqrt{\dfrac{1}{(3\,C_{pm})^2} - (1-C_a)^2}}\right] \tag{7}$$

for $(3C_{pm} - 1)/(3C_{pm}) \le C_a \le 1$. Based on C_{pm}, Ruczinski [7] obtained a lower bound for the yield as: *Yield* $\ge 2\Phi(3C_{pm}) - 1$, or $P(\text{NC}) \le 2\Phi(-3C_{pm})$ for $C_{pm} > \sqrt{3}/3$.

Furthermore, the yield measure formula based on C_{pmk} can be displayed as:

$$P(\text{NC}) = \Phi\left[-\frac{3C_{pmk}(2-C_a)}{\sqrt{C_a^2 - [3C_{pmk}(1-C_a)]^2}}\right] + \Phi\left[-\frac{3C_{pmk}C_a}{\sqrt{C_a^2 - [3C_{pmk}(1-C_a)]^2}}\right] \tag{8}$$

for $3C_{pmk}/(1+3C_{pmk}) \le C_a \le 1$. The corresponding lower bound for the yield is: *Yield* $\ge 2\Phi(3C_{pmk}) - 1$, or $P(\text{NC}) \le 2\Phi(-3C_{pmk})$ for $C_{pmk} > \sqrt{2}/3$.

Hence, $P(\text{NC}) \le 2\Phi(-3C)$ and $0 \le C_a \le 1$, i.e. $LSL \le \mu \le USL$, for $C_{pk} = C$. $P(NC) \le 2\Phi(-3C)$ and $(3C-1)/(3C) \le C_a \le 1$, i.e. $m - d/(3C) \le \mu \le m + d/(3C)$, for $C_{pm} = C$. And, $P(\text{NC}) \le 2\Phi(-3C)$ and $3C/(1+3C) \le C_a \le 1$, i.e. $m - d/(1+3C) \le \mu \le m + d/(1+3C)$, for $C_{pmk} = C$. The three indices C_{pk}, C_{pm}, and C_{pmk} provide the same upper bound $P(\text{NC}) \le 2\Phi(-3C)$ when $C_{pk} = C$, $C_{pm} = C$, or $C_{pmk} = C$, but different information about the process centering measure C_a. For example, if $C_{pk} = 1.0$ we only have the information of process yield through the upper bound $P(\text{NC}) \le 2700$ ppm. But, if $C_{pm} = 1.0$ we have the information of process yield through the upper bound $P(\text{NC}) \le 2700$ ppm and process centering measure $0.667 \le C_a \le 1$. And, if $C_{pmk} = 1.0$ we have the information of process yield through the upper bound $P(\text{NC}) \le 2700$ ppm and process centering measure $0.75 \le C_a \le 1$. Table 1 displays the corresponding upper bound of $P(\text{NC})$ in ppm (nonconforming parts per million – NCPPM) for various $C = 0.90(0.01)2.00$ when $C_{pk} = C$, or $C_{pm} = C$, or $C_{pmk} = C$, where $NCPPM = 10^6 \times P(\text{NC})$. For example, for $C = 1.24$, the number of nonconformities is not greater than 200 ppm.

In general, the process mean, μ, and the process standard deviation, σ, are unknown. But, in practice μ and σ can be estimated by the sample mean, $\bar{X} = \sum_{i=1}^{n} X_i/n$, and the sample standard deviation, $S = \left[\sum_{i=1}^{n}(X_i - \bar{X})^2/(n-1)\right]^{1/2}$ or $S_n = \left[\sum_{i=1}^{n}(X_i - \bar{X})^2/n\right]^{1/2}$, to obtain the natural estimators of the three indices $\hat{C}_{pk}$, $\hat{C}_{pm}$, and $\hat{C}_{pmk}$. In order to calculate these estimators, however, sample data must be collected, and a great degree of uncertainty may be introduced into capability assessments owing to sampling errors. The approach of simply looking at the calculated values of the estimated indices and concluding whether the

Table 1 The corresponding upper bound of P(NC) in ppm (NCPPM) for various $C = 0.90(0.01)2.00$ when $C_{pk} = C$, or $C_{pm} = C$, or $C_{pmk} = C$

C	NCPPM	C	NCPPM	C	NCPPM
0.90	6933.948	1.27	138.967	1.64	0.865
0.91	6333.433	1.28	123.034	1.65	0.742
0.92	5780.136	1.29	108.835	1.66	0.636
0.93	5270.804	1.30	96.193	1.67	0.544
0.94	4802.365	1.31	84.946	1.68	0.466
0.95	4371.923	1.32	74.950	1.69	0.398
0.96	3976.752	1.33	66.073	1.70	0.340
0.97	3614.288	1.34	58.198	1.71	0.290
0.98	3282.122	1.35	51.218	1.72	0.247
0.99	2977.997	1.36	45.036	1.73	0.210
1.00	2699.796	1.37	39.566	1.74	0.179
1.01	2445.537	1.38	34.731	1.75	0.152
1.02	2213.370	1.39	30.460	1.76	0.129
1.03	2001.565	1.40	26.691	1.77	0.110
1.04	1808.510	1.41	23.369	1.78	0.093
1.05	1632.705	1.42	20.443	1.79	0.079
1.06	1472.751	1.43	17.867	1.80	0.067
1.07	1327.350	1.44	15.603	1.81	0.056
1.08	1195.297	1.45	13.614	1.82	0.048
1.09	1075.475	1.46	11.868	1.83	0.040
1.10	966.848	1.47	10.337	1.84	0.034
1.11	868.460	1.48	8.996	1.85	0.029
1.12	779.425	1.49	7.822	1.86	0.024
1.13	698.926	1.50	6.795	1.87	0.020
1.14	626.211	1.51	5.898	1.88	0.017
1.15	560.587	1.52	5.115	1.89	0.014
1.16	501.414	1.53	4.432	1.90	0.012
1.17	448.107	1.54	3.837	1.91	0.010
1.18	400.127	1.55	3.319	1.92	0.008
1.19	356.981	1.56	2.869	1.93	0.007
1.20	318.217	1.57	2.477	1.94	0.006
1.21	283.421	1.58	2.137	1.95	0.005
1.22	252.215	1.59	1.842	1.96	0.004
1.23	224.254	1.60	1.587	1.97	0.003
1.24	199.223	1.61	1.365	1.98	0.003
1.25	176.835	1.62	1.174	1.99	0.002
1.26	156.828	1.63	1.008	2.00	0.002

NCPPM: nonconforming parts per million

given process is capable, is unreliable as the sampling errors are ignored. For normally distributed processes, the testing procedures using the natural estimators $\hat{C}_{\text{pk}}$, $\hat{C}_{pm}$, or $\hat{C}_{\text{pmk}}$ for practitioners have been developed to determine if their processes satisfy the targeted quality condition (see [8–11]).

To obtain an exact measure of the process yield, Boyles [6] considered a yield index, referred to as S_{pk}, for normally distributed processes. The index S_{pk} is defined as:

$$S_{pk} = \frac{1}{3}\Phi^{-1}\left\{\frac{1}{2}[\Phi(3C_{\text{PU}}) + \Phi(3C_{\text{PL}})]\right\} \tag{9}$$

There is a one-to-one relationship between S_{pk} and the process yield because for a process with $S_{pk} = c$, we can obtain $Yield = 2\Phi(3c) - 1$ or $P(\text{NC}) = 2\Phi(-3c)$. Table 2 displays the number of nonconformities (in ppm) for various S_{pk} values $= 0.5(0.1)2.0$.

Unfortunately, the exact distribution of $\hat{S}_{pk}$, the natural estimator of the index S_{pk}, is mathematically intractable. Lee *et al.* [12] derived an approximate distribution of the estimator $\hat{S}_{pk}$ using the Taylor expansion technique. They showed that the estimator $\hat{S}_{pk}$ is approximately normally distributed. Chen and Pearn [13] used the **bootstrap** resampling technique to find the lower confidence bound on S_{pk}, so that practitioners/engineers can use them to perform quality testing and determine whether their processes meet the preset quality requirement.

PCIs for One-Sided Processes

Capability indices C_{PU} and C_{PL} have been designed specifically for assessing processes with one-sided specification limit (which require only *USL* or *LSL*). C_{PU} measures the capability of a smaller-the-better process with an upper specification limit. On the other hand, C_{PL} measures the capability of a larger-the-better process with a lower specification limit.

For a normally distributed process with one-sided specification limit USL, corresponding to C_{PU} value the process yield is:

$$Yield = \Pr(X < USL) = \Phi(3C_{PU}), \text{ or}$$
$$P(\text{NC}) = \Phi(-3C_{PU}) \tag{10}$$

Table 2 The corresponding P(NC) in ppm (NCPPM) for various $S_{pk} = 0.5(0.1)2.0$

S_{pk}	0.5	0.6	0.7	0.8	0.9	1.0	1.1	1.2	1.3	1.4	1.5	1.6	1.7	1.8	1.9	2.0
NCPPM	133614	71861	35729	16395	6934	2700	967	318	96	27	7	2	0.340	0.067	0.012	0.002

NCPPM: nonconforming parts per million

On the other hand, for a normally distributed process with one-sided specification limit LSL, corresponding to C_{PL} value the process yield is:

$$Yield = \Pr(X > LSL) = \Phi(3C_{PL}), \text{ or}$$
$$P(\text{NC}) = \Phi(-3C_{PL}) \quad (11)$$

Given value of C_{PU} or C_{PL}, the corresponding fraction of nonconforming in ppm is

$$\text{NCPPM} = 10^6 \times \Phi(-3C_{PU}) \text{ or}$$
$$\text{NCPPM} = 10^6 \times \Phi(-3C_{PL}) \quad (12)$$

In practice, sample data must be collected in order to calculate these indices. The natural estimators of C_{PU} and C_{PL} are respectively defined as:

$$\hat{C}_{PU} = \frac{USL - \overline{X}}{3S} \text{ and } \hat{C}_{PL} = \frac{\overline{X} - LSL}{3S} \quad (13)$$

For normally distributed processes, Pearn and Chen [14] showed that $\tilde{C}_{PU} = b_f \cdot \hat{C}_{PU}$ and $\tilde{C}_{\text{PL}} = b_f \cdot \hat{C}_{PL}$ with $f = n - 1$ and $b_f = \sqrt{\frac{2}{f}} \cdot \frac{\Gamma[f/2]}{\Gamma[(f-1)/2]}$ are the uniformly minimum variance unbiased estimators (UMVUEs) of C_{PU} and C_{PL}, respectively. Based on the UMVUEs, $\tilde{C}_{PU}$, and $\tilde{C}_{PL}$, Lin and Pearn [15] provided testing procedures and efficient *SAS* computer programs for practitioners to assess processes with one-sided specification limit.

Summary and Related Topics

In this article, PCIs dedicated to the measurement of the yield of normally distributed processes with two-sided specification limits as well as one-sided specification limit are discussed. The yield measure formulas based on $C_{pk}, C_{pm}, C_{pmk}, S_{pk}, C_{PU}$, and C_{PL} are all given. The corresponding upper bounds of NCPPM for various values of C_{pk}, C_{pm}, and C_{pmk} as well as the corresponding NCPPM for various S_{pk} values are tabulated. For non-normal situation, the technique of non-normal quantile estimation is the most common method in use today for modifying PCIs [16, 17].

References

[1] Kane, V.E. (1986). Process capability indices, *Journal of Quality Technology* **18**, 41–52.

[2] Chan, L.K., Cheng, S.W. & Spiring, F.A. (1988). A new measure of process capability: C_{pm}, *Journal of Quality Technology* **20**, 162–175.

[3] Pearn, W.L., Kotz, S. & Johnson, N.L. (1992). Distributional and inferential properties of process capability indices, *Journal of Quality Technology* **24**, 216–233.

[4] Pearn, W.L., Lin, G.H. & Chen, K.S. (1998). Distributional and inferential properties of the process accuracy and process precision indices, *Communications in Statistics: Theory and Methods* **27**, 985–1000.

[5] Pearn, W.L. & Lin, P.C. (2004). Measuring process yield based on capability index C_{pm}, *The International Journal of Advanced Manufacturing Technology* **24**, 503–508.

[6] Boyles, R.A. (1991). The Taguchi capability index, *Journal of Quality Technology* **23**, 17–26.

[7] Ruczinski, I. (1996). *The relation between C_{pm} and the degree of includence*, Ph.D. Thesis, University of Würzberg, Würzberg.

[8] Cheng, S.W. (1994). Practical implementation of the process capability indices, *Quality Engineering* **7**, 239–259.

[9] Pearn, W.L. & Lin, P.C. (2002). Computer program for calculating the p-value in testing process capability index C_{pmk}, *Quality and Reliability Engineering International* **18**, 333–342.

[10] Pearn, W.L. & Lin, P.C. (2004). Testing process performance based on capability index C_{pk} with critical values, *Computers and Industrial Engineering* **47**, 351–369.

[11] Lin, P.C. & Pearn, W.L. (2005). Testing process performance based on the capability index C_{pm}, *The International Journal of Advanced Manufacturing Technology* **27**, 351–358.

[12] Lee, J.C., Hung, H.N., Pearn, W.L. & Kueng, T.L. (2002). On the distribution of the estimated process yield index S_{pk}, *Quality and Reliability Engineering International* **18**, 111–116.

[13] Chen, J.P. & Pearn, W.L. (2002). Testing process performance based on the yield: an application to the liquid-crystal display module, *Microelectronics Reliability* **42**, 1235–1241.

[14] Pearn, W.L. & Chen, K.S. (2002). One-sided capability indices C_{pu} and C_{pl}: decision making with sample information, *International Journal of Quality and Reliability Management* **19**, 221–245.

[15] Lin, P.C. & Pearn, W.L. (2002). Testing process capability for one-sided specification limit with application to the voltage level translator, *Microelectronics Reliability* **42**, 1975–1983.

[16] Pearn, W.L. & Chen, K.S. (1997). Capability indices for non-normal distributions with an application in electrolytic capacitor manufacturing, *Microelectronics Reliability* **37**, 1853–1858.

[17] Kotz, S. & Lovelace, C. (1998). *Process Capability Indices in Theory and Practice*, Arnold, New York.

PI-CHUAN LIN AND WEN LEA PEARN

Processing of Survey Data

Introduction

After the data are collected, a number of processing steps must be performed to convert the **survey data** from its raw and unedited state to a verified and corrected state that is ready for analysis and/or dissemination to the users. If the data were collected by *paper and pencil* (*PAPI*) methods, they will have to be converted into a computer-readable form. Responses to open-ended questions may need to be classified into some number of categories using a coding scheme so that these responses can be tabulated. In addition, a number of operations may be performed on the data to reduce survey errors and missing items.

For example, the data may be "cleaned" by eliminating inconsistencies and addressing unlikely responses (e.g., *outliers*). Survey weights may be computed that account for unequal selection probabilities. These weights may be further refined by a series of "postsurvey adjustments" that are intended to reduce coverage error **bias**, nonresponse bias, and sampling variance. Some survey variables (for example, household income) may have numerous missing values and values may be *imputed* for them. Following these steps, the data contents file should be well documented. Data *masking* and *de-identification* techniques may also be conducted on the file to protect the confidentiality of the respondents. The next section provides a brief overview of these data processing activities.

Overview of the Data Processing Steps

The data processing steps vary depending on mode of **data collection** for the survey and technology available to assist in data processing. The steps involved for processing PAPI questionnaires will be discussed initially. These steps include the following:

- scan editing
- data capture
- data editing
- coding
- file preparation
- data documentation
- analysis.

The steps for *computer-assisted interviewing* (*CAI*) methods are essentially the same except that the data entry step is not necessary since the data are already keyed into the computer by an interviewer.

The PAPI Questionnaire

The PAPI questionnaire is used to collect information for the variables under study. Some questions corresponding to these variables are closed-ended requiring the interviewer (or respondent) to check a box representing a response alternative. For example, marital status may be coded as "1 = single", "2 = married", "3 = divorced", "4 = widow/widower", and "5 = never married". For open-ended questions, a free-format response is written in a blank field on the questionnaire. For example, a response to the question "What is your occupation?" may be "I am a security officer for a bank" or "I am a barber. I cut hair." The questionnaire may also involve branching to skip around questions that do not apply. These legitimately skipped items need to be distinguished from items requiring a response that were not answered by the respondent. The latter type of skip constitutes item-missing data, which is often addressed by the subsequent processing steps.

Scan Editing

Scan editing is an operation involving several steps. First, as questionnaires are received by the survey organization, they are coded as received into the receipt control system and inspected for obvious problems such as blank pages or missing data for key items that must be completed in order for questionnaires to be usable. In *interviewer-assisted* surveys, the questionnaires that are rejected as being incomplete by this process may be sent back to the interviewers for completion. For mail surveys, the questionnaires might be routed to a telephone *follow-up* process for completion. In cases where there is no follow-up of nonrespondents, the questionnaires may be passed on to the next data processing step (i.e., data entry). Ultimately, questionnaires that are not minimally completed will be coded as nonrespondents due to incomplete data. Also as part of the process, questionnaires may be grouped into small batches called *work units* to facilitate the subsequent processing steps.

Data Capture

In *data entry* or *data capture*, the paper questionnaire responses are converted into computer-readable form (i.e., digitized). Data can be entered manually using keying equipment or automatically by using scanning or optical character recognition devices. To use the latter methods, messy questionnaires may have to be copied onto to new, clean forms so that the scanner can read them properly. Keying usually involves some form of **quality control** verification. For example, each questionnaire may be keyed independently by two different keyers. Any discrepancies between the first and second keyings are then rectified by the second keyer. Alternatively, **acceptance sampling** methods (typically, *a* **single sampling** *plan*) may be applied to each work unit. Here, only a sample of questionnaires within each work unit is rekeyed. If the number of discrepancies between the two keyings exceeds some threshold value, the entire work unit would be rekeyed. Because of these verification methods, the error rate associated with keying is usually quite low for closed-ended responses – less than 0.5%. However, for verbal responses such as names and addresses, the error rates are substantially higher – 5% or more [1].

Data Editing

Editing is a process for verifying that the digitized responses are plausible and, if not, modifying them appropriately. Editing rules can be developed for a single variable or for several variables in combination. The editing rules may specify acceptable values for a variable (e.g., an acceptable range of values) or acceptable relationships between two or more variables (e.g., an acceptable range for the ratio of two variables). Typically editing identifies entries that are either definitely in error (called *critical edits*) or are highly suspicious of being in error (called *query edits*). Corrective measures must be taken for all critical edits while various rules may be applied to determine which query edits to address in order to reduce the cost of the editing process. This approach is sometimes referred to as *selective editing* [2].

Some surveys specify that respondents should be recontacted if the number of edit failures is large or if key survey items are flagged as in error or questionable. In this manner, missing, inconsistent, and questionable data can be eliminated by the respondent's input. However, this is not always done either to save costs or due to the impracticality of respondent recontacts. In that case, values may be inserted or changed by means of deducing the correct value based upon other information on the questionnaire or from what is known about the sample unit from prior surveys. Consistency checks, selective editing, deductive editing, and other editing functions can be performed automatically by specially designed computer software.

Coding

Coding is a procedure for classifying open-ended responses into predefined categories that are identified by numeric or alphanumeric code numbers. For example, there may be thousands of different responses to the open-ended question "What is your occupation?" To be able to use this information in subsequent analysis, each response is assigned one of a much smaller number (say 300–400) code numbers, which identify the specific occupation category for the response. To make occupation categories consistent across different surveys and different organizations, a standard occupation classification (SOC) system is used. A typical SOC codebook may contain several hundred occupation titles and/or descriptions with a three-digit code number corresponding to each. In most classification standards, the first digit represents a broad or main category, and the second and third digits represent increasingly detailed categories. Thus, for the response "barber", a coder consults the SOC codebook and looks up the code number for "barber". Suppose the code number is 411. Then the "4" might correspond to the main category "Personal Appearance Workers" 41 might correspond to "Barbers and Cosmetologists", and 411 to "Barber". In *automated coding*, a computer program assigns these code numbers to the majority of the cases while the cases that are too difficult to be accurately coded by computer are coded manually.

File Preparation

The file preparation step results in an analysis file that serves as the output file and the output file is used to produce statistics and analyses. This step consists of a number of activities including *weighting*, ***imputation*** *postsurvey weighting adjustments, statistical disclosure analysis*, and *data suppression*. First, weights have to be computed for the sample units. Often the weights are developed in three steps. First, a base weight is constructed, which is the inverse of the probability of selection for a sample member. Second, the base weight is adjusted to compensate for nonresponse error, and third, further adjustments of the weights might be performed to adjust for frame coverage error depending on availability of external information. These so-called *postsurvey adjustments* are intended to achieve additional improvements in the accuracy of the estimate.

Imputation is a process that replaces missing responses with values that may be treated just like other responses in the subsequent analysis. It has been shown that imputation reduces the biases in analysis due to item nonresponse. A common method is hot-deck imputation, which fills in missing values using corresponding values from similar people in the same dataset. Cold-deck imputation, by contrast, selects donors from another dataset.

Finally, if the data file is to be released to the public, a statistical disclosure analysis may be performed to determine the risk that a sample member on the file could identified by an *intruder* on the basis of a samples member's data. This may involve estimating the probability that an individual in the file can be *reidentified* (i.e., his or her identity can be discovered with certainty) by an intruder. If this probability is unacceptably high, then various techniques are available to suppress or mask some responses to avoid identity disclosure and to protect the confidentiality of the sample member (see [3] for an overview of these methods).

Data Documentation and Analysis

The final processing step is data documentation in which a type of data file users manual is created. This document describes the methods used to collect and process the data and provides detailed information on the variables on the file. For example, each variable on the data file might be linked to one or more questions on the questionnaire. If variables were combined, recoded, or derived, the steps involved in creating these variables is described. The documentation might also include information regarding response rates for the survey, item nonresponse rates, reliability estimates, or other information on the total **survey error** of key variables.

The new technologies have opened up many possibilities to integrate these data processing steps [4]. Therefore, the sequence of steps might be very different from those described above for some surveys. For example, it is possible to integrate data capture and coding into one step; likewise, data capture and editing can be integrated with coding. It is also possible to integrate editing and coding with data collection through the use of CAI technology. The advantage of this type of integration is that inconsistencies in the data or insufficient information for coding can immediately be resolved with the respondent. This reduces follow-up costs and may also result in better information from the respondent. Many other possibilities for combining the various data processing steps exist. The goal of integration is to increase the efficiency of the operations while improving **data quality**.

Data Quality and Data Processing

As is clear from the previous discussion, the data can be modified extensively during data processing. Hence, data processing has the potential to improve data quality for some variables while increasing the error for others. Unfortunately, knowledge about the errors introduced in data processing is very limited in survey organizations and consequently such errors tend to be neglected. Often operations are sometimes run without any particular quality control efforts and the effects of errors on the overall accuracy as measured by the **mean squared error** (MSE) are often unknown except perhaps for national data series of great importance.

As an example, although editing is intended to improve data quality, it misses many errors and can even introduce new ones. Automation can reduce some of the errors made by manual processing, but might introduce new errors. For instance, in optical recognition data capture operations, the recognition errors are not uniformly distributed across digits and other characters, which can introduce systematic errors (i.e., biases). For these reasons, quality control and quality assurance measures should be a standard part of all data processing operations. For more information on data processing steps and their contributions to survey error, see [5].

References

[1] Biemer, P.P. & Lyberg, L.E. (2003). *Introduction to Survey Quality*, John Wiley & Sons, Hoboken, p. 224.
[2] Granquist, L. & Kovar, J. (1997). Editing of survey data: how much is enough? in *Survey Measurement and Process Quality*, L. Lyberg, P. Biemer, M. Collins, E. De Leeuw, C. Dippo, N. Schwartz & D. Trewin, eds, John Wiley & Sons, New York, pp. 415–435.
[3] Willenborg, L. & de Waal, T. (1996). *Statistical Disclosure Control in Practice*, Springer, New York.
[4] Couper, M. & Nicholls, B. (1998). The history and development of computer assisted survey information collection methods, in *Computer Assisted Information Collection*, M. Couper, R. Baker, J. Bethlehem, C. Clark, J. Martin, W. Nicholls & J. O'Reilly, eds, John Wiley & Sons, New York, pp. 1–21.
[5] Biemer, P. & Lyberg, L. (2003). Introduction to Survey Quality, John Wiley & Sons, Hoboken, NJ.

PAUL P. BIEMER

Product Array Designs

Introduction

Robust parameter design or **robust design** is the centerpiece of Taguchi's systematic approach to quality improvement (*see* **Robust Design**), and it consists of three major steps. The first step is to identify parameters (or factors) that affect the performance of a system. Taguchi proposed classification of the factors into three types: control factors (or design factors), noise factors, and signal factors. A system with at least one signal factor is referred to as a *dynamic system*. The primary goal of robust design is to find settings of control factors that simultaneously optimize a system's performance and make it robust to variations caused by noise and signal factors. The second step is to investigate the selected factors through experimentation and to identify optimal control settings to achieve the goal of robust design. Statistical design and analysis of experiments play a critical role in this step. The last step is to conduct confirmatory experiments to ensure that, when the optimal control settings are used, the quality of the system under investigation has indeed been improved.

This article gives a brief review of product arrays, which are among the most popular experimental plans in the second step of robust design. Product arrays were originally proposed by Taguchi and were given a different name (i.e. *inner–outer arrays*). Some researchers and engineers use the term *cross* or *crossed arrays*. The name *product array* is used throughout this article. Though Taguchi's original idea of using robust design for quality improvement is widely considered important, many of his proposed methods have been criticized and more efficient and effective alternatives have been developed by statisticians and other researchers in quality engineering. This is also the case with regard to selecting experimental plans for robust design experiments. In this article, both Taguchi's methods and other new alternatives are discussed.

The organization of the article is as follows. The section titled "Product Arrays" first reviews **factorial** designs and orthogonal arrays, and then defines and discusses product arrays. The section titled "Analysis of Parameter Design Experiments" briefly introduces several popular modeling approaches for analyzing data generated from robust design experiments and discusses the advantage of product arrays from the modeling perspective. The section titled "Optimal Selection of Product Arrays" discusses the optimal selection of product arrays. The section titled "Compound Orthogonal Arrays" introduces compound orthogonal arrays, a generalization of product arrays. The section titled "Product Arrays with Signal Factors" briefly discusses product arrays involving signal factors.

Product Arrays

Factorial Design and Orthogonal Array

Suppose there are k factors with $l_1, l_2, \ldots, l_k$ levels, respectively, in an experiment. These factors have in total $l_1 \cdot l_2 \cdots l_k$ different level combinations or settings. If all the settings are used, the experiment is said to have a full factorial design (or a $l_1 \cdots l_2 \cdots l_k$ design). When $l_1 = l_2 \cdots = l_k = l$, i.e. all the factors have the same number of levels, the full factorial design is then said to be an l^k design. For example, when $l = 2$, we have a full factorial 2^k design. After data are collected from an experiment with a full factorial design, the analysis of variance method is often used to assess the factorial effects of the factors, which include main effects, two-factor interactions, and so on (*see* **Analysis of Variance**). If an effect involves r factors, it is said to be of order r where $1 \leq r \leq k$; a **main effect** only involves one factor, so it is of order 1.

The major advantage of using a full factorial design is that it guarantees the independent **estimation** of all the factorial effects. When the number of factors increases, the full factorial design quickly becomes infeasible. In practice, a subset of the factor settings is usually selected and used in the experiment. The selected subset is referred to as a **fractional factorial** *design*. In a fractional factorial design, factorial effects become entangled with each other and are not independently estimable. Nevertheless, a fractional factorial design is sufficient for the study of factorial effects of lower order such as main effects and two-factor interactions, when it is judiciously selected and effects of higher order are assumed to be negligible. The orthogonal arrays proposed by Rao [1] are an example of such designs. An **orthogonal array** with N rows, k columns (or factors), and strength t is an N by k array whose rows are the factor settings such that for any t columns, all the possible level combinations of the corresponding t factors appear an equal number of times. The orthogonal array is denoted by $OA(N, l_1 l_2 \cdots l_k, t)$. If some of the factors have the same number of levels, they are usually grouped together in the notation. For example, if among the k factors, k_1 factors have s_1 levels, k_2 factors have s_2 levels, and so on, then the orthogonal array is denoted as $OA(N, s_1^{k_1} s_2^{k_2} \ldots, t)$. An $OA(N, s_1^{k_1} s_2^{k_2} \ldots, t)$ is sometimes denoted by $L_N(s_1^{k_1} s_2^{k_2} \ldots)$ when its strength is equal to two. *See* **Orthogonal Arrays** for more details.

Orthogonal arrays of strength two, three, or four are commonly used in practice. Under reasonable assumptions about higher-order factorial effects, these orthogonal arrays are sufficient for investigating main effects and/or two-factor interactions. For example, if interactions are assumed to be negligible, then an orthogonal array of strength two guarantees the independent estimation of main effects; if interactions between three or more factors are assumed to be negligible, an orthogonal array of strength four guarantees that both main effects and two-factor interactions are estimable. The estimability of an orthogonal array of strength three is between those of strengths two and four. For general results regarding the estimability of orthogonal arrays of strength t, the reader can consult Hedayat *et al.* [2] and Dey and Mukerjee [3].

Orthogonal arrays can be classified into symmetrical arrays and asymmetrical arrays, according to whether the factors have the same number of levels or not. They can also be classified into regular arrays and nonregular arrays. In general, regular arrays can be constructed by defining relations between factorial effects. Regular symmetrical arrays are among the most popular experimental designs in practice. Two examples are 2^{k-p} and 3^{k-p} designs, where k is the number of factors, 2 or 3 is the number of levels and p is the fraction index. The resolution of a regular symmetrical array is defined to be the lowest order of the effects that are entangled or aliased with the grand mean. It can be shown that the resolution is equal to the strength of the array plus 1. Hence, a 2^{k-p} or 3^{k-p} design with resolution R is an $OA(2^q, 2^k, R-1)$ or $OA(3^q, 3^k, R-1)$, respectively, where $q = k - p$. The minimum aberration criterion proposed by Fries and Hunter [4] is generally used to select optimal regular symmetrical arrays (*see* **Minimum Aberration**). For a comprehensive account of 2^{k-p} and 3^{k-p} designs and extensive tables of optimal designs, the reader is referred to Wu and Hamada [5]. General orthogonal arrays have also been tabulated; for example, Hedayat *et al.* [2] includes a fairly comprehensive table (i.e. Table 12.7) of orthogonal arrays of strength 2 with up to 100 runs, and Sloane maintains a library of over 200 orthogonal arrays online (http://www.research.att.com/~njas/oadir/index.html).

Product Arrays

As discussed in the introduction, there are three possible types of factors present in a robust design experiment. For ease of discussion, we focus on experiments involving only control and noise factors in the sections titled "Product Arrays", "Analysis of Parameter Design Experiments", Optimal Selection of Product Arrays", and "Compound Orthogonal Arrays", and briefly discuss the dynamic cases (i.e. those involving signal factors) in the section titled "Product Arrays with Signal Factors". Product arrays were originally proposed by Taguchi to accommodate control factors and noise factors in robust design experiments.

Suppose there are k control factors and m noise factors in a robust design experiment. Let

$OA_c(N_c, \ldots, t_c)$ be an orthogonal array for the k control factors. Because the number of levels of each factor can be arbitrary, it is left unspecified in the notation. Denote the ith row (i.e., the ith control setting) of $OA_c(N_c, \ldots, t_c)$ as c_i for $1 \leq i \leq N_c$. Then $OA_c(N_c, \ldots, t_c)$ can be considered a collection of control settings, that is, $OA_c(N_c, \ldots, t_c) = \{c_i : 1 \leq i \leq N_c\}$. Similarly, let $OA_n(N_n, \ldots, t_n)$ be an orthogonal array for the m noise factors. Denote the jth row of $OA_n(N_n, \ldots, t_n)$ as n_j. Then $OA_n(N_n, \ldots, t_n) = \{n_j : 1 \leq j \leq N_n\}$. The two arrays $OA_c(N_c, \ldots, t_c)$ and $OA_n(N_n, \ldots, t_n)$ are referred to as the *control array* and the *noise array*, respectively. Taguchi proposed the use of the combined settings (c_i, n_j) with $1 \leq i \leq N_c$ and $1 \leq j \leq N_n$ for the robust design experiment. Clearly, these combined settings are the Cartesian products of the control settings of OA_c and the noise settings of OA_n. Therefore, the resulting design is named a *product array*. The product array is denoted by $OA_p(N_c \times N_n, \ldots, (t_c, t_n))$, and it is equal to

$$OA_c(N_c, \ldots, t_c) \times OA_n(N_n, \ldots, t_n)$$
$$= \{(c_i, n_j) : 1 \leq i \leq N_c, 1 \leq j \leq N_n\} \quad (1)$$

In Taguchi's original terminology, the control array, noise array, and product array are referred to as the *inner array*, *outer array*, and *inner–outer array*, respectively. The product array OA_p is indeed an orthogonal array with $N_c N_n$ rows and $k + m$ columns.

We present a real robust design experiment as an illustrative example. A robust design experiment for improving an injection-molding process was reported in Engel [6]. There were seven control factors, namely, cycle time (A), mold temperature (B), cavity thickness (C), holding pressure (D), injection speed (E), holding time (F), and gate size (G), and three noise factors, namely, percentage regrind (a), moisture content (b), and ambient temperature (c). Each factor had two levels labeled -1 and $+1$, respectively. The goal of the experiment was to determine process parameter settings at which percent shrinkage would be consistently close to a target value. A regular 2^{7-4} design with resolution III (i.e. $OA_c(2^3, 2^7, 2)$) was selected as the control array and a regular 2^{3-1} design with resolution III (i.e. $OA_n(2^2, 2^3, 2)$) was selected as the noise array. The product array is $OA_p(2^3 \times 2^2, 2^{10}, (2, 2)) = OA_c(2^3, 2^7, 2) \times OA_n(2^2, 2^3, 2)$, whose layout together with the data generated from the experiment is given in Table 1.

The c_i's and n_j's in the above table indicate the control and the noise settings, respectively. The experiment has 32 runs in total and one observation at each combined setting of the control and noise factors.

A robust design experiment using a product array can be conducted in two different ways. First, it can be conducted as a completely randomized experiment in which the run order needs to be completely randomized and the factors need to be reset after each run. Second, it can be conducted as a **split-plot** experiment in which the run order is subject to some restrictions. For example, when readjusting the control factors means rebuilding a different prototype system, which can be both time consuming and costly, it is practically more appealing to build the prototype system according to a fixed control setting, and then test the system across all the noise

Table 1 Product Array and Shrinkage Data, Injection Molding Experiment

									n_1	n_2	n_3	n_4
								a	−1	−1	+1	+1
								b	−1	+1	−1	+1
								c	−1	+1	+1	−1
Runs		A	B	C	D	E	F	G				
1–4	c_1	−1	−1	−1	−1	−1	−1	−1	2.2	2.1	2.3	2.3
5–8	c_2	−1	−1	−1	+1	+1	+1	+1	0.3	2.5	2.7	0.3
9–12	c_3	−1	+1	+1	−1	−1	+1	+1	0.5	3.1	0.4	2.8
13–16	c_4	−1	+1	+1	+1	+1	−1	−1	2.0	1.9	1.8	2.0
17–20	c_5	+1	−1	+1	−1	+1	−1	+1	3.0	3.1	3.0	3.0
21–24	c_6	+1	−1	+1	+1	−1	+1	−1	2.1	4.2	1.0	3.1
25–28	c_7	+1	+1	−1	−1	+1	+1	−1	4.0	1.9	4.6	2.2
29–32	c_8	+1	+1	−1	+1	−1	−1	+1	2.0	1.9	1.9	1.8

settings. Therefore, the control setting will not be reset between these runs. An experiment run in this manner is known as a *split-plot experiment*, and the analysis needs to take account of this structure; for details, *see* **Split-Plot Designs**, Box and Jones [7], and Bisgaard and Steinberg [8]. Bingham and Sitter [9] studied the optimal selection of split-plot designs for robust design experiments. In the remainder of this article, we focus on completely randomized robust design experiments only.

Compounding Noise Factors

A major criticism on using product arrays for robust design experiments is that their size tends to be large. For a product array $OA_\mathrm{p}(N_\mathrm{c} \times N_\mathrm{n}, \ldots, (t_\mathrm{c}, t_\mathrm{n}))$, the total number of runs is $N_\mathrm{c}N_\mathrm{n}$. When the number of control and noise factors increases, $N_\mathrm{c}N_\mathrm{n}$ may become too large for the product array to be practical. Therefore, there is a need to limit the size of a product array. According to Taguchi, robust design needs to investigate as many control factors as possible, so there might not be much room to reduce the size of a control array. Taguchi saw the opportunity of size reduction in the noise array. He proposed to compound noise factors as one hyperfactor and select a small number of representative levels of this hypernoise factor as the noise settings. The success of using compounded noise factors in robust design requires the judicious selection of the hypernoise factor settings as well as domain knowledge and understanding of the noise effects. There has not been much study about this technique in the literature, though it has been occasionally used in practice. One systematic investigation of the advantages and disadvantages of this technique can be found in Hou [10].

Analysis of Parameter Design Experiments

After data are collected from a robust design experiment, statistical methods are used to analyze the data and find the control settings for robust optimization. These methods primarily follow three different modeling approaches, which are the signal-to-noise (SN) ratio modeling approach proposed by Taguchi (*see* **Signal-to-Noise Ratios for Robust Design**), the location-dispersion modeling approach, and the response modeling approach. Statisticians and researchers in quality engineering have shown that while the SN ratio modeling approach is effective under some scenarios, it can be seriously flawed in other cases. Many close followers of Taguchi's quality philosophy and practice however still solely rely on this approach. Product arrays might have been proposed particularly for the SN ratio modeling approach. Nonetheless, they are also well suited for the other two approaches. This section briefly reviews these three approaches and then discusses the major advantage of product arrays from an analysis perspective.

SN Ratio Modeling

The SN ratio summarizes performance at each possible control setting. The particular SN ratio used depends on the nature of the problem; *see* Phadke [11] or **Signal-to-Noise Ratios for Robust Design** for details. A common goal of robust design is to achieve a particular nominal level and in this setting the SN ratio is defined to be $\eta = \ln(\mu^2/\sigma^2)$, where μ and σ^2 are the mean and variance of the response of the system under study. Suppose an $OA_\mathrm{p}(N_\mathrm{c} \times N_\mathrm{n}, \ldots, (t_\mathrm{c}, t_\mathrm{n}))$ is used in a robust design experiment and the obtained data are $\{y_{ij} : 1 \le i \le N_\mathrm{c}, 1 \le j \le N_\mathrm{n}\}$, where y_{ij} is the observed response at (c_i, n_j) for $1 \le i \le N_\mathrm{c}$ and $1 \le j \le N_\mathrm{n}$. From the data, the SN ratio at the control setting c_i $(1 \le i \le N_\mathrm{c})$ can be estimated by $\hat{\eta}_i = \ln(\bar{y}_i^2/s_i^2)$, where $\bar{y}_i$ and s_i^2 are the sample mean and variance of $\{y_{ij}\}_{1 \le j \le N_\mathrm{n}}$. It is clear that the use of the noise array in the experiment is to systematically introduce and evaluate variations attributable to the noise factors at different control settings. The SN ratio $\hat{\eta}_i$ reflects the magnitude of the mean response (i.e. the signal) relative to the variation at the control setting c_i (i.e. the noise). Taguchi proposed to select control settings that maximize the SN ratio to make the system robust to the noise factors.

Once the SN ratios $\{\hat{\eta}_i\}_{1 \le i \le N_\mathrm{c}}$ are available, regression analysis is used to identify the so-called dispersion effects, which are the control effects that significantly affect the SN ratio (*see* **Dispersion Effects**). In a separate regression analysis using $\{\bar{y}_i, 1 \le i \le N_\mathrm{c}\}$ as the responses, the control effects that contribute significantly to the response mean can also be identified. Effects that affect the mean response but do not affect the SN ratio are called the *adjustment effects*. For the nominal-the-best problem, Taguchi suggested the following two-step procedure for robust optimization: (a) Select the settings of the control factors involved in the dispersion effects to maximize the SN ratio $\hat{\eta}$; (b) Select the settings of the factors involved in the adjustment effects to bring the mean response on target. The use of this procedure can be justified by the performance measures independent of adjustment (**PerMIA**), which were proposed by Léon *et al.* [12] and further studied by Léon and Wu [13]; *see* **Performance Measures for Robust Design** for details. The procedure may however fail to identify dispersion or adjustment effects under other scenarios; see Bérubé and Wu [14] for a simulation study. Statisticians and probably the majority of researchers in quality engineering seem to agree that the unquestioning use of the SN ratio for robust design is not desirable; more discussions can be found in Nair [15], Steinberg [16], and Wu and Hamada [5].

Location-Dispersion Modeling

Because the success of robust design hinges on the understanding of how control factors affect the mean μ and variance σ^2 of a system's performance, it is desirable to directly model μ and σ^2 as functions of the control effects. This idea leads to the location-dispersion modeling approach, which fits regression models for $\{\bar{y}_i\}$ and $\{\log s_i^2\}$ separately and uses the resulting models for robust optimization. Modeling of response mean and variance separately, however, suffers from an obvious drawback that the **heteroscedasticity** inherent in the problem has not been taken into account. A more efficient approach is to fit the mean and variance jointly. Such approaches have already been studied in the statistics literature; see Davidian and Carroll [17, 18] and Nelder and Pregibon [19]. Engel [6] proposed a model to jointly fit response mean and variance in robust design, which is $E(y) = \mu = x^\tau\beta$, $\mathrm{Var}(y) = \phi V(\mu, \theta)$, and $\log(\phi) = x^\tau\gamma$, where β and γ are parameter vectors and θ is usually a scalar, and x is the vector of control effects. Engel then proposed the use of either pseudonormal likelihood or extended quasi-likelihood for parameter estimation. The fitted model is then used for robust optimization; see Engel [6] for more details.

Response Modeling

In a robust design experiment, both the control and noise factors are varied systematically and the responses at different combined settings are observed. Therefore, it is possible to directly model the response as a function of all the factorial effects including the control effects, the noise effects, and their interactions. Subsequently, the fitted response model can be used to derive models for the response mean and variance, which can be further used for robust optimization. This approach is referred to as the *response modeling approach*. The response modeling approach was hinted as early as 1985 by Eastering [20], and was investigated in detail by Welch *et al.* [21], Shoemaker *et al.* [22], and Vining and Myers [23].

Let x and z denote the control and noise factors, respectively, and let $f(x)$, $g(z)$, $h(x)$, and $k(z)$ denote vectors of control effects and noise effects accordingly. A general model that relates a response y to the control and noise effects is $y = \alpha_0 + f(x)^\tau\alpha + g(z)^\tau\beta + h(x)^\tau\Gamma k(z) + \epsilon$, where α and β are coefficient vectors, Γ is a coefficient matrix of the interactions between the control and noise factors, and ϵ is a random error. This general model induces the models for the response mean and variance, which are, respectively, $\mu = E(y) = \alpha_0 + f(x)^\tau\alpha$, and $\sigma^2 = \mathrm{Var}(y) = \beta^\tau \mathrm{Var}(g(z))\beta + h(x)^\tau\Gamma \mathrm{Var}(k(z))\Gamma h(x)$. Given the data from a robust design experiment, the coefficients α, β, and Γ can be estimated, using the general response model. Plugging these estimates in μ and σ^2 gives the fitted mean and variance models, which can be further used for robust optimization. The advantage of the response modeling approach is multifold. First, it can reveal how control and noise factors interact with each other to affect a system's performance. Second, more efficient experimental strategies such as combined arrays can be used for robust design. Third, **response surface** methodologies such as sequential design and **central composite** design can be incorporated into robust design. *See* **Robust Design** for more details on this approach.

Modeling Advantage of Product Arrays

As mentioned earlier, the product structure of a product array makes it possible to calculate the response mean and variance at any control setting. Therefore, it is well suited for the SN ratio modeling approach. As a matter of fact, the product structure has its merits for the other two approaches as well.

For the location-dispersion modeling approach, Rosenbaum [24, 25] showed that a product array

$OA_p(N_c \times N_n, \ldots, (t_c, t_n))$ with $t_c = t_n = 2$ ensures that variance calculated at each control setting does not confound control-by-control interactions with any control-by-noise interactions. Hence, when modeling variance as a function of control effects, the dispersion effects can be identified. Furthermore, Rosenbaum extended product arrays to compound arrays that have more **power** in finding location and dispersion effects. Compound orthogonal arrays will be discussed in the section titled "Compound Orthogonal Arrays". For the response modeling approach, the advantage of a product array is stated in the following theorem, which is a modified version of Theorem 10.1 in Wu and Hamada [5]. For ease of presentation, both the control and noise arrays of the product array are two-level regular designs.

Theorem *Suppose a 2^{k-p} design is chosen for the control array OA_c, a 2^{m-q} design is chosen for the noise array OA_n, and the product array is $OA_p = OA_c \times OA_n = 2^{k-p} \times 2^{m-q}$.*

1. *If $\alpha_1, \alpha_2, \ldots, \alpha_u$ are u estimable control effects in OA_c and $\beta_1, \beta_2, \ldots, \beta_v$ are v estimable noise effects in OA_n, then the factorial effects $\alpha_i, \beta_j, \alpha_i\beta_j$ for $1 \le i \le u, 1 \le j \le v$ are estimable in OA_p.*
2. *All the km control-by-noise two-factor interactions are not confounded with other main effects or two-factor interactions.*

Part 2 of the above theorem implies that control-by-noise two-factor interactions can be estimated under the assumption that interactions of order higher than two are negligible. This is considered the major advantage of a product array for the response modeling approach.

Optimal Selection of Product Arrays

Owing to the product structure, the estimation properties of a product array are completely determined by those of its control and noise arrays. Control and noise arrays with good estimation properties lead to good product arrays. Therefore, given the size and the numbers of control and noise factors, an optimal product array is the product of a control array and a noise array, which are optimal in their own rights. For regular orthogonal arrays, the minimum aberration criterion is used to define and select optimal designs; and for nonregular orthogonal arrays, minimum aberration type of criteria have been developed lately (for example, see Xu [26, 27]). In practice, many practitioners are not aware of these better designs and still use standard choices such as those advocated by Taguchi. Extensive tables of minimum aberration two-level, three-level, and mixed-level designs with economical size can be found in Wu and Hamada [5], and some optimal nonregular orthogonal arrays can be found in Xu [26, 27].

Compound Orthogonal Arrays

By relaxing the product structure, Rosenbaum [24, 25] proposed compound orthogonal arrays as a generalization of product arrays. Compared with product arrays, compound orthogonal arrays can have higher strengths for given size and better run-size economy for given strengths. A product array $OA_p(N_c \times N_n, \ldots, (t_c, t_n))$ is a collection of (c_i, n_j) where c_i and n_j are the ith row and jth row of the control array $OA_c(N_c, \ldots, t_c)$ and the noise array $OA_n(N_n, \ldots, t_n)$, respectively. Let $c_i \times OA_n = \{(c_i, n_j), 1 \le j \le N_n\}$. Then $OA_p(N_c \times N_n, \ldots, (t_c, t_n)) = \bigcup_{1 \le i \le N_c} c_i \times OA_n$. Clearly when forming the product array, the same noise array is used for each control setting. Rosenbaum suggested that it should be advantageous to relax this restriction and allow the use of different noise arrays at different control settings. For example, at two different control settings c_i and c_j, two different noise arrays denoted by OA_{n,c_i} and OA_{n,c_j} can be considered, where the subscripts c_i and c_j indicate that these arrays depend on the corresponding control settings. The collection of $c_i \times OA_{n,c_i}$ $(1 \le i \le N_c)$ forms an orthogonal array, which Rosenbaum called a *compound orthogonal array*. Owing to the allowance of using different noise arrays, compound orthogonal arrays provide more flexibility in accommodating control factors and noise factors, which leads to the advantage stated in the beginning of this section.

Rosenbaum proposed various ways to construct compound orthogonal arrays. Hedayat and Stufken [28] and Zhu *et al.* [29] have further studied compound arrays that have optimal estimation properties. Some useful two-level compound orthogonal arrays have been tabulated in these two references. Compound orthogonal arrays have not yet been adopted by practitioners in real robust design experiments.

Product Arrays with Signal Factors

As mentioned in the introduction, a system with at least one signal factor is referred to as a *dynamic system*. Taguchi [30] referred to the robust design of such systems by the term *dynamic problems*. The focus of a dynamic problem is to optimize the relationship between a system performance measure and the signal factor and simultaneously make the relationship robust to the noise factors. Taguchi defined the dynamic SN ratio and proposed its use for robust optimization of dynamic systems. This can be considered a generalization of the SN ratio modeling approach discussed in the section titled "SN Ratio Modeling". The response modeling approach has also been developed for dynamic problems by Miller and Wu [31]. Other analysis methods for dynamic problems can be found in Grove and Davis [32], Lunani *et al.* [33], and McCaskey and Tsui [34]. *See* **Signal–Response Systems** for more details on dynamic systems.

The product arrays discussed in the previous sections do not include signal factors. There are three possible methods to incorporate signal factors in a product array. For ease of discussion, we assume there is only one signal factor, denoted by G. The first method is to cross a product array OA_{p} of control and noise factors with G, which leads to an orthogonal array $OA_{\mathrm{p}} \times G$. In other words, for each combined setting of the control and noise factors, the system is tested across the experimental levels of G. The second method is to add the signal factor to the noise array, which leads to a noise–signal array $OA_{\mathrm{n},g}$; and then it is crossed with a control array to generate a product array $OA_{\mathrm{c}} \times OA_{\mathrm{n},g}$. This way was suggested by Taguchi and was often used in practice. The third method is to add the signal factor to the control array, which results in a control–signal array denoted by $OA_{\mathrm{c},g}$; and then it is crossed with the noise array to generate a product array $OA_{\mathrm{c},g} \times OA_{\mathrm{n}}$. There does not exist much study on optimal selection of product arrays involving signal factors, especially for the last two types. See Wu and Hamada [5] for more discussions.

References

[1] Rao, C.R. (1947). Factorial experiments derivable form combinatorial arrangements of arrays, *Journal of the Royal Statistical Society* **9**, 128–139.

[2] Hedayat, A.S., Sloane, N.J.A. & Stufken, J. (1999). *Orthogonal Arrays: Theory and Applications*, Springer, New York.

[3] Dey, A. & Mukerjee, R. (1999). *Fractional Factorial Plans*, John Wiley & Sons, New York.

[4] Fries, A. & Hunter, W.G. (1980). Minimum aberration 2^{k-p} designs, *Technometrics* **22**, 601–608.

[5] Wu, C.F.J. & Hamada, H. (2000). *Experiments: Planning, Analysis, and Parameter Design Optimization*, John Wiley & Sons, New York.

[6] Engel, J. (1992). Modeling variation in industrial experiments, *Applied Statistics* **41**, 579–593.

[7] Box, G.E.P. & Jones, S. (1992). Split-plot designs for robust product experimentation, *Journal of Applied Statistics* **19**, 3–26.

[8] Bisgaard, S. & Steinberg, D. (1997). The design and analysis of $2^{k-p} \times S$ prototype experiments, *Technometrics* **39**, 52–62.

[9] Bingham, D. & Sitter, R.R. (2003). Fractional factorial split-plot designs for robust parameter experiments, *Technometrics* **45**, 80–89.

[10] Hou, X.S. (2002). On the use of compound noise factor in parameter design experiments, *Applied Stochastic Models in Business and Industry* **18**, 225–243.

[11] Phadke, M.S. (1989). *Quality Engineering Using Robust Design*, Prentice Hall, Eaglewood CLiffs.

[12] Léon, R.V., Shoemaker, A.C. & Kacker, R.N. (1987). Performance measurements independent of adjustment (with discussion), *Technometrics* **29**, 253–285.

[13] Léon, R.V. & Wu, C.F.J. (1992). A theory of performance measures in parameter design, *Statistica Sinica* **2**, 335–358.

[14] Bérubé, J. & Wu, C.F.J. (2000). Signal-to-noise ratio and related measures in parameter design optimization: an overview, *Sankhya, Series B* **62**, 417–432.

[15] Nair, V.N. (1992). Taguchi's parameter design: a panel discussion, *Technometrics* **34**, 127–161.

[16] Steinberg, D.M. (1996). Robust design: experiments for improving quality, *Handbook of Statistics*, S., Ghosh & C.R. Rao, eds, Elsevier Science, Vol. 13, pp. 199–240.

[17] Davidian, M. & Carroll, R.J. (1987). Variance function estimation, *Journal of the American Statistical Association* **82**, 1079–1092.

[18] Davidian, M. & Carroll, R.J. (1988). A note on extended quasilikelihood estimation, *Journal of the Royal Statistical Society. Series B* **50**, 74–82.

[19] Nelder, J.A. & Pregibon, D. (1987). An extended quasi-likelihood function, *Biometrika* **74**, 221–232.

[20] Easterling, R.G. (1985). Discussion of the Paper by Kackar, *Journal of Quality Technology* **17**, 191–192.

[21] Welch, W.J., Yu, T.K., Kang, S.M. & Sacks, J. (1990). Computer experiments for quality control by parameter design, *Journal of Quality Technology* **22**, 15–22.

[22] Shoemaker, A.C., Tsui, K.L. & Wu, C.F.J. (1991). Economical experimentation methods for robust design, *Technometrics* **33**, 415–427.

[23] Vining, G.G. & Myers, R.H. (1990). Combining Taguchi and response surface philosophies: a dual response approach, *Journal of Quality Technology* **22**, 38–45.

[24] Rosenbaum, P.R. (1994). Dispersion effects from fractional factorials in Taguchi's method of quality design, *Journal of the Royal Statistical Society. Series B* **56**, 641–652.

[25] Rosenbaum, P.R. (1996). Some useful compound dispersion experiments in quality design, *Technometrics* **38**, 354–364.

[26] Xu, H. (2002). An algorithm for constructing orthogonal and nearly-orthogonal arrays with mixed levels and small runs, *Technometrics* **44**, 356–368.

[27] Xu, H. (2003). Minimum moment aberration for nonregular designs and supersaturated designs, *Statistica Sinica* **13**, 691–708.

[28] Hedayat, A.S. & Stufken, J. (1999). Compound orthogonal arrays, *Technometrics* **41**, 57–61.

[29] Zhu, Y., Zeng, P. & Jennings, K. (2007). Optimal compound orthogonal arrays and single arrays for robust parameter design experiments, *Technometrics*, to appear.

[30] Taguchi, G. (1987). *System of Experimental Design*, UNIPUB, White Plains, Vols 1 and 2.

[31] Miller, A. & Wu, C.F.J. (1996). Parameter design for signal-response systems: a different look at Taguchi's dynamic parameter design, *Statistical Science* **11**, 122–136.

[32] Grove, D.M. & Davis, T.P. (1992). *Engineering Quality and Experimental Design*, Longman Scientific and Technical, Essex.

[33] Lunani, M., Nair, V. & Wasserman, G. (1997). Graphical methods for robust design with dynamic characteristics *Journal of Quality Technology* **29**, 327–338.

[34] McCaskey, S.D. & Tsui, K.-L. (1997). Analysis of dynamic robust design experiments, *International Journal of Production Research* **35**, 1561–1574.

Further Reading

Taguchi, G. (1991). *Taguchi Methods, Vol. 1: Research and Development; Vol. 3 Signal-to-Noise Ratio for Quality Evaluation*, ASI Press, Dearborn.

YU ZHU

Profile Monitoring

Introduction

Profile monitoring is the use of **control charts** for cases in which the quality of a process or product can be characterized by a functional relationship between a response variable and one or more explanatory variables. These cases appear to be increasingly common in practical applications. For each profile we assume that $n (n > 1)$ values of the response variable (Y) are measured along with the corresponding values of one or more explanatory variables (the Xs). The reader is referred to Woodall *et al.* [1] for a more detailed and comprehensive review of profile monitoring methods. This is a relatively new area of application, so no software is known to be available for its implementation.

Jin and Shi [2] and Zhou *et al.* [3] used the term *waveform signals* to refer to what we call *profiles*. These signals are often collected by sensors during manufacturing processes, e.g., tonnage stamping in stamping, torque signals in tapping, and force signals in welding. Gardner *et al.* [4] used the term *signature* in place of our use of *profile*.

Calibration processes are often characterized by linear functions, as discussed by Mestek *et al.* [5], for example. The profile monitoring framework also includes applications in which numerous measurements of the same variable, e.g., a dimension such as thickness, are made at several locations on each manufactured part.

In many cases it is most efficient to summarize the in-control performance with a parametric model and monitor for shifts in the parameters of this model. The control charts are then based on the estimated parameters of the model from successive profiles observed over time. For nonparametric models one can alternatively monitor metrics that reflect the discrepancies between observed profiles and a baseline profile established using historical data.

Some General Issues

Phase I Applications

In phase I, one analyzes a historical set of process data. The goals in phase I are to understand the variation in a process over time, to evaluate the process stability, and to model the in-control process performance. With profile data collected for phase I, one should examine the fit of the hypothesized model for each profile. One should check, for example, for **outliers**. In some cases it may be more appropriate to delete specific points within a profile dataset rather than to discard the entire profile sample.

The principal component analysis, as described by Jones and Rice [6], can be very useful in summarizing and interpreting a set of phase I profile data with identical, equally spaced values of a independent variable X for each profile. With this approach one treats each profile as a multivariate vector of n response Y values and identifies a few mutually orthogonal linear combinations of the Y variables that explain as much variation in the profiles as possible. If these principal components are interpretable, this approach can be very effective in understanding process performance. Jones and Rice [6] recommended plotting the average profile and the profiles with the largest and smallest principal component scores for aiding in the interpretation of the principal components.

Model Selection

Any model selected for a particular application should be as simple as required to adequately model the profile data. Standard recommendations regarding model selection apply. With any model, however, it must be determined how to design a monitoring procedure to detect profile changes over time most effectively, preferably in a way that allows for the interpretation of out-of-control signals.

It is critically important to assess variation between profiles in phase I as well as variation within profiles. Common cause variation is that variation considered to be characteristic of the process and that cannot be reduced substantially without fundamental process changes. It must be determined how much of the profile-to-profile variation is common cause variation and should be incorporated into the determination of the control chart limits to be used for monitoring in phase II. There is common cause variation between profiles if it is not reasonable to expect each set of observed profile data to be adequately represented using the same set of parameters for the assumed model. Process knowledge is always required in the decision whether or not to include profile-to-profile variation as common cause variation. From our experience, it is often unrealistic to expect that process improvement can be used to remove all profile-to-profile variation.

The Control Chart Statistic(s)

In phase II it is often recommended to monitor the profiles using a separate control chart for each parameter of a parametric model, provided the estimators of the parameters at each sampling stage are independent. If the estimators of the parameters of the model for the profile are dependent, as is more often the case, one can use a T^2 chart (*see* **Hotelling's T^2 Chart**). It is important to use an appropriate estimator of the variance–**covariance** estimator in phase I. In particular, if the vectors of estimators are pooled in order to estimate the variance–covariance matrix, then the resulting T^2 chart will tend to be quite ineffective in detecting outlying profiles or sustained shifts in the profiles over time.

If the approach used is nonparametric, then a reasonable approach is to base control charts on metrics that measure the departures of observed process profiles from a baseline profile model developed in phase I. One such metric is the average deviation of the observed profile from the baseline profile calculated over a grid of X values. Gardner *et al.* [4] recommended this approach.

One should make sure that the choice of statistics to monitor does not result in methods that are unable to detect certain types of profile shifts. For example, if one chooses to monitor only a subset of the principal components, or a subset of wavelet coefficients, that were determined using only in-control data, then certain types of out-of-control profile shifts may be undetectable.

In some cases it may be possible to target control charts to detect specified types of process faults. This can be done by inducing the process faults and observing the effect on several profiles. The control chart statistic can then be determined, perhaps using discriminant analysis, to be most effective in detecting a specific fault. Gardner *et al.* [4], for example, used this type of approach. This strategy reduces the dimensionality of the problem and increases the **power** of the charts, but one must be careful not to overlook important types of process faults.

Analysis of Simple Linear Regression Profiles

In this section we review the work done on the simplest case, the one in which the profile is adequately represented by a straight line.

Phase II Methods

Several researchers have studied phase II control charting methods for monitoring a simple linear regression relationship with assumed known values of the Y intercept, slope, and variance parameters. One approach of Kang and Albin [7] involved a multivariate T^2 control chart based on the successive vectors of the least-squares estimators of the Y intercept and slope. Their second method used statistics based on the successive samples of n deviations from the in-control line with a combination of an exponentially weighted moving average (EWMA) chart (*see* **Exponentially Weighted Moving Average (EWMA) Control Chart**) to monitor the average deviation and a range (R) chart (*see* **Moving Range and R Charts**) to monitor the variation of the deviations.

Kim *et al.* [8] proposed alternative control charts for monitoring in phase II. Since they recommended coding the X values so that the estimators of the Y intercept and slope are independent, they monitored each of the two regression coefficients using a separate EWMA chart. Also, they proposed replacing the R chart of Kang and Albin [7] by one of two EWMA charts for monitoring a process standard deviation. The statistical performance of the combined use of these three EWMA charts was shown to be superior to that of the phase II methods of Kang and Albin [7] for sustained shifts in the regression parameters. Their method is also much more interpretable since each parameter in the model is monitored with a separate control chart. One could use other types of control charts, such as individuals and **moving average** control charts (*see* **Control Charts for the Mean**), with the sequences of estimators of the slope and Y intercept.

The approach of Kim *et al.* [8] is illustrated in Figure 1 with Shewhart charts applied to the data shown in Figure 2. These charts show that the fitted

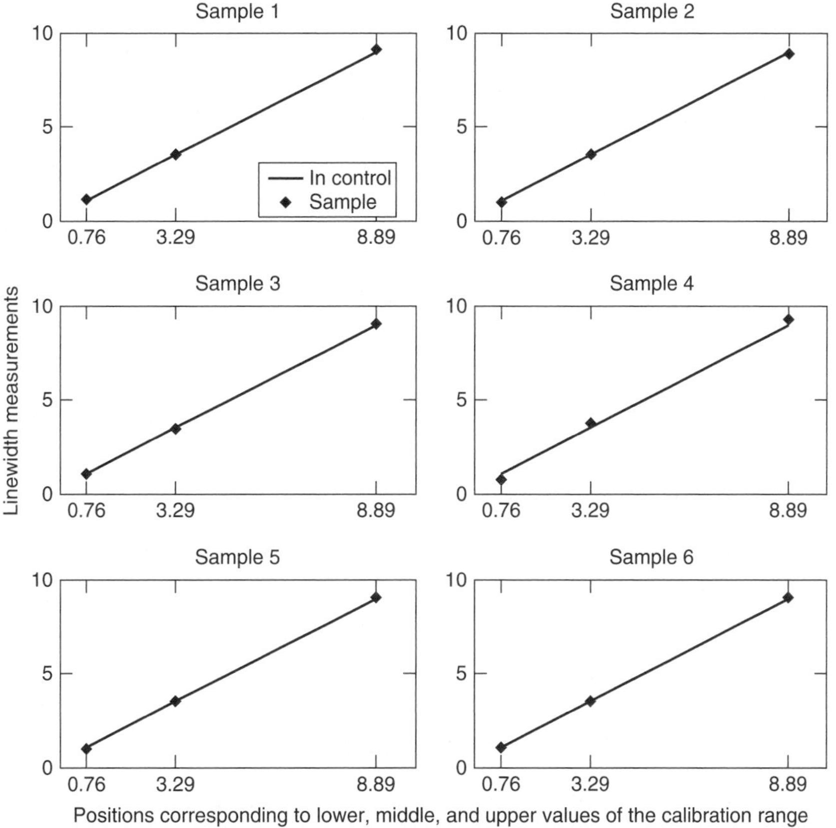

Figure 1 Plot of illustrative phase II calibration data with fitted lines [Reproduced with permission from Gupta *et al.* [9]]

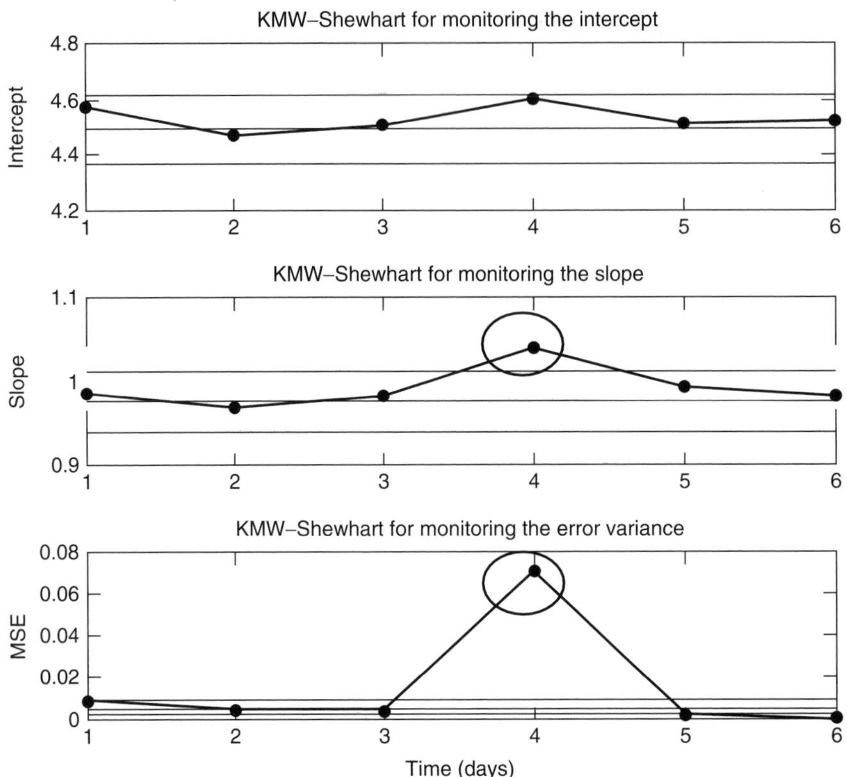

Figure 2 Shewhart control charts using the methods of Kim *et al.* [8] showing a significant deviation from the standard line at sample four [Reproduced with permission from Gupta *et al.* [9]]

line corresponding to sample four was significantly different from the standard line.

Phase I Methods

Kim *et al.* [8] recommended the use of separate Shewhart charts for the *Y* intercept, slope, and the error variance in phase I applications provided one codes the *X* values so that the average *X* value is zero.

Mahmoud and Woodall [10] proposed a phase I method based on the standard approach of testing collinearity using indicator variables for each profile in a multiple regression model. The performance of several methods was compared for different numbers of shifts of a specified size in the regression parameters. The authors showed that some previously proposed phase I methods are ineffective in detecting shifts in the process parameters due to the ways in which the vectors of estimators were pooled to estimate the variance–covariance matrix. Mahmoud and Woodall [10] illustrated several of the phase I methods using a calibration dataset given by Mestek *et al.* [5].

Change-point methods (*see* **Change-Point Methods**) can also be very useful in phase I if one suspects that instability can be modeled adequately by step shifts in the underlying parameter(s). Mahmoud *et al.* [11] developed such a change-point method for the simple linear profile application.

Linear Calibration Applications

As discussed by Kim *et al.* [8], once the parameters are established in a baseline linear calibration study, one would want to detect any change in the *Y* intercept or slope and any increase in the variation about the regression line since such shifts could correspond to greater inaccuracies in the measurement process. The effect of assignable causes in general, however, will vary from application to application. Sometimes, for example, one may wish to detect isolated outliers.

Croarkin and Varner [12] and ISO 5725-6 [13] recommended a method for monitoring a linear calibration line based on the measurement of three standards at each time period. These are at low, medium, and high values, respectively. The three deviations of the measured values from these standards are plotted simultaneously for each sample on a Shewhart-type control chart. Gupta *et al.* [9] showed, however, that the use of control charts based on the estimated regression parameters is more effective. This latter approach is more interpretable, particularly as the number of standards measured at each time period increases. The NIST/SEMATECH *e-Handbook of Statistical Methods*, available online at http://www.itl.nist.gov/div898/handbook/, contains a discussion of the methods proposed by Croarkin and Varner [12].

Other Profile Models and Approaches

As a natural extension of the simple linear regression model, one can consider using control charts to monitor profiles that can be represented by a polynomial regression model or some other multiple linear regression relationship. Jensen *et al.* [14] considered phase II methods for this situation.

In some cases a nonlinear regression model is useful for modeling a profile. A nonlinear model can be used with the vertical board density profile data provided by Walker and Wright [15], where the density of particleboard measured across the diameter of the board is U shaped.

Young *et al.* [16] also considered a vertical density profile application. In their approach, however, the data for each profile were summarized into a top-face, bottom-face, and core average density. Then, a T^2 approach was used for the resulting three-dimensional quality vectors. In general it can be useful to break the range of the X variable into intervals and to model the profile within each interval. Each "subprofile" could be monitored separately or a T^2 method could be used. This approach seems most promising when assignable causes tend to affect the profile only within certain intervals. It is closely related to the use of wavelet models based on the Haar transform.

Jin and Shi [17] used wavelets to model stamping tonnage signals, profiles which can have a rather complicated shape. They recommended Shewhart charts to monitor changes in wavelet coefficients. In some cases a wavelet coefficient can be associated with a specific process fault. Some additional work was reported by Jin and Shi [2], Lada *et al.* [18], and Zhou *et al.* [3].

Nonparametric, or smoothing, methods do not require a specified functional form for the profile. In nonparametric regression methods, one obtains a smoothed curve that can be expressed as a weighted average of the observed responses. Winistorfer *et al.* [19] used this approach to model vertical density profile data for pressed wood panels.

In a semiconductor application, Gardner *et al.* [4] modeled the spatial signatures (i.e., profiles) of wafers using splines and then used various metrics to compare the profiles of new wafers to an established baseline surface. The metrics used were designed to detect specific equipment faults.

Boeing [20, pp. 140–144] also proposed phase II control charting methods using spline fitting. In their application there were multiple gap measurements made at several locations on a manufactured part. Splines were fit to the responses for each part. The average spline value for the first k splines observed or nominal values are used as the baseline. Metrics such as the maximum difference or the average absolute difference was calculated for each spline relative to the baseline. Individuals X charts and moving range charts are then based on the values of the metrics.

References

[1] Woodall, W.H., Spitzner, D.J., Montgomery, D.C. & Gupta, S. (2004). Using control charts to monitor process and product quality profiles, *Journal of Quality Technology* **36**, 309–320.

[2] Jin, J. & Shi, J. (2001). Automatic feature extraction of waveform signals for in-process diagnostic performance improvement, *Journal of Intelligent Manufacturing* **12**, 257–268.

[3] Zhou, S.Y., Sun, B.C. & Shi, J.J. (2006). An SPC monitoring system for cycle-based waveform signals using Haar transform, *IEEE Transactions on Automation Science and Engineering* **3**, 60–72.

[4] Gardner, M.M., Lu, J.C., Gyurcsik, R.S., Wortman, J.J., Hornung, B.E., Heinisch, H.H., Rying, E.A., Rao, S., Davis, J.C. & Mozumder, P.K. (1997). Equipment fault detection using spatial signatures, *IEEE Transactions on Components, Packaging, and Manufacturing Technology – Part C* **20**, 295–304.

[5] Mestek, O., Pavlik, J. & Suchánek, M. (1994). Multivariate control charts: control charts for calibration

curves, *Fresenius' Journal of Analytical Chemistry* **350**, 344–351.
[6] Jones, M.C. & Rice, J.A. (1992). Displaying the important features of large collections of similar curves, *American Statistician* **46**, 140–145.
[7] Kang, L. & Albin, S.L. (2000). On-line monitoring when the process yields a linear profile, *Journal of Quality Technology* **32**, 418–426.
[8] Kim, K., Mahmoud, M.A. & Woodall, W.H. (2003). On the monitoring of linear profiles, *Journal of Quality Technology* **35**, 317–328.
[9] Gupta, S., Montgomery, D.C. & Woodall, W.H. (1996). Performance evaluation of two methods for online monitoring of linear calibration profiles, *International Journal of Production Research* **44**, 1927–1942.
[10] Mahmoud, M.A. & Woodall, W.H. (2003). Phase I analysis of linear profiles with calibration applications, *Technometrics* **46**, 377–391.
[11] Mahmoud, M.A., Parker, P.A., Woodall, W.H. & Hawkins, D.M. (2007). A change point method for linear profile data, *Quality & Reliability Engineering International* **23**, 247–268.
[12] Croarkin, C. & Varner, R. (1982). Measurement assurance for dimensional measurements on integrated-circuit photomasks, NBS Technical Note 1164, U.S. Department of Commerce, Washington, DC.
[13] ISO 5725-6 (1994). *Accuracy (Trueness and Precision) of Measurement Methods and Results – Part 6*. International Organization for Standardization, Geneva.
[14] Jensen, D.R., Hui, Y.V. & Ghare, P.M. (1984). Monitoring an input-output model for production. I. The control charts, *Management Science* **30**, 1197–1206.
[15] Walker, E. & Wright, S.P. (2002). Comparing curves using additive models, *Journal of Quality Technology* **34**, 118–129.
[16] Young, T.M., Winistorfer, P.M. & Wang, S. (1999). Multivariate control charts of MDF and OSB vertical density profile attributes, *Forest Products Journal* **49**, 79–86.
[17] Jin, J. & Shi, J. (1999). Feature-preserving data compression of stamping tonnage information using wavelets, *Technometrics* **41**, 327–339.
[18] Lada, E.K., Lu, J.C. & Wilson, J.R. (2002). A wavelet-based procedure for process fault detection, *IEEE Transactions on Semiconductor Manufacturing* **15**, 79–90.
[19] Winistorfer, P.M., Young, T.M. & Walker, E. (1996). Modeling and comparing vertical density profiles, *Wood and Fiber Science* **28**, 133–141.
[20] Boeing Commercial Airplane Group, Material Division, Procurement Quality Assurance Department (1998). *Advanced Quality System Tools*, AQS D1-9000-1, The Boeing Company, Seattle.

Related Articles

Change-Point Methods; Control Charts, Overview; Exponentially Weighted Moving Average (EWMA) Control Chart; Multivariate Control Charts Overview; Regression Control Charts.

William H. Woodall

Project Definition, Benchmarking Processes for *see* Benchmarking in Project Definition

Project Management, Stage-Gate Approach to

Introduction

The project process is a predetermined set of repeatable steps, or stages, that are used in sequence to plan and implement a variety of project types. The term *stage-gate* has been around for quite sometime and is defined for this article on project process as the criterion or question which must be appropriately addressed prior to proceeding to the next stage in the project.[a]

There is an old saying in the project management world – "Don't work any harder than your project sponsor – or you and your team will surely be frustrated." Keeping the project sponsor fully engaged and working on the right things at the right times before, during, and after completion of the project is one of the key elements necessary for a project's success. As a project manager, you might think they should already know this and it is really not your job to manage the sponsor and the project too. However, many project sponsors are not fully aware of the scope and importance of their role supporting the projects they sponsor and may not even be aware of the appropriate sequence of events necessary to perform the project successfully. As the project manager you may need to become an educator and coach the sponsor on how to effectively perform this key role. After all, why risk your project's

Table 1 Project process phases

Planning phase	Solution design phase	Implementation phase	Operational startup phase
Stage 1 Inception *Stage 2* Feasibility	*Stage 3* Concept design *Stage 4* Scheme design *Stage 5* Detailed design	*Stage 6* Performance *Stage 7* Commissioning and Handover	*Stage 8* Process integration *Stage 9* Closeout and Benefits Realization

success if you can influence the outcome by managing the project sponsor along with the rest of the project?

Every project an organization implements, must be evaluated as an investment of resources to produce a benefit or add value to the organization. Accordingly, a good project manager knows the value of killing a project early to ensure that not a moment of effort is wasted on a project that has no merit. Most organizations are faced with more work than they have the time and resources to complete, so with limited resources even a day spent on a project that will ultimately be canceled, is a day that could have been spent more productively on another project delivering value for the investment being made. As a project manager you also may be thinking, but I do not want to be known as the *project killer* in my organization, what is in it for me? Would it not be better to be leading projects that are hugely successful and adding great value to the organization? This is not necessarily possible if you are working on projects that ultimately will not be implemented, and if they are, do not deliver value to the organization at least several times the investment made in the project.

This article provides an overview of the disciplined approach to project process that engages the right resources at the right time, only when predetermined criteria are met at planned points along the project's timeline. It also ties in the role a project sponsor plays, at each phase in the project process, so the project can proceed in the most efficient and expeditious manner. The project process model applies to a wide variety of projects from Six Sigma and lean projects to laboratory, manufacturing, building construction, and IT projects. Asking the hard questions and responding with honest answers will set the project up for success – or, in some cases, take it off the project list as quickly as possible.

Project Process Model Summary

The project process model is an end-to-end process comprised of four main phases and nine stages within these phases. The key premise is that all projects will be appropriately vetted and prioritized against other opportunities the organization has to invest its resources, and realize the intended benefits. To do this requires that all those involved in the project prioritization and selection effort follow the same disciplined approach for all projects.

The use of a documented process signals to everyone involved, the importance of the process. Any deviation from the process can be flagged and its potential impact on the project's success understood. Table 1 shows the stages in each phase of the project process.

Each project moves forward in a sequence and prior to proceeding to the next stage in the project, a stage-gate criterion or question must be appropriately addressed. Each phase is described further below with an outline of the roles of key players in the project process.

Roles and Responsibilities

Individuals have specific roles and responsibilities assigned during a project because of their position and level of influence within the organization or on a specific project. At a minimum, there are four key roles that need to be established for each project: project sponsor, project manager (or project director), project team, and stakeholders.

Project Sponsor

The project sponsor is accountable for the realization of the project's business benefits and has macrolevel scope control. The project sponsor also ensures the project is appropriately resourced to achieve the intended results and undertakes regular project reviews and makes necessary adjustments. The project sponsor may also hold the executive role as well. Staying connected to the executive or top governing level in the organization is necessary to ensure that the project has the appropriate level of visibility and support in the organization to be successful. If the project sponsor does not hold the executive role, he will need to periodically inform the executive of the progress the project is making against the intended objectives.

Project Manager (or Project Director)

The project manager (or project director) is responsible for leading the planning and execution of a project in a manner consistent with the needs of the organization and the directives of the project sponsor. He is also responsible for providing feedback on the project process in support of continuous quality improvement.

Project Team

The project team is responsible for planning and executing the project in a manner consistent with the needs of the organization, the directives of the project manager, and the approved project plan. They are also responsible for looking for ways of improving the project process and providing feedback in support of continuous quality improvement.

Stakeholders

Stakeholders are individuals and groups who have a stake in the outcome of the project and whose interests must be recognized for the project to be successful. Customers are also considered stakeholders and are members of the project team.

Project Process Phases

Planning Phase

The time spent on planning a project returns many dividends over the life of the project. The planning phase has two stages, Stage 1 – Inception and Stage 2 – Feasibility.

Stage 1 – Inception. The inception stage initiates the project. The project sponsor, working with the senior leadership, determines what is needed by the organization to meet its strategic business objectives and assigns the appropriate resources to begin the inception stage. The project manager should be assigned to work directly with the project sponsor to gather data and begin the analysis to determine the nature of the project opportunity. The initial focus is on asking the right questions to understand why the project is needed and how the project fulfills these needs. At this stage, the unmet business needs will be identified by the customer and strategic alternatives that meet the business needs, generated. This will be summarized as business objectives.

Activities in this stage include determining the nature of the project or business opportunity; identifying strategic alternatives; evaluating potential risks and benefits; estimating costs to a Class I level ±50%; and obtaining preliminary funding. The key deliverable in this stage is a statement of business objectives, which answers the stage-gate question "Is the project needed?"

Stage 2 – Feasibility. In the feasibility stage, the anticipated business benefits gained by executing the project will be identified and quantified where possible. Further, the project description will be written to include how it directly supports the business strategy. The feasibility stage tests the fundamental economic soundness, and business and technical viability of the project. Based on the business needs and anticipated benefits of the project, detailed project objectives will be written.

At this point a clearly defined and agreed scope of work needs to be written, based on the project objectives approved by the project sponsor. Scope definition ensures there are clear boundaries and definitions of processes, facilities, and systems to be delivered. Scope definition is a key component for developing risk analysis, funding requests, and deciding on project controls strategy. Scope will define quality, quantity, and functionality of the end result of the project. Scope will also define work specifically excluded from the project where overlapping activities may occur.

The activities of this stage include defining the project's objective and purpose; defining the general scope of work; addressing the project's impact in relation to facilities, processes, software, and systems; and estimating capital and operational costs to a Class II level ±20–25%. The conclusion of this stage answers the stage-gate question "Is there a business benefit?" prior to proceeding to the next stage.

Solution Design Phase

The solution design phase initiates project design activities based on the upfront planning efforts. In this phase the project team usually expands to include the designer and any subject matter experts necessary to complete the design in a manner consistent with the business and project objectives, and scope of work. Additional project team members may come from both internal and external sources.

During this phase the risk of the scope expanding beyond the original intent is the greatest, and the project sponsor's role is to ensure the design work proceeds according to the agreed scope of work. Any changes to the scope of work places the project at risk. The project sponsor's role during this phase also includes reviewing the complement of project team members and making adjustments as needed, supporting the project in clearing funding hurdles, and keeping the executive leadership updated on the project's progress and informed of any potential risks to the project success.

Stage 3 – Concept Design. The basis of design is written at the beginning of this stage. Sometimes called the *design program*, it is the detailed listing of specifications and design requirements written prior to beginning the solution design. Preliminary design options and the associated costs of each option are then generated and reviewed to determine which designs to pursue further. The project team will seek input from key stakeholders during the review and selection of design options. At the conclusion of this stage a single design option will be selected as the basis of the scheme design stage.

The activities of this stage include developing preferred designs with associated costs; selecting a single option as the basis of the scheme design stage; preparing the project execution plan; and estimating costs to the Class III level ±15–20%. The conclusion of this stage answers the stage-gate question "Is this the right design option?" prior to proceeding to the next stage.

Stage 4 – Scheme Design. During the scheme design stage the designer continues design of the selected option and produces design documentation to 20–35% complete. The project team will involve and seek input from key stakeholders at key points as the scheme design efforts and costs progress. The designer provides other information needed to support a full funding request with cost estimates taken to a Class IV level ±10%. The conclusion of this stage answers the stage-gate question "Should the organization commit full funding to the project?"

Stage 5 – Detailed Design. During the detailed design stage, design documents including flow diagrams, calculations, drawings, specifications, and other information are completed and bidding or pricing packages are produced for the procurement of equipment and services needed during the implementation phase. The project team will continue to involve stakeholders as the detailed design efforts are accomplished. Cost estimates are reviewed at the Class IV level ±10%. The conclusion of this stage answers several stage-gate questions "Are these the right services providers, contractors and suppliers? Are these the right bidding or pricing packages?"

Implementation Phase

During this phase changes to processes, systems, software, and facilities are initiated and completed, and the resulting product is transferred to the operating group. In many cases, this is the most complex and time-consuming portion of the project, requiring constant communication among the project team members, the stakeholders directly affected by all the changes, and those performing the work activities.

Stage 6 – Performance. The performance stage is where changes to the processes, facilities, equipment, software, and systems are underway and completed. The progress and quality of the work during this stage will be tracked according to the design documents, time schedules, and approved costs. Any variances will be tracked and a strategy developed to return to the planned activities.

Activities concluding this stage are focused around review and acceptance of completed and substantially completed processes, facilities, software, equipment, and systems by the customer. The conclusion to this stage answers the stage-gate question: "Is the designed change complete?"

Stage 7 – Commissioning and Handover. Commissioning is the testing of the change (facilities, equipment, software, processes, and systems) to determine if they function as designed. It also includes the documenting of the test data supporting the findings. The commissioning effort begins with a commissioning plan which includes a schedule of activities and is integrated with the overall performance stage schedule.

Handover activities during this stage include training customer operations staff to use the new or changed systems, and then systematically transferring responsibility from the project team to the customer. The conclusion of this stage answers these stage-gate questions "Does the facility, process, software, or system function as designed? Does the business/customer accept operational responsibility?"

Operational Startup

This final phase of the project process has two stages which provide evidence that the designed and implemented changes operate in a consistent manner and that the organization realizes the benefits promised at the beginning of the project, based on its investment in the project.

Stage 8 – Process Integration. The process integration stage, also called *validation in highly regulated environments*, is the implementation of rigorous activities which provide documented evidence that the specific process, facility, software, or system consistently produces results within the design specifications; and ensures product compliance with regulatory standards and full integration with other processes. The conclusion of this stage answers the stage-gate question "Does the facility, process, software, or system produce consistent results?"

Stage 9 – Closeout and Benefits Realization. Closeout and benefits realization stage concludes the project process with activities including final adjustments and repairs, record drawings, contract closeout obligations, financial accounting, customer satisfaction surveys, operation and maintenance lessons learned, and documented evidence that the intended business benefits were realized. This final stage concludes by answering the stage-gate question: "Are the anticipated benefits being realized?" prior to final project closeout.

Summary

Viewing each project as an investment made by your organization to achieve its strategic objectives, underscores the need to gain a significant return on that investment in the form of capability, flexibility, speed, cost, and competitive advantage in the industry. By applying the same disciplined approach to each project, your organization establishes ways of working that sets the organization up to reliably produce projects and have the greatest likelihood of gaining the return on investment promised at the outset.

Keeping the project sponsor connected to the process and aware of the role he plays during each phase, brings visibility of the importance of the project to all levels of the organization and contributes to the likelihood of having the necessary resources to ensure a successful outcome.

End Note

[a] The origin of the stage-gate methodology is based on the experiences, suggestions, and observations of a large number of managers and firms in over 60 cases as observed by Robert Cooper. The term *stage-gate* first appeared in an article by Cooper in the *Journal of Marketing Management*, **3**, 3, Spring 1988. An even earlier version can be found in Cooper's book: *Winning at New Products*, 1988 (http://www.12manage.com/methods_cooper_stage-gate.html).

THOMAS F. TRABERT

Project Management, Team Building in *see* Team Building

Projectivity in Experimental Designs

Concept and Usefulness

Projectivity concerns the properties of a **factorial** design when it is restricted to a subset of experimental factors. Box and Tyssedal [1] defined projectivity of two-level designs as follows: A $n \times k$ design with n runs and k factors each at two levels is said to be of projectivity P if the design contains a complete 2^P factorial in every possible subset of P out of the k factors, possibly with some points replicated. Box and Tyssedal [1] describe such designs as (n, k, P) screens. In general a $n \times k$ design with all factors at s levels is said to be of projectivity P if every subset of P factors out of possible k contains a complete s^P factorial design, possibly with some points replicated. This article will present results for two-level designs only.

The concept of projectivity is related to strength of orthogonal arrays (Rao [2], *see* **Orthogonal Arrays**), but differs in that complete copies of a 2^P factorial are not required in order to have projectivity P. Projectivity is also related to the resolution of a design (Box *et al.* [3], *see* **Factorial Designs, Resolution of**). A design is of resolution R if no p-factor effect is confounded with any other effect containing less than $R - p$ factors. For geometric fractional factorial designs (*see* **Fractional Factorial Designs**); that is a $1/2^s$ fraction of a 2^k factorial it is always the case that $P = R - 1$ [1].

The rationale for considering projection properties of **screening designs** was first pointed out by Box *et al.* [3] and further explained in Box and Tyssedal [4]. In factor screening a model that often sufficiently approximates reality is that out of a larger number k of tested factors only a small subset, typically two or three, maybe four, are expected to really affect a particular response. This is known as *factor sparsity* and tells us that out of the whole k dimensional factor space there is an active subspace of lower dimension within which most of the changes in the measured response occur. When choosing an appropriate design for the experimental investigation, it is then of importance to consider projections of the design on to small subsets of factors.

Statistically, projectivity P implies that all main effects and all interactions of any P factors are estimable with no **bias** if the other factors are inert, but its usefulness for factor screening goes beyond that of **estimation** capacity. If our candidate set under investigation contains the true active factors, replicated runs have the same expected value. Thereby a model independent estimate of the error variance may be obtained regardless of any functional relationship, and the ability of a subset of factors to explain the variation in the response may be evaluated with rather weak assumptions on the underlying model. A pretty common assumption for analyzing a screening experiment is to assume that a model consisting of main effects and interactions gives an adequate approximation of the response. Also the principle of effect heredity (*see* **Plackett–Burman Designs**), is often a precept. These assumptions cannot be relied upon in every situation, Box and Tyssedal [4]. The importance of taking projective properties into account is that it may enable us to identify the active subspace without necessarily imposing restrictive assumptions on the underlying model. A more thorough investigation of the real relationship between design factors and responses can then be performed once the active subspace is identified.

The rest of this article will focus on projective properties of two-level designs, an example to illustrate how such properties can be taken advantage of when a factor screening is performed and some techniques that can be used to obtain designs with good projection properties in various experimental situations.

Projective Properties of Two-Level Orthogonal Arrays

Orthogonal two-level designs may be constructed from Hadamard matrices. A Hadamard matrix is a $n \times n$ matrix with entries ± 1 where rows and columns are pairwise orthogonal. Given a Hadamard matrix, H, an equivalent matrix with all elements in the first column equal to $+1$ can be obtained by multiplying by -1 each element in every row of H whose first element is -1. The remaining $n - 1$ columns must have half 1's and half -1's and constitute an orthogonal two-level design that can accommodate $n - 1$ factors in n runs. Orthogonal designs with $k = n - 1$ are called *saturated*.

When $n = 2^k$ one type of two-level design can be written down as follows. In the first column -1 and $+1$ are alternated. The second column consists of alternating pairs of -1's and $+1$'s. In general the length of the string of -1's and $+1$'s doubles for each column up to column k which consists of 2^{k-1} -1's followed by 2^{k-1} $+1$'s. The rest of the columns are then all possible interaction columns between these k columns. Designs written down in this way are known as the *fractional factorials* (*see* **Fractional Factorial Designs**), sometimes called *geometric designs* or *regular designs*. The rest of the orthogonal two-level arrays are called *nongeometric designs* or sometimes *nonregular designs*. A necessary condition for these designs to exist is that $n \equiv 0$ mod 4.

Now let H_n be a $n \times n$ Hadamard matrix with all elements in the first column equal to $+1$. Then a $2n \times 2n$ Hadamard matrix may be constructed by the following technique: $H_{2n} = \begin{pmatrix} H_n & H_n \\ H_n & -H_n \end{pmatrix}$, called *doubling*. The last $2n-1$ columns constitute a $2n \times 2n-1$ orthogonal array. Another way of obtaining orthogonal arrays is by cyclic generation (*see* **Plackett–Burman Designs**). Box and Tyssedal [1] were able to prove three propositions that are useful for determining the projectivity of an orthogonal two-level array.

- A saturated design obtained from a doubled $n \times n$ Hadamard matrix is always of projectivity $P = 2$ and only 2.
- A saturated design obtained from a cyclic orthogonal array is either a geometric factorial orthogonal array with $P = 2$ and only 2, or else has projectivity at least $P = 3$.
- Any saturated two-level design obtained from an orthogonal array containing $n = 4m$ runs, with m odd, is of projectivity at least $P = 3$.

The last result is also in agreement with a published result in Cheng [5].

All saturated fractional factorials can be obtained from a doubled $n \times n$ Hadamard matrix and are therefore projectivity $P = 2$ and only two designs. However for all of them it is possible to remove $n-1$ columns and obtain a $(2n, n, 3)$ screen [1]. If the saturated design obtained from H_n is of projectivity $P = 3$, the design obtained from doubling H_n and then removing the column with $+1$ for the first n runs and -1 for the last n runs will be a $(2n, 2n-2, 3)$ screen.

It is interesting to note that for $n = 8$ it is only possible to obtain a $(8, 4, 3)$ screen whereas the orthogonal array for $n = 12$ is cyclic generated and provides us with a $(12, 11, 3)$ screen. Hall [6] showed it was possible to construct five 16-run orthogonal arrays. One provides us with the well-known fractional factorial design from which a $(16, 8, 3)$ screen can be constructed. For the four others $(16, 12, 3)$ screens can be obtained from three of them and a $(16, 14, 3)$ screen can be obtained from the fourth [4].

An important class of two-level designs was derived by Plackett and Burman (PB) [7] (*see* **Plackett–Burman Designs**). Using the three propositions above, all PB designs can be classified with respect to being of projectivity $P = 2$ and only 2 or of projectivity at least $P = 3$ (Box and Tyssedal [1], *see* **Plackett–Burman Designs**). Later Samset and Tyssedal [8] discovered by a computer search that PB designs with $n = 68$, 72, 80, and 84 also are of projectivity $P = 4$.

An Example

To show the usefulness of exploiting projective properties in the search for active factors, assume a 12-run PB design (*see* **Plackett–Burman Designs**) in the 11 factors $x_A, x_B, \ldots, x_K$, has been performed and that y is related to the three factors x_A, x_B, and x_C through the functional relationship

$$E(y) = 2x_A + 0.8x_B + \frac{2}{0.4x_B + x_A x_B x_C + 2} \quad (1)$$

Noise from a normally distributed error term with mean 0 and standard deviation $\sigma = 0.3$ was added and the projective-based search given in Tyssedal and Samset [9] and Tyssedal *et al.* [10] (*see also* **Plackett–Burman Designs**) was used to identify the active factors.

The two first columns of Table 1 show the five subspaces of three factors giving the best fit in a projective-based search. The subspace consisting of the three factors x_A, x_B, and x_C is clearly the one that best explains the variation in the data. The two last columns show the result of a search through all possible three-factor models consisting of main effects and two-factor interactions, a rather common precept for a search based on the heredity principles. We notice that only x_A is included in the five candidate sets having the smallest estimated σ. In fact with zero noise and all three-factor interactions assumed inert, x_B and x_C are not included in the 10 subspaces of three factors giving the best fit.

A Bayesian approach to perform a factor based search is given in Box and Meyer [11].

Table 1 A comparison between two searches for identifying three active factors in a 12-run PB design. One is a projective-based search. The other is a search where the three-factor interaction is neglected

Projective-based search		Neglecting interactions with three or more factors	
Factors in model	$\widehat{\sigma}$	Factors in model	$\widehat{\sigma}$
x_A, x_B, x_C	0.23	x_A, x_H, x_I	0.57
x_A, x_F, x_G	0.62	x_A, x_G, x_J	0.73
x_A, x_D, x_J	0.63	x_A, x_E, x_G	0.76
x_A, x_H, x_I	0.64	x_A, x_D, x_F	0.80
x_A, x_G, x_J	0.64	x_A, x_G, x_K	0.85

Some Techniques Used to Construct Designs with Good Projection Properties

The **foldover** of a $n \times k$ two-level design is the design obtained when all mirror images of the original design's runs are added in addition to a column where the first n entries are $+1$ and the last n entries are -1. Let $\mathbf{1}$ be a $n \times 1$ vector of ones. The foldover, $\tilde{\mathbf{X}}$, of a two-level design $\mathbf{X}$ is given by $\tilde{\mathbf{X}} = \begin{bmatrix} \mathbf{X} & \mathbf{1} \\ -\mathbf{X} & -\mathbf{1} \end{bmatrix}$.

Zeiden and Zemach [12] showed that the foldover of a $n \times k$ orthogonal two-level design of strength 2 is a $2n \times (k+1)$ orthogonal two-level design of strength 3 and therefore according to Cheng [5] also a projectivity $P = 4$ design in $k + 1$ factors if n is not a multiple of 8.

Orthogonal two-level arrays have some important hidden projections properties [13, 14]. Assume interactions involving more than two factors can be neglected. The following results are due to Cheng [5] and Cheng [15] respectively.

- If n is not a multiple of 8, then in the projections on to any four factors, all main effects and two-factor interactions are estimable.
- If n is not a multiple of 16, then in the projections on to any five factors in a $P = 4$ design, obtained by foldover, all main effects and two-factor interactions are estimable.

Let $\mathbf{X}$ be a $(n, k, 3)$ screen. The design $\begin{bmatrix} \mathbf{X} & \mathbf{X} \\ \mathbf{X} & -\mathbf{X} \end{bmatrix}$ obtained from doubling $\mathbf{X}$ is then a $(2n, 2k, 3)$ screen. Note that $\mathbf{X}$ does not have to be orthogonal. The 68-run PB design is of projectivity $P = 4$. Lin [16] introduced a new class of supersaturated designs (*see* **Supersaturated Designs**), using half-fractions of Hadamard matrices. Such a construction normally reduces the projectivity by one [8]. A $(34, 66, 3)$ screen can then be constructed using an arbitrary column in the 68-run PB design as a branching column [8]. From this design a $(68, 132, 3)$ screen can be constructed by doubling.

Foldover and doubling have an interesting coupling to split-plot experimentation (*see* **Split-Plot Designs**). Suppose for each level combination of the whole-plot factors, two runs as mirror image pairs are used at the subplot level. The corresponding design can be written as

$$\begin{bmatrix} \mathbf{W} & \mathbf{S} \\ \mathbf{W} & -\mathbf{S} \end{bmatrix}$$

where $\begin{bmatrix} \mathbf{W} \\ \mathbf{W} \end{bmatrix}$ contains the whole-plot factors and $\begin{bmatrix} \mathbf{S} \\ -\mathbf{S} \end{bmatrix}$ the subplot factors. $\mathbf{W}$ and $\mathbf{S}$ may be different or equal. For instance if $\mathbf{X}$ is a $(n, k, 3)$ screen, $\mathbf{W} = \mathbf{S} = \mathbf{X}$ gives us a split-plot design of projectivity $P = 3$ that can accommodate n whole-plot factors and n subplot factors. For more details on these designs (see Tyssedal and Kulahci [17]).

Concluding Remarks

The reason it is important to focus on projective properties is the need to be able to identify active subspaces of factors without necessarily imposing restrictive assumptions on the underlying model. Nongeometric designs have very favorable projective properties compared to geometric ones, and from that perspective seem to be the preferable choice when many factors need to be investigated in few runs and factor sparsity can be assumed. Many of the nongeometric designs have also very favorable hidden projection properties that essentially increase their projectivity by one if the model is adequately approximated by means of main effects and two-factor interactions. Such designs have therefore a great potential for use in screening experimentation and more attention should be paid to their use.

References

[1] Box, G.E.P. & Tyssedal, J. (1996). Projective properties of certain orthogonal arrays, *Biometrika* **83**(4), 950–955.

[2] Rao, C.R. (1947). Factorial experiments derivable from combinatorial arrangements of arrays, *Journal of the Royal Statistical Society, Supplement* **9**, 128–139.

[3] Box, G.E.P., Hunter, J.S. & Hunter, W.G. (1978). *Statistics for Experimenters: An Introduction to Design, Data Analysis, and Model Building*, John Wiley & Sons, New York.

[4] Box, G.E.P. & Tyssedal, J. (2001). Sixteen run designs of high projectivity for screening, *Communications in Statistics: Simulations and Computation* **30**(2), 217–228.

[5] Cheng, C.S. (1995). Some projection properties of orthogonal arrays, *The Annals of Statistics* **23**, 1223–1233.

[6] Hall, M.J. (1961). Hadamard matrices of order 16, *Jet Propulsion Laboratory, Summary* **1**, 21–26.

[7] Plackett, R.L. & Burman, J.P. (1946). The design of optimum multifactorial experiments, *Biometrika* **33**, 305–325.

[8] Tyssedal, J. & Samset, O. (1999). *Two-Level Designs with Good Projection Properties*, Preprint, Statistics No. 12/1999, The Norwegian University of Science and Technology.

[9] Tyssedal, J. & Samset, O. (1997). *Analysis of the 12 Run Plackett-Burman Design*, Preprint, Statistics No. 8/1997, The Norwegian University of Science and Technology.

[10] Tyssedal, J., Grinde, H. & Røstad, C.C. (2006). The use of a 12 run Plackett-Burman design in the injection moulding of a technical plastic component, *Quality and Reliability Engineering* **22**, 651–657.

[11] Box, G.E.P. & Meyer, R.D. (1993). Finding the active factors in fractionated screening experiments, *Journal of Quality Technology* **25**, 94–105.

[12] Seiden, E. & Zemach, R. (1966). On orthogonal arrays, *Annals of Mathematical Statistics* **37**, 1355–1370.

[13] Lin, D.K.J. & Draper, N.R. (1992). Projection properties of Plackett and Burman designs, *Technometrics* **34**, 423–428.

[14] Wang, J.C. & Wu, C.F.J. (1995). A hidden projection property of Plackett-Burman and related designs, *Statistica Sinica* **5**, 235–250.

[15] Cheng, C.S. (1998). Some hidden projection properties of orthogonal arrays with strength three, *Biometrika* **85**(2), 491–495.

[16] Lin, D.K.J. (1993). A new class of supersaturated designs, *Technometrics* **35**, 28–31.

[17] Tyssedal, J. & Kulahci, M. (2005). Analysis of split-plot designs with mirror image pairs as sub-plots, *Quality and Reliability Engineering* **21**(5), 539–551.

JOHN TYSSEDAL

Proportional Hazard Model

The proportional hazard model relates the hazard function (*see* **Hazard Function**) at two different conditions, x and the baseline x_0, by

$$h(t;x) = g(x)h(t;x_0) \tag{1}$$

with $t > 0$ and g a positive function, e.g. $g(x;\beta) = \mathrm{e}^{x\beta}$, where β is a parameter.

The links between the reliability function (*see* **Reliability Function**) and the distribution function and the baselines ones are given, respectively, by $R(t;x) = [R(t;x_0)]^{g(x)}$ and $F(t;x) = 1 - [1 - F(t;x_0)]^{g(x)}$.

Pugh Concept Selection *see* Concept Selection Matrices

QFD *see* Quality Function Deployment

Quadrature and Numerical Integration

Introduction

A numerical integration technique is an algorithm to calculate the numerical value of a definite integral,

$$\int_a^b f(x)\,\mathrm{d}x \tag{1}$$

Numerical integration problems go back at least to Greek antiquity when e.g. the area of a circle was obtained by successively increasing the number of sides of an inscribed polygon. In the seventeenth century, the invention of calculus resulted in a new development of the subject, leading to the basic numerical integration rules. In the following centuries, the field became more sophisticated and, with the introduction of computers in the recent past, many classical and new algorithms have been implemented, leading to very fast and accurate results. Consequently, there is a vast amount of relevant literature on solving numerical integration problems consisting of books, articles, software packages, etc. Some essential references are [1–6].

Numerical integration is sometimes called *quadrature*. A quadrature rule is a numerical approximation of an integral using a weighted sum of some values of the integrand,

$$\int_a^b f(x)\,\mathrm{d}x \approx \sum_{i=1}^{n} \omega_i f(x_i) \tag{2}$$

where $x_1, \ldots, x_n$ are the points or abscissas, usually chosen to be in the interval of integration, and $\omega_1, \ldots, \omega_n$ are the weights associated to these points.

The simplest quadrature rules are the Riemann sums. Given a set of ordered abscissas, $a = x_1 < x_2 < \cdots < x_n = b$, the left-handed Riemann sum is obtained with,

$$\int_a^b f(x)\,\mathrm{d}x \approx \sum_{i=1}^{n-1} h_i f(x_i) \tag{3}$$

where $h_i = (x_{i+1} - x_i)$, the length of the ith interval between points. Observe that the area under the curve in the ith interval is approximated by a rectangle of width h_i and height $f(x_i)$. Analogously, one can define the right-handed and midpoint Riemann sums by evaluating the function on the maximum or midpoint value, respectively, of each ith interval. The Riemann sums are also known as *rectangular rules* because of the use of rectangles to approximate the integral.

The rest of this paper is organized as follows. The section titled "Newton–Cotes Formulas" describes the Newton–Cotes formulas which are based on evaluating the integrand at a number of equally spaced points. The section titled "Gaussian Quadrature" introduces the Gaussian quadrature rules where the abscissas are chosen optimally to give the most accurate approximations possible. This section also includes a numerical example comparing different approaches. Some comments on the extensions of the Gaussian quadrature are also included. The section titled "Discussion and Remarks" concludes with some discussion and remarks.

Newton–Cotes Formulas

Assume that we divide the interval $[a, b]$ into n equally spaced points such that,

$$x_i = x_1 + ih, \quad \text{for } i = 0, 1, \ldots, n-1 \tag{4}$$

where $x_1 = a$, $x_n = b$ and $h = (b-a)/n$. Let us denote $f_i = f(x_i)$, for $i = 1, \ldots, n$. The main idea in Newton–Cotes formulas is to approximate the function $f(x)$ using polynomials that pass through the points (x_i, f_i) and integrate them to approximate the integral from x_1 to x_n. Newton–Cotes formulas are of *closed type* when the endpoints, x_1 and x_n, are included in the set of abscissas and they are called *open type* when the endpoints are not used to approximate the integral.

The basic closed Newton–Cotes formula is the *trapezoidal rule* which uses two points, x_1 and x_2, and approximates the function $f(x)$ on $[x_1, x_2]$ using a straight line joining (x_1, f_1) and (x_2, f_2). Then, the function is approximated by the following first-order polynomial,

$$f(x) \approx \left(\frac{f_2 - f_1}{h}\right) x + \frac{x_2 f_1 - x_1 f_2}{h} \tag{5}$$

This polynomial can be integrated to give the following approximation,

$$\int_{x_1}^{x_2} f(x)\,\mathrm{d}x \approx \int_{x_2}^{x_2} \frac{f_2 - f_1}{h} x + \frac{x_2 f_1 - x_1 f_2}{h} \, \mathrm{d}x = \frac{h}{2}(f_1 + f_2) \tag{6}$$

Observe that this is called the trapezoidal rule because the area under $f(x)$ is approximated by

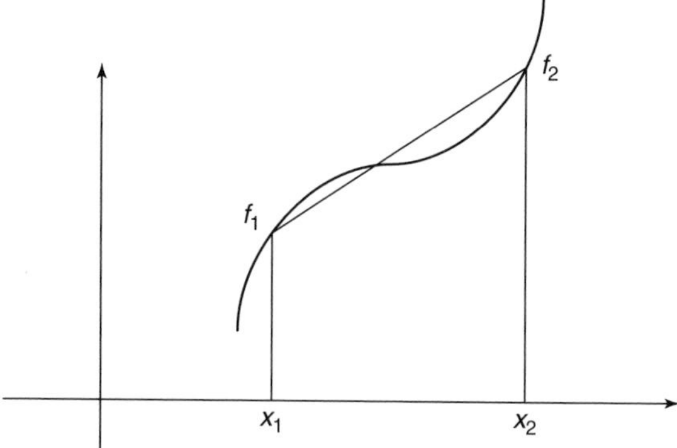

Figure 1 Trapezoidal rule

a trapezoid with bases f_1 and f_2 and height $h = (x_2 - x_1)$, see Figure 1. The error of this approximation can be shown to be given by, see e.g. [2],

$$E_T = -\frac{h^3}{12} f''(\xi) \tag{7}$$

where ξ is some point in the interval $[x_1, x_2]$. Then, the amount of error will not be larger than the maximum value of the second derivative of the function in the interval and it will increase with the curvature of the function. The trapezoidal rule is exact for polynomials up to and including degree 1.

The trapezoidal rule can be used repeatedly to approximate the integral over the whole interval, $[x_1, x_n]$. This is called the extended or composite trapezoidal rule and is given by,

$$\int_{x_1}^{x_n} f(x)\,\mathrm{d}x = \int_{x_1}^{x_2} f(x)\,\mathrm{d}x + \cdots + \int_{x_{n-1}}^{x_n} f(x)\,\mathrm{d}x \approx h\left(\frac{f_1}{2} + f_2 + \cdots + f_{n-1} + \frac{f_n}{2}\right) \tag{8}$$

whose approximation error is equal to $-(n-1)h^3 f''(\xi)/12$. A powerful extension of the trapezoidal rule is the Romberg integration, see e.g. [5], which uses a sequence of refinements of the extended trapezoidal rule by increasing carefully the number of subintervals.

The three-point closed Newton–Cotes formula is called the *Simpson's rule* and uses a quadratic polynomial joining the points (x_1, f_1), (x_2, f_2) and (x_3, f_3) to approximate the function $f(x)$ on the interval $[x_1, x_3]$,

$$f(x) \approx \alpha x^2 + \beta x + \gamma \quad (9)$$

where the coefficients α, β and γ are given by,

$$\alpha = \frac{f_1 - 2f_2 + f_3}{2h^2} \quad (10)$$

$$\beta = \frac{-2x_1(f_1 - 2f_2 + f_3) - h(3f_1 - 4f_2 + f_3)}{2h^2} \quad (11)$$

$$\gamma = \frac{x_1^2(f_1 - 2f_2 + f_3) + hx_1(3f_1 - 4f_2 + f_3) + 2h^2 f_1}{2h^2} \quad (12)$$

Then, we can integrate this second-order polynomial to obtain the Simpson's approximation,

$$\int_{x_1}^{x_3} f(x)\,\mathrm{d}x \approx \frac{h}{3}(f_1 + 4f_2 + f_3) \quad (13)$$

The error in this case can be shown to be given by,

$$E_S = -\frac{h^5}{90} f^{(iv)}(\xi) \quad (14)$$

where $f^{(iv)}(\xi)$ is the fourth derivative of $f(x)$ evaluated at some point ξ such that $x_1 \leq \xi \leq x_3$. Simpson's rule is exact for polynomials up to and including degree 3. This is surprising because it is designed to be exact for quadratic polynomials but, due to the symmetry of the formula, it is also exact for cubic polynomials.

As before, we can use Simpson's rule repeatedly to obtain the following extended or composite Simpson's rule, where the number of points, n, must be odd,

$$\int_{x_1}^{x_n} f(x)\,\mathrm{d}x = \int_{x_1}^{x_3} f(x)\,\mathrm{d}x + \cdots + \int_{x_{n-2}}^{x_n} f(x)\,\mathrm{d}x$$
$$\approx \frac{h}{3}(f_1 + 4f_2 + 2f_3 + 4f_4 + 2f_5 \cdots + 4f_{n-1} + f_n) \quad (15)$$

The approximation error is equal to $-(n-1)h^5 f^{(iv)}(\xi)/180$. Note that Simpson's rule will in general imply better accuracy than the trapezoidal rule if the function $f(x)$ is smooth enough (with finite fourth derivative).

The four-point closed Newton–Cotes formula is called *Simpson's 3/8 rule* and is based on a cubic polynomial approximation,

$$f(x) \approx \alpha x^3 + \beta x^2 + \gamma x + \delta \quad (16)$$

where α, β, γ, and δ are constants such that the polynomial passes through the points $(x_1, f_1), \ldots, (x_4, f_4)$. These coefficients can be calculated using the Lagrange interpolation formulas, see e.g. [2]. Then, the Simpson's 3/8 rule can be obtained to be,

$$\int_{x_1}^{x_3} f(x)\,\mathrm{d}x \approx \frac{3h}{8}(f_1 + 3f_2 + 3f_3 + f_4) \quad (17)$$

whose approximation error is given by,

$$E_{\tilde{S}} = -\frac{3h^5}{80} f^{(iv)}(\xi) \quad (18)$$

As expected, Simpson's 3/8 rule is exact for polynomials up to and including degree 3. Note that the approximation error for Simpson's 3/8 rule is larger than the ordinary Simpson's error given in equation (14). However, Simpson's 3/8 rule has the advantage that the number of points, n, is not required to be odd when it is used repeatedly to approximate the integral over the whole interval, $[x_1, x_n]$. Then, if n is even, one possibility is using the 3/8 Simpson's rule for the first four points and the ordinary Simpson's rule for the remaining points, whose number (including the fourth point) is definitely odd.

The five-point closed Newton–Cotes formula is called the *Boole's rule* and is given by,

$$\int_{x_1}^{x_5} f(x)\,\mathrm{d}x = \frac{2h}{45}(7f_1 + 32f_2 + 12f_3 + 32f_4 + 7f_5) \quad (19)$$

which can be obtained using the fourth-order Lagrange polynomial that passes through $(x_1, f_1), \ldots, (x_5, f_5)$. The error is $-8h^7 f^{(vi)}(\xi)/945$ and it is exact for polynomials of degree 5 or less. Boole's rule is also known as *Bode's rule*, as in e.g. [7], due to an early typo.

Further point closed Newton–Cotes formulas can be obtained by integrating the Lagrange polynomials that pass through the considered points, see e.g. [8]. In general, observe that higher-point formulas will imply higher accuracy only when the function can be well approximated by a polynomial.

Open Newton–Cotes formulas approximate the integral from $a = x_1$ to $b = x_n$ without using the

Table 1 Newton–Cotes formulas of open type

Approximation formula	Error
$\int_{x_1}^{x_3} f(x)\,dx \approx 2hf_2$	$\frac{h^3}{24} f''(\xi)$
$\int_{x_1}^{x_4} f(x)\,dx \approx \frac{3h}{2}(f_2 + f_3)$	$\frac{h^3}{4} f''(\xi)$
$\int_{x_1}^{x_5} f(x)\,dx \approx \frac{4h}{3}(2f_2 - f_3 + 2f_4)$	$\frac{28h^5}{90} f^{(iv)}(\xi)$
$\int_{x_1}^{x_6} f(x)\,dx \approx \frac{5h}{24} \times (11f_2 + f_3 + f_4 + 11f_5)$	$\frac{95h^5}{144} f^{(iv)}(\xi)$
$\int_{x_1}^{x_7} f(x)\,dx \approx \frac{6h}{20}(11f_2 - 14f_3 + 26f_4 - 14f_5 + 11f_6)$	$\frac{-41h^7}{140} f^{(vi)}(\xi)$

interior points, $x_2, x_2, \ldots, x_{n-1}$. Some of these formulas are shown in Table 1. Newton–Cotes formulas of open type can be useful, for example, when the function takes infinite values at the endpoints or when one or both endpoints are infinity. However, these are not very common in practice as it is not reasonable to use them repeatedly to obtain extended open rules. Also, the Gaussian quadrature rules, which will be described in the next section, lead always to more accurate approximations than open Newton–Cotes formulas.

Gaussian Quadrature

Suppose now that we have the freedom to choose the abscissas at which to evaluate the function $f(x)$ rather than being equally spaced points. The Gaussian quadrature rules provide careful choices of these points in order to obtain much more accuracy in approximating the required integral. These are open formulas in the sense that it is not required to evaluate the function at the endpoints. Furthermore, it is possible to define a weighting function, $W(x)$, such that the approximation of the integral will be exact for polynomials times this weight function instead of only for polynomials. Then, the numerical approximation will be as follows,

$$\int_a^b W(x)f(x)\,dx \approx \sum_{i=1}^{n} \omega_i f(x_i) \tag{20}$$

The advantage of incorporating a weighting function, $W(x)$, is that singularities or difficult terms can be removed from the function to be integrated.

An n-point Gaussian quadrature rule is constructed to be exact for polynomials of degree $(2n - 1)$, by a suitable choice of the points, x_i, and weights, ω_i. In fact, it can be shown (see e.g., [5]) that the abscissas used for the n-point Gaussian quadrature formulas are optimal and given precisely by the roots of the nth orthogonal polynomial, $p_n(x)$, for the same interval, $[a, b]$, and weighting function, $W(x)$. A set of polynomials $\{p_n(x)\}$ is orthogonal if they satisfy the condition,

$$\int_a^b W(x)p_n(x)p_m(x)\,dx = 0, \quad \text{for } n \neq m \tag{21}$$

The simplest form of Gaussian quadrature uses uniform weighting, $W(x) = 1$, and the reference interval is $[a, b] = [-1, 1]$. The orthogonal polynomials for this weighting function and interval are the Legendre polynomials, see e.g. [5], which are used to approximate the integrand $f(x)$ over the interval $[-1, +1]$ as follows,

$$\int_{-1}^{1} f(x)\,dx \approx \sum_{i=1}^{n} \omega_i f(x_i) \tag{22}$$

where x_i are the roots of the Legendre polynomials and ω_i are the corresponding weights, which are well known for this case and are given in Table 2. If we are interested in approximating the integral defined on an arbitrary interval, $[a, b]$, we can use a simple change of variable as follows,

$$\int_a^b f(x)\,dx = k\int_{-1}^{1} f(c + kx)\,dx$$

$$\approx k\sum_{i=1}^{n} \omega_i f(c + kx_i) \tag{23}$$

where $c = (b + a)/2$ and $k = (b - a)/2$. This is called the *Gauss–Legendre quadrature rule*. The following example illustrates and compares it with the Simpson's approximation.

Example Assume that we are interested in the approximation of the following integral,

$$\int_0^{\frac{\pi}{2}} \cos(x)\,dx \tag{24}$$

which is known analytically to be one. Firstly, we use the two-point Gauss–Legendre quadrature, given in

Table 2 Gauss–Legendre abscissas and weights

n^o points	2	3		4		5		
x_i	$\pm\sqrt{\frac{1}{3}}$	0.0	±0.7746	±0.3399	±0.8611	0.0	±0.5385	±0.9062
ω_i	1.0	0.88889	0.55556	0.65215	0.34785	0.56889	0.47863	0.23693

equation (23), to obtain the following approximation,

$$\int_0^{\frac{\pi}{2}} \cos(x)\,dx \approx \frac{\pi}{4}\left[1.0 \times \cos\left(\frac{\pi}{4}+\frac{\pi}{4\sqrt{3}}\right) + 1.0 \times \cos\left(\frac{\pi}{4}-\frac{\pi}{4\sqrt{3}}\right)\right] = 0.99847 \quad (25)$$

which is very close to the true value of one and gives an approximation error of 0.00153. Now, we approximate the same integral using the three-point Simpson's quadrature, given in equation (13),

$$\int_0^{\frac{\pi}{2}} \cos(x)\,dx \approx \frac{\pi}{12}\left[\cos(0) + 4\cos\left(\frac{\pi}{4}\right) + \cos\left(\frac{\pi}{2}\right)\right] = 1.0023 \quad (26)$$

This gives an approximation error of 0.0023 which is larger than the error obtained previously with the Gauss–Legendre quadrature that was based only on two points.

The Gaussian quadrature is also superior when it is compared with the composite Simpson's rule. For example, the four-point Gauss–Legendre quadrature gives the following approximation for the same integral,

$$\begin{aligned}\int_0^{\frac{\pi}{2}} \cos(x)\,dx \approx{}& \frac{0.65215\pi}{4} \times \cos\left(\frac{\pi}{4}+\frac{0.3399\pi}{4}\right) + \frac{0.65215\pi}{4} \times \cos\left(\frac{\pi}{4}-\frac{0.3399\pi}{4}\right) \\ &+ \frac{0.34785\pi}{4} \times \cos\left(\frac{\pi}{4}+\frac{0.8611\pi}{4}\right) + \frac{0.34785\pi}{4} \times \cos\left(\frac{\pi}{4}-\frac{0.8611\pi}{4}\right) \\ ={}& 0.999999977 \quad (27)\end{aligned}$$

which is a much better approximation than that obtained with the following composite Simpson's quadrature based on five points, as given in equation (15),

$$\begin{aligned}\int_0^{\frac{\pi}{2}} \cos(x)\,dx \approx{}& \frac{\pi}{24}\cos(0) + 4\cos\left(\frac{\pi}{8}\right) + 2\cos\left(\frac{\pi}{4}\right) + 4\cos\left(\frac{3\pi}{8}\right) + \cos\left(\frac{\pi}{2}\right) \\ ={}& 1.000134585 \quad (28)\end{aligned}$$

Alternatively to the Gauss–Legendre quadrature, other Gaussian rules can be developed using different classical orthogonal polynomials. Table 3 shows some of these polynomials together with their associated intervals and weighting functions. A large amount of information about these polynomials and their properties can be found in [7] and [9]. See also [5] and [10] for numerical procedures on the calculation of the abscissas and weights.

When we have an unusual choice for $W(x)$, we can develop our own Gaussian quadrature although this task usually becomes more difficult. Firstly, we need to obtain the set of orthogonal polynomials for the considered interval and weighting function. This can be done for example using a recurrence relation as described in [5]. Then, we need to calculate the zeros of these polynomials which will be the points, x_i, at which to evaluate the function. These can be obtained using a root-finding algorithm like Newton's method or faster procedures as those described in [5].

Table 3 Classical orthogonal polynomials with their intervals and weighting functions

Interval	$W(x)$	Symbol	Polynomial
$[-1, 1]$	1	$P_n(x)$	Legendre
$[0, \infty)$	e^{-x}	$L_n(x)$	Laguerre
$(-\infty, \infty)$	e^{-x^2}	$H_n(x)$	Hermite
$[-1, 1]$	$(1-x^2)^{-1/2}$	$T_n(x)$	First-kind Chebyshev
$[-1, 1]$	$(1-x^2)^{1/2}$	$U_n(x)$	Second-kind Chebyshev

Finally, we need to calculate the weights ω_i which can be obtained analytically by integrating the orthogonal polynomials for which the approximation is exact. There are also alternative formulas, which are more efficient for the calculation of these weights, see e.g. [5] and [10].

The Gaussian quadrature can be modified in order to incorporate one or both endpoints in the set of abscissas, leading to the Radau and Lobatto quadrature formulas respectively, see e.g. [10]. These formulas use a uniform weighting function $W(x) = 1$ and the free abscissas are the roots of some polynomials related to the Legendre orthogonal polynomials. Radau and Lobatto quadratures are slightly less optimal than the Gaussian quadrature.

Another important extension of the Gaussian quadrature is the Gauss–Kronrod algorithm, see e.g. [5] and [10]. This is an adaptive Gaussian method where the abscissas are suitably selected such that they can be reused in the next iterations reducing the number of function evaluations. This approach is implemented in various software packages such as Mathematica (Wolfram Research Inc.).

Discussion and Remarks

The generalization of the described quadrature rules to the multidimensional case is not straightforward. The main reason is that the number of function evaluations for an n-dimensional integral increases to the power of n and then, it becomes very expensive even for low-dimensional integrals. An alternative approach for these cases is the use of Monte Carlo integration (*see* **Monte Carlo Methods**; **Markov Chain Monte Carlo, Introduction**) which gives reasonable approximations for multidimensional integrals defined over complicated regions.

It is frequent in practice to have a tabulated function at given points, x_i, obtained from experimental data. In these cases, the abscissas are predetermined and cannot be chosen at will. One possibility is to obtain the Lagrange polynomial which interpolates the observed points and integrate it to approximate the integral, see e.g. [10]. One alternative is the use of splines which are piecewise polynomial functions with some finite derivatives that can be arranged to interpolate the data points, see e.g. [2].

Finally, note that the problem of evaluating an integral such as that given in equation (1) is equivalent to solving the differential equation $y'(x) = f(x)$ with $y(a) = 0$. There are many numerical procedures in the literature for solving differential equations that could be applied for this problem, see e.g. [5]. These would be specially well suited when the function to integrate is concentrated around one or various peaks.

References

[1] Davis, P.J. & Rabinowitz, P. (1984). *Methods of Numerical Integration*, 2nd Edition, Academic Press, New York.

[2] Evans, G. (1993). *Practical Numerical Integration*, John Wiley & Sons, Chichester.

[3] Hildebrand, F.B. (1956). *Introduction to Numerical Analysis*, McGraw-Hill, New York.

[4] Krommer, A.R. & Ueberhuber, C.W. (1994). *Numerical Integration on Advanced Computer Systems*, Springer-Verlag, Berlin.

[5] Press, W.H., Teukolsky, S.A., Vetterling, W.T. & Flannery, B.P. (2002). *Numerical Recipes in C++: The Art of Scientific Computing*, 2nd Edition, Cambridge University Press, Cambridge.

[6] Ueberhuber, C.W. (1997). *Numerical Computation: Methods, Software, and Analysis*, Springer-Verlag, Berlin.

[7] Weisstein, E.W. (2007). *Newton-Cotes Formulas. From MathWorld–A Wolfram Web Resource*. http://mathworld.wolfram.com/Newton-CotesFormulas.html.

[8] Abramowitz, M. & Stegun, I.A. (eds) (1972). *Handbook of Mathematical Functions with Formulas, Graphs, and Mathematical Tables*, Dover Publications, New York.

[9] Stroud, A.H. & Secrest, D. (1966). *Gaussian Quadrature Formulas*, Prentice Hall, Englewood Cliffs.

[10] Weisstein, E.W. (2007). *Gaussian Quadrature. From MathWorld–A Wolfram Web Resource*. http://mathworld.wolfram.com/GaussianQuadrature.html.

Related Articles

Convergence and Mixing in Markov Chain Monte Carlo; **Laplace Approximations in Bayesian Lifetime Analysis**; **Monte Carlo Methods**; **Markov Chain Monte Carlo, Introduction**.

M. Concepción Ausín

Quality Control, Computing in

Introduction

The quality movement has benefited tremendously from the development of computer software making important but difficult and/or time-consuming calculations readily available to practitioners. When these calculations were performed by hand, the default was typically to resort to simpler tools such as "rules of thumb" for sample size. **Measurement Sytems analysis** (MSA) called for 10 parts, 3 operators, and 3 replicates with analysis based on the X-bar and range method. Statistical process control (**SPC**) charts were limited to individuals, X-bar and range, or attribute charts. The design and analysis of experiments (DOE) were often simple two-level full or fractional factorials with hand calculations of "average of highs minus average of lows". These simplified approaches are still valuable for training purposes and may work quite well in practice, but can also produce misleading results due to oversimplification or violation of assumptions.

The following is a brief survey of important advances in **quality control** tools that are (or should be) available to practitioners using current statistical quality software tools, as of 2007:

Power and Sample Size

The power of a test $(1 - \beta)$ is the probability of rejecting the null hypothesis, given that the alternative hypothesis is true. Power depends on the type of hypothesis test, and is larger with an increase in sample size (which translates to increased cost), effect size (difference to be detected), **significance level** (α), and decrease in population standard deviation (σ) [1, 2].

Power and sample size calculations should be carried out prior to performing a test of hypothesis or design of experiments. Ideally, when planning an experiment, the power results should be displayed in real time as one selects the design type and number of replicates (using standardized effect sizes). This allows the practitioner to balance the cost of an experiment with expected outcome (*see* **Sample-Size Determination in Experimental Designs**; **Sample-Size Determination**).

The details of power calculations are outside of the scope of this article, but note that the calculations for one-sample z, one proportion, and two proportions are straightforward, with statistical power determined using the same distribution function as that used in the significance test. Power calculations for one-sample t, two-sample t, analysis of variance (**ANOVA**), two-level **factorial**, and χ^2 require a reference distribution function called a *noncentral distribution function* [1]. Modern computer software takes care of the challenges associated with computing these distributions.

Measurement Systems Analysis

MSA studies (also known as *gauge repeatability and reproducibility*, GRR) are often performed with 10 parts, 3 operators/appraisers, and 3 replicates using the Automotive Industry Action Group (AIAG) standard templates ([3], *see* **Gauge Repeatability and Reproducibility (R&R) Studies**; **Measurement Systems Analysis, Overview**; **Measurement Systems Analysis, Capability Measures for**). Unfortunately, these sample sizes typically yield wide **confidence intervals** on GRR statistics such as %GRR (%precision/total) and %precision/tolerance. Without considering the confidence intervals, practitioners run the risk of calling an unacceptable measurement system acceptable (or an acceptable measurement system unacceptable).

Computing confidence intervals for GRR statistics requires the calculation of variance components (*see* **Variance Components**; **Gauge Repeatability and Reproducibility (R&R), Variance Components in**), and the confidence intervals associated with these variance components and ratios (*see* [4]; **Gauge Repeatability and Reproducibility (R&R) Studies, Confidence Intervals for**). Computer software is available to perform these computations.

Statistical Process Control

The SPC chart developed by Walter Shewhart has proved to be an effective and essential tool for quality improvement. As a statistical tool, however, there are times when the **control chart** fails to give proper information about the process. Sometimes there are false alarms leading to expensive and fruitless searches for assignable causes. On the other hand, the Shewhart chart occasionally fails to detect a significant process shift early enough to prevent rework. These problems are particularly acute in process industries.

In order for control charts to function properly, the underlying assumptions must be valid. Specifically, the observations are assumed to be normally distributed, independent, with fixed mean and constant variance. If one is using an X-bar chart, the assumption of normality is usually met because of the central limit theorem – the distribution of sample averages tends to be normal, even if the individual observations are not. If, however, the subgroup size, $n = 1$, and the observations are not normally distributed, a power transformation should be applied to the data. Suggested transformations include $X^2, \sqrt{X}, \ln(X), 1/\sqrt{X}, 1/X, 1/X^2$.

A power transformation resulting in normality should be utilized. The transformed data is then used to calculate the control limits. Computer software simplifies this task by employing the Box–Cox transformation method (*see* **Box and Cox Transformation**). If a suitable power transformation cannot be found, an alternative technique such as the Johnson transformation may be used [5].

The requirement for samples to be statistically independent can be a much more serious problem, since there is no similar mechanism to the central limit theorem assuring reasonable results. The presence of serial autocorrelation in data will adversely affect the performance of a Shewhart chart (*see* **Autocorrelation Function**; **Autocorrelated Data**). In the process industries, autocorrelation is typically positive due to inertial elements, resulting in false alarms. If, however, the autocorrelation is negative, the Shewhart limits will be too wide, and significant process shifts will not be detected.

The key to dealing with this autocorrelation is to directly model it with **time series** analysis and then use that model to remove the autocorrelation from the data. A Shewhart control chart can then be applied to the residual prediction errors in order to detect significant process disturbances. The exponentially weighted moving average (EWMA); *see* **Exponentially Weighted Moving Average (EWMA)**; **Exponentially Weighted Moving Average (EWMA) Control Chart** chart can be used as a one-step-ahead predictor, so it is a valid model to consider when dealing with autocorrelation. The EWMA is an important member of the time series models identified by Box and Jenkins as autoregressive integrated moving average (ARIMA) models (*see* [6, 7]; **Large Autoregressive Integrated Moving Average (ARIMA) Modeling**). It is equivalent to an ARIMA(0, 1, 1). Practically, the EWMA is a good prediction model in the majority of cases. If, however, the autocorrelation is negative or the mean is drifting too quickly, a different model will be required. Other ARIMA models should be considered using computer software.

Design and Analysis of Experiments

Computer software greatly simplifies the design and analysis of experiments, but the use of computer-generated designs is especially important in situations where standard factorial or **fractional factorial** cannot be utilized due to factor constraints. D-optimal designs are the most widely used type of computer-generated designs (*see* **Optimal Design**).

Many designed experiments also require the simultaneous optimization of multiple responses. Typically, an algorithm is used to maximize a desirability function to find the optimum settings of the X factors. However, as the number of factors or responses increases, conventional optimization algorithms tend to locate a local optimum rather than the global optimum.

Recent developments using genetic algorithms bring a new level of design efficiency that works well even with highly constrained factor regions [8]. Genetic algorithms are also very useful in multiple response optimization for locating the global optimum [9]. This translates into more effective experiments run at a lower cost.

References

[1] Thomas, L. & Krebs, C.J. (1997). A review of statistical power analysis software, *Bulletin of the Ecological Society of America* **78**(2), 126–139.

[2] Lenth, R.V. (2001). Some practical guidelines for effective sample size determination, *The American Statistician* **55**(3), 187–193.

[3] Chrysler Corporation, Ford Motor Company, General Motors Corporation. (2002). *Measurement Systems Analysis Reference Manual*, 3rd Edition, Automotive Industry Action Group (AIAG), Detroit.

[4] Burdick, R.K., Borror, C.M. & Montgomery, D.C. (2005). *Design and Analysis of Gauge R&R Studies: Making Decisions with Confidence Intervals in Random and Mixed ANOVA Models, ASA-SIAM Series on Statistics and Applied Probability*, Society for Industrial and Applied Mathematics (SIAM), Philadelphia, American Statistical Association (ASA).
[5] Chou, Y., Polansky, A.M. & Mason, R.L. (1998). Transforming non-normal data to normality in statistical process control, *Journal of Quality Technology* **30**(2), 133–141.
[6] Hunter, J.S. (1986). The exponentially weighted moving average, *Journal of Quality Technology* **18**(4), 203–210.
[7] Box, G.E.P. & Jenkins, G.M. (1976). *Time Series Analysis, Forecasting, and Control*, 2nd edition, Holden-Day.
[8] Heredia-Langner, A., Carlyle, W.M., Montgomery, D.C., Borror, C.M. & Runger, G.C. (2003). Genetic algorithms for the construction of D-optimal designs, *Journal of Quality Technology* **35**(1), 28–46.
[9] Ortiz, F., Jr, Simpson, J.R., Pignatiello, J.J., Jr & Heredia-Langner, A. (2004). A genetic algorithm approach to multiple-response optimization, *Journal of Quality Technology* **36**(4), 432–450.

JOHN NOGUERA

Quality Function Deployment

Introduction

Quality function deployment (QFD) is a planning tool that allows the flow down of high-level customer needs and wants through to design parameters (DPs) and then to process variables (PVs) critical to fulfilling the high-level needs. By following the QFD methodology, relationships are explored between quality characteristics expressed by customers and substitute quality requirements expressed in engineering terms, also known as *critical-to-satisfaction (CTS) characteristics* within the terminology of Design for **Six Sigma**. In the QFD methodology, customers define their wants and needs using their own language, which rarely carries any actionable technical terminology. The voice of the customer (VOC) can be grouped (using affinity diagram) into a list of needs and wants, which can be used as input to a QFD's relationship matrix house of quality (HOQ).

Knowledge of customer's needs and wants is paramount in designing effective products and services with innovative and rapid means. Utilizing the QFD in the context of design for Six Sigma (DFSS) methodology allows the developer to attain the shortest development cycle while ensuring the fulfillment of customer needs and wants as depicted in Figure 1.

History of QFD

QFD was developed in Japan by Dr. Yoji Akao and Shigeru Mizuno in 1966 but was not westernized until the 1980s. Their purpose was to develop a quality assurance method that would design customer satisfaction into a product before it was manufactured. One of the first published applications of QFD happens to be in Bridgestone's Kurume factory by Kiyotaka Oshiumi who followed the steps of Dr Akao. For 6 years the methodology was enhanced from the initial crude usage of Kiyotaka Oshiumi of Bridgestone Tire Corporation. Following the first publication of *Hinshitsu Tenkai*, i.e. quality deployment in Japanese, by Dr. Yoji Akao [1] in 1972 the pivotal development work was conducted at Kobe Shipyards for Mitsubishi Heavy Industry. Stringent government regulations for military vessels coupled with the large capital outlay per ship forced Kobe Shipyard's management to commit to upstream quality assurance. The Kobe engineers drafted the QFD matrix, which relates all the government regulations, critical design requirements, and customer requirements to company technical controlled characteristics of how the company would achieve them. In addition, the matrix also depicted the relative importance of each entry, making it possible for important items to be identified and prioritized to receive a greater share of the available company resources. The Kobe application resembles an application that is very similar to how QFD is defined today.

In 1978 the detailed methodology was published [2, 3] in Japanese and was translated to English in 1994.

QFD Methodology

QFD is accomplished by multidisciplinary teams using a series of matrices, or phases of the HOQs,

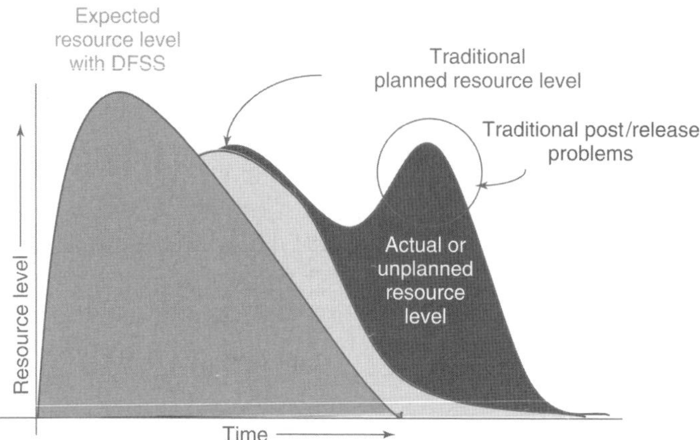

Figure 1 The time-phased effort of QFD *versus* traditional design

to deploy critical customer needs throughout the phases of the design development. Figure 2 depicts the rooms and the generic HOQ. Going room by room, we see that the input is into Room 1 in which we answer the question "what". These "whats" are either the results of VOC synthesis for HOQ 1 or a rotation of the "hows" from Room 3 into the following HOQs. These "whats" are rated in terms of their overall importance and placed in the Importance column. This column contains the importance rating for each "what" or voice listed in Room 1 from the customer perspective.

Next we move to Room 2 and compare our performance and the competitions' performance against these "whats" in the eyes of the customer. This is usually a subjective measure and is generally scaled from 1 to 5. A different symbol is assigned to the different providers so that a graphical representation is depicted in Room 2. Next we must populate Room 3 with the "hows". For each "what" in Room 1 we ask "How can we fulfill this?" We also indicate which direction the improvement is required to satisfy the "what": maximize, minimize, or target. This classification is in alignment with the **robust design** methodology and indicates an optimization direction. In HOQ1, these become "How does the customer measure the what?" In HOQ1, we call these *CTS measures*. In HOQ2, the "hows" are measurable and solution-free functions that are required to fulfill the "whats" of CTSs. In HOQ3 the "hows" become DPs and in HOQ4 the "hows" become PVs. A word of caution: teams involved in designing new services or processes often times jump right to specific solutions

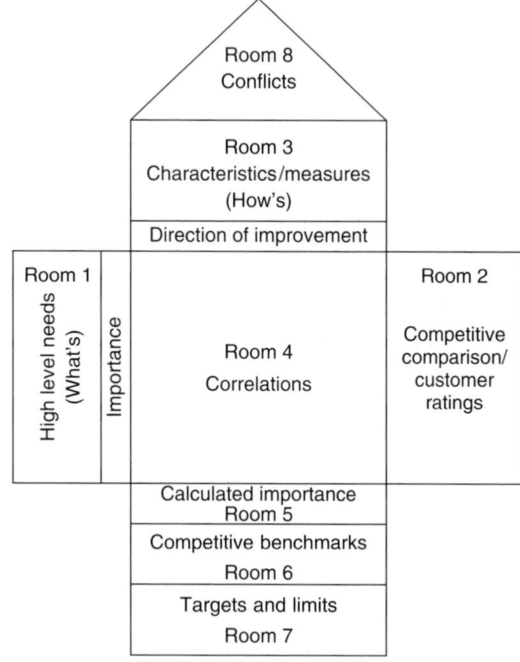

Figure 2 QFD house of quality

in HOQ1. It is a challenge to stay solution free until HOQ3. There are some rare circumstances where the VOC is a specific function that flows straight through each house unchanged.

Within Room 4 we assign the weight of the relationship between each "what" and each "how" using 9 for Strong, 3 for Moderate, and 1 for Weak.

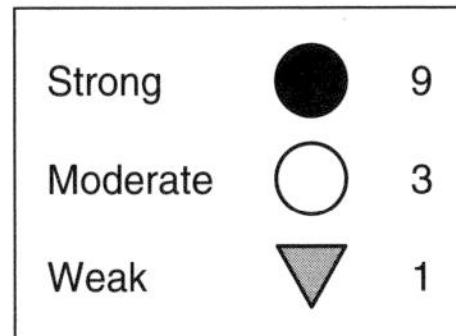

Figure 3 Rating values for the strength of a "what"–"how" relationship

In the actual HOQ these weightings will be depicted with graphical symbols, the most common being the solid circle for Strong, an open circle for Moderate and a triangle for Weak (Figure 3).

Once the relationship assignment is completed, by evaluating the relationship of every "what" to every "how", the calculated importance can be derived by multiplying the weight of the relationship and the Importance of the "what" and summing for each "how" or CTS characteristic. This is the number in Room 5. For each of the "hows" we can also derive quantifiable **benchmark** measures of the competition and our company; in the eyes of industry experts; this is what goes in Room 6. In Room 7, we can state the targets and limits of each of the "hows". Finally, in Room 8, often called the *roof*, we asses the interrelationship of the "hows" to each other. If we were to maximize one of the "hows", what happens to the other "hows"? If both "hows" improve, then we classify their relationship as a synergy, i.e. improving one of the "hows" results in improvement of the other "how", both in their respective direction of goodness, whereas if it were to move away from the direction of improvement, their relationship would be classified as a compromise. Wherever a relationship does not exist it is just left blank. For example, if we wanted to improve customer intimacy by reducing the number of call handlers then the wait time may degrade. This is clearly a compromise. Although it would be ideal to have **correlation** and regression values for these relationships, often they are just based on common sense or business laws. This completes each of the eight rooms in the HOQ. The next step is to sort on the basis of the importance in Room 1 and Room 5 and then evaluate the HOQ for completeness and balance.

The QFD methodology is deployed through a four-phase sequence shown in Figure 4. Each phase is a replica of the QFD HOQ captured in Figure 2 with some incorporated changes due to a structured cascading process. The four phases are

- Phase 1 – critical-to-satisfaction planning – House 1
- Phase 2 – functional requirements planning – House 2
- Phase 3 – DP planning – House 3
- Phase 4 – PV planning – House 4

Each of these four phases deploys the HOQ with the only content variation occurring in Room 1, the VOC room, also known as the "whats", and Room 3, the design substitute quality characteristics that answer the "whats", also known as the "hows".

QFD Example

There was a business that purchased $3 billion of direct materials (materials that are sold to customers) and $113 million of indirect materials and services (materials and services that are consumed internally in order to support the customer facing processes) and yet the business had no formal supply chain organization [4]. The procurement function existed in each functional group so operations purchased what they needed as did human resources, legal, and information systems. This distributed function resulted in a lack of power in dealing with suppliers and mostly supplier favorable contracts and some incestuous relationships (lavish trips and gifts given to the decision makers).

The corporate parent required that all procurement be consolidated into a formal supply-chain organization with sound legal contracts, high integrity, and year-over-year deflation on the total expenses. The design team was a small set of certified Black Belts and a master Black Belt who were being repatriated into a clean-sheet supply-chain organization.

The design team tasked with the project identified the customers and established their wants, needs, delights, and usage profiles. In order to complete the HOQ, the team also needs to understand the competing performance and the environmental aspects that the design will operate within. The design team realized that the actual customer base would include the corporate parent looking for results, a local management team that wanted no increased organizational cost and the operational people who needed materials and services. Through interviews laced with a dose

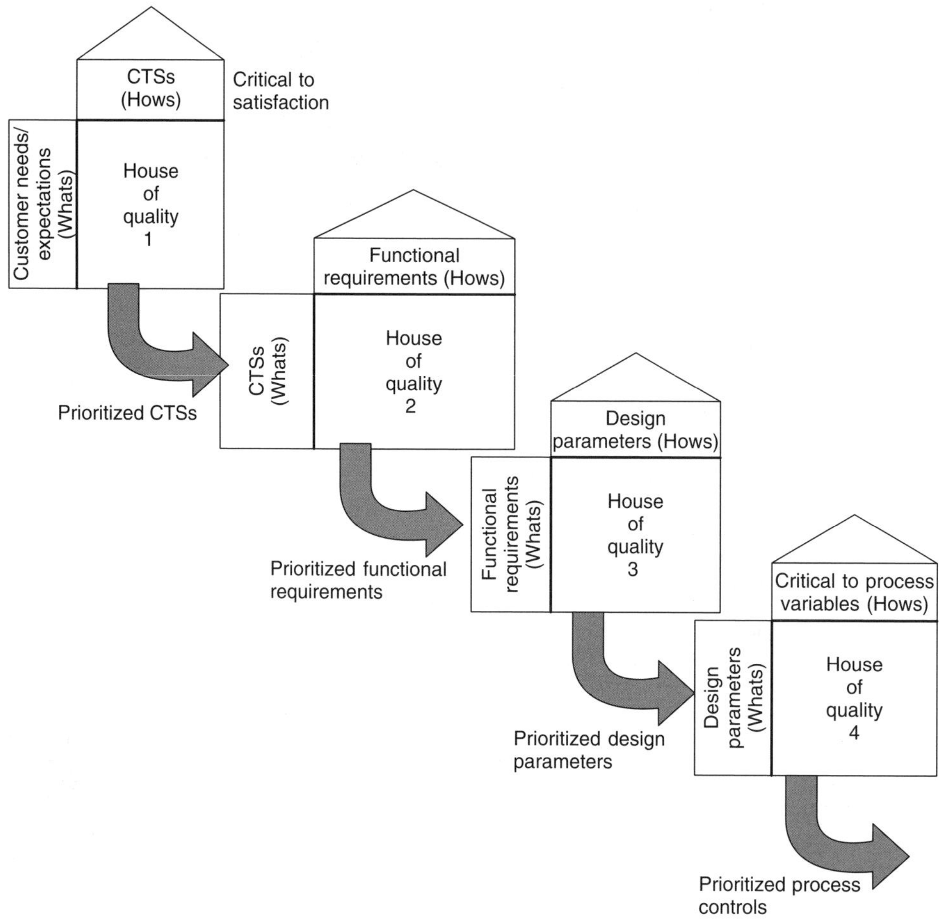

Figure 4 The four phases of QFD

of common sense, the following customer wants and needs (the "whats") were derived:

- Price deflation
- Conforming materials
- Fast process
- Affordable process
- Compliance
- Ease of use

The next step was to determine how these would be measured in the customer's eye. The CTS measures (the "hows") were determined to be

- Year-over-year price deflation (%)
- Defective deliveries (%)
- Late deliveries (%)
- Cost per transaction ($)
- Cost to the organization per order ($)
- Speed of order (hours)
- Number of compliance violations
- Number of process steps

Next the team used the QFD application, Decision Capture,[a] to assign the direction of improvement and importance of the wants/needs, the "whats" or VOC, and the relationship between these and the "hows", or CTS characteristics. Figure 5 shows the results. We can see that price deflation is moderately related to percentage of defective deliveries and number of compliance violations, it is weakly related to percentage of late deliveries and strongly related to year-over-year price deflation. We see that price deflation is extremely important and the direction of improvement is to maximize it.

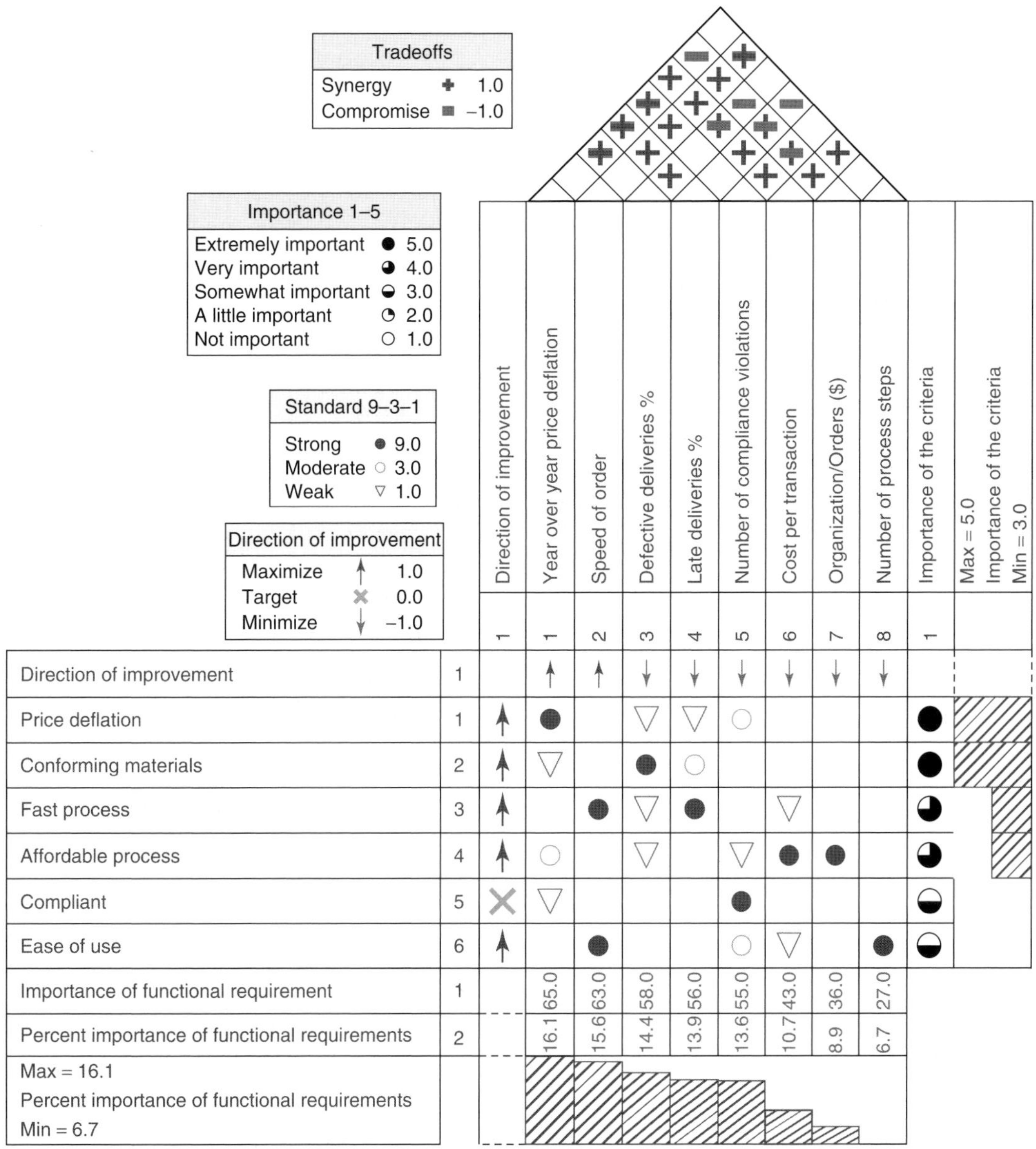

Figure 5 Partially completed phase 1 QFD HOQ

In Figure 5, we can see that the importance of the CTSs has been calculated. The method of calculation is as follows for the top sorted CTS:

Year-over-year price deflation = 9 (strong) × 5 (extremely important) + 1 (weak) × 5 (extremely important) + 3 (moderate) × 4 (very important) + 1 (weak) × 3 (somewhat important) = 65. Thus, the CTS importance score is a product sum of the relationships in Room 4 (in the respective CTS column) by the corresponding "whats" importance rating listed in Room 1. In effect, it is a weighted sum by the importance to the customer. The other columns are similarly calculated.

The upper and lower halves of phase 1 HOQ are shown in Figures 6 and 7, respectively. Figure 6 shows the competitive ratings of the wants/needs

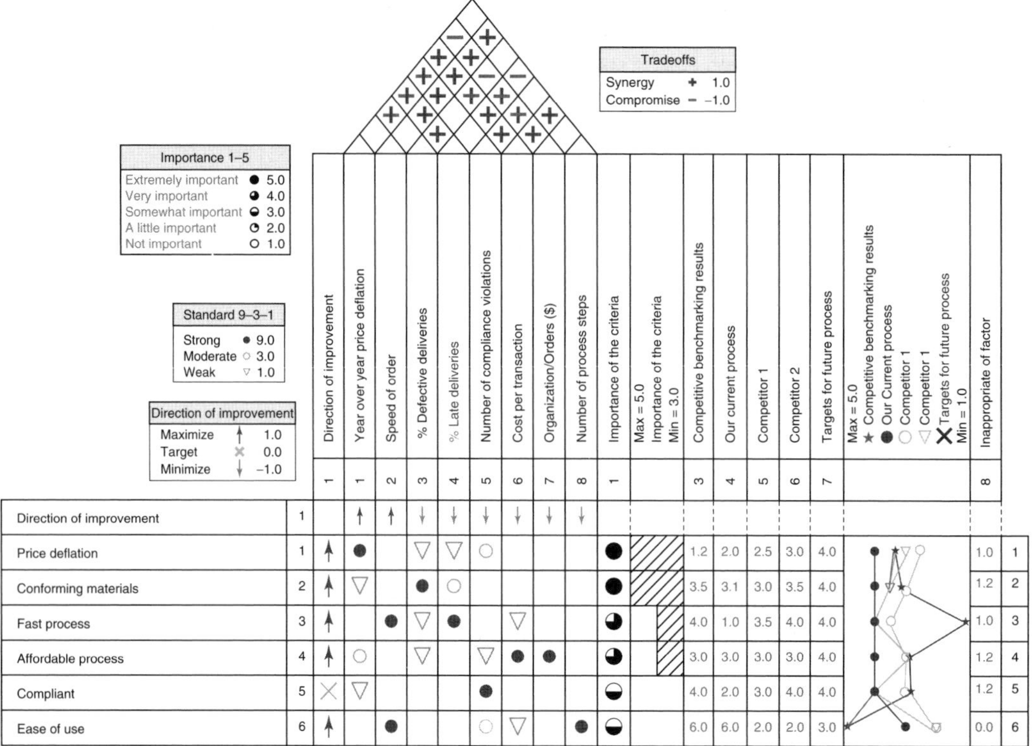

		Direction of improvement	Year over year price deflation	Speed of order	% Defective deliveries	% Late deliveries	Number of compliance violations	Cost per transaction	Organization/Orders ($)	Number of process steps	Importance of the criteria	Competitive benchmarking results	Our current process	Competitor 1	Competitor 2	Targets for future process	Inappropriate of factor	
		1	1	2	3	4	5	6	7	8	1	3	4	5	6	7	8	
Direction of improvement	1		↑	↑	↓	↓	↓	↓	↓	↓								
Price deflation	1	↑	●		▽	▽	○				●	1.2	2.0	2.5	3.0	4.0	1.0	1
Conforming materials	2	↑	▽		●	○					●	3.5	3.1	3.0	3.5	4.0	1.2	2
Fast process	3	↑		●	▽	●		▽			◕	4.0	1.0	3.5	4.0	4.0	1.0	3
Affordable process	4	↑	○		▽		▽	●	●		◕	3.0	3.0	3.0	3.0	4.0	1.2	4
Compliant	5	✕	▽				●				◒	4.0	2.0	3.0	4.0	4.0	1.2	5
Ease of use	6	↑		●			○	▽		●	◒	6.0	6.0	2.0	2.0	3.0	0.0	6

Figure 6 Upper half of phase 1 QFD HOQ

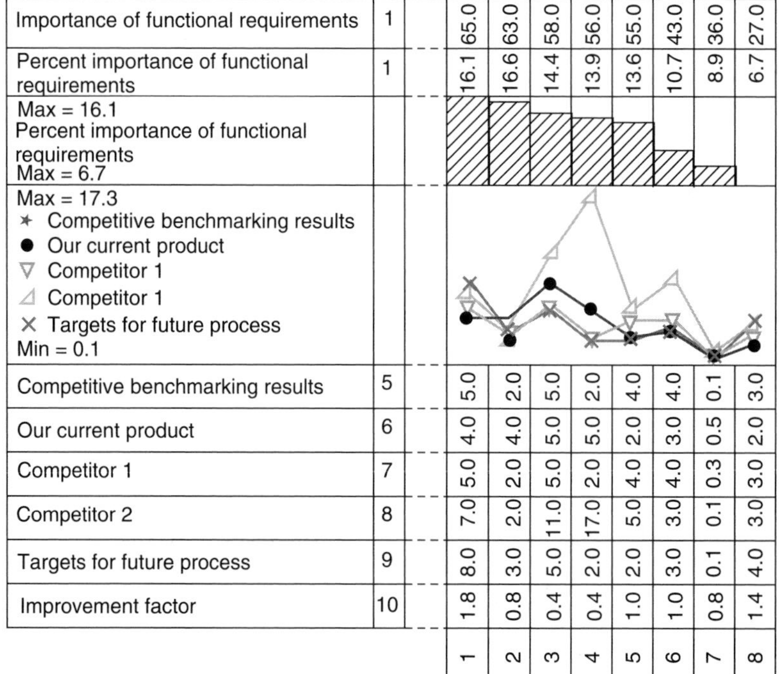

		1	2	3	4	5	6	7	8
Importance of functional requirements	1	65.0	63.0	58.0	56.0	55.0	43.0	36.0	27.0
Percent importance of functional requirements	1	16.1	16.6	14.4	13.9	13.6	10.7	8.9	6.7
Competitive benchmarking results	5	5.0	2.0	5.0	2.0	4.0	4.0	0.1	3.0
Our current product	6	4.0	4.0	5.0	5.0	2.0	3.0	0.5	2.0
Competitor 1	7	5.0	2.0	5.0	2.0	4.0	4.0	0.3	3.0
Competitor 2	8	7.0	2.0	11.0	17.0	5.0	3.0	0.1	3.0
Targets for future process	9	8.0	3.0	5.0	2.0	2.0	3.0	0.1	4.0
Improvement factor	10	1.8	0.8	0.4	0.4	1.0	1.0	0.8	1.4

Figure 7 Lower half of phase 1 QFD HOQ

(Room 2) and the technical ratings of the CTSs (Room 6). For this purpose the team used Competitor 1 as the incumbent process and used a sister business with a fully deployed supply-chain organization for Competitor 2. The synergies and compromises have also been determined.

The next step would be to assess the HOQ for any "whats" with only weak or no relationships. In this example no such case exists. Now we look for "hows" that are blank or weak and also see there are none of these.

Summary

QFD is a planning tool used to translate customer needs and wants into focused design actions. This tool is best accomplished with cross-functional teams and is key in preventing problems from occurring once the design is operationalized. The structured linkage allows for rapid design cycle and effective utilization of resources while achieving Six Sigma levels of performance.

To be successful with the QFD, the team needs to avoid "jumping" right to solutions and process phase 1 QFD HOQ and other phases (Figure 4) thoroughly and properly before performing detailed design.

It is important to have the correct VOC and the appropriate benchmark information. Also a strong cross-functional team willing to think out of the box is required in order to truly obtain Six Sigma–capable products or processes. From this point the QFD is process driven, but it is not the charts that we are trying to complete; it is the total concept of linking the VOC throughout the design effort.

End Note

[a] Decision Capture is a software application from International TechneGroup. www.decisioncapture.com

References

[1] Akao, Y. (1972). New product development and quality assurance – quality deployment system (in Japanese), *Standardization and Quality Control* **25**(4), 7–14.

[2] Shigeru, M. & Akao, Y. (eds) (1978). *Quality Function Deployment: A Company Wide Quality Approach (in Japanese)*, JUSE Press.

[3] Shigeru, M. & Akao, Y. (eds) (1994). *QFD: The Customer-Driven Approach to Quality Planning and Deployment* (Translated by Glenn H. Mazur), Asian Productivity Organization, ISBN 92-833-1122-1.

[4] El-Haik, B. & Roy, D.M. (2005). *Service Design for Six Sigma: A Roadmap For Excellence*, John Wiley & Sons.

BASEM EL-HAIK

Quality Improvement *see* Brainstorming

Quality in Critical Care Medicine

Introduction

Critical care medicine is a growing element of modern hospital medicine. Critically ill patients consume about of 20% of hospital budgets and approximately 0.8% of the GNP in the United States and 0.2% in Canada [1]. It is therefore a very expensive part of hospital medicine.

Intensive care units (ICUs) can be structured in different ways: according to patient population (such as surgical *vs* medical, neuro, and burns) and according to organization of patient care (open *vs* closed units).

The recent interest in quality improvement in healthcare has not skipped the ICU environment and there are many opportunities to improve on the delivery of critical care provided in ICUs. To describe these, a brief look at the current issues that are problematic in caring for critically ill patients is needed.

The patients in the ICU are frequently the most complex patients in the hospital. These are the patients who require intensive monitoring and often complex and multiple therapeutic interventions. Many drug prescriptions are provided to these patients on a daily basis and this has been shown to be a ground for drug errors, which can lead to adverse patient outcomes [2].

Many factors impact the processes and therefore the outcomes of patients in the ICU. The physical design and equipment available, the organizational workflow of various medical services, and the interactions between the different members of the ICU team: physicians, nurses, pharmacists, physiotherapists, respiratory therapists, and social workers are all critical to achieve optimal outcome, and their availability as well as the communication between them all have a critical impact on the function of the ICU.

Determinants of Quality

A look at quality in ICUs requires observation of a number of components. Design, both physical and organizational setup, of the ICU can impact the process and clinical outcome. Process, includes the approach to patients with regard to using protocols, workflow of organization, and guidelines, and outcome primarily deals with the clinical outcome of patients admitted to the ICU, but also pertains to other markers of significance such as cost, length of stay, and resource use. An important component of outcome is patient and family satisfaction with the process in the ICU. An admission to intensive care is frequently a very dramatic episode in a person's life and the impact it has on the patient and family can and should be addressed, measured, and improved upon.

Physical design of the ICU can have significant effect on the workflow, on personal and patient satisfaction, and therefore on clinical outcome. Poorly designed work stations can lead to increased rate of medical errors, may impair communication, and increase the risk for nosocomial infections.

ICU Structure: Open *versus* Closed

Many ICUs are organized in an "open" manner. This means that the patients are admitted to the ICU by a primary physician who continues to carry the responsibility for all aspects of the patients' stay in the ICU. This implies that although consultants are used, the treatment of different patients may be quite diverse with some difficulty in implementing routines and protocols for the different clinical issues that may arise. "Closed" units have a primary ICU team that directs care of the patients when they are admitted to the ICU. These clinicians are "dedicated intensivists" and spend a major part of their time in caring for critically ill patients. Studies have shown differences in outcomes between closed and open units with an advantage to a closed unit [3].

The clinical process in the ICU spans the encounter with the patient from the decision on admission to the ICU and the therapy afforded to patients even before they are admitted to the ICU. Rivers and colleagues have shown that early goal directed therapy in patients with severe infections can dramatically reduce mortality. They applied this approach to critically ill patients who required intensive care while they were still in the emergency room waiting for admission to the ICU [4, 5].

Staffing

Staffing of the ICU by nurses affects the quality of healthcare delivered to the patients and to families. This has been shown in the effect of nursing ratio on nosocomial infections (infections acquired in the ICU). A low nursing ratio led to a 30% increase in nosocomial infections. These infections led to increased morbidity and mortality [6]. Other studies have also shown that staffing by nurses can impact various outcome parameters. The availability of other paramedical personnel in the ICU, such as pharmacists, can also lead to improved quality and cost reduction.

Medical Error and Error Prevention

Medical errors are responsible for a great number of adverse events in critically ill patients. A model of medical error was described by Leape [7]. Most errors are due to a system failure rather than a failure by a particular individual. Errors are difficult to reduce because of the complexity of medical systems, and also because healthcare professionals are often reluctant to report and disclose the details of medical errors because of concern of punishment. A comprehensive approach to reducing medical errors requires an open and blame-free attitude. This will allow reporting of incidents that can be collected and studied, and with the use of information from large data-bases allow conclusions as to root causes of errors.

Errors are often divided into those due to faulty planning and those due to faulty execution, and they can be errors of commission or errors of omission. The usual model for medical errors describes the "Swiss cheese model". The theory behind this model is that all processes have inherent faults in them, the so-called holes, which do not lead to error because each process is composed of many components and the faults in each component need to be aligned to lead to an actual error. To minimize or eliminate errors, efforts must be made to investigate possible errors and near misses, strive to reduce the holes in the system, and allow for backup mechanisms that will decrease the likelihood of a number of faults occurring in the same process in a manner that can lead to error.

Among the various mechanisms designed to reduce errors are the use of computerized systems and the implementation of appropriate communications protocols between staff. Other important issues are the empowerment of all clinical personnel as well as patients and their families to question a process or a decision and to alert and even abort a process if potential error is suspected.

Evidence-Based Medicine

Many activities and clinical procedures in the ICU are based on personal experience and anecdotal information. There is now a growing body of evidence regarding the best policies and procedures that should be used to achieve the best outcome in critically ill patients. This information can direct the appropriate therapy with regard to ventilatory management in patients with acute respiratory distress syndrome (ARDS) [8], weaning from mechanical ventilation [9], the initial approach to hemodynamic management of patients in septic shock [4], and metabolic management of patients with regard to the control of glucose levels and insulin administration [10].

Despite the fact that these issues have significant data to support their implementation, studies have shown that clinicians are not always ready to adopt evidence-based therapies and their implementation is not universal by any means. For example, following the publication of the National Institutes of Health (NIH) study that showed that reduced tidal volumes can decrease mortality in mechanically ventilated patients with ARDS, there was a consistent but moderate application of this information in the treatment of patients with ARDS as seen in a study in a number of ICUs in the New England area [11].

Even established and standard therapies that are uniformly considered to be part of routine care are not universally complied with. Pronovost *et al.* measured compliance with routine procedures such as adequate sedation, elevation of the head of the bed, deep venous thrombosis (DVT) prophylaxis, and appropriate use of blood transfusion in the ICUs of 13 teaching and community hospitals. They found the compliance for these approaches to be between 33 and 89% [12].

It should be noted that although these and many other interventions have substantial levels of evidence that indicate the appropriate therapy to be followed when a number of choices are available, the treatment of critically ill patients is very complex and chaotic. The physiologic response to illness and to therapy is very diverse and general approaches may not always be able to indicate the correct or optimal strategy in each and every case. Probably the most significant determinant of outcome in the treatment of complex critically ill patients is the management by an experienced, dedicated clinician.

Guidelines, Protocols, and Bundles

Although the importance of an experienced intensivist cannot be overstated, it is not possible to provide this service to every critically ill patient. This amplifies the importance of treatment provided using the best available clinical knowledge by following protocols and guidelines. Protocols can provide a framework and suggest the optimal diagnostic and therapeutic interventions required to handle various situations.

A new concept regarding the application of best practice techniques is the bundle concept: the combination of a number of proven therapies to achieve a better outcome. There are a number of clinical issues for which the bundle concept has been applied. The concept includes the combination of a number of practices for which there is substantial evidence regarding their effect on improving outcome. At the same time, not all interventions that have a scientific basis to them are included in the bundle since consideration is given to ease of implementation and cost, in deciding on the components of a bundle. Therefore maximal effort to reduce complications should include the bundle as a minimum, but by no means as the sole, component of the intervention.

The Institute of Healthcare Improvement (IHI) has promoted the ventilator-associated pneumonia (VAP) bundle and the central catheter-related infection bundle among others.

Ventilator-Associated Pneumonia

Pneumonia in ventilated patients is a major clinical problem. Critically ill patients requiring ventilatory support may develop pneumonia owing to the treatment, and this nosocomial infection leads to increases in length of stay in the ICU, to increased time of mechanical ventilation and to increased mortality. It is therefore very important to decrease the incidence of VAP to improve outcome. The VAP bundle includes a number of strategies that have been shown to reduce VAP, and are "bundled" together. There are four components of the ventilator bundle and they are elevation of the head of the bed, daily assessment of patient ability to be extubated by stopping sedation, prophylaxis for upper GI bleeding, and DVT prophylaxis.

This means that when assessing compliance with the approach to reduce VAP, complete application of the bundle is considered as compliance with the bundle. There is no partial application and measurement of complete compliance can be shown to reduce the rate of VAP. It has been shown that compliance with the VAP bundle can be achieved when staff are informed about the importance and scientific validity of the bundle, and are given feedback regarding measured compliance with the bundle. Applying these principles, Cocanour and colleagues have shown a large reduction in the VAP in a trauma ICU [13]. Others have also shown significant reductions in the incidence of VAP in critically ill patients by applying this approach [14].

Central Venous Catheter Infection

Central lines, which are an essential component in patients treated in the ICU, can lead to blood stream infections that can have a significant impact on morbidity and mortality. Almost 50% of patients in the ICU have central lines, which are used for monitoring and delivery of critical drugs. Infections associated with these central lines lead to significant burden in terms of cost, an average cost of a central line associated infection being between \$3700 and \$29 000. There is also a significant attributable increase in length of stay and mortality. The central venous catheter bundle includes the following components: hand hygiene, maximal barrier precautions upon insertion of the central line, chlorhexidine skin antisepsis, optimal catheter site selection, with subclavian vein as the preferred site for nontunneled catheters, and daily review of line necessity with prompt removal of unnecessary lines.

Applying a complex of interventions designed to reduce infections, catheter-related infections could be eliminated in a single institution [15], and in a study of a large number of ICUs it was seen that the application of a group of interventions lead to a significant and sustained reduction in catheter-related infections [16].

Who Regulates? Who Measures?

The increasing concerns over both quality and cost have led many regulatory agencies to try to enforce various process changes into critical care. Not only governmental agencies but privately organized bodies with a large consumer power such as the Leap Frog group have also determined process components that they regard as mandatory to improve outcome. Among these are the implementation of computerized physician order entry, and the staffing of dedicated intensive care physicians 24 h a day all week. The involvement of regulatory agencies is perceived by some clinicians as adding complexity to reporting and regulations without a quality or outcome benefit. However, it is a factor that will probably only increase in magnitude and lead to a growing demand on clinicians and administrators to apply best practice techniques, document, and measure them. The effect of these regulations on quality is as yet unknown.

Summary

Quality improvement in the ICU is an ongoing endeavor that begins with design of the physical aspects of the ICU and workflow structure, entails measurement of different parameters, and impacts outcome of patients, both in terms of requirement of ICU care, length of stay, and mortality.

References

[1] Jacobs, P. & Noseworthy, T.W. (1990). National estimates of intensive care utilization and costs: Canada and the United States, *Critical Care Medicine* **18**(11), 1282–1286.

[2] Donchin, Y., Gopher, D., Olin, M., Badihi, Y., Biesky, M., Sprung, C.L., Pizov, R. & Cotev, S. (1995). A look into the nature and causes of human errors in the intensive care unit, *Critical Care Medicine* **23**(2), 294–300.

[3] Carson, S.S., Stocking, C., Podsadecki, T., Christenson, J., Pohlman, A., MacRae, S., Jordan, J., Humphrey, H., Siegler, M. & Hall, J. (1996). Effects of organizational change in the medical intensive care unit of a teaching hospital: a comparison of 'open' and 'closed' formats, *The Journal of the American Medical Association* **276**(4), 322–328.

[4] Rivers, E. (2006). The outcome of patients presenting to the emergency department with severe sepsis or septic shock, *Critical Care* **10**(4), 154.

[5] Rivers, E., Nguyen, B., Havstad, S., Ressler, J., Muzzin, A., Knoblich, B., Peterson, E. & Tomlanovich, M. (2001). Early goal-directed therapy in the treatment of severe sepsis and septic shock, *The New England Journal of Medicine* **345**(19), 1368–1377.

[6] Hugonnet, S., Chevrolet, J.C. & Pittet, D. (2007). The effect of workload on infection risk in critically ill patients, *Critical Care Medicine* **35**(1), 76–81.

[7] Leape, L.L. (1997). A systems analysis approach to medical error, *Journal of Evaluation in Clinical Practice* **3**(3), 213–222.

[8] The Acute Respiratory Distress Syndrome Network. (2000). Ventilation with lower tidal volumes as compared with traditional tidal volumes for acute lung injury and the acute respiratory distress syndrome. *The New England Journal of Medicine* **342**, 1301–1308.

[9] Esteban, A., Frutos, F., Tobin, M.J., Alia, I., Solsona, J.F., Valverdu, I., Fernandez, R., de la Cal, M.A., Benito, S. & Tomas, R. Spanish Lung Failure Collaborative Group. (1995). A comparison of four methods of weaning patients from mechanical ventilation, *The New England Journal of Medicine* **332**(6), 345–350.

[10] van den Berghe, G., Wouters, P., Weekers, F., Verwaest, C., Bruyninckx, F., Schetz, M., Vlasselaers, D., Ferdinande, P., Lauwers, P. & Bouillon, R. (2001). Intensive insulin therapy in the critically ill patients, *The New England Journal of Medicine* **345**(19), 1359–1367.

[11] Young, M.P., Manning, H.L., Wilson, D.L., Mette, S.A., Riker, R.R., Leiter, J.C., Liu, S.K., Bates, J.T. & Parsons, P.E. (2004). Ventilation of patients with acute lung injury and acute respiratory distress syndrome: has new evidence changed clinical practice? *Critical Care Medicine* **32**(6), 1260–1265.

[12] Pronovost, P.J., Berenholtz, S.M., Ngo, K., McDowell, M., Holzmueller, C., Haraden, C., Resar, R., Rainey, T., Nolan, T. & Dorman T. (2003). Developing and pilot testing quality indicators in the intensive care unit, *Journal of Critical Care* **18**(3), 145–155.

[13] Cocanour, C.S., Peninger, M., Domonoske, B.D., Li, T., Wright, B., Valdivia, A. & Luther, K.M. (2006). Decreasing ventilator-associated pneumonia in a trauma ICU, *The Journal of Trauma* **61**(1), 122–129; discussion 129–130.

[14] Youngquist, P., Carroll, M., Farber, M., Macy, D., Madrid, P., Ronning, J. & Susag, A. (2007). Implementing a ventilator bundle in a community hospital, *Joint Commission Journal on Quality and Patient Safety* **33**(4), 219–225.

[15] Berenholtz, S.M., Pronovost, P.J., Lipsett, P.A., Hobson, D., Earsing, K., Farley, J.E., Milanovich, S., Garrett-Mayer, E., Winters, B.D., Rubin, H.R., Dorman, T. & Perl, T.M. (2004). Eliminating catheter-related bloodstream infections in the intensive care unit, *Critical Care Medicine* **32**(10), 2014–2020.

[16] Pronovost, P., Needham, D., Berenholtz, S., Sinopoli, D., Chu, H., Cosgrove, S., Sexton, B., Hyzy, R., Welsh, R., Roth, G., Bander, J., Kepros, J. & Goeschel, C. (2006). An intervention to decrease catheter-related bloodstream infections in the ICU, *The New England Journal of Medicine* **355**(26), 2725–2732.

ERAN SEGAL

Quality Management, Overview

Introduction

Quality management (QM) is an essential feature in any organization that desires to be considered reliable by its customers – an organization that consistently keeps the promises that it makes. Reliable organizations design compelling promises for customers and consistently perform according to these promises at a high level of service thereby achieving a competitive edge over their rivals. This approach to quality requires two key elements – innovation to develop new products (whether goods or services) and customer care to effectively and efficiently deliver these products – and obtains predictable financial results for business owners. This means that the value proposition offered to consumers must deliver value that is beyond the offerings of competitors, so that consumers will consistently choose their own organization's products/services instead of the alternative offerings. QM is the approach to develop and maintain a "value edge" in competitive differentiation at a level that assures both market leadership and profitable growth. This outcome is achieved through an integrated system of business practices called *quality management* [1].

Quality Defined

To understand QM requires defining *quality* in a way that eliminates any subjective interpretation. In order to provide an operational definition of quality and illustrate how it is applied, three questions will be addressed: what is quality, how is quality created, and how is quality delivered.

First, We Must Ask What is Quality?

Quality is a comparison of expectations for performance outcomes with the perception of their achievement. Technically quality can have two meanings: it can describe either the characteristics of a product or service that give it an ability to satisfy stated or implied customer needs or it can refer to a product or service that is free of deficiencies. But, the most important aspect of quality is the one that creates business competitiveness for an organization. To describe this perspective a theory of **attractive quality** was developed by Dr. Noriaki Kano of Tokyo Science University [2]. The Kano model of attractive quality (Figure 1) illustrates three functions that define the dimensions of competitive quality.

Kano observed that there are three functions that define how quality is perceived by customers as meeting their requirements. When customers describe their quality needs, they define their understood needs; they do not talk about unknown requirements for unanticipated needs or applications nor do they describe their assumptions regarding basic performance that is embedded within their concept of the good or service. The three curves result in different types of customer performance perceptions.

- The lowest curve represents "must be" or standard quality which is characteristic of a commodity product (for instance, there is no competitive value in the standard grade of octane in gasoline – people are likely to change brands for very marginal differences in price). These requirements are not typically described by customers as they are "expected" or "understood" to be part of the package – cars are expected to go and to stop, and most people would feel that these capabilities must not be specified, thus these features deliver no satisfaction to customers but only offer an opportunity to dissatisfy the customer through poor performance.
- The middle curve describes what Kano called *one-dimensional quality* or a product quality feature where companies compete "head to head" to win customer approval (e.g., gasoline consumption of an automobile) and value as perceived by customers is directly proportional to the performance of this feature design. In this category of product feature, quality is directly proportional to the effectiveness of the design to the customer's experience in accomplishing the job they need to accomplish [3].
- The third curve describes the type of quality that always satisfies customers or what Kano refers to as *attractive quality* – quality that anticipates customer needs before the customer knows or understands what capability or feature they are missing. This type of quality comes from innovation – knowing the customer environment and application so intimately that new technology can be applied to create new competence or capabilities in the job that the customers need to get done. This type of quality provides a disrupting influence on markets and can totally change the balance of its competitive structure.

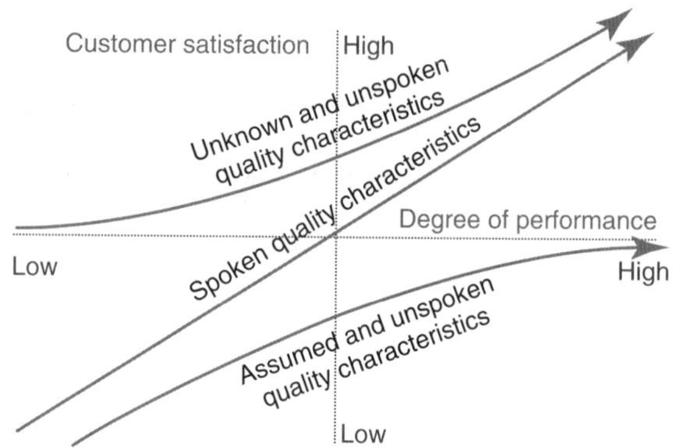

Figure 1 The Kano theory of attractive quality

Thus, according to Kano, quality is directly linked to the ability of a product or service to deliver perceivable value to its targeted customers.

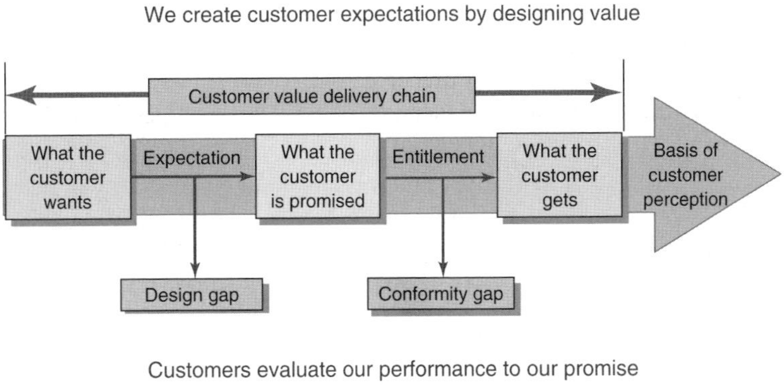

We create customer expectations by designing value

Customer value delivery chain

What the customer wants → Expectation → What the customer is promised → Entitlement → What the customer gets → Basis of customer perception

Design gap

Conformity gap

Customers evaluate our performance to our promise

Figure 2 The Watson model for value delivery

Second, We Must Ask How is Quality Created?

Given this understanding of the importance of the customer's experience, how is this type of quality created? Consider the Watson model for value delivery for a definition of the way quality is created and delivered in the customer experience (Figure 2) [4].

Quality creation is a two-phased process: designing value into products and services and delivering the quality of value that was designed. The design process interprets the customer experience of a Kano analysis to make a commercial promise to the market. This first stage in quality creation establishes the expectations of customers for the value they will receive and defines the offer made to customers. Not understanding or misinterpreting the best value proposition for customers thus results in suboptimization of the organization's potential competitive position. The second stage of quality delivery is in the delivery of the promise that has been defined for customers – a customer care function in which the consistency of promise delivery is evaluated by customers. Managing the daily dynamics of the customer experience must satisfy a quality entitlement – customers are entitled to consistently perceive satisfaction in the delivery of the promised performance. By aggressively executing the delivery of this quality model through an organization's business model, persistent perceptions creating long-term customer loyalty are achieved.

Third, We Must Ask How is Quality Delivered?

Sustained success and profitable growth are achieved when an organization coordinates its activities to deliver quality above its competitors, costs below its competitors, and technology ahead of its competitors. These imperatives must be constructed into a system for quality creation, which is delivered through the business processes of the organization (Figure 3 describes such a basic business model). This model characterizes the three core processes of an organization: product creation, product realization, and the process of management.

Figure 3 overlays the linkage of **Six Sigma** methodology and policy deployment on top of the business model to illustrate the linkage between change management projects (business improvement seeking ***kaizen*** (continuous improvement) or *hoshin kanri* (breakthrough improvement)) to the daily management system (*nichijo kanri*) represented by these core business processes. Note that two types of innovation are required to implement this change management process and that these are expressed by the two different approaches that are taken to resolve issues using Six Sigma business improvement methods:

- DMADV (define–measure–analyze–design–verify) is the Design for Six Sigma (DFSS) process that delivers innovation through business development processes by focusing on value delivery through product creation, service creation, or development of new value delivery work processes. This Six Sigma approach resolves issues where a current performance capability is inadequate to meet its future requirements (the designed capability is unable to meet the customer's need).
- DMAIC (define–measure–analyze–improve–control) is the **lean** Six Sigma process that

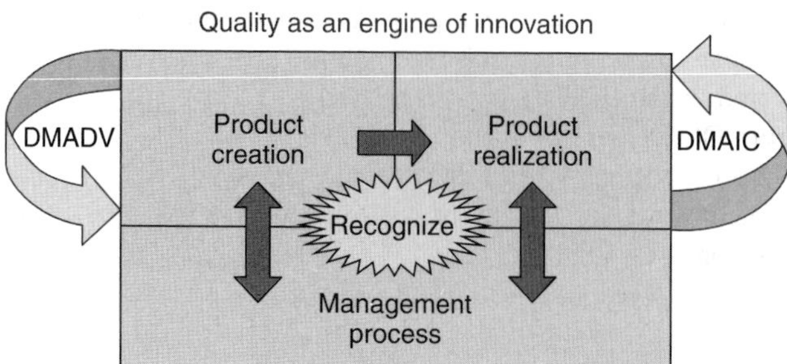

Figure 3 Quality-driven business model

delivers innovation in the product realization process by improving or optimizing the performance of current products, services, and the processes that deliver products or services by applying the methods of statistical problem solving and lean production management. This Six Sigma approach resolves issues where the current level of performance is not achieving its designed capability (achieved process capability is less than the designed process capability).

Quality must be managed over the long term but it is delivered in the short-term actions of the daily management system through a disciplined work process. What is the set of specific components that are contained in a quality management system (QMS) and how do they operate?

The Elements of Quality Management

QM differs from the management of quality. QM delivers excellence through a set of good management practices while the management of quality describes the profession of managing the quality function within an organization – the first is applicable to managers in all organizational functions and at all organizational levels, while the second applies more narrowly to a specific profession and its associated disciplines.

QM is an "unnatural act" of establishing and preserving excellence – QM defies the natural law of entropy whereby everything degenerates into a state of mediocrity before it obsolesces.

Management of quality combines several specific disciplines to assure the consistent delivery of the level of service or product quality intended at the point of the customer experience. In managing quality there are five quality disciplines that have evolved over the past century. These five disciplines for managing for quality relate to practices of engineering, assurance, control, and improvement. Taken together they follow a PDCA cycle (plan–do–check–act) that is often described as the *Deming cycle*.[a]

- *Quality engineering (QE)*
  QE accomplishes Deming's "plan" step for designing an enduring level of product or service quality. QE analyzes the steps of an operational process and gives a statistical basis for measuring and managing process performance so it consistently meets outcomes for customer standards or requirements. QE delivers predictable quality outcomes by maximizing process quality to produce the required quality results in the product or service.
- ***Quality control** (QC)*
  QC describes the methods used to address the Deming "do" step for process management. Control actions include processes for actively testing or monitoring performance, evaluating current performance results against standards for desired performance, and guiding process activities to achieve a state of statistical control by taking corrective action when performance is observed to deviate from this standard. Thus, QC consists of all operational techniques and activities used to fulfill the requirements for quality outcomes by focusing on improving process performance.
- *Quality assurance (QA)*
  QA delivers the "check" step of the quality discipline. QA gives confidence that the customer

requirements will be achieved by demonstrating the product or service capability to fulfill performance results as delivered to the final customer. QA is distinguished from QC in that QA focuses on the final test results or compliance with the external quality requirement while QC focuses on the interim test results or compliance with internal quality requirements, which produce final results.

- *Quality improvement (QI)*
  QI executes the "act" step of the PDCA cycle. The QI process is one of continuously improving or adjusting performance to reduce the cost of poor quality, improve the cycle time of process activities, and eliminate waste from all process steps. QI seeks to obtain and sustain the "ideal" process performance that was initially designed into the process capability. QI activities include the work of both Six Sigma and *kaizen* blitz teams who seek to eliminate waste or streamline process activities.
- *Quality audit and review (QAR)*
  QAR is a systematic, independent assessment of the entire quality system to determine the effectiveness of both the quality plan and its execution through application of the four disciplines of quality in the work processes of the organization. Thus, QAR "checks" on the value of governance in the PDCA activities of an organization's formal QMS.

When these five quality disciplines are combined into a systematic approach for the delivery of quality results, they become the cornerstones of a sound QMS. Thus, the "professional" dimension of QM may be defined as follows:

- *Quality management (QM)*
  The process of designing and deploying a system for managing processes to achieve maximum customer satisfaction with product or service quality at the lowest overall cost to the organization while continuing to improve the performance outcomes of the process.

This professional dimension of QM seeks to instill a systematic way of working in the entire organization for delivering quality outcomes – achieving organization-wide practices that deliver managing for quality. What are the ingredients of this system?

Management System for Delivering Quality

The formal QMS of an organization is a documented process for managing these five dimensions of QM (described above) and QMS describes the context for the design, development, deployment, and delivery of quality activities throughout the organization. The QMS documents and executes the organization's quality policy (strategic direction about how the organization will work to deliver quality) and it can be summarized as the PDCA of the management system: quality planning, quality standards (defines the system of work activities, controls and tests, as well as the control plan to assure that the system operates effectively despite any potential contingencies), QA, QC, and QAR, which must occur supported by the activities of an iterative process for QI. The QMS is a formal system to document the organization structure, management responsibilities, and organization-wide procedures that are required to achieve effective QM as a routine way of working. The ingredients of the QMS documentation include the following:

- *Quality policy*
  An organization's general statement of its beliefs about quality, its vision of how quality will come about and what should be the expected results.
- *Quality plan*
  A document or set of documents that describe the standards, quality practices, resources, and processes pertinent to the assurance of quality for a specific product, service, or project through the application of the four quality disciplines.
- *Quality standards*
  The documented description of accepted way to perform work and quality operating practices so that "best practice" is applied as a discipline in the daily working environment. This documentation includes standard operating procedures as well as control plans to assure proper measurement and interpretation of the critical quality process measures and the judgments made to interpret them regarding quality performance. Standards should also exist to describe how each of the core disciplines of QM will be managed.
- *Quality systems*
  The information systems that manage data for product and service performance, customer and

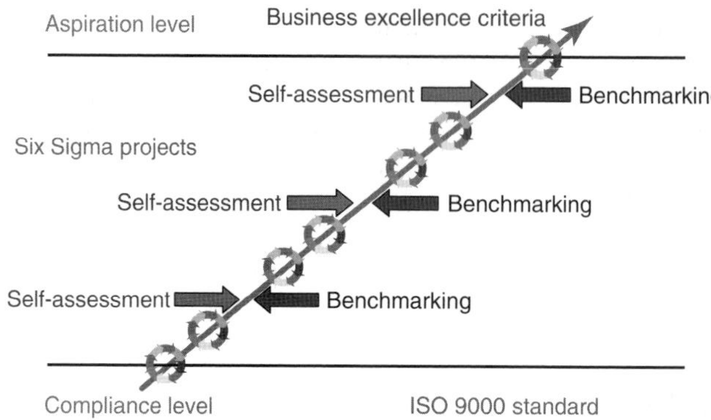

Figure 4 An integrated quality system for total performance management

market information about complaints, failures, warranties, and the record of responses to such claims.

The typical system for delivering quality in a modern company contains a number of proven ingredients for success: ISO 9000, business excellence criteria, and Six Sigma which can be forged into the process management architecture that is shown in Figure 4.

This integrated QMS seeks to maintain a business control system based on the ISO 9000 Quality Management Standard while encouraging the organization to aspire to apply the "best practice" principles of QM, which have been distilled into the award criteria for business excellence prizes such as the Malcolm Baldrige National Quality Award or the European Quality Award. These award criteria address a system of areas for concern in the development of a business management system that were developed as a consensus from many of the world's best companies. No company has ever scored above 800 points using these criteria as a standard, so they provide a way to challenge organizations to initiate self-improvement projects that close the gap in opportunities for improvement that are discovered when the organization conducts a self-assessment or conducts formal benchmarking with an organization that is recognized as superior by their management team. These improvement projects can be conducted using the Six Sigma DMADV or DMAIC methods.

Achievement of exceptional performance using such a management system requires that the organization develop "leadership at all levels" of its structure. What does this mean in terms of the quality actions of the organization's business leaders? We have finished discussing the key dimensions of management of quality; now let us turn back to understanding how QM can be infused into an entire organization as it integrates its QMS into its core business to manage performance in all dimensions of business excellence while ensuring successful results from the perspectives of all the organization's stakeholders (e.g., customers, employees, suppliers, investors, legal and regulatory authorities, etc.).

Pervasive Quality Leadership

It is a tall order to deliver this kind of management system. It requires, as Warren Bennis observed, the kind of situation where: "a leader is a follower is a leader" [5]. Bennis noted that we are facing a "chronic crisis of governance – that is, a pervasive organizational incapacity to cope with the expectations of their constituents – [that] is now an overwhelming factor worldwide. If there was ever a moment in history when a comprehensive strategic view of leadership was needed, not just by a few leaders in high office but by large numbers of leaders in every job, from the factory flow to the executive suite, this is certainly it" [6]. In order to have a system of sound business management there must be good role models of management quality – not just at the top of the organization, but throughout its entire structure so that the people can see this behavior is not just applicable for the "superleaders" but that is also part of the expectation for every ordinary worker – including supervisors – and they are also capable of exhibiting the shared values and behaviors that operationally define quality actions throughout the entire organization. This consistency assures that each person understands that the vision of leadership is not about wishing, hoping, or praying: it is an act of courageous leadership that must be duplicated at all levels of the organization. What does leadership at all levels look like?

Executive Leadership

Senior managers must establish the vision and values that will guide their organization into its future state. One way executives demonstrate leadership is by establishing a framework for action through a management process for cascading values and objectives that define the way the organization will work together and the goals that it will seek to accomplish in pursuit of its long-term mission. A critical aspect of the process of management that remains the active responsibility of the top management team is conducting regular leadership reviews of the business. This activity is both a due diligence responsibility of the business owners and a fundamental approach for exercising leadership. Leadership reviews have at least two emphases: review of the governance structure and "strategic direction of the organization" as well as review of strategic problem areas that lead to business vulnerability through challenging technologies, violation of critical assumptions, or changing "rules of competition" in the market. Leadership reviews offer the top management an opportunity for

- demonstrating personal commitment to their "philosophy of management";
- providing visibility for their "defining moments of quality encouragement";
- mentoring the organization to achieve desired behavioral changes;
- guiding cross-organizational efforts to achieve desired systemic change; and
- encouraging people to "build a desire to win and a will to act".

What else can executives do personally to demonstrate "positive leadership behavior"? A few examples include the following:

- *Executive touch program*
  Program for top executives to get close to the leading targeted external customers. Requires quarterly visits that are intended solely for relationship building to establish a foundation for future business discussions.
- *Executive customer advocate program*
  Program for the senior management team to facilitate significant problems encountered by major customers in each of the primary lines of business.
- *Complaint listening program*
  Program for all levels of management to spend a few hours monthly listening to "real" customer complaints directly on call center lines.
- *Executive escalation program*
  Program for the top level of senior managers to rotate through a monthly "duty day" where they are represent formal escalation "points of last resort" for resolving all customer complaints.
- *Executive compensation program*
  Change the compensation so a significant element of the "reward component" is granted for a "statistically valid" increase in customer satisfaction as measured by a valid external method.

Business leaders must exhibit consistency in all that they do. Lack of consistency is considered by employees to countermand the organizational culture and may cause deterioration in a shared system of values. It is particularly important to demonstrate consistent performance when one of the tenants of the organization is "empowerment" – the ability to make individual choices within a set of boundary conditions. An organization's vision provides direction for empowerment, but its values provide the boundaries for making choices.

Management Leadership

Managers may also become leaders. The definitions of manager and leader are not mutually exclusive. Each person appointed to a management function should at least aspire to becoming a local leader. What can be done to demonstrate leadership at the local level? Several actions can be suggested:

- taking an active role in leading the local implementation of a major business or QI initiative that has strategic value to the organization as a whole;
- showing a personal interest in developing the next generation of leaders through the personal mentoring of high-potential employees;

- "managing by wandering around" and taking time to talk with employees about any issues or concerns they may have, providing brief words of encouragement, and taking action to apply their insights for better management of work or business processes;
- sponsoring and reviewing improvement projects that deliver on the annual continuous improvement objectives of their own management area of responsibility;
- recognizing the improvement efforts of frontline teams and individuals;
- developing the core competence of their organization through team-based on-the-job education and training programs; and
- exercising information transparency by communicating to all employees about business results and briefing them on both the strategic direction and any news that directly concerns them or their livelihood.

Frontline Leadership

Leadership is often required at the frontline in order to coordinate action and encourage common behaviors that reinforce organizational values. Each person can exhibit leadership as a means to encourage fellow employees and provide an example that reinforces the behaviors of the value system. Some of the actions that can be taken at the personal level include

- participate actively in their work group or team process management activities and continuous improvement projects;
- develop their personal problem-solving and statistical analysis skills;
- mentor new employees in the cultural values and historical accomplishments of the organization;
- pursue certification in the core skills involved in the profession and demonstrate mastery of the tools and methods of their own trade;
- provide improvement ideas and suggestions to managers whenever any opportunities for improvement are observed in the work process;
- participate on teams for conducting self-assessments, audits, and cross-functional process improvement;
- take responsibility for their personal development and pursue a combination of both internal and external courses that deliver their career objectives; and
- recognize the contributions of colleagues and team members and encourage their achievements by expressing appreciation for their positive involvement in team improvement projects or personal suggestions made for process improvement.

Personal Leadership

True leadership is not in words, but in deeds: "The authentic test for mastery of learning is not in what a manager [person] says, but in what a manager [person] does" [7]. This requires a consistent practice that is developed from the inside out and exists on both personal and interpersonal levels both within the local work experience and across the whole organization. This effort requires two key focus areas: managers must empower the workforce to exhibit the principles of total quality leadership and frontline workers must become aligned with the strategic direction, improvement objectives, and cultural values of the leadership team. Each employee should aspire to represent a role model of the behavior desired of colleagues in the organization. This principal is also a cornerstone of the lean methods of the Toyota production system where employees are held accountable for the quality of their work and are expected to manage the quality of their work using a self-regulated QC approach. This raises an important question about the responsibility for quality and the role of quality professionals in managing to achieve quality outcomes in contrast to the responsibility of both process owners and workers for the quality of their own work output. What is the role of a quality professional in managing for quality?

Professional Leadership: The Role of a Quality Manager

Responsibility for Quality Performance

Who is responsible for the quality outcome of work? Is it the individual doing the work, the supervisor of the work effort, the owner of the work or business process owner, or the leader of the business area? There is a quandary: if everyone is responsible for the quality of work, then nobody is accountable! The resolution of this quandary lies in the multiple levels of responsibility – which fosters different types of accountability! Responsibility accumulates from bottom up; however, accountability is delegated from higher levels of authority to lower levels in the organizational structure. Thus management delegates responsibility for quality performance of work to the workers, but maintains overall responsibility and accountability. Workers can only be held responsible and accountable in the area of work that has been properly delegated to them. You cannot have accountability without responsibility! What does it take to hold someone properly accountable for executing their responsibility and then delivering the desired level of quality in their work? According to the sage advice of Peter Drucker, there are three conditions that must be satisfied to hold a worker personally accountable for the quality of their work:

1. Employees must have specific knowledge of the job they are asked to perform in terms of specifications, measurements, systems for work performance, and contingency actions that should be performed when work quality is not acceptable according to standards and they must be trained to perform to these work requirements.
2. Employees must know the standard of quality they are expected to deliver (the targets or goal of their performance expectation).
3. Employees must have the ability to monitor their work progress and be delegated the authority to make decisions that allow them to self-regulate their own performance to consistently meet the standards of performance [8, 9].

When these three conditions are not met, then the business managers retain the responsibility and accountability for all aspects of work quality. The job of a professional quality manager is to develop the QMS that assures that responsibility for execution of work practices leading to consistently sustainable quality results is achieved. What behavioral characteristics does it take to be such a "local leader" of the organization's quality practices?

Behavioral Qualities of an Exceptional Quality Manager

On the basis of a behavioral analysis of successful and unsuccessful quality managers, there are some common traits that have been identified as distinguishing between the most successful quality managers and those who were not perceived by their bosses as being successful. Sixteen traits were identified as leading to success [10]. Interestingly, only one trait was observed to detract from managerial performance. The positive behavioral traits that were discovered in this study included customer oriented, customer advocate, organizationally astute, influencing, goal oriented, interpersonally diagnostic, persistent, organizational planning, initiating, mentoring subordinates, collaborative, professional, conceptual, innovative, communicative, and self-confident. The one negative trait that leads to loss of personal influence and ineffective personal interactions was "making fast decisions." On the basis of these characteristics, it is clear that a "local quality leader" must behave like an internal consultant acting without the direct management authority and responsibility for line management (e.g., management of the profit and loss centers of the organization) but must convince others, through persuasive use of data, as to what is the right course to take. In an interview with Peter Drucker he was asked: how can a quality professional convince their business leader to "do" quality? His answer was most educational. Drucker said, "It is not your job to train your CEOs. They are bright people and can understand quickly what needs to be accomplished. However, it is the job of staff to clearly report necessary information that CEOs can easily assimilate and understand, so they may draw their own conclusions" [11]. The "secret weapon" of quality professionals is the use of data as a means to convince management to act. Communicating concerns with clarity and confidence is perhaps the greatest competence that a quality professional can have.

Summary

What is quality? Quality is the never-ending pursuit of excellence in product or service performance as judged according to the standards of customers relative to alternative choices they have for selecting similar goods or services from competitors. These customer standards for quality change over time and are influenced by a wide variety of external forces such as technology and regulatory legislation. As quality improves in one area in life, it often has a positive influence by increasing the requirement for quality in other areas. Thus, managing to improve quality is an essential ingredient to continued success and QM is a business requirement that will continue to be part of the "essential work" of organizations.

End Note

[a] Although PDCA is often called the *Deming cycle* it was first described by Walter Shewhart and the PDCA version was actually developed by the Union of Japanese Scientists and Engineers (JUSE) translators of Dr Deming's lectures under the supervision of Kaoru Ishikawa.

References

[1] Watson, G.H. (2003). Persistent leadership: a key to sustainable quality, in *Quality into the 21st Century: Perspectives on Quality and Competitiveness for Sustained Performance*, T. Conti, Y. Kondo & G.H. Watson, eds, ASQ Quality Press, Milwaukee, Chapter 5.

[2] Kano, N., Seraku, N., Takahashi, F. & Tsuji, S. (1984). Attractive quality and must-be quality, translated by Glenn Mazur, *Quality, The Journal of the Japanese Society for Quality Control* **14**(2), 147–156.

[3] Christensen, C.M. & Raynor, M.E. (2003). *The Innovator's Solution*, Harvard Business School Press, Boston.

[4] Watson, G.H. (2003). Customers, competitors and consistent quality, in *Quality into the 21st Century*, T. Conti, Y. Kondo & G.H. Watson, eds, ASQ Quality Press, 2003, *Ibid.*, Chapter 2.

[5] Bennis, W. (1989, 1994). *On Becoming a Leader*, Perseus Books, Cambridge, p. 39.

[6] Bennis, W. & Nanus, B. (1997). *Leaders: Strategies for Taking Charge*, 2nd Edition, Harper Business, New York, p. 2.

[7] Watson, G.H. (1994). *Business Systems Engineering*, John Wiley & Sons, New York, p. 116.

[8] Drucker, P.F. (1954). *The Practice of Management*, Harper Collins, New York.

[9] Drucker, P.F. (1973). *Management: Tasks, Responsibilities and Practices*, Harper Trade, New York.

[10] Watson, G.H. (1999). Building quality competence: successful management behavior, in *Proceedings of the 53rd Annual Quality Congress*, May 25, 1999.

[11] Watson, G.H. (2002). Selling Six Sigma to CEO's, *Six Sigma Forum Magazine* **2**, 26.

GREGORY H. WATSON

Quantiles

A quantile of a distribution is the value such that a proportion p of the population values are less than or equal to it. In other words, the pth quantile is the value that cuts off a certain proportion p of the area under the probability distribution function (*see* **Cumulative Distribution Function (CDF)**). It is sometimes known as a *theoretical quantile* or a *population quantile* and is denoted by $Q_t(p)$. Quantiles for common distributions are easily obtained using routines for the inverse cumulative distribution function (inverse **cdf**) found in many popular statistical packages.

For example, the 0.5 quantile (which we also recognize as the median or 50th percentile) of a standard **normal distribution** is $P(x \leq x_{0.5}) = \Phi^{-1}(0.5) = 0$, where $\Phi^{-1}(.)$ denotes the inverse cdf of that distribution. Equally easily, we locate the 0.5 quantile for an exponential distribution (*see* **Probability Density Functions**) with mean of 1 as 0.6931. Other useful quantiles such as the lower and upper **quartiles** (0.25 and 0.75 quantiles) can be obtained in a similar way.

However, we are more likely to be concerned with quantiles associated with data. Here, the *empirical quantile* $Q(p)$ is that point on the data scale that splits the data (the empirical distribution; *see* **Estimation**) into two parts so that a proportion p of the observations fall below it and the rest lie above. Although each data point is itself some quantile, obtaining a *specific* quantile requires a more pragmatic approach.

Consider the following 10 observations, ordered from the smallest to the largest.

$$274, 302, 334, 430, 489, 703, 978, 1656, 1697, 2745$$

Imagine that we are interested in the 0.5 and 0.85 quantiles for these data. With 10 data points, the point Q (0.5) on the data scale that has half of the observations below it and half above does not actually correspond to a member of the sample but lies *somewhere* between the middle two values 489 and 703. Also, we have no way of splitting off exactly 0.85 of the data. So, where do we go from here?

Several different remedies have been proposed, each of which essentially entails a modest redefinition of quantile. For most practical purposes, any differences between the estimates obtained from the various versions will be negligible. The only method that we shall consider here is the one that is also used by Minitab and SPSS (but see [1] for a description of an alternative favored by the ***exploratory data analysis*** (EDA) community).

Suppose that we have n ordered observations x_i $(i = 1{-}n)$. These split the data scale into $n + 1$

segments: one below the smallest observation, $n - 1$ between each adjacent pair of values and one above the largest. The proportion p of the distribution that lies below the ith observation is then estimated by $i/(n + 1)$. Setting this equal to p gives $i = p(n + 1)$. If i is an integer, the ith observation is taken to be $Q(p)$. Otherwise, we take the integer part of i, say j, and look for $Q(p)$ between the jth and $j + 1$th observations. Assuming simple linear interpolation, $Q(p)$ lies a fraction $(i - j)$ of the way from x_j to x_{j+1} and is estimated by

$$Q(p) = x_j + (x_{j+1} - x_j) \times (i - j) \qquad (1)$$

A point to note is that we cannot determine quantiles in the tails of the distribution for $p < 1/(n + 1)$ or $p > n/(n + 1)$ since these take us outside the range of the data. If extrapolation is unavoidable, it is safest to define $Q(p)$ for all P values in these two regions as x_1 and x_n, respectively.

Returning to the 10 data points given before, we note that for the 0.5 quantile, $i = 0.5 \times 11 = 5.5$, which is not an integer so $Q(0.5) = 489 + (703 - 489) \times (5.5 - 5) = 596$, which is exactly half way between the fifth and sixth observations. For the 0.85 quantile, $i = 0.85 \times 11 = 9.35$, so the required value will lie 0.35 of the way between the 9th and 10th observations, and is estimated by $Q(0.85) = 1697 + (2745 - 1697) \times (9.35 - 9) = 2063.8$.

From a plot of the empirical quantiles $Q(p)$, that is, the data, against proportion p as in Figure 1, we can get some feeling for the distributional properties of our sample (*see also* **Probability Plots**). For example, we can read off rough values of important quantiles such as the median and quartiles. Moreover, the increasing steepness of the slope on the right-hand side indicates a lower density of data points in that region and thereby alerts us to **skewness** and possible **outliers**. Note that we have also defined $Q(p) = x_1$ when $p < 1/11$ and $Q(p) = x_{10}$ when $p > 10/11$. Clearly, extrapolation would not be advisable, especially at the upper end.

It is possible to obtain ***confidence intervals*** for quantiles, and information on how these are constructed can be found, for example in [2]. Routines for these procedures are available within certain statistical packages such as SAS and S-PLUS.

Quantiles play a central role within *exploratory data analysis*. The median $(Q(0.5))$, upper and lower quartiles $(Q(0.75)$ and $Q(0.25))$, and the maximum and minimum values, which constitute the so-called

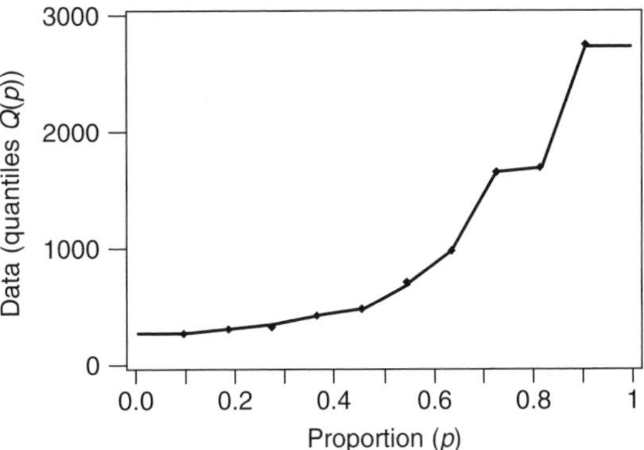

Figure 1 Quantile plot for Necker cube reversal times

five-number summary, are the basic elements of a box plot. An empirical quantile–quantile plot (*see* **Probability Plots**) is a scatterplot of the paired empirical quantiles of two samples of data, while the symmetry plot is a scatterplot of the paired empirical quantiles from a single sample, which has been split and folded at the median. Of course, for these latter plots there is no need for the P values to be explicitly identified.

Finally, it is worth noting that the term *quantile* is a general one referring to the class of statistics, which split a distribution according to a proportion p that can be *any* value between 0 and 1. As we have seen, well-known measures that have specific P values, such as the median, quartile, and percentile (on Galton's original idea of splitting by 100ths [3]), belong to this group. Nowadays, however, quantile and percentile are used almost interchangeably: the only difference is whether p is expressed as a decimal fraction or as a percentage.

References

[1] Chambers, J.M., Cleveland, W.S., Kleiner, B. & Tukey, P.A. (1983). *Graphical Methods for Data Analysis*, Duxbury, Boston.

[2] Conover, W.J. (1999). *Practical Nonparametric Statistics*, 3rd Edition, John Wiley & Sons, New York.

[3] Galton, F. (1885). Anthropometric percentiles, *Nature* **31**, 223–225.

PAT LOVIE

Article originally published in Encyclopedia of Statistics in Behavioral Science *(2005, John Wiley & Sons, Ltd). Minor revisions for this publication by Jeroen de Mast.*

Quartiles

For a continuous variable X, with distribution function F_X, the first, second, and third quartiles are $Q_1 = F^{-1}(0.25)$, $Q_2 = F^{-1}(0.50)$, and $Q_3 = F^{-1}(0.75)$ (that is, the values below which X will be with 25%, 50%, or 75% probability respectively). Note that the second quartile is equal to the median. *See also* **Quantiles**.

Queues in Reliability

Introduction

A queuing system is a mathematical model to characterize situations where certain units, called *customers*, arrive in continuous time in order to receive a service or facility provided by other units, called *servers*. When customers arrive, they are immediately served if there is an empty server. Otherwise, the customers must spend some time waiting in line for service until a server becomes available. Queuing theory is the methodology devoted to the stochastic description of queuing systems and the study of their performance measures, such as the queue length, waiting times, and so on. Some essential references on queuing theory are found in [1–5].

Queuing theory can be very useful in solving many complex reliability problems. Component failures or machine breakdowns can be treated as arrivals of customers and the repair or replacement of failed components may be seen as the service facility in a queuing system. The number of repair facilities or maintenance crews is equivalent to the number of servers. It is useful to find the analogies between reliability and queuing theory because it is frequently found in practice that a given reliability problem has been previously studied with an equivalent problem in queuing theory [6]. For example, the number of failed components corresponds to the *queue length* and the time elapsed from a failure to repair is equivalent to the waiting time in the system or *sojourn time*. Note that the notion of repair may also refer to preventive maintenance activities performed before system failures. For example, a deteriorated machine can be repaired to an upgraded status before it fails.

A queuing model is completely described by six characteristics that, following Kendall's notation [7], are given by $A/B/c/K/m/R$, where A and B describe the arrival and service pattern, respectively, c the number of servers, K the system capacity or maximum number of customers allowed in the system, m the size of the customer population, and R the discipline or the order in which customers are selected for service. Different symbols are traditionally used for the type of distribution of A and B. For example, M denotes exponential (Markovian) random variables, D degenerate (constant) variables, and G general distributions. For instance, the $M/G/1$ queue denotes a single-server system with exponential interarrival times and general service time distribution. Note that when nothing is said about K, m, and R, it is assumed that there is an infinite capacity, an infinite customer population, and a first-come-first-served discipline.

In the case of reliability, the main characteristic of queues is that requests for service are usually generated by a finite customer population, that is, $m < \infty$. This is because, in general, there is a limited number of units (machines, devices, etc.) which can fail, and when they are all in the system (being repaired or waiting for repair), no more can arrive. Then, the arrivals of requests do not form a renewal process as the arrivals may depend on the number of units staying at the system. This is an essential difference from typical queuing systems, where the population of potential arrivals can be considered to be effectively limitless. For example, in telephone operations the number of customers is usually large enough that the probability of having all of them wanting service at once is extremely small. However, in repair of machinery, the whole set of machines may be simultaneously out of order. Queuing systems with finite population are known as *finite source queues*, which have been extensively studied in the queuing literature, see e.g., [8] and the detailed bibliography provided in [9]. Finite-source queuing models are used to describe the classical *machine repairmen problem* which is described in the following section.

The Machine Repairmen Problem

Consider a repairable system (*see* **Repairable Systems Reliability**) consisting of r machines which operate independently of each other and may eventually fail. When a machine breaks down, it has to be repaired by one of the c repairmen, if there are any available. Otherwise, it has to wait for service until a server becomes available. Thus, the operating machines are outside the queuing system and enter the system only when they break down and require repair, as shown in Figure 1. Machine lifetimes and repair times are independent random variables with rates λ and μ, respectively. Then, the so-called mean time between failures (MTBF) and mean time to repair (MTTR) are given by $1/\lambda$ and $1/\mu$, respectively.

This model is known as *the machine repairmen problem* or the *machine interference problem* since the machines interfere with each other when all the servers are busy and new requests for service arrive. It can be used to describe a large variety of real situations, such as manufacturing systems (see e.g., [10]), telecommunication traffic (see e.g., [11]), computer systems (see e.g., [12]), inventory models (see e.g., [13]), and so on. This broad range of applications has motivated a large amount of research on this problem in the literature. For two reviews, see [14, 15].

The simplest machine repairmen model supposes that both lifetimes and service times are exponentially distributed, there is no limit in the waiting room, the repair is carried out by a first-in-first-out discipline and after having been served, each machine starts working again as good as new. The queuing model describing this system is denoted by $M/M/c/r/r$ or, equivalently, $M/M/c//r$. Note that there is an important difference from the notation used for the arrival process in infinite source queues, such as the $M/M/c$ queue, where the symbol M indicates a Poisson arrival process. However, in the $M/M/c/r/r$ queue, the time between failures of each machine is exponential but the arrival of failures from the whole set of machines does not follow a **Poisson process** because, as mentioned before, these arrivals depend on the number of machines waiting for repair. In particular, when the number of machines waiting for service increases, the arrival rate of further service demands decreases. More specifically, when n machines are "down" for repair, there are $(r-n)$ operating machines and the time until the next breakdown is the minimum of $(r-n)$ exponential distributions with rate λ, which is also an exponential distribution of rate $(r-n)\lambda$. Then, given that there are n machines being repaired or waiting for repair, the arrival rate or average number of fails per unit time is given by,

$$\lambda_n = \begin{cases} (r-n)\lambda, & \text{if } 0 \le n \le r \\ 0, & \text{if } n \ge r \end{cases} \quad (1)$$

and the average number of repaired machines per unit time is given by,

$$\mu_n = \begin{cases} n\mu, & \text{if } 0 \le n \le c \\ c\mu, & \text{if } n \ge c \end{cases} \quad (2)$$

Thus, it can be shown that the number of machines in the $M/M/c/r/r$ system follows a birth–death process, in which an arrival of a failure is interpreted as a birth and the completion of a repair is interpreted as a death, see e.g., [3]. Then, the long-run probability of having n machines in the system can be shown to be,

$$p_n = \begin{cases} \binom{r}{n}\left(\dfrac{\lambda}{\mu}\right)^n p_0, & \text{if } 1 \le n \le c \\ \binom{r}{n}\dfrac{n!}{c^{n-c}c!}\left(\dfrac{\lambda}{\mu}\right)^n p_0, & \text{if } c \le n \le r \end{cases} \quad (3)$$

where p_0 is obtained from $p_0 + \cdots + p_r = 1$. Using these probabilities, we can calculate for example the mean number of machines in the system (being repaired or waiting for repair), L, and the mean number of machines waiting in the queue for repair, L_q, using,

$$L = \sum_{n=1}^{r} np_n \quad \text{and} \quad L_q = \sum_{n=c+1}^{r} (n-c)\,p_n \quad (4)$$

As noted in equation (1), the arrival rate, given that the system is in state n, is $(r-n)\lambda$. Then, the

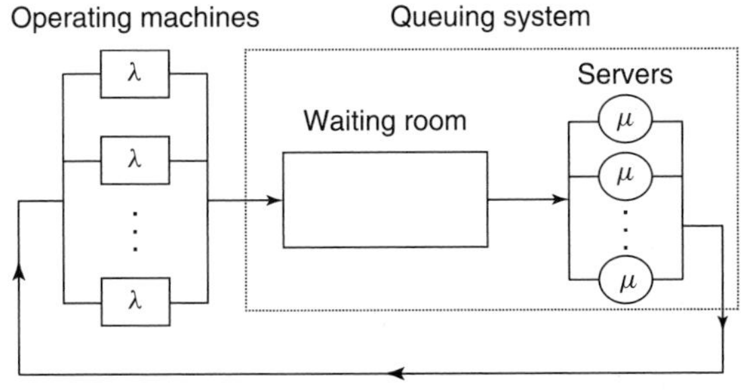

Figure 1 Machine repairmen/interference problem

unconditional average rate at which machines arrive to the system (also called the *effective mean arrival Rate*) is given by,

$$\lambda_{\text{eff}} = \sum_{n=0}^{r-1} (r-n)\,\lambda p_n = \lambda\,(r-L) \tag{5}$$

Thus, we can also obtain the mean waiting time, W, a machine spends in the system (from failure to recovery) and the mean queuing time, W_q, a machine waits in the queue for repair using the well-known Little's formulas,

$$L = \lambda_{\text{eff}} W \quad \text{and} \quad L_q = \lambda_{\text{eff}} W_q \tag{6}$$

Moreover, it is also possible to derive the distribution functions of the waiting time and queuing time in addition to their mean values, W and W_q, see e.g., [2, 3]. Some other performance measures such as the probability that a broken machine must wait for repair can also be obtained. Finally, it is worth observing that these results also hold even though the lifetimes are not exponentially distributed. That is, equation (3) is also valid for the $G/M/c/r/r$ queuing system. Furthermore, equation (3) also holds for systems with exponential lifetimes, general service times, and the same number of repairmen as machines, i.e. for the $M/G/c/c/c$ queue, see [3].

These results have also been extended to more complicated situations where the lifetimes and repair durations follow more general distributions than the exponential one, such as the Erlang distribution, mixtures of exponentials, Coxian and phase-type distributions (see e.g., [16, 17]). Many other features have been introduced to extend the applicability of the machine repairmen problem, including for example *vacation* models where the servers can take a vacation of random duration when the system is empty (see e.g., [18]), *retrial* models where machines that find all the servers busy do not enter the queue but instead reapply for service at random intervals (see e.g., [19]), unreliable servers that may themselves break down (see e.g., [20]), *priority* models where a group of machines are served with priority over the others (see e.g., [21]), *k-out-of-r* systems where at least k machines must be functioning for the whole system to work properly (see e.g., [22]), heterogeneous failure modes (see e.g., [23]), and so on. Another important extension is the use of *spares*, also called *stand-by* or *sparing redundancy* (*see* **Reliability of Redundant Systems**), which is introduced in the following section.

The Machine Repairmen Problem with Spares

Assume now that in addition to the r working machines considered previously, there is another set of s machines standing by as spares such that, when a working machine fails, it is immediately substituted by a spare machine if any is available. When a machine is repaired, it becomes a spare unless the number of working machines is less than r, in which case the repaired machine is restarted immediately.

Consider an $M/M/c/r/r$ with s spares where the lifetimes and repair times are exponentially distributed with rates λ and μ, respectively. Then, given that there are n machines being repaired or waiting for repair, the arrival rate is now given by,

$$\lambda_n = \begin{cases} r\lambda, & \text{if } 0 \le n \le s \\ (r+s-n)\,\lambda, & \text{if } s \le n \le s+r \\ 0, & \text{if } n \ge s+r \end{cases} \tag{7}$$

and μ_n is the same as given in equation (2). Thus, this model can also be described by a birth–death process [3]. And then, it can be shown that the probability of having n machines in the system when $c \le s$ is given by,

$$p_n = \begin{cases} \dfrac{r^n}{n!}\left(\dfrac{\lambda}{\mu}\right)^n p_0, & \text{if } 1 \le n \le c \\ \dfrac{r^n}{c^{n-c}c!}\left(\dfrac{\lambda}{\mu}\right)^n p_0, & \text{if } c \le n \le s \\ \dfrac{r^s r!}{(r+s-n)!c^{n-c}c!}\left(\dfrac{\lambda}{\mu}\right)^n p_0, & \\ \quad \text{if } s \le n \le s+r \end{cases} \tag{8}$$

and when $c > s$,

$$p_n = \begin{cases} \dfrac{r^n}{n!}\left(\dfrac{\lambda}{\mu}\right)^n p_0, & \text{if } 1 \le n \le s \\ \dfrac{r^s r!}{(r+s-n)!n!}\left(\dfrac{\lambda}{\mu}\right)^n p_0, & \text{if } s \le n \le c \\ \dfrac{r^s r!}{(r+s-n)!c^{n-c}c!}\left(\dfrac{\lambda}{\mu}\right)^n p_0, & \\ \quad \text{if } c \le n \le s+r \end{cases} \tag{9}$$

Using these probabilities, we can obtain the mean number of machines in the system and in the queue,